GEOMETRIC FORMULAS

Formulas for area A, perimeter P, circumference C, volume V:

Rectangle

$A = lw$

$P = 2l + 2w$

Box

$V = lwh$

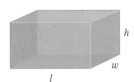

Triangle

$A = \frac{1}{2}bh$

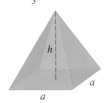

Pyramid

$V = \frac{1}{3}ha^2$

Circle

$A = \pi r^2$

$C = 2\pi r$

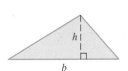

Sphere

$V = \frac{4}{3}\pi r^3$

$A = 4\pi r^2$

Cylinder

$V = \pi r^2 h$

Cone

$V = \frac{1}{3}\pi r^2 h$

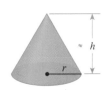

HERON'S FORMULA

$A = \sqrt{s(s-a)(s-b)(s-c)}$

where $s = \dfrac{a+b+c}{2}$

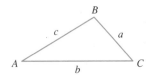

CONIC S

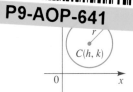

Circles

$(x - h)^2 + (y - k)^2 = r^2$

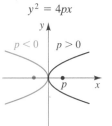

Parabolas

$$x^2 = 4py \qquad\qquad y^2 = 4px$$

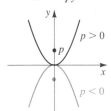

Focus $(0, p)$, directrix $y = -p$ Focus $(p, 0)$, directrix $x = -p$

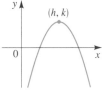

$y = a(x - h)^2 + k,$
$a < 0, \quad h > 0, \quad k > 0$

$y = a(x - h)^2 + k,$
$a > 0, \quad h > 0, \quad k > 0$

Ellipses

$$\frac{x^2}{a^2} + \frac{y^2}{b^2} = 1$$

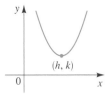

Foci $(\pm c, 0)$, $c^2 = a^2 - b^2$ Foci $(0, \pm c)$, $c^2 = b^2 - a^2$

Hyperbolas

$$\frac{x^2}{a^2} - \frac{y^2}{b^2} = 1 \qquad -\frac{x^2}{a^2} + \frac{y^2}{b^2} = 1$$

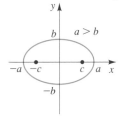

Foci $(\pm c, 0)$, $c^2 = a^2 + b^2$ Foci $(0, \pm c)$, $c^2 = a^2 + b^2$

$$z^n = [r(\cos \theta + i \sin \theta)]^n = r^n(\cos n\theta + i \sin n\theta)$$

$$\sqrt[n]{z} = [r(\cos \theta + i \sin \theta)]^{1/n}$$

$$= r^{1/n}\left(\cos \frac{\theta + 2k\pi}{n} + i \sin \frac{\theta + 2k\pi}{n}\right)$$

where $k = 0, 1, 2, \ldots, n - 1$

ROTATION OF AXES

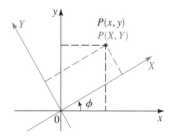

Rotation of axes formulas

$$x = X \cos \phi - Y \sin \phi \qquad y = X \sin \phi + Y \cos \phi$$

Angle-of-rotation formula for conic sections

$$\cot 2\phi = \frac{A - C}{B}$$

POLAR COORDINATES

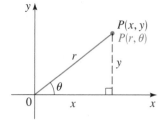

$$x = r \cos \theta$$
$$y = r \sin \theta$$
$$r^2 = x^2 + y^2$$
$$\tan \theta = \frac{y}{x}$$

POLAR EQUATIONS OF CONICS

The graph of a polar equation of the form

$$r = \frac{ed}{1 \pm e \cos \theta} \qquad \text{or} \qquad r = \frac{ed}{1 \pm e \sin \theta}$$

is a conic wth eccentricity e and with one focus at the origin. The conic is

1. a parabola if $e = 1$.

2. an ellipse if $e < 1$.

3. a hyperbola if $e > 1$.

REFERENCES TO CALCULUS

Many students experience difficulty with *precalculus* mathematics in calculus courses. Calculus requires that you understand and remember precalculus topics. For this reason, it may be helpful to retain this text as a reference in your calculus course. It has been written with this purpose in mind.

Some references to calculus topics in this text:

Asymptotes ▪ pages 265, 630

Calculating trigonometric functions ▪ page 363

Components of a vector ▪ page 513

Cycloid ▪ page 669

Definition of the number e ▪ page 294

Difference quotients ▪ pages 40, 134, 135

Extreme values of a function ▪ page 182

Factoring with fractional exponents ▪ page 40

Formulas for lowering powers of sine and cosine ▪ page 477

Fractional expressions ▪ page 40

Half-angle formulas ▪ page 478

Increasing and decreasing functions ▪ page 143

Infinite series ▪ page 704

Limits ▪ pages 265, 705

Local maxima and minima ▪ pages 179, 227

Natural exponential function ▪ page 294

Natural logarithm function ▪ page 309

Newton's Law of Cooling ▪ page 330

Numerical methods ▪ page 363

Parametric curves ▪ page 666

Partial fractions ▪ page 591

Partial sums ▪ page 689

Polar coordinates ▪ page 650

Polar equations of conics ▪ page 661

Radian measure ▪ page 408

Systems of equations ▪ page 531

Trigonometric substitution ▪ page 465

PRECALCULUS
Mathematics for Calculus

ABOUT THE COVER

A precalculus course is an opportunity to learn about the beauty and practical power of mathematics as a prelude to calculus.

The violin, with its sound hole in the shape of an integral sign, has become a symbol of lead author James Stewart's calculus textbook series, which includes *Calculus, Third Edition; Calculus: Early Transcendentals, Third Edition;* and *Calculus: Concepts and Contexts.*

The cover of this edition, which shows violin-making as a "work in progress," reflects both the Stewart authorship and the foundational importance of the precalculus course.

ABOUT THE AUTHORS

James Stewart was educated at the University of Toronto and Stanford University, did research at the University of London, and now teaches at McMaster University. His research field is harmonic analysis.

He is the author of a best-selling calculus textbook series published by Brooks/Cole, including *Calculus, Third Ed.; Calculus: Early Transcendentals, Third Ed.;* and *Calculus: Concepts and Contexts,* as well as a series of high-school mathematics textbooks.

A talented violinist, Stewart was concertmaster of the McMaster Symphony Orchestra for eight years and played professionally in the Hamilton Philharmonic Orchestra. One of his greatest pleasures is playing string quartets.

Lothar Redlin grew up on Vancouver Island, received a Bachelor of Science degree from the University of Victoria, and a Ph.D. from McMaster University in 1978. After completing his education, he did research and taught at the University of Washington, the University of Waterloo, and California State University, Long Beach.

He is currently Professor of Mathematics at The Pennsylvania State University, Abington College. His research field is topology.

Saleem Watson received his Bachelor of Science degree from Andrews University in Michigan. He did his graduate studies at Dalhousie University and McMaster University, where he received his Ph.D. in 1978. After completing his education, he did research at the Mathematics Institute of the University of Warsaw in Poland. He subsequently taught and did research at McMaster University and The Pennsylvania State University.

He is currently Professor of Mathematics at California State University, Long Beach. His research field is functional analysis.

The authors have also published *College Algebra, Second Edition* (Brooks/Cole, 1996).

PRECALCULUS
Mathematics for Calculus
Third Edition

James Stewart
McMaster University

Lothar Redlin
The Pennsylvania State University

Saleem Watson
California State University, Long Beach

BROOKS/COLE PUBLISHING COMPANY
I(T)P® An International Thomson Publishing Company
Pacific Grove ■ Albany ■ Belmont ■ Bonn ■ Boston ■ Cincinnati ■ Detroit ■ Johannesburg ■ London
Madrid ■ Melbourne ■ Mexico City ■ New York ■ Paris ■ Singapore ■ Tokyo ■ Toronto ■ Washington

GWO A GARY W. OSTEDT BOOK

Publisher: *Gary W. Ostedt*
Marketing Team: *Caroline Croley,*
    Christine Davis, and Laura Caldwell
Assistant Editor: *Linda Row*
Editorial Associate: *Carol Ann Benedict*
Production Editor: *Jamie Sue Brooks*
Production Service: *TECH·arts*
Manuscript Editor: *Luana Richards*
Permissions Editor: *Carline Haga*

Interior Design: *Brian Betsill, TECH·arts*
Interior Illustration: *TECH·arts*
Cover Design: *Vernon T. Boes*
Cover Photo: *Mark Tomalty/Masterfile*
Photo Research: *Stephanie Kuhns, TECH·arts*
Typesetting: *Sandy Senter,*
    The Beacon Group
Cover Printing: *Phoenix Color Corp.*
Printing and Binding: *World Color*

For more information, contact:

BROOKS/COLE PUBLISHING COMPANY
511 Forest Lodge Road
Pacific Grove, CA 93950
USA

International Thomson Publishing Europe
Berkshire House 168-173
High Holborn
London WC1V 7AA
England

Thomas Nelson Australia
102 Dodds Street
South Melbourne, 3205
Victoria, Australia

Nelson Canada
1120 Birchmount Road
Scarborough, Ontario
Canada M1K 5G4

International Thomson Editores
Seneca 53
Col. Polanco
México, D. F., México
C. P. 11560

International Thomson Publishing GmbH
Königswinterer Strasse 418
53227 Bonn
Germany

International Thomson Publishing Asia
221 Henderson Road
#05-10 Henderson Building
Singapore 0315

International Thomson Publishing Japan
Hirakawacho Kyowa Building, 3F
2-2-1 Hirakawacho
Chiyoda-ku, Tokyo 102
Japan

Printed in the United States of America

10 9 8 7 6

LIBRARY OF CONGRESS CATALOGING-IN-PUBLICATION DATA
Stewart, James.
    Precalculus : mathematics for calculus / James Stewart, Lothar
Redlin, Saleem Watson.
    Rev. ed. of: Mathematics for calculus. 2nd ed. c1993.
    Includes index.
    ISBN 0-534-34504-2
    1. Mathematics. I. Redlin, L. II. Watson, Saleem.
III. Stewart, James. Mathematics for calculus. 2nd ed.
IV. Title.
QA39.2.S75 1998
510—dc21   97-28428

Photo Credits

xxi ▪ Jim Pickerell/Stock Boston; Lori Adamski Peck/Tony Stone Images xxii ▪ Brian Yarvin/Photo Researchers; Ross M. Horowitz/The Image Bank; Michael Busselle/Tony Stone Images xxiii ▪ Karen Gloria; J. L. Amos, Superstock; Tamara Munzer (Stanford University), Eric Hoffman (Ipsilon Networks), K. Claffy (NLANR), and Bill Fenner (XeroxPARC) xxiv ▪ TSM/Lew Long, 1996; Susan Van Etten/Photo Edit xxv ▪ Will and Denni McIntyre/Photo Researchers 2 ▪ Jim Pickerell/Stock Boston 97 ▪ By courtesy of the National Portrait Gallery, London 123 ▪ Stanford University News Service 126 ▪ The Bettmann Archive 130 ▪ Lori Adamski Peck/Tony Stone Images 143 ▪ Stanford University News Service 165 ▪ The Granger Collection, New York 220 ▪ Brian Yarvin/Photo Researchers 226 ▪ Bill Sanderson/Science Photo Library/Photo Researchers 284 ▪ Michael Ng/NYT Permissions 286 ▪ Ross M. Horowitz/The Image Bank 298 ▪ Garry McMichael/Photo Researchers 350 ▪ Michael Busselle/Tony Stone Images 406 ▪ Karen Gloria 439 ▪ Alan Oddie/Photo Edit 460 ▪ J. L. Amos, Superstock 530 ▪ Tamara Munzer (Stanford University), Eric Hoffman (Ipsilon Networks), K. Claffy (NLANR), and Bill Fenner (XeroxPARC) 561 ▪ The Archives of the National Academy of Sciences 565 ▪ Courtesy of Caltech 571 ▪ The Granger Collection, New York 578 ▪ The Bettmann Archive 580 ▪ The Granger Collection, New York 610 ▪ TSM/Lew Long, 1996 680 ▪ The Granger Collection, New York 684 ▪ Susan Van Etten/Photo Edit 689 ▪ The Granger Collection, New York 691 ▪ The Bettmann Archive 717 ▪ Copyright 1979 National Council of Teachers of Mathematics. Used by permission. Courtesy of Andrejs Dunkels, Sweden. 740 ▪ Will and Denni McIntyre/Photo Researchers 750 ▪ Ken Regen/Camera 5 756 ▪ Stanford University News Service

THIS BOOK IS PRINTED ON ACID-FREE RECYCLED PAPER

To Kathi Townes
for always striving for perfection

PREFACE

The art of teaching is the art of assisting discovery.

MARK VAN DOREN

What does a student really need to know to prepare for calculus? This question has motivated the writing of this text.

To be prepared to study calculus a student needs not only technical skills, but also a clear understanding of *concepts*—in other words, a proper appreciation for what mathematics is all about. We believe that nearly all instructors are committed to the goal of transmitting conceptual understanding. This book was written to provide such instructors and their students with the tools needed to achieve this goal.

The calculus reform movement has obvious implications for the teaching of precalculus. Indeed, the goals of this movement (the primary goal is a focus on conceptual understanding) are best served by implementing its strategies in the precalculus curriculum. In practical implementations of the goals of reform, instructors are using many different approaches. Some use *technology* to help students become active learners; others use the *rule of four,* "topics should be presented geometrically, numerically, algebraically, and verbally," as well as an expanded emphasis on *applications* to promote conceptual reasoning; still others use *group learning* or *writing exercises* as a way of encouraging students to explore their own understanding of a given concept. In this book we have used all these methods of presenting precalculus mathematics as enhancements to a central core of fundamental skills. These methods are tools to be utilized by instructors and their students to chart their own course of action towards the goal of conceptual understanding.

In writing this third edition, our purpose was to sharpen the clarity of the exposition while retaining the main features that have contributed to the success of this book. We continue to present precalculus mathematics as a problem-solving endeavor, with applications to the real world. Throughout the book we encourage the student to organize his or her thoughts and to utilize the appropriate tools for solving a problem. To assist students in understanding the examples, we have included step-by-step comments on solutions; these have been placed in the margin to avoid interrupting the flow of the solution. We have added many graphs and figures to remind students of the geometric meaning behind a calculation and to visually promote insight into formulas and theorems. We have included more numerical examples and exercises. We have integrated the use of the graphing calculator throughout the text in optional subsections, examples, and exercises; these are identified with a special symbol, . We have added new sections on modeling that illustrate how the concepts learned in a given chapter are used to model real-life data. We have also added more mathematical "vignettes" that describe a key application of precalculus or a historical insight in the development of the subject. All these changes have been made with the goal of further enhancing the usability of the book in the classroom.

Many of the changes in the structure of this book are a result of our own experience in teaching precalculus mathematics. We have benefited from the insightful comments of our colleagues who have used the text and critiqued it section by section. We have also benefited greatly from the sharp insight of our reviewers.

SPECIAL FEATURES

Flexible Approach to Trigonometry

The trigonometric functions can be defined in two different, although equivalent, ways—one based on right triangles and the other based on arc length on the unit circle. One way to teach the subject is to begin with the right triangle approach, building on the foundation of a conventional high-school course in trigonometry. The other way is to begin with the unit circle approach, thus emphasizing that the trigonometric functions are functions of real numbers, just like polynomial, logarithmic, and exponential functions. Each way of teaching trigonometry has its merits and so the trigonometry chapters of this text have been written so that *either approach may be studied first.* When each approach is introduced, appropriate applications are also presented. This provides early motivation for studying the trigonometric functions and also makes it very clear why we need to study two different ways of looking at the same functions.

Focus on Modeling

Precalculus mathematics has practical applications in its own right, not just as a prerequisite for calculus. In addition to the applied problems that occur throughout the text, we have concluded several of the chapters with sections entitled *Focus on Modeling.* The first such section, following Chapter 2, introduces the basic idea of modeling a real-life situation using linear functions. Other sections present ways in which polynomial, exponential, logarithmic, and trigonometric functions; systems of inequalities; and probability can all be used to model familiar phenomena from the sciences and from everyday life.

Focus on Problem Solving

We are committed to helping students develop mathematical thinking rather than "memorizing all the rules," an endless and pointless task that many students are not taught to avoid. Thus, an emphasis on problem solving is integrated throughout the text. In addition, we have concluded several chapters with sections entitled *Focus on Problem Solving,* each of which highlights a particular problem-solving technique. The first such section, following Chapter 1, gives a general introduction to the principles of problem solving. Each *Focus* section includes problems of a varied nature that encourage students to apply their problem-solving skills. In selecting these problems we have kept in mind the following advice from David Hilbert: "A mathematical problem should be difficult in order to entice us, yet not inaccessible lest it mock our efforts."

Graphing Calculators and Computers

Calculator and computer technology provides completely new ways of visualizing mathematics and affects not only how a topic is taught but also what is emphasized. We have integrated the use of graphing calculators in numerous subsections, examples, and exercises (see, for instance, pages 182, 378–380, and 303). One example of how technology enhances understanding is that we can use it to draw accurate graphs of *families* of functions, so that students can see how varying a parameter can affect the graph of a function (see pages 230 and 150, Exercises 49–54). These graphing calculator subsections, examples, and exercises, all marked with the special symbol ▦, are optional and may be omitted without loss of continuity.

The availability of graphing calculators makes it not less important but far more important to clearly understand the concepts that underlie the image the calculator produces on its screen. Accordingly, all our calculator-oriented subsections are preceded by sections in which students must sketch graphs by hand and analyze them, so that they can understand precisely what the graphing device is doing when they later use it to simplify the routine, mechanical part of their work. Thus, we treat the calculator as an aid to understanding, an extension of pencil and paper, not as the central feature of the course.

Mathematical Vignettes Throughout this book we make use of the margins to provide short biographies of interesting mathematicians as well as applications of precalculus mathematics to the "real world." The biographies often include a key insight that the mathematician discovered and which is relevant to precalculus. (See, for instance, the vignettes on Viète, page 51; Salt Lake City, page 83; and radiocarbon dating, page 322.) They serve to enliven the material and show that mathematics is an important, vital activity, and that even at this elementary level, it is fundamental in everyday life.

Real-World Applications We have included substantial applied problems that we believe will capture the attention of students. These are integrated throughout the text in both examples and exercises. Applications from engineering, physics, chemistry, business, biology, environmental studies, and other fields show how mathematics is used to model real-life situations. We have carefully chosen applications that show the relevance of precalculus mathematics to our daily lives and its remarkable power as a problem-solving tool. (See, for instance, the examples and exercises on energy expended in bird flight, page 66; determining the optimal shape for a can, page 184; terminal velocity of a skydiver, page 302, Exercise 28; and establishing time of death, page 336, Exercise 30.)

Discovery, Writing, and Group Learning Most exercise sets end with a block of exercises labeled *Discovery • Discussion*. These exercises are designed to encourage the students to experiment, preferably in groups, with the concepts developed in the section, and then to write out what they have learned, rather than simply to look for "the answer."

Check Your Answer The new *Check Your Answer* feature is used wherever possible to emphasize the importance of looking back to check whether an answer is reasonable.

Review Sections and Chapter Tests Each chapter ends with an extensive review section, including a *Chapter Test* designed to help the students gauge their progress. Brief answers to the odd-numbered exercises in each section (including the review exercises) and to all questions in the tests are given in the back of the book.

Concept Checks The review material in each chapter now begins with a *Concept Check* that is designed to get the students to think about and explain in their own words the ideas presented in the chapter. These can be used as writing exercises, in a classroom discussion setting, or for personal study.

MAJOR CHANGES FOR THE THIRD EDITION

- Chapter 1, which is a review chapter, has been restructured so that the material on integer and rational exponents appears together in one section, and the material on algebraic expressions and factoring also appears in one section. Section 1.7 now contains the method of sign diagrams for solving nonlinear inequalities.

- Chapter 3 on polynomial and rational functions has been restructured to begin with the idea of the graph of a polynomial function; the material on the theory of equations now comes later in the chapter. Many of the exercise sets in this chapter have been rewritten to include problems that involve more straightforward factorizations.

- Chapter 4 has been completely rewritten to emphasize the importance of the natural exponential function and its role in applications. Exponential and logarithmic equations are now treated in a separate section, Section 4.5.

- Chapter 7 has been reorganized so that the material on trigonometric equations now appears after the introduction of the inverse trigonometric functions.

- Chapter 8 now begins with the substitution and elimination methods for both linear and nonlinear systems of equations. Many of the Gaussian elimination exercises have been simplified from a computational point of view.

- Chapter 9 has a new section on polar equations of conics.

- Chapter 10 has been reorganized so that the material on arithmetic sequences and series appears together in one section, as does the material on geometric sequences and series.

- Chapter 11 is a new chapter. We have added it because many users indicated that they need material on counting and probability for their courses.

- *Focus on Modeling* sections have been introduced after Chapters 2, 4, 5, 8, 9, and 11.

- The *Focus on Problem Solving* sections include many new problems that are related to either the problem-solving principle discussed in the section or the material of the preceding chapters.

- The emphasis on problem solving has been enhanced throughout the text.

- Numerous margin comments referring to the principles of problem solving have been added, and "guideline" boxes that apply these principles to the topic at hand have been expanded and improved.

- Where possible, prose-style explanations in example solutions have been replaced by author notes to improve readability.

- We have made greater use of summary boxes to clarify the structure of some topics.

- Numerous graphs and figures have been added to enhance understanding through visualization. In particular, figures have been added to summary boxes wherever appropriate.

- Each example has been given a title to clarify its purpose.

- A new *Check Your Answer* feature has been added to many examples.

- Several new mathematical vignettes have been added and existing ones updated.

- This edition employs additional colors to clarify the relationships between different elements of figures.

SECTION 1.8 COORDINATE GEOMETRY   83

sponding angles are equal. It follows that $d(A, P) = d(M, Q)$ and so

$$x - x_1 = x_2 - x$$

Solving for x, we get $2x = x_1 + x_2$, so $x = \dfrac{x_1 + x_2}{2}$. Similarly, $y = \dfrac{y_1 + y_2}{2}$.

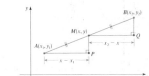

FIGURE 6

MIDPOINT FORMULA

The midpoint of the line segment from $A(x_1, y_1)$ to $B(x_2, y_2)$ is

$$\left(\frac{x_1 + x_2}{2}, \frac{y_1 + y_2}{2} \right)$$

EXAMPLE 3 ■ Finding the Midpoint

The midpoint of the line segment that joins the points $(-2, 5)$ and $(4, 9)$ is

$$\left(\frac{-2 + 4}{2}, \frac{5 + 9}{2} \right) = (1, 7)$$

See Figure 7.

The coordinates of a point in the xy-plane uniquely determine its location. We can think of the coordinates as the "address" of the point. In Salt Lake City, Utah, the addresses of most buildings are in fact expressed as coordinates. The city is divided into quadrants with Main Street as the vertical (North–South) axis and S. Temple Street as the horizontal (East–West) axis. An address such as

1760 W 2100 S

indicates a location 17.6 blocks west of Main Street and 21 blocks south of S. Temple Street. (This is the address of the main post office in Salt Lake City.) With this logical system it is possible for someone unfamiliar with the city to locate any address immediately, as easily as one can locate a point in the coordinate plane.

FIGURE 7

◀ Mathematical vignettes provide applications of algebra to the real world or give short biographies of mathematicians, contemporary as well as historical.

◀ Summary boxes organize and clarify topics and highlight key ideas.

Check Your Answer emphasizes the importance ▶ of looking back at your work to check whether an answer is reasonable.

This *Warning* symbol points out situations ▶ where many students make the same mistake.

Example titles clarify the purpose of examples. ▶

54   CHAPTER 1 FUNDAMENTALS

(b) To eliminate the square root, we first isolate it on one side of the equal sign.

$$x - 1 = -\sqrt{2 - \frac{x}{2}} \qquad \text{Subtract 1}$$

$$(x - 1)^2 = 2 - \frac{x}{2} \qquad \text{Square each side to eliminate square root}$$

$$x^2 - 2x + 1 = 2 - \frac{x}{2} \qquad \text{Expand LHS}$$

$$2x^2 - 4x + 2 = 4 - x \qquad \text{Multiply by 2 to clear denominator}$$

$$2x^2 - 3x - 2 = 0 \qquad \text{Move all terms to LHS}$$

$$(x - 2)(2x + 1) = 0 \qquad \text{Factor}$$

Setting each factor equal to 0 gives us $x = 2$ and $x = -\frac{1}{2}$ as potential solutions. If we substitute these into the original equation (see *Check Your Answers*), we see that $x = -\frac{1}{2}$ is a solution but $x = 2$ is not. The only solution is

$$x = -\tfrac{1}{2} \qquad ■$$

The reason extraneous solutions are often introduced when we square each side of an equation is that the operation of squaring can turn a false equation into a true one. For example, $-1 \neq 1$, but $(-1)^2 = 1^2$. Thus, the squared equation may be true for more values of the variable than the original equation, so we must always check our answers to make sure that each satisfies the original equation.

In some cases, fourth-degree (or higher-degree) polynomial equations can be changed to quadratic equations by performing algebraic substitutions, as shown in the following example.

EXAMPLE 11 ■ An Equation of Quadratic Type

Find the real solutions of $x^4 - 2x^2 - 2 = 0$.

SOLUTION If we set $w = x^2$, then we get a quadratic equation in the new variable w.

$$x^4 - 2x^2 - 2 = (x^2)^2 - 2x^2 - 2$$

$$= w^2 - 2w - 2 = 0 \qquad \text{Set } w = x^2$$

From the quadratic formula, we have

$$w = \frac{2 \pm \sqrt{(-2)^2 + 4 \cdot 2}}{2} = 1 \pm \sqrt{3}$$

Because $x^2 = w$, we have $x = \pm\sqrt{w}$. So the potential solutions are $\pm\sqrt{1 \pm \sqrt{3}}$. Since we can't take the square root of a negative number such as $1 - \sqrt{3}$ in

CHECK YOUR ANSWERS
$x = -\frac{1}{2}$:
   LHS $= -\frac{1}{2}$
   RHS $= 1 - \sqrt{2 - \dfrac{\left(-\frac{1}{2}\right)}{2}}$
     $= 1 - \sqrt{\frac{9}{4}} = 1 - \frac{3}{2} = -\frac{1}{2}$
   LHS = RHS ✓
$x = 2$:
   LHS $= 2$
   RHS $= 1 - \sqrt{2 - \frac{2}{2}}$
     $= 1 - 1 = 0$
   LHS $\neq$ RHS, so $x = 2$ is not a solution. ✗

SECTION 1.10 LINES 115

(b) Find the temperature at which the scales agree. [*Hint:* Suppose that a is the temperature at which the scales agree. Set $F = a$ and $C = a$. Then solve for a.]

63. At the surface of the ocean, the water pressure is the same as the air pressure above the water, 15 lb/in². Below the surface, the water pressure increases by 4.34 lb/in² for every 10 ft of descent.
(a) Find an equation for the relationship between pressure and depth below the ocean surface.
(b) At what depth is the pressure 100 lb/in²?

water pressure increases with depth

64. Jason and Debbie leave Detroit at 2:00 P.M. and drive at a constant speed, traveling west on I-90. They pass Ann Arbor, 40 mi from Detroit, at 2:50 P.M.
(a) Express the distance traveled in terms of the time elapsed.
(b) Draw the graph of the equation in part (a).
(c) What is the slope of this line? What does it represent?

65. The monthly cost of driving a car depends on the number of miles driven. Lynn found that in May her driving cost was $380 for 480 mi and in June her cost was $460 for 800 mi.
(a) Express the monthly cost C in terms of the distance driven d, assuming that a linear relationship gives a suitable model.
(b) Use part (a) to predict the cost of driving 1500 mi per month.
(c) Draw the graph of the linear equation. What does the slope of the line represent?
(d) What does the y-intercept of the graph represent?
(e) Why is a linear relationship a suitable model for this situation?

66. The manager of a furniture factory finds that it costs $2200 to manufacture 100 chairs in one day and $4800 to produce 300 chairs in one day.
(a) Assuming that the relationship between cost and the number of chairs produced is linear, find an equation that expresses this relationship. Then graph this equation.
(b) What is the slope of the line of part (a), and what does it represent?
(c) What is the y-intercept of this line, and what does it represent?

67–68 ■ Equations for supply and demand are given.
(a) Draw the graphs of the two equations in an appropriate viewing rectangle.
(b) Estimate the equilibrium point from the graph.
(c) Estimate the price and the amount of the commodity produced and sold at equilibrium.

67. Supply: $y = 0.45p + 4$
Demand: $y = -0.65p + 28$

68. Supply: $y = 8.5p + 45$
Demand: $y = -0.6p + 300$

DISCOVERY · DISCUSSION

69. What Does the Slope Mean? Suppose that the graph of the outdoor temperature over a certain period of time is a line. How is the weather changing if the slope of the line is positive? If it's negative? If it's zero?

70. Collinear Points Suppose you are given the coordinates of three points in the plane, and you want to see whether they lie on the same line. How can you do this using slopes? Using the Distance Formula? Can you think of another method?

◀ Real-world applications show the relevance of algebra to everyday life and indicate its remarkable problem-solving power.

◀ Graphing calculator exercises are indicated with the graphing calculator logo.

◀ *Discovery·Discussion* exercises encourage students to experiment with, discuss, and write about the concepts they have learned.

Families of curves can be studied using ▶ the graphing calculator, so that students can see how varying a parameter in an equation affects the graph.

CHAPTER 9 TOPICS IN ANALYTIC GEOMETRY

itself? The graph repeats itself when the same value of r is obtained at θ and $\theta + 2n\pi$. Thus, we need to find an integer n, so that

$$\cos\frac{2(\theta + 2n\pi)}{3} = \cos\frac{2\theta}{3}$$

For this equality to hold, $4n\pi/3$ must be a multiple of 2π, and this first happens when $n = 3$. Therefore, we obtain the entire graph if we choose values of θ between $\theta = 0$ and $\theta = 0 + 2(3)\pi = 6\pi$. The graph is shown in Figure 16.

FIGURE 16
$r = \cos(2\theta/3)$

EXAMPLE 10 ■ A Family of Polar Equations

Graph the family of polar equations $r = 1 + c\sin\theta$ for $c = 3, 2.5, 2, 1.5, 1$. How does the shape of the graph change as c changes? (These curves are called **limaçons**, after the French word for snail, because of the shape of the curve for certain values of c.)

SOLUTION Figure 17 shows computer-drawn graphs for the given values of c. For $c > 1$, the graph has an inner loop; the loop decreases in size as c decreases. When $c = 1$, the loop disappears and the graph becomes a cardioid (see Example 7).

$c = 3.0$ $c = 2.5$ $c = 2.0$ $c = 1.5$ $c = 1.0$

FIGURE 17 A family of limaçons $r = 1 + c\sin\theta$ in the viewing rectangle $[-2.5, 2.5]$ by $[-0.5, 4.5]$

The following box gives a summary of some of the basic polar graphs used in calculus.

Express all unknown quantities in terms of the variable

Relate the quantities

Set up an equation

Solve

picture into an equation here is to notice that the total amount of juice on both sides of the equal sign is the same. The orange juice in the first vat is 5% of 900 gal, or 45 gal. The second vat contains x gallons of juice, and the third contains $0.1(900 + x)$ gallons.

Equating the total amounts of pure juice before and after mixing, we get the equation

amount of juice before mixing = amount of juice after mixing

$$45 + x = 0.1(900 + x) \qquad \text{From Figure 3}$$
$$45 + x = 90 + 0.1x \qquad \text{Multiply}$$
$$0.9x = 45 \qquad \text{Subtract } 0.1x \text{ and } 45$$
$$x = \frac{45}{0.9} = 50 \qquad \text{Divide by } 0.9$$

The manufacturer should add 50 gal of pure orange juice to the soda.

CHECK YOUR ANSWER

amount of juice before mixing = 5% of 900 gal + 50 gal pure juice
$$= 45 \text{ gal} + 50 \text{ gal} = 95 \text{ gal}$$
amount of juice after mixing = 10% of 950 gal = 95 gal

Amounts are equal. ✓

EXAMPLE 5 ■ Time Needed to Do a Job

Because of an anticipated heavy rainstorm, the water level in a reservoir must be lowered by 1 ft. Opening spillway A lowers the level by this amount in 4 hours, whereas opening the smaller spillway B does the job in 6 hours. How long will it take to lower the water level by 1 ft if both spillways are opened?

SOLUTION   As usual, we represent the number we are seeking by x.

FIGURE 4

Identify the variable

x = number of hours it takes to lower the water level by 1 ft if both spillways are open

Finding an equation relating x to the other quantities in t[his]
easy. Certainly x is not simply $4 + 6$, since that would m[ean]
ways together require longer to lower the water level than [ ]
Instead, we look at the fraction of the job that can be do[ne by]
each spillway.

Express all quantities in terms of the variable

Spillway A lowers the water by $\frac{1}{4}$ ft in 1[ ]

Spillway B lowers the water by $\frac{1}{6}$ ft in 1[ ]

Both spillways lower the water by $\frac{1}{x}$ ft[ ]

◀ Problem-solving notes in margins apply problem-solving steps to examples in the text.

◀ Author notes provide step-by-step comments on solutions.

◀ *Check Your Answer*

◀ Example title

Each chapter ends with a review section ▶ containing a *Concept Check*, extensive review exercises, and a *Chapter Test*.

---

9   TEST

1. Find the focus and directrix of the parabola $x^2 = -12y$, and sketch its graph.

2. Find the vertices, foci, and the lengths of the major and minor axes for the ellipse $\dfrac{x^2}{16} + \dfrac{y^2}{4} = 1$. Then sketch its graph.

3. Find the vertices, foci, and asymptotes of the hyperbola $\dfrac{y^2}{9} - \dfrac{x^2}{16} = 1$. Then sketch its graph.

4–6 ■ Find an equation for the conic whose graph is shown.

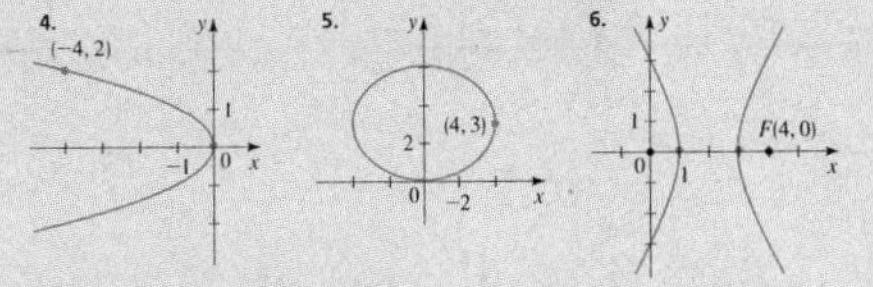

4. $(-4, 2)$

5. $(4, 3)$

6. $F(4, 0)$

7–9 ■ Sketch the graph of the equation.

7. $16x^2 + 36y^2 - 96x + 36y + 9 = 0$ 8. $9x^2 - 8y^2 + 36x + 64y = 92$

9. $2x + y^2 + 8y + 8 = 0$

10. Find an equation for the hyperbola with foci $(0, \pm 5)$ and with asymptotes $y = \pm \frac{3}{4}x$.

11. Find an equation for the parabola with focus $(2, 4)$ and directrix the x-axis.

12. A parabolic reflector for a car headlight forms a bowl shape that is 6 in. wide at its opening and 3 in. deep, as shown in the figure at the left. How far from the vertex should the bulb be placed if it is to be located at the focus?

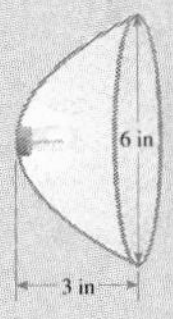

13. (a) Use the discriminant to determine whether the graph of this equation is a parabola, an ellipse, or a hyperbola:
$$5x^2 + 4xy + 2y^2 = 18$$
(b) Use rotation of axes to eliminate the xy-term in the equation.
(c) Sketch the graph of the equation.
(d) Find the coordinates of the vertices of this conic (in the xy-coordinate system).

14. Graph the polar equation $r = 2 + \cos \theta$.

15. Convert the polar equation $r = 2 \cos \theta - 4 \sin \theta$ to rectangular coordinates, and identify the graph.

16. (a) Sketch the graph of the parametric curve
$$x = 3 \sin \theta + 3 \qquad y = 2 \cos \theta \qquad 0 \le \theta \le \pi$$
(b) Eliminate the parameter θ in part (a) to obtain an equation for this curve in rectangular coordinates.

6 in

3 in

PRINCIPLES OF PROBLEM SOLVING

There are no hard and fast rules that will ensure success in solving problems. However, it is possible to outline some general steps in the problem-solving process and to give some principles that may be useful in the solution of certain problems. These steps and principles are just common sense made explicit. They have been adapted from George Polya's book *How To Solve It*.

1 | UNDERSTAND THE PROBLEM

The first step is to read the problem and make sure that you understand it clearly. Ask yourself the following questions:

What is the unknown?

What are the given quantities?

What are the given conditions?

For many problems it is useful to

draw a diagram

and identify the given and required quantities on the diagram.
   Usually it is necessary to

introduce suitable notation

In choosing symbols for the unknown quantities we often use letters such as a, b, c, m, n, x, and y, but in some cases it helps to use initials as suggestive symbols, for instance, V for volume or t for time.

2 | THINK OF A PLAN

Find a connection between the given information and the unknown that will enable you to calculate the unknown. It often helps t̲
"How can I relate the given to the unknown?" If yo̲
immediately, the following ideas may be helpful in dev̲

■ **Try to recognize something familiar**

Relate the given situation to previous knowledge. Look a̲
recall a more familiar problem that has a similar unkn̲

■ **Try to recognize patterns**

Some problems are solved by recognizing that some kir̲
The pattern could be geometric, or numerical, or algeb̲

George Polya (1887–1985) is famous among mathematicians for his ideas on problem solving. His lectures on problem solving at Stanford University attracted overflow crowds whom he held on the edges of their seats, leading them to discover solutions for themselves. He was able to do this because of his deep insight into the psychology of problem solving. His well-known book *How To Solve It* has been translated into 15 languages. He said that Euler (see page 249) was unique among great mathematicians because he explained *how* he found his results. Polya often said to his students and colleagues, "Yes, I see that your proof is correct, but how did you discover it?" In the preface to *How To Solve It*, Polya writes, "A great discovery solves a great problem but there is a grain of discovery in the solution of any problem. Your problem may be modest; but if it challenges your curiosity and brings into play your inventive faculties, and if you solve it by your own means, you may experience the tension and enjoy the triumph of discovery."

◀ *Focus on Problem Solving* sections include many new problems that are related to the problem-solving principle discussed in the section or to the material of the preceding chapters.

◀ Mathematical vignettes provide short biographies of interesting mathematicians, contemporary as well as historical, or give applications of algebra to the real world.

Focus on Modeling sections show how ▶ precalculus mathematics can be used to model important phenomena from the sciences and from everyday life.

FOCUS ON MODELING

Periodic behavior—behavior that repeats over and over again—is common in nature. Perhaps the most familiar example is the daily rising and setting of the sun, which results in the repetitive pattern of day, night, day, night, ... Another example is the daily variation of tide levels at the beach, which results in the repetitive pattern of high tide, low tide, high tide, low tide, ... Certain animal populations increase and decrease in a predictable periodic pattern: a large population exhausts the food supply, which causes the population to dwindle; this in turn results in a more plentiful food supply, which makes it possible for the population to increase; and the pattern then repeats over and over.
   Other common examples of periodic behavior involve motion that is caused by vibration or oscillation. A mass suspended from a spring that has been compressed and then allowed to vibrate vertically is a simple example. This same "back and forth" motion also occurs in such diverse phenomena as sound waves, light waves, alternating electrical current, and pulsating stars, to name a few. In this section we consider the problem of modeling periodic behavior.

MODELING PERIODIC BEHAVIOR

The trigonometric functions are ideally suited for modeling periodic behavior. A glance at the graphs of the sine and cosine functions, for instance, shows that these functions themselves exhibit periodic behavior. Figure 1 shows the graph of $y = \sin t$. If we think of t as time, we see that as time goes on, $y = \sin t$ increases and decreases over and over again. Figure 2 shows that the motion of a vibrating mass on a spring is modeled very accurately by $y = \sin t$.

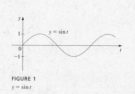

FIGURE 1
$y = \sin t$

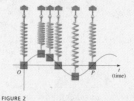

FIGURE 2
Motion of a vibrating spring is modeled by $y = \sin t$.

   Notice that the mass returns to its original position over and over again. A **cycle** is one complete vibration of an object so the mass in Figure 2 completes one cycle of its motion between O and P. Our observations about how the

ANCILLARIES FOR PRECALCULUS, MATHEMATICS FOR CALCULUS, THIRD EDITION

For the Instructor

PRINTED

Instructor's Solutions Manual

by John Banks, *San Jose City College*

- Solutions to all even-numbered text exercises

Test Items

by Andrew Bulman-Fleming

- Over 3000 multiple-choice and short-answer test items
- Computerized version also available

SOFTWARE

Computerized Testing

- Over 3000 multiple-choice and short-answer test items
- Algorithmic test generation
- Print version also available

Computerized Tutorial

- Text-specific items
- Algorithmic quiz generation

Technical support

- Toll-free technical support is available for any Brooks/Cole software product: (800) 327-0325 or E-mail: support@brookscole.com

VIDEO

Text-Specific Video Tutorials

by Barbara Brown, *Anoka-Ramsey Community College*

- Provides additional examples and explanations for each chapter of the text

For the Student

PRINTED

Student Solutions Manual

by John Banks, *San Jose City College*

- Solutions to all odd-numbered text exercises

Study Guide

by John Banks, *San Jose City College*

- Detailed explanations
- Worked-out practice problems

**Precalculus in Context:
Functioning in the Real World, Second Edition**

by Marsha J. Davis, *Eastern Connecticut State University,* Judith Flagg Moran, *Trinity College,* and Mary E. Murphy, *Smith College*

- Thirteen labs and approximately fifty related problems and explorations cover topics traditionally found in a one-semester course
- Real-life situations designed for collaborative group work, including programming a VCR, quilt patterns, moose populations in New England, seasonal affective disorder, and the AIDS pandemic
- A graphing device (calculator or computer) is used as an essential tool for understanding in all projects

Explorations in Precalculus using the TI-82/TI-83

by Deborah Cochener and Bonnie Hodge, *Austin Peay State University*

- Hands-on applications with solutions
- Correlation charts relate course topics to workbook units
- Key charts (specific to the TI-82, TI-83, and TI-85) show which units introduce keys on the calculator
- "Troubleshooting Section" helps you avoid common errors

ACKNOWLEDGMENTS

We thank the following reviewers for their thoughtful and constructive comments on this edition.

Edward Dixon, *Tennessee Technological University*
Richard Dodge, *Jackson Community College*
Floyd Downs, *Arizona State University at Tempe*
Marjorie Kreienbrink, *University of Wisconsin–Wankesha*
Donna Krichiver, *Johnson Community College*
Wayne Lewis, *University of Hawaii* and *Honolulu Community College*
Adam Lutoborski, *Syracuse University*
Keith Oberlander, *Pasadena City College*
Christine Panoff, *University of Michigan at Flint*
Susan Piliero, *Cornell University*
Gregory St. George, *University of Montana*
Gary Stoudt, *Indiana University of Pennsylvania*
Arnold Vobach, *University of Houston*
Tom Walsh, *City College of San Francisco*
Muserref Wiggins, *University of Akron*
Diane Williams, *Northern Kentucky University*
Suzette Wright, *University of California at Davis*
Yisong Yang, *Polytech University*

We have benefited greatly from the suggestions and comments of our colleagues who have used our books in previous editions. We extend special thanks in this regard to Linda Byun, Bruce Chaderjian, David Gau, Daniel Martinez, David McKay, Robert Mena, Kent Merryfield, Marilyn Oba, Robert Valentini, and Derming Wang, of California State University, Long Beach; and to Gloria Dion, Karen Gold, Betsy Huttenlock, Cecilia McVoy, Mike McVoy, Samir Ouzomgi, and Ralph Rush, of The Pennsylvania State University, Abington College.

We thank Kathi Townes, Stephanie Kuhns, and Brian Betsill of TECH·arts for their excellent work and unflagging energy in attending to all the details of design and production. At The Beacon Group, we thank Sandy Senter for her exceptionally speedy and accurate typesetting work. We also thank Phyllis Panman for checking the accuracy of examples, exercises, and answers, and for helping with the proofreading process. At Brooks/Cole, our thanks go to Jamie Sue Brooks, production editor; Vernon Boes, cover designer; Beth Wilbur, assistant editor; Carol Benedict, editorial associate; and Jill Downey, Christine Davis, and Laura Caldwell, marketing team. They have all done an outstanding job.

We are especially grateful to mathematics publisher Gary W. Ostedt. His extensive experience and keen editorial insight were invaluable resources in the writing of this third edition.

TO THE STUDENT

This textbook was written for you to use as a guide to mastering precalculus mathematics. Here are some suggestions to help you get the most out of your course.

First of all, you should read the appropriate section of text *before* you attempt your homework problems. Reading a mathematics text is quite different from reading a novel, a newspaper, or even another textbook. You may find that you have to reread a passage several times before you understand it. Pay special attention to the examples, and work them out yourself with pencil and paper as you read. With this kind of preparation you will be able to do your homework much more quickly and with more understanding.

Don't make the mistake of trying to memorize every single rule or fact you may come across. Mathematics doesn't consist simply of memorization. Mathematics is a *problem-solving art,* not just a collection of facts. To master the subject you must solve problems—lots of problems. Do as many of the exercises as you can. Be sure to write your solutions in a logical, step-by-step fashion. Don't give up on a problem if you can't solve it right away. Try to understand the problem more clearly—reread it thoughtfully and relate it to what you have learned from your teacher and from the examples in the text. Struggle with it until you solve it. Once you have done this a few times you will begin to understand what mathematics is really all about.

Answers to the odd-numbered exercises, as well as all the answers to each chapter test, appear at the back of the book. If your answer differs from the one given, don't immediately assume that you are wrong. There may be a calculation that connects the two answers and makes both correct. For example, if you get $1/(\sqrt{2} - 1)$ but the answer given is $1 + \sqrt{2}$, your answer *is* correct, because you can multiply both numerator and denominator of your answer by $\sqrt{2} + 1$ to change it to the given answer.

The symbol ⊘ is used to warn against committing an error. We have placed this symbol in the margin to point out situations where we have found that many of our students make the same mistake.

CALCULATORS AND CALCULATIONS

Calculators are essential in most mathematics and science subjects. They free us from performing routine tasks, so we can focus more clearly on the concepts we are studying. Calculators are powerful tools but their results need to be interpreted with care. In what follows, we describe the features a calculator suitable for a precalculus course should have, and we give guidelines for interpreting the results of its calculations.

SCIENTIFIC AND GRAPHING CALCULATORS

For this course you will need a *scientific* calculator—one that has, as a minimum, the usual arithmetic operations ($+$, $-$, $\times$, $\div$) and exponential, logarithmic, and trigonometric functions (e^x, 10^x, ln, log, sin, cos, tan). In addition, a memory and at least some degree of programmability will be useful. Many scientific calculators can perform operations on matrices and determinants, which are studied in Chapter 8.

Your instructor may recommend or require that you purchase a *graphing* calculator. This book has optional subsections and exercises that require the use of a graphing calculator or a computer with graphing software. These special subsections and exercises are indicated by the symbol ▱.

It is important to realize that, because of limited resolution, a graphing calculator gives only an *approximation* to the graph of a function. It can plot only a finite number of points and then connect them to form a *representation* of the graph. In many of our examples we point out that we must be careful when interpreting graphs produced by calculators.

CALCULATIONS AND SIGNIFICANT FIGURES

Most of the applied examples and exercises in this book involve approximate values. For example, one exercise states that the moon has a radius of 1074 miles. This does not mean that the moon's radius is exactly 1074 miles but simply that this is the radius rounded to the nearest mile.

One simple method for specifying the accuracy of a number is to state how many **significant digits** it has. The significant digits in a number are the ones from the first nonzero digit to the last nonzero digit (reading from left to right). Thus, 1074 has four significant digits, 1070 has three, 1100 has two, and 1000 has one significant digit. This rule may sometimes lead to ambiguities. For example, if a distance is 200 km to the nearest kilometer, then the number 200 really has three significant digits, not just one. This ambiguity is avoided if we use scientific notation—that is, if we express the number as a multiple of a power of 10:

$$2.00 \times 10^2$$

When working with approximate values, students often make the mistake of giving a final answer with *more* significant digits than the original data. This is incorrect because you cannot "create" precision by using a calculator. The final result can be no more accurate than the measurements given in the problem. For example, suppose we are told that the two shorter sides of a right triangle are measured to be 1.25 and 2.33 inches long. By the Pythagorean Theorem, we find, using a calculator, that the hypotenuse has length

$$\sqrt{1.25^2 + 2.33^2} \approx 2.644125564 \text{ in.}$$

But since the given lengths were expressed to three significant digits, the answer cannot be any more accurate. We can therefore say only that the hypotenuse is 2.64 in. long, rounding to the nearest hundredth.

In general, the final answer should be expressed with the same accuracy as the *least*-accurate measurement given in the statement of the problem. The following rules make this principle more precise.

RULES FOR WORKING WITH APPROXIMATE DATA

1. When multiplying or dividing, round off the final result so that it has as many *significant digits* as the given value with the fewest number of significant digits.

2. When adding or subtracting, round off the final result so that it has its last significant digit in the *decimal place* in which the least-accurate given value has its last significant digit.

3. When taking powers or roots, round off the final result so that it has the same number of *significant digits* as the given value.

As an example, suppose that a rectangular table top is measured to be 122.64 in. by 37.3 in. We express its area and perimeter as follows:

Area = length × width = 122.64 × 37.3 ≈ 4570 in²   Three significant digits

Perimeter = 2(length + width) = 2(122.64 + 37.3) ≈ 319.9 in.   Tenths digit

Note that in the formula for the perimeter, the value 2 is an exact value, not an approximate measurement. It therefore does not affect the accuracy of the final result. In general, if a problem involves only exact values, we may express the final answer with as many significant digits as we wish.

Note also that to make the final result as accurate as possible, *you should wait until the last step to round off your answer.* If necessary, use the memory feature of your calculator to retain the results of intermediate calculations.

ABBREVIATIONS

cm	centimeter		**MHz**	megahertz
dB	decibel		**mi**	mile
F	farad		**min**	minute
ft	foot		**mL**	milliliter
g	gram		**mm**	millimeter
gal	gallon		**N**	Newton
h	hour		**qt**	quart
H	henry		**oz**	ounce
Hz	Hertz		**s**	second
in.	inch		**Ω**	ohm
J	Joule		**V**	volt
kcal	kilocalorie		**W**	watt
kg	kilogram		**yd**	yard
km	kilometer		**yr**	year
kPa	kilopascal		**°C**	degree Celsius
L	liter		**°F**	degree Fahrenheit
lb	pound		**K**	Kelvin
M	mole of solute		$\Rightarrow$	implies
	per liter of solution		$\Leftrightarrow$	is equivalent to
m	meter			

CONTENTS

1 **FUNDAMENTALS**   **2**

1.1   Real Numbers   3
1.2   Exponents and Radicals   14
1.3   Algebraic Expressions   26
1.4   Fractional Expressions   36
1.5   Equations   44
1.6   Problem Solving with Equations   58
1.7   Inequalities   71
1.8   Coordinate Geometry   80
1.9   Graphing Calculators and Computers   95
1.10   Lines   102
          Chapter 1 Review   116
          Chapter 1 Test   120

■   **PRINCIPLES OF PROBLEM SOLVING**   122

2 **FUNCTIONS**   **130**

2.1   What Is a Function?   131
2.2   Graphs of Functions   139
2.3   Applied Functions   152
2.4   Transformations of Functions   162
2.5   Extreme Values of Functions   174
2.6   Combining Functions   187
2.7   One-to-One Functions and Their Inverses   195
          Chapter 2 Review   203
          Chapter 2 Test   208

■   **PRINCIPLES OF MODELING**   210

3 **POLYNOMIALS AND RATIONAL FUNCTIONS   220**

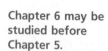

3.1 Polynomial Functions and Their Graphs   221
3.2 Real Zeros of Polynomials   233
3.3 Complex Numbers   248
3.4 Complex Roots and The Fundamental Theorem of Algebra   255
3.5 Rational Functions   262
Chapter 3 Review   277
Chapter 3 Test   280

FOCUS ON PROBLEM SOLVING   281

4 **EXPONENTIAL AND LOGARITHMIC FUNCTIONS   286**

4.1 Exponential Functions   287
4.2 The Natural Exponential Function   294
4.3 Logarithmic Functions   304
4.4 Laws of Logarithms   313
4.5 Exponential and Logarithmic Equations   318
4.6 Applications of Exponential and Logarithmic Functions   325
Chapter 4 Review   338
Chapter 4 Test   341

FOCUS ON MODELING   342

Chapter 6 may be
studied before
Chapter 5.

5 **TRIGONOMETRIC FUNCTIONS OF REAL NUMBERS   350**

5.1 The Unit Circle   351
5.2 Trigonometric Functions of Real Numbers   359
5.3 Trigonometric Graphs   368
5.4 More Trigonometric Graphs   382
Chapter 5 Review   389
Chapter 5 Test   392

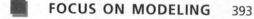

FOCUS ON MODELING   393

Chapter 5 may be
studied after
Chapter 6.

6   TRIGONOMETRIC FUNCTIONS OF ANGLES   406

6.1   Angle Measure   407
6.2   Trigonometry of Right Triangles   415
6.3   Trigonometric Functions of Angles   426
6.4   The Law of Sines   435
6.5   The Law of Cosines   442
      Chapter 6 Review   449
      Chapter 6 Test   453

FOCUS ON PROBLEM SOLVING   455

7   ANALYTIC TRIGONOMETRY   460

7.1   Trigonometric Identities   461
7.2   Addition and Subtraction Formulas   468
7.3   Double-Angle, Half-Angle, and Product-Sum Formulas   475
7.4   Inverse Trigonometric Functions   484
7.5   Trigonometric Equations   493
7.6   Trigonometric Form of Complex Numbers; DeMoivre's Theorem   500
7.7   Vectors   508
      Chapter 7 Review   518
      Chapter 7 Test   522

FOCUS ON PROBLEM SOLVING   524

8   SYSTEMS OF EQUATIONS AND INEQUALITIES   530

8.1   Systems of Equations   531
8.2   Pairs of Lines   539
8.3   Systems of Linear Equations   546
8.4   The Algebra of Matrices   560
8.5   Inverses of Matrices and Matrix Equations   568
8.6   Determinants and Cramer's Rule   577
8.7   Systems of Inequalities   587

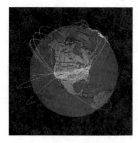

8.8   Partial Fractions   591
          Chapter 8 Review   597
          Chapter 8 Test   600

■   **FOCUS ON MODELING**   602

9 | TOPICS IN ANALYTIC GEOMETRY   610

9.1   Parabolas   611
9.2   Ellipses   619
9.3   Hyperbolas   628
9.4   Shifted Conics   636
9.5   Rotation of Axes   644
9.6   Polar Coordinates   650
9.7   Polar Equations of Conics   661
9.8   Parametric Equations   666
          Chapter 9 Review   675
          Chapter 9 Test   679

■   **FOCUS ON MODELING**   680

10 | SEQUENCES AND SERIES   684

10.1   Sequences and Summation Notation   685
10.2   Arithmetic Sequences   694
10.3   Geometric Sequences   700
10.4   Annuities and Installment Buying   708
10.5   Mathematical Induction   714
10.6   The Binomial Theorem   721
          Chapter 10 Review   731
          Chapter 10 Test   734

■   **FOCUS ON PROBLEM SOLVING**   735

11 COUNTING AND PROBABILITY   740

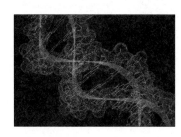

11.1 Counting Principles   741
11.2 Permutations and Combinations   746
11.3 Probability   754
11.4 Expected Value   766
Chapter 11 Review   769
Chapter 11 Test   772

FOCUS ON MODELING   773

ANSWERS TO ODD-NUMBERED EXERCISES AND CHAPTER TESTS   A1

INDEX   A61

PRECALCULUS
Mathematics for Calculus

1 FUNDAMENTALS

For the ancient Egyptian priests, calculating the volume of a pyramid was a complicated process because the language of algebra had not yet been invented. With algebraic notation, however, we can easily find the volume of a pyramid using the formula $V = \frac{1}{3}b^2h$.

One learns by doing the thing; for though you think you know it, you have no certainty until you try.

SOPHOCLES

In this first chapter we review the basic ideas from algebra and coordinate geometry that will be needed throughout this book.

Diophantus lived in Alexandria about 250 A.D. He wrote a book called *Arithmetica,* which is considered the first book on algebra. It contains methods for finding integer solutions of algebraic equations. The *Arithmetica* continued to be read for more than a thousand years. Fermat (see page 458) made some of his most important discoveries while studying this book. Diophantus' major contribution is the use of symbols to stand for the unknowns in a problem. Although his symbolism is not as simple as that used today, it was a major advance over writing everything in words. In Diophantus' notation the equation

$$x^5 - 7x^2 + 8x - 5 = 24$$

is written

$$\Delta K^\gamma \alpha \zeta \theta \;\; \Delta^\gamma \eta \dot{M} \varepsilon \iota \kappa \delta$$

Our modern algebraic notation did not come into common use until the 17th century.

1.1 REAL NUMBERS

Let's recall the types of numbers that make up the real number system. We start with the **natural numbers**:

$$1, 2, 3, 4, \ldots$$

The **integers** consist of the natural numbers together with their negatives and 0:

$$\ldots, -3, -2, -1, 0, 1, 2, 3, 4, \ldots$$

We construct the **rational numbers** by taking ratios of integers. Thus, any rational number r can be expressed as

$$r = \frac{m}{n} \qquad \text{where } m \text{ and } n \text{ are integers and } n \neq 0$$

Examples are

$$\tfrac{1}{2} \qquad -\tfrac{3}{7} \qquad 46 = \tfrac{46}{1} \qquad 0.17 = \tfrac{17}{100}$$

(Recall that division by 0 is always ruled out, so expressions like $\frac{3}{0}$ and $\frac{0}{0}$ are undefined.) There are also real numbers, such as $\sqrt{2}$, that cannot be expressed as a ratio of integers and are therefore called **irrational numbers**. It can be shown, with varying degrees of difficulty, that each of the following numbers is also an irrational number:

$$\sqrt{3} \qquad \sqrt{5} \qquad \sqrt[3]{2} \qquad \pi \qquad \frac{3}{\pi^2}$$

The set of all real numbers is usually denoted by the symbol $\mathbb{R}$. When we use the word *number* without qualification, we will mean "real number." Figure 1 is a diagram of the types of real numbers that we work with in this book.

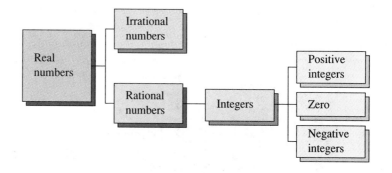

FIGURE 1

The real number system

Every real number has a decimal representation. If the number is rational, then its corresponding decimal is repeating. For example,

$$\tfrac{1}{2} = 0.5000\ldots = 0.5\overline{0} \qquad\qquad \tfrac{2}{3} = 0.66666\ldots = 0.\overline{6}$$

$$\tfrac{157}{495} = 0.3171717\ldots = 0.3\overline{17} \qquad \tfrac{9}{7} = 1.285714285714\ldots = 1.\overline{285714}$$

A repeating decimal such as

$$x = 3.5474747\ldots$$

is a rational number. To convert it to a ratio of two integers, we write:

$$1000x = 3547.47474747\ldots$$
$$\underline{10x = 35.47474747\ldots}$$
$$990x = 3512.0$$

Thus $x = \tfrac{3512}{990}$. (The idea is to multiply x by appropriate powers of 10, and then subtract to eliminate the repeating part.)

(The bar indicates that the sequence of digits repeats forever.) If the number is irrational, the decimal representation is nonrepeating:

$$\sqrt{2} = 1.414213562373095\ldots \qquad \pi = 3.141592653589793\ldots$$

If we stop the decimal expansion of any number at a certain place, we get an approximation to the number. For instance, we can write

$$\pi \approx 3.14159265$$

where the symbol $\approx$ is read "is approximately equal to." The more decimal places we retain, the better the approximation we get.

The real numbers can be represented by points on a line, as shown in Figure 2. The positive direction (toward the right) is indicated by an arrow. We choose an arbitrary reference point O, called the **origin**, which corresponds to the real number 0. Given any convenient unit of measurement, each positive number x is represented by the point on the line a distance of x units to the right of the origin, and each negative number $-x$ is represented by the point x units to the left of the origin. Thus, every real number is represented by a point on the line, and every point P on the line corresponds to exactly one real number. The number associated with the point P is called the coordinate of P, and the line is then called a **coordinate line**, or a **real number line**, or simply a **real line**. Often we identify the point with its coordinate and think of a number as being a point on the real line.

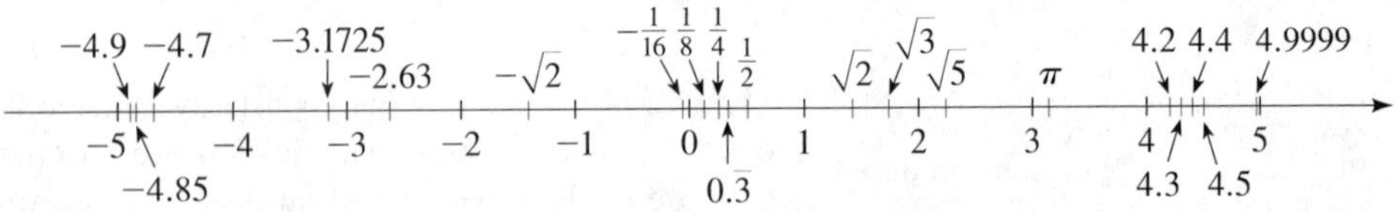

FIGURE 2
The real line

The real numbers are *ordered*. We say that **a is less than b** and write $a < b$ if $b - a$ is a positive number. Geometrically, this means that a lies to the left of b on the number line. (Equivalently, we can say that b is greater than a and write $b > a$.) The symbol $a \le b$ (or $b \ge a$) means that either $a < b$ or $a = b$ and is read "a is less than or equal to b." For instance, the following are true inequalities (see Figure 3):

$$7 < 7.4 < 7.5 \qquad -\pi < -3 \qquad \sqrt{2} < 2 \qquad 2 \le 2$$

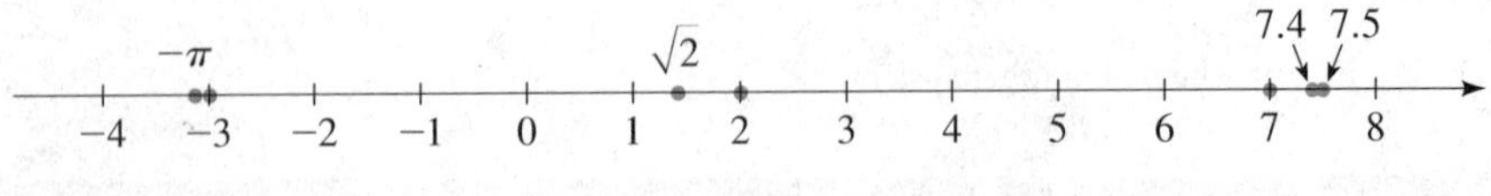

FIGURE 3

■ PROPERTIES OF REAL NUMBERS

In combining real numbers using the familiar operations of addition and multiplication, we use the following properties of real numbers.

PROPERTIES OF REAL NUMBERS

Property	Example	Name and Description
$a + b = b + a$	$7 + 3 = 3 + 7$	**Commutative Property for addition** When we add two numbers, order doesn't matter.
$ab = ba$	$3 \cdot 5 = 5 \cdot 3$	**Commutative Property for multiplication** When we multiply two numbers, order doesn't matter.
$(a + b) + c = a + (b + c)$	$(2 + 4) + 7 = 2 + (4 + 7)$	**Associative Property for addition** When we add three numbers, it doesn't matter which two we add first.
$(ab)c = a(bc)$	$(3 \cdot 7) \cdot 5 = 3 \cdot (7 \cdot 5)$	**Associative Property for multiplication** When we multiply three numbers, it doesn't matter which two we multiply first.
$a(b + c) = ab + ac$ $(b + c)a = ab + ac$	$2 \cdot (3 + 5) = 2 \cdot 3 + 2 \cdot 5$ $(3 + 5) \cdot 2 = 2 \cdot 3 + 2 \cdot 5$	**Distributive Property** When we multiply a number by a sum of two numbers, we get the same result as multiplying the number by each of the terms and then adding the results.

From our experience with numbers we know intuitively that these properties are valid. To reacquaint yourself with these properties, calculate the expressions on both sides of the equal sign in each of the examples in the table. The operations in the parentheses are to be performed first. So, for example,

$$(2 + 4) + 7 = 6 + 7 = 13 \quad \text{and} \quad 2 + (4 + 7) = 2 + 11 = 13$$

The Distributive Property is crucial in algebra because it describes the way addition and multiplication interact with each other. It applies whenever we multiply a number by a sum. Figure 4 explains why this property works for the case in which all the numbers are positive integers, but the property is true for any real numbers a, b, and c.

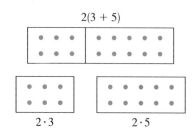

$2(3 + 5)$

$2 \cdot 3$ $2 \cdot 5$

FIGURE 4

The Distributive Property

EXAMPLE 1 ■ Using the Properties of the Real Numbers

Let x, y, z, and w be real numbers.

(a) $(x + y)(2zw) = (2zw)(x + y)$ Commutative Property for multiplication

(b) $(x + y)(z + w) = (x + y)z + (x + y)w$ Distributive Property
(with $a = x + y$)

$$= (zx + zy) + (wx + wy) \quad \text{Distributive Property}$$

$$= zx + zy + wx + wy \quad \text{Associative Property of addition}$$

In the last step we removed the parentheses because, according to the Associative Property, the order of addition doesn't matter. ■

The number 0 is special for addition; it is called the **additive identity** because $a + 0 = a$ for any real number a. Every real number a has a **negative**, $-a$, that satisfies $a + (-a) = 0$. **Subtraction** is the operation that undoes addition; to subtract a number from another, we simply add the negative of that number. By definition,

$$a - b = a + (-b)$$

To combine real numbers involving negatives, we use the following properties.

⊘ Don't make the mistake of assuming that $-a$ is a negative number. Whether $-a$ is negative or positive depends on the value of a. For example, if $a = 5$, then $-a = -5$, a negative number; but if $a = -5$, then $-a = -(-5) = 5$ (Property 2), a positive number.

PROPERTIES OF NEGATIVES

Property	Example
1. $(-1)a = -a$	$(-1)5 = -5$
2. $-(-a) = a$	$-(-5) = 5$
3. $(-a)b = a(-b) = -(ab)$	$(-5)7 = 5(-7) = -(5 \cdot 7)$
4. $(-a)(-b) = ab$	$(-4)(-3) = 4 \cdot 3$
5. $-(a + b) = -a - b$	$-(3 + 5) = -3 - 5$
6. $-(a - b) = b - a$	$-(5 - 8) = 8 - 5$

Property 6 states the intuitive fact that $a - b$ and $b - a$ are negatives of each other. Property 5 is often used with more than two terms:

$$-(a + b + c) = -a - b - c$$

EXAMPLE 2 ■ Using Properties of Negatives

Let x, y, and z be real numbers.

(a) $-(x + 2) = -x - 2$ Property 5

(b) $-(x + y - z) = -x - y - (-z)$ Property 5

$$= -x - y + z \quad \text{Property 2} \qquad ■$$

The number 1 is special for multiplication; it is called the **multiplicative identity** because $a \cdot 1 = a$ for any real number a. Every nonzero real number a has an **inverse**, $1/a$, that satisfies $a \cdot (1/a) = 1$. **Division** is the operation that undoes multiplication; to divide by a number, we multiply by the inverse of that number.

If $b \neq 0$, then, by definition,

$$a \div b = a \cdot \frac{1}{b}$$

We write $a \cdot (1/b)$ as simply a/b. We refer to a/b as the **quotient** of a and b or as the **fraction** a over b; a is the **numerator** and b is the **denominator** (or **divisor**). To combine real numbers using the operation of division, we use the following properties.

PROPERTIES OF FRACTIONS		
Property	Example	Description
1. $\dfrac{a}{b} \cdot \dfrac{c}{d} = \dfrac{ac}{bd}$	$\dfrac{2}{3} \cdot \dfrac{5}{7} = \dfrac{2 \cdot 5}{3 \cdot 7} = \dfrac{10}{21}$	To multiply fractions, multiply numerators and denominators.
2. $\dfrac{a}{b} \div \dfrac{c}{d} = \dfrac{a}{b} \cdot \dfrac{d}{c}$	$\dfrac{2}{3} \div \dfrac{5}{7} = \dfrac{2}{3} \cdot \dfrac{7}{5} = \dfrac{14}{15}$	To divide fractions, invert the divisor and multiply.
3. $\dfrac{a}{c} + \dfrac{b}{c} = \dfrac{a + b}{c}$	$\dfrac{2}{5} + \dfrac{7}{5} = \dfrac{2 + 7}{5} = \dfrac{9}{5}$	To add fractions having the same denominator, add the numerators.
4. $\dfrac{a}{b} + \dfrac{c}{d} = \dfrac{ad + bc}{bd}$	$\dfrac{2}{5} + \dfrac{3}{7} = \dfrac{2 \cdot 7 + 3 \cdot 5}{35} = \dfrac{29}{35}$	To add fractions having different denominators, find a common denominator. Then add the numerators.
5. $\dfrac{ac}{bc} = \dfrac{a}{b}$	$\dfrac{2 \cdot 5}{3 \cdot 5} = \dfrac{2}{3}$	Cancel numbers that are common factors in numerator and denominator.
6. If $\dfrac{a}{b} = \dfrac{c}{d}$, then $ad = bc$.	$\dfrac{2}{3} = \dfrac{6}{9}$, so $2 \cdot 9 = 3 \cdot 6$	Cross multiply.

When adding fractions with different denominators, we don't usually use Property 4. Instead we rewrite the fractions so that they have a common denominator (often smaller than the product of the denominators), and then we use Property 3. This denominator is the **Least Common Denominator** (LCD) described in the next example.

EXAMPLE 3 ■ Using the LCD to Add Fractions

Evaluate: $\dfrac{5}{36} + \dfrac{7}{120}$

SOLUTION   Factoring each denominator into prime factors gives

$$36 = 2^2 \cdot 3^2 \qquad \text{and} \qquad 120 = 2^3 \cdot 3 \cdot 5$$

We find the least common denominator (LCD) by forming the product of all the factors that occur in these factorizations, using the highest power of each factor. Thus, the LCD is $2^3 \cdot 3^2 \cdot 5 = 360$. So

$$\frac{5}{36} + \frac{7}{120} = \frac{5 \cdot 10}{36 \cdot 10} + \frac{7 \cdot 3}{120 \cdot 3} \qquad \text{Property 5}$$

$$= \frac{50}{360} + \frac{21}{360}$$

$$= \frac{71}{360} \qquad \text{Property 3} \qquad \blacksquare$$

SETS AND INTERVALS

In the discussion that follows, we need to use set notation. A **set** is a collection of objects, and these objects are called the **elements** of the set. If S is a set, the notation $a \in S$ means that a is an element of S, and $b \notin S$ means that b is not an element of S. For example, if Z represents the set of integers, then $-3 \in Z$ but $\pi \notin Z$.

Some sets can be described by listing their elements within braces. For instance, the set A that consists of all positive integers less than 7 can be written as

$$A = \{1, 2, 3, 4, 5, 6\}$$

We could also write A in **set-builder notation** as

$$A = \{x \mid x \text{ is an integer and } 0 < x < 7\}$$

which is read "A is the set of all x such that x is an integer and $0 < x < 7$."

If S and T are sets, then their **union** $S \cup T$ is the set that consists of all elements that are in S *or* T (or in both). The **intersection** of S and T is the set $S \cap T$ consisting of all elements that are in both S *and* T. In other words, $S \cap T$ is the common part of S and T. The **empty set**, denoted by $\varnothing$, is the set that contains no element.

EXAMPLE 4 ■ Union and Intersection of Sets

If $S = \{1, 2, 3, 4, 5\}$, $T = \{4, 5, 6, 7\}$, and $V = \{6, 7, 8\}$, find the sets $S \cup T$, $S \cap T$, and $S \cap V$.

SOLUTION

$$S \cup T = \{1, 2, 3, 4, 5, 6, 7\} \qquad \text{All elements in } S \text{ or } T$$

$$S \cap T = \{4, 5\} \qquad \text{Elements common to both } S \text{ and } T$$

$$S \cap V = \varnothing \qquad S \text{ and } V \text{ have no element in common} \qquad \blacksquare$$

$$\overbrace{1, 2, 3, \underbrace{4, 5, }_{S} \underbrace{6, 7, }^{T} 8}_{V}$$

Certain sets of real numbers, called **intervals**, occur frequently in calculus and correspond geometrically to line segments. For example, if $a < b$, then the **open interval** from a to b consists of all numbers between a and b and is denoted

FIGURE 5

The open interval (a, b)

FIGURE 6

The closed interval $[a, b]$

NO SMALLEST OR LARGEST NUMBER IN AN OPEN INTERVAL

Any interval contains infinitely many numbers—every point on the graph of an interval corresponds to a real number. In the closed interval $[0, 1]$, the smallest number is 0 and the largest is 1, but the open interval $(0, 1)$ contains no smallest or largest number. To see this, note that 0.01 is close to zero, but 0.001 is closer, 0.0001 closer yet, and so on. So we can always find a number in the interval $(0, 1)$ closer to zero than any given number. Since 0 itself is not in the interval, the interval contains no smallest number. Similarly, 0.99 is close to 1, but 0.999 is closer, 0.9999 closer yet, and so on. Since 1 itself is not in the interval, the interval has no largest number.

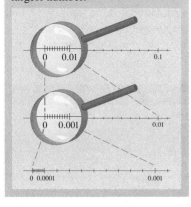

---

by the symbol (a, b). Using set-builder notation, we can write

$$(a, b) = \{x \mid a < x < b\}$$

Note that the endpoints, a and b, are excluded from this interval. This fact is indicated by the parentheses ( ) in the interval notation and the open circles on the graph of the interval in Figure 5.

The **closed interval** from a to b is the set

$$[a, b] = \{x \mid a \leq x \leq b\}$$

Here the endpoints of the interval are included. This is indicated by the square brackets [ ] in the interval notation and the solid circles on the graph of the interval in Figure 6. It is also possible to include only one endpoint in an interval, as shown in the table of intervals below.

We also need to consider infinite intervals, such as

$$(a, \infty) = \{x \mid x > a\}$$

This does not mean that ∞ ("infinity") is a number. The notation (a, ∞) stands for the set of all numbers that are greater than a, so the symbol ∞ simply indicates that the interval extends indefinitely far in the positive direction.

The following table lists the nine possible types of intervals. When these intervals are discussed, we will always assume that $a < b$.

Notation	Set description	Graph
(a, b)	$\{x \mid a < x < b\}$	
$[a, b]$	$\{x \mid a \leq x \leq b\}$	
$[a, b)$	$\{x \mid a \leq x < b\}$	
$(a, b]$	$\{x \mid a < x \leq b\}$	
(a, ∞)	$\{x \mid a < x\}$	
$[a, \infty)$	$\{x \mid a \leq x\}$	
$(-\infty, b)$	$\{x \mid x < b\}$	
$(-\infty, b]$	$\{x \mid x \leq b\}$	
$(-\infty, \infty)$	$\mathbb{R}$ (set of all real numbers)	

EXAMPLE 5 ■ Graphing Intervals

Express each interval in terms of inequalities, and then graph the interval.

(a) $[-1, 2) = \{x \mid -1 \leq x < 2\}$

(b) $[1.5, 4] = \{x \mid 1.5 \leq x \leq 4\}$

(c) $(-3, \infty) = \{x \mid -3 < x\}$

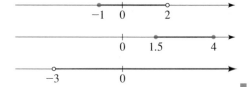

EXAMPLE 6 ■ Finding Unions and Intersections of Intervals

Graph each set.

(a) $(1, 3) \cap [2, 7]$ (b) $(-2, -1) \cup (1, 2)$

SOLUTION

(a) The intersection of two intervals consists of the numbers that are in both intervals. Therefore

$$(1, 3) \cap [2, 7] = \{x \mid 1 < x < 3 \quad \text{and} \quad 2 \leqslant x \leqslant 7\}$$
$$= \{x \mid 2 \leqslant x < 3\}$$
$$= [2, 3)$$

This set is illustrated in Figure 7.

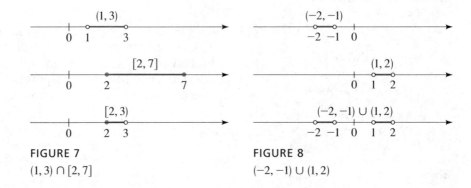

FIGURE 7
$(1, 3) \cap [2, 7]$

FIGURE 8
$(-2, -1) \cup (1, 2)$

(b) The union of the intervals $(-2, -1)$ and $(1, 2)$ consists of the numbers that are in either $(-2, -1)$ or $(1, 2)$, so

$$(-2, -1) \cup (1, 2) = \{x \mid -2 < x < -1 \quad \text{or} \quad 1 < x < 2\}$$

This set is illustrated in Figure 8. ■

ABSOLUTE VALUE AND DISTANCE

The **absolute value** of a number a, denoted by $|a|$, is the distance from a to 0 on the real number line (see Figure 9). Distance is always positive or zero, so we have $|a| \geqslant 0$ for every number a. Remembering that $-a$ is positive when a is negative, we have the following definition.

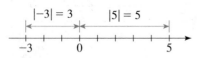

FIGURE 9

> **DEFINITION OF ABSOLUTE VALUE**
>
> If a is a real number, then the **absolute value** of a is
>
> $$|a| = \begin{cases} a & \text{if } a \geqslant 0 \\ -a & \text{if } a < 0 \end{cases}$$

EXAMPLE 7 ■ Evaluating Absolute Values

(a) $|3| = 3$

(b) $|-3| = -(-3) = 3$

(c) $|3 - \pi| = -(3 - \pi) = \pi - 3$ (since $3 < \pi \Rightarrow 3 - \pi < 0$)

(d) If $x \geq 4$, then $|x - 4| = x - 4$ (since in this case $x - 4 \geq 0$)

(e) If $x < 4$, then $|x - 4| = -(x - 4) = 4 - x$ (since in this case
$$x - 4 < 0)$$ ■

The following properties can be proved using the definition of absolute value.

PROPERTIES OF ABSOLUTE VALUE	
Property	Description
1. $|a| = |-a|$	A number and its negative have the same absolute value.
2. $|ab| = |a||b|$	The absolute value of a product is the product of the absolute values.
3. $\left| \dfrac{a}{b} \right| = \dfrac{|a|}{|b|}$	The absolute value of a quotient is the quotient of the absolute values.
4. $|a^n| = |a|^n$	The absolute value of a power is the power of the absolute value.

EXAMPLE 8 ■ Simplifying the Absolute Value of an Expression

(a) $|-4 - x^2| = |-(4 + x^2)| = |4 + x^2| = 4 + x^2$ Property 1 and the fact that $4 + x^2 > 0$

(b) $|2x - 6| = |2(x - 3)| = 2|x - 3|$ Property 2   ■

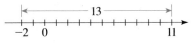

−2   0   11

FIGURE 10

What is the distance on the real line between the numbers -2 and 11? From Figure 10 we see that the distance is 13. We arrive at this by finding either $|11 - (-2)| = 13$ or $|(-2) - 11| = 13$. From this observation we make the following definition (see Figure 11).

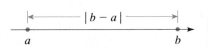

a b

FIGURE 11

Length of a line segment $= |b - a|$

DISTANCE BETWEEN POINTS ON THE REAL LINE
If a and b are real numbers, then the **distance** between the points a and b on the real line is $$d(a, b) = |b - a|$$

From Property 6 of negatives it follows that

$$|b - a| = |a - b|$$

This confirms that, as we would expect, the distance from a to b is the same as the distance from b to a.

EXAMPLE 9 ■ Distance between Points on the Real Line

The distance between the numbers -8 and 2 is

$$d(a, b) = |-8 - 2| = |-10| = 10$$

We can check this calculation geometrically, as shown in Figure 12. ■

FIGURE 12

| 1.1 | **EXERCISES** |

1–8 ■ State the property of real numbers being used.

1. $2x + y = y + 2x$

2. $c(a + b) = (a + b)c$

3. $(x + y) + 5z = x + (y + 5z)$

4. $2(w + x) = 2w + 2x$

5. $3(5x + 1) = 15x + 3$

6. $(xy)S = x(yS)$

7. $(x + a)(x + b) = (x + a)x + (x + a)b$

8. $a(x + y + z) = ax + ay + az$

9–16 ■ Use properties of real numbers to write the expression without parentheses.

9. $3(x + y)$

10. $8(a - b)$

11. $4(2m)$

12. $\frac{1}{2}(10z)$

13. $\frac{4}{3}(-6y)$

14. $-2(r + s)$

15. $-\frac{5}{2}(2x - 4y)$

16. $(3a)(b + c - 2d)$

17–20 ■ Perform the indicated operations.

17. (a) $\frac{4}{13} + \frac{3}{13}$

(b) $\frac{3}{10} \cdot \frac{25}{27}$

18. (a) $\frac{7}{45} \cdot \frac{9}{10}$

(b) $\frac{5}{14} - \frac{1}{21} + 1$

19. (a) $\frac{2}{5} \div \frac{9}{10}$

(b) $\left(4 \div \frac{1}{2}\right) - \frac{1}{2}$

20. (a) $\left(\frac{1}{8} - \frac{1}{9}\right) \div \frac{1}{72}$

(b) $\left(2 \div \frac{2}{3}\right) - \left(\frac{2}{3} \div 2\right)$

21–24 ■ State whether each inequality is true or false.

21. (a) $-6 < -10$

(b) $\sqrt{2} > 1.41$

22. (a) $\frac{10}{11} < \frac{12}{13}$

(b) $8 \leq 8$

23. (a) $-\pi > -3$

(b) $8 \leq 9$

24. (a) $1.1 > 1.\overline{1}$

(b) $-\frac{1}{2} < -1$

25–26 ■ Write each statement in terms of inequalities.

25. (a) x is positive
(b) t is less than 4
(c) a is greater than or equal to π
(d) x is less than $\frac{1}{3}$ and is greater than -5
(e) The distance from p to 3 is at most 5

26. (a) y is negative
(b) z is greater than 1
(c) b is at most 8
(d) w is positive and is less than or equal to 17
(e) y is at least 2 units from π

27–30 ■ Find the indicated set if

$$A = \{1, 2, 3, 4, 5, 6\} \qquad B = \{2, 4, 6, 8\}$$
$$C = \{7, 8, 9, 10\}$$

27. (a) $A \cup B$

(b) $A \cap B$

28. (a) $B \cup C$

(b) $B \cap C$

29. (a) $A \cup C$

(b) $A \cap C$

30. (a) $A \cup B \cup C$

(b) $A \cap B \cap C$

31–32 ■ Find the indicated set if

$$A = \{x \mid x \geq -2\} \qquad B = \{x \mid x < 4\}$$
$$C = \{x \mid -1 < x \leq 5\}$$

31. (a) $B \cup C$ (b) $B \cap C$

32. (a) $A \cap C$ (b) $A \cap B$

33–38 ■ Express the interval in terms of inequalities, and then graph the interval.

33. $(-3, 0)$ **34.** $(2, 8]$

35. $[2, 8)$ **36.** $\left[-6, -\frac{1}{2}\right]$

37. $[2, \infty)$ **38.** $(-\infty, 1)$

39–44 ■ Express the inequality in interval notation, and then graph the corresponding interval.

39. $x \leq 1$ **40.** $1 \leq x \leq 2$

41. $-2 < x \leq 1$ **42.** $x \geq -5$

43. $x > -1$ **44.** $-5 < x < 2$

45–50 ■ Graph the set.

45. $(-2, 0) \cup (-1, 1)$ **46.** $(-2, 0) \cap (-1, 1)$

47. $[-4, 6] \cap [0, 8)$ **48.** $[-4, 6) \cup [0, 8)$

49. $(-\infty, -4) \cup (4, \infty)$ **50.** $(-\infty, 6] \cap (2, 10)$

51–54 ■ Evaluate each expression.

51. (a) $|100|$ (b) $|-73|$

52. (a) $|-8 - (-2)|$ (b) $|\pi - 10|$

53. (a) $||-6| - |-4||$ (b) $\dfrac{-1}{|-1|}$

54. (a) $|2 - |-12||$ (b) $-1 - |1 - |-1||$

55–60 ■ Use the properties of absolute value (as in Example 8) to simplify the expression.

55. $|3x + 9|$ **56.** $|4x - 16|$

57. $\left|\frac{1}{2}x - \frac{5}{2}\right|$ **58.** $|-2x - 10|$

59. $|-x^2 - 9|$ **60.** $\left|\dfrac{x - 1}{1 - x}\right|$

61–62 ■ Find the distance between the given numbers.

61. (a) 2 and 17   **62.** (a) $\frac{7}{15}$ and $-\frac{1}{21}$
    (b) -3 and 21   (b) -38 and -57
    (c) $\frac{11}{8}$ and $-\frac{3}{10}$ (c) -2.6 and -1.8

63–64 ■ Express each repeating decimal as a fraction. (See the margin note on page 4.)

63. (a) $0.\overline{7}$ (b) $0.\overline{28}$ (c) $0.\overline{57}$

64. (a) $5.\overline{23}$ (b) $1.3\overline{7}$ (c) $2.1\overline{35}$

● DISCOVERY · DISCUSSION

65. Sums and Products of Rational and Irrational Numbers Explain why the sum, the difference, and the product of two rational numbers are again rational numbers. Is the product of two irrational numbers again irrational? What about the sum?

66. Limiting Behavior of Reciprocals Complete the tables. What happens to the size of the fraction $\dfrac{1}{x}$ as x gets large? As x gets small?

x	$\dfrac{1}{x}$		x	$\dfrac{1}{x}$
1			1.0	
2			0.5	
10			0.1	
100			0.01	
1000			0.001	

67. Irrational Numbers and Geometry Using the following figure, explain how to locate the point $\sqrt{2}$ on a number line. Can you locate $\sqrt{5}$ by a similar method? What about $\sqrt{6}$? List some other irrational numbers that can be located this way.

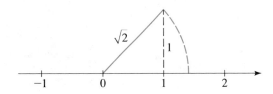

1.2 EXPONENTS AND RADICALS

In this section we give meaning to expressions such as $a^{m/n}$ in which the exponent m/n is a rational number. In order to do this, we need to recall some facts about integer exponents, radicals, and nth roots.

INTEGER EXPONENTS

A product of identical numbers is usually written in exponential notation. For example, $5 \cdot 5 \cdot 5$ is written as 5^3. In general, we have the following definition.

EXPONENTIAL NOTATION

If a is any real number and n is a positive integer, then the **nth power of a** is

$$a^n = \underbrace{a \cdot a \cdot \cdots \cdot a}_{n \text{ factors}}$$

The number a is called the **base** and n is called the **exponent**.

EXAMPLE 1 ■ Exponential Notation

(a) $\left(\frac{1}{2}\right)^5 = \left(\frac{1}{2}\right)\left(\frac{1}{2}\right)\left(\frac{1}{2}\right)\left(\frac{1}{2}\right)\left(\frac{1}{2}\right) = \frac{1}{32}$

(b) $(-3)^4 = (-3) \cdot (-3) \cdot (-3) \cdot (-3) = 81$

(c) $-3^4 = -(3 \cdot 3 \cdot 3 \cdot 3) = -81$ ■

⊘ Note the distinction between $(-3)^4$ and -3^4. In $(-3)^4$ the exponent applies to -3, but in -3^4 the exponent applies only to 3.

We can state several useful rules for working with exponential notation. To discover the rule for multiplication, we multiply 5^4 by 5^2:

$$5^4 \cdot 5^2 = \underbrace{(5 \cdot 5 \cdot 5 \cdot 5)}_{4 \text{ factors}}\underbrace{(5 \cdot 5)}_{2 \text{ factors}} = \underbrace{5 \cdot 5 \cdot 5 \cdot 5 \cdot 5 \cdot 5}_{6 \text{ factors}} = 5^6 = 5^{4+2}$$

It appears that *to multiply two powers of the same base, we add their exponents.* In general, we can confirm this by considering any real number a and any positive integers m and n:

$$a^m a^n = \underbrace{(a \cdot a \cdot \cdots \cdot a)}_{m \text{ factors}}\underbrace{(a \cdot a \cdot \cdots \cdot a)}_{n \text{ factors}} = \underbrace{a \cdot a \cdot a \cdot \cdots \cdot a}_{m + n \text{ factors}} = a^{m+n}$$

Thus, we have shown that

$$a^m a^n = a^{m+n}$$

whenever m and n are positive integers.

We would like this rule to be true even when m and n are 0 or negative integers. Thus, for instance, we must have

$$2^0 \cdot 2^3 = 2^{0+3} = 2^3$$

But this can happen only if $2^0 = 1$. Likewise, we want to have

$$5^4 \cdot 5^{-4} = 5^{4+(-4)} = 5^{4-4} = 5^0 = 1$$

and this will be true if $5^{-4} = 1/5^4$. These observations lead to the following definition.

ZERO AND NEGATIVE EXPONENTS

If $a \neq 0$ is any real number and n is a positive integer, then

$$\text{(a) } a^0 = 1 \qquad \text{(b) } a^{-n} = \frac{1}{a^n}$$

EXAMPLE 2 ■ **Zero and Negative Exponents**

(a) $\left(\frac{4}{7}\right)^0 = 1$

(b) $x^{-1} = \frac{1}{x^1} = \frac{1}{x}$

(c) $(-2)^{-3} = \frac{1}{(-2)^3} = \frac{1}{-8} = -\frac{1}{8}$ ■

Familiarity with the following rules is essential for our work with exponents and bases. In the table the bases a and b are real numbers, and the exponents m and n are integers.

LAWS OF EXPONENTS

Law	Description
1. $a^m a^n = a^{m+n}$	To multiply two powers of the same number, add the exponents.
2. $\dfrac{a^m}{a^n} = a^{m-n}$	To divide two powers of the same number, subtract the exponents.
3. $(a^m)^n = a^{mn}$	To raise a power to a new power, multiply the exponents.
4. $(ab)^n = a^n b^n$	To raise a product to a power, raise each factor to the power.
5. $\left(\dfrac{a}{b}\right)^n = \dfrac{a^n}{b^n}$	To raise a quotient to a power, raise both numerator and denominator to the power.

We have already seen how to prove Law 1 whenever m and n are positive integers, but it can be proved for any integer exponents by using the definition of negative exponents. We now give the proofs of Laws 3 and 4. The proofs of Laws 2 and 5 are requested in Exercise 77.

■ **Proof of Law 3** If m and n are positive integers, we have

$$(a^m)^n = \underbrace{(a \cdot a \cdot \cdots \cdot a)}_{m \text{ factors}}{}^n$$

$$= \underbrace{\underbrace{(a \cdot a \cdot \cdots \cdot a)}_{m \text{ factors}} \underbrace{(a \cdot a \cdot \cdots \cdot a)}_{m \text{ factors}} \cdots \underbrace{(a \cdot a \cdot \cdots \cdot a)}_{m \text{ factors}}}_{n \text{ groups of factors}}$$

$$= \underbrace{a \cdot a \cdot \cdots \cdot a}_{mn \text{ factors}} = a^{mn}$$

The cases for which $m \le 0$ or $n \le 0$ can be proved using the definition of negative exponents. □

■ **Proof of Law 4** For the case in which m and n are positive integers, we have

$$(ab)^n = \underbrace{(ab)(ab)\cdots(ab)}_{n \text{ factors}} = \underbrace{(a \cdot a \cdot \cdots \cdot a)}_{n \text{ factors}} \cdot \underbrace{(b \cdot b \cdot \cdots \cdot b)}_{n \text{ factors}} = a^n b^n$$

Here we have used the Commutative and Associative Properties repeatedly. If $m \le 0$ or $n \le 0$, Law 4 can be proved using the definition of negative exponents. □

EXAMPLE 3 ■ **Using Laws of Exponents**

(a) $x^4 x^7 = x^{4+7} = x^{11}$ Law 1

(b) $y^4 y^{-7} = y^{4-7} = y^{-3} = \dfrac{1}{y^3}$ Law 1

(c) $\dfrac{c^9}{c^5} = c^{9-5} = c^4$ Law 2

(d) $(b^4)^5 = b^{4 \cdot 5} = b^{20}$ Law 3

(e) $(3x)^3 = 3^3 x^3 = 27x^3$ Law 4

(f) $\left(\dfrac{x}{2}\right)^5 = \dfrac{x^5}{2^5} = \dfrac{x^5}{32}$ Law 5   ■

EXAMPLE 4 ■ **Simplifying Expressions with Exponents**

Simplify: (a) $(2a^3 b^2)(3ab^4)^3$ (b) $\left(\dfrac{x}{y}\right)^3 \left(\dfrac{y^2 x}{z}\right)^4$

SOLUTION

(a) $(2a^3b^2)(3ab^4)^3 = (2a^3b^2)[3^3a^3(b^4)^3]$ Law 4

$= (2a^3b^2)(27a^3b^{12})$ Law 3

$= (2)(27)a^3a^3b^2b^{12}$ Group together factors with the same base

$= 54a^6b^{14}$ Law 1

(b) $\left(\dfrac{x}{y}\right)^3\left(\dfrac{y^2x}{z}\right)^4 = \dfrac{x^3}{y^3}\dfrac{(y^2)^4x^4}{z^4}$ Laws 5 and 4

$= \dfrac{x^3}{y^3}\dfrac{y^8x^4}{z^4}$ Law 3

$= (x^3x^4)\left(\dfrac{y^8}{y^3}\right)\dfrac{1}{z^4}$ Group together factors with the same base

$= \dfrac{x^7y^5}{z^4}$ Laws 1 and 2   ■

When simplifying an expression, you will find that many different methods will lead to the same result; you should feel free to use any of the rules of exponents to arrive at your own method. We now give two additional laws that are useful in simplifying expressions with negative exponents.

LAWS OF EXPONENTS

6. $\left(\dfrac{a}{b}\right)^{-n} = \left(\dfrac{b}{a}\right)^n$ To raise a fraction to a negative power, invert the fraction and change the sign of the exponent.

7. $\dfrac{a^{-n}}{b^{-m}} = \dfrac{b^m}{a^n}$ To move a number raised to a power from numerator to denominator or from denominator to numerator, change the sign of the exponent.

■ **Proof of Law 7** Using first the definition of negative exponents and then Property 2 of fractions, we have

$$\frac{a^{-n}}{b^{-m}} = \frac{1/a^n}{1/b^m} = \frac{1}{a^n}\cdot\frac{b^m}{1} = \frac{b^m}{a^n}$$
□

The proof of Law 6 is requested in Exercise 77.

EXAMPLE 5 ■ Simplifying Expressions with Negative Exponents

Eliminate negative exponents and simplify each expression.

(a) $\dfrac{6st^{-4}}{2s^{-2}t^2}$ (b) $\left(\dfrac{y}{3z^2}\right)^{-2}$

SOLUTION

(a) We use Law 7, which allows us to move a number raised to a power from the numerator to the denominator (or vice versa) by changing the sign of the exponent.

$$\frac{6st^{-4}}{2s^{-2}t^2} = \frac{6ss^2}{2t^4t^2} \qquad \text{Law 7}$$

$$= \frac{3s^3}{t^6} \qquad \text{Law 1}$$

(b) We use Law 6, which allows us to change the sign of the exponent of a fraction by inverting the fraction.

$$\left(\frac{y}{3z^2}\right)^{-2} = \left(\frac{3z^2}{y}\right)^2 \qquad \text{Law 6}$$

$$= \frac{9z^4}{y^2} \qquad \text{Laws 5 and 4}$$

■

SCIENTIFIC NOTATION

Exponential notation is used by scientists as a compact way of writing very large numbers and very small numbers. For example, the nearest star beyond the sun, Proxima Centauri, is approximately 40,000,000,000,000 km away. The mass of a hydrogen atom is about 0.00000000000000000000000166 g. Scientists usually write such numbers in a more convenient way, called *scientific notation*. A positive number x is said to be written in **scientific notation** if it is expressed as follows:

$$x = a \times 10^n \qquad \text{where } 1 \le a < 10 \text{ and } n \text{ is an integer}$$

For instance, when we state that the distance to the star Proxima Centauri is 4×10^{13} km, the positive exponent 13 indicates that the decimal point should be moved 13 places to the *right*:

$$4 \times 10^{13} = 40,\underbrace{000,000,000,000}_{13 \text{ places}}$$

When we state that the mass of a hydrogen atom is 1.66×10^{-24} g, the exponent -24 indicates that the decimal point should be moved 24 places to the *left*:

$$1.66 \times 10^{-24} = 0.\underbrace{00000000000000000000000166}_{24 \text{ places}}$$

EXAMPLE 6 ■ **Writing Numbers in Scientific Notation**

(a) $56920 = 5.692 \times 10^4$ (b) $0.000093 = 9.3 \times 10^{-5}$ ■

To use scientific notation on a calculator, use a key labeled $\boxed{\text{EE}}$ or $\boxed{\text{EXP}}$ or $\boxed{\text{EEX}}$ to enter the exponent. For example, to enter the number 3.629×10^{15} on a TI-82 calculator, we enter

$$3.629 \quad \boxed{\text{2nd}} \quad \boxed{\text{EE}} \quad 15$$

and the display reads

$$\boxed{3.629\text{E}15}$$

Scientific notation is often used on a calculator to display a very large or very small number. For instance, if we use a calculator to square the number 1,111,111, the display panel may show (depending on the calculator model) the approximation

$$\boxed{1.234568 \quad 12} \quad \text{or} \quad \boxed{1.234568 \quad \text{E}12}$$

Here the final digits indicate the power of 10, and we interpret the result as

$$1.234568 \times 10^{12}$$

EXAMPLE 7 ■ **Using Scientific Notation**

If $a \approx 0.00046$, $b \approx 1.697 \times 10^{22}$, and $c \approx 2.91 \times 10^{-18}$, use a calculator to approximate the quotient ab/c.

SOLUTION   We could enter the data using scientific notation, or we could use laws of exponents as follows:

$$\frac{ab}{c} \approx \frac{(4.6 \times 10^{-4})(1.697 \times 10^{22})}{2.91 \times 10^{-18}}$$

$$= \frac{(4.6)(1.697)}{2.91} \times 10^{-4+22+18}$$

$$\approx 2.7 \times 10^{36}$$

We state the answer correct to two significant figures because the least accurate of the given numbers is stated to two significant figures.   ■

RADICALS

We know what 2^n means whenever n is an integer. To give meaning to a power, such as $2^{4/5}$, whose exponent is a rational number, we need to discuss radicals.

The symbol $\sqrt{}$ means "the positive square root of." Thus

$$\boxed{\sqrt{a} = b \quad \text{means} \quad b^2 = a \text{ and } b \geq 0}$$

It is true that the number 9 has two square roots, 3 and -3, but the notation $\sqrt{9}$ is reserved for the *positive* square root of 9 (sometimes called the *principal square root* of 9). If we want the negative root, we *must* write $-\sqrt{9}$, which is -3.

Since $a = b^2 \geq 0$, the symbol $\sqrt{a}$ makes sense only when $a \geq 0$. For instance,

$$\sqrt{9} = 3 \quad \text{because} \quad 3^2 = 9 \text{ and } 3 \geq 0$$

Square roots are special cases of nth roots. The nth root of x is the number that, when raised to the nth power, gives x.

DEFINITION OF _n_th ROOT

If n is any positive integer, then the **principal _n_th root** of a is defined as follows:

$$\sqrt[n]{a} = b \quad \text{means} \quad b^n = a$$

If n is even, we must have $a \geq 0$ and $b \geq 0$.

Thus

$$\sqrt[4]{81} = 3 \qquad \text{because} \qquad 3^4 = 81 \quad \text{and} \quad 3 \geq 0$$

$$\sqrt[3]{-8} = -2 \qquad \text{because} \qquad (-2)^3 = -8$$

But $\sqrt{-8}$, $\sqrt[4]{-8}$, and $\sqrt[6]{-8}$ are not defined. (For instance, $\sqrt{-8}$ is not defined because the square of every real number is nonnegative.) Odd roots are unique, but even roots are not.

The equation $x^5 = 31$ has only one real solution: $x = \sqrt[5]{31}$.

The equation $x^4 = 31$ has two real solutions: $x = \pm\sqrt[4]{31}$.

Notice that

$$\sqrt{4^2} = \sqrt{16} = 4 \qquad \text{but} \qquad \sqrt{(-4)^2} = \sqrt{16} = 4 = |-4|$$

Thus, the equation $\sqrt{a^2} = a$ is not always true; it is true only when $a \geq 0$. However, we can always write $\sqrt{a^2} = |a|$. This last equation is true not only for square roots, but for any even root. This and other rules used in working with nth roots are listed in the following box. In each property we assume that all the given roots exist.

PROPERTIES OF _n_th ROOTS

Property	Example
1. $\sqrt[n]{ab} = \sqrt[n]{a}\,\sqrt[n]{b}$	$\sqrt[3]{-8 \cdot 27} = \sqrt[3]{-8}\,\sqrt[3]{27} = (-2)(3) = 6$
2. $\sqrt[n]{\dfrac{a}{b}} = \dfrac{\sqrt[n]{a}}{\sqrt[n]{b}}$	$\sqrt[4]{\dfrac{16}{81}} = \dfrac{\sqrt[4]{16}}{\sqrt[4]{81}} = \dfrac{2}{3}$
3. $\sqrt[m]{\sqrt[n]{a}} = \sqrt[mn]{a}$	$\sqrt{\sqrt[3]{729}} = \sqrt[6]{729} = 3$
4. $\sqrt[n]{a^n} = a$ if n is odd	$\sqrt[3]{(-5)^3} = -5, \quad \sqrt[5]{2^5} = 2$
5. $\sqrt[n]{a^n} = |a|$ if n is even	$\sqrt[4]{(-3)^4} = |-3| = 3$

EXAMPLE 8 ■ **Using Properties of nth Roots**

(a) $\sqrt{20} \cdot \sqrt{5} = \sqrt{20 \cdot 5} = \sqrt{100} = 10$ Property 1

(b) $\dfrac{\sqrt[5]{64}}{\sqrt[5]{2}} = \sqrt[5]{\dfrac{64}{2}} = \sqrt[5]{32} = 2$ Property 2

(c) $\sqrt[6]{64} = \sqrt[3]{\sqrt{64}} = \sqrt[3]{8} = 2$ Property 3   ■

EXAMPLE 9 ■ **Simplifying Expressions Involving nth Roots**

(a) $\sqrt[3]{x^4} = \sqrt[3]{x^3 x}$ Factor out the largest cube

       $= \sqrt[3]{x^3} \sqrt[3]{x}$ Property 1

       $= x \sqrt[3]{x}$ Property 4

(b) $\sqrt[4]{81 x^8 y^4} = \sqrt[4]{81} \, \sqrt[4]{x^8} \, \sqrt[4]{y^4}$ Property 1

       $= 3 \sqrt[4]{(x^2)^4} \, |y|$ Property 5

       $= 3x^2 |y|$ Property 5   ■

It is frequently useful to combine like radicals in an expression such as $2\sqrt{3} + 5\sqrt{3}$. This can be done by using the Distributive Property. Thus

$$2\sqrt{3} + 5\sqrt{3} = (2 + 5)\sqrt{3} = 7\sqrt{3}$$

The next example further illustrates this process.

⊘ Avoid making the following error:
$$\sqrt{a + b} \neq \sqrt{a} + \sqrt{b}$$
For instance, if we let $a = 9$ and $b = 16$, then we see the error:
$$\sqrt{9 + 16} \overset{?}{=} \sqrt{9} + \sqrt{16}$$
$$\sqrt{25} \overset{?}{=} 3 + 4$$
$$5 \overset{?}{=} 7 \quad \text{Wrong!}$$

EXAMPLE 10 ■ **Combining Radicals**

(a) $10\sqrt[3]{4} + 7\sqrt[3]{4} - 2\sqrt[3]{4} = (10 + 7 - 2)\sqrt[3]{4}$ Distributive Property

       $= 15\sqrt[3]{4}$

(b) $\sqrt{32} + \sqrt{200} = \sqrt{16 \cdot 2} + \sqrt{100 \cdot 2}$ Factor out the largest squares

       $= \sqrt{16} \, \sqrt{2} + \sqrt{100} \, \sqrt{2}$ Property 1

       $= 4\sqrt{2} + 10\sqrt{2} = 14\sqrt{2}$ Distributive Property

(c) If $b > 0$, then

$$b\sqrt{25b} - \sqrt{b^3} = b\sqrt{25} \, \sqrt{b} - \sqrt{b^2} \, \sqrt{b} \quad \text{Property 1}$$

       $= 5b\sqrt{b} - b\sqrt{b}$ Property 5

       $= (5b - b)\sqrt{b}$ Distributive Property

       $= 4b\sqrt{b}$

(d) $\sqrt[3]{8a^8} + \sqrt[3]{27a^5} = \sqrt[3]{(8a^6)a^2} + \sqrt[3]{(27a^3)a^2}$ Factor out the largest cubes

       $= \sqrt[3]{8a^6} \, \sqrt[3]{a^2} + \sqrt[3]{27a^3} \, \sqrt[3]{a^2}$ Property 1

       $= 2a^2 \sqrt[3]{a^2} + 3a \sqrt[3]{a^2}$ Find the cube roots

       $= a\sqrt[3]{a^2} \, (2a + 3)$ Distributive Property   ■

RATIONAL EXPONENTS

To define what is meant by a *rational exponent* or, equivalently, a *fractional exponent* such as $a^{1/3}$, we need to use radicals. In order to give meaning to the symbol $a^{1/n}$ in a way that is consistent with the Laws of Exponents, we would have to have

$$(a^{1/n})^n = a^{(1/n)n} = a^1 = a$$

So, by the definition of nth root,

$$a^{1/n} = \sqrt[n]{a}$$

In general, we define rational exponents as follows.

DEFINITION OF RATIONAL EXPONENTS

For any rational exponent m/n in lowest terms, where m and n are integers and $n > 0$, we define

$$a^{m/n} = \left(\sqrt[n]{a}\right)^m$$

or, equivalently, $a^{m/n} = \sqrt[n]{a^m}$

If n is even, then we require that $a \geq 0$.

With this definition it can be proved that *the Laws of Exponents also hold for rational exponents.*

EXAMPLE 11 ■ Using the Definition of Rational Exponents

(a) $4^{1/2} = \sqrt{4} = 2$

(b) $64^{1/3} = \sqrt[3]{64} = 4$

(c) $4^{3/2} = (\sqrt{4})^3 = 2^3 = 8$ Alternate solution: $4^{3/2} = \sqrt{4^3} = \sqrt{64} = 8$

(d) $(125)^{-1/3} = \dfrac{1}{125^{1/3}} = \dfrac{1}{\sqrt[3]{125}} = \dfrac{1}{5}$

(e) $\dfrac{1}{\sqrt[3]{x^4}} = \dfrac{1}{x^{4/3}} = x^{-4/3}$ ■

EXAMPLE 12 ■ Using the Laws of Exponents with Rational Exponents

(a) $a^{1/3}a^{7/3} = a^{8/3}$ Law 1

(b) $\dfrac{a^{2/5}a^{7/5}}{a^{3/5}} = a^{2/5+7/5-3/5} = a^{6/5}$ Laws 1 and 2

(c) $(2a^3b^4)^{3/2} = 2^{3/2}(a^3)^{3/2}(b^4)^{3/2}$ Law 4

$\qquad\qquad = (\sqrt{2})^3 a^{3(3/2)} b^{4(3/2)}$ Law 3

$\qquad\qquad = 2\sqrt{2}\, a^{9/2} b^6$

(d) $\left(\dfrac{2x^{3/4}}{y^{1/3}}\right)^3 \left(\dfrac{y^4}{x^{-1/2}}\right) = \dfrac{2^3(x^{3/4})^3}{(y^{1/3})^3} \cdot (y^4 x^{1/2})$ Laws 5, 4, and 7

$\qquad\qquad\qquad\qquad = \dfrac{8x^{9/4}}{y} \cdot y^4 x^{1/2}$ Law 3

$\qquad\qquad\qquad\qquad = 8x^{11/4} y^3$ Laws 1 and 2   ∎

EXAMPLE 13 ∎ **Simplifying by Writing Radicals as Rational Exponents**

(a) $(2\sqrt{x})(3\sqrt[3]{x}) = (2x^{1/2})(3x^{1/3})$ Definition of rational exponents

$\qquad\qquad\qquad = 6x^{1/2+1/3} = 6x^{5/6}$ Law 1

(b) $\sqrt{x\sqrt{x}} = (xx^{1/2})^{1/2}$ Definition of rational exponents

$\qquad\qquad = (x^{3/2})^{1/2}$ Law 1

$\qquad\qquad = x^{3/4}$ Law 3   ∎

It is often useful for us to eliminate the radical in a denominator by multiplying both numerator and denominator by an appropriate expression. This procedure is called **rationalizing the denominator**. If the denominator is of the form $\sqrt{a}$, we multiply numerator and denominator by $\sqrt{a}$. In doing so we multiply the given quantity by 1, so we do not change its value. For instance,

$$\frac{1}{\sqrt{a}} = \frac{1}{\sqrt{a}} \cdot 1 = \frac{1}{\sqrt{a}} \cdot \frac{\sqrt{a}}{\sqrt{a}} = \frac{\sqrt{a}}{a}$$

Note that the denominator in the last fraction contains no radical. In general, if the denominator is of the form $\sqrt[n]{a^m}$ with $m < n$, then multiplying numerator and denominator by $\sqrt[n]{a^{n-m}}$ will rationalize the denominator, because (for $a > 0$)

$$\sqrt[n]{a^m}\,\sqrt[n]{a^{n-m}} = \sqrt[n]{a^{m+n-m}} = \sqrt[n]{a^n} = a$$

EXAMPLE 14 ∎ **Rationalizing Denominators**

(a) $\dfrac{2}{\sqrt{3}} = \dfrac{2}{\sqrt{3}} \cdot \dfrac{\sqrt{3}}{\sqrt{3}} = \dfrac{2\sqrt{3}}{3}$

(b) $\dfrac{1}{\sqrt[3]{x^2}} = \dfrac{1}{\sqrt[3]{x^2}} \dfrac{\sqrt[3]{x}}{\sqrt[3]{x}} = \dfrac{\sqrt[3]{x}}{\sqrt[3]{x^3}} = \dfrac{\sqrt[3]{x}}{x}$

(c) $\sqrt[7]{\dfrac{1}{a^2}} = \dfrac{1}{\sqrt[7]{a^2}} = \dfrac{1}{\sqrt[7]{a^2}} \dfrac{\sqrt[7]{a^5}}{\sqrt[7]{a^5}} = \dfrac{\sqrt[7]{a^5}}{\sqrt[7]{a^7}} = \dfrac{\sqrt[7]{a^5}}{a}$ ∎

1.2 EXERCISES

1–10 ■ Evaluate each of the given numbers.

1. (a) $(-2)^4$ (b) -2^4 (c) π^0

2. (a) $\left(\frac{1}{3}\right)^4 4^{-3}$ (b) $\left(\frac{1}{4}\right)^{-2}$ (c) $\left(\frac{4}{9}\right)^0 \cdot 2^{-3}$

3. (a) $2^{-3}5^4$ (b) $\dfrac{10^9}{10^4}$ (c) $(2^4 \cdot 2^2)^2$

4. (a) $\left(\frac{2}{3}\right)^{-1}$ (b) $\sqrt[3]{-64}$ (c) $\sqrt[5]{-32}$

5. (a) $\sqrt{\frac{4}{9}}$ (b) $\sqrt[4]{256}$ (c) $\sqrt[6]{\frac{1}{64}}$

6. (a) $\sqrt{7}\sqrt{28}$ (b) $\sqrt[3]{3}\sqrt[3]{9}$ (c) $\sqrt[4]{24}\sqrt[4]{54}$

7. (a) $\dfrac{\sqrt{72}}{\sqrt{2}}$ (b) $\dfrac{\sqrt{48}}{\sqrt{3}}$ (c) $\sqrt{\frac{9}{25}}$

8. (a) $9^{7/2}$ (b) $(-32)^{2/5}$ (c) $(-125)^{-1/3}$

9. (a) $\left(\frac{4}{9}\right)^{-1/2}$ (b) $\left(-\frac{27}{8}\right)^{2/3}$ (c) $\left(\frac{25}{64}\right)^{3/2}$

10. (a) $1024^{-0.1}$ (b) $3^{2/7}3^{5/7}$ (c) $3^{1/2}9^{1/4}$

11–14 ■ Simplify the expression.

11. $\sqrt[3]{108} - \sqrt[3]{32}$ **12.** $\sqrt{8} + \sqrt{50}$

13. $\sqrt{245} - \sqrt{125}$ **14.** $\sqrt[3]{54} - \sqrt[3]{16}$

15–32 ■ Simplify the expression and eliminate any negative exponent(s).

15. $t^7 t^{-2}$ **16.** $(4x^2)(6x^7)$

17. $(12x^2y^4)\left(\frac{1}{2}x^5y\right)$ **18.** $(6y)^3$

19. $\dfrac{x^9(2x)^4}{x^3}$ **20.** $\dfrac{a^{-3}b^4}{a^{-5}b^5}$

21. $b^4\left(\frac{1}{3}b^2\right)(12b^{-8})$ **22.** $(2s^3t^{-1})\left(\frac{1}{4}s^6\right)(16t^4)$

23. $(rs)^3(2s)^{-2}(4r)^4$ **24.** $(2u^2v^3)^3(3u^3v)^{-2}$

25. $\dfrac{(6y^3)^4}{2y^5}$ **26.** $\dfrac{(2x^3)^2(3x^4)}{(x^3)^4}$

27. $\dfrac{(x^2y^3)^4(xy^4)^{-3}}{x^2y}$ **28.** $\left(\dfrac{c^4d^3}{cd^2}\right)\left(\dfrac{d^2}{c^3}\right)^3$

29. $\dfrac{(xy^2z^3)^4}{(x^3y^2z)^3}$ **30.** $\left(\dfrac{xy^{-2}z^{-3}}{x^2y^3z^{-4}}\right)^{-3}$

31. $\left(\dfrac{q^{-1}rs^{-2}}{r^{-5}sq^{-8}}\right)^{-1}$ **32.** $(3ab^2c)\left(\dfrac{2a^2b}{c^3}\right)^{-2}$

33–48 ■ Simplify the expression and eliminate any negative exponent(s). Assume that all letters denote positive numbers.

33. $x^{2/3}x^{1/5}$ **34.** $(-2a^{3/4})(5a^{3/2})$

35. $(4b)^{1/2}(8b^{2/5})$ **36.** $(8x^6)^{-2/3}$

37. $(c^2d^3)^{-1/3}$ **38.** $(4x^6y^8)^{3/2}$

39. $(y^{3/4})^{2/3}$ **40.** $(a^{2/5})^{-3/4}$

41. $(2x^4y^{-4/5})^3(8y^2)^{2/3}$ **42.** $(x^{-5}y^3z^{10})^{-3/5}$

43. $\left(\dfrac{x^6y}{y^4}\right)^{5/2}$ **44.** $\left(\dfrac{-2x^{1/3}}{y^{1/2}z^{1/6}}\right)^4$

45. $\left(\dfrac{3a^{-2}}{4b^{-1/3}}\right)^{-1}$ **46.** $\dfrac{(y^{10}z^{-5})^{1/5}}{(y^{-2}z^3)^{1/3}}$

47. $\dfrac{(9st)^{3/2}}{(27s^3t^{-4})^{2/3}}$ **48.** $\left(\dfrac{a^2b^{-3}}{x^{-1}y^2}\right)^3\left(\dfrac{x^{-2}b^{-1}}{a^{3/2}y^{1/3}}\right)$

49–56 ■ Simplify the expression. Assume the letters denote any real numbers.

49. $\sqrt[4]{x^4}$ **50.** $\sqrt[3]{x^3y^6}$

51. $\sqrt[3]{x^3y}$ **52.** $\sqrt{x^4y^4}$

53. $\sqrt[5]{a^6b^7}$ **54.** $\sqrt[3]{a^2b}\sqrt[3]{a^4b}$

55. $\sqrt[3]{\sqrt{64x^6}}$ **56.** $\sqrt[4]{x^4y^2z^2}$

57–60 ■ Rationalize the denominator.

57. (a) $\dfrac{1}{\sqrt{6}}$ (b) $\sqrt{\dfrac{x}{3y}}$ (c) $\sqrt{\dfrac{3}{20}}$

58. (a) $\sqrt{\dfrac{x^5}{2}}$ (b) $\sqrt{\dfrac{2}{3}}$ (c) $\sqrt{\dfrac{1}{2x^3y^5}}$

59. (a) $\dfrac{1}{\sqrt[3]{x}}$ (b) $\dfrac{1}{\sqrt[5]{x^2}}$ (c) $\dfrac{1}{\sqrt[7]{x^3}}$

60. (a) $\dfrac{1}{\sqrt[3]{x^2}}$ (b) $\dfrac{1}{\sqrt[4]{x^3}}$ (c) $\dfrac{1}{\sqrt[3]{x^4}}$

61–62 ■ Write each number in scientific notation.

61. (a) 69,300,000
(b) 0.000028536
(c) 129,540,000

62. (a) 7,259,000,000
(b) 0.0000000014
(c) 0.0007029

63–64 ■ Write each number in ordinary decimal notation.

63. (a) 3.19×10^5
(b) 2.670×10^{-8}
(c) 7.1×10^{14}

64. (a) 8.55×10^{-3}
(b) 6×10^{12}
(c) 6.257×10^{-10}

65–66 ■ Write the number indicated in each statement in scientific notation.

65. (a) A light-year, the distance that light travels in one year, is about 5,900,000,000,000 mi.
(b) The diameter of an electron is about 0.0000000000004 cm.
(c) A drop of water contains more than 33 billion billion molecules.

66. (a) The distance from the earth to the sun is about 93 million miles.
(b) The mass of an oxygen molecule is about 0.000000000000000000000053 g.
(c) The mass of the earth is about 5,970,000,000,000,000,000,000,000 kg.

67–72 ■ Use scientific notation, Laws of Exponents, and a calculator to perform the indicated operations. State your answer correct to the number of significant digits indicated by the given data.

67. $(7.2 \times 10^{-9})(1.806 \times 10^{-12})$

68. $(1.062 \times 10^{24})(8.61 \times 10^{19})$

69. $\dfrac{1.295643 \times 10^9}{(3.610 \times 10^{-17})(2.511 \times 10^6)}$

70. $\dfrac{(73.1)(1.6341 \times 10^{28})}{0.0000000019}$

71. $\dfrac{(0.0000162)(0.01582)}{(594621000)(0.0058)}$

72. $\dfrac{(3.542 \times 10^{-6})^9}{(5.05 \times 10^4)^{12}}$

73. The speed of light is about 186,000 mi/s. Use the information in Exercise 66(a) to find how long it takes for a light ray from the sun to reach the earth.

74. Without using a calculator, determine which number is larger in each pair of numbers.
(a) $7^{1/4}$ and $4^{1/3}$ (b) $\sqrt[3]{5}$ and $\sqrt{3}$

75. Due to the curvature of the earth, the maximum distance D that you can see from the top of a tall building of height h is estimated by the formula

$$D = \sqrt{2rh + h^2}$$

where $r = 3960$ mi is the radius of the earth and D and h are also measured in miles. How far can you see from the observation deck of the Toronto CN Tower, 1135 ft above the ground?

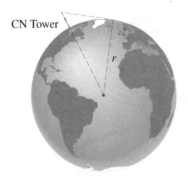

CN Tower

r

76. Police use the formula $s = \sqrt{30fd}$ to estimate the speed s (in mi/h) at which a car is traveling if it skids d feet after the brakes are applied suddenly. The number f is the coefficient of friction of the road, which is a measure of the "slipperiness" of the road. The table gives some typical estimates for f.

	Tar	Concrete	Gravel
Dry	1.0	0.8	0.2
Wet	0.5	0.4	0.1

(a) If a car skids 65 ft on wet concrete, how fast was it moving when the brakes were applied?
(b) If a car is traveling at 50 mi/h, how far will it skid on wet tar?

77. Prove each of the given Laws of Exponents for the case in which m and n are positive integers and $m > n$.

    (a) Law 2   **(b)** Law 5   **(c)** Law 6

DISCOVERY · DISCUSSION

78. Limiting Behavior of Powers   Complete the following tables. What happens to the nth root of 2 as n gets large? What about the nth root of $\frac{1}{2}$?

n	$2^{1/n}$
1	
2	
5	
10	
100	

n	$\left(\frac{1}{2}\right)^{1/n}$
1	
2	
5	
10	
100	

79. Distances between Powers   Which of the following pairs of numbers is closer together?

$$10^{10} \text{ and } 10^{50} \quad \text{or} \quad 10^{100} \text{ and } 10^{101}$$

80. Relativity   The theory of relativity states that as an object travels with velocity v, its rest mass m_0 changes to a mass m given by the formula

$$m = \frac{m_0}{\sqrt{1 - \dfrac{v^2}{c^2}}}$$

where $c \approx 3.0 \times 10^8$ m/s is the speed of light. By what factor is the rest mass of a spaceship multiplied if the ship travels at one-tenth the speed of light? At one-half the speed of light? At 90% of the speed of light? How does the mass of the spaceship change as it travels at a speed very close to the speed of light? Do we need to know the actual value of the speed of light to answer these questions?

1.3   ALGEBRAIC EXPRESSIONS

Algebraic expressions such as

$$2x^2 - 3x + 4 \qquad ax + b$$

$$\frac{y - 1}{y^2 + 2} \qquad \frac{cx^2y + dy^2z}{\sqrt{x^2 + y^2 + z^2}}$$

are obtained by starting with variables such as x, y, and z and constants such as 2, -3, a, b, c, and d, and combining them using addition, subtraction, multiplication, division, and roots. A **variable** is a letter that can represent any number in a given set of numbers, whereas a **constant** represents a fixed (or specific) number. The **domain** of a variable is the set of numbers that the variable is permitted to have. For instance, in the expression $\sqrt{x}$ the domain of x is $\{x \mid x \geq 0\}$, whereas in the expression $2/(x - 3)$ the domain of x is $\{x \mid x \neq 3\}$.

    The simplest types of algebraic expressions use only addition, subtraction, and multiplication. Such expressions are called **polynomials**. The general form of a polynomial of degree n (where n is a nonnegative integer) in the variable x is

$$a_n x^n + a_{n-1} x^{n-1} + \cdots + a_1 x + a_0$$

where $a_0, a_1, \ldots, a_n$ are constants and $a_n \neq 0$. The **degree** of a polynomial is the highest power of the variable. Any polynomial is a sum of **terms** of the form ax^k, called **monomials**, where a is a constant and k is a nonnegative integer. A **binomial** is a sum of two monomials, a **trinomial** is the sum of three monomials, and so on. Thus, $2x^2 - 3x + 4$, $ax + b$, and $x^4 + 2x^3$ are polynomials of degree 2, 1, and 4, respectively; the first is a trinomial, the other two are binomials.

We **add** and **subtract** polynomials using the properties of real numbers that were discussed in Section 1.1. The idea is to combine **like terms** (that is, terms with the same variables raised to the same powers) using the Distributive Property. For instance,

$ac + bc = (a + b)c$

$$5x^7 + 3x^7 = (5 + 3)x^7 = 8x^7$$

In subtracting polynomials we have to remember that if a minus sign precedes an expression in parentheses, then the sign of every term within the parentheses is changed when we remove the parentheses:

$$-(b + c) = -b - c$$

[This is simply a case of the Distributive Property, $a(b + c) = ab + ac$, with $a = -1$.]

EXAMPLE 1 ■ Adding and Subtracting Polynomials
(a) Find the sum $(x^3 - 6x^2 + 2x + 4) + (3x^3 + 5x^2 - 4x)$.
(b) Find the difference $(x^3 - 6x^2 + 2x + 4) - (3x^3 + 5x^2 - 4x)$.

SOLUTION

(a) $(x^3 - 6x^2 + 2x + 4) + (3x^3 + 5x^2 - 4x)$

$\qquad = (x^3 + 3x^3) + (-6x^2 + 5x^2) + (2x - 4x) + 4$ Group like terms

$\qquad = 4x^3 - x^2 - 2x + 4$ Combine like terms

(b) $(x^3 - 6x^2 + 2x + 4) - (3x^3 + 5x^2 - 4x)$

$\qquad = x^3 - 6x^2 + 2x + 4 - 3x^3 - 5x^2 + 4x$ Distributive Property

$\qquad = (x^3 - 3x^3) + (-6x^2 - 5x^2) + (2x + 4x) + 4$ Group like terms

$\qquad = -2x^3 - 11x^2 + 6x + 4$ Combine like terms

■

To find the **product** of polynomials or other algebraic expressions, we need to use the Distributive Property repeatedly. In particular, using it three times on the product of two binomials, we get

$$(a + b)(c + d) = a(c + d) + b(c + d) = ac + ad + bc + bd$$

This says that we multiply the two factors by multiplying each term in one factor by each term in the other factor and adding these products. Schematically, we have

The acronym **FOIL** helps us remember that the product of two binomials is the sum of the products of the **F**irst terms, the **O**uter terms, the **I**nner terms, and the **L**ast terms.

$$(a + b)(c + d) = ac + ad + bc + bd$$
$$\text{F} \quad \text{O} \quad \text{I} \quad \text{L}$$

EXAMPLE 2 ■ Multiplying Binomials

(a) $(2x + 1)(3x - 5) = 6x^2 - 10x + 3x - 5$ Distributive Property

$= 6x^2 - 7x - 5$ Combine like terms

(b) $3(x - 1)(4x + 3) = 3(4x^2 + 3x - 4x - 3)$ Distributive Property

$= 3(4x^2 - x - 3)$ Combine like terms

$= 12x^2 - 3x - 9$ Distributive Property   ■

In general, we can multiply any two polynomials by using the Distributive Property and the Laws of Exponents.

EXAMPLE 3 ■ Multiplying Polynomials

Find the product $(x^2 - 3)(x^3 + 2x + 1)$.

SOLUTION We start by regarding the second factor as a single number and using the Distributive Property:

$(x^2 - 3)(x^3 + 2x + 1) = x^2(x^3 + 2x + 1) - 3(x^3 + 2x + 1)$ Distributive Property

$= x^5 + 2x^3 + x^2 - 3x^3 - 6x - 3$ Distributive Property

$= x^5 - x^3 + x^2 - 6x - 3$ Combine like terms   ■

The next example shows that the methods we have used for multiplying polynomials also apply to other algebraic expressions.

EXAMPLE 4 ■ Multiplying Algebraic Expressions

(a) $\sqrt{x}\,(x^2 + 2x + \sqrt{x}\,) = x^2\sqrt{x} + 2x\sqrt{x} + \sqrt{x}\,\sqrt{x}$ Distributive Property

$= x^{5/2} + 2x^{3/2} + x$ Laws of Exponents

(b) $(1 + \sqrt{x}\,)(2 - 3\sqrt{x}\,) = 2 - 3\sqrt{x} + 2\sqrt{x} - 3(\sqrt{x}\,)^2$ Distributive Property

$= 2 - \sqrt{x} - 3x$ Combine like terms

■

We can also consider polynomials in two or more variables. For instance,

$$4x^2y^3 - 2xy + 6$$

is a polynomial in the variables x and y. The techniques for combining such polynomials are similar to those for polynomials in one variable.

EXAMPLE 5 ■ Multiplying Polynomials with More Than One Variable

Find the product $(x^2 - xy + y^2)(x - y)$.

SOLUTION

$$(x^2 - xy + y^2)(x - y) = (x^2 - xy + y^2)x - (x^2 - xy + y^2)y \qquad \text{Distributive Property}$$

$$= x^3 - x^2y + xy^2 - x^2y + xy^2 - y^3 \qquad \text{Distributive Property}$$

$$= x^3 - 2x^2y + 2xy^2 - y^3 \qquad \text{Combine like terms} \ \blacksquare$$

Certain types of products occur so frequently that you should memorize them. You can verify each of the following formulas by performing the multiplications.

The figures give a geometric interpretation of the formula

$$(A + B)^2 = A^2 + 2AB + B^2$$

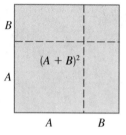

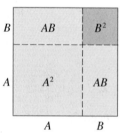

SPECIAL PRODUCT FORMULAS

1. $(A - B)(A + B) = A^2 - B^2$

2. $(A + B)^2 = A^2 + 2AB + B^2$

3. $(A - B)^2 = A^2 - 2AB + B^2$

4. $(A + B)^3 = A^3 + 3A^2B + 3AB^2 + B^3$

5. $(A - B)^3 = A^3 - 3A^2B + 3AB^2 - B^3$

The key idea in using these formulas (or any other formula in algebra) is the **Principle of Substitution**: We may substitute any algebraic expression for any letter in a formula. For example, to find $(x^2 + y^3)^2$ we use Product Formula 2:

$$(A + B)^2 = A^2 + 2AB + B^2$$

We substitute x^2 for A and y^2 for B to get

$$(x^2 + y^3)^2 = (x^2)^2 + 2(x^2)(y^3) + (y^3)^2$$

This type of substitution is valid because every algebraic expression (in this case, x^2 or y^3) represents a number.

EXAMPLE 6 ■ Using the Special Product Formulas

Use the Special Product Formulas to find each product.

(a) $(3x + 5)^2$ (b) $(2x - \sqrt{y})(2x + \sqrt{y})$ (c) $(x^2 - 2)^3$

SOLUTION

(a) Product Formula 2, with $A = 3x$ and $B = 5$, gives

$$(3x + 5)^2 = (3x)^2 + 2(3x)(5) + 5^2 = 9x^2 + 30x + 25$$

(b) Using Product Formula 1 with $A = 2x$ and $B = \sqrt{y}$, we have

$$(2x - \sqrt{y})(2x + \sqrt{y}) = (2x)^2 - (\sqrt{y})^2 = 4x^2 - y$$

(c) Substituting $A = x^2$ and $B = 2$ in Product Formula 5, we get

$$(x^2 - 2)^3 = (x^2)^3 - 3(x^2)^2(2) + 3(x^2)(2)^2 - 2^3$$
$$= x^6 - 6x^4 + 12x^2 - 8$$

■

FACTORING

We used the Distributive Property to expand algebraic expressions. We some-times need to reverse this process (again using the Distributive Property) by **factoring** an expression as a product of simpler ones. For example, we can write

■■■ FACTORING ➡

$$x^2 - 4 = (x - 2)(x + 2)$$

⬅ EXPANDING ■■■

and we say that $x - 2$ and $x + 2$ are **factors** of $x^2 - 4$. The easiest type of fac-toring occurs when the terms have a common factor.

EXAMPLE 7 ■ **Factoring Out Common Factors**

Factor each expression.

(a) $3x^2 - 6x$ (b) $8x^4y^2 + 6x^3y^3 - 2xy^4$

CHECK YOUR ANSWER

Multiplying gives

$$3x(x - 2) = 3x^2 - 6x \quad ✓$$

SOLUTION

(a) The greatest common factor of the terms $3x^2$ and $-6x$ is $3x$, so we have

$$3x^2 - 6x = 3x(x - 2)$$

(b) We note that

8, 6, and -2 have the greatest common factor 2

x^4, x^3, and x have the greatest common factor x

y^2, y^3, and y^4 have the greatest common factor y^2

CHECK YOUR ANSWER

Multiplying gives

$$2xy^2(4x^3 + 3x^2y - y^2) =$$
$$8x^4y^2 + 6x^3y^3 - 2xy^4 \quad ✓$$

So the greatest common factor of the three terms in the polynomial is $2xy^2$, and we have

$$8x^4y^2 + 6x^3y^3 - 2xy^4 = (2xy^2)(4x^3) + (2xy^2)(3x^2y) + (2xy^2)(-y^2)$$
$$= 2xy^2(4x^3 + 3x^2y - y^2)$$

■

To factor a second-degree polynomial, or **quadratic**, of the form $x^2 + bx + c$, we note that

$$(x + r)(x + s) = x^2 + (r + s)x + rs$$

so we need to choose numbers r and s so that $r + s = b$ and $rs = c$.

EXAMPLE 8 ■ Factoring $x^2 + bx + c$ by Trial and Error

Factor: $x^2 + 7x + 12$

SOLUTION   In this case, $rs = 12$ and so r and s must be factors of 12 and their sum must be 7. We enumerate the trial factors of 12:

r	1	2	3
s	12	6	4
Sum	13	8	7

Thus, $r = 3$ and $s = 4$ are the factors of 12 whose sum is 7. The factorization is

$$x^2 + 7x + 12 = (x + 3)(x + 4)$$ ■

To factor a quadratic of the form $ax^2 + bx + c$ with $a \neq 1$, we look for factors of the form $px + r$ and $qx + s$:

$$ax^2 + bx + c = (px + r)(qx + s) = pqx^2 + (ps + qr)x + rs$$

Therefore, we try to find numbers p, q, r, and s such that

$$pq = a \qquad rs = c \qquad ps + qr = b$$

If these numbers are all integers, then we will have a limited number of possibilities to try for p, q, r, and s.

EXAMPLE 9 ■ Factoring $ax^2 + bx + c$ by Trial and Error

Factor: $6x^2 + 7x - 5$

SOLUTION   We can factor 6 as $6 \cdot 1$ or $3 \cdot 2$, and -5 as $-5 \cdot 1$ or $5 \cdot (-1)$. By trying these possibilities, we arrive at the factorization

$$6x^2 + 7x - 5 = (3x + 5)(2x - 1)$$ ■

Some special algebraic expressions can be factored using the following formulas. The first three are simply the Special Product Formulas written backward.

FACTORING FORMULAS

Formula	Name
1. $A^2 - B^2 = (A - B)(A + B)$	Difference of squares
2. $A^2 + 2AB + B^2 = (A + B)^2$	Perfect square
3. $A^2 - 2AB + B^2 = (A - B)^2$	Perfect square
4. $A^3 - B^3 = (A - B)(A^2 + AB + B^2)$	Difference of cubes
5. $A^3 + B^3 = (A + B)(A^2 - AB + B^2)$	Sum of cubes

CHECK YOUR ANSWER

Multiplying gives

$$(x + 3)(x + 4) = x^2 + 7x + 12$$ ✓

factors of a
$$ax^2 + bx + c = (px + r)(qx + s)$$
factors of c

CHECK YOUR ANSWER

Multiplying gives

$$(3x + 5)(2x - 1) = 6x^2 + 7x - 5$$ ✓

EXAMPLE 10 ■ Factoring Differences of Squares

Factor each polynomial.

(a) $4x^2 - 25$ (b) $9x^4 - y^6$ (c) $(x + y)^2 - x^2$

SOLUTION

(a) Using the formula for a difference of squares with $A = 2x$ and $B = 5$, we have

$$4x^2 - 25 = (2x)^2 - 5^2$$
$$= (2x - 5)(2x + 5)$$

(b) We recognize that $9x^4 = (3x^2)^2$ and $y^6 = (y^3)^2$ and so the expression is a difference of squares. We may use the formula for the difference of squares with $A = 3x^2$ and $B = y^3$. We have

$$9x^4 - y^6 = (3x^2)^2 - (y^3)^2$$
$$= (3x^2 - y^3)(3x^2 + y^3)$$

(c) We use the formula for the difference of squares with $A = x + y$ and $B = x$.

$$(x + y)^2 - x^2 = [(x + y) - x][(x + y) + x]$$
$$= y(2x + y)$$ ■

EXAMPLE 11 ■ Factoring Perfect Squares

Factor each trinomial: (a) $x^2 + 6x + 9$ (b) $4x^2 - 4xy + y^2$

SOLUTION

(a) Using Formula 2 with $a = x$ and $b = 3$, we get

$$x^2 + 6x + 9 = x^2 + 2(3x) + 3^2 = (x + 3)^2$$

(b) Here we use $A = 2x$ and $B = y$ in Formula 3.

$$4x^2 - 4xy + y^2 = (2x)^2 - 2(2x)y + y^2$$
$$= (2x - y)^2$$ ■

EXAMPLE 12 ■ Factoring Differences and Sums of Cubes

Factor each polynomial: (a) $27x^3 - 1$ (b) $x^6 + 8$

SOLUTION

(a) Using the formula for a difference of cubes with $A = 3x$ and $B = 1$, we get

$$27x^3 - 1 = (3x)^3 - 1^3 = (3x - 1)[(3x)^2 + (3x)(1) + 1^2]$$
$$= (3x - 1)(9x^2 + 3x + 1)$$

FACTORING NUMBERS

To factor a whole number completely means to write it as a product of smaller whole numbers that cannot themselves be factored, that is, as a product of primes. For example, $420 = 2 \cdot 2 \cdot 3 \cdot 5 \cdot 7$. Factoring a very large number can be a difficult task. High-speed computers employing the fastest known methods would take about a day to factor an arbitrary 30-digit number and about a million years to factor a 40-digit number. Ted Rivest, Adi Shamir, and Leonard Adleman used this fact in the 1970s to devise the RSA code for sending secret messages. This code uses an extremely large number to encode a message, but requires knowledge of the factors to decode it. Since multiplying numbers is easy but factoring the result is hard, this code is very difficult to break. It was at first thought that a carefully selected 80-digit number would provide an unbreakable code, but recent advances in the study of factoring have made numbers with many more digits necessary to assure complete security.

(b) Using the formula for a sum of cubes with $A = x^2$ and $B = 2$, we have

$$x^6 + 8 = (x^2)^3 + 2^3 = (x^2 + 2)(x^4 - 2x^2 + 4)$$ ∎

When we factor an expression, the result can sometimes be factored further. In general, we first factor out common factors, then inspect the result to see if it can be factored by any of the other methods of this section. We repeat this process until we have factored the expression completely.

EXAMPLE 13 ■ Factoring an Expression Completely
Factor completely each expression.

(a) $2x^4 - 8x^2$ (b) $x^5y^2 - xy^6$

SOLUTION

(a) We first factor out the power of x with the smallest exponent.

$$2x^4 - 8x^2 = 2x^2(x^2 - 4) \qquad \text{Common factor is } 2x^2$$
$$= 2x^2(x - 2)(x + 2) \qquad \text{Factor } x^2 - 4 \text{ as a difference of squares}$$

(b) We first factor out the powers of x and y with the smallest exponents.

$$x^5y^2 - xy^6 = xy^2(x^4 - y^4) \qquad \text{Common factor is } xy^2$$
$$= xy^2(x^2 + y^2)(x^2 - y^2) \qquad \text{Factor } x^4 - y^4 \text{ as a difference of squares}$$
$$= xy^2(x^2 + y^2)(x + y)(x - y) \qquad \text{Factor } x^2 - y^2 \text{ as a difference of squares}$$ ∎

In the next example we factor out variables with fractional exponents. This type of factoring occurs in calculus.

EXAMPLE 14 ■ Factoring Variables with Fractional Exponents
Factor each expression.

(a) $3x^{3/2} - 9x^{1/2} + 6x^{-1/2}$ (b) $(1 + x)^{-2/3}x + (1 + x)^{1/3}$

SOLUTION

(a) Factor out the power of x with the *smallest exponent*, that is, $x^{-1/2}$.

$$3x^{3/2} - 9x^{1/2} + 6x^{-1/2} = 3x^{-1/2}(x^2 - 3x + 2) \qquad \text{Factor out } 3x^{-1/2}$$
$$= 3x^{-1/2}(x - 1)(x - 2) \qquad \text{Factor the quadratic } x^2 - 3x + 2$$

(b) Factor out the power of $1 + x$ with the smallest exponent, that is, $(1 + x)^{-2/3}$.

$$(1 + x)^{-2/3}x + (1 + x)^{1/3} = (1 + x)^{-2/3}[x + (1 + x)] \qquad \text{Factor out } (1 + x)^{-2/3}$$
$$= (1 + x)^{-2/3}(1 + 2x)$$ ∎

CHECK YOUR ANSWER

To see that you have factored correctly, multiply using the Laws of Exponents.

(a) $3x^{-1/2}(x^2 - 3x + 2)$
$= 3x^{3/2} - 9x^{1/2} + 6x^{-1/2}$ ✓

(b) $(1 + x)^{-2/3}[x + (1 + x)]$
$= (1 + x)^{-2/3}x + (1 + x)^{1/3}$ ✓

Polynomials with at least four terms can sometimes be factored by grouping terms. The following example illustrates the idea.

EXAMPLE 15 ■ Factoring by Grouping

Factor each polynomial.

(a) $x^3 + x^2 + 4x + 4$ 　　　　　　　　(b) $x^3 - 2x^2 - 3x + 6$

SOLUTION

(a) $x^3 + x^2 + 4x + 4 = (x^3 + x^2) + (4x + 4)$ 　　Group terms

$\qquad\qquad\qquad\quad = x^2(x + 1) + 4(x + 1)$ 　　Factor out common factors

$\qquad\qquad\qquad\quad = (x^2 + 4)(x + 1)$ 　　Factor out $x + 1$ from each term

(b) $x^3 - 2x^2 - 3x + 6 = (x^3 - 2x^2) - (3x - 6)$ 　　Group terms

$\qquad\qquad\qquad\quad = x^2(x - 2) - 3(x - 2)$ 　　Factor out common factors

$\qquad\qquad\qquad\quad = (x^2 - 3)(x - 2)$ 　　Factor out $x - 2$ from each term ■

1.3 EXERCISES

1–40 ■ Expand and simplify.

1. $2(x - 1) + 4(x + 2)$

2. $5(2x + 3) - 7(2x - 3)$

3. $(2x^2 + x + 1) + (x^2 - 3x + 5)$

4. $(2x^2 + x + 1) - (x^2 - 3x + 5)$

5. $(x^3 + 6x^2 - 4x + 7) - (3x^2 + 2x - 4)$

6. $4(x^2 - x + 2) - 5(x^2 - 2x + 1)$

7. $2(2 - 5t) + t^2(t - 1) - (t^4 - 1)$

8. $5(3t - 4) - (t^2 + 2) - 2t(t - 3)$

9. $\sqrt{x}\,(x - \sqrt{x})$ 　　**10.** $x^{3/2}(\sqrt{x} - 1/\sqrt{x})$

11. $\sqrt[3]{y}\,(y^2 - 1)$ 　　**12.** $(4x - 1)(3x + 7)$

13. $(3t - 2)(7t - 5)$ 　　**14.** $(t + 6)(t + 5) - 3(t + 4)$

15. $(x + 2y)(3x - y)$ 　　**16.** $(4x - 3y)(2x + 5y)$

17. $(1 - 2y)^2$ 　　**18.** $(3x + 4)^2$

19. $(2x - 5)(x^2 - x + 1)$ 　　**20.** $(x^2 + 3)(5x - 6)$

21. $x(x - 1)(x + 2)$ 　　**22.** $(1 + 2x)(x^2 - 3x + 1)$

23. $y^4(6 - y)(5 + y)$

24. $(t - 5)^2 - 2(t + 3)(8t - 1)$

25. $(2x^2 + 3y^2)^2$

26. $(x^{1/2} + y^{1/2})(x^{1/2} - y^{1/2})$

27. $(x^2 - a^2)(x^2 + a^2)$

28. $(\sqrt{h^2 + 1} + 1)(\sqrt{h^2 + 1} - 1)$

29. $(1 + a^3)^3$

30. $(x - 1)(x^2 + x + 1)$

31. $\left(\sqrt{a} - \dfrac{1}{b}\right)\left(\sqrt{a} + \dfrac{1}{b}\right)$

32. $\left(c + \dfrac{1}{c}\right)^2$

33. $(x^2 + x - 2)(x^3 - x + 1)$

34. $(1 + x + x^2)(1 - x + x^2)$

35. $(1 + x^{4/3})(1 - x^{2/3})$

36. $(x^{3/2} - x + 1)(x^2 + x^{1/2} - 2)$

37. $(1 - b)^2(1 + b)^2$

38. $(1 + x - x^2)^2$

39. $(3x^2y + 7xy^2)(x^2y^3 - 2y^2)$

40. $(x^4y - y^5)(x^2 + xy + y^2)$

41–90 ■ Factor the expression completely.

41. $2x + 12x^3$

42. $8x^5 + 4x^3$

43. $6y^4 - 15y^3$

44. $5ab - 8abc$

45. $x^2 + 7x + 6$

46. $x^2 - x - 6$

47. $x^2 - 2x - 8$

48. $x^2 - 14x + 48$

49. $y^2 - 8y + 15$

50. $z^2 + 6z - 16$

51. $2x^2 + 5x + 3$

52. $2x^2 + 7x - 4$

53. $9x^2 - 36$

54. $8x^2 + 10x + 3$

55. $6x^2 - 5x - 6$

56. $6 + 5t - 6t^2$

57. $(x - 1)(x + 2)^2 - (x - 1)^2(x + 2)$

58. $(x + 1)^3 x - 2(x + 1)^2 x^2 + x^3(x + 1)$

59. $y^4(y + 2)^3 + y^5(y + 2)^4$

60. $n(x - y) + (n - 1)(y - x)$

61. $(a^2 - 1)b^2 - 4(a^2 - 1)$

62. $(a + b)^2 - (a - b)^2$

63. $t^3 + 1$

64. $4t^2 - 9s^2$

65. $4t^2 - 12t + 9$

66. $x^3 - 27$

67. $x^3 + 2x^2 + x$

68. $3x^3 - 27x$

69. $4x^2 + 4xy + y^2$

70. $4r^2 - 12rs + 9s^2$

71. $x^4 + 2x^3 - 3x^2$

72. $x^6 + 64$

73. $8x^3 - 125$

74. $x^4 + 2x^2 + 1$

75. $x^4 + x^2 - 2$

76. $x^3 + 3x^2 - x - 3$

77. $y^3 - 3y^2 - 4y + 12$

78. $y^3 - y^2 + y - 1$

79. $2x^3 + 4x^2 + x + 2$

80. $3x^3 + 5x^2 - 6x - 10$

81. $x^6 - y^6$

82. $x^8 - 1$

83. $x^{5/2} - x^{1/2}$

84. $3x^{-1/2} + 4x^{1/2} + x^{3/2}$

85. $x^{-3/2} + 2x^{-1/2} + x^{1/2}$

86. $(x - 1)^{7/2} - (x - 1)^{3/2}$

87. $(x^2 + 1)^{1/2} + 2(x^2 + 1)^{-1/2}$

88. $x^{-1/2}(x + 1)^{1/2} + x^{1/2}(x + 1)^{-1/2}$

89. $(a^2 + 1)^2 - 7(a^2 + 1) + 10$

90. $(a^2 + 2a)^2 - 2(a^2 + 2a) - 3$

91. Factor $x^4 + 3x^2 + 4$.   [*Hint:* Write the expression as $(x^4 + 4x^2 + 4) - x^2$ and note that this is a difference of squares.]

92. Show that $ab = \frac{1}{2}[(a + b)^2 - (a^2 + b^2)]$.

93. Show that

$$(a^2 + b^2)(c^2 + d^2) = (ac + bd)^2 + (ad - bc)^2$$

94. Show that $(a^2 + b^2)^2 - (a^2 - b^2)^2$ is a perfect square.

95. Factor the expression completely:

$$4a^2c^2 - (c^2 - b^2 + a^2)^2$$

96. Verify each of the following formulas algebraically.
(a) Special Product Formulas 1 and 2
(b) Special Product Formulas 3 and 4
(c) The formula for a difference of cubes
(d) The formula for a sum of cubes

▲ **DISCOVERY · DISCUSSION**

97. Geometric Interpretations of Algebraic Formulas
Explain how each figure verifies the given formula. In each case $a > b > 0$.

$$a^2 - b^2 = (a + b)(a - b)$$

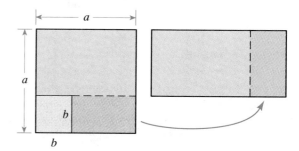

$$(a + b)^3 = a^3 + 3a^2b + 3ab^2 + b^3$$

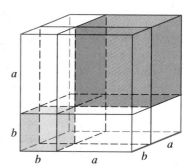

Is it possible to create a figure such as these to illustrate the product formula for $(a + b)^4$? Explain your answer. (Think about the dimensions of the objects in the figures.)

98. Volume of a Cylindrical Shell Using the formula for the volume of a cylinder given on the inside of the front cover of this book, explain why the volume of the cylindrical shell shown at the right is

$$V = \pi R^2 h - \pi r^2 h$$

Factor this to show that

$$V = 2\pi \cdot \text{average radius} \cdot \text{height} \cdot \text{thickness}$$

Use the "unrolled" diagram to explain why this makes sense geometrically.

1.4 FRACTIONAL EXPRESSIONS

A quotient of two algebraic expressions is called a **fractional expression**. We assume that all fractions are defined; that is, *we deal only with values of the variables such that the denominators are not zero.*

A common type of fractional expression occurs when both the numerator and denominator are polynomials. This is called a **rational expression**. For instance,

$$\frac{4x^3 + 2x + 5}{x + 3}$$

is a rational expression whose denominator is 0 when $x = -3$. So, in dealing with this expression, we implicitly assume that $x \neq -3$.

In simplifying rational expressions we factor both numerator and denominator and use the following property of fractions:

$$\frac{AC}{BC} = \frac{A}{B}$$

This says that we can **cancel** common factors from numerator and denominator.

EXAMPLE 1 ■ Simplifying Fractional Expressions by Cancellation

Simplify: (a) $\dfrac{x^2 - 1}{x^2 + x - 2}$ (b) $\dfrac{2x^3 + 5x^2 - 3x}{6 - x - x^2}$

SOLUTION

⊘ We can't cancel the x^2's in $\dfrac{x^2 - 1}{x^2 + x - 2}$ because x^2 is not a factor.

(a) $\dfrac{x^2 - 1}{x^2 + x - 2} = \dfrac{(x - 1)(x + 1)}{(x - 1)(x + 2)}$ Factor

$= \dfrac{x + 1}{x + 2}$ Cancel common factors

(b) $\dfrac{2x^3 + 5x^2 - 3x}{6 - x - x^2} = \dfrac{x(2x^2 + 5x - 3)}{-(x^2 + x - 6)} = \dfrac{x(2x - 1)(x + 3)}{-(x - 2)(x + 3)}$ Factor

$= -\dfrac{x(2x - 1)}{x - 2}$ Cancel common factors

∎

When multiplying fractional expressions, we use the following property of fractions:

$$\dfrac{A}{B} \cdot \dfrac{C}{D} = \dfrac{AC}{BD}$$

This says that to multiply two fractions we multiply their numerators and multiply their denominators.

EXAMPLE 2 ■ Multiplying Fractional Expressions

Perform the indicated multiplication and simplify: $\dfrac{x^2 + 2x - 3}{x^2 + 8x + 16} \cdot \dfrac{3x + 12}{x - 1}$

SOLUTION We first factor.

$\dfrac{x^2 + 2x - 3}{x^2 + 8x + 16} \cdot \dfrac{3x + 12}{x - 1} = \dfrac{(x - 1)(x + 3)}{(x + 4)^2} \cdot \dfrac{3(x + 4)}{x - 1}$ Factor

$= \dfrac{3(x - 1)(x + 3)(x + 4)}{(x - 1)(x + 4)^2}$ Property of fractions

$= \dfrac{3(x + 3)}{x + 4}$ Cancel common factors

∎

When dividing fractional expressions, we use the following property of fractions:

$$\dfrac{A}{B} \div \dfrac{C}{D} = \dfrac{A}{B} \cdot \dfrac{D}{C}$$

This says that to divide a fraction by another fraction we invert the divisor and multiply.

EXAMPLE 3 ■ Dividing Fractional Expressions

Perform the indicated division and simplify: $\dfrac{x-4}{x^2-4} \div \dfrac{x^2-3x-4}{x^2+5x+6}$

SOLUTION

$$\frac{x-4}{x^2-4} \div \frac{x^2-3x-4}{x^2+5x+6} = \frac{x-4}{x^2-4} \cdot \frac{x^2+5x+6}{x^2-3x-4} \qquad \text{Invert and multiply}$$

$$= \frac{(x-4)(x+2)(x+3)}{(x-2)(x+2)(x-4)(x+1)} \qquad \text{Factor}$$

$$= \frac{x+3}{(x-2)(x+1)} \qquad \text{Cancel common factors}$$

■

⊘ Avoid making the following error:

$$\frac{A}{B+C} \ne \frac{A}{B} + \frac{A}{C}$$

For instance, if we let $A = 2$, $B = 1$, and $C = 1$, then we see the error:

$$\frac{2}{1+1} \stackrel{?}{=} \frac{2}{1} + \frac{2}{1}$$

$$\frac{2}{2} \stackrel{?}{=} 2 + 2$$

$$1 \stackrel{?}{=} 4 \quad \text{Wrong!}$$

In adding and subtracting rational expressions, we first find a common denominator and then use the following property of fractions:

$$\boxed{\frac{A}{C} + \frac{B}{C} = \frac{A+B}{C}}$$

Although any common denominator will work, it is best to use the **least common denominator** (LCD) as explained in Section 1.1. The LCD is found by factoring each denominator and taking the product of the distinct factors, using the highest power that appears in any of the factors.

EXAMPLE 4 ■ Adding and Subtracting Fractional Expressions

Perform the indicated operations and simplify.

(a) $\dfrac{3}{x-1} + \dfrac{x}{x+2}$ (b) $\dfrac{1}{x^2-1} - \dfrac{2}{(x+1)^2}$

SOLUTION

(a) Here the LCD is simply the product $(x-1)(x+2)$, so we have

$$\frac{3}{x-1} + \frac{x}{x+2} = \frac{3(x+2)}{(x-1)(x+2)} + \frac{x(x-1)}{(x-1)(x+2)} \qquad \text{Write fractions using LCD}$$

$$= \frac{3x+6+x^2-x}{(x-1)(x+2)} \qquad \text{Add fractions}$$

$$= \frac{x^2+2x+6}{(x-1)(x+2)} \qquad \text{Combine terms in numerator}$$

(b) The LCD of $x^2 - 1 = (x - 1)(x + 1)$ and $(x + 1)^2$ is $(x - 1)(x + 1)^2$, so we have

$$\frac{1}{x^2 - 1} - \frac{2}{(x + 1)^2} = \frac{1}{(x - 1)(x + 1)} - \frac{2}{(x + 1)^2} \qquad \text{Factor}$$

$$= \frac{(x + 1) - 2(x - 1)}{(x - 1)(x + 1)^2} \qquad \begin{array}{l}\text{Combine fractions}\\ \text{using LCD}\end{array}$$

$$= \frac{x + 1 - 2x + 2}{(x - 1)(x + 1)^2} \qquad \text{Distributive Property}$$

$$= \frac{3 - x}{(x - 1)(x + 1)^2} \qquad \begin{array}{l}\text{Combine terms}\\ \text{in numerator}\end{array} \quad \blacksquare$$

In the next example we simplify a **compound fraction**, which contains a fraction in both the numerator and the denominator.

EXAMPLE 5 ■ **Simplifying a Compound Fraction**

Simplify: $\dfrac{\dfrac{x}{y} + 1}{1 - \dfrac{y}{x}}$

SOLUTION 1 We combine the terms in the numerator into a single fraction. We do the same in the denominator. Then we invert and multiply.

$$\frac{\dfrac{x}{y} + 1}{1 - \dfrac{y}{x}} = \frac{\dfrac{x + y}{y}}{\dfrac{x - y}{x}} = \frac{x + y}{y} \cdot \frac{x}{x - y}$$

$$= \frac{x(x + y)}{y(x - y)}$$

SOLUTION 2 We find the least common denominator (LCD) of all the fractions in the expression, then multiply numerator and denominator by it. In this example the LCD of all the fractions is xy. Thus

$$\frac{\dfrac{x}{y} + 1}{1 - \dfrac{y}{x}} = \frac{\dfrac{x}{y} + 1}{1 - \dfrac{y}{x}} \cdot \frac{xy}{xy} \qquad \begin{array}{l}\text{Multiply numerator and}\\ \text{denominator by } xy\end{array}$$

$$= \frac{x^2 + xy}{xy - y^2} \qquad \text{Simplify}$$

$$= \frac{x(x + y)}{y(x - y)} \qquad \text{Factor} \qquad \blacksquare$$

The remaining examples show situations in calculus that require the ability to work with fractional expressions.

EXAMPLE 6 ■ Simplifying a Compound Fraction

Simplify: $\dfrac{\dfrac{1}{(a + h)^2} - \dfrac{1}{a^2}}{h}$

SOLUTION As in the first solution to Example 5, we begin by combining the fractions in the numerator using a common denominator.

$$\frac{\dfrac{1}{(a + h)^2} - \dfrac{1}{a^2}}{h} = \frac{\dfrac{a^2 - (a + h)^2}{(a + h)^2 a^2}}{h} \qquad \text{Combine fractions in the numerator}$$

$$= \frac{a^2 - (a + h)^2}{(a + h)^2 a^2} \cdot \frac{1}{h} \qquad \text{Property 2 of fractions (invert divisor and multiply)}$$

$$= \frac{a^2 - (a^2 + 2ah + h^2)}{(a + h)^2 a^2} \cdot \frac{1}{h} \qquad \text{Product Formula 2}$$

$$= \frac{-2ah - h^2}{h(a + h)^2 a^2} \qquad \text{Subtract}$$

$$= \frac{h(-2a - h)}{h(a + h)^2 a^2} \qquad \text{Factor out } h$$

$$= -\frac{2a + h}{(a + h)^2 a^2} \qquad \text{Property 5 of fractions (cancel common factors)} \qquad ■$$

EXAMPLE 7 ■ Simplifying a Compound Fraction

Simplify: $\dfrac{(1 + x^2)^{1/2} - x^2(1 + x^2)^{-1/2}}{1 + x^2}$

SOLUTION 1 Factor $(1 + x^2)^{-1/2}$ from the numerator.

$$\frac{(1 + x^2)^{1/2} - x^2(1 + x^2)^{-1/2}}{1 + x^2} = \frac{(1 + x^2)^{-1/2}[(1 + x^2) - x^2]}{1 + x^2}$$

$$= \frac{(1 + x^2)^{-1/2}}{1 + x^2} = \frac{1}{(1 + x^2)^{3/2}}$$

SOLUTION 2 Since $(1 + x^2)^{-1/2} = 1/(1 + x^2)^{1/2}$ is a fraction, we may clear all fractions by multiplying numerator and denominator by $(1 + x^2)^{1/2}$.

$$\frac{(1 + x^2)^{1/2} - x^2(1 + x^2)^{-1/2}}{1 + x^2} = \frac{(1 + x^2)^{1/2} - x^2(1 + x^2)^{-1/2}}{1 + x^2} \cdot \frac{(1 + x^2)^{1/2}}{(1 + x^2)^{1/2}}$$

$$= \frac{(1 + x^2) - x^2}{(1 + x^2)^{3/2}} = \frac{1}{(1 + x^2)^{3/2}} \qquad ■$$

If a fraction has a denominator of the form $A + B\sqrt{C}$, we may rationalize the denominator by multiplying numerator and denominator by the **conjugate radical** $A - B\sqrt{C}$. This is effective because, by Product Formula 1 in Section 1.3, the product of the denominator and its conjugate radical does not contain a radical:

$$(A + B\sqrt{C})(A - B\sqrt{C}) = A^2 - B^2C$$

EXAMPLE 8 ■ Rationalizing the Denominator

Rationalize the denominator: $\dfrac{1}{1 + \sqrt{2}}$

SOLUTION We multiply both the numerator and the denominator by the conjugate radical of $1 + \sqrt{2}$, which is $1 - \sqrt{2}$.

$$\frac{1}{1 + \sqrt{2}} = \frac{1}{1 + \sqrt{2}} \cdot \frac{1 - \sqrt{2}}{1 - \sqrt{2}} \qquad \text{Multiply numerator and denominator by the conjugate radical}$$

$$= \frac{1 - \sqrt{2}}{1^2 - (\sqrt{2})^2} \qquad \text{Product Formula 1}$$

$$= \frac{1 - \sqrt{2}}{1 - 2} = \frac{1 - \sqrt{2}}{-1} = \sqrt{2} - 1 \qquad ■$$

Product Formula 1
$(a + b)(a - b) = a^2 - b^2$

EXAMPLE 9 ■ Rationalizing a Numerator

Rationalize the numerator: $\dfrac{\sqrt{4 + h} - 2}{h}$

SOLUTION We multiply numerator and denominator by the conjugate radical $\sqrt{4 + h} + 2$. The advantage of doing this is that we can use Product Formula 1, and the radicals then disappear from the numerator.

$$\frac{\sqrt{4 + h} - 2}{h} = \frac{\sqrt{4 + h} - 2}{h} \cdot \frac{\sqrt{4 + h} + 2}{\sqrt{4 + h} + 2} \qquad \text{Multiply numerator and denominator by the conjugate radical}$$

$$= \frac{(\sqrt{4 + h})^2 - 2^2}{h(\sqrt{4 + h} + 2)} \qquad \text{Product Formula 1}$$

$$= \frac{4 + h - 4}{h(\sqrt{4 + h} + 2)}$$

$$= \frac{h}{h(\sqrt{4 + h} + 2)} = \frac{1}{\sqrt{4 + h} + 2} \qquad \text{Property 5 of fractions (cancel common factors)}$$

■

 Don't make the mistake of applying properties of multiplication to the operation of addition. Many of the common errors in algebra involve doing just that. The table on page 42 states several properties of multiplication and illustrates the error in applying them to addition.

Correct multiplication property	Common error with addition
$(a \cdot b)^2 = a^2 \cdot b^2$	$(a + b)^2 \neq a^2 + b^2$
$\sqrt{a \cdot b} = \sqrt{a}\,\sqrt{b}$ $(a, b \geq 0)$	$\sqrt{a + b} \neq \sqrt{a} + \sqrt{b}$
$\sqrt{a^2 \cdot b^2} = a \cdot b$ $(a, b \geq 0)$	$\sqrt{a^2 + b^2} \neq a + b$
$\dfrac{1}{a} \cdot \dfrac{1}{b} = \dfrac{1}{a \cdot b}$	$\dfrac{1}{a} + \dfrac{1}{b} \neq \dfrac{1}{a + b}$
$\dfrac{ab}{a} = b$	$\dfrac{a + b}{a} \neq b$
$a^{-1} \cdot b^{-1} = (a \cdot b)^{-1}$	$a^{-1} + b^{-1} \neq (a \cdot b)^{-1}$

To verify that the formulas in the right-hand column are wrong, simply substitute numbers for a and b and calculate each side. For example, if we take $a = 2$ and $b = 2$ in the fourth property, we find that the left-hand side is

$$\frac{1}{a} + \frac{1}{b} = \frac{1}{2} + \frac{1}{2} = 1$$

whereas the right-hand side is

$$\frac{1}{a + b} = \frac{1}{2 + 2} = \frac{1}{4}$$

Since $1 \neq \frac{1}{4}$, the stated formula is wrong. You should similarly convince yourself of the error in each of the other formulas.

1.4   EXERCISES

1–48 ■ Simplify the expression.

1. $\dfrac{x^2 + 3x + 2}{x^2 + 5x + 6}$

2. $\dfrac{x^2 + x - 6}{x^2 - 4}$

3. $\dfrac{y - y^2}{y^2 - 1}$

4. $\dfrac{2y^2 - 9y - 18}{4y^2 + 16y + 15}$

5. $\dfrac{2x^3 - x^2 - 6x}{2x^2 - 7x + 6}$

6. $\dfrac{1 - x^2}{x^3 - 1}$

7. $\dfrac{t - 3}{t^2 + 9} \cdot \dfrac{t + 3}{t^2 - 9}$

8. $\dfrac{x^2 - x - 6}{x^2 + 2x} \cdot \dfrac{x^3 + x^2}{x^2 - 2x - 3}$

9. $\dfrac{x^2 + 7x + 12}{x^2 + 3x + 2} \cdot \dfrac{x^2 + 5x + 6}{x^2 + 6x + 9}$

10. $\dfrac{x^2 + 2xy + y^2}{x^2 - y^2} \cdot \dfrac{2x^2 - xy - y^2}{x^2 - xy - 2y^2}$

11. $\dfrac{2x^2 + 3x + 1}{x^2 + 2x - 15} \div \dfrac{x^2 + 6x + 5}{2x^2 - 7x + 3}$

12. $\dfrac{4y^2 - 9}{2y^2 + 9y - 18} \div \dfrac{2y^2 + y - 3}{y^2 + 5y - 6}$

13. $\dfrac{\dfrac{x^3}{x + 1}}{\dfrac{x}{x^2 + 2x + 1}}$

14. $\dfrac{\dfrac{2x^2 - 3x - 2}{x^2 - 1}}{\dfrac{2x^2 + 5x + 2}{x^2 + x - 2}}$

15. $\dfrac{x/y}{z}$

16. $\dfrac{x}{y/z}$

17. $\dfrac{1}{x + 5} + \dfrac{2}{x - 3}$

18. $\dfrac{1}{x + 1} + \dfrac{1}{x - 1}$

19. $\dfrac{1}{x + 1} - \dfrac{1}{x + 2}$

20. $\dfrac{x}{x - 4} - \dfrac{3}{x + 6}$

21. $\dfrac{x}{(x + 1)^2} + \dfrac{2}{x + 1}$

22. $\dfrac{5}{2x - 3} - \dfrac{3}{(2x - 3)^2}$

23. $u + 1 + \dfrac{u}{u + 1}$

24. $\dfrac{2}{a^2} - \dfrac{3}{ab} + \dfrac{4}{b^2}$

25. $\dfrac{1}{x^2} + \dfrac{1}{x^2 + x}$

26. $\dfrac{1}{x} + \dfrac{1}{x^2} + \dfrac{1}{x^3}$

27. $\dfrac{2}{x + 3} - \dfrac{1}{x^2 + 7x + 12}$

28. $\dfrac{x}{x^2 - 4} + \dfrac{1}{x - 2}$

29. $\dfrac{1}{x + 3} + \dfrac{1}{x^2 - 9}$

30. $\dfrac{x}{x^2 + x - 2} - \dfrac{2}{x^2 - 5x + 4}$

31. $\dfrac{2}{x} + \dfrac{3}{x - 1} - \dfrac{4}{x^2 - x}$

32. $\dfrac{x}{x^2 - x - 6} - \dfrac{1}{x + 2} - \dfrac{2}{x - 3}$

33. $\dfrac{1}{x^2 + 3x + 2} - \dfrac{1}{x^2 - 2x - 3}$

34. $\dfrac{1}{x + 1} - \dfrac{2}{(x + 1)^2} + \dfrac{3}{x^2 - 1}$

35. $\dfrac{\dfrac{x}{y} - \dfrac{y}{x}}{\dfrac{1}{x^2} - \dfrac{1}{y^2}}$

36. $x - \dfrac{y}{\dfrac{x}{y} + \dfrac{y}{x}}$

37. $\dfrac{1 + \dfrac{1}{c - 1}}{1 - \dfrac{1}{c - 1}}$

38. $1 + \dfrac{1}{1 + \dfrac{1}{1 + x}}$

39. $\dfrac{\dfrac{5}{x - 1} - \dfrac{2}{x + 1}}{\dfrac{x}{x - 1} + \dfrac{1}{x + 1}}$

40. $\dfrac{\dfrac{a - b}{a} - \dfrac{a + b}{b}}{\dfrac{a - b}{b} + \dfrac{a + b}{a}}$

41. $\dfrac{x^{-2} - y^{-2}}{x^{-1} + y^{-1}}$

42. $\dfrac{x^{-1} + y^{-1}}{(x + y)^{-1}}$

43. $\dfrac{\dfrac{1}{a + h} - \dfrac{1}{a}}{h}$

44. $\dfrac{(x + h)^{-3} - x^{-3}}{h}$

45. $\dfrac{\dfrac{1 - (x + h)}{2 + (x + h)} - \dfrac{1 - x}{2 + x}}{h}$

46. $\dfrac{(x + h)^3 - 7(x + h) - (x^3 - 7x)}{h}$

47. $\sqrt{1 + \left(\dfrac{x}{\sqrt{1 - x^2}}\right)^2}$

48. $\sqrt{1 + \left(x^3 - \dfrac{1}{4x^3}\right)^2}$

49–52 ■ Simplify the expression.

49. $\dfrac{2(1 + x)^{1/2} - x(1 + x)^{-1/2}}{x + 1}$

50. $\dfrac{(1 - x^2)^{1/2} + x^2(1 - x^2)^{-1/2}}{1 - x^2}$

51. $\dfrac{3(1 + x)^{1/3} - x(1 + x)^{-2/3}}{(1 + x)^{2/3}}$

52. $\dfrac{(7 - 3x)^{1/2} + \frac{3}{2}x(7 - 3x)^{-1/2}}{7 - 3x}$

53–56 ■ Rationalize the denominator.

53. $\dfrac{2}{3 + \sqrt{5}}$

54. $\dfrac{1}{\sqrt{x} + 1}$

55. $\dfrac{2}{\sqrt{2} + \sqrt{7}}$

56. $\dfrac{y}{\sqrt{3} + \sqrt{y}}$

57–62 ■ Rationalize the numerator.

57. $\dfrac{1 - \sqrt{5}}{3}$

58. $\dfrac{\sqrt{3} + \sqrt{5}}{2}$

59. $\dfrac{\sqrt{r} + \sqrt{2}}{5}$

60. $\dfrac{\sqrt{x} - \sqrt{x + h}}{h\sqrt{x}\sqrt{x + h}}$

61. $\sqrt{x^2 + 1} - x$

62. $\sqrt{x + 1} - \sqrt{x}$

63–72 ■ State whether the given equation is true for all values of the variables. (Disregard any value that makes a denominator zero.)

63. $\dfrac{16 + a}{16} = 1 + \dfrac{a}{16}$ **64.** $\dfrac{b}{b - c} = 1 - \dfrac{b}{c}$

65. $\dfrac{2}{4 + x} = \dfrac{1}{2} + \dfrac{2}{x}$ **66.** $\dfrac{x + 1}{y + 1} = \dfrac{x}{y}$

67. $\dfrac{x}{x + y} = \dfrac{1}{1 + y}$ **68.** $2\left(\dfrac{a}{b}\right) = \dfrac{2a}{2b}$

69. $\dfrac{-a}{b} = -\dfrac{a}{b}$ **70.** $\dfrac{1 + x + x^2}{x} = \dfrac{1}{x} + 1 + x$

71. $\dfrac{x^2 + 1}{x^2 + x - 1} = \dfrac{1}{x - 1}$ **72.** $\dfrac{x^2 - 1}{x - 1} = x + 1$

73. If two electrical resistors with resistances R_1 and R_2 are connected in parallel (see the figure), then the total resistance R is given by

$$R = \dfrac{1}{\dfrac{1}{R_1} + \dfrac{1}{R_2}}$$

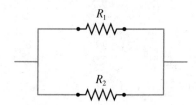

R_1

R_2

(a) Simplify the expression for R.
(b) If $R_1 = 10$ ohms and $R_2 = 20$ ohms, what is the total resistance R?

 DISCOVERY · DISCUSSION

74. Limiting Behavior of a Rational Expression The rational expression

$$\frac{x^2 - 9}{x - 3}$$

is not defined for $x = 3$. Complete the tables and determine what value the expression approaches as x gets closer and closer to 3. Why is this reasonable? (Factor the numerator of the expression and simplify to see why.)

x	$\dfrac{x^2 - 9}{x - 3}$
2.80	
2.90	
2.95	
2.99	
2.999	

x	$\dfrac{x^2 - 9}{x - 3}$
3.20	
3.10	
3.05	
3.01	
3.001	

75. Is This Rationalization? In the expression

$$\frac{2}{\sqrt{x}}$$

we would eliminate the radical if we were to square both numerator and denominator. Is this the same thing as rationalizing the denominator?

1.5   **EQUATIONS**

An equation is a statement that two mathematical expressions are equal. For example,

$$3 + 5 = 8$$

is an equation. But it is not a very interesting one—it just states a simple arithmetic fact. Most equations that we study in algebra contain **variables**, which are symbols (usually letters) that stand for numbers. In the equations

$$(w - 4)(w + 4) = w^2 - 16 \qquad \text{and} \qquad 4x + 7 = 19$$

the letters w and x are variables. In the first of these equations, the equation is true no matter what value the variable w stands for. This equation is the "difference of squares" formula from Section 1.3; it is true for all w, so we say it is an **identity**. The second equation is *not* true for all values of the variable x. The values of x that make the equation true are called the **solutions** or **roots** of the equation, and the process of finding the solutions is called **solving the equation**.

Two equations with exactly the same solutions are called **equivalent equations**. To solve an equation, we try to find a simpler, equivalent equation in which the variable stands alone on one side of the "equal" sign. Here are the properties that we use to solve an equation. (In these properties, A, B, and c stand for any algebraic expressions, and the symbol $\iff$ means "is equivalent to.")

PROPERTIES OF EQUALITY

Property	Description
1. $A = B \iff A + c = B + c$	Adding the same quantity to both sides of an equation gives an equivalent equation.
2. $A = B \iff cA = cB \quad (c \neq 0)$	Multiplying both sides of an equation by the same nonzero quantity gives an equivalent equation.

These properties require that you *perform the same operation on both sides of an equation* when solving it. Thus, if we say "*add* -7" when solving an equation, this is just a short way of saying "*add* -7 to each side of the equation."

This is how we use the properties of equality to solve the equation $4x + 7 = 19$:

$$4x + 7 + (-7) = 19 + (-7) \qquad \text{Add } -7$$

$$4x = 12 \qquad \text{Simplify}$$

$$\tfrac{1}{4} \cdot 4x = \tfrac{1}{4} \cdot 12 \qquad \text{Multiply by } \tfrac{1}{4}$$

$$x = 3 \qquad \text{Simplify}$$

So the solution of this equation is $x = 3$. To verify this, we check our answer by substituting $x = 3$ to make sure that this value of x does indeed make the equation true:

$$\boxed{x = 3}$$
$$\downarrow$$
$$4(3) + 7 \overset{?}{=} 19$$

$$19 = 19 \qquad \text{Correct!}$$

LINEAR EQUATIONS

Linear Equations

$$4x - 5 = 3$$
$$2x = \tfrac{1}{2}x - 5$$

Nonlinear Equations

$$x^2 + 2x = 8$$
$$\sqrt{x} - \frac{3}{x} = 6x - 1$$

The simplest type of equation is a **linear equation**, or first-degree equation; this is an equation in which each term is either a constant or a nonzero multiple of the variable. Thus, a linear equation is equivalent to an equation of the form $ax + b = 0$. Here a and b represent real numbers with $a \neq 0$, and x is the unknown variable that we are solving for. The equation in the following example is linear.

EXAMPLE 1 ■ Solving a Linear Equation

Solve the equation $7x - 4 = 3x + 8$.

SOLUTION   We solve this by changing it to an equivalent equation with all terms that have the variable x on one side and all constant terms on the other.

$$7x - 4 = 3x + 8$$

$$(7x - 4) + 4 = (3x + 8) + 4 \qquad \text{Add 4}$$

$$7x = 3x + 12 \qquad \text{Simplify}$$

$$7x - 3x = (3x + 12) - 3x \qquad \text{Subtract } 3x$$

$$4x = 12 \qquad \text{Simplify}$$

$$\tfrac{1}{4} \cdot 4x = \tfrac{1}{4} \cdot 12 \qquad \text{Multiply by } \tfrac{1}{4}$$

$$x = 3 \qquad \text{Simplify}$$

Since all of these equations are equivalent, the solution is 3. To verify this answer we *look back* (one of the principles of problem solving introduced on pages 122–124) by substituting $x = 3$ into the original equation.

$$7(3) - 4 \overset{?}{=} 3(3) + 8$$

$$17 = 17$$

The last statement is true, so $x = 3$ is the solution.   ■

The word *algebra* comes from the ninth century Arabic book *Hisâb al-Jabr w'al-Muqabala*, written by al-Khowarizmi. The title refers to transposing and combining terms, two processes used in solving equations. In Latin translations the title was shortened to *Aljabr*, from which we get the word *algebra*. The author's name itself made its way into the English language in the form of our word *algorithm*.

Because checking the answer is so important, we do this frequently in the remaining examples. In these checks, LHS stands for "left-hand side" of the original equation, and RHS stands for "right-hand side."

In the next example we solve an equation that doesn't look like a linear equation, but it does simplify to an equivalent linear equation.

EXAMPLE 2 ■ An Equation that Reduces to a Linear Equation

Solve the equation $\dfrac{x}{x + 1} = \dfrac{2x + 1}{2x - 3}$.

SOLUTION If $x \neq -1$ and $x \neq \frac{3}{2}$, then the denominators of the fractions in this equation are not 0, so we can multiply each side of the equation by the LCD, which is $(x + 1)(2x - 3)$.

$$(x + 1)(2x - 3)\left(\frac{x}{x + 1}\right) = (x + 1)(2x - 3)\left(\frac{2x + 1}{2x - 3}\right) \qquad \text{Multiply by LCD}$$

$$(2x - 3)x = (x + 1)(2x + 1) \qquad \text{Simplify}$$

$$2x^2 - 3x = 2x^2 + 3x + 1 \qquad \text{Expand}$$

$$-3x = 3x + 1 \qquad \text{Subtract } 2x^2$$

$$-6x = 1 \qquad \text{Subtract } 3x$$

$$x = -\frac{1}{6} \qquad \text{Divide by } -6$$

The solution is $-\frac{1}{6}$.

CHECK YOUR ANSWER

$x = -\frac{1}{6}$:

$$\text{LHS} = \frac{-\frac{1}{6}}{(-\frac{1}{6}) + 1} = \frac{-\frac{1}{6}}{\frac{5}{6}} = -\frac{1}{5} \qquad\qquad \text{RHS} = \frac{2(-\frac{1}{6}) + 1}{2(-\frac{1}{6}) - 3} = \frac{\frac{4}{6}}{-\frac{20}{6}} = -\frac{1}{5}$$

$\text{LHS} = \text{RHS} \;\; \checkmark$

EXAMPLE 3 ■ **Solving for One Variable in Terms of Others**

Solve for the variable M in the equation

This is Newton's Law of Gravity.

$$F = G\frac{mM}{r^2}$$

SOLUTION Although this equation involves more than one variable, we solve it as usual by isolating M on one side and treating the other variables as we would numbers.

$$F = \left(\frac{Gm}{r^2}\right)M \qquad \text{Factor } M \text{ from RHS}$$

$$\left(\frac{r^2}{Gm}\right)F = \left(\frac{r^2}{Gm}\right)\left(\frac{Gm}{r^2}\right)M \qquad \text{Multiply by reciprocal of } \frac{Gm}{r^2}$$

$$\frac{r^2 F}{Gm} = M \qquad \text{Simplify}$$

The solution is $M = \dfrac{r^2 F}{Gm}$.

■

QUADRATIC EQUATIONS

Linear equations are first-degree equations of the form $ax + b = 0$. Quadratic equations are second-degree equations; they contain an additional term involving the square of the variable.

QUADRATIC EQUATIONS

A **quadratic equation** is an equation equivalent to one of the form

$$ax^2 + bx + c = 0$$

where a, b, and c are real numbers with $a \neq 0$.

Some quadratic equations can be solved by factoring and using the following basic property of real numbers.

ZERO-PRODUCT PROPERTY

$$AB = 0 \quad \text{if and only if} \quad A = 0 \text{ or } B = 0$$

This means that if we can factor the left-hand side of a quadratic (or other) equation, then we can solve it by setting each factor equal to 0 in turn. This method works only when the right-hand side of the equation is 0.

EXAMPLE 4 ■ Solving a Quadratic Equation by Factoring

Solve the equation $x^2 + 5x = 24$.

SOLUTION   We must first rewrite the equation so that the right-hand side is 0.

$$x^2 + 5x = 24$$

$$x^2 + 5x - 24 = 0 \qquad \text{Subtract 24}$$

$$(x - 3)(x + 8) = 0 \qquad \text{Factor}$$

$$x - 3 = 0 \quad \text{or} \quad x + 8 = 0 \qquad \text{Set each factor equal to 0}$$

$$x = 3 \qquad\qquad x = -8 \qquad \text{Solve}$$

The solutions are $x = 3$ and $x = -8$. ∎

CHECK YOUR ANSWERS

$x = 3$:
$$(3)^2 + 5(3) = 9 + 15 = 24 \;\checkmark$$
$x = -8$:
$$(-8)^2 + 5(-8) = 64 - 40 = 24 \;\checkmark$$

A quadratic equation of the form $x^2 - c = 0$, where c is a positive constant, factors as $(x - \sqrt{c})(x + \sqrt{c}) = 0$, and so the solutions are $x = \sqrt{c}$ and $x = -\sqrt{c}$. We often abbreviate this as $x = \pm\sqrt{c}$.

SOLVING A SIMPLE QUADRATIC EQUATION

The solutions of the equation $x^2 = c$ are $x = \sqrt{c}$ and $x = -\sqrt{c}$.

Euclid (circa 300 B.C.) taught in Alexandria. His *Elements* is the most widely influential scientific book in history. For 2000 years it was the standard introduction to geometry in schools, and for many generations it was considered the best way to develop logical reasoning. Abraham Lincoln, for instance, studied the *Elements* as a way to sharpen his mind. The story is told that King Ptolemy once asked Euclid if there was a faster way to learn geometry than through the *Elements*. Euclid replied that there is "no royal road to geometry"—meaning by this that mathematics does not respect wealth or social status. Euclid was revered in his own time and was referred to by the title "The Geometer" or "The Writer of the *Elements*." The greatness of the *Elements* stems from its precise, logical, and systematic treatment of geometry. For dealing with equality, Euclid listed the following rules, which he called "common notions."

1. Things that are equal to the same thing are equal to each other.

2. If equals are added to equals, the sums are equal.

3. If equals are subtracted from equals, the remainders are equal.

4. Things that coincide with one another are equal.

5. The whole is greater than the part.

EXAMPLE 5 ■ **Solving Simple Quadratics**

Solve each equation: (a) $x^2 = 5$ (b) $(x - 4)^2 = 5$

SOLUTION

(a) From the principle in the preceding box, we get $x = \pm\sqrt{5}$.

(b) We can take the square root of each side of this equation as well.

$$(x - 4)^2 = 5$$

$$x - 4 = \pm\sqrt{5} \qquad \text{Take the square root}$$

$$x = 4 \pm \sqrt{5} \qquad \text{Add 4}$$

The solutions are $x = 4 + \sqrt{5}$ and $x = 4 - \sqrt{5}$.   ■

As we saw in Example 5, if a quadratic equation is of the form $(x + a)^2 = c$, then we can solve it by taking the square root of each side. In an equation of this form, the left-hand side is a *perfect square*: the square of a linear expression in x. So, if a quadratic equation doesn't factor readily, then we can solve it using the technique of **completing the square**. This means that we add a constant to an expression to make it a perfect square. For example, to make $x^2 - 6x$ a perfect square we must add 9, since $x^2 - 6x + 9 = (x - 3)^2$. In general, from the identity

$$x^2 + bx + \left(\frac{b}{2}\right)^2 = \left(x + \frac{b}{2}\right)^2$$

it follows that to make $x^2 + bx$ a perfect square, we must add the square of half the coefficient of x. (Note that this works whether b is positive or negative.)

COMPLETING THE SQUARE

To make $x^2 + bx$ a perfect square, add $\left(\frac{b}{2}\right)^2$.

EXAMPLE 6 ■ **Solving Quadratic Equations by Completing the Square**

Solve each equation.

(a) $x^2 - 8x + 13 = 0$ (b) $3x^2 - 12x + 4 = 0$

Completing the Square

Area of blue region is

$$x^2 + 2\left(\frac{b}{2}\right)x = x^2 + bx$$

Add a small square of area $\left(\dfrac{b}{2}\right)^2$ to "complete" the square.

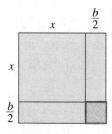

SOLUTION

(a) $x^2 - 8x + 13 = 0$

$$x^2 - 8x \qquad = -13 \qquad \text{Subtract 13}$$

$$x^2 - 8x + 16 = -13 + 16 \qquad \text{Complete the square: add } \left(\frac{-8}{2}\right)^2 = 16$$

$$(x - 4)^2 = 3 \qquad \text{Perfect square}$$

$$x - 4 = \pm\sqrt{3} \qquad \text{Take the square root}$$

$$x = 4 \pm \sqrt{3} \qquad \text{Add 4}$$

(b) After subtracting 4 from each side of the equation, we must factor the coefficient of x^2 (the 3) from the left side to put the equation in the correct form for completing the square.

$$3x^2 - 12x + 4 = 0$$

$$3x^2 - 12x \qquad = -4 \qquad \text{Subtract 4}$$

$$3(x^2 - 4x \quad) = -4 \qquad \text{Factor 3 from LHS}$$

Now we complete the square by adding $(-2)^2 = 4$ inside the parentheses. Since everything inside the parentheses is multiplied by 3, this means that we are actually adding $3 \cdot 4 = 12$ to the left side of the equation. Thus, we must add 12 to the right side as well.

$$3(x^2 - 4x + 4) = -4 + 3 \cdot 4 \qquad \text{Complete the square}$$

$$3(x - 2)^2 = 8 \qquad \text{Perfect square}$$

$$(x - 2)^2 = \frac{8}{3} \qquad \text{Divide by 3}$$

$$x - 2 = \pm\sqrt{\frac{8}{3}} = \pm\frac{2\sqrt{6}}{3} \qquad \begin{array}{l}\text{Take the square root}\\\text{Rationalize the denominator}\end{array}$$

$$x = 2 \pm \frac{2\sqrt{6}}{3} \qquad \text{Add 2} \qquad \blacksquare$$

We can use the technique of completing the square to derive a formula for the roots of the general quadratic equation $ax^2 + bx + c = 0$. First, we divide each side of the equation by a and move the constant to the right side, giving

$$x^2 + \frac{b}{a}x = -\frac{c}{a}$$

We now complete the square by adding $[b/(2a)]^2$ to each side of the equation:

$$x^2 + \frac{b}{a}x + \left(\frac{b}{2a}\right)^2 = -\frac{c}{a} + \left(\frac{b}{2a}\right)^2$$

$$\left(x + \frac{b}{2a}\right)^2 = \frac{-4ac + b^2}{4a^2} \qquad \text{Perfect square}$$

$$x + \frac{b}{2a} = \pm\sqrt{\frac{-4ac + b^2}{4a^2}} = \pm\frac{\sqrt{b^2 - 4ac}}{2a} \qquad \text{Take square root}$$

$$x = \frac{-b \pm \sqrt{b^2 - 4ac}}{2a} \qquad \text{Subtract } \frac{b}{2a}$$

This is the quadratic formula.

François Viète (1540–1603) was a French mathematician, sometimes known by the Latin form of his name, Vieta. He introduced a new level of abstraction in algebra by using letters to stand for *known* quantities in an equation. Before Viète's time, each equation had to be solved on its own. For instance, the quadratic equations

$$3x^2 + 2x + 8 = 0$$
$$5x^2 - 6x + 4 = 0$$

had to be solved separately for the unknown x. Viète's idea was to consider all quadratic equations at once by writing

$$ax^2 + bx + c = 0$$

where a, b, and c are known quantities. Thus, it is possible to write a *formula* (in this case, the quadratic formula) involving a, b, and c that can be used to solve all such equations in one fell swoop.

THE QUADRATIC FORMULA

The roots of the quadratic equation $ax^2 + bx + c = 0$, where $a \neq 0$, are

$$x = \frac{-b \pm \sqrt{b^2 - 4ac}}{2a}$$

The quadratic formula could be used to solve the equations in Examples 4, 5, and 6. You should carry out the details of these calculations.

EXAMPLE 7 ■ Using the Quadratic Formula

Find all solutions of each equation.

(a) $3x^2 - 5x - 1 = 0$ (b) $4x^2 + 12x + 9 = 0$ (c) $x^2 + 2x + 2 = 0$

SOLUTION

(a) Using the quadratic formula with $a = 3$, $b = -5$, and $c = -1$, we get

$$x = \frac{-(-5) \pm \sqrt{(-5)^2 - 4(3)(-1)}}{2(3)} = \frac{5 \pm \sqrt{37}}{6}$$

If approximations are desired, we use a calculator and obtain

$$x = \frac{5 + \sqrt{37}}{6} \approx 1.8471 \qquad \text{and} \qquad x = \frac{5 - \sqrt{37}}{6} \approx -0.1805$$

Another Method

$$4x^2 + 12x + 9 = 0$$
$$(2x + 3)^2 = 0$$
$$2x + 3 = 0$$
$$x = -\tfrac{3}{2}$$

(b) Using the quadratic formula with $a = 4$, $b = 12$, and $c = 9$ gives

$$x = \frac{-12 \pm \sqrt{(12)^2 - 4 \cdot 4 \cdot 9}}{2 \cdot 4} = \frac{-12 \pm 0}{8} = -\frac{3}{2}$$

This equation has only one solution, $x = -\tfrac{3}{2}$.

(c) Using the quadratic formula with $a = 1$, $b = 2$, and $c = 2$ gives

$$x = \frac{-2 \pm \sqrt{2^2 - 4 \cdot 2}}{2} = \frac{-2 \pm \sqrt{-4}}{2} = \frac{-2 \pm 2\sqrt{-1}}{2} = -1 \pm \sqrt{-1}$$

Since the square of any real number is nonnegative, $\sqrt{-1}$ is undefined in the real number system. The equation has no real solution. ∎

In Section 3.3 we will study the complex number system, in which the square roots of negative numbers do exist. The equation in Example 7(c) does have solutions in the complex number system.

The quantity $b^2 - 4ac$ that appears under the square root sign in the quadratic formula is called the **discriminant** of the equation $ax^2 + bx + c = 0$ and is given the symbol D. If $D < 0$, then $\sqrt{b^2 - 4ac}$ is undefined, so the quadratic equation has no real solution, as in Example 7(c). If $D = 0$, then the equation has only one real solution, as in Example 7(b). Finally, if $D > 0$, then the equation has two distinct real solutions, as in Example 7(a).

The following box summarizes what we have observed about the discriminant.

THE DISCRIMINANT

The discriminant of the general quadratic equation $ax^2 + bx + c = 0$ ($a \neq 0$) is $D = b^2 - 4ac$.

1. If $D > 0$, then the equation has two distinct real solutions.

2. If $D = 0$, then the equation has exactly one real solution.

3. If $D < 0$, then the equation has no real solution.

EXAMPLE 8 ■ **Using the Discriminant**

Use the discriminant to determine how many real solutions each equation has.

(a) $x^2 + 2x + 8 = 0$ (b) $3x^2 - 5x + \frac{3}{2} = 0$

SOLUTION

(a) The discriminant is

$$D = 2^2 - 4 \cdot 1 \cdot 8 = -28 < 0$$

Thus, the equation has no real solution.

(b) The discriminant is $D = (-5)^2 - 4 \cdot 3 \cdot \frac{3}{2} = 25 - 18 = 7 > 0$. Thus, the equation has two distinct real solutions. ∎

OTHER EQUATIONS

We now consider other types of equations, including those that involve higher powers or radicals.

EXAMPLE 9 ■ **Solving Equations Using nth Roots**

Find all real solutions of each of the following equations.

(a) $x^3 = -8$ (b) $16x^4 = 81$

SOLUTION

(a) Since every real number has exactly one real cube root, we can solve this equation by taking the cube root of each side.

$$(x^3)^{1/3} = (-8)^{1/3}$$

$$x = -2$$

(b) Here we must remember that if n is even, then every positive real number has *two* real nth roots, a positive one and a negative one.

$$x^4 = \tfrac{81}{16} \qquad \text{Divide by 16}$$

$$(x^4)^{1/4} = \pm\left(\tfrac{81}{16}\right)^{1/4} \qquad \text{Take the fourth root}$$

$$x = \pm\tfrac{3}{2}$$

   ■

Some equations that at first glance may not appear to be quadratic can be changed into quadratic equations by performing simple algebraic operations on them, such as multiplying each side by a common denominator or squaring each side. Special care must be taken in solving such equations to avoid keeping *extraneous solutions* that may be introduced when we manipulate the equations.

EXAMPLE 10 ■ **Taking Care of Extraneous Solutions**

Solve each equation.

(a) $x + 3 = \dfrac{-2x^2 + 7x - 3}{x - 3}$ (b) $x = 1 - \sqrt{2 - \dfrac{x}{2}}$

SOLUTION

(a) We can multiply each side of the equation by $x - 3$ to clear the denominator as long as $x \neq 3$.

$$(x + 3)(x - 3) = \left(\frac{-2x^2 + 7x - 3}{x - 3}\right)(x - 3)$$

$$x^2 - 9 = -2x^2 + 7x - 3 \qquad \text{Expand and cancel}$$

$$3x^2 - 7x - 6 = 0 \qquad \begin{array}{l}\text{Move all terms to LHS}\\\text{and simplify}\end{array}$$

$$(3x + 2)(x - 3) = 0 \qquad \text{Factor}$$

$$x = -\tfrac{2}{3} \quad \text{or} \quad x = 3 \qquad \text{Set each factor equal to 0}$$

Thus, $x = -\tfrac{2}{3}$ or $x = 3$. But $x = 3$ does not satisfy the original equation (since division by 0 is impossible), so the only solution is $x = -\tfrac{2}{3}$.

CHECK YOUR ANSWERS

$x = -\tfrac{2}{3}$:

  LHS $= \left(-\tfrac{2}{3}\right) + 3 = \tfrac{7}{3}$

  RHS $= \dfrac{-2\left(-\tfrac{2}{3}\right)^2 + 7\left(-\tfrac{2}{3}\right) - 3}{-\tfrac{2}{3} - 3}$

    $= \dfrac{-\tfrac{8}{9} - \tfrac{14}{3} - 3}{-\tfrac{11}{3}} = \dfrac{\tfrac{77}{9}}{\tfrac{11}{3}} = \tfrac{7}{3}$

  LHS = RHS   ✓

$x = 3$:

  LHS $= 3 + 3 = 6$

  RHS $= \dfrac{-2(3)^2 + 7(3) - 3}{3 - 3} = \dfrac{0}{0}$

  RHS is undefined, so $x = 3$ is not a solution.   ✗

(b) To eliminate the square root, we first isolate it on one side of the equal sign.

$$x - 1 = -\sqrt{2 - \frac{x}{2}} \qquad \text{Subtract 1}$$

$$(x - 1)^2 = 2 - \frac{x}{2} \qquad \text{Square each side to elimate square root}$$

$$x^2 - 2x + 1 = 2 - \frac{x}{2} \qquad \text{Expand LHS}$$

$$2x^2 - 4x + 2 = 4 - x \qquad \text{Multiply by 2 to clear denominator}$$

$$2x^2 - 3x - 2 = 0 \qquad \text{Move all terms to LHS}$$

$$(x - 2)(2x + 1) = 0 \qquad \text{Factor}$$

Setting each factor equal to 0 gives us $x = 2$ and $x = -\frac{1}{2}$ as potential solutions. If we substitute these into the original equation (see *Check Your Answers*), we see that $x = -\frac{1}{2}$ is a solution but $x = 2$ is not. The only solution is

$$x = -\frac{1}{2} \qquad \blacksquare$$

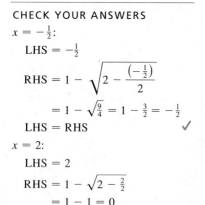

CHECK YOUR ANSWERS

$x = -\frac{1}{2}$:

   LHS $= -\frac{1}{2}$

   RHS $= 1 - \sqrt{2 - \dfrac{(-\frac{1}{2})}{2}}$

   $= 1 - \sqrt{\frac{9}{4}} = 1 - \frac{3}{2} = -\frac{1}{2}$

   LHS $=$ RHS   ✓

$x = 2$:

   LHS $= 2$

   RHS $= 1 - \sqrt{2 - \frac{2}{2}}$

   $= 1 - 1 = 0$

   LHS $\neq$ RHS, so $x = 2$ is not a solution.   ✗

The reason extraneous solutions are often introduced when we square each side of an equation is that the operation of squaring can turn a false equation into a true one. For example, $-1 \neq 1$, but $(-1)^2 = 1^2$. Thus, the squared equation may be true for more values of the variable than the original equation, so we must always check our answers to make sure that each satisfies the original equation.

In some cases, fourth-degree (or higher-degree) polynomial equations can be changed to quadratic equations by performing algebraic substitutions, as shown in the following example.

EXAMPLE 11 ■ An Equation of Quadratic Type

Find the real solutions of $x^4 - 2x^2 - 2 = 0$.

SOLUTION   If we set $w = x^2$, then we get a quadratic equation in the new variable w.

$$x^4 - 2x^2 - 2 = (x^2)^2 - 2x^2 - 2$$

$$= w^2 - 2w - 2 = 0 \qquad \text{Set } w = x^2$$

From the quadratic formula, we have

$$w = \frac{2 \pm \sqrt{(-2)^2 + 4 \cdot 2}}{2} = 1 \pm \sqrt{3}$$

Because $x^2 = w$, we have $x = \pm\sqrt{w}$. So the potential solutions are $\pm\sqrt{1 \pm \sqrt{3}}$. Since we can't take the square root of a negative number such as $1 - \sqrt{3}$ in

the real number system, we have

$$x = \pm\sqrt{1 + \sqrt{3}}$$

as the only real solutions of the original equation.   ■

EXAMPLE 12 ■ **Solving an Equation Involving Fractional Exponents**

Find all solutions of the equation $x^{5/6} + x^{2/3} = 2x^{1/2}$.

SOLUTION

$$x^{1/6} = w$$
$$x^{1/3} = w^2$$
$$x^{1/2} = w^3$$

$x^{5/6} + x^{2/3} - 2x^{1/2} = 0$	Move all terms to LHS
$x^{1/2}(x^{1/3} + x^{1/6} - 2) = 0$	Factor $x^{1/2}$ (lowest power of x)
$w^3(w^2 + w - 2) = 0$	Substitute $w = x^{1/6}$
$w^3(w - 1)(w + 2) = 0$	Factor the quadratic

So $w = 0$ or $w = 1$ or $w = -2$

$\quad\quad\quad x^{1/6} = 0 \quad\quad\quad x^{1/6} = 1 \quad\quad\quad x^{1/6} = -2$

$\quad\quad\quad\quad x = 0 \quad\quad\quad x = 1^6 = 1 \quad\quad x = (-2)^6 = 64$

Checking these answers, we see that $x = 0$ and $x = 1$ are solutions, but $x = 64$ is not (see *Check Your Answers*). The only solutions are 0 and 1.

CHECK YOUR ANSWERS

$x = 0$:
   LHS $= 0^{5/6} + 0^{2/3}$
      $= 0$
   RHS $= 2 \cdot 0^{1/2}$
      $= 0$
   LHS $=$ RHS ✓

$x = 1$:
   LHS $= 1^{5/6} + 1^{2/3}$
      $= 2$
   RHS $= 2 \cdot 1^{1/2}$
      $= 2$
   LHS $=$ RHS ✓

$x = 64$:
   LHS $= 64^{5/6} + 64^{2/3}$
      $= 32 + 16 = 48$
   RHS $= 2 \cdot 64^{1/2}$
      $= 2 \cdot 8 = 16$
   LHS $\neq$ RHS ✗   ■

To divide each side of the original equation in Example 12 by the common factor $x^{1/2}$ would have been wrong, because we would have lost the solution $x = 0$ by doing so. Never divide both sides of an equation by an expression containing the variable (unless you know that expression can never equal 0).

EXAMPLE 13 ■ **An Absolute-Value Equation**

Solve the equation $|2x - 5| = 3$.

SOLUTION   By the definition of absolute value, $|2x - 5| = 3$ is equivalent to

$$2x - 5 = 3 \quad\quad \text{or} \quad\quad 2x - 5 = -3$$
$$2x = 8 \quad\quad\quad\quad\quad\quad 2x = 2$$
$$x = 4 \quad\quad\quad\quad\quad\quad\quad x = 1$$

The solutions are $x = 1$, $x = 4$.   ■

| 1.5 | **EXERCISES** |

1–4 ■ Determine whether each of the given values of the variable is a solution of the equation.

1. $2x - 3 = x + 1$
  (a) $x = 4$ (b) $x = \frac{3}{2}$

2. $4(x - 1) - (2 - x) = 5(x - 2) + 4$
  (a) $x = -3$ (b) $x = 0$

3. $\dfrac{1}{x} - \dfrac{1}{x + 3} = \dfrac{1}{6}$
  (a) $x = -3$ (b) $x = 3$

4. $\sqrt{x^2 - 7} = x - 1$
  (a) $x = 4$ (b) $x = -4$

5–28 ■ Solve the equation.

5. $3x - 5 = 7$ **6.** $4x + 7 = 9x - 13$

7. $-7w = 15 - 2w$ **8.** $5t - 13 = 12 - 5t$

9. $\frac{1}{2}y - 2 = \frac{1}{3}y$ **10.** $\dfrac{z}{5} = \dfrac{3}{10}z + 7$

11. $2(1 - x) = 3(1 + 2x) + 5$

12. $5(x + 3) + 9 = -2(x - 2) - 1$

13. $\dfrac{1}{x} = \dfrac{4}{3x} + 1$ **14.** $\dfrac{2}{t + 6} = \dfrac{3}{t - 1}$

15. $\dfrac{1}{t - 1} + \dfrac{t}{3t - 2} = \dfrac{1}{3}$

16. $r - 2[1 - 3(2r + 4)] = 61$

17. $(t - 4)^2 = (t + 4)^2 + 32$

18. $\frac{2}{3}x - \frac{1}{4} = \frac{1}{6}x - \frac{1}{9}$

19. $\dfrac{2}{x} - 5 = \dfrac{6}{x} + 4$

20. $\dfrac{4}{x - 1} + \dfrac{2}{x + 1} = \dfrac{35}{x^2 - 1}$

21. $\dfrac{2x - 7}{2x + 4} = \dfrac{2}{3}$

22. $\dfrac{1}{z} - \dfrac{1}{2z} - \dfrac{1}{5z} = \dfrac{10}{z + 1}$

23. $\sqrt{x - 4} = \sqrt{2x}$

24. $\sqrt{2x + 8} = \sqrt{6x}$

25. $\dfrac{1}{x + 3} + \dfrac{5}{x^2 - 9} = \dfrac{2}{x - 3}$

26. $x^2 = 49$

27. $x^2 = 18$ **28.** $x^2 - 24 = 0$

29–32 ■ Solve the equation by factoring.

29. $x^2 - x - 6 = 0$ **30.** $x^2 + 2x = 8$

31. $x^2 + 4 = 4x$ **32.** $2y^2 + 7y + 3 = 0$

33–36 ■ Solve the equation by completing the square.

33. $x^2 - 4x + 2 = 0$ **34.** $x^2 - 6x - 9 = 0$

35. $x^2 + x - \frac{3}{4} = 0$ **36.** $2x^2 + 8x + 1 = 0$

37–74 ■ Find all real solutions of the equation.

37. $x^2 - 2x - 8 = 0$ **38.** $2x^2 + x - 3 = 0$

39. $x^2 + 12x - 27 = 0$ **40.** $8x^2 - 6x - 9 = 0$

41. $3x^2 + 6x - 5 = 0$ **42.** $x^2 - 6x + 1 = 0$

43. $4x^2 + 16x - 9 = 0$ **44.** $0 = x^2 - 4x + 1$

45. $3 + 5z + z^2 = 0$ **46.** $w^2 = 3(w - 1)$

47. $x^2 - \sqrt{5}\,x + 1 = 0$

48. $\sqrt{6}\,x^2 + 2x - \sqrt{\frac{3}{2}} = 0$

49. $\dfrac{x^2}{x + 100} = 50$

50. $1 + \dfrac{2x}{(x + 3)(x + 4)} = \dfrac{2}{x + 3} + \dfrac{4}{x + 4}$

51. $\dfrac{x + 5}{x - 2} = \dfrac{5}{x + 2} + \dfrac{28}{x^2 - 4}$

52. $\dfrac{1}{x - 1} - \dfrac{2}{x^2} = 0$

53. $x^4 - 16 = 0$ **54.** $64x^6 = 27$

55. $x^4 - x^3 - 2x^2 = 0$

56. $(x - 2)^5 - 9(x - 2)^3 = 0$

57. $\sqrt{2x + 1} + 1 = x$ **58.** $x - \sqrt{9 - 3x} = 0$

59. $\sqrt{5 - x} + 1 = x - 2$ **60.** $2x + \sqrt{x + 1} = 8$

61. $\sqrt{\sqrt{x - 5} + x} = 5$

62. $(x + 5)^2 - 3(x + 5) - 10 = 0$

63. $4(x + 1)^{1/2} - 5(x + 1)^{3/2} + (x + 1)^{5/2} = 0$

64. $x^{1/2} + 3x^{-1/2} = 10x^{-3/2}$

65. $x^{4/3} - 5x^{2/3} + 6 = 0$

66. $\sqrt{x} - 3\sqrt[4]{x} - 4 = 0$

67. $x^{1/2} - 3x^{1/3} = 3x^{1/6} - 9$

68. $x - 5\sqrt{x} + 6 = 0$

69. $\dfrac{1}{x^3} + \dfrac{4}{x^2} + \dfrac{4}{x} = 0$

70. $4x^{-4} - 16x^{-2} + 4 = 0$

71. $|2x| = 3$ **72.** $|3x + 5| = 1$

73. $|x - 4| = 0.01$ **74.** $|x - 6| = -1$

75–78 ■ Find the solution(s) of the equation correct to two decimal places.

75. $2.15x - 4.63 = x + 1.19$

76. $3.16(x + 4.63) = 4.19(x - 7.24)$

77. $x^2 - 2.45x + 1.50 = 0$

78. $x^2 - 2.45x + 1.51 = 0$

79–90 ■ Solve the equation for the indicated variable.

79. $PV = nRT$;   for R

80. $P = 2l + 2w$;   for w

81. $A = 2lw + 2wh + 2lh$;   for h

82. $\dfrac{1}{R} = \dfrac{1}{R_1} + \dfrac{1}{R_2}$;   for R_1

83. $\dfrac{ax + b}{cx + d} = 2$;   for x

84. $a - 2[b - 3(c - x)] = 6$;   for x

85. $V = \frac{1}{3}\pi r^2 h$;   for r

86. $F = G\dfrac{mM}{r^2}$;   for r

87. $a^2 + b^2 = c^2$;   for b

88. $S = \dfrac{n(n + 1)}{2}$;   for n

89. $A = P\left(1 + \dfrac{i}{100}\right)^2$;   for i

90. $\dfrac{1}{s + a} + \dfrac{1}{s + b} = \dfrac{1}{c}$;   for s

91–94 ■ Use the discriminant to determine the number of real solutions of the equation. Do not solve the equation.

91. $x^2 - 6x + 1 = 0$

92. $x^2 = 6x - 9$

93. $x^2 + 2.20x + 1.21 = 0$

94. $x^2 + 2.21x + 1.21 = 0$

95–96 ■ Find all values for k that ensure that the given equation has exactly one solution.

95. $4x^2 + kx + 25 = 0$

96. $kx^2 + 36x + k = 0$

⬤ DISCOVERY · DISCUSSION

97. Find the Error   Find the mistake in the following solution, and then solve the equation correctly:

$$x^2 + 6x + 5 = x^2 - 1$$

$$(x + 5)(x + 1) = (x - 1)(x + 1) \qquad \text{Factor}$$

$$x + 5 = x - 1 \qquad \text{Divide by } x + 1$$

$$5 \overset{?}{=} -1 \qquad \text{Subtract } x$$

98. Volumes of Solids   The sphere, cylinder, and cone shown here all have the same radius r and the same volume V. Express the heights of the cylinder and cone in terms of r. (Use the volume formulas given on the inside of the front cover of this book.)

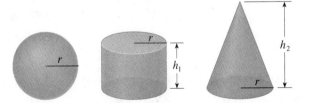

99. Types of Equations   Give an example of each of the following types of equations, all containing the variable x: an identity, a linear equation, a quadratic equation, and an equation that has no solution. Describe the solution set of each equation.

1.6 PROBLEM SOLVING WITH EQUATIONS

Many problems in the sciences, economics, finance, medicine, and numerous other fields can be translated into algebra problems; this is one reason that algebra is so useful. In this section we study some problem-solving principles to help us tackle applied problems.

EXAMPLE 1 ■ Interest on an Investment

Mary inherits $100,000 and invests it in two certificates of deposit. One certificate pays 6% and the other pays $4\frac{1}{2}\%$ simple interest annually. If Mary's total interest is $5025 per year, how much money is invested at each rate?

SOLUTION The key to solving any applied problem is to translate the information given into the language of algebra, that is, into an equation. First, we need to identify the variables. This can usually be done by a careful reading of the question asked in the problem. Here we are asked to find out how much money is invested at each rate. So we let

Identify the variable

$$x = \text{amount invested at } 6\%$$

Express all unknown quantities in terms of the variable

The rest of Mary's money is invested at $4\frac{1}{2}\%$, so

$$100{,}000 - x = \text{amount invested at } 4\frac{1}{2}\%$$

Now we translate the fact that her total annual interest is $5025 into an equation:

Relate the quantities

$$(\text{interest at } 6\%) + (\text{interest at } 4\tfrac{1}{2}\%) = 5025$$

Since x dollars are invested at 6%, the annual interest paid by this certificate is 6% of x, or $0.06x$. Similarly, the interest received from the $4\frac{1}{2}\%$ certificate will be $0.045(100{,}000 - x)$, and so the interest equation is expressed in terms of x as

Set up an equation

$$0.06x + 0.045(100{,}000 - x) = 5025$$

Solve

Now we solve for x:

$$0.06x + 4500 - 0.045x = 5025 \qquad \text{Multiply}$$

$$0.015x + 4500 = 5025 \qquad \text{Combine the } x \text{ terms}$$

$$0.015x = 525 \qquad \text{Subtract 4500}$$

$$x = \frac{525}{0.015} = 35{,}000$$

So Mary has invested $35,000 at 6% and the remaining $65,000 at $4\frac{1}{2}\%$. ■

CHECK YOUR ANSWER

total interest = 6% of $35,000
+
$4\frac{1}{2}\%$ of $65,000
= $2100 + $2925
= $5025   ✓

We now adapt the problem-solving principles given on pages 122–124 to the process of translating a "word problem" from English into algebra. The following guidelines should help you set up the equation that expresses the English statement of a problem in the language of algebra. These guidelines are referred to in the margin notes for the examples in this section.

GUIDELINES FOR SOLVING WORD PROBLEMS

1. IDENTIFY THE VARIABLE. Identify the quantity that the problem asks you to find. This quantity can usually be determined by a careful reading of the question posed at the end of the problem. **Introduce notation** for the variable (call it x or some other letter). Make sure to write down precisely what the variable represents.

2. EXPRESS ALL UNKNOWN QUANTITIES IN TERMS OF THE VARIABLE. Read each sentence in the problem again, and express all the quantities mentioned in the problem in terms of the variable you defined in Step 1. To organize this information, it is sometimes helpful to **draw a diagram** or **make a table** (see Examples 2 and 6 in this section).

3. RELATE THE QUANTITIES. Find the crucial fact in the problem that relates two or more of the expressions you listed in Step 2. A statement that one quantity "is," "equals," or "is the same as" another often signals the type of relationship we are looking for.

4. SET UP AN EQUATION. Set up an equation that expresses the crucial fact you found in Step 3 in algebraic form. You will often need to use a formula to translate from English into algebra. (For instance, in Example 1 we needed to know that *interest = rate × principal* to set up the equation.)

5. SOLVE THE PROBLEM AND CHECK YOUR ANSWER. Solve the equation, check to make sure your solution satisfies the original problem, and express your final answer as an English sentence that answers the question posed in the problem.

Pythagoras (circa 580–500 B.C.) founded a school in Croton in southern Italy, devoted to the study of arithmetic, geometry, music, and astronomy. The Pythagoreans, as they were called, were a secret society with peculiar rules and initiation rites. They wrote nothing down, and were not to reveal to anyone what they had learned from the Master. Although women were barred by law from attending public meetings, Pythagoras allowed women in his school, and his most famous student was Theano (whom he later married).

According to Aristotle, the Pythagoreans were convinced that "the principles of mathematics are the principles of all things." Their motto was "Everything is Number," by which they meant *whole* numbers. The outstanding contribution of Pythagoras is the theorem that bears his name: In a right triangle the area of the square on the hypotenuse is equal to the sum of the areas of the squares on the other two sides.

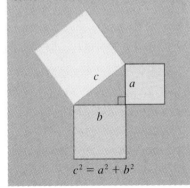

$$c^2 = a^2 + b^2$$

EXAMPLE 2 ■ Dimensions of a Poster

A poster has on it a rectangular printed area 100 cm by 140 cm, with a blank strip of uniform width around the four edges. The perimeter of the poster is $1\frac{1}{2}$ times the perimeter of the printed area. What is the width of the blank strip, and what are the dimensions of the poster?

SOLUTION   In a problem such as this, which involves geometry, it is essential to draw a diagram like the one shown in Figure 1 on page 60.

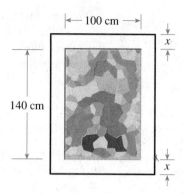

FIGURE 1

Let

| Identify the variable |

$$x = \text{width of the blank strip}$$

From the figure we see that the poster is $(100 + 2x)$ cm by $(140 + 2x)$ cm, so its perimeter is $2(100 + 2x) + 2(140 + 2x)$. The perimeter of the printed area is $2(100) + 2(140) = 480$ cm. We are told that

| Express all unknown quantities in terms of the variable |

| Relate the quantities |

$$(\text{perimeter of poster}) = \tfrac{3}{2} \times (\text{perimeter of printed part})$$

so

| Set up an equation |

| Solve |

$$2(100 + 2x) + 2(140 + 2x) = \tfrac{3}{2} \cdot 480$$

$$480 + 8x = 720 \qquad \text{Expand and combine like terms on LHS}$$

$$8x = 240 \qquad \text{Subtract 480}$$

$$x = 30 \qquad \text{Divide by 8}$$

The blank strip is 30 cm wide, so the dimensions of the poster are

$$100 + 30 + 30 = 160 \text{ cm wide}$$

by

$$140 + 30 + 30 = 200 \text{ cm high} \qquad\blacksquare$$

EXAMPLE 3 ■ Determining the Height of a Building Using Similar Triangles

A man 6 ft tall wishes to find the height of a certain four-story building. He measures its shadow and finds it to be 28 ft long, while his own shadow is $3\frac{1}{2}$ ft long. How tall is the building?

| Identify the variable |

| Relate the quantities |

| Set up an equation |

| Solve |

SOLUTION   First we let h represent the height of the building. We can use the fact that the triangles in Figure 2 are similar to relate the sides to each other. Since the ratios of corresponding sides are the same in pairs of similar triangles, we get the equation

$$\frac{h}{28} = \frac{6}{3.5}$$

$$h = \frac{6 \cdot 28}{3.5} = 48$$

Thus, the building is 48 ft tall.

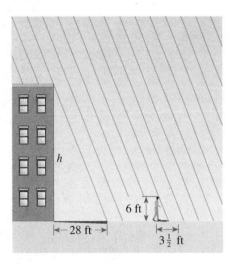

FIGURE 2

EXAMPLE 4 ■ Mixtures and Concentration

A manufacturer of soft drinks makes a type of orange soda that is advertised as "naturally flavored," although it contains only 5% orange juice. A new federal regulation stipulates that to be called "natural" a drink must contain at least 10% fruit juice. How much pure orange juice must this manufacturer add to 900 gal of orange soda to conform to the new regulation?

SOLUTION   In any problem of this type—in which two different substances are to be mixed—a diagram helps us set up the required equation (see Figure 3).

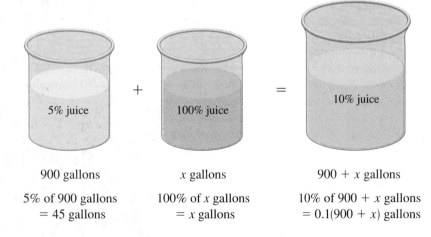

	5% juice	100% juice	10% juice
Volume	900 gallons	x gallons	$900 + x$ gallons
Amount of orange juice	5% of 900 gallons = 45 gallons	100% of x gallons = x gallons	10% of $900 + x$ gallons = $0.1(900 + x)$ gallons

FIGURE 3

Identify the variable

Let x be the amount (in gallons) of pure juice to be added. Then $900 + x$ gallons of 10% orange juice mixture will result. The key idea to turn the picture into an equation here is to notice that the total amount of juice on both sides of the equal sign is the same. The orange juice in the first vat is 5% of 900 gal, or 45 gal. The second vat contains x gallons of juice, and the third contains $0.1(900 + x)$ gallons.

Equating the total amounts of pure juice before and after mixing, we get the equation

| | Express all unknown quantities in terms of the variable |

amount of juice before mixing = amount of juice after mixing

	Relate the quantities
	Set up an equation
	Solve

$$45 + x = 0.1(900 + x) \qquad \text{From Figure 3}$$

$$45 + x = 90 + 0.1x \qquad \text{Multiply}$$

$$0.9x = 45 \qquad \text{Subtract } 0.1x \text{ and } 45$$

$$x = \frac{45}{0.9} = 50 \qquad \text{Divide by } 0.9$$

The manufacturer should add 50 gal of pure orange juice to the soda.

CHECK YOUR ANSWER

amount of juice before mixing = 5% of 900 gal + 50 gal pure juice

= 45 gal + 50 gal = 95 gal

amount of juice after mixing = 10% of 950 gal = 95 gal

Amounts are equal. ✓

EXAMPLE 5 ■ Time Needed to Do a Job

Because of an anticipated heavy rainstorm, the water level in a reservoir must be lowered by 1 ft. Opening spillway A lowers the level by this amount in 4 hours, whereas opening the smaller spillway B does the job in 6 hours. How long will it take to lower the water level by 1 ft if both spillways are opened?

FIGURE 4

| | Identify the variable |

SOLUTION As usual, we represent the number we are seeking by x.

x = number of hours it takes to lower the water level by 1 ft if both spillways are open

Finding an equation relating x to the other quantities in this problem is not easy. Certainly x is not simply $4 + 6$, since that would mean that both spillways together require longer to lower the water level than either spillway alone. Instead, *we look at the fraction of the job that can be done in one hour by each spillway.*

| | Express all quantities in terms of the variable |

Spillway A lowers the water by $\frac{1}{4}$ ft in 1 h

Spillway B lowers the water by $\frac{1}{6}$ ft in 1 h

Both spillways lower the water by $\dfrac{1}{x}$ ft in 1 h

Therefore, we get the equation

| Relate the quantities |

(fraction done by A) + (fraction done by B) = (fraction done by both)

| Set up an equation |

$$\frac{1}{4} + \frac{1}{6} = \frac{1}{x}$$

| Solve |

$$3x + 2x = 12 \quad \text{Multiply by the LCD, } 12x$$

$$5x = 12 \quad \text{Add}$$

$$x = \frac{12}{5} \quad \text{Divide by 5}$$

It will take $2\frac{2}{5}$ hours, or 2 h 24 min to lower the water level by 1 ft if both spillways are open. ■

The next example deals with distance, rate (speed), and time. The formula to keep in mind here is

$$\text{distance} = \text{rate} \times \text{time}$$

where the rate is either the constant speed or average speed of a moving object. For example, driving at 60 mi/h for 4 hours takes you a distance of $60 \cdot 4 = 240$ mi.

EXAMPLE 6 ■ A Distance-Rate-Time Problem

A jet flew from New York to Los Angeles, a distance of 4200 km. The speed for the return trip was 100 km/h faster than the outbound speed. If the total trip took 13 hours, what was the speed from New York to Los Angeles?

SOLUTION

| Identify the variable |

Let $\qquad s = $ speed from New York to Los Angeles

Then $\qquad s + 100 = $ speed from Los Angeles to New York

In problems involving motion it is often helpful to organize the information in a table, as shown on page 64. First we fill in the "Distance" column, since we know that the cities are 4200 km apart. Then we fill in the "Rate" column, since we have expressed both speeds (rates) in terms of the variable s. Finally, we calculate the entries for the "Time" column, using

$$\text{time} = \frac{\text{distance}}{\text{rate}}$$

	Distance (km)	Rate (km/h)	Time (h)
N.Y. to L.A.	4200	s	$\dfrac{4200}{s}$
L.A. to N.Y.	4200	$s + 100$	$\dfrac{4200}{s + 100}$

Express all unknown quantities in terms of the variable

Relate the quantities

Set up an equation

The total trip took 13 hours, so we have the equation

$$\frac{4200}{s} + \frac{4200}{s + 100} = 13$$

Multiplying by the common denominator, $s(s + 100)$, we get

$$4200(s + 100) + 4200s = 13s(s + 100)$$

$$8400s + 420{,}000 = 13s^2 + 1300s$$

$$0 = 13s^2 - 7100s - 420{,}000$$

Although this equation does factor, with numbers this large it is probably quickest to use the quadratic formula and a calculator.

Solve

$$s = \frac{7100 \pm \sqrt{(7100)^2 - 4(13)(-420{,}000)}}{2(13)}$$

$$= \frac{7100 \pm 8500}{26}$$

$$s = 600 \qquad \text{or} \qquad s = \frac{-700}{13} \approx -53.8$$

Since s represents speed, we reject the negative answer and conclude that the jet's speed from New York to Los Angeles was 600 km/h.   ∎

EXAMPLE 7 ■ The Path of a Projectile

An object thrown or fired straight upward at an initial speed of v_0 ft/s will reach a height of h feet after t seconds, where h and t are related by the formula

$$h = -16t^2 + v_0 t$$

(This formula is derived in elementary physics courses and depends on the fact that the acceleration due to gravity is constant near the surface of the earth. Here we are neglecting the effect of air resistance.)

Suppose that a bullet is shot straight upward with an initial speed of 800 ft/s. Its path is shown in Figure 5.

(a) When does the bullet fall back to ground level?

(b) When does it reach a height of 6400 ft?

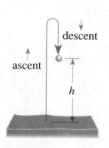

FIGURE 5

(c) When does it reach a height of 2 mi?

(d) How high is the highest point the bullet reaches?

SOLUTION Since the initial speed in this case is $v_0 = 800$ ft/s, the formula is $h = -16t^2 + 800t$.

(a) Ground level corresponds to $h = 0$, so we must solve the equation

$$0 = -16t^2 + 800t$$

$$0 = -16t(t - 50)$$

Thus, $t = 0$ or $t = 50$. This means the bullet starts ($t = 0$) at ground level and returns to ground level after 50 s.

(b) Setting $h = 6400$ gives the equation

$$6400 = -16t^2 + 800t$$

$$16t^2 - 800t + 6400 = 0$$

$$t^2 - 50t + 400 = 0 \qquad \text{Divide by 16}$$

$$(t - 10)(t - 40) = 0 \qquad \text{Factor}$$

$$t = 10 \quad \text{or} \quad t = 40$$

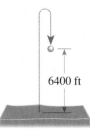

6400 ft

The bullet reaches 6400 ft after 10 s (on its ascent) and again after 40 s (on its descent to earth).

(c) Two miles is $2 \times 5280 = 10{,}560$ ft.

$$10{,}560 = -16t^2 + 800t$$

$$16t^2 - 800t + 10{,}560 = 0$$

$$t^2 - 50t + 660 = 0 \qquad \text{Divide by 16}$$

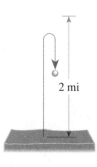

2 mi

The discriminant of this equation is $D = (-50)^2 - 4(660) = -140$, which is negative. Thus the equation has no real solution. The bullet never reaches a height of 2 mi.

(d) Each height the bullet reaches is attained twice, once on its ascent and once on its descent. The only exception is the highest point of its path, which is reached only once. This means that for the highest value of h, the equation

$$h = -16t^2 + 800t$$

10,000 ft

or

$$16t^2 - 800t + h = 0$$

has only one solution for t. This in turn means that the discriminant of the equation is 0, and so

$$D = (-800)^2 - 4(16)h = 0$$

$$640{,}000 - 64h = 0$$

$$h = 10{,}000$$

The maximum height reached is 10,000 ft.   ∎

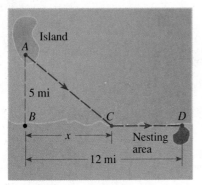

FIGURE 6

Identify the variable

Express all unknown quantities
in terms of the variable

Relate the quantities

Set up an equation

Solve

EXAMPLE 8 ■ Energy Expended in Bird Flight

Ornithologists have determined that some species of birds tend to avoid flights over large bodies of water during daylight hours, because air generally rises over land and falls over water in the daytime, so flying over water requires more energy. A bird is released from point A on an island, 5 mi from the nearest point B on a straight shoreline. The bird flies to a point C on the shoreline and then flies along the shoreline to its nesting area D, as shown in Figure 6. Suppose the bird has 170 kcal of energy reserves. It uses 10 kcal/mi flying over land and 14 kcal/mi flying over water.

(a) Where should the point C be located so that the bird uses exactly 170 kcal of energy during its flight?

(b) Does the bird have enough energy reserves to fly directly from A to D?

SOLUTION

(a) Let x be the distance in miles from B to C. Then

$$\text{distance flown over water} = \sqrt{x^2 + 25} \qquad \text{Pythagorean Theorem}$$

$$\text{distance flown over land} = 12 - x$$

Since

$$\text{energy used} = (\text{energy per mile}) \times (\text{miles flown})$$

and since

$$\text{total energy} = (\text{energy over water}) + (\text{energy over land})$$

we set up the following equation for the energy that the bird uses on its trip:

$$170 = 14\sqrt{x^2 + 25} + 10(12 - x)$$

To solve this equation we eliminate the square root by first bringing all other terms to the left of the equal sign and then squaring each side.

$$170 - 10(12 - x) = 14\sqrt{x^2 + 25} \qquad \text{Isolate the square-root term on RHS}$$

$$50 + 10x = 14\sqrt{x^2 + 25} \qquad \text{Simplify LHS}$$

$$(50 + 10x)^2 = (14)^2(x^2 + 25) \qquad \text{Square each side}$$

$$2500 + 1000x + 100x^2 = 196x^2 + 4900 \qquad \text{Expand}$$

$$0 = 96x^2 - 1000x + 2400 \qquad \text{Move all terms to RHS}$$

This equation could be factored, but because the numbers are so large it is easier to use the quadratic formula:

$$x = \frac{1000 \pm \sqrt{(-1000)^2 - 4(96)(2400)}}{2(96)}$$

$$= \frac{1000 \pm 280}{192} = 6\tfrac{2}{3} \quad \text{or} \quad 3\tfrac{3}{4}$$

Point C should be either $6\frac{2}{3}$ mi or $3\frac{3}{4}$ mi from B so that the bird uses exactly 170 kcal of energy during its flight.

(b) By the Pythagorean Theorem, the length of the route directly from A to D is $\sqrt{5^2 + 12^2} = 13$ mi, so the energy the bird requires for that route is $14 \times 13 = 182$ kcal. This is more energy than the bird has available, so that route can't be taken by this bird. ∎

1.6 EXERCISES

1–6 ■ Express the given quantity in terms of the indicated variable.

1. The interest obtained after 2 years on an investment at 7% simple interest per year; x = the number of dollars invested

2. The perimeter (in cm) of a rectangle that is 5 cm longer than it is wide; w = width of the rectangle (in cm)

3. The area (in square inches) of a rectangle that is 50 in. long; w = width of the rectangle (in inches)

4. The time (in hours) it takes to travel a given distance at 55 mi/h; d = the given distance (in miles)

5. The distance (in miles) traveled when driving at a certain speed for 2 hours, then driving 15 mi/h faster for another hour; s = initial speed (in mi/h)

6. The concentration (in oz/gal) of salt in a mixture of 3 gal of brine containing 25 oz of salt, to which has been added some pure water; x = volume of pure water added (in gallons)

7–64 ■ Use the problem-solving principles described in this section to answer the question posed.

7. During their major league careers, Hank Aaron hit 31 more home runs than Babe Ruth. Together they hit 1459 home runs. How many home runs did Babe Ruth hit?

8. A student has scores of 79, 81, and 72 on his first three tests. He needs an average of at least 80 to earn a grade of B. What score does he need to make on his fourth test to raise his average test score to 80?

9. Find three consecutive integers whose sum is 336.

10. Find four consecutive odd integers whose sum is 272.

11. Find two numbers whose sum is 55 and whose product is 684.

12. The sum of the squares of two consecutive even integers is 1252. Find the integers.

13. Phyllis invested $12,000, a portion earning a simple interest rate of $4\frac{1}{2}$% per year and the rest earning a rate of 4% per year. After one year the total interest earned on these investments was $525. How much money did she invest at each rate?

14. If Ben invests $4000 at 4% interest per year, how much additional money must he invest at $5\frac{1}{2}$% annual interest to ensure that the interest he receives each year is $4\frac{1}{2}$% of the total amount invested?

15. What annual rate of interest would you have to earn on an investment of $3500 to ensure receiving $262.50 interest after one year?

16. Jack invests $1000 at a certain annual interest rate, and he invests another $2000 at an annual rate that is one-half percent higher. If he receives a total of $190 interest in one year, at what rate is the $1000 invested?

17. A rectangular garden is 25 ft wide. If its area is 1125 ft², what is the length of the garden?

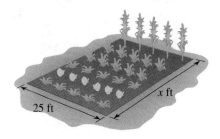

18. A room is $1\frac{1}{2}$ times as long as it is wide. Its perimeter is 80 ft. How wide is the room?

19. Al paints with watercolors on a sheet of paper 20 in. wide by 15 in. high. He then places this sheet on a mat so that a uniformly wide strip of the mat shows all around the picture. The perimeter of the mat is 102 in. How wide is the strip of the mat showing around the picture?

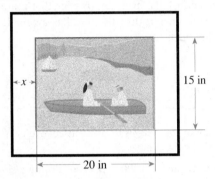

15 in

20 in

20. A factory is to be built on a lot measuring 180 ft by 240 ft. A local building code specifies that a lawn of uniform width and equal in area to the factory must surround the factory. What must the width of this lawn be, and what are the dimensions of the factory?

21. A rectangular garden is 10 ft longer than it is wide. Its area is 875 ft². What are its dimensions?

22. A parcel of land is 6 ft longer than it is wide. Each diagonal from one corner to the opposite corner is 174 ft long. What are the dimensions of the parcel?

23. A box with a square base and no top is to be made from a square piece of cardboard by cutting 4-inch squares from each corner and folding up the sides, as shown in the figure. The box is to hold 100 in³. How big a piece of cardboard is needed?

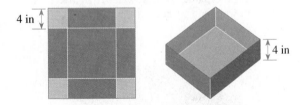

4 in

4 in

24. A cylindrical can has a volume of 40π cm³ and is 10 cm tall. What is its diameter?   [*Hint:* Use the volume formula listed on the inside of the front cover of this book.]

10 cm

25. A commercial jet took off from Kansas City for San Francisco, traveling a distance of 2550 km at a speed of 800 km/h. At the same time a private jet left San Francisco, traveling at 900 km/h and bound for Kansas City. How long after takeoff will the jets pass each other?

26. After robbing a bank in Dodge City, the robber gallops off at 14 mi/h. Ten minutes later the marshal leaves to pursue him, riding at 16 mi/h. How long does it take the marshal to catch up with the bank robber?

27. Wilma drove at an average speed of 50 mi/h from her home in Boston to visit her sister in Buffalo. She stayed in Buffalo 10 hours and, on the trip back, averaged 45 mi/h. She returned home 29 hours after leaving. How many miles is Buffalo from Boston?

28. A coast guard boat that patrols a river separating two countries cruises at 20 knots (nautical miles per hour) in still water. The river flows at 4 knots. For each patrol the captain of the boat is ordered to travel upstream and then return. His trip takes exactly 6 hours. How far upstream does the boat travel?

29. Wendy took a trip from Davenport to Omaha, a distance of 300 mi. Traveling the first part of the way by bus, Wendy arrived at the train station just in time to complete her journey by train. The bus averaged 40 mi/h and the train 60 mi/h. The entire trip took $5\frac{1}{2}$ hours. How long did Wendy spend on the train?

30. It took a crew 2 h 40 min to row 6 km upstream and back again. If the rate of flow of the stream was 3 km/h, what was the rowing rate of the crew in still water?

31. A salesman drove from Ajax to Barrington, a distance of 120 mi, at a steady speed. He then increased his speed by 10 mi/h to drive the 150 mi from Barrington to Collins. If the second leg of his trip took 6 minutes more time than the first leg, how fast was he driving between Ajax and Barrington?

32. Kiran drove from Tortula to Cactus, a distance of 250 mi. She increased her speed by 10 mi/h for the 360-mile trip from Cactus to Dry Junction. If the total trip took 11 hours, what was her speed from Tortula to Cactus?

33. What quantity of a 60% acid solution must be mixed with a 30% solution to produce 300 mL of a 50% solution?

34. A pot contains 6 L of brine at a concentration of 120 g/L. How much of the water should be boiled off to increase the concentration to 200 g/L?

35. A jeweler has five rings, each weighing 18 g, made of an alloy of 10% silver and 90% gold. He wishes to melt down the rings and add enough silver to reduce the gold content to 75%. How much silver should he add?

36. A health clinic uses a solution of bleach to sterilize petri dishes in which cultures are grown. The sterilization tank contains 100 gal of a solution of 2% ordinary household bleach mixed with pure distilled water. New research indicates that the concentration of bleach should be 5% for complete sterilization. How much of the solution should be drained and replaced with bleach to increase the bleach content to the recommended level?

37. The radiator in a car is filled with a solution of 60% antifreeze and 40% water. The manufacturer of the antifreeze suggests that, for summer driving, optimal cooling of the engine is obtained with only 50% antifreeze. If the capacity of the radiator is 3.6 L, how much coolant should be drained and replaced with water to reduce the antifreeze concentration to the recommended level?

38. A bottle contains 750 mL of fruit punch with a concentration of 50% pure fruit juice. Jill drinks 100 mL of the punch and then refills the bottle with an equal amount of a cheaper brand of punch. If the concentration of juice in the bottle is now reduced to 48%, what was the concentration of the punch that Jill added?

39. Candy and Tim share a paper route. It takes Candy 70 min to deliver all the papers, whereas Tim takes 80 min. How long will it take them if they work together?

40. Stan and Hilda can mow the lawn in 40 min if they work together. If Hilda works twice as fast as Stan, how long would it take Stan to mow the lawn alone?

41. Betty and Karen have been hired to paint the houses in a new development. Working together the women can paint a house in two-thirds the time that it takes Karen working alone. Betty takes 6 hours to paint a house alone. How long does it take Karen to paint a house working alone?

42. Henry and Irene working together can wash all the windows of their house in 1 h 48 min. Working alone, it takes Henry $1\frac{1}{2}$ hours more than Irene to do the job.

How long does it take each person working alone to wash all the windows?

43. Jack, Kay, and Lynn deliver advertising flyers in a small town. If each person works alone, it takes Jack 4 hours to deliver all the flyers, and it takes Lynn 1 hour longer than it takes Kay. Working together, they can deliver all the flyers in 40% of the time it takes Kay working alone. How long does it take Kay to deliver all the flyers alone?

44. A man is walking away from a lamppost with a light source 6 m above the ground. The man is 2 m tall. How far from the lamppost is the man when his shadow is 5 m long? [*Hint:* Use similar triangles.]

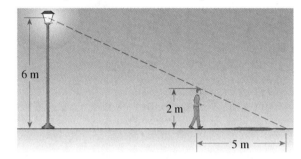

45. At 3:00 P.M. a man 165 cm tall casts a shadow 132 cm long. At the same time, a tall building nearby casts a shadow 160 m long. How tall is the building?

46. A woodcutter determines the height of a tall tree by first measuring a smaller one 125 ft away, then moving so that his eyes are in the line of sight along the tops of the trees. He then measures how far he is standing from the small tree (see the figure). Suppose the small tree is 20 ft tall, the man is 25 ft from the small tree, and his eye level is 5 ft above the ground. How tall is the taller tree?

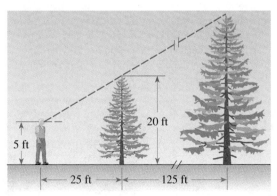

47. A ball is thrown straight upward at an initial speed of 40 ft/s. Use the formula $h = -16t^2 + v_0t$ discussed in Example 7 to answer the following questions.
 (a) When does the ball reach a height of 24 ft?
 (b) When does it reach a height of 48 ft?
 (c) What is the greatest height reached by the ball?
 (d) When does the ball reach the highest point of its path?
 (e) When does the ball hit the ground?

48. How fast would a ball have to be thrown upward to reach a maximum height of 100 ft?   [*Hint:* Refer to the formula in Example 7 and use the discriminant of the equation $16t^2 - v_0t + h = 0$.]

49. The fish population in a certain lake rises and falls according to the formula

$$F = 1000(30 + 17t - t^2)$$

Here F is the number of fish at time t, where t is measured in years since January 1, 1997, when the fish population was first estimated.
 (a) On what date will the fish population again be the same as on January 1, 1997?
 (b) By what date will all the fish in the lake have died?

50. A small-appliance manufacturer finds that the profit P (in dollars) generated by producing x microwave ovens per week is given by the formula $P = \frac{1}{10}x(300 - x)$ provided that $0 \leqslant x \leqslant 200$. How many ovens must be manufactured in a given week to generate a profit of $1250?

51. If an imaginary line segment is drawn between the centers of the earth and the moon, then the net gravitational force F acting on an object situated on this line segment is

$$F = \frac{-K}{x^2} + \frac{0.012K}{(239 - x)^2}$$

where $K > 0$ is a constant and x is the distance of the object from the center of the earth, measured in thousands of miles. How far from the center of the earth is the "dead spot" where no net gravitational force acts upon the object? (Express your answer to the nearest thousand miles.)

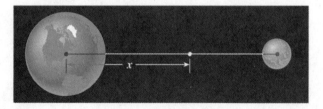

52. A spherical tank has a capacity of 750 gal. Using the fact that one gallon is about 0.1337 ft³, find the radius of the tank (to the nearest hundredth of a foot).

53. A city lot has the shape of a right triangle whose hypotenuse is 7 ft longer than one of the other sides. The perimeter of the lot is 392 ft. How long is each of the three sides of the lot?

54. A boardwalk is parallel to and 210 ft inland from a straight shoreline. A sandy beach lies between the boardwalk and the shoreline. A man is standing on the boardwalk, exactly 750 ft across the sand from his beach umbrella, which is right at the shoreline. The man walks 4 ft/s on the boardwalk and 2 ft/s on the sand. How far should he walk on the boardwalk before veering off onto the sand if he wishes to reach his umbrella in exactly 4 min 45 s?

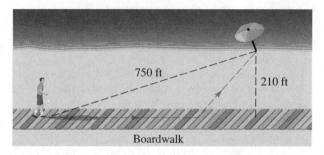

750 ft   210 ft

Boardwalk

55. Helen earns $7.50 per hour at her job, but if she works more than 35 hours in a week she is paid $1\frac{1}{2}$ times her regular salary for the overtime hours worked. One week her gross pay was $352.50. How many overtime hours did she work that week?

56. A change purse contains an equal number of pennies, nickels, and dimes. The total value of the coins is $1.44. How many coins of each type does the purse contain?

57. The oldest child in a family of four children is twice as old as the youngest. The two middle children are 10 and 11 years old. If the average age of the children is $10\frac{1}{2}$ years, how old is the youngest child?

58. Rochelle's psychology professor gives a grade of A for an average of 90 or better and a grade of B for an average of 80 or better in her course. Rochelle has obtained scores of 82, 75, and 71 on her midterm examinations. If the final exam counts twice as much as a midterm, what score must she make on her final exam to earn a grade of B? To earn an A? (Assume that the maximum possible score on each test is 100.)

59. The fuel consumption for William's car is 30 mi/gal on the highway and 25 mi/gal in the city. On a vacation trip of 400 mi he used 14 gal of gasoline. How many highway miles did he drive on this trip?

60. A rectangular parcel of land is 50 ft wide. The length of a diagonal between opposite corners is 10 ft more than the length of the parcel. What is the length of the parcel?

61. A running track has the shape shown in the figure, with straight sides and semicircular ends. If the length of the track is 440 yd and the two straight parts are each 110 yd long, what is the radius of the semicircular parts (to the nearest yard)?

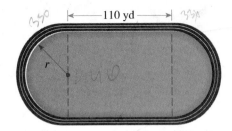

62. Next-door neighbors Bob and Jim use hoses from both houses to fill Bob's swimming pool. They know it takes 18 h using both hoses. They also know that Bob's hose, used alone, takes 20% less time than Jim's hose alone. How much time is required to fill the pool by each hose alone?

63. This problem is taken from a Chinese mathematics text-book called *Chui-chang suan-shu,* or *Nine Chapters on the Mathematical Art,* which was written about 250 B.C.

A 10-ft-long stem of bamboo is broken in such a way that its tip touches the ground 3 ft from the base of the stem, as shown in the figure. What is the height of the break?

[*Hint:* Use the Pythagorean Theorem.]

64. A man wishes to determine the water level in a deep well. He drops a stone into the well and hears the stone hit the water 3 s later. If the stone drops $16t^2$ feet after t seconds, and the speed of sound is 1090 ft/s, how far below the opening of the well is the surface of the water (to the nearest foot)?

 DISCOVERY · DISCUSSION

65. Historical Research Read the biographical notes on Pythagoras (page 59), Euclid (page 49), and Viète (page 51). Choose one of these mathematicians and find out more about him from the library. Write a short essay on your findings. Include both biographical information and a description of the mathematics for which he is famous.

1.7 **INEQUALITIES**

Some problems in algebra lead to **inequalities** instead of equations. An inequality looks just like an equation, except that in the place of the equal sign is one of the symbols $<$, $>$, $\leq$, or $\geq$. Here is an example of an inequality:

$$4x + 7 \leq 19$$

To **solve** an inequality that contains a variable means to find all values of the variable that make the inequality true. Unlike an equation, an inequality generally has infinitely many solutions, which form an interval or a union of intervals on the real line. (See page 9 for a table that describes intervals.) The following illustration indicates this difference for one inequality:

Solution

Equation: $4x + 7 = 19$ $x = 3$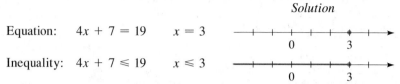

Inequality: $4x + 7 \le 19$ $x \le 3$

To solve inequalities we use the following rules to isolate the variable on one side of the inequality sign. These rules tell us when two inequalities are *equivalent* (the symbol $\Longleftrightarrow$ means "is equivalent to"). In these rules the symbols A, B, and C stand for real numbers or algebraic expressions. Here we state the rules for inequalities involving the symbol $\le$, but they apply to each of the four inequality symbols.

RULES FOR INEQUALITIES

Rule	Description
1. $A \le B \iff A + C \le B + C$	**Adding** the same quantity to each side of an inequality gives an equivalent inequality.
2. $A \le B \iff A - C \le B - C$	**Subtracting** the same quantity from each side of an inequality gives an equivalent inequality.
3. If $C > 0$, then $A \le B \iff CA \le CB$	**Multiplying** each side of an inequality by the same *positive* quantity gives an equivalent inequality.
4. If $C < 0$, then $A \le B \iff CA \ge CB$	**Multiplying** each side of an inequality by the same *negative* quantity *reverses the direction* of the inequality.
5. $0 < A \le B \iff \dfrac{1}{A} \ge \dfrac{1}{B} > 0$	**Taking reciprocals** of each side of an inequality involving *positive* quantities *reverses the direction* of the inequality.
6. If $A \le B$ and $C \le D$, then $A + C \le B + D$	Inequalities can be added.

Pay special attention to Rules 3 and 4. Rule 3 says that we can multiply (or divide) each side of an inequality by a *positive* number, but Rule 4 says that if we multiply each side of an inequality by a *negative* number, then we reverse the direction of the inequality. For example, if we start with the inequality

$$3 < 5$$

and multiply by 2, we get

$$6 < 10$$

but if we multiply by -2, we get

$$-6 > -10$$

EXAMPLE 1 ■ **Solving a Linear Inequality**

Solve the inequality $3x < 9x + 4$ and sketch the solution set.

SOLUTION

$$3x < 9x + 4$$

$$3x - 9x < 9x + 4 - 9x \qquad \text{Subtract } 9x$$

$$-6x < 4 \qquad \text{Simplify}$$

$$\left(-\tfrac{1}{6}\right)(-6x) > -\tfrac{1}{6}(4) \qquad \text{Multiply by } -\tfrac{1}{6} \text{ (or divide by } -6)$$

$$x > -\tfrac{2}{3} \qquad \text{Simplify}$$

The solution set consists of all numbers greater than $-\tfrac{2}{3}$. In other words, the solution of the inequality is the interval $\left(-\tfrac{2}{3}, \infty\right)$. It is graphed in Figure 1. ■

FIGURE 1

EXAMPLE 2 ■ **Relationship between Fahrenheit and Celsius Scales**

The instructions on a box of film indicate that the box should be stored at a temperature between $5\,°C$ and $30\,°C$. What range of temperatures does this correspond to on the Fahrenheit scale?

SOLUTION The relationship between degrees Celsius (C) and degrees Fahrenheit (F) is given by the equation $C = \tfrac{5}{9}(F - 32)$. Expressing the statement on the box in terms of inequalities, we have

$$5 < C < 30$$

so the corresponding Fahrenheit temperatures satisfy the inequalities

$$5 < \tfrac{5}{9}(F - 32) < 30$$

$$\tfrac{9}{5} \cdot 5 < F - 32 < \tfrac{9}{5} \cdot 30 \qquad \text{Multiply by } \tfrac{9}{5}$$

$$9 < F - 32 < 54 \qquad \text{Simplify}$$

$$9 + 32 < F < 54 + 32 \qquad \text{Add 32}$$

$$41 < F < 86 \qquad \text{Simplify}$$

The film should be stored at a temperature between $41\,°F$ and $86\,°F$. ■

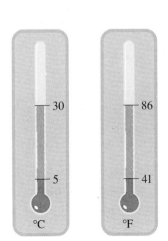

In Examples 1 and 2 the inequalities involved were **linear**; that is, they were equivalent to an inequality of the form $ax + b > 0$, where a and b are real numbers and the symbol $>$ can be replaced by an inequality symbol ($<$, $\leq$, or $\geq$). We now consider nonlinear inequalities.

NONLINEAR INEQUALITIES

To solve inequalities involving squares and other powers of the variable, we use factoring, together with the following principle.

> ## THE SIGN OF A PRODUCT OR QUOTIENT
>
> If a product or a quotient has an *even* number of *negative* factors, then its value is *positive*.
>
> If a product or quotient has an *odd* number of *negative* factors, then its value is *negative*.

EXAMPLE 3 ■ A Quadratic Inequality

Solve the inequality $x^2 - 5x + 6 \leq 0$.

SOLUTION   First we factor the left side.

$$(x - 2)(x - 3) \leq 0$$

We know that the corresponding equation $(x - 2)(x - 3) = 0$ has the solutions 2 and 3. As shown in Figure 2, the numbers 2 and 3 divide the real line into three intervals: $(-\infty, 2)$, $(2, 3)$, and $(3, \infty)$. On each of these intervals we determine the signs of the factors using **test values**. We choose a number inside each interval and check the sign of the factors $x - 2$ and $x - 3$ at the value selected. For instance, if we use the test value $x = 1$ for the interval $(-\infty, 2)$ (see Figure 3), then substitution in the factors $x - 2$ and $x - 3$ gives

$$x - 2 = 1 - 2 = -1 < 0$$

and

$$x - 3 = 1 - 3 = -2 < 0$$

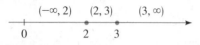

FIGURE 2

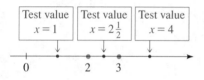

FIGURE 3

So both factors are negative on this interval. (The factors $x - 2$ and $x - 3$ change sign only at 2 and 3, respectively, so they maintain their signs over the length of each interval. That is why using a single test value on each interval is sufficient.)

Using the test values $x = 2\frac{1}{2}$ and $x = 4$ for the intervals $(2, 3)$ and $(3, \infty)$ (see Figure 3), respectively, we construct the following sign table. The final row of the table is obtained from the fact that the expression in the last row is the product of the two factors.

Interval	$(-\infty, 2)$	$(2, 3)$	$(3, \infty)$
Sign of $x - 2$	$-$	$+$	$+$
Sign of $x - 3$	$-$	$-$	$+$
Sign of $(x - 2)(x - 3)$	$+$	$-$	$+$

If you prefer, you can represent this information on a real number line, as in the following sign diagram. The vertical lines indicate the points at which the real line is divided into intervals:

	2		3	
Sign of $x - 2$	$-$		$+$	$+$
Sign of $x - 3$	$-$		$-$	$+$
Sign of $(x - 2)(x - 3)$	$+$		$-$	$+$

We read from the table or the diagram that $(x - 2)(x - 3)$ is negative on the interval $(2, 3)$. Thus, the solution of the inequality $(x - 2)(x - 3) \leqslant 0$ is

$$\{x \mid 2 \leqslant x \leqslant 3\} = [2, 3]$$

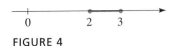

FIGURE 4

We have included the endpoints 2 and 3 because we seek values of x such that the product is either less than *or equal to* zero. The solution is illustrated in Figure 4.

We use the following steps to solve an inequality that can be factored.

GUIDELINES FOR SOLVING NONLINEAR INEQUALITIES

1. If necessary, rewrite the inequality so that all nonzero terms appear on one side of the inequality sign.

2. If the nonzero side of the inequality involves quotients, bring them to a common denominator.

3. Factor the nonzero side of the inequality.

4. List the intervals determined by the factorization.

5. Use test values to make a table or diagram of the signs of each factor on each interval. In the last row of the table, determine the sign of the product (or quotient) of these factors.

6. Determine the solution set from the last row of the sign table. Be sure to check whether the inequality is satisfied by some or all of the endpoints of the intervals (this may happen if the inequality involves $\geqslant$ or $\leqslant$).

 The factoring technique described in these steps works only if all nonzero terms appear on one side of the inequality symbol. If the inequality is not written in this form, first rewrite it, as indicated in Step 1. This technique is illustrated in the examples that follow.

EXAMPLE 4 ■ An Inequality Involving a Quotient

Solve $\dfrac{1 + x}{1 - x} \geq 1$.

SOLUTION   First we move all nonzero terms to the left side, and then we simplify using a common denominator.

$$\frac{1 + x}{1 - x} \geq 1$$

$$\frac{1 + x}{1 - x} - 1 \geq 0 \qquad \text{Subtract 1}$$

$$\frac{1 + x}{1 - x} - \frac{1 - x}{1 - x} \geq 0 \qquad \text{Common denominator } 1 - x$$

$$\frac{1 + x - 1 + x}{1 - x} \geq 0 \qquad \text{Combine the fractions}$$

$$\frac{2x}{1 - x} \geq 0 \qquad \text{Simplify}$$

The numerator is zero when $x = 0$ and the denominator is zero when $x = 1$, so we construct the following sign table using these values to define intervals on the real line.

Interval	$(-\infty, 0)$	$(0, 1)$	$(1, \infty)$
Sign of $2x$	$-$	$+$	$+$
Sign of $1 - x$	$+$	$+$	$-$
Sign of $\dfrac{2x}{1 - x}$	$-$	$+$	$-$

From the table we see that the solution set is $\{x \mid 0 \leq x < 1\} = [0, 1)$. We include the endpoint 0 because the original inequality requires the quotient to be greater than *or equal to* 1. However, we do not include the other endpoint 1, since the quotient in the inequality is not defined at 1. Always check the endpoints of solution intervals to determine whether they satisfy the original inequality.

The solution set $[0, 1)$ is illustrated in Figure 5.   ■

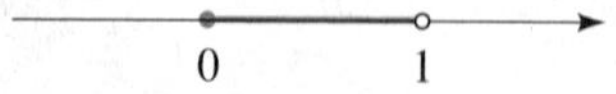

$$\overset{\displaystyle 0 \qquad\quad 1}{\longrightarrow}$$

FIGURE 5

EXAMPLE 5 ■ Solving an Inequality with Three Factors

Solve the inequality $x < \dfrac{2}{x - 1}$.

SOLUTION   After moving all nonzero terms to one side of the inequality, we use a common denominator to combine the terms.

$$x - \frac{2}{x-1} < 0 \qquad \text{Subtract } \frac{2}{x-1}$$

$$\frac{x(x-1)}{x-1} - \frac{2}{x-1} < 0 \qquad \text{Common denominator } x-1$$

$$\frac{x^2 - x - 2}{x-1} < 0 \qquad \text{Combine the fractions}$$

$$\frac{(x+1)(x-2)}{x-1} < 0 \qquad \text{Factor the numerator}$$

The factors in this quotient change sign at -1, 1, and 2, so we must examine the intervals $(-\infty, -1)$, $(-1, 1)$, $(1, 2)$, and $(2, \infty)$. Using test values, we get the following sign table.

Interval	$(-\infty, -1)$	$(-1, 1)$	$(1, 2)$	$(2, \infty)$
Sign of $x + 1$	$-$	$+$	$+$	$+$
Sign of $x - 2$	$-$	$-$	$-$	$+$
Sign of $x - 1$	$-$	$-$	$+$	$+$
Sign of $\dfrac{(x+1)(x-2)}{x-1}$	$-$	$+$	$-$	$+$

Since the quotient must be negative, the solution is

$$(-\infty, -1) \cup (1, 2)$$

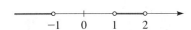

FIGURE 6

as illustrated in Figure 6. ■

ABSOLUTE-VALUE INEQUALITIES

The following properties are useful in solving inequalities that involve absolute values. They can be proved using the definition of absolute value (Section 1.1).

PROPERTIES OF ABSOLUTE-VALUE INEQUALITIES

Inequality	Equivalent form	Graph
1. $\lvert x \rvert < c$	$-c < x < c$	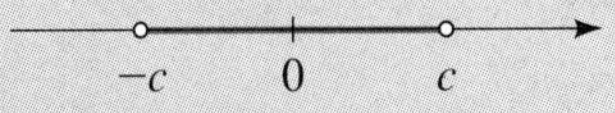 $-c \quad 0 \quad c$
2. $\lvert x \rvert \leq c$	$-c \leq x \leq c$	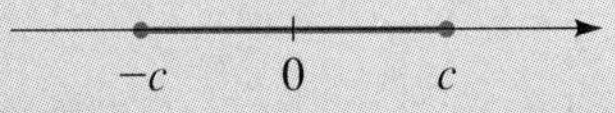 $-c \quad 0 \quad c$
3. $\lvert x \rvert > c$	$x < -c \;$ or $\; c < x$	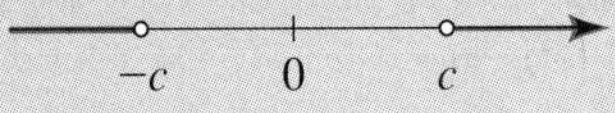 $-c \quad 0 \quad c$
4. $\lvert x \rvert \geq c$	$x \leq -c \;$ or $\; c \leq x$	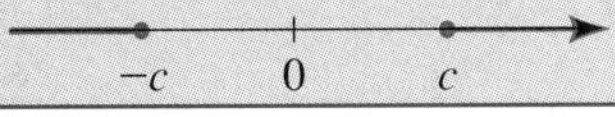 $-c \quad 0 \quad c$

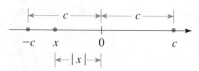

FIGURE 7

To prove Property 1, for example, we see that the inequality $|x| < c$ says that the distance from x to 0 is less than c, and from Figure 7 you can see that this is true if and only if x is between c and $-c$.

EXAMPLE 6 ■ **Solving an Inequality Involving Absolute Value**

Solve the inequality $|x - 5| < 2$.

SOLUTION 1 The inequality $|x - 5| < 2$ is equivalent to

$$-2 < x - 5 < 2 \qquad \text{Property 1}$$

$$3 < x < 7 \qquad \text{Add 5}$$

The solution set is the open interval $(3, 7)$.

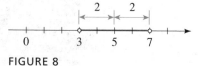

FIGURE 8

SOLUTION 2 Geometrically, the solution set consists of all numbers x whose distance from 5 is less than 2. From Figure 8 we see that this is the interval $(3, 7)$.
■

EXAMPLE 7 ■ **Solving an Inequality Involving Absolute Value**

Solve the inequality $|3x + 2| \geq 4$.

SOLUTION By Property 4, the inequality $|3x + 2| \geq 4$ is equivalent to

$$3x + 2 \geq 4 \qquad \text{or} \qquad 3x + 2 \leq -4$$

$$3x \geq 2 \qquad\qquad\qquad 3x \leq -6$$

$$x \geq \tfrac{2}{3} \qquad\qquad\qquad x \leq -2$$

So the solution set is

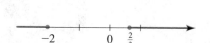

FIGURE 9

$$\left\{ x \mid x \leq -2 \quad \text{or} \quad x \geq \tfrac{2}{3} \right\} = (-\infty, -2] \cup \left[\tfrac{2}{3}, \infty\right)$$

The set is graphed in Figure 9.
■

1.7 | **EXERCISES**

1–4 ■ Let $S = \left\{-1, 0, \tfrac{1}{2}, \sqrt{2}, 2\right\}$. Use substitution to determine which of the elements of S satisfy the given inequality.

1. $x + 1 \geq 0$

2. $x - 2 < 0$

3. $\dfrac{1}{x} \leq \dfrac{1}{2}$

4. $|x - 1| > 1$

5–64 ■ Solve the inequality. Express the solution in interval form and illustrate the solution set on the real number line.

5. $3x \leq 12$

6. $-4x > 16$

7. $20 < -4x$

8. $-1 \geq 7x$

9. $2x - 5 > 3$

10. $3x + 11 < 5$

11. $7 - x \geq 5$

12. $5 - 3x \leq -16$

13. $2x + 1 < 0$

14. $0 < 5 - 2x$

15. $3x + 11 \leq 6x + 8$

16. $6 - x \geq 2x + 9$

17. $1 - x \leq 2$

18. $4 - 3x \geq 6$

19. $4 - 3x \leq -(1 + 8x)$

20. $2(7x - 3) \leq 12x + 16$

21. $(x - 2)(x - 5) > 0$

22. $(3x + 1)(x - 1) \geq 0$

23. $x^2 - 3x - 18 \leq 0$

24. $x^2 + 5x + 6 > 0$

25. $2x^2 + x \geq 1$

26. $x^2 < x + 2$

27. $x^2 > 3(x + 6)$

28. $x^2 + 2x > 3$

29. $x^2 < 4$

30. $(x + 2)(x - 1)(x - 3) \leq 0$

31. $x(x^2 - 4) \geq 0$

32. $x^3 > x$

33. $\dfrac{x - 3}{x + 1} \geq 0$

34. $\dfrac{2x + 6}{x - 2} < 0$

35. $\dfrac{4x}{2x + 3} > 2$

36. $-2 < \dfrac{x + 1}{x - 3}$

37. $\dfrac{2x + 1}{x - 5} \leq 3$

38. $\dfrac{3 + x}{3 - x} \geq 1$

39. $\dfrac{4}{x} < x$

40. $\dfrac{x}{x + 1} > 3x$

41. $\dfrac{x^2 - 4}{x^2 + 4} \geq 0$

42. $\dfrac{x}{x + 2} \leq \dfrac{1}{x}$

43. $\dfrac{1}{1 - x} \leq \dfrac{3}{x}$

44. $\dfrac{(x - 1)^2}{(x + 1)(x + 2)} > 0$

45. $\dfrac{x - 3}{2x + 5} \geq 1$

46. $\dfrac{1}{x} + \dfrac{1}{x + 1} < \dfrac{2}{x + 2}$

47. $|x| < 2$

48. $|x| \geq 4$

49. $|x - 5| \leq 3$

50. $|x - 9| > 9$

51. $|x + 5| < 2$

52. $|x + 1| \geq 3$

53. $|2x - 3| \leq 0.4$

54. $|5x - 2| < 6$

55. $|x + 6| < 0.001$

56. $\dfrac{1}{|x + 7|} > 2$

57. $-1 < 2x - 5 < 7$

58. $1 < 3x + 4 \leq 16$

59. $0 \leq 1 - x < 1$

60. $-5 \leq 3 - 2x \leq 9$

61. $\dfrac{1}{x} < 4$

62. $\dfrac{2}{3} \leq \dfrac{1}{x - 2} < 1$

63. $x^4 > x^2$

64. $x^5 > x^2$

65. Use the relationship $C = \frac{5}{9}(F - 32)$ given in Example 2 to find the interval on the Fahrenheit scale corresponding to the temperature range $20 \leq C \leq 30$.

66. What interval on the Celsius scale corresponds to the temperature range $50 \leq F \leq 95$?

67. As dry air moves upward, it expands and in so doing cools at a rate of about 1°C for each 100-meter rise, up to about 12 km.

(a) If the ground temperature is 20 °C, write a formula for the temperature at height h.

(b) What range of temperatures can be expected if a plane takes off and reaches a maximum height of 5 km?

68. It is estimated that the annual cost of driving a new Destini is given by the formula

$$C = 0.35m + 2200$$

where m represents the number of miles driven per year and C is the cost in dollars. Jane has purchased a Destini and decides to budget between $6400 and $7100 for next year's driving costs. What is the corresponding range of miles that she can drive her new Destini?

69. Using calculus it can be shown that if a ball is thrown upward with an initial velocity of 16 ft/s from the top of a building 128 ft high, then its height h above the ground t seconds later will be

$$h = 128 + 16t - 16t^2$$

During what time interval will the ball be at least 32 ft above the ground?

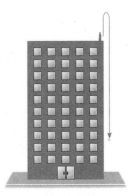

70. The gravitational force F exerted by the earth on an object having a mass of 100 kg is given by the equation

$$F = \dfrac{4,000,000}{d^2}$$

where d is the distance (in km) of the object from the center of the earth, and the force F is measured in newtons (N). For what distances will the gravitational force exerted by the earth on this object be between 0.0004 N and 0.01 N?

71. In the vicinity of a bonfire, the temperature T in °C at a distance of x meters from the center of the fire was given by $T = \dfrac{600{,}000}{x^2 + 300}$. At what range of distances from the fire center was the temperature less than 500 °C?

72. The gas mileage g (measured in mi/gal) for a particular vehicle, driven at v mi/h, is given by the formula $g = 10 + 0.9v - 0.01v^2$, as long as v is between 10 mi/h and 75 mi/h. For what range of speeds is the vehicle's mileage 30 mi/gal or better?

73–74 ■ Solve the inequality for x, assuming that $0 < a < b < c$.

73. $a(bx - c) \geq bc$
74. $\dfrac{x^2 + (a - b)x - ab}{x + c} \leq 0$

75. Show that if $a < b$, then $a < \dfrac{a + b}{2} < b$.

76. Suppose that a, b, c, and d are positive numbers such that

$$\frac{a}{b} < \frac{c}{d}$$

Show that

$$\frac{a}{b} < \frac{a + c}{b + d} < \frac{c}{d}$$

▲ **DISCOVERY · DISCUSSION**

77. Do Powers Preserve Order? If $a < b$, is $a^2 < b^2$? (Check both positive and negative values for a and b.) If $a < b$, is $a^3 < b^3$? Based on your observations, state a general rule about the relationship between a^n and b^n when $a < b$ and n is a positive integer.

78. Using Distances to Solve Absolute-Value Inequalities Recall that $|a - b|$ is the distance between a and b on the number line. For any number x, what do $|x - 1|$ and $|x - 3|$ represent? Use this interpretation to solve the inequality $|x - 1| < |x - 3|$ geometrically. In general, if $a < b$, what is the solution of the inequality $|x - a| < |x - b|$?

1.8   COORDINATE GEOMETRY

FIGURE 1

Points on a line can be identified with real numbers by assigning them coordinates, as described in Section 1.1. In a similar way, points in a plane can be identified with ordered pairs of real numbers. We start by drawing two perpendicular coordinate lines that intersect at the **origin** O on each line. Usually one line is horizontal with positive direction to the right and is called the **x-axis**; the other line is vertical with positive direction upward and is called the **y-axis**.

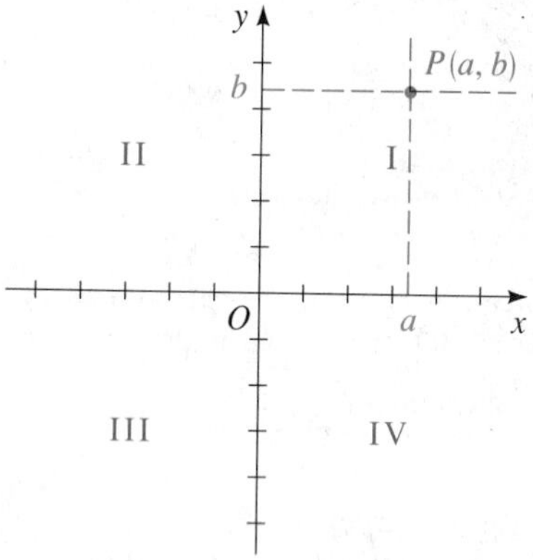

Any point P in the plane can be located by a unique ordered pair of numbers as follows. Draw lines through P perpendicular to the x- and y-axes. These lines will intersect the axes in points with coordinates a and b as shown in Figure 1. Then the point P is assigned the ordered pair (a, b). The first number a is called the **x-coordinate** (or **abscissa**) of P; the second number is called the **y-coordinate** (or **ordinate**) of P. We say that P is the point with coordinates (a, b), and we denote the point by the symbol $P(a, b)$. We can think of the coordinates of a point as its "address" in the plane, since the coordinates specify the location of the point. Several points are labeled with their coordinates in Figure 2.

By reversing the preceding process, starting with an ordered pair (a, b), we can arrive at the corresponding point P. Often we identify the point P with the ordered pair (a, b) and refer to "the point (a, b)." [Although the notation used for an open interval (a, b) is the same as the notation used for a point (a, b), you will be able to tell from the context which meaning is intended.]

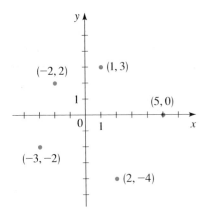

FIGURE 2

This coordinate system is called the **rectangular coordinate system** or the **Cartesian coordinate system** in honor of the French mathematician René Descartes (1596–1650), although another Frenchman, Pierre Fermat (1601–1665), also invented the principles of analytic geometry at about the same time as Descartes. The plane used with this coordinate system is called the **coordinate plane** or the **Cartesian plane** and is denoted by $\mathbb{R}^2$.

The x- and y-axes, or the **coordinate axes**, divide the Cartesian plane into four quadrants, which are labeled I, II, III, and IV in Figure 1. Notice that the first quadrant consists of those points whose x- and y-coordinates are both positive. (The points *on* the coordinate axes are not assigned to any quadrant.)

EXAMPLE 1 ■ Graphing Regions in the Coordinate Plane

Describe and sketch the region given by each of the following sets.

(a) $\{(x, y) \mid x \geqslant 0\}$ (b) $\{(x, y) \mid y = 1\}$ (c) $\{(x, y) \mid |y| < 1\}$

SOLUTION

(a) The points whose x-coordinates are 0 or positive lie on the y-axis or to the right of it, as shown in Figure 3(a).

(b) The set of all points with y-coordinate 1 is a horizontal line 1 unit above the x-axis, as in Figure 3(b).

(c) Recall from Section 1.7 that

$$|y| < 1 \qquad \text{if and only if} \qquad -1 < y < 1$$

The given region consists of those points in the plane whose y-coordinates lie between -1 and 1. Thus, the region consists of all points that lie between (but not on) the horizontal lines $y = 1$ and $y = -1$. These lines are shown as dashed lines in Figure 3(c) to indicate that the points on these lines do not lie in the set.

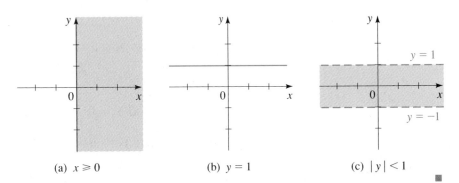

FIGURE 3   (a) $x \geqslant 0$ (b) $y = 1$ (c) $|y| < 1$

We now find a formula for the distance $d(A, B)$ between two points $A(x_1, y_1)$ and $B(x_2, y_2)$ in the plane. Recall from Section 1.1 that the distance between points a and b on a number line is $d(a, b) = |b - a|$. So, from Figure 4 we see that the distance between the points $A(x_1, y_1)$ and $C(x_2, y_1)$ on a horizontal line must be $|x_2 - x_1|$ and the distance between $B(x_2, y_2)$ and $C(x_2, y_1)$ on a vertical line must be $|y_2 - y_1|$. Since triangle ABC is a right triangle, the Pythagorean Theorem gives

$$d(A, B) = \sqrt{|x_2 - x_1|^2 + |y_2 - y_1|^2}$$

$$= \sqrt{(x_2 - x_1)^2 + (y_2 - y_1)^2}$$

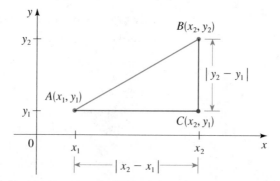

FIGURE 4

DISTANCE FORMULA

The distance between the points $A(x_1, y_1)$ and $B(x_2, y_2)$ in the plane is

$$d(A, B) = \sqrt{(x_2 - x_1)^2 + (y_2 - y_1)^2}$$

EXAMPLE 2 ■ Applying the Distance Formula

Which of the points $P(1, -2)$ or $Q(8, 9)$ is closer to the point $A(5, 3)$?

SOLUTION   By the Distance Formula we have

$$d(P, A) = \sqrt{(5 - 1)^2 + [3 - (-2)]^2} = \sqrt{4^2 + 5^2} = \sqrt{41}$$

$$d(Q, A) = \sqrt{(5 - 8)^2 + (3 - 9)^2} = \sqrt{(-3)^2 + (-6)^2} = \sqrt{45}$$

This shows that $d(P, A) < d(Q, A)$, so P is closer to A (see Figure 5).   ■

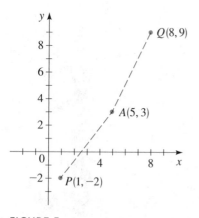

FIGURE 5

Now let's now find the coordinates (x, y) of the midpoint M of the line segment that joins the point $A(x_1, y_1)$ to the point $B(x_2, y_2)$. In Figure 6 notice that triangles APM and MQB are congruent because $d(A, M) = d(M, B)$ and corresponding angles are equal. It follows that $d(A, P) = d(M, Q)$ and so

$$x - x_1 = x_2 - x$$

Solving for x, we get $2x = x_1 + x_2$, so $x = \dfrac{x_1 + x_2}{2}$. Similarly, $y = \dfrac{y_1 + y_2}{2}$.

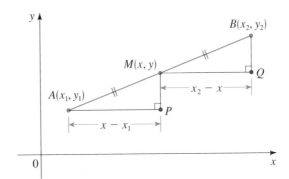

FIGURE 6

MIDPOINT FORMULA

The midpoint of the line segment from $A(x_1, y_1)$ to $B(x_2, y_2)$ is

$$\left(\frac{x_1 + x_2}{2}, \frac{y_1 + y_2}{2} \right)$$

EXAMPLE 3 ■ Finding the Midpoint

The midpoint of the line segment that joins the points $(-2, 5)$ and $(4, 9)$ is

$$\left(\frac{-2 + 4}{2}, \frac{5 + 9}{2} \right) = (1, 7)$$

See Figure 7.

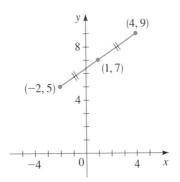

FIGURE 7

The coordinates of a point in the xy-plane uniquely determine its location. We can think of the coordinates as the "address" of the point. In Salt Lake City, Utah, the addresses of most buildings are in fact expressed as coordinates. The city is divided into quadrants with Main Street as the vertical (North-South) axis and S. Temple Street as the horizontal (East-West) axis. An address such as

    1760 W   2100 S

indicates a location 17.6 blocks west of Main Street and 21 blocks south of S. Temple Street. (This is the address of the main post office in Salt Lake City.) With this logical system it is possible for someone unfamiliar with the city to locate any address immediately, as easily as one can locate a point in the coordinate plane.

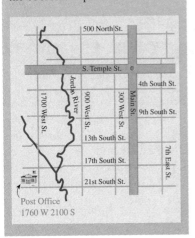

EXAMPLE 4 ■ Applying the Midpoint Formula

Show that the quadrilateral with vertices $P(1, 2)$, $Q(4, 4)$, $R(5, 9)$, and $S(2, 7)$ is a parallelogram by proving that its two diagonals bisect each other.

SOLUTION   If the two diagonals have the same midpoint, then they must bisect each other. The midpoint of the diagonal PR is

$$\left(\frac{1 + 5}{2}, \frac{2 + 9}{2} \right) = \left(3, \tfrac{11}{2} \right)$$

and the midpoint of the diagonal QS is

$$\left(\frac{4 + 2}{2}, \frac{4 + 7}{2} \right) = \left(3, \tfrac{11}{2} \right)$$

so each diagonal bisects the other, as shown in Figure 8. (A theorem from elementary geometry states that the quadrilateral is therefore a parallelogram.)

■

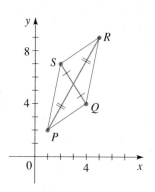

FIGURE 8

GRAPHS OF EQUATIONS

Suppose we have an equation involving the variables x and y, such as

$$x^2 + y^2 = 25 \qquad \text{or} \qquad x = y^2 \qquad \text{or} \qquad y = \frac{2}{x}$$

Fundamental Principle of Analytic Geometry

A point (x, y) lies on the graph of an equation if and only if its coordinates satisfy the equation.

A point (x, y) **satisfies** the equation if the equation is true when the coordinates of the point are substituted into the equation. For example, the point $(3, 4)$ satisfies the first equation, since $3^2 + 4^2 = 25$, but the point $(2, -3)$ does not, since $2^2 + (-3)^2 = 13 \neq 25$.

THE GRAPH OF AN EQUATION

The **graph** of an equation in x and y is the set of all points (x, y) in the coordinate plane that satisfy the equation.

The graphs of most of the equations that we will encounter are curves. For example, we will see later in this section that the first equation, $x^2 + y^2 = 25$, represents a circle; we will determine the shapes represented by the other equations later in this chapter. Graphs are important in the study of equations because they give visual representations of equations.

EXAMPLE 5 ■ Sketching a Graph by Plotting Points

Sketch the graph of the equation $2x - y = 3$.

SOLUTION   We first solve the given equation for y to get

$$y = 2x - 3$$

This helps us calculate the *y*-coordinates in the following table.

x	*y = 2x − 3*	*(x, y)*
−1	−5	(−1, −5)
0	−3	(0, −3)
1	−1	(1, −1)
2	1	(2, 1)
3	3	(3, 3)
4	5	(4, 5)

Of course, there are infinitely many points on the graph, and it is impossible to plot all of them. But the more points we plot, the better we can imagine what the graph represented by the equation looks like. We plot the points we found in the table in Figure 9; they appear to lie on a line. So, we complete the graph by joining the points by a line. (In Section 1.10 we verify that the graph of an equation of this type is indeed a line.)

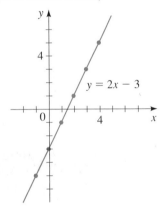

FIGURE 9

A detailed discussion of parabolas and their geometric properties is presented in Chapter 9.

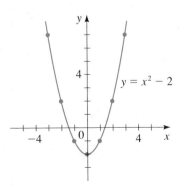

FIGURE 10

EXAMPLE 6 ■ Sketching a Graph by Plotting Points

Sketch the graph of the equation $y = x^2 - 2$.

SOLUTION   We find some of the points that satisfy the equation in the following table. In Figure 10 we plot these points and then connect them by a smooth curve. A curve with this shape is called a *parabola*.

x	$y = x^2 - 2$	*(x, y)*
−3	7	(−3, 7)
−2	2	(−2, 2)
−1	−1	(−1, −1)
0	−2	(0, −2)
1	−1	(1, −1)
2	2	(2, 2)
3	7	(3, 7)

EXAMPLE 7 ■ **Sketching the Graph of an Equation Involving Absolute Value**

Sketch the graph of the equation $y = |x|$.

SOLUTION   We make a table of values.

x	$y = \lvert x \rvert$	(x, y)
-3	3	$(-3, 3)$
-2	2	$(-2, 2)$
-1	1	$(-1, 1)$
0	0	$(0, 0)$
1	1	$(1, 1)$
2	2	$(2, 2)$
3	3	$(3, 3)$

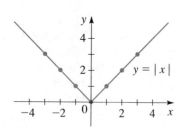

FIGURE 11

In Figure 11 we plot the points with the coordinates given by the table and use these to sketch the graph of the equation.   ■

The x-coordinates of the points where a graph intersects the x-axis are called the **x-intercepts** of the graph and are obtained by setting $y = 0$ in the equation of the graph. The y-coordinates of the points where a graph intersects the y-axis are called the **y-intercepts** of the graph and are obtained by setting $x = 0$ in the equation of the graph.

DEFINITION OF INTERCEPTS

Intercepts	How to find them	Where they are on the graph
x-intercepts:		
The x-coordinates of points where the graph of an equation intersects the x-axis	Set $y = 0$ and solve for x	
y-intercepts:		
The y-coordinates of points where the graph of an equation intersects the y-axis	Set $x = 0$ and solve for y	

EXAMPLE 8 ■ Finding Intercepts

Find the x- and y-intercepts of the graph of the equation $y = x^2 - 2$.

SOLUTION   To find the x-intercepts we set $y = 0$ and solve for x.

$$0 = x^2 - 2 \qquad \text{Set } y = 0$$

$$x^2 = 2 \qquad \text{Add 2 to each side}$$

$$x = \pm\sqrt{2} \qquad \text{Take the square root}$$

Thus, the x-intercepts are $\sqrt{2}$ and $-\sqrt{2}$.

To find the y-intercepts, we set $x = 0$ and solve for y.

$$y = 0^2 - 2 \qquad \text{Set } x = 0$$

$$y = -2$$

Thus, the y-intercept is -2.

The graph of this equation was sketched in Example 6. We sketch it again in Figure 12 with the x- and y-intercepts labeled.   ■

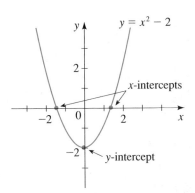

FIGURE 12

CIRCLES

So far we have discussed how to find the graph of an equation in x and y. The converse problem is to find an equation of a graph, that is, an equation that represents a given curve in the xy-plane. Such an equation is satisfied by the coordinates of the points on the curve and by no other point. This is the other half of the basic principle of analytic geometry as formulated by Descartes and Fermat. The idea is that if a geometric curve can be represented by an algebraic equation, then the rules of algebra can be used to analyze the curve.

As an example of this type of problem, let's find the equation of a circle with radius r and center (h, k). By definition, the circle is the set of all points $P(x, y)$ whose distance from the center $C(h, k)$ is r (see Figure 13). Thus, P is on the circle if and only if $d(P, C) = r$. From the Distance Formula we have

$$\sqrt{(x - h)^2 + (y - k)^2} = r$$

$$(x - h)^2 + (y - k)^2 = r^2 \qquad \text{Square each side}$$

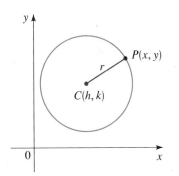

FIGURE 13

This is the desired equation.

EQUATION OF A CIRCLE

An equation of the circle with center (h, k) and radius r is

$$(x - h)^2 + (y - k)^2 = r^2$$

In particular, if the center is the origin $(0, 0)$, then the equation is

$$x^2 + y^2 = r^2$$

René Descartes (1596–1650) was born in the town of La Haye in southern France. From an early age Descartes liked mathematics because of "the certainty of its results and the clarity of its reasoning." He believed that in order to arrive at truth, one must begin by doubting everything, including one's own existence; this led him to formulate perhaps the most well known sentence in all of philosophy: "I think, therefore I am." In his book *Discourse on Method*, he described what is now called the Cartesian plane. This idea of combining algebra and geometry allowed mathematicians for the first time to "see" the equations they were studying. The philosopher John Stuart Mill called this invention "the greatest single step ever made in the progress of the exact sciences." Descartes liked to get up late and spend the morning in bed thinking and writing. He invented the coordinate plane while lying in bed watching a fly crawl on the ceiling, and reasoning that he could describe the exact location of the fly by knowing its distance from two perpendicular walls. In 1649 Descartes became the tutor of Queen Christina of Sweden. She liked her lessons at 5 o'clock in the morning when, she

(continued)

EXAMPLE 9 ■ **Graphing a Circle**

Graph each equation.

(a) $x^2 + y^2 = 25$ (b) $(x - 2)^2 + (y + 1)^2 = 25$

SOLUTION

(a) Rewriting the equation as $x^2 + y^2 = 5^2$, we see that this is an equation of the circle of radius 5 centered at the origin. Its graph is sketched in Figure 14.

(b) Rewriting the equation as $(x - 2)^2 + (y + 1)^2 = 5^2$, we see that this is an equation of the circle of radius 5 centered at $(2, -1)$. Its graph is sketched in Figure 15.

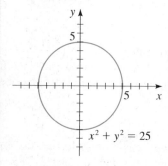

FIGURE 14

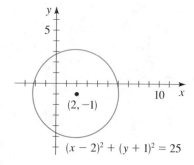

FIGURE 15

EXAMPLE 10 ■ **Finding the Equation of a Circle**

(a) Find an equation of the circle with radius 3 and center $(2, -5)$.

(b) Find an equation of the circle that has the points $P(1, 8)$ and $Q(5, -6)$ as the endpoints of a diameter.

SOLUTION

(a) Using the equation of a circle with $r = 3$, $h = 2$, and $k = -5$, we obtain

$$(x - 2)^2 + (y + 5)^2 = 9$$

Its graph is shown in Figure 16.

(b) We first observe that the center is the midpoint of the diameter PQ, so by the Midpoint Formula the center is

$$\left(\frac{1 + 5}{2}, \frac{8 - 6}{2}\right) = (3, 1)$$

The radius r is the distance from P to the center, so

$$r^2 = (3 - 1)^2 + (1 - 8)^2 = 2^2 + (-7)^2 = 53$$

Therefore, the equation of the circle is

$$(x - 3)^2 + (y - 1)^2 = 53$$

said, her mind was sharpest. However, the change from his usual habits and the ice-cold library where they studied proved too much for him. In February 1650, after just two months of this, he caught pneumonia and died.

Its graph is shown in Figure 17.

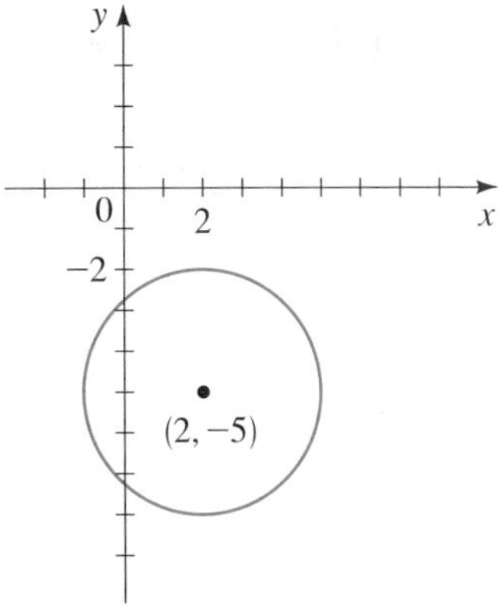

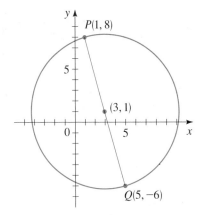

FIGURE 16
$(x - 2)^2 + (y + 5)^2 = 9$

FIGURE 17
$(x - 3)^2 + (y - 1)^2 = 53$

■

EXAMPLE 11 ■ Identifying an Equation of a Circle

Sketch the graph of the equation $x^2 + y^2 + 2x - 6y + 7 = 0$ by first showing that it represents a circle and then finding its center and radius.

SOLUTION We first group the x-terms and y-terms. Then we complete the square within each grouping (by adding the square of half the coefficient of x and the square of half the coefficient of y to each side of the equation).

$$(x^2 + 2x \qquad) + (y^2 - 6y \qquad) = -7 \qquad \text{Group terms}$$

$$(x^2 + 2x + 1) + (y^2 - 6y + 9) = -7 + 1 + 9 \qquad \text{Complete the squares by adding 1 and 9 to each side}$$

$$(x + 1)^2 + (y - 3)^2 = 3$$

Comparing this equation with the standard equation of a circle, we see that $h = -1$, $k = 3$, and $r = \sqrt{3}$, so the given equation represents a circle with center $(-1, 3)$ and radius $\sqrt{3}$. The circle is sketched in Figure 18. ■

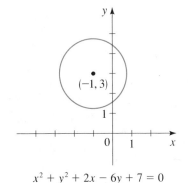

$x^2 + y^2 + 2x - 6y + 7 = 0$

FIGURE 18

SYMMETRY

Figure 19 shows the graph of $y = x^2$. Notice that the part of the graph to the left of the y-axis is the mirror image of the part to the right of the y-axis. The reason is that if the point (x, y) is on the graph, then so is $(-x, y)$, and these points are reflections of each other about the y-axis. In this situation we say the graph is **symmetric with respect to the y-axis**. Similarly, we say a graph is **symmetric with respect to the x-axis** if whenever the point (x, y) is on the graph, then so is $(x, -y)$. A graph is **symmetric with respect to the origin** if whenever (x, y) is on the graph, so is $(-x, -y)$. The following table gives definitions of the three

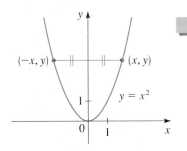

FIGURE 19

types of symmetry and explains how to determine whether the graph of an equation exhibits each of these symmetries.

DEFINITION OF SYMMETRY

Type of symmetry	How to test for symmetry	What the graph looks like (figures in this section)	Geometric meaning
Symmetry with respect to the x-axis	The equation is unchanged when y is replaced by $-y$	(Figures 14, 21)	Graph is unchanged when reflected in the x-axis.
Symmetry with respect to the y-axis	The equation is unchanged when x is replaced by $-x$	(Figures 10, 11, 12, 14, 19, 21)	Graph is unchanged when reflected in the y-axis.
Symmetry with respect to the origin	The equation is unchanged when x is replaced by $-x$ and y by $-y$	(Figures 14, 20, 21)	Graph is unchanged when rotated 180° about the origin. This is the same as a reflection in the x-axis followed by a reflection in the y-axis.

One reason that symmetry is important is that we can use it to help us sketch the graphs of equations. The remaining examples in this section illustrate the use of symmetry in graphing.

EXAMPLE 12 ■ Using Symmetry and Intercepts to Sketch a Graph

Find the x- and y-intercepts of the graph of the equation $y = x^3 - 9x$. Test the equation for symmetry and sketch its graph.

SOLUTION Setting $x = 0$ in the equation gives $y = 0$, so the y-intercept is 0. Setting $y = 0$ gives $x^3 - 9x = 0$, or $x(x^2 - 9) = x(x - 3)(x + 3) = 0$. Thus, the x-intercepts are 0, 3, and -3.

If we replace x by $-x$ and y by $-y$ in the equation, we get

$$-y = (-x)^3 - 9(-x)$$

$$-y = -x^3 + 9x$$

$$y = x^3 - 9x$$

and so the equation is unchanged. This means that the graph is symmetric with respect to the origin. We sketch it by first plotting points for $x > 0$ and then using symmetry about the origin (see Figure 20).

x	$y = x^3 - 9x$	(x, y)
0	0	$(0, 0)$
1	-8	$(1, -8)$
1.5	-10.125	$(1.5, -10.125)$
2	-10	$(2, -10)$
2.5	-6.875	$(2.5, -6.875)$
3	0	$(3, 0)$
4	28	$(4, 28)$

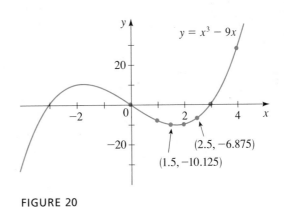

FIGURE 20

EXAMPLE 13 ■ **A Circle that Has All Three Types of Symmetry**

Test the equation of the circle $x^2 + y^2 = 4$ for symmetry.

SOLUTION   The equation $x^2 + y^2 = 4$ remains unchanged when x is replaced by $-x$ and y is replaced by $-y$, since $(-x)^2 = x^2$ and $(-y)^2 = y^2$, so the circle exhibits all three types of symmetry. It is symmetric with respect to the x-axis, the y-axis, and the origin. See Figure 21.

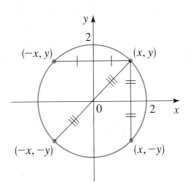

FIGURE 21
$x^2 + y^2 = 4$

1.8 EXERCISES

1. Plot the given points in a coordinate plane:

$(2,3)$, $(-2,3)$, $(4,5)$, $(4,-5)$, $(-4,5)$, $(-4,-5)$

2. Find the coordinates of the points shown in the figure.

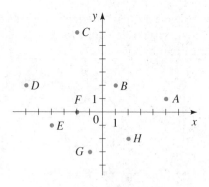

3–8 ■ A pair of points is given.
(a) Plot the points in a coordinate plane.
(b) Find the distance between them.
(c) Find the midpoint of the segment that joins them.

3. $(2,3)$, $(5,2)$ **4.** $(2,-1)$, $(4,3)$

5. $(6,-2)$, $(-1,3)$ **6.** $(1,-6)$, $(-1,-3)$

7. $(3,4)$, $(-3,-4)$ **8.** $(5,0)$, $(0,6)$

9. Draw the rectangle with vertices $A(1,3)$, $B(5,3)$, $C(1,-3)$, and $D(5,-3)$ on a coordinate plane. Find the area of the rectangle.

10. Draw the parallelogram with vertices $A(1,2)$, $B(5,2)$, $C(3,6)$, and $D(7,6)$ on a coordinate plane. Find the area of the parallelogram.

11. Plot the points $A(1,0)$, $B(5,0)$, $C(4,3)$, and $D(2,3)$ on a coordinate plane. Draw the segments AB, BC, CD, and DA. What kind of quadrilateral is $ABCD$, and what is its area?

12. Plot the points $P(5,1)$, $Q(0,6)$, and $R(-5,1)$ on a coordinate plane. Where must the point S be located so that the quadrilateral $PQRS$ is a square? Find the area of this square.

13–24 ■ Sketch the region given by the set.

13. $\{(x,y) \mid x \le 0\}$ **14.** $\{(x,y) \mid y \ge 0\}$

15. $\{(x,y) \mid x = 3\}$ **16.** $\{(x,y) \mid y = -2\}$

17. $\{(x,y) \mid 1 < x < 2\}$ **18.** $\{(x,y) \mid 0 \le y \le 4\}$

19. $\{(x,y) \mid xy < 0\}$ **20.** $\{(x,y) \mid xy > 0\}$

21. $\{(x,y) \mid x \ge 1 \text{ and } y < 3\}$ **22.** $\{(x,y) \mid |y| > 1\}$

23. $\{(x,y) \mid |x| \le 2\}$

24. $\{(x,y) \mid |x| < 3 \text{ and } |y| > 2\}$

25. Which of the points $A(6,7)$ or $B(-5,8)$ is closer to the origin?

26. Which of the points $C(-6,3)$ or $D(3,0)$ is closer to the point $E(-2,1)$?

27. Show that the triangle with vertices $A(0,2)$, $B(-3,-1)$, and $C(-4,3)$ is isosceles.

28. Find the area of the triangle shown in the figure.

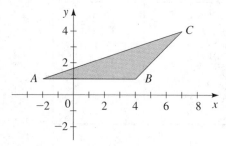

29. Refer to triangle ABC in the figure.
(a) Show that triangle ABC is a right triangle by using the converse of the Pythagorean Theorem.
(b) Find the area of triangle ABC.

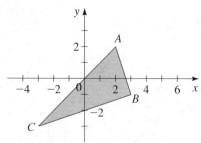

30. Find a point on the y-axis that is equidistant from the points $(5,-5)$ and $(1,1)$.

31. (a) Sketch the parallelogram with vertices $A(-2,-1)$, $B(4,2)$, $C(7,7)$, and $D(1,4)$.
(b) Find the midpoints of the diagonals of this parallelogram.
(c) Conclude from part (b) that the diagonals bisect each other.

32. The point M in the figure is the midpoint of the line segment AB. Show that M is equidistant from the vertices of triangle ABC.

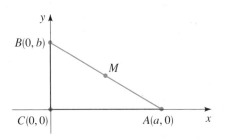

33. Suppose that each point in the coordinate plane is shifted to the right 3 units and upward 2 units.
(a) The point $(5, 3)$ is shifted to what new point?
(b) The point (a, b) is shifted to what new point?
(c) Triangle ABC in the figure has been shifted to triangle $A'B'C'$. Find the coordinates of the points A', B', and C'.

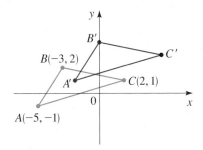

34. Suppose that the y-axis acts as a mirror that reflects each point to the right of it into a point to the left of it.
(a) The point $(3, 7)$ is reflected into what point?
(b) The point (a, b) is reflected into what point?
(c) Triangle ABC in the figure is reflected into triangle $A'B'C'$. Find the coordinates of the points A', B', and C'.

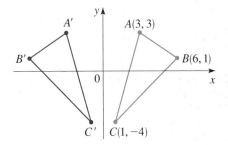

35–52 ■ Make a table of values and sketch the graph of the equation. Find x- and y-intercepts, and test for symmetry.

35. $y = x$ **36.** $y = -x$

37. $y = x - 1$ **38.** $y = 2x + 5$

39. $3x - y = 5$ **40.** $x + y = 3$

41. $y = 1 - x^2$ **42.** $y = x^2 + 2x$

43. $y = x^2 - 9$ **44.** $y = 9 - x^2$

45. $y = \sqrt{x}$ **46.** $y = \sqrt{4 - x^2}$

47. $y = |x|$ **48.** $x = |y|$

49. $y = 4 - |x|$ **50.** $y = |4 - x|$

51. $x = y^3$ **52.** $y = x^3 - 4x$

53–58 ■ Test the equation for symmetry.

53. $y = x^4 + x^2$ **54.** $x = y^4 - y^2$

55. $x^2 y^2 + xy = 1$ **56.** $x^4 y^4 + x^2 y^2 = 1$

57. $y = x^3 + 10x$ **58.** $y = x^2 + |x|$

59–62 ■ Complete the graph using the given symmetry property.

59. Symmetric with respect to the y-axis

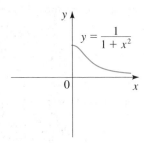

60. Symmetric with respect to the x-axis

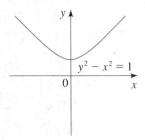

61. Symmetric with respect to the origin

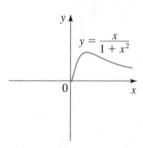

62. Symmetric with respect to the origin

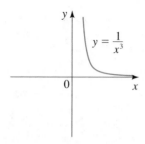

63–68 ■ Find an equation of the circle that satisfies the given conditions.

63. Center $(2, -1)$; radius 3

64. Center $(-1, -4)$; radius 8

65. Center at the origin; passes through $(4, 7)$

66. Center $(-1, 5)$; passes through $(-4, -6)$

67. Endpoints of a diameter are $P(-1, 3)$ and $Q(7, -5)$

68. Center $(7, -3)$; tangent to the x-axis

69–70 ■ Find the equation of the circle shown in the figure.

69.   **70.**

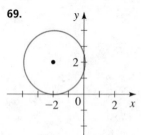
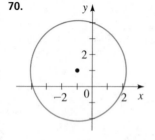

71–76 ■ Show that the equation represents a circle, and find the center and radius of the circle.

71. $x^2 + y^2 - 2x + 4y + 1 = 0$

72. $x^2 + y^2 - 2x - 2y = 2$

73. $x^2 + y^2 - 4x + 10y + 13 = 0$

74. $x^2 + y^2 + 6y + 2 = 0$

75. $2x^2 + 2y^2 + x = 0$

76. $16x^2 + 16y^2 + 8x + 32y + 1 = 0$

77–80 ■ Sketch the graph of the equation.

77. $x^2 + y^2 + 4x - 10y = 21$

78. $4x^2 + 4y^2 + 2x = 0$

79. $x^2 + y^2 + 6x - 12y + 45 = 0$

80. $x^2 + y^2 - 16x + 12y + 200 = 0$

81–82 ■ Sketch the region given by the set.

81. $\{(x, y) \mid x^2 + y^2 \le 1\}$

82. $\{(x, y) \mid 1 \le x^2 + y^2 < 9\}$

83. Find the area of the region that lies outside the circle $x^2 + y^2 = 4$ but inside the circle
$$x^2 + y^2 - 4y - 12 = 0$$

84. Sketch the region in the coordinate plane that satisfies both the inequalities $x^2 + y^2 \le 9$ and $y \ge |x|$. What is the area of this region?

85. Under what conditions on the coefficients a, b, and c does the equation $x^2 + y^2 + ax + by + c = 0$ represent a circle? When that condition is satisfied, find the center and radius of the circle. [*Hint:* Complete the squares.]

■ **DISCOVERY · DISCUSSION**

86. Completing a Line Segment Plot the points $M(6, 8)$ and $A(2, 3)$ on a coordinate plane. If M is the midpoint of the line segment AB, find the coordinates of B. Write a brief description of the steps you took in finding B, and your reasons for taking them.

87. Completing a Parallelogram Plot the points $P(-1, -4)$, $Q(1, 1)$, and $R(4, 2)$ on a coordinate plane. Where should the point S be located so that the figure $PQRS$ is a parallelogram? Write a brief description of the steps you took and your reasons for taking them.

1.9 GRAPHING CALCULATORS AND COMPUTERS

In this section we make use of graphing calculators or computers to graph more complicated equations and to solve more complex problems than we would otherwise be able to solve. We begin by describing how these devices work and how to recognize and avoid some common pitfalls of using such devices.

HOW TO USE GRAPHING DEVICES

A graphing calculator or computer displays a rectangular portion of the g   `/
an equation in a **display window** or **viewing screen**, which we call a **viewin** **rectangle**. The default screen often gives an incomplete or misleading picture, so it is important to choose the viewing rectangle with care. If we choose the x-values to range from a minimum value of $\text{Xmin} = a$ to a maximum of $\text{Xmax} = b$ and the y-values to range from a minimum of $\text{Ymin} = c$ to a maximum of $\text{Ymax} = d$, then the portion of the graph displayed lies in the rectangle

$$[a, b] \times [c, d] = \{(x, y) \mid a \leqslant x \leqslant b, c \leqslant y \leqslant d\}$$

(a, d) $y = d$ (b, d)

$x = a$ $x = b$

(a, c) $y = c$ (b, c)

FIGURE 1
The viewing rectangle $[a, b]$ by $[c, d]$

shown in Figure 1. We refer to this as the $[a, b]$ *by* $[c, d]$ *viewing rectangle*.

The graphing device draws the graph of an equation much as you would. It plots points of the form (x, y) for a certain number of values of x, equally spaced between a and b. If the equation is not defined for an x-value, or if the corresponding y-value lies outside the viewing rectangle, the device ignores this value and moves on to the next x-value. The machine connects each point to the preceding plotted point to form a representation of the graph of the equation.

EXAMPLE 1 ■ Drawing Graphs in Different Viewing Rectangles

Draw the graph of the equation $y = x^2 + 3$ in each viewing rectangle.

(a) $[-2, 2]$ by $[-2, 2]$

(b) $[-4, 4]$ by $[-4, 4]$

(c) $[-10, 10]$ by $[-5, 30]$

(d) $[-50, 50]$ by $[-100, 1000]$

SOLUTION For part (a) select the range by setting $\text{Xmin} = -2$, $\text{Xmax} = 2$, $\text{Ymin} = -2$, and $\text{Ymax} = 2$. The resulting graph is shown in Figure 2(a). The

FIGURE 2
Graphs of $y = x^2 + 3$

(a)

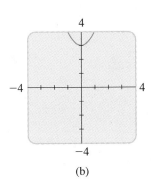

(b)

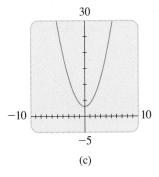

(c)

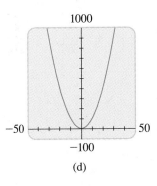

(d)

display window is blank! A moment's thought provides the explanation. Notice that $x^2 \geqslant 0$ for all x, so $y = x^2 + 3 \geqslant 3$ for all x. This means that the graph of the equation lies totally outside the viewing rectangle $[-2, 2]$ by $[-2, 2]$.

The graphs for the viewing rectangles in parts (b), (c), and (d) are shown in Figure 2, parts (b), (c), and (d). Observe that we get more complete pictures in parts (c) and (d), but the graph in part (d) doesn't show clearly that the y-intercept is 3. ■

EXAMPLE 2 ■ Graphing a Cubic Equation

Graph the equation $y = x^3 - 49x$.

SOLUTION   Let's experiment with different viewing rectangles. If we start with the viewing rectangle $[-5, 5]$ by $[-5, 5]$, we get the graph in Figure 3. On most calculators the screen appears to be blank, but it is not quite blank because the point $(0, 0)$ has been plotted. It turns out that for all the other x-values that the calculator chooses between -5 and 5, the value of y is greater than 5 or less than -5, so the corresponding point on the graph lies outside the viewing rectangle.

If we use the zoom-out feature of a graphing calculator to change the viewing rectangle to $[-10, 10]$ by $[-10, 10]$, then we get the picture shown in Figure 4(a). The graph appears to consist of vertical lines, but we know that it can't be correct. If we look carefully while the graph is being drawn, we see that the graph leaves the screen and reappears during the graphing process. This indicates that we need to see more of the graph in the vertical direction, so we change the viewing rectangle to $[-10, 10]$ by $[-100, 100]$. The resulting graph is shown in Figure 4(b). It still doesn't reveal all the main features of the equation, so we try $[-10, 10]$ by $[-200, 200]$ in Figure 4(c). Now we are more confident that we have arrived at an appropriate viewing rectangle. In Chapter 3, when we discuss third-degree polynomials, we will see that the graph shown in Figure 4(c) does indeed reveal all the main features of the equation.

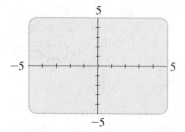

FIGURE 3

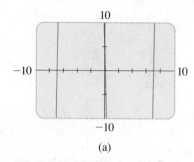

(a)

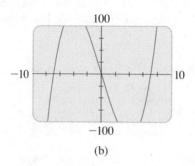

(b)

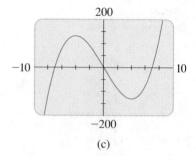

(c)

FIGURE 4

Graphs of $y = x^3 - 49x$ ■

Alan Turing (1912–1954) was at the center of two pivotal events of the 20th century—World War II and the invention of computers. At the age of 23 Turing made his mark on mathematics by solving Hilbert's third problem (see page 578). In this research he invented a theoretical machine, now called a Turing machine, which was the inspiration for modern digital computers. During World War II Turing was in charge of the British effort to decipher secret German codes. His complete success in this endeavor played a decisive role in the Allies' victory. To carry out the numerous logical steps required to break a coded message, Turing developed decision procedures similar to modern computer programs. After the war he helped develop the first electronic computers in Britain. He also did pioneering work on artificial intelligence and computer models of biological processes. At the age of 42 Turing died of cyanide poisoning under mysterious circumstances.

We see from Examples 1 and 2 that the choice of a viewing rectangle can make a big difference in the appearance of a graph. Sometimes it is necessary to change to a larger viewing rectangle to obtain a more complete picture—a more global view—of the graph. In the next example we see how the equation itself can provide us with enough information to select a good viewing rectangle.

EXAMPLE 3 ■ Choosing an Appropriate Viewing Rectangle

Determine an appropriate viewing rectangle for the equation $y = \sqrt{8 - 2x^2}$ and graph the equation.

SOLUTION   The expression for y is defined only when the expression under the square root sign is not negative. So we must have

$$0 \le 8 - 2x^2 \qquad \text{Expression under the root sign must be nonnegative}$$

$$2x^2 \le 8 \qquad \text{Add } 2x^2$$

$$x^2 \le 4 \qquad \text{Divide by 2}$$

$$|x| \le 2 \qquad \text{Take the square root}$$

$$-2 \le x \le 2 \qquad \text{Meaning of } |x| \le 2 \text{ (see page 77)}$$

Therefore, the values of x for which y is defined lie in the interval $[-2, 2]$. Also

$$0 \le \sqrt{8 - 2x^2} \le \sqrt{8} = 2\sqrt{2} \approx 2.83$$

so the corresponding values of y lie in the interval $[0, 2.83]$.

We choose the viewing rectangle so that the x- and y-intervals are somewhat larger than those we found. Taking the viewing rectangle to be $[-3, 3]$ by $[-1, 4]$, we get the graph shown in Figure 5.

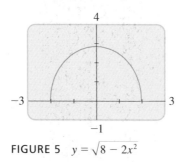

FIGURE 5 $y = \sqrt{8 - 2x^2}$

Graphing calculators are able to easily graph equations in which y can be isolated on one side of the equal sign. The next example shows how to graph equations that don't have this property.

EXAMPLE 4 ■ Graphing a Circle

Graph the circle $x^2 + y^2 = 1$.

SOLUTION We know from Section 1.8 how to graph this circle by hand; it is a circle with center the origin and radius 1. But graphing a circle with a graphing calculator is not a straightforward task, because we cannot isolate y on one side of the equal sign. If we try to solve the equation of the circle for y, we have

$$y^2 = 1 - x^2$$
$$y = \pm\sqrt{1 - x^2}$$

Therefore, we can regard the circle as being described by the graphs of *two* equations:

$$y = \sqrt{1 - x^2} \qquad \text{and} \qquad y = -\sqrt{1 - x^2}$$

The first equation represents the top half of the circle (because $y \geq 0$), and the second represents the bottom half of the circle (because $y \leq 0$). If we graph the first equation in the viewing rectangle $[-2, 2]$ by $[-2, 2]$, we get the semicircle shown in Figure 6(a). The graph of the second equation is the semicircle in Figure 6(b). Graphing these semicircles together on the same viewing screen, we get the full circle in Figure 6(c).

The graph in Figure 6(c) looks somewhat flattened. Most graphing calculators allow you to set the scales on the axes so that circles really look like circles. On the TI-82, for example, from the Zoom menu, choose "ZSquare" to set the scales appropriately. (On the TI-85 the command is "Zsq.")

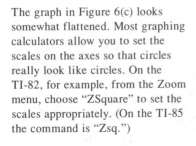

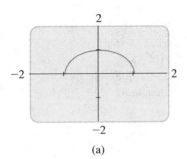

(a)

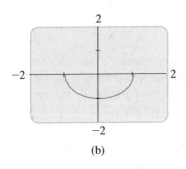

(b)

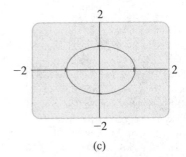

(c)

FIGURE 6

Graphing the equation $x^2 + y^2 = 1$

We have already discovered one of the pitfalls of using graphing calculators and computers: Examples 1 and 2 showed that the use of an inappropriate viewing rectangle can give a misleading representation of the graph of an equation. We also saw how to remedy the situation—we included the crucial parts of the graph by changing to a larger viewing rectangle. Another pitfall is illustrated in the next example.

 ### EXAMPLE 5 ■ Avoiding Extraneous Lines in Graphs

Draw the graph of the equation $y = \dfrac{1}{1 - x}$.

SOLUTION   Figure 7(a) shows the graph produced by a graphing calculator with viewing rectangle $[-9, 9]$ by $[-9, 9]$. In connecting successive points on the graph, the calculator produced a steep line segment from the top to the bottom of the screen. That line segment should not be part of the graph. Notice that the right side of the equation $y = 1/(1 - x)$ is not defined for $x = 1$. Sometimes we can get rid of the extraneous near-vertical line by experimenting with a change of scale. Here, for example, when we change to the smaller viewing rectangle $[-5, 5]$ by $[-5, 5]$, we obtain the much better graph in Figure 7(b).

Another way to avoid the extraneous line is to change the graphing mode on the calculator so that the dots are not connected.

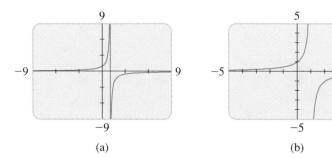

FIGURE 7

Graphs of $y = \dfrac{1}{1 - x}$

(a)   (b)

SOLVING EQUATIONS AND INEQUALITIES GRAPHICALLY

In Sections 1.5 and 1.7 we solved equations and inequalities algebraically. The ability to sketch graphs of equations quickly and accurately gives us another method of finding approximate solutions to equations and inequalities. For example, consider the equation

$$x^2 - 3x - 10 = 0$$

To solve this equation we first graph

$$y = x^2 - 3x - 10$$

as shown in Figure 8(a). We are interested in those values of x for which $y = 0$. But these are precisely the x-intercepts of the graph. By moving the cursor near the x-intercepts, we see that they appear to be at $x = -2$ and $x = 5$. To confirm this we may use the zoom-in feature on the graphing device to get a closer look

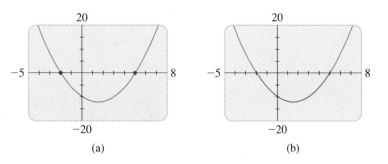

FIGURE 8

Graphs of $y = x^2 - 3x - 10$

(a)   (b)

at each of the intercepts. In this particular case, we can prove that these are the roots by substituting them directly into the equation.

To solve the inequality

$$x^2 - 3x - 10 > 0$$

we again sketch the graph of $y = x^2 - 3x - 10$ and look for the x-coordinates of those points on the graph with y-coordinate greater than 0. These are simply the x-values for which the graph lies above the x-axis. So, from Figure 8(b) we see that the solution of the inequality is $(-\infty, -2) \cup (5, \infty)$. Similarly, the solution of the inequality

$$x^2 - 3x - 10 < 0$$

consists of the x-coordinates of those points on the graph with y-coordinate less than 0, that is, points where the graph lies below the x-axis. So, the solution is the interval $(-2, 5)$.

EXAMPLE 6 ■ Solving an Equation Graphically

Solve the equation $2x^2 + 3x - 7 = 0$.

SOLUTION We graph the equation

$$y = 2x^2 + 3x - 7$$

in the viewing rectangle $[-5, 5]$ by $[-15, 15]$ (see Figure 9). The solutions occur at the x-intercepts. Moving the cursor to the x-intercepts, we find that $x \approx -2.8$ and $x \approx 1.3$. Using the zoom-in feature to get a closer look at the x-intercepts, we read the more accurate approximations

$$x \approx -2.77 \qquad \text{and} \qquad x \approx 1.27 \qquad \blacksquare$$

EXAMPLE 7 ■ Solving an Inequality Graphically

Solve the inequality $x^3 - 5x^2 \geq -8$.

SOLUTION We write the inequality as

$$x^3 - 5x^2 + 8 \geq 0$$

and then graph the equation

$$y = x^3 - 5x^2 + 8$$

in the viewing rectangle $[-6, 6]$ by $[-15, 15]$, as shown in Figure 10. The solution of the inequality consists of those intervals on which the graph lies on or above the x-axis. By moving the cursor to the x-intercepts we find that, correct to one decimal place, the solution is $[-1.1, 1.5] \cup [4.6, \infty)$. $\blacksquare$

In Example 6 we can use the quadratic formula to get

$$x = \frac{-3 \pm \sqrt{3^2 - 4 \cdot 2 \cdot (-7)}}{2 \cdot 2}$$

$$= \frac{-3 \pm \sqrt{65}}{4}$$

You can check that, correct to two decimal places, these are the solutions we obtained using the graphing calculator.

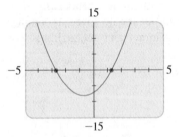

FIGURE 9
$y = 2x^2 + 3x - 7$

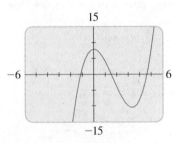

FIGURE 10
$y = x^3 - 5x^2 + 8$

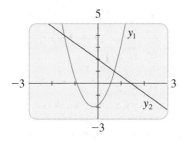

FIGURE 11

$y_1 = 3.7x^2 + 1.3x - 1.9$

$y_2 = 2.0 - 1.4x$

EXAMPLE 8 ■ Solving an Inequality Graphically

Solve the inequality $3.7x^2 + 1.3x - 1.9 \leq 2.0 - 1.4x$.

SOLUTION We graph the equations

$$y_1 = 3.7x^2 + 1.3x - 1.9$$

$$y_2 = 2.0 - 1.4x$$

in the same viewing rectangle in Figure 11. We are interested in those values of x for which $y_1 \leq y_2$; these are points for which the graph of y_2 lies on or above the graph of y_1. To determine the appropriate interval we look for the x-coordinates of points where the graphs intersect. We conclude that the solution is (approximately) the interval $[-1.45, 0.72]$. ■

1.9 EXERCISES

1–6 ■ Use a graphing calculator or computer to decide which one of the viewing rectangles (a)–(d) produces the most appropriate graph of the equation.

1. $y = x^4 + 2$
(a) $[-2, 2]$ by $[-2, 2]$
(b) $[0, 4]$ by $[0, 4]$
(c) $[-8, 8]$ by $[-4, 40]$
(d) $[-40, 40]$ by $[-80, 800]$

2. $y = x^2 + 7x + 6$
(a) $[-5, 5]$ by $[-5, 5]$
(b) $[0, 10]$ by $[-20, 100]$
(c) $[-15, 8]$ by $[-20, 100]$
(d) $[-10, 3]$ by $[-100, 20]$

3. $y = 100 - x^2$
(a) $[-4, 4]$ by $[-4, 4]$
(b) $[-10, 10]$ by $[-10, 10]$
(c) $[-15, 15]$ by $[-30, 110]$
(d) $[-4, 4]$ by $[-30, 110]$

4. $y = 2x^2 - 1000$
(a) $[-10, 10]$ by $[-10, 10]$
(b) $[-10, 10]$ by $[-100, 100]$
(c) $[-10, 10]$ by $[-1000, 1000]$
(d) $[-25, 25]$ by $[-1200, 200]$

5. $y = 10 + 25x - x^3$
(a) $[-4, 4]$ by $[-4, 4]$
(b) $[-10, 10]$ by $[-10, 10]$
(c) $[-20, 20]$ by $[-100, 100]$
(d) $[-100, 100]$ by $[-200, 200]$

6. $y = \sqrt{8x - x^2}$
(a) $[-4, 4]$ by $[-4, 4]$
(b) $[-5, 5]$ by $[0, 100]$
(c) $[-10, 10]$ by $[-10, 40]$
(d) $[-2, 10]$ by $[-2, 6]$

7–20 ■ Determine an appropriate viewing rectangle for the equation and use it to draw the graph.

7. $y = 100x^2$

8. $y = -100x^2$

9. $y = 4 + 6x - x^2$

10. $y = 0.3x^2 + 1.7x - 3$

11. $y = \sqrt[4]{256 - x^2}$

12. $y = \sqrt{12x - 17}$

13. $y = 0.01x^3 - x^2 + 5$

14. $y = x(x + 6)(x - 9)$

15. $y = \dfrac{1}{x^2 + 25}$

16. $y = \dfrac{x}{x^2 + 25}$

17. $y = x^4 - 4x^3$

18. $y = x^3 + \dfrac{1}{x}$

19. $y = 1 + |x - 1|$

20. $y = 2x - |x^2 - 5|$

21. Graph the circle $x^2 + y^2 = 9$ by solving for y and graphing two equations as in Example 4.

22. Graph the circle $(y - 1)^2 + x^2 = 1$ by solving for y and graphing two equations as in Example 4.

23. Graph the equation $4x^2 + 2y^2 = 1$ by solving for y and graphing two equations corresponding to the negative and positive square roots. (The graph is called an *ellipse*.)

24. Graph the equation $y^2 - 9x^2 = 1$ by solving for y and graphing the two equations corresponding to the positive and negative square roots. (The graph is called a *hyperbola*.)

25–32 ■ Find the solutions of the equation in the given interval by drawing a graph in an appropriate viewing rectangle. State each answer correct to two decimal places.

25. $x^2 - 7x + 12 = 0$; $[0, 6]$

26. $x^2 - 0.75x + 0.125 = 0$; $[-2, 2]$

27. $x^3 - 6x^2 + 11x - 6 = 0;$ $[-1, 4]$

28. $16x^3 + 16x^2 = x + 1;$ $[-2, 2]$

29. $x - \sqrt{x + 1} = 0;$ $[-1, 5]$

30. $1 + \sqrt{x} = \sqrt{1 + x^2};$ $[-1, 5]$

31. $x^{1/3} - x = 0;$ $[-3, 3]$

32. $x^{1/2} + x^{1/3} - x = 0;$ $[-1, 5]$

33–40 ■ Find the solutions of the inequality by drawing appropriate graphs. State each answer correct to two decimal places.

33. $x^2 - 3x - 10 \le 0$

34. $0.5x^2 + 0.875x \le 0.25$

35. $x^3 + 11x \le 6x^2 + 6$

36. $16x^3 + 24x^2 > -9x - 1$

37. $x^{1/3} < x$

38. $\sqrt{0.5x^2 + 1} \le 2|x|$

39. $(x + 1)^2 < (x - 1)^2$

40. $(x + 1)^2 \le x^3$

 DISCOVERY · DISCUSSION

41. Equation Notation on Graphing Calculators   When you enter the following equations into your calculator, how does what you see on the screen differ from the usual way of writing the equations? (Check your user's manual if you're not sure.)

(a) $y = |x|$

(b) $y = \sqrt[5]{x}$

(c) $y = \dfrac{x}{x - 1}$

1.10   LINES

In this section we find equations for straight lines lying in a coordinate plane. We first need a way of measuring the "steepness" of a line. From Figure 1 we see that the steepness is determined by how quickly the line rises (or falls) as we move from left to right. Thus, we define the slope of a line as the ratio of the change in the y-coordinate (that is, the distance the line rises or falls) to the change in the x-coordinate (that is, the distance we move to the right). The change in the y-coordinate between two points on a line is called the **rise**, and the change in the x-coordinate is called the **run**. So, to find the slope of a line we choose two points on the line and find

$$\text{slope} = \frac{\text{rise}}{\text{run}}$$

More precisely, we have the following definition.

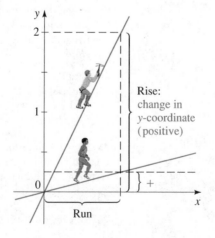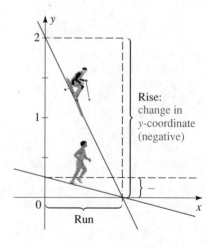

FIGURE 1

SLOPE OF A LINE

The **slope** m of a nonvertical line that passes through the points $A(x_1, y_1)$ and $B(x_2, y_2)$ is

$$m = \frac{y_2 - y_1}{x_2 - x_1}$$

The slope of a vertical line is not defined.

The slope is independent of which two points are chosen on the line. We can see that this is true from the similar triangles in Figure 2:

$$\frac{y_2 - y_1}{x_2 - x_1} = \frac{y_2' - y_1'}{x_2' - x_1'}$$

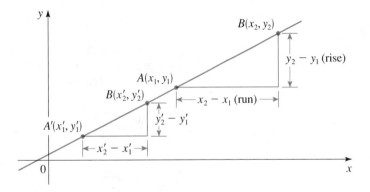

FIGURE 2

Figure 3 shows several lines labeled with their slopes. Notice that lines with positive slope slant upward to the right, whereas lines with negative slope slant downward to the right. The steepest lines are those for which the absolute value of the slope is the largest; a horizontal line has slope zero.

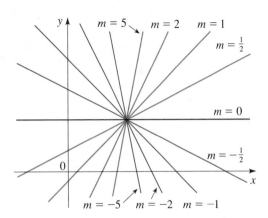

FIGURE 3

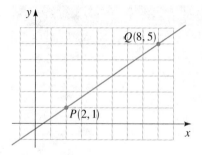

FIGURE 4

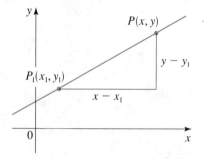

FIGURE 5

EXAMPLE 1 ■ Finding the Slope of a Line through Two Points

Find the slope of the line that passes through the points $P(2, 1)$ and $Q(8, 5)$.

SOLUTION   Since any two different points determine a line, there is only one line that passes through these two points. From the definition, the slope is

$$m = \frac{y_2 - y_1}{x_2 - x_1} = \frac{5 - 1}{8 - 2} = \frac{4}{6} = \frac{2}{3}$$

This says that for every 3 units we move to the right, the line rises 2 units. The line is drawn in Figure 4.   ■

Now let's find the equation of the line that passes through a given point $P(x_1, y_1)$ and has slope m. A point $P(x, y)$ with $x \neq x_1$ lies on this line if and only if the slope of the line through P_1 and P is equal to m (see Figure 5), that is,

$$\frac{y - y_1}{x - x_1} = m$$

This equation can be rewritten in the form $y - y_1 = m(x - x_1)$, and we observe that this equation is also satisfied when $x = x_1$ and $y = y_1$. Therefore, it is an equation of the given line.

POINT-SLOPE FORM OF THE EQUATION OF A LINE

An equation of the line that passes through the point (x_1, y_1) and has slope m is

$$y - y_1 = m(x - x_1)$$

EXAMPLE 2 ■ Finding the Equation of a Line with Given Point and Slope

(a) Find an equation of the line through $(1, -3)$ with slope $-\frac{1}{2}$.
(b) Sketch the line.

SOLUTION

(a) Using the point-slope form with $m = -\frac{1}{2}$, $x_1 = 1$, and $y_1 = -3$, we obtain an equation of the line as

$$y + 3 = -\tfrac{1}{2}(x - 1) \qquad \text{From point-slope equation}$$

$$2y + 6 = -x + 1 \qquad \text{Multiply by 2}$$

$$x + 2y + 5 = 0 \qquad \text{Rearrange}$$

(b) The fact that the slope is $-\frac{1}{2}$ tells us that when we move to the right 2 units, the line drops 1 unit. This enables us to sketch the line in Figure 6.   ■

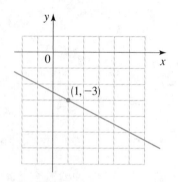

FIGURE 6

EXAMPLE 3 ■ Finding the Equation of a Line through Two Given Points

Find an equation of the line through the points $(-1, 2)$ and $(3, -4)$.

SOLUTION   The slope of the line is

$$m = \frac{-4 - 2}{3 - (-1)} = -\frac{6}{4} = -\frac{3}{2}$$

Using the point-slope form with $x_1 = -1$ and $y_1 = 2$, we obtain

$$y - 2 = -\tfrac{3}{2}(x + 1) \qquad \text{From point-slope equation}$$

$$2y - 4 = -3x - 3 \qquad \text{Multiply by 2}$$

$$3x + 2y - 1 = 0 \qquad \text{Rearrange} \qquad \blacksquare$$

Suppose a nonvertical line has slope m and y-intercept b (see Figure 7). This means the line intersects the y-axis at the point $(0, b)$, so the point-slope form of the equation of the line, with $x = 0$ and $y = b$, becomes

$$y - b = m(x - 0)$$

This simplifies to $y = mx + b$, which is called the **slope-intercept form** of the equation of a line.

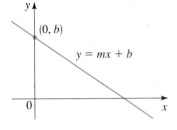

FIGURE 7

SLOPE-INTERCEPT FORM OF THE EQUATION OF A LINE

An equation of the line that has slope m and y-intercept b is

$$y = mx + b$$

EXAMPLE 4 ■ Finding the Slope of a Line from Its Equation

Find the slope and y-intercept of each line.

(a) $y = 3x - 2$ 　　　　　　　　　　(b) $3y - 2x = 1$

SOLUTION

(a) The equation $y = 3x - 2$ is in the form $y = mx + b$, where $m = 3$ and $b = -2$. So, from the slope-intercept form of the equation, we see that the slope is 3 and the y-intercept is -2.

(b) We first write the equation in the form $y = mx + b$:

$$3y - 2x = 1$$

$$3y = 2x + 1 \qquad \text{Add } 2x$$

$$y = \tfrac{2}{3}x + \tfrac{1}{3} \qquad \text{Divide by 3}$$

From the slope-intercept form of the equation of a line, we see that the slope is $m = \tfrac{2}{3}$ and the y-intercept is $b = \tfrac{1}{3}$. 　　　　　■

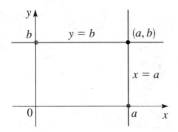

FIGURE 8

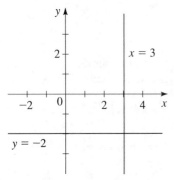

FIGURE 9

If a line is horizontal, its slope is $m = 0$, so its equation is $y = b$, where b is the y-intercept (see Figure 8). A vertical line does not have a slope, but we can write its equation as $x = a$, where a is the x-intercept, because the x-coordinate of every point on the line is a.

VERTICAL AND HORIZONTAL LINES

The equation of a vertical line through (a, b) is $x = a$.

The equation of a horizontal line through (a, b) is $y = b$.

EXAMPLE 5 ■ Vertical and Horizontal Lines

(a) The graph of the equation $x = 3$ is a vertical line with x-intercept 3.
(b) The graph of the equation $y = -2$ is a horizontal line with y-intercept -2.

The lines are graphed in Figure 9. ■

A **linear equation** is an equation of the form

$$Ax + By + C = 0$$

where A, B, and C are constants and A and B are not both 0. The equation of a line is a linear equation:

■ A nonvertical line has the equation $y = mx + b$ or $-mx + y - b = 0$, which is a linear equation with $A = -m$, $B = 1$, and $C = -b$.

■ A vertical line has the equation $x = a$ or $x - a = 0$, which is a linear equation with $A = 1$, $B = 0$, and $C = -a$.

Conversely, the graph of a linear equation is a line:

■ If $B \neq 0$, the equation becomes

$$y = -\frac{A}{B}x - \frac{C}{B}$$

and this is the slope-intercept form of the equation of a line (with $m = -A/B$ and $b = -C/B$).

■ If $B = 0$, the equation becomes $Ax + C = 0$, or $x = -C/A$, which represents a vertical line.

We have proved the following.

GENERAL EQUATION OF A LINE

The graph of every **linear equation**

$$Ax + By + C = 0 \qquad (A, B \text{ not both zero})$$

is a line. Conversely, every line is the graph of a linear equation.

EXAMPLE 6 ■ Graphing a Linear Equation

Sketch the graph of the equation $3x - 5y = 15$.

SOLUTION Since the equation $3x - 5y = 15$ is linear, its graph is a line. To draw the graph, it is enough to find just two points on the line. The intercepts are the easiest points to find.

x-intercept: Substitute $y = 0$, to get $3x = 15$, so $x = 5$.

y-intercept: Substitute $x = 0$, to get $-5y = 15$, so $y = -3$.

With these points we can sketch the graph shown in Figure 10.

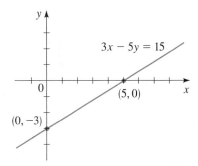

FIGURE 10

Another method is to write the equation in slope-intercept form:

$$3x - 5y = 15$$
$$5y = 3x - 15$$
$$y = \tfrac{3}{5}x - 3$$

This equation is in the form $y = mx + b$, so the slope is $m = \tfrac{3}{5}$ and the y-intercept is $b = -3$. This information can be used to sketch the line. ■

PARALLEL AND PERPENDICULAR LINES

Since slope measures the steepness of a line, it seems reasonable that parallel lines should have the same slope. In fact, we can prove this.

PARALLEL LINES

Two nonvertical lines are parallel if and only if they have the same slope.

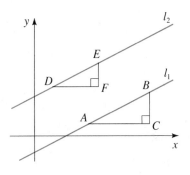

FIGURE 11

■ **Proof** Let the lines l_1 and l_2 in Figure 11 have slopes m_1 and m_2. If the lines are parallel, then the right triangles ABC and DEF are similar, so

$$m_1 = \frac{d(B,C)}{d(A,C)} = \frac{d(E,F)}{d(D,F)} = m_2$$

Conversely, if the slopes are equal, then the triangles will be similar, so $\angle BAC = \angle EDF$ and the lines are parallel.   □

EXAMPLE 7 ■ Finding the Equation of a Line Parallel to a Given Line

Find an equation of the line through the point $(5, 2)$ that is parallel to the line $4x + 6y + 5 = 0$.

SOLUTION   First we write the equation of the given line in slope-intercept form.

$$4x + 6y + 5 = 0$$

$$6y = -4x - 5 \qquad \text{Subtract } 4x + 5$$

$$y = -\tfrac{2}{3}x - \tfrac{5}{6} \qquad \text{Divide by } 6$$

So the line has slope $m = -\tfrac{2}{3}$. Since the required line is parallel to the given line, it also has slope $m = -\tfrac{2}{3}$. From the point-slope form of the equation of a line, we get

$$y - 2 = -\tfrac{2}{3}(x - 5) \qquad \text{Slope } m = -\tfrac{2}{3}, \text{ point } (5, 2)$$

$$3y - 6 = -2x + 10 \qquad \text{Multiply by } 3$$

$$2x + 3y - 16 = 0 \qquad \text{Rearrange}$$

Thus, the equation of the required line is $2x + 3y - 16 = 0$.   ■

The condition for perpendicular lines is not as obvious as that for parallel lines.

PERPENDICULAR LINES

Two lines with slopes m_1 and m_2 are perpendicular if and only if $m_1 m_2 = -1$, that is, their slopes are negative reciprocals:

$$m_2 = -\frac{1}{m_1}$$

Also, a horizontal line (slope 0) is perpendicular to a vertical line (no slope).

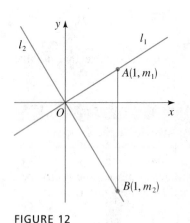

FIGURE 12

■ **Proof**   In Figure 12 we show two lines intersecting at the origin. (If the lines intersect at some other point, we consider lines parallel to these that intersect at the origin. These lines have the same slopes as the original lines.) If the lines l_1 and l_2 have slopes m_1 and m_2, then their equations are $y = m_1 x$ and $y = m_2 x$. Notice that $A(1, m_1)$ lies on l_1 and $B(1, m_2)$ lies on l_2. By the Pythagorean Theorem and its converse, $OA \perp OB$ if and only if

$$[d(O, A)]^2 + [d(O, B)]^2 = [d(A, B)]^2$$

By the Distance Formula, this becomes

$$(1^2 + m_1^2) + (1^2 + m_2^2) = (1 - 1)^2 + (m_2 - m_1)^2$$

$$2 + m_1^2 + m_2^2 = m_2^2 - 2m_1 m_2 + m_1^2$$

$$2 = -2m_1 m_2$$

$$m_1 m_2 = -1 \qquad \square$$

EXAMPLE 8 ■ Perpendicular Lines

Show that the points $P(3, 3)$, $Q(8, 17)$, and $R(11, 5)$ are the vertices of a right triangle.

SOLUTION   The slopes of the lines containing PR and QR are, respectively,

$$m_1 = \frac{5 - 3}{11 - 3} = \frac{1}{4} \qquad \text{and} \qquad m_2 = \frac{5 - 17}{11 - 8} = -4$$

Since $m_1 m_2 = -1$, these lines are perpendicular and so PQR is a right triangle. It is sketched in Figure 13.

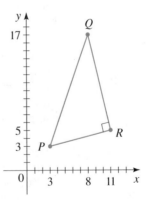

FIGURE 13

EXAMPLE 9 ■ Finding the Equation of a Line Perpendicular to a Given Line

Find the equation of a line that is perpendicular to the line $4x + 6y + 5 = 0$ and passes through the origin.

SOLUTION   In Example 7 we found that the slope of the line $4x + 6y + 5 = 0$ is $-\frac{2}{3}$. Thus, the slope of a perpendicular line is the negative reciprocal, that is, $\frac{3}{2}$. Since the required line passes through $(0, 0)$, the point-slope form gives

$$y - 0 = \tfrac{3}{2}(x - 0)$$

$$y = \tfrac{3}{2}x$$

APPLICATIONS OF LINEAR EQUATIONS

The simplest equations relating two variables are linear equations. We now consider how such equations are used in applications to model the relationship between two quantities.

EXAMPLE 10 ■ Linear Relationship between Fahrenheit and Celsius Scales

The relationship between the Fahrenheit (F) and Celsius (C) temperature scales is given by the equation $F = \frac{9}{5}C + 32$.

(a) Sketch a graph of this equation, letting the horizontal axis be the C-axis and the vertical axis be the F-axis.

(b) What is the slope of this graph and what does it represent? What is the F-intercept and what does it represent?

SOLUTION

(a) This is a linear equation, so its graph is a line. (We are calling the coordinates C and F instead of x and y, but this makes no difference in the shape of the graph.) Since two points determine a line, we first find two points that satisfy the equation, and plot those points. Then we draw a straight line through them. When $C = 0$, then $F = \frac{9}{5}(0) + 32 = 32$, and when $C = 5$, then $F = \frac{9}{5}(5) + 32 = 41$. Thus, the points $(0, 32)$ and $(5, 41)$ lie on the line. These points and the line containing them are shown in Figure 14.

(b) Since the equation is given in slope-intercept form, we see that the slope is $\frac{9}{5}$ and the F-intercept is 32. The slope represents the change in °F for every 1 °C change; thus, a 9 °F increase in temperature corresponds to a 5 °C increase. The F-intercept is the F-coordinate of the point on the graph whose C-coordinate is 0. Thus, 32 °F is the same as 0 °C (the freezing point of water). ■

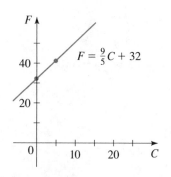

FIGURE 14

EXAMPLE 11 ■ Linear Relationship between Temperature and Elevation

(a) As dry air moves upward, it expands and cools. If the ground temperature is 20 °C and the temperature at a height of 1 km is 10 °C, express the temperature T (in °C) in terms of the height h (in kilometers). (Assume that the expression is linear.)

(b) Draw the graph of the linear equation. What does its slope represent?

(c) What is the temperature at a height of 2.5 km?

SOLUTION

(a) Because we are assuming a linear relationship between h and T, the equation must be of the form

$$T = mh + b$$

where m and b are constants. When $h = 0$, we are given that $T = 20$, so

$$20 = m(0) + b$$
$$b = 20$$

Thus, we have

$$T = mh + 20$$

When $h = 1$, we have $T = 10$ and so

$$10 = m(1) + 20$$

$$m = 10 - 20 = -10$$

The required expression is

$$T = -10h + 20$$

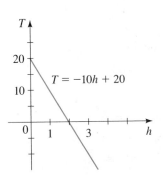

FIGURE 15

(b) The graph is sketched in Figure 15. The slope is $m = -10$ °C/km, and this represents the rate of change of temperature with respect to distance above the ground.

(c) At a height of $h = 2.5$ km, the temperature is

$$T = -10(2.5) + 20 = -25 + 20 = -5 \, ^\circ\text{C} \qquad \blacksquare$$

Economists model supply and demand for a commodity by linear equations. For example, for a certain commodity we might have

Supply equation $y = 8p - 10$

Demand equation $y = -3p + 15$

where p is the price of the commodity. In the supply equation y (the *amount produced*) increases as the price increases because if the price is high, more people will produce the commodity. The demand equation indicates that y (the *amount sold*) decreases as the price increases. The **equilibrium point** is the point of intersection of the graphs of the supply and demand equations; at that point the amount produced equals the amount sold.

In the supply and demand equations, most economists express p, the price, in terms of y, the amount produced or sold. For clarity, we have chosen to express y in terms of p.

EXAMPLE 12 ■ Supply and Demand for Wheat

An economist models the market for wheat by the following equations:

Supply equation $y = 8.33p - 14.58$

Demand equation $y = -1.39p + 23.35$

Here p is the price per bushel (in dollars), and y is the number of bushels produced and sold (in millions).

(a) At what point is the price so low that no wheat is produced?

(b) At what point is the price so high that no wheat is sold?

(c) Draw the graph of the supply and demand lines in the same viewing rectangle and find the equilibrium point. Estimate the equilibrium price and the quantities produced and sold at equilibrium.

SOLUTION

(a) If no wheat is produced, then $y = 0$ in the supply equation.

$$0 = 8.33p - 14.58 \qquad \text{Set } y = 0 \text{ in the supply equation}$$

$$p = 1.75 \qquad\qquad \text{Solve for } p$$

So, at the low price of $1.75 per bushel, the production of wheat halts completely.

(b) If no wheat is sold, then $y = 0$ in the demand equation.

$$0 = -1.39p + 23.35 \qquad \text{Set } y = 0 \text{ in the demand equation}$$

$$p = 16.80 \qquad\qquad \text{Solve for } p$$

So, at the high price of $16.80 per bushel, no wheat is sold.

(c) In Figure 16 we draw the graphs of both the supply and demand equations in the same viewing rectangle. By moving the cursor to their point of intersection, we find that the lines intersect (approximately) at the point (3.9, 17.9). Thus, the equilibrium point occurs when the price is $3.90 per bushel and 17.9 million bushels are produced and sold. ∎

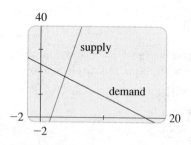

FIGURE 16

1.10 EXERCISES

1–6 ■ Find the slope of the line through P and Q.

1. $P(2, 4)$, $Q(4, 12)$ **2.** $P(-2, 5)$, $Q(4, -3)$

3. $P(-1, 3)$, $Q(1, -6)$ **4.** $P(-1, -4)$, $Q(6, 0)$

5. $P(2, 2)$, $Q(0, 0)$ **6.** $P(-2, 3)$, $Q(5, 5)$

7. Find the slopes of the lines l_1, l_2, l_3, and l_4 in the figure.

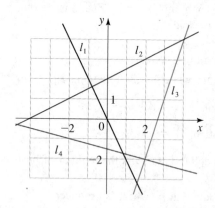

8. (a) Sketch lines through $(0, 0)$ with slopes $1, 0, \frac{1}{2}, 2$, and -1.

(b) Sketch lines through $(0, 0)$ with slopes $\frac{1}{3}, \frac{1}{2}, -\frac{1}{3}$, and 3.

9–12 ■ Find an equation for the line whose graph is sketched.

9.

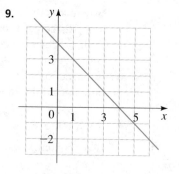

10.

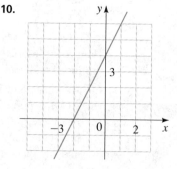

11.

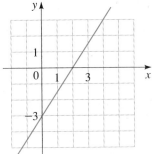

12.

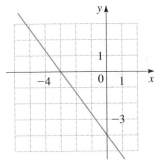

13–32 ■ Find an equation of the line that satisfies the given conditions.

13. Through $(2, 3)$;   slope 1

14. Through $(-2, 4)$;   slope -1

15. Through $(1, 7)$;   slope $\frac{2}{3}$

16. Through $(-3, -5)$;   slope $-\frac{7}{2}$

17. Through $(2, 1)$ and $(1, 6)$

18. Through $(-1, -2)$ and $(4, 3)$

19. Slope 3; y-intercept -2

20. Slope $\frac{2}{5}$; y-intercept 4

21. x-intercept 1; y-intercept -3

22. x-intercept -8; y-intercept 6

23. Through $(4, 5)$;   parallel to the x-axis

24. Through $(4, 5)$;   parallel to the y-axis

25. Through $(1, -6)$;   parallel to the line $x + 2y = 6$

26. y-intercept 6;   parallel to the line $2x + 3y + 4 = 0$

27. Through $(-1, 2)$;   parallel to the line $x = 5$

28. Through $(2, 6)$;   perpendicular to the line $y = 1$

29. Through $(-1, -2)$;   perpendicular to the line $2x + 5y + 8 = 0$

30. Through $(\frac{1}{2}, -\frac{2}{3})$;   perpendicular to the line $4x - 8y = 1$

31. Through $(1, 7)$;   parallel to the line passing through $(2, 5)$ and $(-2, 1)$

32. Through $(-2, -11)$;   perpendicular to the line passing through $(1, 1)$ and $(5, -1)$

33. (a) Sketch the line with slope $\frac{3}{2}$ that passes through the point $(-2, 1)$.
    (b) Find an equation for this line.

34. (a) Sketch the line with slope -2 that passes through the point $(4, -1)$.
    (b) Find an equation for this line.

35–36 ■ Use a graphing device to graph the given family of lines on a common screen. What do the lines have in common?

35. $y = 2 + m(x - 3)$ for $m = 0, \pm0.5, \pm1, \pm2, \pm10$

36. $y = 1.3x + b$ for $b = 0, \pm1, \pm2.8$

37–48 ■ Find the slope and y-intercept of the line and draw its graph.

37. $x + y = 3$

38. $3x - 2y = 12$

39. $x + 3y = 0$

40. $2x - 5y = 0$

41. $\frac{1}{2}x - \frac{1}{3}y + 1 = 0$

42. $-3x - 5y + 30 = 0$

43. $y = 4$

44. $4y + 8 = 0$

45. $3x - 4y = 12$

46. $x = -5$

47. $3x + 4y - 1 = 0$

48. $4x + 5y = 10$

49. Show that $A(1, 1)$, $B(7, 4)$, $C(5, 10)$, and $D(-1, 7)$ are vertices of a parallelogram.

50. Show that $A(-3, -1)$, $B(3, 3)$, and $C(-9, 8)$ are vertices of a right triangle.

51. Show that $A(1, 1)$, $B(11, 3)$, $C(10, 8)$, and $D(0, 6)$ are vertices of a rectangle.

52. Use slopes to determine whether the given points are collinear (lie on a line).
    (a) $(1, 1)$, $(3, 9)$, $(6, 21)$
    (b) $(-1, 3)$, $(1, 7)$, $(4, 15)$

53. Find the equation of the perpendicular bisector of the line segment joining the points $A(1, 4)$ and $B(7, -2)$.

54. Find the area of the triangle formed by the coordinate axes and the line $2y + 3x - 6 = 0$.

55. (a) Show that if the x- and y-intercepts of a line are nonzero numbers a and b, then the equation of the line can be written in the form

$$\frac{x}{a} + \frac{y}{b} = 1$$

This is called the **two-intercept form** of the equation of a line.

(b) Use part (a) to find an equation of the line whose x-intercept is 6 and whose y-intercept is -8.

56. (a) Find an equation for the line tangent to the circle $x^2 + y^2 = 25$ at the point $(3, -4)$. (See the figure.)

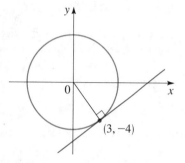

(b) At what other point on the circle will a tangent line be parallel to the tangent line in part (a)?

57. West of Albuquerque, New Mexico, Route 40 eastbound is straight and makes a steep descent towards the city. The highway has a 6% grade, which means that its slope is $-\frac{6}{100}$. Driving on this road you notice from elevation signs that you have descended a distance of 1000 ft. What is the change in your horizontal distance?

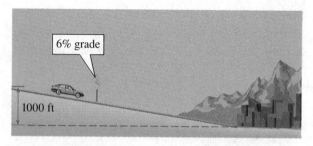

58. A small business buys a computer for \$4000. After 4 years the value of the computer is expected to be \$200. For accounting purposes, the business uses *linear depreciation* to assess the value of the computer at a given time. This means that if V is the value of the computer at time t, then a linear equation is used to relate V and t.

(a) Find a linear equation that relates V and t.

(b) Find the depreciated value of the computer 3 years from the date of purchase.

59. A small-appliance manufacturer finds that if he produces x toaster ovens in a month his production cost is given by the equation $y = 6x + 3000$ (where y is measured in dollars).

(a) Sketch a graph of this linear equation.

(b) What do the slope and y-intercept of the graph represent?

60. The manager of a weekend flea market knows from past experience that if he charges x dollars for a rental space at the flea market, then the number y of spaces he can rent is given by the equation $y = 200 - 4x$.

(a) Sketch a graph of this linear equation. (Remember that the rental charge per space and the number of spaces rented must both be nonnegative quantities.)

(b) What do the slope, the y-intercept, and the x-intercept of the graph represent?

61. Biologists have observed that the chirping rate of crickets of a certain species is related to temperature, and the relationship appears to be very nearly linear. A cricket produces 120 chirps per minute at 70 °F and 168 chirps per minute at 80 °F.

(a) Find the linear equation that relates the temperature t and the number of chirps per minute n.

(b) If the crickets are chirping at 150 chirps per minute, estimate the temperature.

62. The relationship between the Fahrenheit (F) and Celsius (C) temperature scales is given by the equation $F = \frac{9}{5}C + 32$ (see Example 10).

(a) Complete the table to compare the two scales at the given values.

C	F
$-30°$	
$-20°$	
$-10°$	
$0°$	
	$50°$
	$68°$
	$86°$

(b) Find the temperature at which the scales agree. [*Hint:* Suppose that *a* is the temperature at which the scales agree. Set $F = a$ and $C = a$. Then solve for *a*.]

63. At the surface of the ocean, the water pressure is the same as the air pressure above the water, 15 lb/in^2. Below the surface, the water pressure increases by 4.34 lb/in^2 for every 10 ft of descent.
   (a) Find an equation for the relationship between pressure and depth below the ocean surface.
   (b) At what depth is the pressure 100 lb/in^2?

water pressure increases with depth

64. Jason and Debbie leave Detroit at 2:00 P.M. and drive at a constant speed, traveling west on I-90. They pass Ann Arbor, 40 mi from Detroit, at 2:50 P.M.
   (a) Express the distance traveled in terms of the time elapsed.
   (b) Draw the graph of the equation in part (a).
   (c) What is the slope of this line? What does it represent?

65. The monthly cost of driving a car depends on the number of miles driven. Lynn found that in May her driving cost was $380 for 480 mi and in June her cost was $460 for 800 mi.
   (a) Express the monthly cost *C* in terms of the distance driven *d*, assuming that a linear relationship gives a suitable model.
   (b) Use part (a) to predict the cost of driving 1500 mi per month.
   (c) Draw the graph of the linear equation. What does the slope of the line represent?
   (d) What does the *y*-intercept of the graph represent?
   (e) Why is a linear relationship a suitable model for this situation?

66. The manager of a furniture factory finds that it costs $2200 to manufacture 100 chairs in one day and $4800 to produce 300 chairs in one day.
   (a) Assuming that the relationship between cost and the number of chairs produced is linear, find an equation that expresses this relationship. Then graph this equation.
   (b) What is the slope of the line of part (a), and what does it represent?
   (c) What is the *y*-intercept of this line, and what does it represent?

67–68 ■ Equations for supply and demand are given.
   (a) Draw the graphs of the two equations in an appropriate viewing rectangle.
   (b) Estimate the equilibrium point from the graph.
   (c) Estimate the price and the amount of the commodity produced and sold at equilibrium.

67. Supply: $y = 0.45p + 4$
   Demand: $y = -0.65p + 28$

68. Supply: $y = 8.5p + 45$
   Demand: $y = -0.6p + 300$

DISCOVERY · DISCUSSION

69. What Does the Slope Mean? Suppose that the graph of the outdoor temperature over a certain period of time is a line. How is the weather changing if the slope of the line is positive? If it's negative? If it's zero?

70. Collinear Points Suppose you are given the coordinates of three points in the plane, and you want to see whether they lie on the same line. How can you do this using slopes? Using the Distance Formula? Can you think of another method?

1 REVIEW

CONCEPT CHECK

1. Define each term in your own words. (Check by referring to the definition in the text.)
   (a) An integer
   (b) A rational number
   (c) An irrational number
   (d) A real number

2. State each of these properties of real numbers.
   (a) Commutative Property
   (b) Associative Property
   (c) Distributive Property

3. Explain what is meant by the union and intersection of sets. What notation is used for these ideas?

4. What is an open interval? What is a closed interval? What notation is used for these intervals?

5. What is the absolute value of a number?

6. (a) In the expression a^x, which is the base and which is the exponent?
   (b) What does a^x mean if $x = n$, a positive integer?
   (c) What if $x = 0$?
   (d) What if x is a negative integer: $x = -n$, where n is a positive integer?
   (e) What if $x = m/n$, a rational number?
   (f) State the Laws of Exponents.

7. (a) What does $\sqrt[n]{a} = b$ mean?
   (b) Why is $\sqrt{a^2} = |a|$?
   (c) How many real nth roots does a positive real number have if n is odd? If n is even?

8. Explain how the procedure of rationalizing the denominator works.

9. What is the difference between a variable and a constant in an algebraic expression?

10. State the Special Product Formulas for $(a + b)^2$, $(a - b)^2$, $(a + b)^3$, and $(a - b)^3$.

11. State each factoring formula.
    (a) Difference of squares
    (b) Difference of cubes
    (c) Sum of cubes

12. What is the difference between an identity and an equation?

13. What is a root of an equation?

14. Write the general form of each type of equation.
    (a) A linear equation
    (b) A quadratic equation

15. What is the Zero-Product Property?

16. Describe the process of completing the square.

17. State the quadratic formula.

18. What is the discriminant of a quadratic equation?

19. State the rules for working with inequalities.

20. How do you solve a nonlinear inequality?

21. State each formula.
    (a) The Distance Formula
    (b) The Midpoint Formula

22. Given an equation, what is its graph?

23. How do you find the x-intercepts and y-intercepts of a graph?

24. Write an equation of the circle with center (h, k) and radius r.

25. Explain the meaning of each type of symmetry. How do you test for it?
    (a) Symmetry with respect to the x-axis
    (b) Symmetry with respect to the y-axis
    (c) Symmetry with respect to the origin

26. Define the slope of a line.

27. Write each form of the equation of a line.
    (a) The point-slope form
    (b) The slope-intercept form

28. Given lines with slopes m_1 and m_2, explain how you can tell if the lines are
    (a) parallel.
    (b) perpendicular.

EXERCISES

1–4 ■ State the property of real numbers being used.

1. $x + 5 = 5 + x$

2. $(a + b)(a - b) = (a - b)(a + b)$

3. $A(x + y) = Ax + Ay$

4. $(A + 1)(x + y) = (A + 1)x + (A + 1)y$

5–6 ■ Express the interval in terms of inequalities, and then graph the interval.

5. $(-1, 3]$

6. $(-\infty, 4]$

7–8 ■ Express the inequality in interval notation, and then graph the corresponding interval.

7. $x > 2$

8. $1 \leqslant x \leqslant 6$

9–16 ■ Evaluate the expression.

9. $\big| 3 - |-9| \big|$

10. $\sqrt[3]{-125}$

11. $216^{-1/3}$

12. $64^{2/3}$

13. $\dfrac{\sqrt{242}}{\sqrt{2}}$

14. $\sqrt[4]{4}\,\sqrt[4]{324}$

15. $2^{1/2} 8^{1/2}$

16. $\sqrt{2}\,\sqrt{50}$

17–26 ■ Simplify the expression.

17. $(2x^3 y)^2 (3x^{-1} y^2)$

18. $(a^2)^{-3}(a^3 b)^2 (b^3)^4$

19. $\dfrac{x^4 (3x)^2}{x^3}$

20. $\left(\dfrac{r^2 s^{4/3}}{r^{1/3} s}\right)^6$

21. $\sqrt[3]{(x^3 y)^2 y^4}$

22. $\sqrt{x^2 y^4}$

23. $\dfrac{x}{2 + \sqrt{x}}$

24. $\dfrac{\sqrt{x} + 1}{\sqrt{x} - 1}$

25. $\dfrac{8r^{1/2} s^{-3}}{2r^{-2} s^4}$

26. $\left(\dfrac{ab^2 c^{-3}}{2a^3 b^{-4}}\right)^{-2}$

27. Write the number 78,250,000,000 in scientific notation.

28. Write the number 2.08×10^{-8} in ordinary decimal notation.

29. If $a \approx 0.00000293$, $b \approx 1.582 \times 10^{-14}$, and $c \approx 2.8064 \times 10^{12}$, use a calculator to approximate the number ab/c.

30. If your heart beats 80 times per minute and you live to be 90 years old, estimate the number of times your heart beats during your lifetime. State your answer in scientific notation.

31–44 ■ Factor the expression completely.

31. $12x^2 y^4 - 3xy^5 + 9x^3 y^2$

32. $x^2 - 9x + 18$

33. $x^2 + 3x - 10$

34. $6x^2 + x - 12$

35. $4t^2 - 13t - 12$

36. $x^4 - 2x^2 + 1$

37. $25 - 16t^2$

38. $y^3 - 2y^2 - y + 2$

39. $x^{-1/2} - 2x^{1/2} + x^{3/2}$

40. $8x^3 + y^6$

41. $a^2 y - b^2 y$

42. $ax^2 + bx^2 - a - b$

43. $x^3 (x - 6)^2 + x^4 (x - 6)$

44. $x^2 (x - 2) + x(x - 2)^2$

45–60 ■ Perform the indicated operations and simplify.

45. $(2x + 1)(3x - 2) - 5(4x - 1)$

46. $(2y - 7)(2y + 7)$

47. $(2a^2 - b)^2$

48. $(2x + 1)^3$

49. $\sqrt{x}\,(\sqrt{x} + 1)(2\sqrt{x} - 1)$

50. $\dfrac{x^3 + 2x^2 + 3x}{x}$

51. $\dfrac{x^2 - 2x - 3}{2x^2 + 5x + 3}$

52. $\dfrac{x^3/(x - 1)}{x^2/(x^3 - 1)}$

53. $\dfrac{x^2 - 2x - 15}{x^2 - 6x + 5} \div \dfrac{x^2 - x - 12}{x^2 - 1}$

54. $x - \dfrac{1}{x + 1}$

55. $\dfrac{1}{x - 1} - \dfrac{2}{x^2 - 1}$

56. $\dfrac{1}{x + 2} + \dfrac{1}{x^2 - 4} - \dfrac{2}{x^2 - x - 2}$

57. $\dfrac{\dfrac{1}{x} - \dfrac{1}{2}}{x - 2}$

58. $\dfrac{\dfrac{1}{x} - \dfrac{1}{x + 1}}{\dfrac{1}{x} + \dfrac{1}{x + 1}}$

59. $\dfrac{\sqrt{x+h}-\sqrt{x}}{h}$ (rationalize the numerator)

60. $\dfrac{\sqrt{6}}{\sqrt{3}-\sqrt{2}}$ (rationalize the denominator)

61–74 ■ Find all real solutions of the equation.

61. $3x+12=24$ **62.** $5x-7=42$

63. $\frac{1}{3}x-\frac{1}{2}=2$

64. $2(x+3)-4(x-5)=8-5x$

65. $\dfrac{x-5}{2}-\dfrac{2x+5}{3}=\dfrac{5}{6}$ **66.** $\dfrac{x+1}{x-1}=\dfrac{2x-1}{2x+1}$

67. $x^2-9x+14=0$ **68.** $x^2+24x+144=0$

69. $2x^2+x=1$ **70.** $3x^2+5x-2=0$

71. $3x^2+4x-1=0$ **72.** $x^2-3x+9=0$

73. $\dfrac{1}{x}+\dfrac{2}{x-1}=3$ **74.** $x-4\sqrt{x}=32$

75. The owner of a store sells raisins for \$3.20 per pound and nuts for \$2.40 per pound. He decides to mix the raisins and nuts and sell 50 lb of the mixture for \$2.72 per pound. What quantities of raisins and nuts should he use?

76. The approximate distance d (in feet) that a driver travels after noticing that he or she must come to a sudden stop is given by the following formula, where x is the speed of the car (in mi/h):

$$d=x+\frac{x^2}{20}$$

If a car travels 75 ft before stopping, what must its speed have been before the brakes were applied?

77. Anthony leaves Kingstown at 2:00 P.M. and drives to Queensville, 160 mi distant, at 45 mi/h. At 2:15 P.M. Helen leaves Queensville and drives to Kingstown at 40 mi/h. At what time do they pass each other on the road?

78. A woman cycles 8 mi/h faster than she runs. Every morning she cycles 4 mi and runs $2\frac{1}{2}$ mi, for a total of one hour of exercise. How fast does she run?

79. Abbie paints twice as fast as Beth and three times as fast as Cathie. If it takes them 60 min to paint a living room with all three working together, how long would it take Abbie if she works alone?

80. A rectangular swimming pool is 8 ft deep everywhere and twice as long as it is wide. If the pool holds 8464 ft^3 of water, what are its dimensions?

81–89 ■ Solve the inequality. Express the solution using interval notation and graph the solution set on the real number line.

81. $3x-2>-11$ **82.** $12-x\geqslant 7x$

83. $3-x\leqslant 2x-7$ **84.** $x^2+4x-12>0$

85. $\dfrac{2x+5}{x+1}\leqslant 1$ **86.** $2x^2\geqslant x+3$

87. $\dfrac{x-4}{x^2-4}\leqslant 0$ **88.** $|x-4|<0.02$

89. $|2x+1|\geqslant 1$

90. The volume of a sphere is given by $V=\frac{4}{3}\pi r^3$, where r is the radius. Find the interval of values of the radius so that the volume is between 8 ft^3 and 12 ft^3, inclusive.

91–92 ■ Two points P and Q are given.
(a) Plot P and Q on a coordinate plane.
(b) Find the distance from P to Q.
(c) Find the midpoint of the segment PQ.
(d) Sketch the line determined by P and Q, and find its equation in slope-intercept form.
(e) Sketch the circle that passes through Q and has center P, and find the equation of this circle.

91. $P(2,0),\quad Q(-5,12)$

92. $P(7,-1),\quad Q(2,-11)$

93–94 ■ Sketch the region given by the set.

93. $\{(x,y)\mid |x|<4 \text{ and } |y|<2\}$

94. $\{(x,y)\mid |x|\geqslant 4 \text{ or } |y|\geqslant 2\}$

95. Which of the points $A(4,4)$ or $B(5,3)$ is closer to the point $C(-1,-3)$?

96. Find an equation of the circle that has center $(2,-5)$ and radius $\sqrt{2}$.

97. Find an equation of the circle that has center $(-5,-1)$ and passes through the origin.

98. Find an equation of the circle that contains the points $P(2,3)$ and $Q(-1,8)$ and has the midpoint of the segment PQ as its center.

99–102 ■ Determine whether the equation represents a circle, a point, or has no graph. If the equation is that of a circle, find its center and radius.

99. $x^2 + y^2 + 2x - 6y + 9 = 0$

100. $2x^2 + 2y^2 - 2x + 8y = \frac{1}{2}$

101. $x^2 + y^2 + 72 = 12x$

102. $x^2 + y^2 - 6x - 10y + 34 = 0$

103. Find an equation for the line that passes through the points $(-1, -6)$ and $(2, -4)$.

104. Find an equation for the line that passes through the point $(6, -3)$ and has slope $-\frac{1}{2}$.

105. Find an equation for the line that has x-intercept 4 and y-intercept 12.

106. Find an equation for the line that passes through the point $(1, 7)$ and is perpendicular to the line $x - 3y + 16 = 0$.

107. Find an equation for the line that passes through the origin and is parallel to the line $3x + 15y = 22$.

108. Find an equation for the line that passes through the point $(5, 2)$ and is parallel to the line passing through $(-1, -3)$ and $(3, 2)$.

109–110 ■ Find equations for the circle and the line in the figure.

109.

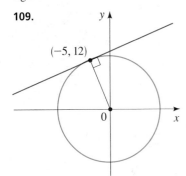

110.

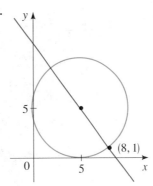

111–120 ■ Test the equation for symmetry and sketch its graph.

111. $y = 2 - 3x$

112. $2x - y + 1 = 0$

113. $x + 3y = 21$

114. $x = 2y + 12$

115. $\dfrac{x}{2} - \dfrac{y}{7} = 1$

116. $\dfrac{x}{4} + \dfrac{y}{5} = 0$

117. $y = 16 - x^2$

118. $8x + y^2 = 0$

119. $x = \sqrt{y}$

120. $y = -\sqrt{1 - x^2}$

121–124 ■ Use a graphing device to solve the equation or inequality correct to two decimal places.

121. $x^3 - 3x - 1 = 0$

122. $x^2 = \sqrt{x + 4}$

123. $0.1x + 1 \geqslant x^3$

124. $4 - x^2 \leqslant 3 - \sqrt{x + 3}$

1. (a) Graph the intervals $[-3, 2]$ and $(4, \infty)$ on a real number line.
   (b) Express the inequalities $x < 5$ and $-2 \leq x \leq 1$ in interval notation.
   (c) Find the distance between -22 and 31 on the number line.

2. Evaluate each expression.

   (a) $(-3)^4$ (b) $\dfrac{5^{18}}{5^{12}}$ (c) $\dfrac{\sqrt{32}}{\sqrt{8}}$ (d) $16^{-3/4}$

3. Simplify each expression.

   (a) $\sqrt{200} - \sqrt{8}$ (b) $(2a^3b^2)(3ab^4)^3$ (c) $\left(\dfrac{2x^{1/4}}{y^{1/3}x^{1/6}}\right)^3$

   (d) $\dfrac{x^2 + 3x + 2}{x^2 - x - 2}$ (e) $\dfrac{x^2}{x^2 - 4} - \dfrac{x + 1}{x + 2}$

4. Write each number in scientific notation.
   (a) 325,000,000,000   (b) 0.000008931

5. Perform the indicated operations and simplify.

   (a) $(x - 5)(2x + 3)$ (b) $(\sqrt{x} + \sqrt{y})(\sqrt{x} - \sqrt{y})$ (c) $(3t + 4)^2$

6. Factor completely each expression.

   (a) $9x^2 - 25$ (b) $6x^2 + 7x - 5$ (c) $x^3 - 4x^2 - 3x + 12$
   (d) $x^4 + 27x$ (e) $3x^{3/2} - 9x^{1/2} + 6x^{-1/2}$

7. Rationalize the denominator: $\dfrac{x}{\sqrt{x} - 2}$

8. Solve for w in the equation

$$V = 4wh^2 + 2wh$$

9. Find all real solutions of each equation.
   (a) $2x + 7 = 12 + \frac{5}{2}x$
   (b) $\dfrac{2x}{x + 1} = \dfrac{2x - 1}{x}$
   (c) $x^2 - x - 12 = 0$
   (d) $2x^2 + 4x + 3 = 0$
   (e) $x^{1/2} - 3x^{3/2} + 2x^{5/2} = 0$
   (f) $x^4 + 27x = 0$

10. Bill drove from Ajax to Bixby at an average speed of 50 mi/h. On the way back, he drove at 60 mi/h. The total trip took $4\frac{2}{3}$ hours of driving time. Find the distance between these two cities.

11. A rectangular parcel of land is 70 ft longer than it is wide. Each diagonal between opposite corners is 130 ft. What are the dimensions of the parcel?

12. Solve each inequality. Sketch the solution on a real number line, and write the answer using interval notation.

(a) $-1 \leqslant 5 - 2x$

(b) $x(x - 1)(x - 2) > 0$

(c) $|x - 3| < 2$

(d) $\dfrac{2x + 5}{x + 1} \leqslant 1$

13. A bottle of medicine is to be stored at a temperature between $5\,°C$ and $10\,°C$. What range does this correspond to on a Fahrenheit scale? [*Note:* The Fahrenheit (F) and Celsius (C) scales satisfy the relation $C = \frac{5}{9}(F - 32)$.]

14. Let $A(-7, 4)$ and $B(5, -12)$ be points in the plane.

(a) Find the length of the segment AB.

(b) Find the midpoint of the segment AB.

(c) Find an equation of the line that passes through the points A and B.

(d) Find an equation of the perpendicular bisector of AB.

(e) Find an equation of the circle for which AB is a diameter.

15. Sketch the graph of each equation.

(a) $x^2 + y^2 - 6x + 10y + 9 = 0$

(b) $2x^2 + 2y^2 + 6x + 10y + 17 = 0$

16. Find an equation for the line with the given property.

(a) Passing through $(-2, 3)$ and parallel to the line $2x + 6y = 17$

(b) Having x-intercept -3 and y-intercept 12

17. Sketch the graph of each equation.

(a) $-3x + 4y = 24$ (b) $x = y^2 - 1$

PRINCIPLES OF PROBLEM SOLVING

There are no hard and fast rules that will ensure success in solving problems. However, it is possible to outline some general steps in the problem-solving process and to give some principles that may be useful in the solution of certain problems. These steps and principles are just common sense made explicit. They have been adapted from George Polya's book *How To Solve It*.

1 │ UNDERSTAND THE PROBLEM

The first step is to read the problem and make sure that you understand it clearly. Ask yourself the following questions:

> *What is the unknown?*
> *What are the given quantities?*
> *What are the given conditions?*

For many problems it is useful to

> *draw a diagram*

and identify the given and required quantities on the diagram.
   Usually it is necessary to

> *introduce suitable notation*

In choosing symbols for the unknown quantities we often use letters such as a, b, c, m, n, x, and y, but in some cases it helps to use initials as suggestive symbols, for instance, V for volume or t for time.

2 │ THINK OF A PLAN

Find a connection between the given information and the unknown that will enable you to calculate the unknown. It often helps to ask yourself explicitly: "How can I relate the given to the unknown?" If you do not see a connection immediately, the following ideas may be helpful in devising a plan.

■ Try to recognize something familiar

Relate the given situation to previous knowledge. Look at the unknown and try to recall a more familiar problem that has a similar unknown.

■ Try to recognize patterns

Some problems are solved by recognizing that some kind of pattern is occurring. The pattern could be geometric, or numerical, or algebraic. If you can see regu-

George Polya (1887–1985) is famous among mathematicians for his ideas on problem solving. His lectures on problem solving at Stanford University attracted overflow crowds whom he held on the edges of their seats, leading them to discover solutions for themselves. He was able to do this because of his deep insight into the psychology of problem solving. His well-known book *How To Solve It* has been translated into 15 languages. He said that Euler (see page 249) was unique among great mathematicians because he explained *how* he found his results. Polya often said to his students and colleagues, "Yes, I see that your proof is correct, but how did you discover it?" In the preface to *How To Solve It,* Polya writes, "A great discovery solves a great problem but there is a grain of discovery in the solution of any problem. Your problem may be modest; but if it challenges your curiosity and brings into play your inventive faculties, and if you solve it by your own means, you may experience the tension and enjoy the triumph of discovery."

larity or repetition in a problem, then you might be able to guess what the continuing pattern is and then prove it.

■ Use analogy

Try to think of an analogous problem, that is, a similar problem, a related problem, but one that is easier than the original problem. If you can solve the similar, simpler problem, then it might give you the clues you need to solve the original, more difficult problem. For instance, if a problem involves very large numbers, you could first try a similar problem with smaller numbers. Or if the problem is in three-dimensional geometry, you could look for a similar problem in two-dimensional geometry. Or if the problem you start with is a general one, you could first try a special case.

■ Introduce something extra

You may sometimes need to introduce something new—an auxiliary aid—to help make the connection between the given and the unknown. For instance, in a problem for which a diagram is useful, the auxiliary aid could be a new line drawn in the diagram. In a more algebraic problem the aid could be a new unknown that is related to the original unknown.

■ Take cases

You may sometimes have to split a problem into several cases and give a different argument for each of the cases. For instance, we often have to use this strategy in dealing with absolute value.

■ Work backward

Sometimes it is useful to imagine that your problem is solved and work backward, step by step, until you arrive at the given data. Then you may be able to reverse your steps and thereby construct a solution to the original problem. This procedure is commonly used in solving equations. For instance, in solving the equation $3x - 5 = 7$, we suppose that x is a number that satisfies $3x - 5 = 7$ and work backward. We add 5 to each side of the equation and then divide each side by 3 to get $x = 4$. Since each of these steps can be reversed, we have solved the problem.

■ Establish subgoals

In a complex problem it is often useful to set subgoals (in which the desired situation is only partially fulfilled). If you can first reach these subgoals, then you may be able to build on them to reach your final goal.

■ Indirect reasoning

Sometimes it is appropriate to attack a problem indirectly. In using **proof by contradiction** to prove that P implies Q, we assume that P is true and Q is false and try to see why this cannot happen. Somehow we have to use this information and arrive at a contradiction to what we absolutely know is true.

■ Mathematical induction

In proving statements that involve a positive integer n, it is frequently helpful to use the Principle of Mathematical Induction, which is discussed in Section 10.5.

3 CARRY OUT THE PLAN

In Step 2 a plan was devised. In carrying out that plan, you must check each stage of the plan and write the details that prove that each stage is correct.

4 LOOK BACK

Having completed your solution, it is wise to look back over it, partly to see if errors have been made in the solution and partly to see if you can discover an easier way to solve the problem. Another reason for looking back is that it will familiarize you with the method of solution, and this may be useful for solving a future problem. Descartes said, "Every problem that I solved became a rule which served afterwards to solve other problems."

We illustrate some of these principles of problem solving in an example. Further illustrations of these principles will be presented at the end of selected chapters.

PROBLEM ■ Average Speed

A driver sets out on a journey. For the first half of the distance she drives at the leisurely pace of 30 mi/h; during the second half she drives 60 mi/h. What is her average speed on this trip?

PRELIMINARY THOUGHTS It is tempting to take the average of the speeds and say that the average speed for the entire trip is

$$\frac{30 + 60}{2} = 45 \text{ mi/h}$$

But is this simple-minded approach really correct?

Let's look at an easily calculated special case. Suppose that the total distance traveled is 120 mi. Since the first 60 mi is traveled at 30 mi/h, it takes 2 h. The second 60 mi is traveled at 60 mi/h, so it takes one hour. Thus,

Try a special case

the total time is $2 + 1 = 3$ hours and the average speed is

$$\frac{120}{3} = 40 \text{ mi/h}$$

So our guess of 45 mi/h was wrong.

Understand the problem

SOLUTION   We need to look more carefully at the meaning of average speed. It is defined as

$$\text{average speed} = \frac{\text{distance traveled}}{\text{time elapsed}}$$

Introduce notation

Let d be the distance traveled on each half of the trip. Let t_1 and t_2 be the times taken for the first and second halves of the trip. Now we can write down the information that we have been given. For the first half of the trip, we have

State what is given

$$(1) \qquad 30 = \frac{d}{t_1}$$

and, for the second half, we have

$$(2) \qquad 60 = \frac{d}{t_2}$$

Identify the unknown

Now we identify the quantity we are asked to find:

$$\text{average speed for entire trip} = \frac{\text{total distance}}{\text{total time}} = \frac{2d}{t_1 + t_2}$$

Connect the given with the unknown

To calculate this quantity we need to know t_1 and t_2, so we solve Equations 1 and 2 for these times:

$$t_1 = \frac{d}{30} \qquad t_2 = \frac{d}{60}$$

Now we have the ingredients needed to calculate the desired quantity:

$$\text{average speed} = \frac{2d}{t_1 + t_2} = \frac{2d}{\dfrac{d}{30} + \dfrac{d}{60}}$$

$$= \frac{60(2d)}{60\left(\dfrac{d}{30} + \dfrac{d}{60}\right)} \qquad \text{Multiply numerator and denominator by 60}$$

$$= \frac{120d}{2d + d} = \frac{120d}{3d} = 40$$

So, the average speed for the entire trip is 40 mi/h.

PROBLEMS

1. A man drives from home to work at a speed of 50 mi/h. The return trip from work to home is traveled at the more leisurely pace of 30 mi/h. What is the man's average speed for the round-trip?

2. An old car has to travel a 2-mile route, uphill and down. Because it is so old, the car can climb the first mile—the ascent—no faster than an average speed of 15 mi/h. How fast does the car have to travel the second mile—on the descent it can go faster, of course—in order to achieve an average speed of 30 mi/h for the trip?

3. A car and a van are parked 120 mi apart on a straight road. The drivers start driving toward each other at noon, each at a speed of 40 mi/h. A fly starts from the front bumper of the van at noon and flies to the bumper of the car, then immediately back to the bumper of the van, back to the car, and so on, until the car and the van meet. If the fly flies at a speed of 100 mi/h, what is the total distance it travels?

4. Which price is better for the buyer, a 40% discount or two successive discounts of 20%?

5. Use a calculator to find the value of the expression

$$\sqrt{3 + 2\sqrt{2}} - \sqrt{3 - 2\sqrt{2}}$$

   The number looks very simple. Show that the calculated value is correct.

6. Use a calculator to evaluate

$$\frac{\sqrt{2} + \sqrt{6}}{\sqrt{2} + \sqrt{3}}$$

   Show that the calculated value is correct.

7. An amoeba propagates by simple division; each split takes three minutes to complete. When such an amoeba is put into a glass container with a nutrient fluid, the container is full of amoebas in one hour. How long would it take for the container to be filled if we start with not one amoeba, but two?

8. Two runners start running laps at the same time, from the same starting position. George runs a lap in 50 s; Sue runs a lap in 30 s. When will the runners next be side by side?

9. Player A has a higher batting average than Player B for the first half of the baseball season. Player A also has a higher batting average than Player B for the second half of the season. Is it true that Player A has a higher batting average than Player B for the entire season?

Don't feel bad if you don't solve these problems right away. Problems 2 and 7 were sent to Albert Einstein by his friend Wertheimer. Einstein (and his friend Bucky) enjoyed the problems and wrote back to Wertheimer. Here is part of his reply:

Your letter gave us a lot of amusement. The first intelligence test fooled both of us (Bucky and me). Only on working it out did I notice that no time is available for the downhill run! Mr. Bucky was also taken in by the second example, but I was not. Such drolleries show us how stupid we are!

(See *Mathematical Intelligencer*, Spring 1990, page 41)

10. A woman starts at a point P on the earth's surface and walks 1 mi south, then 1 mi east, then 1 mi north, and finds herself back at P, the starting point. Describe all points P for which this is possible (there are infinitely many).

11. A spoonful of cream is taken from a cup of cream and put into a cup of coffee. The coffee is then stirred. Then a spoonful of this mixture is put into the cup of cream. Is there now more cream in the coffee cup or more coffee in the cup of cream?

12. An ice cube is floating in a cup of water, full to the brim, as shown in the sketch. As the ice melts, what happens? Does the cup overflow, or does the water level drop, or does it remain the same? (You need to know Archimedes' principle: A floating object displaces a volume of water whose weight equals the weight of the object.)

13. An extended family consists of the members

{Baby, Son, Daughter, Mother, Father, Uncle, Aunt, Grandma, Grandpa}

listed in increasing order of age. If x and y are members of this family, define $x + y$ to be the older of x and y, and define $x \cdot y$ to be the younger of x and y.
(a) Find (Baby + Uncle) + Mother.
(b) Find Father · (Grandpa + Aunt).
(c) Show that the operations $+$ and $\cdot$ that we have defined satisfy the same Commutative, Associative, and Distributive Properties as do the real numbers with regular addition and multiplication.
(d) Show that the operations also satisfy the property

$$x + (y \cdot z) = (x + y) \cdot (x + z)$$

Do the operations of regular addition and multiplication with real numbers also satisfy this property?

14. The ancient Egyptians, as a result of their pyramid-building, knew that the volume of a pyramid with height h and square base of side length a is $V = \frac{1}{3}ha^2$. They were able to use this fact to prove that the volume of a truncated pyramid is $V = \frac{1}{3}h(a^2 + ab + b^2)$, where h is the height and b and a are the lengths of the sides of the square top and bottom, as shown in the figure. Prove the truncated pyramid volume formula.

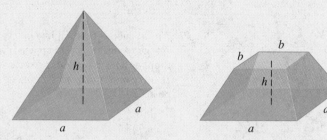

15. The ancient Babylonians developed the following process for finding the square root of a number N. First they made a guess at the square root—let's call this first guess r_1. Noting that

$$r_1 \cdot \left(\frac{N}{r_1}\right) = N$$

they concluded that the actual square root must be somewhere between r_1 and N/r_1, so their next guess for the square root, r_2, was the average of these two numbers:

$$r_2 = \frac{1}{2}\left(r_1 + \frac{N}{r_1}\right)$$

Continuing in this way, their next approximation was given by

$$r_3 = \frac{1}{2}\left(r_2 + \frac{N}{r_2}\right)$$

and so on. In general, once we have the nth approximation to the square root of N, we find the $(n + 1)$st using

$$r_{n+1} = \frac{1}{2}\left(r_n + \frac{N}{r_n}\right)$$

Use this procedure to find $\sqrt{72}$, correct to two decimal places.

16. This is a problem adapted from a tenth-century Arab manuscript. In his will, a nobleman leaves half of his horses to his oldest son, a third to his second son, and a ninth to his youngest. At his death he has 17 horses, so none of these bequests results in a whole number of horses for any of the sons. The executor of the will solves this dilemma by adding one of his own horses to the estate, making a total of 18. Now, according to the provisions of the will, he gives 9 to the oldest son, 6 to the second, and 2 to the third. Thus, each son inherits a little more than he was originally entitled to, and one horse is left over, which the executor takes back again. Everybody is happy, but something seems wrong. What, in fact, *is* wrong?

17. Suppose that each point in the coordinate plane is colored either red or blue. Show that there must always be two points of the same color that are exactly one unit apart.

18. Suppose that each point (x, y) in the plane, both of whose coordinates are rational numbers, represents a tree. If you are standing at the point $(0, 0)$, how far could you see in this forest?

19. A thousand points are graphed in the coordinate plane. Explain why it is possible to draw a straight line in the plane so that half of the points are on one side of the line and half are on the other. [*Hint:* Consider the slopes of the lines determined by each *pair* of points.]

20. Sketch the region in the plane consisting of all points (x, y) such that

$$|x| + |y| \leqslant 1$$

21. Find the last digit in the number 3^{459}. [*Hint:* Calculate the first few powers of 3, and look for a pattern.]

22. Draw the graph of the equation

$$x^2 y - y^3 - 5x^2 + 5y^2 = 0$$

[*Hint:* Factor.]

23. Suppose $x + y = 1$ and $x^2 + y^2 = 4$. Find $x^3 + y^3$.

2

FUNCTIONS

A function is a rule that describes how one quantity depends on another; for example, when studying motion, the distance traveled is a function of time.

That flower of modern mathematical thought—the notion of a function.

THOMAS J. McCORMACK

One of the most basic and important ideas in all of mathematics is that of a *function*. In this chapter we study functions, their graphs, and some of their applications.

WHAT IS A FUNCTION?

In nearly every physical phenomenon, we observe that one quantity depends on another. For example, your height depends on your age, the temperature depends on the date, the cost of mailing a package depends on its weight. These are all examples of functions. We say that your height is a function of your age, the temperature is a function of the date, and the cost of mailing a package is a function of its weight. (See Figure 1.) Although there is no simple rule relating height to age or temperature to date, there is a definite rule (one that the post office actually uses) that relates the cost of mailing a package to its weight.

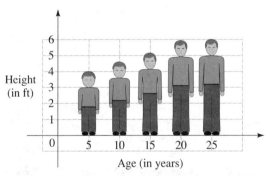

Height is a function of age.

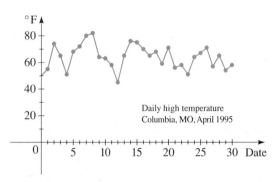

Temperature is a function of date.

FIGURE 1

Can you think of other examples of functions? Here are two more examples. The area A of a circle depends on its radius r. The rule that connects A and r is given by the formula $A = \pi r^2$. To each positive number r there is associated one value of A, so we say that A is a function of r. If a bacteria culture starts with 5000 bacteria and the population doubles every hour, then the number N of bacteria depends on the time t. The rule that connects N and t is given by the formula $N = (5000)2^t$. For each value of t there is a corresponding value of N, so we say that N is a function of t.

From these examples you can see why functions are important in science. For instance, a biologist observes that the number of bacteria in a culture increases with time. The biologist then tries to find the rule or function that relates the number of bacteria to the time. A physicist observes that the weight of an astronaut depends on her elevation. The physicist then tries to discover the rule or function that relates the astronaut's weight to her elevation. An economist observes that the price for a certain commodity depends on the demand for that commodity. The economist then tries to discover the rule or function that relates these quantities.

In the following list the numbers on the right are related to the numbers on the left.

$$
\begin{array}{rcl}
1 & \to & 1 \\
2 & \to & 4 \\
3 & \to & 9 \\
4 & \to & 16 \\
& \vdots &
\end{array}
$$

Can you discover the rule that relates these numbers? The numbers on the right are the squares of those on the left; so the rule is "square the number." In general, we have $x \to x^2$. In order to talk about this rule, we need to give it a name; we will call it f. Thus, in this example,

$$f \quad \text{is the rule} \quad \text{"square the number"}$$

We can express this rule in one of two ways, using either arrow notation or function notation:

$$x \xrightarrow{f} x^2 \quad \text{or} \quad f(x) = x^2$$

We have previously used letters to stand for numbers. Here we do something quite different. We use letters to represent *rules*. If the letter f represents a function, then the notation $f(x)$ means "apply the rule f to the number x."

When we write $f(2)$, we mean "apply the rule f to 2"; applying the rule to 2 gives $2^2 = 4$. So we write $f(2) = 4$. Similarly, $f(3) = 9$ and $f(4) = 16$.

The following definition concisely captures these ideas.

DEFINITION OF FUNCTION

A **function** f is a rule that assigns to each element x in a set A exactly one element, called $f(x)$, in a set B.

We usually consider functions for which the sets A and B are sets of real numbers. The set A is called the **domain** of the function. The symbol $f(x)$ is read "f of x" or "f at x" and is called the **value of f at x**, or the **image of x under f**. The **range** of f is the set of all possible values of $f(x)$ as x varies throughout the domain, that is,

$$\{f(x) \mid x \in A\}$$

The symbol that represents an arbitrary number in the domain of a function f is called an **independent variable**. The symbol that represents a number in the range of f is called a **dependent variable**. For instance, in the bacteria example, t is the independent variable and N is the dependent variable.

It is helpful to think of a function as a **machine** (see Figure 2). If x is in the domain of the function f, then when x enters the machine, it is accepted as an input and the machine produces an output $f(x)$ according to the rule of the function. Thus, we can think of the domain as the set of all possible inputs and the range as the set of all possible outputs.

FIGURE 2

Machine diagram of f

The preprogrammed functions in a calculator are good examples of a function as a machine. For example, the $\sqrt{x}$ key on your calculator is such a function. First you input x into the display. Then you press the key labeled $\sqrt{x}$. If $x < 0$, then x is not in the domain of this function; that is, x is not an acceptable input and the calculator will indicate an error. If $x > 0$, then an approximation to $\sqrt{x}$ will appear in the display, correct to a certain number of decimal places. [Thus, the $\sqrt{x}$ key on your calculator is not quite the same as the exact mathematical function f defined by $f(x) = \sqrt{x}$.]

Another way to picture a function is by an **arrow diagram** as in Figure 3. Each arrow connects an element of A to an element of B. The arrow indicates that $f(x)$ is associated with x, $f(a)$ is associated with a, and so on.

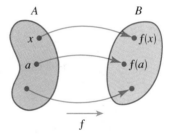

FIGURE 3

Arrow diagram of f

EXAMPLE 1 ■ The Squaring Function

The squaring function assigns to each real number x its square x^2. It is defined by

$$f(x) = x^2$$

(a) Evaluate $f(3)$, $f(-2)$, and $f(\sqrt{5})$.
(b) Find the domain and range of f.
(c) Draw a machine diagram and an arrow diagram for f.

SOLUTION

(a) The values of f are found by substituting for x in $f(x) = x^2$.

$$f(3) = 3^2 = 9 \qquad f(-2) = (-2)^2 = 4 \qquad f(\sqrt{5}) = (\sqrt{5})^2 = 5$$

(b) The domain of f is the set $\mathbb{R}$ of all real numbers. The range of f consists of all values of $f(x)$, that is, all numbers of the form x^2. But $x^2 \geqslant 0$ for all numbers x, and any nonnegative number c is a square, since $c = (\sqrt{c})^2 = f(\sqrt{c})$. Therefore, the range of f is $\{y \mid y \geqslant 0\} = [0, \infty)$.

(c) Machine and arrow diagrams for this function are shown in Figure 4.

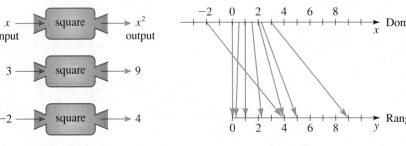

FIGURE 4   Machine diagram   Arrow diagram   ■

EXAMPLE 2 ■ The Domain and Range of a Function

If we define a function g by

$$g(x) = x^2, \qquad 0 \leqslant x \leqslant 3$$

then the domain of g is given as the closed interval $[0, 3]$. The function g is different from the function f in Example 1, because in considering g we are restricting our attention to those values of x between 0 and 3. The range of g is

$$\{x^2 \mid 0 \leqslant x \leqslant 3\} = \{y \mid 0 \leqslant y \leqslant 9\} = [0, 9] \qquad ■$$

Consider the function f defined by the rule

$$f(x) = x^3 + \sqrt{x^2 + 1}$$

To evaluate f at a number we simply substitute that number for x in the definition of f. For example, we have

$$f(5) = 5^3 + \sqrt{5^2 + 1}$$
$$f(a) = a^3 + \sqrt{a^2 + 1}$$
$$f(a + h) = (a + h)^3 + \sqrt{(a + h)^2 + 1}$$

To find $f(a + h)$ we substituted the expression $a + h$ for each occurrence of x because the rule f applies to the number $a + h$.

EXAMPLE 3 ■ Evaluating a Function

If $f(x) = 2x^2 + 3x - 1$, evaluate each function value.

Expressions like the one in part (d) of Example 3 occur frequently in calculus; they are called *difference quotients.*

(a) $f(a)$

(b) $f(-a)$

(c) $f(a + h)$

(d) $\dfrac{f(a + h) - f(a)}{h}$, $\quad h \neq 0$

SOLUTION

(a) $f(a) = 2a^2 + 3a - 1$

(b) $f(-a) = 2(-a)^2 + 3(-a) - 1 = 2a^2 - 3a - 1$

(c) $f(a + h) = 2(a + h)^2 + 3(a + h) - 1$
$$= 2(a^2 + 2ah + h^2) + 3(a + h) - 1$$
$$= 2a^2 + 4ah + 2h^2 + 3a + 3h - 1$$

The *difference quotient* in part (d) represents the average change in the value of f between $x = a$ and $x = a + h$.

(d) Using the results from parts (c) and (a), we have

$$\frac{f(a + h) - f(a)}{h} = \frac{(2a^2 + 4ah + 2h^2 + 3a + 3h - 1) - (2a^2 + 3a - 1)}{h}$$

$$= \frac{4ah + 2h^2 + 3h}{h} = 4a + 2h + 3 \quad \blacksquare$$

We should distinguish between a function f (which is a rule) and the number $f(x)$, which is the value of f at x. Nonetheless, it is common to abbreviate an expression such as

$$\text{the function } f \text{ defined by } f(x) = x^2 + x$$

to

$$\text{the function } f(x) = x^2 + x$$

In Examples 1 and 2 the domain of the function was given explicitly. But if a function is given by a formula and the domain is not stated explicitly, *the convention is that the domain is the set of all real numbers for which the formula makes sense and defines a real number.* In the next example we find the domain from the formula of a function.

EXAMPLE 4 ■ Finding Domains of Functions

Find the domain of each function.

(a) $f(x) = \dfrac{1}{x^2 - x}$ (b) $g(t) = \dfrac{t}{\sqrt{t + 1}}$ (c) $h(x) = \sqrt{2 - x - x^2}$

SOLUTION

(a) The function is not defined when the denominator is 0. Since

$$f(x) = \frac{1}{x^2 - x} = \frac{1}{x(x - 1)}$$

we see that $f(x)$ is not defined when $x = 0$ or $x = 1$. Thus, the domain of f is

$$\{x \mid x \neq 0, x \neq 1\}$$

The domain may also be written in interval notation as

$$(-\infty, 0) \cup (0, 1) \cup (1, \infty)$$

(b) The square root of a negative number is not defined (as a real number), so we require that $t + 1 \geqslant 0$. Also, the denominator cannot be 0, that is, $\sqrt{t + 1} \neq 0$, so we also require that $t + 1 \neq 0$. Thus, $g(t)$ exists when $t + 1 > 0$, that is, $t > -1$. So, the domain of g is

$$\{t \mid t > -1\} = (-1, \infty)$$

(c) Since the square root of a negative number is not defined (as a real number), the domain of h consists of all values of x such that

$$2 - x - x^2 \geqslant 0$$

We solve this inequality using the methods of Section 1.7. Since $2 - x - x^2 = (2 + x)(1 - x)$, the product will change sign when $x = -2$ or 1, as indicated in the following sign table.

Interval	$x < -2$	$-2 < x < 1$	$1 < x$
Sign of $2 + x$	−	+	+
Sign of $1 - x$	+	+	−
Sign of $(2 + x)(1 - x)$	−	+	−

Therefore, the domain of h is

$$\{x \mid -2 \leqslant x \leqslant 1\} = [-2, 1] \qquad \blacksquare$$

FOUR WAYS TO REPRESENT A FUNCTION

To help us understand what a function is we have used machine and arrow diagrams. To describe a specific function we can use the following four ways:

- verbally (by a description in words)
- algebraically (by an explicit formula)
- visually (by a graph)
- numerically (by a table of values)

A single function may be represented in all four ways, and it is often useful to go from one representation to another to gain insight into the function. However, certain functions are described more naturally by one method than by another. An example of a verbal description is

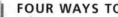

$$P(t) \quad \text{is} \quad \text{"the population of the world at time } t\text{"}$$

The function P can also be described numerically by giving a table of values (see page 342). A useful representation of the area of a circle as a function of its radius is the algebraic formula

$$A(r) = \pi r^2$$

The graph produced by a seismograph (see the following box) is a visual representation of the vertical acceleration function $a(t)$ of the ground during an earthquake. As a final example, consider the function $C(w)$, which is described verbally as "the cost of mailing a first-class letter with weight w." The most convenient way of describing this function numerically is by a table of values.

We will be using all four representations of functions throughout this book. We summarize them in the following box.

FOUR WAYS TO REPRESENT A FUNCTION

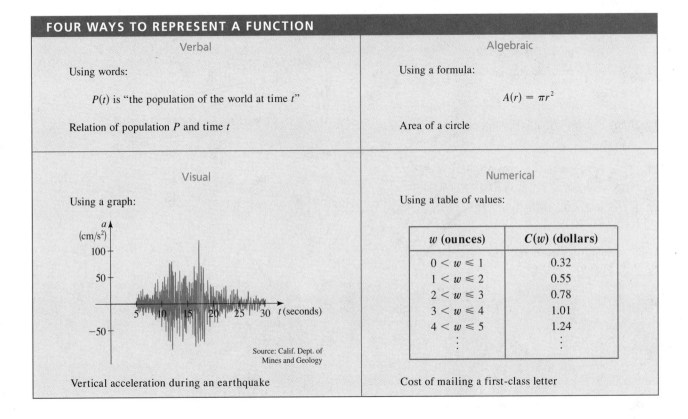

Verbal

Using words:

$P(t)$ is "the population of the world at time t"

Relation of population P and time t

Algebraic

Using a formula:

$$A(r) = \pi r^2$$

Area of a circle

Visual

Using a graph:

Vertical acceleration during an earthquake

Source: Calif. Dept. of Mines and Geology

Numerical

Using a table of values:

w (ounces)	$C(w)$ (dollars)
$0 < w \leq 1$	0.32
$1 < w \leq 2$	0.55
$2 < w \leq 3$	0.78
$3 < w \leq 4$	1.01
$4 < w \leq 5$	1.24
$\vdots$	$\vdots$

Cost of mailing a first-class letter

2.1 EXERCISES

1–4 ■ Express the rule in function notation. [For example, the rule "square, then subtract 5" is expressed as the function $f(x) = x^2 - 5$.]

1. Multiply by 3, then add 1

2. Subtract 5, then divide by 7

3. Add 2, then square

4. Square, add 1, then take the square root

5–8 ■ Express the function (or rule) in words.

5. $f(x) = \dfrac{x}{3} - 5$

6. $g(x) = \dfrac{x - 5}{3}$

7. $h(x) = 2x^2 - 3$

8. $k(x) = \sqrt{2x + 1}$

9–10 ■ Draw a machine diagram and an arrow diagram for the function.

9. $f(x) = \sqrt{x}, \quad 0 \leqslant x \leqslant 4$

10. $f(x) = \dfrac{2}{x}, \quad 1 \leqslant x \leqslant 4$

11. If $f(x) = 2x + 1$, find $f(1), f(-2), f(\frac{1}{2}), f(\sqrt{5}), f(a),$
$f(-a)$, and $f(a + b)$.

12. If $f(x) = x^3 + 2x^2 - 3$, find $f(0), f(3), f(-3), f(-x),$
and $f(1/a)$.

13. If $g(x) = \dfrac{1 - x}{1 + x}$, find $g(2), g(-2), g(\pi), g(a), g(a - 1),$
and $g(-a)$.

14. If $h(t) = t + \dfrac{1}{t}$, find $h(1), h(\pi), h(t + 1), h(t) + h(1),$
and $h(x)$.

15. If $f(x) = 2x^2 + 3x - 4$, find $f(0), f(2), f(\sqrt{2}),$
$f(1 + \sqrt{2}), f(-x), f(x + 1), 2f(x),$ and $f(2x)$.

16. If $f(x) = 2 - 3x$, find $f(1), f(-1), f(\frac{1}{3}), f(x/3), f(3x),$
$f(x^2)$, and $[f(x)]^2$.

17–22 ■ Find $f(a), f(a) + f(h), f(a + h)$, and
$\dfrac{f(a + h) - f(a)}{h}$, where $h \neq 0$.

17. $f(x) = 3x + 2$ **18.** $f(x) = x^2 + 1$

19. $f(x) = 5$ **20.** $f(x) = \dfrac{1}{x + 1}$

21. $f(x) = 3 - 5x + 4x^2$ **22.** $f(x) = x^3 + x + 1$

23–26 ■ Find $f(2 + h), f(x + h)$, and $\dfrac{f(x + h) - f(x)}{h}$,
where $h \neq 0$.

23. $f(x) = 8x - 1$ **24.** $f(x) = x - x^2$

25. $f(x) = \dfrac{1}{x}$ **26.** $f(x) = \dfrac{x}{x + 1}$

27–34 ■ Find the domain and range of the function.

27. $f(x) = 2x$

28. $f(x) = x^2 + 1$

29. $f(x) = 2x, \quad -1 \leqslant x \leqslant 5$

30. $f(x) = x^2 + 1, \quad 0 \leqslant x \leqslant 5$

31. $f(x) = 2 - x^2$ **32.** $g(x) = \sqrt{7 - 3x}$

33. $h(x) = \sqrt{2x - 5}$ **34.** $G(x) = \sqrt{x^2 - 9}$

35–54 ■ Find the domain of the function.

35. $f(x) = \dfrac{1}{x - 3}$ **36.** $f(x) = \dfrac{1}{3x - 6}$

37. $f(x) = \dfrac{x + 2}{x^2 - 1}$ **38.** $f(x) = \dfrac{x^4}{x^2 + x - 6}$

39. $f(x) = \dfrac{x + 2}{x^2 + 1}$ **40.** $f(x) = \dfrac{x^3 + 1}{x^3 - x}$

41. $f(x) = \sqrt{x - 5}$ **42.** $f(x) = \sqrt[4]{x + 9}$

43. $f(t) = \sqrt[3]{t - 1}$ **44.** $f(t) = \sqrt{t^2 + 1}$

45. $g(x) = \dfrac{\sqrt{2 + x}}{3 - x}$ **46.** $g(x) = \dfrac{\sqrt{x}}{2x^2 + x - 1}$

47. $F(x) = \dfrac{x}{\sqrt{x - 10}}$ **48.** $F(x) = x^2 - \dfrac{x}{\sqrt{9 - 2x}}$

49. $G(x) = \sqrt{x} + \sqrt{1 - x}$

50. $G(x) = \sqrt{x + 2} - 2\sqrt{x - 3}$

51. $g(x) = \sqrt[4]{x^2 - 6x}$

52. $g(x) = \sqrt{x^2 - 2x - 8}$

53. $\phi(x) = \sqrt{\dfrac{x}{\pi - x}}$

54. $\phi(x) = \sqrt{\dfrac{x^2 - 2x}{x - 1}}$

⬤ **DISCOVERY · DISCUSSION**

55. Examples of Functions   At the beginning of this section we discussed three examples of everyday, ordinary functions: Height is a function of age, temperature is a function of date, and postage cost is a function of weight. Give three other examples of functions from everyday life.

56. Four Ways to Represent a Function   In the table on page 137 we represented four different functions verbally, algebraically, visually, and numerically. Think of a function that can be represented all four ways, and write down the four representations.

2.2 | GRAPHS OF FUNCTIONS

In the preceding section we saw how to picture functions using machine diagrams and arrow diagrams. A third method for visualizing a function is its graph.

> ### THE GRAPH OF A FUNCTION
>
> If f is a function with domain A, then the **graph** of f is the set of ordered pairs
>
> $$\{(x, f(x)) \mid x \in A\}$$
>
> In other words, the graph of f is the set of all points (x, y) such that $y = f(x)$; that is, the graph of f is the graph of the equation $y = f(x)$.

The graph of a function f gives us a useful picture of the behavior or "life history" of a function. Since the y-coordinate of any point (x, y) on the graph is $y = f(x)$, we can read the value of $f(x)$ from the graph as being the height of the graph above the point x (see Figure 1). The graph of f also allows us to picture the domain and range of f on the x-axis and y-axis as in Figure 2.

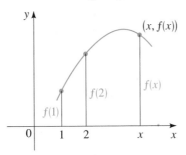

FIGURE 1

The graph of f

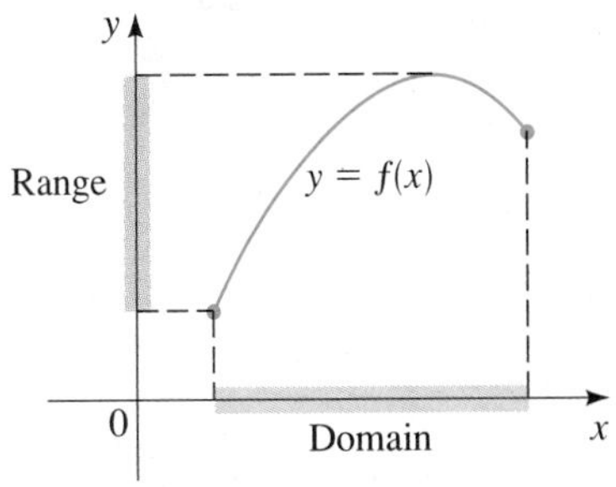

FIGURE 2

Domain and range of f

A function f of the form $f(x) = mx + b$ is called a **linear function** because its graph is the graph of the equation $y = mx + b$, which represents a line with slope m and y-intercept b. A special case of the linear function occurs when the slope is $m = 0$. The function $f(x) = b$, where b is a given number, is called a **constant function** because all its values are the same number, namely, b. Its graph is the horizontal line $y = b$. Figure 3 shows the graph of the constant function $f(x) = 3$.

FIGURE 3

The constant function $f(x) = 3$

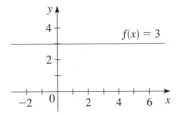

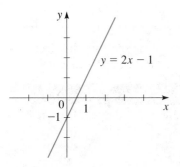

FIGURE 4

EXAMPLE 1 ■ **The Graph of a Linear Function**

Sketch the graph of the function $f(x) = 2x - 1$.

SOLUTION   To graph this linear function we graph the equation $y = 2x - 1$. We recognize this equation as being that of a line with slope 2 and y-intercept -1. This enables us to sketch the graph of f in Figure 4.   ■

EXAMPLE 2 ■ **The Graph of the Squaring Function**

Sketch the graph of $f(x) = x^2$.

SOLUTION   To graph this function we graph the equation $y = x^2$. We draw it in Figure 5 by setting up a table of values and plotting points. Recall from Example 1 of Section 2.1 that the domain is $\mathbb{R}$ and the range is $[0, \infty)$. The graph is called a *parabola*.

x	$f(x) = x^2$
0	0
$\pm\frac{1}{2}$	$\frac{1}{4}$
± 1	1
± 2	4
± 3	9

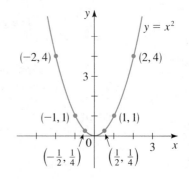

FIGURE 5

■

EXAMPLE 3 ■ **Graphing a Function by Plotting Points**

Sketch the graph of $f(x) = x^3$.

SOLUTION   We first make a table of values for the function. Then we plot the points given by the table and join them by a smooth curve to obtain the graph shown in Figure 6.

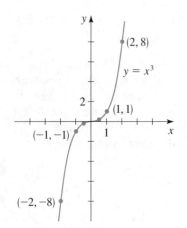

FIGURE 6

x	$f(x) = x^3$
0	0
$\frac{1}{2}$	$\frac{1}{8}$
1	1
2	8
$-\frac{1}{2}$	$-\frac{1}{8}$
-1	-1
-2	-8

■

EXAMPLE 4 ■ **Finding the Domain and Range from a Graph**

(a) Sketch the graph of $f(x) = \sqrt{4 - x^2}$.

(b) Find the domain and range of f.

SOLUTION

(a) We must graph the equation $y = \sqrt{4 - x^2}$. Because we are taking the positive square root, we know that $y \geqslant 0$. Squaring each side of the equation, we get

$$y^2 = 4 - x^2 \qquad \text{Square both sides}$$

$$x^2 + y^2 = 4 \qquad \text{Add } x^2$$

which we recognize as the equation of a circle with center the origin and radius 2. But, since $y \geqslant 0$, the graph of f consists of just the upper half of this circle. See Figure 7.

(b) From the graph in Figure 7 we see that the domain is the closed interval $[-2, 2]$ and the range is $[0, 2]$.

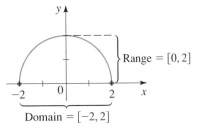

FIGURE 7

Graph of $f(x) = \sqrt{4 - x^2}$

The graph of a function is a curve in the xy-plane. But the question arises: Which curves in the xy-plane are graphs of functions? This is answered by the following test.

THE VERTICAL LINE TEST

A curve in the coordinate plane is the graph of a function if and only if no vertical line intersects the curve more than once.

We can see from Figure 8 why the Vertical Line Test is true. If each vertical line $x = a$ intersects a curve only once at (a, b), then exactly one functional value is defined by $f(a) = b$. But if a line $x = a$ intersects the curve twice, at (a, b) and

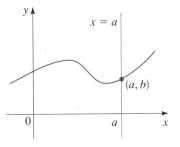 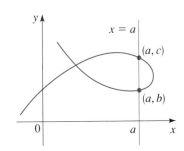

FIGURE 8

Vertical Line Test

Graph of a function   Not a graph of a function

at (a, c), then the curve can't represent a function because a function cannot assign two different values to a.

EXAMPLE 5 ■ Using the Vertical Line Test

Using the Vertical Line Test, we see that the curves in Figures 9(b) and (c) represent functions, whereas those in parts (a) and (d) do not.

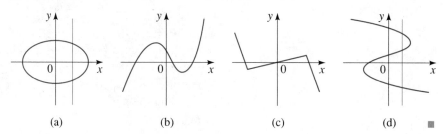

FIGURE 9   (a)   (b)   (c)   (d)   ■

EXAMPLE 6 ■ Graphing the Square Root Function

Sketch the graph of $f(x) = \sqrt{x}$.

SOLUTION   First we note that the domain is $\{x \mid x \geqslant 0\} = [0, \infty)$. Then we plot the points given by the following table and use them to sketch the graph in Figure 10.

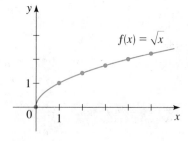

FIGURE 10

x	$f(x) = \sqrt{x}$
0	0
1	1
2	$\sqrt{2}$
3	$\sqrt{3}$
4	2
5	$\sqrt{5}$

■

It is very useful to know where the graph of a function rises and where it falls. The graph shown in Figure 11 rises, falls, then rises again as we move from left to right: It rises from A to B, falls from B to C, and rises again from C to D. The function f is said to be *increasing* when its graph rises and *decreasing* when its graph falls. We have the following definition.

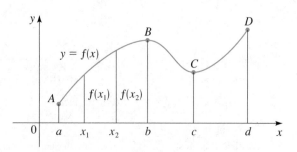

FIGURE 11

Donald Knuth was born in Milwaukee in 1938 and is now Professor of Computer Science at Stanford University. While still a graduate student at Caltech, he started writing a monumental series of books entitled *The Art of Computer Programming*. President Carter awarded him the National Medal of Science in 1979. When Knuth was a high school student, he became fascinated with graphs of functions and laboriously drew many hundreds of them because he wanted to see the behavior of a great variety of functions. (Now, with progress in computers and programming, it is far easier to use computers and graphing calculators to do this.) Knuth is also famous for his invention of TEX, a system of computer-assisted typesetting. This system was used in the preparation of the manuscript for this textbook. He has written a novel entitled *Surreal Numbers: How Two Ex-Students Turned On to Pure Mathematics and Found Total Happiness.*

DEFINITION OF INCREASING AND DECREASING FUNCTIONS

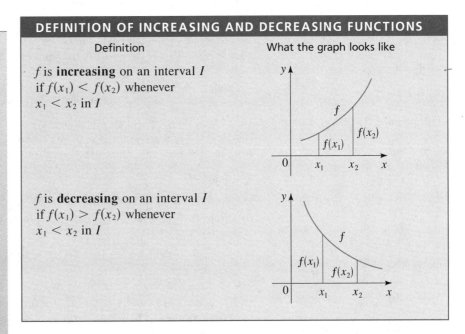

For instance, the functions in Examples 1, 3, and 6 are all increasing on their domains. In Example 2, f is decreasing on $(-\infty, 0]$ and increasing on $[0, \infty)$. In Example 4, f is increasing on $[-2, 0]$ and decreasing on $[0, 2]$.

EXAMPLE 7 ■ **Intervals on which a Function Increases or Decreases**

State the intervals on which the function whose graph is shown in Figure 12 is increasing or decreasing.

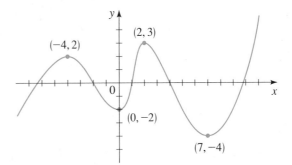

FIGURE 12

SOLUTION The function is increasing on $(-\infty, -4]$, $[0, 2]$, and $[7, \infty)$. It is decreasing on $[-4, 0]$ and $[2, 7]$. ■

Graphing calculators and computers are ideal devices to use for graphing functions because many graphing devices give exactly one value of y for each value of x, and this is the property that functions have. In Section 1.9 we observed that in order to graph an equation on a graphing calculator, we need to isolate y and write the equation as $y = $ *an expression involving only x*. We see now that such an equation actually represents a function $y = f(x)$.

EXAMPLE 8 ■ Using a Graph to Find Intervals where a Function Increases and Decreases

(a) Sketch the graph of the function $f(x) = x^{2/3}$.
(b) Find the domain and range of the function.
(c) Find the intervals on which f increases and decreases.

SOLUTION

(a) We use a graphing calculator to sketch the graph in Figure 13.

(b) From the graph we observe that the domain of f is $\mathbb{R}$ and the range is $[0, \infty)$.

(c) From the graph we see that f is decreasing on $(-\infty, 0]$ and increasing on $[0, \infty)$. ■

Some graphing calculators, such as the TI-82, do not evaluate $x^{2/3}$ [entered as $x^{\wedge}(2/3)$] for negative x. To graph a function like $f(x) = x^{2/3}$, we enter it as $y_1 = (x^{\wedge}(1/3))^{\wedge}2$ because these calculators correctly evaluate powers of the form $x^{\wedge}(1/n)$.

FIGURE 13
Graph of $f(x) = x^{2/3}$

To understand how the equation of a function relates to its graph, it is helpful to graph a **family of functions**, that is, a collection of functions whose equations are related. Usually, the equations differ only in a single constant called a *parameter*. In the next example we graph a family of **power functions** $f(x) = x^n$, where n is a positive integer. (In this case, n is the parameter.)

EXAMPLE 9 ■ A Family of Power Functions

(a) Graph the functions $f(x) = x^n$ for $n = 2, 4$, and 6 in the viewing rectangle $[-2, 2]$ by $[-1, 3]$.
(b) Graph the functions $f(x) = x^n$ for $n = 1, 3$, and 5 in the viewing rectangle $[-2, 2]$ by $[-2, 2]$.
(c) Graph all of the functions from parts (a) and (b) in the viewing rectangle $[-1, 3]$ by $[-1, 3]$.
(d) What conclusions can you draw from these graphs?

SOLUTION   The graphs for parts (a), (b), and (c) are shown in Figure 14.

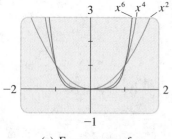

(a) Even powers of x

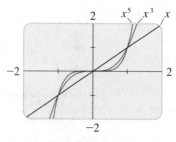

(b) Odd powers of x

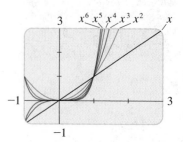

(c) Even and odd powers

FIGURE 14   A family of power functions $f(x) = x^n$

(d) We see that the general shape of $f(x) = x^n$ depends on whether n is even or odd.

From part (a) we see that if n is even, then the graph of $f(x) = x^n$ is similar to the parabola $y = x^2$. These functions are decreasing on $(-\infty, 0]$ and increasing on $[0, \infty)$.

From part (b) we see that if n is odd, then the graph of $f(x) = x^n$ is similar to that of $y = x^3$. These functions are increasing on all of $\mathbb{R}$.

Notice from part (c) that as n increases the graph of $y = x^n$ becomes flatter near 0 and steeper when $x > 1$. When $0 < x < 1$, the lower powers of x are the "bigger" functions. But when $x > 1$, the higher powers of x are the dominant functions. ■

PIECEWISE-DEFINED FUNCTIONS

Each function we have graphed so far has been defined by just one equation. But the rule defining a function is often given by more than one equation. A function may be defined by different equations on different parts of its domain. Some examples from everyday experience are the cost of mailing a first-class letter as a function of its weight, the population of New York City as a function of time, and the cost of a taxi ride as a function of distance. The following examples show how to graph such functions.

EXAMPLE 10 ■ The Graph of a Piecewise-Defined Function

Sketch the graph of the function f defined by

$$f(x) = \begin{cases} 1 - x & \text{if } x \leq 1 \\ x^2 & \text{if } x > 1 \end{cases}$$

SOLUTION Remember that a function is a rule. Here is how we apply the rule for this particular function: First look at the value of the input x. If it happens that $x \leq 1$, then the value of $f(x)$ is $1 - x$. On the other hand, if $x > 1$, then the value of $f(x)$ is x^2.

Since $0 \leq 1$, we have $f(0) = 1 - 0 = 1$.

Since $1 \leq 1$, we have $f(1) = 1 - 1 = 0$.

Since $2 > 1$, we have $f(2) = 2^2 = 4$.

Since $-3 \leq 1$, we have $f(-3) = 1 - (-3) = 4$.

How do we draw the graph of f? We observe that if $x \leq 1$, then $f(x) = 1 - x$, so the part of the graph of f that lies to the left of the vertical line $x = 1$ must coincide with the line $y = 1 - x$, which has slope -1 and y-intercept 1. If $x > 1$, then $f(x) = x^2$, so the part of the graph of f that lies to the right of the line $x = 1$ must coincide with the graph of $y = x^2$, which we sketched in Example 2. This enables us to sketch the graph in Figure 15. The solid dot at $(1, 0)$ indicates that this point is included on the graph; the open dot at $(1, 1)$ indicates that this point is excluded from the graph. ■

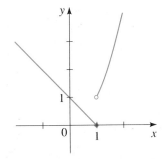

FIGURE 15

$$f(x) = \begin{cases} 1 - x & \text{if } x \leq 1 \\ x^2 & \text{if } x > 1 \end{cases}$$

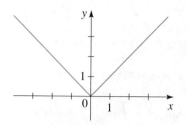

FIGURE 16

Graph of $f(x) = |x|$

EXAMPLE 11 ■ Graph of the Absolute-Value Function

Sketch the graph of the absolute-value function $f(x) = |x|$.

SOLUTION   Recall that

$$|x| = \begin{cases} x & \text{if } x \geq 0 \\ -x & \text{if } x < 0 \end{cases}$$

Using the same method as in Example 10, we note that the graph of f coincides with the line $y = x$ to the right of the y-axis and coincides with the line $y = -x$ to the left of the y-axis (see Figure 16). ■

Most piecewise-defined functions have graphs that consist of several "pieces" that may or may not be connected to each other. One important function with this property that crops up frequently in applications is the **greatest integer function**. The *greatest integer function* of any number x is the largest integer less than or equal to x. The greatest integer of x is denoted $[\![x]\!]$. Thus, $[\![2]\!] = 2$, $[\![2.3]\!] = 2$, $[\![1.999]\!] = 1$, $[\![0.002]\!] = 0$, $[\![-0.002]\!] = -1$, $[\![-3]\!] = -3$, and so on. More precisely, we have

$$[\![x]\!] = \begin{cases} \vdots & & \\ -1 & \text{if} & -1 \leq x < 0 \\ 0 & \text{if} & 0 \leq x < 1 \\ 1 & \text{if} & 1 \leq x < 2 \\ 2 & \text{if} & 2 \leq x < 3 \\ 3 & \text{if} & 3 \leq x < 4 \\ \vdots & & \end{cases}$$

A graph of the greatest integer function is shown in Figure 17. In the next example we study a real-world function whose values can be expressed using the greatest integer function.

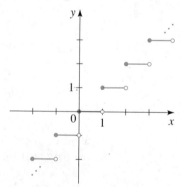

FIGURE 17

The greatest integer function,
$y = [\![x]\!]$

EXAMPLE 12 ■ The Cost Function for Long-Distance Phone Calls

The cost of a long-distance daytime phone call from Toronto to New York City is 69 cents for the first minute and 58 cents for each additional minute (or part of a minute). Draw the graph of the cost C (in dollars) of the phone call as a function of time t (in minutes).

SOLUTION Let $C(t)$ be the cost for t minutes. Since $t > 0$, the domain of the function is $(0, \infty)$. From the given information, we have

$$
\begin{aligned}
C(t) &= 0.69 & &\text{if } 0 < t \leq 1 \\
C(t) &= 0.69 + 0.58 = 1.27 & &\text{if } 1 < t \leq 2 \\
C(t) &= 0.69 + 2(0.58) = 1.85 & &\text{if } 2 < t \leq 3 \\
C(t) &= 0.69 + 3(0.58) = 2.43 & &\text{if } 3 < t \leq 4
\end{aligned}
$$

and so on. The graph is shown in Figure 18. ∎

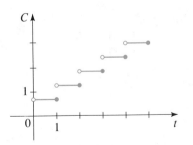

FIGURE 18
Cost of a long-distance call

The following example shows how graphing calculators can be used to help draw piecewise-defined functions.

EXAMPLE 13 ■ Drawing a Piecewise-Defined Function Using a Graphing Device

Use a graphing device to draw the graph of the function

$$
f(x) = \begin{cases} x^3 & \text{if } x \leq 0 \\ x + 2 & \text{if } x > 0 \end{cases}
$$

SOLUTION First we use a calculator to draw the graphs of the functions

$$
g(x) = x^3 \qquad \text{and} \qquad h(x) = x + 2
$$

in the viewing rectangle $[-3, 3]$ by $[-5, 5]$, as shown in Figure 19(a). Then we draw the graph of f (by hand) in Figure 19(b) by taking the part of the graph of g to the left of $x = 0$ and combining it with the part of the graph of h to the right of $x = 0$. Note that $f(0) = 0^3 = 0$, so we place a solid dot at $(0, 0)$ and an open dot at $(0, 2)$.

On many graphing calculators the graph in Figure 19(b) can be produced by using the logical functions in the calculator. For example, on the TI-82 the following equation gives the required graph:

$y = (x \leq 0)x^\wedge 3 + (x > 0)(x + 2)$

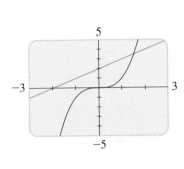

(a)

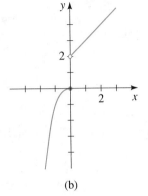

(b)

FIGURE 19

2.2 | EXERCISES

1. The graph of a function f is given.
   (a) State the values of $f(-1)$, $f(0)$, $f(1)$, and $f(3)$.
   (b) State the domain and range of f.
   (c) State the intervals on which f is increasing and on which f is decreasing.

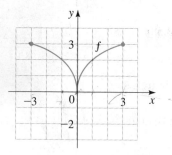

2. The graph of a function h is given.
   (a) State the values of $h(-2)$, $h(0)$, $h(2)$, and $h(3)$.
   (b) State the domain and range of h.
   (c) State the intervals on which h is increasing and on which h is decreasing.

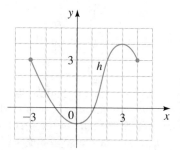

3. The graph of a function g is given.
   (a) State the values of $g(-4)$, $g(-2)$, $g(0)$, $g(2)$, and $g(4)$.
   (b) State the domain and range of g.
   (c) State the intervals on which g is increasing and on which g is decreasing.

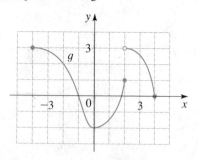

4. The graph of a function f is given.
   (a) State the values of $f(-3)$, $f(1)$, $f(2)$, and $f(3)$.
   (b) State the domain and range of f.

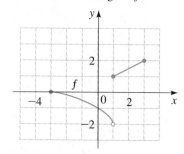

5. Graphs of the functions f and g are given.
   (a) Which is larger, $f(0)$ or $g(0)$?
   (b) Which is larger, $f(-3)$ or $g(-3)$?
   (c) For which values of x is $f(x) = g(x)$?

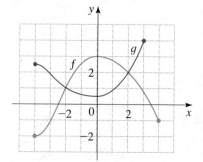

6. The graph of a function f is given.
   (a) Estimate $f(0.5)$ to the nearest tenth.
   (b) Estimate $f(3)$ to the nearest tenth.
   (c) Find all the numbers x in the domain of f so that $f(x) = 1$.

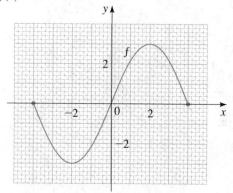

7–8 ■ Determine whether each curve is the graph of a function of *x*.

7. (a)

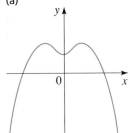

(b)

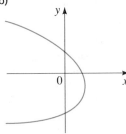

(c)

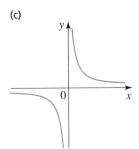

(d)

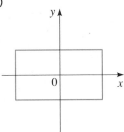

8. (a)

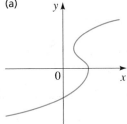

(b)

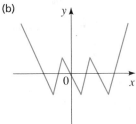

(c)

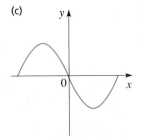

(d)

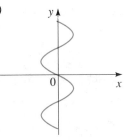

9–12 ■ Determine whether the curve is the graph of a function of *x*. If it is, state the domain and range of the function.

9.

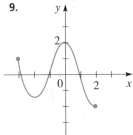

10.

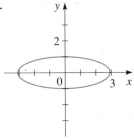

11.

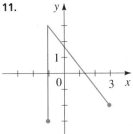

12.

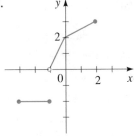

13–20 ■ A function *f* is given.
(a) Sketch the graph of *f*.
(b) Find the domain of *f*.
(c) Find the intervals on which *f* is increasing and on which *f* is decreasing.

13. $f(x) = 1 - x$ **14.** $f(x) = \frac{1}{2}(x + 1)$

15. $f(x) = x^2 - 4x$ **16.** $f(x) = x^2 - 4x + 4$

17. $f(x) = \sqrt{9 - x}$ **18.** $f(x) = \sqrt{2x + 6}$

19. $f(x) = \sqrt{16 - x^2}$ **20.** $f(x) = -\sqrt{25 - x^2}$

21–44 ■ Sketch the graph of the function.

21. $f(x) = 3$ **22.** $f(x) = -5$

23. $f(x) = 2x + 3$ **24.** $f(x) = 6 - 3x$

25. $f(x) = -x + 4, \quad -1 \leqslant x \leqslant 4$

26. $f(x) = \dfrac{x + 3}{2}, \quad -2 \leqslant x \leqslant 2$

27. $f(x) = -x^2$ **28.** $f(x) = x^2 - 4$

29. $f(x) = x^2 + 2x + 1$ **30.** $f(x) = x^2 + 6x - 7$

31. $g(x) = x^3 - 8$ **32.** $g(x) = 4x^2 - x^4$

33. $g(x) = \sqrt{-x}$

34. $g(x) = \sqrt{6 - 2x}$

35. $F(x) = \dfrac{1}{x}$

36. $F(x) = \dfrac{2}{x + 4}$

37. $H(x) = |2x|$

38. $H(x) = |x + 1|$

39. $G(x) = |x| + x$

40. $G(x) = |x| - x$

41. $f(x) = |2x - 2|$

42. $f(x) = \dfrac{x}{|x|}$

43. $f(x) = \dfrac{x^2 - 1}{x - 1}$

44. $f(x) = \dfrac{x^2 + 5x + 6}{x + 2}$

 45–48 ■ A function f is given.
(a) Use a graphing device to draw the graph of f.
(b) State approximately the intervals on which f is increasing and on which f is decreasing.

45. $f(x) = x^{2/5}$

46. $f(x) = 4 - x^{2/3}$

47. $f(x) = x^3 + 2x^2 - x - 2$

48. $f(x) = x^4 - 4x^3 + 2x^2 + 4x - 3$

 49. In this exercise we consider the family of root functions $f(x) = x^c$, where $0 < c \le 1$.
(a) Draw the graphs of f for $c = \frac{1}{2}, \frac{1}{4}$, and $\frac{1}{6}$ on the same screen using the viewing rectangle $[-1, 4]$ by $[-1, 3]$.
(b) Draw the graphs of f for $c = 1, \frac{1}{3}$, and $\frac{1}{5}$ on the same screen using the viewing rectangle $[-3, 3]$ by $[-2, 2]$.
(c) Draw the graphs of f for $c = \frac{1}{2}, \frac{1}{3}, \frac{1}{4}$ and $\frac{1}{5}$ on the same screen using the viewing rectangle $[-1, 4]$ by $[-1, 3]$.
(d) What conclusions can you make from these graphs?

 50. In this exercise we consider the family of functions $f(x) = 1/x^n$.
(a) Draw the graphs of f for $n = 1$ and $n = 3$ on the same screen using the viewing rectangle $[-3, 3]$ by $[-3, 3]$.
(b) Draw the graphs of f for $n = 2$ and $n = 4$ on the same screen using the same viewing rectangle.
(c) Draw the graphs of all the functions in parts (a) and (b) on the same screen using the viewing rectangle $[-1, 3]$ by $[-1, 3]$.
(d) What conclusions can you make from these graphs?

 51. In this exercise we consider the family of functions $f(x) = x^2 + c$.
(a) Draw the graphs of f for $c = 0, 2, 4$, and 6 on the same screen using the viewing rectangle $[-5, 5]$ by $[-10, 10]$.

(b) Draw the graphs of f for $c = 0, -2, -4$, and -6 on the same screen using the viewing rectangle $[-5, 5]$ by $[-10, 10]$.
(c) What conclusions can you make from these graphs?

 52. In this exercise we consider the family of functions $f(x) = (x - c)^2$.
(a) Draw the graphs of f for $c = 0, 1, 2$, and 3 on the same screen using the viewing rectangle $[-5, 5]$ by $[-10, 10]$.
(b) Draw the graphs of f for $c = 0, -1, -2$, and -3 on the same screen using the viewing rectangle $[-5, 5]$ by $[-10, 10]$.
(c) What conclusions can you make from these graphs?

 53. In this exercise we consider the family of functions $f(x) = (x - c)^3$.
(a) Draw the graphs of f for $c = 0, 2, 4$, and 6 on the same screen using the viewing rectangle $[-10, 10]$ by $[-10, 10]$.
(b) Draw the graphs of f for $c = 0, -2, -4$, and -6 on the same screen using the viewing rectangle $[-10, 10]$ by $[-10, 10]$.
(c) What conclusions can you make from these graphs?

 54. In this exercise we consider the family of functions $f(x) = cx^2$.
(a) Draw the graphs of f for $c = 1, 2, 3$, and 4 on the same screen using the viewing rectangle $[-5, 5]$ by $[-10, 10]$.
(b) Draw the graphs of f for $c = 1, \frac{1}{2}, \frac{1}{4}$, and $\frac{1}{10}$ on the same screen using the viewing rectangle $[-5, 5]$ by $[-10, 10]$.
(c) Draw the graphs of f for $c = 1, -1, -\frac{1}{2}$, and -2 on the same screen using the viewing rectangle $[-5, 5]$ by $[-10, 10]$.
(d) What conclusions can you make from these graphs?

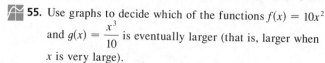 **55.** Use graphs to decide which of the functions $f(x) = 10x^2$ and $g(x) = \dfrac{x^3}{10}$ is eventually larger (that is, larger when x is very large).

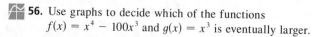 **56.** Use graphs to decide which of the functions $f(x) = x^4 - 100x^3$ and $g(x) = x^3$ is eventually larger.

 57. (a) Draw the graphs of the functions $f(x) = x^2 + x - 6$ and $g(x) = |x^2 + x - 6|$.
(b) How are the graphs of f and g related?

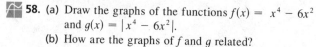 **58.** (a) Draw the graphs of the functions $f(x) = x^4 - 6x^2$ and $g(x) = |x^4 - 6x^2|$.
(b) How are the graphs of f and g related?

59–72 ■ Sketch the graph of the piecewise-defined function.

59. $f(x) = \begin{cases} 0 & \text{if } x < 2 \\ 1 & \text{if } x \geq 2 \end{cases}$

60. $f(x) = \begin{cases} 1 & \text{if } x \leq 1 \\ x + 1 & \text{if } x > 1 \end{cases}$

61. $f(x) = \begin{cases} 3 & \text{if } x < 2 \\ x - 1 & \text{if } x \geq 2 \end{cases}$

62. $f(x) = \begin{cases} 1 - x & \text{if } x < -2 \\ 5 & \text{if } x \geq -2 \end{cases}$

63. $f(x) = \begin{cases} x & \text{if } x \leq 0 \\ x + 1 & \text{if } x > 0 \end{cases}$

64. $f(x) = \begin{cases} 2x + 3 & \text{if } x < -1 \\ 3 - x & \text{if } x \geq -1 \end{cases}$

65. $f(x) = \begin{cases} x + 1 & \text{if } x \neq 1 \\ 1 & \text{if } x = 1 \end{cases}$

66. $f(x) = \begin{cases} x + 3 & \text{if } x \neq -2 \\ 4 & \text{if } x = -2 \end{cases}$

67. $f(x) = \begin{cases} -1 & \text{if } x < -1 \\ 1 & \text{if } -1 \leq x \leq 1 \\ -1 & \text{if } x > 1 \end{cases}$

68. $f(x) = \begin{cases} -1 & \text{if } x < -1 \\ x & \text{if } -1 \leq x \leq 1 \\ 1 & \text{if } x > 1 \end{cases}$

69. $f(x) = \begin{cases} 2 & \text{if } x \leq -1 \\ x^2 & \text{if } x > -1 \end{cases}$

70. $f(x) = \begin{cases} 1 - x^2 & \text{if } x \leq 2 \\ x & \text{if } x > 2 \end{cases}$

71. $f(x) = \begin{cases} 0 & \text{if } |x| \leq 2 \\ 3 & \text{if } |x| > 2 \end{cases}$

72. $f(x) = \begin{cases} x^2 & \text{if } |x| \leq 1 \\ 1 & \text{if } |x| > 1 \end{cases}$

73–76 ■ Use a graphing device to draw the graph of the piecewise-defined function.

73. $f(x) = \begin{cases} x + 2 & \text{if } x \leq -1 \\ x^2 & \text{if } x > -1 \end{cases}$

74. $f(x) = \begin{cases} 2x - x^2 & \text{if } x > 1 \\ (x - 1)^3 & \text{if } x \leq 1 \end{cases}$

75. $f(x) = \begin{cases} x^3 - 2x + 1 & \text{if } x \leq 0 \\ x - x^2 & \text{if } 0 < x < 1 \\ \sqrt[4]{x - 1} & \text{if } x \geq 1 \end{cases}$

76. $f(x) = \begin{cases} \sqrt{-x} & \text{if } x < 0 \\ \sqrt{2x - x^2} & \text{if } 0 \leq x \leq 2 \\ \sqrt{x - 2} & \text{if } x > 2 \end{cases}$

77. A taxi company charges $2.00 for the first mile (or part of a mile) and 20 cents for each succeeding tenth of a mile (or part). Express the cost C (in dollars) of a ride as a function of the distance x traveled (in miles) for $0 < x < 2$, and sketch the graph of this function.

78. The domestic postage rate for first-class letters weighing 12 oz or less is 32 cents for a letter weighing 1 oz or less and 23 cents for each additional ounce (or part of an ounce). Express the postage P as a function of the weight x of a letter, with $0 < x \leq 12$.

79–82 ■ Find a function whose graph is the given curve.

79. The line segment joining the points $(-2, 1)$ and $(4, -6)$

80. The line segment joining the points $(-3, -2)$ and $(6, 3)$

81. The bottom half of the parabola $x + (y - 1)^2 = 0$

82. The top half of the circle $(x - 1)^2 + y^2 = 1$

▨ **DISCOVERY · DISCUSSION**

83. When Does a Graph Represent a Function? For every integer n, the graph of the equation $y = x^n$ is the graph of a function, namely, $f(x) = x^n$. Explain why the graph of $x = y^2$ is *not* the graph of a function. Is the graph of $x = y^3$ the graph of a function? If so, of what function of x is it the graph? Determine for what integers n the graph of $x = y^n$ is the graph of a function of x.

84. Step Functions In Example 12 and Exercises 77 and 78 we are given functions whose graphs consist of horizontal line segments. Such functions are often called *step functions,* because their graphs look like stairs. Give some other examples of step functions that arise in everyday life.

85. Stretched Step Functions Sketch the functions $f(x) = [\![x]\!]$, $g(x) = [\![2x]\!]$, and $h(x) = [\![3x]\!]$ on separate graphs. How are the graphs related? If n is a positive integer, what does the graph of $k(x) = [\![nx]\!]$ look like?

2.3　APPLIED FUNCTIONS

Mathematical models are discussed in more detail in the special section *Principles of Modeling* that begins on page 210.

When scientists or applied mathematicians talk about a *mathematical model* for a real-world phenomenon, they mean a function that describes, at least approximately, the dependence of one physical quantity on another. For instance, the model may describe the population of an animal species as a function of time, or the pressure of a gas as a function of its volume. In this section we study such applied functions.

Our first example shows that even when a precise formula for functional dependence is not available, it is still possible to visualize the situation from a graph of the function.

EXAMPLE 1　■　A Graph of Water Temperature

When you turn on a hot water faucet, the temperature T of the water depends on how long the water has been running. Draw a rough graph of T as a function of the time t that has elapsed since the faucet was turned on.

SOLUTION　The initial temperature of the water is close to room temperature because that water has been standing in the pipes. When the water from the hot water tank starts coming out, T increases quickly. In the next phase, T is constant at the temperature of the water in the tank. When the tank is drained, T decreases to the temperature of the water supply. This enables us to make the rough sketch of T as a function of t in Figure 1.　　■

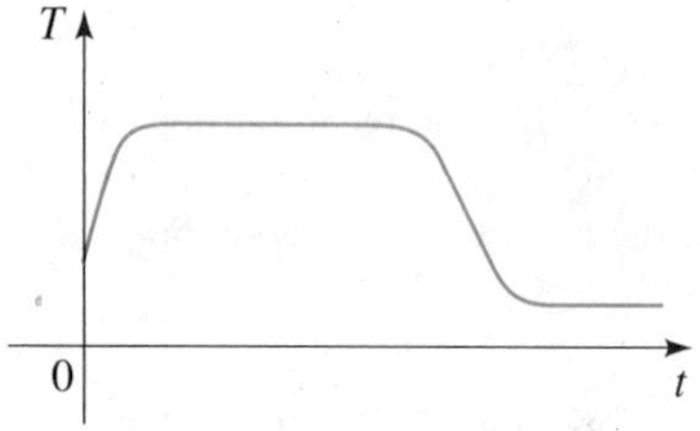

FIGURE 1

Graph of water temperature T as a function of time t

A more accurate graph of the function in Example 1 could be obtained by using a thermometer to measure the temperature of the water at 10-second intervals. In general, scientists collect experimental data and use them to sketch the graphs of functions; they then try to discover a formula that fits the graph. Once a defining formula for a function is found, scientists are then able to predict how a certain process will behave. In the next examples we find explicit formulas for applied functions. In discovering these functions we will rely on the problem-solving principles introduced on pages 122–124.

EXAMPLE 2　■　Perimeter as a Function of Area

Express the perimeter of a square as a function of its area.

SOLUTION　Let P, A, and s represent the perimeter, area, and side length of the square in Figure 2, respectively.

Introduce notation

Draw a diagram

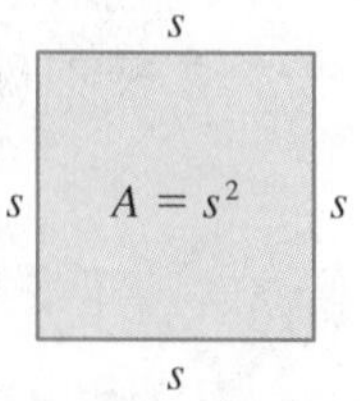

$$P = s + s + s + s = 4s$$

FIGURE 2

Relate the quantities P and A

Our goal is to find the relationship between P and A. We start with the known formulas for perimeter and area:

$$P = 4s \quad \text{and} \quad A = s^2$$

Eliminate s

The second formula gives $s = \sqrt{A}$. To express P as a function of A, we substitute this into the first formula to get

$$P = 4s \qquad \text{Formula for perimeter}$$
$$= 4\sqrt{A} \qquad \text{Substitute } \sqrt{A} \text{ for } s$$

So, the function that relates P and A is

$$P = 4\sqrt{A}$$

Since A must be positive, the domain is given by $A > 0$ or $(0, \infty)$. ■

The function $P = 4\sqrt{A}$ in Example 2 gives the precise relationship between the perimeter and the area of a square. The graph in Figure 3 allows us to visualize how fast the perimeter increases as the area increases.

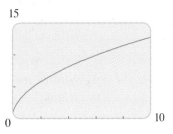

FIGURE 3
Graph of $P = 4\sqrt{A}$

In Example 2 the margin notes indicate the steps we used to solve the problem. These are explained in the following box.

GUIDELINES FOR SETTING UP APPLIED FUNCTIONS

1. INTRODUCE NOTATION. Assign a symbol to the quantity the problem asks for (for now, let's call it Q) and then select symbols for the other unknown quantities in the problem. It helps to use initials as suggestive variables—for example, A for area, h for height, t for time, and so on.

2. RELATE THE QUANTITIES. Find equations that relate Q and the other variables from Step 1, using the given information. It is often helpful to **draw a diagram** and identify the variables and given quantities on the diagram. You may need to use formulas for volume, area, perimeter, or other quantities, as we did in Example 2.

3. ELIMINATE UNNEEDED VARIABLES. If Q has been expressed as a function of more than one variable in Step 2, use any relationships that exist between the variables to eliminate all but one of them in the expression for Q. Then Q will be expressed as a function of one variable only.

EXAMPLE 3 ■ Surface Area as a Function of Radius

A can holds 1 L (liter) of oil. Express the surface area of the can as a function of its radius.

SOLUTION   Let r be the radius and h be the height of the can (in centimeters) shown in Figure 4.

Introduce notation

Draw a diagram

FIGURE 4

Relate the quantities S, r, and k

Then the area of the top is πr^2, the area of the bottom is also πr^2, and the area of the sides is the circumference times the height, that is, $2\pi rh$. So the total surface area is

$$S = 2\pi r^2 + 2\pi rh$$

Eliminate h

To express S as a function of r we need to eliminate h, and we do this by using the fact that the volume is given as 1 L, which is eqivalent to 1000 cm³. Thus, the volume is

$$\pi r^2 h = 1000$$

The formula for the volume of a cylinder is given on the inside of the front cover of this book.

Substituting $h = 1000/(\pi r^2)$ into the expression for S, we have

$$S = 2\pi r^2 + 2\pi r\left(\frac{1000}{\pi r^2}\right) = 2\pi r^2 + \frac{2000}{r}$$

Therefore, the equation

$$S = 2\pi r^2 + \frac{2000}{r}, \quad r > 0$$

expresses S as a function of r.   ■

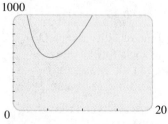

FIGURE 5

Graph of $S = 2\pi r^2 + \dfrac{2000}{r}$

The graph of the function $S = 2\pi r^2 + \dfrac{2000}{r}$ for $r > 0$ gives us a picture of how the surface area of the can changes as the radius changes (see Figure 5): As r increases, the surface area S first decreases, then increases.

DIRECT AND INVERSE VARIATION

Two types of mathematical models occur so often in various sciences that they are given special names. The first is called *direct variation* and occurs when one quantity is a constant multiple of another.

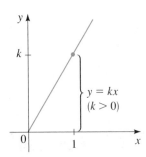

FIGURE 6
Direct variation

> ## DIRECT VARIATION
>
> If the quantities x and y are related by an equation
>
> $$y = kx$$
>
> for some constant $k \neq 0$, we say that y **varies directly as** x, or y **is directly proportional to** x, or simply y **is proportional to** x. The constant k is called the **constant of proportionality**.

Recall that a linear function is a function of the form $f(x) = mx + b$ and its graph is a line with slope m and y-intercept b. So if y varies directly as x, then the equation $y = kx$ or, equivalently, $f(x) = kx$, shows that y is a linear function of x. The graph of this function is a line with slope k, the constant of proportionality (see Figure 6). The y-intercept is $b = 0$, so this line passes through the origin.

EXAMPLE 4 ■ Direct Variation

During a thunderstorm you see the lightning before you hear the thunder because light travels much faster than sound. The distance between you and the center of the storm varies directly as the time interval between the lightning and the thunder.

(a) Suppose that the thunder from a storm whose center is 5400 ft away takes 5 s to reach you. Determine the constant of proportionality and write the equation for the variation.

(b) Sketch the graph of this equation. What does the constant of proportionality represent?

(c) If the time interval between the lightning and thunder is now 8 s, how far away is the center of the storm?

SOLUTION

Introduce notation

Relate the quantities d and t

Eliminate k

(a) Let d be the distance from you to the storm and let t be the length of the time interval. We are given that d varies directly as t, so

$$d = kt$$

where k is a constant. To find k, we use the fact that $t = 5$ when $d = 5400$. Substituting these values in the equation, we get

$$5400 = k(5) \qquad \text{Substitute}$$

$$k = \frac{5400}{5} = 1080 \qquad \text{Solve for } k$$

Substituting this value of k in the equation for d, we obtain

$$d = 1080t$$

as the equation for d as a function of t.

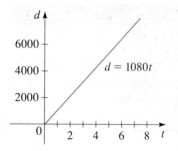

FIGURE 7

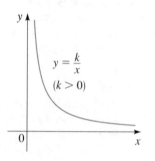

FIGURE 8
Inverse variation

(b) The graph of the equation $d = 1080t$ is a line through the origin with slope 1080 and is shown in Figure 7. The constant $k = 1080$ is the approximate speed of sound (in ft/s).

(c) When $t = 8$, we have

$$d = 1080 \cdot 8 = 8640$$

So, the storm center is 8640 ft $\approx$ 1.6 mi away.   ∎

Another function that is used frequently in mathematical modeling is the function $f(x) = k/x$, where k is a constant.

INVERSE VARIATION

If the quantities x and y are related by the equation

$$y = \frac{k}{x}$$

for some constant $k \neq 0$, we say that y **is inversely proportional to** x, or y **varies inversely as** x.

The graph of the function $f(x) = k/x$ for $x > 0$ is shown in Figure 8 for the case $k > 0$. It gives a picture of what happens when y is inversely proportional to x.

EXAMPLE 5 ■ Inverse Variation

Boyle's law states that when a sample of gas is compressed at a constant temperature, the pressure of the gas is inversely proportional to the volume of the gas.

(a) Suppose that the pressure of a sample of air that occupies 0.106 m³ at 25 °C is 50 kPa. Find the constant of proportionality, and write the equation that expresses the inverse proportionality.

(b) If the sample expands to a volume of 0.3 m³, find the new pressure.

SOLUTION

Introduce notation

(a) Let P be the pressure of the sample of gas and let V be its volume. Then, by the definition of inverse proportionality, we have

Relate the quantities P and V

$$P = \frac{k}{V}$$

Eliminate k

where k is a constant. To find k we use the fact that $P = 50$ when $V = 0.106$. Substituting these values in the equation, we get

$$50 = \frac{k}{0.106} \qquad \text{Substitute}$$

$$k = (50)(0.106) = 5.3 \qquad \text{Solve for } k$$

Putting this value of k in the equation for P, we have

$$P = \frac{5.3}{V}$$

(b) When $V = 0.3$, we have

$$P = \frac{5.3}{0.3} \approx 17.7$$

So, the new pressure is about 17.7 kPa. ■

A physical quantity often depends on more than one other quantity. For instance, if the quantities x, y, and z are related by the equation

$$z = kxy$$

where k is a nonzero constant, then we say that z **varies jointly as** x and y, or z **is jointly proportional to** x and y. If

$$z = k\frac{x}{y}$$

we say that z **is proportional to** x and **inversely proportional to** y.

EXAMPLE 6 ■ Newton's Law of Gravitation

Newton's Law of Gravitation says that two objects with masses m_1 and m_2 attract each other with a force F that is jointly proportional to their masses and is inversely proportional to the square of the distance r between the objects. Express Newton's Law of Gravitation as an equation.

SOLUTION   Using the definitions of joint and inverse proportionality, and using the traditional notation G for the constant of proportionality, we have

$$F = G\frac{m_1 m_2}{r^2}$$

■

If m_1 and m_2 are fixed masses, then the gravitational force between them is $F = C/r^2$ (where $C = Gm_1m_2$ is a constant). Figure 9 shows the graph of this function for $r > 0$ with $C = 1$. Observe how the gravitational attraction decreases with increasing distance.

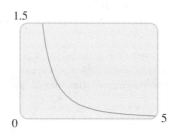

FIGURE 9

Graph of $F = \dfrac{1}{r^2}$

2.3 EXERCISES

1. The graph gives the weight of a certain person as a function of age. Describe in words how this person's weight has varied over time. What do you think happened when this person was 30 years old?

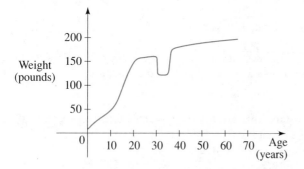

2. The graph gives a salesman's distance from his home as a function of time on a certain day. Describe in words what the graph indicates about his travels on this day.

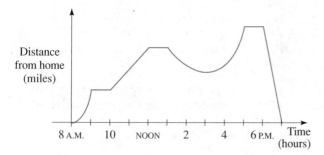

3. You put some ice cubes in a glass, fill the glass with cold water, and then let the glass sit on a table. Sketch a rough graph of the temperature of the water as a function of the elapsed time.

4. A home owner mows the lawn every Wednesday afternoon. Sketch a rough graph of the height of the grass as a function of time over the course of a four-week period beginning on a Sunday.

5. When a football is kicked from a tee, its height depends on the time elapsed since the kickoff. Sketch a rough graph of the height of the football as a function of time.

6. Sketch a rough graph of the number of hours of daylight as a function of the time of year in the Northern Hemisphere.

7. Sketch a rough graph of the outdoor temperature as a function of time during a typical spring day.

8. You place a frozen pie in an oven and bake it for an hour. Then you take it out and let it cool before eating it. Sketch a rough graph of the temperature of the pie as a function of time.

9. The figure shows the power consumption in San Francisco for September 19, 1996 (P is measured in megawatts; t is measured in hours starting at midnight).
 (a) What was the power consumption at 6 A.M.? At 6 P.M.?
 (b) When was the power consumption the lowest?
 (c) When was the power consumption the highest?

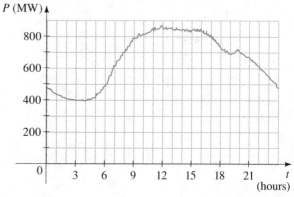

Source: Pacific Gas & Electric

10. The graph shows the vertical acceleration of the ground from the 1994 Northridge earthquake in Los Angeles, as measured by a seismograph. (Here t represents the time in seconds.)

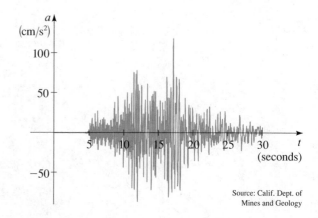

Source: Calif. Dept. of Mines and Geology

(a) At what time t did the earthquake first make noticeable movements of the earth?

(b) At what time t did the earthquake seem to end?

(c) At what time t was the maximum intensity of the earthquake reached?

11. Temperature readings T (in °F) were recorded every 2 hours from midnight to noon in Atlanta, Georgia, on March 18, 1996. The time t was measured in hours from midnight.

t	T
0	58
2	57
4	53
6	50
8	51
10	57
12	61

Use the readings to sketch a rough graph of T as a function of t.

12. The population P (in thousands) of San Jose, California, from 1984 to 1994 is shown in the table. (Midyear estimates are given.)

t	P
1984	695
1986	716
1988	733
1990	782
1992	800
1994	817

Draw a graph of P as a function of time t.

13–26 ■ Find a formula for the described function and state its domain.

13. A rectangle has a perimeter of 20 ft. Express the area A of the rectangle as a function of the length x of one of its sides.

14. A rectangle has an area of 16 m². Express the perimeter P of the rectangle as a function of the length x of one of its sides.

15. Express the area A of an equilateral triangle as a function of the length x of a side.

16. Express the surface area A of a cube as a function of its volume V.

17. Express the radius r of a circle as a function of its area A.

18. Express the area A of a circle as a function of the circumference C.

19. An open rectangular box with a volume of 12 ft³ has a square base. Express the surface area A of the box as a function of the length x of a side of the base.

20. A woman 5 ft tall is standing near a street lamp that is 12 ft tall, as shown in the figure. Express the length L of her shadow as a function of her distance d from the base of the lamp.

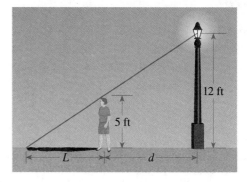

21. A Norman window has the shape of a rectangle surmounted by a semicircle, as shown in the figure. If the perimeter of the window is 30 ft, express the area A of the window as a function of the width x of the window.

22. A box with an open top is to be constructed from a rectangular piece of cardboard with dimensions

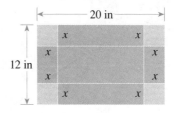

12 in. by 20 in. by cutting out equal squares of side x at each corner and then folding up the sides, as shown in the figure. Express the volume V of the box as a function of x.

23. A farmer has 2400 ft of fencing and wants to fence off a rectangular field that borders a straight river, as shown in the figure. He needs no fence along the river. Express the area A of the field in terms of the width x of the field.

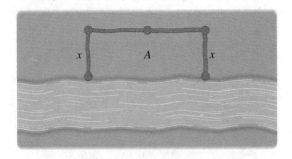

24. A rectangle is inscribed in a semicircle of radius r, as shown in the figure. Express the area A of the rectangle as a function of the height h of the rectangle.

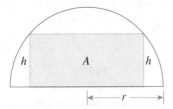

25. Two ships leave a port at the same time. One sails south at 15 mi/h and the other sails east at 20 mi/h. Express the distance d between the ships as a function of t, the time (in hours) elapsed since their departure.

26. A man is standing at point A on the bank of a straight river, 2 mi wide, and he wants to reach point B, 7 mi

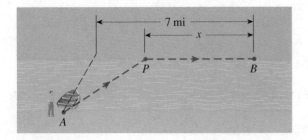

downstream on the opposite bank, by first rowing his boat to a point P on the opposite bank and then walking the remaining distance x to B. He can row at 2 mi/h and walk at 5 mi/h. Express the total time T that he takes to go from A to B as a function of x.

27. The weight w of an object at a height h above the surface of the earth is given by

$$w = \left(\frac{R}{R + h}\right)^2 w_0$$

where w_0 is the weight of the object at sea level and $R = 6400$ km is the radius of the earth. Suppose the weight of an object at sea level is 80 N. Using an appropriate viewing rectangle, sketch the graph of the weight of the object as a function of its height.

28. A highway engineer wants to estimate the maximum number of cars that can safely travel a particular highway at a given speed. She assumes that each car is 17 ft long, travels at a speed s, and follows the car in front of it at the "safe following distance" for that speed. The safe following distance at 20 mi/h is one car length. Using the fact that the distance required to stop is proportional to the square of the speed and the fact that 1 mi/h = 88 ft/s, she finds that the number N of cars per minute that pass a given point is a function of the speed given by

$$N(s) = \frac{88s}{17 + 17\left(\dfrac{s}{20}\right)^2}$$

Sketch a graph of this function in an appropriate viewing rectangle.

29. For a fish swimming at a speed v relative to the water, the energy expenditure per unit time is proportional to v^3. If the fish is swimming against a current with speed u miles per hour, where $u < v$, then the time required to travel a distance of L miles is $L/(v - u)$ and the total energy required to travel the distance is

$$E(v) = 2.73v^3 \frac{L}{v - u}$$

Suppose the speed of the current is $u = 5$ mi/h. Sketch the graph of the energy $E(v)$ needed to swim a distance of 10 mi as a function of the speed v of the fish.

30. In the theory of relativity, the mass of an object changes as its speed changes. If m_0 is the rest mass of the object, then its mass m is a function of its speed v given by

$$m(v) = \frac{m_0}{\sqrt{1 - \dfrac{v^2}{c^2}}}$$

where $c = 3.0 \times 10^5$ km/s is the speed of light. Sketch the graph of the mass m as a function of the speed v for an object with rest mass 1000 kg.

31–38 ■ Write an equation that expresses the statement.

31. R varies directly as t.

32. P is directly proportional to u.

33. v is inversely proportional to z.

34. w is jointly proportional to m and n.

35. y is proportional to s and inversely proportional to t.

36. P varies inversely as T.

37. z is proportional to the square root of y.

38. A is proportional to the square of t and inversely proportional to the cube of x.

39–44 ■ Express the statement as a formula. Use the given information to find the constant of proportionality.

39. y is directly proportional to x. If $x = 4$, then $y = 72$.

40. z varies inversely as t. If $t = 3$, then $z = 5$.

41. M varies directly as x and inversely as y. If $x = 2$ and $y = 6$, then $M = 5$.

42. S varies jointly as p and q. If $p = 4$ and $q = 5$, then $S = 180$.

43. W is inversely proportional to the square of r. If $r = 6$, then $W = 10$.

44. t is jointly proportional to x and y and inversely proportional to r. If $x = 2$, $y = 3$, and $r = 12$, then $t = 25$.

45. Hooke's law states that the force needed to keep a spring stretched x units beyond its natural length is directly proportional to x. Here the constant of proportionality is called the **spring constant**.
(a) Write Hooke's law as an equation.
(b) If a spring has a natural length of 10 cm and a force of 40 N is required to maintain the spring stretched to a length of 15 cm, find the spring constant.

(c) What force is needed to keep the spring stretched to a length of 14 cm?

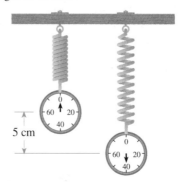

46. The period of a pendulum (the time elapsed during one complete swing of the pendulum) varies directly with the square root of the length of the pendulum.
(a) Express this relationship by writing an equation.
(b) In order to double the period, how would we have to change the length l?

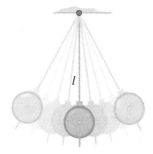

47 The cost of printing a magazine is jointly proportional to the number of pages in the magazine and the number of magazines printed.
(a) Write an equation for this joint variation if the printing cost is $60,000 for 4000 copies of a 120-page magazine.
(b) How much would the printing cost be for 5000 copies of a 92-page magazine?

48. The pressure P of a sample of gas is directly proportional to the temperature T and inversely proportional to the volume V.
(a) Write an equation that expresses this fact if 100 L of gas exerts a pressure of 33.2 kPa at a temperature of 400 K (absolute temperature measured on the Kelvin scale).
(b) If the temperature is increased to 500 K and the volume is decreased to 80 L, what is the pressure of the gas?

49. The resistance R of a wire varies directly as its length L and inversely as the square of its diameter d.
    (a) A wire 1.2 m long and 0.005 m in diameter has a resistance of 140 ohms. Write an equation for this variation and find the constant of proportionality.
    (b) Find the resistance of a wire made of the same material that is 3 m long and has a diameter of 0.008 m.

50. Kepler's Third Law of planetary motion states that the square of the period T of a planet (the time it takes for the planet to make a complete revolution about the sun) is directly proportional to the cube of the average distance d from the sun.

    (a) Express Kepler's Third Law as an equation.

    (b) Find the constant of proportionality by using the fact that for our planet the period is about 365 days and the average distance is about 93 million miles.

    (c) The planet Neptune is about 2.79×10^9 mi from the sun. Find the period of Neptune.

2.4 TRANSFORMATIONS OF FUNCTIONS

In this section we study how certain transformations of a function affect its graph. This will give us a better understanding of how to graph functions. The transformations we study are shifting, reflecting, and stretching.

EXAMPLE 1 ■ Vertical Shifts of Graphs

Sketch the graph of each function.
(a) $f(x) = x^2 + 3$ (b) $g(x) = x^2 - 2$

SOLUTION

Recall that the graph of the function f is the same as the graph of the equation $y = f(x)$. It is often convenient to express a function in equation form, particularly when discussing its graph. For instance, we may refer to the function $f(x) = x^2$ by the equation $y = x^2$, as we do in Example 1.

(a) We start with the graph of the function $y = x^2$ (from Example 2 in Section 2.2). The equation $y = x^2 + 3$ indicates that the y-coordinate of a point on the graph of f is 3 more than the y-coordinate of the corresponding point on the curve $y = x^2$. This means that we obtain the graph of $f(x) = x^2 + 3$ simply by shifting the graph of $y = x^2$ upward 3 units, as shown in Figure 1.

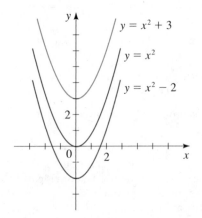

FIGURE 1

(b) Similarly, we get the graph of $g(x) = x^2 - 2$ by shifting the parabola $y = x^2$ downward 2 units (see Figure 1). ■

In general, suppose we know the graph of $y = f(x)$. How do we obtain from it the graphs of

$$y = f(x) + c \quad\text{and}\quad y = f(x) - c \quad (c > 0)$$

The equation $y = f(x) + c$ tells us that the y-coordinate of each point on its graph is c units above the y-coordinate of the corresponding point on the graph of $y = f(x)$. So, we obtain the graph of $y = f(x) + c$ simply by shifting the graph of $y = f(x)$ upward c units. We summarize these observations in the following box.

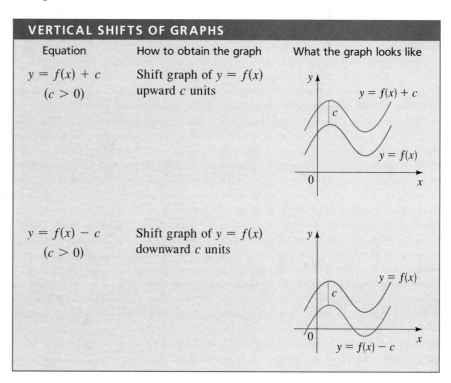

VERTICAL SHIFTS OF GRAPHS

Equation	How to obtain the graph	What the graph looks like
$y = f(x) + c$ $(c > 0)$	Shift graph of $y = f(x)$ upward c units	
$y = f(x) - c$ $(c > 0)$	Shift graph of $y = f(x)$ downward c units	

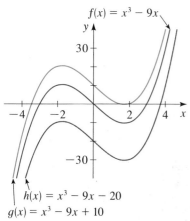

$f(x) = x^3 - 9x$

$h(x) = x^3 - 9x - 20$
$g(x) = x^3 - 9x + 10$

FIGURE 2

EXAMPLE 2 ■ Vertical Shifts of Graphs

Sketch the graph of each function.

(a) $f(x) = x^3 - 9x$

(b) $g(x) = x^3 - 9x + 10$

(c) $h(x) = x^3 - 9x - 20$

SOLUTION

(a) The graph of f was sketched in Example 12 in Section 1.8. It is sketched again in Figure 2.

(b) To obtain the graph of g, we shift the graph of f upward 10 units.

(c) To obtain the graph of h, we shift the graph of f downward 20 units.

The graphs of f, g, and h are sketched in Figure 2. ■

In the next example we consider transformations that shift the graph of a function horizontally.

EXAMPLE 3 ■ Horizontal Shifts of Graphs

The graph of $y = f(x)$ is sketched in Figure 3. Sketch each of the following graphs.

(a) $y = f(x - 5)$ (b) $y = f(x + 8)$

SOLUTION

(a) The value of $f(x - 5)$ at x is the same as the value of $f(x)$ at $x - 5$. So the graph of $y = f(x - 5)$ is simply the graph of $y = f(x)$ shifted to the right 5 units (see Figure 4).

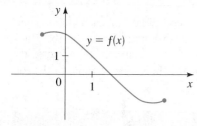

FIGURE 3

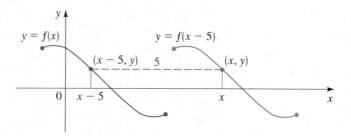

FIGURE 4

(b) Similar reasoning shows that the graph of $y = f(x + 8)$ is the graph of $y = f(x)$ shifted to the left 8 units (see Figure 5).

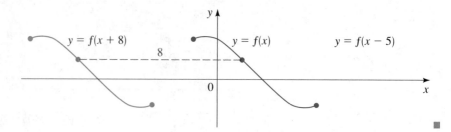

FIGURE 5   ■

In general, suppose we know the graph of $y = f(x)$. How do we use it to obtain the graphs of

$$y = f(x + c) \quad \text{and} \quad y = f(x - c) \qquad (c > 0)$$

The value of $f(x - c)$ at x is the same as the value of $f(x)$ at $x - c$. Since $x - c$ is c units to the left of x, it follows that the graph of $y = f(x - c)$ is just the graph of $y = f(x)$ shifted to the right c units. Similar reasoning shows that the graph of $y = f(x + c)$ is the graph of $y = f(x)$ shifted to the left c units. The following box summarizes these facts.

HORIZONTAL SHIFTS OF GRAPHS

Equation	How to obtain the graph	What the graph looks like
$y = f(x - c)$ $(c > 0)$	Shift graph of $y = f(x)$ to the right c units	
$y = f(x + c)$ $(c > 0)$	Shift graph of $y = f(x)$ to the left c units	

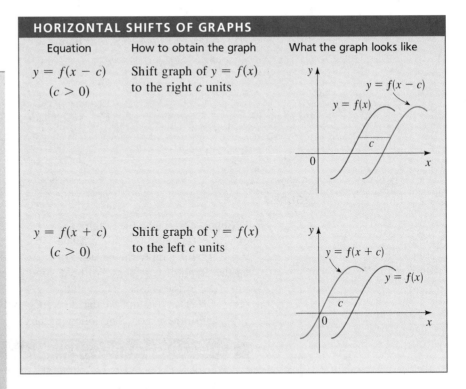

Sonya Kovalevsky (1850–1891) is considered the most important woman mathematician of the 19th century. She was born in Moscow to an aristocratic family. While a child, she was exposed to the principles of calculus in a very unusual fashion—her bedroom was temporarily wallpapered with the pages of a calculus book. She later wrote that she "spent many hours in front of that wall, trying to understand it." Since Russian law forbade women from studying in universities, she entered a marriage of convenience, which allowed her to travel to Germany and obtain a doctorate in mathematics from the University of Göttingen. She eventually was awarded a full professorship at the University of Stockholm, where she taught for eight years before dying in an influenza epidemic at the age of 41. Her research was instrumental in helping put the ideas and applications of functions and calculus on a sound and logical foundation. She received many accolades and prizes for her research work.

EXAMPLE 4 ■ Horizontal Shifts of Graphs

Sketch the graph of each function.

(a) $f(x) = (x + 4)^2$ (b) $g(x) = (x - 2)^2$

SOLUTION We start with the graph of $y = x^2$, then move it to the left 4 units to get the graph of f and to the right 2 units to obtain the graph of g. See Figure 6.

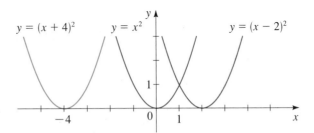

FIGURE 6

EXAMPLE 5 ■ Combining Horizontal and Vertical Shifts

Sketch the graph of the function $f(x) = \sqrt{x - 3} + 4$.

SOLUTION We start with the graph of the square root function $y = \sqrt{x}$ (Example 6 in Section 2.2). We move it to the right 3 units to get the graph of $y = \sqrt{x-3}$. Then we move it upward 4 units to obtain the graph of $f(x) = \sqrt{x-3} + 4$ shown in Figure 7.

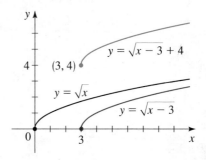

FIGURE 7

Suppose we know the graph of $y = f(x)$. How do we obtain from it the graphs of $y = -f(x)$ and $y = f(-x)$? The y-coordinate of each point on the graph of $y = -f(x)$ is simply the negative of the y-coordinate of the corresponding point on the graph of $y = f(x)$. So the desired graph is the reflection of the graph of $y = f(x)$ in the x-axis. The value of $y = f(-x)$ at x is the same as the value of $y = f(x)$ at $-x$ and so the desired graph is the reflection of the graph of $y = f(x)$ in the y-axis. The following box summarizes these observations.

REFLECTING GRAPHS

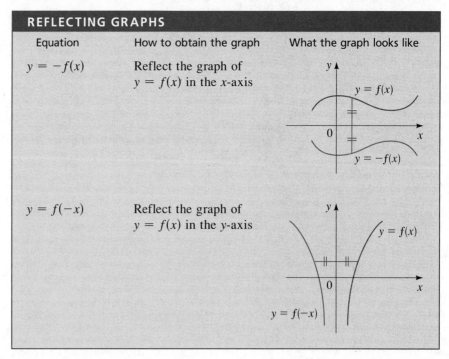

Equation	How to obtain the graph	What the graph looks like
$y = -f(x)$	Reflect the graph of $y = f(x)$ in the x-axis	
$y = f(-x)$	Reflect the graph of $y = f(x)$ in the y-axis	

EXAMPLE 6 ■ Reflecting Graphs

Sketch the graph of each function.

(a) $f(x) = -x^2$ (b) $g(x) = \sqrt{-x}$

SOLUTION

(a) We start with the graph of $y = x^2$. The graph of $f(x) = -x^2$ is the graph of $y = x^2$ reflected in the x-axis (see Figure 8).

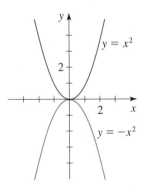

FIGURE 8

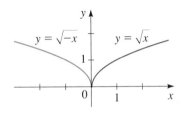

FIGURE 9

(b) We start with the graph of $y = \sqrt{x}$ (Example 6 in Section 2.2). The graph of $g(x) = \sqrt{-x}$ is the graph of $y = \sqrt{x}$ reflected in the y-axis (see Figure 9). Note that the domain of the function $g(x) = \sqrt{-x}$ is $\{x \mid x \leq 0\}$. ∎

Suppose we know the graph of $y = f(x)$. How do we obtain from it the graph of $y = af(x)$? The y-coordinate of $y = af(x)$ at x is the same as the corresponding y-coordinate of $y = f(x)$ multiplied by a. Multiplying the y-coordinates by a has the effect of vertically stretching or shrinking (depending on whether $a > 1$ or $0 < a < 1$) the graph by a factor of a.

VERTICAL STRETCHING AND SHRINKING OF GRAPHS

Equation	How to obtain the graph	What the graph looks like
$y = af(x)$ $(a > 1)$	Stretch the graph of $y = f(x)$ vertically by a factor of a	
$y = af(x)$ $(0 < a < 1)$	Shrink the graph of $y = f(x)$ vertically by a factor of a	

EXAMPLE 7 ■ Vertical Stretching and Shrinking of Graphs

Sketch the graph of each function.

(a) $f(x) = 3x^2$ (b) $g(x) = \frac{1}{3}x^2$

SOLUTION

(a) We start with the parabola $y = x^2$. The graph of $f(x) = 3x^2$ is the graph of $y = x^2$ stretched vertically by a factor of 3. The result is a narrower parabola as shown in Figure 10(a). To obtain the desired graph, we have multiplied the y-coordinate of each point on the graph of $y = x^2$ by 3.

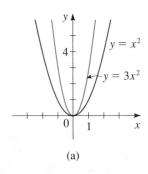

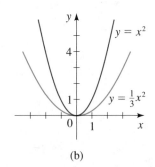

FIGURE 10   (a)   (b)

(b) The graph of $g(x) = \frac{1}{3}x^2$ is the graph of $y = x^2$ shrunk vertically by a factor of $\frac{1}{3}$. The result is a wider parabola as in Figure 10(b). To get the desired graph, we have multiplied the y-coordinate of each point on the graph of $y = x^2$ by $\frac{1}{3}$.   ■

FIGURE 11

EXAMPLE 8 ■ Vertical Stretching, Shrinking, and Reflecting

Given the graph of f in Figure 11, draw the graphs of the following functions.

(a) $y = 2f(x)$ (b) $y = \frac{1}{2}f(x)$

(c) $y = -f(x)$ (d) $y = -2f(x)$

(e) $y = -\frac{1}{2}f(x)$

SOLUTION The graphs are shown in Figure 12. Notice, for instance, that the graph of $y = -2f(x)$ is obtained by stretching the graph of f by a factor of 2 and then reflecting in the x-axis.   ■

We illustrate the effect of combining shifts, reflections, and stretching in the following example.

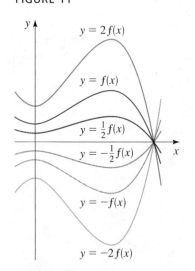

FIGURE 12

EXAMPLE 9 ■ Combining Shifting, Stretching, and Reflecting

Sketch the graph of the function $f(x) = 1 - 2(x - 3)^2$.

SOLUTION Starting with the graph of $y = x^2$, we first shift this graph to the right 3 units to get the graph of $y = (x - 3)^2$. Then we reflect in the x-axis and stretch by a factor of 2 to get the graph of $y = -2(x - 3)^2$. Finally, we shift upward 1 unit to get the graph of $f(x) = 1 - 2(x - 3)^2$ shown in Figure 13.

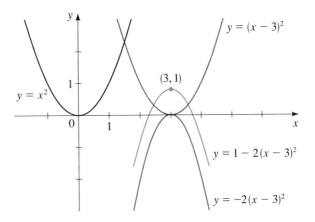

FIGURE 13

Now we consider horizontal shrinking and stretching of graphs. If we know the graph of $y = f(x)$, then how is the graph of $y = f(ax)$ related to it? The y-coordinate of $y = f(ax)$ at x is the same as the y-coordinate of $y = f(x)$ at ax. Thus, the x-coordinates in the graph of $y = f(x)$ correspond to the x-coordinates in the graph of $y = f(ax)$ multiplied by a. Looking at this the other way around, we see that the x-coordinates in the graph of $y = f(ax)$ are the x-coordinates in the graph of $y = f(x)$ multiplied by $1/a$. In other words, to change the graph of $y = f(x)$ to the graph of $y = f(ax)$, we must shrink (or stretch) the graph horizontally by a factor of $1/a$, as summarized in the following box.

HORIZONTAL SHRINKING AND STRETCHING OF GRAPHS

Equation	How to obtain the graph	What the graph looks like
$y = f(ax)$ $(a > 1)$	Shrink the graph of $y = f(x)$ horizontally by a factor of $\dfrac{1}{a}$	
$y = f(ax)$ $(0 < a < 1)$	Stretch the graph of $y = f(x)$ horizontally by a factor of $\dfrac{1}{a}$	

EXAMPLE 10 ■ Horizontal Stretching and Shrinking of Graphs

The graph of $y = f(x)$ is shown in Figure 14. Sketch the graph of each function.

(a) $y = f(2x)$ 　　　　　　　　　　　　(b) $y = f(\frac{1}{2}x)$

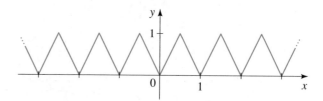

FIGURE 14

$y = f(x)$

SOLUTION　Using the principles described in the preceding box, we obtain the graphs shown in Figures 15 and 16.

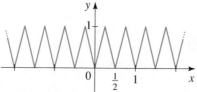

FIGURE 15

$y = f(2x)$

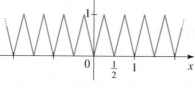

FIGURE 16

$y = f(\frac{1}{2}x)$ ■

EVEN AND ODD FUNCTIONS

If a function f satisfies $f(-x) = f(x)$ for every number x in its domain, then f is called an **even function**. For instance, the function $f(x) = x^2$ is even because

$$f(-x) = (-x)^2 = (-1)^2 x^2 = x^2 = f(x)$$

The geometric significance of an even function is that its graph is symmetric with respect to the y-axis (see Figure 17). This means that if we have plotted the graph of f for $x \geq 0$, then we can obtain the entire graph simply by reflecting this portion in the y-axis.

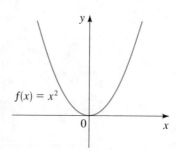

FIGURE 17

$f(x) = x^2$ is an even function.

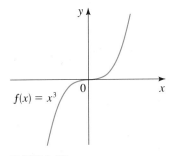

FIGURE 18

$f(x) = x^3$ is an odd function.

If f satisfies $f(-x) = -f(x)$ for every number x in its domain, then f is called an **odd function**. For example, the function $f(x) = x^3$ is odd because

$$f(-x) = (-x)^3 = (-1)^3 x^3 = -x^3 = -f(x)$$

The graph of an odd function is symmetric about the origin (see Figure 18). If we have plotted the graph of f for $x \geq 0$, then we can obtain the entire graph by rotating this portion through $180°$ about the origin. (This is equivalent to reflecting first in the x-axis and then in the y-axis.)

EVEN AND ODD FUNCTIONS

Definition	Symmetry of graph of f	What the graph looks like
f is **even** if $f(-x) = f(x)$ for all x in the domain of f	Graph of f is symmetric with respect to the y-axis	$f(-x) = = f(x)$
f is **odd** if $f(-x) = -f(x)$ for all x in the domain of f	Graph of f is symmetric with respect to the origin	$f(x)$, $f(-x) =$

EXAMPLE 11 ■ **Even and Odd Functions**

Determine whether each of the following functions is even, odd, or neither even nor odd.

(a) $f(x) = x^5 + x$ (b) $g(x) = 1 - x^4$

(c) $h(x) = 2x - x^2$

SOLUTION

(a) $f(-x) = (-x)^5 + (-x)$

$$= -x^5 - x = -(x^5 + x)$$

$$= -f(x)$$

Therefore, f is an odd function.

(b) $g(-x) = 1 - (-x)^4 = 1 - x^4 = g(x)$

So g is even.

(c) $h(-x) = 2(-x) - (-x)^2 = -2x - x^2$

Since $h(-x) \neq h(x)$ and $h(-x) \neq -h(x)$, we conclude that h is neither even nor odd. ■

The graphs of the functions in Example 11 are shown in Figure 19. The graph of f was drawn by plotting points for $x \geq 0$ and rotating $180°$ about the origin. The graph of g was drawn by plotting points for $x \geq 0$ and reflecting in the y-axis. Notice that the graph of h is symmetric neither about the y-axis nor about the origin.

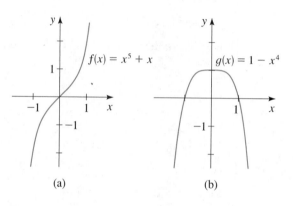

FIGURE 19　　　(a)　　　　　　　　(b)　　　　　　　　(c)

2.4　EXERCISES

1–8 ■ Suppose that the graph of f is given. Describe how the graph of each of the following functions can be obtained from the graph of f.

1. (a) $y = f(x) - 4$　　　(b) $y = f(x - 4)$

2. (a) $y = f(x + 5)$　　　(b) $y = f(x) + 5$

3. (a) $y = 3f(x)$　　　(b) $y = \frac{1}{3}f(x)$

4. (a) $y = -f(x)$　　　(b) $y = f(-x)$

5. (a) $y = -f(x) + 5$　　　(b) $y = f(-x) + 5$

6. (a) $y = -4f(x)$　　　(b) $y = -\frac{1}{4}f(x)$

7. (a) $y = f(x - 2) - 3$　　　(b) $y = 2f(x - 3)$

8. (a) $y = \frac{1}{2}f(x) + 10$　　　(b) $y = \frac{1}{2}f(x + 10)$

9. The graph of f is given. Sketch the graph of each of the following functions.

(a) $y = f(x - 2)$　　　(b) $y = f(x) - 2$

(c) $y = 2f(x)$　　　(d) $y = -f(x) + 3$

(e) $y = f(-x)$　　　(f) $y = \frac{1}{2}f(x - 1)$

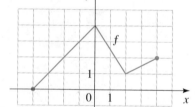

10. The graph of $y = f(x)$ is given. Match each equation with its graph.

(a) $y = f(x - 4)$ (b) $y = f(x) + 3$

(c) $y = \frac{1}{3} f(x)$ (d) $y = -f(x + 4)$

(e) $y = 2f(x + 6)$

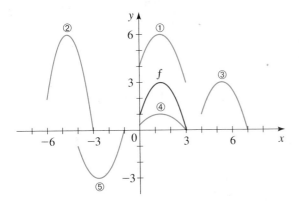

11. (a) Sketch the graph of $f(x) = \dfrac{1}{x}$ by plotting points.

(b) Use the graph of f to sketch the graph of each of the following functions.

(i) $y = -\dfrac{1}{x}$ (ii) $y = \dfrac{1}{x - 1}$

(iii) $y = \dfrac{2}{x + 2}$ (iv) $y = 1 + \dfrac{1}{x - 3}$

12. (a) Sketch the graph of $g(x) = \sqrt[3]{x}$ by plotting points.

(b) Use the graph of g to sketch the graph of each of the following functions.

(i) $y = \sqrt[3]{x - 2}$ (ii) $y = \sqrt[3]{x + 2} + 2$

(iii) $y = 1 - \sqrt[3]{x}$

13–28 ■ Sketch the graph of the function, not by plotting points, but by starting with the graph of a standard function and applying transformations.

13. $f(x) = (x - 2)^2$ **14.** $f(x) = (x + 7)^2$

15. $f(x) = -(x + 1)^2$ **16.** $f(x) = 1 - x^2$

17. $f(x) = x^3 + 2$ **18.** $f(x) = -x^3$

19. $y = 1 + \sqrt{x}$ **20.** $y = 2 - \sqrt{x + 1}$

21. $y = \frac{1}{2}\sqrt{x + 4} - 3$ **22.** $y = 3 - 2(x - 1)^2$

23. $y = 5 + (x + 3)^2$ **24.** $y = \frac{1}{3}x^3 - 1$

25. $y = |x| - 1$ **26.** $y = |x - 1|$

27. $y = |x + 2| + 2$ **28.** $y = 2 - |x|$

29–32 ■ Graph the functions on the same screen using the given viewing rectangle. How is each graph related to the graph in part (a)?

29. Viewing rectangle $[-8, 8]$ by $[-2, 8]$

(a) $y = \sqrt[4]{x}$ (b) $y = \sqrt[4]{x + 5}$

(c) $y = 2\sqrt[4]{x + 5}$ (d) $y = 4 + 2\sqrt[4]{x + 5}$

30. Viewing rectangle $[-8, 8]$ by $[-6, 6]$

(a) $y = |x|$ (b) $y = -|x|$

(c) $y = -3|x|$ (d) $y = -3|x - 5|$

31. Viewing rectangle $[-4, 6]$ by $[-4, 4]$

(a) $y = x^6$ (b) $y = \frac{1}{3}x^6$

(c) $y = -\frac{1}{3}x^6$ (d) $y = -\frac{1}{3}(x - 4)^6$

32. Viewing rectangle $[-6, 6]$ by $[-4, 4]$

(a) $y = \dfrac{1}{\sqrt{x}}$ (b) $y = \dfrac{1}{\sqrt{x + 3}}$

(c) $y = \dfrac{1}{2\sqrt{x + 3}}$ (d) $y = \dfrac{1}{2\sqrt{x + 3}} - 3$

33. The graphs of $f(x) = x^2 - 4$ and $g(x) = |x^2 - 4|$ are shown. Explain how the graph of g is obtained from the graph of f.

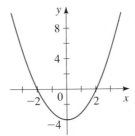
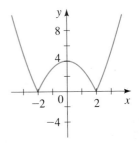

$f(x) = x^2 - 4$ $g(x) = |x^2 - 4|$

34. The graph of $f(x) = x^4 - 4x^2$ is shown. Use this graph to sketch the graph of $g(x) = |x^4 - 4x^2|$.

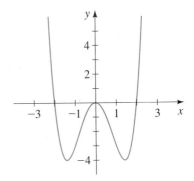

35–36 ■ Sketch the graph of each function.

35. (a) $f(x) = 4x - x^2$ (b) $g(x) = |4x - x^2|$

36. (a) $f(x) = x^3$ (b) $g(x) = |x^3|$

37. The graph of f is given. Use it to graph each of the following functions.

(a) $y = f(2x)$ (b) $y = f\left(\frac{1}{2}x\right)$

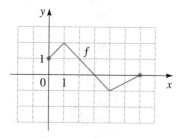

38. The graph of f is given. Use it to graph each of the following functions.

(a) $y = f(2x)$ (b) $y = f\left(\frac{1}{2}x\right)$

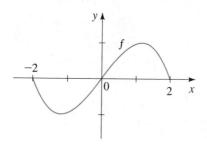

39–40 ■ Using the graph of $f(x) = [\![x]\!]$ described in Section 2.2, graph the indicated function.

39. $y = [\![2x]\!]$ **40.** $y = \left[\!\left[\frac{1}{4}x\right]\!\right]$

 41. If $f(x) = \sqrt{2x - x^2}$, graph the following functions in the viewing rectangle $[-5, 5]$ by $[-4, 4]$. How is each graph related to the graph in part (a)?

(a) $y = f(x)$ (b) $y = f(2x)$ (c) $y = f\left(\frac{1}{2}x\right)$

42. If $f(x) = \sqrt{2x - x^2}$, graph the following functions in the viewing rectangle $[-5, 5]$ by $[-4, 4]$. How is each graph related to the graph in part (a)?

(a) $y = f(x)$ (b) $y = f(-x)$ (c) $y = -f(-x)$

(d) $y = f(-2x)$ (e) $y = f\left(-\frac{1}{2}x\right)$

43–50 ■ Determine whether the function f is even, odd, or neither. If f is even or odd, use symmetry to sketch its graph.

43. $f(x) = x^{-2}$ **44.** $f(x) = x^{-3}$

45. $f(x) = x^2 + x$ **46.** $f(x) = x^4 - 4x^2$

47. $f(x) = x^3 - x$ **48.** $f(x) = 3x^3 + 2x^2 + 1$

49. $f(x) = 1 - \sqrt[3]{x}$ **50.** $f(x) = x + \dfrac{1}{x}$

DISCOVERY · DISCUSSION

51. Sums of Even and Odd Functions   If f and g are both even functions, is $f + g$ necessarily even? If both are odd, is their sum necessarily odd? What can you say about the sum if one is odd and one is even? In each case, prove your answer.

52. Products of Even and Odd Functions   Answer the same questions as in Exercise 51, except this time consider the *product* of f and g instead of the sum.

53. Even and Odd Power Functions   What must be true about the integer n if the function $f(x) = x^n$ is an even function? If it is an odd function? Why do you think the names "even" and "odd" were chosen for these function properties?

2.5   EXTREME VALUES OF FUNCTIONS

An extreme value of a function is the largest or smallest value of the function on some interval. Finding points where functions achieve extreme values is important in applications. For example, if the function represents the profit in a business, we would be interested in the maximum values; if a function represents the amount of material to be used in a manufacturing process, we would be interested in the minimum values. We begin by giving a method of finding extreme values of quadratic functions and then use graphing devices to find extreme values of other functions.

EXTREME VALUES OF QUADRATIC FUNCTIONS

A **quadratic function** is a function f of the form

$$f(x) = ax^2 + bx + c$$

where a, b, and c are real numbers and $a \neq 0$.

In particular, if we take $a = 1$ and $b = c = 0$, we get the simple quadratic function $f(x) = x^2$ whose graph is the parabola that we drew in Example 2 of Section 2.2. In fact, the graph of any quadratic function is a **parabola**; it can be obtained from the graph of $y = x^2$ by the transformations given in Section 2.4.

EXAMPLE 1 ■ Graphing a Quadratic Function

Sketch the graph of the quadratic function $f(x) = -2x^2 + 3$.

SOLUTION As in Section 2.4, we start with the graph of $y = x^2$. We stretch vertically by a factor of 2 to get the narrower parabola $y = 2x^2$. Then we reflect in the x-axis to get the graph of $y = -2x^2$ shown in Figure 1(a). Finally, we shift this graph upward 3 units to obtain the graph of $f(x) = -2x^2 + 3$ shown in Figure 1(b).

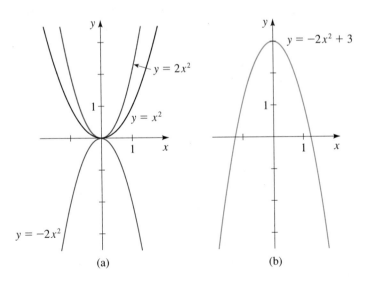

FIGURE 1

Steps in graphing $y = -2x^2 + 3$

(a)   (b)

If $f(x) = ax^2 + bx + c$ with $b \neq 0$, we first have to complete the square to put the function in a more convenient form for graphing.

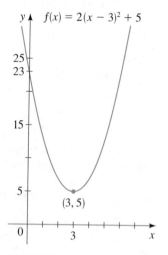

$f(x) = 2(x - 3)^2 + 5$

(3, 5)

FIGURE 2

EXAMPLE 2 ■ Graphing a Quadratic Function by Completing the Square

Sketch the graph of the quadratic function $f(x) = 2x^2 - 12x + 23$.

SOLUTION   Since the coefficient of x^2 is not 1, we must factor this coefficient from the terms involving x before we complete the square.

$$f(x) = 2x^2 - 12x + 23$$
$$= 2(x^2 - 6x) + 23 \qquad \text{Factor 2 from the } x\text{-terms}$$
$$= 2(x^2 - 6x + 9) + 23 - 18 \qquad \begin{array}{l}\text{Complete the square}\\\text{inside the parentheses}\end{array}$$
$$= 2(x - 3)^2 + 5$$

(When we add 9 inside the parentheses, we actually add $2 \cdot 9 = 18$ because every term inside the parentheses is multiplied by 2. So, to preserve equality we subtract 18 outside the parentheses.) This form of the function tells us that we get the graph of f by taking the parabola $y = x^2$, shifting it to the right 3 units, stretching it by a factor of 2, and moving it upward 5 units. Notice that the vertex is at $(3, 5)$ and the parabola opens upward. We sketch the graph in Figure 2 after noting that the y-intercept is $f(0) = 23$.   ■

If we start with a general quadratic function $f(x) = ax^2 + bx + c$ and complete the square as in Example 2, we arrive at an expression in the standard form $f(x) = a(x - h)^2 + k$. We know from Section 2.4 that the graph of this function is obtained from the graph of $y = x^2$ by a horizontal shift, a vertical stretch, and then a vertical shift. The vertex of the resulting parabola is the point (h, k). If $a > 0$, the parabola opens upward, as shown in Figure 3. But if $a < 0$, then a reflection in the x-axis is also involved, so the parabola opens downward as in Figure 4.

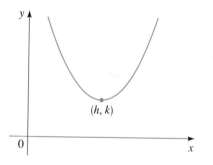

FIGURE 3
$f(x) = a(x - h)^2 + k$,
$a > 0, h > 0, k > 0$

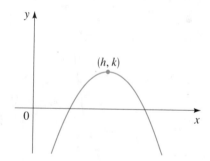

FIGURE 4
$f(x) = a(x - h)^2 + k$,
$a < 0, h > 0, k > 0$

Observe from Figure 3 that if $a > 0$, then the lowest point on the parabola is the vertex (h, k), so the minimum value of the function occurs when $x = h$ and this **minimum value** is $f(h) = k$. Even without the picture we could note that

$(x - h)^2 \geq 0$ for all x, so $a(x - h)^2 \geq 0$ (since $a > 0$) and therefore

$$f(x) = a(x - h)^2 + k \geq k \qquad \text{for all } x$$

and $f(h) = k$. Similarly, if $a < 0$, then the highest point on the parabola is (h, k), so the **maximum value** of f is $f(h) = k$. We summarize this discussion in the following box.

STANDARD FORM OF A QUADRATIC FUNCTION

A quadratic function $f(x) = ax^2 + bx + c$ can be expressed in the **standard form**

$$f(x) = a(x - h)^2 + k$$

by completing the square. The graph of f is a parabola with vertex (h, k); the parabola opens upward if $a > 0$ or opens downward if $a < 0$.

If $a > 0$, then the **minimum value** of f occurs at $x = h$ and this value is $f(h) = k$.

If $a < 0$, then the **maximum value** of f occurs at $x = h$ and this value is $f(h) = k$.

EXAMPLE 3 ■ Expressing a Quadratic Function in Standard Form

Consider the quadratic function $f(x) = 5x^2 - 30x + 49$.
(a) Express f in standard form.
(b) Sketch the graph of f.
(c) Find the minimum value of f.

SOLUTION

(a) Completing the square, we have

$$f(x) = 5x^2 - 30x + 49$$

$$= 5(x^2 - 6x) + 49 \qquad \text{Factor 5 from the } x\text{-terms}$$

$$= 5(x^2 - 6x + 9) + 49 - 45 \qquad \text{Complete the square inside the parentheses}$$

$$= 5(x - 3)^2 + 4$$

(When we add 9 inside the parentheses, we actually add $5 \cdot 9 = 45$ because every term inside the parentheses is multiplied by 5. So, we subtract 45 outside the parentheses to preserve equality.)

(b) The graph is a parabola that has its vertex at $(3, 4)$ and opens upward, as sketched in Figure 5.

(c) Since the coefficient of x^2 is positive, f has a minimum value. The minimum value is $f(3) = 4$. ■

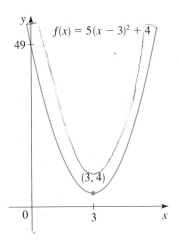

$f(x) = 5(x - 3)^2 + 4$

$(3, 4)$

FIGURE 5

EXAMPLE 4 ■ **Expressing a Quadratic Function in Standard Form**

Consider the quadratic function $f(x) = -x^2 + x + 2$.
(a) Express f in standard form.
(b) Sketch the graph of f.
(c) Find the maximum value of f.

SOLUTION

(a) To express this quadratic function in standard form, we complete the square.

$$y = -x^2 + x + 2$$

$$= -(x^2 - x) + 2 \qquad \text{Factor } -1 \text{ from the } x\text{-terms}$$

$$= -(x^2 - x + \tfrac{1}{4}) + 2 + \tfrac{1}{4} \qquad \text{Complete the square inside the parentheses}$$

$$= -(x - \tfrac{1}{2})^2 + \tfrac{9}{4}$$

[To complete the square for $x^2 - x$, we add $(-\tfrac{1}{2})^2 = \tfrac{1}{4}$; but adding $\tfrac{1}{4}$ inside the parentheses is actually adding $-\tfrac{1}{4}$ because every term inside the parentheses is multiplied by -1. So, we must add $\tfrac{1}{4}$ outside the parentheses to preserve equality.]

(b) From the standard form we see that the graph is a parabola that opens downward and has vertex $(\tfrac{1}{2}, \tfrac{9}{4})$. As an aid to sketching the graph, we find the intercepts. The y-intercept is $f(0) = 2$. To find the x-intercepts, we set $f(x) = 0$ and factor the resulting equation.

$$-x^2 + x + 2 = 0$$

$$-(x^2 - x - 2) = 0$$

$$-(x - 2)(x + 1) = 0$$

Thus, the x-intercepts are $x = 2$ and $x = -1$. The graph of f is sketched in Figure 6.

(c) Since the coefficient of x^2 is negative, f has a maximum value, which is $f(\tfrac{1}{2}) = \tfrac{9}{4}$. ■

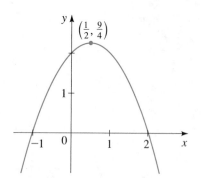

FIGURE 6
Graph of $f(x) = -x^2 + x + 2$

Expressing a quadratic function in standard form helps us sketch its graph as well as find its maximum or minimum value. If we are interested only in finding the maximum or minimum value, then a formula is available for doing so. This formula is obtained by completing the square for the general quadratic function as follows:

$$f(x) = ax^2 + bx + c$$

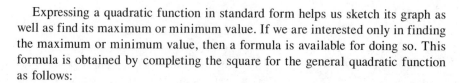

$$= a\left(x^2 + \frac{b}{a}x\right) + c \qquad \text{Factor } a \text{ from the } x\text{-terms}$$

$$= a\left(x^2 + \frac{b}{a}x + \frac{b^2}{4a^2}\right) + c - \frac{b^2}{4a} \qquad \text{Complete the square}$$
$$\text{inside the parentheses}$$

$$= a\left(x + \frac{b}{2a}\right)^2 + c - \frac{b^2}{4a}$$

This equation is in standard form with $h = -b/(2a)$ and $k = c - b^2/(4a)$. Since the maximum or minimum value occurs at $x = h$, we have the following result.

MAXIMUM OR MINIMUM VALUE OF A QUADRATIC FUNCTION

The maximum or minimum value of a quadratic function
$f(x) = ax^2 + bx + c$ occurs at

$$x = -\frac{b}{2a}$$

If $a > 0$, then the **minimum value** is $f\left(-\frac{b}{2a}\right)$.

If $a < 0$, then the **maximum value** is $f\left(-\frac{b}{2a}\right)$.

EXAMPLE 5 ■ **Finding Maximum and Minimum Values of Quadratic Functions**

Find the maximum or minimum value of each quadratic function.

(a) $f(x) = x^2 + 4x$　　　　　　　(b) $g(x) = -2x^2 + 4x - 5$

SOLUTION

(a) This is a quadratic function with $a = 1$ and $b = 4$. Thus, the maximum or minimum value occurs at

$$x = -\frac{b}{2a} = -\frac{4}{2 \cdot 1} = -2$$

Since $a > 0$, this quadratic function has the minimum value

$$f(-2) = (-2)^2 + 4(-2) = -4$$

(b) This is a quadratic function with $a = -2$ and $b = 4$. Thus, the maximum or minimum value occurs at

$$x = -\frac{b}{2a} = -\frac{4}{2 \cdot (-2)} = 1$$

Since $a < 0$, the function has the maximum value

$$f(1) = -2(1)^2 + 4(1) - 5 = -3$$

■

APPLIED MAXIMUM AND MINIMUM PROBLEMS

See the Guidelines for Setting Up
Applied Functions on page 153.

In solving applied maximum and minimum problems, we use the steps for setting up applied problems that we discussed in Section 2.3.

EXAMPLE 6 ■ The Largest Product of Two Numbers with a Given Sum

Among all pairs of numbers whose sum is 100, find a pair whose product is as large as possible.

SOLUTION   Let the numbers be x and y. We know that $x + y = 100$, so $y = 100 - x$. Thus, their product is

> Introduce notation
> Relate the quantities x, y, and P
> Eliminate y

$$P = xy = x(100 - x) = -x^2 + 100x$$

We must therefore find the maximum value of the function

$$P(x) = -x^2 + 100x$$

This is a quadratic function with $a = -1$ and $b = 100$, so the maximum occurs when

$$x = -\frac{b}{2a} = -\frac{100}{2(-1)} = 50$$

When $x = 50$, then $y = 100 - 50 = 50$, so the two numbers are 50 and 50. [Note that the maximum value of P is $P(50) = 50(100 - 50) = 2500$.]   ■

EXAMPLE 7 ■ Maximizing the Area of a Fenced Field

A farmer wants to enclose a rectangular field by a fence and divide it into two smaller rectangular fields by constructing another fence parallel to one side of the field. He has 3000 yards of fencing. Find the dimensions of the field so that the total enclosed area is a maximum.

> Introduce notation

SOLUTION   Let x and y be the dimensions of the field (in yards) and draw a diagram as in Figure 7.

> Draw a diagram

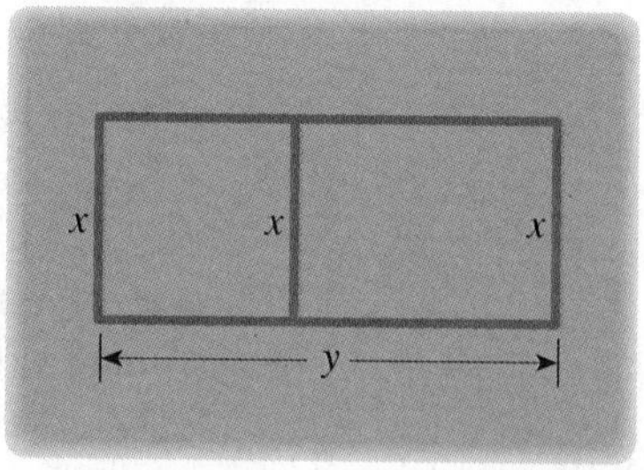

FIGURE 7

Then the area of the field is

> Relate the quantities x, y, and A
> Eliminate y

$$A = xy$$

We eliminate y by using the fact that the total length of the fencing is 3000.

Therefore

$$x + x + x + y + y = 3000$$

$$3x + 2y = 3000$$

$$2y = 3000 - 3x$$

$$y = 1500 - \tfrac{3}{2}x$$

This enables us to express A in terms of x alone:

$$A = xy$$

$$= x\left(1500 - \tfrac{3}{2}x\right)$$

$$= 1500x - \tfrac{3}{2}x^2$$

This is a quadratic function with $a = -\tfrac{3}{2}$ and $b = 1500$, so its maximum value occurs when

$$x = -\frac{b}{2a} = -\frac{1500}{2\left(-\frac{3}{2}\right)} = 500$$

When $x = 500$, then $y = 1500 - \tfrac{3}{2}(500) = 750$. Thus, the field should be 500 yd by 750 yd. ■

EXAMPLE 8 ■ Maximizing Revenue from Ticket Sales

A hockey team plays in an arena with a seating capacity of 15,000 spectators. With the ticket price set at $12, average attendance at recent games has been 11,000. A market survey indicates that for each dollar the ticket price is lowered, the average attendance will increase by 1000. What price should the owners of the team set as the ticket price to maximize their revenue from ticket sales?

Introduce notation

SOLUTION Let x be the selling price of the ticket. Then $12 - x$ is the amount the ticket price has been lowered, so the number of tickets sold is

$$11{,}000 + 1000(12 - x) = 23{,}000 - 1000x$$

Relate the quantities R and x

The revenue is

$$R(x) = x(23{,}000 - 1000x)$$

$$= -1000x^2 + 23{,}000x$$

This is a quadratic function with $a = -1000$ and $b = 23{,}000$, so the maximum value occurs when

$$x = -\frac{b}{2a} = -\frac{23{,}000}{2(-1000)} = 11.5$$

This shows that the revenue is maximized when $x = 11.5$. The owners should set the ticket price at $11.50. ■

USING GRAPHING DEVICES TO FIND EXTREME VALUES

The methods we have discussed apply to finding extreme values of quadratic functions only. We now show how to locate extreme values of any function that can be graphed with a calculator or computer.

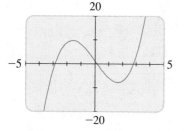

20

−5   **5**

−20

FIGURE 8

Graph of $f(x) = x^3 - 8x + 1$

EXAMPLE 9 ■ Peaks and Valleys of a Graph

Draw the graph of the function $f(x) = x^3 - 8x + 1$ in the viewing rectangle $[-5, 5]$ by $[-20, 20]$. Comment on the peaks and valleys of the graph.

SOLUTION   The graph is shown in Figure 8. A high point of the graph occurs for a value of x between -3 and 0. In fact, if we restrict our attention to the viewing rectangle $[-3, 0]$ by $[0, 15]$, as in Figure 9(a), we see that there is a *highest* point on the restricted graph. Similarly, when we confine our attention to the viewing rectangle $[0, 3]$ by $[-15, 0]$ in Figure 9(b), we see that there is also a *lowest* point on the restricted graph.

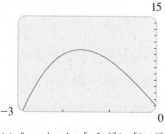

15

−3

0

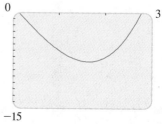

0   **3**

−15

FIGURE 9   (a) f restricted to $[-3, 0]$ by $[0, 15]$ (b) f restricted to $[0, 3]$ by $[-15, 0]$ ■

In general, if there is a viewing rectangle such that the point $(a, f(a))$ is the highest point on the graph of f *within* the viewing rectangle (not on the edge), then the number $f(a)$ is called a **local maximum value** of f (see Figure 10). Notice that $f(a) \geq f(x)$ for all numbers x that are close to a.

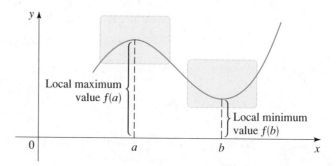

FIGURE 10

Similarly, if there is a viewing rectangle such that the point $(b, f(b))$ is the lowest point on the graph of f within the viewing rectangle, then the number $f(b)$ is called a **local minimum value** of f. In this case, $f(b) \leq f(x)$ for all numbers x that are close to b.

EXAMPLE 10 ■ Finding Local Maxima and Minima from a Graph

Find the approximate local maximum and minimum values of the function $f(x) = x^3 - 8x + 1$ of Example 9.

SOLUTION   Let's first find the local maximum value. Figure 9(a) shows that the maximum point lies in the rectangle $[-2, -1]$ by $[9, 10]$, so we draw the graph restricted to this viewing rectangle in Figure 11(a). By moving the cursor close to the maximum point, we see that the y-coordinates don't change very much in the vicinity of the maximum. The maximum value is about 9.7, and since the distance between the scale marks is 0.1, this estimate for the maximum value is accurate to within 0.1.

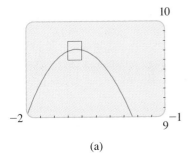
(a)

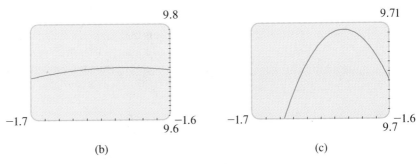
(b)   (c)

FIGURE 11

Steps in finding the local maximum of $f(x) = x^3 - 8x + 1$

If we want an even more accurate estimate, we can zoom in to a smaller rectangle containing the maximum point. Figure 11(b) shows the graph restricted to the viewing rectangle $[-1.7, -1.6]$ by $[9.6, 9.8]$. The curve in Figure 11(b) looks quite flat. It is easier to pinpoint the maximum value if we use a viewing rectangle that is much wider than tall. So we use the viewing rectangle $[-1.7, -1.6]$ by $[9.7, 9.71]$ in Figure 11(c). By moving the cursor along the curve and observing how the y-coordinates change, we can estimate the maximum value quite accurately. It appears that the local maximum value is about 9.709, and this value occurs when x is about -1.633.

We locate the local minimum in a similar fashion. By zooming in to the rectangles shown in Figure 12, we find that the local minimum value is about -7.709, and this value occurs when $x \approx 1.633$.

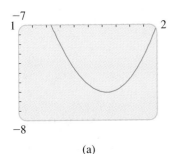
(a)

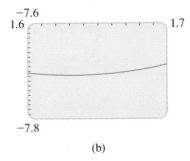
(b)

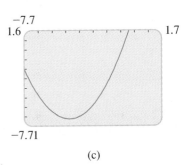
(c)

FIGURE 12 Steps in finding the local minimum of $f(x) = x^3 - 8x + 1$ ■

Example 11 shows how to make a 1-L can with the smallest amount of metal. This knowledge may not be very useful if you are making just one can. But a manufacturer making millions of cans may save a substantial sum of money by using the optimal shape.

EXAMPLE 11 ■ Minimizing Material Used in Manufacturing a Can

A can is to be made that will hold 1 L of oil. Find the value of the radius that minimizes the amount of metal required to manufacture the can.

SOLUTION   In order to minimize the amount of the metal, we minimize the total surface area of the can. In Example 3 in Section 2.3 we found the surface area to be

$$S = 2\pi r^2 + \frac{2000}{r}, \qquad r > 0$$

where r is the radius of the can (in cm). The graph of S as a function of r is shown in Figure 13(a). We see that the minimum point occurs in the rectangle $[5, 6]$ by $[500, 600]$, so we draw the graph in that viewing rectangle in Figure 13(b). But that graph is very flat, so we change the view to $[5, 6]$ by $[550, 560]$ in Figure 13(c). Moving the cursor near the minimum point and observing the y-coordinates (or S-coordinates in this case), we see that the minimum value of S is about 554, and this value occurs when the radius is about 5.4 cm.

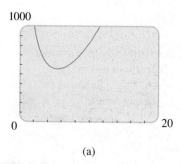

(a)

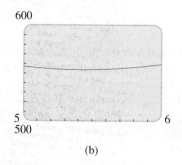

(b)

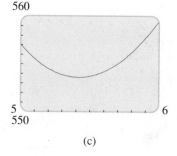

(c)

FIGURE 13

2.5 EXERCISES

1–12 ■ Sketch the graph of the given parabola and state the coordinates of its vertex and its intercepts.

1. $y = x^2 - 8$

2. $y = 4 - x^2$

3. $y = -x^2 - 2$

4. $y = x^2 + 6x$

5. $y = 2x^2 - 6x$

6. $y = x^2 - 2x + 2$

7. $y = x^2 + 6x + 8$

8. $y = -x^2 - 4x + 4$

9. $y = 2x^2 + 4x + 3$

10. $y = -3x^2 + 6x - 2$

11. $y = 2x^2 - 20x + 57$

12. $y = 2x^2 + x - 6$

13–22 ■ A quadratic function is given.
(a) Express the quadratic function in standard form.
(b) Sketch its graph.
(c) Find its maximum or minimum value.

13. $f(x) = 2x - x^2$

14. $f(x) = x + x^2$

15. $f(x) = x^2 + 2x - 1$

16. $f(x) = x^2 - 8x + 8$

17. $f(x) = -x^2 - 3x + 3$

18. $f(x) = 1 - 6x - x^2$

19. $g(x) = 3x^2 - 12x + 13$

20. $g(x) = 2x^2 + 8x + 11$

21. $h(x) = 1 - x - x^2$ **22.** $h(x) = 3 - 4x - 4x^2$

23–30 ■ Find the maximum or minimum value of the function.

23. $f(x) = x^2 + x + 1$ **24.** $f(x) = 1 + 3x - x^2$

25. $f(t) = 100 - 49t - 7t^2$ **26.** $f(t) = 10t^2 + 40t + 113$

27. $f(s) = s^2 - 1.2s + 16$ **28.** $g(x) = 100x^2 - 1500x$

29. $h(x) = \frac{1}{2}x^2 + 2x - 6$ **30.** $f(x) = -\dfrac{x^2}{3} + 2x + 7$

31. Find a function whose graph is a parabola with vertex $(1, -2)$ and that passes through the point $(4, 16)$.

32. Find a function whose graph is a parabola with vertex $(3, 4)$ and that passes through the point $(1, -8)$.

33–34 ■ Find the domain and range of the function.

33. $f(x) = -x^2 + 4x - 3$ **34.** $f(x) = x^2 - 2x - 3$

35. If a ball is thrown directly upward into the air with a velocity of 40 ft/s, its height (in feet) after t seconds is given by $y = 40t - 16t^2$. What is the maximum height attained by the ball?

36. The effectiveness of a television commercial depends on how many times a viewer watches it. After some experiments, an advertising agency found that if the effectiveness E is measured on a scale of 0 to 10, then

$$E(n) = \tfrac{2}{3}n - \tfrac{1}{90}n^2$$

where n is the number of times a viewer watches a given commercial. For a commercial to have maximum effectiveness, how many times should a viewer watch it?

37. Find two numbers whose difference is 100 and whose product is as small as possible.

38. Find two positive numbers whose sum is 100 and the sum of whose squares is a minimum.

39. Find two numbers whose sum is -24 and whose product is a maximum.

40. Among all rectangles that have a perimeter of 20 ft, find the dimensions of the one with the largest area.

41. A farmer has 2400 ft of fencing and wants to fence off a rectangular field that borders a straight river. He needs no fence along the river. (See the figure for Exercise 23 in Section 2.3.) What are the dimensions of the field that has the largest area?

42. Find the area of the largest rectangle that can be inscribed in a right triangle with legs of lengths 3 cm and 4 cm if two sides of the rectangle lie along the legs as in the figure.

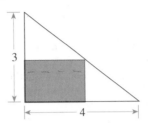

43. A farmer with 750 ft of fencing wants to enclose a rectangular area and then divide it into four pens with fencing parallel to one side of the rectangle (see the figure). What is the largest possible total area of the four pens?

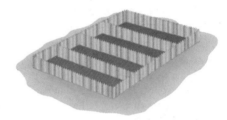

44. A student makes and sells necklaces at the beach during the summer months. The material for each necklace costs her $6 and she has been selling about 20 per day at $10 each. She has been wondering whether or not to raise the price, so she takes a survey and finds that for every dollar increase she would lose two sales a day. What price should she set for the necklaces to maximize her profit?

45–46 ■ A quadratic function is given.
(a) Use a graphing calculator or computer to find the maximum or minimum value of the quadratic function f, correct to two decimal places.
(b) Find the exact maximum or minimum value of f, and compare with your answer to part (a).

45. $f(x) = x^2 + 1.79x - 3.21$

46. $f(x) = 1 + x - \sqrt{2}x^2$

47–54 ■ Find the local maximum and minimum values of the function and the value of x at which each occurs. State each answer correct to two decimal places.

47. $f(x) = x^3 - x$ **48.** $f(x) = 3 + x + x^2 - x^3$

49. $g(x) = x^4 - 2x^3 - 11x^2$

50. $g(x) = x^5 - 8x^3 + 20x$

51. $U(x) = x\sqrt{6 - x}$ **52.** $U(x) = x\sqrt{x - x^2}$

53. $V(x) = \dfrac{1 - x^2}{x^3}$ **54.** $V(x) = \dfrac{1}{x^2 + x + 1}$

55. For a fish swimming at a speed v relative to the water, the energy expenditure per unit of time is proportional to v^3. Exercise 29 in Section 2.3 stated that if a fish is swimming against a current of 5 mi/h, then the total energy required to travel a distance of 10 mi is

$$E(v) = 2.73v^3 \frac{10}{v - 5}$$

Biologists believe that migrating fish try to minimize the total energy required to swim a fixed distance. Find the value of v that minimizes energy required.

NOTE: This result has been verified; migrating fish swim against a current at a speed 50% greater than the speed of the current.

56. A highway engineer wants to estimate the maximum number of cars that can safely travel a particular highway at a given speed. She assumes that each car is 17 ft long, travels at a speed s, and follows the car in front of it at the "safe following distance" for that speed. We saw in Exercise 28 in Section 2.3 that the number N of cars per minute that pass a given point is a function of the speed given by

$$N(s) = \frac{88s}{17 + 17\left(\dfrac{s}{20}\right)^2}$$

At what speed can the greatest number of cars travel the highway safely?

57. An open box with a square base must have a volume of 12 ft³. Find the dimensions of the box that minimize the amount of material used. (See Exercise 19 in Section 2.3.)

58. An open box is to be constructed from a rectangular piece of cardboard with dimensions 12 in. by 20 in. by cutting out equal squares of side x at each corner and then folding up the sides. (See the figure in Exercise 22 in Section 2.3.) Find the largest volume that such a box can have.

59. A Norman window has the shape of a rectangle surmounted by a semicircle. (See the figure in Exercise 21 in Section 2.3.) If the perimeter of the window is 30 ft,

find the dimensions of the window so that the greatest possible amount of light is admitted.

60. Find the area of the largest rectangle that can be inscribed in a semicircle of radius 1 ft. (See the figure in Exercise 24 in Section 2.3.)

61. Find the dimensions of the rectangle of largest area that has its base on the x-axis and its other two vertices above the x-axis, lying on the parabola $y = 8 - x^2$.

62. A man is standing at a point A on the bank of a straight river, 2 mi wide, and he wants to reach point B, 7 mi downstream on the opposite bank, by first rowing his boat to a point P on the opposite bank and then walking the remaining distance x to B. (See the figure in Exercise 26 in Section 2.3.) He can row at a speed of 2 mi/h and walk at a speed of 5 mi/h. Where should he land in order to reach B as soon as possible?

63. A kite frame is to be made from six pieces of wood. The four pieces that form the border of the kite have been cut to the lengths indicated in the figure. Let x be as in the figure.

  (a) Show that the area of the kite is given by the function $A(x) = x\left(\sqrt{25 - x^2} + \sqrt{144 - x^2}\right)$.

  (b) How long should each of the two crosspieces be to maximize the area of the kite?

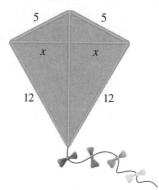

64. Find the maximum value of the function
$f(x) = 3 + 4x^2 - x^4$.   [*Hint:* Let $t = x^2$.]

65. Find a function whose graph is a parabola that passes through the points $(1, -1)$, $(-1, -3)$, and $(3, 9)$.

◆ **DISCOVERY · DISCUSSION**

66. Maxima and Minima   In Example 11 we saw a real-world situation in which it was important to be able to find the minimum value of a function. Think of several other everyday situations in which a maximum value or a minimum value is important.

2.6 | COMBINING FUNCTIONS

Two functions f and g can be combined to form new functions $f + g, f - g, fg,$ and f/g in a manner similar to the way we add, subtract, multiply, and divide real numbers. For example, we define the function $f + g$ by

$$(f + g)(x) = f(x) + g(x)$$

The sum of f and g is defined by

$$(f + g)(x) = f(x) + g(x)$$

The name of the new function is "$f + g$." So this $+$ sign stands for the operation of addition of *functions*. The $+$ sign on the right side, however, stands for addition of the *numbers* $f(x)$ and $g(x)$.

The new function $f + g$ is called the **sum** of the functions f and g; its value at x is $f(x) + g(x)$. Of course, the sum on the right-hand side makes sense only if both $f(x)$ and $g(x)$ are defined, that is, if x belongs to the domain of f and also to the domain of g. So, if the domain of f is A and the domain of g is B, then the domain of $f + g$ is the intersection of these domains, that is, $A \cap B$. Similarly, we can define the **difference** $f - g$, the **product** fg, and the **quotient** f/g of the functions f and g. Their domains are $A \cap B$, but in the case of the quotient we must remember not to divide by 0.

ALGEBRA OF FUNCTIONS

Let f and g be functions with domains A and B. Then the functions $f + g, f - g, fg,$ and f/g are defined as follows.

$$(f + g)(x) = f(x) + g(x) \qquad \text{Domain } A \cap B$$

$$(f - g)(x) = f(x) - g(x) \qquad \text{Domain } A \cap B$$

$$(fg)(x) = f(x)g(x) \qquad \text{Domain } A \cap B$$

$$\left(\frac{f}{g}\right)(x) = \frac{f(x)}{g(x)} \qquad \text{Domain } \{x \in A \cap B \mid g(x) \neq 0\}$$

EXAMPLE 1 ■ Combining Functions

Let $f(x) = x^3$ and $g(x) = \sqrt{x}$.
(a) Find the functions $f + g, f - g, fg,$ and f/g and their domains.
(b) Find $(f + g)(4)$, $(f - g)(4)$, $(fg)(3)$, and $(f/g)(1)$.

SOLUTION

(a) The domain of f is $\mathbb{R}$ and the domain of g is $\{x \mid x \geq 0\}$. We have

$$(f + g)(x) = f(x) + g(x) = x^3 + \sqrt{x} \qquad \text{Domain } \{x \mid x \geq 0\}$$

$$(f - g)(x) = f(x) - g(x) = x^3 - \sqrt{x} \qquad \text{Domain } \{x \mid x \geq 0\}$$

$$(fg)(x) = f(x)g(x) = x^3\sqrt{x} \qquad \text{Domain } \{x \mid x \geq 0\}$$

$$\left(\frac{f}{g}\right)(x) = \frac{f(x)}{g(x)} = \frac{x^3}{\sqrt{x}} \qquad \text{Domain } \{x \mid x > 0\}$$

Note that in the domain of f/g we exclude 0 because $g(0) = 0$.

(b) $(f + g)(4) = f(4) + g(4) = 4^3 + \sqrt{4} = 66$

$(f - g)(4) = f(4) - g(4) = 4^3 - \sqrt{4} = 62$

$(fg)(3) = f(3)g(3) = 3^3\sqrt{3} = 27\sqrt{3}$

$\left(\dfrac{f}{g}\right)(1) = \dfrac{f(1)}{g(1)} = \dfrac{1^3}{\sqrt{1}} = 1$ ∎

EXAMPLE 2 ∎ Finding Domains of Combinations of Functions

If $f(x) = \sqrt{x}$ and $g(x) = \sqrt{4 - x^2}$, find the functions $f + g, f - g, fg$, and f/g and their domains.

SOLUTION   The domain of $f(x) = \sqrt{x}$ is $[0, \infty)$. The domain of $g(x) = \sqrt{4 - x^2}$ is the interval $[-2, 2]$ (from Example 4 in Section 2.2). The intersection of the domains of f and g is

$$[0, \infty) \cap [-2, 2] = [0, 2]$$

Thus, we have

$(f + g)(x) = \sqrt{x} + \sqrt{4 - x^2}$ Domain $\{x \mid 0 \leqslant x \leqslant 2\}$

$(f - g)(x) = \sqrt{x} - \sqrt{4 - x^2}$ Domain $\{x \mid 0 \leqslant x \leqslant 2\}$

$(fg)(x) = \sqrt{x}\sqrt{4 - x^2} = \sqrt{4x - x^3}$ Domain $\{x \mid 0 \leqslant x \leqslant 2\}$

$\left(\dfrac{f}{g}\right)(x) = \dfrac{\sqrt{x}}{\sqrt{4 - x^2}} = \sqrt{\dfrac{x}{4 - x^2}}$ Domain $\{x \mid 0 \leqslant x < 2\}$

Notice that the domain of f/g is the interval $[0, 2)$ because we must exclude the points where $g(x) = 0$, that is, $x = \pm 2$.   ∎

The graph of the function $f + g$ is obtained from the graphs of f and g by **graphical addition**. This means that we add corresponding y-coordinates as in Figure 1.

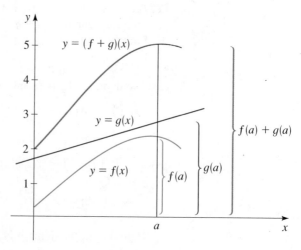

FIGURE 1

Graphical addition

EXAMPLE 3 ■ Using Graphical Addition

Use graphical addition to graph the function $f + g$ from Example 2.

SOLUTION   To graph the function $(f + g)(x) = \sqrt{x} + \sqrt{4 - x^2}$ (from Example 2), we first graph the function $f(x) = \sqrt{x}$ and the function $g(x) = \sqrt{4 - x^2}$ (the top half of the circle $x^2 + y^2 = 4$) in Figure 2. To obtain the graph of $f + g$, we add the y-coordinates of the points on the graphs of f and g. Given the points $Q(x, f(x))$ and $R(x, g(x))$, the point S has coordinates $(x, f(x) + g(x))$. Thus, $d(P, Q) = d(R, S)$.

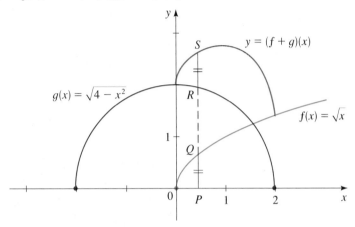

FIGURE 2

COMPOSITION OF FUNCTIONS

We now consider a very important way of combining two functions to get a new function. Suppose $f(x) = \sqrt{x}$ and $g(x) = x^2 + 1$. We may define a function h as

$$h(x) = f(g(x)) = f(x^2 + 1) = \sqrt{x^2 + 1}$$

The function h is made up of the functions f and g in an interesting way: Given a number x we first apply to it the function g, then apply f to the result. In this case, f is the rule "take the square root," g is the rule "square, then add 1," and h is the rule "square, then add 1, then take the square root." In other words, we get the rule h by applying the rule g and then the rule f.

In general, given any two functions f and g, we start with a number x in the domain of g and find its image $g(x)$. If this number $g(x)$ is in the domain of f, then we can calculate the value of $f(g(x))$. The result is a new function $h(x) = f(g(x))$ obtained by substituting g into f. It is called the *composition* (or *composite*) of f and g and is denoted by $f \circ g$ ("f composed with g").

COMPOSITION OF FUNCTIONS

Given two functions f and g, the **composite function** $f \circ g$ (also called the **composition** of f and g) is defined by

$$(f \circ g)(x) = f(g(x))$$

The domain of $f \circ g$ is the set of all x in the domain of g such that $g(x)$ is in the domain of f. In other words, $(f \circ g)(x)$ is defined whenever both $g(x)$ and $f(g(x))$ are defined. The best way to picture $f \circ g$ is by a machine diagram (Figure 3) or an arrow diagram (Figure 4).

FIGURE 3

The $f \circ g$ machine is composed of the g machine (first) and then the f machine.

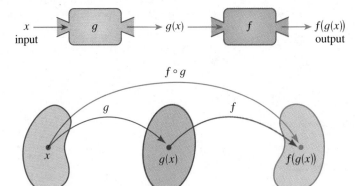

FIGURE 4

Arrow diagram for $f \circ g$

EXAMPLE 4 ■ Finding the Composition of Functions

Let $f(x) = x^2$ and $g(x) = x - 3$.
(a) Find the functions $f \circ g$ and $g \circ f$ and their domains.
(b) Find $(f \circ g)(5)$ and $(g \circ f)(7)$.

SOLUTION

In Example 4, f is the rule "square" and g is the rule "subtract 3." The function $f \circ g$ *first* subtracts 3 and *then* squares; the function $g \circ f$ *first* squares and *then* subtracts 3.

(a) We have

$$(f \circ g)(x) = f(g(x)) \qquad \text{Definition of } f \circ g$$
$$= f(x - 3) \qquad \text{Definition of } g$$
$$= (x - 3)^2 \qquad \text{Definition of } f$$

and

$$(g \circ f)(x) = g(f(x)) \qquad \text{Definition of } g \circ f$$
$$= g(x^2) \qquad \text{Definition of } f$$
$$= x^2 - 3 \qquad \text{Definition of } g$$

The domains of both $f \circ g$ and $g \circ f$ are $\mathbb{R}$.

(b) We have

$$(f \circ g)(5) = f(g(5)) = f(2) = 2^2 = 4$$
$$(g \circ f)(7) = g(f(7)) = g(49) = 49 - 3 = 46$$ ■

You can see from Example 4 that, in general, $f \circ g \neq g \circ f$. Remember that the notation $f \circ g$ means that the function g is applied first and then f is applied second.

EXAMPLE 5 ■ Finding the Composition of Functions

If $f(x) = \sqrt{x}$ and $g(x) = \sqrt{2-x}$, find the following functions and their domains.

(a) $f \circ g$　　(b) $g \circ f$　　(c) $f \circ f$　　(d) $g \circ g$

The graphs of f and g of Example 5, as well as $f \circ g$, $g \circ f$, $f \circ f$, and $g \circ g$ are shown. These graphs indicate that the operation of composition can produce functions quite different from the original functions.

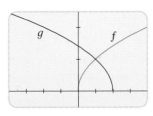

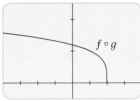

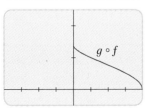

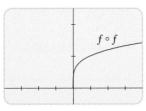

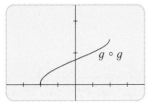

SOLUTION

(a)
$$(f \circ g)(x) = f(g(x)) \qquad \text{Definition of } f \circ g$$
$$= f(\sqrt{2-x}) \qquad \text{Definition of } g$$
$$= \sqrt{\sqrt{2-x}} \qquad \text{Definition of } f$$
$$= \sqrt[4]{2-x}$$

The domain of $f \circ g$ is $\{x \mid 2 - x \geq 0\} = \{x \mid x \leq 2\} = (-\infty, 2]$.

(b)
$$(g \circ f)(x) = g(f(x)) \qquad \text{Definition of } g \circ f$$
$$= g(\sqrt{x}) \qquad \text{Definition of } f$$
$$= \sqrt{2 - \sqrt{x}} \qquad \text{Definition of } g$$

For $\sqrt{x}$ to be defined, we must have $x \geq 0$. For $\sqrt{2 - \sqrt{x}}$ to be defined, we must have $2 - \sqrt{x} \geq 0$, that is, $\sqrt{x} \leq 2$, or $x \leq 4$. Thus, we have $0 \leq x \leq 4$, so the domain of $g \circ f$ is the closed interval $[0, 4]$.

(c)
$$(f \circ f)(x) = f(f(x)) \qquad \text{Definition of } f \circ f$$
$$= f(\sqrt{x}) \qquad \text{Definition of } f$$
$$= \sqrt{\sqrt{x}} \qquad \text{Definition of } f$$
$$= \sqrt[4]{x}$$

The domain of $f \circ f$ is $[0, \infty)$.

(d)
$$(g \circ g)(x) = g(g(x)) \qquad \text{Definition of } g \circ g$$
$$= g(\sqrt{2-x}) \qquad \text{Definition of } g$$
$$= \sqrt{2 - \sqrt{2-x}} \qquad \text{Definition of } g$$

This expression is defined when both $2 - x \geq 0$ and $2 - \sqrt{2-x} \geq 0$. The first inequality means $x \leq 2$, and the second is equivalent to $\sqrt{2-x} \leq 2$, or $2 - x \leq 4$, that is, $x \geq -2$. Thus, $-2 \leq x \leq 2$, so the domain of $g \circ g$ is $[-2, 2]$. ■

It is possible to take the composition of three or more functions. For instance, the composite function $f \circ g \circ h$ is found by first applying h, then g, and then f as follows:

$$(f \circ g \circ h)(x) = f(g(h(x)))$$

EXAMPLE 6 ■ A Composition of Three Functions

Find $f \circ g \circ h$ if $f(x) = x/(x + 1)$, $g(x) = x^{10}$, and $h(x) = x + 3$.

SOLUTION

$$(f \circ g \circ h)(x) = f(g(h(x))) \qquad \text{Definition of } f \circ g \circ h$$

$$= f(g(x + 3)) \qquad \text{Definition of } h$$

$$= f((x + 3)^{10}) \qquad \text{Definition of } g$$

$$= \frac{(x + 3)^{10}}{(x + 3)^{10} + 1} \qquad \text{Definition of } f$$

■

So far we have used composition to build complicated functions from simpler ones. But in calculus it is useful to be able to "decompose" a complicated function into simpler ones, as shown in the following example.

EXAMPLE 7 ■ Recognizing a Composition of Functions

Given $F(x) = \sqrt[4]{x + 9}$, find functions f and g such that $F = f \circ g$.

SOLUTION
Since the formula for F says to first add 9 and then take the fourth root, we let

$$g(x) = x + 9 \qquad \text{and} \qquad f(x) = \sqrt[4]{x}$$

Then $$(f \circ g)(x) = f(g(x))$$

$$= f(x + 9)$$

$$= \sqrt[4]{x + 9}$$

$$= F(x)$$

■

EXAMPLE 8 ■ An Application of Composition of Functions

A ship is traveling at 20 mi/h parallel to a straight shoreline. The ship is 5 mi from shore. It passes a lighthouse at noon.

(a) Express the distance s between the lighthouse and the ship as a function of d, the distance the ship has traveled since noon; that is, find f so that $s = f(d)$.

(b) Express d as a function of t, the time elapsed since noon; that is, find g so that $d = g(t)$.

(c) Find $f \circ g$. What does this function represent?

SOLUTION
We first draw a diagram as in Figure 5.

(a) We can relate the distances s and d by the Pythagorean Theorem. Thus, s can be expressed as a function of d by

$$s = f(d) = \sqrt{25 + d^2}$$

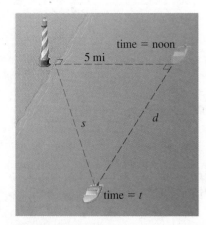

FIGURE 5

(b) Since the ship is traveling at 20 mi/h, the distance d it has traveled is a function of t as follows:

distance = rate × time

$$d = g(t) = 20t$$

(c) We have

$$(f \circ g)(t) = f(g(t)) \qquad \text{Definition of } f \circ g$$

$$= f(20t) \qquad \text{Definition of } g$$

$$= \sqrt{25 + (20t)^2} \qquad \text{Definition of } f$$

The function $f \circ g$ gives the distance of the ship from the lighthouse as a function of time. ■

2.6 EXERCISES

1–6 ■ Find $f + g, f - g, fg,$ and f/g and their domains.

1. $f(x) = x^2 - x, \quad g(x) = x + 5$

2. $f(x) = x^3 + 2x^2, \quad g(x) = 3x^2 - 1$

3. $f(x) = \sqrt{1 + x}, \quad g(x) = \sqrt{1 - x}$

4. $f(x) = \sqrt{9 - x^2}, \quad g(x) = \sqrt{x^2 - 1}$

5. $f(x) = \dfrac{2}{x}, \quad g(x) = -\dfrac{2}{x + 4}$

6. $f(x) = \dfrac{1}{x + 1}, \quad g(x) = \dfrac{x}{x + 1}$

7–8 ■ Find the domain of the function.

7. $F(x) = \dfrac{\sqrt{4 - x} + \sqrt{3 + x}}{x^2 - 1}$

8. $F(x) = \sqrt{1 - x} + \sqrt{x - 2}$

9–10 ■ Use graphical addition to sketch the graph of $f + g$.

9.

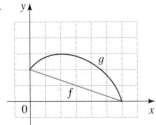

10.

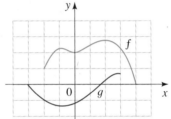

11–14 ■ Draw the graphs of $f, g,$ and $f + g$ on a common screen to illustrate graphical addition.

11. $f(x) = \sqrt{1 + x}, \quad g(x) = \sqrt{1 - x}$

12. $f(x) = x^2, \quad g(x) = \sqrt{x}$

13. $f(x) = x^2, \quad g(x) = x^3$

14. $f(x) = \sqrt[4]{1 - x}, \quad g(x) = \sqrt{1 - \dfrac{x^2}{9}}$

15–20 ■ Use $f(x) = 3x - 5$ and $g(x) = 2 - x^2$ to evaluate the expression.

15. (a) $f(g(0))$ (b) $g(f(0))$

16. (a) $f(f(4))$ (b) $g(g(3))$

17. (a) $(f \circ g)(-2)$ (b) $(g \circ f)(-2)$

18. (a) $(f \circ f)(-1)$ (b) $(g \circ g)(2)$

19. (a) $(f \circ g)(x)$ (b) $(g \circ f)(x)$

20. (a) $(f \circ f)(x)$ (b) $(g \circ g)(x)$

21–26 ■ Use the given graphs of f and g to evaluate the expression.

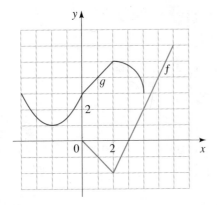

21. $f(g(2))$

22. $g(f(0))$

23. $(g \circ f)(4)$

24. $(f \circ g)(0)$

25. $(g \circ g)(-2)$

26. $(f \circ f)(4)$

27–38 ■ Find the functions $f \circ g$, $g \circ f$, $f \circ f$, and $g \circ g$ and their domains.

27. $f(x) = 2x + 3$, $g(x) = 4x - 1$

28. $f(x) = 6x - 5$, $g(x) = \dfrac{x}{2}$

29. $f(x) = 2x^2 - x$, $g(x) = 3x + 2$

30. $f(x) = \dfrac{1}{x}$, $g(x) = x^3 + 2x$

31. $f(x) = \sqrt{x - 1}$, $g(x) = x^2$

32. $f(x) = x + 2$, $g(x) = \dfrac{1}{x - 2}$

33. $f(x) = \dfrac{1}{x - 1}$, $g(x) = \dfrac{x - 1}{x + 1}$

34. $f(x) = \dfrac{2}{x}$, $g(x) = \dfrac{x}{x + 2}$

35. $f(x) = \sqrt[3]{x}$, $g(x) = 1 - \sqrt{x}$

36. $f(x) = \sqrt{x^2 - 1}$, $g(x) = \sqrt{1 - x}$

37. $f(x) = \dfrac{x + 2}{2x + 1}$, $g(x) = \dfrac{x}{x - 2}$

38. $f(x) = \dfrac{1}{\sqrt{x}}$, $g(x) = x^2 - 4x$

39–42 ■ Find $f \circ g \circ h$.

39. $f(x) = x - 1$, $g(x) = \sqrt{x}$, $h(x) = x - 1$

40. $f(x) = \dfrac{1}{x}$, $g(x) = x^3$, $h(x) = x^2 + 2$

41. $f(x) = x^4 + 1$, $g(x) = x - 5$, $h(x) = \sqrt{x}$

42. $f(x) = \sqrt{x}$, $g(x) = \dfrac{x}{x - 1}$, $h(x) = \sqrt[3]{x}$

43–48 ■ Express the function in the form $f \circ g$.

43. $F(x) = (x - 9)^5$

44. $F(x) = \sqrt{x} + 1$

45. $G(x) = \dfrac{x^2}{x^2 + 4}$

46. $G(x) = \dfrac{1}{x + 3}$

47. $H(x) = |1 - x^3|$

48. $H(x) = \sqrt{1 + \sqrt{x}}$

49–52 ■ Express the function in the form $f \circ g \circ h$.

49. $F(x) = \dfrac{1}{x^2 + 1}$

50. $F(x) = \sqrt[3]{\sqrt{x} - 1}$

51. $G(x) = (4 + \sqrt[3]{x})^9$

52. $G(x) = \dfrac{2}{(3 + \sqrt{x})^2}$

53. A stone is dropped into a lake, creating a circular ripple that travels outward at a speed of 60 cm/s. Express the area of this circle as a function of time t (in seconds).

54. A spherical balloon is being inflated. If the radius of the balloon is increasing at a rate of 1 cm/s, express the volume of the balloon as a function of time t (in seconds).

55. An airplane is flying at a speed of 350 mi/h at an altitude of one mile. The plane passes directly above a radar station at time $t = 0$.

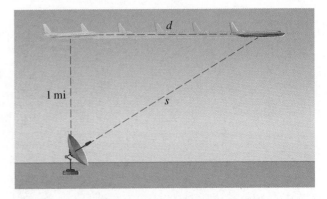

(a) Express the distance s between the plane and the radar station as a function of d.

(b) Express the horizontal distance d (in miles) that the plane has flown as a function of t (in hours).

(c) Use composition to express s as a function of t.

DISCOVERY · DISCUSSION

56. Compound Interest   A savings account earns 5% interest compounded annually. If you invest x dollars in such an account, then the amount $A(x)$ of the investment after one year is the initial investment plus 5%; that is, $A(x) = x + 0.05x = 1.05x$. Find $A \circ A$, $A \circ A \circ A$, and $A \circ A \circ A \circ A$. What do these compositions represent? Find a formula for what you get when you compose n copies of A.

57. Solving an Equation for an Unknown Function   If $g(x) = 2x + 1$ and $h(x) = 4x^2 + 4x + 7$, find a function f such that $f \circ g = h$. (Think about what operations you would have to perform on the formula for g to end up with the formula for h.)

Now suppose that

$$f(x) = 3x + 5 \quad \text{and} \quad h(x) = 3x^2 + 3x + 2$$

Use the same sort of reasoning to find a function g such that $f \circ g = h$.

58. Compositions of Odd and Even Functions   Suppose that $h = f \circ g$. If g is an even function, is h necessarily even? If g is odd, is h odd? What if g is odd and f is odd? What if g is odd and f is even?

2.7 ONE-TO-ONE FUNCTIONS AND THEIR INVERSES

Let's compare the functions f and g whose arrow diagrams are shown in Figure 1. Note that f never takes on the same value twice (any two numbers in A have different images), whereas g does take on the same value twice (both 2 and 3 have the same image, 4). In symbols, $g(2) = g(3)$ but $f(x_1) \neq f(x_2)$ whenever $x_1 \neq x_2$. Functions that have this latter property are called *one-to-one*.

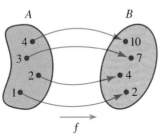
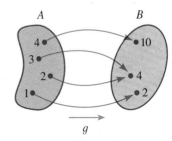

FIGURE 1 f is one-to-one g is not one-to-one

> **DEFINITION OF A ONE-TO-ONE FUNCTION**
>
> A function with domain A is called a **one-to-one function** if no two elements of A have the same image, that is,
>
> $$f(x_1) \neq f(x_2) \quad \text{whenever } x_1 \neq x_2$$

An equivalent way of writing the condition for a one-to-one function is this:

If $f(x_1) = f(x_2)$, then $x_1 = x_2$.

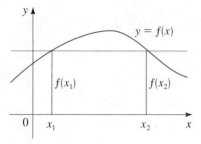

FIGURE 2

This function is not one-to-one because $f(x_1) = f(x_2)$.

If a horizontal line intersects the graph of f in more than one point, then we see from Figure 2 that there are numbers $x_1 \neq x_2$ such that $f(x_1) = f(x_2)$. This means that f is not one-to-one. Therefore, we have the following geometric method for determining whether a function is one-to-one.

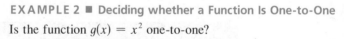

HORIZONTAL LINE TEST

A function is one-to-one if and only if no horizontal line intersects its graph more than once.

EXAMPLE 1 ■ Deciding whether a Function Is One-to-One

Is the function $f(x) = x^3$ one-to-one?

SOLUTION 1 If $x_1 \neq x_2$, then $x_1^3 \neq x_2^3$ (two different numbers cannot have the same cube). Therefore, $f(x) = x^3$ is one-to-one.

SOLUTION 2 From Figure 3 we see that no horizontal line intersects the graph of $f(x) = x^3$ more than once. Therefore, by the Horizontal Line Test, f is one-to-one.

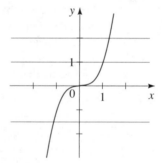

FIGURE 3

$f(x) = x^3$ is one-to-one.

Notice that the function f of Example 1 is increasing and is also one-to-one. In fact, it can be proved that every increasing function and every decreasing function is one-to-one (see Exercise 49).

EXAMPLE 2 ■ Deciding whether a Function Is One-to-One

Is the function $g(x) = x^2$ one-to-one?

SOLUTION 1 This function is not one-to-one because, for instance,

$$g(1) = 1 = g(-1)$$

and so 1 and -1 have the same image.

SOLUTION 2 From Figure 4 we see that there are horizontal lines that intersect the graph of g more than once. Therefore, by the Horizontal Line Test, g is not one-to-one.

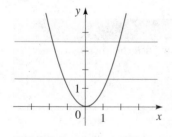

FIGURE 4

$g(x) = x^2$ is not one-to-one.

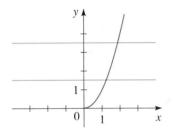

FIGURE 5
$h(x) = x^2 \ (x \geq 0)$ is one-to-one.

Although the function g in Example 2 is not one-to-one, it is possible to restrict its domain so that the resulting function is one-to-one. In fact, if we define

$$h(x) = x^2, \qquad x \geq 0$$

then h is one-to-one, as you can see from Figure 5 and the Horizontal Line Test.

EXAMPLE 3 ■ **Showing that a Function Is One-to-One**

Show that the function $f(x) = 3x + 4$ is one-to-one.

SOLUTION   Suppose there are numbers x_1 and x_2 such that $f(x_1) = f(x_2)$. Then

$$3x_1 + 4 = 3x_2 + 4$$

$$3x_1 = 3x_2$$

$$x_1 = x_2$$

Therefore, f is one-to-one.   ■

INVERSE FUNCTIONS

One-to-one functions are important because they are precisely the functions that possess inverse functions according to the following definition.

⊘ Don't mistake the -1 in f^{-1} for an exponent.

$$f^{-1} \quad does \ not \ mean \quad \frac{1}{f(x)}$$

The reciprocal $1/f(x)$ is written as $[f(x)]^{-1}$.

> **DEFINITION OF INVERSE FUNCTION**
>
> Let f be a one-to-one function with domain A and range B. Then its **inverse function** f^{-1} has domain B and range A and is defined by
>
> $$f^{-1}(y) = x \iff f(x) = y$$
>
> for any y in B.

This definition says that if f takes x into y, then f^{-1} takes y back into x. (If f were not one-to-one, then f^{-1} would not be defined uniquely.) The arrow diagram in Figure 6 indicates that f^{-1} reverses the effect of f. From the definition, we have

$$\text{domain of } f^{-1} = \text{range of } f$$

$$\text{range of } f^{-1} = \text{domain of } f$$

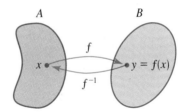

FIGURE 6

EXAMPLE 4 ■ Finding f^{-1} for Specific Values

If $f(1) = 5$, $f(3) = 7$, and $f(8) = -10$, find $f^{-1}(5)$, $f^{-1}(7)$, and $f^{-1}(-10)$.

SOLUTION From the definition of f^{-1}, we have

$$f^{-1}(5) = 1 \qquad \text{because} \qquad f(1) = 5$$
$$f^{-1}(7) = 3 \qquad \text{because} \qquad f(3) = 7$$
$$f^{-1}(-10) = 8 \qquad \text{because} \qquad f(8) = -10$$

The diagram in Figure 7 makes it clear how f^{-1} reverses the effect of f in this case.

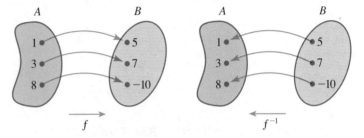

FIGURE 7

By its definition, the inverse function f^{-1} undoes what f does: If we start with x, apply f, and then apply f^{-1}, we arrive back at x, where we started. Similarly, f undoes what f^{-1} does. In general, any function that reverses the effect of f in this way must be the inverse of f. These observations are expressed precisely as follows.

PROPERTY OF INVERSE FUNCTIONS

Let f be a one-to-one function with domain A and range B. The inverse function f^{-1} satisfies the following cancellation properties.

$$f^{-1}(f(x)) = x \qquad \text{for every } x \text{ in } A$$
$$f(f^{-1}(x)) = x \qquad \text{for every } x \text{ in } B$$

Conversely, any function f^{-1} satisfying these equations is the inverse of f.

These properties indicate that f is the inverse function of f^{-1}, so we often say that f and f^{-1} are *inverses of each other*.

EXAMPLE 5 ■ Verifying that Two Functions Are Inverses

Show that $f(x) = x^3$ and $g(x) = x^{1/3}$ are inverses of each other.

SOLUTION Note that the domain and range of both f and g is $\mathbb{R}$. We have

$$g(f(x)) = g(x^3) = (x^3)^{1/3} = x$$
$$f(g(x)) = f(x^{1/3}) = (x^{1/3})^3 = x$$

This shows that f and g are inverses of each other. These equations simply say that the cube function and the cube root function, when composed, cancel each other.   ∎

We now explain how to compute inverse functions. We first observe from the definition of f^{-1} that

$$y = f(x) \iff f^{-1}(y) = x$$

So, if $y = f(x)$ and if we are able to solve this equation for x in terms of y, then we must have $x = f^{-1}(y)$. If we then interchange x and y, we have $y = f^{-1}(x)$, which is the desired equation.

> **HOW TO FIND THE INVERSE FUNCTION OF A ONE-TO-ONE FUNCTION**
>
> 1. Write $y = f(x)$.
>
> 2. Solve this equation for x in terms of y (if possible).
>
> 3. Interchange x and y. The resulting equation is $y = f^{-1}(x)$.

Note that Steps 2 and 3 can be reversed. In other words, it is possible to interchange x and y first and then solve for y in terms of x.

In Example 6, note how f^{-1} reverses the effect of f. The function f is the rule "multiply by 3, then subtract 2," whereas f^{-1} is the rule "add 2, then divide by 3."

EXAMPLE 6 ∎ Finding an Inverse Function

Find the inverse function of $f(x) = 3x - 2$.

SOLUTION   First we write $y = f(x)$.

$$y = 3x - 2$$

Then we solve this equation for x:

$$3x = y + 2 \qquad \text{Add 2}$$

$$x = \frac{y + 2}{3} \qquad \text{Divide by 3}$$

Finally, we interchange x and y:

$$y = \frac{x + 2}{3}$$

Therefore, the inverse function is $f^{-1}(x) = \dfrac{x + 2}{3}$.   ∎

CHECK YOUR ANSWER

We can check our answer by using the Inverse Function Property.

$$f^{-1}(f(x)) = f^{-1}(3x - 2)$$
$$= \frac{(3x - 2) + 2}{3}$$
$$= \frac{3x}{3} = x$$

$$f(f^{-1}(x)) = f\left(\frac{x + 2}{3}\right)$$
$$= 3\left(\frac{x + 2}{3}\right) - 2$$
$$= x + 2 - 2 = x \quad \checkmark$$

In Example 7, note how f^{-1} reverses the effect of f. The function f is the rule "take the fifth power, subtract 3, then divide by 2," whereas f^{-1} is the rule "multiply by 2, add 3, then take the fifth root."

CHECK YOUR ANSWER

We can check our answer by using the Inverse Function Property.

$$f^{-1}(f(x)) = f^{-1}\left(\frac{x^5 - 3}{2}\right)$$

$$= \left[2\left(\frac{x^5 - 3}{2}\right) + 3\right]^{1/5}$$

$$= (x^5 - 3 + 3)^{1/5}$$

$$= (x^5)^{1/5} = x$$

$$f(f^{-1}(x)) = f((2x + 3)^{1/5})$$

$$= \frac{[(2x + 3)^{1/5}]^5 - 3}{2}$$

$$= \frac{2x + 3 - 3}{2}$$

$$= \frac{2x}{2} = x \qquad ✓$$

EXAMPLE 7 ■ Finding an Inverse Function

Find the inverse function of $f(x) = \dfrac{x^5 - 3}{2}$.

SOLUTION   We first write $y = (x^5 - 3)/2$ and solve for x.

$$y = \frac{x^5 - 3}{2} \qquad \text{Equation defining function}$$

$$2y = x^5 - 3 \qquad \text{Multiply by 2}$$

$$x^5 = 2y + 3 \qquad \text{Add 3}$$

$$x = (2y + 3)^{1/5} \qquad \text{Take fifth roots}$$

Then we interchange x and y to get $y = (2x + 3)^{1/5}$. Therefore, the inverse function is $f^{-1}(x) = (2x + 3)^{1/5}$.   ■

The principle of interchanging x and y to find the inverse function also gives us the method for obtaining the graph of f^{-1} from the graph of f. If $f(a) = b$, then $f^{-1}(b) = a$. Thus, the point (a, b) is on the graph of f if and only if the point (b, a) is on the graph of f^{-1}. But we get the point (b, a) from the point (a, b) by reflecting in the line $y = x$ (see Figure 8).

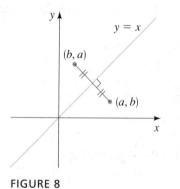

FIGURE 8

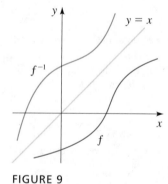

FIGURE 9

Therefore, as illustrated in Figure 9, the following is true.

> The graph of f^{-1} is obtained by reflecting the graph of f in the line $y = x$.

EXAMPLE 8 ■ Finding an Inverse Function

(a) Sketch the graph of $f(x) = \sqrt{x - 2}$.

(b) Use the graph of f to sketch the graph of f^{-1}.

(c) Find an equation for f^{-1}.

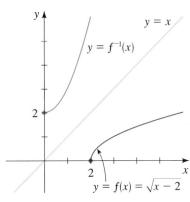

FIGURE 10

In Example 8, note how f^{-1} reverses the effect of f. The function f is the rule "subtract 2, then take the square root"; f^{-1} is the rule "square, then add 2."

SOLUTION

(a) Using the transformations from Section 2.4, we sketch the graph of $y = \sqrt{x - 2}$ by plotting the graph of the function $y = \sqrt{x}$ (Example 6 in Section 2.2) and moving it to the right 2 units.

(b) The graph of f^{-1} is obtained from the graph of f in part (a) by reflecting it in the line $y = x$, as shown in Figure 10.

(c) Solve $y = \sqrt{x - 2}$ for x, noting that $y \geq 0$.

$$\sqrt{x - 2} = y$$

$$x - 2 = y^2 \qquad \text{Square each side}$$

$$x = y^2 + 2, \quad y \geq 0 \qquad \text{Add 2}$$

Interchange x and y:

$$y = x^2 + 2, \qquad x \geq 0$$

Thus $\qquad f^{-1}(x) = x^2 + 2, \qquad x \geq 0$

This expression shows that the graph of f^{-1} is the right half of the parabola $y = x^2 + 2$ and, from the graph shown in Figure 10, this seems reasonable. ∎

2.7 EXERCISES

1–6 ■ The graph of a function f is given. Determine whether f is one-to-one.

1.

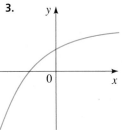

2.

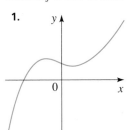

3.

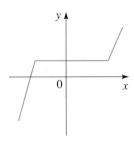

4.

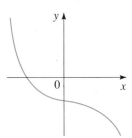

5.

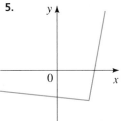

6.

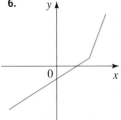

7–12 ■ Determine whether the function is one-to-one.

7. $f(x) = 7x - 3$ **8.** $f(x) = x^2 - 2x + 5$

9. $g(x) = \sqrt{x}$ **10.** $g(x) = |x|$

11. $h(x) = x^4 + 5$

12. $h(x) = x^4 + 5, \quad 0 \leq x \leq 2$

13–16 ■ Assume f is a one-to-one function.

13. (a) If $f(2) = 7$, find $f^{-1}(7)$.
(b) If $f^{-1}(3) = -1$, find $f(-1)$.

14. (a) If $f(5) = 18$, find $f^{-1}(18)$.
(b) If $f^{-1}(4) = 2$, find $f(2)$.

15. If $f(x) = 5 - 2x$, find $f^{-1}(3)$.

16. If $g(x) = x^2 + 4x$ with $x \geq -2$, find $g^{-1}(5)$.

17–22 ■ Use the Property of Inverse Functions to show that f and g are inverses of each other.

17. $f(x) = 2x - 5$; $g(x) = \dfrac{x + 5}{2}$

18. $f(x) = \dfrac{3 - x}{4}$; $g(x) = 3 - 4x$

19. $f(x) = x^2 - 4$, $x \geq 0$;
$g(x) = \sqrt{x + 4}$, $x \geq -4$

20. $f(x) = x^3 + 1$; $g(x) = (x - 1)^{1/3}$

21. $f(x) = \dfrac{1}{x - 1}$, $x \neq 1$;

$g(x) = \dfrac{1}{x} + 1$, $x \neq 0$

22. $f(x) = \sqrt{4 - x^2}$, $0 \leq x \leq 2$;
$g(x) = \sqrt{4 - x^2}$, $0 \leq x \leq 2$

23–40 ■ Find the inverse function of f.

23. $f(x) = 2x + 1$

24. $f(x) = 6 - x$

25. $f(x) = 4x + 7$

26. $f(x) = 3 - 5x$

27. $f(x) = \dfrac{1}{x + 2}$, $x > -2$

28. $f(x) = \dfrac{x - 2}{x + 2}$

29. $f(x) = \dfrac{1 + 3x}{5 - 2x}$

30. $f(x) = 5 - 4x^3$

31. $f(x) = \sqrt{2 + 5x}$

32. $f(x) = x^2 + x$, $x \geq -\frac{1}{2}$

33. $f(x) = 4 - x^2$, $x \geq 0$

34. $f(x) = \sqrt{2x - 1}$

35. $f(x) = 4 + \sqrt[3]{x}$

36. $f(x) = (2 - x^3)^5$

37. $f(x) = 1 + \sqrt{1 + x}$

38. $f(x) = \sqrt{9 - x^2}$, $0 \leq x \leq 3$

39. $f(x) = x^4$, $x \geq 0$

40. $f(x) = 1 - x^3$

41–44 ■ A function f is given.
(a) Sketch the graph of f.
(b) Use the graph of f to sketch the graph of f^{-1}.
(c) Find f^{-1}.

41. $f(x) = 3x - 6$

42. $f(x) = 16 - x^2$, $x \geq 0$

43. $f(x) = \sqrt{x + 1}$

44. $f(x) = x^3 - 1$

 45–48 ■ Draw the graph of f and use it to determine whether the function is one-to-one.

45. $f(x) = x^3 - x$

46. $f(x) = x^3 + x$

47. $f(x) = \dfrac{x + 12}{x - 6}$

48. $f(x) = \sqrt{x^3 - 4x + 1}$

49. Prove that if f is an increasing function on its entire domain, then f is one-to-one.

50. Show that it is possible to restrict the domain of the function $f(x) = x^2 + x + 1$ in such a way that the resulting function is one-to-one.

 DISCOVERY · DISCUSSION

51. The Inverse of a Price Function Marcello's Pizza charges a base price of \$7 for a large pizza, plus \$2 for each topping. Thus, if you order a large pizza with x toppings, the price of your pizza is given by the function $f(x) = 7 + 2x$. Find f^{-1}. What does the function f^{-1} represent?

52. Determining when a Linear Function Has an Inverse In order for the linear function $f(x) = mx + b$ to be one-to-one, what must be true about its slope? If it is one-to-one, what is its inverse?

53. Finding an Inverse "In Your Head" In the margin notes in this section we have pointed out that the inverse of a function can be found simply by reversing the operations that make up the function. For instance, in Example 6 we saw that the inverse of

$$f(x) = 3x - 2 \quad \text{is} \quad f^{-1}(x) = \frac{x + 2}{3}$$

because the "reverse" of "multiply by 3 and subtract 2" is "add 2 and divide by 3." Consider another function:

$$f(x) = \frac{3x - 2}{x + 7}$$

Is it possible to use the same sort of simple reversal of operations to find the inverse of this function? If so, do it. If not, explain what is different about this function that makes this task difficult.

2 REVIEW

CONCEPT CHECK

1. Define each concept in your own words. (Check by referring to the definition in the text.)
   (a) Function
   (b) Domain and range of a function
   (c) Graph of a function
   (d) Independent and dependent variables
   (e) Increasing function

2. Give an example of each type of function.
   (a) Constant function
   (b) Linear function
   (c) Quadratic function

3. Sketch by hand, on the same axes, the graphs of the following functions.
   (a) $f(x) = x$ (b) $g(x) = x^2$
   (c) $h(x) = x^3$ (d) $j(x) = x^4$

4. (a) State the Vertical Line Test.
   (b) State the Horizontal Line Test.

5. Write an equation that expresses each relationship.
   (a) y is directly proportional to x.
   (b) y is inversely proportional to x.
   (c) z is jointly proportional to x and y.

6. Suppose the graph of f is given. Write an equation for each of the graphs that are obtained from the graph of f as follows.
   (a) Shift 3 units upward.
   (b) Shift 3 units downward.
   (c) Shift 3 units to the right.
   (d) Shift 3 units to the left.
   (e) Reflect in the x-axis.

   (f) Reflect in the y-axis.
   (g) Stretch vertically by a factor of 3.
   (h) Shrink vertically by a factor of $\frac{1}{3}$.
   (i) Stretch horizontally by a factor of 3.
   (j) Stretch horizontally by a factor of $\frac{1}{3}$.

7. (a) What is an even function? What symmetry does its graph possess? Give an example of an even function.
   (b) What is an odd function? What symmetry does its graph possess? Give an example of an odd function.

8. Write the standard form of a quadratic function.

9. What does it mean to say that $f(3)$ is a local maximum value of f?

10. Suppose that f has domain A and g has domain B.
    (a) What is the domain of $f + g$?
    (b) What is the domain of fg?
    (c) What is the domain of f/g?

11. How is the composite function $f \circ g$ defined?

12. (a) What is a one-to-one function?
    (b) How can you tell from the graph of a function whether it is one-to-one?
    (c) Suppose f is a one-to-one function with domain A and range B. How is the inverse function f^{-1} defined? What is the domain of f^{-1}? What is the range of f^{-1}?
    (d) If you are given a formula for f, how do you find a formula for f^{-1}?
    (e) If you are given the graph of f, how do you find the graph of f^{-1}?

EXERCISES

1. If $f(x) = x^2 - x + 1$, find $f(0), f(2), f(-2), f(a), f(-a)$, $f(x + 1), f(2x)$, and $2f(x) - 2$.

2. If $f(x) = 1 + \sqrt{x - 1}$, find $f(5), f(9), f(a + 1), f(-x)$, $f(x^2)$, and $[f(x)]^2$.

3. The graph of a function is given.
(a) Find $f(-2)$ and $f(2)$.
(b) Find the domain of f.
(c) Find the range of f.
(d) On what intervals is f increasing? On what intervals is f decreasing?
(e) Is f one-to-one?

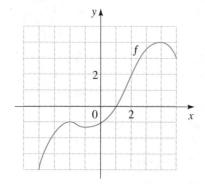

4. Which of the following figures are graphs of functions? Which of the functions are one-to-one?

(a)
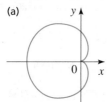

(b)
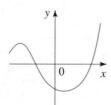

(c)
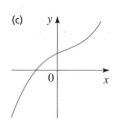

(d)
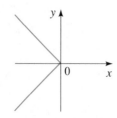

5–6 ■ Find the domain and range of the function.

5. $f(x) = \sqrt{x + 3}$ **6.** $F(t) = t^2 + 2t + 5$

7–14 ■ Find the domain of the function.

7. $f(x) = 7x + 15$

8. $f(x) = \dfrac{2x + 1}{2x - 1}$

9. $f(x) = \sqrt{x + 4}$

10. $f(x) = 3x - \dfrac{2}{\sqrt{x + 1}}$

11. $f(x) = \dfrac{1}{x} + \dfrac{1}{x + 1} + \dfrac{1}{x + 2}$

12. $g(x) = \dfrac{2x^2 + 5x + 3}{2x^2 - 5x - 3}$

13. $h(x) = \sqrt{4 - x} + \sqrt{x^2 - 1}$

14. $f(x) = \dfrac{\sqrt[3]{2x + 1}}{\sqrt[3]{2x + 2}}$

15–32 ■ Sketch the graph of the function.

15. $f(x) = 1 - 2x$

16. $f(x) = \frac{1}{3}(x - 5), \quad 2 \le x \le 8$

17. $f(t) = 1 - \frac{1}{2}t^2$ **18.** $g(t) = t^2 - 2t$

19. $f(x) = x^2 - 6x + 6$ **20.** $f(x) = 3 - 8x - 2x^2$

21. $y = 1 - \sqrt{x}$ **22.** $y = -|x|$

23. $y = \frac{1}{2}x^3$ **24.** $y = \sqrt{x + 3}$

25. $h(x) = \sqrt[3]{x}$ **26.** $H(x) = x^3 - 3x^2$

27. $g(x) = \dfrac{1}{x^2}$ **28.** $G(x) = \dfrac{1}{(x - 3)^2}$

29. $f(x) = \begin{cases} 1 - x & \text{if } x < 0 \\ 1 & \text{if } x \ge 0 \end{cases}$

30. $f(x) = \begin{cases} 1 - 2x & \text{if } x \le 0 \\ 2x - 1 & \text{if } x > 0 \end{cases}$

31. $f(x) = \begin{cases} x + 6 & \text{if } x < -2 \\ x^2 & \text{if } x \ge -2 \end{cases}$

32. $f(x) = \begin{cases} -x & \text{if } x < 0 \\ x^2 & \text{if } 0 \le x < 2 \\ 1 & \text{if } x \ge 2 \end{cases}$

33. Determine which of the following viewing rectangles produces the most appropriate graph of the function $f(x) = 6x^3 - 15x^2 + 4x - 1$.
   (i) $[-2, 2]$ by $[-2, 2]$
   (ii) $[-8, 8]$ by $[-8, 8]$
   (iii) $[-4, 4]$ by $[-12, 12]$
   (iv) $[-100, 100]$ by $[-100, 100]$

 34. Determine which of the following viewing rectangles produces the most appropriate graph of the function $f(x) = \sqrt{100 - x^3}$.
   (i) $[-4, 4]$ by $[-4, 4]$
   (ii) $[-10, 10]$ by $[-10, 10]$
   (iii) $[-10, 10]$ by $[-10, 40]$
   (iv) $[-100, 100]$ by $[-100, 100]$

 35–38 ■ Draw the graph of the function in an appropriate viewing rectangle.

35. $f(x) = x^2 + 25x + 173$

36. $f(x) = 1.1x^3 - 9.6x^2 - 1.4x + 3.2$

37. $y = \dfrac{x}{\sqrt{x^2 + 16}}$

38. $y = |x(x + 2)(x + 4)|$

 39. Find, approximately, the domain of the function $f(x) = \sqrt{x^3 - 4x + 1}$.

 40. Find, approximately, the range of the function $f(x) = x^4 - x^3 + x^2 + 3x - 6$.

41. Suppose that the graph of f is given. Describe how the graph of each of the following functions can be obtained from the graph of f.
   (a) $y = f(x) + 8$
   (b) $y = f(x + 8)$
   (c) $y = 1 + 2f(x)$
   (d) $y = f(x - 2) - 2$
   (e) $y = f(-x)$
   (f) $y = -f(-x)$
   (g) $y = -f(x)$
   (h) $y = f^{-1}(x)$

42. The graph of f is given. Draw the graph of each of the following functions.

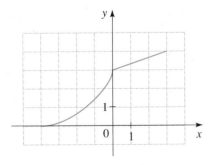

(a) $y = f(x - 2)$
(b) $y = -f(x)$
(c) $y = 3 - f(x)$
(d) $y = \frac{1}{2}f(x) - 1$
(e) $y = f^{-1}(x)$
(f) $y = f(-x)$

43. Determine whether f is even, odd, or neither even nor odd.
   (a) $f(x) = 2x^5 - 3x^2 + 2$
   (b) $f(x) = x^3 - x^7$
   (c) $f(x) = \dfrac{1 - x^2}{1 + x^2}$
   (d) $f(x) = \dfrac{1}{x + 2}$

44. Determine whether the function in the figure is even, odd, or neither.

(a)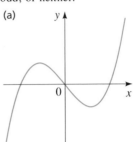

(b)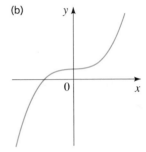

(c)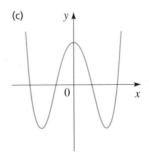

45. Express the quadratic function $f(x) = x^2 + 4x + 1$ in standard form.

46. Express the quadratic function $f(x) = -2x^2 + 12x + 12$ in standard form.

47. Find the minimum value of the function $g(x) = 2x^2 + 4x - 5$.

48. Find the maximum value of the function $f(x) = 1 - x - x^2$.

 49–50 ■ Find the local maximum and minimum values of the function and the values of x at which they occur. State each answer correct to two decimal places.

49. $f(x) = 3.3 + 1.6x - 2.5x^3$

50. $f(x) = x^{2/3}(6 - x)^{1/3}$

51. If $f(x) = x^2 - 3x + 2$ and $g(x) = 4 - 3x$, find each of the following functions.
    (a) $f + g$ (b) $f - g$ (c) fg
    (d) f/g (e) $f \circ g$ (f) $g \circ f$

52. If $f(x) = 1 + x^2$ and $g(x) = \sqrt{x - 1}$, find each of the following functions.
    (a) $f \circ g$ (b) $g \circ f$ (c) $(f \circ g)(2)$
    (d) $(f \circ f)(2)$ (e) $f \circ g \circ f$ (f) $g \circ f \circ g$

53–54 ■ Find the functions $f \circ g$, $g \circ f$, $f \circ f$, and $g \circ g$ and their domains.

53. $f(x) = 3x - 1$, $g(x) = 2x - x^2$

54. $f(x) = \sqrt{x}$, $g(x) = \dfrac{2}{x - 4}$

55. Find $f \circ g \circ h$, where $f(x) = \sqrt{1 - x}$, $g(x) = 1 - x^2$, and $h(x) = 1 + \sqrt{x}$.

56. If $T(x) = \dfrac{1}{\sqrt{1 + \sqrt{x}}}$, find functions f, g, and h such that $f \circ g \circ h = T$.

57–62 ■ Determine whether the function is one-to-one.

57. $f(x) = 3 + x^3$

58. $g(x) = 2 - 2x + x^2$

59. $h(x) = \dfrac{1}{x^4}$

60. $r(x) = 2 + \sqrt{x + 3}$

 61. $p(x) = 3.3 + 1.6x - 2.5x^3$

 **62.** $q(x) = 3.3 + 1.6x + 2.5x^3$

63. Show that $f(x) = 3x - 2$ is a one-to-one function and find its inverse function.

64. If $f(x) = 1 + \sqrt[5]{x - 2}$, find f^{-1}.

65. (a) Sketch the graph of the function
$$f(x) = x^2 - 4, \quad x \geq 0$$
    (b) Use part (a) to sketch the graph of f^{-1}.
    (c) Find an equation for f^{-1}.

66. (a) Show that the function $f(x) = 1 + \sqrt[4]{x}$ is one-to-one.
    (b) Sketch the graph of f.
    (c) Use part (b) to sketch the graph of f^{-1}.
    (d) Find an equation for f^{-1}.

67. Suppose that M varies directly as z, and $M = 120$ when $z = 15$. Write an equation that expresses this variation.

68. Suppose that z is inversely proportional to y. If $y = 16$, then $z = 12$. Use this information to write an equation that expresses z in terms of y.

69. The intensity of illumination I from a light varies inversely as the square of the distance d from the light.
    (a) Write this statement as an equation.
    (b) Determine the constant of proportionality if it is known that a lamp has an intensity of 1000 candles at a distance of 8 m.
    (c) What is the intensity of this lamp at a distance of 20 m?

70. The number of Christmas cards sold by a greeting-card store depends on the time of year. Sketch a rough graph of the number of Christmas cards sold as a function of the time of year.

71. An isosceles triangle has a perimeter of 8 cm. Express the area A of the triangle as a function of the length b of the base of the triangle.

72. A rectangle is inscribed in an equilateral triangle with a perimeter of 30 cm as in the figure.
    (a) Express the area A of the rectangle as a function of the length x shown in the figure.
    (b) Find the dimensions of the rectangle with the largest area.

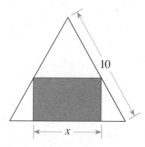

73. A piece of wire 10 m long is cut into two pieces. One piece, of length x, is bent into the shape of a square. The other piece is bent into the shape of an equilateral triangle.
    (a) Express the total area enclosed as a function of x.
    (b) For what value of x is this total area a minimum?

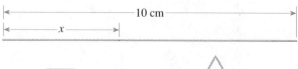

74. A baseball team plays in a stadium that holds 55,000 spectators. With the ticket price set at $10, the average attendance at recent games has been 27,000. A market survey indicates that for every dollar the ticket price is lowered, attendance will increase by 3000. How should the ticket price be set to maximize revenue?

75. A farmer wants to enclose a rectangular pen so that it has an area of 100 m². Find the dimensions of the pen that will require using the minimum amount of fencing.

76. A bird is released from point A on an island, 5 mi from the nearest point B on a straight shoreline. The bird flies to a point C on the shoreline, and then flies along the shoreline to its nesting area D (see the figure). Suppose the bird requires 10 kcal/mi of energy to fly over land and 14 kcal/mi to fly over water (see Example 8 in Section 1.6). If the bird instinctively chooses a path that minimizes its energy expenditure, to what point does it fly?

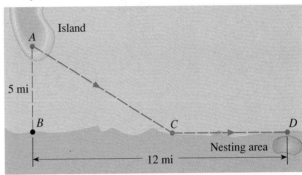

77. Under what condition on the constants a, b, c, and d is the following function one-to-one?

$$f(x) = \frac{ax + b}{cx + d}$$

1. Which of the following are graphs of functions? If the graph is that of a function, is it one-to-one?

(a)

(b)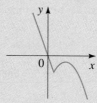

(c)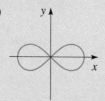

(d)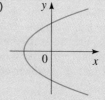

2. Find the domain of the function $f(x) = \dfrac{\sqrt{x}}{x - 1}$.

3. (a) Sketch the graph of the function $f(x) = x^3$.
   (b) Use part (a) to graph the function $g(x) = (x - 1)^3 - 2$.

4. (a) How is the graph of $y = 2 - f(x + 3)$ obtained from the graph of f?
   (b) How is the graph of $y = -f(-x)$ obtained from the graph of f?

5. (a) Sketch the graph of the function $f(x) = 2x^2 - 8x + 13$.
   (b) What is the minimum value of f?

6. Let $f(x) = \begin{cases} 1 - x^2 & \text{if } x \leqslant 0 \\ 2x + 1 & \text{if } x > 0 \end{cases}$

   (a) Evaluate $f(-2)$ and $f(1)$.
   (b) Sketch the graph of f.

7. If $f(x) = x^2 + 2x - 1$ and $g(x) = 2x - 3$, find each of the following functions.
   (a) $f \circ g$ (b) $g \circ f$
   (c) $f(g(2))$ (d) $g(f(2))$
   (e) $g \circ g \circ g$

8. (a) If $f(x) = \sqrt{3 - x}$, find the inverse function f^{-1}.
   (b) Sketch the graphs of f and f^{-1} on the same coordinate axes.

9. (a) If 800 ft of fencing is available to build three adjacent pens, as shown in the diagram, express the total area of the pens as a function of x.

(b) What value of x will maximize the total area?

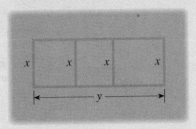

10. Let $f(x) = 3x^4 - 14x^2 + 5x - 3$.

(a) Draw the graph of f in an appropriate viewing rectangle.

(b) Is f one-to-one?

(c) Find the local maximum and minimum values of f and the values of x at which they occur. State each answer correct to two decimal places.

(d) Find the approximate range of f.

(e) Find the approximate intervals on which f is increasing and on which f is decreasing.

PRINCIPLES OF MODELING

A model is a representation of an object or phenomenon. For example, a toy Ferrari is a *model* of the actual car; a road map is a model of the streets and highways in a city. A model usually represents just one aspect of the original thing. The toy Ferrari is not an actual car, but it does represent what that particular type of car looks like; a road map does not contain the actual streets in a city, but it does represent the relationship of the streets to each other.

A **mathematical model** is a mathematical representation of an object or phenomenon. Often a mathematical model is a function that describes a certain process. We saw in Example 7 of Section 1.6 that the function $h = -16t^2 + 800t$ models the height h of a bullet t seconds after it has been fired straight up. In that example we used the model to predict the height of the bullet at certain times. Figure 1 illustrates the process of mathematical modeling.

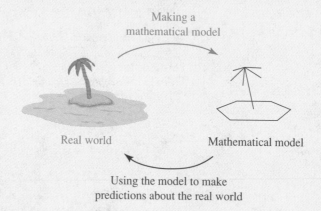

Making a
mathematical model

Real world

Mathematical model

Using the model to make
predictions about the real world

FIGURE 1

Mathematical models are useful because they allow us to isolate critical aspects of the thing we are studying and then to predict how it will behave. Models are used extensively in engineering, industry, and manufacturing. For example, engineers use computer models of skyscrapers to predict their strength and how they would behave in an earthquake. Aircraft manufacturers use elaborate mathematical models to predict the aerodynamic properties of a new design before the aircraft is actually built.

How are mathematical models developed? How are they used to predict the behavior of a process? In the next few pages and in subsequent *Focus on Modeling* sections, we explain how mathematical models can be obtained from real-world data, and we describe some of their applications.

LINEAR FUNCTIONS AS MODELS

Consider the following data, which gives the winning Olympic pole vaults in this century.

Year	Gold Medalist(s)	Height (ft)	Year	Gold Medalist(s)	Height (ft)
1900	Irving Baxter, USA	10.83	1956	Robert Richards, USA	14.96
1904	Charles Dvorak, USA	11.48	1960	Don Bragg, USA	15.42
1908	A. Gilbert, E. Cook, USA	12.17	1964	Fred Hansen, USA	16.73
1912	Harry Babcock, USA	12.96	1968	Bob Seagren, USA	17.71
1920	Frank Foss, USA	13.42	1972	W. Nordwig, E. Germany	18.04
1924	Lee Barnes, USA	12.96	1976	Tadeusz Slusarski, Poland	18.04
1928	Sabin Carr, USA	13.77	1980	W. Kozakiewicz, Poland	18.96
1932	William Miller, USA	14.15	1984	Pierre Quinon, France	18.85
1936	Earle Meadows, USA	14.27	1988	Sergei Bubka, USSR	19.77
1948	Guinn Smith, USA	14.10	1992	M. Tarassob, Unified Team	19.02
1952	Robert Richards, USA	14.92	1996	Jean Jalfione, France	19.42

From the table we see that athletes generally pole-vault higher as time goes on. To see this trend better we make a **scatter plot**, as shown in Figure 2. How can we model this data mathematically? Since it appears that the points lie more or less along a line, we can try to fit a line visually to approximate the points in the scatter plot. In other words, we may be able to use a linear function as our model.

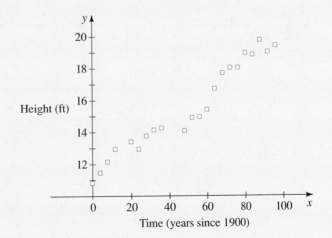

FIGURE 2
Scatter plot of pole-vault data

WHAT IS THE LINE OF BEST FIT?

How do we find a line that best fits the data? It seems reasonable to choose the line that is as close as possible to all the data points in the scatter plot. So, we

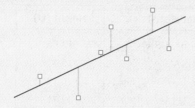

FIGURE 3

Distances from the points to the line

want the line for which the sum of the distances from the data points to the line is smallest (see Figure 3).

For technical reasons it is better to find the line where the sum of the squares of these distances is smallest. The resulting line is called the **regression line** or the **least squares line**. The formula for the regression line is found using calculus.* Fortunately, this formula is programmed into most graphing calculators (such as the TI-82, TI-83, or TI-85). Using a calculator we find that the regression line for the pole-vault data is

$$y = 0.0891x + 11.09$$

In Figure 4 the regression line is graphed, together with the scatter plot.

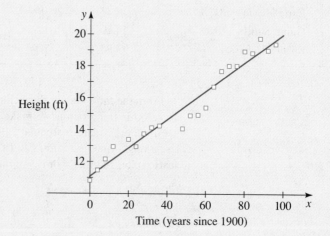

FIGURE 4

Scatter plot and regression line

What do you think the Olympic pole-vault record will be in the year 2000? We can use the regression line to make a reasonable prediction. In the year 2000, $x = 100$ so the regression line gives $y = 0.0891(100) + 11.09 \approx 19.99$ ft. Such predictions are reasonable for points close to our measured data, but we can't predict too far away from the measured data. Is it reasonable to use this model to predict the record 100 years from now?

ANOTHER EXAMPLE

When laboratory rats are exposed to asbestos fibers, some of them develop lung tumors. The following table lists the results of several such experiments by different scientists. Using a graphing calculator we get the regression line $y = 0.0177x + 0.54047$. The scatter plot and the regression line are graphed in Figure 5. Is it reasonable to use a linear model for this data?

*This formula is discussed in Exercise 51 on page 546.

Asbestos exposure (fibers/mL)	Percent that develop lung tumors
50	2
400	6
500	5
900	10
1100	26
1600	42
1800	37
2000	28
3000	50

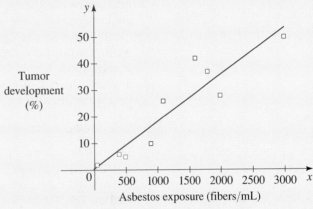

FIGURE 5 Scatter plot and regression line for asbestos-tumor data

HOW GOOD IS THE FIT?

For any given data it is always possible to find the regression line, even if the data do not tend to lie along a line. Consider the three scatter plots in Figure 6.

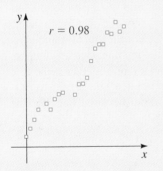

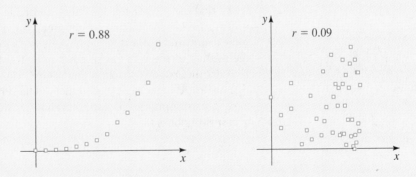

FIGURE 6

The data in the first scatter plot appear to lie along a line, in the second they appear to lie along a parabola, and in the third there appears to be no discernible trend. We can easily find the regression lines for each scatter plot using a graphing calculator. But how well do these lines represent the data? The calculator gives a **correlation coefficient** r, which is a statistical measure of how well the data lie along the regression line, or how well the two variables are correlated. The correlation coefficient is a number between -1 and 1. A correlation coefficient r close to 1 or -1 indicates strong correlation and a coefficient close to 0

indicates very little correlation; the slope of the line determines whether the correlation coefficient is positive or negative. Also, the more data points we have, the more meaningful the correlation coefficient will be. Using a calculator we find that the correlation coefficient between asbestos fibers and lung tumors in rats is $r = 0.92$. We can reasonably conclude that the presence of asbestos increases the risk of lung tumors in rats. Can we conclude that asbestos *causes* lung tumors in rats?

THE MEDIAN-MEDIAN LINE

Different factors enter into the choice of the line that best fits the data. One consideration is that points, called *outliers,* that are too far away from the general trend should not be given too much weight. In this case, we may use the **median-median line**. Most graphing calculators are programmed to find the median-median line. For the asbestos-tumor data, a graphing calculator gives the median-median line

$$y = 0.02x - 0.7$$

The **median** of a set of data points is the middle number, when the numbers are placed in increasing order. (If there is an even number of data points in a set, the median is the average of the middle pair.) The median of a set of data points is different from their average; it is not influenced by outliers.

To find the median-median line, we divide the data into three groups—a left group, a middle group, and a right group (see Figure 5). For each group we determine a **summary point** (x^*, y^*), where x^* is the median of the x-values and y^* is the median of the y-values in the group. To obtain the median-median line, we find the line passing through the left and right summary points, then slide this line one-third of the way toward the middle summary point, while keeping it parallel to its original position (see Figure 7). We now describe these steps for the asbestos-tumor data.

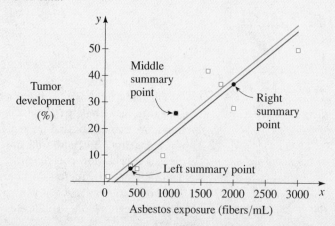

FIGURE 7

Moving the line up toward the middle summary point

Step 1 Find the summary points.   First divide the data into three groups based on their x-coordinates. For the left group we have (in increasing order)

$$x\text{-values:}\quad 50,\ 400\,,\ 500$$

$$y\text{-values:}\quad 2,\ 5\,,\ 6$$

So the summary point for the left group is $(400, 5)$. The summary points for the other groups are found similarly.

Group	Points in group	Summary point
Left	$(50, 2)$, $(400, 6)$, $(500, 5)$	$(400, 5)$
Middle	$(900, 10)$, $(1100, 26)$, $(1600, 42)$	$(1100, 26)$
Right	$(1800, 37)$, $(2000, 28)$, $(3000, 50)$	$(2000, 37)$

Step 2 Find the equation of the line passing through the left and right summary points.   The slope determined by the left and right summary points is

$$m = \frac{37 - 5}{2000 - 400} = 0.02$$

So the equation of the line passing through these points is

$$y - 5 = 0.02(x - 400) \qquad \text{or} \qquad y = 0.02x - 3$$

Step 3 Find the equation of the median-median line.   We need to slide the line found in Step 2 one-third of the way toward the middle summary point. So we first determine the distance and direction that we need to move the line found in Step 2. The y-value of the point on this line corresponding to the x-coordinate of the middle summary point is

$$y = 0.02(1100) - 3 = 19.0$$

Since the y-value of the middle summary point is 26, the difference is $26 - 19 = 7$, and one-third of this value is $\frac{7}{3} \approx 2.3$. Since the middle summary point is above the line, we need to move the line *upward* 2.3 units. Thus, the equation of the median-median line is

$$y = 0.02x - 3 + 2.3 \qquad \text{or} \qquad y = 0.02x - 0.7$$

The scatter plot and the median-median line are graphed in Figure 8. (Compare the median-median line with the regression line in Figure 5.)

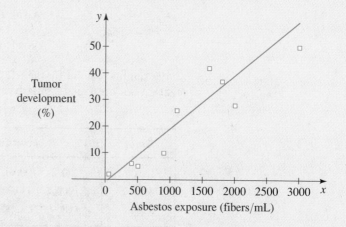

FIGURE 8
Scatter plot and median-median line

 PROBLEMS

1. The table lists average carbon dioxide (CO_2) levels in the atmosphere, measured in parts per million (ppm) at Mauna Loa Observatory from 1972 to 1988.
   (a) Make a scatter plot of the data.
   (b) Find and graph the regression line.
   (c) Use the linear model in part (b) to estimate the CO_2 level in the atmosphere in 1990. Compare your answer with the actual CO_2 level of 354.0 measured in 1990.

Year	CO_2 level (ppm)
1972	327.3
1974	330.0
1976	332.0
1978	335.3
1980	338.5
1982	341.0
1984	344.3
1986	347.0
1988	351.3

2. Biologists have observed that the chirping rate of crickets of a certain species appears to be related to temperature. The table shows the chirping rates for various temperatures.

Temperature (°F)	Chirping rate (chirps/min)
50	20
55	46
60	79
65	91
70	113
75	140
80	173
85	198
90	211

(a) Make a scatter plot of the data.
(b) Find and graph the regression line.
(c) Use the linear model in part (b) to estimate the chirping rate at 100°F.

Income	Ulcer rate
$4,000	14.1
$6,000	13.0
$8,000	13.4
$12,000	12.4
$16,000	12.0
$20,000	12.5
$30,000	10.5
$45,000	9.4
$60,000	8.2

Flow rate (%)	Mosquito positive rate (%)
0	22
10	16
40	12
60	11
90	6
100	2

3. The table shows (lifetime) peptic ulcer rates (per 100 population) for various family incomes as reported by the 1989 National Health Interview Survey.
(a) Make a scatter plot of the data.
(b) Find and graph the regression line.
(c) Find and graph the median-median line.
(d) Estimate the peptic ulcer rate for an income level of $25,000 according to the linear model in part (b).
(e) Estimate the peptic ulcer rate for an income level of $80,000 according to the linear model in part (c).

4. The table lists the relative abundance of mosquitoes (as measured by the mosquito positive rate) versus the flow rate (measured as a percentage of maximum flow) of canal networks in Saga City, Japan.
(a) Make a scatter plot of the data.
(b) Find and graph the regression line.
(c) Use the linear model in part (b) to estimate the mosquito positive rate if the canal flow is 70% of maximum.
(d) Find and graph the median-median line.
(e) Use the linear model in part (d) to estimate the mosquito positive rate if the canal flow is 70% of maximum.

5. The tables show the results of experiments giving the average number of mosquito bites per 15 minutes versus temperature and wind speed.

Temperature (°C)	Bites/15 min
10	0.03
15	7.2
20	46.2
25	65.5
30	37.8

Wind speed (km/h)	Bites/15 min
0.5	59.3
2	35.7
4	24.8
6	21.9
8	12.0
14	4.8

(a) Make a scatter plot for each set of the data.
(b) Find and graph the regression line for the temperature-bites data.
(c) Find and graph the median-median line for the wind-bites data.
(d) Find the correlation coefficient for each set of data. Are linear models appropriate for these data?

6. The table shows (lifetime) peptic ulcer rates (per 100 population) for various education levels as reported by the 1989 National Health Interview Survey.

Education level (years)	Ulcer rate
8	12
11	12.5
12	11
15	10
16	8
18	9

(a) Make a scatter plot of the data.
(b) Find and graph the regression line.
(c) Use the linear model in part (b) to estimate the ulcer rate for a person with 20 years of education.

7. Audiologists study the intelligibility of spoken sentences under different noise levels. Intelligibility, the MRT score, is measured as the percent of a spoken sentence that the listener can decipher at a certain noise level in decibels (dB). The table shows the results of one such test.

Noise level (dB)	MRT score (%)
80	99
84	91
88	84
92	70
96	47
100	23
104	11

(a) Make a scatter plot of the data.
(b) Find and graph the regression line.
(c) Find the correlation coefficient. Is a linear model appropriate?
(d) Use the linear model in part (b) to estimate the intelligibility of a sentence at a 94-decibel noise level.

Year	IQ index
1940	85
1950	89
1960	90
1970	95
1980	97
1990	100

8. The IQ scores in the United States have risen dramatically since the tests were first introduced, as shown in the table at the left.
(a) Make a scatter plot of the data.
(b) Find and graph the regression line.
(c) Use the linear model you found in part (b) to predict the IQ index in the year 2000.

9. The average life expectancy in the United States has been rising steadily over the past few decades, as shown in the table.

Year	Life expectancy
1920	54.1
1930	59.7
1940	62.9
1950	68.2
1960	69.7
1970	70.8
1980	73.7
1990	75.4

(a) Make a scatter plot of the data.

(b) Find and graph the regression line.

(c) Use the linear model you found in part (b) to predict the life expectancy in the year 2000.

10. Do you think that shoe size and height are correlated? Find out by surveying the shoe sizes and heights of people in your class. (Of course, the data for men and women should be separate.) Find the correlation coefficient.

3

POLYNOMIALS AND RATIONAL FUNCTIONS

In applications of mathematics, many situations can be modeled by polynomial functions. For example, the volume of a silo of fixed height is a polynomial function of its radius.

Each problem that I solved became a rule which served afterwards to solve other problems.

RENÉ DESCARTES

We have previously studied constant, linear, and quadratic functions, which have the equations $f(x) = c$, $f(x) = mx + b$, and $f(x) = ax^2 + bx + c$, respectively. All these functions are special cases of an important class of functions called polynomials. A **polynomial** P **of degree** n is a function of the form

$$P(x) = a_n x^n + a_{n-1} x^{n-1} + \cdots + a_1 x + a_0$$

where $a_n \neq 0$. Polynomials are constructed using the operations of addition, subtraction, and multiplication. If we introduce division as well, we obtain the set of rational functions. A **rational function** r is a function of the form

$$r(x) = \frac{P(x)}{Q(x)}$$

where P and Q are polynomials.

Virtually all the functions used in mathematics and the sciences are evaluated numerically by using polynomial approximations. In this chapter we study this important class of functions by first learning about the graphs of polynomial functions. We then learn how to find rational, irrational, and complex solutions of polynomial equations. Finally, we study rational functions and their graphs.

Working with polynomials and rational functions has numerical, algebraic, and graphical aspects, and we use all three approaches to give us a deeper understanding of these functions.

3.1   POLYNOMIAL FUNCTIONS AND THEIR GRAPHS

Before we learn to work with polynomials, we must agree on some terminology.

POLYNOMIAL FUNCTIONS

A function of the form

$$P(x) = a_n x^n + a_{n-1} x^{n-1} + \cdots + a_1 x + a_0$$

where $a_n \neq 0$, is a **polynomial of degree** n. The numbers $a_0, a_1, a_2, \ldots, a_n$ are called the **coefficients** of the polynomial. The number a_0 is the **constant coefficient**. The number a_n, the coefficient of the highest power, is the **leading coefficient**.

If a polynomial consists of just a single term, then it is called a **monomial**. For example, $P(x) = x^3$ and $Q(x) = -6x^5$ are monomials.

The graphs of polynomials of degree 0 or 1 are lines (Section 1.10), and the graphs of polynomials of degree 2 are parabolas (Section 2.5). The greater the degree of a polynomial, the more complicated its graph. In this section we study the general features of polynomial graphs. We begin with the simplest polynomial functions, those of the form $y = x^n$.

EXAMPLE 1 ■ Graphs of Simple Monomials

Graph each of the following functions.

(a) $y = x$ (b) $y = x^2$ (c) $y = x^3$ (d) $y = x^4$ (e) $y = x^5$

SOLUTION   We are already familiar from our previous work with some of these graphs, but we include them for completeness.

x	$y = x$	$y = x^2$	$y = x^3$	$y = x^4$	$y = x^5$
0.1	0.1	0.01	0.001	0.0001	0.00001
0.2	0.2	0.04	0.008	0.0016	0.00032
0.5	0.5	0.25	0.125	0.0625	0.03125
0.7	0.7	0.49	0.343	0.2401	0.16807
1.0	1.0	1.00	1.000	1.0000	1.00000
1.2	1.2	1.44	1.728	2.0736	2.48832
1.5	1.5	2.25	3.375	5.0625	7.59375
2.0	2.0	4.00	8.000	16.0000	32.00000

If n is odd, then $(-x)^n = -x^n$, so $y = x^n$ is an odd function. If n is even, then $(-x)^n = x^n$, so $y = x^n$ is an even function (see Section 2.4). We use these facts to find the values for negative x from the table. Plotting the points leads to the graphs in Figure 1.

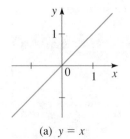
(a) $y = x$

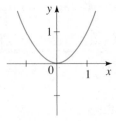
(b) $y = x^2$

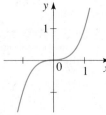
(c) $y = x^3$

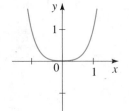
(d) $y = x^4$

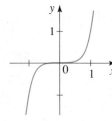
(e) $y = x^5$

FIGURE 1   ■

As Example 1 suggests, when n is odd, the graph of $y = x^n$ has the same general shape as $y = x^3$, and when n is even, the graph of $y = x^n$ has more or less the same U-shape as $y = x^2$. However, note that as the degree n becomes larger, the graphs become flatter around the origin and steeper elsewhere.

EXAMPLE 2 ■ Transformations of Monomials

Sketch the graph of each of the following functions.

(a) $y = -x^3$ (b) $y = (x - 2)^4$ (c) $y = -2x^5 + 4$

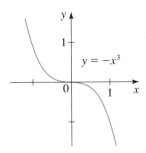

FIGURE 2

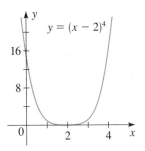

FIGURE 3

SOLUTION We use the graphs in Example 1 and transform them using the techniques of Section 2.4.

(a) The function $y = -x^3$ is the negative of $y = x^3$, so we simply reflect the graph in Figure 1(c) in the x-axis to obtain Figure 2.

(b) The graph of the function $y = (x - 2)^4$ has the same shape as $y = x^4$. Replacing x by $x - 2$ shifts the graph of $y = x^4$ in Figure 1(d) to the right 2 units (see Figure 3).

(c) We begin with the graph of $y = x^5$ in Figure 1(e). Multiplying the function by 2 stretches the graph vertically. The negative sign reflects the graph about the x-axis, so at this stage we have the dashed blue graph in Figure 4. Finally, adding 4 to the function shifts the graph upward 4 units. Since $-2x^5 + 4 = 0$ when $x^5 = 2$, the graph crosses the x-axis at $x = \sqrt[5]{2}$.

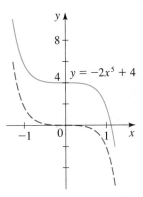

FIGURE 4   ■

If $y = P(x)$ is a polynomial and if c is a number such that $P(c) = 0$, then we say that c is a **zero** of P. In other words, the zeros of P are the solutions of the polynomial equation $P(x) = 0$. When we are dealing with polynomial equations, we often refer to solutions as **roots**. To find the roots of polynomial equations, we factor the polynomial and then use the Zero-Product Principle, as in Section 1.5. Note that if $P(c) = 0$, then the graph of $y = P(x)$ has an x-intercept at $x = c$, so the x-intercepts of the graph are the zeros of the function.

> ### ZEROS OF POLYNOMIALS
>
> If P is a polynomial and if c is a number such that $P(c) = 0$, then we say that c is a **zero** of P. The following are equivalent ways of saying the same thing.
>
> 1. c is a zero of P
>
> 2. $x = c$ is a root of the equation $P(x) = 0$
>
> 3. $x - c$ is a factor of $P(x)$
>
> 4. $x = c$ is an x-intercept of the graph of P

Between any two successive zeros of the polynomial, the values of the polynomial will be either all positive or all negative. Thus, the graph will lie entirely above or entirely below the x-axis between any two successive zeros (or x-intercepts). In the next example we use the zeros of a polynomial to help us sketch its graph.

EXAMPLE 3 ■ Graph of a Third-Degree Polynomial

Sketch the graph of the function $P(x) = (x + 2)(x - 2)(x - 1)$.

SOLUTION The zeros of P occur where the factors of P equal 0. Thus, $P(x) = 0$ when $x + 2 = 0$, $x - 2 = 0$, and when $x - 1 = 0$. It follows that the zeros of P are -2, 2, and 1, so these are the x-intercepts of its graph. Since $P(0) = 4$, the y-intercept is 4.

To sketch the graph, we must calculate $P(x)$ for other values of x as well. We choose values between (and to the right and left of) successive zeros to determine whether $P(x)$ is positive or negative on each of the intervals determined by the zeros. If $P(x)$ is positive at any x between successive zeros, then the graph of $y = P(x)$ lies above the x-axis on the interval between those zeros, and if $P(x)$ is negative on such an interval, then the graph lies below the x-axis on that interval. Plotting the points given in the table and completing the sketch gives us the graph in Figure 5.

x	$P(x)$
-3	-20
-2	0
-1	6
0	4
1	0
$\frac{3}{2}$	$-\frac{7}{8}$
2	0
3	10

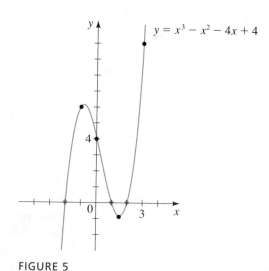

FIGURE 5

To see why P and Q have the same end behavior, factor P as follows:

$$P(x) = 3x^5\left(1 - \frac{5}{3x^2} + \frac{2}{3x^4}\right)$$

$$Q(x) = 3x^5$$

When $|x|$ is large, the second and third terms within the parentheses are very small (see Exercise 66 on page 13), so for large $|x|$,

$$P(x) \approx 3x^5(1 - 0 + 0)$$
$$= 3x^5 = Q(x)$$

In Example 3 we graphed a polynomial in the vicinity of its zeros—that is, for relatively small values of x. The **end behavior** of a polynomial is a description of what happens to the values of the polynomial as $|x|$ becomes large. *The end behavior is determined by the term in the polynomial containing the highest power of x, because when x is large, the other terms are relatively insignificant in size.* For example, the polynomials $P(x) = 3x^5 - 5x^3 + 2x$ and $Q(x) = 3x^5$

x	$P(x)$	$Q(x)$
15	2,261,280	2,278,125
30	72,765,060	72,900,000
50	936,875,100	937,500,000
−15	−2,261,280	−2,278,125
−30	−72,765,060	−72,900,000
−50	−936,875,100	−937,500,000

have the same end behavior. This is because when x becomes large in the positive direction, the values of both P and Q become very large and positive, and when x becomes large in the negative direction, the values of P and Q are very large and negative (see the table in the margin). We describe this by writing

$$y \to \infty \quad \text{as} \quad x \to \infty \quad \text{and} \quad y \to -\infty \quad \text{as} \quad x \to -\infty$$

The following box shows the four possible types of end behavior, based on the degree of the highest power and the sign of its coefficient.

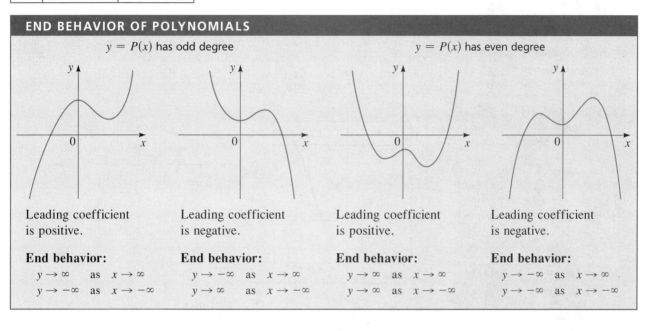

END BEHAVIOR OF POLYNOMIALS

$y = P(x)$ has odd degree

Leading coefficient is positive.

End behavior:

$y \to \infty \quad$ as $\quad x \to \infty$
$y \to -\infty \quad$ as $\quad x \to -\infty$

Leading coefficient is negative.

End behavior:

$y \to -\infty \quad$ as $\quad x \to \infty$
$y \to \infty \quad$ as $\quad x \to -\infty$

$y = P(x)$ has even degree

Leading coefficient is positive.

End behavior:

$y \to \infty \quad$ as $\quad x \to \infty$
$y \to \infty \quad$ as $\quad x \to -\infty$

Leading coefficient is negative.

End behavior:

$y \to -\infty \quad$ as $\quad x \to \infty$
$y \to -\infty \quad$ as $\quad x \to -\infty$

To sketch the graph of a polynomial, we use the zeros and end behavior of the polynomial, as indicated in the following box.

GUIDELINES FOR GRAPHING A POLYNOMIAL

1. Factor the polynomial to find all its real zeros; these are the x-intercepts of the graph.

2. Make a table of values for the polynomial, using values of x between and to the left and right of the zeros found in Step 1. Include the y-intercept in the table.

3. Plot the intercepts and the points determined in Step 2.

4. Determine the end behavior.

5. Sketch a smooth curve that passes through the points plotted in Step 3 and exhibits the required end behavior.

Sir Isaac Newton (1642–1727) continues to be universally regarded, more than 270 years after his death, as one of the giants of physics and mathematics. He is most well known for discovering the laws of motion and gravity and for inventing the calculus, but his contributions also include the Binomial Theorem, the laws of optics, and methods for solving polynomial equations to any desired accuracy. He was born on Christmas Day, a sickly infant whose father had died a few months before. After an unhappy childhood, he entered Cambridge University, where he learned mathematics by studying the writings of Euclid and Descartes.

During the plague years of 1665 and 1666, the university was closed, and Newton spent his time thinking and writing about his ideas that were later to revolutionize the sciences. Because of a pathological fear of criticism, he did not publish these writings until much later, partly as a result of the encouragement of Edmund Halley (who discovered the now-famous comet). After his work was published, it brought Newton enormous fame and prestige in his own lifetime. Even the poets were

(continued)

EXAMPLE 4 ■ Graph of a Fourth-Degree Polynomial

Sketch the graph of the function $P(x) = -2x^4 - x^3 + 3x^2$.

SOLUTION   We factor to obtain

$$P(x) = -x^2(2x^2 + x - 3)$$

$$= -x^2(2x + 3)(x - 1)$$

so the zeros of P are 0, $-\frac{3}{2}$, and 1. The y-intercept is $P(0) = 0$. In the following table we give the values of P at a number of other points. (Such values are most easily obtained using a programmable calculator.) Remember that we must choose points between successive zeros to determine whether the polynomial is positive or negative between these zeros. We plot these points as shown in Figure 6.

x	$P(x)$
-2	-12
-1.5	0
-1	2
-0.5	0.75
0	0
0.5	0.5
1	0
1.5	-6.75

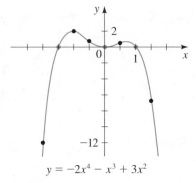

$$y = -2x^4 - x^3 + 3x^2$$

FIGURE 6

Because P has even degree and its leading coefficient is negative, it has the following end behavior:

$$y \to -\infty \quad \text{as} \quad x \to \infty \qquad \text{and} \qquad y \to -\infty \quad \text{as} \quad x \to -\infty$$

Completing the graph with a smooth curve, we obtain the graph shown in Figure 6.   ■

Note that although $x = 0$ is a zero of the polynomial in Example 4, the graph does not cross the x-axis at the x-intercept 0. It just touches the x-axis there and then moves upward again. We say that the graph is *tangent* to the x-axis at $x = 0$.

 GRAPHING DEVICES AND POLYNOMIALS

We now use graphing devices to explore further features of the graphs of polynomials. In the next example we use a graphing calculator to help us visualize the end behavior of a fifth-degree polynomial.

EXAMPLE 5 ■ **End Behavior of a Fifth-Degree Polynomial**

Graph the polynomials

$$P(x) = 3x^5 - 5x^3 + 2x \qquad \text{and} \qquad Q(x) = 3x^5$$

on the same screen, first using the viewing rectangle $[-2, 2]$ by $[-2, 2]$ and
then changing the rectangle to $[-10, 10]$ by $[-10{,}000, 10{,}000]$. Describe and
compare the end behavior of the two functions.

SOLUTION   Figure 7(a) shows that the two functions look quite different on
the smaller viewing rectangle. The polynomial P has four local extrema, but Q
has none. On the large viewing rectangle shown in Figure 7(b), however, the
two functions look almost the same. This means that P and Q have the same
end behavior:

$$y \to \infty \quad \text{as} \quad x \to \infty \qquad \text{and} \qquad y \to -\infty \quad \text{as} \quad x \to -\infty$$

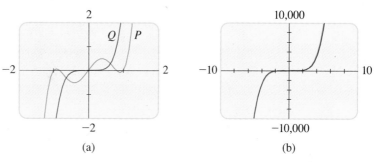

(a)   (b)

FIGURE 7 $P(x) = 3x^5 - 5x^3 + 2x$
$Q(x) = 3x^5$ ■

The margin note on page 224
explains why P and Q have the
same end behavior.

Recall from Section 2.5 that if the point $(a, f(a))$ is the highest point on the
graph of f within some viewing rectangle, then $f(a)$ is a local maximum value of
f, and if $(b, f(b))$ is the lowest point on the graph of f within a viewing rectangle,
then $f(b)$ is a local minimum value (see Figure 8). We say that such a point
$(a, f(a))$ is a **local maximum point** on the graph and that $(b, f(b))$ is a **local**

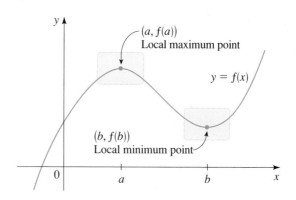

FIGURE 8

minimum point. The set of all local maximum and minimum points on the graph of a function is called its **local extrema**.

EXAMPLE 6 ■ Graphing a Third-Degree Polynomial

Graph the polynomial function

$$P(x) = x^3 - 3x^2 - x + 15$$

Determine the x- and y-intercepts and the coordinates of the local extrema of this graph.

SOLUTION   By setting $x = 0$, we see that the y-intercept of the graph is 15, so we must choose a viewing rectangle that extends at least this far upward. The graph of P in the viewing rectangle $[-5, 5]$ by $[-10, 20]$ is shown in Figure 9. The graph has one x-intercept, $x \approx -1.86$, and two extrema, a local maximum and a local minimum. By zooming in on each of these and tracing along the graph with the cursor (as described in Section 2.5), we find that the local maximum point is $(-0.16, 15.08)$ and the local minimum point is $(2.15, 8.92)$, rounded to two decimal places.   ■

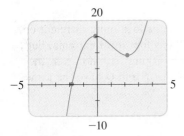

FIGURE 9

$P(x) = x^3 - 3x^2 - x + 15$

EXAMPLE 7 ■ Polynomials of Fourth and Fifth Degree

Using the viewing rectangle $[-5, 5]$ by $[-100, 100]$, graph each of the following polynomials, and determine how many local extrema each function has.

(a) $P_1(x) \doteq x^4 + x^3 - 16x^2 - 4x + 48$

(b) $P_2(x) = x^5 + 3x^4 - 5x^3 - 15x^2 + 4x - 15$

SOLUTION   The graphs are shown in Figure 10. We see that P_1 has two local minimum points and one local maximum point, for a total of three local extrema. The polynomial P_2 has two local minimum and two local maximum points—a total of four local extrema.

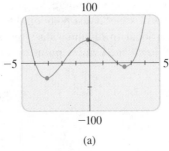

(a)

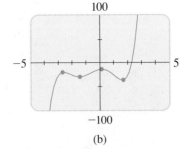

(b)

FIGURE 10 $P_1(x) = x^4 + x^3 - 16x^2 - 4x + 48$ $P_2(x) = x^5 + 3x^4 - 5x^3 - 15x^2 + 4x - 15$ ■

The polynomial P in Example 6 is of degree 3 and its graph has two local extrema. In Example 7, P_1 is a fourth-degree polynomial and has three local extrema, whereas P_2 is of degree 5 and has four local extrema. The number of extrema in each case is one less than the degree. This is not a coincidence, as the principle in the following box indicates. (A proof of this principle requires the methods of calculus.)

LOCAL EXTREMA OF POLYNOMIALS

If $P(x) = a_n x^n + a_{n-1} x^{n-1} + \cdots + a_1 x + a_0$ is a polynomial of degree n, then the graph of P has at most $n - 1$ local extrema.

 A polynomial of degree n may in fact have less than $n - 1$ local extrema, as we see in Example 8. The preceding principle tells us only that a polynomial of degree n can have *no more* than $n - 1$ local extrema.

EXAMPLE 8 ■ Local Extrema of Polynomials

Graph each of the following polynomials using a graphing device, and determine the approximate coordinates of all local extrema.

(a) $F(x) = -2x^3 + 3x^2 - 5x + 2$

(b) $G(x) = 7x^4 + 3x^2 - 10x$

SOLUTION

(a) Using the viewing rectangle $[-5, 5]$ by $[-100, 100]$ gives us the graph of F shown in Figure 11(a). As the graph is plotted, we see that the values of $F(x)$ appear to decline continuously, so the function has no local maximum or minimum. If we select other viewing rectangles, we observe the same behavior, so the function has no local extrema.

(b) In the viewing rectangle $[-5, 5]$ by $[-100, 100]$, the graph of G has the shape shown in Figure 11(b). It appears to have just a single local minimum in quadrant IV. To confirm this, we zoom in on the portion of the graph near this minimum point. The viewing rectangle $[0, 1]$ by $[-5, 0]$ gives us the graph in Figure 11(c). Using the cursor, we locate the local minimum point at $(0.61, -4.01)$, correct to two decimal places.

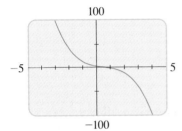

(a) $F(x) = -2x^3 + 3x^2 - 5x + 2$

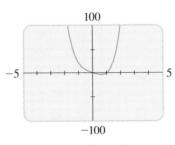

(b) $G(x) = 7x^4 + 3x^2 - 10x$

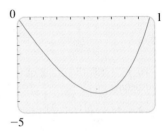

(c) $G(x) = 7x^4 + 3x^2 - 10x$

FIGURE 11 ■

FAMILIES OF POLYNOMIALS

A graphing calculator enables us to quickly draw the graphs of many functions at once, on the same viewing screen. This enables us to see how changing a value in the definition of the functions affects the shape of its graph. In the next example we apply this principle to a family of third-degree polynomials.

EXAMPLE 9 ■ A Family of Polynomials

Sketch the family of polynomials $P(x) = x^3 - cx^2$ for $c = 0, 1, 2$, and 3. How does changing the value of c affect the graph?

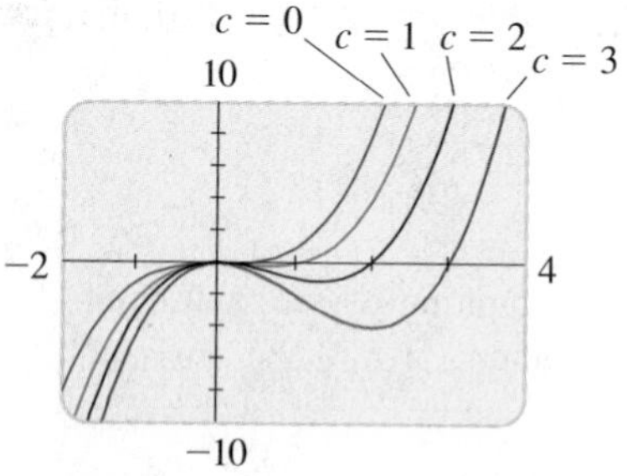

SOLUTION   The polynomials

$$P_0(x) = x^3 \qquad\qquad P_1(x) = x^3 - x^2$$
$$P_2(x) = x^3 - 2x^2 \qquad\qquad P_3(x) = x^3 - 3x^2$$

are graphed in Figure 12. We see that increasing the value of c causes the graph to develop an increasingly deep "valley" to the right of the y-axis, creating a local maximum at the origin and a local minimum at a point in quadrant IV. This local minimum moves lower and further to the right as c increases. To see why this happens, factor $P(x) = x^2(x - c)$. The polynomial P has zeros at 0 and c, and the larger c gets, the further to the right the minimum between 0 and c will be.   ■

FIGURE 12

A family of polynomials
$P(x) = x^3 - cx^2$

3.1   EXERCISES

1–8 ■ Sketch the graph of the function by transforming the graph of an appropriate function of the form $y = x^n$. Indicate all x- and y-intercepts on each graph.

1. $y = x^3 - 8$

2. $y = (x - 2)^3$

3. $y = -x^4 + 16$

4. $y = -2(x - 1)^3$

5. $y = -(x - 1)^4 + 1$

6. $y = 3x^5 - 9$

7. $y = 4(x - 2)^5 - 4$

8. $y = 3x^4 - 27$

9–26 ■ Sketch the graph of the function, and describe its end behavior.

9. $y = (x - 3)(x + 1)$

10. $y = (x - 1)(x + 1)(x - 2)$

11. $y = x(x - 2)(x + 1)$

12. $y = \frac{1}{5}x(x - 5)^2$

13. $y = (x - 1)^2(x - 3)$

14. $y = \frac{1}{4}(x + 1)^3(x - 3)$

15. $y = \frac{1}{12}(x + 2)^2(x - 3)^2$

16. $y = x^3 + x^2 - x - 1$

17. $y = x^3 + 3x^2 - 4x - 12$

18. $y = 2x^3 - x^2 - 18x + 9$

19. $y = x^3 - x^2 - 6x$

20. $y = (x^2 - 2x - 3)^2$

21. $y = \frac{1}{8}(2x^4 + 3x^3 - 16x - 24)$

22. $y = x^4 - 3x^2 - 4$

23. $y = x^4 - 2x^3 - 8x + 16$

24. $y = x^4 - 2x^3 + 8x - 16$

25. $y = x^5 - 9x^3$

26. $y = x^6 - 2x^3 + 1$

27–30 ■ Match the polynomial function with one of the graphs I–IV. Give reasons for your choice.

27. $P(x) = x(x^2 - 4)$

28. $Q(x) = -x^2(x^2 - 4)$

29. $R(x) = -x^5 + 5x^3 - 4x$

30. $S(x) = \frac{1}{2}x^6 - 2x^4$

I
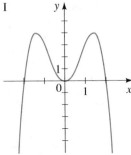

II
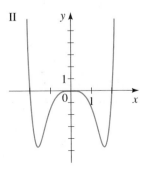

III
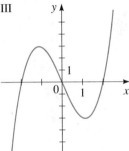

IV
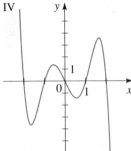

31–36 ■ Use a graphing device to graph the polynomial, and describe its end behavior.

31. $y = 3x^3 - x^2 + 5x + 1$

32. $y = -\frac{1}{8}x^3 + \frac{1}{4}x^2 + 12x$

33. $y = x^4 - 7x^2 + 5x + 5$

34. $y = (1 - x)^5$

35. $y = x^{11} - 9x^9$

36. $y = 2x^2 - x^{12}$

37–44 ■ Graph the polynomial in the given viewing rectangle. Find the x-intercept(s), if any, the y-intercept, and the coordinates of all local extrema. State each answer correct to two decimal places.

37. $y = -x^2 + 8x$, $[-4, 12]$ by $[-50, 30]$

38. $y = x^3 - 3x^2$, $[-2, 5]$ by $[-10, 10]$

39. $y = x^3 - 12x + 9$, $[-5, 5]$ by $[-30, 30]$

40. $y = 2x^3 - 3x^2 - 12x - 32$, $[-5, 5]$ by $[-60, 30]$

41. $y = x^4 + 4x^3$, $[-5, 5]$ by $[-30, 30]$

42. $y = x^4 - 18x^2 + 32$, $[-5, 5]$ by $[-100, 100]$

43. $y = 3x^5 - 5x^3 + 3$, $[-3, 3]$ by $[-5, 10]$

44. $y = x^5 - 5x^2 + 6$, $[-3, 3]$ by $[-5, 10]$

45–54 ■ Find all local extrema of the polynomial, correct to two decimal places.

45. $y = -2x^2 + 3x + 5$

46. $y = x^3 + 12x$

47. $y = x^3 - x^2 - x$

48. $y = 6x^3 + 3x + 1$

49. $y = x^4 - 5x^2 + 4$

50. $y = 1.2x^5 + 3.75x^4 - 7x^3 - 15x^2 + 18x$

51. $y = (x - 2)^5 + 32$

52. $y = (x^2 - 2)^3$

53. $y = x^8 - 3x^4 + x$

54. $y = \frac{1}{3}x^7 - 17x^2 + 7$

55–60 ■ Graph the family of polynomials in the same viewing rectangle, using the given values of c. Explain how changing the value of c affects the graph.

55. $P(x) = cx^3$; $c = 1, 2, 5, \frac{1}{2}$

56. $P(x) = (x - c)^4$; $c = -1, 0, 1, 2$

57. $P(x) = x^4 + c$; $c = -1, 0, 1, 2$

58. $P(x) = x^3 + cx$; $c = 2, 0, -2, -4$

59. $P(x) = x^4 - cx$; $c = 0, 1, 8, 27$

60. $P(x) = x^c$; $c = 1, 3, 5, 7$

61. (a) On the same coordinate axes, sketch graphs (as accurately as possible) of the functions

$$y = x^3 - 2x^2 - x + 2 \quad \text{and} \quad y = -x^2 + 5x + 2$$

(b) Based on your sketch in part (a), at how many points do the two graphs appear to intersect?

(c) Find the coordinates of all the intersection points.

62. Portions of the graphs of $y = x^2$, $y = x^3$, $y = x^4$, $y = x^5$, and $y = x^6$ are plotted in the figures. Determine which function belongs to each graph.

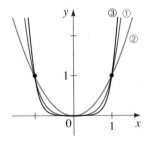

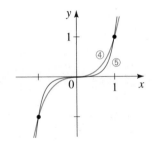

63. Recall that a function f is *odd* if $f(-x) = -f(x)$ or *even* if $f(-x) = f(x)$ for all real x.

(a) Show that if f and g are both odd, then so is the function $f + g$.

(b) Show that if f and g are both even, then so is the function $f + g$.

(c) Show that if f is odd and g is even, and neither has constant value 0, then the function $f + g$ is neither even nor odd.

(d) Show that a polynomial $P(x)$ that contains only odd powers of x is an odd function.

(e) Show that a polynomial $P(x)$ that contains only even powers of x is an even function.

(f) Show that if a polynomial $P(x)$ contains both odd and even powers of x, then it is neither an odd nor an even function.

(g) Express the function

$$P(x) = x^5 + 6x^3 - x^2 - 2x + 5$$

as the sum of an odd function and an even function.

64. A cardboard box has a square base, with each of the four edges of the base having length x inches, as shown in the figure. The total length of all 12 edges of the box is 144 in.

(a) Express the volume V of the box as a function of x.

(b) Sketch the graph of the function V.

(c) Since both x and V represent positive quantities (length and volume, respectively), what is the domain of V?

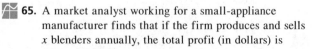

65. A market analyst working for a small-appliance manufacturer finds that if the firm produces and sells x blenders annually, the total profit (in dollars) is

$$P(x) = 8x + 0.3x^2 - 0.0013x^3 - 372$$

Graph the function P in an appropriate viewing

rectangle and use the graph to answer the following questions.

(a) When just a few blenders are manufactured, the firm loses money (profit is negative). [For example, $P(10) = -263.3$, so the firm loses \$263.30 if it produces and sells only 10 blenders.] How many blenders must the firm produce to break even?

(b) Does profit increase indefinitely as more blenders are produced and sold? If not, what is the largest possible profit the firm could have?

66. The rabbit population on a small island is observed to be given by the function

$$P(t) = 120t - 0.4t^4 + 1000$$

where t is the time (in months) since observations of the island began.

(a) When is the maximum population attained, and what is that maximum population?

(b) When does the rabbit population disappear from the island?

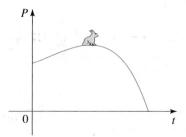

67. (a) Graph the function $P(x) = (x - 1)(x - 3)(x - 4)$ and find all local extrema, correct to the nearest tenth.

(b) Graph the function

$$Q(x) = (x - 1)(x - 3)(x - 4) + 5$$

and use your answers to part (a) to find all local extrema, correct to the nearest tenth.

(c) If $a < b < c$, explain why the function

$$P(x) = (x - a)(x - b)(x - c)$$

must have two local extrema.

(d) If $a < b < c$ and d is any real number, explain why the function $Q(x) = (x - a)(x - b)(x - c) + d$ must have two local extrema.

68. (a) How many x-intercepts and how many local extrema
does the polynomial $P(x) = x^3 - 4x$ have?

(b) How many x-intercepts and how many local extrema
does the polynomial $Q(x) = x^3 + 4x$ have?

(c) If $a > 0$, how many x-intercepts and how many
local extrema does each of the polynomials
$P(x) = x^3 - ax$ and $Q(x) = x^3 + ax$ have? Explain
your answer.

DISCOVERY · DISCUSSION

69. Graphs of Large Powers Graph the functions $y = x^2$,
$y = x^3$, $y = x^4$, and $y = x^5$, for $-1 \le x \le 1$, on the
same coordinate axes. What do you think the graph of
$y = x^{100}$ would look like on this same interval? What
about $y = x^{101}$? Make a table of values to confirm your
answers.

70. Maximum Number of Local Extrema What is the

smallest possible degree that the polynomial whose
graph is shown can have? Explain.

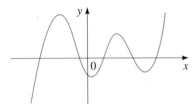

71. Possible Number of Local Extrema Is it possible for a
third-degree polynomial to have exactly one local extre-
mum? Can a fourth-degree polynomial have exactly two
local extrema? How many local extrema can polynomials of third, fourth, fifth, and sixth degree have? (Think
about the end behavior of such polynomials.) Now give
an example of a polynomial that has six local extrema.

72. Impossible Situation? Is it possible for a polynomial to
have two local maxima and no local minimum? Explain.

3.2 REAL ZEROS OF POLYNOMIALS

So far in this chapter we have been studying polynomial functions *graphically*. In
this section we begin to study polynomials *algebraically*. Most of our work will
be concerned with finding the solutions of polynomial equations, but first we consider division of polynomials.

DIVIDING POLYNOMIALS

Long division for polynomials is very much like the familiar process of long
division for numbers. For example, to divide $6x^2 - 26x + 12$ (the **dividend**) by
$x - 4$ (the **divisor**), we arrange our work as follows.

$$
\begin{array}{r}
\boxed{\text{quotient}} \\
\downarrow \\
6x - 2 \\
\boxed{\text{divisor}} \rightarrow \quad x - 4 \overline{)6x^2 - 26x + 12} \leftarrow \boxed{\text{dividend}} \\
\underline{6x^2 - 24x} \\
-2x + 12 \\
\underline{-2x + 8} \\
4 \leftarrow \boxed{\text{remainder}}
\end{array}
$$

Multiply divisor by $6x$

Subtract and "bring down" 12

Multiply divisor by -2

Subtract

Evariste Galois (1811–1832) is one of the very few mathematicians to have an entire theory named in his honor. Though he died while not yet 21, he completely settled the central problem in the theory of equations by describing a criterion that reveals whether any given equation can be solved by algebraic operations. Although Galois was one of the greatest mathematicians in the world at that time, no one knew it but him. He repeatedly sent his work to the eminent mathematicians Cauchy and Poisson, who either lost his letters or did not understand his ideas. Galois wrote in a terse style and included few details, which probably played a role in his failure to pass the entrance exams at the Ecole Polytechnique in Paris. Galois was a political radical and spent several months in prison for his revolutionary activities. His brief life came to a tragic end when he was killed in a duel over a love affair. The night before his duel, fearing that he would be killed, Galois wrote down the essence of his ideas and entrusted them to his friend Auguste Chevalier. He concluded by writing "... there will, I *(continued)*

The division process ends when the last line is of lesser degree than the divisor. The last line then contains the **remainder**, and the top line contains the **quotient**. The result of the division can be interpreted in either of two ways.

$$\frac{6x^2 - 26x + 12}{x - 4} = 6x - 2 + \frac{4}{x - 4}$$

or

$$6x^2 - 26x + 12 = (x - 4)(6x - 2) + 4$$

We summarize what happens in this or any such long division problem with the following theorem.

DIVISION ALGORITHM

If $P(x)$ and $D(x)$ are polynomials, with $D(x) \neq 0$, then there exist unique polynomials $Q(x)$ and $R(x)$ such that

$$P(x) = D(x) \cdot Q(x) + R(x)$$

where $R(x)$ is either 0 or of degree less than the degree of $D(x)$. The polynomials $P(x)$ and $D(x)$ are called the **dividend** and **divisor**, respectively, $Q(x)$ is the **quotient**, and $R(x)$ is the **remainder**.

EXAMPLE 1 ■ Long Division of Polynomials

Let $P(x) = 8x^4 + 6x^2 - 3x + 1$ and $D(x) = 2x^2 - x + 2$. Find polynomials $Q(x)$ and $R(x)$ such that $P(x) = D(x) \cdot Q(x) + R(x)$.

SOLUTION   We use long division after first inserting the term $0x^3$ into the dividend to ensure that the columns will line up correctly in the long division process.

$$
\begin{array}{r}
4x^2 + 2x \\
2x^2 - x + 2 \overline{\smash{)}\,8x^4 + 0x^3 + 6x^2 - 3x + 1} \\
\underline{8x^4 - 4x^3 + 8x^2 } \\
4x^3 - 2x^2 - 3x \\
\underline{4x^3 - 2x^2 + 4x } \\
-7x + 1
\end{array}
$$

The process is completed at this point because $-7x + 1$ is of lesser degree than the divisor $2x^2 - x + 2$. From the long division table, we see that $Q(x) = 4x^2 + 2x$ and $R(x) = -7x + 1$, so

$$8x^4 + 6x^2 - 3x + 1 = (2x^2 - x + 2)(4x^2 + 2x) + (-7x + 1) \qquad ■$$

Synthetic division is a quick method of dividing polynomials; it can be used when the divisor is of the form $x - c$. In synthetic division we write down only the essential parts of the long division table. Compare these long division and synthetic division tables, in which we divide $2x^3 - 7x^2 + 5$ by $x - 3$:

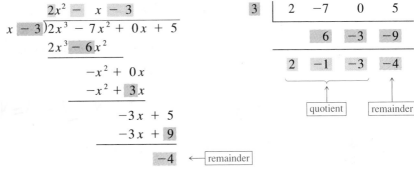

Long Division

$$
\begin{array}{r}
2x^2 - x - 3 \\
x - 3 \overline{\smash{)}2x^3 - 7x^2 + 0x + 5} \\
\underline{2x^3 - 6x^2} \\
-x^2 + 0x \\
\underline{-x^2 + 3x} \\
-3x + 5 \\
\underline{-3x + 9} \\
-4 \quad \leftarrow \boxed{\text{remainder}}
\end{array}
$$

In synthetic division we write down only the essential parts of the long division table. We abbreviate $2x^3 - 7x^2 + 5$ by writing only the coefficients: 2 −7 0 5, and instead of $x - 3$, we simply write 3. (Writing 3 instead of −3 allows us to add instead of subtract in the division process, but this changes the sign of all the numbers that appear in the gold boxes.)

Here is how you get the synthetic division table in practice. Start by writing the divisor and dividend:

$$
3 \, \big|\; 2 \;\; -7 \;\;\; 0 \;\;\; 5
$$

Bring down the 2, multiply $3 \cdot 2 = 6$, and write the result in the middle row. Then add:

$$
\begin{array}{c|cccc}
3 & 2 & -7 & 0 & 5 \\
  &   & 6 &   &   \\
\hline
  & 2 & -1 &   &
\end{array}
$$

Repeat this process of multiplying and then adding until the table is complete.

$$
\begin{array}{c|cccc}
3 & 2 & -7 & 0 & 5 \\
  &   & 6 & -3 & -9 \\
\hline
  & 2 & -1 & -3 & -4
\end{array}
$$

SYNTHETIC DIVISION

To divide $a_n x^n + a_{n-1} x^{n-1} + \cdots + a_1 x + a_0$ by $x - c$, we proceed as follows:

$$
\begin{array}{c|cccccccc}
c & a_n & a_{n-1} & a_{n-2} & a_{n-3} & \cdots & a_2 & a_1 & a_0 \\
  &   & cb_{n-1} & cb_{n-2} & cb_{n-3} & \cdots & cb_2 & cb_1 & cb_0 \\
\hline
  & b_{n-1} & b_{n-2} & b_{n-3} & b_{n-4} & \cdots & b_1 & b_0 & r
\end{array}
$$

Here $b_{n-1} = a_n$ and each number in the bottom row is obtained by adding the numbers above it. The remainder is r and the quotient is

$$b_{n-1} x^{n-1} + b_{n-2} x^{n-2} + \cdots + b_1 x + b_0$$

THE REMAINDER AND FACTOR THEOREMS

If the divisor in the Division Algorithm is of the form $x - c$ for some real number c, then the remainder must be a constant (since the degree of the remainder is less than the degree of the divisor). If we call this constant r, then

$$P(x) = (x - c) \cdot Q(x) + r$$

Setting $x = c$ in this equation, we get $P(c) = (c - c) \cdot Q(x) + r = 0 + r = r$. This proves the next theorem.

REMAINDER THEOREM

If the polynomial $P(x)$ is divided by $x - c$, then the remainder is the value $P(c)$.

EXAMPLE 2 ■ Using the Remainder Theorem to Find
              The Value of a Polynomial

Let $P(x) = 3x^5 + 5x^4 - 4x^3 + 7x + 3$.

(a) Find the quotient and remainder when $P(x)$ is divided by $x + 2$.
(b) Use the Remainder Theorem to find $P(-2)$.

SOLUTION

(a) Since $x + 2 = x - (-2)$, the synthetic division table for this problem takes the following form.

$$
\begin{array}{c|cccccc}
-2 & 3 & 5 & -4 & 0 & 7 & 3 \\
   &   & -6 & 2 & 4 & -8 & 2 \\
\hline
   & 3 & -1 & -2 & 4 & -1 & 5
\end{array}
$$

The quotient is $3x^4 - x^3 - 2x^2 + 4x - 1$ and the remainder is 5. Thus

$$3x^5 + 5x^4 - 4x^3 + 7x + 3 = (x + 2)(3x^4 - x^3 - 2x^2 + 4x - 1) + 5$$

or $\qquad \dfrac{3x^5 + 5x^4 - 4x^3 + 7x + 3}{x + 2} = 3x^4 - x^3 - 2x^2 + 4x - 1 + \dfrac{5}{x + 2}$

(b) By the Remainder Theorem, $P(-2)$ is the remainder when $P(x)$ is divided by $x - (-2) = x + 2$. From part (a), the remainder is 5, so $P(-2) = 5$. ∎

If the polynomial $P(x)$ factors as $P(x) = (x - c) \cdot Q(x)$, then

$$P(c) = (c - c) \cdot Q(c) = 0 \cdot Q(c) = 0$$

Conversely, if $P(c) = 0$, then by the Remainder Theorem

$$P(x) = (x - c) \cdot Q(x) + 0 = (x - c) \cdot Q(x)$$

so $x - c$ is a factor of $P(x)$. Thus, we have proved the following theorem.

FACTOR THEOREM

$P(c) = 0$ if and only if $x - c$ is a factor of $P(x)$.

EXAMPLE 3 ■ Factoring a Polynomial Using the Factor Theorem

Let $P(x) = x^3 - 7x + 6$. Show that $P(1) = 0$, and use this fact to factor $P(x)$ completely.

SOLUTION   Substituting, we see that $P(1) = 1^3 - 7 \cdot 1 + 6 = 0$. By the Factor Theorem, this means that $x - 1$ is a factor of $P(x)$. Using synthetic or long division (see the margin), we see that

$$P(x) = (x - 1)(x^2 + x - 6)$$

$$= (x - 1)(x - 2)(x + 3) \qquad \text{Factor the quadratic } x^2 + x - 6$$ ∎

$$
\begin{array}{r|rrrr}
1 & 1 & 0 & -7 & 6 \\
  &   & 1 & 1 & -6 \\
\hline
  & 1 & 1 & -6 & 0
\end{array}
$$

$$
\begin{array}{r}
x^2 + x - 6 \\
x - 1\overline{)x^3 + 0x^2 - 7x + 6} \\
\underline{x^3 - x^2} \\
x^2 - 7x \\
\underline{x^2 - x} \\
-6x + 6 \\
\underline{-6x + 6} \\
0
\end{array}
$$

Recall that if $P(c) = 0$, then the number c is a **zero** of the polynomial P. We also express this by saying that c is a **root** of the polynomial equation $P(x) = 0$. Thus, the zeros of the polynomial of Example 3 are 1, 2, and -3. With this terminology the Factor Theorem tells us that $x - c$ is a factor of $P(x)$ if and only if c is a zero of $P(x)$.

EXAMPLE 4 ■ Finding a Polynomial with Specified Zeros

Find a polynomial $F(x)$ of degree 4 that has zeros -3, 0, 1, and 5.

SOLUTION   By the Factor Theorem, $x - (-3)$, $x - 0$, $x - 1$, and $x - 5$ must all be factors of the desired polynomial, so let

$$F(x) = (x + 3)(x - 0)(x - 1)(x - 5) = x^4 - 3x^3 - 13x^2 + 15x$$

Since $F(x)$ is to have degree 4, any other solution of the problem must be a constant multiple of the polynomial we have chosen (because multiplication by any polynomial other than a constant will increase the degree).   ■

 ## RATIONAL ZEROS OF POLYNOMIALS

The Factor Theorem tells us that finding the zeros of a polynomial is really the same thing as factoring it into linear factors. We now study a method for finding all the *rational* zeros of a polynomial.

To help us understand the next theorem, consider the polynomial

$$P(x) = (x - 2)(x - 3)(x + 4) \qquad \text{Factored form}$$
$$= x^3 - x^2 - 14x + 24 \qquad \text{Expanded form}$$

From the factored form we see that the zeros of P are 2, 3, and -4. When the polynomial is expanded, the constant 24 is obtained by multiplying $(-2) \times (-3) \times 4$. This means that the zeros of the polynomial are all factors of the constant term. The next theorem generalizes this observation.

RATIONAL ZEROS THEOREM

If the polynomial $P(x) = a_n x^n + a_{n-1} x^{n-1} + \cdots + a_1 x + a_0$ has integer coefficients, then every rational zero of P is of the form

$$\frac{p}{q}$$

where p is a factor of the constant coefficient a_0

and q is a factor of the leading coefficient a_n.

■ **Proof** If p/q is a rational zero, in lowest terms, of the polynomial P, then we have

$$a_n \left(\frac{p}{q}\right)^n + a_{n-1}\left(\frac{p}{q}\right)^{n-1} + \cdots + a_1\left(\frac{p}{q}\right) + a_0 = 0$$

$$a_n p^n + a_{n-1} p^{n-1} q + \cdots + a_1 p q^{n-1} + a_0 q^n = 0 \qquad \text{Multiply by } q^n$$

$$p(a_n p^{n-1} + a_{n-1} p^{n-2} q + \cdots + a_1 q^{n-1}) = -a_0 q^n \qquad \begin{array}{l}\text{Subtract } a_0 q^n \\ \text{and factor LHS}\end{array}$$

Now p is a factor of the left side, so it must be a factor of the right as well. Since p/q is in lowest terms, p and q have no factor in common, and so p must be a factor of a_0. A similar proof shows that q is a factor of a_n.   □

If $a_n = 1$ in the Rational Zeros Theorem, then q must be 1 or -1, so in this case, any rational zero p/q is in fact an *integer* factor of a_0. Thus, if the highest-power term in a polynomial is just x^n, all its rational zeros are integers, not fractions.

EXAMPLE 5 ■ **Using the Rational Zeros Theorem to Factor a Polynomial**

Factor the polynomial $P(x) = 2x^3 + x^2 - 13x + 6$.

SOLUTION   By the Rational Zeros Theorem, if p/q is a zero of $P(x)$, then p divides 6 and q divides 2, so p/q is of the form

$$\frac{\text{factor of } 6}{\text{factor of } 2}$$

The factors of 6 are ± 1, ± 2, ± 3, ± 6, and the factors of 2 are ± 1, ± 2. Thus, the possible values of p/q are

$$\pm \tfrac{1}{1}, \quad \pm \tfrac{2}{1}, \quad \pm \tfrac{3}{1}, \quad \pm \tfrac{6}{1}, \quad \pm \tfrac{1}{2}, \quad \pm \tfrac{2}{2}, \quad \pm \tfrac{3}{2}, \quad \pm \tfrac{6}{2}$$

Simplifying the fractions and eliminating duplicates, we get the following list of possible values of p/q.

$$\pm 1, \quad \pm 2, \quad \pm 3, \quad \pm 6, \quad \pm \tfrac{1}{2}, \quad \pm \tfrac{3}{2}$$

Now we check to see which of these *possible* zeros actually *are* zeros by substituting them, one at a time, into the polynomial P until we find one that makes $P(x) = 0$. We have

$$P(1) = 2(1)^3 + (1)^2 - 13(1) + 6 = -4 \qquad \text{1 is } \textit{not} \text{ a zero of } P$$

$$P(2) = 2(2)^3 + (2)^2 - 13(2) + 6 = 0 \qquad \text{2 is a zero of } P$$

Since $x = 2$ is a zero of P, it follows that $x - 2$ is a factor of $P(x)$. Using synthetic division (as shown in the margin), we obtain the following factorization:

$$
\begin{array}{r|rrrr}
2 & 2 & 1 & -13 & 6 \\
  &   & 4 & 10 & -6 \\
\hline
  & 2 & 5 & -3 & 0
\end{array}
$$

$$
\begin{aligned}
P(x) &= 2x^3 + x^2 - 13x + 6 \\
&= (x - 2)(2x^2 + 5x - 3) \qquad \text{See margin} \\
&= (x - 2)(2x - 1)(x + 3) \qquad \text{Factor } 2x^2 + 5x - 3
\end{aligned}
$$

■

The following box explains how to use the Rational Zeros Theorem with synthetic division to factor a polynomial.

FINDING THE RATIONAL ZEROS OF A POLYNOMIAL

1. List all possible rational zeros using the Rational Zeros Theorem.

2. Use synthetic division to evaluate the polynomial at each of the candidates for rational zeros that you found in Step 1. When the remainder is 0, note the quotient you have obtained.

3. Repeat Steps 1 and 2 for the quotient. Stop when you reach a quotient that is quadratic or factors easily, and use the quadratic formula or factor to find the remaining zeros.

EXAMPLE 6 ■ Using the Rational Zeros Theorem and the Quadratic Formula

Find all solutions of the equation $x^4 - 5x^3 - 5x^2 + 23x + 10 = 0$.

SOLUTION   Let $P(x) = x^4 - 5x^3 - 5x^2 + 23x + 10$. The leading coefficient of P is 1, so all the rational zeros are integers: They are divisors of the constant term 10. Thus, the possible candidates are

$$\pm 1, \quad \pm 2, \quad \pm 5, \quad \pm 10$$

Using synthetic division (see the margin) we find that 1 and 2 are not zeros, but that 5 is a zero and that P factors as

$$x^4 - 5x^3 - 5x^2 + 23x + 10 = (x - 5)(x^3 - 5x - 2)$$

We now try to factor the quotient $x^3 - 5x - 2$. Its possible zeros are the divisors of -2, namely,

$$\pm 1, \quad \pm 2$$

Since we have already seen that 1 and 2 are not zeros of the original polynomial P, we don't need to try them again. Checking the remaining candidates -1 and -2, we see that -2 is a zero, and P factors as

$$x^4 - 5x^3 - 5x^2 + 23x + 10 = (x - 5)(x^3 - 5x - 2)$$

$$= (x - 5)(x + 2)(x^2 - 2x - 1)$$

Now we use the quadratic formula to obtain the two remaining zeros of P:

$$x = \frac{2 \pm \sqrt{(-2)^2 - 4(1)(-1)}}{2} = 1 \pm \sqrt{2}$$

The solutions of the equation are 5, -2, $1 + \sqrt{2}$, and $1 - \sqrt{2}$.   ■

Synthetic division (margin):

```
1 | 1  -5   -5   23   10
  |      1   -4   -9   14
  ------------------------
    1  -4   -9   14   24

2 | 1  -5   -5   23   10
  |      2   -6  -22    2
  ------------------------
    1  -3  -11    1   12

5 | 1  -5   -5   23   10
  |      5    0  -25  -10
  ------------------------
    1   0   -5   -2    0

-2 | 1   0   -5   -2
   |     -2    4    2
   ----------------------
     1  -2   -1    0
```

DESCARTES' RULE OF SIGNS AND UPPER AND LOWER BOUNDS FOR ROOTS

In some cases, the following rule—discovered by the French philosopher and mathematician René Descartes around 1637 (see page 88)—is helpful in eliminating candidates from lengthy lists of possible rational roots. To describe this rule, we need the concept of *variation in sign*. If $P(x)$ is a polynomial with real coefficients, written with descending powers of x (and omitting powers with coefficient 0), then a **variation in sign** occurs whenever adjacent coefficients have opposite signs. For example,

$$P(x) = 5x^7 - 3x^5 - x^4 + 2x^2 + x - 3$$

has three variations in sign.

> ### DESCARTES' RULE OF SIGNS
>
> Let $P(x)$ be a polynomial with real coefficients.
>
> 1. The number of positive real zeros of $P(x)$ is either equal to the number of variations in sign in $P(x)$ or is less than that by an even whole number.
>
> 2. The number of negative real zeros of $P(x)$ is either equal to the number of variations in sign in $P(-x)$ or is less than that by an even whole number.

EXAMPLE 7 ■ Using Descartes' Rule

Use Descartes' Rule of Signs to determine the possible number of positive and negative real zeros of the polynomial

$$P(x) = 3x^6 + 4x^5 + 3x^3 - x - 3$$

SOLUTION The polynomial has one variation in sign and so it has one positive zero. Now

$$P(-x) = 3(-x)^6 + 4(-x)^5 + 3(-x)^3 - (-x) - 3$$
$$= 3x^6 - 4x^5 - 3x^3 + x - 3$$

So, $P(-x)$ has three variations in sign. Thus, $P(x)$ has either three or one negative root(s), making a total of either two or four real roots. The remaining roots are imaginary numbers, which we will study in Section 3.3. ■

We say that a is a **lower bound** and b is an **upper bound** for the roots of a polynomial equation if every real root c of the equation satisfies $a \le c \le b$. The next theorem helps us find such bounds for any polynomial equation.

> ### THE UPPER AND LOWER BOUNDS THEOREM
>
> Let $P(x)$ be a polynomial with real coefficients.
>
> 1. If we divide $P(x)$ by $x - b$ (with $b > 0$) using synthetic division, and if the row that contains the quotient and remainder has no negative entry, then b is an upper bound for the real roots of $P(x) = 0$.
>
> 2. If we divide $P(x)$ by $x - a$ (with $a < 0$) using synthetic division, and if the row that contains the quotient and remainder has entries that are alternately nonpositive and nonnegative, then a is a lower bound for the real roots of $P(x) = 0$.

A proof of this theorem is suggested in Exercises 90 and 91. The phrase "alternately nonpositive and nonnegative" simply means that the signs of the numbers alternate, with zero considered to be positive or negative as required.

EXAMPLE 8 ■ Upper and Lower Bounds for Zeros of a Polynomial

Show that all the real roots of the equation $x^4 - 3x^2 + 2x - 5 = 0$ lie between -3 and 2.

SOLUTION   We divide the polynomial by $x - 2$ and $x + 3$ using synthetic division.

$$
\begin{array}{r|rrrrr}
2 & 1 & 0 & -3 & 2 & -5 \\
  &   & 2 & 4 & 2 & 8 \\
\hline
  & 1 & 2 & 1 & 4 & 3
\end{array}
\qquad \leftarrow \text{All entries positive}
$$

$$
\begin{array}{r|rrrrr}
-3 & 1 & 0 & -3 & 2 & -5 \\
   &   & -3 & 9 & -18 & 48 \\
\hline
   & 1 & -3 & 6 & -16 & 43
\end{array}
\qquad \leftarrow \text{Entries alternate in sign}
$$

By the Upper and Lower Bounds Theorem, -3 is a lower bound and 2 is an upper bound for the roots. Since neither -3 nor 2 is a root (the remainders are not 0 in the division table), all the real roots lie between these numbers.   ■

EXAMPLE 9 ■ Factoring a Fifth-Degree Polynomial

Factor completely the polynomial

$$P(x) = 2x^5 + 5x^4 - 8x^3 - 14x^2 + 6x + 9$$

SOLUTION   The possible rational zeros of $P(x)$ are $\pm\frac{1}{2}$, ± 1, $\pm\frac{3}{2}$, ± 3, $\pm\frac{9}{2}$, and ± 9. We check the positive candidates first, beginning with the smallest.

$$
\begin{array}{r|rrrrrr}
\frac{1}{2} & 2 & 5 & -8 & -14 & 6 & 9 \\
 & & 1 & 3 & -\frac{5}{2} & -\frac{33}{4} & -\frac{9}{8} \\
\hline
 & 2 & 6 & -5 & -\frac{33}{2} & -\frac{9}{4} & \frac{63}{8}
\end{array}
\qquad \leftarrow \frac{1}{2} \text{ is not a zero}
$$

$$
\begin{array}{r|rrrrrr}
1 & 2 & 5 & -8 & -14 & 6 & 9 \\
 & & 2 & 7 & -1 & -15 & -9 \\
\hline
 & 2 & 7 & -1 & -15 & -9 & 0
\end{array}
\qquad \leftarrow P(1) = 0
$$

So 1 is a zero, and $P(x) = (x - 1)(2x^4 + 7x^3 - x^2 - 15x - 9)$. We continue by factoring the quotient.

$$
\begin{array}{r|rrrrr}
1 & 2 & 7 & -1 & -15 & -9 \\
 & & 2 & 9 & 8 & -7 \\
\hline
 & 2 & 9 & 8 & -7 & -16
\end{array}
\qquad \leftarrow 1 \text{ is not a zero}
$$

$$
\begin{array}{r|rrrrr}
\frac{3}{2} & 2 & 7 & -1 & -15 & -9 \\
 & & 3 & 15 & 21 & 9 \\
\hline
 & 2 & 10 & 14 & 6 & 0
\end{array}
\qquad \leftarrow P\!\left(\frac{3}{2}\right) = 0, \text{ all entries nonnegative}
$$

We see that $\frac{3}{2}$ is both a zero and an upper bound for the zeros of $P(x)$, so we don't need to check any further for positive zeros, because all the remaining candidates are greater than $\frac{3}{2}$.

$$P(x) = (x - 1)\left(x - \frac{3}{2}\right)(2x^3 + 10x^2 + 14x + 6)$$

$$= (x - 1)(2x - 3)(x^3 + 5x^2 + 7x + 3)$$

Factor 2 from last factor, multiply into second factor

By Descartes' Rule of Signs, $x^3 + 5x^2 + 7x + 3$ has no positive zero, so the only possible rational zeros are -1 and -3.

$$
\begin{array}{r|rrrr}
-1 & 1 & 5 & 7 & 3 \\
   &   & -1 & -4 & -3 \\
\hline
   & 1 & 4 & 3 & 0 \quad \leftarrow P(-1) = 0
\end{array}
$$

Therefore
$$P(x) = (x - 1)(2x - 3)(x + 1)(x^2 + 4x + 3)$$
$$= (x - 1)(2x - 3)(x + 1)^2(x + 3)$$

This means that the zeros of P are $1, \frac{3}{2}, -1$, and -3.   ■

USING ALGEBRA AND GRAPHING DEVICES TO SOLVE POLYNOMIAL EQUATIONS

In Section 1.9 we used graphing devices to solve equations graphically. We can now use the algebraic techniques we've learned to help us select an appropriate viewing rectangle when solving polynomial equations.

EXAMPLE 10 ■ Solving a Fourth-Degree Equation

Find all real solutions of the following equation, correct to the nearest tenth.

$$3x^4 + 4x^3 - 7x^2 - 2x - 3 = 0$$

SOLUTION First we use the Upper and Lower Bounds Theorem to find two numbers between which all the solutions must lie. This will allow us to choose a viewing rectangle that is certain to contain all the x-intercepts of the polynomial function. We use synthetic division and proceed by trial and error.

> We use the Upper and Lower Bounds Theorem to see where the roots can be found.

To find an upper bound, we try the whole numbers $1, 2, 3, \ldots$ as potential candidates. We see that 2 is an upper bound for the roots.

$$
\begin{array}{r|rrrrr}
1 & 3 & 4 & -7 & -2 & -3 \\
  &   & 3 & 7 & 0 & -2 \\
\hline
  & 3 & 7 & 0 & -2 & -5
\end{array}
\qquad
\begin{array}{r|rrrrr}
2 & 3 & 4 & -7 & -2 & -3 \\
  &   & 6 & 20 & 26 & 48 \\
\hline
  & 3 & 10 & 13 & 24 & 45 \quad \leftarrow \text{All} \\
  &   &   &   &   &   \text{positive}
\end{array}
$$

Now we look for a lower bound, trying the numbers $-1, -2$, and -3 as potential candidates.

$$
\begin{array}{r|rrrrr}
-1 & 3 & 4 & -7 & -2 & -3 \\
   &   & -3 & -1 & 8 & -6 \\
\hline
   & 3 & 1 & -8 & 6 & -9
\end{array}
\qquad
\begin{array}{r|rrrrr}
-2 & 3 & 4 & -7 & -2 & -3 \\
   &   & -6 & 4 & 6 & -8 \\
\hline
   & 3 & -2 & -3 & 4 & -11
\end{array}
\qquad
\begin{array}{r|rrrrr}
-3 & 3 & 4 & -7 & -2 & -3 \\
   &   & -9 & 15 & -24 & 78 \\
\hline
   & 3 & -5 & 8 & -26 & 75 \quad \leftarrow \text{Entries} \\
   &   &   &   &   &   \text{alternate} \\
   &   &   &   &   &   \text{in sign}
\end{array}
$$

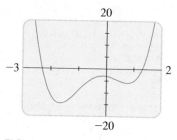

FIGURE 1
$y = 3x^4 + 4x^3 - 7x^2 - 2x - 3$

Thus, -3 is a lower bound for the roots. So the viewing rectangle $[-3, 2]$ by $[-20, 20]$ contains all the x-intercepts of $y = 3x^4 + 4x^3 - 7x^2 - 2x - 3$. The graph in Figure 1 has two x-intercepts, one between -3 and -2 and the other between 1 and 2. Zooming in, we find that the solutions of the equation, to the nearest tenth, are -2.3 and 1.3.   ■

EXAMPLE 11 ■ Determining the Size of a Fuel Tank

A fuel tank consists of a cylindrical center section 4 ft long and two hemispherical end sections, as shown in Figure 2. If the tank has a volume of 100 ft^3, what is the radius r shown in the figure, correct to the nearest hundredth of a foot?

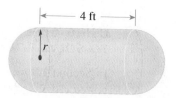

FIGURE 2

SOLUTION   Using the volume formula listed on the inside of the front cover of this book, we see that the volume of the cylindrical section of the tank is

$$\pi \cdot r^2 \cdot 4$$

Volume of a cylinder: $V = \pi r^2 h$

The two hemispherical parts together form a complete sphere whose volume is

$$\tfrac{4}{3}\pi r^3$$

Volume of a sphere: $V = \tfrac{4}{3}\pi r^3$

Because the total volume of the tank is 100 ft^3, we get the following equation:

$$\tfrac{4}{3}\pi r^3 + 4\pi r^2 = 100$$

A negative solution for r would be meaningless in this physical situation, and by substitution we can verify that $r = 3$ leads to a tank that is over 226 ft^3 in volume, much larger than the required 100 ft^3. Thus, we know the correct radius lies somewhere between 0 and 3 ft, and so we use a viewing rectangle of $[0, 3]$ by $[50, 150]$ to graph the function $y = \tfrac{4}{3}\pi x^3 + 4\pi x^2$, as shown in Figure 3. Since we want the value of this function to be 100, we also graph the horizontal line $y = 100$ in the same viewing rectangle. The correct radius will be the x-coordinate of the point of intersection of the curve and the line. Using the cursor and zooming in, we see that at the point of intersection $x \approx 2.15$, correct to two decimal places. Thus, the tank has a radius of about 2.15 ft.   ■

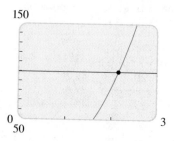

FIGURE 3
$y = \tfrac{4}{3}\pi x^3 + 4\pi x^2$ and $y = 100$

Note that we also could have solved the equation in Example 11 by first writing it as

$$\tfrac{4}{3}\pi r^3 + 4\pi r^2 - 100 = 0$$

and then finding the x-intercept of the function $y = \tfrac{4}{3}\pi x^3 + 4\pi x^2 - 100$.

1–12 ■ Find the quotient and remainder.

1. $\dfrac{x^3 + 2x^2 + 2x + 1}{x + 2}$

2. $\dfrac{3x^3 - 12x^2 - 9x + 1}{x - 5}$

3. $\dfrac{x^3 - 8x + 2}{x + 3}$

4. $\dfrac{x^4 - x^3 + x^2 - x + 2}{x - 2}$

5. $\dfrac{x^5 + 3x^3 - 6}{x - 1}$

6. $\dfrac{x^3 - 9x^2 + 27x - 27}{x - 3}$

7. $\dfrac{x^3 + 6x + 3}{x^2 - 2x + 2}$

8. $\dfrac{3x^4 - 5x^3 - 20x - 5}{x^2 + x + 3}$

9. $\dfrac{6x^3 + 2x^2 + 22x}{2x^2 + 5}$

10. $\dfrac{9x^2 - x + 5}{3x^2 - 7x}$

11. $\dfrac{2x^3 + 3x^2 - 2x + 1}{x - \frac{1}{2}}$

12. $\dfrac{6x^4 + 10x^3 + 5x^2 + x + 1}{x + \frac{2}{3}}$

13–19 ■ Use synthetic division and the Remainder Theorem to evaluate $P(c)$.

13. $P(x) = 4x^2 + 12x + 5, \quad c = -1$

14. $P(x) = 2x^2 + 9x + 1, \quad c = \frac{1}{2}$

15. $P(x) = x^3 + 3x^2 - 7x + 6, \quad c = 2$

16. $P(x) = 2x^3 - 21x^2 + 9x - 200, \quad c = 11$

17. $P(x) = 5x^4 + 30x^3 - 40x^2 + 36x + 14, \quad c = -7$

18. $P(x) = 6x^5 + 10x^3 + x + 1, \quad c = -2$

19. $P(x) = 3x^3 + 4x^2 - 2x + 1, \quad c = \frac{2}{3}$

20. Let

$$P(x) = 6x^7 - 40x^6 + 16x^5 - 200x^4$$
$$- 60x^3 - 69x^2 + 13x - 139$$

Calculate $P(7)$ by (a) using synthetic division, and (b) substituting $x = 7$ into the polynomial and evaluating directly.

21–24 ■ List all possible rational zeros given by the Rational Zeros Theorem (but don't check to see which values actually are zeros).

21. $P(x) = x^3 - 2x^2 - 5x + 3$

22. $Q(x) = x^3 - 4x^2 + 6x + 20$

23. $R(x) = 2x^4 - 3x^3 - x + 6$

24. $S(x) = 6x^4 - x^3 + x^2 - 12$

25–44 ■ Find all rational roots of the equation, and then use the quadratic formula to find the irrational roots, if any.

25. $x^3 - 3x^2 - 4x + 12 = 0$

26. $x^3 - x^2 - 8x + 12 = 0$

27. $x^3 - 4x^2 + x + 6 = 0$

28. $x^3 + 3x^2 - 4 = 0$

29. $x^3 - 7x^2 + 14x - 8 = 0$

30. $x^3 - 4x^2 - 7x + 10 = 0$

31. $x^3 + 3x^2 + 6x + 4 = 0$

32. $x^3 - 2x^2 - 2x - 3 = 0$

33. $x^4 - 5x^2 + 4 = 0$

34. $x^4 - 2x^3 - 3x^2 + 8x - 4 = 0$

35. $x^4 + 6x^3 + 7x^2 - 6x - 8 = 0$

36. $x^4 - x^3 - 23x^2 - 3x + 90 = 0$

37. $4x^4 - 25x^2 + 36 = 0$

38. $x^4 - x^3 - 5x^2 + 3x + 6 = 0$

39. $4x^3 - 7x + 3 = 0$

40. $2x^3 - 3x^2 - 2x + 3 = 0$

41. $4x^3 + 4x^2 - x - 1 = 0$

42. $8x^3 + 10x^2 - x - 3 = 0$

43. $x^5 - 4x^4 - 3x^3 + 22x^2 - 4x - 24 = 0$

44. $3x^5 - 14x^4 - 14x^3 + 36x^2 + 43x + 10 = 0$

45–46 ■ Show that the polynomial does not have any rational zeros.

45. $x^3 - x - 2$

46. $2x^4 - x^3 + x + 2$

47–50 ■ Find a polynomial of the specified degree that has the given zeros.

47. Degree 3; zeros $-1, 1, 3$

48. Degree 4; zeros $-2, 0, 2, 4$

49. Degree 4; zeros $-1, 1, 3, 5$

50. Degree 5; zeros $-2, -1, 0, 1, 2$

51–56 ■ Use Descartes' Rule of Signs to determine how many positive and how many negative real zeros the polynomial can have. Then determine the possible total number of real zeros.

51. $x^3 - x^2 - x - 3$

52. $2x^3 - x^2 + 4x - 7$

53. $x^4 + x^3 + x^2 + x + 12$

54. $x^5 + 4x^3 - x^2 + 6x$

55. $4x^7 - 3x^5 + 5x^4 + x^3 - 3x^2 + 2x - 5$

56. $x^8 - x^5 + x^4 - x^3 + x^2 - x + 1$

57–60 ■ Show that the given values for a and b are lower and upper bounds, respectively, for the real roots of the equation.

57. $2x^3 + 5x^2 + x - 2 = 0$; $a = -3, b = 1$

58. $x^4 - 2x^3 - 9x^2 + 2x + 8 = 0$; $a = -3, b = 5$

59. $8x^3 + 10x^2 - 39x + 9 = 0$; $a = -3, b = 2$

60. $3x^4 - 17x^3 + 24x^2 - 9x + 1 = 0$; $a = 0, b = 6$

61–64 ■ Find integers that are upper and lower bounds for the real roots of the equation.

61. $x^3 - 3x^2 + 4 = 0$

62. $2x^3 - 3x^2 - 8x + 12 = 0$

63. $x^4 - 2x^3 + x^2 - 9x + 2 = 0$

64. $x^5 - x^4 + 1 = 0$

65–70 ■ Find all rational roots of the equation, and then find the irrational roots, if any. Whenever appropriate, use the Rational Zeros Theorem, the Upper and Lower Bounds Theorem, Descartes' Rule of Signs, the quadratic formula, or other factoring techniques.

65. $2x^4 + 3x^3 - 4x^2 - 3x + 2 = 0$

66. $2x^4 + 15x^3 + 31x^2 + 20x + 4 = 0$

67. $4x^4 - 21x^2 + 5 = 0$

68. $6x^4 - 7x^3 - 8x^2 + 5x = 0$

69. $x^5 - 7x^4 + 9x^3 + 23x^2 - 50x + 24 = 0$

70. $8x^5 - 14x^4 - 22x^3 + 57x^2 - 35x + 6 = 0$

71–74 ■ All the real solutions of the given equation are rational. List all possible rational roots using the Rational Zeros Theorem, and then graph the polynomial in the given viewing rectangle to determine which values are actually the solutions of the equation. (All solutions can be seen in the given viewing rectangle.)

71. $x^3 - 3x^2 - 4x + 12 = 0$; $[-4, 4]$ by $[-15, 15]$

72. $x^4 - 5x^2 + 4 = 0$; $[-4, 4]$ by $[-30, 30]$

73. $2x^4 - 5x^3 - 14x^2 + 5x + 12 = 0$;
$[-2, 5]$ by $[-40, 40]$

74. $3x^3 + 8x^2 + 5x + 2 = 0$; $[-3, 3]$ by $[-10, 10]$

75–82 ■ Use a graphing device to find all real solutions of the equation, correct to two decimal places.

75. $x^3 - 5x^2 - 4 = 0$

76. $x^4 - x - 4 = 0$

77. $3x^3 + x^2 + x - 2 = 0$

78. $2x^3 - 8x^2 + 9x - 9 = 0$

79. $10x^4 - 9x^3 - 11x^2 + 5x - 3 = 0$

80. $3x^4 + 8x^3 + 2x^2 + 5x + 2 = 0$

81. $4.00x^4 + 4.00x^3 - 10.96x^2 - 5.88x + 9.09 = 0$

82. $x^5 + 2.00x^4 + 0.96x^3 + 5.00x^2 + 10.00x + 4.80 = 0$

83. A grain silo consists of a cylindrical main section and a hemispherical roof. If the total volume of the silo (including the part inside the roof section) is 15,000 ft^3 and the cylindrical part is 30 ft tall, what is the radius of the silo, correct to the nearest tenth of a foot?

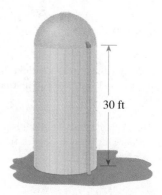

30 ft

84. A rectangular parcel of land has an area of 5000 ft². A diagonal between opposite corners is measured to be 10 ft longer than one side of the parcel. What are the dimensions of the land, correct to the nearest foot?

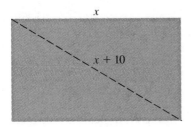

 85. Snow began falling at noon on Sunday. The amount of snow on the ground at a certain location at time t was given by the function

$$h(t) = 11.60t - 12.41t^2 + 6.20t^3$$
$$- 1.58t^4 + 0.20t^5 - 0.01t^6$$

where t is measured in days from the start of the snowfall and $h(t)$ is the depth of snow in inches. Draw a graph of this function and use your graph to answer the following questions.

(a) What happened shortly after noon on Tuesday?

(b) Was there ever more than 5 in. of snow on the ground? If so, on what day(s)?

(c) On what day and at what time (to the nearest hour) did the snow disappear completely?

 86. An open box with a volume of 1500 cm³ is to be constructed by taking a piece of cardboard 20 cm by 40 cm, cutting squares of side length x cm from each corner, and folding up the sides. Show that this can be done in two different ways, and find the exact dimensions of the box in each case.

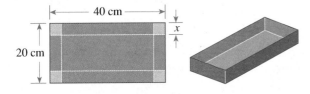

 87. A rocket consists of a right circular cylinder of height 20 m surmounted by a cone whose height and diameter are equal and whose radius is the same as that of the cylindrical section. What should this radius be (correct to two decimal places) if the total volume is to be $500\pi/3$ m³?

 88. A rectangular box with a volume of $2\sqrt{2}$ ft³ has a square base. The diagonal of the box (between a pair of opposite corners) is 1 ft longer than each side of the base.

(a) If the base has sides of length x feet, show that

$$x^6 - 2x^5 - x^4 + 8 = 0$$

(b) Show that there are two different boxes that satisfy the given conditions. Find the dimensions in each case, correct to the nearest hundredth of a foot.

 89. A box with a square base has length plus girth of 108 in. (Girth is the distance "around" the box.) What is the length of the box if its volume is 2200 in³?

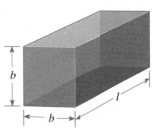

90. Let $P(x)$ be a polynomial with real coefficients and let $b > 0$. Use the Division Algorithm to write

$$P(x) = (x - b) \cdot Q(x) + r$$

Suppose that $r \geq 0$ and that all the coefficients in $Q(x)$ are nonnegative. Let $z > b$.
(a) Show that $P(z) > 0$.
(b) Prove the first part of the Upper and Lower Bounds Theorem.

91. Use the first part of the Upper and Lower Bounds Theorem to prove the second part.   [*Hint:* Show that if $P(x)$ satisfies the second part of the theorem, then $P(-x)$ satisfies the first part.]

92. Show that the equation

$$x^5 - x^4 - x^3 - 5x^2 - 12x - 6 = 0$$

has exactly one rational root, and then prove that it must have either two or four irrational roots.

■ **DISCOVERY · DISCUSSION**

93. Impossible Division?   Suppose you were asked to solve the following two problems on a test:

Find the remainder when $6x^{1000} - 17x^{562} + 12x + 26$ is divided by $x + 1$.

Is $x - 1$ a factor of $x^{567} - 3x^{400} + x^9 + 2$?

Obviously, it's impossible to solve these problems by dividing, because the polynomials are of such large degree. Use one or more of the theorems in this section to solve these problems *without* actually dividing.

94. How Many Real Zeros Can a Polynomial Have?   Give examples of polynomials that have the following properties, or explain why it is impossible to find such a polynomial.
(a) A polynomial of degree 3 that has no real zeros
(b) A polynomial of degree 4 that has no real zeros
(c) A polynomial of degree 3 that has three real zeros, only one of which is rational
(d) A polynomial of degree 4 that has four real zeros, none of which is rational
What must be true about the degree of a polynomial with integer coefficients if it has no real zeros?

95. Nested Form of a Polynomial   Expand Q to prove that the polynomials P and Q are the same.

$$P(x) = 3x^4 - 5x^3 + x^2 - 3x + 5$$

$$Q(x) = (((3x - 5)x + 1)x - 3)x + 5$$

Try to evaluate $P(2)$ and $Q(2)$ in your head, using the forms given. Which is easier? Now write the polynomial $R(x) = x^5 - 2x^4 + 3x^3 - 2x^2 + 3x + 4$ in "nested" form, like the polynomial Q. Use the nested form to find $R(3)$ in your head.

3.3   COMPLEX NUMBERS

In Section 1.5, we saw that if the discriminant of a quadratic equation is negative, the equation has no real solution. For example, the equation

$$x^2 + 4 = 0$$

has no real solution. If we try to solve this equation, we get $x^2 = -4$, so

$$x = \pm\sqrt{-4}$$

But this is impossible, since the square of any real number is positive. [For example, $(-2)^2 = 4$, a positive number.] Thus, negative numbers don't have real square roots.

To make it possible to solve *all* quadratic equations, mathematicians invented an expanded number system, called the *complex number system*. First they defined the new number

$$i = \sqrt{-1}$$

This means $i^2 = -1$. A complex number is then a number of the form $a + bi$, where a and b are real numbers.

Leonhard Euler (1707–1783) was born in Basel, Switzerland, the son of a pastor. At age 13 his father sent him to the University at Basel to study theology, but Euler soon decided to devote himself to the sciences. Besides theology he studied mathematics, medicine, astronomy, physics, and oriental languages. It is said that Euler could calculate as effortlessly as "men breathe or as eagles fly." One hundred years before Euler, Fermat (see page 458) had conjectured that $2^{2^n} + 1$ is a prime number for all n. The first five of these numbers are 5, 17, 257, 65537, and 4,294,967,297. It is easy to show that the first four are prime. The fifth was also thought to be prime until Euler, with his phenomenal calculating ability, showed that it is the product $641 \times 6,700,417$ and so is not prime. Euler published more than any other mathematician in history. His collected works comprise 75 large volumes. Although he was blind for the last 17 years of his life, he continued to work and publish. In his writings he popularized the use of the symbols π, e, and i, which you will find in this textbook. One of Euler's most lasting contributions is the development of complex numbers.

DEFINITION OF COMPLEX NUMBERS

A **complex number** is an expression of the form

$$a + bi$$

where a and b are real numbers and $i^2 = -1$. The **real part** of this complex number is a and the **imaginary part** is b. Two complex numbers are **equal** if and only if their real parts are equal and their imaginary parts are equal.

Note that both the real and imaginary parts of a complex number are real numbers.

EXAMPLE 1 ■ Complex Numbers

The following are examples of complex numbers.

$3 + 4i$	Real part 3, imaginary part 4
$\frac{1}{2} - \frac{2}{3}i$	Real part $\frac{1}{2}$, imaginary part $-\frac{2}{3}$
$6i$	Real part 0, imaginary part 6
-7	Real part -7, imaginary part 0

A number such as $6i$, which has real part 0, is called a **pure imaginary number**. A real number like -7 can be thought of as a complex number with imaginary part 0.

In the complex number system every quadratic equation has solutions. The numbers $2i$ and $-2i$ are solutions of $x^2 = -4$ because

$$(2i)^2 = 2^2 i^2 = 4(-1) = -4 \qquad \text{and} \qquad (-2i)^2 = (-2)^2 i^2 = 4(-1) = -4$$

Although we use the term *imaginary* in this context, imaginary numbers should not be thought of as any less "real" (in the ordinary rather than the mathematical sense of that word) than negative numbers or irrational numbers. All numbers (except possibly the positive integers) are creations of the human mind—the numbers -1 and $\sqrt{2}$ as well as the number i. We study complex numbers because they complete, in a useful and elegant fashion, the study of the solutions of polynomial equations. (See Cardano, page 256.) In fact, imaginary numbers are useful not only in algebra and mathematics, but in the other sciences as well. To give just one example, in electrical theory the *reactance* of a circuit is a quantity whose measure is an imaginary number.

ARITHMETIC OPERATIONS ON COMPLEX NUMBERS

Complex numbers are added, subtracted, multiplied, and divided just as we would any number of the form $a + b\sqrt{c}$. The only difference we must keep in mind is that $i^2 = -1$. Thus, the following calculations are valid.

$$(a + bi)(c + di) = ac + (ad + bc)i + bdi^2 \qquad \text{Multiply and collect like terms}$$

$$= ac + (ad + bc)i + bd(-1) \qquad i^2 = -1$$

$$= (ac - bd) + (ad + bc)i \qquad \text{Combine real and imaginary parts}$$

We therefore define the sum, difference, and product of complex numbers as follows.

ADDING, SUBTRACTING, AND MULTIPLYING COMPLEX NUMBERS

Definition	Description
Addition	
$(a + bi) + (c + di) = (a + c) + (b + d)i$	To add complex numbers, add the real parts and the imaginary parts.
Subtraction	
$(a + bi) - (c + di) = (a - c) + (b - d)i$	To subtract complex numbers, subtract the real parts and the imaginary parts.
Multiplication	
$(a + bi) \cdot (c + di) = (ac - bd) + (ad + bc)i$	Multiply complex numbers like binomials, using $i^2 = -1$.

EXAMPLE 2 ■ Adding, Subtracting, and Multiplying Complex Numbers

Express each of the following in the form $a + bi$.

(a) $(3 + 5i) + (4 - 2i)$ (b) $(3 + 5i) - (4 - 2i)$

(c) $(3 + 5i)(4 - 2i)$ (d) i^{23}

SOLUTION

(a) According to the definition, we add the real parts and we add the imaginary parts.

$$(3 + 5i) + (4 - 2i) = (3 + 4) + (5 - 2)i = 7 + 3i$$

(b) $(3 + 5i) - (4 - 2i) = (3 - 4) + [5 - (-2)]i = -1 + 7i$

(c) $(3 + 5i)(4 - 2i) = [3 \cdot 4 - 5(-2)] + [3(-2) + 5 \cdot 4]i = 22 + 14i$

(d) $i^{23} = i^{20+3} = (i^2)^{10}i^3 = (-1)^{10}i^2i = (1)(-1)i = -i$ ■

Complex Conjugates

Number	Conjugate
$3 + 2i$	$3 - 2i$
$1 - i$	$1 + i$
$4i$	$-4i$
5	5

Division of complex numbers is much like rationalizing the denominator of a radical expression, which we considered in Section 1.4. For the complex number $z = a + bi$ we define its **complex conjugate** to be $\bar{z} = a - bi$. Note that $z \cdot \bar{z} = (a + bi)(a - bi) = a^2 + b^2$. So the product of a complex number and its conjugate is always a nonnegative real number. We use this property to divide complex numbers.

DIVIDING COMPLEX NUMBERS

To simplify the quotient $\dfrac{a + bi}{c + di}$, multiply the numerator and the denominator by the complex conjugate of the denominator:

$$\frac{a + bi}{c + di} = \left(\frac{a + bi}{c + di} \right) \left(\frac{c - di}{c - di} \right) = \frac{(ac + bd) + (bc - ad)i}{c^2 + d^2}$$

Rather than memorize this entire formula, it's best just to remember the first step and then multiply out the numerator and the denominator as usual.

EXAMPLE 3 ■ Dividing Complex Numbers

Express each of the following in the form $a + bi$.

(a) $\dfrac{3 + 5i}{1 - 2i}$

(b) $\dfrac{7 + 3i}{4i}$

SOLUTION We multiply both numerator and denominator by the complex conjugate of the denominator to make the new denominator a real number.

(a) The complex conjugate of $1 - 2i$ is $\overline{1 - 2i} = 1 + 2i$.

$$\frac{3 + 5i}{1 - 2i} = \left(\frac{3 + 5i}{1 - 2i} \right) \left(\frac{1 + 2i}{1 + 2i} \right) = \frac{-7 + 11i}{5} = -\frac{7}{5} + \frac{11}{5}i$$

(b) The complex conjugate of $4i$ is $-4i$. Therefore

$$\frac{7 + 3i}{4i} = \left(\frac{7 + 3i}{4i} \right) \left(\frac{-4i}{-4i} \right) = \frac{12 - 28i}{16} = \frac{3}{4} - \frac{7}{4}i \qquad \blacksquare$$

Just as every positive real number r has two square roots ($\sqrt{r}$ and $-\sqrt{r}$), every negative number has two square roots as well. Both square roots are pure imaginary numbers, for if $r > 0$ is real, then

$$(i\sqrt{r})^2 = i^2 r = -r \qquad \text{and} \qquad (-i\sqrt{r})^2 = (-1)^2 i^2 r = -r$$

We call $i\sqrt{r}$ the *principal square root* of $-r$, and we will use the symbol $\sqrt{-r}$ to denote this principal square root. The other square root will then be $-\sqrt{-r} = -i\sqrt{r}$. Note that the two square roots of a negative real number are complex conjugates of each other.

> ### SQUARE ROOTS OF NEGATIVE NUMBERS
>
> If $-r < 0$, then the square roots of $-r$ are
>
> $$i\sqrt{r} \quad \text{and} \quad -i\sqrt{r}$$
>
> The **principal square root** of $-r$ is $i\sqrt{r}$.

We usually write $i\sqrt{b}$ instead of $\sqrt{b}\,i$ to avoid confusion with $\sqrt{bi}$.

EXAMPLE 4 ■ Square Roots of Negative Numbers

(a) $\sqrt{-1} = i\sqrt{1} = i$ (b) $\sqrt{-16} = i\sqrt{16} = 4i$ (c) $\sqrt{-3} = i\sqrt{3}$ ■

Special care must be taken when performing calculations involving square roots of negative numbers. Although $\sqrt{a} \cdot \sqrt{b} = \sqrt{ab}$ when a and b are positive, this is *not* true when both are negative. For example,

$$\sqrt{-2} \cdot \sqrt{-3} = i\sqrt{2} \cdot i\sqrt{3} = i^2\sqrt{6} = -\sqrt{6}$$

but $$\sqrt{(-2)(-3)} = \sqrt{6}$$

so $$\sqrt{-2} \cdot \sqrt{-3} \ne \sqrt{(-2)(-3)}$$

When multiplying radicals of negative numbers, express them first in the form $i\sqrt{r}$ (where $r > 0$) to avoid possible errors of this type.

EXAMPLE 5 ■ Using Square Roots of Negative Numbers

Evaluate $\left(\sqrt{12} - \sqrt{-3}\right)\left(3 + \sqrt{-4}\right)$ and express in the form $a + bi$.

SOLUTION

$$\left(\sqrt{12} - \sqrt{-3}\right)\left(3 + \sqrt{-4}\right) = \left(\sqrt{12} - i\sqrt{3}\right)\left(3 + i\sqrt{4}\right)$$
$$= \left(2\sqrt{3} - i\sqrt{3}\right)\left(3 + 2i\right)$$
$$= \left(6\sqrt{3} + 2\sqrt{3}\right) + i\left(2 \cdot 2\sqrt{3} - 3\sqrt{3}\right)$$
$$= 8\sqrt{3} + i\sqrt{3}$$ ■

GRAPHING COMPLEX NUMBERS

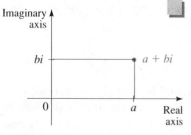

FIGURE 1

To graph real numbers or sets of real numbers, we have been using the number line, which has just one dimension. Complex numbers, however, have two components: the real part and the imaginary part. This suggests that we need two axes to graph complex numbers: one for the real part and one for the imaginary part. We call these the **real axis** and the **imaginary axis**, respectively. The plane determined by these two axes is called the **complex plane**. To graph the complex number $a + bi$, we plot the ordered pair of numbers (a, b) in this plane, as indicated in Figure 1.

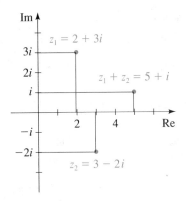

FIGURE 2

EXAMPLE 6 ■ Graphing Complex Numbers

Graph the complex numbers $z_1 = 2 + 3i$, $z_2 = 3 - 2i$, and $z_1 + z_2$.

SOLUTION   We have $z_1 + z_2 = (2 + 3i) + (3 - 2i) = 5 + i$. The graph is shown in Figure 2. ■

EXAMPLE 7 ■ Graphing Sets of Complex Numbers

Graph each of the following sets of complex numbers.

(a) $S = \{a + bi \mid a \geq 0\}$ (b) $T = \{a + bi \mid a < 1, b \geq 0\}$

SOLUTION

(a) S is the set of complex numbers whose real part is nonnegative. The graph is shown in Figure 3(a).

(b) T is the set of complex numbers for which the real part is less than 1 and the imaginary part is nonnegative. The graph is shown in Figure 3(b).

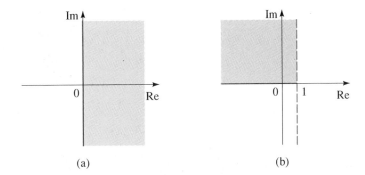

FIGURE 3   (a)   (b)   ■

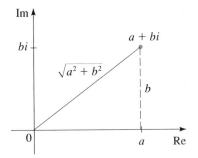

FIGURE 4

The plural of *modulus* is *moduli*.

Recall from Section 1.1 that the absolute value of a real number can be thought of as its distance from the origin on the real number line. We define absolute value for complex numbers in a similar fashion. From Figure 4 we can see, using the Pythagorean Theorem, that the distance between $a + bi$ and the origin in the complex plane is $\sqrt{a^2 + b^2}$. This leads to the following definition.

> The **modulus** (or **absolute value**) of the complex number $z = a + bi$ is
> $$|z| = \sqrt{a^2 + b^2}$$

EXAMPLE 8 ■ Calculating the Modulus

Find the moduli of the complex numbers $3 + 4i$ and $8 - 5i$.

SOLUTION

$$|3 + 4i| = \sqrt{3^2 + 4^2} = \sqrt{25} = 5$$
$$|8 - 5i| = \sqrt{8^2 + (-5)^2} = \sqrt{89}$$

■

EXAMPLE 9 ■ Absolute Value of Complex Numbers

Graph each of the following sets of complex numbers.

(a) $C = \{z \mid |z| = 1\}$ (b) $D = \{z \mid |z| \leq 1\}$

SOLUTION

(a) C is the set of complex numbers whose distance from the origin is 1. Thus, C is a circle of radius 1 with center at the origin.

(b) D is the set of complex numbers whose distance from the origin is less than or equal to 1. Thus, D is the disk that consists of all complex numbers on and inside the circle C of part (a).

The graphs of C and D are shown in Figure 5.

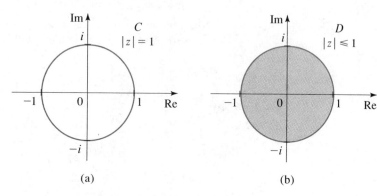

FIGURE 5   (a)   (b)   ■

1–6 ■ Find the real and imaginary parts of the complex number.

1. $3 - 5i$

2. $\dfrac{2 + 4i}{3}$

3. $6i$

4. $\frac{1}{2}$

5. $\sqrt{2} + \sqrt{-3}$

6. $\dfrac{2 + 4i}{\sqrt{-16}}$

7–40 ■ Evaluate the expression and write the result in the form $a + bi$.

7. $(4 + 3i) + (5 - 2i)$

8. $(7 - 6i) + (-3 + 7i)$

9. $\left(7 - \frac{1}{2}i\right) + \left(5 + \frac{3}{2}i\right)$

10. $(-4 + i) - (2 - 5i)$

11. $(-12 + 8i) - (7 + 4i)$

12. $6i - (4 - i)$

13. $4(-1 + 2i)$

14. $2i\left(\frac{1}{2} - i\right)$

15. $(7 - i)(4 + 2i)$

16. $(5 - 3i)(1 + i)$

17. $(3 - 4i)(5 - 12i)$

18. $\left(\frac{2}{3} + 12i\right)\left(\frac{1}{6} + 24i\right)$

19. $\dfrac{1}{i}$

20. $\dfrac{1}{1 + i}$

21. $\dfrac{2 - 3i}{1 - 2i}$

22. $\dfrac{5 - i}{3 + 4i}$

23. $\dfrac{26 + 39i}{2 - 3i}$

24. $\dfrac{25}{4 - 3i}$

25. $\dfrac{10i}{1 - 2i}$

26. $(2 - 3i)^{-1}$

27. i^3

28. $(2i)^4$

29. i^{100}

30. i^{1002}

31. $\sqrt{-25}$

32. $\sqrt{\dfrac{-9}{4}}$

33. $\sqrt{-3}\,\sqrt{-12}$

34. $\sqrt{\frac{1}{3}}\,\sqrt{-27}$

35. $(3 - \sqrt{-5})(1 + \sqrt{-1})$ **36.** $\dfrac{1 - \sqrt{-1}}{1 + \sqrt{-1}}$

37. $\dfrac{2 + \sqrt{-8}}{1 + \sqrt{-2}}$

38. $(\sqrt{3} - \sqrt{-4})(\sqrt{6} - \sqrt{-8})$

39. $\dfrac{\sqrt{-36}}{\sqrt{-2}\sqrt{-9}}$ **40.** $\dfrac{\sqrt{-7}\sqrt{-49}}{\sqrt{28}}$

41–48 ■ Graph the complex number and find its modulus.

41. $3i$ **42.** -3

43. $5 + 2i$ **44.** $7 - 3i$

45. $\sqrt{3} + i$ **46.** $-1 - \dfrac{\sqrt{3}}{3}i$

47. $\dfrac{3 + 4i}{5}$ **48.** $\dfrac{-\sqrt{2} + i\sqrt{2}}{2}$

49–50 ■ Sketch the complex number z, and also sketch $2z$, $-z$, and $\frac{1}{2}z$ on the same complex plane.

49. $z = 1 + i$ **50.** $z = 2 - 3i$

51–52 ■ Sketch the complex number z and its complex conjugate $\bar{z}$ on the same complex plane.

51. $z = 8 + 2i$ **52.** $z = -5 + 6i$

53–54 ■ Sketch z_1, z_2, and $z_1 + z_2$ on the same complex plane.

53. $z_1 = 2 - i$, $z_2 = 2 + i$

54. $z_1 = -1 + i$, $z_2 = 2 - 3i$

55–62 ■ Sketch the set in the complex plane.

55. $\{z = a + bi \mid a \leqslant 0, b \geqslant 0\}$

56. $\{z = a + bi \mid a > 1, b > 1\}$

57. $\{z \mid |z| = 3\}$ **58.** $\{z \mid |z| \geqslant 1\}$

59. $\{z \mid |z| < 2\}$ **60.** $\{z \mid 2 \leqslant |z| \leqslant 5\}$

61. $\{z = a + bi \mid a + b < 2\}$

62. $\{z = a + bi \mid a \geqslant b\}$

63–70 ■ Prove each statement. Assume that $z = a + bi$ and $w = c + di$.

63. $\bar{z} + \bar{w} = \overline{z + w}$ **64.** $\overline{zw} = \bar{z} \cdot \bar{w}$

65. $(\bar{z})^2 = \overline{z^2}$ **66.** $\bar{\bar{z}} = z$

67. $z + \bar{z}$ is a real number

68. $z = \bar{z}$ if and only if z is real

69. $\dfrac{z}{\bar{z}} + \dfrac{\bar{z}}{z} = \dfrac{2(a^2 - b^2)}{a^2 + b^2}$ **70.** $\dfrac{z}{\bar{z}} - \dfrac{\bar{z}}{z} = \dfrac{4abi}{a^2 + b^2}$

▲ DISCOVERY · DISCUSSION

71. Complex Conjugate Roots   Suppose that the equation $ax^2 + bx + c = 0$ has real coefficients and complex roots. Why must the roots be complex conjugates of each other? (Think about how you would find the roots.)

72. Powers of i   Calculate the first 12 powers of i, that is, $i, i^2, i^3, \ldots, i^{12}$. Do you notice a pattern? Explain how you would calculate any whole number power of i, using the pattern you have discovered. Use this procedure to calculate i^{4446}.

COMPLEX ROOTS AND THE FUNDAMENTAL THEOREM OF ALGEBRA

3.4

We've already seen that the quadratic equation $ax^2 + bx + c = 0$, with $a \neq 0$, has the solutions

$$x = \frac{-b \pm \sqrt{b^2 - 4ac}}{2a}$$

If $b^2 - 4ac < 0$, then the equation has no real solution. But in the complex number system, this equation will always have solutions because negative numbers have square roots in this expanded setting.

Gerolamo Cardano (1501–1576) is certainly one of the most colorful figures in the history of mathematics. He was the most renowned physician in Europe in his day, yet he was plagued throughout his life by numerous maladies, including ruptures, hemorrhoids, and the irrational fear of encountering rabid dogs. His beloved sons broke his heart—his favorite was eventually beheaded for murdering his wife. He was a compulsive gambler, but he made the best of this vice by writing the *Book on Games of Chance*, the first study of probability from a sound mathematical point of view.

His major mathematical work was the *Ars Magna,* in which he detailed the solution of the general third- and fourth-degree polynomial equations. At the time of its publication, mathematicians were uncomfortable even with negative numbers, but Cardano's formulas paved the way for the acceptance not just of negative numbers, but also of imaginary numbers, because they occurred naturally in solving polynomial equations. For example, one of his formulas gives the solution

$$x = \sqrt[3]{2 + \sqrt{-121}}$$
$$- \sqrt[3]{-2 + \sqrt{-121}}$$

for the cubic equation

$$x^3 - 15x - 4 = 0$$

This value for x actually turns out to be the *integer* 4, yet to find it Cardano had to use the imaginary number $\sqrt{-121} = 11i$.

EXAMPLE 1 ■ **Using the Quadratic Formula to Find Complex Roots**

Solve each of the following equations.

(a) $x^2 + 9 = 0$ 　　　　　　　　　　　(b) $x^2 + 4x + 5 = 0$

SOLUTION

(a) $x^2 + 9 = 0$ means $x^2 = -9$, so $x = \pm\sqrt{-9} = \pm i \sqrt{9} = \pm 3i$. The solutions are $3i$ and $-3i$.

(b) By the quadratic formula

$$x = \frac{-4 \pm \sqrt{4^2 - 4 \cdot 5}}{2}$$

$$= \frac{-4 \pm \sqrt{-4}}{2}$$

$$= \frac{-4 \pm 2i}{2} = -2 \pm i$$

so the solutions are $-2 + i$ and $-2 - i$.　　　　　　　　　■

EXAMPLE 2 ■ **Factoring and Using the Quadratic Formula To Find Complex Roots**

Find all the roots of the equation $x^6 - 64 = 0$.

SOLUTION

$$x^6 - 64 = (x^3)^2 - 8^2$$

$$= (x^3 - 8)(x^3 + 8) \qquad \text{Difference of squares formula}$$

$$= (x - 2)(x^2 + 2x + 4)(x + 2)(x^2 - 2x + 4) \qquad \text{Difference of cubes and sum of cubes formulas}$$

This expression will equal 0 when any factor is 0, so we find the solutions as follows.

$$x - 2 = 0 \qquad \text{means} \qquad x = 2$$

$$x^2 + 2x + 4 = 0 \qquad \text{means} \qquad x = \frac{-2 \pm \sqrt{2^2 - 4 \cdot 4}}{2}$$

$$= \frac{-2 \pm \sqrt{-12}}{2} = -1 \pm i\sqrt{3}$$

$$x + 2 = 0 \qquad \text{means} \qquad x = -2$$

$$x^2 - 2x + 4 = 0 \qquad \text{means} \qquad x = \frac{2 \pm \sqrt{(-2)^2 - 4 \cdot 4}}{2}$$

$$= \frac{2 \pm \sqrt{-12}}{2} = 1 \pm i\sqrt{3}$$

The roots of the equation are

$$2, \quad -2, \quad -1 + i\sqrt{3}, \quad -1 - i\sqrt{3}, \quad 1 + i\sqrt{3}, \quad \text{and} \quad 1 - i\sqrt{3} \quad \blacksquare$$

THE FUNDAMENTAL THEOREM OF ALGEBRA

It is a remarkable fact that adding just the number $\sqrt{-1}$ and its real multiples to the real number system is sufficient to provide a number system in which *every* polynomial equation has a root. Although we will not prove this fact (a proof requires mathematical expertise well beyond the scope of this book), it nevertheless forms the basis for much of our work in solving polynomial equations. This theorem was proved by the German mathematician C. F. Gauss in 1799.

FUNDAMENTAL THEOREM OF ALGEBRA

Every polynomial

$$P(x) = a_n x^n + a_{n-1} x^{n-1} + \cdots + a_1 x + a_0 \qquad (n \geq 1, a_n \neq 0)$$

with complex coefficients has at least one complex zero.

Because any real number is also a complex number, the theorem applies to polynomials with real coefficients as well.

Since every zero of a polynomial corresponds to a linear factor (by the Factor Theorem), the Fundamental Theorem of Algebra ensures that we can factor any polynomial $P(x)$ of degree n as

$$P(x) = (x - c_1) \cdot Q_1(x)$$

where $Q_1(x)$ is of degree $n - 1$ and c_1 is a zero of $P(x)$. But now applying the Fundamental Theorem to the quotient $Q_1(x)$ gives us the factorization

$$P(x) = (x - c_1) \cdot (x - c_2) \cdot Q_2(x)$$

where $Q_2(x)$ is of degree $n - 2$ and c_2 is a zero of $Q_1(x)$. Continuing this process for n steps, we will get a final quotient $Q_n(x)$ of degree 0, which is therefore a nonzero constant that we will call a. This proves the following corollary of the Fundamental Theorem of Algebra.

COMPLETE FACTORIZATION THEOREM

If $P(x)$ is a polynomial of degree $n > 0$, then there exist complex numbers $a, c_1, c_2, \ldots, c_n$ (with $a \neq 0$) such that

$$P(x) = a(x - c_1)(x - c_2)\cdots(x - c_n)$$

The number a is clearly the coefficient of x^n in $P(x)$. The numbers c_1, $c_2, \ldots, c_n$ are the zeros of $P(x)$ (by the Factor Theorem). These need not all be different: If the factor $x - c$ appears k times in the complete factorization of $P(x)$, then we say that c is a zero of **multiplicity** k.

It follows that $P(x)$ can have no zero other than $c_1, c_2, \ldots, c_n$, because if

$$P(c) = a(c - c_1)(c - c_2) \cdots (c - c_n) = 0$$

then at least one of the factors $c - c_i$ must be zero, so $c = c_i$ for some $i \in \{1, 2, \ldots, n\}$. We have thus proved the following theorem.

ZEROS THEOREM

Every polynomial of degree $n \geq 1$ has exactly n zeros, provided that a zero of multiplicity k is counted k times.

In Example 2, for instance, the polynomial $x^6 - 64$ is of degree 6, and it has exactly six zeros.

EXAMPLE 3 ■ Factoring a Polynomial with Complex Zeros

Find the complete factorization and all five zeros of the polynomial

$$P(x) = 3x^5 + 24x^3 + 48x$$

SOLUTION   The terms of P have $3x$ as a common factor, so we get the following factorization:

$$P(x) = 3x(x^4 + 8x^2 + 16)$$
$$= 3x(x^2 + 4)^2$$

To factor $x^2 + 4$, note that $2i$ and $-2i$ are zeros of this polynomial. Thus, $x^2 + 4 = (x - 2i)(x + 2i)$, and so

$$P(x) = 3x[(x - 2i)(x + 2i)]^2$$
$$= 3x(x - 2i)(x - 2i)(x + 2i)(x + 2i)$$

Setting each factor equal to zero in turn, we see that the zeros of P are 0, $2i$, and $-2i$. However, $2i$ and $-2i$ are each counted twice, since the factor of P that corresponds to each occurs twice in the factorization of P. Each of these is a zero of multiplicity 2 (or a *double zero*), so the total number of zeros, counting multiplicity, is five.   ■

EXAMPLE 4 ■ Finding Polynomials with Specified Zeros

Find a polynomial that satisfies the given description.

(a) A polynomial $P(x)$ of degree 4, with zeros i, $-i$, 2, and -2 and with $P(3) = 25$

(b) A polynomial $Q(x)$ of degree 4, with zeros -2 and 0, where -2 is a zero of multiplicity three

SOLUTION

(a) The required polynomial has the form

$$P(x) = a(x - i)(x - (-i))(x - 2)(x - (-2))$$
$$= a(x^2 + 1)(x^2 - 4) \qquad \text{Difference of squares}$$
$$= a(x^4 - 3x^2 - 4) \qquad \text{Multiply}$$

We know that $P(3) = a(3^4 - 3 \cdot 3^2 - 4) = 50a = 25$, so $a = \frac{1}{2}$. Thus

$$P(x) = \tfrac{1}{2}x^4 - \tfrac{3}{2}x^2 - 2$$

(b) We require

$$Q(x) = a(x - (-2))^3(x - 0)$$
$$= a(x + 2)^3 x$$
$$= a(x^3 + 6x^2 + 12x + 8)x \qquad \text{Product Formula 4 (Section 1.3)}$$
$$= a(x^4 + 6x^3 + 12x^2 + 8x)$$

Since we are given no information about Q other than its zeros and their multiplicity, we can choose any number for a. If we use $a = 1$, we get

$$Q(x) = x^4 + 6x^3 + 12x^2 + 8x \qquad \blacksquare$$

EXAMPLE 5 ■ **Finding All the Zeros of a Polynomial**

Find all four zeros of $P(x) = 3x^4 - 2x^3 - x^2 - 12x - 4$.

SOLUTION Using the Rational Zeros Theorem from Section 3.2, we obtain the following list of possible rational zeros: $\pm 1, \pm 2, \pm 4, \pm\frac{1}{3}, \pm\frac{2}{3}, \pm\frac{4}{3}$. Checking these using synthetic division, we find that 2 and $-\frac{1}{3}$ are zeros, and we get the following factorization.

$$P(x) = 3x^4 - 2x^3 - x^2 - 12x - 4$$
$$= (x - 2)(3x^3 + 4x^2 + 7x + 2) \qquad \text{Factor } x - 2$$
$$= (x - 2)(x + \tfrac{1}{3})(3x^2 + 3x + 6) \qquad \text{Factor } x + \tfrac{1}{3}$$
$$= (x - 2)(x + \tfrac{1}{3})(3)(x^2 + x + 2) \qquad \text{Factor 3}$$
$$= (x - 2)(3x + 1)(x^2 + x + 2) \qquad \text{Multiply}$$

The zeros of the quadratic factor are

$$x = \frac{-1 \pm \sqrt{1 - 8}}{2} = -\frac{1}{2} \pm i\frac{\sqrt{7}}{2} \qquad \text{Quadratic formula}$$

so the zeros of $P(x)$ are

$$2, \quad -\frac{1}{3}, \quad -\frac{1}{2} + i\frac{\sqrt{7}}{2}, \quad \text{and} \quad -\frac{1}{2} - i\frac{\sqrt{7}}{2} \qquad \blacksquare$$

Carl Friedrich Gauss (1777–1855) is considered the greatest mathematician of modern times. He was referred to by his contemporaries as "The Prince of Mathematics." Gauss was born into a poor family; his father made a living as a mason. As a very small child he found a calculation error in his father's accounts. This was the first of many incidents that gave evidence of his mathematical precocity. (See also page 696.) At the age of 19 Gauss demonstrated that the regular 17-sided polygon can be constructed with straightedge and compass alone. This was remarkable because, since the time of Euclid, it was thought that the only regular polygons constructible in this way were the triangle and pentagon. Because of this discovery Gauss decided to pursue a career in mathematics instead of languages, his other passion. In his doctoral dissertation, written at the age of 22, Gauss proved the Fundamental Theorem of Algebra: A polynomial of degree n with complex coefficients has n roots. His other accomplishments range over every branch of mathematics, as well as physics and astronomy.

As you may have noticed from the examples so far, the complex roots of polynomial equations with real coefficients come in pairs. Whenever $a + bi$ is a root, so is its complex conjugate $a - bi$. This is always the case, as the following theorem states.

CONJUGATE ROOTS THEOREM

If the polynomial $P(x)$ of degree $n > 0$ has real coefficients, and if the complex number z is a root of the equation $P(x) = 0$, then its complex conjugate $\bar{z}$ is also a root.

■ **Proof**   Let

$$P(x) = a_n x^n + a_{n-1} x^{n-1} + \cdots + a_1 x + a_0$$

where each coefficient is real. Suppose that $P(z) = 0$. We must prove that $P(\bar{z}) = 0$. We use the facts that the complex conjugate of a sum of two complex numbers is the sum of the conjugates and that the conjugate of a product is the product of the conjugates (see Exercises 63 and 64 in Section 3.3).

$$P(\bar{z}) = a_n(\bar{z})^n + a_{n-1}(\bar{z})^{n-1} + \cdots + a_1\bar{z} + a_0$$
$$= \overline{a_n}\,\overline{z^n} + \overline{a_{n-1}}\,\overline{z^{n-1}} + \cdots + \overline{a_1}\,\overline{z} + \overline{a_0} \qquad \text{Because the coefficients are real}$$
$$= \overline{a_n z^n} + \overline{a_{n-1}z^{n-1}} + \cdots + \overline{a_1 z} + \overline{a_0}$$
$$= \overline{a_n z^n + a_{n-1}z^{n-1} + \cdots + a_1 z + a_0}$$
$$= \overline{P(z)} = \overline{0} = 0$$

This derivation shows that the conjugate $\bar{z}$ is also a zero of $P(x)$, and we have proved the theorem. □

EXAMPLE 6 ■ A Polynomial with a Specified Complex Zero

Find a polynomial $P(x)$ of degree 3 that has integer coefficients and zeros $\frac{1}{2}$ and $3 - i$.

SOLUTION   Since $3 - i$ is a zero, then so is $3 + i$ by the Conjugate Roots Theorem. This means that $P(x)$ has the form

$$P(x) = a\left(x - \tfrac{1}{2}\right)[x - (3 - i)][x - (3 + i)]$$
$$= a\left(x - \tfrac{1}{2}\right)[(x - 3) + i][(x - 3) - i] \qquad \text{Regroup}$$
$$= a\left(x - \tfrac{1}{2}\right)[(x - 3)^2 - i^2] \qquad \text{Difference of squares formula}$$
$$= a\left(x - \tfrac{1}{2}\right)(x^2 - 6x + 10) \qquad \text{Expand}$$
$$= a\left(x^3 - \tfrac{13}{2}x^2 + 13x - 5\right) \qquad \text{Expand}$$

To make all coefficients integers, we set $a = 2$ and get

$$P(x) = 2x^3 - 13x^2 + 26x - 10$$

Any other polynomial that satisfies the given requirements must be an integer multiple of this one. ∎

EXAMPLE 7 ■ Using Descartes' Rule to Count Real and Imaginary Zeros

Without actually factoring, determine how many positive real zeros, negative real zeros, and imaginary zeros the following polynomial could have:

$$P(x) = x^4 + 6x^3 - 12x^2 - 14x - 24$$

SOLUTION   Since there is one change of sign, by Descartes' Rule of Signs, P has one positive real zero. Also, $P(-x) = x^4 - 6x^3 - 12x^2 + 14x - 24$ has three changes of sign, so there are either three or one negative real zero(s). So P has a total of either four or two real zeros. Since P is of degree 4, it has four zeros in all, which gives the following possibilities.

Positive real zeros	Negative real zeros	Imaginary zeros
1	3	0
1	1	2

∎

3.4 EXERCISES

1–22 ■ Find all solutions of the equation.

1. $x^2 + 16 = 0$

2. $4x^2 + 25 = 0$

3. $x^2 + 2x + 2 = 0$

4. $x^2 - x + 1 = 0$

5. $x^2 + 4x + 8 = 0$

6. $2x^2 + 2x + 1 = 0$

7. $3x^2 - 5x + 4 = 0$

8. $2x^2 - 3x + 2 = 0$

9. $x^2 - 8x + 17 = 0$

10. $3x^2 - 4x + 2 = 0$

11. $t + 3 + \dfrac{3}{t} = 0$

12. $\theta^3 + \theta^2 + \theta = 0$

13. $x^4 - 1 = 0$

14. $x^3 - 64 = 0$

15. $x^3 + 8 = 0$

16. $x^4 + 4 = 0$

17. $x^4 - 16 = 0$

18. $16x^4 - 81 = 0$

19. $x^6 - 729 = 0$

20. $x^4 + 2x^2 + 1 = 0$

21. $x^4 + 10x^2 + 25 = 0$

22. $x^6 + 7x^3 - 8 = 0$

23–28 ■ Find a polynomial with integer coefficients that satisfies the given conditions.

23. $P(x)$ has degree 3, and zeros 2 and i.

24. $Q(x)$ has degree 3, and zeros -3 and $1 + i$.

25. $R(x)$ has degree 4, and zeros $1 - 2i$ and 1, with 1 a zero of multiplicity 2.

26. $S(x)$ has degree 4, and zeros $2i$ and $3i$.

27. $T(x)$ has degree 4, zeros i and $1 + i$, and constant coefficient 12.

28. $U(x)$ has degree 5, zeros $\frac{1}{2}$, -1, and $-i$, and leading coefficient 4; the zero -1 has multiplicity 2.

29–36 ■ Find all solutions of the equation.

29. $x^3 + 2x^2 + 4x + 8 = 0$

30. $x^3 - 7x^2 + 17x - 15 = 0$

31. $x^3 - 2x^2 + 2x - 1 = 0$

32. $x^3 + 7x^2 + 18x + 18 = 0$

33. $x^3 - 3x^2 + 3x - 2 = 0$

34. $2x^3 - 8x^2 + 9x - 9 = 0$

35. $x^4 + x^3 + 7x^2 + 9x - 18 = 0$

36. $x^5 + x^3 + 8x^2 + 8 = 0$ [*Hint:* Factor by grouping.]

37–44 ■ Find the complete factorization of the polynomial.

37. $P(x) = x^3 + 27$ **38.** $P(x) = x^4 - 625$

39. $P(x) = x^6 + 7x^3 - 8$ **40.** $P(x) = x^5 + 3x^3 + 2x$

41. $P(x) = x^3 - x - 6$

42. $P(x) = 2x^3 + 7x^2 + 12x + 9$

43. $P(x) = x^4 - x^3 + 7x^2 - 9x - 18$

44. $P(x) = x^4 - 2x^3 - 2x^2 - 2x - 3$

45–48 ■ Use Descartes' Rule of Signs to determine how many positive real zeros, negative real zeros, and imaginary zeros the polynomial can have.

45. $P(x) = 3x^4 + 2x^3 + x^2 + 5$

46. $Q(x) = 2x^4 - 4x^3 + x^2 - 5x + 12$

47. $R(x) = 6x^5 - x^4 - 5x - 2$

48. $S(x) = x^6 + x^4 - 3x^3 - x^2 + 10$

 49. By the Zeros Theorem, every nth-degree polynomial equation has exactly n solutions (including possibly some that are repeated). Some of these may be real and some may be imaginary. Use a graphing calculator to determine how many real and how many imaginary solutions each of the following equations has.
(a) $x^4 - 2x^3 - 11x^2 + 12x = 0$
(b) $x^4 - 2x^3 - 11x^2 + 12x - 5 = 0$
(c) $x^4 - 2x^3 - 11x^2 + 12x + 40 = 0$

50–52 ■ So far we have worked only with polynomials that have real coefficients. These exercises involve polynomials with real and imaginary coefficients.

50. Find all solutions of the equation.
(a) $2x + 4i = 1$ (b) $x^2 - ix = 0$
(c) $x^2 + 2ix - 1 = 0$ (d) $ix^2 - 2x + i = 0$

51. (a) Show that $2i$ and $1 - i$ are both solutions of the equation
$$x^2 - (1 + i)x + (2 + 2i) = 0$$
but that their complex conjugates $-2i$ and $1 + i$ are not.
(b) Explain why the result of part (a) does not violate the Conjugate Roots Theorem.

52. (a) Find the polynomial with *real* coefficients of the smallest possible degree for which i and $1 + i$ are zeros and in which the coefficient of the highest power is 1.
(b) Find the polynomial with *complex* coefficients of the smallest possible degree for which i and $1 + i$ are zeros and in which the coefficient of the highest power is 1.

 DISCOVERY • DISCUSSION

53. Polynomials of Odd Degree   The Conjugate Roots Theorem says that the complex zeros of a polynomial with real coefficients occur in complex conjugate pairs. Explain how this fact proves that a polynomial with real coefficients and odd degree has at least one real zero.

54. Roots of Unity   There are two square roots of 1, namely 1 and -1. These are the solutions of $x^2 = 1$. The fourth roots of 1 are the solutions of the equation $x^4 = 1$ or $x^4 - 1 = 0$. How many fourth roots of 1 are there? Find them. The cube roots of 1 are the solutions of the equation $x^3 = 1$ or $x^3 - 1 = 0$. How many cube roots of 1 are there? Find them. How would you find the sixth roots of 1? How many are there? Make a conjecture about the number of nth roots of 1.

3.5 RATIONAL FUNCTIONS

A **rational function** is a function of the form

$$r(x) = \frac{P(x)}{Q(x)}$$

where P and Q are polynomials. We assume that $P(x)$ and $Q(x)$ have no factor in common. Rational functions are not defined for those values of x for which the denominator $Q(x)$ is zero. The graph of a rational function is largely determined by its shape near these x-values. We begin by graphing a very simple rational function.

EXAMPLE 1 ■ A Simple Rational Function

Sketch a graph of the rational function $r(x) = \dfrac{1}{x}$.

SOLUTION The function r is not defined for $x = 0$. The following tables show that when x is close to zero, the value of $|r(x)|$ is large, and the closer x gets to zero, the larger $|r(x)|$ gets.

For positive real numbers,

$$\dfrac{1}{\text{BIG NUMBER}} = \text{small number}$$

$$\dfrac{1}{\text{small number}} = \text{BIG NUMBER}$$

x	$r(x)$
-0.1	-10
-0.01	-100
-0.00001	$-100{,}000$

x	$r(x)$
0.1	10
0.01	100
0.00001	$100{,}000$

We describe this behavior by saying "$r(x)$ approaches negative infinity as x approaches zero from the left" and "$r(x)$ approaches infinity as x approaches zero from the right." In symbols, we write

$$r(x) \to -\infty \quad \text{as} \quad x \to 0^- \qquad \text{and} \qquad r(x) \to \infty \quad \text{as} \quad x \to 0^+$$

We say that $x = 0$ is a **vertical asymptote** of the function r.

The next two tables show the behavior of the function r as $|x|$ becomes large.

x	$r(x)$
-10	-0.1
-100	-0.01
$-100{,}000$	-0.00001

x	$r(x)$
10	0.1
100	0.01
$100{,}000$	0.00001

These tables show that as $|x|$ becomes large, the value of $r(x)$ gets closer and closer to zero. We describe this situation in symbols by writing

$$r(x) \to 0 \quad \text{as} \quad x \to -\infty \qquad \text{and} \qquad r(x) \to 0 \quad \text{as} \quad x \to \infty$$

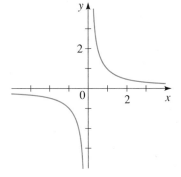

FIGURE 1

$r(x) = \dfrac{1}{x}$

We say that the line $y = 0$ is a **horizontal asymptote** of the function r. Using the information given by these tables and plotting a few additional points, we obtain the graph shown in Figure 1. ■

EXAMPLE 2 ■ Graphing a Rational Function

Sketch a graph of the rational function $s(x) = \dfrac{x - 1}{x - 2}$.

SOLUTION 1 The function is not defined for $x = 2$, so we first determine the graph of the function for x-values near 2.

$x \to 2^-$		$x \to 2^+$	
x	y	x	y
1	0	3	2
1.5	-1	2.5	3
1.9	-9	2.1	11
1.95	-19	2.05	21
1.99	-99	2.01	101
1.999	-999	2.001	1001

We see from the first table that as x approaches 2 from the left, the y-values decrease without bound. In symbols,

$$y \to -\infty \quad \text{as} \quad x \to 2^- \qquad \text{"} y \text{ approaches negative infinity}$$
$$\text{as } x \text{ approaches 2 from the left"}$$

The second table shows that as x approaches 2 from the right, the y-values increase without bound. In symbols,

$$y \to \infty \quad \text{as} \quad x \to 2^+ \qquad \text{"} y \text{ approaches infinity}$$
$$\text{as } x \text{ approaches 2 from the right"}$$

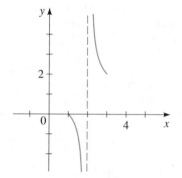

FIGURE 2

The graph of $y = r(x)$ therefore has the shape near $x = 2$ shown in Figure 2.

Now we examine the behavior of the function as x becomes progressively larger in absolute value (for both negative and positive x).

$x \to -\infty$		$x \to \infty$	
x	y	x	y
-10	0.8750	10	1.0833
-100	0.9898	100	1.0098
-1000	0.9990	1000	1.0010
$-10,000$	0.9999	10,000	1.0001

As $|x|$ becomes larger and larger, the values of y get progressively closer to 1. This means that the graph of $y = r(x)$ will approach the horizontal line $y = 1$ as x increases or decreases without bound. We express this by saying "y approaches 1 as x approaches infinity or negative infinity" and we write

$$y \to 1 \quad \text{as} \quad x \to \infty \qquad \text{and} \qquad y \to 1 \quad \text{as} \quad x \to -\infty$$

This means we can complete the graph of $y = r(x)$ as shown in Figure 3.

SOLUTION 2 Using long division, we see that

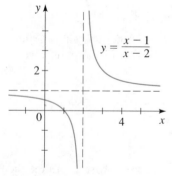

$$y = \frac{x-1}{x-2}$$

FIGURE 3

$$s(x) = 1 + \frac{1}{x - 2}$$

Shifting functions is discussed in Section 2.4.

This means that the graph of s is just the graph of the function r of Example 1 shifted upward 1 unit and to the right 2 units. Thus, we can obtain the graph in Figure 3 from the graph in Figure 1 simply by shifting horizontally and then vertically. ∎

In Example 2 we introduced the notation described in the following box.

ARROW NOTATION	
Symbol	Meaning
$x \to a^-$	x approaches a from the left
$x \to a^+$	x approaches a from the right
$x \to -\infty$	x goes to negative infinity; that is, x decreases without bound
$x \to \infty$	x goes to infinity; that is, x increases without bound

The line $x = 2$ is called a *vertical asymptote* of the graph in Figure 3, and the line $y = 1$ is a *horizontal asymptote*. Informally speaking, an asymptote of a function is a line that the graph of the function gets closer and closer to as one travels along that line in either direction. More formally, we make the following definitions.

ASYMPTOTES

1. The line $x = a$ is a **vertical asymptote** of the function $y = f(x)$ if

$$y \to \infty \text{ or } y \to -\infty \quad \text{as} \quad x \to a^+ \text{ or } x \to a^-$$

2. The line $y = b$ is a **horizontal asymptote** of the function $y = f(x)$ if

$$y \to b \quad \text{as} \quad x \to \infty \text{ or } x \to -\infty$$

In the next example we show a general procedure for finding horizontal and vertical asymptotes and graphing rational functions.

EXAMPLE 3 ■ Graphing a Rational Function

Sketch a graph of the rational function $r(x) = \dfrac{2x^2 + 7x - 4}{x^2 + x - 2}$.

SOLUTION We factor the numerator and denominator, find the intercepts and asymptotes, and sketch the graph.

Factor: $r(x) = \dfrac{(2x - 1)(x + 4)}{(x - 1)(x + 2)}$

A fraction is 0 if and only if its numerator is 0.

x-intercepts: The *x*-intercepts are the zeros of the numerator, $x = \frac{1}{2}$ and $x = -4$.

y-intercept: To find the *y*-intercept, we substitute $x = 0$ into the original form of the function:

$$r(0) = \frac{2(0)^2 + 7(0) - 4}{(0)^2 + (0) - 2} = \frac{-4}{-2} = 2$$

The *y*-intercept is 2.

Horizontal asymptote: The horizontal asymptote (if it exists) is the value that *y* approaches as $x \to \pm\infty$. To find this value we divide both the numerator and the denominator by the highest power of *x* that appears in the denominator (in this case, x^2).

$$y = r(x) = \frac{2x^2 + 7x - 4}{x^2 + x - 2} \cdot \frac{\dfrac{1}{x^2}}{\dfrac{1}{x^2}} = \frac{2 + \dfrac{7}{x} - \dfrac{4}{x^2}}{1 + \dfrac{1}{x} - \dfrac{2}{x^2}}$$

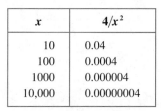

x	$4/x^2$
10	0.04
100	0.0004
1000	0.000004
10,000	0.00000004

$\uparrow$

approaching 0

Any expression of the form c/x^n approaches 0 as $x \to \pm\infty$ (if $n > 0$). The table in the margin illustrates this for the term $4/x^2$. So as $x \to \pm\infty$, we have

$$y = \frac{2 + \dfrac{7}{x} - \dfrac{4}{x^2}}{1 + \dfrac{1}{x} - \dfrac{2}{x^2}} \to \frac{2 + 0 - 0}{1 + 0 - 0} = 2$$

Thus, the horizontal asymptote is $y = 2$.

Vertical asymptotes: The vertical asymptotes occur where the denominator is 0, that is, where the function is undefined. Thus, from the factored form the vertical asymptotes are $x = 1$ and $x = -2$.

Behavior near vertical asymptotes: We need to know whether $y \to \infty$ or $y \to -\infty$ on each side of each vertical asymptote. To determine the sign of *y* for *x*-values near the vertical asymptotes, we use test values. For instance, as $x \to 1^-$, we use a test value close to and to the left of 1 ($x = 0.9$, say) to check whether *y* is positive or negative to the left of $x = 1$:

$$y = \frac{(2(0.9) - 1)((0.9) + 4)}{((0.9) - 1)((0.9) + 2)} \qquad \text{whose sign is} \qquad \frac{(+)(+)}{(-)(+)} \quad \text{(negative)}$$

So $y \to -\infty$ as $x \to 1^-$. On the other hand, as $x \to 1^+$, we use a test value close to and to the right of 1 ($x = 1.1$, say), to get

$$y = \frac{(2(1.1) - 1)((1.1) + 4)}{((1.1) - 1)((1.1) + 2)} \qquad \text{whose sign is} \qquad \frac{(+)(+)}{(+)(+)} \quad \text{(positive)}$$

So $y \to \infty$ as $x \to 1^+$. The other entries in the following table are calculated similarly.

As $x \to$	-2^-	-2^+	1^-	1^+
the sign of $y = \dfrac{(2x - 1)(x + 4)}{(x - 1)(x + 2)}$ is	$\dfrac{(-)(+)}{(-)(-)}$	$\dfrac{(-)(+)}{(-)(+)}$	$\dfrac{(+)(+)}{(-)(+)}$	$\dfrac{(+)(+)}{(+)(+)}$
so $y \to$	$-\infty$	∞	$-\infty$	∞

Additional values: **Graph:**

x	y
-6	0.93
-3	-1.75
-1	4.50
1.5	6.29
2	4.50
3	3.50

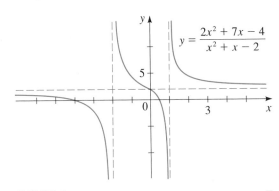

$$y = \frac{2x^2 + 7x - 4}{x^2 + x - 2}$$

FIGURE 4   ■

Here is a summary of the procedure to be followed when graphing rational functions.

SKETCHING GRAPHS OF RATIONAL FUNCTIONS

1. FACTOR. Factor the numerator and denominator.

2. INTERCEPTS. Find the x-intercepts by determining the zeros of the numerator, and the y-intercept from the value of the function at $x = 0$.

3. VERTICAL ASYMPTOTES. Find the vertical asymptotes by determining the zeros of the denominator, and then see if $y \to \infty$ or $y \to -\infty$ on each side of every vertical asymptote.

4. HORIZONTAL ASYMPTOTE. Find the horizontal asymptote (if any) by dividing both numerator and denominator by the highest power of x that appears in the denominator, and then letting $x \to \pm\infty$.

5. SKETCH THE GRAPH. Graph the information provided by the first four steps. Then plot as many additional points as needed to fill in the rest of the graph of the function.

The Three Forms of A Rational Function

To find the **y-intercept** of a rational function, we use the *original form* of the function.

To find the **x-intercepts** and **vertical asymptotes**, we use the *factored form* of the function.

To find the **horizontal asymptote**, we use a *compound-fraction form* of the function.

Notice that we use three forms of the equation defining the rational function when we perform this analysis. In Example 3 the *original form*

$$r(x) = \frac{2x^2 + 7x - 4}{x^2 + x - 2}$$

easily gave us the y-intercept $r(0) = (-4)/(-2) = 2$. The *factored form*

$$r(x) = \frac{(2x - 1)(x + 4)}{(x - 1)(x + 2)}$$

gave us the x-intercepts $\frac{1}{2}$ and -4, and the vertical asymptotes $x = 1$ and $x = -2$. Finally, the *compound-fraction form*

$$r(x) = \frac{2 + \dfrac{7}{x} - \dfrac{4}{x^2}}{1 + \dfrac{1}{x} - \dfrac{2}{x^2}}$$

tells us that the horizontal asymptote is $y = \frac{2}{1} = 2$.

We can determine whether a rational function $r(x) = P(x)/Q(x)$ has a horizontal asymptote by considering the degrees of the numerator and denominator. If the degrees of P and Q are the same (both n, say), then dividing numerator and denominator by x^n shows that the horizontal asymptote is

$$y = \frac{\text{leading coefficient of } P}{\text{leading coefficient of } Q}$$

as we saw in Example 3. The following box summarizes the procedure for finding asymptotes.

ASYMPTOTES OF RATIONAL FUNCTIONS

Let

$$r(x) = \frac{a_n x^n + a_{n-1} x^{n-1} + \cdots + a_1 x + a_0}{b_m x^m + b_{m-1} x^{m-1} + \cdots + b_1 x + b_0}$$

be a rational function.

1. The vertical asymptotes are the lines $x = a$, where a is a zero of the denominator.

2. (a) If $n < m$, then r has horizontal asymptote $y = 0$.

   (b) If $n = m$, then r has horizontal asymptote $y = \dfrac{a_n}{b_m}$.

   (c) If $n > m$, then r has no horizontal asymptote.

EXAMPLE 4 ■ Graphing a Rational Function

Sketch a graph of the rational function $r(x) = \dfrac{x-2}{x^2-1}$.

SOLUTION

Factor: $y = \dfrac{x-2}{(x-1)(x+1)}$

x-intercept: 2, from $x - 2 = 0$

y-intercept: 2, because $r(0) = \dfrac{0-2}{0^2-1} = 2$

Horizontal asymptote: $y = 0$, since degree of numerator < degree of denominator

Vertical asymptotes: $x = 1$ and $x = -1$, from the zeros of the denominator

Behavior near vertical asymptotes:

As $x \rightarrow$	1^+	1^-	-1^+	-1^-
the sign of $y = \dfrac{x-2}{(x-1)(x+1)}$ is	$\dfrac{(-)}{(+)(+)}$	$\dfrac{(-)}{(-)(+)}$	$\dfrac{(-)}{(-)(+)}$	$\dfrac{(-)}{(-)(-)}$
so $y \rightarrow$	$-\infty$	∞	∞	$-\infty$

Additional values: **Graph:**

x	y
-2	-1.33
-0.5	3.33
0.5	2
1.5	-0.4
3	0.125
4	0.133
5	0.125

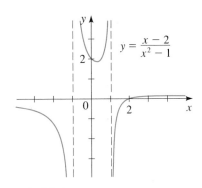

FIGURE 5   ■

From the graph in Figure 5, we see that, contrary to common misconception, a graph may cross a horizontal asymptote.

EXAMPLE 5 ■ Graphing a Rational Function

Sketch a graph of the rational function $r(x) = \dfrac{x^2 - 3x - 4}{2x^2 + 4x}$.

SOLUTION

Factor: $\quad y = \dfrac{(x+1)(x-4)}{2x(x+2)}$

x-intercepts: -1 and 4, from $x + 1 = 0$ and $x - 4 = 0$

y-intercept: None, because $r(0)$ is undefined

Horizontal asymptote: $y = \frac{1}{2}$, because degrees of numerator and denominator are the same and

$$\frac{\text{leading coefficient of numerator}}{\text{leading coefficient of denominator}} = \frac{1}{2}$$

Vertical asymptotes: $x = 0$ and $x = -2$, from the zeros of the denominator

Behavior near vertical asymptotes:

As $x \to$	-2^-	-2^+	0^-	0^+
the sign of $y = \dfrac{(x+1)(x-4)}{2x(x+2)}$ is	$\dfrac{(-)(-)}{(-)(-)}$	$\dfrac{(-)(-)}{(-)(+)}$	$\dfrac{(+)(-)}{(-)(+)}$	$\dfrac{(+)(-)}{(+)(+)}$
so $y \to$	∞	$-\infty$	∞	$-\infty$

Additional values:

x	y
-3	2.33
-2.5	3.90
-0.5	1.50
1	-1.00
3	-0.13
5	0.09

Graph:

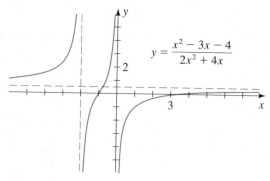

$$y = \frac{x^2 - 3x - 4}{2x^2 + 4x}$$

FIGURE 6

SLANT ASYMPTOTES

If $r(x) = P(x)/Q(x)$ is a rational function in which the degree of the numerator is one more than the degree of the denominator, we can use the Division Algorithm to express the function in the form

$$r(x) = ax + b + \frac{R(x)}{Q(x)}$$

where the degree of R is less than the degree of Q and $a \neq 0$. This means that as $x \to \pm\infty$, $R(x)/Q(x) \to 0$, so for large values of $|x|$, the graph of $y = r(x)$

approaches the graph of the line $y = ax + b$. In this situation we say that $y = ax + b$ is a **slant asymptote**, or an **oblique asymptote**.

EXAMPLE 6 ■ A Rational Function with a Slant Asymptote

Sketch a graph of the rational function $r(x) = \dfrac{x^2 - 4x - 5}{x - 3}$.

SOLUTION   Since the degree of the numerator is one more than the degree of the denominator, the function has a slant asymptote. By dividing $x - 3$ into $x^2 - 4x - 5$, we obtain

$$r(x) = x - 1 - \frac{8}{x - 3}$$

$$
\begin{array}{r}
x - 1 \\
x - 3 \overline{) x^2 - 4x - 5} \\
\underline{x^2 - 3x} \\
-x - 5 \\
\underline{-x + 3} \\
-8
\end{array}
$$

so the slant asymptote is the line $y = x - 1$. The line $x = 3$ is a vertical asymptote, and it's easy to see that $r(x) \to -\infty$ as $x \to 3^{+}$ and $r(x) \to \infty$ as $x \to 3^{-}$.

Factoring the numerator in the original form for r gives

$$r(x) = \frac{(x + 1)(x - 5)}{x - 3}$$

so the x-intercepts are -1 and 5, and the y-intercept is $\frac{5}{3}$. Plotting the asymptotes, intercepts, and the additional points listed in the table, we can complete the graph as shown in Figure 7.

x	y
-2	-1.4
1	4
2	9
4	-5
6	2.33

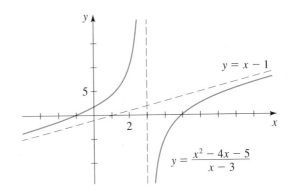

FIGURE 7

USING GRAPHING DEVICES TO GRAPH RATIONAL FUNCTIONS

So far we have considered only horizontal and slant asymptotes as end behaviors for rational functions. In the next example we graph a function that behaves like a parabola for large values of $|x|$.

EXAMPLE 7 ■ End Behavior of a Rational Function

Graph the function

$$f(x) = \frac{x^3 - 2x^2 + 3}{x - 2}$$

in appropriate viewing rectangles to show the vertical asymptote and to determine its end behavior.

SOLUTION   First we graph the function in a narrow viewing rectangle to see the vertical asymptote. The function is undefined when $x = 2$, so we choose the viewing rectangle $[-4, 4]$ by $[-20, 20]$ and obtain the graph in Figure 8(a). The function has x-intercept -1, vertical asymptote $x = 2$, and a local minimum point with approximate coordinates $(2.74, 11.56)$. To determine the end behavior, we try a larger viewing rectangle—in this case, $[-30, 30]$ by $[-200, 200]$. In the graph in Figure 8(b) the vertical asymptote has all but disappeared, and the graph looks like a parabola. To see why this is the case, we divide the denominator of f into the numerator and write the result in quotient-remainder form:

$$\begin{array}{r} x^2 \\ x - 2 \overline{)x^3 - 2x^2 + 0x + 3} \\ \underline{x^3 - 2x^2 } \\ 3 \end{array}$$

$$f(x) = x^2 + \frac{3}{x - 2}$$

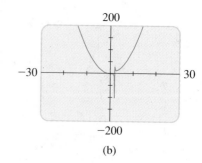

FIGURE 8
$$f(x) = \frac{x^3 - 2x^2 + 3}{x - 2}$$

(a)   (b)

When $|x|$ is large, $3/(x - 2)$ is small; that is, $3/(x - 2) \to 0$ as $x \to \pm\infty$. This means that for large $|x|$, the graph of f will be close to the graph of $y = x^2$. Thus, the end behavior of the function f is like that of the parabola $y = x^2$.

In Figure 9 the graphs of $y = (x^3 - 2x^2 + 3)/(x - 2)$ and $y = x^2$ are displayed in the viewing rectangle $[-8, 8]$ by $[-5, 20]$. From the figure we see that the graphs of the two functions are very close to each other everywhere except near the vertical asymptote.   ■

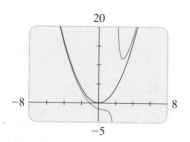

FIGURE 9
$$y = \frac{x^3 - 2x^2 + 3}{x - 2} \quad \text{and} \quad y = x^2$$

Rational functions occur frequently in scientific applications of algebra. In the next example we analyze the graph of a function from the theory of electricity.

EXAMPLE 8 ■ Electrical Resistance

When two resistors with resistances R_1 and R_2 are connected in parallel, their combined resistance R is given by the formula

$$R = \frac{R_1 R_2}{R_1 + R_2}$$

Suppose that a fixed 8-ohm resistor is connected in parallel with a variable resistor, as shown in Figure 10. If the resistance of the variable resistor is denoted by x, then the combined resistance R is a function of x. Graph R and give a physical interpretation of the graph.

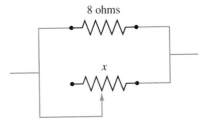

8 ohms

x

FIGURE 10

SOLUTION Substituting $R_1 = 8$ and $R_2 = x$ into the formula gives the function

$$R(x) = \frac{8x}{8 + x}$$

Since resistance cannot be negative, this function has physical meaning only when $x > 0$. The function is graphed in Figure 11(a) using the viewing rectangle $[0, 20]$ by $[0, 10]$. The function has no vertical asymptote when x is restricted to positive values. The combined resistance R increases as the variable resistance x increases. If we widen the viewing rectangle to $[0, 100]$ by $[0, 10]$, we obtain the graph in Figure 11(b). For large x, the combined resistance R levels off, getting closer and closer to the horizontal asymptote $R = 8$. No matter how large the variable resistance x, the combined resistance is never greater than 8 ohms.

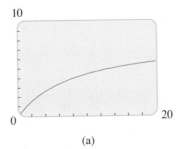

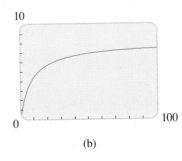

FIGURE 11

$R(x) = \dfrac{8x}{8 + x}$

(a)   (b)

■

3.5 | EXERCISES

1–4 ■ Find the x- and y-intercepts of the function.

1. $y = \dfrac{x - 6}{x + 1}$

2. $y = \dfrac{2}{x - 2}$

3. $y = \dfrac{x}{x^2 - 2x - 15}$

4. $y = \dfrac{x^2 + 10}{2x}$

5–6 ■ From the graph, determine the x- and y-intercepts and the vertical and horizontal asymptotes.

5.

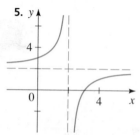

6.

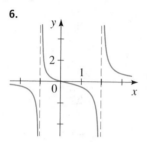

7–16 ■ Find all asymptotes (including vertical, horizontal, and slant).

7. $y = \dfrac{5}{x + 3}$

8. $y = \dfrac{3x + 3}{x - 3}$

9. $y = \dfrac{x^2}{x^2 - x - 6}$

10. $y = \dfrac{2x - 4}{x^2 + 2x + 1}$

11. $y = \dfrac{6}{x^2 + 2}$

12. $y = \dfrac{(x - 1)(x - 2)}{(x - 3)(x - 4)}$

13. $y = \dfrac{x^2 + 2}{x - 1}$

14. $y = \dfrac{x^3 + 3x^2}{x^2 - 4}$

15. $y = \dfrac{2x^3 - x^2 - 8x + 4}{x + 3}$

16. $y = \dfrac{6x^4}{x^2 - 3}$

17–38 ■ Find the intercepts and asymptotes, and then sketch a graph of the rational function.

17. $y = \dfrac{4}{x - 2}$

18. $y = \dfrac{9}{x + 3}$

19. $y = \dfrac{x - 1}{x - 2}$

20. $y = \dfrac{x + 9}{x - 3}$

21. $y = \dfrac{4x - 4}{x + 2}$

22. $y = \dfrac{2x + 6}{-6x + 3}$

23. $y = \dfrac{2x - 4}{x}$

24. $y = \dfrac{x}{2x - 4}$

25. $y = \dfrac{18}{(x - 3)^2}$

26. $y = \dfrac{x - 2}{(x + 1)^2}$

27. $y = \dfrac{4x + 8}{(x - 4)(x + 1)}$

28. $y = \dfrac{x - 9}{(x + 3)(x - 1)}$

29. $y = \dfrac{(x - 1)(x + 2)}{(x + 1)(x - 3)}$

30. $y = \dfrac{2x(x + 4)}{(x - 1)(x - 2)}$

31. $y = \dfrac{x^2 - 2x + 1}{x^2 + 2x + 1}$

32. $y = \dfrac{4x^2}{x^2 - 2x - 3}$

33. $y = \dfrac{2x^2 + 10x - 12}{x^2 + x - 6}$

34. $y = \dfrac{2x^2 + 2x - 4}{x^2 + x}$

35. $y = \dfrac{x^2 - x - 6}{x^2 + 3x}$

36. $y = \dfrac{x^2 + 3x}{x^2 - x - 6}$

37. $y = \dfrac{3x^2 + 6}{x^2 - 2x - 3}$

38. $y = \dfrac{5x^2 + 5}{x^2 + 4x + 4}$

39–46 ■ Find the slant asymptote, the vertical asymptotes, and sketch a graph of the function.

39. $y = \dfrac{x^2}{x - 2}$

40. $y = \dfrac{x^2 + 2x}{x - 1}$

41. $y = \dfrac{x^2 - 2x - 8}{x}$

42. $y = \dfrac{3x - x^2}{2x - 2}$

43. $y = \dfrac{x^2 + 5x + 4}{x - 3}$

44. $y = \dfrac{x^3 + 4}{2x^2 + x - 1}$

45. $y = \dfrac{x^3 + x^2}{x^2 - 4}$

46. $y = \dfrac{2x^3 + 2x}{x^2 - 1}$

47–50 ■ Graph the rational function f and determine all vertical asymptotes from your graph. Then graph f and g in a sufficiently large viewing rectangle to show that they have the same end behavior.

47. $f(x) = \dfrac{2x^2 + 6x + 6}{x + 3}$, $g(x) = 2x$

48. $f(x) = \dfrac{-x^3 + 6x^2 - 5}{x^2 - 2x}$, $g(x) = -x + 4$

49. $f(x) = \dfrac{x^3 - 2x^2 + 16}{x - 2}$, $g(x) = x^2$

50. $f(x) = \dfrac{-x^4 + 2x^3 - 2x}{(x - 1)^2}$, $g(x) = 1 - x^2$

51–54 ■ Graph the rational function and find all vertical asymptotes, x- and y-intercepts, and local extrema, correct to the nearest decimal. Then use long division to find a polynomial that has the same end behavior as the rational function, and graph both functions in a sufficiently large viewing rectangle to verify that the end behaviors of the polynomial and the rational function are the same.

51. $y = \dfrac{2x^2 - 5x}{2x + 3}$

52. $y = \dfrac{x^4 - 3x^3 + x^2 - 3x + 3}{x^2 - 3x}$

53. $y = \dfrac{x^5}{x^3 - 1}$

54. $y = \dfrac{x^4}{x^2 - 2}$

55. In this chapter we have adopted the convention that in rational functions, the numerator and denominator don't share a common factor. In this exercise we consider the graph of a rational function that doesn't satisfy this rule. Show that the graph of

$$r(x) = \frac{3x^2 - 3x - 6}{x - 2}$$

is the line $y = 3x + 3$ with the point $(2, 9)$ removed. [*Hint:* Factor. What is the domain of r?]

56–59 ■ Graph the rational function, using the method of Exercise 55.

56. $y = \dfrac{x^2 + x - 20}{x + 5}$

57. $y = \dfrac{2x^2 - x - 1}{x - 1}$

58. $y = \dfrac{x^2 - 3x + 2}{x^2 - 4x + 4}$

59. $y = \dfrac{2x^2 - 5x - 3}{x^2 - 2x - 3}$

60. For a camera with a lens of fixed focal length F to focus on an object located a distance x from the lens, the film must be placed a distance y behind the lens, where F, x, and y are related by

$$\frac{1}{x} + \frac{1}{y} = \frac{1}{F}$$

(See the figure.)

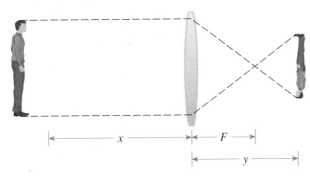

Suppose the camera has a 55-millimeter lens ($F = 55$).
(a) Express y as a function of x and graph the function.
(b) What happens to the focusing distance y as the object moves far away from the lens?
(c) What happens to the focusing distance y as the object moves close to the lens?

61. The rabbit population on Mr. Jenkins' farm follows the formula

$$p(t) = \frac{3000t}{t + 1}$$

where $t \geq 0$ is the time (in months) since the beginning of the year.
(a) Sketch a graph of the rabbit population.
(b) What eventually happens to the rabbit population?

62. After a certain drug is injected into a patient, the concentration c of the drug in the blood is monitored. At time $t \geq 0$ (in minutes since the injection), the concentration (in mg/L) is given by

$$c(t) = \frac{30t}{t^2 + 2}$$

(a) Sketch a graph of the drug concentration.
(b) What eventually happens to the concentration of drug in the blood?

63. A drug is administered to a patient and the concentration of the drug in the bloodstream is monitored. At time $t \geq 0$ (in hours since giving the drug), the concentration (in mg/L) is given by

$$c(t) = \frac{5t}{t^2 + 1}$$

Graph the function c with a graphing device.
(a) What is the highest concentration of drug that is reached in the patient's bloodstream?
(b) What happens to the drug concentration after a long period of time?

(c) How long does it take for the concentration to drop below 0.3 mg/L?

64. Suppose a rocket is fired upward from the surface of the earth with an initial velocity v (measured in m/s). Then the maximum height h (in meters) reached by the rocket is given by the function

$$h(v) = \frac{Rv^2}{2gR - v^2}$$

where $R = 6.4 \times 10^6$ m is the radius of the earth and $g = 9.8$ m/s^2 is the acceleration due to gravity. Use a graphing device to sketch a graph of the function h. (Note that h and v must both be positive, so the viewing rectangle need not contain negative values.) What does the vertical asymptote represent physically?

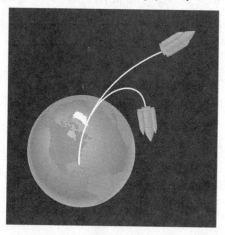

65. As a train moves towards an observer (see the figure), the pitch of its whistle sounds higher to the observer

than it would if the train were at rest, because the crests of the sound waves are compressed closer together. This phenomenon is called the *Doppler effect*. The observed pitch P is a function of the speed v of the train and is given by

$$P(v) = P_0\left(\frac{s_0}{s_0 - v}\right)$$

where P_0 is the actual pitch of the whistle at the source and $s_0 = 332$ m/s is the speed of sound in air. Suppose that a train has a whistle pitched at $P_0 = 440$ Hz. Graph the function $y = P(v)$ using a graphing device. How can the vertical asymptote of this function be interpreted physically?

⬤ DISCOVERY · DISCUSSION

66. Constructing a Rational Function from Its Asymptotes
Give an example of a rational function that has vertical asymptote $x = 3$. Now give an example of one that has vertical asymptote $x = 3$ *and* horizontal asymptote $y = 2$. Now give an example of a rational function with vertical asymptotes $x = 1$ and $x = -1$, horizontal asymptote $y = 0$, and x-intercept 4.

67. A Rational Function with No Asymptote   Explain how you can tell (without graphing it) that the function

$$r(x) = \frac{x^6 + 10}{x^4 + 8x^2 + 15}$$

has no x-intercept and no horizontal, vertical, or slant asymptote. What is its end behavior?

3 | REVIEW

CONCEPT CHECK

1. (a) Write the defining equation for a polynomial P of degree n.
(b) What does it mean to say that c is a zero of P?

2. Sketch graphs showing the possible end behaviors of polynomials of odd degree and of even degree.

3. What steps would you follow to graph a polynomial by hand?

4. (a) What is meant by a local maximum point or local minimum point of a polynomial?
(b) How many local extrema can a polynomial of degree n have?

5. State the Division Algorithm and identify the dividend, divisor, quotient, and remainder.

6. How does synthetic division work?

7. (a) State the Remainder Theorem.
(b) State the Factor Theorem.

8. (a) State the Rational Zeros Theorem.
(b) What steps would you take to find the rational zeros of a polynomial?

9. State Descartes' Rule of Signs.

10. (a) What does it mean to say that a is a lower bound and b is an upper bound for the zeros of a polynomial?
(b) State the Upper and Lower Bounds Theorem.

11. (a) What is a complex number?
(b) What are the real and imaginary parts of a complex number?
(c) What is the complex conjugate of a complex number?
(d) What is the modulus of a complex number?
(e) How do you add, subtract, multiply, and divide complex numbers?

12. (a) State the Fundamental Theorem of Algebra.
(b) State the Complete Factorization Theorem.
(c) What does it mean to say that c is a zero of multiplicity k of a polynomial P?
(d) State the Zeros Theorem.
(e) State the Conjugate Roots Theorem.

13. (a) What is a rational function?
(b) What does it mean to say that $x = a$ is a vertical asymptote of $y = f(x)$?
(c) How do you locate a vertical asymptote?
(d) What does it mean to say that $y = b$ is a horizontal asymptote of $y = f(x)$?
(e) How do you locate a horizontal asymptote?
(f) What steps do you follow to sketch the graph of a rational function by hand?
(g) Under what circumstances does a rational function have a slant asymptote? If one exists, how do you find it?

EXERCISES

1–6 ■ Graph the polynomial. Show clearly all x- and y-intercepts.

1. $y = (x - 2)^3 + 8$

2. $y = 32 - 2x^4$

3. $y = x^3 - 9x$

4. $y = x^3 - 5x^2 - 6x$

5. $y = x^3 - 5x^2 - 4x + 20$

6. $y = x^4 - 9x^2$

7–10 ■ Use a graphing device to graph the polynomial. Find the x- and y-intercepts and the coordinates of all local extrema, correct to the nearest decimal. Describe the end behavior of the function.

7. $y = 2x^3 + x^2 - 18x - 9$

8. $y = x^4 - 8x^2 + 16$

9. $y = x^5 + x^2 - 5$ **10.** $y = 3x^5 + x^4 - 4x$

11–18 ■ Find the quotient and remainder.

11. $\dfrac{x^3 - x^2 + x - 11}{x - 3}$ **12.** $\dfrac{x^4 + 30x + 12}{2x + 6}$

13. $\dfrac{x^3 - x^2 - 11x + 6}{x^2 + 2x - 5}$

14. $\dfrac{x^5 - 3x^4 + 3x^3 + 20x - 6}{x^2 + 2x - 6}$

15. $\dfrac{x^4 - 25x^2 + 4x + 15}{x + 5}$

16. $\dfrac{2x^3 - x^2 - 5}{x - \frac{3}{2}}$

17. $\dfrac{x^4 + x^3 - 2x^2 - 3x - 1}{x - \sqrt{3}}$

18. $\dfrac{15x - 7}{5x + 12}$

19–20 ■ Find the indicated value of the polynomial using the Remainder Theorem.

19. $P(x) = 2x^3 - 9x^2 - 7x + 13;$ find $P(5)$

20. $Q(x) = x^4 + 4x^3 + 7x^2 + 10x + 15;$ find $Q(-3)$

21. Show that $\frac{1}{2}$ is a zero of the polynomial
$$2x^4 + x^3 - 5x^2 + 10x - 4$$

22. Use the Factor Theorem to show that $x + 4$ is a factor of the polynomial
$$x^5 + 4x^4 - 7x^3 - 23x^2 + 23x + 12$$

23. What is the remainder when the polynomial $x^{500} + 6x^{201} - x^2 - 2x + 4$ is divided by $x - 1$?

24. What is the remainder when $x^{101} - x^4 + 2$ is divided by $x + 1$?

25–26 ■ List all possible rational roots (without testing to see if they actually are roots), and then determine the possible number of positive and negative real roots using Descartes' Rule of Signs.

25. $x^5 - 6x^3 - x^2 + 2x + 18 = 0$

26. $6x^4 + 3x^3 + x^2 + 3x + 4 = 0$

27–36 ■ Evaluate the expression and write the result in the form $a + bi$.

27. $(3 - 5i) - (6 + 4i)$

28. $(-2 + 3i) + \left(\frac{1}{2} - i\right)$

29. $(2 + 7i)(6 - i)$

30. $3(5 - 2i)\dfrac{i}{5}$

31. $\dfrac{2 - 3i}{2 + 3i}$

32. $\dfrac{2 + i}{4 - 3i}$

33. i^{45}

34. $(3 - i)^3$

35. $\left(1 - \sqrt{-3}\right)\left(2 + \sqrt{-4}\right)$

36. $\sqrt{-5} \cdot \sqrt{-20}$

37. Find a polynomial of degree 3 with constant coefficient 12 and zeros $-\frac{1}{2}$, 2, and 3.

38. Find a polynomial of degree 4 having integer coefficients and zeros $3i$ and 4, with 4 a double zero.

39. Does there exist a polynomial of degree 4 with integer coefficients that has zeros i, $2i$, $3i$, and $4i$? If so, find it. If not, explain why.

40. Prove that the equation $3x^4 + 5x^2 + 2 = 0$ has no real root.

41–50 ■ Find all rational, irrational, and imaginary roots (and state their multiplicities). Use Descartes' Rule of Signs, the Upper and Lower Bounds Theorem, the quadratic formula, or other factoring techniques to help you whenever possible.

41. $x^3 - 3x^2 - 13x + 15 = 0$

42. $2x^3 + 5x^2 - 6x - 9 = 0$

43. $x^4 + 6x^3 + 17x^2 + 28x + 20 = 0$

44. $x^4 + 7x^3 + 9x^2 - 17x - 20 = 0$

45. $x^5 - 3x^4 - x^3 + 11x^2 - 12x + 4 = 0$

46. $x^4 = 81$

47. $x^6 = 64$

48. $18x^3 + 3x^2 - 4x - 1 = 0$

49. $6x^4 - 18x^3 + 6x^2 - 30x + 36 = 0$

50. $x^4 + 15x^2 + 54 = 0$

51–54 ■ Use a graphing device to find all real solutions of the equation.

51. $2x^2 = 5x + 3$

52. $x^3 + x^2 - 14x - 24 = 0$

53. $x^4 - 3x^3 - 3x^2 - 9x - 2 = 0$

54. $x^5 = x + 3$

55–60 ■ Graph the rational function. Show clearly all x- and y-intercepts and asymptotes.

55. $y = \dfrac{3x - 12}{x + 1}$

56. $y = \dfrac{1}{(x + 2)^2}$

57. $y = \dfrac{x - 2}{x^2 - 2x - 8}$

58. $y = \dfrac{2x^2 - 6x - 7}{x - 4}$

59. $y = \dfrac{x^2 - 9}{2x^2 + 1}$

60. $y = \dfrac{x^3 + 27}{x + 4}$

 61–64 ■ Use a graphing device to analyze the graph of the rational function. Find all x- and y-intercepts, vertical, horizontal, and slant asymptotes, and the coordinates of local extrema. If the function has no horizontal or slant asymptote, find a polynomial that has the same end behavior as the rational function.

61. $y = \dfrac{x - 3}{2x + 6}$

62. $y = \dfrac{2x - 7}{x^2 + 9}$

63. $y = \dfrac{x^3 + 8}{x^2 - x - 2}$

64. $y = \dfrac{2x^3 - x^2}{x + 1}$

65. (a) Show that -1 is a root of the equation
$$2x^4 + 5x^3 + x + 4 = 0$$

(b) Use the information from part (a) to show that $2x^3 + 3x^2 - 3x + 4 = 0$ has no positive real root. [*Hint:* Compare the coefficients of this polynomial to your synthetic division table from part (a).]

66. Find the coordinates of all points of intersection of the graphs of
$$y = x^4 + x^2 + 24x \quad \text{and} \quad y = 6x^3 + 20$$

1. Graph the function $P(x) = x^3 - x^2 - 9x + 9$, showing clearly all x- and y-intercepts.

2. Use synthetic division to find the quotient and remainder when $x^4 - 4x^2 + 2x + 5$ is divided by $x - 2$.

3. Let $P(x) = 2x^4 - 7x^3 + x^2 + 7x - 3$.
   (a) List all possible rational zeros of P.
   (b) Find the complete factorization of P.
   (c) What are the zeros of P?

4. Evaluate and write your answer in the form $a + bi$.
   (a) $\dfrac{6 - 2i}{2 + 3i}$ (b) $(2 - i)^3$ (c) i^{13}

5. Find all real and complex roots of the equation $x^4 + x^3 - 2x^2 - 6x - 4 = 0$.

6. Let
$$P(x) = x^{23} - 5x^{12} + 8x - 1 \qquad Q(x) = 3x^4 + x^2 - x - 15$$
$$R(x) = 4x^6 + x^4 + 2x^2 + 16$$

   (a) Explain why an even integer could not possibly be a zero of any of these three polynomials.
   (b) Does R have any real zeros? Why or why not?
   (c) How many real zeros does Q have? Why?
   (d) Show that P has no rational zero.

7. Find a fourth-degree polynomial with integer coefficients that has zeros $1 + 2i$ and -1, with -1 a zero of multiplicity 2.

8. Let $P(x) = 2x^4 - 17x^3 + 53x^2 - 72x + 36$.
   (a) Use Descartes' Rule of Signs to determine how many positive and how many negative real roots the equation $P(x) = 0$ may have.
   (b) Show that 9 is an upper bound for the real roots of $P(x) = 0$ but is not itself a root.

9. Consider the following four rational functions:

$$r(x) = \frac{2x - 1}{x^2 - x - 2} \qquad s(x) = \frac{x^3 + 27}{x^2 + 4} \qquad t(x) = \frac{x^3 - 9x}{x + 2} \qquad u(x) = \frac{x^2 + x - 6}{x^2 - 25}$$

   (a) Which of these four rational functions has a horizontal asymptote?
   (b) Which of these functions has a slant asymptote?
   (c) Which of these functions has no vertical asymptote?
   (d) Graph $y = u(x)$, showing clearly any asymptotes and x- and y-intercepts the function may have.

10. Choose an appropriate viewing rectangle and graph the following function. Find all its x-intercepts and local extrema, correct to two decimal places:
$$P(x) = x^4 - 4x^3 + 8x$$

FOCUS ON PROBLEM SOLVING

It is often necessary to combine different problem-solving principles. The principles of **drawing a diagram** and **introducing something extra** are used in the two problems that follow. Euler's solution of the "Königsberg bridge problem" is so brilliant that it is the basis for a whole field of mathematics called *network theory*, which has practical applications to electric circuits and economics.

THE KÖNIGSBERG BRIDGE PROBLEM

The old city of Königsberg, now called Kaliningrad, is situated on both banks of the Pregel River and on two islands in the river. Seven bridges connect the parts of the city, as shown in Figure 1. After the last bridge was built, the Königsberg townspeople posed the problem of finding a route through the city that would cross all seven bridges in a continuous walk, without recrossing any bridge. No one was able to find such a route. When Euler heard of the problem, he realized that an interesting mathematical principle must be at work. To discover the principle Euler first stripped away the nonessential parts of the problem by drawing a simpler diagram of the city so that the land masses were represented by points and the bridges by lines connecting the points (Figure 2).

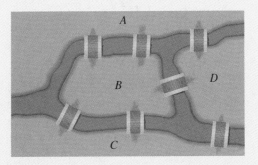

FIGURE 1

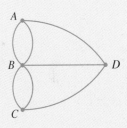

FIGURE 2

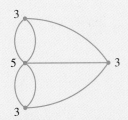

FIGURE 3

The problem now was to draw the diagram without lifting the pencil and without retracing any line. To show that this was impossible, Euler introduced something extra by labeling each point with the number of line segments that end at the point (Figure 3). He then argued that every pass through a point accounts for two line ends (one entering and one leaving). Thus each point, except possibly the starting point and the ending point, must be labeled by an even number in order for the walk to be possible. Since more than two points are labeled by odd numbers, the walk is impossible!

The beauty of Euler's solution is that it is completely general. In modern language a diagram consisting of points and connecting lines is called a *network*, each point is a *vertex*, each line an *edge*, and the number of edges ending at a vertex is the *order* of the vertex. A network is *traversable* if it can be drawn in one stroke without retracing any edge. Euler's reasoning about the Königsberg bridge problem proved the following principle: *If a network has more than two vertices with odd order, then it is not traversable.*

Here is another clever use of the principle of **introducing something extra**.

THE IMPOSSIBLE MUSEUM TOUR

The floor plan of a museum is in the shape of a square with six square rooms to a side. Each room is connected to each adjacent room by a door. The museum entrance and exit are at diagonally opposite corners, as shown in Figure 4.

A visitor making a tour of this museum would like to visit each room exactly once and then exit. Can you find such a path? Here are examples of attempts that failed:

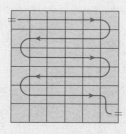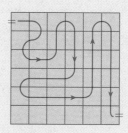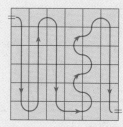

Trial and error will make it plausible to you that such a path is in fact not possible. But trial and error does not *prove* that our proposed tour of the museum is impossible. How do we give a convincing argument for this? Here is a clever idea: Imagine that the rooms are colored alternately black and white, like a checkerboard, as shown in Figure 5. So there are 18 black rooms and 18 white rooms. By coloring the rooms we've introduced something extra that at first does not seem relevant to the problem. Indeed, the tour either can or cannot be done, regardless of the colors of the rooms. But notice how coloring the rooms in this fashion allows us to give a convincing argument that explains why this tour is impossible. We reason as follows.

Entrance ⊨ ▢▢▢▢▢
▢▢▢▢▢▢
▢▢▢▢▢▢
▢▢▢▢▢▢
▢▢▢▢▢▢
▢▢▢▢▢ ⊨ Exit

FIGURE 4

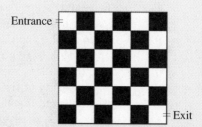

FIGURE 5

If a visitor is in a white room, he must next go to a black room, and if he is in a black room, he must next go to a white room. The entrance is in a white room, so the visitor's path, in terms of colors of rooms, is

W   B   W   B   W   B   ...

Since the museum contains an even number of rooms and the visitor starts in a white room, he must end his tour in a black room. But the exit is in a white room, so the proposed tour is impossible!

PROBLEMS

1–4 ■ Determine whether each network is traversable. If it is, find a path that traverses it.

1.

2.

3.

4.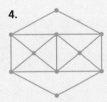

5. Is the museum tour possible if there are five rooms to a side instead of six? Find a general result about when the tour is possible for an $n \times n$ museum.

6. A Tibetan monk leaves the monastery at 7:00 A.M. and walks along his usual path to the top of the mountain, arriving at 7:00 P.M. The following morning, he starts at 7:00 A.M. at the top and takes the same path back, arriving at the monastery at 7:00 P.M. Show that there is exactly one point on the path that the monk will cross at exactly the same time of day on both trips.

7. A curve that starts and ends at the same point and does not intersect itself is called a **Jordan curve**. Such a curve always has an "inside" and an "outside." Is the point A in the figure inside or outside this Jordan curve? Find an easy method of determining whether a point is inside or outside a Jordan curve. [*Hint:* The point B is clearly outside the curve. Add a line segment connecting A and B.]

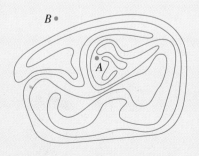

Srinivasa Ramanujan (1887–1920) was born into a poor family in the small town of Kumbakonam in India. He was self-taught in mathematics and worked in virtual isolation from other mathematicians. At the age of 25 he wrote a letter to G. H. Hardy, the leading British mathematician at the time, listing some of his discoveries. Hardy immediately recognized Ramanujan's genius and for the next six years the two worked together in London until Ramanujan fell ill and went back to his hometown in India, where he died a year later. Ramanujan was a genius with phenomenal ability to see hidden patterns in the properties of numbers. Most of his discoveries were written as complicated infinite series, the importance of which was not recognized until many years after his death. In the last year of his life he wrote 130 pages of mysterious formulas, many of which still defy proof. Hardy tells the story that when he visited Ramanujan in a hospital and arrived in a taxi, he remarked to Ramanujan that the cab's num-
(continued)

8. Consider the following sequence of polynomials:

$$P_1(x) = (x - a)$$
$$P_2(x) = (x - a)(x - b)$$
$$P_3(x) = (x - a)(x - b)(x - c)$$

$$\vdots \qquad \vdots \qquad \qquad \vdots$$

(a) Expand the polynomials P_2 and P_3.
(b) Using the pattern you observe from part (a), write the expansions of P_4 and P_5 without actually multiplying out these polynomials. The patterns you have found are called **Viète's relations** (see page 51).
(c) For a polynomial of degree n, express the coefficient of x^{n-1} and the constant coefficient in terms of the zeros of the polynomial.

9. If the equation $x^4 + ax^2 + bx + c = 0$ has roots 1, 2, and 3, find c.

10. Find a polynomial of degree 3 with integer coefficients that has $2 - \sqrt[3]{2}$ as one of its zeros. How many other real zeros does the polynomial have?

11. You have an 8×8 grid with squares removed from two diagonally opposite corners, as shown in the figure. You also have a set of dominoes, each of which covers exactly two squares of the grid. Is it possible to cover all the remaining squares of the grid with dominoes, without overlapping any dominoes? [*Hint:* Color the grid as a checkerboard.]

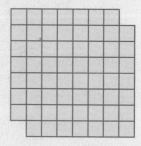

12. The number 1729 is the smallest positive integer that can be represented in two different ways as the sum of two cubes. What are the two ways?

13. If $f_0(x) = \dfrac{1}{1 - x}$ and $f_{n+1} = f_0 \circ f_n$ for $n = 0, 1, 2, \ldots$, find $f_{100}(3)$.

14. Sketch a graph of the function $g(x) = |x^2 - 1| - |x^2 - 4|$.

ber, 1729, was uninteresting. Ramanujan replied "No, it is a very interesting number. It is the smallest number expressible as the sum of two cubes in two different ways." (See Problem 12.)

15. Sketch a graph of the function $f(x) = x - [\![x]\!]$, where $[\![x]\!]$ is the greatest integer function defined on page 146.

16. Sketch the region in the plane defined by the equation

$$[\![x]\!]^2 + [\![y]\!]^2 = 1$$

17. (a) Show that if you multiply four consecutive integers and then add 1 to the result, you get a perfect square.
   (b) Show that if you multiply three consecutive integers and then add the middle integer to the result, you get a perfect cube.

18. The equation $x^3 = 1$ has only one real solution, so it must have two imaginary solutions. Let ω be one of the imaginary solutions. Show that the expression

$$(1 - \omega + \omega^2)(1 + \omega - \omega^2)$$

is a real number, and find its value.

19. (a) If w and z are complex numbers, show that $|wz| = |w||z|$.
   (b) Show that for any integers a and b, one can find integers c and d such that $c^2 + d^2 = 2(a^2 + b^2)$. [*Hint:* Let $w = 1 + i$, $z = a + bi$; evaluate $|wz|^2$ in two ways, using part (a).]

   Remark: This problem shows how complex numbers can be used to find properties of integers.

20. A rectangle is divided into smaller rectangles in such a way that each of the smaller rectangles has at least one side of integer length. Show that one of the sides of the original, large rectangle must have integer length. [*Hint:* Imagine the rectangle lying in the coordinate plane, which has been tiled into a checkerboard pattern of black and white squares, each $\frac{1}{2}$ unit by $\frac{1}{2}$ unit. Observe that a rectangle has at least one integer side if and only if its area is composed of equal amounts of black and white.]

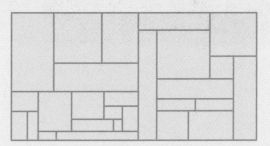

4

EXPONENTIAL AND LOGARITHMIC FUNCTIONS

Population growth—such as the growth of a school of fish in the ocean—is modeled by exponential functions. These same functions describe compound interest, radioactive decay, and temperature changes.

Mathematics compares the most diverse phenomena and discovers the secret analogies which unite them.

JOSEPH FOURIER

So far we have studied relatively simple functions such as polynomials and rational functions. We now turn our attention to two of the most important functions in mathematics, the *exponential function* and its inverse function, the *logarithmic function*. We use these functions to describe exponential growth in biology and economics and radioactive decay in physics and chemistry.

4.1 EXPONENTIAL FUNCTIONS

In Section 1.2 we defined a^x if $a > 0$ and x is a rational number, but we have not yet defined irrational powers. For instance, what is meant by $2^{\sqrt{3}}$ or 5^{π}? To help us answer these questions we first look at the graph of the function $f(x) = 2^x$, where x is rational. A very crude representation of this graph is shown in Figure 1. Holes occur throughout the graph wherever x is irrational, and we want to fill in these holes with a smooth curve. To do this we enlarge the domain of $f(x) = 2^x$ to all of $\mathbb{R}$ by defining irrational powers of 2 appropriately. For example, to define $2^{\sqrt{3}}$ we use rational approximations of $\sqrt{3}$. Since

$$\sqrt{3} \approx 1.73205\ldots$$

we successively approximate $2^{\sqrt{3}}$ by the following rational powers:

$$2^{1.7}, \quad 2^{1.73}, \quad 2^{1.732}, \quad 2^{1.7320}, \quad 2^{1.73205}, \quad \ldots$$

Using advanced mathematics it can be shown that there is exactly one number that these powers approach. We define $2^{\sqrt{3}}$ to be this number. Intuitively, these rational powers of 2 are getting closer and closer to $2^{\sqrt{3}}$. By this approximation process we can compute $2^{\sqrt{3}}$ correct to six decimal places:

$$2^{\sqrt{3}} \approx 3.321997$$

Similarly, we can define 2^x (or a^x if $a > 0$) where x is any irrational number.

Using this definition of irrational powers, we can graph $f(x) = 2^x$ on all of $\mathbb{R}$, as shown in Figure 2. It is an increasing function and, in fact, it increases very rapidly when $x > 0$ (see the margin note).

To demonstrate just how quickly $f(x) = 2^x$ increases, let's perform the following thought experiment. Suppose we start with a piece of paper a thousandth of an inch thick, and we fold it in half 50 times. Each time we fold the paper, the thickness of the paper stack doubles, so the thickness of the resulting stack would be $2^{50}/1000$ inches. How thick do you think that is? It works out to be more than 17 million miles!

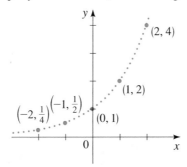

FIGURE 1
Representation of $f(x) = 2^x$ for x rational

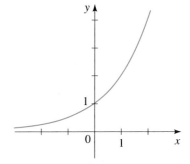

FIGURE 2
$f(x) = 2^x$ for x real

It can be proved that *the Laws of Exponents are still true when the exponents are real numbers.*

EXAMPLE 1 ■ Graphing Exponential Functions by Plotting Points

Draw the graph of each of the following functions.

(a) $f(x) = 3^x$ 　　　　　　　　　　(b) $g(x) = \left(\dfrac{1}{3}\right)^x$

SOLUTION　We calculate values of $f(x)$ and $g(x)$ and plot points to sketch the graphs in Figure 3.

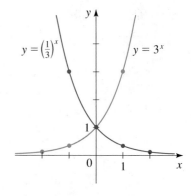

x	$f(x) = 3^x$	$g(x) = (\frac{1}{3})^x$
-3	$\frac{1}{27}$	27
-2	$\frac{1}{9}$	9
-1	$\frac{1}{3}$	3
0	1	1
1	3	$\frac{1}{3}$
2	9	$\frac{1}{9}$
3	27	$\frac{1}{27}$

FIGURE 3

Notice that

$$g(x) = \left(\frac{1}{3}\right)^x = \frac{1}{3^x} = 3^{-x} = f(-x)$$

and so we could have obtained the graph of g from the graph of f by reflecting in the y-axis.　■

Figure 4 shows the graphs of the family of exponential functions $f(x) = a^x$ for various values of the base a. All of these graphs pass through the point $(0, 1)$

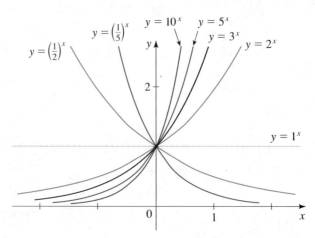

FIGURE 4

A family of exponential functions

because $a^0 = 1$ for $a \neq 0$. You can see from Figure 4 that there are three kinds of exponential functions $y = a^x$: If $0 < a < 1$, the exponential function decreases rapidly. If $a = 1$, it is constant. If $a > 1$, the function increases rapidly, and the larger the base, the more rapid the increase.

If $a > 1$, then the graph of $y = a^x$ approaches $y = 0$ as x decreases through negative values, and so the x-axis is a horizontal asymptote. If $0 < a < 1$, the graph approaches $y = 0$ as x increases indefinitely and, again, the x-axis is a horizontal asymptote. In either case the graph never touches the x-axis because $a^x > 0$ for all x. Thus, for $a \neq 1$, the exponential function $f(x) = a^x$ has domain $\mathbb{R}$ and range $(0, \infty)$. We summarize the preceding discussion.

EXPONENTIAL FUNCTIONS

For $a > 0$, the **exponential function with base a** is defined by

$$f(x) = a^x$$

For $a \neq 1$, the domain of f is $\mathbb{R}$, the range of f is $(0, \infty)$, and the graph of f has one of the following shapes:

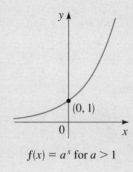

$f(x) = a^x$ for $a > 1$

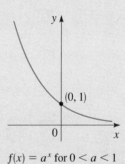

$f(x) = a^x$ for $0 < a < 1$

EXAMPLE 2 ■ **Identifying the Graphs of Exponential Functions**

Find the exponential function $f(x) = a^x$ whose graph is given.

(a)

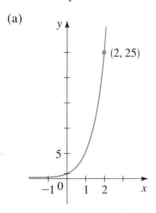

(b)

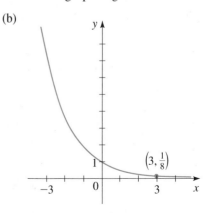

SOLUTION

(a) Since $f(2) = a^2 = 25$, we see by inspection that the base is $a = 5$. So this is the graph of the exponential function $f(x) = 5^x$.

(b) Since $f(3) = a^3 = \frac{1}{8}$, we see by inspection that the base is $a = \frac{1}{2}$. So this is the graph of the exponential function $f(x) = \left(\frac{1}{2}\right)^x$. ■

In the next two examples we see how to graph certain functions, not by plotting points, but by taking the basic graphs of the exponential functions in Figure 4 and applying the shifting and reflecting transformations of Section 2.4.

EXAMPLE 3 ■ Transformations of Exponential Functions

Vertical shifting and reflecting of graphs is explained in Section 2.4.

Use the graph of $f(x) = 2^x$ to sketch the graph of each of the following functions.

(a) $g(x) = 1 + 2^x$ (b) $h(x) = -2^x$

SOLUTION

(a) The graph of $g(x) = 1 + 2^x$ is obtained by starting with the graph of $f(x) = 2^x$ in Figure 5(a) and shifting it upward 1 unit. Notice from Figure 5(b) that the line $y = 1$ is now a horizontal asymptote.

(b) Again we start with the graph of $f(x) = 2^x$, but here we reflect in the x-axis to get the graph of $h(x) = -2^x$ shown in Figure 5(c).

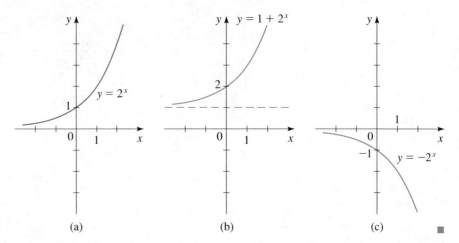

FIGURE 5   (a)   (b)   (c)   ■

EXAMPLE 4 ■ Transformations of Exponential Functions

Vertical and horizontal shifting of graphs is explained in Section 2.4.

(a) Use the graph of $f(x) = 10^x$ to sketch the graph of $g(x) = 10^{x-1} - 2$.

(b) State the asymptote, the domain, and the range of this function.

SOLUTION

(a) Recall from Section 2.4 that we get the graph of $y = f(x - 1)$ from the graph of $y = f(x)$ by shifting to the right 1 unit. Thus, we get the graph of

$y = 10^{x-1} - 2$ by shifting the graph of $y = 10^x$ to the right 1 unit and downward 2 units, as shown in Figure 6.

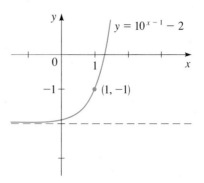

FIGURE 6

(b) The horizontal asymptote is $y = -2$, the domain is $\mathbb{R}$, and the range is

$$\{y \mid y > -2\} = (-2, \infty) \qquad \blacksquare$$

EXAMPLE 5 ■ Comparing Exponential and Power Functions

Compare the rates of growth of the exponential function $f(x) = 2^x$ and the power function $g(x) = x^2$ by drawing the graphs of both functions in each of the following viewing rectangles.

(a) $[0, 3]$ by $[0, 8]$ (b) $[0, 6]$ by $[0, 25]$ (c) $[0, 20]$ by $[0, 1000]$

SOLUTION

(a) Figure 7(a) shows that the graph of $g(x) = x^2$ catches up with, and becomes higher than, the graph of $f(x) = 2^x$ at $x = 2$.

(b) The larger viewing rectangle in Figure 7(b) shows that the graph of $f(x) = 2^x$ overtakes that of $g(x) = x^2$ when $x = 4$.

(c) Figure 7(c) gives a more global view and shows that, when x is large, $f(x) = 2^x$ is much larger than $g(x) = x^2$.

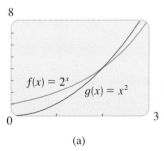

(a)

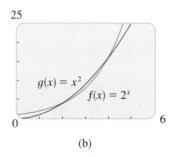

(b)

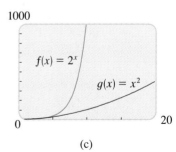

(c)

FIGURE 7   ■

4.1   **EXERCISES**

1–8 ■ Sketch the graph of the function by making a table of values. Use a calculator if necessary.

1. $f(x) = 2^x$

2. $g(x) = 8^x$

3. $h(x) = 6^x$

4. $h(x) = (0.8)^x$

5. $f(x) = \left(\frac{1}{3}\right)^x$

6. $h(x) = (1.1)^x$

7. $g(x) = \left(\frac{1}{4}\right)^x$

8. $f(x) = \left(\frac{3}{2}\right)^x$

9–10 ■ Graph both functions on one set of axes.

9. $y = 4^x$ and $y = 7^x$

10. $y = \left(\frac{2}{3}\right)^x$ and $y = \left(\frac{4}{3}\right)^x$

11–14 ■ Find the exponential function $f(x) = a^x$ whose graph is given.

11.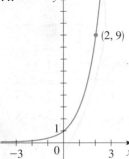

12.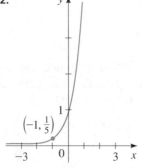

13.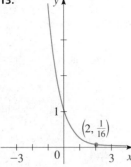

14.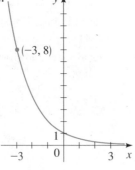

15–20 ■ Match the exponential function with one of the graphs labeled I–VI.

15. $f(x) = 5^x$

16. $f(x) = -5^x$

17. $f(x) = 5^{-x}$

18. $f(x) = 5^x + 3$

19. $f(x) = 5^{x-3}$

20. $f(x) = 5^{x+1} - 4$

I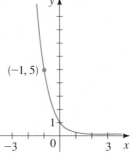

II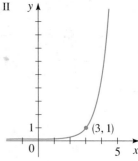

III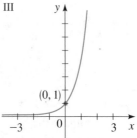

IV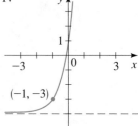

V

VI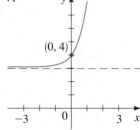

21–36 ■ Graph the function, not by plotting points, but by starting from the graphs in Figure 4. State the domain, range, and asymptote.

21. $f(x) = -3^x$

22. $f(x) = 10^{-x}$

23. $g(x) = 2^x - 3$

24. $g(x) = 2^{x-3}$

25. $h(x) = 4 + \left(\frac{1}{2}\right)^x$

26. $h(x) = 6 - 3^x$

27. $f(x) = 10^{x+3}$

28. $f(x) = -\left(\frac{1}{5}\right)^x$

29. $f(x) = -3^{-x}$

30. $f(x) = 10^{-x} - 4$

31. $y = 5^{-2x}$

32. $y = 1 + 2^{x+1}$

33. $f(x) = 5 - 2^{x-1}$

34. $f(x) = 1 - 2^{-x}$

35. $y = 2^{|x|}$

36. $y = 2^{-|x|}$

37–38 ■ Find the function of the form $f(x) = Ca^x$ whose graph is given.

37.

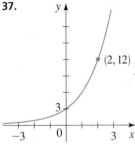

38.

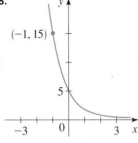

39. (a) Sketch the graphs of $f(x) = 2^x$ and $g(x) = 3(2^x)$.
 (b) How are the graphs related?

40. (a) Sketch the graphs of $f(x) = 9^{x/2}$ and $g(x) = 3^x$.
 (b) Use the Laws of Exponents to explain the relationship between these graphs.

41. If $f(x) = 10^x$, show that

$$\frac{f(x + h) - f(x)}{h} = 10^x \left(\frac{10^h - 1}{h} \right)$$

42. Compare the functions $f(x) = x^3$ and $g(x) = 3^x$ by evaluating both of them for $x = 0, 1, 2, 3, 4, 5, 6, 7, 8, 9,$ 10, 15, and 20. Then draw the graphs of f and g on the same set of axes.

 43. (a) Compare the rates of growth of the functions $f(x) = 2^x$ and $g(x) = x^5$ by drawing the graphs of both functions in each of the following viewing rectangles.
  (i) $[0, 5]$ by $[0, 20]$
  (ii) $[0, 25]$ by $[0, 10^7]$
  (iii) $[0, 50]$ by $[0, 10^8]$
 (b) Find the solutions of the equation $2^x = x^5$, correct to one decimal place.

 44. (a) Compare the rates of growth of the functions $f(x) = 3^x$ and $g(x) = x^4$ by drawing the graphs of

both functions in each of the the following viewing rectangles:
  (i) $[-4, 4]$ by $[0, 20]$
  (ii) $[0, 10]$ by $[0, 5000]$
  (iii) $[0, 20]$ by $[0, 10^5]$
 (b) Find the solutions of the equation $3^x = x^4$, correct to two decimal places.

45–46 ■ Draw graphs of the given family of functions for $c = 0.25, 0.5, 1, 2, 4$. How are the graphs related?

45. $f(x) = c2^x$

46. $f(x) = 2^{cx}$

47–48 ■ Graph the function and comment on vertical and horizontal asymptotes.

47. $y = 2^{1/x}$

48. $y = \dfrac{2^x}{x}$

49–50 ■ Find the local maximum and minimum values of the function and the value of x at which each occurs. State each answer correct to two decimal places.

49. $g(x) = x^x$ $(x > 0)$

50. $g(x) = \sqrt[x]{x}$ $(x > 0)$

51–52 ■ Find, correct to two decimal places, (a) the intervals on which the function is increasing or decreasing and (b) the range of the function.

51. $y = 10^{x-x^2}$

52. $y = x2^x$

 DISCOVERY · DISCUSSION

53. Growth of an Exponential Function Suppose you are offered a job that lasts one month, and you are to be very well paid. Which of the following methods of payment is more profitable for you?
 (a) One million dollars at the end of the month
 (b) Two cents on the first day of the month, 4 cents on the second day, 8 cents on the third day, and, in general, 2^n cents on the nth day

54. The Height of the Graph of an Exponential Function Your mathematics instructor asks you to sketch a graph of the exponential function $f(x) = 2^x$ for x between 0 and 40, using a scale of 10 units to one inch. What are the dimensions of the sheet of paper you will need to sketch this graph?

n	$\left(1 + \dfrac{1}{n}\right)^n$
1	2.00000
5	2.48832
10	2.59374
100	2.70481
1000	2.71692
10,000	2.71815
100,000	2.71827
1,000,000	2.71828
10,000,000	2.71828

4.2   THE NATURAL EXPONENTIAL FUNCTION

Any positive number can be used as the base for an exponential function, but some bases are used more frequently than others. We will see in the remaining sections of this chapter that the bases 2 and 10 are convenient for certain applications, but the most important base is the number denoted by the letter e.

The number e is defined as the value that $(1 + 1/n)^n$ approaches as n becomes large. (In calculus this idea is made more precise through the concept of a limit. See Exercise 41.) The table in the margin shows the values of the expression $(1 + 1/n)^n$ for increasingly large values of n. It appears that, correct to five decimal places, $e \approx 2.71828$; in fact, the approximate value to 20 decimal places is

$$e \approx 2.71828182845904523536$$

It can be shown that e is an irrational number, so we cannot write its exact value.

Why use such a strange base for an exponential function? It may seem at first that a base such as 10 is easier to work with. We will see, however, that in certain applications the number e is the best possible base. In this section we study how e occurs in the description of compound interest and population growth.

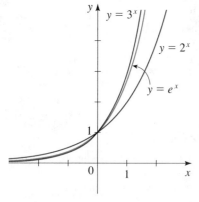

FIGURE 1

Graph of the natural exponential function

THE NATURAL EXPONENTIAL FUNCTION

The **natural exponential function** is the exponential function

$$f(x) = e^x$$

with base e. It is often referred to as *the* exponential function.

Since $2 < e < 3$, the graph of the natural exponential function lies between the graphs of $y = 2^x$ and $y = 3^x$, as shown in Figure 1.

Scientific calculators have a special key for the function $f(x) = e^x$. We use this key in the next example.

EXAMPLE 1 ■ Evaluating the Exponential Function

Evaluate each expression correct to five decimal places.

(a) e^3 (b) $2e^{-0.53}$ (c) $e^{4.8}$

SOLUTION   We use the $\boxed{e^x}$ key on a calculator to evaluate the exponential function.

(a) $e^3 \approx 20.08554$

(b) $2e^{-0.53} \approx 1.17721$

(c) $e^{4.8} \approx 121.51042$ ∎

The notation e for the base of the natural exponential function was chosen by the Swiss mathematician Leonhard Euler (see page 249), probably because it is the first letter of the word *exponential*.

EXAMPLE 2 ■ Transformations of the Exponential Function

Sketch the graph of each function.

(a) $f(x) = e^{-x}$ (b) $g(x) = 3e^{0.5x}$

SOLUTION

Reflecting graphs is explained in
Section 2.4.

(a) We start with the graph of $y = e^x$ and reflect in the y-axis to obtain the
graph of $y = e^{-x}$ as in Figure 2.

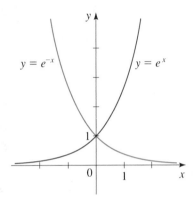

FIGURE 2

(b) We calculate several values, plot the resulting points, then connect the
points with a smooth curve. The graph is shown in Figure 3.

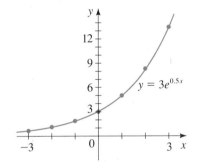

FIGURE 3

x	$f(x) = 3e^{0.5x}$
-3	0.67
-2	1.10
-1	1.82
0	3.00
1	4.95
2	8.15
3	13.45

■

COMPOUND INTEREST

The natural exponential function occurs in calculating *continuously* compounded
interest. We begin by explaining compound interest.

If an amount of money P, called the **principal**, is invested at a simple interest
rate r, then after one time period the interest is Pr, and the amount A of money is

$$A = P + Pr = P(1 + r)$$

If the interest is reinvested, then the new principal is now $P(1 + r)$, and the
amount after another time period is $A = P(1 + r)(1 + r) = P(1 + r)^2$. Similarly, after a third time period the amount is $A = P(1 + r)^3$. In general, after k

periods the amount is

$$A = P(1 + r)^k$$

Notice that this is an exponential function with base $1 + r$.

If interest is compounded n times per year, then in each time period the interest rate is r/n, and there are nt time periods in t years. This leads to the following formula for the amount after t years.

r is often referred to as the nominal annual interest rate.

COMPOUND INTEREST

Compound interest is calculated by the formula

$$A = P\left(1 + \frac{r}{n}\right)^{nt}$$

where
A = amount after t years

P = principal

r = interest rate

n = number of times interest is compounded per year

t = number of years

EXAMPLE 3 ■ Calculating Compound Interest

A sum of $1000 is invested at an interest rate of 12% per year. Find the amounts in the account after 3 years if interest is compounded annually, semiannually, quarterly, monthly, and daily.

SOLUTION We use the compound interest formula with $P = \$1000$, $r = 0.12$, and $t = 3$.

Compounding	n	Amount after 3 years
Annual	1	$1000\left(1 + \dfrac{0.12}{1}\right)^{1(3)} = \$1404.93$
Semiannual	2	$1000\left(1 + \dfrac{0.12}{2}\right)^{2(3)} = \$1418.52$
Quarterly	4	$1000\left(1 + \dfrac{0.12}{4}\right)^{4(3)} = \$1425.76$
Monthly	12	$1000\left(1 + \dfrac{0.12}{12}\right)^{12(3)} = \$1430.77$
Daily	365	$1000\left(1 + \dfrac{0.12}{365}\right)^{365(3)} = \$1433.24$

■

We see from Example 3 that the interest paid increases as the number of compounding periods n increases. Let's see what happens as n increases indefinitely. If we let $m = n/r$, then

$$A = P\left(1 + \frac{r}{n}\right)^{nt} = P\left[\left(1 + \frac{r}{n}\right)^{n/r}\right]^{rt} = P\left[\left(1 + \frac{1}{m}\right)^{m}\right]^{rt}$$

Recall that as m becomes large, the quantity $(1 + 1/m)^{m}$ approaches the number e. Thus, the amount approaches $A = Pe^{rt}$. This expression gives the amount when the interest is compounded at "every instant."

CONTINUOUSLY COMPOUNDED INTEREST

Continuously compounded interest is calculated by the formula

$$A = Pe^{rt}$$

where A = amount after t years

P = principal

r = interest rate

t = number of years

EXAMPLE 4 ■ Calculating Continuously Compounded Interest

Find the amount after 3 years if $1000 is invested at an interest rate of 12% per year, compounded continuously.

SOLUTION   We use the formula for continuously compounded interest with $P = \$1000$, $r = 0.12$, and $t = 3$ to get

$$A = 1000e^{(0.12)3} = 1000e^{0.36} = \$1433.33$$

Compare this amount with the amounts in Example 3.   ■

EXPONENTIAL GROWTH

Biologists have observed that the population of a species always doubles its size in a fixed period of time. For example, under ideal conditions a certain population of bacteria doubles in size every 3 hours. If the culture is started with 1000 bacteria, then after 3 hours there will be 2000 bacteria, after another 3 hours there will be 4000, and so on. If we let $n = n(t)$ be the number of bacteria after t hours, then

$$n(0) = 1000$$
$$n(3) = 1000 \cdot 2$$
$$n(6) = (1000 \cdot 2) \cdot 2 = 1000 \cdot 2^{2}$$
$$n(9) = (1000 \cdot 2^{2}) \cdot 2 = 1000 \cdot 2^{3}$$
$$n(12) = (1000 \cdot 2^{3}) \cdot 2 = 1000 \cdot 2^{4}$$

The **Gateway Arch** in St. Louis, Missouri, is shaped in the form of the graph of a combination of exponential functions (*not* a parabola, as it might first appear). Specifically, it is a **catenary**, which is the graph of an equation of the form

$$y = a(e^{bx} + e^{-bx})$$

(see Exercise 43). This shape was chosen because it is optimal for distributing the internal structural forces of the arch. Chains and cables suspended between two points (for example, the stretches of cable between pairs of telephone poles) hang in the shape of a catenary.

From this pattern it appears that the number of bacteria after t hours is

$$n(t) = 1000 \cdot 2^{t/3}$$

In general, suppose that the initial size of a population is n_0 and the doubling period is a. Then the size of the population at time t is

$$n(t) = n_0 2^{ct}$$

where $c = 1/a$. If we knew the tripling time b, then the formula would be $n(t) = n_0 3^{ct}$ where $c = 1/b$. These formulas indicate that the growth of the bacteria is exponential. But what base should we use in our formula? The answer is e, because then it can be shown (using calculus) that the formula is

$$n(t) = n_0 e^{rt}$$

where r is the *relative rate of growth of population, expressed as a proportion of the population at any time*. For instance, if $r = 0.02$, then at any time t the growth rate is 2% of the population at time t.

Notice that the formula for population growth is the same as that for continuously compounded interest. In fact, the same principle is at work in both cases: The growth of a population (or an investment) per time period is proportional to the size of the population (or the amount of the investment). A population of 1,000,000 will increase more in one year than a population of 1000; in exactly the same way, an investment of $1,000,000 will increase more in one year than an investment of $1000.

EXPONENTIAL GROWTH

A population that experiences **exponential growth** increases according to the formula

$$n(t) = n_0 e^{rt}$$

where $n(t)$ = population at time t

   n_0 = initial size of the population

   r = relative rate of growth (expressed as a proportion of the population)

   t = time

EXAMPLE 5 ■ Predicting the Size of a Population

The initial bacterium count in a culture is 500. A biologist later makes a sample count of bacteria in the culture and finds that the relative rate of growth is 40% per hour.

(a) Find a formula for the number of bacteria $n(t)$ after t hours.
(b) What is the estimated count after 10 hours?
(c) Sketch the graph of the function $n(t)$.

SOLUTION

(a) We use the formula for exponential population growth with $n_0 = 500$ and $r = 0.4$ to get

$$n(t) = 500e^{0.4t}$$

where t is measured in hours.

(b) Using the formula in part (a) we find that the bacterium count after 10 hours is

$$n(10) = 500e^{0.4(10)} = 500e^4 \approx 27{,}300$$

(c) We plot several points and sketch the graph in Figure 4.

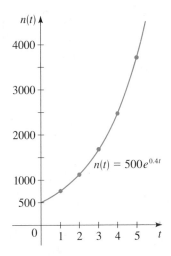

FIGURE 4

 EXAMPLE 6 ■ Comparing Effects of Different Rates of Population Growth

In 1995 the population of the world was 5.7 billion and the relative rate of growth was 2% per year. It is claimed that a rate of 1.6% per year would make a significant difference in the total population in just a few decades. Test this claim by estimating the population of the world in the year 2035 using a relative rate of growth of (a) 2% per year and (b) 1.6% per year.

Graph the population functions for the next 100 years for the two relative growth rates in the same viewing rectangle.

SOLUTION

(a) By the formula for exponential population growth, we have

$$n(t) = 5.7e^{0.02t}$$

where $n(t)$ is measured in billions and t is measured in years since 1995. Because the year 2035 is 40 years after 1995, we find

$$n(40) = 5.7e^{0.02(40)} = 5.7e^{0.8} \approx 12.7$$

Thus, the estimated population in the year 2035 is about 13 billion.

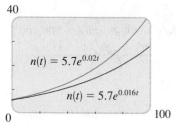

FIGURE 5

(b) We use the formula

$$n(t) = 5.7e^{0.016t}$$

and find $n(40) = 5.7e^{0.016(40)} = 5.7e^{0.64} \approx 10.8$

So, the estimated population in the year 2035 is about 11 billion.

The graphs in Figure 5 show that a small change in the relative rate of growth can, over time, make a large difference in population size. ∎

EXAMPLE 7 ■ Finding the Initial Population

A certain breed of rabbit was introduced onto a small island about 8 years ago. The current rabbit population on the island is estimated to be 4100, with a relative growth rate of 55% per year.

(a) What was the initial size of the rabbit population?

(b) Estimate the population 12 years from now.

SOLUTION

(a) From the formula for exponential population growth, we have

$$n(t) = n_0 e^{0.55t}$$

and we know that the population at time $t = 8$ is $n(8) = 4100$. We substitute what we know into the equation and solve for n_0:

$$4100 = n_0 e^{0.55(8)}$$

$$n_0 = \frac{4100}{e^{0.55(8)}} \approx \frac{4100}{81.45} \approx 50$$

Thus, we estimate that 50 rabbits were introduced onto the island.

Another way to solve part (b) is to let t be the number of years from *now*. In this case, $n_0 = 4100$ (the current population). So, the population 12 years from now will be

$$n(12) = 4100e^{0.55(12)} \approx 3 \text{ million}$$

(b) Now that we know n_0, we can write a formula for population growth:

$$n(t) = 50e^{0.55t}$$

Twelve years from now, $t = 20$ and

$$n(20) = 50e^{0.55(20)} \approx 2{,}993{,}707$$

So, we estimate that the rabbit population on the island 12 years from now will be about 3 million. ∎

Can the rabbit population in Example 7(b) actually reach such a high number? In reality, as the island becomes overpopulated with rabbits, the rabbit population growth will be slowed due to food shortage and other factors. One model that takes into account such factors is the *logistic growth formula* described in Exercises 31 and 32.

| 4.2 | EXERCISES |

1–2 ■ Complete the table and use it to graph the function.

1. $f(x) = 3e^x$

x	$f(x) = 3e^x$
-2	
-1.5	
-1	
-0.5	
0	
0.5	
1	
1.5	
2	

2. $g(x) = 2e^{-0.5x}$

x	$g(x) = 2e^{-0.5x}$
-4	
-3	
-2	
-1	
0	
1	
2	
3	
4	

3–8 ■ Graph the function, not by plotting points, but by starting from the graph of $y = e^x$ in Figure 1. State the domain, range, and asymptote.

3. $y = -e^x$ **4.** $y = 1 - e^x$

5. $y = e^{-x} - 1$ **6.** $y = -e^{-x}$

7. $y = e^{x-2}$ **8.** $y = e^{x-3} + 4$

9. If \$10,000 is invested at an interest rate of 10% per year, compounded semiannually, find the value of the investment after the given number of years.
(a) 5 years   (b) 10 years   (c) 15 years

10. If \$4000 is borrowed at a rate of 16% interest per year, compounded quarterly, find the amount due at the end of the given number of years.
(a) 4 years   (b) 6 years   (c) 8 years

11. If \$3000 is invested at an interest rate of 9% per year, find the amount of the investment at the end of 5 years for each of the following compounding methods.
(a) Annual   (b) Semiannual
(c) Monthly   (d) Weekly
(e) Daily   (f) Hourly
(g) Continuously

12. If \$4000 is invested in an account for which interest is compounded quarterly, find the amount of the investment at the end of 5 years for each of the following interest rates.
(a) 6%   (b) $6\frac{1}{2}$%   (c) 7%   (d) 8%

13. Which of the given interest rates and compounding periods would provide the best investment?
(i) $8\frac{1}{2}$% per year, compounded semiannually
(ii) $8\frac{1}{4}$% per year, compounded quarterly
(iii) 8% per year, compounded continuously

14. Which of the given interest rates and compounding periods would provide the better investment?
(i) $9\frac{1}{4}$% per year, compounded semiannually
(ii) 9% per year, compounded continuously

15–16 ■ The **present value** of a sum of money is the amount that must be invested now, at a given rate of interest, to produce the desired sum at a later date.

15. Find the present value of \$10,000 if interest is paid at a rate of 9% per year, compounded semiannually, for 3 years.

16. Find the present value of \$100,000 if interest is paid at a rate of 8% per year, compounded monthly, for 5 years.

17. The number of bacteria in a culture is given by the formula

$$n(t) = 500e^{0.45t}$$

where t is measured in hours.
(a) What is the relative rate of growth of this bacterium population? Express your answer as a percentage.
(b) What is the initial population of the culture (at $t = 0$)?
(c) How many bacteria will the culture contain at time $t = 5$?

18. The rat population in New York City is given by the formula

$$n(t) = 54e^{0.12t}$$

where t is measured in years since 1990 and $n(t)$ is measured in millions.
(a) What is the relative rate of growth of the rat population? Express your answer as a percentage.
(b) What was the rat population in 1990?
(c) What is the rat population expected to be in the year 2000?
(d) Sketch a graph of the population function.

19. The fox population in a certain region has a relative growth rate of 8% per year. It is estimated that the population in 1992 was 18,000.
(a) Find a formula for the population t years after 1992.
(b) Use the formula from part (a) to estimate the fox population in the year 2000.
(c) Sketch the graph of the fox population function for the years 1992–2000.

20. The population of a certain city has a relative growth rate of 5% per year. The population in 1988 was 421,000. Find the projected population of the city for each of the given years.
(a) 2000   (b) 2030

21. The population of a country has relative growth rate of 3% per year. The government is trying to reduce the growth rate to 2%. The population in 1995 was approximately 110 million. Find the projected population for the year 2020 for each of the following conditions.
(a) The relative growth rate remains at 3% per year.
(b) The relative growth rate is reduced to 2% per year.

22. The population of a certain city was 680,000 in 1992 and is growing at the relative growth rate of 12% per year.
(a) Find a formula for the population of this city t years after 1992.
(b) Estimate the population in the year 2000.

23. The population of the world in 1987 was 5 billion and the relative growth rate was estimated at 2% per year. Assuming that the world population follows an exponential growth model, find the projected world population in 1995. (Compare this with the actual world population of 5.7 billion in 1995.)

24. The relative growth rate for a certain bacteria population is 80% per hour. A small culture is formed, and 3 hours later a count shows approximately 21,500 bacteria in the culture.
(a) Find the initial number of bacteria in the culture.
(b) Estimate the number of bacteria 5 hours from the time the culture was started.

25. Under ideal conditions, a certain type of bacteria has a relative growth rate of 220% per hour. A number of these bacteria are introduced accidentally into a food product. Two hours after contamination, a bacterium count shows that there are about 40,000 bacteria in the food.
(a) Find the initial number of bacteria introduced into the food.
(b) Estimate the number of bacteria in the food 3 hours after contamination.

26. A radioactive substance decays in such a way that the amount of mass remaining after t days is given by the function
$$m(t) = 13e^{-0.015t}$$
where $m(t)$ is measured in kilograms.
(a) Find the mass at time $t = 0$.
(b) How much of the mass remains after 45 days?

27. Radioactive iodine is used by doctors as a tracer in diagnosing certain thyroid gland disorders. This type of iodine decays in such a way that the mass remaining after t days is given by the function
$$m(t) = 6e^{-0.087t}$$
where $m(t)$ is measured in grams.
(a) Find the mass at time $t = 0$.
(b) How much of the mass remains after 20 days?

28. A sky diver jumps from a reasonable height above the ground. The air resistance she experiences is proportional to her velocity, and the constant of proportionality is 0.2. It can be shown that the velocity of the sky diver at time t is given by
$$v(t) = 80(e^{-0.2t} - 1)$$
where t is measured in seconds and $v(t)$ is measured in feet per second (ft/s).

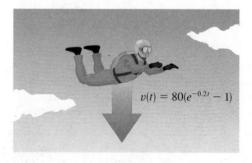

$v(t) = 80(e^{-0.2t} - 1)$

(a) Find the initial velocity of the sky diver.
(b) Find the velocity after 5 s and after 10 s.

(c) Draw a graph of the velocity function $v(t)$.
(d) The maximum velocity of a falling object with wind resistance is called its *terminal velocity*. From the graph in part (c), find the terminal velocity of this sky diver.

29. A 50-gallon barrel is filled completely with pure water. Salt water with a concentration of 0.3 lb/gal is then pumped into the barrel, and the resulting mixture overflows at the same rate. The amount of salt in the barrel at time t is given by

$$Q(t) = 15(1 - e^{-0.04t})$$

where t is measured in minutes and $Q(t)$ is measured in pounds.
(a) How much salt is in the barrel after 5 min?
(b) How much salt is in the barrel after 10 min?
(c) Draw a graph of the function $Q(t)$.
(d) Use the graph in part (c) to determine the value that the amount of salt in the barrel approaches as t becomes large. Is this what you would expect?

$$Q(t) = 15(1 - e^{-0.04t})$$

30. A sum of $5000 is invested at an interest rate of 9% per year, compounded semiannually.
(a) Find the value $A(t)$ of the investment after t years.
(b) Draw the graph of $A(t)$.
(c) Use the graph of $A(t)$ to determine when this investment will amount to $25,000.

31. Assume that the rabbit population in Example 7 behaves according to the *logistic growth formula*

$$n(t) = \frac{300}{0.05 + \left(\dfrac{300}{n_0} - 0.05 \right) e^{-0.55t}}$$

where n_0 is the initial rabbit population.
(a) If the initial population is 50 rabbits, what will the population be after 12 years?

(b) Draw the graphs of the function $n(t)$ for $n_0 = 50$, 500, 2000, 8000, and 12,000 in the viewing rectangle $[0, 15]$ by $[0, 12,000]$.
(c) From the graphs in part (b), observe that, regardless of the initial population, the rabbit population seems to approach a certain number as time goes on. What is that number? (This is the number of rabbits that the island can support.)

32. The population of a certain species of bird is limited by the type of habitat required for nesting. The population is modeled by the *logistic growth formula*

$$n(t) = \frac{5600}{0.5 + 27.5e^{-0.044t}}$$

where t is measured in years.
(a) Find the initial bird population.
(b) Draw a graph of the function $n(t)$.
(c) What size does the population approach as time goes on?

33. Use a calculator to help graph the function $f(x) = e^{-x^2}$ for $x \geq 0$. Then use the fact that f is an even function to draw the rest of the graph.

34. (a) Compare the functions $f(x) = e^x$ and $g(x) = x^3$ by drawing their graphs in each of the following viewing rectangles:
   (i) $[0, 3]$ by $[0, 15]$
   (ii) $[0, 6]$ by $[0, 120]$
   (iii) $[0, 20]$ by $[0, 10,000]$
(b) Find the solutions of the equation $e^x = x^3$, correct to two decimal places.

35. The *hyperbolic cosine function* is defined by

$$\cosh(x) = \frac{e^x + e^{-x}}{2}$$

Sketch the graphs of the functions $y = \frac{1}{2}e^x$ and $y = \frac{1}{2}e^{-x}$ on the same axes and use graphical addition (see Section 2.6) to sketch the graph of $y = \cosh(x)$.

36. The *hyperbolic sine function* is defined by

$$\sinh(x) = \frac{e^x - e^{-x}}{2}$$

Sketch the graph of this function using graphical addition as in Exercise 35.

37–40 ■ Use the definitions in Exercises 35 and 36 to prove the identity.

37. $\cosh(-x) = \cosh(x)$ **38.** $\sinh(-x) = -\sinh(x)$

39. $[\cosh(x)]^2 - [\sinh(x)]^2 = 1$

40. $\sinh(x + y) = \sinh(x)\cosh(y) + \cosh(x)\sinh(y)$

 41. Illustrate the definition of the number e by graphing the curve $y = (1 + 1/x)^x$ and the line $y = e$ on the same screen using the viewing rectangle $[0, 40]$ by $[0, 4]$.

 42. Investigate the behavior of the function

$$f(x) = \left(1 - \frac{1}{x}\right)^x$$

as $x \to \infty$ by graphing f and the line $y = 1/e$ on the same screen using the viewing rectangle $[0, 20]$ by $[0, 1]$.

 43. (a) Draw the graphs of the family of functions

$$f(x) = \frac{a}{2}(e^{x/a} + e^{-x/a})$$

for $a = 0.5, 1, 1.5$, and 2.

(b) How does a larger value of a affect the graph?

 44. Graph the function $y = e^x/x$ and comment on vertical and horizontal asymptotes.

45–46 ■ Find the local maximum and minimum values of the function and the value of x at which each occurs. State each answer correct to two decimal places.

45. $f(x) = xe^{-x}$ **46.** $f(x) = e^x + e^{-3x}$

4.3 LOGARITHMIC FUNCTIONS

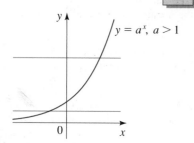

y

$y = a^x, \ a > 1$

0

x

FIGURE 1

$f(x) = a^x$ is one-to-one

Every exponential function $f(x) = a^x$, with $a > 0$ and $a \neq 1$, is a one-to-one function by the Horizontal Line Test (see Figure 1 for the case $a > 1$) and therefore has an inverse function. The inverse function f^{-1} is called the *logarithmic function with base a* and is denoted by $\log_a$. Recall from Section 2.7 that f^{-1} is defined by

$$f^{-1}(x) = y \iff f(y) = x$$

This leads to the following definition of the logarithmic function.

We read $\log_a x = y$ as "log base a of x is y."

> **DEFINITION OF THE LOGARITHMIC FUNCTION**
>
> Let a be a positive number with $a \neq 1$. The **logarithmic function with base a**, denoted by $\log_a$, is defined by
>
> $$\log_a x = y \iff a^y = x$$

In words, this says that

By tradition, the name of the logarithmic function is $\log_a$, not just a single letter. Also, we usually omit the parentheses in the function notation and write

$$\log_a(x) = \log_a x$$

> $\log_a x$ is the exponent to which the base a must be raised to give x.

When we use the definition of logarithms to switch back and forth between the **logarithmic form** $\log_a x = y$ and the **exponential form** $a^y = x$, it's helpful to notice that, in both forms, the base is the same:

Logarithmic form   Exponential form

exponent   exponent

↓   ↓

$\log_a x = y$ $a^y = x$

↑   ↑

base   base

The following examples illustrate how to change an equation from one of these forms to the other.

EXAMPLE 1 ■ Logarithmic and Exponential Forms

The logarithmic and exponential forms are equivalent equations—if one is true, then so is the other. So, we can switch from one form to the other as in the following illustrations.

Logarithmic form	Exponential form
$\log_{10} 100{,}000 = 5$	$10^5 = 100{,}000$
$\log_2 8 = 3$	$2^3 = 8$
$\log_2\left(\frac{1}{8}\right) = -3$	$2^{-3} = \frac{1}{8}$
$\log_5 s = r$	$5^r = s$

■

It is important to understand that $\log_a x$ is an *exponent*. For example, the numbers in the right column of the table in the margin are the logarithms (base 10) of the numbers in the left column. This is the case for all bases, as the following example illustrates.

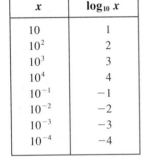

x	$\log_{10} x$
10	1
10^2	2
10^3	3
10^4	4
10^{-1}	-1
10^{-2}	-2
10^{-3}	-3
10^{-4}	-4

EXAMPLE 2 ■ Evaluating Logarithms

(a) $\log_{10} 1000 = 3$ because $10^3 = 1000$

(b) $\log_2 32 = 5$ because $2^5 = 32$

(c) $\log_{10} 0.1 = -1$ because $10^{-1} = 0.1$

(d) $\log_{16} 4 = \frac{1}{2}$ because $16^{1/2} = 4$

■

GRAPHS OF LOGARITHMIC FUNCTIONS

Recall that if a one-to-one function f has domain A and range B, then its inverse function f^{-1} has domain B and range A. Since the exponential function $f(x) = a^x$ with $a \neq 1$ has domain $\mathbb{R}$ and range $(0, \infty)$, we conclude that its inverse function, $f^{-1}(x) = \log_a x$, has domain $(0, \infty)$ and range $\mathbb{R}$.

The graph of $f^{-1}(x) = \log_a x$ is obtained by reflecting the graph of $f(x) = a^x$ in the line $y = x$. Figure 2 shows the case $a > 1$. The fact that $y = a^x$ (for $a > 1$) is a very rapidly increasing function for $x > 0$ implies that $y = \log_a x$ is a very slowly increasing function for $x > 1$ (see Exercise 74.)

Notice that since $a^0 = 1$, we have

$$\log_a 1 = 0$$

and so the x-intercept of the function $y = \log_a x$ is 1. Notice also that because the x-axis is a horizontal asymptote of $y = a^x$, the y-axis is a vertical asymptote of $y = \log_a x$. In fact, using the notation of Section 3.5, we can write

$$\log_a x \to -\infty \quad \text{as} \quad x \to 0^+$$

for the case $a > 1$.

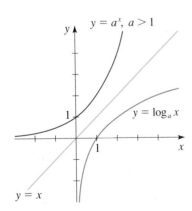

FIGURE 2

Graph of the logarithmic function $f(x) = \log_a x$

EXAMPLE 3 ■ Graphing a Logarithmic Function by Plotting Points

Sketch the graph of $f(x) = \log_2 x$.

SOLUTION   To make a table of values, we choose the x-values to be powers of 2 so that we can easily find their logarithms. We plot these points and connect them with a smooth curve as in Figure 3.

x	$\log_2 x$
1	0
2	1
2^2	2
2^3	3
2^4	4
2^{-1}	-1
2^{-2}	-2
2^{-3}	-3
2^{-4}	-4

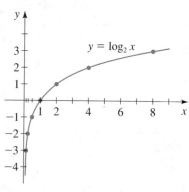

FIGURE 3   ■

Figure 4 shows the graphs of the family of logarithmic functions with bases 2, 3, 5, and 10. These graphs are drawn by reflecting the graphs of $y = 2^x$, $y = 3^x$, $y = 5^x$, and $y = 10^x$ (see Figure 4 in Section 4.1) in the line $y = x$. We can also plot points as an aid to sketching these graphs, as illustrated in Example 3.

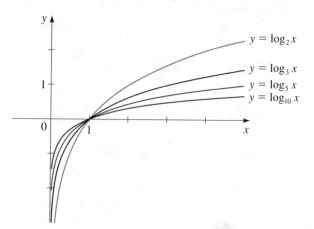

FIGURE 4

A family of logarithmic functions

In the next two examples we graph logarithmic functions by starting with the basic graphs in Figure 4 and using the transformations of Section 2.4.

EXAMPLE 4 ■ Reflecting Graphs of Logarithmic Functions

Sketch the graph of each function.

(a) $f(x) = -\log_2 x$ (b) $g(x) = \log_2(-x)$

SOLUTION

Reflecting graphs is explained in Section 2.4.

(a) We start with the graph of $y = \log_2 x$ in Figure 5(a) and reflect in the x-axis to get the graph of $f(x) = -\log_2 x$ in Figure 5(b).

(b) To obtain the graph of $g(x) = \log_2(-x)$, we reflect the graph of $y = \log_2 x$ in the y-axis. See Figure 5(c).

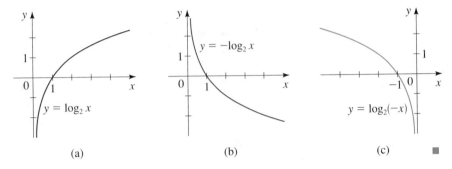

FIGURE 5   (a)   (b)   (c)   ■

EXAMPLE 5 ■ **Shifting Graphs of Logarithmic Functions**

Find the domain of each function, and sketch the graph.

(a) $f(x) = 2 + \log_5 x$ (b) $g(x) = \log_{10}(x - 3)$

SOLUTION

(a) The graph of f is obtained from the graph of $y = \log_5 x$ (Figure 4) by shifting upward 2 units (see Figure 6). The domain of f is $(0, \infty)$.

(b) The graph of g is obtained from the graph of $y = \log_{10} x$ (Figure 4) by shifting to the right 3 units (see Figure 7). The line $x = 3$ is a vertical asymptote. The domain of $y = \log_a x$ is the interval $(0, \infty)$, so $\log_{10} x$ is defined only when $x > 0$. Therefore, the domain of $g(x) = \log_{10}(x - 3)$ is

$$\{x \mid x - 3 > 0\} = \{x \mid x > 3\} = (3, \infty)$$

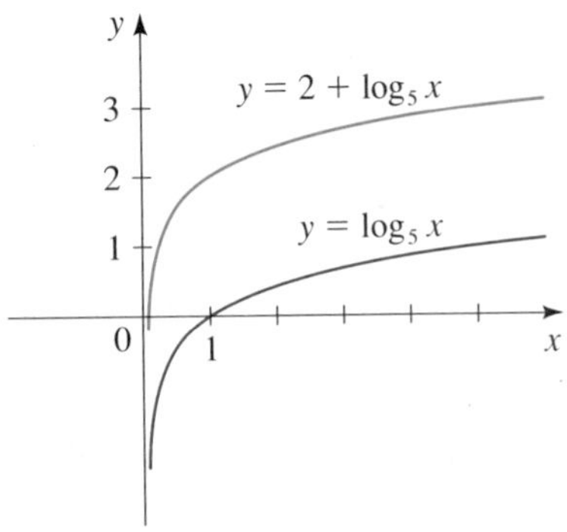

FIGURE 6

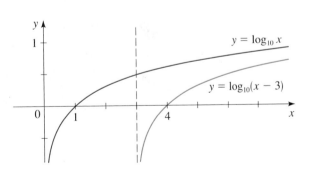

FIGURE 7   ■

In Section 2.7 we saw that a function f and its inverse function f^{-1} satisfy the equations

$$f^{-1}(f(x)) = x \qquad \text{for } x \text{ in the domain of } f$$
$$f(f^{-1}(x)) = x \qquad \text{for } x \text{ in the domain of } f^{-1}$$

When applied to $f(x) = a^x$ and $f^{-1}(x) = \log_a x$, these equations become

$$\log_a(a^x) = x \qquad x \in \mathbb{R}$$

$$a^{\log_a x} = x \qquad x > 0$$

We list these and other properties of logarithms discussed in this section.

PROPERTIES OF LOGARITHMS

Property	Reason
1. $\log_a 1 = 0$	We must raise a to the power 0 to get 1.
2. $\log_a a = 1$	We must raise a to the power 1 to get a.
3. $\log_a a^x = x$	We must raise a to the power x to get a^x.
4. $a^{\log_a x} = x$	$\log_a x$ is the power to which a must be raised to get x.

EXAMPLE 6 ■ Applying Properties of Logarithms

We illustrate the properties of logarithms when the base is 5.

$\log_5 1 = 0$ *Property 1* $\log_5 5 = 1$ *Property 2*

$\log_5 5^8 = 8$ *Property 3* $5^{\log_5 12} = 12$ *Property 4*   ■

COMMON AND NATURAL LOGARITHMS

We now use a calculator to evaluate logarithms with base 10 (common logarithms) and base e (natural logarithms). In the next section we will see that we can use a calculator to find logarithms with *any* base.

COMMON LOGARITHM

The logarithm with base 10 is called the **common logarithm** and is denoted by omitting the base:

$$\log x = \log_{10} x$$

From the definition of logarithms we can easily find that

$$\log 10 = 1 \qquad \text{and} \qquad \log 100 = 2$$

But how can we find $\log 50$? We need to find the exponent y such that $10^y = 50$. Clearly, 1 is too small and 2 is too large. So

$$1 < \log 50 < 2$$

To find a better approximation we can experiment to find a power of 10 closer to 50. Fortunately, scientific calculators are equipped with a $\boxed{\log}$ key that directly gives the values of common logarithms.

EXAMPLE 7 ■ Evaluating Common Logarithms

Use a calculator to find appropriate values of $f(x) = \log x$ and use the values to sketch the graph.

SOLUTION   We make a table of values, using a calculator to evaluate the function at those values of x that are not powers of 10. We plot those points and connect them by a smooth curve as in Figure 8.

x	$\log x$
0.01	-2
0.1	-1
0.5	-0.301
1	0
4	0.602
5	0.699
10	1
15	1.176

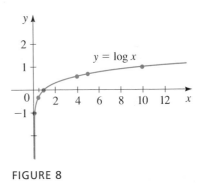

FIGURE 8   ■

Of all possible bases a for logarithms, it turns out that the most convenient choice for the purposes of calculus is the number e, which we defined in Section 4.2.

The notation ln is an abbreviation for *logarithmus naturalis*.

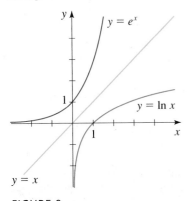

FIGURE 9

Graph of the natural logarithmic function

NATURAL LOGARITHM

The logarithm with base e is called the **natural logarithm** and is denoted by **ln**:

$$\ln x = \log_e x$$

The natural logarithmic function $y = \ln x$ is the inverse function of the exponential function $y = e^x$. Both functions are graphed in Figure 9. By the definition of inverse functions we have

$$\ln x = y \iff e^y = x$$

If we substitute $a = e$ and write "ln" for "$\log_e$" in the properties of logarithms mentioned earlier, we obtain the following properties of natural logarithms.

PROPERTIES OF NATURAL LOGARITHMS	
Property	Reason
1. $\ln 1 = 0$	We must raise e to the power 0 to get 1.
2. $\ln e = 1$	We must raise e to the power 1 to get e.
3. $\ln e^x = x$	We must raise e to the power x to get e^x.
4. $e^{\ln x} = x$	$\ln x$ is the power to which e must be raised to get x.

Calculators are equipped with an $\boxed{\ln}$ key that directly gives the values of natural logarithms.

EXAMPLE 8 ■ Evaluating the Natural Logarithm Function

(a) $\ln e^8 = 8$ Definition of natural logarithm

(b) $\ln\left(\dfrac{1}{e^2}\right) = \ln e^{-2} = -2$ Definition of natural logarithm

(c) $\ln 5 \approx 1.609$ Use the $\boxed{\ln}$ key on a calculator   ■

EXAMPLE 9 ■ Finding the Domain of a Logarithmic Function

Find the domain of the function $f(x) = \ln(4 - x^2)$.

SOLUTION   As with any logarithmic function, $\ln x$ is defined when $x > 0$. Thus, the domain of f is

$$\{x \mid 4 - x^2 > 0\} = \{x \mid x^2 < 4\} = \{x \mid |x| < 2\}$$

$$= \{x \mid -2 < x < 2\} = (-2, 2)$$ ■

EXAMPLE 10 ■ Drawing the Graph of a Logarithmic Function

Draw the graph of the function $y = x\ln(4 - x^2)$ and use it to find the asymptotes and local maximum and minimum values.

SOLUTION   As in Example 9 the domain of this function is the interval $(-2, 2)$, so we choose the viewing rectangle $[-3, 3]$ by $[-3, 3]$. The graph is shown in Figure 10, and from it we see that the lines $x = -2$ and $x = 2$ are vertical asymptotes.

The function has a local maximum point to the right of $x = 1$ and a local minimum point to the left of $x = -1$. By zooming in and tracing along the graph with the cursor, we find that the local maximum value is approximately 1.13 and this occurs when $x \approx 1.15$. Similarly (or by noticing that the function is odd), we find that the local minimum value is about -1.13, and it occurs when $x \approx -1.15$.   ■

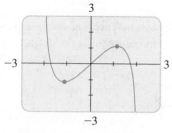

FIGURE 10

$y = x\ln(4 - x^2)$

4.3 EXERCISES

1–6 ■ Express the equation in exponential form.

1. (a) $\log_2 32 = 5$ (b) $\log_5 1 = 0$

2. (a) $\log_{10} 0.1 = -1$ (b) $\log_8 512 = 3$

3. (a) $\log_4 2 = \frac{1}{2}$ (b) $\log_2\left(\frac{1}{16}\right) = -4$

4. (a) $\log_3 81 = 4$ (b) $\log_8 4 = \frac{2}{3}$

5. (a) $\ln 5 = x$ (b) $\ln y = 5$

6. (a) $\ln(x + 1) = 2$ (b) $\ln(x - 1) = 4$

7–12 ■ Express the equation in logarithmic form.

7. (a) $2^3 = 8$ (b) $10^{-3} = 0.001$

8. (a) $10^4 = 10,000$ (b) $81^{1/2} = 9$

9. (a) $4^{-3/2} = 0.125$ (b) $7^3 = 343$

10. (a) $8^{-1} = \frac{1}{8}$ (b) $10^m = n$

11. (a) $e^x = 2$ (b) $e^3 = y$

12. (a) $e^{x+1} = 0.5$ (b) $e^{0.5x} = t$

13–22 ■ Evaluate the expression.

13. (a) $\log_5 5^4$ (b) $\log_4 64$ (c) $\log_9 9$

14. (a) $\log_3 3$ (b) $\log_3 1$ (c) $\log_3 3^2$

15. (a) $\log_8 64$ (b) $\log_7 49$ (c) $\log_7 7^{10}$

16. (a) $\log_2 32$ (b) $\log_8 8^{17}$ (c) $\log_6 1$

17. (a) $\log_3\left(\frac{1}{27}\right)$ (b) $\log_{10} \sqrt{10}$ (c) $\log_5 0.2$

18. (a) $\log_5 125$ (b) $\log_{49} 7$ (c) $\log_9 \sqrt{3}$

19. (a) $2^{\log_2 37}$ (b) $3^{\log_3 8}$ (c) $e^{\ln \sqrt{5}}$

20. (a) $e^{\ln \pi}$ (b) $10^{\log 5}$ (c) $10^{\log 87}$

21. (a) $\log_8 0.25$ (b) $\ln e^4$ (c) $\ln(1/e)$

22. (a) $\log_4 \sqrt{2}$ (b) $\log_4\left(\frac{1}{2}\right)$ (c) $\log_4 8$

23–28 ■ Use the definition of the logarithmic function to find x.

23. (a) $\log_2 x = 5$ (b) $\log_2 16 = x$

24. (a) $\log_5 x = 4$ (b) $\log_{10} 0.1 = x$

25. (a) $\log_{10} x = 2$ (b) $\log_5 x = 2$

26. (a) $\log_x 1000 = 3$ (b) $\log_x 25 = 2$

27. (a) $\log_x 16 = 4$ (b) $\log_x 8 = \frac{3}{2}$

28. (a) $\log_x 6 = \frac{1}{2}$ (b) $\log_x 3 = \frac{1}{3}$

29–32 ■ Use a calculator to evaluate the expression, correct to four decimal places.

29. (a) $\log 2$ (b) $\log 35.2$ (c) $\log\left(\frac{2}{3}\right)$

30. (a) $\log 50$ (b) $\log \sqrt{2}$ (c) $\log(3\sqrt{2})$

31. (a) $\ln 5$ (b) $\ln 25.3$ (c) $\ln(1 + \sqrt{3})$

32. (a) $\ln 27$ (b) $\ln 7.39$ (c) $\ln 54.6$

33–36 ■ Find the function of the form $y = \log_a x$ whose graph is given.

33.   **34.**

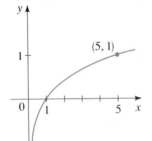

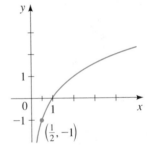

35.   **36.**

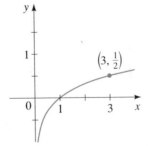

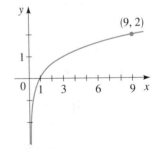

37–42 ■ Match the logarithmic function with one of the graphs labeled I–VI on page 312.

37. $f(x) = -\ln x$ **38.** $f(x) = \ln(x - 2)$

39. $f(x) = 2 + \ln x$ **40.** $f(x) = \ln(-x)$

41. $f(x) = \ln(2 - x)$ **42.** $f(x) = -\ln(-x)$

I
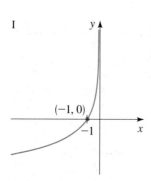

II
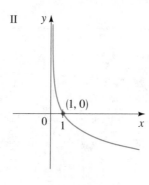

III
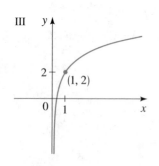

IV
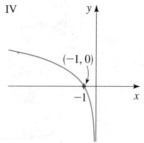

V
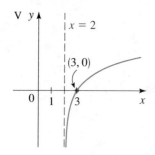

VI
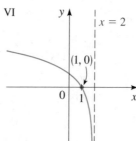

43. Draw the graph of $y = 4^x$, then use it to draw the graph of $y = \log_4 x$.

44. Draw the graph of $y = 3^x$, then use it to draw the graph of $y = \log_3 x$.

45–54 ■ Graph the function, not by plotting points, but by starting from the graphs in Figures 4 and 9. State the domain, range, and asymptote.

45. $f(x) = \log_2(x - 4)$ **46.** $f(x) = -\log_{10} x$

47. $g(x) = \log_5(-x)$ **48.** $g(x) = \ln(x + 2)$

49. $y = 2 + \log_3 x$ **50.** $y = \log_3(x - 1) - 2$

51. $y = 1 - \log_{10} x$ **52.** $y = 1 + \ln(-x)$

53. $y = |\ln x|$ **54.** $y = \ln|x|$

55–60 ■ Find the domain of the function.

55. $f(x) = \log_{10}(2 + 5x)$ **56.** $f(x) = \log_2(10 - 3x)$

57. $g(x) = \log_3(x^2 - 1)$ **58.** $g(x) = \ln(x - x^2)$

59. $h(x) = \ln x + \ln(2 - x)$

60. $h(x) = \sqrt{x - 2} - \log_5(10 - x)$

61–66 ■ Draw the graph of the function in a suitable viewing rectangle and use it to find the domain, the asymptotes, and the local maximum and minimum values.

61. $y = \log_{10}(1 - x^2)$ **62.** $y = \ln(x^2 - x)$

63. $y = x + \ln x$ **64.** $y = x(\ln x)^2$

65. $y = \dfrac{\ln x}{x}$ **66.** $y = x \log_{10}(x + 10)$

67. Compare the rates of growth of the functions $f(x) = \ln x$ and $g(x) = \sqrt{x}$ by drawing their graphs on a common screen using the viewing rectangle $[-1, 30]$ by $[-1, 6]$.

68. (a) By drawing the graphs of the functions
$$f(x) = 1 + \ln(1 + x) \quad \text{and} \quad g(x) = \sqrt{x}$$
in a suitable viewing rectangle, show that even when a logarithmic function starts out higher than a root function, it is ultimately overtaken by the root function.

    (b) Find, correct to two decimal places, the solutions of the equation $\sqrt{x} = 1 + \ln(1 + x)$.

69–70 ■ A family of functions is given.
(a) Draw graphs of the family for $c = 1, 2, 3,$ and 4.
(b) How are the graphs in part (a) related?

69. $f(x) = \log(cx)$ **70.** $f(x) = c \log x$

71–72 ■ A function $f(x)$ is given.
(a) Find the domain of the function f.
(b) Find the inverse function of f.

71. $f(x) = \log_2(\log_{10} x)$ **72.** $f(x) = \ln(\ln(\ln x))$

73. (a) Find the inverse of the function $f(x) = \dfrac{2^x}{1 + 2^x}$.

    (b) What is the domain of the inverse function?

DISCOVERY · DISCUSSION

74. The Height of the Graph of a Logarithmic Function Suppose that the graph of $y = 2^x$ is drawn on a coordinate plane where the unit of measurement is an inch.

(a) Show that at a distance 2 ft to the right of the origin the height of the graph is about 265 mi.

(b) If the graph of $y = \log_2 x$ is drawn on the same set of axes, how far to the right of the origin do we have to go before the height of the curve reaches 2 ft?

75. The Googolplex A **googol** is 10^{100}, and a **googolplex** is 10^{googol}. Find $\log(\log(\text{googol}))$ and $\log(\log(\log(\text{googolplex})))$.

76. Comparing Logarithms Which is larger, $\log_4 17$ or $\log_5 24$? Explain your reasoning.

77. The Number of Digits in an Integer Compare $\log 1000$ to the number of digits in 1000. Do the same for 10,000. How many digits does any number between 1000 and 10,000 have? Between what two values must the common logarithm of such a number lie? Use your observations to explain why the number of digits in any positive integer x is $[\![\log x]\!] + 1$. (The symbol $[\![n]\!]$ is the greatest integer function defined in Section 2.2.)

4.4 LAWS OF LOGARITHMS

Since logarithms are exponents, the Laws of Exponents give rise to the Laws of Logarithms. These properties give logarithmic functions a wide range of applications, as we will see in Section 4.6.

LAWS OF LOGARITHMS

Let a be a positive number, with $a \neq 1$. Let $A > 0$, $B > 0$, and C be any real numbers.

Law	Description
1. $\log_a(AB) = \log_a A + \log_a B$	The logarithm of a product of numbers is the sum of the logarithms of the numbers.
2. $\log_a\left(\dfrac{A}{B}\right) = \log_a A - \log_a B$	The logarithm of a quotient of numbers is the difference of the logarithms of the numbers.
3. $\log_a(A^C) = C \log_a A$	The logarithm of a power of a number is the exponent times the logarithm of the number.

■ **Proof** We make use of the property $\log_a a^x = x$ from Section 4.3.

1. Let
$$\log_a A = u \quad \text{and} \quad \log_a B = v$$

When written in exponential form, these equations become
$$a^u = A \quad \text{and} \quad a^v = B$$

Thus
$$\log_a(AB) = \log_a(a^u a^v) = \log_a(a^{u+v})$$
$$= u + v = \log_a A + \log_a B$$

John Napier (1550–1617) was a Scottish landowner for whom mathematics was a hobby. Today he is best known as the inventor of logarithms. He published his invention in 1614 under the title *A Description of the Marvelous Rule of Logarithms*. In Napier's time logarithms were used exclusively for simplifying complicated calculations. For example, to multiply two large numbers we write them as powers of 10. The exponents are simply the logarithms of the numbers. For instance,

$$4532 \times 57783 \approx 10^{3.65629} \times 10^{4.76180}$$
$$= 10^{8.41809}$$
$$\approx 261{,}872{,}564$$

The idea is that multiplying powers of 10 is easy (we simply add their exponents). Napier produced extensive tables giving the logarithms (or exponents) of numbers. Since the advent of calculators and computers, logarithms are no longer used for this purpose. The logarithmic functions, however, have found many applications, some of which are described in this chapter.

Napier wrote on many other topics. One of his most colorful works is a book entitled *A Plaine Discovery of the Whole Revelation of Saint John*, in which he predicted that the world would end in the year 1700.

2. Using Law 1, we have

$$\log_a A = \log_a\left[\left(\frac{A}{B}\right)B\right] = \log_a\left(\frac{A}{B}\right) + \log_a B$$

so

$$\log_a\left(\frac{A}{B}\right) = \log_a A - \log_a B$$

3. Let $\log_a A = u$. Then $a^u = A$, so

$$\log_a(A^C) = \log_a(a^u)^C = \log_a(a^{uC}) = uC = C\log_a A \qquad \square$$

As the following examples illustrate, these laws are used in both directions. Since the domain of any logarithmic function is the interval $(0, \infty)$, we assume that all quantities whose logarithms occur are positive.

EXAMPLE 1 ■ Using the Laws of Logarithms to Expand Expressions

Use the Laws of Logarithms to rewrite each expression.

(a) $\log_2(6x)$

(b) $\log\sqrt{5}$

(c) $\log_5(x^3 y^6)$

(d) $\ln\left(\dfrac{ab}{\sqrt[3]{c}}\right)$

SOLUTION

(a) $\log_2(6x) = \log_2 6 + \log_2 x$ Law 1

(b) $\log\sqrt{5} = \log 5^{1/2} = \frac{1}{2}\log 5$ Law 3

(c) $\log_5(x^3 y^6) = \log_5 x^3 + \log_5 y^6$ Law 1

$\qquad\qquad\quad = 3\log_5 x + 6\log_5 y$ Law 3

(d) $\ln\left(\dfrac{ab}{\sqrt[3]{c}}\right) = \ln(ab) - \ln\sqrt[3]{c}$ Law 2

$\qquad\qquad\quad = \ln a + \ln b - \ln c^{1/3}$ Law 1

$\qquad\qquad\quad = \ln a + \ln b - \frac{1}{3}\ln c$ Law 3   ■

EXAMPLE 2 ■ Using the Laws of Logarithms to Evaluate Expressions

Evaluate each expression.

(a) $\log_4 2 + \log_4 32$ (b) $\log_2 80 - \log_2 5$ (c) $-\frac{1}{3}\log 8$

SOLUTION

(a) $\log_4 2 + \log_4 32 = \log_4(2 \cdot 32)$ Law 1

$\qquad\qquad\qquad = \log_4 64 = 3$ Because $4^3 = 64$

(b) $\log_2 80 - \log_2 5 = \log_2\left(\frac{80}{5}\right)$ Law 2

$\qquad\qquad\qquad = \log_2 16 = 4$ Because $2^4 = 16$

(c) $-\frac{1}{3} \log 8 = \log 8^{-1/3}$ Law 3

$\qquad\qquad = \log\left(\frac{1}{2}\right)$ Property of negative exponents

$\qquad\qquad \approx -0.301$ Use a calculator   ■

EXAMPLE 3 ■ Writing an Expression as a Single Logarithm

Express $3 \log x + \frac{1}{2} \log(x + 1)$ as a single logarithm.

SOLUTION

$$3 \log x + \tfrac{1}{2} \log(x + 1) = \log x^3 + \log(x + 1)^{1/2} \qquad \text{Law 3}$$

$$= \log(x^3(x + 1)^{1/2}) \qquad \text{Law 1} \qquad ■$$

EXAMPLE 4 ■ Writing an Expression as a Single Logarithm

Express $3 \ln s + \frac{1}{2} \ln t - 4 \ln(t^2 + 1)$ as a single logarithm.

SOLUTION

$$3 \ln s + \tfrac{1}{2} \ln t - 4 \ln(t^2 + 1) = \ln s^3 + \ln t^{1/2} - \ln(t^2 + 1)^4 \qquad \text{Law 3}$$

$$= \ln(s^3 t^{1/2}) - \ln(t^2 + 1)^4 \qquad \text{Law 1}$$

$$= \ln\left(\frac{s^3 \sqrt{t}}{(t^2 + 1)^4}\right) \qquad \text{Law 2} \qquad ■$$

WARNING Although the Laws of Logarithms tell us how to compute the logarithm of a product or a quotient, *there is no corresponding rule for the logarithm of a sum or a difference*. For instance,

$$\log_a(x + y) \neq \log_a x + \log_a y$$

In fact, we know that the right side is equal to $\log_a(xy)$.

Also, don't improperly simplify quotients or powers of logarithms. For instance,

$$\frac{\log 6}{\log 2} \neq \log\left(\frac{6}{2}\right)$$

$$(\log_2 x)^3 \neq 3 \log_2 x$$

CHANGE OF BASE

For some purposes, it's useful to be able to change from logarithms in one base to logarithms in another base. Suppose we are given $\log_a x$ and want to find $\log_b x$. Let

$$y = \log_b x$$

We write this in exponential form and take the logarithm, with base a, of each side.

$$b^y = x \qquad \text{Exponential form}$$

$$\log_a(b^y) = \log_a x \qquad \text{Take } \log_a \text{ of each side}$$

$$y \log_a b = \log_a x \qquad \text{Law 3}$$

$$y = \frac{\log_a x}{\log_a b} \qquad \text{Divide by } \log_a b$$

This proves the following formula.

CHANGE OF BASE FORMULA

$$\log_b x = \frac{\log_a x}{\log_a b}$$

We may write the change of base formula as

$$\log_b x = \left(\frac{1}{\log_a b} \right) \log_a x$$

So, $\log_b x$ is just a constant multiple of $\log_a x$; the constant is $\dfrac{1}{\log_a b}$.

In particular if we put $x = a$, then $\log_a a = 1$ and this formula becomes

$$\log_b a = \frac{1}{\log_a b}$$

We can now evaluate a logarithm to *any* base by using the Change of Base Formula to express the logarithm in terms of common logarithms or natural logarithms and then using a calculator.

EXAMPLE 5 ■ Using the Change of Base Formula to Evaluate Logarithms

Use the Change of Base Formula and common or natural logarithms to evaluate each logarithm, correct to five decimal places.

(a) $\log_8 5$ (b) $\log_9 20$

SOLUTION

(a) We use the Change of Base Formula with $b = 8$ and $a = 10$ to convert to common logarithms:

$$\log_8 5 = \frac{\log_{10} 5}{\log_{10} 8} \approx 0.77398$$

(b) We use the Change of Base Formula with $b = 9$ and $a = e$ to convert to natural logarithms:

$$\log_9 20 = \frac{\ln 20}{\ln 9} \approx 1.36342$$

■

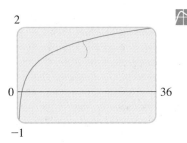

FIGURE 1

$$f(x) = \log_6 x = \frac{\ln x}{\ln 6}$$

EXAMPLE 6 ■ Using the Change of Base Formula To Graph a Logarithmic Function

Use a graphing calculator to graph $f(x) = \log_6 x$.

SOLUTION Calculators do not have a key for $\log_6$, so we use the Change of Base Formula to write

$$f(x) = \log_6 x = \frac{\ln x}{\ln 6}$$

Since calculators do have an $\boxed{\text{ln}}$ key, we can enter this new form of the function and graph it. The graph is shown in Figure 1. ■

4.4 EXERCISES

1–24 ■ Use the Laws of Logarithms to rewrite the expression in a form with no logarithm of a product, quotient, or power.

1. $\log_2(x(x-1))$

2. $\log_5\left(\dfrac{x}{2}\right)$

3. $\log 7^{23}$

4. $\ln(\pi x)$

5. $\log_2(AB^2)$

6. $\log_6 \sqrt[4]{17}$

7. $\log_3(x\sqrt{y})$

8. $\log_2(xy)^{10}$

9. $\log_5 \sqrt[3]{x^2+1}$

10. $\log_a\left(\dfrac{x^2}{yz^3}\right)$

11. $\ln\sqrt{ab}$

12. $\ln\sqrt[3]{3r^2s}$

13. $\log\left(\dfrac{x^3y^4}{z^6}\right)$

14. $\log\left(\dfrac{a^2}{b^4\sqrt{c}}\right)$

15. $\log_2\left(\dfrac{x(x^2+1)}{\sqrt{x^2-1}}\right)$

16. $\log_5\sqrt{\dfrac{x-1}{x+1}}$

17. $\ln\left(x\sqrt{\dfrac{y}{z}}\right)$

18. $\ln\dfrac{3x^2}{(x+1)^{10}}$

19. $\log\sqrt[4]{x^2+y^2}$

20. $\log\left(\dfrac{x}{\sqrt[3]{1-x}}\right)$

21. $\log\sqrt{\dfrac{x^2+4}{(x^2+1)(x^3-7)^2}}$

22. $\log\sqrt{x\sqrt{y\sqrt{z}}}$

23. $\ln\left(\dfrac{z^4\sqrt{x}}{\sqrt[3]{y^2+6y+17}}\right)$

24. $\log\left(\dfrac{10^x}{x(x^2+1)(x^4+2)}\right)$

25–34 ■ Evaluate the expression.

25. $\log_5\sqrt{125}$

26. $\log_2 112 - \log_2 7$

27. $\log 2 + \log 5$

28. $\log\sqrt{0.1}$

29. $\log_4 192 - \log_4 3$

30. $\log_{12} 9 + \log_{12} 16$

31. $\ln 6 - \ln 15 + \ln 20$

32. $e^{3\ln 5}$

33. $10^{2\log 4}$

34. $\log_2 8^{33}$

35–44 ■ Rewrite the expression as a single logarithm.

35. $\log_3 5 + 5\log_3 2$

36. $\log 12 + \frac{1}{2}\log 7 - \log 2$

37. $\log_2 A + \log_2 B - 2\log_2 C$

38. $\log_5(x^2-1) - \log_5(x-1)$

39. $4\log x - \frac{1}{3}\log(x^2+1) + 2\log(x-1)$

40. $\ln(a+b) + \ln(a-b) - 2\ln c$

41. $\ln 5 + 2\ln x + 3\ln(x^2+5)$

42. $2(\log_5 x + 2\log_5 y - 3\log_5 z)$

43. $\frac{1}{3}\log(2x+1) + \frac{1}{2}[\log(x-4) - \log(x^4-x^2-1)]$

44. $\log_a b + c\log_a d - r\log_a s$

45–52 ■ Use the Change of Base Formula and a calculator to evaluate the logarithm, correct to six decimal places. Use either natural or common logarithms.

45. $\log_2 7$

46. $\log_5 2$

47. $\log_3 11$

48. $\log_6 92$

49. $\log_7 3.58$

50. $\log_6 532$

51. $\log_4 322$

52. $\log_{12} 2.5$

 53. Use the Change of Base Formula to show that

$$\log_3 x = \frac{\ln x}{\ln 3}$$

Then use this fact to draw the graph of the function $f(x) = \log_3 x$.

 54. Draw the graph of the family of functions $y = \log_a x$ for $a = 2$, e, 5, and 10 on the same screen, using the viewing rectangle $[0, 5]$ by $[-3, 3]$. How are these graphs related?

55. Use the Change of Base Formula to show that

$$\log e = \frac{1}{\ln 10}$$

56. Simplify: $(\log_2 5)(\log_5 7)$

57. Show that $-\ln(x - \sqrt{x^2 - 1}) = \ln(x + \sqrt{x^2 - 1})$.

 DISCOVERY · DISCUSSION

58. Is the Equation an Identity? Discuss each equation and determine whether it is true for all possible values of the variables. (Ignore values of the variables for which any term is undefined.)

(a) $\log\left(\dfrac{x}{y}\right) = \dfrac{\log x}{\log y}$

(b) $\log_2(x - y) = \log_2 x - \log_2 y$

(c) $\log_5\left(\dfrac{a}{b^2}\right) = \log_5 a - 2 \log_5 b$

(d) $\log 2^z = z \log 2$

(e) $(\log P)(\log Q) = \log P + \log Q$

(f) $\dfrac{\log a}{\log b} = \log a - \log b$

(g) $(\log_2 7)^x = x \log_2 7$

(h) $\log_a a^a = a$

(i) $\log(x - y) = \dfrac{\log x}{\log y}$

(j) $-\ln\left(\dfrac{1}{A}\right) = \ln A$

59. Find the Error What is wrong with the following argument?

$$\log 0.1 < 2 \log 0.1$$
$$= \log(0.1)^2$$
$$= \log 0.01$$
$$\log 0.1 < \log 0.01$$
$$0.1 < 0.01$$

60. Shifting, Shrinking, and Stretching Graphs of Functions Let $f(x) = x^2$. Show that $f(2x) = 4f(x)$, and explain how this shows that shrinking the graph of f horizontally has the same effect as stretching it vertically. Then use the identities $e^{2+x} = e^2 e^x$ and $\ln(2x) = \ln 2 + \ln x$ to show that for $g(x) = e^x$, a horizontal shift is the same as a vertical stretch and for $h(x) = \ln x$, a horizontal shrinking is the same as a vertical shift.

4.5 EXPONENTIAL AND LOGARITHMIC EQUATIONS

In this section we solve equations that involve exponential or logarithmic functions. The techniques we develop here will be used in the next section for solving applied problems.

EXPONENTIAL EQUATIONS

An exponential equation is one in which the variable occurs in the exponent. For example,

$$2^x = 7$$

The variable x presents a difficulty because it is in the exponent. To deal with this difficulty we take the logarithm of each side and then use the Laws of Logarithms to "bring down x" from the exponent.

$$2^x = 7$$

$$\ln 2^x = \ln 7 \qquad \text{Take ln of each side}$$

$$x \ln 2 = \ln 7 \qquad \text{Law 3 (bring down the exponent)}$$

$$x = \frac{\ln 7}{\ln 2} \qquad \text{Solve for } x$$

$$\approx 2.807 \qquad \text{Use a calculator}$$

Recall that Law 3 of the Laws of Logarithms says that $\log_a A^C = C \log_a A$.

The method we used to solve $2^x = 7$ is typical of the methods we use to solve all exponential equations, and it can be summarized in the following three steps.

GUIDELINES FOR SOLVING EXPONENTIAL EQUATIONS

1. Isolate the exponential expression on one side of the equation.

2. Take the logarithm of each side, then use the Laws of Logarithms to "bring down the exponent."

3. Solve for the variable.

EXAMPLE 1 ■ Solving an Exponential Equation

Find the solution of the equation $3^{x+2} = 7$, correct to six decimal places.

SOLUTION   We take the common logarithm of each side and use Law 3.

$$3^{x+2} = 7$$

$$\log(3^{x+2}) = \log 7 \qquad \text{Take log of each side}$$

$$(x + 2)\log 3 = \log 7 \qquad \text{Law 3 (bring down the exponent)}$$

$$x + 2 = \frac{\log 7}{\log 3} \qquad \text{Divide by } \log 3$$

$$x = \frac{\log 7}{\log 3} - 2 \qquad \text{Subtract 2}$$

$$\approx -0.228756 \qquad \text{Use a calculator}$$

We could have used natural logarithms instead of common logarithms. In fact, using the same steps we get

$$x = \frac{\ln 7}{\ln 3} - 2 \approx -0.228756$$

CHECK YOUR ANSWER ■ Substituting $x = -0.228756$ into the original equation and using a calculator, we get

$$3^{(-0.228756)+2} \approx 7 \quad \checkmark$$

EXAMPLE 2 ■ Solving an Exponential Equation

Solve the equation $8e^{2x} = 20$.

SOLUTION We first divide by 8 in order to isolate the exponential term on one side of the equation.

$$8e^{2x} = 20$$

$$e^{2x} = \frac{20}{8} \qquad \text{Divide by 8}$$

$$\ln e^{2x} = \ln 2.5 \qquad \text{Take ln of each side}$$

$$2x = \ln 2.5 \qquad \text{Property of ln}$$

$$x = \frac{\ln 2.5}{2} \qquad \text{Divide by 2}$$

$$\approx 0.458 \qquad \text{Use a calculator} \qquad ■$$

CHECK YOUR ANSWER

Substituting $x = 0.458$ into the original equation and using a calculator, we get $8e^{2(0.458)} \approx 20$. ✓

An alternative method is to solve graphically (see Section 1.9). For the equation in Example 3 we draw the graphs of $y = e^{3-2x}$ and $y = 4$ in the same viewing rectangle.

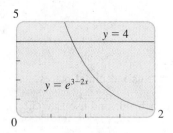

Zooming in on the point of intersection of the two graphs, we see that the solution is $x \approx 0.807$.

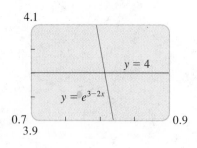

EXAMPLE 3 ■ Solving an Exponential Equation

Solve the equation $e^{3-2x} = 4$.

SOLUTION Since the base of the exponential term is e, we use natural logarithms to solve this equation.

$$\ln(e^{3-2x}) = \ln 4 \qquad \text{Take ln of each side}$$

$$3 - 2x = \ln 4 \qquad \text{Property of ln}$$

$$2x = 3 - \ln 4$$

$$x = \tfrac{1}{2}(3 - \ln 4) \approx 0.807$$

Check that this answer satisfies the original equation. ■

EXAMPLE 4 ■ Solving an Exponential Equation

Solve the equation $e^{2x} - e^x - 6 = 0$.

SOLUTION To isolate the exponential term, we factor.

$$e^{2x} - e^x - 6 = 0$$

$$(e^x)^2 - e^x - 6 = 0 \qquad \text{Law of Exponents}$$

$$(e^x - 3)(e^x + 2) = 0 \qquad \text{Factor (a quadratic in } e^x\text{)}$$

$$e^x - 3 = 0 \quad \text{or} \quad e^x + 2 = 0 \qquad \text{Zero-Product Property}$$

$$e^x = 3 \qquad\qquad e^x = -2$$

The equation $e^x = 3$ leads to $x = \ln 3$. But the equation $e^x = -2$ has no solution because $e^x > 0$ for all x. Thus, $x = \ln 3 \approx 1.0986$ is the only solution. Check that this answer satisfies the original equation. ■

EXAMPLE 5 ■ Solving an Exponential Equation

Solve the equation $3x^2e^x + x^3e^x = 0$.

SOLUTION   First we factor the left side of the equation.

$$3x^2e^x + x^3e^x = 0$$

$$(3x^2 + x^3)e^x = 0 \qquad \text{Factor out } e^x$$

$$x^2(3 + x)e^x = 0 \qquad \text{Factor out } x^2$$

$$x^2 = 0 \quad \text{or} \quad 3 + x = 0 \quad \text{or} \quad e^x = 0 \qquad \text{Zero-Product Property}$$

$$x = 0 \qquad\qquad x = -3$$

The equation $e^x = 0$ has no solution because $e^x > 0$ for all x. Thus, $x = 0$ and $x = -3$ are the only solutions.   ■

CHECK YOUR ANSWERS

$x = 0$:

$$3(0)^2e^0 + 0^3e^0 = 0 \quad \checkmark$$

$x = -3$:

$$3(-3)^2e^{-3} + (-3)^3e^{-3}$$

$$= 27e^{-3} - 27e^{-3} = 0 \quad \checkmark$$

LOGARITHMIC EQUATIONS

A *logarithmic equation* is one in which a logarithm of the variable occurs. For example,

$$\log_2(x + 2) = 5$$

To solve for x we write the equation in exponential form.

$$x + 2 = 2^5 \qquad \text{Exponential form}$$

$$x = 32 - 2 = 30 \qquad \text{Solve for } x$$

Another way of looking at the first step is to raise the base, 2, to each side of the equation.

$$2^{\log_2(x+2)} = 2^5 \qquad \text{Raise 2 to each side}$$

$$x + 2 = 2^5 \qquad \text{Property of logarithms}$$

$$x = 32 - 2 = 30 \qquad \text{Solve for } x$$

The methods used to solve this simple problem are typical. We summarize the steps as follows.

GUIDELINES FOR SOLVING LOGARITHMIC EQUATIONS

1. Isolate the logarithmic term on one side of the equation; you may need to first combine the logarithmic terms.

2. Write the equation in exponential form (or raise the base to each side of the equation).

3. Solve for the variable.

Radiocarbon dating is a method archeologists use to determine the age of ancient objects. The carbon dioxide in the atmosphere always contains a fixed fraction of radioactive carbon, carbon-14 (^{14}C), with a half-life of about 5730 years. Plants absorb carbon dioxide from the atmosphere, which then makes its way to animals through the food chain. Thus, all living creatures contain the same fixed proportions of ^{14}C to nonradioactive ^{12}C as the atmosphere.

After an organism dies, it stops assimilating ^{14}C, and the amount of ^{14}C in it begins to decay exponentially. We can then determine the time elapsed since the death of the organism by measuring the amount of ^{14}C left in it.

For example, if a donkey bone contains 73% as much ^{14}C as a living donkey and it died t years ago, then by the formula for radioactive decay (Section 4.6),

$$0.73 = (1.00)e^{-(t\ln 2)/5730}$$

We solve this exponential equation to find $t \approx 2600$, so the bone is about 2600 years old.

CHECK YOUR ANSWER

If $x = 5000$, we get

$$4 + 3\log 2(5000) = 4 + 3\log 10000$$
$$= 4 + 3(4)$$
$$= 16 \qquad \checkmark$$

EXAMPLE 6 ■ Solving Logarithmic Equations

Solve each of the following equations for x.

(a) $\ln x = 8$

(b) $\log_2(25 - x) = 3$

SOLUTION

(a)
$$\ln x = 8$$
$$x = e^8 \qquad \text{Exponential form}$$

Therefore, $x = e^8 \approx 2981$.

Another way to solve this problem is as follows.

$$\ln x = 8$$
$$e^{\ln x} = e^8 \qquad \text{Raise } e \text{ to each side}$$
$$x = e^8 \qquad \text{Property of ln}$$

(b) The first step is to rewrite the equation in exponential form.

$$\log_2(25 - x) = 3$$
$$25 - x = 2^3 \qquad \text{Exponential form (or raise 2 to each side)}$$
$$25 - x = 8$$
$$x = 25 - 8 = 17$$

CHECK YOUR ANSWER ■ If $x = 17$, we get

$$\log_2(25 - 17) = \log_2 8 = 3 \qquad \checkmark$$

EXAMPLE 7 ■ Solving Logarithmic Equations

Solve the equation $4 + 3\log(2x) = 16$.

SOLUTION   We first isolate the logarithmic term. This will allow us to write the equation in exponential form.

$$4 + 3\log(2x) = 16$$
$$3\log(2x) = 12 \qquad \text{Subtract 4}$$
$$\log(2x) = 4 \qquad \text{Divide by 3}$$
$$2x = 10^4 \qquad \text{Exponential form (or raise 10 to each side)}$$
$$x = 5000 \qquad \text{Divide by 2}$$

EXAMPLE 8 ■ Solving Logarithmic Equations

Solve the equation $\log(x + 2) + \log(x - 1) = 1$.

SOLUTION   We first combine the logarithmic terms using the Laws of Logarithms.

$$\log[(x + 2)(x - 1)] = 1 \qquad \text{Law 1}$$

$$(x + 2)(x - 1) = 10 \qquad \text{Exponential form (or raise 10 to each side)}$$

$$x^2 + x - 2 = 10 \qquad \text{Expand left side}$$

$$x^2 + x - 12 = 0 \qquad \text{Subtract 10}$$

$$(x + 4)(x - 3) = 0 \qquad \text{Factor}$$

$$x = -4 \qquad \text{or} \qquad x = 3$$

We check these potential solutions in the original equation and find that $x = -4$ is not a solution (because logarithms of negative numbers are undefined), but $x = 3$ is a solution. (See *Check Your Answers*.)

CHECK YOUR ANSWERS

$x = -4$:

$$\log(-4 + 2) + \log(-4 - 1) = \log(-2) + \log(-5) \quad \text{undefined} \qquad \times$$

$x = 3$:

$$\log(3 + 2) + \log(3 - 1) = \log 5 + \log 2 = \log(5 \cdot 2) = \log 10 = 1 \qquad \checkmark$$

4.5 EXERCISES

1–24 ■ Find the solution of the exponential equation, correct to four decimal places.

1. $5^x = 16$

2. $10^{-x} = 2$

3. $2^{1-x} = 3$

4. $3^{2x-1} = 5$

5. $3e^x = 10$

6. $2e^{12x} = 17$

7. $e^{1-4x} = 2$

8. $4(1 + 10^{5x}) = 9$

9. $4 + 3^{5x} = 8$

10. $2^{3x} = 34$

11. $8^{0.4x} = 5$

12. $3^{x/14} = 0.1$

13. $5^{-x/100} = 2$

14. $e^{3-5x} = 16$

15. $e^{2x+1} = 200$

16. $\left(\frac{1}{4}\right)^x = 75$

17. $5^x = 4^{x+1}$

18. $10^{1-x} = 6^x$

19. $2^{3x+1} = 3^{x-2}$

20. $7^{x/2} = 5^{1-x}$

21. $\dfrac{50}{1 + e^{-x}} = 4$

22. $\dfrac{10}{1 + e^{-x}} = 2$

23. $100(1.04)^{2t} = 300$

24. $(1.00625)^{12t} = 2$

25–32 ■ Solve the equation.

25. $x^2 2^x - 2^x = 0$

26. $x^2 10^x - x10^x = 2(10^x)$

27. $4x^3 e^{-3x} - 3x^4 e^{-3x} = 0$

28. $x^2 e^x + xe^x - e^x = 0$

29. $e^{2x} - 3e^x + 2 = 0$

30. $e^{2x} - e^x - 6 = 0$

31. $e^{4x} + 4e^{2x} - 21 = 0$

32. $e^x - 12e^{-x} - 1 = 0$

33–48 ■ Solve the logarithmic equation for x.

33. $\ln x = 10$

34. $\ln(2 + x) = 1$

35. $\log x = -2$

36. $\log(x - 4) = 3$

37. $\log(3x + 5) = 2$

38. $\log_3(2 - x) = 3$

39. $2 - \ln(3 - x) = 0$

40. $\log_2(x^2 - x - 2) = 2$

41. $\log_2 3 + \log_2 x = \log_2 5 + \log_2(x - 2)$

42. $2 \log x = \log 2 + \log(3x - 4)$

43. $\log x + \log(x - 1) = \log(4x)$

44. $\log_5 x + \log_5(x + 1) = \log_5 20$

45. $\log_5(x + 1) - \log_5(x - 1) = 2$

46. $\log x + \log(x - 3) = 1$

47. $\log_9(x - 5) + \log_9(x + 3) = 1$

48. $\ln(x - 1) + \ln(x + 2) = 1$

49. For what value of x is the following true?

$$\log(x + 3) = \log x + \log 3$$

50. For what value of x is it true that $(\log x)^3 = 3 \log x$?

51. Solve for x: $2^{2/\log_5 x} = \frac{1}{16}$

52. Solve for x: $\log_2(\log_3 x) = 4$

53. A 15-gram sample of radioactive iodine decays in such a way that the mass remaining after t days is given by $m(t) = 15e^{-0.087t}$ where $m(t)$ is measured in grams. After how many days is there only 5 g remaining?

54. The velocity of a sky diver t seconds after jumping is given by $v(t) = 80(e^{-0.2t} - 1)$. After how many seconds is the velocity 70 ft/s?

55. The figure shows an electric circuit containing a battery producing a voltage of 60 volts, a resistor with a resistance of 13 ohms, and an inductor with an inductance of 5 henrys. Using calculus it can be shown that the current $I = I(t)$ (in amperes, A) t seconds after the switch is closed is $I = \frac{60}{13}(1 - e^{-13t/5})$.
(a) Use this equation to express the time t as a function of the current I.
(b) After how many seconds is the current 2 amperes?

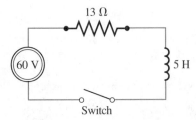

13 Ω

60 V

5 H

Switch

56. A *learning curve* is a graph of a function $P(t)$ that measures the performance of someone learning a skill as a function of the training time t. At first, the rate of learning is rapid. Then, as performance increases and approaches a maximal value M, the rate of learning

decreases. It has been found that the function $P(t) = M - Ce^{-kt}$, where k and C are positive constants and $C < M$, is a reasonable model for learning.
(a) Express the learning time t as a function of the performance level P.
(b) For a pole-vaulter in training, the learning curve is given by $P(t) = 20 - 14e^{-0.024t}$, where $P(t)$ is the height he is able to pole-vault after t months. After how many months of training is he able to vault 12 ft?
(c) Draw a graph of the learning curve in part (b).

57–64 ■ Use a graphing calculator to find all solutions of the equation, correct to two decimal places.

57. $\ln x = 3 - x$

58. $\log x = x^2 - 2$

59. $x^3 - x = \log(x + 1)$

60. $x = \ln(4 - x^2)$

61. $e^x = -x$

62. $2^{-x} = x - 1$

63. $4^{-x} = \sqrt{x}$

64. $e^{x^2} - 2 = x^3 - x$

65. Solve the inequality $\log(x - 2) + \log(9 - x) < 1$.

66. Solve the inequality $3 \leq \log_2 x \leq 4$.

67. Solve the inequality $2 < 10^x < 5$.

68. Solve the inequality $x^2 e^x - 2e^x < 0$.

69. Solve the equation $(x - 1)^{\log(x-1)} = 100(x - 1)$.

70. Solve the equation $\log_2 x + \log_4 x + \log_8 x = 11$.

71. Solve the equation $4^x - 2^{x+1} = 3$.
[*Hint:* First write the equation as a quadratic equation in 2^x.]

DISCOVERY · DISCUSSION

72. Estimating a Solution Without actually solving the equation, find two whole numbers between which the solution of $9^x = 20$ must lie. Do the same for $9^x = 100$. Explain how you reached your conclusions.

73. A Surprising Equation Take logarithms to show that the equation $x^{1/\log x} = 5$ has no solution. For what values of k does the equation $x^{1/\log x} = k$ have a solution? What does this tell us about the graph of the function $f(x) = x^{1/\log x}$? Confirm your answer using a graphing calculator.

4.6
APPLICATIONS OF EXPONENTIAL AND LOGARITHMIC FUNCTIONS

Logarithms were invented by John Napier (page 314) to eliminate the tedious calculations involved in multiplying, dividing, and taking powers and roots of the large numbers that occur in astronomy and other sciences. With the advent of computers and calculators, logarithms are no longer important for such calculations. However, logarithms arise in problems of exponential growth and decay because they are inverses of exponential functions. Because of the Laws of Logarithms, they also turn out to be useful in the measurement of the loudness of sounds, the intensity of earthquakes, and many other phenomena. In this section we study some of these applications.

COMPOUND INTEREST

Recall the formulas for interest that we found in Section 4.2.

> If a principal P is invested at an interest rate r for a period of t years, then the amount A of the investment is given by
>
> $A = P(1 + r)$ Simple interest (for one year)
>
> $A = P\left(1 + \dfrac{r}{n}\right)^{nt}$ Interest compounded n times per year
>
> $A = Pe^{rt}$ Interest compounded continuously

We can use logarithms to determine the time it takes for the principal to increase to a given amount.

EXAMPLE 1 ■ Finding the Time for an Investment to Double

A sum of $5000 is invested at an interest rate of 9% per year. Find the time required for the money to double if the interest is compounded according to the following method.

(a) Semiannual   (b) Continuous

SOLUTION

(a) We use the formula for compound interest with $P = \$5000$, $A = \$10,000$, $r = 0.09$, $n = 2$, and solve the resulting exponential equation for t.

$$5000\left(1 + \frac{0.09}{2}\right)^{2t} = 10,000$$

$$(1.045)^{2t} = 2 \qquad \text{Divide by 5000}$$

$$\log 1.045^{2t} = \log 2 \qquad \text{Take log of each side}$$

$$2t \log 1.045 = \log 2 \qquad \text{Law 3}$$

$$t = \frac{\log 2}{2 \log 1.045} \qquad \text{Divide by } 2 \log 1.045$$

$$\approx 7.9 \qquad \text{Use a calculator}$$

The money will double in 7.9 years.

(b) We use the formula for continuously compounded interest with $P = \$5000$, $A = \$10,000$, $r = 0.09$, and solve the resulting exponential equation.

$$5000e^{0.09t} = 10,000$$

$$e^{0.09t} = 2 \qquad \text{Divide by 5000}$$

$$\ln e^{0.09t} = \ln 2 \qquad \text{Take ln of each side}$$

$$0.09t = \ln 2 \qquad \text{Property of ln}$$

$$t = \frac{\ln 2}{0.09} \qquad \text{Divide by 0.09}$$

$$t \approx 7.702 \qquad \text{Use a calculator}$$

The money will double in 7.7 years. ■

If an investment earns compound interest, then the **annual percentage yield** (APY) is the *simple* interest rate that yields the same amount at the end of one year.

EXAMPLE 2 ■ Calculating the Annual Percentage Yield

Find the annual percentage yield for an investment that earns interest at a rate of 6% per year, compounded daily.

SOLUTION   After one year, a principal P will grow to the amount

$$A = P\left(1 + \frac{0.06}{365}\right)^{365} = P(1.06183)$$

The formula for simple interest is

$$A = P(1 + r)$$

Comparing, we see that the annual percentage yield is 6.183%. ■

EXPONENTIAL GROWTH

In Section 4.2 we studied the formula for exponential growth, which describes the growth of an animal or bacterium population.

> If n_0 is the initial size of the population, then the population $n(t)$ at time t is given by
>
> $$n(t) = n_0 e^{rt}$$
>
> where r is the relative rate of growth expressed as a fraction of the population.

Now that we are equipped with logarithms, we can answer questions concerning the time at which the population reaches a certain size.

EXAMPLE 3 ■ World Population Projections

The population of the world in 1995 was 5.7 billion, and the estimated relative growth rate is 2% per year. If the population continues to grow at this rate, when will it reach 57 billion?

SOLUTION We use the formula for population growth with $n_0 = 5.7$ billion, $r = 0.02$, and $n(t) = 57$ billion. This leads to an exponential equation, which we solve for t.

$$5.7e^{0.02t} = 57$$

$$e^{0.02t} = 10 \qquad \text{Divide by 5.7}$$

$$\ln e^{0.02t} = \ln 10 \qquad \text{Take ln of each side}$$

$$0.02t = \ln 10 \qquad \text{Property of ln}$$

$$t = \frac{\ln 10}{0.02} \qquad \text{Divide by 0.02}$$

$$t \approx 115.129 \qquad \text{Use a calculator}$$

Thus, the population will reach 57 billion in approximately 115 years, that is, in the year $1995 + 115 = 2110$. ■

STANDING ROOM ONLY

The population of the world was about 5.7 billion in 1995. The recently observed rate of increase is 2% per year. Using the exponential model for population growth, and assuming that each person occupies an average of 4 ft^2 of the surface of the earth, we find that by the year 2559 there will be standing room only! (The total land surface area of the world is about 1.8×10^{15} ft^2.)

EXAMPLE 4 ■ The Number of Bacteria in a Culture

A culture starts with 10,000 bacteria, and the number doubles every 40 min.

(a) Find a formula for the number of bacteria at time t.
(b) Find the number of bacteria after one hour.
(c) After how many minutes will there be 50,000 bacteria?

(d) Sketch the graph of the number of bacteria at time t.

SOLUTION

(a) To find the formula for population growth, we need to find the rate r. To do this we use the formula with $n_0 = 10{,}000$, $t = 40$, and $n(t) = 20{,}000$, and then solve for r.

$$10{,}000 \cdot e^{r(40)} = 20{,}000$$

$$e^{40r} = 2 \qquad \text{Divide by 10,000}$$

$$\ln e^{40r} = \ln 2 \qquad \text{Take ln of each side}$$

$$40r = \ln 2 \qquad \text{Property of ln}$$

$$r = \frac{\ln 2}{40} \qquad \text{Divide by 40}$$

$$r \approx 0.01733 \qquad \text{Use a calculator}$$

Now that we know $r \approx 0.01733$, we can write the formula for the population growth:

$$n(t) = 10{,}000 \cdot e^{0.01733t}$$

(b) Using the formula we found in part (a) with $t = 60$ min (one hour), we get

$$n(60) = 10{,}000 \cdot e^{0.01733(60)} \approx 28{,}287$$

Thus, the number of bacteria after one hour is approximately 28,000.

(c) We use the formula we found in part (a) with $n(t) = 50{,}000$ and solve the resulting exponential equation for t.

$$10{,}000 \cdot e^{0.01733t} = 50{,}000$$

$$e^{0.01733t} = 5 \qquad \text{Divide by 10,000}$$

$$\ln e^{0.01733t} = \ln 5 \qquad \text{Take ln of each side}$$

$$0.01733t = \ln 5 \qquad \text{Property of ln}$$

$$t = \frac{\ln 5}{0.01733} \qquad \text{Divide by 0.01733}$$

$$t \approx 92.9 \qquad \text{Use a calculator}$$

The bacterium count will reach 50,000 in approximately 93 min.

(d) We sketch a graph of the function $n(t) = 10{,}000 \cdot e^{0.01733t}$ in Figure 1. ∎

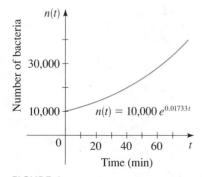

FIGURE 1

RADIOACTIVE DECAY

Radioactive substances decay by spontaneously emitting radiation. The rate of decay is directly proportional to the mass of the substance. This is analogous to population growth, except that the mass of radioactive material *decreases*. It can be shown that the formula for the mass $m(t)$ remaining at time t is given by

$$m(t) = m_0 e^{-rt}$$

where r is the rate of decay expressed as a proportion of the mass and m_0 is the initial mass. Physicists express the rate of decay in terms of **half-life**, the time required for half the mass to decay. We can obtain the rate r from this as follows. If h is the half-life, then a mass of 1 unit will become $\frac{1}{2}$ unit when $t = h$. Substituting this into the formula, we get

$$\frac{1}{2} = 1 \cdot e^{-rh}$$

$$\ln\left(\tfrac{1}{2}\right) = -rh \qquad \text{Take ln of each side}$$

$$r = -\frac{1}{h}\ln(2^{-1}) \qquad \text{Solve for } r$$

$$r = \frac{\ln 2}{h} \qquad \ln 2^{-1} = -\ln 2 \text{ by Law 3}$$

This last equation allows us to find the rate r from the half-life h.

If m_0 is the initial mass of a radioactive substance with half-life h, then the mass $m(t)$ remaining at time t is given by

$$m(t) = m_0 e^{-rt}$$

where $r = \dfrac{\ln 2}{h}$.

EXAMPLE 5 ■ Radioactive Decay

Polonium-210 (^{210}Po) has a half-life of 140 days. Suppose a sample of this substance has a mass of 300 mg.

(a) Find a formula for the amount of the sample remaining at time t.
(b) Find the mass remaining after one year.
(c) How long will it take for the sample to decay to a mass of 200 mg?
(d) Draw a graph of the sample mass as a function of time.

SOLUTION

(a) Using the formula for radioactive decay with $m_0 = 300$ and $r = (\ln 2/140) \approx 0.00495$, we have

$$m(t) = 300e^{-0.00495t}$$

(b) We use the formula we found in part (a) with $t = 365$ (one year).

$$m(365) = 300e^{-0.00495(365)} \approx 49.256$$

Thus, approximately 49 mg of ^{210}Po remains after one year.

(c) We use the formula we found in part (a) with $m(t) = 200$ and solve the resulting exponential equation for t.

$$300e^{-0.00495t} = 200$$

$$e^{-0.00495t} = \tfrac{2}{3} \qquad \text{Divide by 300}$$

$$\ln e^{-0.00495t} = \ln\left(\tfrac{2}{3}\right) \qquad \text{Take ln of each side}$$

$$-0.00495t = \ln\left(\tfrac{2}{3}\right) \qquad \text{Property of ln}$$

$$t = -\frac{\ln\left(\tfrac{2}{3}\right)}{0.00495} \qquad \text{Divide by } -0.00495$$

$$t \approx 81.9 \qquad \text{Use a calculator}$$

The time required for the sample to decay is about 82 days.

(d) A graph of the function $m(t) = 300e^{-0.00495t}$ is shown in Figure 2. ∎

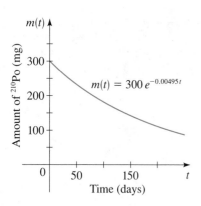

FIGURE 2

NEWTON'S LAW OF COOLING

Newton's Law of Cooling states that the rate of cooling of an object is proportional to the temperature difference between the object and its surroundings, provided that the temperature difference is not too large. Using calculus, the following formula can be deduced from this law.

> If D_0 is the initial temperature difference between an object and its surroundings, and if its surroundings have temperature T_s, then the temperature of the object at time t is given by
>
> $$T(t) = T_s + D_0 e^{-kt}$$
>
> where k is a positive constant that depends on the type of object.

EXAMPLE 6 ■ Newton's Law of Cooling

A cup of coffee has a temperature of $200\,°\text{F}$ and is placed in a room that has a temperature of $70\,°\text{F}$. After 10 min the temperature of the coffee is $150\,°\text{F}$.

(a) Find a formula for the temperature of the coffee at time t.
(b) Find the temperature of the coffee after 15 min.
(c) When will the coffee have cooled to $100\,°\text{F}$?
(d) Illustrate by drawing a graph of the temperature function.

SOLUTION

(a) The temperature of the room is $T_s = 70\,°\text{F}$, and the initial temperature difference is

$$D_0 = 200 - 70 = 130\,°\text{F}$$

Half-lives of **radioactive elements** vary from very long to very short. Here are some examples.

Element	Half-life
Thorium-232	14.5 billion years
Uranium-235	4.5 billion years
Thorium-230	80,000 years
Plutonium-239	24,360 years
Carbon-14	5,730 years
Radium-226	1,600 years
Cesium-137	30 years
Strontium-90	28 years
Polonium-210	140 days
Thorium-234	25 days
Iodine-135	8 days
Radon-222	3.8 days
Lead-211	3.6 minutes
Krypton-91	10 seconds

So, by Newton's Law of Cooling, the temperature after t minutes is

$$T(t) = 70 + 130e^{-kt}$$

We need to find the constant k associated with this cup of coffee. To do this we use the fact that when $t = 10$, the temperature is $T(10) = 150$. So, we have

$$70 + 130e^{-10k} = 150$$

$$130e^{-10k} = 80 \qquad \text{Subtract 70}$$

$$e^{-10k} = \tfrac{8}{13} \qquad \text{Divide by 130}$$

$$-10k = \ln\left(\tfrac{8}{13}\right) \qquad \text{Take ln of each side}$$

$$k = -\tfrac{1}{10}\ln\left(\tfrac{8}{13}\right) \qquad \text{Divide by } -10$$

$$k \approx 0.04855 \qquad \text{Use a calculator}$$

Substituting this value of k into the expression for $T(t)$, we get

$$T(t) = 70 + 130e^{-0.04855t}$$

(b) We use the formula we found in part (a) with $t = 15$.

$$T(15) = 70 + 130e^{-0.04855(15)} \approx 133\,°\text{F}$$

(c) We use the formula we found in part (a) with $T(t) = 100$ and solve the resulting exponential equation for t.

$$70 + 130e^{-0.04855t} = 100$$

$$130e^{-0.04855t} = 30 \qquad \text{Subtract 70}$$

$$e^{-0.04855t} = \tfrac{3}{13} \qquad \text{Divide by 130}$$

$$-0.04855t = \ln\left(\tfrac{3}{13}\right) \qquad \text{Take ln of each side}$$

$$t = -\frac{\ln\left(\tfrac{3}{13}\right)}{0.04855} \qquad \text{Divide by } -0.04855$$

$$t \approx 30.2 \qquad \text{Use a calculator}$$

The coffee will have cooled to $100\,°\text{F}$ after about half an hour.

(d) The graph of the temperature function is sketched in Figure 3. Notice that the line $T = 70$ is a horizontal asymptote. (Why?) ■

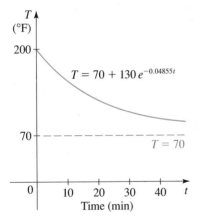

FIGURE 3

Temperature of coffee after t minutes

LOGARITHMIC SCALES

When a physical quantity can vary over a very large range, it is often convenient to take its logarithm in order to have a more manageable set of numbers. We discuss three such situations: the pH scale, which measures acidity; the Richter scale, which measures the intensity of earthquakes; and the decibel scale, which measures the loudness of sounds. Other quantities that are measured on logarithmic scales are light intensity, information capacity, and radiation.

THE pH SCALE   Chemists measured the acidity of a solution by giving its hydrogen ion concentration until Sorensen, in 1909, proposed a more convenient measure. He defined

$$pH = -\log[H^+]$$

where $[H^+]$ is the concentration of hydrogen ions measured in moles per liter (M). He did this to avoid very small numbers and negative exponents. For instance,

$$\text{if} \quad [H^+] = 10^{-4}\text{ M}, \quad \text{then} \quad pH = -\log_{10}(10^{-4}) = -(-4) = 4$$

Solutions with a pH of 7 are defined as *neutral*, those with pH < 7 are *acidic*, and those with pH > 7 are *basic*. Notice that when the pH increases by one unit, $[H^+]$ decreases by a factor of 10.

EXAMPLE 7 ■ pH Scale and Hydrogen Ion Concentration

(a) The hydrogen ion concentration of a sample of human blood was measured to be $[H^+] = 3.16 \times 10^{-8}$ M. Find the pH and classify the blood as acidic or basic.
(b) The most acidic rainfall ever measured occurred in Scotland in 1974; its pH was 2.4. Find the hydrogen ion concentration.

SOLUTION

(a) A calculator gives

$$pH = -\log[H^+] = -\log(3.16 \times 10^{-8}) \approx 7.5$$

Since this is greater than 7, the blood is basic.

(b) To find the hydrogen ion concentration, we need to solve for $[H^+]$ in the logarithmic equation

$$\log[H^+] = -pH$$

So, we write it in exponential form.

$$[H^+] = 10^{-pH}$$

In this case, pH = 2.4, so

$$[H^+] = 10^{-2.4} \approx 4.0 \times 10^{-3}\text{ M} \qquad \blacksquare$$

THE RICHTER SCALE   In 1935 the American geologist Charles Richter (1900–1984) defined the magnitude of an earthquake to be

$$M = \log \frac{I}{S}$$

where I is the intensity of the earthquake (measured by the amplitude of a seismograph reading taken 100 km from the epicenter of the earthquake) and S

is the intensity of a "standard" earthquake (whose amplitude is 1 micron = 10^{-4} cm). The magnitude of a standard earthquake is

$$M = \log \frac{S}{S} = \log 1 = 0$$

Richter studied many earthquakes that occurred between 1900 and 1950. The largest had magnitude 8.9 on the Richter scale, and the smallest had magnitude 0. This corresponds to a ratio of intensities of 800,000,000, so the Richter scale provides more manageable numbers to work with. For instance, an earthquake of magnitude 6 is ten times stronger than an earthquake of magnitude 5.

EXAMPLE 8 ■ Magnitude of Earthquakes

The 1906 earthquake in San Francisco had an estimated magnitude of 8.3 on the Richter scale. In the same year the strongest earthquake ever recorded occurred on the Colombia-Ecuador border and was four times as intense. What was the magnitude of the Colombia-Ecuador earthquake on the Richter scale?

SOLUTION   If I is the intensity of the San Francisco earthquake, then from the definition of magnitude we have

$$M = \log \frac{I}{S} = 8.3$$

The intensity of the Colombia-Ecuador earthquake was $4I$, so its magnitude was

$$M = \log \frac{4I}{S} = \log 4 + \log \frac{I}{S} = \log 4 + 8.3 \approx 8.9$$ ■

EXAMPLE 9 ■ Intensity of Earthquakes

The 1989 Loma Prieta earthquake that shook San Francisco had a magnitude of 7.1 on the Richter scale. How many times more intense was the 1906 earthquake (see Example 8) than the 1989 event?

SOLUTION   If I_1 and I_2 are the intensities of the 1906 and 1989 earthquakes, then we are required to find I_1/I_2. To relate this to the definition of magnitude, we divide numerator and denominator by S.

$$\log \frac{I_1}{I_2} = \log \frac{I_1/S}{I_2/S} \qquad \text{Divide numerator and denominator by } S$$

$$= \log \frac{I_1}{S} - \log \frac{I_2}{S} \qquad \text{Law 2 of Logarithms}$$

$$= 8.3 - 7.1 = 1.2 \qquad \text{Definition of earthquake magnitude}$$

Therefore

$$\frac{I_1}{I_2} = 10^{\log(I_1/I_2)} = 10^{1.2} \approx 16$$

The 1906 earthquake was about 16 times as intense as the 1989 earthquake.   ■

THE DECIBEL SCALE   The ear is sensitive to an extremely wide range of sound intensities. We take as a reference intensity $I_0 = 10^{-12}$ watts/m^2 at a frequency of 1000 hertz, which measures a sound that is just barely audible (the threshold of hearing). The psychological sensation of loudness varies with the logarithm of the intensity (the Weber-Fechner Law) and so the **intensity level** β, measured in decibels (dB), is defined as

$$\beta = 10 \log \frac{I}{I_0}$$

The intensity level of the barely audible reference sound is

$$\beta = 10 \log \frac{I_0}{I_0} = 10 \log 1 = 0 \text{ dB}$$

Source of Sound	β (dB)
Jet takeoff	140
Jackhammer	130
Rock concert	120
Subway	100
Heavy traffic	80
Ordinary traffic	70
Normal conversation	50
Whisper	30
Rustling leaves	10–20
Threshold of hearing	0

EXAMPLE 10 ■ Sound Intensity of a Jet Takeoff

Find the decibel intensity level of a jet engine during takeoff if the intensity was measured at 100 W/m^2 (watts per square meter).

SOLUTION   From the definition of intensity level, we see that

$$\beta = 10 \log \frac{I}{I_0} = 10 \log \frac{10^2}{10^{-12}} = 10 \log 10^{14} = 140 \text{ dB}$$

Thus, the intensity level is 140 dB.   ■

The table in the margin lists decibel intensity levels for some common sounds ranging from the threshold of hearing to the jet takeoff of Example 10. The threshold of pain is about 120 dB.

4.6   EXERCISES

1. A man invests $10,000 in an account that pays 8.5% per year, compounded quarterly.
   (a) Find the amount after 3 years.
   (b) How long will it take for the investment to double?

2. A man invests $6500 in an account that pays 6% per year, compounded continuously.
   (a) What is the amount after 2 years?
   (b) How long will it take for the amount to be $8000?

3. Find the time required for an investment of $5000 to grow to $8000 at an interest rate of 9.5% per year, compounded quarterly.

4. Nancy wants to invest $4000 in saving certificates that bear an interest rate of 9.75% per year, compounded

semiannually. How long a time period should she choose in order to save an amount of $5000?

5. How long will it take for an investment of $1000 to double in value if the interest rate is 8.5% per year, compounded continuously?

6. A sum of $1000 was invested for 4 years, and the interest was compounded semiannually. If this sum amounted to $1435.77 in the given time, what was the interest rate?

7. Find the annual percentage yield for an investment that earns 8% per year, compounded monthly.

8. Find the annual percentage yield for an investment that earns $5\frac{1}{2}$% per year, compounded continuously.

9. The number of bacteria in a culture is given by the formula

$$n(t) = 500e^{0.45t}$$

where t is measured in hours.
   (a) What is the initial number of bacteria?
   (b) What is the relative rate of growth of this bacterium population? Express your answer as a percentage.
   (c) How many bacteria are in the culture after 3 hours?
   (d) After how many hours will the number of bacteria reach 10,000?

10. The number of a certain species of fish is given by the formula $n(t) = 12e^{0.012t}$ where t is measured in years and $n(t)$ is measured in millions.
   (a) What is the relative rate of growth of the fish population? Express your answer as a percentage.
   (b) What will the fish population be after 5 years?
   (c) After how many years will the number of fish reach 30 million?
   (d) Sketch a graph of the fish population function $n(t)$.

11. The population of a certain city was 112,000 in 1994, and the observed relative growth rate is 4% per year.
   (a) Find a formula for the population $n(t)$ after t years.
   (b) Find the projected population in the year 2000.
   (c) In what year will the population reach 200,000?

12. The frog population in a small pond grows exponentially. The current population is 85 frogs, and the relative growth rate is 18% per year.
   (a) Find a formula for the population $n(t)$ after t years.
   (b) Find the projected population after 3 years.
   (c) Find the number of years required for the frog population to reach 600.

13. The graph shows the deer population in a Pennsylvania county between 1990 and 1994. Assume that the population grows exponentially.

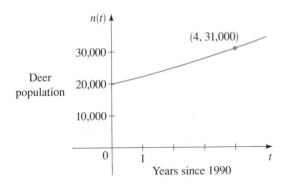

(a) What was the deer population in 1990?

(b) Find a formula for the deer population t years after 1990.
   (c) What is the projected deer population in 1998?
   (d) In what year will the deer population reach 100,000?

14. A culture contains 1500 bacteria initially and doubles every 30 min.
   (a) Find a formula for the number of bacteria $n(t)$ after t minutes.
   (b) Find the number of bacteria after 2 hours.
   (c) After how many minutes will the culture contain 4000 bacteria?

15. A culture starts with 8600 bacteria. After one hour the count is 10,000.
   (a) Find a formula for the number of bacteria $n(t)$ after t hours.
   (b) Find the number of bacteria after 2 hours.
   (c) After how many hours will the number of bacteria double?

16. The count in a culture of bacteria was 400 after 2 hours and 25,600 after 6 hours.
   (a) What is the relative rate of growth of the bacterium population? Express your answer as a percentage.
   (b) What was the initial size of the culture?
   (c) Find a formula for the number of bacteria $n(t)$ after t hours.
   (d) Find the number of bacteria after 4.5 hours.
   (e) When will the number of bacteria be 50,000?

17. The population of the world was 5.7 billion in 1995 and the observed relative growth rate was 2% per year.
   (a) By what year will the population have doubled?
   (b) By what year will the population have tripled?

18. The population of California was 10,586,223 in 1950 and 23,668,562 in 1980. Assume the population grows exponentially.
   (a) Find a formula for the population t years after 1950.
   (b) Find the time required for the population to double.
   (c) Use the data to predict the population of California in the year 2000.

19. An infectious strain of bacteria increases in number at a relative growth rate of 200% per hour. When a certain critical number of bacteria are present in the bloodstream, a person becomes ill. If a single bacterium infects a person, the critical level is reached in 24 hours. How long will it take for the critical level to be reached if the same person is infected with 10 bacteria?

20. The half-life of radium-226 is 1600 years. Suppose we have a 22-mg sample.
    (a) Find a formula for the mass remaining after t years.
    (b) How much of the sample will remain after 4000 years?
    (c) After how long will only 18 mg of the sample remain?

21. The half-life of cesium-137 is 30 years. Suppose we have a 10-g sample.
    (a) Find a formula for the mass remaining after t years.
    (b) How much of the sample will remain after 80 years?
    (c) After how long will only 2 g of the sample remain?

22. The mass $m(t)$ remaining after t days from a 40-g sample of thorium-234 is given by

$$m(t) = 40e^{-0.0277t}$$

    (a) How much of the sample will remain after 60 days?
    (b) After how long will only 10 g of the sample remain?
    (c) Find the half-life of thorium-234.

23. The half-life of strontium-90 is 25 years. How long will it take a 50-mg sample to decay to a mass of 32 mg?

24. Radium-221 has a half-life of 30 s. How long will it take for 95% of a sample to decompose?

25. If 250 mg of a radioactive element decays to 200 mg in 48 hours, find the half-life of the element.

26. After three days a sample of radon-222 has decayed to 58% of its original amount.
    (a) What is the half-life of radon-222?
    (b) How long will it take the sample to decay to 20% of its original amount?

27. A wooden artifact from an ancient tomb contains 65% of the carbon-14 that is present in living trees. How long ago was the artifact made? (The half-life of carbon-14 is 5730 years.)

28. The burial cloth of an Egyptian mummy is estimated to contain 59% of the carbon-14 it contained originally. How long ago was the mummy buried? (The half-life of carbon-14 is 5730 years.)

29. A hot bowl of soup is served at at a dinner party. It starts to cool according to Newton's Law of Cooling so that its temperature at time t is given by

$$T(t) = 65 + 145e^{-0.05t}$$

where t is measured in minutes and T is measured in °F.
    (a) What is the initial temperature of the soup?
    (b) What is the temperature after 10 min?
    (c) After how long will the temperature be 100°F?

30. Newton's Law of Cooling is used in homicide investigations to determine the time of death. The normal body temperature is 98.6°F. Immediately following death the body begins to cool. It has been determined experimentally that the constant in Newton's Law of Cooling is approximately $k = 0.1947$. If the temperature of the surroundings is 60°F and the temperature of the body is now 72°F, how long ago was the time of death?

31. A roasted turkey is taken from an oven when its temperature has reached 185°F and is placed on a table in a room where the temperature is 75°F.
    (a) If the temperature of the turkey is 150°F after half an hour, what is its temperature after 45 min?
    (b) When will the turkey cool to 100°F?

32. A kettle full of water is brought to a boil in a room with temperature 20°C. After 15 min the temperature of the water has decreased from 100°C to 75°C. Find the temperature after another 10 min. Illustrate by sketching the graph of the temperature function.

33. The hydrogen ion concentration of a sample of each substance is given. Calculate the pH of the substance.
    (a) Lemon juice: $[H^+] = 5.0 \times 10^{-3}$ M
    (b) Tomato juice: $[H^+] = 3.2 \times 10^{-4}$ M
    (c) Seawater: $[H^+] = 5.0 \times 10^{-9}$ M

34. An unknown substance has a hydrogen ion concentration of $[H^+] = 3.1 \times 10^{-8}$ M. Find the pH and classify the substance as acidic or basic.

35. The pH reading of a sample of each substance is given. Calculate the hydrogen ion concentration of the substance.
    (a) Vinegar: pH = 3.0
    (b) Milk: pH = 6.5

36. The pH reading of a glass of liquid is given. Find the hydrogen ion concentration of the liquid.
    (a) Beer: pH = 4.6
    (b) Water: pH = 7.3

37. The hydrogen ion concentrations in cheeses range from 4.0×10^{-7} M to 1.6×10^{-5} M. Find the corresponding range of pH readings.

38. The pH readings for wines vary from 2.8 to 3.8. Find the corresponding range of hydrogen ion concentrations.

39. If one earthquake is 20 times as intense as another, how much larger is its magnitude on the Richter scale?

40. The 1906 earthquake in San Francisco had a magnitude of 8.3 on the Richter scale. At the same time in Japan there was an earthquake with magnitude 4.9 that caused only minor damage. How many times more intense was the San Francisco earthquake than the Japanese earthquake?

41. The Alaska earthquake of 1964 had a magnitude of 8.6 on the Richter scale. How many times more intense was this than the 1906 San Francisco earthquake?

42. The Northridge, California, earthquake of 1994 had a magnitude of 6.8 on the Richter scale. A year later, a 7.2-magnitude earthquake struck Kobe, Japan. How many times more intense was the Kobe earthquake than the Northridge earthquake?

43. The 1985 Mexico City earthquake had a magnitude of 8.1 on the Richter scale. The 1976 earthquake in Tangshan, China, was 1.26 times as intense. What was the magnitude of the Tangshan earthquake?

44. The intensity of the sound of traffic at a busy intersection was measured at 2.0×10^{-5} W/m^2. Find the intensity level in decibels.

45. The intensity level of the sound of a subway train was measured at 98 dB. Find the intensity in W/m^2.

46. The noise from a power mower was measured at 106 dB. The noise level at a rock concert was measured at 120 dB. Find the ratio of the intensity of the rock music to that of the power mower.

47. It is a law of physics that the intensity of sound is inversely proportional to the square of the distance d from the source:

$$I = \frac{k}{d^2}$$

(a) Use this and the equation

$$\beta = 10 \log \frac{I}{I_0}$$

(described in this section) to show that the decibel levels β_1 and β_2 at distances d_1 and d_2 from a sound source are related by the equation

$$\beta_2 = \beta_1 + 20 \log \frac{d_1}{d_2}$$

(b) The intensity level at a rock concert is 120 dB at a distance 2 m from the speakers. Find the intensity level at a distance of 10 m.

| 4 | REVIEW |

CONCEPT CHECK

1. (a) Write an equation that defines the exponential function with base a.
   (b) What is the domain of this function?
   (c) If $a \neq 1$, what is the range of this function?
   (d) Sketch the general shape of the graph of the exponential function for each of the following cases.
   　(i) $a > 1$　　　　(ii) $a = 1$
   　(iii) $a < 1$

2. If x is large, which function grows faster, $y = 2^x$ or $y = x^2$?

3. (a) How is the number e defined?
   (b) What is the natural exponential function?

4. (a) How is the logarithmic function $y = \log_a x$ defined?
   (b) What is the domain of this function?
   (c) What is the range of this function?
   (d) Sketch the general shape of the graph of the function $y = \log_a x$ if $a > 1$.
   (e) What is the natural logarithm?
   (f) What is the common logarithm?

5. State the three Laws of Logarithms.

6. State the Change of Base Formula.

7. (a) How do you solve an exponential equation?
   (b) How do you solve a logarithmic equation?

8. Suppose an amount P is invested at an interest rate r and A is the amount after t years.
   (a) Write an expression for A if the interest is compounded n times per year.
   (b) Write an expression for A if the interest is compounded continuously.

9. If the initial size of a population is n_0 and the population grows exponentially with relative growth rate r, write an expression for the population $n(t)$ at time t.

10. (a) What is the half-life of a radioactive substance?
    (b) If a radioactive substance has initial mass m_0 and half-life h, write an expression for the mass $m(t)$ remaining at time t.

11. What does Newton's Law of Cooling say?

12. What do the pH scale, the Richter scale, and the decibel scale have in common? What do they measure?

EXERCISES

1–12 ■ Sketch the graph of the function. State the domain, range, and asymptote.

1. $f(x) = \dfrac{1}{2^x}$

2. $g(x) = 3^{x-2}$

3. $y = 5 - 10^x$

4. $y = 1 + 5^{-x}$

5. $f(x) = \log_3(x - 1)$

6. $g(x) = \log(-x)$

7. $y = 2 - \log_2 x$

8. $y = 3 + \log_5(x + 4)$

9. $F(x) = e^x - 1$

10. $G(x) = \tfrac{1}{2} e^{x-1}$

11. $y = 2 \ln x$

12. $y = \ln(x^2)$

13–14 ■ Find the domain of the function.

13. $f(x) = 10^{x^2} + \log(1 - 2x)$　　**14.** $g(x) = \ln(2 + x - x^2)$

15–18 ■ Write the equation in exponential form.

15. $\log_2 1024 = 10$

16. $\log_6 37 = x$

17. $\log x = y$

18. $\ln c = 17$

19–22 ■ Write the equation in logarithmic form.

19. $2^6 = 64$

20. $49^{-1/2} = \tfrac{1}{7}$

21. $10^x = 74$

22. $e^k = m$

23–38 ■ Evaluate the expression without using a calculator.

23. $\log_2 128$

24. $\log_8 1$

25. $10^{\log 45}$

26. $\log 0.000001$

27. $\ln(e^6)$

28. $\log_4 8$

29. $\log_3\left(\tfrac{1}{27}\right)$

30. $2^{\log_2 13}$

31. $\log_5 \sqrt{5}$

32. $e^{2\ln 7}$

33. $\log 25 + \log 4$

34. $\log_3 \sqrt{243}$

35. $\log_2 16^{23}$

36. $\log_5 250 - \log_5 2$

37. $\log_8 6 - \log_8 3 + \log_8 2$ **38.** $\log \log 10^{100}$

39–44 ■ Rewrite the expression in a form with no logarithms of products, quotients, or powers.

39. $\log(AB^2C^3)$ **40.** $\log_2(x\sqrt{x^2+1}\,)$

41. $\ln\sqrt{\dfrac{x^2-1}{x^2+1}}$ **42.** $\log\left(\dfrac{4x^3}{y^2(x-1)^5}\right)$

43. $\log_5\left(\dfrac{x^2(1-5x)^{3/2}}{\sqrt{x^3-x}}\right)$ **44.** $\ln\left(\dfrac{\sqrt[3]{x^4+12}}{(x+16)\sqrt{x-3}}\right)$

45–50 ■ Rewrite the expression as a single logarithm.

45. $\log 6 + 4\log 2$

46. $\log x + \log(x^2 y) + 3\log y$

47. $\frac{3}{2}\log_2(x-y) - 2\log_2(x^2+y^2)$

48. $\log_5 2 + \log_5(x+1) - \frac{1}{3}\log_5(3x+7)$

49. $\log(x-2) + \log(x+2) - \frac{1}{2}\log(x^2+4)$

50. $\frac{1}{2}[\ln(x-4) + 5\ln(x^2+4x)]$

51–60 ■ Use a calculator to find the solution of the equation, correct to two decimal places.

51. $\log_2(1-x) = 4$ **52.** $2^{3x-5} = 7$

53. $5^{5-3x} = 26$ **54.** $\ln(2x-3) = 14$

55. $e^{3x/4} = 10$ **56.** $2^{1-x} = 3^{2x+5}$

57. $\log x + \log(x+1) = \log 12$

58. $\log_8(x+5) - \log_8(x-2) = 1$

59. $x^2 e^{2x} + 2xe^{2x} = 8e^{2x}$

60. $2^{3^x} = 5$

61–64 ■ Use a calculator to find the solution of the equation, correct to six decimal places.

61. $5^{-2x/3} = 0.63$ **62.** $2^{3x-5} = 7$

63. $5^{2x+1} = 3^{4x-1}$ **64.** $e^{-15k} = 10{,}000$

 65–68 ■ Draw the graph of the function and use it to determine the asymptotes and the local maximum and minimum values.

65. $y = e^{x/(x+2)}$ **66.** $y = 2x^2 - \ln x$

67. $y = \log(x^3 - x)$ **68.** $y = 10^x - 5^x$

69–70 ■ Find the solutions of the equations, correct to two decimal places.

69. $3\log x = 6 - 2x$ **70.** $4 - x^2 = e^{-2x}$

71–72 ■ Solve the inequality graphically.

71. $\ln x > x - 2$ **72.** $e^x < 4x^2$

73. Use a graph of $f(x) = e^x - 3e^{-x} - 4x$ to find, approximately, the intervals on which f is increasing and on which f is decreasing.

74. Find an equation of the line shown in the figure.

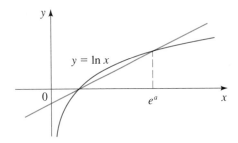

75. Evaluate $\log_4 15$, correct to six decimal places.

76. Solve the inequality: $0.2 \le \log x < 2$

77. Which is larger, $\log_4 258$ or $\log_5 620$?

78. Find the inverse function of the function $f(x) = 2^{3^x}$ and state its domain and range.

79. If \$12,000 is invested at an interest rate of 10% per year, find the amount of the investment at the end of 3 years for each compounding method.
(a) Semiannual   (b) Monthly
(c) Daily   (d) Continuously

80. A sum of \$5000 is invested at an interest rate of $8\frac{1}{2}\%$ per year, compounded semiannually.
(a) Find the amount of the investment after $1\frac{1}{2}$ years.
(b) After what period of time will the investment amount to \$7000?

81. The stray-cat population in a small town grows exponentially. In 1994, the town had 30 stray cats and the relative growth rate was 15% per year.
(a) Find a formula for the stray-cat population $n(t)$ after t years.
(b) Find the projected population after 4 years.
(c) Find the number of years required for the stray-cat population to reach 500.

82. A culture contains 10,000 bacteria initially. After an hour the bacteria count is 25,000.
  (a) Find the doubling period.
  (b) Find the population after 3 hours.

83. Uranium-234 has a half-life of 2.7×10^5 years.
  (a) Find the amount remaining from a 10-mg sample after a thousand years.
  (b) How long will it take this sample to decompose until its mass is 7 mg?

84. A sample of bismuth-210 decayed to 33% of its original mass after 8 days.
  (a) Find the half-life of this element.
  (b) Find the mass remaining after 12 days.

85. The half-life of radium-226 is 1590 years.
  (a) If a sample has a mass of 150 mg, find a formula for the mass that remains after t years.
  (b) Find the mass that will remain after 1000 years.
  (c) After how many years will only 50 mg remain?

86. The half-life of palladium-100 is 4 days. After 20 days a sample has been reduced to a mass of 0.375 g.
  (a) What was the initial mass of the sample?
  (b) Find a formula for the mass remaining after t days.
  (c) What is the mass after 3 days?
  (d) After how many days will only 0.15 g remain?

87. The graph shows the population of a rare species of bird, where t represents years since 1988 and $n(t)$ is measured in thousands.
  (a) Find a formula for the bird population at time t in the form $n(t) = n_0 e^{rt}$.
  (b) What is the bird population expected to be in the year 1999?

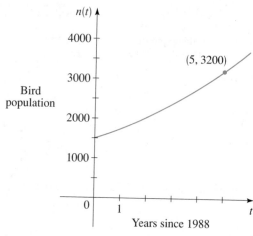

Bird population

Years since 1988

88. A car engine runs at a temperature of 190°F. When the engine is turned off, it cools according to Newton's Law of Cooling with constant $k = 0.341$. Find the time needed for the engine to cool to 90°F if the surrounding temperature is 60°F.

89. The hydrogen ion concentration of fresh egg whites was measured as
$$[H^+] = 1.3 \times 10^{-8} \text{ M}$$
Find the pH, and classify the substance as acidic or basic.

90. The pH of lime juice is 1.9. Find the hydrogen ion concentration.

91. If one earthquake has magnitude 6.5 on the Richter scale, what is the magnitude of another quake that is 35 times as intense?

92. The drilling of a jackhammer was measured at 132 dB. The sound of whispering was measured at 28 dB. Find the ratio of the intensity of the drilling to that of the whispering.

1. Graph the functions $y = 4^x$ and $y = \log_4 x$ on the same axes.

2. Sketch the graph of the function $f(x) = \log(x + 2)$ and state the domain, range, and asymptote.

3. Evaluate each logarithmic expression.
   (a) $\log_3 \sqrt{27}$ (b) $\log_2 56 - \log_2 7$
   (c) $\log_8 4$ (d) $\log_6 4 + \log_6 9$

4. Use the Laws of Logarithms to rewrite the expression

$$\log \sqrt{\frac{x^2 - 1}{x^3(y^2 + 1)^5}}$$

   without logarithms of products, quotients, powers, or roots.

5. Write as a single logarithm: $\ln x - 2\ln(x^2 + 1) + \frac{1}{2}\ln(3 - x^4)$

6. Find the solution of the equation, correct to two decimal places.
   (a) $2^{x-1} = 10$ (b) $5\ln(3 - x) = 4$
   (c) $10^{x+3} = 6^{2x}$ (d) $\log_2(x + 2) + \log_2(x - 1) = 2$

7. The initial size of a culture of bacteria is 1000. After one hour the bacterium count is 8000.
   (a) Find a formula for the population after t hours.
   (b) Find the population after 1.5 hours.
   (c) When will the population reach 15,000?
   (d) Sketch the graph of the population function.

8. How long will it take for an investment to double in value if the interest rate is 8.5% per year, compounded semiannually?

9. Find the domain of the function $f(x) = \log(x + 4) + \log(8 - 5x)$.

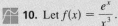 10. Let $f(x) = \dfrac{e^x}{x^3}$.
   (a) Graph f in an appropriate viewing rectangle.
   (b) State the asymptotes of f.
   (c) Find, correct to two decimal places, the local minimum value of f and the value of x at which it occurs.
   (d) Find the range of f.
   (e) Solve the equation $\dfrac{e^x}{x^3} = 2x + 1$. State each solution correct to two decimal places.

FOCUS ON MODELING

In *Principles of Modeling* (page 210), we learned how to construct linear models of data. What do we do if the data we are studying are not linear? In that case, our model would be some other type of function that best fits the data. The type of function we choose is determined by the shape of the scatter plot or by some physical principle that underlies the data. In this section we learn how to construct exponential, power, and polynomial models.

EXPONENTIAL FUNCTIONS AS MODELS

Table 1 gives the population of the world in the 20th century. The scatter plot in Figure 1 shows that the population grows too quickly for a linear model to be appropriate. In fact, as we saw in Section 4.2, population tends to increase exponentially. So, we should seek an *exponential model* for population growth, that is, a function like

$$P = Ce^{kt}$$

Table 1 World population

Year (t)	World population (P, in millions)
1900	1650
1910	1750
1920	1860
1930	2070
1940	2300
1950	2520
1960	3020
1970	3700
1980	4450
1990	5300
1996	5770

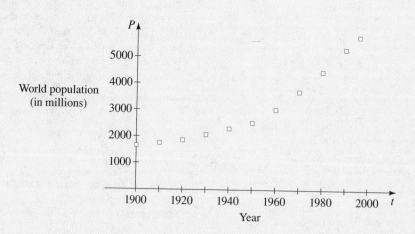

FIGURE 1
Scatter plot of world population

To find an appropriate exponential model of the data points

$$(t, P)$$

we first linearize or "straighten" the data by calculating ln P for each data point. Table 2 gives the linearized data

$$(t, \ln P)$$

[Because the population is given in millions, the first entry in Table 2 is $\ln(1,650,000,000) = 21.224$; subsequent entries are calculated in the same way.]

Table 2 World population data

t	P (in millions)	ln P
1900	1650	21.224
1910	1750	21.283
1920	1860	21.344
1930	2070	21.451
1940	2300	21.556
1950	2520	21.648
1960	3020	21.829
1970	3700	22.032
1980	4450	22.216
1990	5300	22.391
1996	5770	22.476

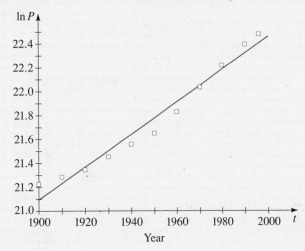

FIGURE 2

We expect the data $(t, \ln P)$ to be nearly linear, so we find a linear model. A calculator gives the regression line

$$\ln P = 0.013623t - 4.7911$$

with correlation coefficient 0.98. Figure 2 shows the scatter plot and the regression line of the linearized data.

To find a model for the population P, we take the exponential of each side of the preceding equation:

$$e^{\ln P} = e^{0.013623t - 4.7911} \qquad \text{Take exponential of each side}$$

$$P = e^{0.013623t - 4.7911} \qquad \text{Property of logarithms}$$

$$= e^{0.013623t} e^{-4.7911} \qquad \text{Law of exponentials}$$

$$= 0.008303 e^{0.013623t} \qquad \text{Use a calculator}$$

Thus, the exponential model is

$$(1) \qquad\qquad P = 0.008303 e^{0.013623t}$$

Figure 3 (on page 344) shows the exponential model for population growth together with the original scatter plot. We see that the exponential curve fits the data reasonably well. The period of relatively slow population growth is explained by the depression of the 1930s and the two world wars.

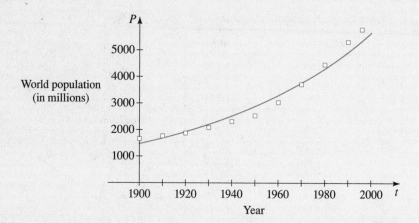

FIGURE 3

Exponential model
for world population

We obtained the exponential model for population growth by taking logarithms, applying linear regression to the resulting data, then solving for P by applying the exponential function. Most graphing calculators, however, are capable of applying the method of least squares *directly* to the data to obtain an exponential model. One such calculator applies exponential regression to the data in Table 1 and obtains the function

(2) $$P = (0.008306) \cdot (1.013716)^t$$

To compare the relative growth rates of different exponential functions, we need to use the same base. So, we convert to base e by writing

$$1.013716 = e^k$$

$$k = \ln 1.013716 \approx 0.013623 \quad \text{Solve for } k$$

So $$P = 0.008306e^{0.013623t}$$

The slight discrepancy between Equations 1 and 2 is explained by the fact that applying least squares to the transformed data doesn't quite correspond to the least squares fit of the original data.

POWER FUNCTIONS AS MODELS

If the scatter plot of the data we are studying appears to resemble the graph of $y = ax^2$, $y = ax^3$, or some other power function, then we seek a *power model,* that is, a function of the form

$$y = ax^n$$

where a is a positive constant and n is any real number. If we take the logarithm of each side of the preceding equation, we get

$$\ln y = \ln ax^n \qquad \text{Take ln of each side}$$

$$= \ln a + \ln x^n \qquad \text{Property of logarithms}$$

$$= \ln a + n \ln x \qquad \text{Property of logarithms}$$

The last equation shows that $\ln y$ is a linear function of $\ln x$. To see this, write $X = \ln x$, $Y = \ln y$, and $A = \ln a$. The last equation becomes

$$Y = nX + A$$

and this is a linear equation. So, if we suspect that a power function is an appropriate model, we may linearize or "straighten" the data by graphing the points $(\ln x, \ln y)$. The graph of the points $(\ln x, \ln y)$ is called a **log-log plot**.

Let's apply the preceding observations to the problem of finding a relationship between the distance of a planet from the sun and the period of the planet, that is, the time it takes the planet to make a complete revolution around the sun. Table 3 shows the mean distance d of each planet from the sun in astronomical units and its period T in years. An astronomical unit (AU) is the mean distance from the sun to the earth. We seek a power model of this data, so we calculate the points $(\ln d, \ln T)$ in Table 4. The resulting log-log plot in Figure 4 is nearly linear.

Table 3 Distances and periods of the planets

Planet	d	T
Mercury	0.387	0.241
Venus	0.723	0.615
Earth	1.000	1.000
Mars	1.523	1.881
Jupiter	5.203	11.861
Saturn	9.541	29.457
Uranus	19.190	84.008
Neptune	30.086	164.784
Pluto	39.507	248.350

Table 4 Log-log table

$\ln d$	$\ln T$
−0.94933	−1.4230
−0.32435	−0.48613
0	0
0.42068	0.6318
1.6492	2.4733
2.2556	3.3829
2.9544	4.4309
3.4041	5.1046
3.6765	5.5148

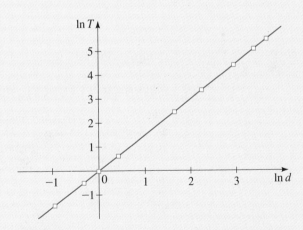

FIGURE 4

Log-log plot and regression line of data in Table 4

Using a calculator we find the regression line of the log-log data:

$$\ln T = 1.5 \ln d + 0.0004$$

(See Figure 4.) This is a linear model for the data in Table 4. To find a model for the original data, we take exponentials of each side.

$$e^{\ln T} = e^{1.5 \ln d + 0.0004} \qquad \text{Take exponentials of each side}$$

$$T = e^{1.5 \ln d} e^{0.0004} \qquad \text{Law of Exponents}$$

$$= e^{\ln d^{1.5}} e^{0.0004} \qquad \text{Property of logarithms}$$

$$= 1.0004 d^{1.5} \qquad \text{Property of logarithms}$$

Since $1.0004 \approx 1$, the power model for this data is

$$T = d^{1.5}$$

This is the remarkable relationship discovered by Johannes Kepler (1571–1630). We can now use this model to calculate the period of an asteroid whose mean distance is five times as far from the sun as that of the earth. In this case, $d = 5$ AU, and so the period is $T = 5^{1.5} \approx 11.18$ years.

Graphing calculators are capable of applying the method of least squares *directly* to obtain a power model of data. One such calculator applies power regression to the data in Table 2 and gives the model

$$y = ax^n$$

where $\qquad a = 1.0039 \qquad$ and $\qquad n = 1.49966$

Compare this with the model obtained above.

POLYNOMIAL FUNCTIONS AS MODELS

Polynomial functions can also be used to model data. Most graphing calculators use the method of least squares to fit a polynomial of specified degree to the given data. Let's try, for instance, to fit a cubic polynomial to the world population data in Table 1. A graphing calculator using the method of least squares gives the cubic model

$$P = at^3 + bt^2 + ct + d$$

where $\qquad a = 2325.67 \qquad\qquad b = -1.306488 \times 10^7$

$\qquad\qquad c = 2.44631 \times 10^{10} \qquad d = -1.52658 \times 10^{13}$

In Figure 5 we graph this cubic function and the data points of Table 1.

We see from the graph that the cubic function models the world population of the 20th century very well. Perhaps surprisingly, it's quite a bit better than the exponential model and would be useful for estimating the world population in

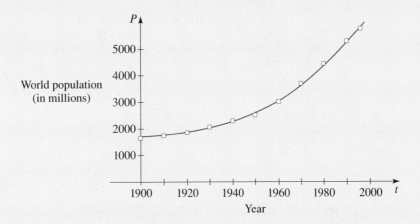

FIGURE 5
Cubic model for world population

1925 or 1985, for instance. For the purpose of predicting the population in 2050 or 2100, however, the cubic model would probably not be nearly as accurate, because we know that the physical mechanism of population growth is inherently exponential.

 PROBLEMS

1. The U.S. Constitution requires a census every 10 years. The census data for 1790–1990 is given in the table.

Year	Population (in millions)	Year	Population (in millions)	Year	Population (in millions)
1790	3.9	1860	31.4	1930	123.2
1800	5.3	1870	38.6	1940	132.2
1810	7.2	1880	50.2	1950	151.3
1820	9.6	1890	63.0	1960	179.3
1830	12.9	1900	76.2	1970	203.3
1840	17.1	1910	92.2	1980	226.5
1850	23.2	1920	106.0	1990	248.7

(a) Make a scatter plot of the data.
(b) Use a calculator to find an exponential model for the data.
(c) Use your model to predict the population at the next census (in the year 2000).
(d) Use your model to estimate the population in 1965.
(e) Compare your answers from parts (c) and (d) to the values in the table. Do you think an exponential model is appropriate for these data?

2. In a physics experiment a lead ball is dropped from a height of 5 m. The students record the distance the ball has fallen every one-tenth of a second. (This can be done using a camera and a strobe light.)

Time (s)	Distance (m)
0.1	0.048
0.2	0.197
0.3	0.441
0.4	0.882
0.5	1.227
0.6	1.765
0.7	2.401
0.8	3.136
0.9	3.969
1.0	4.902

(a) Make a scatter plot of the data.
(b) Use a calculator to find a power model.
(c) Use the model you found to predict how far a dropped ball would fall in 3 s.

3. The U.S. health-care expenditures for 1960–1993 are given in the table, and a scatter plot of the data is shown in the figure.

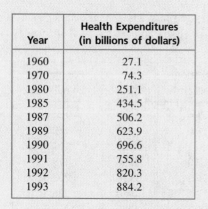

Year	Health Expenditures (in billions of dollars)
1960	27.1
1970	74.3
1980	251.1
1985	434.5
1987	506.2
1989	623.9
1990	696.6
1991	755.8
1992	820.3
1993	884.2

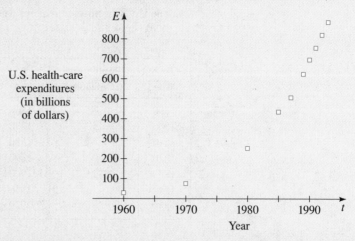

(a) Does the scatter plot shown suggest an exponential model?
(b) Make a table of the values $(t, \ln E)$ and a scatter plot. Does the scatter plot appear to be linear?
(c) Find the regression line for the data in part (b).
(d) Use the results of part (b) to find an exponential model for the growth of health-care expenditures.

(e) Use the model you found in part (d) to predict the total health-care expenditures in 1996.

4. A student is trying to determine the half-life of radioactive iodine-131. He measures the amount of iodine-131 in a sample solution every 8 hours. His data are shown in the table.

Time (h)	Amount (g)
0	4.80
8	4.66
16	4.51
24	4.39
32	4.29
40	4.14
48	4.04

(a) Make a scatter plot of the data.
(b) Use a calculator to find an exponential model.
(c) Use your model to find the half-life of iodine-131.

5. The table gives U.S. lead emissions into the environment in millions of metric tons for 1970–1992.
(a) Find an exponential model for these data.
(b) Find a fourth-degree polynomial model for these data.
(c) Which of these curves gives a better model for the data? Use graphs of the two models to decide.
(d) Use each model to estimate the lead emissions in 1972 and 1982.

Year	Lead Emissions
1970	199.1
1975	143.8
1980	68.0
1985	18.3
1988	5.9
1989	5.5
1990	5.1
1991	4.5
1992	4.7

6. A study of the U.S. Office of Science and Technology in 1972 estimated the cost of reducing automobile emissions by certain percentages.

Reduction in emissions (%)	Cost per car ($)
50	45
55	55
60	62
65	70
70	80
75	90
80	100
85	200
90	375
95	600

Find a model that captures the "diminishing returns" trend of these data.

5

TRIGONOMETRIC FUNCTIONS OF REAL NUMBERS

Trigonometric functions are used to describe periodic phenomena such as the daily ebb and flow of tides.

Trigonometry contains the science of continually undulating magnitude...

AUGUSTUS DE MORGAN

Trigonometry is one of the most versatile branches of mathematics. Ever since its invention in the ancient world, it has been important in both theoretical and practical applications. In modern times it has found applications in such diverse fields as signal processing in the telephone industry, coding of music on compact disc players, finding distances to stars, designing guidance systems in the space shuttle, producing CAT scans for medical use, and many others. It's an indispensable tool for electrical engineers, physicists, computer scientists, and for practically all of the sciences. In Chapters 5, 6, and 7 we will see how some of these applications are possible.

The power and versatility of trigonometry stem from the fact that it can be viewed in two distinct ways. One of these sees trigonometry as the study of *functions of real numbers,* the other as the study of *functions of angles.* The trigonometric functions defined in these two different ways are identical—they assign the same value to a given real number (in the second case, the real number is the measure of an angle). The difference is one of point of view. This difference is most clearly apparent when we consider the applications of trigonometry. One view lends itself to applications involving dynamic processes such as harmonic motion, the study of sound waves, and the description of other periodic phenomena, whereas the other view lends itself to static applications such as distance measurement, force, velocity, and, in general, applications involving measurements of length and direction. To fully appreciate its uses, we must study both approaches to trigonometry. In this chapter we study trigonometric functions of real numbers, sketch their graphs, and describe their applications to harmonic motion. The angle approach is treated in Chapter 6.

The two approaches to trigonometry are independent of each other, so either Chapter 5 or Chapter 6 may be studied first.

5.1 THE UNIT CIRCLE

In this section we explore some properties of the circle of radius 1 centered at the origin. These properties are used in the next section to define the trigonometric functions.

THE UNIT CIRCLE

The set of points at a distance 1 from the origin is a circle of radius 1 (see Figure 1). In Section 1.8 we learned that the equation of this circle is $x^2 + y^2 = 1$.

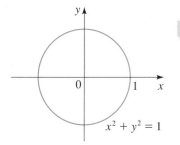

$x^2 + y^2 = 1$

FIGURE 1
The unit circle

THE UNIT CIRCLE
The **unit circle** is the circle of radius 1 centered at the origin in the xy-plane. Its equation is $$x^2 + y^2 = 1$$

EXAMPLE 1 ■ A Point on the Unit Circle

Show that the point $P\left(\dfrac{\sqrt{3}}{3}, \dfrac{\sqrt{2}}{\sqrt{3}}\right)$ is on the unit circle.

SOLUTION   We need to show that this point satisfies the equation of the unit circle, that is, $x^2 + y^2 = 1$. Since

$$\left(\frac{\sqrt{3}}{3}\right)^2 + \left(\frac{\sqrt{2}}{\sqrt{3}}\right)^2 = \frac{3}{9} + \frac{2}{3} = \frac{1}{3} + \frac{2}{3} = 1$$

P is on the unit circle.   ■

EXAMPLE 2 ■ Locating a Point on the Unit Circle

The point $P(\sqrt{3}/2, y)$ is on the unit circle in quadrant IV. Find its y-coordinate.

SOLUTION   Since the point is on the unit circle, we have

$$\left(\frac{\sqrt{3}}{2}\right)^2 + y^2 = 1$$

$$y^2 = 1 - \frac{3}{4} = \frac{1}{4}$$

$$y = \pm\frac{1}{2}$$

Since the point is in quadrant IV, its y-coordinate must be negative, so $y = -\frac{1}{2}$.   ■

TERMINAL POINTS ON THE UNIT CIRCLE

Suppose t is a real number. Let's mark off a distance t along the unit circle, starting at the point $(1, 0)$ and moving in a counterclockwise direction if t is positive or in a clockwise direction if t is negative (Figure 2). In this way we arrive at a

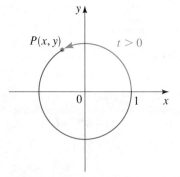

(a) Terminal point $P(x, y)$ determined
    by $t > 0$

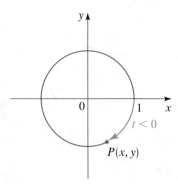

(b) Terminal point $P(x, y)$ determined
    by $t < 0$

FIGURE 2

point $P(x, y)$ on the unit circle. The point $P(x, y)$ obtained in this way is called the **terminal point** determined by the real number t.

The circumference of the unit circle is $C = 2\pi(1) = 2\pi$. So, if a point starts at $(1, 0)$ and moves counterclockwise all the way around the unit circle and returns to $(1, 0)$, it travels a distance of 2π. To move halfway around the circle, it travels a distance of $\frac{1}{2}(2\pi) = \pi$. To move a quarter of the distance around the circle, it travels a distance of $\frac{1}{4}(2\pi) = \pi/2$. Where does the point end up when it travels these distances along the circle? From Figure 3 we see, for example, that when it travels a distance of π starting at $(1, 0)$, its terminal point is $(-1, 0)$.

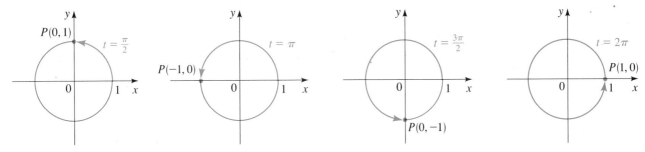

FIGURE 3
Terminal points determined by $t = \frac{\pi}{2}, \pi, \frac{3\pi}{2}$, and 2π

EXAMPLE 3 ■ Finding Terminal Points

Find the terminal point on the unit circle determined by each of the following real numbers t.

(a) $t = 3\pi$ (b) $t = -\pi$ (c) $t = -\dfrac{\pi}{2}$

SOLUTION   From Figure 4 we get the following.

(a) The terminal point determined by 3π is $(-1, 0)$.

(b) The terminal point determined by $-\pi$ is $(-1, 0)$.

(c) The terminal point determined by $-\pi/2$ is $(0, -1)$.

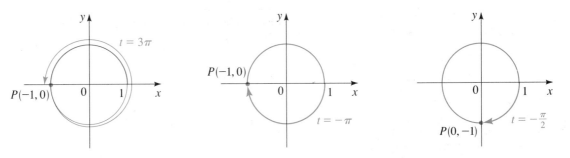

FIGURE 4

Notice that different values of t can determine the same terminal point. ■

The terminal point $P(x, y)$ determined by $t = \pi/4$ is the same distance from $(1,0)$ as from $(0,1)$ along the unit circle (see Figure 5).

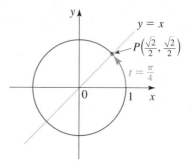

FIGURE 5

Since the unit circle is symmetric with respect to the line $y = x$, it follows that P lies on the line $y = x$. So P is the point of intersection (in the first quadrant) of the circle $x^2 + y^2 = 1$ and the line $y = x$. Substituting x for y in the equation of the circle, we get

$$x^2 + x^2 = 1$$

$$2x^2 = 1 \qquad \text{Combine like terms}$$

$$x^2 = \frac{1}{2} \qquad \text{Divide by 2}$$

$$x = \pm\frac{1}{\sqrt{2}} \qquad \text{Take the square roots}$$

Since P is in the first quadrant, $x = 1/\sqrt{2}$ and since $y = x$, we have $y = 1/\sqrt{2}$ also. Thus, the terminal point determined by $\pi/4$ is

$$P\left(\frac{1}{\sqrt{2}}, \frac{1}{\sqrt{2}}\right) = P\left(\frac{\sqrt{2}}{2}, \frac{\sqrt{2}}{2}\right)$$

Similar methods can be used to find the terminal points determined by $t = \pi/6$ and $t = \pi/3$ (see Exercises 49 and 50). Table 1 and Figure 6 give the terminal points for some special values of t.

Table 1

t	Terminal point determined by t
0	$(1, 0)$
$\dfrac{\pi}{6}$	$\left(\dfrac{\sqrt{3}}{2}, \dfrac{1}{2}\right)$
$\dfrac{\pi}{4}$	$\left(\dfrac{\sqrt{2}}{2}, \dfrac{\sqrt{2}}{2}\right)$
$\dfrac{\pi}{3}$	$\left(\dfrac{1}{2}, \dfrac{\sqrt{3}}{2}\right)$
$\dfrac{\pi}{2}$	$(0, 1)$

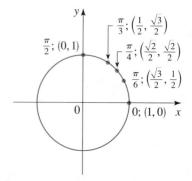

FIGURE 6

EXAMPLE 4 ■ Finding Terminal Points

Find the terminal point determined by each of the following values of t.

(a) $t = -\dfrac{\pi}{4}$ 　　　 (b) $t = \dfrac{3\pi}{4}$ 　　　 (c) $t = -\dfrac{5\pi}{6}$

SOLUTION

(a) Let P be the terminal point determined by $-\pi/4$, and let Q be the terminal point determined by $\pi/4$. From Figure 7(a) we see that the point P has the same coordinates as Q except for sign. Since P is in quadrant IV, its x-coordinate is positive and its y-coordinate is negative. Thus, the terminal point is $P(\sqrt{2}/2, -\sqrt{2}/2)$.

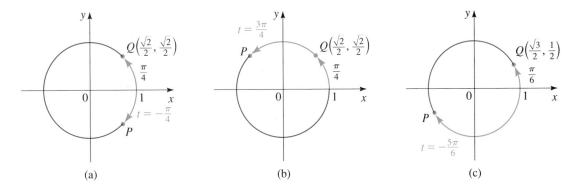

FIGURE 7 　　　(a)　　　　　　　　　　(b)　　　　　　　　　　(c)

(b) Let P be the terminal point determined by $3\pi/4$, and let Q be the terminal point determined by $\pi/4$. From Figure 7(b) we see that the point P has the same coordinates as Q except for sign. Since P is in quadrant II, its x-coordinate is negative and its y-coordinate is positive. Thus, the terminal point is $P(-\sqrt{2}/2, \sqrt{2}/2)$.

(c) Let P be the terminal point determined by $-5\pi/6$, and let Q be the terminal point determined by $\pi/6$. From Figure 7(c) we see that the point P has the same coordinates as Q except for sign. Since P is in quadrant III, its coordinates are both negative. Thus, the terminal point is $P(-\sqrt{3}/2, -\frac{1}{2})$. ■

THE REFERENCE NUMBER

From Examples 3 and 4, we see that to find a terminal point in any quadrant we need only know the "corresponding" terminal point in the first quadrant. We give a procedure for finding such terminal points using the idea of the *reference number*.

REFERENCE NUMBER

Let t be a real number. The **reference number** $\bar{t}$ associated with t is the shortest distance along the unit circle between the terminal point determined by t and the x-axis.

Figure 8 shows that to find the reference number $\bar{t}$ it's helpful to know the quadrant in which the terminal point determined by t lies.

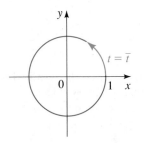
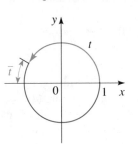
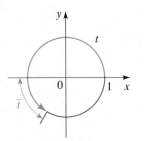
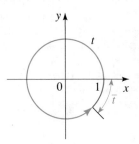

FIGURE 8
The reference number $\bar{t}$ for t

EXAMPLE 5 ■ Finding Reference Numbers

Find the reference number determined by each of the following real numbers t.

(a) $t = \dfrac{5\pi}{6}$ (b) $t = \dfrac{7\pi}{4}$ (c) $t = -\dfrac{2\pi}{3}$ (d) $t = 5.80$

SOLUTION   From Figure 9 we find the reference numbers as follows.

(a) $\bar{t} = \pi - \dfrac{5\pi}{6} = \dfrac{\pi}{6}$

(b) $\bar{t} = 2\pi - \dfrac{7\pi}{4} = \dfrac{\pi}{4}$

(c) $\bar{t} = \pi - \dfrac{2\pi}{3} = \dfrac{\pi}{3}$

(d) $\bar{t} = 2\pi - 5.80 \approx 0.48$

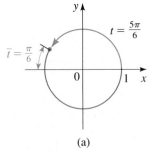
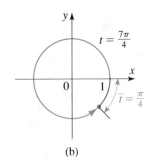
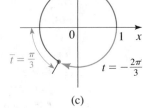
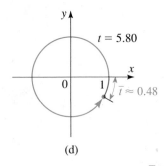

(a)   (b)   (c)   (d)

FIGURE 9   ■

USING REFERENCE NUMBERS TO FIND TERMINAL POINTS

To find the terminal point P determined by any value of t, we use the following steps:

1. Find the reference number $\bar{t}$.

2. Find the terminal point $Q(a, b)$ determined by $\bar{t}$.

3. The terminal point determined by t is $P(\pm a, \pm b)$, where the signs are chosen according to the quadrant in which this terminal point lies.

EXAMPLE 6 ■ Using Reference Numbers to Find Terminal Points

Find the terminal point determined by each of the following values of t.

(a) $t = \dfrac{5\pi}{6}$ (b) $t = \dfrac{7\pi}{4}$ (c) $t = -\dfrac{2\pi}{3}$

SOLUTION   The reference numbers associated with these values of t were found in Example 5.

(a) The reference number is $\bar{t} = \pi/6$, which determines the terminal point $\left(\sqrt{3}/2, \tfrac{1}{2}\right)$ from Table 1. Since the terminal point determined by t is in quadrant II, its x-coordinate is negative and its y-coordinate is positive. Thus, the desired terminal point is

$$\left(-\frac{\sqrt{3}}{2}, \frac{1}{2}\right)$$

(b) The reference number is $\bar{t} = \pi/4$, which determines the terminal point $\left(\sqrt{2}/2, \sqrt{2}/2\right)$ from Table 1. Since the terminal point is in quadrant IV, its x-coordinate is positive and its y-coordinate is negative. Thus, the desired terminal point is

$$\left(\frac{\sqrt{2}}{2}, -\frac{\sqrt{2}}{2}\right)$$

(c) The reference number is $\bar{t} = \pi/3$, which determines the terminal point $\left(\tfrac{1}{2}, \sqrt{3}/2\right)$ from Table 1. Since the terminal point determined by t is in quadrant III, its coordinates are both negative. Thus, the desired terminal point is

$$\left(-\frac{1}{2}, -\frac{\sqrt{3}}{2}\right)$$

■

Since the circumference of the unit circle is 2π, the terminal point determined by t is the same as that determined by $t + 2\pi$ or $t - 2\pi$. In general, we can add or subtract 2π any number of times without changing the terminal point determined by t. We use this observation in the next example to find terminal points for large t.

EXAMPLE 7 ■ Finding the Terminal Point for Large t

Find the terminal point determined by $t = \dfrac{29\pi}{6}$.

SOLUTION   Since

$$t = \frac{29\pi}{6} = 4\pi + \frac{5\pi}{6}$$

we see that the terminal point of t is the same as that of $5\pi/6$ (that is, we subtract 4π). So by Example 6(a) the terminal point is $\left(-\sqrt{3}/2, \tfrac{1}{2}\right)$.

■

5.1 EXERCISES

1–4 ■ Show that the point is on the unit circle.

1. $\left(\frac{3}{5}, \frac{4}{5}\right)$

2. $\left(\frac{40}{41}, \frac{9}{41}\right)$

3. $\left(-\frac{1}{3}, 2\sqrt{2}/3\right)$

4. $\left(-\frac{5}{13}, -\frac{12}{13}\right)$

5–10 ■ The point P is on the unit circle. Find $P(x, y)$ from the given information.

5. The x-coordinate of P is $\frac{3}{5}$ and P is in quadrant I.

6. The y-coordinate of P is $-\frac{1}{3}$ and P is in quadrant III.

7. The x-coordinate of P is $\frac{2}{3}$ and the y-coordinate is negative.

8. The x-coordinate of P is positive and the y-coordinate of P is $-\sqrt{5}/5$.

9. The x-coordinate of P is $\sqrt{2}/3$ and P is in quadrant IV.

10. The x-coordinate of P is $-\frac{2}{5}$ and P is in quadrant II.

11–16 ■ Sketch on the unit circle the approximate location of the terminal point of the arc whose length is the given real number t [as always, the initial point of the arc is $(1, 0)$]. Determine the quadrant in which this terminal point lies.

11. $t = \dfrac{\pi}{8}$

12. $t = \dfrac{3\pi}{7}$

13. $t = -\dfrac{11\pi}{9}$

14. $t = 4$

15. $t = -1.7$

16. $t = 7$

17–26 ■ Find the terminal point $P(x, y)$ on the unit circle determined by the given value of t.

17. $t = \dfrac{\pi}{4}$

18. $t = \dfrac{\pi}{2}$

19. $t = -\dfrac{\pi}{3}$

20. $t = \dfrac{7\pi}{6}$

21. $t = \pi$

22. $t = -\dfrac{5\pi}{6}$

23. $t = \dfrac{2\pi}{3}$

24. $t = -\dfrac{\pi}{2}$

25. $t = -\dfrac{3\pi}{4}$

26. $t = \dfrac{11\pi}{6}$

27. Suppose that the terminal point determined by t is the point $\left(\frac{3}{5}, \frac{4}{5}\right)$ on the unit circle. Find the terminal point determined by each of the following.
(a) $\pi - t$ (b) $-t$ (c) $\pi + t$ (d) $2\pi + t$

28. Suppose that the terminal point determined by t is the point $\left(\frac{1}{3}, 2\sqrt{2}/3\right)$ on the unit circle. Find the terminal point determined by each of the following.
(a) $-t$ (b) $4\pi + t$ (c) $\pi - t$ (d) $t - \pi$

29–32 ■ Find the reference number for each value of t.

29. (a) $t = \dfrac{5\pi}{4}$

(b) $t = \dfrac{7\pi}{3}$

(c) $t = -\dfrac{4\pi}{3}$

(d) $t = \dfrac{\pi}{6}$

30. (a) $t = \dfrac{5\pi}{7}$

(b) $t = -\dfrac{9\pi}{8}$

(c) $t = 3.55$

(d) $t = -2.9$

31. (a) $t = -\dfrac{11\pi}{5}$

(b) $t = \dfrac{13\pi}{6}$

(c) $t = \dfrac{7\pi}{3}$

(d) $t = -\dfrac{5\pi}{6}$

32. (a) $t = 3$

(b) $t = 6$

(c) $t = -3$

(d) $t = -6$

33–44 ■ Find (a) the reference number for each value of t, and (b) the terminal point determined by t.

33. $t = \dfrac{3\pi}{4}$

34. $t = \dfrac{7\pi}{3}$

35. $t = -\dfrac{2\pi}{3}$

36. $t = -\dfrac{7\pi}{6}$

37. $t = \dfrac{13\pi}{4}$

38. $t = \dfrac{13\pi}{6}$

39. $t = \dfrac{7\pi}{6}$

40. $t = \dfrac{17\pi}{4}$

41. $t = -\dfrac{11\pi}{3}$

42. $t = \dfrac{31\pi}{6}$

43. $t = \dfrac{16\pi}{3}$

44. $t = -\dfrac{41\pi}{4}$

45–48 ■ Use the figure to find the terminal point determined by the real number t, with coordinates correct to one decimal place.

45. $t = 1$ **46.** $t = 2.5$

47. $t = -1.1$ **48.** $t = 4.2$

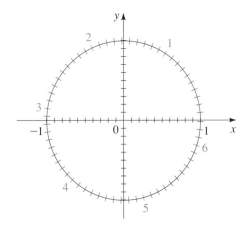

DISCOVERY · DISCUSSION

49. Finding the Terminal Point for $\pi/6$ Suppose the terminal point determined by $t = \pi/6$ is $P(x, y)$ and that the points Q and R are as shown in the figure. Why are the distances PQ and PR the same? Use this fact, together with the Distance Formula, to show that the coordinates of P satisfy the equation

$2y = \sqrt{x^2 + (y - 1)^2}$. Simplify this equation using the fact that $x^2 + y^2 = 1$. Solve the simplified equation to find $P(x, y)$.

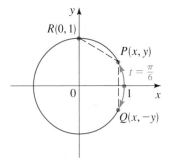

50. Finding the Terminal Point for $\pi/3$ Now that you know the terminal point determined by $t = \pi/6$, use symmetry to find the terminal point determined by $t = \pi/3$ (see the figure). Explain your reasoning.

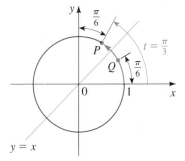

5.2 TRIGONOMETRIC FUNCTIONS OF REAL NUMBERS

A function is a rule that assigns to each real number another real number. In this section we use properties of the unit circle from the preceding section to define certain functions of real numbers, the trigonometric functions.

THE TRIGONOMETRIC FUNCTIONS

Recall that to find the terminal point $P(x, y)$ for a given real number t, we move a distance t along the unit circle, starting at the point $(1, 0)$. We move in a counterclockwise direction if t is positive and in a clockwise direction if t is negative

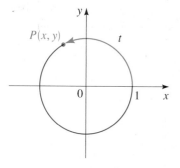

FIGURE 1

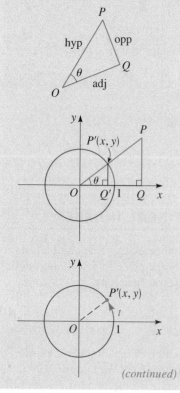

Relationship to Trigonometry of Right Triangles

If you have previously studied trigonometry of right triangles, you are probably wondering how the sine and cosine of an *angle* relate to those of this section. The following figures show how.

(continued)

(see Figure 1). We now use the *x*- and *y*-coordinates of the point $P(x, y)$ to define several functions. For instance, we define the function called *sine* by assigning to each real number *t* the *y*-coordinate of the terminal point $P(x, y)$ determined by *t*. The functions *cosine, tangent, cosecant, secant,* and *cotangent* are also defined using the coordinates of $P(x, y)$.

DEFINITION OF THE TRIGONOMETRIC FUNCTIONS

Let *t* be any real number and let $P(x, y)$ be the terminal point on the unit circle determined by *t*. We define

$$\sin t = y \qquad \cos t = x \qquad \tan t = \frac{y}{x} \quad (x \neq 0)$$

$$\csc t = \frac{1}{y} \quad (y \neq 0) \qquad \sec t = \frac{1}{x} \quad (x \neq 0) \qquad \cot t = \frac{x}{y} \quad (y \neq 0)$$

Because the trigonometric functions can be defined in terms of the unit circle, they are sometimes called the **circular functions**.

EXAMPLE 1 ■ Evaluating Trigonometric Functions

Find the six trigonometric functions of each of the following values of *t*.

(a) $t = \dfrac{\pi}{3}$ (b) $t = \dfrac{\pi}{2}$

SOLUTION

(a) The terminal point determined by $t = \pi/3$ is $P\left(\frac{1}{2}, \sqrt{3}/2\right)$. Since the coordinates are $x = \frac{1}{2}$ and $y = \sqrt{3}/2$, we have

$$\sin \frac{\pi}{3} = \frac{\sqrt{3}}{2} \qquad \cos \frac{\pi}{3} = \frac{1}{2} \qquad \tan \frac{\pi}{3} = \frac{\sqrt{3}/2}{1/2} = \sqrt{3}$$

$$\csc \frac{\pi}{3} = \frac{2\sqrt{3}}{3} \qquad \sec \frac{\pi}{3} = 2 \qquad \cot \frac{\pi}{4} = \frac{1/2}{\sqrt{3}/2} = \frac{\sqrt{3}}{3}$$

(b) The terminal point determined by $\pi/2$ is $P(0, 1)$. So

$$\sin \frac{\pi}{2} = 1 \qquad \cos \frac{\pi}{2} = 0 \qquad \csc \frac{\pi}{2} = \frac{1}{1} = 1 \qquad \cot \frac{\pi}{2} = \frac{0}{1} = 0$$

But $\tan \pi/2$ and $\sec \pi/2$ are undefined because $x = 0$ appears in the denominator in each of their definitions.   ■

Some special values of the trigonometric functions are listed in Table 1. This table is easily obtained from Table 1 of Section 5.1, together with the definitions of the trigonometric functions.

By the triangle definition

$$\sin\theta = \frac{\text{opp}}{\text{hyp}} = \frac{PQ}{OP} = \frac{P'Q'}{OP'}$$

$$= \frac{y}{1} = y$$

$$\cos\theta = \frac{\text{adj}}{\text{hyp}} = \frac{OQ}{OP} = \frac{OQ'}{OP'}$$

$$= \frac{x}{1} = x$$

If θ is measured in radians, then $\theta = t$ and we can write

$$\sin t = y$$

$$\cos t = x$$

These are precisely the definitions of this section, so the two definitions give identical values.

Why then study trigonometry in two different ways? Because different applications require that we view the trigonometric functions differently (see *Focus on Modeling,* page 393, and Section 6.2).

Table 1 Special values of the trigonometric functions

t	$\sin t$	$\cos t$	$\tan t$	$\csc t$	$\sec t$	$\cot t$
0	0	1	0	—	1	—
$\dfrac{\pi}{6}$	$\dfrac{1}{2}$	$\dfrac{\sqrt{3}}{2}$	$\dfrac{\sqrt{3}}{3}$	2	$\dfrac{2\sqrt{3}}{3}$	$\sqrt{3}$
$\dfrac{\pi}{4}$	$\dfrac{\sqrt{2}}{2}$	$\dfrac{\sqrt{2}}{2}$	1	$\sqrt{2}$	$\sqrt{2}$	1
$\dfrac{\pi}{3}$	$\dfrac{\sqrt{3}}{2}$	$\dfrac{1}{2}$	$\sqrt{3}$	$\dfrac{2\sqrt{3}}{3}$	2	$\dfrac{\sqrt{3}}{3}$
$\dfrac{\pi}{2}$	1	0	—	1	—	0

Example 1 shows that some of the trigonometric functions fail to be defined for certain real numbers. So we need to determine their domains. The functions sine and cosine are defined for all values of t. Since the functions cotangent and cosecant have y in the denominator of their definitions, they are not defined whenever the y-coordinate of the terminal point $P(x, y)$ determined by t is 0. This happens when $t = n\pi$ for any integer n, so their domains do not include these points. The functions tangent and secant have x in the denominator in their definitions, so they are not defined whenever $x = 0$. This happens when $t = (\pi/2) + n\pi$ for any integer n.

DOMAINS OF THE TRIGONOMETRIC FUNCTIONS

Function	Domain
sin, cos	All real numbers
tan, sec	All real numbers other than $\dfrac{\pi}{2} + n\pi$ for any integer n
cot, csc	All real numbers other than $n\pi$ for any integer n

---

VALUES OF THE TRIGONOMETRIC FUNCTIONS

To compute other values of the trigonometric functions, we first determine their signs. The signs of the trigonometric functions depend on the quadrant in which the terminal point of t lies. For example, if the terminal point $P(x, y)$ determined by t lies in quadrant III, then its coordinates are both negative. So $\sin t$, $\cos t$, $\csc t$, and $\sec t$ are all negative, while $\tan t$ and $\cot t$ are positive. You can check the other entries in the following box.

The following mnemonic device can be used to remember which trigonometric functions are positive in each quadrant: **A**ll of them, **S**ine, **T**angent, or **C**osine.

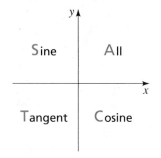

You can remember this as "All Students Take Calculus."

SIGNS OF THE TRIGONOMETRIC FUNCTIONS		
Quadrant	Positive functions	Negative functions
I	all	none
II	sin, csc	cos, sec, tan, cot
III	tan, cot	sin, csc, cos, sec
IV	cos, sec	sin, csc, tan, cot

EXAMPLE 2 ■ Determining the Sign of a Trigonometric Function

(a) $\cos \dfrac{\pi}{3} > 0$, since the terminal point of $t = \dfrac{\pi}{3}$ is in quadrant I.

(b) $\tan 4 > 0$, since the terminal point of $t = 4$ is in quadrant III.

(c) If $\cos t < 0$ and $\sin t > 0$, then t must be in quadrant II.   ■

In Section 5.1 we used the reference number to find the terminal point determined by a real number t. Since the trigonometric functions are defined in terms of the coordinates of terminal points, we can use the reference number to find values of the trigonometric functions. Suppose that $\bar{t}$ is the reference number for t. Then the terminal point of $\bar{t}$ has the same coordinates, except possibly for sign, as the terminal point of t. So the values of the trigonometric functions at t are the same, except possibly for sign, as their values at $\bar{t}$. We illustrate the procedure in the next example.

EXAMPLE 3 ■ Evaluating Trigonometric Functions

Find each of the following values.

(a) $\cos \dfrac{2\pi}{3}$ (b) $\tan\left(-\dfrac{\pi}{3}\right)$ (c) $\sin \dfrac{19\pi}{4}$

SOLUTION

(a) The reference number for $2\pi/3$ is $\pi/3$. Since $2\pi/3$ is in quadrant II, $\cos(2\pi/3)$ is negative. Thus

$$\cos \frac{2\pi}{3} = -\cos \frac{\pi}{3} = -\frac{1}{2}$$

$$\underset{\substack{\uparrow \\ \text{sign}}}{} \quad \underset{\substack{\uparrow \\ \text{reference} \\ \text{number}}}{} \quad \underset{\substack{\uparrow \\ \text{from} \\ \text{Table 1}}}{}$$

(b) The reference number for $-\pi/3$ is $\pi/3$. Since $-\pi/3$ is in quadrant IV, $\tan(-\pi/3)$ is negative. Thus

$$\tan\left(-\frac{\pi}{3}\right) = -\tan \frac{\pi}{3} = -\sqrt{3}$$

$$\underset{\substack{\uparrow \\ \text{sign}}}{} \quad \underset{\substack{\uparrow \\ \text{reference} \\ \text{number}}}{} \quad \underset{\substack{\uparrow \\ \text{from} \\ \text{Table 1}}}{}$$

(c) Since $(19\pi/4) - 4\pi = 3\pi/4$, the terminal points determined by $19\pi/4$ and $3\pi/4$ are the same. The reference number for $3\pi/4$ is $\pi/4$. Since $3\pi/4$ is in quadrant II, $\sin(3\pi/4)$ is positive. Thus

$$\sin \frac{19\pi}{4} = \sin \frac{3\pi}{4} = +\sin \frac{\pi}{4} = \frac{\sqrt{2}}{2}$$

$$\uparrow \qquad\qquad \uparrow \quad \uparrow \quad \uparrow$$
$$\text{subtract } 4\pi \qquad \text{sign reference from}$$
$$\text{number Table 1}$$

So far we have been able to compute the values of the trigonometric functions only for certain values of t. In fact, we can compute the values of the trigonometric functions whenever t is a multiple of $\pi/6$, $\pi/4$, $\pi/3$, and $\pi/2$. How can we compute the trigonometric functions for other values of t? For example, how can we find $\sin 1.5$? One way is to carefully sketch a diagram and read the value (see Exercises 33–40); however, this method is not very accurate. Fortunately, programmed directly into scientific calculators are mathematical procedures (called *numerical methods*) that find the values of *sine, cosine,* and *tangent* correct to the number of digits in the display. The calculator must be put in *radian mode* to evaluate these functions. To find values of cosecant, secant, and cotangent using a calculator, we need to use the following *reciprocal relations:*

$$\csc t = \frac{1}{\sin t} \qquad \sec t = \frac{1}{\cos t} \qquad \cot t = \frac{1}{\tan t}$$

These identities follow from the definitions of the trigonometric functions. For instance, since $\sin t = y$ and $\csc t = 1/y$, we have $\csc t = 1/y = 1/(\sin t)$. The others follow similarly.

EXAMPLE 4 ■ Using a Calculator to Evaluate Trigonometric Functions

Making sure our calculator is set to radian mode and rounding the results to six decimal places, we get

(a) $\sin 2.2 \approx 0.808496$

(b) $\cos 1.1 \approx 0.453596$

(c) $\cot 28 = \dfrac{1}{\tan 28} \approx -3.553286$

(d) $\csc 0.98 = \dfrac{1}{\sin 0.98} \approx 1.204098$

Numerical methods are ways of finding approximations for the values of functions. Since the values of polynomials are easy to compute, it's helpful to express functions that we wish to approximate in terms of polynomials. For example, using calculus it can be shown that

$$\sin t = t - \frac{t^3}{3!} + \frac{t^5}{5!} - \frac{t^7}{7!} + \cdots$$

$$\cos t = 1 - \frac{t^2}{2!} + \frac{t^4}{4!} - \frac{t^6}{6!} + \cdots$$

where $n! = 1 \cdot 2 \cdot 3 \cdots n$ (for example, $4! = 1 \cdot 2 \cdot 3 \cdot 4 = 24$). These remarkable formulas were proved by the British mathematician Brook Taylor (1685–1731). The more terms we use from the series on the right-hand side of these formulas, the more accurate the values obtained for $\sin t$ and $\cos t$. For instance, if we use the first three terms of Taylor's series to find $\sin 0.9$, we get

$$\sin 0.9 \approx 0.9 - \frac{(0.9)^3}{3!} + \frac{(0.9)^5}{5!}$$

$$\approx 0.78342075$$

(Compare this with the value obtained using your calculator.) Formulas like these are programmed into your calculator, which uses them to compute functions like $\sin t$, $\cos t$, e^t, and $\log t$. The calculator uses enough terms to ensure that all displayed digits are accurate.

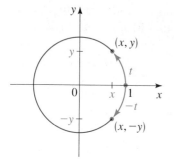

FIGURE 2

Let's consider the relationship between the trigonometric functions of t and those of $-t$. From Figure 2 we see that

$$\sin(-t) = -y = -\sin t$$

$$\cos(-t) = x = \cos t$$

$$\tan(-t) = \frac{-y}{x} = -\frac{y}{x} = -\tan t$$

These equations show that sine and tangent are odd functions, whereas cosine is an even function. It's easy to see that the reciprocal of an even function is even and the reciprocal of an odd function is odd (see Section 2.4). This fact, together with the reciprocal relations, completes our knowledge of the even-odd properties for all the trigonometric functions.

EVEN-ODD PROPERTIES

Sine, cosecant, tangent, and cotangent are odd functions; cosine and secant are even functions.

$$\sin(-t) = -\sin t \qquad \cos(-t) = \cos t \qquad \tan(-t) = -\tan t$$

$$\csc(-t) = -\csc t \qquad \sec(-t) = \sec t \qquad \cot(-t) = -\cot t$$

EXAMPLE 5 ■ Even and Odd Trigonometric Functions

Use the even-odd properties of the trigonometric functions to determine each of the following values.

(a) $\sin\left(-\dfrac{\pi}{6}\right)$
 (b) $\cos\left(-\dfrac{\pi}{4}\right)$

SOLUTION   By the even-odd properties and Table 1, we have

(a) $\sin\left(-\dfrac{\pi}{6}\right) = -\sin\dfrac{\pi}{6} = -\dfrac{1}{2}$

(b) $\cos\left(-\dfrac{\pi}{4}\right) = \cos\dfrac{\pi}{4} = \dfrac{\sqrt{2}}{2}$ ·
 ■

FUNDAMENTAL IDENTITIES

The trigonometric functions are related to each other through equations called **trigonometric identities**. We give the most important ones in the following box.*

---

*We follow the usual convention of writing $\sin^2 t$ for $(\sin t)^2$. In general, we write $\sin^n t$ for $(\sin t)^n$ for all integers n except $n = -1$. The exponent $n = -1$ will be assigned another meaning in Section 7.4. Of course, the same convention applies to the other five trigonometric functions.

> **FUNDAMENTAL IDENTITIES**
>
> Reciprocal Identities
>
> $$\csc t = \frac{1}{\sin t} \qquad \sec t = \frac{1}{\cos t} \qquad \cot t = \frac{1}{\tan t}$$
>
> $$\tan t = \frac{\sin t}{\cos t} \qquad \cot t = \frac{\cos t}{\sin t}$$
>
> Pythagorean Identities
>
> $$\sin^2 t + \cos^2 t = 1 \qquad \tan^2 t + 1 = \sec^2 t \qquad 1 + \cot^2 t = \csc^2 t$$

■ **Proof** The reciprocal identities follow immediately from the definition on page 360. We now prove the Pythagoream identities. By definition, $\cos t = x$ and $\sin t = y$, where x and y are the coordinates of a point $P(x, y)$ on the unit circle. Since $P(x, y)$ is on the unit circle, we have $x^2 + y^2 = 1$. Thus

$$\sin^2 t + \cos^2 t = 1$$

Dividing both sides by $\cos^2 t$ (provided $\cos t \neq 0$), we get

$$\frac{\sin^2 t}{\cos^2 t} + \frac{\cos^2 t}{\cos^2 t} = \frac{1}{\cos^2 t}$$

$$\left(\frac{\sin t}{\cos t} \right)^2 + 1 = \left(\frac{1}{\cos t} \right)^2$$

$$\tan^2 t + 1 = \sec^2 t$$

We have used the reciprocal identities $\sin t / \cos t = \tan t$ and $1/\cos t = \sec t$. Similarly, dividing both sides of the first Pythagorean identity by $\sin^2 t$ (provided $\sin t \neq 0$) gives us $1 + \cot^2 t = \csc^2 t$. □

As their name indicates, the fundamental identities play a central role in trigonometry. This is because they can be used to relate any trigonometric function to any other. So, if we know the value of any one of the trigonometric functions at t, then we can find the values of all the others at t.

EXAMPLE 6 ■ **Finding All Trigonometric Functions from the Value of One**

If $\cos t = \frac{3}{5}$ and t is in quadrant IV, find the values of all the trigonometric functions at t.

SOLUTION From the Pythagorean identities we have

$$\sin^2 t + \cos^2 t = 1$$

$$\sin^2 t + \left(\tfrac{3}{5} \right)^2 = 1 \qquad \text{Substitute } \cos t = \tfrac{3}{5}$$

$$\sin^2 t = 1 - \tfrac{9}{25} = \tfrac{16}{25} \qquad \text{Solve for } \sin^2 t$$

$$\sin t = \pm \tfrac{4}{5} \qquad \text{Take the square roots}$$

Since this point is in quadrant IV, $\sin t$ is negative, so $\sin t = -\frac{4}{5}$. Now that we know both $\sin t$ and $\cos t$, we can find the values of the other trigonometric functions using the reciprocal identities:

$$\sin t = -\frac{4}{5} \qquad \cos t = \frac{3}{5} \qquad \tan t = \frac{\sin t}{\cos t} = \frac{-\frac{4}{5}}{\frac{3}{5}} = -\frac{4}{3}$$

$$\csc t = \frac{1}{\sin t} = -\frac{5}{4} \qquad \sec t = \frac{1}{\cos t} = \frac{5}{3} \qquad \cot t = \frac{1}{\tan t} = -\frac{3}{4} \qquad \blacksquare$$

EXAMPLE 7 ■ **Writing One Trigonometric Function in Terms of Another**

Write $\tan t$ in terms of $\cos t$, where t is in quadrant III.

SOLUTION Since $\tan t = \sin t / \cos t$, we need to write $\sin t$ in terms of $\cos t$. By the Pythagorean identities we have

$$\sin^2 t + \cos^2 t = 1$$

$$\sin^2 t = 1 - \cos^2 t \qquad \text{Solve for } \sin^2 t$$

$$\sin t = \pm\sqrt{1 - \cos^2 t} \qquad \text{Take the square roots}$$

Since $\sin t$ is negative in quadrant III, the negative sign applies here. Thus

$$\tan t = \frac{\sin t}{\cos t} = \frac{-\sqrt{1 - \cos^2 t}}{\cos t} \qquad \blacksquare$$

5.2 ■ EXERCISES

1–20 ■ Find the exact value of the trigonometric function at the given real number.

1. (a) $\sin 0$
(b) $\cos 0$

2. (a) $\sin \pi$
(b) $\cos \pi$

3. (a) $\sin(-\pi)$
(b) $\cos(-\pi)$

4. (a) $\cos \dfrac{\pi}{6}$
(b) $\cos \dfrac{5\pi}{6}$

5. (a) $\sin \dfrac{\pi}{2}$
(b) $\sin \dfrac{3\pi}{2}$

6. (a) $\sin \dfrac{7\pi}{6}$
(b) $\cos \dfrac{7\pi}{6}$

7. (a) $\cos \dfrac{\pi}{2}$
(b) $\cos \dfrac{5\pi}{2}$

8. (a) $\sin \dfrac{5\pi}{6}$
(b) $\sec \dfrac{5\pi}{6}$

9. (a) $\cos \dfrac{7\pi}{3}$
(b) $\sec \dfrac{7\pi}{3}$

10. (a) $\sin \dfrac{3\pi}{4}$
(b) $\cos \dfrac{3\pi}{4}$

11. (a) $\cos \dfrac{\pi}{3}$
(b) $\cos\left(-\dfrac{\pi}{3}\right)$

12. (a) $\sin \dfrac{\pi}{6}$
(b) $\sin\left(-\dfrac{\pi}{6}\right)$

13. (a) $\tan \dfrac{\pi}{6}$
(b) $\tan\left(-\dfrac{\pi}{6}\right)$

14. (a) $\tan \dfrac{\pi}{3}$
(b) $\cot \dfrac{\pi}{3}$

15. (a) $\sec \dfrac{11\pi}{3}$
(b) $\csc \dfrac{11\pi}{3}$

16. (a) $\sec \dfrac{13\pi}{6}$
(b) $\sec\left(-\dfrac{13\pi}{6}\right)$

17. (a) $\sin \dfrac{9\pi}{4}$
(b) $\csc \dfrac{9\pi}{4}$

18. (a) $\sec \pi$
(b) $\csc \dfrac{\pi}{2}$

19. (a) $\tan\left(-\dfrac{\pi}{4}\right)$
(b) $\cot\left(-\dfrac{\pi}{4}\right)$

20. (a) $\tan \dfrac{3\pi}{4}$
(b) $\tan \dfrac{11\pi}{4}$

21–24 ■ Find the value of each of the six trigonometric functions (if it is defined) at the given real number t. Use your answers to complete the table.

21. $t = 0$

22. $t = \dfrac{\pi}{2}$

23. $t = \pi$

24. $t = \dfrac{3\pi}{2}$

t	$\sin t$	$\cos t$	$\tan t$	$\csc t$	$\sec t$	$\cot t$
0	0	1		undefined		
$\dfrac{\pi}{2}$						
π			0			undefined
$\dfrac{3\pi}{2}$						

25–32 ■ The terminal point $P(x, y)$ determined by t is given. Find $\sin t$, $\cos t$, and $\tan t$.

25. $\left(\dfrac{3}{5}, \dfrac{4}{5} \right)$

26. $\left(-\dfrac{3}{5}, \dfrac{4}{5} \right)$

27. $\left(\dfrac{6}{7}, -\dfrac{\sqrt{13}}{7} \right)$

28. $\left(-\dfrac{1}{3}, -\dfrac{2\sqrt{2}}{3} \right)$

29. $\left(\dfrac{40}{41}, \dfrac{9}{41} \right)$

30. $\left(-\dfrac{3}{5}, -\dfrac{4}{5} \right)$

31. $\left(-\dfrac{5}{13}, -\dfrac{12}{13} \right)$

32. $\left(\dfrac{\sqrt{5}}{5}, \dfrac{2\sqrt{5}}{5} \right)$

33–40 ■ Find the approximate value of the given trigonometric function by using **(a)** the figure and **(b)** a calculator. Compare the two values.

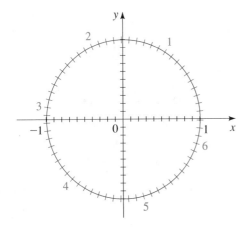

33. $\sin 1$

34. $\cos 0.8$

35. $\sin 1.2$

36. $\cos 5$

37. $\tan 0.8$

38. $\tan(-1.3)$

39. $\cos 4.1$

40. $\sin(-5.2)$

41–44 ■ Find the sign of the expression if the terminal point determined by t is in the given quadrant.

41. $\tan t \cos t$, t in quadrant II

42. $\sin^2 t \cos t$, t in quadrant IV

43. $\dfrac{\tan t \sin t}{\cot t}$, t in quadrant III

44. $\cos t \sec t$, t in any quadrant

45–48 ■ From the information given, find the quadrant in which the terminal point determined by t lies.

45. $\sin t > 0$ and $\cos t < 0$ **46.** $\tan t > 0$ and $\sin t < 0$

47. $\csc t > 0$ and $\sec t < 0$ **48.** $\cos t < 0$ and $\cot t < 0$

49–58 ■ Write the first expression in terms of the second if the terminal point determined by t is in the given quadrant.

49. $\sin t$, $\cos t$; t in quadrant II

50. $\cos t$, $\sin t$; t in quadrant IV

51. $\tan t$, $\sin t$; t in quadrant IV

52. $\tan t$, $\cos t$; t in quadrant III

53. $\sec t$, $\tan t$; t in quadrant II

54. $\csc t$, $\cot t$; t in quadrant III

55. $\tan t$, $\sec t$; t in quadrant III

56. $\sin t$, $\sec t$; t in quadrant IV

57. $\tan^2 t$, $\sin t$; t in any quadrant

58. $\sec^2 t \sin^2 t$, $\cos t$; t in any quadrant

59–66 ■ Find the values of the trigonometric functions of t from the given information.

59. $\sin t = \dfrac{3}{5}$, t in quadrant II

60. $\cos t = -\dfrac{4}{5}$, t in quadrant III

61. $\tan t = -\dfrac{3}{4}$, $\cos t > 0$

62. $\sec t = 3$, t in quadrant IV

63. $\sec t = 2$, $\sin t < 0$

64. $\tan t = \dfrac{1}{4}$, t in quadrant III

65. $\sin t = -\frac{1}{4}, \quad \sec t < 0$

66. $\tan t = -4, \quad t$ in quadrant II

67–74 ■ Determine whether the function is even, odd, or neither.

67. $f(x) = x^2 \sin x$

68. $f(x) = x^2 \cos 2x$

69. $f(x) = \sin x \cos x$

70. $f(x) = e^x \sin x$

71. $f(x) = |x| \cos x$

72. $f(x) = x \sin^3 x$

73. $f(x) = x^3 + \cos x$

74. $f(x) = \cos(\sin x)$

▣ DISCOVERY · DISCUSSION

75. Reduction Formulas   Explain how the figure shows that the following "reduction formulas" are valid:

$$\sin(t + \pi) = -\sin t \qquad \cos(t + \pi) = -\cos t$$

$$\tan(t + \pi) = \tan t$$

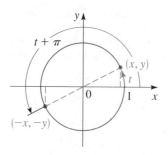

76. More Reduction Formulas   By the "Angle-Side-Angle" theorem from elementary geometry, triangles CDO and AOB in the figure are congruent. Explain how this proves that if B has coordinates (x, y), then D has coordinates $(-y, x)$. Then explain how the figure shows that the following "reduction formulas" are valid:

$$\sin\left(t + \frac{\pi}{2}\right) = \cos t$$

$$\cos\left(t + \frac{\pi}{2}\right) = -\sin t$$

$$\tan\left(t + \frac{\pi}{2}\right) = -\cot t$$

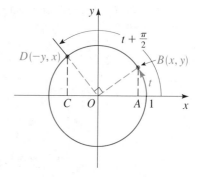

5.3   TRIGONOMETRIC GRAPHS

The graph of a function helps us get a better idea of its behavior. So, in this section we sketch graphs of the sine and cosine functions and certain transformations of these functions. The other trigonometric functions are sketched in the next section.

▮ GRAPHS OF THE SINE AND COSINE FUNCTIONS

To help us sketch the graphs of the sine and cosine functions, we first observe that these functions repeat their values in a regular fashion. To see exactly how this happens, recall that the circumference of the unit circle is 2π. It follows that the terminal point $P(x, y)$ determined by the real number t is the same as that determined by $t + 2\pi$. Since the sine and cosine functions are defined in terms of the coordinates of $P(x, y)$, it follows that their values are unchanged by the addition of any integer multiple of 2π. In other words,

$$\sin(t + 2n\pi) = \sin t \qquad \text{for any integer } n$$

$$\cos(t + 2n\pi) = \cos t \qquad \text{for any integer } n$$

Thus, the sine and cosine functions are *periodic* according to the following definition: A function f is **periodic** if there is a positive number p such that $f(t + p) = f(t)$ for every t. The least such positive number (if it exists) is the **period** of f. If f has period p, then the graph of f on any interval of length p is called **one complete period** of f.

PERIODIC PROPERTIES OF SINE AND COSINE

The function sine has period 2π: $\sin(t + 2\pi) = \sin t$

The function cosine has period 2π: $\cos(t + 2\pi) = \cos t$

So the sine and cosine functions repeat their values in any interval of length 2π. To sketch their graphs we first sketch the graph of one period. To sketch the graphs on the interval $0 \le t \le 2\pi$, we could try to make a table of values and use those points to draw the graph. Since no such table can be complete, let's look more closely at the definitions of these functions.

Recall that $\sin t$ is the y-coordinate of the terminal point $P(x, y)$ on the unit circle determined by the real number t. How does the y-coordinate of this point vary as t increases? It's easy to see that the y-coordinate of $P(x, y)$ increases to 1, then decreases to -1 repeatedly as the point $P(x, y)$ travels around the unit circle. (See Figure 1.) In fact, as t increases from 0 to $\pi/2$, $y = \sin t$ increases from 0 to 1. As t increases from $\pi/2$ to π, the value of $y = \sin t$ decreases from 1 to 0. Table 1 shows the variation of the sine and cosine functions for t between 0 and 2π.

Table 1

t	$\sin t$	$\cos t$
$0 \to \dfrac{\pi}{2}$	$0 \to 1$	$1 \to 0$
$\dfrac{\pi}{2} \to \pi$	$1 \to 0$	$0 \to -1$
$\pi \to \dfrac{3\pi}{2}$	$0 \to -1$	$-1 \to 0$
$\dfrac{3\pi}{2} \to 2\pi$	$-1 \to 0$	$0 \to 1$

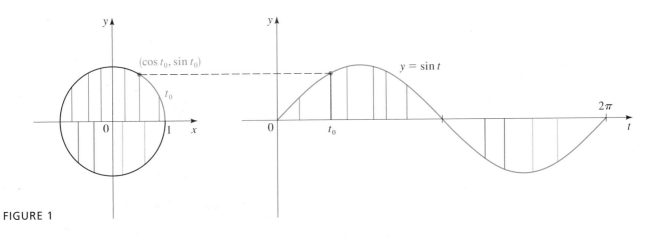

FIGURE 1

To draw the graphs more accurately, we find a few other values of $\sin t$ and $\cos t$ in Table 2. We could find still other values with the aid of a calculator.

Table 2

t	0	$\dfrac{\pi}{6}$	$\dfrac{\pi}{3}$	$\dfrac{\pi}{2}$	$\dfrac{2\pi}{3}$	$\dfrac{5\pi}{6}$	π	$\dfrac{7\pi}{6}$	$\dfrac{4\pi}{3}$	$\dfrac{3\pi}{2}$	$\dfrac{5\pi}{3}$	$\dfrac{11\pi}{6}$	2π
$\sin t$	0	$\dfrac{1}{2}$	$\dfrac{\sqrt{3}}{2}$	1	$\dfrac{\sqrt{3}}{2}$	$\dfrac{1}{2}$	0	$-\dfrac{1}{2}$	$-\dfrac{\sqrt{3}}{2}$	-1	$-\dfrac{\sqrt{3}}{2}$	$-\dfrac{1}{2}$	0
$\cos t$	1	$\dfrac{\sqrt{3}}{2}$	$\dfrac{1}{2}$	0	$-\dfrac{1}{2}$	$-\dfrac{\sqrt{3}}{2}$	-1	$-\dfrac{\sqrt{3}}{2}$	$-\dfrac{1}{2}$	0	$\dfrac{1}{2}$	$\dfrac{\sqrt{3}}{2}$	1

Now we use this information to sketch the graphs of the functions $\sin t$ and $\cos t$ for t between 0 and 2π in Figures 2 and 3. These are the graphs of one period. Using the fact that these functions are periodic with period 2π, we get their complete graphs by continuing the same pattern to the left and to the right in every successive interval of length 2π.

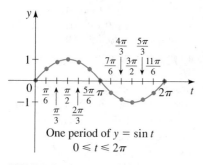

One period of $y = \sin t$
$0 \le t \le 2\pi$

FIGURE 2 Graph of $\sin t$

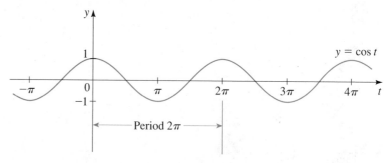

FIGURE 2

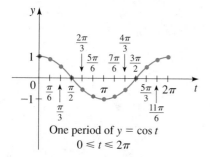

One period of $y = \cos t$
$0 \le t \le 2\pi$

FIGURE 3 Graph of $\cos t$

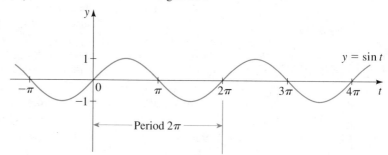

The graph of the sine function is symmetric with respect to the origin. This is as expected, since sine is an odd function. Since the cosine function is an even function, its graph is symmetric with respect to the y-axis.

GRAPHS OF TRANSFORMATIONS OF SINE AND COSINE

We now consider graphs of functions that are transformations of the sine and cosine functions. Thus, the graphing techniques of Section 2.4 are very useful here. The graphs we obtain are important for understanding applications to physical situations such as harmonic motion (see *Focus on Modeling,* page 393), but some of them are beautiful graphs that are interesting in their own right.

It's traditional to use the letter x to denote the variable in the domain of a function. So, from here on we use the letter x and write $y = \sin x$, $y = \cos x$, $y = \tan x$, and so on to denote these functions.

EXAMPLE 1 ■ Cosine Curves

Sketch the graph of each of the following functions.

(a) $f(x) = 2 + \cos x$ (b) $g(x) = -\cos x$

SOLUTION

(a) The graph of $y = 2 + \cos x$ is the same as the graph of $y = \cos x$, but shifted up 2 units [see Figure 4(a)].

(b) The graph of $y = -\cos x$ in Figure 4(b) is the reflection of the graph of $y = \cos x$ in the x-axis.

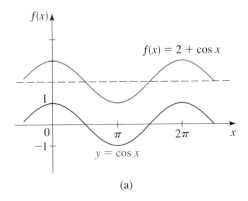

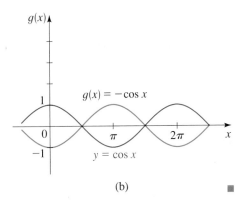

FIGURE 4   (a)   (b)   ■

Vertical stretching and shrinking of graphs is discussed in Section 2.4.

Let's sketch the graph of $y = 2 \sin x$. We start with the graph of $y = \sin x$ and multiply the y-coordinate of each point by 2. This has the effect of stretching the graph vertically by a factor of 2. To graph $y = \frac{1}{2} \sin x$, we start with the graph of $y = \sin x$ and multiply the y-coordinate of each point by $\frac{1}{2}$. This has the effect of shrinking the graph vertically by a factor of $\frac{1}{2}$ (see Figure 5).

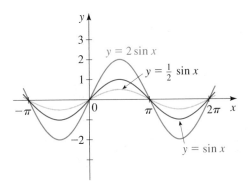

FIGURE 5

In general, for the functions

$$y = a \sin x \qquad \text{and} \qquad y = a \cos x$$

the number $|a|$ is called the **amplitude** and is the largest value these functions attain. Graphs of $y = a \sin x$ for several values of a are sketched in Figure 6.

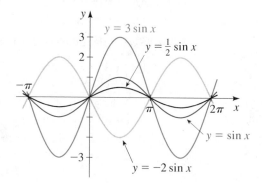

FIGURE 6

EXAMPLE 2 ■ Stretching a Cosine Curve

Find the amplitude of $y = -3 \cos x$ and sketch its graph.

SOLUTION   The amplitude is $|-3| = 3$, so the largest value the graph attains is 3 and the smallest value is -3. To sketch the graph we begin with the graph of $y = \cos x$, stretch the graph vertically by a factor of 3, and reflect in the x-axis, arriving at the graph in Figure 7.

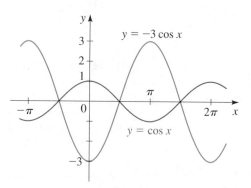

FIGURE 7   ■

Since the sine and cosine functions have period 2π, the functions

$$y = a \sin kx \qquad \text{and} \qquad y = a \cos kx \qquad (k > 0)$$

complete one period as kx varies from 0 to 2π, that is, for $0 \le kx \le 2\pi$ or for $0 \le x \le 2\pi/k$. So these functions complete one period as x varies between 0 and $2\pi/k$ and thus have period $2\pi/k$. The graphs of these functions are called **sine curves** and **cosine curves**, respectively.

> ## SINE AND COSINE CURVES
>
> The sine and cosine curves
>
> $$y = a \sin kx \quad \text{and} \quad y = a \cos kx \quad (k > 0)$$
>
> have amplitude $|a|$ and period $2\pi/k$.
>
> An appropriate interval on which to sketch one complete period is $[0, 2\pi/k]$.

To see how the value of k affects the graph of $y = \sin kx$, let's sketch the sine curve $y = \sin 2x$. Since the period is $2\pi/2 = \pi$, the graph completes one period in the interval $0 \le x \le \pi$ [see Figure 8(a)]. For the sine curve $y = \sin \frac{1}{2}x$, the period is $2\pi \div \frac{1}{2} = 4\pi$, and so the graph completes one period in the interval $0 \le x \le 4\pi$ [see Figure 8(b)]. We see that the effect is to shrink the graph if $k > 1$ or to stretch the graph if $k < 1$.

Horizontal stretching and shrinking of graphs is discussed in Section 2.4.

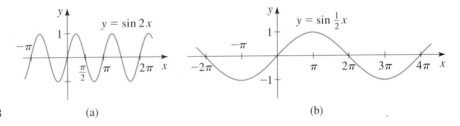

FIGURE 8   (a)   (b)

For comparison we sketch in Figure 9 the graphs of one period of the sine curve $y = a \sin kx$ for several values of k.

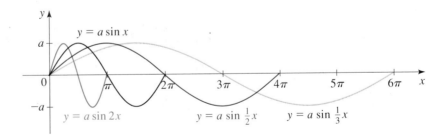

FIGURE 9

EXAMPLE 3 ■ Amplitude and Period

Find the amplitude and period of each of the following functions, and sketch its graph.

(a) $y = 4 \cos 3x$ (b) $y = -2 \sin \frac{1}{2}x$

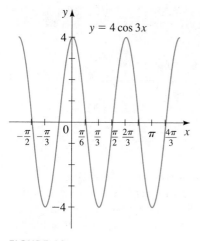

FIGURE 10

SOLUTION

(a) For $y = 4 \cos 3x$,

$$\text{amplitude} = |a| = 4$$

$$\text{period} = \frac{2\pi}{3}$$

The graph is sketched in Figure 10.

(b) For $y = -2 \sin \frac{1}{2}x$,

$$\text{amplitude} = |a| = |-2| = 2$$

$$\text{period} = \frac{2\pi}{\frac{1}{2}} = 4\pi$$

The graph is sketched in Figure 11.

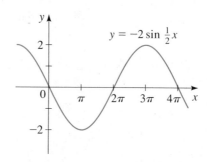

FIGURE 11

The graphs of functions of the form $y = a \sin k(x - b)$ and $y = a \cos k(x - b)$ are simply sine and cosine curves that are shifted horizontally by an amount $|b|$. They are shifted to the right if $b > 0$ or to the left if $b < 0$. The number b is the **phase shift**. We summarize the properties of these functions in the following box.

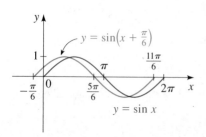

FIGURE 12

SHIFTED SINE AND COSINE CURVES

The sine and cosine curves

$$y = a \sin k(x - b) \qquad \text{and} \qquad y = a \cos k(x - b) \qquad (k > 0)$$

have amplitude $|a|$, period $2\pi/k$, and phase shift b.

An appropriate interval on which to sketch one complete period is $[b, b + (2\pi/k)]$.

The graphs of

$$y = \sin\left(x - \frac{\pi}{3}\right) \qquad \text{and} \qquad y = \sin\left(x + \frac{\pi}{6}\right)$$

are sketched in Figure 12.

EXAMPLE 4 ■ A Shifted Sine Curve

Find the amplitude, period, and phase shift of $y = 3\sin 2\left(x - \dfrac{\pi}{4}\right)$, and sketch the graph of one complete period.

SOLUTION We have

$$\text{amplitude} = |a| = 3$$

$$\text{period} = \frac{2\pi}{2} = \pi$$

$$\text{phase shift} = \frac{\pi}{4} \qquad \text{Shift } \tfrac{\pi}{4} \text{ to the } right$$

Since the phase shift is $\pi/4$ and the period is π, one complete period occurs on the interval

$$\left[\frac{\pi}{4}, \frac{\pi}{4} + \pi\right] = \left[\frac{\pi}{4}, \frac{5\pi}{4}\right]$$

As an aid in sketching the graph we divide this interval into four equal parts, then sketch a sine curve with amplitude 3 as in Figure 13.

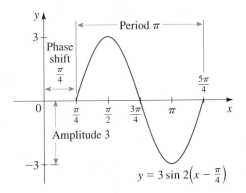

FIGURE 13

EXAMPLE 5 ■ A Shifted Cosine Curve

Find the amplitude, period, and phase shift of

$$y = \frac{3}{4}\cos\left(2x + \frac{2\pi}{3}\right)$$

and sketch the graph of one complete period.

SOLUTION We first write this function in the form $y = a\cos k(x - b)$. To do this we factor 2 from the expression $2x + \dfrac{2\pi}{3}$ to get

$$y = \frac{3}{4}\cos 2\left[x - \left(-\frac{\pi}{3}\right)\right]$$

Thus, we have

$$\text{amplitude} = |a| = \frac{3}{4}$$

$$\text{period} = \frac{2\pi}{k} = \frac{2\pi}{2} = \pi$$

$$\text{phase shift} = b = -\frac{\pi}{3} \qquad \text{Shift } \tfrac{\pi}{3} \text{ to the } \textit{left}$$

From this information it follows that one period of this cosine curve begins at $-\pi/3$ and ends at $-\pi/3 + \pi = 2\pi/3$. To sketch the graph over the interval $[-\pi/3, 2\pi/3]$, we divide this interval into four equal parts and sketch a cosine curve with amplitude $\frac{3}{4}$ as shown in Figure 14.

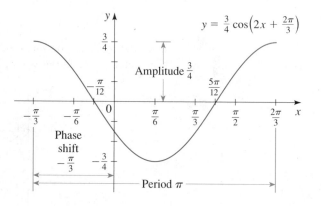

FIGURE 14

 USING GRAPHING DEVICES TO GRAPH TRIGONOMETRIC FUNCTIONS

In Section 1.9 we discussed the use of graphing calculators and computers, and we saw that it's important to choose a viewing rectangle carefully in order to produce a reasonable graph of a function. This is especially true for trigonometric functions; Example 6 shows that, if care is not taken, it's easy to produce a very misleading graph of a trigonometric function.

EXAMPLE 6 ■ Choosing the Viewing Rectangle

Graph the function $f(x) = \sin 50x$ in an appropriate viewing rectangle.

SOLUTION   Figure 15(a) shows the graph of f produced by a graphing calculator using the viewing rectangle $[-12, 12]$ by $[-1.5, 1.5]$. At first glance the graph appears to be reasonable. But if we change the viewing rectangle to the ones shown in the following parts of Figure 15, the graphs look very different. Something strange is happening.

1.5   −12   12   −1.5   (a)

1.5   −10   10   −1.5   (b)

The appearance of the graphs in Figure 15 depends on the machine used. The graphs you get with your own graphing device might not look like these figures, but they will also be quite inaccurate.

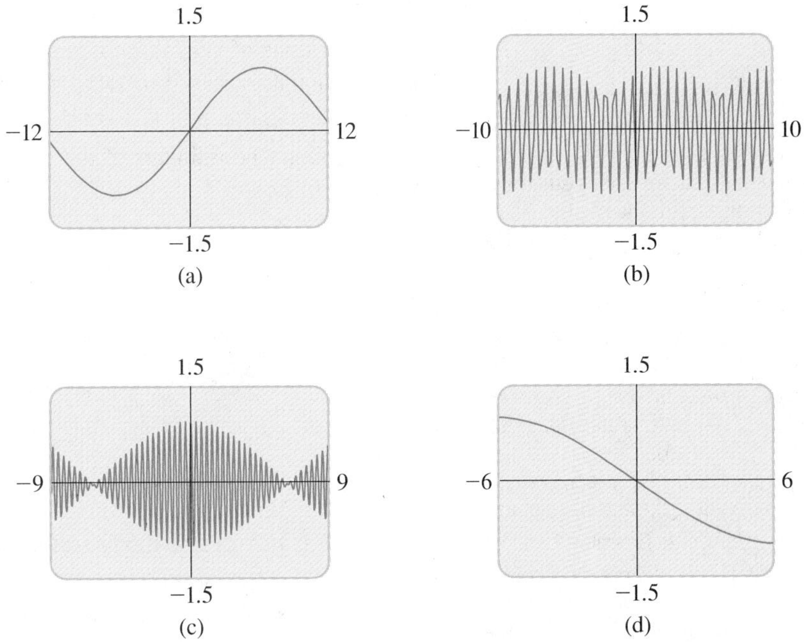

1.5   −9   9   −1.5   (c)

1.5   −6   6   −1.5   (d)

FIGURE 15

Graphs of $f(x) = \sin 50x$ in different viewing rectangles

To explain the big differences in appearance of these graphs and to find an appropriate viewing rectangle, we need to find the period of the function $y = \sin 50x$:

$$\text{period} = \frac{2\pi}{50} = \frac{\pi}{25} \approx 0.126$$

This suggests that we should deal only with small values of x in order to show just a few oscillations of the graph. If we choose the viewing rectangle $[-0.25, 0.25]$ by $[-1.5, 1.5]$, we get the graph shown in Figure 16.

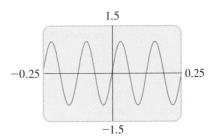

FIGURE 16

$f(x) = \sin 50x$

Now we see what went wrong in Figure 15. The oscillations of $y = \sin 50x$ are so rapid that when the calculator plots points and joins them, it misses most of the maximum and minimum points and therefore gives a very misleading impression of the graph. ■

The function in Example 7 is **periodic** with period 2π. In general, functions that are sums of functions from the following list

1, $\cos kx$, $\cos 2kx$, $\cos 3kx$, ...

$\sin kx$, $\sin 2kx$, $\sin 3kx$, ...

are periodic. Although these functions appear to be special, they are actually fundamental to describing all periodic functions that arise in practice. The French mathematician J. B. J. Fourier (see page 476) discovered that nearly every periodic function can be written as a sum (usually an infinite sum) of these functions (see Section 10.3). This is remarkable because it means that any situation in which periodic variation occurs can be described mathematically using the functions sine and cosine. A modern application of Fourier's discovery is the digital encoding of sound on compact discs.

EXAMPLE 7 ■ A Sum of Sine and Cosine Curves

Draw the graphs of $f(x) = 2\cos x$, $g(x) = \sin 2x$, and $h(x) = 2\cos x + \sin 2x$ on a common screen to illustrate the method of graphical addition.

SOLUTION   Notice that $h = f + g$, so its graph is obtained by adding the corresponding y-coordinates of the graphs of f and g. The graphs of f, g, and h are shown in Figure 17.

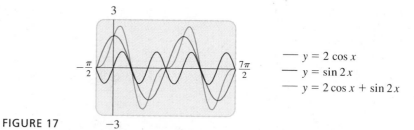

— $y = 2\cos x$
— $y = \sin 2x$
— $y = 2\cos x + \sin 2x$

FIGURE 17

EXAMPLE 8 ■ A Cosine Curve with Exponential Amplitude

Draw the graphs of the functions $y = e^{-x}$, $y = -e^{-x}$, and $y = e^{-x}\cos 6\pi x$ on a common screen. Comment on and explain the relationship among the graphs.

SOLUTION   Figure 18 shows all three graphs in the viewing rectangle $[-1, 2]$ by $[-3, 3]$. It appears that the graph of $y = e^{-x}\cos 6\pi x$ lies between the graphs of the exponential functions $y = e^{-x}$ and $y = -e^{-x}$.

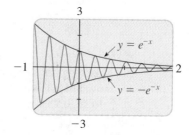

FIGURE 18

$y = e^{-x}\cos 6\pi x$

To understand this, recall that the values of $\cos 6\pi x$ lie between -1 and 1, that is,

$$-1 \le \cos 6\pi x \le 1$$

for all values of x. Multiplying the inequalities by e^{-x}, and noting that $e^{-x} \ge 0$, we get

$$-e^{-x} \le e^{-x}\cos 6\pi x \le e^{-x}$$

This explains why the exponential functions form a boundary for the graph of $y = e^{-x}\cos 6\pi x$. (Note that the graphs touch when $\cos 6\pi x = \pm 1$.)   ■

Example 8 shows that the function $y = e^{-x}$ controls the amplitude of the graph of $y = e^{-x}\cos 6\pi x$. In general, if $f(x) = a(x)\sin kx$ or $a(x)\cos kx$, the function a determines how the amplitude of f varies, and the graph of f lies between the graphs of $y = -a(x)$ and $y = a(x)$. Here is another example.

EXAMPLE 9 ■ **A Cosine Curve with Variable Amplitude**

Graph the function $f(x) = \cos 2\pi x \cos 16\pi x$.

SOLUTION The graph is shown in Figure 19. Although it was drawn by a computer, we could have drawn it by hand, by first sketching the boundary curves $y = \cos 2\pi x$ and $y = -\cos 2\pi x$. The graph of f is a cosine curve that lies between the graphs of these two functions.

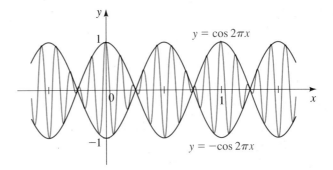

FIGURE 19

$f(x) = \cos 2\pi x \cos 16\pi x$

EXAMPLE 10 ■ **A Sine Curve with Decaying Amplitude**

The function $f(x) = \dfrac{\sin x}{x}$ is important in calculus. Graph this function and comment on its behavior when x is close to 0.

SOLUTION The viewing rectangle $[-15, 15]$ by $[-0.5, 1.5]$ shown in Figure 20(a) gives a good global view of the graph of f. The viewing rectangle $[-1, 1]$ by $[-0.5, 1.5]$ in Figure 20(b) focuses on the behavior of f when $x \approx 0$. Notice that although $f(x)$ is not defined when $x = 0$ (in other words, 0 is not in the domain of f), the values of f seem to approach 1 when x gets close to 0. This fact is crucial in calculus.

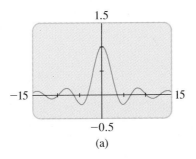

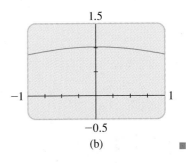

FIGURE 20

$f(x) = \dfrac{\sin x}{x}$

(a)

(b)

The function in Example 10 can be written as

$$f(x) = \left(\frac{1}{x}\right) \cdot \sin x$$

and may thus be viewed as a sine function whose amplitude is controlled by the function $a(x) = 1/x$.

5.3 EXERCISES

1–10 ■ Sketch the graph of the function.

1. $y = 2 + \sin x$ **2.** $y = -\sin x$

3. $y = 1 - \cos x$ **4.** $y = -1 + \cos x$

5. $y = 2\cos x$ **6.** $y = -3\sin x$

7. $y = 3 + 3\cos x$ **8.** $y = 4 - 2\sin x$

9. $y = |\sin x|$ **10.** $y = |\cos x|$

11–18 ■ Find the amplitude and period of the function, and sketch its graph.

11. $y = 3\sin 3x$ **12.** $y = -2\sin 2\pi x$

13. $y = 10\sin \frac{1}{2}x$ **14.** $y = \cos 10\pi x$

15. $y = -\cos \frac{1}{3}x$ **16.** $y = \sin(-2x)$

17. $y = 3\cos 3\pi x$ **18.** $y = 5 - 2\sin 2x$

19–32 ■ Find the amplitude, period, and phase shift of the function, and sketch the graph of one complete period.

19. $y = \cos\left(x - \frac{\pi}{2}\right)$ **20.** $y = 2\sin\left(x - \frac{\pi}{3}\right)$

21. $y = -2\sin\left(x - \frac{\pi}{6}\right)$ **22.** $y = 3\cos\left(x + \frac{\pi}{4}\right)$

23. $y = 5\cos\left(3x - \frac{\pi}{4}\right)$ **24.** $y = -4\sin 2\left(x + \frac{\pi}{2}\right)$

25. $y = 2\sin\left(\frac{2}{3}x - \frac{\pi}{6}\right)$ **26.** $y = \sin\frac{1}{2}\left(x + \frac{\pi}{4}\right)$

27. $y = 3\cos\pi\left(x + \frac{1}{2}\right)$ **28.** $y = 1 + \cos\left(3x + \frac{\pi}{2}\right)$

29. $y = -\frac{1}{2}\cos\left(2x - \frac{\pi}{3}\right)$ **30.** $y = 3 + 2\sin 3(x + 1)$

31. $y = \sin(3x + \pi)$ **32.** $y = \cos\left(\frac{\pi}{2} - x\right)$

33–38 ■ The graph of one complete period of a sine or cosine curve is given.
(a) Find the amplitude, period, and phase shift.
(b) Write an equation that represents the curve in the form

$$y = a\sin k(x - b) \quad \text{or} \quad y = a\cos k(x - b)$$

33.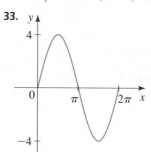

34.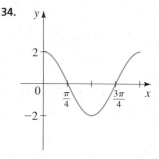

35.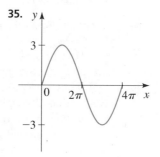

36.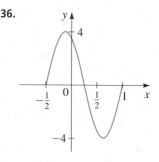

37.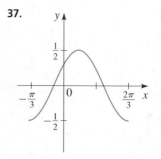

38.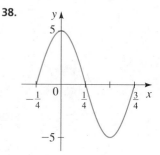

39–46 ■ Determine an appropriate viewing rectangle for each function, and use it to draw the graph.

39. $f(x) = \cos 100x$

40. $f(x) = 3 \sin 120x$

41. $f(x) = \sin(x/40)$

42. $f(x) = \cos(x/80)$

43. $y = \tan 25x$

44. $y = \csc 40x$

45. $y = e^{\sin 20x}$

46. $y = \sqrt{\tan 10\pi x}$

47–48 ■ Draw the graphs of f, g, and $f + g$ on a common screen to illustrate graphical addition.

47. $f(x) = x$, $g(x) = \sin x$

48. $f(x) = \sin x$, $g(x) = \sin 2x$

49–54 ■ Graph the three functions on a common screen. How are the graphs related?

49. $y = x^2$, $y = -x^2$, $y = x^2 \sin x$

50. $y = x$, $y = -x$, $y = x \cos x$

51. $y = e^x$, $y = -e^x$, $y = e^x \sin 5\pi x$

52. $y = \dfrac{1}{1 + x^2}$, $y = -\dfrac{1}{1 + x^2}$, $y = \dfrac{\cos 2\pi x}{1 + x^2}$

53. $y = \cos 3\pi x$, $y = -\cos 3\pi x$, $y = \cos 3\pi x \cos 21\pi x$

54. $y = \sin 2\pi x$, $y = -\sin 2\pi x$, $y = \sin 2\pi x \sin 10\pi x$

55–60 ■ (a) Use a graphing device to graph the function.
(b) Determine from the graph whether the function is periodic and, if so, determine the period.
(c) Determine from the graph whether the function is odd, even, or neither.

55. $y = |\sin x|$

56. $y = \sin|x|$

57. $y = e^{\sin x}$

58. $y = 2^{\cos x}$

59. $y = \sin^2 x$

60. $y = \sin(x^2)$

61–64 ■ Find the maximum and minimum values of the function.

61. $y = \sin x + \sin 2x$

62. $y = x - 2 \sin x$, $0 \le x \le 2\pi$

63. $y = 2 \sin x + \sin^2 x$

64. $y = \dfrac{\cos x}{2 + \sin x}$

65–68 ■ Find all solutions of the equation that lie in the interval $[0, \pi]$. State each answer correct to two decimal places.

65. $\cos x = 0.4$

66. $\tan x = 2$

67. $\csc x = 3$

68. $\cos x = x$

69. Let $f(x) = \dfrac{1 - \cos x}{x}$.

(a) Is the function f even, odd, or neither?
(b) Find the x-intercepts of the graph of f.
(c) Draw the graph of f in an appropriate viewing rectangle.
(d) Describe the behavior of the function as $x \to \infty$ and as $x \to -\infty$.
(e) Notice that $f(x)$ is not defined when $x = 0$. What happens as $x \to 0$?

DISCOVERY · DISCUSSION

70. Compositions Involving Trigonometric Functions   This exercise explores the effect of the inner function g on a composite function $y = f(g(x))$.

(a) Graph the function $y = \sin(\sqrt{x})$ using the viewing rectangle $[0, 400]$ by $[-1.5, 1.5]$. In what ways does this graph differ from the graph of the sine function?
(b) Graph the function $y = \sin(x^2)$ using the viewing rectangle $[-5, 5]$ by $[-1.5, 1.5]$. In what ways does this graph differ from the graph of the sine function?

71. Sinusoidal Curves   The graph of $y = \sin x$ is the same as the graph of $y = \cos x$ shifted to the right $\pi/2$ units. So the sine curve $y = \sin x$ is also at the same time a cosine curve: $y = \cos(x - \pi/2)$. In fact, any sine curve is also a cosine curve with a different phase shift, and any cosine curve is also a sine curve. Sine and cosine curves are collectively referred to as *sinusoidal*. Find all possible ways of expressing the curve whose graph is shown as a sine curve $y = a \sin(x - b)$ and as a cosine curve $y = a \cos(x - b)$. Explain why you think you have found all possible choices for a and b in each case.

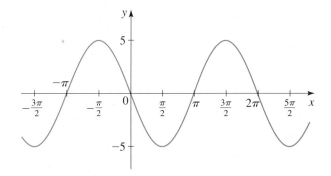

| 5.4 | **MORE TRIGONOMETRIC GRAPHS** |

In this section we sketch the graphs of the tangent, cotangent, secant, and cosecant functions, and transformations of these functions.

GRAPHS OF THE TANGENT, COTANGENT, SECANT, AND COSECANT FUNCTION

We begin by stating the periodic properties of these functions. Recall that sine and cosine have period 2π. Since cosecant and secant are the reciprocals of sine and cosine, respectively, they also have period 2π (see Exercise 47). Tangent and cotangent, however, have period π (see Exercise 75 of Section 5.2).

> **PERIODIC PROPERTIES**
>
> The functions tangent and cotangent have period π:
>
> $$\tan(x + \pi) = \tan x \qquad \cot(x + \pi) = \cot x$$
>
> The functions cosecant and secant have period 2π:
>
> $$\csc(x + 2\pi) = \csc x \qquad \sec(x + 2\pi) = \sec x$$

x	$\tan x$
0	0
$\dfrac{\pi}{6}$	0.58
$\dfrac{\pi}{4}$	1.00
$\dfrac{\pi}{3}$	1.73
1.4	5.80
1.5	14.10
1.55	48.08
1.57	1255.77
1.5707	$10{,}381.33$

We first sketch the graph of tangent. Since it has period π, we need only sketch the graph on any interval of length π and then repeat the pattern to the left and to the right. We sketch the graph on the interval $(-\pi/2, \pi/2)$. Since $\tan \pi/2$ and $\tan(-\pi/2)$ aren't defined, we need to be careful in sketching the graph at points near $\pi/2$ and $-\pi/2$. As x gets near $\pi/2$ through values less than $\pi/2$, the value of $\tan x$ becomes large. To see this, notice that as x gets close to $\pi/2$, $\cos x$ approaches 0 and $\sin x$ approaches 1 and so $\tan x = \sin x/\cos x$ is large. A table of values of $\tan x$ for x close to $\pi/2$ (≈ 1.570796) is shown in the margin.

Thus, by choosing x close enough to $\pi/2$ through values less than $\pi/2$, we can make the value of $\tan x$ larger than any given positive number. In a similar way, by choosing x close to $-\pi/2$ through values greater than $-\pi/2$, we can make $\tan x$ smaller than any given negative number. In the notation of Section 3.5 we have

$$\tan x \to \infty \qquad \text{as} \quad x \to \frac{\pi^-}{2}$$

$$\tan x \to -\infty \qquad \text{as} \quad x \to -\frac{\pi^+}{2}$$

Thus, $x = \pi/2$ and $x = -\pi/2$ are vertical asymptotes (see Section 3.5). With the information we have so far, we sketch the graph of $\tan x$ for $-\pi/2 < x < \pi/2$ in

Figure 1. The complete graph of tangent [see Figure 5(a) on page 384] is now obtained using the fact that tangent is periodic with period π.

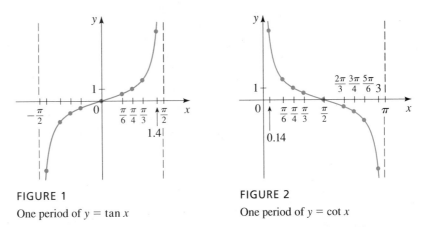

FIGURE 1

One period of $y = \tan x$

FIGURE 2

One period of $y = \cot x$

The graph of $y = \cot x$ is sketched on the interval $(0, \pi)$ by a similar analysis (see Figure 2). Since $\cot x$ is undefined for $x = n\pi$ with n an integer, its complete graph [in Figure 5(b)] has vertical asymptotes at these values.

To sketch the graphs of the cosecant and secant functions, we use the reciprocal identities

$$\csc x = \frac{1}{\sin x} \quad \text{and} \quad \sec x = \frac{1}{\cos x}$$

So, to graph $y = \csc x$ we take the reciprocals of the y-coordinates of the points of the graph of $y = \sin x$ (see Figure 3). Similarly, to graph $y = \sec x$ we take the reciprocals of the y-coordinates of the points of the graph of $y = \cos x$ (see Figure 4).

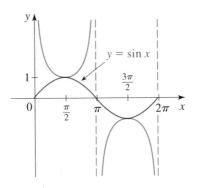

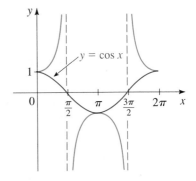

FIGURE 3

One period of $y = \csc x$

FIGURE 4

One period of $y = \sec x$

Let's consider more closely the graph of the function $y = \csc x$ on the interval $0 < x < \pi$. We need to examine the values of the function near 0 and π since at these values $\sin x = 0$, and $\csc x$ is thus undefined. We see that

$$\csc x \to \infty \quad \text{as} \quad x \to 0^+$$

$$\csc x \to \infty \quad \text{as} \quad x \to \pi^-$$

Thus, the lines $x = 0$ and $x = \pi$ are vertical asymptotes. In the interval $\pi < x < 2\pi$ the graph is sketched in the same way. The values of $\csc x$ in that interval are the same as those in the interval $0 < x < \pi$ except for sign (see Figure 3). The complete graph in Figure 5(c) is now obtained from the fact that the function cosecant is periodic with period 2π. Note that the graph has vertical asymptotes at the points where $\sin x = 0$, that is, at $x = n\pi$, for n an integer.

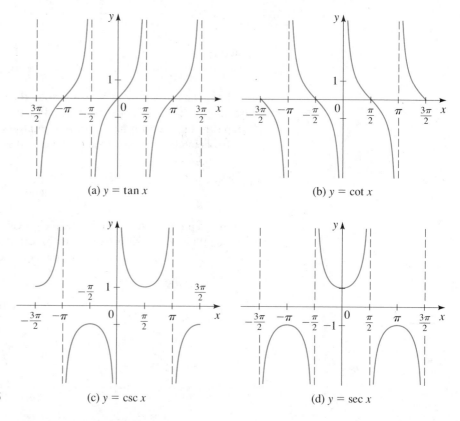

(a) $y = \tan x$

(b) $y = \cot x$

(c) $y = \csc x$

(d) $y = \sec x$

FIGURE 5

The graph of $y = \sec x$ is sketched in a similar manner. Observe that the domain of $\sec x$ is the set of all real numbers other than $x = (\pi/2) + n\pi$, for n an integer, so the graph has vertical asymptotes at those points. The complete graph is shown in Figure 5(d).

It's apparent that the graphs of $y = \tan x$, $y = \cot x$, and $y = \csc x$ are symmetric about the origin, whereas that of $y = \sec x$ is symmetric about the y-axis. This is because tangent, cotangent, and cosecant are odd funtions, whereas secant is an even function.

GRAPHS INVOLVING TANGENT AND COTANGENT FUNCTIONS

We now consider graphs of transformations of the tangent and cotangent functions.

EXAMPLE 1 ■ Graphing Tangent Curves

Sketch the graph of each of the following functions.

(a) $y = 2 \tan x$ (b) $y = -\tan x$

SOLUTION   We first sketch the graph of $y = \tan x$ and then transform it as required.

(a) To graph $y = 2 \tan x$, we multiply the y-coordinate of each point on the graph of $y = \tan x$ by 2. The resulting graph is sketched in Figure 6(a).

(b) The graph of $y = -\tan x$ in Figure 6(b) is obtained from that of $y = \tan x$ by reflecting in the x-axis.

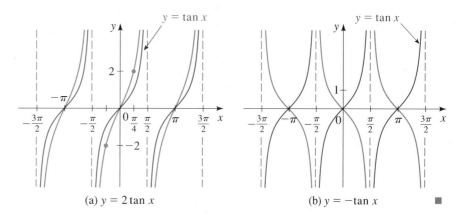

FIGURE 6   (a) $y = 2 \tan x$ (b) $y = -\tan x$ ■

Since the tangent and cotangent functions have period π, the functions

$$y = a \tan kx \qquad \text{and} \qquad y = a \cot kx \qquad (k > 0)$$

complete one period as kx varies from 0 to π, that is, for $0 \le kx \le \pi$ or for $0 \le x \le \pi/k$. So they each have period π/k.

TANGENT AND COTANGENT CURVES

The functions

$$y = a \tan kx \qquad \text{and} \qquad y = a \cot kx \qquad (k > 0)$$

have period π/k.

Thus, one complete period of the graphs of these functions occurs on any interval of length π/k. To sketch a complete period of these graphs, it's convenient to select an interval between vertical asymptotes:

To graph one period of $y = a \tan kx$, an appropriate interval is $\left(-\dfrac{\pi}{2k}, \dfrac{\pi}{2k}\right)$.

To graph one period of $y = a \cot kx$, an appropriate interval is $\left(0, \dfrac{\pi}{k}\right)$.

EXAMPLE 2 ■ Graphing Tangent Curves

Sketch the graph of each of the following functions.

(a) $y = \tan 2x$

(b) $y = \tan 2\left(x - \dfrac{\pi}{4}\right)$

SOLUTION

(a) The period is $\pi/2$ and an appropriate interval is $(-\pi/4, \pi/4)$. The endpoints $x = -\pi/4$ and $x = \pi/4$ are vertical asymptotes. Thus, we sketch on $(-\pi/4, \pi/4)$ one complete period of the function. The graph has the same shape as that of the tangent function, but is shrunk horizontally by a factor of $\frac{1}{2}$. We then repeat that portion of the graph to the left and to the right. See Figure 7(a).

(b) The graph is the same as that in part (a), but it is shifted to the right $\pi/4$, as shown in Figure 7(b).

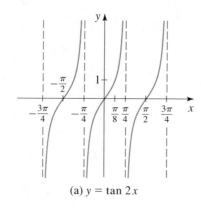

(a) $y = \tan 2x$

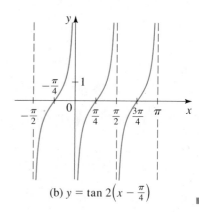

(b) $y = \tan 2\left(x - \frac{\pi}{4}\right)$

FIGURE 7

EXAMPLE 3 ■ A Shifted Cotangent Curve

Sketch the graph of $y = 2 \cot\left(3x - \dfrac{\pi}{2}\right)$.

SOLUTION   We first put this in the form $y = a \cot k(x - b)$ by factoring 3 from the expression $3x - \dfrac{\pi}{2}$:

$$y = 2\cot\left(3x - \frac{\pi}{2}\right) = 2\cot 3\left(x - \frac{\pi}{6}\right)$$

Thus, the graph is the same as that of $y = 2 \cot 3x$, but is shifted to the right $\pi/6$. The period of $y = 2 \cot 3x$ is $\pi/3$, and an appropriate interval is

$(0, \pi/3)$. To get the corresponding interval for the desired graph, we shift this interval to the right $\pi/6$. This gives

$$\left(0 + \frac{\pi}{6}, \frac{\pi}{3} + \frac{\pi}{6}\right) = \left(\frac{\pi}{6}, \frac{\pi}{2}\right)$$

Finally, we graph one period in the shape of cotangent on the interval $(\pi/6, \pi/2)$ and repeat that portion of the graph to the left and to the right. See Figure 8.

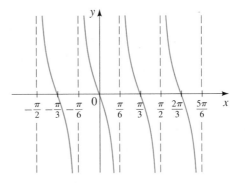

FIGURE 8
$y = 2 \cot\left(3x - \frac{\pi}{2}\right)$

GRAPHS INVOLVING THE COSECANT AND SECANT FUNCTIONS

We have already observed that the cosecant and secant functions are the reciprocals of the sine and cosine functions. Thus, the following result is the counterpart of the result for sine and cosine curves in Section 5.3.

COSECANT AND SECANT CURVES

The functions

$$y = a \csc kx \qquad \text{and} \qquad y = a \sec kx \qquad (k > 0)$$

have period $2\pi/k$.

An appropriate interval on which to sketch one complete period is $[0, 2\pi/k]$.

EXAMPLE 4 ■ Graphing Cosecant Curves

Sketch the graph of each of the following functions.

(a) $y = \frac{1}{2} \csc 2x$ \qquad\qquad (b) $y = \frac{1}{2} \csc\left(2x + \frac{\pi}{2}\right)$

SOLUTION

(a) The period is $2\pi/2 = \pi$. An appropriate interval is $[0, \pi]$, and the asymptotes occur in this interval whenever $\sin 2x = 0$. So the asymptotes in this interval are $x = 0$, $x = \pi/2$, and $x = \pi$. With this information we sketch on the interval $[0, \pi]$ a graph with the same general shape as that of one period of the cosecant function. The complete graph in Figure 9(a) is obtained by repeating this portion of the graph to the left and to the right.

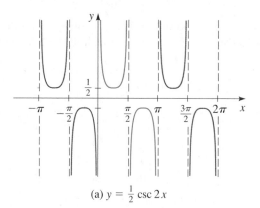

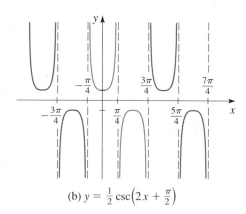

FIGURE 9   (a) $y = \frac{1}{2} \csc 2x$ (b) $y = \frac{1}{2} \csc\left(2x + \frac{\pi}{2}\right)$

(b) We first write

$$y = \frac{1}{2} \csc\left(2x + \frac{\pi}{2}\right) = \frac{1}{2} \csc 2\left(x + \frac{\pi}{4}\right)$$

From this we see that the graph is the same as that in part (a), but shifted to the left $\pi/4$. The graph is sketched in Figure 9(b). ■

EXAMPLE 5 ■ Graphing a Secant Curve

Sketch the graph of $y = 3 \sec \frac{1}{2} x$.

SOLUTION   The period is $2\pi \div \frac{1}{2} = 4\pi$. An appropriate interval is $[0, 4\pi]$, and the asymptotes occur in this interval wherever $\cos \frac{1}{2} x = 0$. Thus, the asymptotes in this interval are $x = \pi, x = 3\pi$. With this information we sketch on the interval $[0, 4\pi]$ a graph with the same general shape as that of one period of the secant function. The complete graph in Figure 10 is obtained by repeating this portion of the graph to the left and to the right. ■

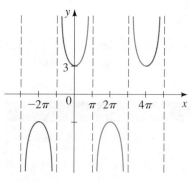

FIGURE 10
$y = 3 \sec \frac{1}{2} x$

5.4   EXERCISES

1–46 ■ Find the period and sketch the graph of the function.

1. $y = 3 \tan x$

2. $y = -3 \tan x$

3. $y = \frac{1}{2} \tan x$

4. $y = -\frac{1}{2} \tan x$

5. $y = 4 \cot x$

6. $y = \frac{1}{4} \cot x$

7. $y = 2 \csc x$

8. $y = \frac{1}{2} \csc x$

9. $y = 4 \sec x$

10. $y = \frac{1}{4} \sec x$

11. $y = \tan\left(x + \frac{\pi}{2}\right)$

12. $y = \tan\left(x - \frac{\pi}{4}\right)$

13. $y = \csc\left(x + \dfrac{\pi}{2}\right)$ **14.** $y = \sec\left(x - \dfrac{\pi}{4}\right)$

15. $y = \cot\left(x + \dfrac{\pi}{4}\right)$ **16.** $y = 2\csc\left(x - \dfrac{\pi}{3}\right)$

17. $y = \frac{1}{2}\sec\left(x - \dfrac{\pi}{6}\right)$ **18.** $y = 3\csc\left(x - \dfrac{\pi}{2}\right)$

19. $y = \tan 2x$ **20.** $y = \tan \frac{1}{2}x$

21. $y = \tan \pi x$ **22.** $y = \cot \dfrac{\pi}{2} x$

23. $y = \sec 2x$ **24.** $y = 5\csc 3x$

25. $y = \csc 2x$ **26.** $y = \csc \frac{1}{2}x$

27. $y = 2\tan 3x$ **28.** $y = 2\tan \dfrac{\pi}{2}x$

29. $y = 5\csc 3x$ **30.** $y = 5\sec 2\pi x$

31. $y = \tan 2\left(x + \dfrac{\pi}{2}\right)$ **32.** $y = \csc 2\left(x + \dfrac{\pi}{2}\right)$

33. $y = \tan 2(x - \pi)$ **34.** $y = \sec 2\left(x - \dfrac{\pi}{2}\right)$

35. $y = \cot\left(2x - \dfrac{\pi}{2}\right)$ **36.** $y = \frac{1}{2}\tan(\pi x - \pi)$

37. $y = 2\csc\left(\pi x - \dfrac{\pi}{3}\right)$ **38.** $y = 2\sec\left(\dfrac{1}{2}x - \dfrac{\pi}{3}\right)$

39. $y = 5\sec\left(3x - \dfrac{\pi}{2}\right)$ **40.** $y = \frac{1}{2}\sec(2\pi x - \pi)$

41. $y = \tan\left(\dfrac{2}{3}x - \dfrac{\pi}{6}\right)$ **42.** $y = \tan \dfrac{1}{2}\left(x + \dfrac{\pi}{4}\right)$

43. $y = 3\sec \pi\left(x + \dfrac{1}{2}\right)$ **44.** $y = \sec\left(3x + \dfrac{\pi}{2}\right)$

45. $y = -2\tan\left(2x - \dfrac{\pi}{3}\right)$ **46.** $y = 2\csc(3x + 3)$

47. (a) Prove that if f is periodic with period p, then $1/f$ is also periodic with period p.
   (b) Prove that cosecant and secant each have period 2π.

DISCOVERY · DISCUSSION

48. Reduction Formulas   Use the graphs in Figure 5 to explain why the following formulas are true.

$$\tan\left(x - \dfrac{\pi}{2}\right) = -\cot x$$

$$\sec\left(x - \dfrac{\pi}{2}\right) = \csc x$$

---

5   REVIEW

CONCEPT CHECK

1. (a) What is the unit circle?
   (b) Use a diagram to explain what is meant by the terminal point determined by a real number t.
   (c) What is the reference number $\bar{t}$ associated with t?
   (d) If t is a real number and $P(x, y)$ is the terminal point determined by t, write equations that define $\sin t$, $\cos t$, $\tan t$, $\cot t$, $\sec t$, and $\csc t$.
   (e) What are the domains of the six functions that you defined in part (d)?
   (f) Which trigonometric functions are positive in quadrants I, II, III, and IV?

2. (a) What is an even function?
   (b) Which trigonometric functions are even?
   (c) What is an odd function?
   (d) Which trigonometric functions are odd?

3. (a) State the reciprocal identities.
   (b) State the Pythagorean identities.

4. (a) What is a periodic function?
   (b) What are the periods of the six trigonometric functions?

5. Sketch the graphs of the sine and cosine functions. How is the graph of cosine related to the graph of sine?

6. Write expressions for the amplitude, period, and phase shift of the sine curve $y = a \sin k(x - b)$ and the cosine curve $y = a \cos k(x - b)$.

7. (a) Sketch the graphs of the tangent and cotangent functions.

(b) State the periods of the tangent curve $y = a \tan kx$ and the cotangent curve $y = a \cot kx$.

8. (a) Sketch the graphs of the secant and cosecant functions.

(b) State the periods of the secant curve $y = a \sec kx$ and the cosecant curve $y = a \csc kx$.

EXERCISES

1–2 ■ A point $P(x, y)$ is given.
(a) Show that P is on the unit circle.
(b) Suppose that P is the terminal point determined by t. Find $\sin t$, $\cos t$, and $\tan t$.

1. $P\left(\dfrac{1}{2}, \dfrac{\sqrt{3}}{2}\right)$

2. $P\left(\dfrac{3}{5}, -\dfrac{4}{5}\right)$

3–6 ■ A real number t is given.
(a) Find the reference number for t.
(b) Find the terminal point $P(x, y)$ on the unit circle determined by t.
(c) Find the six trigonometric functions of t.

3. $t = \dfrac{\pi}{3}$

4. $t = \dfrac{5\pi}{3}$

5. $t = \dfrac{11\pi}{4}$

6. $t = -\dfrac{7\pi}{6}$

7–16 ■ Find the value of the trigonometric function. If possible, give the exact value; otherwise, use a calculator to find an approximate value correct to five decimal places.

7. (a) $\sin \dfrac{3\pi}{4}$ (b) $\cos \dfrac{3\pi}{4}$

8. (a) $\tan \dfrac{\pi}{3}$ (b) $\tan\left(-\dfrac{\pi}{3}\right)$

9. (a) $\sin 1.1$ (b) $\cos 1.1$

10. (a) $\cos \dfrac{\pi}{5}$ (b) $\cos\left(-\dfrac{\pi}{5}\right)$

11. (a) $\cos \dfrac{9\pi}{2}$ (b) $\sec \dfrac{9\pi}{2}$

12. (a) $\sin \dfrac{\pi}{7}$ (b) $\csc \dfrac{\pi}{7}$

13. (a) $\tan \dfrac{5\pi}{2}$ (b) $\cot \dfrac{5\pi}{2}$

14. (a) $\sin 2\pi$ (b) $\csc 2\pi$

15. (a) $\tan \dfrac{\pi}{8}$ (b) $\cot \dfrac{\pi}{8}$

16. (a) $\cos \dfrac{\pi}{3}$ (b) $\sin \dfrac{\pi}{6}$

17–20 ■ Use the fundamental identities to write the first expression in terms of the second.

17. $\dfrac{\tan t}{\cos t}$, $\sin t$

18. $\tan^2 t \sec t$, $\cos t$

19. $\tan t$, $\sin t$; t in quadrant IV

20. $\sec t$, $\sin t$; t in quadrant II

21–24 ■ Find the values of the remaining trigonometric functions at t from the given information.

21. $\sin t = \frac{5}{13}$, $\cos t = -\frac{12}{13}$

22. $\sin t = -\frac{1}{2}$, $\cos t > 0$

23. $\cot t = -\frac{1}{2}$, $\csc t = \sqrt{5}/2$

24. $\cos t = -\frac{3}{5}$, $\tan t < 0$

25. If $\tan t = \frac{1}{4}$ and t is in quadrant III, find $\sec t + \cot t$.

26. If $\sin t = -\frac{8}{17}$ and t is in quadrant IV, find $\csc t + \sec t$.

27. If $\cos t = \frac{3}{5}$ and t is in quadrant I, find $\tan t + \sec t$.

28. If $\sec t = -5$ and t is in quadrant II, find $\sin^2 t + \cos^2 t$.

29–36 ■ (a) Find the amplitude, period, and phase shift of the function, and (b) sketch the graph.

29. $y = 10 \cos \frac{1}{2} x$

30. $y = 4 \sin 2\pi x$

31. $y = -\sin \frac{1}{2} x$

32. $y = 2 \sin\left(x - \dfrac{\pi}{4}\right)$

33. $y = 3 \sin(2x - 2)$

34. $y = \cos 2\left(x - \dfrac{\pi}{2}\right)$

35. $y = -\cos\left(\dfrac{\pi}{2}x + \dfrac{\pi}{6}\right)$

36. $y = 10 \sin\left(2x - \dfrac{\pi}{2}\right)$

37–40 ■ The graph of one period of a function of the form $y = a \sin k(x - b)$ or $y = a \cos k(x - b)$ is shown. Determine the function.

37.

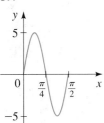

38.

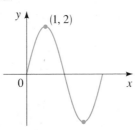

(1, 2)

39.

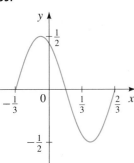

40.

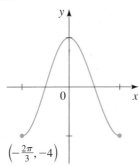

$\left(-\dfrac{2\pi}{3}, -4\right)$

41–48 ■ Find the period, and sketch the graph.

41. $y = 3 \tan x$

42. $y = \tan \pi x$

43. $y = 2 \cot\left(x - \dfrac{\pi}{2}\right)$

44. $y = \sec\left(\dfrac{1}{2}x - \dfrac{\pi}{2}\right)$

45. $y = 4 \csc(2x - \pi)$

46. $y = \tan\left(x + \dfrac{\pi}{6}\right)$

47. $y = \tan\left(\dfrac{1}{2}x - \dfrac{\pi}{8}\right)$

48. $y = -4 \sec 4\pi x$

49–54 ■

(a) Use a graphing device to graph the function.
(b) Determine from the graph whether the function is periodic and, if so, determine the period.
(c) Determine from the graph whether the function is odd, even, or neither.

49. $y = |\cos x|$

50. $y = \sin(\cos x)$

51. $y = \cos(2^{0.1x})$

52. $y = 1 + 2^{\cos x}$

53. $y = e^{-|x|} \cos 3x$

54. $y = \ln x \sin 3x \quad (x > 0)$

55–58 ■ Graph the three functions on a common screen. How are the graphs related?

55. $y = x, \quad y = -x, \quad y = x \sin x$

56. $y = 2^{-x}, \quad y = -2^{-x}, \quad y = 2^{-x} \cos 4\pi x$

57. $y = x, \quad y = \sin 4x, \quad y = x + \sin 4x$

58. $y = \sin^2 x, \quad y = \cos^2 x, \quad y = \sin^2 x + \cos^2 x$

59–60 ■ Find the maximum and minimum values of the function.

59. $y = \cos x + \sin 2x$

60. $y = \cos x + \sin^2 x$

61. Find the solutions of $\sin x = 0.3$ in the interval $[0, 2\pi]$.

62. Find the solutions of $\cos 3x = x$ in the interval $[0, \pi]$.

63. Let $f(x) = \dfrac{\sin^2 x}{x}$.

(a) Is the function f even, odd, or neither?
(b) Find the x-intercepts of the graph of f.
(c) Draw the graph of f in an appropriate viewing rectangle.
(d) Describe the behavior of the function as $x \to \infty$ and as $x \to -\infty$.
(e) Notice that $f(x)$ is not defined when $x = 0$. What happens as $x \to 0$?

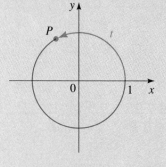

1. The point $P(x, y)$ is on the unit circle in quadrant IV. If $x = \frac{12}{13}$, find y.

2. The point P in the figure at the left has y-coordinate $\frac{4}{5}$. Find each of the following.
   (a) $\sin t$ (b) $\cos(-t)$
   (c) $\tan(\pi - t)$ (d) $\sec(\pi + t)$

3. Express $\tan t$ in terms of $\sin t$ for t in quadrant II.

4. If $\cos t = -\frac{8}{17}$ and t is in quadrant III, find $\tan t \cot t + \csc t$.

5–6 ■ (a) Find the amplitude, period, and phase shift of the function, and (b) sketch the graph.

5. $y = 2 \cos 3x$

6. $y = \sin\left(\dfrac{1}{2}x - \dfrac{\pi}{6}\right)$

7–8 ■ Find the period, and sketch the graph of the function.

7. $y = -\csc 2x$

8. $y = \tan\left(2x - \dfrac{\pi}{2}\right)$

9. The graph shown is one period of a function of the form $y = a \sin k(x - b)$. Determine the function.

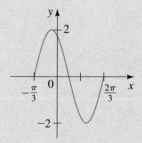

10–11 ■
(a) Use a graphing device to graph the function in an appropriate viewing rectangle.
(b) Determine from the graph if the function is even, odd, or neither.
(c) Find the minimum and maximum values of the function.

10. $y = \dfrac{\cos x}{1 + x^2}$

11. $y = e^{-\sin 3x}$

12. Let $y_1 = \cos(\sin x)$ and $y_2 = \sin(\cos x)$.
    (a) Sketch the graphs of y_1 and y_2 in the same viewing rectangle.
    (b) Determine the period of each of these functions from its graph.
    (c) Find an inequality between $\sin(\cos x)$ and $\cos(\sin x)$ that is valid for all x.

FOCUS ON MODELING

Periodic behavior—behavior that repeats over and over again—is common in nature. Perhaps the most familiar example is the daily rising and setting of the sun, which results in the repetitive pattern of day, night, day, night, ... Another example is the daily variation of tide levels at the beach, which results in the repetitive pattern of high tide, low tide, high tide, low tide, ... Certain animal populations increase and decrease in a predictable periodic pattern: a large population exhausts the food supply, which causes the population to dwindle; this in turn results in a more plentiful food supply, which makes it possible for the population to increase; and the pattern then repeats over and over.

Other common examples of periodic behavior involve motion that is caused by vibration or oscillation. A mass suspended from a spring that has been compressed and then allowed to vibrate vertically is a simple example. This same "back and forth" motion also occurs in such diverse phenomena as sound waves, light waves, alternating electrical current, and pulsating stars, to name a few. In this section we consider the problem of modeling periodic behavior.

MODELING PERIODIC BEHAVIOR

The trigonometric functions are ideally suited for modeling periodic behavior. A glance at the graphs of the sine and cosine functions, for instance, shows that these functions themselves exhibit periodic behavior. Figure 1 shows the graph of $y = \sin t$. If we think of t as time, we see that as time goes on, $y = \sin t$ increases and decreases over and over again. Figure 2 shows that the motion of a vibrating mass on a spring is modeled very accurately by $y = \sin t$.

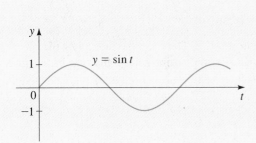

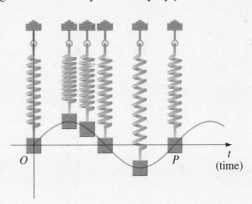

FIGURE 1
$y = \sin t$

FIGURE 2
Motion of a vibrating spring is modeled by $y = \sin t$.

Notice that the mass returns to its original position over and over again. A **cycle** is one complete vibration of an object so the mass in Figure 2 completes one cycle of its motion between O and P. Our observations about how the sine and cosine functions model periodic behavior are summarized in the following box.

SIMPLE HARMONIC MOTION

If the equation describing the displacement y of an object at time t is

$$y = a \sin \omega t \qquad \text{or} \qquad y = a \cos \omega t$$

then the object is in **simple harmonic motion**. In this case,

amplitude $= |a|$ Maximum displacement of the object

period $= \dfrac{2\pi}{\omega}$ Time required to complete one cycle

frequency $= \dfrac{\omega}{2\pi}$ Number of cycles per unit of time

---

Notice that the functions

$$y = a \sin 2\pi\nu t \qquad \text{and} \qquad y = a \cos 2\pi\nu t$$

have frequency ν, because $2\pi\nu/(2\pi) = \nu$. Since we can immediately read the frequency from these equations, we often write equations of simple harmonic motion in this form.

EXAMPLE 1 ■ A Vibrating Spring

The displacement of the mass in Figure 3 is given by

$$y = 10 \sin 4\pi t$$

where y is measured in inches and t in seconds. Describe the motion of the mass. (We assume here that there is no friction, so the spring and weight will oscillate in the same way indefinitely.)

SOLUTION   The mass is in simple harmonic motion. The amplitude, period, and frequency of the motion are

$$\text{amplitude} = |a| = 10$$

$$\text{period} = \frac{2\pi}{\omega} = \frac{2\pi}{4\pi} = \frac{1}{2}$$

$$\text{frequency} = \frac{4\pi}{2\pi} = 2$$

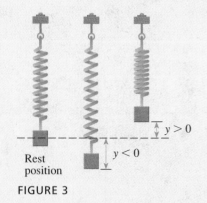

Rest position

$y > 0$

$y < 0$

FIGURE 3

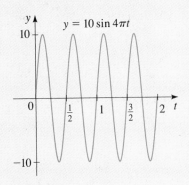

$y = 10 \sin 4\pi t$

FIGURE 4

Thus, the maximum displacement of the mass is 10 cm, it takes $\frac{1}{2}$ s to complete one cycle, and it goes through two complete cycles every second. At $t = 0$, the displacement is $y = 10 \sin 4\pi 0 = 0$, so the mass is initially at its rest position. A nice description of the motion is provided by the graph of its displacement at time t in Figure 4. Notice that y is the displacement of the mass above its rest position. So when $y > 0$, the mass is above its rest position and when $y < 0$, it's below its rest position. ■

The main difference between the two equations describing simple harmonic motion is the starting point. At $t = 0$, we get

$$y = a \sin \omega \cdot 0 = 0 \qquad \text{or} \qquad y = a \cos \omega \cdot 0 = a$$

In other words, in the first case the motion "starts" with zero displacement, while in the second case the motion "starts" with the displacement at maximum (at the amplitude a). Of course, we may think of the motion as "starting" at any point in its cycle. In this case, the equations for simple harmonic motion take the form

$$y = a \sin \omega(t + \theta) \qquad \text{and} \qquad y = a \cos \omega(t + \theta)$$

where θ determines the starting point.

EXAMPLE 2 ■ A Vibrating Spring

A mass is suspended from a spring as shown in Figure 5. The spring is compressed a distance of 4 cm and then released. It is observed that the mass returns to the compressed position after $\frac{1}{3}$ s. Describe the motion of the mass.

SOLUTION The motion of the mass is given by one of the equations for simple harmonic motion. The amplitude of the motion is 4 cm. Since this amplitude is reached at time $t = 0$ when the mass is released, the appropriate equation to use is

$$y = a \cos \omega t$$

The period $p = \frac{1}{3}$, so the frequency is $1/p = 3$. The motion of the mass is described by the equation

$$y = 4 \cos(2\pi)3t$$

$$= 4 \cos 6\pi t$$

where y is the displacement from the rest position at time t. Notice that when $t = 0$, the displacement is $y = 4$, as we expect. ■

An important situation where simple harmonic motion occurs is in the production of sound. Sound is produced by a regular variation in air pressure from the

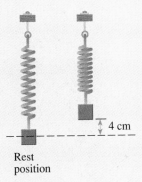

4 cm

Rest
position

FIGURE 5

normal pressure. If the pressure varies in simple harmonic motion, then a pure sound is produced. The tone of the sound depends on the frequency and the loudness depends on the amplitude.

EXAMPLE 3 ■ Sound Production

A tuba player plays the note E and sustains the sound for some time. Assuming that the sound is a pure E, its equation is given by

$$V(t) = 0.2 \sin 80\pi t$$

where $V(t)$ is the variation in pressure from normal pressure at time t, measured in pounds per square inch. Find the frequency, amplitude, and period of the sound. If the tuba player increases the loudness of the note, how does the equation for $V(t)$ change? If the player is playing the note incorrectly and it is a little flat, how does this affect $V(t)$?

SOLUTION   The frequency is $80\pi/(2\pi) = 40$, the period is $\frac{1}{40}$, and the amplitude is 0.2 (see Figure 6). If the player increases the loudness, the amplitude of the wave increases. In other words, the number 0.2 is replaced by a larger number. If the note is flat, then the frequency is less than 40. In this case, the coefficient of t is less than 80π.

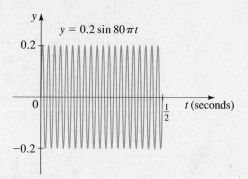

FIGURE 6

EXAMPLE 4 ■ Variable Stars

A variable star is one whose brightness alternately increases and decreases. For the most visible variable star, Delta Cephei, the time between periods of maximum brightness is 5.4 days. The average brightness (or magnitude) of the star is 4.0, and its brightness varies by ±0.35 magnitude. Express the brightness as a function of time.

SOLUTION   The brightness increases and decreases from the average brightness in simple harmonic fashion. Let's measure the time in days, taking $t = 0$

to be a time when the star is at its average brightness. Since the period of the star is 5.4 days,

$$\frac{2\pi}{\omega} = 5.4 \qquad \text{or} \qquad \omega = \frac{2\pi}{5.4}$$

Thus, the variation $V(t)$ from average brightness is

$$V(t) = a \sin \omega t = 0.35 \sin \frac{2\pi}{5.4} t$$

The graph of $V(t)$ in Figure 7 shows that at time $t = 1.35$ days, the star is at maximum brightness (0.35 magnitude above its average brightness) and at $t = 4.05$ days, it is at its minimum brightness (0.35 magnitude below its average brightness). The brightness $B(t)$ at time t is given by

$$B(t) = 4.0 + 0.35 \sin \frac{2\pi}{5.4} t$$

The graph of $B(t)$ is sketched in Figure 8.

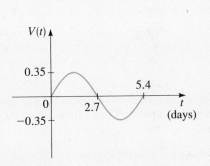

FIGURE 7

Variation from average brightness

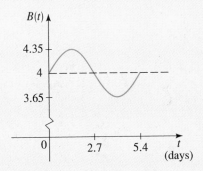

FIGURE 8

Brightness

The number of hours of daylight varies throughout the course of a year. The longest day in the Northern Hemisphere is June 21, and the shortest is December 21. The average length of daylight is 12 h, and the variation from this average depends on the latitude. (For example, Fairbanks, Alaska, experiences more than 20 h of daylight on the longest day and less than 4 h on the shortest day!). The graph in Figure 9 shows the number of hours of daylight at different times of the year for various latitudes. It's apparent from the graph that the variation in hours of daylight is simple harmonic.

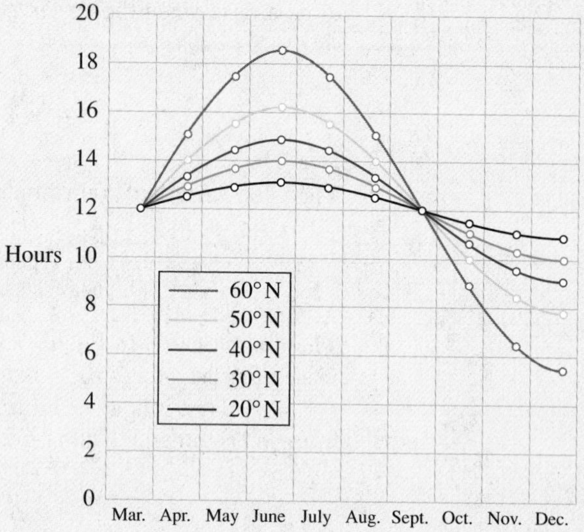

20
18
16
14
12
Hours 10
8   ——— 60° N
6   ——— 50° N
     ——— 40° N
4   ——— 30° N
     ——— 20° N
2
0
Mar. Apr. May June July Aug. Sept. Oct. Nov. Dec.

FIGURE 9

Graph of the length of daylight
from March 21 through December 21
at various latitudes

Source: Lucia C. Harrison, *Daylight, Twilight, Darkness and Time*
(New York: Silver, Burdett, 1935) page 40.

EXAMPLE 5 ■ Variation in the Length of Daylight

The length of daylight L in Philadelphia is modeled by the function

$$L(t) = 12 + 2.83 \sin\left[\frac{2\pi}{365}(t - 80)\right]$$

where t is the number of days after January 1. Find the number of hours of daylight in Philadelphia on each of the following days.

(a) March 21   (b) June 21

SOLUTION

(a) March 21 is the 80th day of the year, so the length of daylight is

$$L(80) = 12 + 2.83 \sin\left[\frac{2\pi}{365}(80 - 80)\right] = 12 \text{ h}$$

(b) June 21 is the 172nd day of the year, so the length of daylight is

$$L(172) = 12 + 2.83 \sin\left[\frac{2\pi}{365}(172 - 80)\right] \approx 14.83 \text{ h}$$

We now explain how the function in Example 5 was obtained. Philadelphia is located approximately 40° N latitude. So we need to find a sine function that represents the 40° N latitude curve in Figure 9. In Philadelphia, daylight lasts 14 h 50 min on June 21 and 9 h 10 min on December 21, so the amplitude of the curve is $14\frac{5}{6} - 12 = \frac{17}{6} \approx 2.83$. Also, since there are approximately 365 days in a year, the period is 365, so we have $2\pi/\omega = 365$, or $\omega = 2\pi/365$. The graph is shifted upward 12 units, and since the sine curve begins its cycle on March 21 (the 80th day of the year), the curve is shifted right 80 units. This information leads to the function L in the example.

DAMPED HARMONIC MOTION

The spring in Figure 2 is assumed to oscillate in a frictionless environment. In this hypothetical case, the amplitude of the oscillation will not change. In the presence of friction, however, the motion of the spring will eventually "die down"; that is, the amplitude of the motion will decrease with time. Motion of this type is called *damped harmonic motion*.

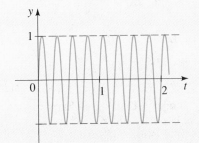

(a) Harmonic motion: $y = \sin 8\pi t$

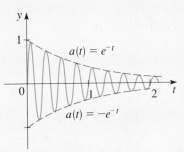

(b) Damped harmonic motion:
$y = e^{-t} \sin 8\pi t$

FIGURE 10

> ### DAMPED HARMONIC MOTION
>
> If the equation describing the displacement y of an object at time t is
>
> $$y = ke^{-ct} \sin \omega t \qquad \text{or} \qquad y = ke^{-ct} \cos \omega t \qquad (c > 0)$$
>
> then the object is in **damped harmonic motion**. The constant c is the **damping constant**.

Damped harmonic motion is simply harmonic motion for which the amplitude is governed by the function $a(t) = ke^{-ct}$. Figure 10 shows the difference between harmonic motion and damped harmonic motion.

The larger the damping constant c, the quicker the oscillation dies down. When a guitar string is plucked and then allowed to vibrate freely, a point on that string undergoes damped harmonic motion. We can hear the damping of the motion as the sound produced by the vibration of the string fades. How fast the damping of the string occurs (as measured by the size of the constant c) is a property of the size of the string and the material it is made of. Another example of damped harmonic motion is the motion that a shock absorber on a car undergoes when the car hits a bump in the road. In this case, the shock absorber is engineered to damp the motion as quickly as possible (large c) and to have the frequency as small as possible (small ω). On the other hand, the sound produced by a tuba player playing a note is undamped as long as the player can maintain the

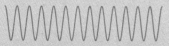

loudness of the note. The electromagnetic waves that produce light move in simple harmonic motion that is not damped.

EXAMPLE 6 ■ A Vibrating Violin String

The G-string on a violin is pulled a distance of 0.5 cm above its rest position, then released and allowed to vibrate. The damping constant c for this string is determined to be 1.4. Suppose that the note produced is a pure G (frequency = 200 cycles per second). Find the equation that describes the motion of the point at which the string was plucked.

SOLUTION   Let P be the point at which the string was plucked. We will find a function $f(t)$ that gives the distance at time t of the point P from its original rest position. Since the maximum displacement occurs at $t = 0$, we find an equation in the form

$$y = ke^{-ct}\cos \omega t$$

From this equation, we see that $f(0) = k$. But we know that the original displacement of the string is 0.5 cm. Thus, $k = 0.5$. Since the frequency of the vibration is 200, we have $\omega/(2\pi) = 200$ or $\omega = (200)(2\pi)$. Finally, since we know that the damping constant is 1.4, we get

$$f(t) = 0.5e^{-1.4t}\cos 400\pi t$$

EXAMPLE 7 ■ Ripples on a Pond

A stone is dropped in a calm lake, causing waves to form. The up-and-down motion of a point on the surface of the water is modeled by damped harmonic motion. At some time the amplitude of the wave is measured, and 20 s later it is found that the amplitude had dropped to $\frac{1}{10}$ of this value. Find the damping constant c.

SOLUTION   The amplitude is governed by the coefficient ke^{-ct} in the equations for damped harmonic motion. Thus, the amplitude at time t is ke^{-ct}, and 20 s later, it is $ke^{-c(t+20)}$. So, from the given information

$$\frac{ke^{-ct}}{ke^{-c(t+20)}} = 10$$

$$ke^{-c(t+20)} = \tfrac{1}{10}ke^{-ct}$$

We now solve this equation for c. Canceling k and using the Laws of Exponents, we get

$$e^{-ct} \cdot e^{-20c} = \tfrac{1}{10}e^{-ct}$$

$$e^{-20c} = \tfrac{1}{10} \qquad \text{Cancel } e^{-ct}$$

$$e^{20c} = 10 \qquad \text{Take reciprocals}$$

Taking the natural logarithm for each side gives

$$20c = \ln(10)$$

$$c = \tfrac{1}{20}\ln(10) \approx \tfrac{1}{20}(2.30) \approx 0.12$$

Thus, the damping constant is $c \approx 0.12$.

PROBLEMS

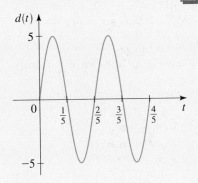

1. The graph at the left gives the displacement $d(t)$ at time t of an object in simple harmonic motion. Find a formula for the function $d(t)$.

2. The graph shows the variation of the water level relative to mean sea level in Commencement Bay at Tacoma, Washington, for a particular 24-hour period. Assuming that this variation is modeled by simple harmonic motion, find an equation of the form $y = a \sin \omega t$ that describes the variation in water level as a function of the number of hours after midnight.

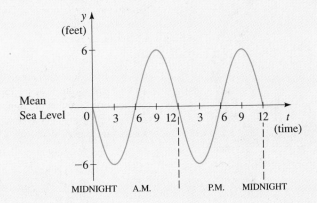

3. The Bay of Fundy in Nova Scotia has the highest tides in the world. In one 12-hour period, the water starts at mean sea level, rises to 21 ft above, drops to 21 ft below, and then returns to mean sea level. Assuming that the motion of the tides is simple harmonic, find an equation that describes the height of the tide in the Bay of Fundy above mean sea level. Sketch a graph that shows the level of the tides over a 24-hour period.

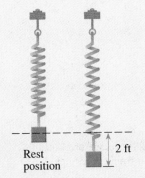

4. A mass suspended from a spring is pulled down a distance of 2 ft from its rest position, as shown in the figure at the left. The mass is released at time $t = 0$ and allowed to oscillate. If the mass returns to this position after 1 s, find an equation that describes its motion.

5. A mass is suspended on a spring. The spring is compressed so that the mass is located 5 cm above its rest position. The mass is released at time $t = 0$ and allowed to oscillate. It is observed that the mass reaches its lowest point $\frac{1}{2}$ s after it is released. Find an equation that describes the motion of the mass.

6. The frequency of oscillation of an object suspended on a spring depends on the stiffness k of the spring (called the *spring constant*) and the mass m of the object. If the spring is compressed a distance a and then allowed to oscillate, its displacement is given by

$$f(t) = a \cos \sqrt{k/m}\ t$$

(a) A 10-g mass is suspended from a spring with stiffness $k = 3$. If the spring is compressed a distance 5 cm and then released, find the equation that describes the oscillation of the spring.

(b) Find a general formula for the frequency (in terms of k and m).

(c) How is the frequency affected if the mass is increased? Is the oscillation faster or slower?

(d) How is the frequency affected if a stiffer spring is used (larger k)? Is the oscillation faster or slower?

7. A ferris wheel has a radius of 10 m, and the bottom of the wheel passes 1 m above the ground. If the ferris wheel makes one complete revolution every 20 s, find an equation that gives the height above the ground of a person on the ferris wheel as a function of time.

8. The pendulum in a grandfather clock makes one complete swing every 2 s. The maximum angle that the pendulum makes with respect to its rest position is 10°. We know from physical principles that the angle θ between the pendulum and its rest position changes in simple harmonic fashion. Find an equation that describes the size of the angle θ as a function of time. (Take $t = 0$ to be a time when the pendulum is vertical.)

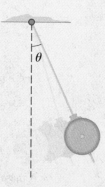

9. The variable star Zeta Gemini has a period of 10 days. The average brightness of the star is 3.8 magnitudes, and the maximum variation from the average is

0.2 magnitude. Assuming that the variation in brightness is simple harmonic, find an equation that gives the brightness of the star as a function of time.

10. Astronomers believe that the radius of a variable star increases and decreases with the brightness of the star. The variable star Delta Cephei (Example 4) has an average radius of 20 million miles and changes by a maximum of 1.5 million miles from this average during a single pulsation. Find an equation that describes the radius of this star as a function of time.

11. An *alternator,* or alternating current (ac) generator, produces an electrical current by rotating an *armature* (or coil) in a magnetic field. The figure represents a simple version of such a generator. As the wire passes through the magnetic field, a voltage E is generated in the wire. It can be shown that the voltage generated is given by

$$E(t) = E_0 \cos \omega t$$

where E_0 is the maximum voltage produced (which depends on the strength of the magnetic field) and $\omega/(2\pi)$ is the number of revolutions per second of the armature (the frequency). Ordinary 110-V household alternating current varies from $+155$ V to -155 V with a frequency of 60 Hz (cycles per second). Find an equation that describes this variation in voltage.

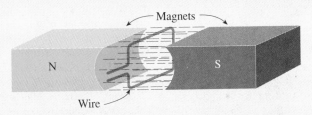

12. The armature in an electric generator is rotating at the rate of 100 revolutions per second (rps). If the maximum voltage produced is 310 V (see the marginal note), find an equation that describes this variation in voltage. What is the rms voltage?

13. The graph shows an oscilloscope reading of the variation in voltage of an ac current produced by a simple generator.

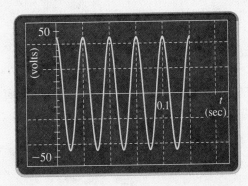

Why do we say that household current is 110 V when the maximum voltage produced is 155 V? From the symmetry of the cosine function, we see that the average voltage produced is zero. This average value would be the same for all ac generators and so gives no information about the voltage generated. To obtain a more informative measure of voltage, engineers use the **root-mean-square** (rms) method. It can be shown that the rms voltage is $1/\sqrt{2}$ times the maximum voltage. So, for household current the rms voltage is

$$155 \times \frac{1}{\sqrt{2}} \approx 110 \text{ V}$$

(a) Find the maximum voltage produced.
(b) Find the frequency (cycles per second) of the generator.
(c) How many revolutions per second does the armature in the generator make?
(d) Find a formula that describes the variation in voltage as a function of time.

14. A train whistle produces a sharp, pure tone. While the train is moving toward us, the pitch of the whistle increases; it then decreases as the train moves away from us. This phenomenon is known as the *Doppler effect*. Suppose that the maximum change in pressure caused by the whistle is 10 N/m² and the frequency is 20 cycles per second.

   (a) Find an equation that describes the variation in pressure as a function of time, and sketch the graph.

   (b) How does the graph in part (a) change as the train moves toward us? Away from us?

15. A strong gust of wind strikes a tall building, causing it to sway back and forth in damped harmonic motion. The frequency of the oscillation is 0.5 cycle per second and the damping constant is $c = 0.9$. Find an equation that describes the motion of the building. (Assume $k = 1$ and take $t = 0$ to be the instant when the gust of wind strikes the building.)

16. When a car hits a certain bump on the road, a shock absorber on the car is compressed a distance of 6 in., then released (see the figure). The shock absorber vibrates in damped harmonic motion with a frequency of 2 cycles per second. The damping constant for this particular shock absorber is 2.8.

   (a) Find an equation that describes the displacement of the shock absorber from its rest position as a function of time. Take $t = 0$ to be the instant that the shock absorber is released.

   (b) How long does it take for the amplitude of the vibration to decrease to 0.5 in.?

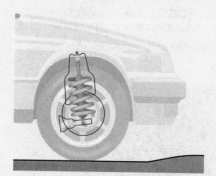 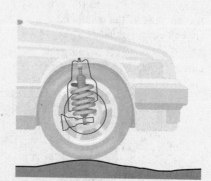 

17. A tuning fork is struck and oscillates in damped harmonic motion. The amplitude of the motion is measured, and 3 s later it is found that the amplitude has dropped to $\frac{1}{4}$ of this value. Find the damping constant c for this tuning fork.

18. A guitar string is pulled at the point P a distance of 3 cm above its rest position. It is then released and vibrates in damped harmonic motion with a frequency of 165 cycles per second. After 2 s, it is observed that the amplitude of the vibration at the point P is 0.6 cm.

(a) Find the damping constant c.

(b) Find an equation that describes the position of the point P above its rest position as a function of time. Take $t = 0$ to be the instant that the string is released.

6

TRIGONOMETRIC FUNCTIONS OF ANGLES

Trigonometric functions of angles are important in surveying, navigation, and astronomy, where they are used to find distances to nearby stars.

The greatest mathematicians, as Archimedes, Newton, and Gauss, always united theory and applications in equal measure.

FELIX KLEIN

In this chapter we present the trigonometric functions as functions of angles, that is, functions that associate to each *angle* a real number. For example, the function sine is defined here as one that assigns to each angle θ a real number $\sin\theta$. Of course, the measure of the angle θ is a real number, so we may view the trigonometric functions of angles as functions of real numbers. Viewed in this way, they are identical with the trigonometric functions defined in Chapter 5. Thus, the difference is one of point of view, but the resulting trigonometric functions are the same.

Why do we study trigonometry in two different ways? Because different applications require that we view these functions differently. For example, when we apply the trigonometric functions to the study of harmonic motion, we must view them as functions of time, a real number. On the other hand, in many applications it's imperative to view them as functions of angles. For example, to find the distance to the sun, we can first measure the angle formed by the sun, earth, and moon. Using this information and the distance to the moon, together with a knowledge of the trigonometric functions of *angles,* we can easily solve this interesting problem (see Exercise 49 of Section 6.2). Put simply, the reason the trigonometric functions are so useful and important is that they can be viewed in these different ways.

The two approaches to trigonometry are independent of each other, so either Chapter 5 or Chapter 6 may be studied first.

6.1 ANGLE MEASURE

An **angle** AOB consists of two rays R_1 and R_2 with a common vertex O (see Figure 1). We often interpret an angle as a rotation of the ray R_1 onto R_2. In this case, R_1 is called the **initial side**, and R_2 is called the **terminal side** of the angle. If the rotation is counterclockwise, the angle is considered **positive**, and if the rotation is clockwise, the angle is considered **negative**.

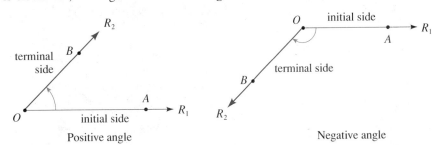

FIGURE 1   Positive angle   Negative angle

ANGLE MEASURE

The **measure** of an angle is the amount of rotation about the vertex required to move R_1 onto R_2. Intuitively, this is how much the angle "opens." One unit of measurement for angles is the **degree**. An angle of measure 1 degree is formed by rotating the initial side $\frac{1}{360}$ of a complete revolution. In calculus and other branches of mathematics, a more natural method of measuring angles is used—*radian measure.* The amount an angle opens is measured along the arc of a circle of radius 1 with center at the vertex of the angle.

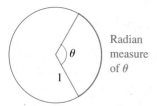

FIGURE 2

> ### DEFINITION OF RADIAN MEASURE
>
> If a circle of radius 1 is drawn with the vertex of an angle at its center, then the measure of this angle in **radians** (abbreviated rad) is the length of the arc that subtends the angle (see Figure 2).

The circumference of the circle of radius 1 is 2π and so a complete revolution has measure 2π rad, a straight angle has measure π rad, and a right angle has measure $\pi/2$ rad. An angle that is subtended by an arc of length 2 along the unit circle has radian measure 2 (see Figure 3).

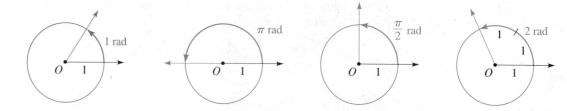

FIGURE 3
Radian measure

Since a complete revolution measured in degrees is 360° and measured in radians is 2π rad, we get the following simple relationship between these two methods of angle measurement.

> ### RELATIONSHIP BETWEEN DEGREES AND RADIANS
>
> $$180° = \pi \text{ rad} \qquad 1 \text{ rad} = \left(\frac{180}{\pi}\right)° \qquad 1° = \frac{\pi}{180} \text{ rad}$$
>
> **1.** To convert degrees to radians, multiply by $\dfrac{\pi}{180}$.
>
> **2.** To convert radians to degrees, multiply by $\dfrac{180}{\pi}$.

To get some idea of the size of a radian, notice that

$$1 \text{ rad} \approx 57.296° \qquad \text{and} \qquad 1° \approx 0.01745 \text{ rad}$$

An angle θ of measure 1 rad is sketched in Figure 4.

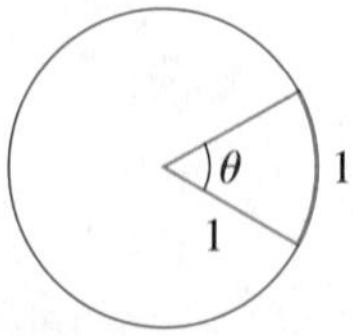

Measure of $\theta = 1$ rad
Measure of $\theta \approx 57.296°$

FIGURE 4

EXAMPLE 1 ■ Converting between Radians and Degrees

(a) Express $60°$ in radians.

(b) Express $\dfrac{\pi}{6}$ rad in degrees.

SOLUTION The relationship between degrees and radians gives

(a) $60° = 60\left(\dfrac{\pi}{180}\right)$ rad $= \dfrac{\pi}{3}$ rad

(b) $\dfrac{\pi}{6}$ rad $= \left(\dfrac{\pi}{6}\right)\left(\dfrac{180}{\pi}\right) = 30°$ ■

A note on terminology: We often use a phrase such as "a $30°$ angle" to mean *an angle whose measure is* $30°$. Also, for an angle θ, we write $\theta = 30°$ or $\theta = \pi/6$ to mean *the measure of θ is $30°$ or $\pi/6$ rad*. When no unit is given, the angle is assumed to be measured in radians.

ANGLES IN STANDARD POSITION

An angle is in **standard position** if it is drawn in the xy-plane with its vertex at the origin and its initial side on the positive x-axis. Figure 5 gives examples of angles in standard position.

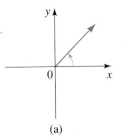

(a)

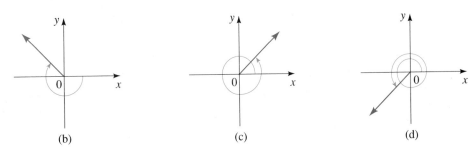

(b) (c) (d)

FIGURE 5
Angles in standard position

Two angles in standard position are **coterminal** if their sides coincide. In Figure 5 the angles in (a) and (c) are coterminal.

EXAMPLE 2 ■ Coterminal Angles

(a) Find angles that are coterminal with the angle $\theta = 30°$ in standard position.

(b) Find angles that are coterminal with the angle $\theta = \dfrac{\pi}{3}$ in standard position.

SOLUTION

(a) To find positive angles that are coterminal with θ, we add any multiple of $360°$. Thus

$$30° + 360° = 390° \qquad \text{and} \qquad 30° + 720° = 750°$$

are coterminal with $\theta = 30°$. To find negative angles that are coterminal with θ, we subtract any multiple of 360°. Thus

$$30° - 360° = -330° \qquad \text{and} \qquad 30° - 720° = -690°$$

are coterminal with θ. See Figure 6.

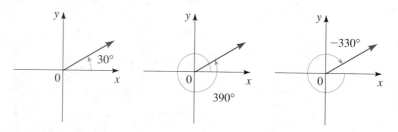

FIGURE 6

(b) To find positive angles that are coterminal with θ, we add any multiple of 2π. Thus

$$\frac{\pi}{3} + 2\pi = \frac{7\pi}{3} \qquad \text{and} \qquad \frac{\pi}{3} + 4\pi = \frac{13\pi}{3}$$

are coterminal with $\theta = \pi/3$. To find negative angles that are coterminal with θ, we subtract any multiple of 2π. Thus

$$\frac{\pi}{3} - 2\pi = -\frac{5\pi}{3} \qquad \text{and} \qquad \frac{\pi}{3} - 4\pi = -\frac{11\pi}{3}$$

are coterminal with θ. See Figure 7.

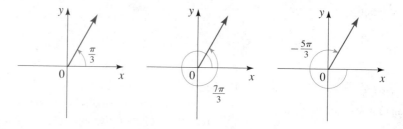

FIGURE 7

EXAMPLE 3 ■ Coterminal Angles

Find an angle with measure between 0° and 360° that is coterminal with the angle of measure 1290° in standard position.

SOLUTION We can subtract 360° as many times as we wish from 1290°, and the resulting angle will be coterminal with 1290°. Thus, $1290° - 360° = 930°$ is coterminal with 1290°, and so is the angle $1290° - 2(360°) = 570°$.

To find the angle that we want between 0° and 360°, we need to subtract 360° from 1290° as many times as necessary. An efficient way to do this is to determine how many times 360° goes into 1290°, that is, to divide 1290 by 360, and the remainder will be the angle we are looking for. We see that 360 goes into 1290 three times with a remainder of 210. Thus, 210° is the desired angle (see Figure 8).

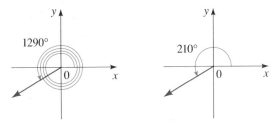

FIGURE 8

LENGTH OF A CIRCULAR ARC

An angle whose radian measure is θ is subtended by an arc that is the fraction $\theta/(2\pi)$ of the circumference of a circle. Thus, in a circle of radius r, the length s of an arc that subtends the angle θ (see Figure 9) is

$$s = \frac{\theta}{2\pi} \times \text{(circumference of circle)}$$

$$= \frac{\theta}{2\pi}(2\pi r) = \theta r$$

FIGURE 9
$s = \theta r$

LENGTH OF A CIRCULAR ARC

In a circle of radius r, the length s of an arc that subtends a central angle of θ radians is

$$s = r\theta$$

Solving for θ, we get the important formula

$$\theta = \frac{s}{r}$$

This formula allows us to define radian measure using a circle of any radius r: The radian measure of an angle θ is s/r, where s is the length of the circular arc that subtends θ in a circle of radius r (see Figure 10).

FIGURE 10

The radian measure of θ is the number of "radiuses" that can fit in the arc that subtends θ; hence the term *radian*.

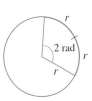

EXAMPLE 4 ■ Arc Length and Angle Measure

(a) Find the length of an arc of a circle with radius 10 m that subtends a central angle of 30°.

(b) A central angle θ in a circle of radius 4 m is subtended by an arc of length 6 m. Find the measure of θ in radians.

SOLUTION

(a) From Example 1(b) we see that $30° = \pi/6$ rad. So the length of the arc is

$$s = r\theta = (10)\frac{\pi}{6} = \frac{5\pi}{3} \text{ m}$$

⊘ The formula $s = r\theta$ is true only when θ is measured in radians.

(b) By the formula $\theta = s/r$, we have

$$\theta = \frac{s}{r} = \frac{6}{4} = \frac{3}{2} \text{ rad}$$ ■

AREA OF A CIRCULAR SECTOR

The area of a circle of radius r is $A = \pi r^2$. A sector of this circle with central angle θ has an area that is the fraction $\theta/(2\pi)$ of the area of the entire circle (see Figure 11). So the area of this sector is

$$A = \frac{\theta}{2\pi} \times \text{(area of circle)}$$

$$= \frac{\theta}{2\pi}(\pi r^2) = \frac{1}{2}r^2\theta$$

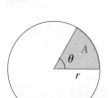

FIGURE 11
$A = \frac{1}{2}r^2\theta$

> **AREA OF A CIRCULAR SECTOR**
>
> In a circle of radius r, the area A of a sector with central angle of θ radians is
>
> $$A = \tfrac{1}{2}r^2\theta$$

EXAMPLE 5 ■ Area of a Sector

Find the area of a sector of a circle with central angle 60° if the radius of the circle is 3 m.

SOLUTION To use the formula for the area of a circular sector, we must find the central angle of the sector in radians: $60° = 60(\pi/180)$ rad $= \pi/3$ rad. Thus, the area of the sector is

⊘ The formula $A = \frac{1}{2}r^2\theta$ is true only when θ is measured in radians.

$$A = \frac{1}{2}r^2\theta = \frac{1}{2}(3)^2\left(\frac{\pi}{3}\right) = \frac{3\pi}{2} \text{ m}^2$$ ■

| 6.1 | **EXERCISES** |

1–9 ■ Find the radian measure of the angle with the given degree measure.

1. 40°   **2.** 330°   **3.** 72°

4. −30°   **5.** 45°   **6.** −80°

7. 765°   **8.** −150°   **9.** 36°

10–18 ■ Find the degree measure of the angle with the given radian measure.

10. $\dfrac{3\pi}{4}$ **11.** $-\dfrac{7\pi}{2}$ **12.** $\dfrac{5\pi}{6}$

13. 2   **14.** 1.5   **15.** $\dfrac{2\pi}{9}$

16. $-\dfrac{\pi}{12}$ **17.** $\dfrac{\pi}{5}$ **18.** $\dfrac{\pi}{18}$

19–24 ■ The measure of an angle in standard position is given. Find two positive angles and two negative angles that are coterminal with the given angle.

19. 300°   **20.** 135°   **21.** $\dfrac{3\pi}{4}$

22. $\dfrac{11\pi}{6}$ **23.** $-\dfrac{\pi}{4}$ **24.** −50°

25–30 ■ The measures of two angles in standard position are given. Determine whether the angles are coterminal.

25. 70°, 430°   **26.** −30°, 330°

27. $\dfrac{5\pi}{6}$, $\dfrac{17\pi}{6}$ **28.** $\dfrac{32\pi}{3}$, $\dfrac{11\pi}{3}$

29. 155°, 875°   **30.** 50°, 340°

31–36 ■ Find an angle between 0° and 360° that is coterminal with the given angle.

31. 733°   **32.** 361°

33. 2223°   **34.** −100°

35. −800°   **36.** 1270°

37–42 ■ Find an angle between 0 and 2π that is coterminal with the given angle.

37. $\dfrac{12\pi}{5}$ **38.** $-\dfrac{7\pi}{3}$ **39.** 87π

40. 10   **41.** $\dfrac{17\pi}{4}$ **42.** $\dfrac{51\pi}{2}$

43. Find the length of the arc s in the figure.

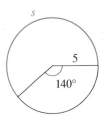

44. Find the angle θ in the figure.

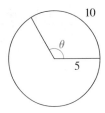

45. Find the radius r of the circle in the figure.

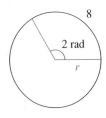

46. Find the length of an arc that subtends a central angle of 45° in a circle of radius 10 m.

47. Find the length of an arc that subtends a central angle of 2 rad in a circle of radius 2 mi.

48. A central angle θ in a circle of radius 5 m is subtended by an arc of length 6 m. Find the measure of θ in degrees and in radians.

49. An arc of length 100 m subtends a central angle θ in a circle of radius 50 m. Find the measure of θ in degrees and in radians.

50. A circular arc of length 3 ft subtends a central angle of 25°. Find the radius of the circle.

51. Find the radius of the circle if an arc of length 6 m on the circle subtends a central angle of $\pi/6$ rad.

52. How many revolutions will a car wheel of diameter 30 in. make as the car travels a distance of one mile?

53. Pittsburgh, Pennsylvania, and Miami, Florida, lie approximately on the same meridian. Pittsburgh has a latitude of 40.5° N and Miami, 25.5° N. Find the distance between these two cities. (The radius of the earth is 3960 mi.)

54. Memphis, Tennessee, and New Orleans, Louisiana, lie approximately on the same meridian. Memphis has latitude 30° N and New Orleans, 35° N. Find the distance between these two cities. (The radius of the earth is 3960 mi.)

55. Find the distance that the earth travels in one day in its path around the sun. Assume that a year has 365 days and that the path of the earth around the sun is a circle of radius 93 million miles. [The path of the earth around the sun is actually an *ellipse* with the sun at one focus (see Section 9.2). This ellipse, however, has very small eccentricity, so it is nearly circular.]

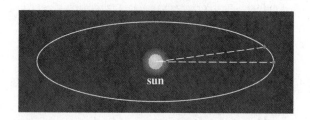

56. The Greek mathematician Eratosthenes (ca. 276–195 B.C.) measured the circumference of the earth from the following observations. He noticed that on a certain day the sun shone directly down a deep well in Syene (modern Aswan). At the same time in Alexandria,

500 miles north (on the same meridian), the rays of the sun shone at an angle of 7.2° to the zenith. Use this information and the figure to find the radius and circumference of the earth. (The data used in this problem are more accurate than those available to Eratosthenes.)

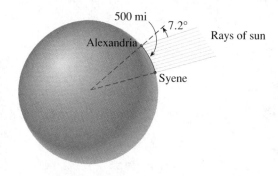

57. Find the distance along an arc on the surface of the earth that subtends a central angle of 1 minute (1 minute = $\frac{1}{60}$ degree). This distance is called a *nautical mile*. (The radius of the earth is 3960 mi.)

58. Find the area of the sector shown in the figure.

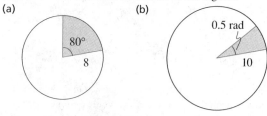

59. Find the area of a sector with central angle 1 rad in a circle of radius 10 m.

60. A sector of a circle has a central angle of 60°. Find the area of the sector if the radius of the circle is 3 mi.

61. The area of a sector of a circle with a central angle of 2 rad is 16 m². Find the radius of the circle.

62. A sector of a circle of radius 24 mi has an area of 288 mi². Find the central angle of the sector.

63. The area of a circle is 72 cm². Find the area of a sector of this circle that subtends a central angle of $\pi/6$ rad.

64. Three circles with radii 1, 2, and 3 ft are externally tangent to one another, as shown in the figure. Find the area of the sector of the circle of radius 1 that is cut off by the line segments joining the center of that circle to the centers of the other two circles.

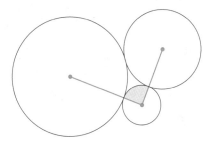

DISCOVERY · DISCUSSION

65. Different Ways of Measuring Angles The custom of measuring angles using degrees, with 360° in a circle, dates back to the ancient Babylonians, who used a number system based on groups of 60. Another system of measuring angles divides the circle into 400 units, called *grads*. In this system a right angle is 100 grad, so this fits in with our base 10 number system.

Write a short essay comparing the advantages and disadvantages of these two systems and the radian system of measuring angles. Which system do you prefer?

66. Clocks and Angles In one hour, the minute hand on a clock moves through a complete circle, and the hour hand moves through $\frac{1}{12}$ of a circle. Through how many radians do the minute and the hour hand move between 1:00 P.M. and 6:45 P.M. (on the same day)?

6.2 TRIGONOMETRY OF RIGHT TRIANGLES

In this section we study certain ratios of the sides of right triangles, called trigonometric ratios, and give several applications.

TRIGONOMETRIC RATIOS

Consider a right triangle with θ as one of its acute angles. The trigonometric ratios are defined as follows (see Figure 1).

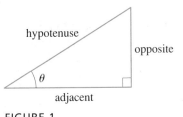

FIGURE 1

THE TRIGONOMETRIC RATIOS		
$\sin\theta = \dfrac{\text{opposite}}{\text{hypotenuse}}$	$\cos\theta = \dfrac{\text{adjacent}}{\text{hypotenuse}}$	$\tan\theta = \dfrac{\text{opposite}}{\text{adjacent}}$
$\csc\theta = \dfrac{\text{hypotenuse}}{\text{opposite}}$	$\sec\theta = \dfrac{\text{hypotenuse}}{\text{adjacent}}$	$\cot\theta = \dfrac{\text{adjacent}}{\text{opposite}}$

The symbols we use for these ratios are abbreviations for their full names: **sine, cosine, tangent, cosecant, secant, cotangent**. Since any two right triangles with angle θ are similar, these ratios are the same, regardless of the size of the triangle; they depend only on the angle θ (see Figure 2).

FIGURE 2

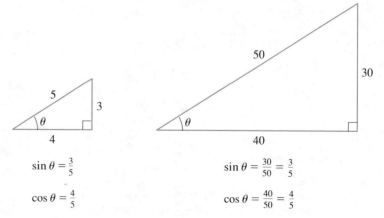

$$\sin \theta = \tfrac{3}{5} \qquad\qquad\qquad \sin \theta = \tfrac{30}{50} = \tfrac{3}{5}$$

$$\cos \theta = \tfrac{4}{5} \qquad\qquad\qquad \cos \theta = \tfrac{40}{50} = \tfrac{4}{5}$$

EXAMPLE 1 ■ Finding Trigonometric Ratios

Find the six trigonometric ratios of the angle θ in Figure 3.

SOLUTION

FIGURE 3

$$\sin \theta = \frac{2}{3} \qquad \cos \theta = \frac{\sqrt{5}}{3} \qquad \tan \theta = \frac{2}{\sqrt{5}}$$

$$\csc \theta = \frac{3}{2} \qquad \sec \theta = \frac{3}{\sqrt{5}} \qquad \cot \theta = \frac{\sqrt{5}}{2}$$ ■

EXAMPLE 2 ■ Finding Trigonometric Ratios

If $\cos \alpha = \tfrac{3}{4}$, sketch a right triangle with acute angle α, and find the other five trigonometric ratios of α.

SOLUTION   Since $\cos \alpha$ is defined as the ratio of the adjacent side to the hypotenuse, we sketch a triangle with hypotenuse of length 4 and a side of length 3 adjacent to α. If the opposite side is x, then by the Pythagorean Theorem, $3^2 + x^2 = 4^2$ or $x^2 = 7$, so $x = \sqrt{7}$. We then use the triangle in Figure 4 to find the ratios.

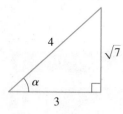

FIGURE 4

$$\sin \alpha = \frac{\sqrt{7}}{4} \qquad \cos \alpha = \frac{3}{4} \qquad \tan \alpha = \frac{\sqrt{7}}{3}$$

$$\csc \alpha = \frac{4}{\sqrt{7}} \qquad \sec \alpha = \frac{4}{3} \qquad \cot \alpha = \frac{3}{\sqrt{7}}$$ ■

SPECIAL TRIANGLES

Certain right triangles have ratios that can be calculated easily from the Pythagorean Theorem. Since they are used frequently, we mention them here.

Hipparchus (circa 140 B.C.) is considered the founder of trigonometry. He constructed tables for a function closely related to the modern sine function and evaluated for angles at half-degree intervals. These are considered the first trigonometric tables. He used his tables mainly to calculate the paths of the planets through the heavens.

The first triangle is obtained by drawing a diagonal in a square of side 1 (see Figure 5). By the Pythagorean Theorem this diagonal has length $\sqrt{2}$. The resulting triangle has angles 45°, 45°, and 90° (or $\pi/4$, $\pi/4$, and $\pi/2$). To get the second triangle, we start with an equilateral triangle ABC of side 2 and draw the perpendicular bisector DB of the base, as in Figure 6. By the Pythagorean Theorem the length of DB is $\sqrt{3}$. Since DB also bisects angle ABC, we obtain a triangle with angles 30°, 60°, and 90° (or $\pi/6$, $\pi/3$, and $\pi/2$).

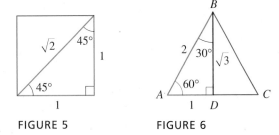

FIGURE 5   FIGURE 6

We can now use the special triangles in Figures 5 and 6 to calculate the trigonometric ratios for angles with measures 30°, 45°, and 60° (or $\pi/6$, $\pi/4$, and $\pi/3$). These are listed in Table 1.

Table 1 Values of the trigonometric ratios for special angles

θ in degrees	θ in radians	$\sin \theta$	$\cos \theta$	$\tan \theta$	$\csc \theta$	$\sec \theta$	$\cot \theta$
30°	$\dfrac{\pi}{6}$	$\dfrac{1}{2}$	$\dfrac{\sqrt{3}}{2}$	$\dfrac{\sqrt{3}}{3}$	2	$\dfrac{2\sqrt{3}}{3}$	$\sqrt{3}$
45°	$\dfrac{\pi}{4}$	$\dfrac{\sqrt{2}}{2}$	$\dfrac{\sqrt{2}}{2}$	1	$\sqrt{2}$	$\sqrt{2}$	1
60°	$\dfrac{\pi}{3}$	$\dfrac{\sqrt{3}}{2}$	$\dfrac{1}{2}$	$\sqrt{3}$	$\dfrac{2\sqrt{3}}{3}$	2	$\dfrac{\sqrt{3}}{3}$

It's useful to remember these special trigonometric ratios because they occur often. Of course, they can be recalled easily if we remember the triangles from which they are obtained.

To find the values of the trigonometric ratios for other angles, we use a calculator. Mathematical methods (called *numerical methods*) used in finding the trigonometric ratios are programmed directly into scientific calculators. For instance, when the $\boxed{\text{SIN}}$ key is pressed, the calculator computes an approximation to the value of the sine of the given angle. Calculators give the values of sine, cosine, and tangent; the other ratios can be easily calculated from these using the following *reciprocal relations:*

For an explanation of numerical methods, see the marginal note on page 363.

$$\csc t = \frac{1}{\sin t} \qquad \sec t = \frac{1}{\cos t} \qquad \cot t = \frac{1}{\tan t}$$

You should check that these relations follow immediately from the definitions of the trigonometric ratios.

We follow the convention that when we write sin *t, we mean the sine of the angle whose radian measure is t.* For instance, sin 1 means the sine of the angle whose radian measure is 1. When using a calculator to find an approximate value for this number, set your calculator to radian mode; you will find that

$$\sin 1 \approx 0.841471$$

If you want to find the sine of the angle whose measure is 1°, set your calculator to degree mode; you will find that

$$\sin 1° \approx 0.0174524$$

EXAMPLE 3 ■ Using a Calculator to Find Trigonometric Ratios

With our calculator in degree mode, and writing the results correct to five decimal places, we find

$$\sin 17° \approx 0.29237 \qquad \sec 88° = \frac{1}{\cos 88°} \approx 28.65371$$

With our calculator in radian mode, and writing the results correct to five decimal places, we find

$$\cos 1.2 \approx 0.36236 \qquad \cot 1.54 = \frac{1}{\tan 1.54} \approx 0.03081$$ ■

APPLICATIONS OF TRIGONOMETRY OF RIGHT TRIANGLES

A triangle has six parts: three angles and three sides. To **solve a triangle** means to determine all of its parts from the information known about the triangle, that is, to determine the lengths of the three sides and the measures of the three angles.

EXAMPLE 4 ■ Solving a Right Triangle

Solve triangle *ABC,* shown in Figure 7.

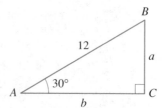

FIGURE 7

SOLUTION It's clear that $\angle B = 60°$. To find a, we look for an equation that relates a to the lengths and angles we already know. In this case, we have

$\sin 30° = a/12$, so

$$a = 12 \sin 30° = 12(\tfrac{1}{2}) = 6$$

Similarly, $\cos 30° = b/12$, so

$$b = 12 \cos 30° = 12\left(\frac{\sqrt{3}}{2}\right) = 6\sqrt{3}$$

∎

It's very useful to know that, using the information given in Figure 8, the lengths of the legs of a right triangle are

$$a = r \sin \theta \qquad \text{and} \qquad b = r \cos \theta$$

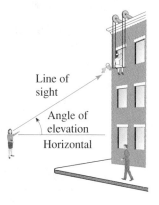

$a = r \sin \theta$

$b = r \cos \theta$

FIGURE 8

The ability to solve right triangles using the trigonometric ratios is fundamental to many problems in navigation, surveying, astronomy, and the measurement of distances. The applications we consider in this section always involve right triangles but, as we will see in the next three sections, trigonometry is also useful in solving triangles that are not right triangles.

To discuss the next examples we need some terminology: If an observer is looking at an object, then the line from the observer's eye to the object is called the **line of sight** (Figure 9). If the object being observed is above the horizontal, then the angle between the line of sight and the horizontal is called the **angle of elevation**. If the object is below the horizontal, then the angle between the line of sight and the horizontal is called the **angle of depression**. In many of the examples and exercises in this chapter, angles of elevation and depression will be given for a hypothetical observer at ground level. If the line of sight follows a physical object, such as an inclined plane or a hillside, we use the term **angle of inclination**.

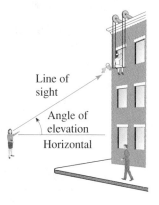
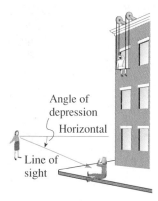

FIGURE 9

The next example gives an important application of trigonometry to the problem of measurement: We measure the height of a tall tree without having to climb it! Although the example is simple, the result is fundamental to the method of applying the trigonometric ratios to such problems.

EXAMPLE 5 ■ Finding the Height of a Tree

A giant redwood tree casts a shadow 532 ft long. Find the height of the tree if the angle of elevation of the sun is 25.7°.

SOLUTION   Let the height of the tree be h. From Figure 10 we see that

$$\frac{h}{532} = \tan 25.7°$$ Definition of tan

$$h = 532 \tan 25.7°$$ Multiply by 532

$$\approx 532(0.48127) \approx 256$$ Use a calculator

Therefore, the height of the tree is about 256 ft.

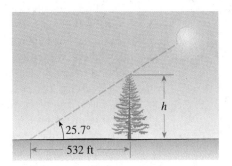

FIGURE 10

EXAMPLE 6 ■ A Problem Involving Right Triangles

From a point on the ground 500 ft from the base of a building, it is observed that the angle of elevation to the top of the building is 24° and the angle of elevation to the top of a flagpole atop the building is 27°. Find the height of the building and the length of the flagpole.

SOLUTION   Figure 11 illustrates the situation. The height of the building is found in the same way that we found the height of the tree in Example 5.

$$\frac{h}{500} = \tan 24°$$ Definition of tan

$$h = 500 \tan 24°$$ Multiply by 500

$$\approx 500(0.4452) \approx 223$$ Use a calculator

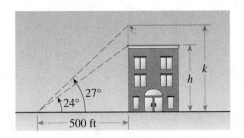

FIGURE 11

The height of the building is approximately 223 ft. To find the length of the flagpole, let's first find the height from the ground to the top of the pole:

$$\frac{k}{500} = \tan 27°$$

$$k = 500 \tan 27° \approx 500(0.5095) \approx 255$$

To find the length of the flagpole, we subtract h from k. So the length of the pole is approximately $255 - 223 = 32$ ft.   ∎

In some problems we need to find an angle in a right triangle whose sides are given. To do this we use Table 1 (page 417) "backward"; that is, we find the *angle* with the specified trigonometric ratio. For example, if $\sin \theta = \frac{1}{2}$, what is the angle θ? From Table 1 we can tell that $\theta = 30°$. To find an angle whose sine is not given in the table, we use the $\boxed{\text{SIN}^{-1}}$ or $\boxed{\text{INV}}\ \boxed{\text{SIN}}$ or $\boxed{\text{ARCSIN}}$ keys on a calculator. For example, if $\sin \theta = 0.8$, we apply the $\boxed{\text{SIN}^{-1}}$ key to get $\theta = 53.13°$ or 0.927 rad. The calculator also gives angles whose cosine or tangent are known, using the $\boxed{\text{COS}^{-1}}$ or $\boxed{\text{TAN}^{-1}}$ key.

The key labels $\boxed{\text{SIN}^{-1}}$ or $\boxed{\text{INV}}\ \boxed{\text{SIN}}$ stand for "inverse sine." The inverse trigonometric functions are studied in Section 7.4.

EXAMPLE 7 ■ Solving for an Angle in a Right Triangle

A 40-ft ladder leans against a building. If the base of the ladder is 6 ft from the base of the building, what is the angle formed by the ladder and the building?

SOLUTION   First we sketch a diagram as in Figure 12. If θ is the angle between the ladder and the building, then

$$\sin \theta = \tfrac{6}{40} = 0.15$$

So θ is the angle whose sine is 0.15. To find the angle θ, we use a calculator and use the $\boxed{\text{SIN}^{-1}}$ key. With our calculator in degree mode, we get

$$\theta \approx 8.6°$$

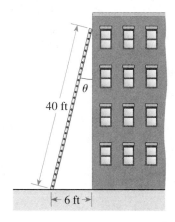

FIGURE 12

6.2   EXERCISES

1–4 ■ Find the values of the six trigonometric ratios of the angle θ in the triangle shown.

1.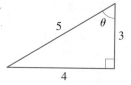

2.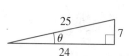

3.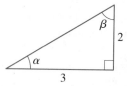

4.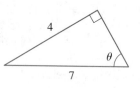

5–6 ■ Find (a) $\sin\alpha$ and $\cos\beta$, (b) $\tan\alpha$ and $\cot\beta$, and (c) $\sec\alpha$ and $\csc\beta$.

5.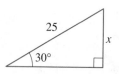

6.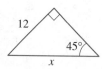

7–12 ■ Find the side labeled x. In Exercises 11 and 12 state your answer correct to five decimal places.

7.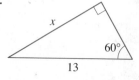

8.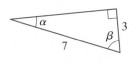

9.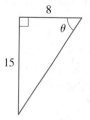

10.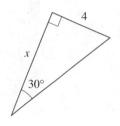

11.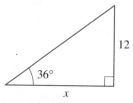

12.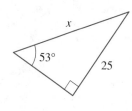

13–14 ■ Express x and y in terms of trigonometric ratios of θ.

13.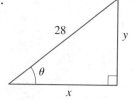

14.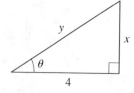

15–20 ■ Sketch a triangle that has acute angle θ, and find the other five trigonometric ratios of θ.

15. $\sin\theta = \frac{3}{5}$

16. $\cos\theta = \frac{2}{7}$

17. $\cot\theta = 1$

18. $\tan\theta = \sqrt{3}$

19. $\sec\theta = 7$

20. $\csc\theta = \frac{13}{12}$

21–26 ■ Evaluate the expression.

21. $\sin\dfrac{\pi}{6} + \cos\dfrac{\pi}{6}$

22. $\sin 30° \csc 30°$

23. $\sin 30° \cos 60° + \sin 60° \cos 30°$

24. $(\sin 60°)^2 + (\cos 60°)^2$

25. $(\cos 30°)^2 - (\sin 30°)^2$

26. $\left(\sin\dfrac{\pi}{3} \cos\dfrac{\pi}{4} - \sin\dfrac{\pi}{4} \cos\dfrac{\pi}{3} \right)^2$

27–30 ■ Solve the right triangle.

27.

28.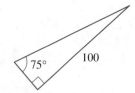

29.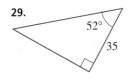

30.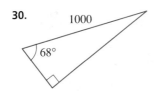

31. Use a ruler to carefully measure the sides of the triangle, and then use your measurements to estimate the six trigonometric ratios of θ.

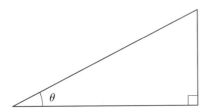

32. Using a protractor, sketch a right triangle that has the acute angle 40°. Measure the sides carefully and use your results to estimate the six trigonometric ratios of 40°.

33. The angle of elevation to the top of the Empire State Building in New York is found to be 11° from the ground at a distance of 1 mi from the base of the building. Using this information, find the height of the Empire State Building.

34. A plane is flying within sight of the Gateway Arch in St. Louis, Missouri, at an elevation of 35,000 ft. The pilot would like to estimate her distance from the Gateway Arch. She finds that the angle of depression to a point on the ground below the arch is 22°.
(a) What is the distance between the plane and the arch?
(b) What is the distance between a point on the ground directly below the plane and the arch?

35. A laser beam is to be directed toward the center of the moon, but the beam strays 0.5° from its intended path.
(a) How far has the beam diverged from its assigned target when it reaches the moon? (The distance from the earth to the moon is 240,000 mi.)
(b) The radius of the moon is about 1000 mi. Will the beam strike the moon?

36. From the top of a 200-ft lighthouse, the angle of depression to a ship in the ocean is 23°. How far is the ship from the base of the lighthouse?

37. A 20-ft ladder leans against a building so that the angle between the ground and the ladder is 72°. How high does the ladder reach on the building?

38. A 20-ft ladder is leaning against a building. If the base of the ladder is 6 ft from the base of the building, what is the angle of elevation of the ladder? How high does the ladder reach on the building?

39. A 96-ft tree casts a shadow that is 120 ft long. What is the angle of elevation of the sun?

40. A 600-ft guy wire is attached to the top of a communication tower. If the wire makes an angle of 65° with the ground, how tall is the communication tower?

41. A man is lying on the beach, flying a kite. He holds the end of the kite string at ground level, and estimates the angle of elevation of the kite to be 50°. If the string is 450 ft long, how high is the kite above the ground?

42. The height of a steep cliff is to be measured from a point on the opposite side of the river. Find the height of the cliff from the information given in the figure.

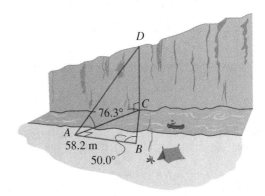

43. A water tower is located 325 ft from a building (see the figure). From a window in the building it is observed that the angle of elevation to the top of the tower is 39°

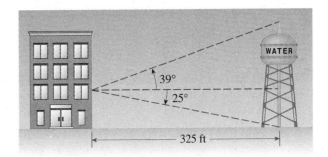

and the angle of depression to the bottom of the tower is 25°. How tall is the tower? How high is the window?

44. An airplane is flying at an elevation of 5150 ft, directly above a straight highway. Two motorists are driving cars on the highway on opposite sides of the plane, and the angle of depression to one car is 35° and to the other is 52°. How far apart are the cars?

45. If both cars in Exercise 44 are on one side of the plane and if the angle of depression to one car is 38° and to the other car is 52°, how far apart are the cars?

46. A hot-air balloon is floating above a straight road. To estimate their height above the ground, the balloonists simultaneously measure the angle of depression to two consecutive mileposts on the road on the same side of the balloon. The angles of depression are found to be 20° and 22°. How high is the balloon?

47. To estimate the height of a mountain above a level plain, the angle of elevation to the top of the mountain is measured to be 32°. One thousand feet closer to the mountain along the plain, it is found that the angle of elevation is 35°. Estimate the height of the mountain.

48. (a) Show that the height h of the mountain in the figure is given by

$$h = d \frac{\tan \beta \tan \alpha}{\tan \beta - \tan \alpha} = \frac{d}{\cot \alpha - \cot \beta}$$

(b) Use the formula in part (a) to find the height h of the mountain if $\alpha \approx 25°$, $\beta \approx 29°$, and $d \approx 800$ ft.

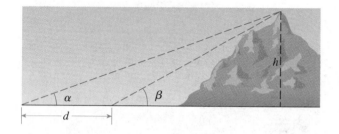

49. When the moon is exactly half full, the earth, moon, and sun form a right angle (see the figure). At that time the angle formed by the sun, earth, and moon is measured to be 89.85°. If the distance from the earth to the moon is 240,000 mi, estimate the distance from the earth to the sun.

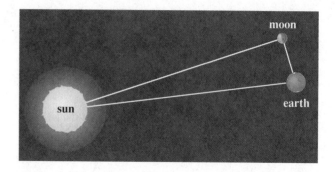

50. To find the distance to the sun as in Exercise 49, we needed to know the distance to the moon. Here is a way of estimating that distance: When the moon is seen at its zenith at a point A on the earth, it is observed to be at the horizon from point B (see the figure). Points A and B are 6155 mi apart, and the radius of the earth is 3960 mi.
   (a) Find the angle θ in degrees.
   (b) Estimate the distance from point A to the moon.

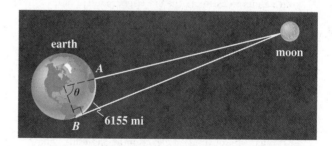

51. In Exercise 56 of Section 6.1 a method was given for finding the radius of the earth. Here is a more modern method of doing this: From a satellite 600 mi above the earth, it is observed that the angle formed by the vertical and the line of sight to the horizon is 60.276°. Use this information to find the radius of the earth.

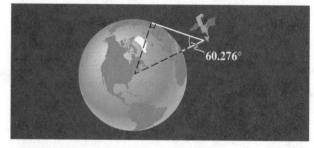

52. To find the distance to nearby stars, the method of parallax is used. The idea is to find a triangle with the star at one vertex and with a base as large as possible. To do this the star is observed at two different times exactly 6 months apart, and its apparent change in position is recorded. From these two observations, $\angle E_1 S E_2$ can be calculated. (The times are chosen so that $\angle E_1 S E_2$ is as large as possible, which guarantees that $\angle E_1 O S$ is 90°.) The angle $E_1 SO$ is called the *parallax* of the star. Alpha Centauri, the star nearest the earth, has a parallax of 0.000211°. Estimate the distance to this star. (Take the distance from the earth to the sun to be 9.3×10^7 mi.)

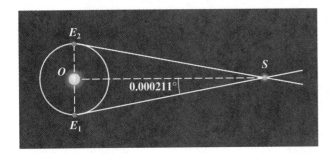

53–56 ■ Find x correct to one decimal place.

53.

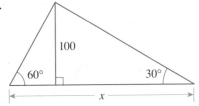

54.

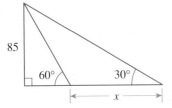

55.

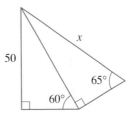

56.

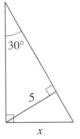

57. Express the lengths a, b, c, and d in the figure in terms of the trigonometric ratios of θ.

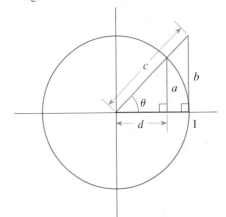

DISCOVERY · DISCUSSION

58. Similar Triangles If two triangles are similar, what properties do they share? Explain how these properties make it possible to define the trigonometric ratios without regard to the size of the triangle.

6.3   TRIGONOMETRIC FUNCTIONS OF ANGLES

In the preceding section we defined the trigonometric ratios for acute angles. Here we extend the trigonometric ratios to all angles by defining the trigonometric functions of angles. With these functions we can solve practical problems that involve angles which are not necessarily acute.

TRIGONOMETRIC FUNCTIONS OF ANGLES

Relationship to the Trigonometric Functions of Real Numbers

You may already have studied the trigonometric functions defined using the unit circle. Here is how they are related to the trigonometric functions of angles.

Let θ be an angle in standard position and choose $P(x, y)$ on the terminal side so that its distance r to the origin is 1, as shown.

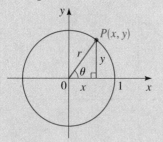

By the definition of the trigonometric functions of the *angle* θ, we have

$$\sin\theta = \frac{y}{1} = y \quad \cos\theta = \frac{x}{1} = x$$

On the other hand, $P(x, y)$ is also the terminal point of an arc of length t.

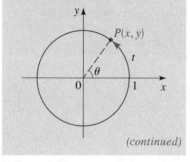

(continued)

Let POQ be a right triangle with acute angle θ as shown in Figure 1(a). Place θ in standard position as shown in Figure 1(b).

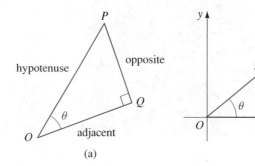

FIGURE 1   (a)   (b)

Then $P = P(x, y)$ is a point on the terminal side of θ. In triangle POQ, the opposite side has length y and the adjacent side has length x. Using the Pythagorean Theorem we see that the hypotenuse has length $r = \sqrt{x^2 + y^2}$. So

$$\sin\theta = \frac{y}{r}, \qquad \cos\theta = \frac{x}{r},$$

$$\tan\theta = \frac{y}{x}$$

The other trigonometric ratios can be found in the same way.

These observations allow us to extend the definition of the trigonometric ratios to any angle. We begin by defining the trigonometric functions of angles as follows (see Figure 2).

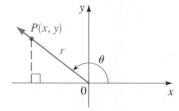

FIGURE 2

By the definition of the trigonometric functions of the *real number* t, we have

$$\sin t = y \quad \text{and} \quad \cos t = x$$

If θ is measured in radians, then $\theta = t$. Comparing the two ways of defining the trigonometric functions, we see that they give identical values. In other words, viewed as functions they assign identical values to a given real number (the real number is the radian measure of θ in one case or the length t of a circular arc in the other).

DEFINITION OF THE TRIGONOMETRIC FUNCTIONS OF ANGLES

Let θ be an angle in standard position and let $P(x, y)$ be a point on the terminal side. If $r = \sqrt{x^2 + y^2}$ is the distance from the origin to the point $P(x, y)$, then

$$\sin \theta = \frac{y}{r} \qquad \cos \theta = \frac{x}{r} \qquad \tan \theta = \frac{y}{x} \quad (x \neq 0)$$

$$\csc \theta = \frac{r}{y} \quad (y \neq 0) \qquad \sec \theta = \frac{r}{x} \quad (x \neq 0) \qquad \cot \theta = \frac{x}{y} \quad (y \neq 0)$$

It is a crucial fact that the values of the trigonometric functions do *not* depend on the choice of the point $P(x, y)$. This is because if $P'(x', y')$ is any other point on the terminal side, as in Figure 3, then triangles POQ and $P'OQ'$ are similar.

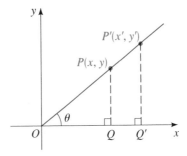

FIGURE 3

EVALUATING TRIGONOMETRIC FUNCTIONS AT ANY ANGLE

The following mnemonic device can be used to remember which trigonometric functions are positive in each quadrant: All of them, Sine, Tangent, or Cosine.

You can remember this as "All Students Take Calculus."

From the definition we see that the values of the trigonometric functions are all positive if the angle θ has its terminal side in quadrant I. This is because x and y are positive in this quadrant. [Of course, r is always positive, since it is simply the distance from the origin to the point $P(x, y)$.] If the terminal side of θ is in quadrant II, however, then x is negative and y is positive. Thus, in quadrant II the functions $\sin \theta$ and $\csc \theta$ are positive, and all the other trigonometric functions have negative values. You can check the other entries in the following table.

SIGNS OF THE TRIGONOMETRIC FUNCTIONS

Quadrant	Positive functions	Negative functions
I	all	none
II	sin, csc	cos, sec, tan, cot
III	tan, cot	sin, csc, cos, sec
IV	cos, sec	sin, csc, tan, cot

Since division by 0 is an undefined operation, certain trigonometric functions are not defined for certain angles. For example, $\tan 90° = y/x$ is undefined because $x = 0$. The angles for which the trigonometric functions may be undefined are the angles for which either the x- or y-coordinate of a point on the terminal side of the angle is 0. These are **quadrantal angles**—angles that are coterminal with the coordinate axes.

We now turn our attention to finding the values of the trigonometric functions for angles that are not acute.

EXAMPLE 1 ■ Finding Trigonometric Functions of Angles

Find (a) $\cos 135°$ and (b) $\tan 390°$.

SOLUTION

(a) From Figure 4 we see that $\cos 135° = -x/r$. But $\cos 45° = x/r$, and since $\cos 45° = \sqrt{2}/2$, we have

$$\cos 135° = -\frac{\sqrt{2}}{2}$$

(b) The angles $390°$ and $30°$ are coterminal. From Figure 5 it's clear that $\tan 390° = \tan 30°$ and, since $\tan 30° = \sqrt{3}/3$, we have

$$\tan 390° = \frac{\sqrt{3}}{3}$$ ∎

From Example 1 we see that the trigonometric functions for angles that aren't acute have the same value, except possibly for sign, as the corresponding trigonometric functions of an acute angle. That acute angle will be called the *reference angle*.

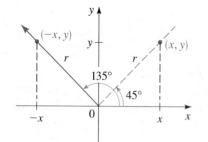

FIGURE 4

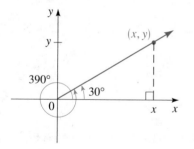

FIGURE 5

REFERENCE ANGLE

Let θ be an angle in standard position. The **reference angle** $\bar{\theta}$ associated with θ is the acute angle formed by the terminal side of θ and the x-axis.

Figure 6 shows that to find a reference angle it's useful to know the quadrant in which the terminal side of the angle lies.

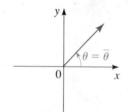 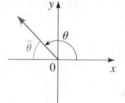 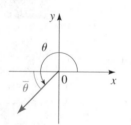 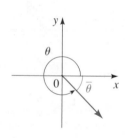

FIGURE 6
The reference angle $\bar{\theta}$
for an angle θ

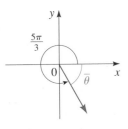

FIGURE 7

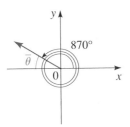

FIGURE 8

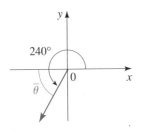

FIGURE 9

$\frac{S|A}{T|C}$ sin 240° *is negative.*

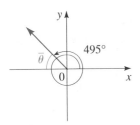

FIGURE 10

$\frac{S|A}{T|C}$ tan 495° *is negative,*
so cot 495° *is negative.*

EXAMPLE 2 ■ Finding Reference Angles

Find the reference angle for (a) $\theta = \dfrac{5\pi}{3}$ and (b) $\theta = 870°$.

SOLUTION

(a) The reference angle is the acute angle formed by the terminal side of the angle $5\pi/3$ and the x-axis (see Figure 7). Since the terminal side of this angle is in quadrant IV, the reference angle is

$$2\pi - \frac{5\pi}{3} = \frac{\pi}{3}$$

(b) The angles 870° and 150° are coterminal [because $870 - 2(360) = 150$]. Thus, the terminal side of this angle is in quadrant II (see Figure 8). So the reference angle is

$$180° - 150° = 30° \qquad ■$$

EVALUATING TRIGONOMETRIC FUNCTIONS AT ANY ANGLE

To find the values of the trigonometric functions for any angle θ, we carry out the following steps.

1. Find the reference angle $\bar{\theta}$ associated with the angle θ.

2. Determine the sign of the trigonometric function of θ.

3. The value of the trigonometric function of θ is the same, except possibly for sign, as the value of the trigonometric function of $\bar{\theta}$.

EXAMPLE 3 ■ Using the Reference Angle to
Evaluate Trigonometric Functions

Find (a) sin 240° and (b) cot 495°.

SOLUTION

(a) This angle has its terminal side in quadrant III, as shown in Figure 9. The reference angle is therefore $240° - 180° = 60°$, and the value of sin 240° is negative. Thus

$$\sin 240° = -\sin 60° = -\frac{\sqrt{3}}{2}$$

↑   ↑
sign   reference angle

(b) The angle 495° is coterminal with the angle 135°, and the terminal side of this angle is in quadrant II, as shown in Figure 10. So the reference angle is $180° - 135° = 45°$, and the value of cot 495° is negative. We have

$$\cot 495° = \cot 135° = -\cot 45° = -1$$

↑   ↑   ↑
coterminal angles   sign   reference angle

■

EXAMPLE 4 ■ Using the Reference Angle to Evaluate Trigonometric Functions

Find (a) $\sin \dfrac{16\pi}{3}$ and (b) $\sec\left(-\dfrac{\pi}{4}\right)$.

SOLUTION

(a) The angle $16\pi/3$ is coterminal with $4\pi/3$, and these angles are in quadrant III (see Figure 11). Thus, the reference angle is $(4\pi/3) - \pi = \pi/3$. Since the value of sine is negative in quadrant III, we have

$$\sin \frac{16\pi}{3} = \sin \frac{4\pi}{3} = -\sin \frac{\pi}{3} = -\frac{\sqrt{3}}{2}$$

↑   ↑   ↑
coterminal angles   sign   reference angle

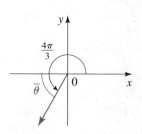

FIGURE 11

$\dfrac{S\,|\,A}{T\,|\,C}$ $\sin \dfrac{16\pi}{3}$ *is negative.*

(b) The angle $-\pi/4$ is in quadrant IV, and its reference angle is $\pi/4$ (see Figure 12). Since secant is positive in this quadrant, we get

$$\sec\left(-\frac{\pi}{4}\right) = +\sec \frac{\pi}{4} = \frac{\sqrt{2}}{2}$$

↑   ↑
sign   reference angle

FIGURE 12

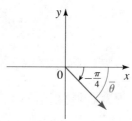

$\dfrac{S\,|\,A}{T\,|\,C}$ $\cos\left(-\dfrac{\pi}{4}\right)$ *is positive, so* $\sec\left(-\dfrac{\pi}{4}\right)$ *is positive.*   ■

TRIGONOMETRIC IDENTITIES

The trigonometric functions of angles are related to each other through several important equations called **trigonometric identities**. We've already encountered the reciprocal identities. These identities continue to hold for any angle θ, provided both sides of the equation are defined. The Pythagorean identities are a consequence of the Pythagorean Theorem.*

---

*We follow the usual convention of writing $\sin^2\theta$ for $(\sin\theta)^2$. In general, we write $\sin^n\theta$ for $(\sin\theta)^n$ for all integers n except $n = -1$. The exponent $n = -1$ will be assigned another meaning in Section 7.4. Of course, the same convention applies to the other five trigonometric functions.

FUNDAMENTAL IDENTITIES

Reciprocal Identities

$$\csc \theta = \frac{1}{\sin \theta} \qquad \sec \theta = \frac{1}{\cos \theta} \qquad \cot \theta = \frac{1}{\tan \theta}$$

$$\tan \theta = \frac{\sin \theta}{\cos \theta} \qquad \cot \theta = \frac{\cos \theta}{\sin \theta}$$

Pythagorean Identities

$$\sin^2 \theta + \cos^2 \theta = 1 \qquad \tan^2 \theta + 1 = \sec^2 \theta \qquad 1 + \cot^2 \theta = \csc^2 \theta$$

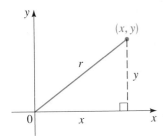

FIGURE 13

■ **Proof** Let's prove the first Pythagorean identity. Applying the Pythagorean Theorem to the triangle in Figure 13 gives

$$\sin^2 \theta + \cos^2 \theta = \left(\frac{y}{r}\right)^2 + \left(\frac{x}{r}\right)^2 = \frac{x^2 + y^2}{r^2} = \frac{r^2}{r^2} = 1$$

Thus, $\sin^2 \theta + \cos^2 \theta = 1$. (Although the figure indicates an acute angle, you should check that the proof holds for all angles θ.)

See Exercises 55 and 56 for the proofs of the other two Pythagorean identities.   □

EXAMPLE 5 ■ Expressing One Trigonometric Function in Terms of Another

(a) Express $\sin \theta$ in terms of $\cos \theta$.
(b) Express $\tan \theta$ in terms of $\sin \theta$, where θ is in quadrant II.

SOLUTION

(a) From the first Pythagorean identity we get

$$\sin \theta = \pm \sqrt{1 - \cos^2 \theta}$$

where the sign depends on the quadrant. If θ is in quadrant I or II, then $\sin \theta$ is positive, and hence

$$\sin \theta = \sqrt{1 - \cos^2 \theta}$$

whereas if θ is in quadrant III or IV, $\sin \theta$ is negative and so

$$\sin \theta = -\sqrt{1 - \cos^2 \theta}$$

(b) Since $\tan \theta = \sin \theta / \cos \theta$, we need to write $\cos \theta$ in terms of $\sin \theta$. By part (a)

$$\cos \theta = \pm \sqrt{1 - \sin^2 \theta}$$

and since $\cos \theta$ is negative in quadrant II, the negative sign applies here. Thus

$$\tan \theta = \frac{\sin \theta}{\cos \theta} = \frac{\sin \theta}{-\sqrt{1 - \sin^2 \theta}}$$ ■

EXAMPLE 6 ■ Evaluating a Trigonometric Function

If $\tan \theta = \frac{2}{3}$ and θ is in quadrant III, find $\cos \theta$.

SOLUTION 1   We need to write $\cos \theta$ in terms of $\tan \theta$. From the identity $\tan^2 \theta + 1 = \sec^2 \theta$, we get $\sec \theta = \pm \sqrt{\tan^2 \theta + 1}$. In quadrant III, $\sec \theta$ is negative, so

$$\sec \theta = -\sqrt{\tan^2 \theta + 1}$$

Thus
$$\cos \theta = \frac{1}{\sec \theta} = \frac{1}{-\sqrt{\tan^2 \theta + 1}}$$

$$= \frac{1}{-\sqrt{\left(\frac{2}{3}\right)^2 + 1}} = \frac{1}{-\sqrt{\frac{13}{9}}} = -\frac{3}{\sqrt{13}}$$

SOLUTION 2   This problem can be solved more easily using the method of Example 2 of Section 6.2. Recall that, except for sign, the values of the trigonometric functions of any angle are the same as those of an acute angle (the reference angle). So, ignoring the sign for the moment, let's sketch a right triangle with an acute angle $\overline{\theta}$ satisfying $\tan \overline{\theta} = \frac{2}{3}$ (see Figure 14). By the Pythagorean Theorem the hypotenuse of this triangle has length $\sqrt{13}$. From the triangle in Figure 14 we immediately see that $\cos \overline{\theta} = 3/\sqrt{13}$. Since θ is in quadrant III, $\cos \theta$ is negative and so

$$\cos \theta = -\frac{3}{\sqrt{13}}$$

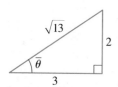

FIGURE 14

EXAMPLE 7 ■ Evaluating Trigonometric Functions

If $\sec \theta = 2$ and θ is in quadrant IV, find the other five trigonometric functions of θ.

SOLUTION   We sketch a triangle as in Figure 15 so that $\sec \overline{\theta} = 2$. Taking into account the fact that θ is in quadrant IV, we get

$$\sin \theta = -\frac{\sqrt{3}}{2} \qquad \cos \theta = \frac{1}{2} \qquad \tan \theta = -\sqrt{3}$$

$$\csc \theta = -\frac{2}{\sqrt{3}} \qquad \sec \theta = 2 \qquad \cot \theta = -\frac{1}{\sqrt{3}}$$

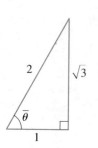

FIGURE 15

AREAS OF TRIANGLES

We end this section by giving an application of the trigonometric functions that involves angles that are not necessarily acute. More extensive applications appear in the next two sections.

The area of a triangle is $\mathcal{A} = \frac{1}{2}(\text{base}) \times (\text{height})$. If we know two sides and the included angle of a triangle, then we can find the height using the trigonometric functions, and from this we can find the area.

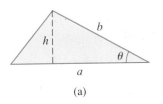

(a)

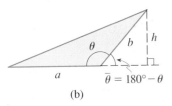

(b)

FIGURE 16

If θ is an acute angle, then the height of the triangle in Figure 16(a) is given by $h = b \sin \theta$. Thus, the area is

$$\mathcal{A} = \tfrac{1}{2} \times \text{base} \times \text{height} = \tfrac{1}{2} ab \sin \theta$$

If the angle θ is not acute, then from Figure 16(b) we see that the height of the triangle is

$$h = b \sin(180° - \theta) = b \sin \theta$$

This is so because the reference angle of θ is the angle $180° - \theta$. Thus, in this case also, the area of the triangle is

$$\mathcal{A} = \tfrac{1}{2} \times \text{base} \times \text{height} = \tfrac{1}{2} ab \sin \theta$$

AREA OF A TRIANGLE

The area $\mathcal{A}$ of a triangle with sides of lengths a and b and with included angle θ is

$$\mathcal{A} = \tfrac{1}{2} ab \sin \theta$$

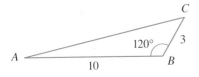

FIGURE 17

EXAMPLE 8 ■ **Finding the Area of a Triangle**

Find the area of triangle ABC shown in Figure 17.

SOLUTION   The triangle has sides of length 10 and 3, with included angle $120°$. Therefore

$$\mathcal{A} = \tfrac{1}{2} ab \sin \theta$$
$$= \tfrac{1}{2} (10)(3) \sin(120°)$$
$$= 15 \sin(60°) \qquad \text{Reference angle}$$
$$= 15 \frac{\sqrt{3}}{2} \qquad \qquad ■$$

6.3   **EXERCISES**

1–6 ■ Find the reference angle for the given angle.

1. (a) $225°$ (b) $-35°$ (c) $181°$

2. (a) $290°$ (b) $750°$ (c) $570°$

3. (a) $335°$ (b) $-95°$ (c) $165°$

4. (a) $\dfrac{3\pi}{5}$ (b) $\dfrac{7\pi}{6}$ (c) $-\dfrac{2\pi}{3}$

5. (a) $\dfrac{17\pi}{3}$ (b) $-\dfrac{\pi}{4}$ (c) 3

6. (a) $\dfrac{23\pi}{11}$ (b) $\dfrac{23}{11}$ (c) $\dfrac{17\pi}{7}$

7–30 ■ Find the exact value of the trigonometric function.

7. $\sin 150°$ **8.** $\cos 225°$ **9.** $\sin 135°$

10. $\tan 330°$ **11.** $\sin(-60°)$ **12.** $\sec(-60°)$

13. $\csc(-630°)$ **14.** $\cot 210°$ **15.** $\cos 570°$

16. $\sec 120°$ **17.** $\tan 750°$ **18.** $\cos 660°$

19. $\sin \dfrac{2\pi}{3}$ **20.** $\sin \dfrac{5\pi}{3}$ **21.** $\sin \dfrac{3\pi}{2}$

22. $\cos \dfrac{7\pi}{3}$ **23.** $\cos\left(-\dfrac{7\pi}{3}\right)$ **24.** $\tan \dfrac{5\pi}{6}$

25. $\sec \dfrac{17\pi}{3}$ **26.** $\csc \dfrac{5\pi}{4}$ **27.** $\cot\left(-\dfrac{\pi}{4}\right)$

28. $\cos \dfrac{7\pi}{4}$ **29.** $\tan \dfrac{5\pi}{2}$ **30.** $\sin \dfrac{11\pi}{6}$

31–34 ■ Find the quadrant in which θ lies from the information given.

31. $\sin \theta < 0$ and $\cos \theta < 0$

32. $\tan \theta < 0$ and $\sin \theta < 0$

33. $\sec \theta > 0$ and $\tan \theta < 0$

34. $\csc \theta > 0$ and $\cos \theta < 0$

35–40 ■ Write the first trigonometric function in terms of the second for θ in the given quadrant.

35. $\tan \theta$, $\cos \theta$; θ in quadrant III

36. $\cot \theta$, $\sin \theta$; θ in quadrant II

37. $\cos \theta$, $\sin \theta$; θ in quadrant IV

38. $\sec \theta$, $\sin \theta$; θ in quadrant I

39. $\sec \theta$, $\tan \theta$; θ in quadrant II

40. $\csc \theta$, $\cot \theta$; θ in quadrant III

41–48 ■ Find the values of the trigonometric functions of θ from the information given.

41. $\sin \theta = \frac{3}{5}$, θ in quadrant II

42. $\cos \theta = -\frac{7}{12}$, θ in quadrant III

43. $\tan \theta = -\frac{3}{4}$, $\cos \theta > 0$

44. $\sec \theta = 5$, $\sin \theta < 0$

45. $\csc \theta = 2$, θ in quadrant I

46. $\cot \theta = \frac{1}{4}$, $\sin \theta < 0$

47. $\cos \theta = -\frac{2}{7}$, $\tan \theta < 0$

48. $\tan \theta = -4$, $\sin \theta > 0$

49. If $\theta = \pi/3$, find the value of each of the following expressions.
(a) $\sin 2\theta$, $2 \sin \theta$

(b) $\sin \frac{1}{2}\theta$, $\dfrac{\sin \theta}{2}$

(c) $\sin^2\theta$, $\sin(\theta^2)$

50. Find the area of a triangle with sides of length 7 and 9 and included angle 72°.

51. Find the area of a triangle with sides of length 10 and 22 and included angle 10°.

52. Find the area of an equilateral triangle with side of length 10.

53. A triangle has an area of 16 in², and two of the sides of the triangle have lengths 5 in. and 7 in. Find the angle included by these two sides.

54. Find the area of the shaded region in the figure.

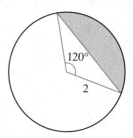

55. Use the first Pythagorean identity to prove the second. [*Hint:* Divide by $\cos^2\theta$.]

56. Use the first Pythagorean identity to prove the third.

▲ DISCOVERY · DISCUSSION

57. Using a Calculator To solve a certain problem you need to find the sine of 4 rad. Your study partner uses his calculator and tells you that sin 4 is 0.0697564737. On your calculator you get −0.7568024953. What is wrong? What mistake did your partner make?

THE LAW OF SINES

In Section 6.2 we used the trigonometric ratios to solve right triangles. The trigonometric functions can also be used to solve *oblique triangles,* that is, triangles with no right angles. To do this we first study the Law of Sines here and then the Law of Cosines in the next section. To state these laws (or formulas) more easily, we follow the convention of labeling the angles of a triangle as A, B, C, and the lengths of the corresponding opposite sides as a, b, c, as in Figure 1.

The **Law of Sines** says that in any triangle the lengths of the sides are proportional to the sines of the corresponding opposite angles.

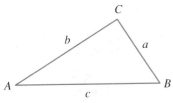

FIGURE 1

THE LAW OF SINES

In triangle ABC we have

$$\frac{\sin A}{a} = \frac{\sin B}{b} = \frac{\sin C}{c}$$

■ **Proof** To see why the Law of Sines is true, refer to Figure 2. By the formula in Section 6.3 the area of triangle ABC is $\frac{1}{2}ab\sin C$. By the same formula the area of this triangle is also $\frac{1}{2}ac\sin B$ and $\frac{1}{2}bc\sin A$. Thus

$$\tfrac{1}{2}bc\sin A = \tfrac{1}{2}ac\sin B = \tfrac{1}{2}ab\sin C$$

Multiplying by $2/(abc)$ gives the Law of Sines. ☐

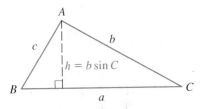

FIGURE 2

EXAMPLE 1 ■ Tracking a Satellite

A satellite orbiting the earth passes directly overhead at observation stations in Phoenix and Los Angeles, 340 mi apart. At an instant when the satellite is between these two stations, its angle of elevation is simultaneously observed to be 60° at Phoenix and 75° at Los Angeles. How far is the satellite from Los Angeles? In other words, find the distance AC in Figure 3.

SOLUTION Whenever two angles in a triangle are known, the third angle can be determined immediately because the sum of the angles of a triangle is 180°. In this case, $\angle C = 180° - (75° + 60°) = 45°$ (see Figure 3), so we have

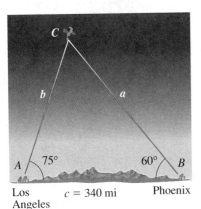

FIGURE 3

$$\frac{\sin B}{b} = \frac{\sin C}{c} \qquad \text{Law of Sines}$$

$$\frac{\sin 60°}{b} = \frac{\sin 45°}{340} \qquad \text{Substitute}$$

$$b = \frac{340\sin 60°}{\sin 45°} \approx 416 \qquad \text{Solve for } b$$

The distance of the satellite from Los Angeles is approximately 416 mi. ■

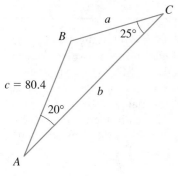

FIGURE 4

EXAMPLE 2 ■ Solving a Triangle

Solve the triangle in Figure 4.

SOLUTION   First, $\angle B = 180° - (20° + 25°) = 135°$. Since side c is known, to find side a we use the relation

$$\frac{\sin A}{a} = \frac{\sin C}{c} \qquad \text{Law of Sines}$$

$$a = \frac{c \sin A}{\sin C} = \frac{80.4 \sin 20°}{\sin 25°} \approx 65.1 \qquad \text{Solve for } a$$

Similarly, to find b we use

$$\frac{\sin B}{b} = \frac{\sin C}{c} \qquad \text{Law of Sines}$$

$$b = \frac{c \sin B}{\sin C} = \frac{80.4 \sin 135°}{\sin 25°} \approx 134.5 \qquad \text{Solve for } b$$

■

THE AMBIGUOUS CASE

To solve a triangle we need to know certain information about its sides and angles. To decide whether we've got enough information, it's often helpful to make a sketch. For instance, if we are given two angles and the included side, then it's clear that one and only one triangle can be formed [see Figure 5(a)]. Similarly, if two sides and the included angle are known, then a unique triangle is determined [Figure 5(b)]. But if we know all three angles and no sides, we cannot uniquely determine the triangle because many triangles can have the same three angles. (All these triangles would be similar, of course.) So we won't consider this last case.

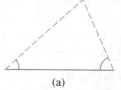 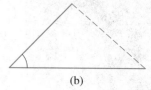

FIGURE 5   (a)   (b)

In general, a triangle is determined by three of its six parts (angles and sides) as long as at least one of these three parts is a side. So, the possibilities are as follows:

Case 1   SAA   One side and two angles
Case 2   SSA   Two sides and the angle opposite one of those sides
Case 3   SAS   Two sides and the included angle
Case 4   SSS   Three sides

In each of these cases, except Case 2, a unique triangle is determined by the information given. Case 2 presents a situation where there may be two triangles, one triangle, or no triangle with the given properties. For this reason, Case 2 is sometimes called the **ambiguous case**. To see why this is so, we show in Figure 6 the possibilities when angle A and sides a and b are given. In part (a) no solution is possible, since side a is too short to complete the triangle. In part (b) the solution is a right triangle. In part (c) two solutions are possible, and in part (d) there is a unique triangle with the given properties. We illustrate the possibilities of Case 2 in the following examples.

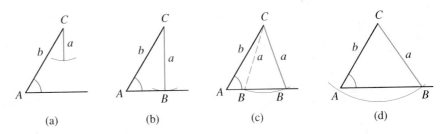

FIGURE 6
The ambiguous case

(a)   (b)   (c)   (d)

EXAMPLE 3 ■ SSA, the One-Solution Case

Solve triangle ABC, where $\angle A = 45°$, $a = 7\sqrt{2}$, and $b = 7$.

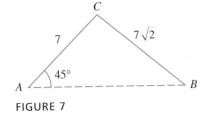

FIGURE 7

SOLUTION   We first sketch the triangle with the information we have (see Figure 7). Our sketch is necessarily tentative, since we don't yet know the other angles. Nevertheless, we can now see the possibilities.

We first find $\angle B$.

$$\frac{\sin A}{a} = \frac{\sin B}{b} \qquad \text{Law of Sines}$$

$$\sin B = \frac{b \sin A}{a} = \frac{7}{7\sqrt{2}} \sin 45° = \left(\frac{1}{\sqrt{2}}\right)\left(\frac{\sqrt{2}}{2}\right) = \frac{1}{2} \qquad \text{Solve for } \sin B$$

We consider only angles smaller than 180°, since no triangle can contain an angle of 180° or larger.

Which angles B have $\sin B = \frac{1}{2}$? From the preceding section we know that there are two such angles smaller than 180° (they are 30° and 150°). Which of these angles is compatible with what we know about triangle ABC? Since $\angle A = 45°$, we cannot have $\angle B = 150°$, because $45° + 150° > 180°$. So $\angle B = 30°$, and the remaining angle is $\angle C = 180° - (30° + 45°) = 105°$.

Now we can find side c.

$$\frac{\sin B}{b} = \frac{\sin C}{c} \qquad \text{Law of Sines}$$

$$c = \frac{b \sin C}{\sin B} = \frac{7 \sin 105°}{\sin 30°} = \frac{7 \sin 105°}{\frac{1}{2}} \approx 13.5 \qquad \text{Solve for } c \qquad ■$$

In Example 3 there were two possibilities for angle B, and one of these was not compatible with the rest of the information. In general, if $\sin A < 1$, we must check the angle and its supplement as possibilities, because any angle smaller than 180° can be in the triangle. To decide whether either possibility works, we check to see whether the resulting sum of the angles exceeds 180°. It can happen,

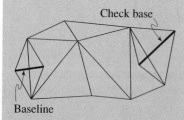
as in Figure 6(c), that both possibilities are compatible with the given information. In that case, two different triangles are solutions to the problem.

EXAMPLE 4 ■ SSA, the Two-Solution Case

Solve triangle ABC if $\angle A = 43.1°$, $a = 186.2$, and $b = 248.6$.

SOLUTION From the given information we sketch the triangle shown in Figure 8. Note that side a may be drawn in two possible positions to complete the triangle. From the Law of Sines

$$\sin B = \frac{b \sin A}{a} = \frac{248.6 \sin 43.1°}{186.2} \approx 0.91225$$

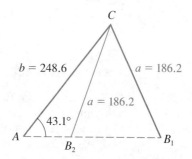

FIGURE 8

There are two possible angles B between $0°$ and $180°$ such that $\sin B = 0.91225$. Using the $\boxed{\text{SIN}^{-1}}$ key on a calculator (or $\boxed{\text{INV}}$ $\boxed{\text{SIN}}$ or $\boxed{\text{ARCSIN}}$), we find that one of these angles is approximately $65.8°$. The other is approximately $180° - 65.8° = 114.2°$. We denote these two angles by B_1 and B_2 so that

$$\angle B_1 \approx 65.8° \qquad \text{and} \qquad \angle B_2 \approx 114.2°$$

Thus, two triangles satisfy the given conditions: triangle $A_1 B_1 C_1$ and triangle $A_2 B_2 C_2$.

Solve triangle $A_1 B_1 C_1$:

$$\angle C_1 \approx 180° - (43.1° + 65.8°) = 71.1° \qquad \text{Find } \angle C_1$$

Thus, $\quad c_1 = \dfrac{a_1 \sin C_1}{\sin A_1} \approx \dfrac{186.2 \sin 71.1°}{\sin 43.1°} \approx 257.8 \qquad$ Law of Sines

Solve triangle $A_2 B_2 C_2$:

$$\angle C_2 \approx 180° - (43.1° + 114.2°) = 22.7° \qquad \text{Find } \angle C_2$$

Thus $\quad c_2 = \dfrac{a_2 \sin C_2}{\sin A_2} \approx \dfrac{186.2 \sin 22.7°}{\sin 43.1°} \approx 105.2 \qquad$ Law of Sines

Triangles $A_1 B_1 C_1$ and $A_2 B_2 C_2$ are shown in Figure 9.

expedition reached the foothills of the Himalayas. A later expedition, using triangulation, calculated the height of the highest peak of the Himalayas to be 29,002 ft. The peak was named in honor of Sir George Everest. Today, using satellites, the height of Mt. Everest is estimated to be 29,028 ft. The very close agreement of these two estimates shows the great accuracy of the trigonometric method.

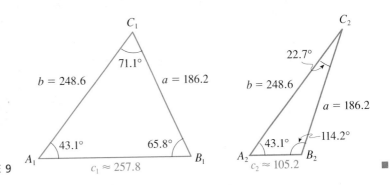

FIGURE 9

The next example presents a situation for which no triangle is compatible with the given data.

EXAMPLE 5 ■ SSA, the No-Solution Case

Solve triangle ABC, where $\angle A = 42°$, $a = 70$, and $b = 122$.

SOLUTION First, let's try to find $\angle B$. We have

$$\frac{\sin A}{a} = \frac{\sin B}{b} \qquad \text{Law of Sines}$$

$$\sin B = \frac{b \sin A}{a} = \frac{122 \sin 42°}{70} \approx 1.17 \qquad \text{Solve for } \sin B$$

Since the sine of an angle is never greater than 1, we conclude that no triangle satisfies the conditions given in this problem. ■

6.4 EXERCISES

1–6 ■ Use the Law of Sines to find the indicated side x or angle θ.

1.

2.

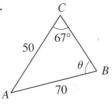

3.

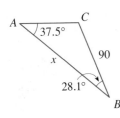

4.

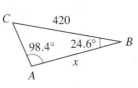

5.

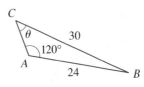

6.

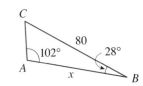

7–8 ■ Solve the triangle using the Law of Sines.

7.

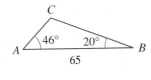

8.

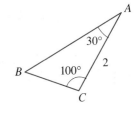

9–14 ■ Sketch each triangle and then solve the triangle using the Law of Sines.

9. $\angle A = 50°$, $\angle B = 68°$, $c = 230$

10. $\angle A = 23°$, $\angle B = 110°$, $c = 50$

11. $\angle A = 30°$, $\angle C = 65°$, $b = 10$

12. $\angle A = 22°$, $\angle B = 95°$, $a = 420$

13. $\angle B = 29°$, $\angle C = 51°$, $b = 44$

14. $\angle B = 10°$, $\angle C = 100°$, $c = 115$

15–22 ■ Use the Law of Sines to solve for all possible triangles that satisfy the given conditions.

15. $a = 28$, $b = 15$, $\angle A = 110°$

16. $a = 30$, $c = 40$, $\angle A = 37°$

17. $a = 20$, $c = 45$, $\angle A = 125°$

18. $b = 45$, $c = 42$, $\angle C = 38°$

19. $b = 25$, $c = 30$, $\angle B = 25°$

20. $a = 75$, $b = 100$, $\angle A = 30°$

21. $a = 50$, $b = 100$, $\angle A = 50°$

22. $a = 100$, $b = 80$, $\angle A = 135°$

23. To find the distance across a river, a surveyor chooses points A and B, which are 200 ft apart on one side of the river (see the figure). She then chooses a reference point C on the opposite side of the river and finds that $\angle BAC \approx 82°$ and $\angle ABC \approx 52°$. Approximate the distance from A to C.

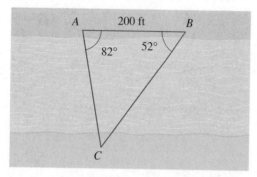

24. A pilot is flying over a straight highway. He determines the angles of depression to two mileposts, 5 mi apart, to be 32° and 48°, as shown in the figure.
(a) Find the distance of the plane from point A.
(b) Find the elevation of the plane.

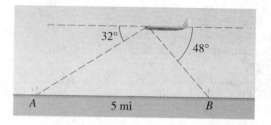

25. The path of a satellite orbiting the earth causes it to pass directly over two tracking stations A and B, which are 50 mi apart. When the satellite is on one side of the two stations, the angles of elevation at A and B are measured to be 87.0° and 84.2°, respectively.
(a) How far is the satellite from station A?
(b) How high is the satellite above the ground?

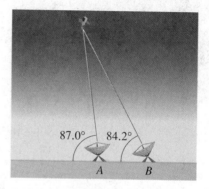

26. A tree on a hillside casts a shadow 215 ft down the hill. If the angle of inclination of the hillside is 22° to the horizontal and the angle of elevation of the sun is 52°, find the height of the tree.

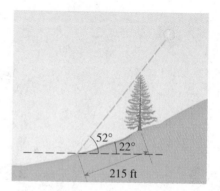

27. A communication tower is located at the top of a steep hill, as shown. The angle of inclination of the hill is

58°. A guy wire is to be attached to the top of the tower and to the ground, 100 m downhill from the base of the tower. The angle α in the figure is determined to be 12°. Find the length of cable required for the guy wire.

28. Points A and B are separated by a lake. To find the distance between them, a surveyor locates a point C on land such that $\angle CAB = 48.6°$. He also measures CA as 312 ft and CB as 527 ft. Find the distance between A and B.

29. To calculate the height of a mountain, angles α, β, and distance d are determined, as shown.
  (a) Find the length of BC in terms of α, β, and d.
  (b) Show that the height h of the mountain is given by the formula

$$h = d \frac{\sin \alpha \sin \beta}{\sin(\beta - \alpha)}$$

  (c) Use the formula in part (b) to find the height of a mountain if $\alpha \approx 25°$, $\beta \approx 29°$, and $d \approx 800$ ft.

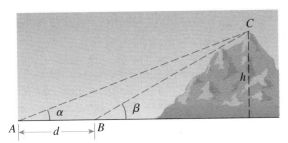

30. Observers at P and Q are located on the side of a hill that is inclined 32° to the horizontal, as shown. The observer at P determines the angle of elevation to a hot-air balloon to be 62°. At the same instant, the observer at Q measures the angle of elevation to the balloon to

be 71°. If P is 60 m down the hill from Q, find the distance from Q to the balloon.

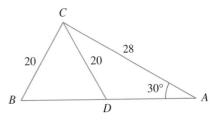

31. A water tower 30 m tall is located at the top of a hill. From a distance of 120 m down the hill, it is observed that the angle formed between the top and base of the tower is 8°. Find the angle of inclination of the hill.

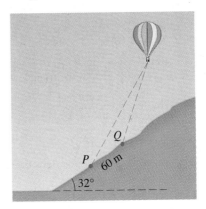

32. For the triangle shown, find (a) $\angle BCD$ and (b) $\angle DCA$.

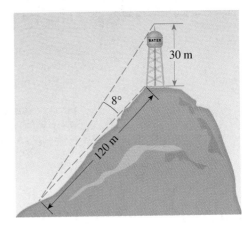

33. In triangle ABC, $\angle A = 40°$, $a = 15$, and $b = 20$.
  (a) Show that there are two triangles, ABC and A'B'C', that satisfy these conditions.

(b) Show that the areas of the triangles in part (a) are proportional to the sines of the angles C and C', that is,

$$\frac{\text{area of } \triangle ABC}{\text{area of } \triangle A'B'C'} = \frac{\sin C}{\sin C'}$$

34. Show that, given the three angles A, B, C of a triangle and one side, say a, the area of the triangle is

$$\text{area} = \frac{a^2 \sin B \sin C}{2 \sin A}$$

  DISCOVERY · DISCUSSION

35. Number of Solutions in the Ambiguous Case   We have seen that when using the Law of Sines to solve a triangle in the SSA case, there may be two, one, or no

solution(s). Sketch triangles like those in Figure 6 to verify the criteria in the table for the number of solutions if you are given $\angle A$ and sides a and b.

Criterion	Number of solutions
$a \geq b$	1
$b > a > b \sin A$	2
$a = b \sin A$	1
$a < b \sin A$	0

If $\angle A = 30°$ and $b = 100$, use these criteria to find the range of values of a for which the triangle ABC has two solutions, one solution, or no solution.

6.5 THE LAW OF COSINES

The Law of Sines cannot be used directly to solve triangles if we know two sides and the angle between them or if we know all three sides (these are Cases 3 and 4 of the preceding section). In these two cases, the **Law of Cosines** applies.

> **THE LAW OF COSINES**
>
> In any triangle ABC, we have
>
> $$a^2 = b^2 + c^2 - 2bc \cos A$$
> $$b^2 = a^2 + c^2 - 2ac \cos B$$
> $$c^2 = a^2 + b^2 - 2ab \cos C$$

■ **Proof**   To prove the Law of Cosines, place triangle ABC so that $\angle A$ is at the origin, as shown in Figure 1. The coordinates of the vertices B and C are $(c, 0)$ and $(b \cos A, b \sin A)$, respectively. (You should check that the coordinates of these points would be the same if we had drawn angle A as an acute angle.) Using the Distance Formula, we get

$$a^2 = (b \cos A - c)^2 + (b \sin A - 0)^2$$
$$= b^2 \cos^2 A - 2bc \cos A + c^2 + b^2 \sin^2 A$$
$$= b^2 (\cos^2 A + \sin^2 A) - 2bc \cos A + c^2$$
$$= b^2 + c^2 - 2bc \cos A \quad \text{Because } \sin^2 A + \cos^2 A = 1$$

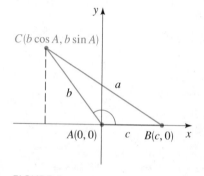

FIGURE 1

This proves the first formula. The other two formulas are obtained in the same way by placing each of the other vertices of the triangle at the origin and repeating the preceding argument.   □

In words, the Law of Cosines says that the square of any side of a triangle is equal to the sum of the squares of the other two sides minus twice the product of those two sides times the cosine of the included angle.

If one of the angles of a triangle, say $\angle C$, is a right angle, then $\cos C = 0$ and the Law of Cosines reduces to the Pythagorean Theorem, $c^2 = a^2 + b^2$. Thus, the Pythagorean Theorem is a special case of the Law of Cosines.

EXAMPLE 1 ■ Length of a Tunnel

A tunnel is to be built through a mountain. To estimate the length of the tunnel, a surveyor makes the measurements shown in Figure 2. Use the surveyor's data to approximate the length of the tunnel.

SOLUTION   To approximate the length c of the tunnel, we use the Law of Cosines:

$$c^2 = a^2 + b^2 - 2ab\cos C \qquad \text{Law of Cosines}$$
$$= 388^2 + 212^2 - 2(388)(212)\cos 82.4° \qquad \text{Substitute}$$
$$\approx 173730.2367 \qquad \text{Use a calculator}$$
$$c \approx \sqrt{173730.2367} \approx 416.8 \qquad \text{Take square roots}$$

Thus, the tunnel will be approximately 417 ft long.   ■

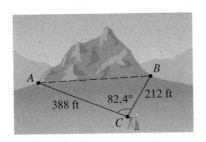

FIGURE 2

EXAMPLE 2 ■ SSS, the Law of Cosines

The sides of a triangle are $a = 5$, $b = 8$, and $c = 12$ (see Figure 3). Find the angles of the triangle.

SOLUTION   We first find $\angle A$. From the Law of Cosines, we have $a^2 = b^2 + c^2 - 2bc\cos A$. Solving for $\cos A$, we get

$$\cos A = \frac{b^2 + c^2 - a^2}{2bc} = \frac{8^2 + 12^2 - 5^2}{2(8)(12)} = \frac{183}{192} = 0.953125$$

Using a calculator, we find that $\angle A \approx 18°$. In the same way the equations

$$\cos B = \frac{a^2 + c^2 - b^2}{2ac} = \frac{5^2 + 12^2 - 8^2}{2(5)(12)} = 0.875$$

$$\cos C = \frac{a^2 + b^2 - c^2}{2ab} = \frac{5^2 + 8^2 - 12^2}{2(5)(8)} = -0.6875$$

give $\angle B \approx 29°$ and $\angle C \approx 133°$. Of course, once two angles are calculated, the third can more easily be found from the fact that the sum of the angles of a triangle is 180°. However, it's a good idea to calculate all three angles using the Law of Cosines and add the three angles as a check on your computations.   ■

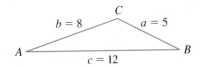

FIGURE 3

$$\boxed{\text{COS}^{-1}}$$
or
$$\boxed{\text{INV}}\;\boxed{\text{COS}}$$
or
$$\boxed{\text{ARC}}\;\boxed{\text{COS}}$$

EXAMPLE 3 ■ SAS, the Law of Cosines and the Law of Sines

Solve triangle ABC, where $\angle A = 46.5°$, $b = 10.5$, and $c = 18.0$.

SOLUTION We can find a using the Law of Cosines.

$$a^2 = b^2 + c^2 - 2bc \cos A$$

$$= (10.5)^2 + (18.0)^2 - 2(10.5)(18.0)(\cos 46.5°) \approx 174.05$$

Thus, $a \approx \sqrt{174.05} \approx 13.2$. The two remaining angles can now be found using the Law of Sines. We have

$$\sin B = \frac{b \sin A}{a} \approx \frac{10.5 \sin 46.5°}{13.2} \approx 0.577$$

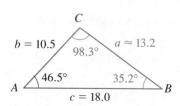

FIGURE 4

So B is the angle whose sine is 0.577. For this, we use our calculator to get $\angle B \approx 35.2°$. Since angle B can have measure between 0° and 180°, another possibility for angle B is $\angle B = 180° - 35.2° = 144.8°$. It's a simple matter to choose between these two possibilities, since the largest angle in a triangle must be opposite the longest side. So the correct choice is $\angle B \approx 35.2°$. In this case, $\angle C \approx 180° - (46.5° + 35.2°) = 98.3°$, and indeed the largest angle, $\angle C$, is opposite the longest side, $c = 18.0$.

To summarize: $\angle B \approx 35.2°$, $\angle C \approx 98.3°$, and $a \approx 13.2$. (See Figure 4.)   ■

As we saw in Example 3, when using both the Law of Cosines and the Law of Sines to solve a triangle in the SAS case, we must be careful to choose the correct measure for the remaining angle. In any triangle, the longest side is opposite the largest angle, and the shortest side is opposite the smallest angle. Thus, when using the Law of Sines, we must choose the angle so that this condition is satisfied. That's why it's a good idea to sketch the triangle so you can check your final answer.

AREA OF TRIANGLES

An interesting application of the Law of Cosines involves a formula for finding the area of a triangle from the lengths of its three sides (see Figure 5).

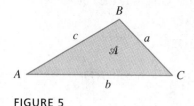

FIGURE 5

HERON'S FORMULA

The area $\mathcal{A}$ of triangle ABC is given by

$$\mathcal{A} = \sqrt{s(s-a)(s-b)(s-c)}$$

where $s = \frac{1}{2}(a + b + c)$ is the **semiperimeter** of the triangle; that is, s is half the perimeter.

■ **Proof** We start with the formula $\mathcal{A} = \frac{1}{2}ab\sin C$ from Section 6.3. Thus

$$\mathcal{A}^2 = \tfrac{1}{4}a^2b^2\sin^2 C$$

$$= \tfrac{1}{4}a^2b^2(1 - \cos^2 C) \qquad \text{Pythagorean identity}$$

$$= \tfrac{1}{4}a^2b^2(1 - \cos C)(1 + \cos C) \qquad \text{Factor}$$

Next, we write the expressions $1 - \cos C$ and $1 + \cos C$ in terms of a, b, and c. By the Law of Cosines we have

$$\cos C = \frac{a^2 + b^2 - c^2}{2ab} \qquad \text{Law of Cosines}$$

$$1 + \cos C = 1 + \frac{a^2 + b^2 - c^2}{2ab} \qquad \text{Add 1}$$

$$= \frac{2ab + a^2 + b^2 - c^2}{2ab} \qquad \text{Common denominator}$$

$$= \frac{(a + b)^2 - c^2}{2ab} \qquad \text{Factor}$$

$$= \frac{(a + b + c)(a + b - c)}{2ab} \qquad \text{Difference of squares}$$

Similarly

$$1 - \cos C = \frac{(c + a - b)(c - a + b)}{2ab}$$

Substituting these expressions in the formula we obtained for $\mathcal{A}^2$ gives

$$\mathcal{A}^2 = \tfrac{1}{4}a^2b^2\frac{(a + b + c)(a + b - c)}{2ab}\frac{(c + a - b)(c - a + b)}{2ab}$$

$$= \frac{(a + b + c)}{2}\frac{(a + b - c)}{2}\frac{(c + a - b)}{2}\frac{(c - a + b)}{2}$$

$$= s(s - c)(s - b)(s - a)$$

Showing that each factor in the last expression equals the corresponding factor in the preceding expression is left as an exercise. Heron's Formula now follows by taking the square root of each side. □

EXAMPLE 4 ■ Area of a Lot

A businessman wishes to buy a triangular lot in a busy downtown location (see Figure 6). The lot frontages on the three adjacent streets are 125, 280, and 315 ft. Find the area of the lot.

SOLUTION   The semiperimeter of the lot is

$$s = \frac{125 + 280 + 315}{2} = 360$$

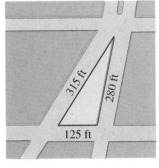

FIGURE 6

By Heron's Formula the area is

$$\mathcal{A} = \sqrt{360(360 - 125)(360 - 280)(360 - 315)} \approx 17{,}451.6$$

Thus, the area is approximately $17{,}452 \text{ ft}^2$.   ■

| 6.5 | **EXERCISES** |

1–6 ■ Use the Law of Cosines to determine the indicated side x or angle θ.

1.
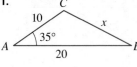

2.
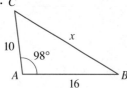

3.
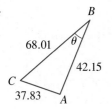

4.
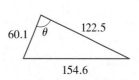

5.
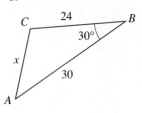

6.
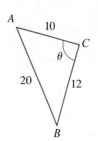

7–16 ■ Solve triangle ABC.

7.
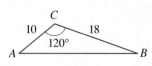

8.
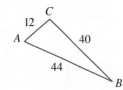

9. $a = 3.0$, $b = 4.0$, $\angle C = 53°$

10. $b = 60$, $c = 30$, $\angle A = 70°$

11. $a = 20$, $b = 25$, $c = 22$

12. $a = 10$, $b = 12$, $c = 16$

13. $b = 125$, $c = 162$, $\angle B = 40°$

14. $a = 65$, $c = 50$, $\angle C = 52°$

15. $a = 50$, $b = 65$, $\angle A = 55°$

16. $a = 73.5$, $\angle B = 61°$, $\angle C = 83°$

17–24 ■ Find the indicated side x or angle θ. (Use either the Law of Sines or the Law of Cosines, as appropriate.)

17.
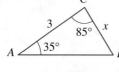

18.
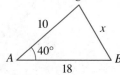

19.
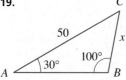

20.
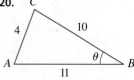

21.
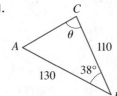

22.
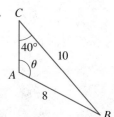

23.
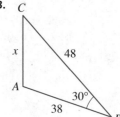

24.
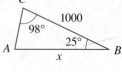

25. To find the distance across a small lake, a surveyor has taken the measurements shown. Find the distance across the lake using this information.

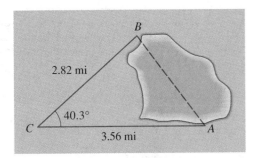

26. A parallelogram has sides of lengths 3 and 5, and one angle is 50°. Find the length of the diagonals.

27. Two straight roads diverge at an angle of 65°. Two cars leave the intersection at 2:00 P.M., one traveling at 50 mi/h and the other at 30 mi/h. How far apart are the cars at 2:30 P.M.?

28. A car travels along a straight road, heading east for 1 h, then traveling for 30 min on another road that leads northeast. If the car has maintained a constant speed of 40 mi/h, how far is it from its starting position?

29. A pilot flies in a straight path for 1 h 30 min. She then makes a course correction, heading 10° to the right of her original course, and flies 2 h in the new direction. If she maintains a constant speed of 625 mi/h, how far is she from her starting position?

30. Two boats leave the same port at the same time. One travels at a speed of 30 mi/h in the direction N 50° E and the other travels at a speed of 26 mi/h in a direction S 70° E (see the figure). How far apart are the two boats after one hour?

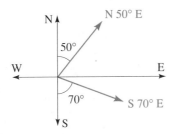

31. A triangular field has sides of lengths 22, 36, and 44 yd. Find the largest angle.

32. Two tugboats that are 120 ft apart pull a barge, as shown. If the length of one cable is 212 ft and the length of the other is 230 ft, find the angle formed by the two cables.

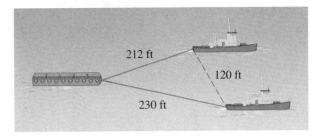

33. A boy is flying two kites at the same time. He has 380 ft of line out to one kite and 420 ft to the other. He estimates the angle between the two lines to be 30°. Approximate the distance between the kites.

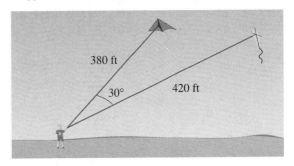

34. A 125-ft tower is located on the side of a mountain that is inclined 32° to the horizontal. A guy wire is to be attached to the top of the tower and anchored at a point 55 ft downhill from the base of the tower. Find the shortest length of wire needed.

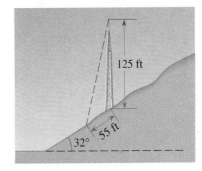

35. A steep mountain is inclined 74° to the horizontal and rises 3400 ft above the surrounding plain. A cable car is to be installed from a point 800 ft from the base to the top of the mountain, as shown. Find the shortest length of cable needed.

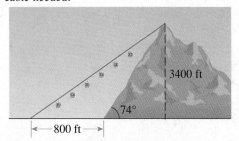

36. The CN Tower in Toronto, Canada, is the highest tower in the world. A woman on the observation deck, 1150 ft above the ground, wants to determine the distance between two landmarks on the ground below. She observes that the angle formed by the lines of sight to these two landmarks is 43°. She also observes that the angle between the vertical and the line of sight to one of the landmarks is 62° and to the other landmark is 54°. Find the distance between the two landmarks.

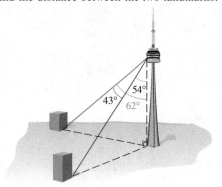

37. Three circles of radii 4, 5, and 6 cm are mutually tangent. Find the shaded area enclosed between the circles.

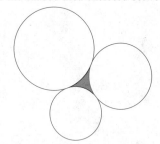

38. Prove that in triangle ABC

$$a = b \cos C + c \cos B$$
$$b = c \cos A + a \cos C$$
$$c = a \cos B + b \cos A$$

These are called the *Projection Laws*. [*Hint:* To get the first equation, add together the second and third equations in the Law of Cosines and solve for a.]

39. A surveyor wishes to find the distance between two points A and B on the opposite side of a river. On her side of the river, she chooses two points C and D that are 20 m apart and measures the angles shown. Find the distance between A and B.

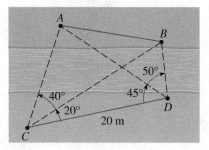

40. Find the area of a triangle with sides of lengths 12, 18, and 24 m.

41. Land in downtown Columbia is valued at $20 a square foot. What is the value of a triangular lot with sides of lengths 112, 148, and 190 ft?

42. Find the area of the quadrilateral in the figure, correct to two decimal places.

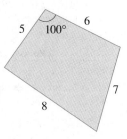

DISCOVERY · DISCUSSION

43. Solving for the Angles in a Triangle When we solved for $\angle B$ in Example 3 using the Law of Sines, we had to be careful because there were two possibilities, 35.2° and 144.8°. That is because these are the *two* angles between 0° and 180° whose sine is 0.577. What if we had used the Law of Cosines to find $\angle B$? Would the problem of two possible measures for $\angle B$ have arisen? Which method do you prefer here: the Law of Sines or the Law of Cosines?

CONCEPT CHECK

1. (a) Explain the difference between a positive angle and a negative angle.
   (b) How is an angle of measure 1 degree formed?
   (c) How is an angle of measure 1 radian formed?
   (d) How is the radian measure of an angle θ defined?
   (e) How do you convert from degrees to radians?
   (f) How do you convert from radians to degrees?

2. (a) When is an angle in standard position?
   (b) When are two angles coterminal?

3. (a) What is the length s of an arc of a circle with radius r that subtends a central angle of θ radians?
   (b) What is the area A of a sector of a circle with radius r and central angle θ radians?

4. If θ is an acute angle in a right triangle, define the six trigonometric ratios in terms of the adjacent and opposite sides and the hypotenuse.

5. What does it mean to solve a triangle?

6. If θ is an angle in standard position, $P(x, y)$ is a point on the terminal side, and r is the distance from the origin to P, write expressions for the six trigonometric functions of θ.

7. Which trigonometric functions are positive in quadrants I, II, III, and IV?

8. If θ is an angle in standard position, what is its reference angle $\overline{\theta}$?

9. (a) State the reciprocal identities.
   (b) State the Pythagorean identities.

10. (a) What is the area of a triangle with sides of length a and b and with included angle θ?
    (b) What is the area of a triangle with sides of length a, b, and c?

11. (a) State the Law of Sines.
    (b) State the Law of Cosines.

12. Explain the ambiguous case in the Law of Sines.

EXERCISES

1–2 ■ Find the radian measure that corresponds to the given degree measure.

1. (a) 70°   (b) 420°   (c) −240°   (d) −40°

2. (a) 24°   (b) −330°   (c) 750°   (d) 5°

3–4 ■ Find the degree measure that corresponds to the given radian measure.

3. (a) $\dfrac{7\pi}{2}$ (b) $-\dfrac{\pi}{3}$
   (c) $\dfrac{7\pi}{4}$ (d) 2.1

4. (a) 8   (b) $-\dfrac{5}{2}$
   (c) $\dfrac{11\pi}{6}$ (d) $\dfrac{3\pi}{5}$

5. Find the length of an arc of a circle of radius 8 m if the arc subtends a central angle of 1 rad.

6. Find the measure of a central angle θ in a circle of radius 5 ft if the angle is subtended by an arc of length 7 ft.

7. A circular arc of length 100 ft subtends a central angle of 70°. Find the radius of the circle.

8. How many revolutions will a car wheel of diameter 28 in. make over a period of half an hour if the car is traveling at 60 mi/h?

9. New York and Los Angeles are 2450 mi apart. Find the angle that the arc between these two cities subtends at the center of the earth. (The radius of the earth is 3960 mi.)

10. Find the area of a sector with central angle 2 rad in a circle of radius 5 m.

11. Find the area of a sector with central angle 52° in a circle of radius 200 ft.

12. A sector in a circle of radius 25 ft has an area of 125 ft^2. Find the central angle of the sector.

13–14 ■ Find the values of the six trigonometric ratios of θ.

13.

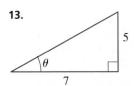

14.

15–18 ■ Find the sides labeled x and y, correct to two decimal places.

15.

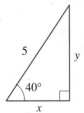

16.

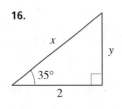

17.

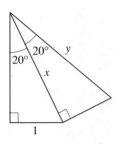

18.

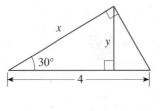

19–20 ■ Solve the triangle.

19.

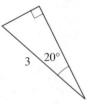

20.

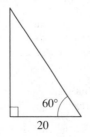

21. Express the lengths a and b in the figure in terms of the trigonometric ratios of θ.

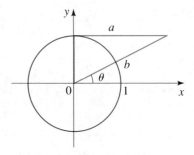

22. The highest tower in the world is the CN Tower in Toronto, Canada. From a distance of 1 km from its base, the angle of elevation to the top of the tower is 28.81°. Find the height of the tower.

23. Find the perimeter of a regular hexagon that is inscribed in a circle of radius 8 m.

24. The pistons in a car engine move up and down repeatedly to turn the crankshaft, as shown. Find the height of the point P above the center O of the crankshaft in terms of the angle θ.

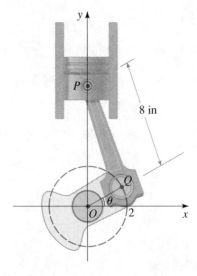

25. As viewed from the earth, the angle subtended by the full moon is 0.518°. Use this information and the fact that the distance AB from the earth to the moon is 236,900 mi to find the radius of the moon.

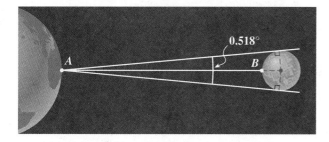

26. A pilot measures the angles of depression to two ships to be 40° and 52° (see the figure). If the pilot is flying at an elevation of 35,000 ft, find the distance between the two ships.

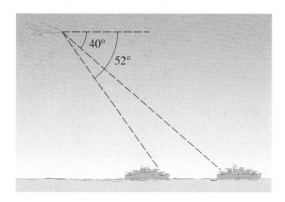

27–38 ■ Find the exact value.

27. $\sin 315°$ **28.** $\csc \dfrac{9\pi}{4}$ **29.** $\tan(-135°)$

30. $\cos \dfrac{5\pi}{6}$ **31.** $\cot\left(-\dfrac{22\pi}{3}\right)$ **32.** $\sin 405°$

33. $\cos 585°$ **34.** $\sec \dfrac{22\pi}{3}$ **35.** $\csc \dfrac{8\pi}{3}$

36. $\sec \dfrac{13\pi}{6}$ **37.** $\cot(-390°)$ **38.** $\tan \dfrac{23\pi}{4}$

39. Find the values of the six trigonometric ratios of the angle θ in standard position if the point $(-5, 12)$ is on the terminal side of θ.

40. Find $\sin \theta$ if θ is in standard position and its terminal side intersects the circle of radius 1 centered at the origin at the point $\left(-\sqrt{3}/2, \frac{1}{2}\right)$.

41. Find the acute angle that is formed by the line $y - \sqrt{3}\,x + 1 = 0$ and the x-axis.

42. Find the six trigonometric ratios of the angle θ in standard position if its terminal side is in quadrant III and is parallel to the line $4y - 2x - 1 = 0$.

43–46 ■ Write the first expression in terms of the second for θ in the given quadrant.

43. $\tan \theta$, $\cos \theta$; θ in quadrant II

44. $\sec \theta$, $\sin \theta$; θ in quadrant III

45. $\tan^2\theta$, $\sin \theta$; θ in any quadrant

46. $\csc^2\theta \cos^2\theta$, $\sin \theta$; θ in any quadrant

47–50 ■ Find the values of the six trigonometric functions of θ from the information given.

47. $\tan \theta = \sqrt{7}/3$, $\sec \theta = \frac{4}{3}$

48. $\sec \theta = \frac{41}{40}$, $\csc \theta = -\frac{41}{9}$

49. $\sin \theta = \frac{3}{5}$, $\cos \theta < 0$

50. $\sec \theta = -\frac{13}{5}$, $\tan \theta > 0$

51. If $\tan \theta = -\frac{1}{2}$ for θ in quadrant II, find $\sin \theta + \cos \theta$.

52. If $\sin \theta = \frac{1}{2}$ for θ in quadrant I, find $\tan \theta + \sec \theta$.

53. If $\tan \theta = -1$, find $\sin^2\theta + \cos^2\theta$.

54. If $\cos \theta = -\sqrt{3}/2$ and $\pi/2 < \theta < \pi$, find $\sin 2\theta$.

55–60 ■ Find the side labeled x.

55.

56.

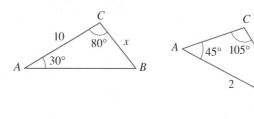

57.

58.

59.

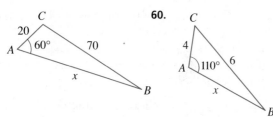

60.

61. Two ships leave a port at the same time. One travels at 20 mi/h in a direction N 32° E, and the other travels at 28 mi/h in a direction S 42° E (see the figure). How far apart are the two ships after 2 h?

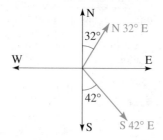

62. From a point A on the ground, the angle of elevation to the top of a tall building is 24.1°. From a point B, which is 600 ft closer to the building, the angle of elevation is measured to be 30.2°. Find the height of the building.

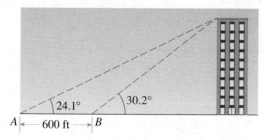

63. Find the distance between the points A and B on opposite sides of a lake from the information shown.

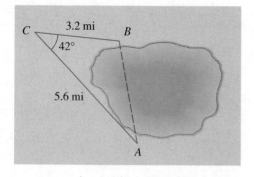

64. A boat is cruising the ocean off a straight shoreline. Points A and B are 120 mi apart on the shore, as shown. It is found that $\angle A = 42.3°$ and $\angle B = 68.9°$. Find the shortest distance from the boat to the shore.

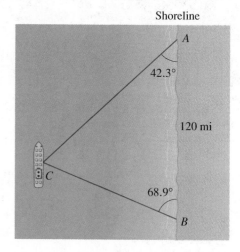

65. To measure the height of an inaccessible cliff on the opposite side of a river, a surveyor makes the measurements shown. Find the height of the cliff.

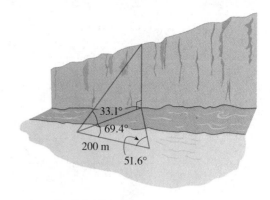

66. Find the area of a triangle with sides of length 8 and 14 and included angle 35°.

67. Find the area of a triangle with sides of length 5, 6, and 8.

1. Find the radian measures that correspond to the degree measures $300°$ and $-18°$.

2. Find the degree measures that correspond to the radian measures $\dfrac{5\pi}{6}$ and 2.4.

3. Find the exact value of each of the following.
(a) $\sin 405°$ (b) $\tan(-150°)$
(c) $\sec \dfrac{5\pi}{3}$ (d) $\csc \dfrac{5\pi}{2}$

4. Find $\tan\theta + \sin\theta$ for the angle θ shown.

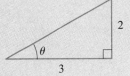

5. Find the lengths a and b shown in the figure in terms of θ.

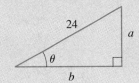

6. If $\cos\theta = -\frac{1}{3}$ and θ is in quadrant III, find $\tan\theta\cot\theta + \csc\theta$.

7. If $\sin\theta = \frac{5}{13}$ and $\tan\theta = -\frac{5}{12}$, find $\sec\theta$.

8. Express $\tan\theta$ in terms of $\sec\theta$ for θ in quadrant II.

9. The base of the ladder in the figure is 6 ft from the building, and the angle formed by the ladder and the ground is $73°$. How high up the building does the ladder touch?

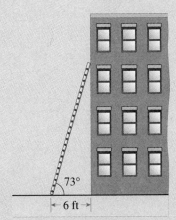

10–13 ■ Find the side labeled x.

10.

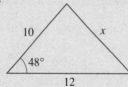

11.

12.

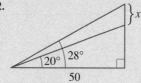

13.

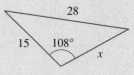

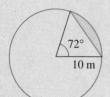

14–15 ■ Refer to the figure at the left.

14. Find the area of the shaded region.

15. Find the perimeter of the shaded region.

16–17 ■ Refer to the figure at the right.

16. Find the angle opposite the longest side.

17. Find the area of the triangle.

18. Two wires tether a balloon to the ground, as shown. How high is the balloon above the ground?

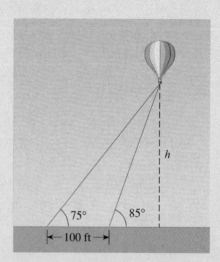

FOCUS ON PROBLEM SOLVING

One important problem-solving technique is the strategy of **taking cases**. A classic use of this strategy is in the classification of the regular polyhedra. These are called *Platonic solids* because they were mentioned in the writings of Plato.

A *regular polygon* is one in which all sides and all angles are equal. There are infinitely many regular polygons, as indicated in Figure 1.

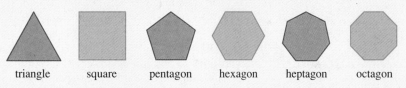

FIGURE 1
The regular polygons

triangle   square   pentagon   hexagon   heptagon   octagon

For three-dimensional shapes, the analogous concept is that of regular polyhedra. A *regular polyhedron* is a solid in which all faces are congruent regular polygons, and the same number of polygons meet at each corner, or *vertex*. We would like to find all possible regular polyhedra. It might seem at first that there should be infinitely many, just as for regular polygons. But we will see that there are just finitely many polyhedra.

CLASSIFYING THE REGULAR POLYHEDRA

We now prove that there are exactly five regular polyhedra (see Figure 2). To show this, we consider all possible cases.

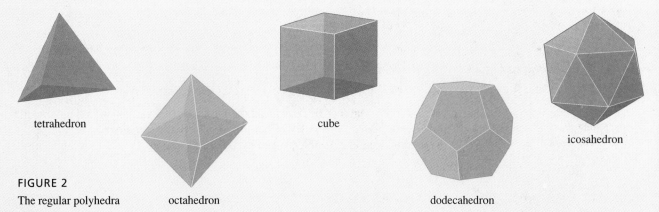

tetrahedron

cube

icosahedron

FIGURE 2
The regular polyhedra   octahedron

dodecahedron

Three, four, and five equilateral triangles can be folded up to make a corner, but six such triangles lie flat.

■ **CASE 1** Suppose the faces of a regular polyhedron are equilateral triangles. How many such triangles can meet at a corner? To make a corner there must be at least three triangles. We can also have four or five. But six equilateral triangles cannot meet at a point to make a corner. (Why?) If three triangles meet at each vertex, we can complete the polyhedron by adding one more triangle to make a *tetrahedron*. If four triangles meet at each vertex, we have an *octahedron*. If five triangles meet at each vertex, the resulting regular polyhedron is an *icosahedron*. Thus, we have found all the regular polyhedra with triangular faces.

■ **CASE 2** Suppose the faces of a regular polygon are squares. If three squares meet at each point, then the polyhedron is a *cube*. It's impossible for four or more squares to meet at a point to make a corner. (Why?) Thus, the only regular polyhedron with square faces is the cube.

■ **CASE 3** Suppose the faces of a regular polygon are pentagons. If three pentagons meet at each vertex, the resulting polyhedron is a *dodecahedron*. Since the angles of a regular pentagon are 108°, it's impossible for more than three regular pentagons to meet at a vertex. Thus, the only regular polyhedron with pentagonal faces is the dodecahedron.

■ **CASE 4** Is it possible for the faces of a regular polygon to be regular hexagons? Since the angle of a regular hexagon is 120°, when three such hexagons meet at a point they do not form a corner, so it's impossible for a regular polyhedron to have hexagonal faces. The same reasoning shows that no other regular polygon can be the face of a regular polyhedron.

Since these four cases account for all the possibilities, we have shown that there are exactly five regular polyhedra.

■ EULER'S FORMULA

How many faces, edges, and vertices does a regular polyhedron have? In the 18th century, Euler observed that

$$F - E + V = 2 \qquad \text{Euler's Formula}$$

where F is the number of faces, E is the number of edges, and V is the number of vertices. We can use Euler's Formula to answer the question. For example, the icosahedron is assembled from F equilateral triangles:

In these triangles the total number of sides is $3F$, and the total number of angles is also $3F$. In an icosahedron, five angles of these triangles meet to form a vertex, so the total number of vertices must be

$$V = \frac{3F}{5}$$

Since two sides of adjacent triangles meet to form one edge of the polyhedron,

It's interesting to note that there are infinitely many 2-dimensional regular polygons but only five 3-dimensional regular polyhedra. Although it's impossible to draw the 4-dimensional regular polyhedra, mathematicians have shown that there are six of them. Surprisingly, in all higher dimensions there are exactly three regular polyhedra—the n-dimensional cube, tetrahedron, and octahedron.

the number of edges must be

$$E = \frac{3F}{2}$$

Substituting into Euler's Formula gives

$$F - \frac{3F}{2} + \frac{3F}{5} = 2$$

Solving gives $F = 20$. Substituting this value of F into the formulas for edges and faces gives $E = 30$ and $V = 12$. Thus, for the icosahedron, $F = 20$, $E = 30$, and $V = 12$. Using similar reasoning we can find the number of faces, edges, and vertices for the other regular polyhedra.

 PROBLEMS

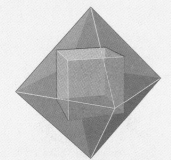

1. Find the number of faces, edges, and vertices for each of the regular polyhedra, not by counting them, but by using Euler's Formula.

2. Describe the polyhedron whose edges are the line segments joining the centers of the faces of an octahedron, as shown in the figure at the left. Do the same for the other Platonic solids.

3. As the following figures indicate, it's possible to *tile* the plane (that is, completely cover it) with equilateral triangles and with squares. Find all other regular polygons that tile the plane. Prove your answer.

4. (a) Find a formula for the area A_n of the regular polygon with n sides inscribed in a circle of radius 1 (see the figure). Express your answer as a trigonometric function of n.

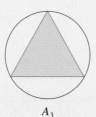 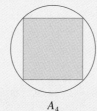 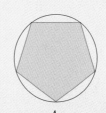 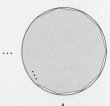

A_3 A_4 A_5 ... A_n

(b) Find A_3, A_4, A_{100}, and A_{1000}, correct to six decimal places. Notice that the values get closer and closer to π. Why?

5. A group of n pulleys, all of radius 1, is fixed so that their centers form a convex n-gon of perimeter P. (The figure shows the case $n = 4$.) Find the length of the belt that fits around the pulleys. [*Hint:* Try fitting together the sectors of the pulleys that touch the belt.]

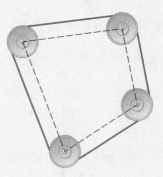

6. A pair of pulleys is connected by a belt, as shown in each figure. Find the length of the belt.

(a)   (b)

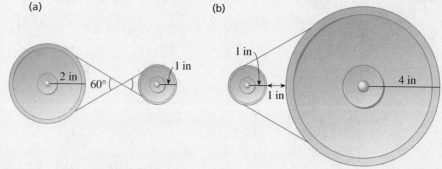

7. Two circles of radius 1 are placed so that their centers are one unit apart. Find the area of the region common to both circles.

8. Suppose that the lengths of the sides of a triangle are rational numbers. Prove that the cosine of each angle is a rational number.

9. Find the exact value of

$$\sin \frac{\pi}{100} + \sin \frac{2\pi}{100} + \sin \frac{3\pi}{100} + \sin \frac{4\pi}{100} + \cdots + \sin \frac{199\pi}{100}$$

where the sum contains the sines of all numbers of the form $k\pi/100$, with k varying from 1 to 199.

Pierre de Fermat (1601–1665) was a French lawyer who became interested in mathematics at the age of 30. Because of his job as a magistrate, Fermat had little time to write complete proofs of his discoveries and often wrote them in the margin of whatever book he was reading at the time. After his death, his copy of Diophantus' *Arithmetica* (see page 3) was found to contain a particularly tantalizing comment. Where Diophantus discusses the solutions of $x^2 + y^2 = z^2$ (for example, $x = 3$, $y = 4$, $z = 5$), Fermat states in the margin that for $n \geq 3$ there are no natural number solutions to the equation $x^n + y^n = z^n$. In other words, it's impossible for a cube to equal the sum of two cubes, a fourth power to equal the sum of two fourth powers, and so on. Fermat writes "I have discovered a truly wonderful proof for this but the margin is too small to contain it." All the other margin comments in Fermat's copy of *Arithmetica* have been proved. This one, however, remained unproved, and it came to be known as "Fermat's Last Theorem."

(continued)

10. In order to draw a map of the spherical earth on a flat sheet of paper, several different ingenious methods have been developed. One of these is the *Mercator projection*. In this method, each point on the spherical earth is projected onto a circumscribed cylinder (tangent to the sphere on the equator) by a line through the center of the earth. By what factor are each of the following distances distorted using this method of projection? That is, what is the ratio of the projected distance on the cylinder to the actual distance on the sphere?

(a) The distance between 20° and 21° N latitude along a meridian
(b) The distance between 40° and 41° N latitude along a meridian
(c) The distance between 80° and 81° N latitude along a meridian
(d) The distance between two points that are 1° apart on the 20th parallel
(e) The distance between two points that are 1° apart on the 40th parallel
(f) The distance between two points that are 1° apart on the 80th parallel

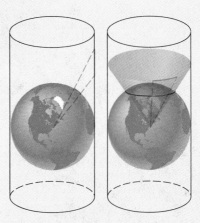

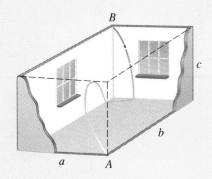

11. A bug is sitting at point *A* in one corner of a room and wants to crawl to point *B* in the corner diagonally opposite (see the figure at the left). Find the shortest path for the bug.

12. Prove that every prime number is the leg of exactly one right triangle with integer sides. (This problem was first stated by Fermat.)

13. Show that the equation $x^2 + y^2 = 4z + 3$ has no solution in integers.
[*Hint:* Recall that an even number is of the form $2n$ and an odd number is of the form $2n + 1$. Consider all possible cases for x and y even or odd.]

14. (a) Find all prime numbers p such that $2p + 1$ is a perfect square.
[*Hint:* Write the equation $2p + 1 = n^2$ as $2p = n^2 - 1$ and factor. Then consider cases.]

(b) Find all prime numbers p such that $2p + 1$ is a perfect cube.

15. (a) Write 13 as the sum of two squares. Then do the same for 41.
(b) Verify that $(a^2 + b^2)(c^2 + d^2) = (ac + bd)^2 + (ad - bc)^2$.
(c) Express 533 as the sum of two squares in two different ways.
[*Hint:* Factor 533 and use parts (a) and (b).]

7

ANALYTIC TRIGONOMETRY

Trigonometric identities are used in analyzing directed quantities, or vectors. The path of a sailboat is determined by resolving the vector forces of wind and current that act on it.

The function of mathematics in providing answers is often less important than its function in providing understanding.

BEN NOBLE

In Chapters 5 and 6 we studied the graphical and geometric properties of trigonometric functions. In this chapter we study the algebraic aspects of trigonometry, that is, simplifying and factoring expressions and solving equations that involve trigonometric functions. The basic tools in the algebra of trigonometry are trigonometric identities. We will find identities for trigonometric functions of sums and differences of real numbers, multiple-angle formulas, and other related identities. These identities are used in the study of complex numbers, vectors, and analytic geometry.

7.1 TRIGONOMETRIC IDENTITIES

We begin this section by reviewing some of the basic trigonometric identities that we studied in Chapters 5 and 6.

FUNDAMENTAL TRIGONOMETRIC IDENTITIES

Reciprocal Identities

$$\csc x = \frac{1}{\sin x} \qquad \sec x = \frac{1}{\cos x} \qquad \cot x = \frac{1}{\tan x}$$

$$\tan x = \frac{\sin x}{\cos x} \qquad \cot x = \frac{\cos x}{\sin x}$$

Pythagorean Identities

$$\sin^2 x + \cos^2 x = 1 \qquad \tan^2 x + 1 = \sec^2 x \qquad 1 + \cot^2 x = \csc^2 x$$

Even-Odd Identities

$$\sin(-x) = -\sin x \qquad \cos(-x) = \cos x \qquad \tan(-x) = -\tan x$$

Cofunction Identities

$$\sin\left(\frac{\pi}{2} - u\right) = \cos u \qquad \tan\left(\frac{\pi}{2} - u\right) = \cot u \qquad \sec\left(\frac{\pi}{2} - u\right) = \csc u$$

$$\cos\left(\frac{\pi}{2} - u\right) = \sin u \qquad \cot\left(\frac{\pi}{2} - u\right) = \tan u \qquad \csc\left(\frac{\pi}{2} - u\right) = \sec u$$

SIMPLIFYING TRIGONOMETRIC EXPRESSIONS

Identities enable us to write the same expression in different ways. It's often possible to rewrite a complicated-looking expression as a much simpler one, as the next examples show.

EXAMPLE 1 ■ **Simplifying a Trigonometric Expression**

Simplify the expression $(1 + \sin x)(\sec x - \tan x)$.

SOLUTION We use the fundamental identities.

$$(1 + \sin x)(\sec x - \tan x) = (1 + \sin x)\left(\frac{1}{\cos x} - \frac{\sin x}{\cos x}\right) \qquad \text{Reciprocal identities}$$

$$= (1 + \sin x)\left(\frac{1 - \sin x}{\cos x}\right) \qquad \text{Common denominator}$$

$$= \frac{1 - \sin^2 x}{\cos x} \qquad \text{Difference of squares}$$

$$= \frac{\cos^2 x}{\cos x} \qquad \text{Pythagorean identity}$$

$$= \cos x \qquad \blacksquare$$

EXAMPLE 2 ■ **Simplifying by Combining Fractions**

Simplify the expression $\dfrac{\sin \theta}{\cos \theta} + \dfrac{\cos \theta}{1 + \sin \theta}$.

SOLUTION We combine the fractions by using a common denominator.

$$\frac{\sin \theta}{\cos \theta} + \frac{\cos \theta}{1 + \sin \theta} = \frac{\sin \theta (1 + \sin \theta) + \cos^2 \theta}{\cos \theta (1 + \sin \theta)} \qquad \text{Common denominator}$$

$$= \frac{\sin \theta + \sin^2 \theta + \cos^2 \theta}{\cos \theta (1 + \sin \theta)} \qquad \text{Expand}$$

$$= \frac{\sin \theta + 1}{\cos \theta (1 + \sin \theta)} \qquad \text{Pythagorean identity}$$

$$= \frac{1}{\cos \theta} = \sec \theta \qquad \begin{array}{l}\text{Cancel and use} \\ \text{reciprocal identity}\end{array} \qquad \blacksquare$$

PROVING TRIGONOMETRIC IDENTITIES

Many identities follow from the fundamental identities. In the examples that follow we learn how to prove that a given trigonometric equation is an identity, and in the process we will see how to discover new identities.

First, it's easy to decide when a given equation is *not* an identity. All we need to do is to show that the equation does not hold for some value of the variable (or variables). Thus, the equation

$$\sin x + \cos x = 1$$

is not an identity, because when $x = \pi/4$, we have

$$\sin \frac{\pi}{4} + \cos \frac{\pi}{4} = \frac{\sqrt{2}}{2} + \frac{\sqrt{2}}{2} = \sqrt{2} \neq 1$$

To verify that a trigonometric equation is an identity, we transform one side of the equation into the other side by a series of steps, each of which is itself an identity.

GUIDELINES FOR PROVING TRIGONOMETRIC IDENTITIES

1. Pick one side of the equation and write it down. Your goal is to transform it into the other side. It's usually easier to start with the more complicated side.

2. Use algebra and the identities you know to change the side you started with. Bring fractional expressions to a common denominator, factor, and use the fundamental identities to simplify expressions.

3. It's sometimes helpful to rewrite all functions in terms of sines and cosines.

EXAMPLE 3 ■ Proving a Trigonometric Identity

Verify the identity $\cos\theta\,(\sec\theta - \cos\theta) = \sin^2\theta$.

SOLUTION   The left-hand side looks more complicated, so we start with it and try to transform it into the right-hand side.

$$\begin{aligned}
\text{LHS} &= \cos\theta\,(\sec\theta - \cos\theta) \\[6pt]
&= \cos\theta\left(\frac{1}{\cos\theta} - \cos\theta\right) &&\text{Reciprocal identity} \\[6pt]
&= 1 - \cos^2\theta &&\text{Expand} \\[6pt]
&= \sin^2\theta = \text{RHS} &&\text{Pythagorean identity} \quad\blacksquare
\end{aligned}$$

In Example 3 it isn't easy to see how to change the right-hand side into the left-hand side, but it's definitely possible. Simply notice that each of the steps is reversible. In other words, if we start with the last expression in the proof and work backward through the steps, the right side is transformed into the left side. You will probably agree, however, that it's more difficult to prove the identity this way. That's why it's better to change the more complicated side of the identity into the simpler side.

EXAMPLE 4 ■ Proving an Identity by Combining Fractions

Verify the identity

$$2\tan x \sec x = \frac{1}{1 - \sin x} - \frac{1}{1 + \sin x}$$

SOLUTION Finding a common denominator and combining the fractions on the right-hand side of this equation, we get

$$\text{RHS} = \frac{1}{1 - \sin x} - \frac{1}{1 + \sin x}$$

$$= \frac{(1 + \sin x) - (1 - \sin x)}{(1 - \sin x)(1 + \sin x)} \qquad \text{Common denominator}$$

$$= \frac{2 \sin x}{1 - \sin^2 x} \qquad \text{Simplify}$$

$$= \frac{2 \sin x}{\cos^2 x} \qquad \text{Pythagorean identity}$$

$$= 2 \frac{\sin x}{\cos x} \left(\frac{1}{\cos x} \right) \qquad \text{Factor}$$

$$= 2 \tan x \sec x = \text{LHS} \qquad \text{Reciprocal identities} \qquad \blacksquare$$

See the *Principles of Problem Solving* on pages 122–124.

In Example 5 we introduce "something extra" to the problem by multiplying the numerator and the denominator by a trigonometric expression, chosen so that we can simplify the result.

EXAMPLE 5 ■ Proving an Identity by Introducing Something Extra

Prove that this equation is an identity:

$$\frac{1}{1 - \sin z} = \sec^2 z + \tan z \sec z$$

SOLUTION We start with the left-hand side.

$$\text{LHS} = \frac{1}{1 - \sin z}$$

$$= \frac{1}{1 - \sin z} \cdot \frac{1 + \sin z}{1 + \sin z} \qquad \begin{array}{l}\text{Multiply numerator and} \\ \text{denominator by } 1 + \sin z\end{array}$$

$$= \frac{1 + \sin z}{1 - \sin^2 z} \qquad \text{Expand}$$

$$= \frac{1 + \sin z}{\cos^2 z} \qquad \text{Pythagorean identity}$$

$$= \frac{1}{\cos^2 z} + \frac{\sin z}{\cos^2 z} \qquad \text{Separate into two fractions}$$

$$= \frac{1}{\cos^2 z} + \frac{\sin z}{\cos z} \frac{1}{\cos z} \qquad \text{Factor}$$

$$= \sec^2 z + \tan z \sec z = \text{RHS} \qquad \text{Reciprocal identities} \qquad \blacksquare$$

Here's another method for proving that an equation is an identity. If we can transform each side of the equation separately, by way of identities, to arrive at the same result, then the equation is an identity. Example 6 illustrates this procedure.

EXAMPLE 6 ■ **Proving an Identity by Working with Both Sides**

Verify the identity $\dfrac{1 + \cos\theta}{\cos\theta} = \dfrac{\tan^2\theta}{\sec\theta - 1}$.

SOLUTION We prove the identity by changing each side separately into the same expression. Can you supply the reasons for each of these steps?

$$\text{LHS} = \frac{1 + \cos\theta}{\cos\theta} = \frac{1}{\cos\theta} + \frac{\cos\theta}{\cos\theta} = \sec\theta + 1$$

$$\text{RHS} = \frac{\tan^2\theta}{\sec\theta - 1} = \frac{\sec^2\theta - 1}{\sec\theta - 1} = \frac{(\sec\theta - 1)(\sec\theta + 1)}{\sec\theta - 1} = \sec\theta + 1$$

It follows that LHS = RHS, so the equation is an identity. ■

⊘ *Warning:* To prove an identity we do *not* just perform the same operations on both sides of the equation. For example, if we start with an equation that is not an identity, such as

(1) $\sin x = -\sin x$

and square both sides, we get the equation

(2) $\sin^2 x = \sin^2 x$

which is clearly an identity. Does this mean that the original equation is an identity? Of course not. The problem here is that the operation of squaring is not **reversible** in the sense that we cannot arrive back at (1) from (2) by taking square roots (reversing the procedure). Only operations that are reversible will necessarily transform an identity into an identity.

We end this section by describing the technique of *trigonometric substitution*, which we use to convert algebraic expressions to trigonometric ones. This is often useful in calculus, for instance, in finding the area of a circle or an ellipse.

EXAMPLE 7 ■ **Trigonometric Substitution**

Substitute $\sin\theta$ for x in the expression $\sqrt{1 - x^2}$ and simplify. Assume that $0 \leqslant \theta \leqslant \pi/2$.

SOLUTION Setting $x = \sin\theta$, we have

$$\sqrt{1 - x^2} = \sqrt{1 - \sin^2\theta} \qquad \text{Substitute } x = \sin\theta$$

$$= \sqrt{\cos^2\theta} \qquad \text{Pythagorean identity}$$

$$= \cos\theta \qquad \text{Take square root}$$

The last equality is true because $\cos\theta \geqslant 0$ for the values of θ in question. ■

7.1 EXERCISES

1–6 ■ Write the trigonometric expression in terms of sine and cosine, and then simplify.

1. $\cos x \tan x$

2. $\sin \theta \cos \theta \csc \theta$

3. $\sec^2 x - \tan^2 x$

4. $\dfrac{\tan x + \cot x}{\sec x \csc x}$

5. $\cos u + \tan u \sin u$

6. $\cos^2 x (1 + \tan^2 x)$

7–20 ■ Simplify the trigonometric expression.

7. $\dfrac{\cos x \sec x}{\cot x}$

8. $\cos^3 x + \sin^2 x \cos x$

9. $\dfrac{1 + \sin y}{1 + \csc y}$

10. $\dfrac{\tan x}{\sec(-x)}$

11. $\dfrac{\sec^2 x - 1}{\sec^2 x}$

12. $\dfrac{\sec x - \cos x}{\tan x}$

13. $\dfrac{1 + \csc x}{\cos x + \cot x}$

14. $\dfrac{\sin x}{\csc x} + \dfrac{\cos x}{\sec x}$

15. $\dfrac{1 + \sin u}{\cos u} + \dfrac{\cos u}{1 + \sin u}$

16. $\tan x \cos x \csc x$

17. $\dfrac{2 + \tan^2 x}{\sec^2 x} - 1$

18. $\dfrac{1 + \cot A}{\csc A}$

19. $\tan \theta + \cos(-\theta) + \tan(-\theta)$

20. $\dfrac{\cos x}{\sec x + \tan x}$

21–82 ■ Verify the identity.

21. $\sin \theta \cot \theta = \cos \theta$

22. $\dfrac{\tan x}{\sec x} = \sin x$

23. $\dfrac{\cos u \sec u}{\tan u} = \cot u$

24. $\dfrac{\cot x \sec x}{\csc x} = 1$

25. $\dfrac{\tan y}{\csc y} = \sec y - \cos y$

26. $\dfrac{\cos v}{\sec v \sin v} = \csc v - \sin v$

27. $\sin B + \cos B \cot B = \csc B$

28. $\cos(-x) - \sin(-x) = \cos x + \sin x$

29. $\cot(-\alpha) \cos(-\alpha) + \sin(-\alpha) = -\csc \alpha$

30. $\csc x [\csc x + \sin(-x)] = \cot^2 x$

31. $(1 - \sin x)(1 + \sin x) = \cos^2 x$

32. $(\sin x + \cos x)^2 = 1 + 2 \sin x \cos x$

33. $(1 - \cos \beta)(1 + \cos \beta) = \dfrac{1}{\csc^2 \beta}$

34. $\dfrac{\cos x}{\sec x} + \dfrac{\sin x}{\csc x} = 1$

35. $\dfrac{(\sin x + \cos x)^2}{\sin^2 x - \cos^2 x} = \dfrac{\sin^2 x - \cos^2 x}{(\sin x - \cos x)^2}$

36. $(\sin x + \cos x)^4 = (1 + 2 \sin x \cos x)^2$

37. $\dfrac{\sec t - \cos t}{\sec t} = \sin^2 t$

38. $\dfrac{1 - \sin x}{1 + \sin x} = (\sec x - \tan x)^2$

39. $\dfrac{1}{1 - \sin^2 y} = 1 + \tan^2 y$

40. $\csc x - \sin x = \cos x \cot x$

41. $(\cot x - \csc x)(\cos x + 1) = -\sin x$

42. $\sin^4 \theta - \cos^4 \theta = \sin^2 \theta - \cos^2 \theta$

43. $(1 - \cos^2 x)(1 + \cot^2 x) = 1$

44. $\cos^2 x - \sin^2 x = 2 \cos^2 x - 1$

45. $2 \cos^2 x - 1 = 1 - 2 \sin^2 x$

46. $\tan y + \cot y = \sec y \csc y$

47. $\dfrac{1 - \cos \alpha}{\sin \alpha} = \dfrac{\sin \alpha}{1 + \cos \alpha}$

48. $\sin^2 \alpha + \cos^2 \alpha + \tan^2 \alpha = \sec^2 \alpha$

49. $\dfrac{\sin x - 1}{\sin x + 1} = \dfrac{-\cos^2 x}{(\sin x + 1)^2}$

50. $\dfrac{\sin w}{\sin w + \cos w} = \dfrac{\tan w}{1 + \tan w}$

51. $\dfrac{(\sin t + \cos t)^2}{\sin t \cos t} = 2 + \sec t \csc t$

52. $\sec t \csc t (\tan t + \cot t) = \sec^2 t + \csc^2 t$

53. $\dfrac{1 + \tan^2 u}{1 - \tan^2 u} = \dfrac{1}{\cos^2 u - \sin^2 u}$

54. $\dfrac{1 + \sec^2 x}{1 + \tan^2 x} = 1 + \cos^2 x$

55. $\dfrac{\sec x}{\sec x - \tan x} = \sec x (\sec x + \tan x)$

56. $\dfrac{\sec x + \csc x}{\tan x + \cot x} = \sin x + \cos x$

57. $\sec v - \tan v = \dfrac{1}{\sec v + \tan v}$

58. $\dfrac{\sin A}{1 - \cos A} - \cot A = \csc A$

59. $\dfrac{\sin x + \cos x}{\sec x + \csc x} = \sin x \cos x$

60. $\dfrac{1 - \cos x}{\sin x} + \dfrac{\sin x}{1 - \cos x} = 2 \csc x$

61. $\dfrac{\csc x - \cot x}{\sec x - 1} = \cot x$ **62.** $\dfrac{\csc^2 x - \cot^2 x}{\sec^2 x} = \cos^2 x$

63. $\tan^2 u - \sin^2 u = \tan^2 u \sin^2 u$

64. $\dfrac{\tan v \sin v}{\tan v + \sin v} = \dfrac{\tan v - \sin v}{\tan v \sin v}$

65. $\sec^4 x - \tan^4 x = \sec^2 x + \tan^2 x$

66. $\dfrac{\cos \theta}{1 - \sin \theta} = \sec \theta + \tan \theta$

67. $\dfrac{\cos \theta}{\sin \theta - 1} = \dfrac{\sin \theta - \csc \theta}{\cos \theta - \cot \theta}$

68. $\dfrac{1 + \tan x}{1 - \tan x} = \dfrac{\cos x + \sin x}{\cos x - \sin x}$

69. $\dfrac{\cos^2 t + \tan^2 t - 1}{\sin^2 t} = \tan^2 t$

70. $\dfrac{1}{1 - \sin x} - \dfrac{1}{1 + \sin x} = 2 \sec x \tan x$

71. $\dfrac{1}{\sec x + \tan x} + \dfrac{1}{\sec x - \tan x} = 2 \sec x$

72. $\dfrac{1 + \sin x}{1 - \sin x} - \dfrac{1 - \sin x}{1 + \sin x} = 4 \tan x \sec x$

73. $(\tan x + \cot x)^2 = \sec^2 x + \csc^2 x$

74. $\tan^2 x - \cot^2 x = \sec^2 x - \csc^2 x$

75. $\dfrac{\sec u - 1}{\sec u + 1} = \dfrac{1 - \cos u}{1 + \cos u}$

76. $\dfrac{\cot x + 1}{\cot x - 1} = \dfrac{1 + \tan x}{1 - \tan x}$

77. $\dfrac{\sin^3 x + \cos^3 x}{\sin x + \cos x} = 1 - \sin x \cos x$

78. $\dfrac{\tan v - \cot v}{\tan^2 v - \cot^2 v} = \sin v \cos v$

79. $\dfrac{1 + \sin x}{1 - \sin x} = (\tan x + \sec x)^2$

80. $\dfrac{\tan x + \tan y}{\cot x + \cot y} = \tan x \tan y$

81. $(\tan x + \cot x)^4 = \csc^4 x \sec^4 x$

82. $(\sin \alpha - \tan \alpha)(\cos \alpha - \cot \alpha) = (\cos \alpha - 1)(\sin \alpha - 1)$

83–88 ■ Make the indicated trigonometric substitution in the given algebraic expression and simplify (see Example 7). Assume $0 \leqslant \theta < \pi/2$.

83. $\dfrac{x}{\sqrt{1 - x^2}}$, $x = \sin \theta$

84. $\sqrt{1 + x^2}$, $x = \tan \theta$

85. $\sqrt{x^2 - 1}$, $x = \sec \theta$

86. $\dfrac{1}{x^2 \sqrt{4 + x^2}}$, $x = 2 \tan \theta$

87. $\sqrt{9 - x^2}$, $x = 3 \sin \theta$

88. $\dfrac{\sqrt{x^2 - 25}}{x}$, $x = 5 \sec \theta$

89–92 ■ Show that the equation is not an identity.

89. $\sin 2x = 2 \sin x$

90. $\sin(x + y) = \sin x + \sin y$

91. $\sec^2 x + \csc^2 x = 1$

92. $\dfrac{1}{\sin x + \cos x} = \csc x + \sec x$

93–96 ■ Graph f and g in the same viewing rectangle. Do the graphs suggest that the equation $f(x) = g(x)$ is an identity? Prove your answer.

93. $f(x) = \cos^2 x - \sin^2 x$, $g(x) = 1 - 2 \sin^2 x$

94. $f(x) = \tan x (1 + \sin x)$, $g(x) = \dfrac{\sin x \cos x}{1 + \sin x}$

95. $f(x) = (\sin x + \cos x)^2$, $g(x) = 1$

96. $f(x) = \cos^4 x - \sin^4 x$, $g(x) = 2\cos^2 x - 1$

🔘 **DISCOVERY · DISCUSSION**

97. Cofunction Identities   In the right triangle shown, explain why

$$v = \frac{\pi}{2} - u$$

Explain how you can obtain all six cofunction identities from this triangle, for $0 < u < \pi/2$.

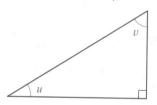

98. Graphs and Identities   Suppose you graph two functions, f and g, on a graphing device, and their graphs appear identical in the viewing rectangle. Does this prove that the equation $f(x) = g(x)$ is an identity? Explain.

7.2   ADDITION AND SUBTRACTION FORMULAS

We now derive identities for trigonometric functions of sums and differences.

ADDITION AND SUBTRACTION FORMULAS

$$\sin(s + t) = \sin s \cos t + \cos s \sin t \qquad \sin(s - t) = \sin s \cos t - \cos s \sin t$$

$$\cos(s + t) = \cos s \cos t - \sin s \sin t \qquad \cos(s - t) = \cos s \cos t + \sin s \sin t$$

$$\tan(s + t) = \frac{\tan s + \tan t}{1 - \tan s \tan t} \qquad \tan(s - t) = \frac{\tan s - \tan t}{1 + \tan s \tan t}$$

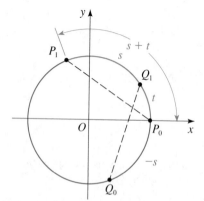

FIGURE 1

■ **Proof**   To prove the formula $\cos(s + t) = \cos s \cos t - \sin s \sin t$, we use Figure 1. In the figure, the distances t, $s + t$, and $-s$ have been marked on the unit circle, starting at $P_0(1, 0)$ and terminating at Q_1, P_1, and Q_0, respectively. The coordinates of these points are

$$P_0(1, 0) \qquad\qquad Q_0(\cos(-s), \sin(-s))$$

$$P_1(\cos(s + t), \sin(s + t)) \qquad Q_1(\cos t, \sin t)$$

Since $\cos(-s) = \cos s$ and $\sin(-s) = -\sin s$, it follows that the point Q_0 has the coordinates $Q_0(\cos s, -\sin s)$. Notice that the distances between P_0 and P_1 and between Q_0 and Q_1 measured along the arc of the circle are equal. Since equal arcs are subtended by equal chords, it follows that $d(P_0, P_1) = d(Q_0, Q_1)$.

Using the Distance Formula, we get

$$\sqrt{[\cos(s + t) - 1]^2 + [\sin(s + t) - 0]^2} = \sqrt{(\cos t - \cos s)^2 + (\sin t + \sin s)^2}$$

Squaring both sides and expanding, we have

these add to 1

$$\cos^2(s + t) - 2\cos(s + t) + 1 + \sin^2(s + t)$$

$$= \cos^2 t - 2\cos s \cos t + \cos^2 s + \sin^2 t + 2\sin s \sin t + \sin^2 s$$

these add to 1   these add to 1

Using the Pythagorean identity $\sin^2 z + \cos^2 z = 1$ three times gives

$$2 - 2\cos(s + t) = 2 - 2\cos s \cos t + 2\sin s \sin t$$

Finally, subtracting 2 from each side and dividing both sides by -2, we get

$$\cos(s + t) = \cos s \cos t - \sin s \sin t$$

which proves the addition formula for cosine.

The subtraction formula for cosine is obtained by replacing t by $-t$ in the addition formula as follows:

$$\cos(s - t) = \cos(s + (-t))$$

$$= \cos s \cos(-t) - \sin s \sin(-t) \qquad \text{Addition formula for cosine}$$

$$= \cos s \cos t + \sin s \sin t \qquad \text{Even-odd identities}$$

This proves the subtraction formula for cosine. See Exercises 46 and 47 for proofs of the other addition formulas. ☐

EXAMPLE 1 ■ Using the Addition and Subtraction Formulas

Find the exact value of each expression: (a) $\cos 75°$ (b) $\cos \dfrac{\pi}{12}$

SOLUTION

(a) Notice that $75° = 45° + 30°$. Since we know the exact values of sine and cosine at $45°$ and $30°$, we use the addition formula for cosine to get

$$\cos 75° = \cos(45° + 30°)$$

$$= \cos 45° \cos 30° - \sin 45° \sin 30°$$

$$= \frac{\sqrt{2}}{2} \frac{\sqrt{3}}{2} - \frac{\sqrt{2}}{2} \frac{1}{2} = \frac{\sqrt{2}\sqrt{3} - \sqrt{2}}{4} = \frac{\sqrt{6} - \sqrt{2}}{4}$$

(b) Since $\dfrac{\pi}{12} = \dfrac{\pi}{4} - \dfrac{\pi}{6}$, the subtraction formula for cosine gives

$$\cos \frac{\pi}{12} = \cos\left(\frac{\pi}{4} - \frac{\pi}{6}\right)$$

$$= \cos \frac{\pi}{4} \cos \frac{\pi}{6} + \sin \frac{\pi}{4} \sin \frac{\pi}{6}$$

$$= \frac{\sqrt{2}}{2} \frac{\sqrt{3}}{2} + \frac{\sqrt{2}}{2} \frac{1}{2} = \frac{\sqrt{6} + \sqrt{2}}{4}$$

■

EXAMPLE 2 ■ Using the Addition Formula for Sine

Find the exact value of the expression $\sin 20° \cos 40° + \cos 20° \sin 40°$.

SOLUTION   We recognize the expression as the right-hand side of the addition formula for sine with $s = 20°$ and $t = 40°$. So we have

$$\sin 20° \cos 40° + \cos 20° \sin 40° = \sin(20° + 40°) = \sin 60° = \frac{\sqrt{3}}{2}$$

■

EXAMPLE 3 ■ Proving an Identity

Prove the cofunction identity: $\cos\left(\dfrac{\pi}{2} - u\right) = \sin u$

SOLUTION   By the subtraction formula for cosine,

$$\cos\left(\frac{\pi}{2} - u\right) = \cos \frac{\pi}{2} \cos u + \sin \frac{\pi}{2} \sin u$$

$$= 0 \cdot \cos u + 1 \cdot \sin u = \sin u$$

■

EXAMPLE 4 ■ Proving an Identity

Verify the identity: $\dfrac{1 + \tan x}{1 - \tan x} = \tan\left(\dfrac{\pi}{4} + x\right)$

SOLUTION   Starting with the right-hand side and using the addition formula for tangent, we get

$$\text{RHS} = \tan\left(\frac{\pi}{4} + x\right) = \frac{\tan \dfrac{\pi}{4} + \tan x}{1 - \tan \dfrac{\pi}{4} \tan x}$$

$$= \frac{1 + \tan x}{1 - \tan x} = \text{LHS}$$

■

The next example is a typical use of the addition and subtraction formulas in calculus.

EXAMPLE 5 ■ An Identity from Calculus

If $f(x) = \sin x$, show that

$$\frac{f(x + h) - f(x)}{h} = \sin x \left(\frac{\cos h - 1}{h} \right) + \cos x \left(\frac{\sin h}{h} \right)$$

SOLUTION

$$\frac{f(x + h) - f(x)}{h} = \frac{\sin(x + h) - \sin x}{h}$$

$$= \frac{\sin x \cos h + \cos x \sin h - \sin x}{h}$$

$$= \frac{\sin x (\cos h - 1) + \cos x \sin h}{h}$$

$$= \sin x \left(\frac{\cos h - 1}{h} \right) + \cos x \left(\frac{\sin h}{h} \right) \qquad ■$$

EXPRESSIONS OF THE FORM $A \sin x + B \cos x$

We can write expressions of the form $A \sin x + B \cos x$ in terms of a single trigonometric function using the addition formula for sine. For example, consider the expression

$$\frac{1}{2} \sin x + \frac{\sqrt{3}}{2} \cos x$$

If we set $\phi = \pi/3$, then $\cos \phi = \frac{1}{2}$ and $\sin \phi = \sqrt{3}/2$, and we can write

$$\frac{1}{2} \sin x + \frac{\sqrt{3}}{2} \cos x = \cos \phi \sin x + \sin \phi \cos x$$

$$= \sin(x + \phi) = \sin\left(x + \frac{\pi}{3} \right)$$

We are able to do this because the coefficients $\frac{1}{2}$ and $\sqrt{3}/2$ are precisely the cosine and sine of a particular number, in this case, $\pi/3$. We can use this same idea in general to write $A \sin x + B \cos x$ in the form $k \sin(x + \phi)$. We start by multiplying the numerator and denominator by $\sqrt{A^2 + B^2}$ to get

$$A \sin x + B \cos x = \sqrt{A^2 + B^2} \left(\frac{A}{\sqrt{A^2 + B^2}} \sin x + \frac{B}{\sqrt{A^2 + B^2}} \cos x \right)$$

We need a number ϕ with the property that

$$\cos \phi = \frac{A}{\sqrt{A^2 + B^2}} \qquad \text{and} \qquad \sin \phi = \frac{B}{\sqrt{A^2 + B^2}}$$

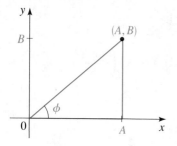

FIGURE 2

Figure 2 shows that the point (A, B) in the plane determines a number ϕ with precisely this property. With this ϕ, we have

$$A \sin x + B \cos x = \sqrt{A^2 + B^2}\,(\cos\phi \sin x + \sin\phi \cos x)$$
$$= \sqrt{A^2 + B^2}\,\sin(x + \phi)$$

We have proved the following theorem.

SUMS OF SINES AND COSINES

If A and B are real numbers, then

$$A \sin x + B \cos x = k \sin(x + \phi)$$

where $k = \sqrt{A^2 + B^2}$ and ϕ satisfies

$$\cos\phi = \frac{A}{\sqrt{A^2 + B^2}} \qquad \text{and} \qquad \sin\phi = \frac{B}{\sqrt{A^2 + B^2}}$$

EXAMPLE 6 ■ A Sum of Sine and Cosine Terms

Express $3 \sin x + 4 \cos x$ in the form $k \sin(x + \phi)$.

SOLUTION   By the preceding theorem, $k = \sqrt{A^2 + B^2} = \sqrt{3^2 + 4^2} = 5$. The angle ϕ has the property that $\sin\phi = \frac{4}{5}$ and $\cos\phi = \frac{3}{5}$. Using a calculator we find $\phi \approx 53.1°$. Thus

$$3 \sin x + 4 \cos x \approx 5 \sin(x + 53.1°) \qquad\qquad ■$$

EXAMPLE 7 ■ Graphing a Trigonometric Function

Write the function $f(x) = -\sin 2x + \sqrt{3} \cos 2x$ in the form $k \sin(2x + \phi)$ and use the new form to sketch a graph of the function.

SOLUTION   Since $A = -1$ and $B = \sqrt{3}$, we have $k = \sqrt{A^2 + B^2} = \sqrt{1 + 3} = 2$. The angle ϕ satisfies $\cos\phi = -\frac{1}{2}$ and $\sin\phi = \sqrt{3}/2$. From the signs of these quantities we conclude that ϕ is in quadrant II. Thus, $\phi = 2\pi/3$. By the preceding theorem we can write

$$f(x) = -\sin 2x + \sqrt{3} \cos 2x = 2 \sin\left(2x + \frac{2\pi}{3}\right)$$

Using the form

$$f(x) = 2 \sin 2\left(x + \frac{\pi}{3}\right)$$

we see that the graph is a sine curve with amplitude 2, period $2\pi/2 = \pi$, and phase shift $-\pi/3$. The graph is shown in Figure 3. ■

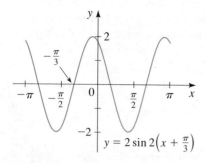

$$y = 2 \sin 2\left(x + \frac{\pi}{3}\right)$$

FIGURE 3

7.2 EXERCISES

1–6 ■ Use an addition or subtraction formula to find the exact value of the expression, as demonstrated in Example 1.

1. $\sin 15°$ **2.** $\cos 165°$ **3.** $\tan 105°$

4. $\cos \dfrac{\pi}{12}$ **5.** $\sin \dfrac{11\pi}{12}$ **6.** $\sin 75°$

7–10 ■ Use an addition or subtraction formula to write the expression as a trigonometric function of one number, and find its exact value.

7. $\sin 18° \cos 27° + \cos 18° \sin 27°$

8. $\cos \dfrac{3\pi}{7} \cos \dfrac{2\pi}{21} + \sin \dfrac{3\pi}{7} \sin \dfrac{2\pi}{21}$

9. $\dfrac{\tan 73° - \tan 13°}{1 + \tan 73° \tan 13°}$

10. $\cos \dfrac{13\pi}{15} \cos\left(-\dfrac{\pi}{5}\right) - \sin \dfrac{13\pi}{15} \sin\left(-\dfrac{\pi}{5}\right)$

11–14 ■ Prove the cofunction identity using the addition and subtraction formulas.

11. $\tan\left(\dfrac{\pi}{2} - u\right) = \cot u$ **12.** $\cot\left(\dfrac{\pi}{2} - u\right) = \tan u$

13. $\sec\left(\dfrac{\pi}{2} - u\right) = \csc u$ **14.** $\csc\left(\dfrac{\pi}{2} - u\right) = \sec u$

15–32 ■ Prove the identity.

15. $\sin\left(x - \dfrac{\pi}{2}\right) = -\cos x$ **16.** $\cos\left(x - \dfrac{\pi}{2}\right) = \sin x$

17. $\sin(x - \pi) = -\sin x$ **18.** $\cos(x - \pi) = -\cos x$

19. $\tan(x - \pi) = \tan x$

20. $\sin\left(\dfrac{\pi}{2} - x\right) = \sin\left(\dfrac{\pi}{2} + x\right)$

21. $\cos\left(x + \dfrac{\pi}{6}\right) + \sin\left(x - \dfrac{\pi}{3}\right) = 0$

22. $\tan\left(x - \dfrac{\pi}{4}\right) = \dfrac{\tan x - 1}{\tan x + 1}$

23. $\sin(x + y) - \sin(x - y) = 2 \cos x \sin y$

24. $\cos(x + y) + \cos(x - y) = 2 \cos x \cos y$

25. $\cot(x - y) = \dfrac{\cot x \cot y + 1}{\cot y - \cot x}$

26. $\cot(x + y) = \dfrac{\cot x \cot y - 1}{\cot x + \cot y}$

27. $\tan x - \tan y = \dfrac{\sin(x - y)}{\cos x \cos y}$

28. $1 - \tan x \tan y = \dfrac{\cos(x + y)}{\cos x \cos y}$

29. $\dfrac{\sin(x + y) - \sin(x - y)}{\cos(x + y) + \cos(x - y)} = \tan y$

30. $\cos(x + y) \cos(x - y) = \cos^2 x - \sin^2 y$

31. $\sin(x + y + z) = \sin x \cos y \cos z + \cos x \sin y \cos z$ $+ \cos x \cos y \sin z - \sin x \sin y \sin z$

32. $\tan(x - y) + \tan(y - z) + \tan(z - x)$ $= \tan(x - y) \tan(y - z) \tan(z - x)$

33–36 ■ Write the expression in terms of sine only.

33. $-\sqrt{3} \sin x + \cos x$ **34.** $\sin x + \cos x$

35. $5(\sin 2x - \cos 2x)$ **36.** $3 \sin \pi x + 3\sqrt{3} \cos \pi x$

37–38 ■ Express the function in terms of sine only, and sketch a graph of the function.

37. $f(x) = \sin x + \cos x$

38. $g(x) = \cos 2x + \sqrt{3} \sin 2x$

39. Show that if $\beta - \alpha = \pi/2$, then

$$\sin(x + \alpha) + \cos(x + \beta) = 0$$

40. Let $g(x) = \cos x$. Show that

$$\dfrac{g(x + h) - g(x)}{h} = -\cos x \left(\dfrac{1 - \cos h}{h}\right) - \sin x \left(\dfrac{\sin h}{h}\right)$$

41. Refer to the figure. Show that $\alpha + \beta = \gamma$, and find $\tan \gamma$.

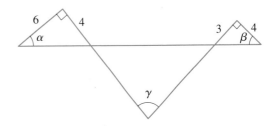

42. (a) If L is a line in the plane and θ is the angle formed by the line and the x-axis as shown in the figure, show that the slope m of the line is given by

$$m = \tan \theta$$

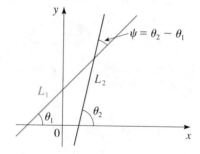

(b) Let L_1 and L_2 be two nonparallel lines in the plane with slopes m_1 and m_2, respectively. Let ψ be the acute angle formed by the two lines (see the figure). Show that

$$\tan \psi = \frac{m_2 - m_1}{1 + m_1 m_2}$$

(c) Find the acute angle formed by the two lines $y = \frac{1}{3}x + 1$ and $y = -\frac{1}{2}x - 3$.

(d) Show that if two lines are perpendicular, then the slope of one is the negative reciprocal of the slope of the other. [*Hint:* First find an expression for $\cot \psi$.]

 43–44 ■ (a) Graph the function and make a conjecture, and (b) prove that your conjecture is true.

43. $y = \sin^2\left(x + \frac{\pi}{4}\right) + \sin^2\left(x - \frac{\pi}{4}\right)$

44. $y = -\frac{1}{2}[\cos(x + \pi) + \cos(x - \pi)]$

45. A digital delay-device echoes an input signal by repeating it a fixed length of time after it is received. If such a device receives the pure note $f_1(t) = 5 \sin t$ and echoes the pure note $f_2(t) = 5 \cos t$, then the combined sound is $f(t) = f_1(t) + f_2(t)$.
(a) Graph $y = f(t)$ and observe that the graph has the form of a sine curve $y = k \sin(t + \phi)$.
(b) Find k and ϕ.

 DISCOVERY · DISCUSSION

46. Addition Formula for Sine   In the text we proved only the addition and subtraction formulas for cosine. Use these formulas and the cofunction identities

$$\sin x = \cos\left(\frac{\pi}{2} - x\right)$$

$$\cos x = \sin\left(\frac{\pi}{2} - x\right)$$

to prove the addition formula for sine. [*Hint:* To get started, use the first cofunction identity to write

$$\sin(s + t) = \cos\left(\frac{\pi}{2} - (s + t)\right)$$

$$= \cos\left(\left(\frac{\pi}{2} - s\right) - t\right)$$

and use the subtraction formula for cosine.]

47. Addition Formula for Tangent   Use the addition formulas for cosine and sine to prove the addition formula for tangent. [*Hint:* Use

$$\tan(s + t) = \frac{\sin(s + t)}{\cos(s + t)}$$

and divide the numerator and denominator by $\cos s \cos t$.]

7.3 DOUBLE-ANGLE, HALF-ANGLE, AND PRODUCT-SUM FORMULAS

The identities we consider in this section are consequences of the addition formulas. The **double-angle formulas** allow us to find the values of the trigonometric functions at $2x$ from their values at x. The **half-angle formulas** relate the values of the trigonometric functions at $\frac{1}{2}x$ to their values at x. The **product-sum formulas** relate products of sines and cosines to sums of sines and cosines.

DOUBLE-ANGLE FORMULAS

The formulas in the following box are consequences of the addition formulas, which we proved in the preceding section.

DOUBLE-ANGLE FORMULAS

$$\sin 2x = 2 \sin x \cos x$$

$$\cos 2x = \cos^2 x - \sin^2 x = 1 - 2\sin^2 x = 2\cos^2 x - 1$$

$$\tan 2x = \frac{2 \tan x}{1 - \tan^2 x}$$

■ **Proof** Setting $x = y$ in the addition formula for sine, we get

$$\sin 2x = \sin(x + x)$$
$$= \sin x \cos x + \sin x \cos x$$
$$= 2 \sin x \cos x$$

Setting $x = y$ in the addition formula for cosine gives

$$\cos 2x = \cos(x + x)$$
$$= \cos x \cos x - \sin x \sin x$$
$$= \cos^2 x - \sin^2 x$$

The second and third formulas for $\cos 2x$ are obtained from the formula we just proved and the Pythagorean identity. Substituting $\cos^2 x = 1 - \sin^2 x$ gives

$$\cos 2x = \cos^2 x - \sin^2 x$$
$$= (1 - \sin^2 x) - \sin^2 x$$
$$= 1 - 2 \sin^2 x$$

The third formula is obtained in the same way, by substituting $\sin^2 x = 1 - \cos^2 x$.

Finally, we obtain the double-angle formula for tangent by setting $x = y$ in the addition formula for tangent.   □

EXAMPLE 1 ■ Using the Double-Angle Formulas

If $\cos x = -\frac{2}{3}$ and x is in quadrant II, find $\sin 2x$ and $\cos 2x$.

SOLUTION   Using one of the double-angle formulas for cosine, we get

$$\cos 2x = 2\cos^2 x - 1$$

$$= 2\left(-\frac{2}{3}\right)^2 - 1 = \frac{8}{9} - 1 = -\frac{1}{9}$$

To use the formula $\sin 2x = 2\sin x \cos x$, we need to find $\sin x$ first. We have

$$\sin x = \sqrt{1 - \cos^2 x} = \sqrt{1 - (-\frac{2}{3})^2} = \frac{\sqrt{5}}{3}$$

where we have used the positive square root because $\sin x$ is positive in quadrant II. Thus

$$\sin 2x = 2\sin x \cos x$$

$$= 2\left(\frac{\sqrt{5}}{3}\right)\left(-\frac{2}{3}\right) = -\frac{4\sqrt{5}}{9}$$
■

EXAMPLE 2 ■ A Triple-Angle Formula

Write $\cos 3x$ in terms of $\cos x$.

SOLUTION

$$\cos 3x = \cos(2x + x)$$

$= \cos 2x \cos x - \sin 2x \sin x$	Addition formula
$= (2\cos^2 x - 1)\cos x - (2\sin x \cos x)\sin x$	Double-angle formulas
$= 2\cos^3 x - \cos x - 2\sin^2 x \cos x$	Expand
$= 2\cos^3 x - \cos x - 2\cos x\,(1 - \cos^2 x)$	Pythagorean identity
$= 2\cos^3 x - \cos x - 2\cos x + 2\cos^3 x$	Expand
$= 4\cos^3 x - 3\cos x$	Simplify

■

Example 2 shows that $\cos 3x$ can be written as a polynomial of degree 3 in $\cos x$. The identity $\cos 2x = 2\cos^2 x - 1$ shows that $\cos 2x$ is a polynomial of degree 2 in $\cos x$. In fact, for any natural number n, we can write $\cos nx$ as a polynomial in $\cos x$ of degree n (see Exercise 75). The analogous result for $\sin nx$ is not true in general.

EXAMPLE 3 ■ Proving an Identity

Prove the identity: $\dfrac{\sin 3x}{\sin x \cos x} = 4 \cos x - \sec x$

SOLUTION   We start with the left-hand side.

$$\frac{\sin 3x}{\sin x \cos x} = \frac{\sin(x + 2x)}{\sin x \cos x}$$

$$= \frac{\sin x \cos 2x + \cos x \sin 2x}{\sin x \cos x} \qquad \text{Addition formula}$$

$$= \frac{\sin x \left(2 \cos^2 x - 1\right) + \cos x \left(2 \sin x \cos x\right)}{\sin x \cos x} \qquad \text{Double-angle formulas}$$

$$= \frac{\sin x \left(2 \cos^2 x - 1\right)}{\sin x \cos x} + \frac{\cos x \left(2 \sin x \cos x\right)}{\sin x \cos x} \qquad \text{Separate fraction}$$

$$= \frac{2 \cos^2 x - 1}{\cos x} + 2 \cos x \qquad \text{Cancel}$$

$$= 2 \cos x - \frac{1}{\cos x} + 2 \cos x \qquad \text{Separate fraction}$$

$$= 4 \cos x - \sec x \qquad \text{Reciprocal identity} \quad ■$$

HALF-ANGLE FORMULAS

The following formulas allow us to write any trigonometric expression involving even powers of sine and cosine in terms of the first power of cosine only. This technique is important in calculus. The half-angle formulas are immediate consequences of these formulas.

FORMULAS FOR LOWERING POWERS

$$\sin^2 x = \frac{1 - \cos 2x}{2} \qquad \cos^2 x = \frac{1 + \cos 2x}{2}$$

$$\tan^2 x = \frac{1 - \cos 2x}{1 + \cos 2x}$$

■ **Proof**   The first formula is obtained by solving for $\sin^2 x$ in the double-angle formula $\cos 2x = 1 - 2 \sin^2 x$. Similarly, the second formula is obtained by solving for $\cos^2 x$ in the double-angle formula $\cos 2x = 2 \cos^2 x - 1$.

The last formula follows from the first two and the reciprocal identities as follows:

$$\tan^2 x = \frac{\sin^2 x}{\cos^2 x} = \frac{\dfrac{1 - \cos 2x}{2}}{\dfrac{1 + \cos 2x}{2}} = \frac{1 - \cos 2x}{1 + \cos 2x}$$

☐

EXAMPLE 4 ■ Lowering Powers in a Trigonometric Expression

Express $\sin^2 x \cos^2 x$ in terms of the first power of cosine.

SOLUTION   We use the formulas for lowering powers repeatedly.

$$\sin^2 x \cos^2 x = \left(\frac{1 - \cos 2x}{2} \right) \left(\frac{1 + \cos 2x}{2} \right)$$

$$= \frac{1 - \cos^2 2x}{4} = \frac{1}{4} - \frac{1}{4} \cos^2 2x$$

$$= \frac{1}{4} - \frac{1}{4} \left(\frac{1 + \cos 4x}{2} \right) = \frac{1}{4} - \frac{1}{8} - \frac{\cos 4x}{8}$$

$$= \tfrac{1}{8} - \tfrac{1}{8} \cos 4x = \tfrac{1}{8}(1 - \cos 4x)$$

Another way to obtain this identity is to use the double-angle formula for sine in the form $\sin x \cos x = \frac{1}{2} \sin 2x$. Thus

$$\sin^2 x \cos^2 x = \frac{1}{4} \sin^2 2x = \frac{1}{4} \left(\frac{1 - \cos 4x}{2} \right) = \frac{1}{8}(1 - \cos 4x)$$

■

HALF-ANGLE FORMULAS

$$\sin \frac{u}{2} = \pm \sqrt{\frac{1 - \cos u}{2}} \qquad \cos \frac{u}{2} = \pm \sqrt{\frac{1 + \cos u}{2}}$$

$$\tan \frac{u}{2} = \frac{1 - \cos u}{\sin u} = \frac{\sin u}{1 + \cos u}$$

The choice of the $+$ or $-$ sign depends on the quadrant in which $u/2$ lies.

■ **Proof**   We substitute $x = u/2$ in the formulas for lowering powers and take the square root of each side. This yields the first two half-angle formulas.

In the case of the half-angle formula for tangent, we get

$$\tan\frac{u}{2} = \pm\sqrt{\frac{1-\cos u}{1+\cos u}}$$

$$= \pm\sqrt{\left(\frac{1-\cos u}{1+\cos u}\right)\left(\frac{1-\cos u}{1-\cos u}\right)} \qquad \text{Multiply numerator and denominator by } 1-\cos u$$

$$= \pm\sqrt{\frac{(1-\cos u)^2}{1-\cos^2 u}} \qquad \text{Simplify}$$

$$= \pm\frac{|1-\cos u|}{|\sin u|} \qquad \sqrt{A^2} = |A|$$

Now, $1 - \cos u$ is nonnegative for all values of u. It is also true that $\sin u$ and $\tan(u/2)$ always have the same sign. (Verify this.) It follows that

$$\tan\frac{u}{2} = \frac{1-\cos u}{\sin u}$$

The other half-angle formula for tangent is derived from this by multiplying the numerator and denominator by $1 + \cos u$. ☐

EXAMPLE 5 ■ Using a Half-Angle Formula

Find the exact value of $\sin 22.5°$.

SOLUTION Since $22.5°$ is half of $45°$, we use the half-angle formula for sine with $u = 45°$. We choose the $+$ sign because $22.5°$ is in the first quadrant.

$$\sin\frac{45°}{2} = \sqrt{\frac{1-\cos 45°}{2}} \qquad \text{Half-angle formula}$$

$$= \sqrt{\frac{1-\sqrt{2}/2}{2}} \qquad \cos 45° = \sqrt{2}/2$$

$$= \sqrt{\frac{2-\sqrt{2}}{4}} \qquad \text{Common denominator}$$

$$= \tfrac{1}{2}\sqrt{2-\sqrt{2}} \qquad \text{Simplify} \qquad ■$$

EXAMPLE 6 ■ Using a Half-Angle Formula

Find $\tan(u/2)$ if $\sin u = \frac{2}{5}$ and u is in quadrant II.

SOLUTION To use the half-angle formulas for tangent, we first need to find $\cos u$. Since cosine is negative in quadrant II, we have

$$\cos u = -\sqrt{1-\sin^2 u}$$

$$= -\sqrt{1-\left(\tfrac{2}{5}\right)^2} = -\frac{\sqrt{21}}{5}$$

Thus
$$\tan \frac{u}{2} = \frac{1 - \cos u}{\sin u}$$

$$= \frac{1 + \sqrt{21}/5}{\frac{2}{5}} = \frac{5 + \sqrt{21}}{2}$$ ∎

PRODUCT-SUM FORMULAS

It is possible to write the product $\sin u \cos v$ as a sum of trigonometric functions. To see this, consider the addition and subtraction formulas for the sine function:

$$\sin(u + v) = \sin u \cos v + \cos u \sin v$$

$$\sin(u - v) = \sin u \cos v - \cos u \sin v$$

Adding the left- and right-hand sides of these formulas gives

$$\sin(u + v) + \sin(u - v) = 2 \sin u \cos v$$

Dividing by 2 yields the formula

$$\sin u \cos v = \tfrac{1}{2}[\sin(u + v) + \sin(u - v)]$$

The other three **product-to-sum formulas** follow from the addition formulas in a similar way.

PRODUCT-TO-SUM FORMULAS

$$\sin u \cos v = \tfrac{1}{2}[\sin(u + v) + \sin(u - v)]$$

$$\cos u \sin v = \tfrac{1}{2}[\sin(u + v) - \sin(u - v)]$$

$$\cos u \cos v = \tfrac{1}{2}[\cos(u + v) + \cos(u - v)]$$

$$\sin u \sin v = \tfrac{1}{2}[\cos(u - v) - \cos(u + v)]$$

EXAMPLE 7 ■ Expressing a Trigonometric Product as a Sum

Express $\sin 3x \sin 5x$ as a sum of trigonometric functions.

SOLUTION   Using the fourth product-to-sum formula with $u = 3x$ and $v = 5x$ and the fact that cosine is an even function, we get

$$\sin 3x \sin 5x = \tfrac{1}{2}[\cos(3x - 5x) - \cos(3x + 5x)]$$

$$= \tfrac{1}{2}\cos(-2x) - \tfrac{1}{2}\cos 8x$$

$$= \tfrac{1}{2}\cos 2x - \tfrac{1}{2}\cos 8x$$ ∎

The product-to-sum formulas can also be used as sum-to-product formulas. This is possible because the right-hand side of each of the product-to-sum formulas is a sum and the left side is a product. For example, if we let

$$u = \frac{x + y}{2} \quad \text{and} \quad v = \frac{x - y}{2}$$

in the first product-to-sum formula, we get

$$\sin \frac{x + y}{2} \cos \frac{x - y}{2} = \tfrac{1}{2}[\sin x + \sin y]$$

so

$$\sin x + \sin y = 2 \sin \frac{x + y}{2} \cos \frac{x - y}{2}$$

The remaining three of the following **sum-to-product formulas** are obtained in a similar manner.

SUM-TO-PRODUCT FORMULAS

$$\sin x + \sin y = 2 \sin \frac{x + y}{2} \cos \frac{x - y}{2}$$

$$\sin x - \sin y = 2 \cos \frac{x + y}{2} \sin \frac{x - y}{2}$$

$$\cos x + \cos y = 2 \cos \frac{x + y}{2} \cos \frac{x - y}{2}$$

$$\cos x - \cos y = -2 \sin \frac{x + y}{2} \sin \frac{x - y}{2}$$

EXAMPLE 8 ■ **Expressing a Trigonometric Sum as a Product**

Write $\sin 7x + \sin 3x$ as a product.

SOLUTION The first sum-to-product formula gives

$$\sin 7x + \sin 3x = 2 \sin \frac{7x + 3x}{2} \cos \frac{7x - 3x}{2}$$

$$= 2 \sin 5x \cos 2x$$ ■

EXAMPLE 9 ■ **Proving an Identity**

Verify the identity: $\dfrac{\sin 3x - \sin x}{\cos 3x + \cos x} = \tan x$

SOLUTION We apply the second sum-to-product formula to the numerator and the third formula to the denominator.

$$\text{LHS} = \frac{\sin 3x - \sin x}{\cos 3x + \cos x} = \frac{2 \cos \dfrac{3x + x}{2} \sin \dfrac{3x - x}{2}}{2 \cos \dfrac{3x + x}{2} \cos \dfrac{3x - x}{2}} \qquad \text{Sum-to-product formulas}$$

$$= \frac{2 \cos 2x \sin x}{2 \cos 2x \cos x} \qquad \text{Simplify}$$

$$= \frac{\sin x}{\cos x} = \tan x = \text{RHS} \qquad \text{Cancel} \qquad \blacksquare$$

7.3 EXERCISES

1–6 ■ Find $\sin 2x$, $\cos 2x$, and $\tan 2x$ from the given information.

1. $\sin x = \frac{5}{13}$, x in quadrant I

2. $\cos x = \frac{4}{5}$, $\csc x < 0$

3. $\tan x = -\frac{4}{3}$, x in quadrant II

4. $\csc x = 4$, $\tan x < 0$

5. $\sin x = -\frac{3}{5}$, x in quadrant III

6. $\cot x = \frac{2}{3}$, $\sin x > 0$

7–12 ■ Use the formulas for lowering powers to rewrite the expression in terms of the first power of cosine, as in Example 4.

7. $\sin^4 x$ **8.** $\cos^4 x$

9. $\cos^4 x \sin^4 x$ **10.** $\cos^4 x \sin^2 x$

11. $\cos^2 x \sin^4 x$ **12.** $\cos^6 x$

13–18 ■ Use an appropriate half-angle formula to find the exact value of the expression.

13. $\sin 15°$ **14.** $\tan 15°$

15. $\cos 22.5°$ **16.** $\tan \dfrac{\pi}{8}$

17. $\sin \dfrac{\pi}{12}$ **18.** $\cos \dfrac{5\pi}{12}$

19–24 ■ Simplify the expression by using a double-angle formula or a half-angle formula.

19. (a) $2 \sin 18° \cos 18°$ (b) $2 \sin 3\theta \cos 3\theta$

20. (a) $\dfrac{2 \tan 7°}{1 - \tan^2 7°}$ (b) $\dfrac{2 \tan 7\theta}{1 - \tan^2 7\theta}$

21. (a) $\cos^2 34° - \sin^2 34°$ (b) $\cos^2 5\theta - \sin^2 5\theta$

22. (a) $\cos^2 \dfrac{\theta}{2} - \sin^2 \dfrac{\theta}{2}$ (b) $2 \sin \dfrac{\theta}{2} \cos \dfrac{\theta}{2}$

23. (a) $\dfrac{\sin 8°}{1 + \cos 8°}$ (b) $\dfrac{1 - \cos 4\theta}{\sin 4\theta}$

24. (a) $\sqrt{\dfrac{1 - \cos 30°}{2}}$ (b) $\sqrt{\dfrac{1 - \cos 8\theta}{2}}$

25–30 ■ Find $\sin \dfrac{x}{2}$, $\cos \dfrac{x}{2}$, and $\tan \dfrac{x}{2}$ from the given information.

25. $\sin x = \frac{3}{5}$, $0° < x < 90°$

26. $\cos x = -\frac{4}{5}$, $180° < x < 270°$

27. $\csc x = 3$, $90° < x < 180°$

28. $\tan x = 1$, $0° < x < 90°$

29. $\sec x = \frac{3}{2}$, $270° < x < 360°$

30. $\cot x = 5$, $180° < x < 270°$

31–34 ■ Write the product as a sum.

31. $\sin 2x \cos 3x$ **32.** $\sin x \sin 5x$

33. $3 \cos 4x \cos 7x$ **34.** $11 \sin \dfrac{x}{2} \cos \dfrac{x}{4}$

35–40 ■ Write the sum as a product.

35. $\sin 5x + \sin 3x$

36. $\sin x - \sin 4x$

37. $\cos 4x - \cos 6x$

38. $\cos 9x + \cos 2x$

39. $\sin 2x - \sin 7x$

40. $\sin 3x + \sin 4x$

41–46 ■ Find the value of the product or sum.

41. $2 \sin 52.5° \sin 97.5°$

42. $3 \cos 37.5° \cos 7.5°$

43. $\cos 37.5° \sin 7.5°$

44. $\sin 75° + \sin 15°$

45. $\cos 255° - \cos 195°$

46. $\cos \dfrac{\pi}{12} + \cos \dfrac{5\pi}{12}$

47–64 ■ Prove the identity.

47. $\cos^2 5x - \sin^2 5x = \cos 10x$

48. $\sin 8x = 2 \sin 4x \cos 4x$

49. $(\sin x + \cos x)^2 = 1 + \sin 2x$

50. $\dfrac{2 \tan x}{1 + \tan^2 x} = \sin 2x$

51. $\dfrac{\sin 4x}{\sin x} = 4 \cos x \cos 2x$

52. $\dfrac{1 + \sin 2x}{\sin 2x} = 1 + \frac{1}{2} \sec x \csc x$

53. $\dfrac{2(\tan x - \cot x)}{\tan^2 x - \cot^2 x} = \sin 2x$

54. $\cot 2x = \dfrac{1 - \tan^2 x}{2 \tan x}$

55. $\tan 3x = \dfrac{3 \tan x - \tan^3 x}{1 - 3 \tan^2 x}$

56. $4(\sin^6 x + \cos^6 x) = 4 - 3 \sin^2 2x$

57. $\cos^4 x - \sin^4 x = \cos 2x$

58. $\tan^2\!\left(\dfrac{x}{2} + \dfrac{\pi}{4}\right) = \dfrac{1 + \sin x}{1 - \sin x}$

59. $\dfrac{\sin x + \sin 5x}{\cos x + \cos 5x} = \tan 3x$

60. $\dfrac{\sin 3x + \sin 7x}{\cos 3x - \cos 7x} = \cot 2x$

61. $\dfrac{\sin 10x}{\sin 9x + \sin x} = \dfrac{\cos 5x}{\cos 4x}$

62. $\dfrac{\sin x + \sin 3x + \sin 5x}{\cos x + \cos 3x + \cos 5x} = \tan 3x$

63. $\dfrac{\sin x + \sin y}{\cos x + \cos y} = \tan\!\left(\dfrac{x + y}{2}\right)$

64. $\tan y = \dfrac{\sin(x + y) - \sin(x - y)}{\cos(x + y) + \cos(x - y)}$

65. Show that $\sin 45° + \sin 15° = \sin 75°$.

66. Show that $\cos 87° + \cos 33° = \sin 63°$.

67. Prove the identity

$$\frac{\sin x + \sin 2x + \sin 3x + \sin 4x + \sin 5x}{\cos x + \cos 2x + \cos 3x + \cos 4x + \cos 5x} = \tan 3x$$

68. Use the identity

$$\sin 2x = 2 \sin x \cos x$$

n times to show that

$$\sin(2^n x) = 2^n \sin x \cos x \cos 2x \cos 4x \cdots \cos 2^{n-1}x$$

69. (a) Draw the graph of $f(x) = \dfrac{\sin 3x}{\sin x} - \dfrac{\cos 3x}{\cos x}$ and make a conjecture.
(b) Prove the conjecture you made in part (a).

70. (a) Draw the graph of $f(x) = \cos 2x + 2 \sin^2 x$ and make a conjecture.
(b) Prove the conjecture you made in part (a).

71. Let $f(x) = \sin 6x + \sin 7x$.
(a) Graph $y = f(x)$.
(b) Verify that $f(x) = 2 \cos \frac{1}{2}x \sin \frac{13}{2}x$.
(c) Graph $y = 2 \cos \frac{1}{2}x$ and $y = -2 \cos \frac{1}{2}x$, together with the graph in part (a), in the same viewing rectangle. How are these graphs related to the graph of f?

72. When two pure notes that are close in frequency are played together, their sounds interfere to produce *beats*; that is, the loudness (or amplitude) of the sound alternately increases and decreases. If the two notes are given by

$$f_1(t) = \cos 11t \qquad \text{and} \qquad f_2(t) = \cos 13t$$

the resulting sound is $f(t) = f_1(t) + f_2(t)$.
(a) Graph the function $y = f(t)$.
(b) Verify that $f(t) = 2 \cos t \cos 12t$.
(c) Graph $y = 2 \cos t$ and $y = -2 \cos t$, together with the graph in part (a), in the same viewing rectangle. How do these graphs describe the variation in the loudness of the sound?

73. The lower right-hand corner of a long piece of paper 6 in. wide is folded over to the left-hand edge as shown. The length L of the fold depends on the angle θ. Show that

$$L = \frac{3}{\sin\theta\cos^2\theta}$$

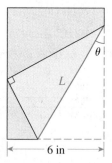

74. Let $3x = \pi/3$ and let $y = \cos x$. Use the result of Example 2 to show that y satisfies the equation

$$8y^3 - 6y - 1 = 0$$

NOTE: This equation has roots of a certain kind that are used in showing that the angle $\pi/3$ cannot be trisected using a ruler and compass only.

75. (a) Show that there is a polynomial $P(t)$ of degree 4 such that $\cos 4x = P(\cos x)$ (see Example 2).
(b) Show that there is a polynomial $Q(t)$ of degree 5 such that $\cos 5x = Q(\cos x)$.

NOTE: In general, there is a polynomial $P_n(t)$ of degree n such that $\cos nx = P_n(\cos x)$. These polynomials are called Tchebycheff polynomials, after the Russian mathematician P. L. Tchebycheff (1821–1894).

76. In triangle ABC (see the figure) the line segment s bisects angle C. Show that the length of s is given by

$$s = \frac{2ab\cos x}{a + b}$$

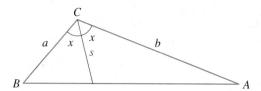

![triangle icon] **DISCOVERY · DISCUSSION**

77. Geometric Proof of a Double-Angle Formula Use the figure to prove that $\sin 2\theta = 2\sin\theta\cos\theta$.

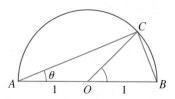

Hint: Find the area of triangle ABC in two different ways. You will need the following facts from geometry:

An angle inscribed in a semicircle is a right angle, so $\angle ACB$ is a right angle.

The central angle subtended by the chord of a circle is twice the angle subtended by the chord on the circle, so $\angle BOC$ is 2θ.

7.4   INVERSE TRIGONOMETRIC FUNCTIONS

If f is a one-to-one function with domain A and range B, then its inverse f^{-1} is the function with domain B and range A defined by

$$f^{-1}(x) = y \iff f(y) = x$$

(See Section 2.7.) In other words, f^{-1} is the rule that reverses the action of f. Figure 1 represents the actions of f and f^{-1} graphically.

For a function to have an inverse, it must be one-to-one. Since the trigonometric functions are not one-to-one, they do not have inverses. It is possible, however, to restrict the domains of the trigonometric functions in such a way that the resulting functions are one-to-one.

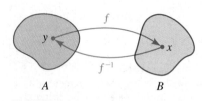

FIGURE 1
$f^{-1}(x) = y \Leftrightarrow f(y) = x$

THE INVERSE SINE FUNCTION

Let's first consider the sine function. There are many ways to restrict the domain of sine so that the new function is one-to-one. A natural way to do this is to restrict the domain to the interval $[-\pi/2, \pi/2]$. The reason for this choice is that sine attains each of its values exactly once on this interval. We write Sin x (with a capital S) for the new function, which has the domain $[-\pi/2, \pi/2]$ and the same values as sin x on this interval. The graphs of sin x and Sin x are shown in Figure 2. The function Sin x is one-to-one (by the Horizontal Line Test), and so has an inverse.

FIGURE 2

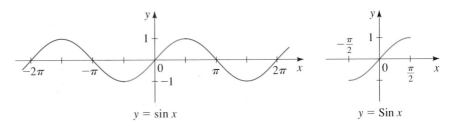

$y = \sin x$ $y = \operatorname{Sin} x$

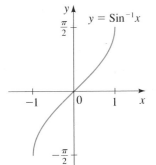

FIGURE 3

The inverse of the function Sin x is the function $\operatorname{Sin}^{-1} x$ defined by

$$\operatorname{Sin}^{-1} x = y \iff \operatorname{Sin} y = x$$

for $-1 \le x \le 1$ and $-\pi/2 \le y \le \pi/2$. The graph of $y = \operatorname{Sin}^{-1} x$ is shown in Figure 3; it is obtained by reflecting the graph of $y = \operatorname{Sin} x$ in the line $y = x$. It is customary to write $\operatorname{Sin}^{-1} x$ simply as $\sin^{-1} x$.

DEFINITION OF THE INVERSE SINE FUNCTION

The **inverse sine function** is the function $\sin^{-1}$ with domain $[-1, 1]$ and range $[-\pi/2, \pi/2]$ defined by

$$\sin^{-1} x = y \iff \sin y = x$$

The inverse sine function is also called **arcsine** and is denoted by **arcsin**.

Thus, $\sin^{-1} x$ *is the number in the interval* $[-\pi/2, \pi/2]$ *whose sine is* x. In other words, $\sin(\sin^{-1} x) = x$. In fact, from the general properties of inverse functions studied in Section 2.7, we have the following relations.

$$\sin(\sin^{-1} x) = x \qquad \text{for } -1 \le x \le 1$$

$$\sin^{-1}(\sin x) = x \qquad \text{for } -\frac{\pi}{2} \le x \le \frac{\pi}{2}$$

EXAMPLE 1 ■ Evaluating the Inverse Sine Function

Find (a) $\sin^{-1}\frac{1}{2}$, (b) $\sin^{-1}(-\frac{1}{2})$, and (c) $\sin^{-1}\frac{3}{2}$.

SOLUTION

(a) The number in the interval $[-\pi/2, \pi/2]$ whose sine is $\frac{1}{2}$ is $\pi/6$. Thus, $\sin^{-1}\frac{1}{2} = \pi/6$.

(b) Again, $\sin^{-1}(-\frac{1}{2})$ is the number in the interval $[-\pi/2, \pi/2]$ whose sine is $-\frac{1}{2}$. Since $\sin(-\pi/6) = -\frac{1}{2}$, we have $\sin^{-1}(-\frac{1}{2}) = -\pi/6$.

(c) Since $\frac{3}{2} > 1$, it is not in the domain of $\sin^{-1}x$, so $\sin^{-1}\frac{3}{2}$ is not defined. ■

EXAMPLE 2 ■ Using a Calculator to Evaluate Inverse Sine

Find approximate values for (a) $\sin^{-1}(0.82)$ and (b) $\sin^{-1}\frac{1}{3}$.

SOLUTION Since no rational multiple of π has a sine of 0.82 or $\frac{1}{3}$, we use a calculator to approximate these values. Using the $\boxed{\text{INV}}$ $\boxed{\text{SIN}}$, or $\boxed{\text{SIN}^{-1}}$, or $\boxed{\text{ARCSIN}}$ key(s) on the calculator (with the calculator in radian mode), we get

$$\text{(a) } \sin^{-1}(0.82) \approx 0.96141 \qquad \text{(b) } \sin^{-1}\frac{1}{3} \approx 0.33984$$ ■

EXAMPLE 3 ■ Composing Trigonometric Functions and Their Inverses

Find $\cos(\sin^{-1}\frac{3}{5})$.

SOLUTION 1 It is easy to find $\sin(\sin^{-1}\frac{3}{5})$. In fact, by the properties of inverse functions, this value is exactly $\frac{3}{5}$. To find $\cos(\sin^{-1}\frac{3}{5})$, we reduce this to the easier problem by writing the cosine function in terms of the sine function. Let $u = \sin^{-1}\frac{3}{5}$. Since $-\pi/2 \le u \le \pi/2$, $\cos u$ is positive and we can write

$$\cos u = +\sqrt{1 - \sin^2 u}$$

Thus $\cos(\sin^{-1}\frac{3}{5}) = \sqrt{1 - \sin^2(\sin^{-1}\frac{3}{5})}$

$$= \sqrt{1 - (\tfrac{3}{5})^2} = \sqrt{1 - \tfrac{9}{25}} = \sqrt{\tfrac{16}{25}} = \tfrac{4}{5}$$

SOLUTION 2 Let $\theta = \sin^{-1}\frac{3}{5}$. Then θ is the number in the interval $[-\pi/2, \pi/2]$ whose sine is $\frac{3}{5}$. Let's interpret θ as an angle and draw a right triangle with θ as one of its acute angles, with opposite side 3 and hypotenuse 5 (see Figure 4). The remaining leg of the triangle is found by the Pythagorean Theorem to be 4. From the figure we get

$$\cos(\sin^{-1}\tfrac{3}{5}) = \cos\theta = \tfrac{4}{5}$$ ■

From Solution 2 of Example 3 we can immediately find the values of the other trigonometric functions of $\theta = \sin^{-1}\frac{3}{5}$ from the triangle. Thus,

$$\tan(\sin^{-1}\tfrac{3}{5}) = \tfrac{3}{4} \qquad \sec(\sin^{-1}\tfrac{3}{5}) = \tfrac{5}{4} \qquad \csc(\sin^{-1}\tfrac{3}{5}) = \tfrac{5}{3}$$

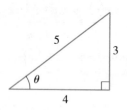

FIGURE 4

THE INVERSE COSINE FUNCTION

If the domain of the cosine function is restricted to the interval $[0, \pi]$, the resulting function is one-to-one and so has an inverse. We choose this interval because on it, cosine attains each of its values exactly once (see Figure 5).

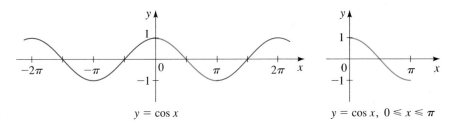

FIGURE 5

$y = \cos x$ $y = \cos x,\ 0 \leq x \leq \pi$

DEFINITION OF THE INVERSE COSINE FUNCTION

The **inverse cosine function** is the function $\cos^{-1}$ with domain $[-1, 1]$ and range $[0, \pi]$ defined by

$$\cos^{-1}x = y \iff \cos y = x$$

The inverse cosine function is also called **arccosine** and is denoted by **arccos**.

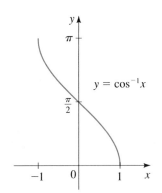

FIGURE 6

Thus, $y = \cos^{-1}x$ *is the number in the interval* $[0, \pi]$ *whose cosine is* x. The following relations follow from the inverse function properties.

$$\cos(\cos^{-1}x) = x \qquad \text{for } -1 \leq x \leq 1$$
$$\cos^{-1}(\cos x) = x \qquad \text{for } 0 \leq x \leq \pi$$

The graph of $y = \cos^{-1}x$ is sketched in Figure 6; it is obtained by reflecting the graph of $y = \cos x,\ 0 \leq x \leq \pi$, in the line $y = x$.

EXAMPLE 4 ■ Evaluating the Inverse Cosine Function

Find (a) $\cos^{-1}(\sqrt{3}/2)$, (b) $\cos^{-1}0$, and (c) $\cos^{-1}\frac{5}{7}$.

SOLUTION

(a) The number in the interval $[0, \pi]$ whose cosine is $\sqrt{3}/2$ is $\pi/6$. Thus, $\cos^{-1}(\sqrt{3}/2) = \pi/6$.

(b) Since $\cos(\pi/2) = 0$, it follows that $\cos^{-1}0 = \pi/2$.

(c) Since no rational multiple of π has cosine $\frac{5}{7}$, we use a calculator to find this value approximately: $\cos^{-1}\frac{5}{7} \approx 0.77519$. ■

EXAMPLE 5 ■ Composing Trigonometric Functions and Their Inverses

Write $\sin(\cos^{-1}x)$ and $\tan(\cos^{-1}x)$ as algebraic expressions in x for $-1 \le x \le 1$.

SOLUTION 1   Let $u = \cos^{-1}x$. We need to find $\sin u$ and $\tan u$ in terms of x. As in Example 3 the idea here is to write sine and tangent in terms of cosine. We have

$$\sin u = \pm\sqrt{1 - \cos^2 u} \qquad \text{and} \qquad \tan u = \frac{\sin u}{\cos u} = \frac{\pm\sqrt{1 - \cos^2 u}}{\cos u}$$

To choose the proper signs, note that u lies in the interval $[0, \pi]$ because $u = \cos^{-1}x$. Since $\sin u$ is positive on this interval, the $+$ sign is the correct choice. Now substituting $u = \cos^{-1}x$ in the displayed equations and using the relation $\cos(\cos^{-1}x) = x$ gives

$$\sin(\cos^{-1}x) = \sqrt{1 - x^2} \qquad \text{and} \qquad \tan(\cos^{-1}x) = \frac{\sqrt{1 - x^2}}{x}$$

SOLUTION 2   Let $\theta = \cos^{-1}x$, so $\cos\theta = x$. In Figure 7 we draw a right triangle with an acute angle θ, adjacent side x, and hypotenuse 1. By the Pythagorean Theorem, the remaining leg is $\sqrt{1 - x^2}$. From the figure,

$$\sin(\cos^{-1}x) = \sin\theta = \sqrt{1 - x^2} \qquad \text{and} \qquad \tan(\cos^{-1}x) = \tan\theta = \frac{\sqrt{1 - x^2}}{x}$$

■

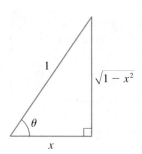

FIGURE 7

$\cos\theta = \dfrac{x}{1} = x$

NOTE   In Solution 2 of Example 5, it may seem that because we are sketching a triangle, the angle $\theta = \cos^{-1}x$ must be acute. But it turns out that the triangle method works for any θ and for any x. The domains and ranges of all six inverse trigonometric functions have been chosen in such a way that we can always use a triangle to find $S(T^{-1}(x))$, where S and T are any trigonometric functions.

EXAMPLE 6 ■ Composing a Trigonometric Function and an Inverse

Write $\sin(2\cos^{-1}x)$ as an algebraic expression in x for $-1 \le x \le 1$.

SOLUTION   Let $\theta = \cos^{-1}x$ and sketch a triangle as shown in Figure 8. We need to find $\sin 2\theta$, but from the triangle we can find the trigonometric function only of θ, not of 2θ. The double-angle identity for sine is useful here. We have

$$
\begin{aligned}
\sin(2\cos^{-1}x) &= \sin 2\theta \\
&= 2\sin\theta\cos\theta && \text{Double-angle formula} \\
&= 2\sqrt{1 - x^2}\, x && \text{From triangle} \\
&= 2x\sqrt{1 - x^2}
\end{aligned}
$$

■

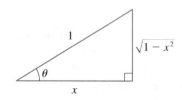

FIGURE 8

$\cos\theta = \dfrac{x}{1} = x$

THE INVERSE TANGENT FUNCTION

We restrict the domain of the tangent function to the interval $(-\pi/2, \pi/2)$ in order to obtain a one-to-one function.

DEFINITION OF THE INVERSE TANGENT FUNCTION

The **inverse tangent function** is the function $\tan^{-1}$ with domain $\mathbb{R}$ and range $(-\pi/2, \pi/2)$ defined by

$$\tan^{-1}x = y \iff \tan y = x$$

The inverse tangent function is also called **arctangent** and is denoted by **arctan**.

Thus, $\tan^{-1}x$ *is the number in the interval* $(-\pi/2, \pi/2)$ *whose tangent is* x. The following relations follow from the inverse function properties.

$$\tan(\tan^{-1}x) = x \qquad \text{for } x \in \mathbb{R}$$

$$\tan^{-1}(\tan x) = x \qquad \text{for } -\frac{\pi}{2} < x < \frac{\pi}{2}$$

Figure 9 shows the graph of $y = \tan x$ on the interval $(-\pi/2, \pi/2)$ and the graph of its inverse function, $y = \tan^{-1}x$.

FIGURE 9

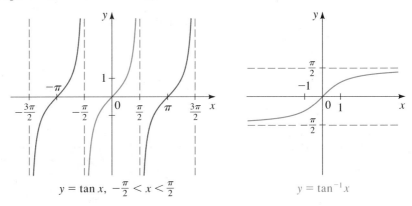

$y = \tan x, \; -\frac{\pi}{2} < x < \frac{\pi}{2}$ $\qquad\qquad$ $y = \tan^{-1}x$

EXAMPLE 7 ■ Evaluating the Inverse Tangent Function

Find (a) $\tan^{-1} 1$, (b) $\tan^{-1} \sqrt{3}$, and (c) $\tan^{-1}(-20)$.

SOLUTION

(a) The number in the interval $(-\pi/2, \pi/2)$ with tangent 1 is $\pi/4$. Thus, $\tan^{-1} 1 = \pi/4$.

(b) Since $\tan(\pi/3) = \sqrt{3}$, we have $\tan^{-1} \sqrt{3} = \pi/3$.

(c) We use a calculator to find that $\tan^{-1}(-20) \approx -1.52084$. ■

EXAMPLE 8 ■ The Angle of a Beam of Light

A lighthouse is located on an island that is 2 mi off a straight shoreline (see Figure 10). Express the angle formed by the beam of light and the shoreline in terms of the distance d in the figure.

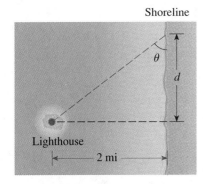

FIGURE 10

SOLUTION   From the figure, we see that $\tan \theta = 2/d$. Taking the inverse tangent of both sides, we get

$$\tan^{-1}(\tan \theta) = \tan^{-1}\left(\frac{2}{d}\right)$$

$$\theta = \tan^{-1}\left(\frac{2}{d}\right) \qquad \text{Cancellation property}$$

■

THE INVERSE SECANT, COSECANT, AND COTANGENT FUNCTIONS

To define the inverse functions of the secant, cosecant, and cotangent functions, we restrict the domain of each function to a set on which it is one-to-one and on which it attains all its values. Although any interval satisfying these criteria is appropriate, we choose to restrict the domains in a way that simplifies the choice of sign in computations involving inverse trigonometric functions. The choices we make are also appropriate for calculus. This explains the seemingly strange restriction for the domains of the secant and cosecant functions. We end this section by displaying the graphs of the secant, cosecant, and cotangent functions with their restricted domains and the graphs of their inverse functions (Figures 11–13).

FIGURE 11

The inverse secant function

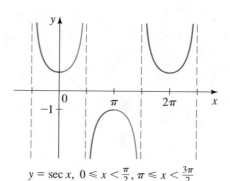

$y = \sec x, \ 0 \le x < \frac{\pi}{2}, \ \pi \le x < \frac{3\pi}{2}$

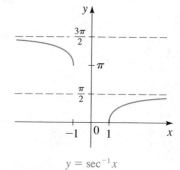

$y = \sec^{-1} x$

FIGURE 12

The inverse cosecant function

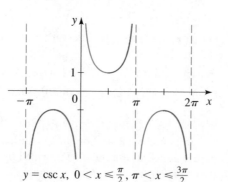

$y = \csc x, \ 0 < x \le \frac{\pi}{2}, \ \pi < x \le \frac{3\pi}{2}$

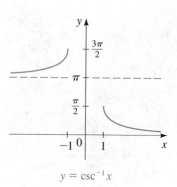

$y = \csc^{-1} x$

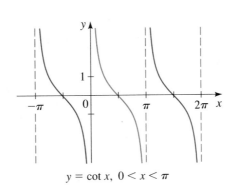

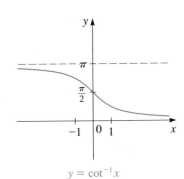

FIGURE 13

The inverse cotangent function

$y = \cot x,\ 0 < x < \pi$

$y = \cot^{-1} x$

7.4   EXERCISES

1–8 ■ Find the exact value of each expression, if it is defined.

1. (a) $\sin^{-1}\frac{1}{2}$ (b) $\cos^{-1}\frac{1}{2}$ (c) $\cos^{-1}2$

2. (a) $\sin^{-1}\frac{\sqrt{3}}{2}$ (b) $\cos^{-1}\frac{\sqrt{3}}{2}$ (c) $\cos^{-1}\left(-\frac{\sqrt{3}}{2}\right)$

3. (a) $\sin^{-1}\frac{\sqrt{2}}{2}$ (b) $\cos^{-1}\frac{\sqrt{2}}{2}$ (c) $\sin^{-1}\left(-\frac{\sqrt{2}}{2}\right)$

4. (a) $\tan^{-1}\sqrt{3}$ (b) $\tan^{-1}(-\sqrt{3})$ (c) $\sin^{-1}\sqrt{3}$

5. (a) $\sin^{-1}1$ (b) $\cos^{-1}1$ (c) $\cos^{-1}(-1)$

6. (a) $\tan^{-1}1$ (b) $\tan^{-1}(-1)$ (c) $\tan^{-1}0$

7. (a) $\tan^{-1}\frac{\sqrt{3}}{3}$ (b) $\tan^{-1}\left(-\frac{\sqrt{3}}{3}\right)$ (c) $\sin^{-1}(-2)$

8. (a) $\sin^{-1}0$ (b) $\cos^{-1}0$ (c) $\cos^{-1}\left(-\frac{1}{2}\right)$

9–10 ■ Use a calculator to find an approximate value of each expression correct to five decimal places.

9. (a) $\sin^{-1}(0.7688)$ (b) $\cos^{-1}(-0.5014)$

10. (a) $\cos^{-1}(0.3388)$ (b) $\tan^{-1}(15.2000)$

11–26 ■ Find the exact value of the expression, if it is defined.

11. $\sin(\sin^{-1}\frac{1}{3})$ **12.** $\cos(\cos^{-1}\frac{3}{4})$

13. $\tan(\tan^{-1}10)$ **14.** $\sin(\sin^{-1}10)$

15. $\cos^{-1}\left(\cos\frac{\pi}{3}\right)$ **16.** $\tan^{-1}\left(\tan\frac{\pi}{6}\right)$

17. $\sin^{-1}\left[\sin\left(-\frac{\pi}{6}\right)\right]$ **18.** $\sin^{-1}\left(\sin\frac{5\pi}{6}\right)$

19. $\tan^{-1}\left(\tan\frac{2\pi}{3}\right)$

20. $\cos^{-1}\left[\cos\left(-\frac{\pi}{4}\right)\right]$

21. $\tan(\sin^{-1}\frac{1}{2})$

22. $\sin(\sin^{-1}0)$

23. $\cos\left(\sin^{-1}\frac{\sqrt{3}}{2}\right)$

24. $\tan\left(\sin^{-1}\frac{\sqrt{2}}{2}\right)$

25. $\tan^{-1}\left(2\sin\frac{\pi}{3}\right)$

26. $\cos^{-1}\left(\sqrt{3}\sin\frac{\pi}{6}\right)$

27–38 ■ Evaluate the expression by sketching a triangle, as in Solution 2 of Example 3.

27. $\sin(\cos^{-1}\frac{3}{5})$ **28.** $\tan(\sin^{-1}\frac{4}{5})$

29. $\sin(\tan^{-1}\frac{12}{5})$ **30.** $\cos(\tan^{-1}5)$

31. $\sec(\sin^{-1}\frac{12}{13})$ **32.** $\csc(\cos^{-1}\frac{7}{25})$

33. $\cos(\tan^{-1}2)$ **34.** $\cot(\sin^{-1}\frac{2}{3})$

35. $\sin(2\cos^{-1}\frac{3}{5})$ **36.** $\tan(2\tan^{-1}\frac{5}{13})$

37. $\sin(\sin^{-1}\frac{1}{2} + \cos^{-1}\frac{1}{2})$ **38.** $\cos(\sin^{-1}\frac{3}{5} - \cos^{-1}\frac{3}{5})$

39–46 ■ Rewrite the expression as an algebraic expression in x.

39. $\cos(\sin^{-1}x)$

40. $\sin(\tan^{-1}x)$

41. $\tan(\sin^{-1}x)$

42. $\cos(\tan^{-1}x)$

43. $\cos(2\tan^{-1}x)$

44. $\sin(2\sin^{-1}x)$

45. $\cos(\cos^{-1}x + \sin^{-1}x)$

46. $\sin(\tan^{-1}x - \sin^{-1}x)$

47. A 50-ft pole casts a shadow of length s as shown. Express the angle θ of elevation of the sun in terms of the length s of the shadow.

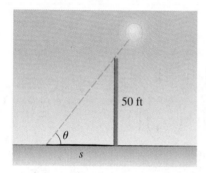

48. A 680-ft rope anchors a hot-air balloon. Express the angle θ in the figure as a function of the height h.

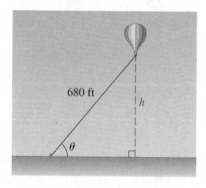

49. A painting that is 2 m high hangs in a museum with its bottom edge 3 m above the floor. A person whose eye level is h meters above the floor stands at a distance of x meters directly in front of the painting. The size of the painting as it appears to the viewer is determined by the size of the angle θ that the painting subtends at the viewer's eyes (see the figure). The larger θ is, the larger the apparent size of the painting. The angle θ depends on the distance x; in other words, the angle θ is a function of x.

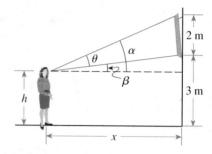

(a) Show that

$$\theta = \tan^{-1}\left(\frac{2x}{x^2 + (3-h)(5-h)}\right)$$

[*Hint:* Use the subtraction formula for tangent and the fact that $\theta = \alpha - \beta$.]

(b) Assume the viewer's eye level is $h = 2$ m. Find the angle θ (in degrees) subtended by the picture for $x = 0.5$, 2, and 5 m.

50–51 ■ (a) Graph the function and make a conjecture, and (b) prove that your conjecture is true.

50. $y = \tan^{-1}x + \tan^{-1}\dfrac{1}{x}$

51. $y = \sin^{-1}x + \cos^{-1}x$

52–53 ■ (a) Use a graphing device to find all solutions to the equation, correct to two decimal places, and (b) find the exact solution.

52. $\sin^{-1}x - \cos^{-1}x = 0$

53. $\tan^{-1}x + \tan^{-1}2x = \dfrac{\pi}{4}$

DISCOVERY · DISCUSSION

54. Two Different Compositions The functions

$$f(x) = \sin(\sin^{-1}x) \qquad \text{and} \qquad g(x) = \sin^{-1}(\sin x)$$

both simplify to just x for suitable values of x. But these functions are not the same for all x. Sketch graphs of both f and g to show how the functions differ. (Think carefully about the domain and range of $\sin^{-1}$.)

TRIGONOMETRIC EQUATIONS

A *trigonometric equation* is an equation that contains trigonometric functions. For example,

$$\sin^2 x + \cos^2 x = 1 \qquad \text{and} \qquad 2 \sin x - 1 = 0$$

are both trigonometric equations. The first equation is an identity—that means it is true for every value of the variable x. The second equation is true only for certain values of x. In this section we are interested in solving trigonometric equations, that is, in finding all values of the variable that make the equation true.

EXAMPLE 1 ■ Solving a Trigonometric Equation

Solve the equation $2 \sin x - 1 = 0$.

SOLUTION   We first solve the equation for $\sin x$.

$$2 \sin x - 1 = 0 \qquad \text{Given equation}$$

$$2 \sin x = 1 \qquad \text{Add 1}$$

$$\sin x = \tfrac{1}{2} \qquad \text{Divide by 2}$$

In the interval $[0, 2\pi)$, the values of x for which this equation is true are $x = \pi/6$ and $x = 5\pi/6$. But since the sine function is periodic with period 2π, adding any integer multiple of 2π to these values gives another solution. Thus, all the solutions are of the form

$$x = \frac{\pi}{6} + 2k\pi \qquad \text{or} \qquad x = \frac{5\pi}{6} + 2k\pi$$

for any integer k. Figure 1 shows a graphical representation of the solutions.

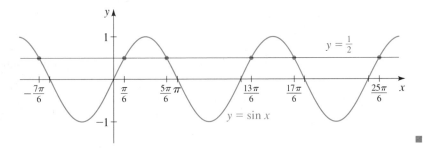

FIGURE 1

In general, as in Example 1, if a trigonometric equation has one solution, then it has infinitely many solutions. To find all the solutions of such an equation, we need only find the solutions in an appropriate interval and then use the fact that the trigonometric functions are periodic.

EXAMPLE 2 ■ Solving by Factoring

Solve the equation $2\cos^2 x - 7\cos x + 3 = 0$.

SOLUTION   We factor the left-hand side of the equation.

$$2\cos^2 x - 7\cos x + 3 = 0 \qquad \text{Given equation}$$

$$(2\cos x - 1)(\cos x - 3) = 0 \qquad \text{Factor}$$

$$2\cos x - 1 = 0 \qquad \text{or} \qquad \cos x - 3 = 0$$

$$\cos x = \tfrac{1}{2} \qquad\qquad\qquad \cos x = 3$$

$2C^2 - 7C + 3 = 0$

$(2C - 1)(C - 3) = 0$

Since $\cos x$ is never greater than 1, we see that $\cos x = 3$ has no solution. In the interval $[0, 2\pi)$, the equation $\cos x = \tfrac{1}{2}$ has solutions $x = \pi/3$ and $x = 5\pi/3$. Because the cosine function is periodic with period 2π, all the solutions are of the form

$$x = \frac{\pi}{3} + 2k\pi \qquad \text{or} \qquad x = \frac{5\pi}{3} + 2k\pi$$

for any integer k. ■

EXAMPLE 3 ■ An Equation Involving a Double Argument

Solve the trigonometric equation $\tan^4 2x - 9 = 0$.

SOLUTION

$$\tan^4 2x - 9 = 0 \qquad \text{Given equation}$$

$$\tan^4 2x = 9 \qquad \text{Add 9}$$

$$\tan 2x = \sqrt{3} \qquad \text{or} \qquad \tan 2x = -\sqrt{3} \qquad \text{Take fourth roots}$$

The interval $(-\pi/2, \pi/2)$ contains one complete period of the tangent function. In this interval, we have

$$2x = \frac{\pi}{3} \qquad \text{or} \qquad 2x = -\frac{\pi}{3}$$

Since the tangent function is periodic with period π, all the solutions are given by

$$2x = \frac{\pi}{3} + k\pi \qquad \text{or} \qquad 2x = -\frac{\pi}{3} + k\pi$$

$$x = \frac{\pi}{6} + \frac{k\pi}{2} \qquad\qquad x = -\frac{\pi}{6} + \frac{k\pi}{2} \qquad \text{Divide by 2}$$

for any integer k. ■

Trigonometric identities are useful tools for solving trigonometric equations. They can be used to transform an equation into an equivalent equation that's simpler to solve.

EXAMPLE 4 ■ Using a Trigonometric Identity

Solve the equation $3 \sin x = 2 \cos^2 x$ in the interval $[0, 2\pi)$.

SOLUTION   Using the identity $\cos^2 x = 1 - \sin^2 x$, we get an equivalent equation that involves only the sine function.

$$3 \sin x = 2 \cos^2 x \qquad \text{Given equation}$$

$$3 \sin x = 2(1 - \sin^2 x) \qquad \text{Pythagorean identity}$$

$$2 \sin^2 x + 3 \sin x - 2 = 0 \qquad \text{Simplify}$$

$$(2 \sin x - 1)(\sin x + 2) = 0 \qquad \text{Factor}$$

$$2 \sin x - 1 = 0 \qquad \text{or} \qquad \sin x + 2 = 0$$

$$\sin x = \tfrac{1}{2} \qquad\qquad\qquad \sin x = -2$$

Since $-1 \le \sin x \le 1$, the equation $\sin x = -2$ has no solution. The solutions of the given equation are thus the solutions of $\sin x = \tfrac{1}{2}$, that is, $x = \pi/6, 5\pi/6$. ■

EXAMPLE 5 ■ Finding Intersection Points

Find the points of intersection of the graphs of $f(x) = \sin x$ and $g(x) = \cos x$.

SOLUTION   The graphs of f and g are shown in Figure 2.

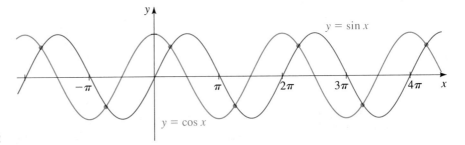

FIGURE 2

The graphs intersect at the points where $f(x) = g(x)$. So we need to find the solutions of the equation

$$\sin x = \cos x \qquad \text{Equate functions}$$

Notice that the numbers x for which $\cos x = 0$ are not solutions of this equation. For $\cos x \ne 0$, we can divide both sides of the equation by $\cos x$ to get

$$\frac{\sin x}{\cos x} = 1 \qquad \text{Divide by } \cos x$$

$$\tan x = 1 \qquad \text{Reciprocal identity}$$

The solution of this last equation in the interval $(-\pi/2, \pi/2)$ is $x = \pi/4$. Since the tangent function is periodic with period π, the solutions are

$$x = \frac{\pi}{4} + k\pi$$

for any integer k.   ∎

EXAMPLE 6 ■ Finding Solutions in an Interval

Find the solutions of $\cos 3x \sec x = 2 \cos 3x$ in the interval $[0, 2\pi)$.

SOLUTION

$$\cos 3x \sec x = 2 \cos 3x \qquad \text{Given equation}$$
$$\cos 3x \sec x - 2 \cos 3x = 0 \qquad \text{Subtract } 2 \cos 3x$$
$$\cos 3x(\sec x - 2) = 0 \qquad \text{Factor}$$
$$\cos 3x = 0 \qquad \text{or} \qquad \sec x = 2$$

We begin by solving $\cos 3x = 0$. Since we are seeking solutions in the interval $0 \le x \le 2\pi$, we have $0 \le 3x \le 6\pi$. In this interval, $\cos 3x = 0$ has the solutions

$$3x = \frac{\pi}{2}, \quad \frac{3\pi}{2}, \quad \frac{5\pi}{2}, \quad \frac{7\pi}{2}, \quad \frac{9\pi}{2}, \quad \frac{11\pi}{2}$$

so

$$x = \frac{\pi}{6}, \quad \frac{\pi}{2}, \quad \frac{5\pi}{6}, \quad \frac{7\pi}{6}, \quad \frac{3\pi}{2}, \quad \frac{11\pi}{6}$$

Now we solve $\sec x = 2$. In the interval $[0, 2\pi]$, the solutions of $\sec x = 2$ are $x = \pi/3$ and $x = 5\pi/3$. Thus, all the solutions of the given equation are

$$x = \frac{\pi}{6}, \quad \frac{\pi}{3}, \quad \frac{\pi}{2}, \quad \frac{5\pi}{6}, \quad \frac{7\pi}{6}, \quad \frac{3\pi}{2}, \quad \frac{5\pi}{3}, \quad \frac{11\pi}{6}$$ ∎

⊘   Notice that in Example 6 we did not divide both sides by $\cos 3x$. It would have been wrong to do so, since we could be dividing by 0. Indeed, if we divide both sides by $\cos 3x$, then we lose all solutions of the given equation that are solutions of $\cos 3x = 0$.

EXAMPLE 7 ■ Taking Care of Extraneous Solutions

Solve the equation $\cos x + 1 = \sin x$ in the interval $[0, 2\pi)$.

SOLUTION   To get an equation that involves either sine only or cosine only, we square both sides and use the Pythagorean identities.

$$\cos x + 1 = \sin x \qquad \text{Given equation}$$
$$(\cos x + 1)^2 = \sin^2 x \qquad \text{Square both sides}$$
$$\cos^2 x + 2 \cos x + 1 = \sin^2 x \qquad \text{Expand}$$
$$\cos^2 x + 2 \cos x + 1 = 1 - \cos^2 x \qquad \text{Pythagorean identity}$$

$$2\cos^2 x + 2\cos x = 0 \qquad \text{Simplify}$$

$$2\cos x(\cos x + 1) = 0 \qquad \text{Factor}$$

$$\cos x = 0 \quad \text{or} \quad \cos x + 1 = 0$$

From these equations we get the possible solutions

$$x = \frac{\pi}{2}, \quad \frac{3\pi}{2}, \quad \pi \qquad \text{Potential solutions}$$

Since we may have introduced extraneous roots by squaring both sides of the equation, we must check to see whether each of these values for x satisfies the original equation. From *Check Your Answers,* we see that the solutions of the given equation in the interval $[0, 2\pi)$ are $\pi/2$ and π.

CHECK YOUR ANSWERS

$x = \dfrac{\pi}{2}$:

$$\cos\frac{\pi}{2} + 1 \overset{?}{=} \sin\frac{\pi}{2}$$

$$0 + 1 = 1 \quad \checkmark$$

$x = \dfrac{3\pi}{2}$:

$$\cos\frac{3\pi}{2} + 1 \overset{?}{=} \sin\frac{3\pi}{2}$$

$$0 + 1 \overset{?}{=} -1 \quad \times$$

$x = \pi$:

$$\cos\pi + 1 \overset{?}{=} \sin\pi$$

$$-1 + 1 = 0 \quad \checkmark \qquad \blacksquare$$

 If we perform an operation on an equation that may introduce new roots, then we must check that the solutions obtained are not extraneous; that is, we must verify that they satisfy the original equation, as in Example 7.

EXAMPLE 8 ■ **Using a Trigonometric Identity**

Solve the equation: $\sin 2x - \cos x = 0$

SOLUTION   Since the argument in the two terms is different, we first use the double-angle formula for sine.

$$\sin 2x - \cos x = 0 \qquad \text{Given equation}$$

$$2\sin x\cos x - \cos x = 0 \qquad \text{Double-angle formula}$$

$$\cos x(2\sin x - 1) = 0 \qquad \text{Factor}$$

$$\cos x = 0 \quad \text{or} \quad 2\sin x - 1 = 0$$

$$\sin x = \tfrac{1}{2}$$

For x in the interval $[0, 2\pi)$, the first of these equations has solutions $x = \pi/2, 3\pi/2$, and the second has solutions $x = \pi/6, 5\pi/6$. Thus, the solutions of the original equation are

$$x = \frac{\pi}{2} + 2k\pi, \quad \frac{3\pi}{2} + 2k\pi, \quad \frac{\pi}{6} + 2k\pi, \quad \frac{5\pi}{6} + 2k\pi$$

for any integer k.   ■

USING A CALCULATOR TO SOLVE TRIGONOMETRIC EQUATIONS

So far all the equations we've solved have had solutions like $\pi/4$, $\pi/3$, $5\pi/6$, $3\pi/2$, and so on. We were able to find these solutions from the special values of the trigonometric functions that we've memorized. Now let's consider examples whose solutions require us to use the inverse trigonometric functions on a calculator.

EXAMPLE 9 ■ Using the Inverse Trigonometric Functions and a Calculator

Solve the equation $\sin x = \frac{2}{3}$ in the interval $[0, 2\pi]$.

SOLUTION   The solutions of this equation are the numbers x whose sine is $\frac{2}{3}$. From Figure 3 we see that there are two such numbers, x_1 and x_2, in the interval $[0, 2\pi]$. Using a calculator in radian mode, we can get x_1:

$$\sin x = \frac{2}{3} \qquad \text{Given equation}$$

$$x_1 = \sin^{-1} \frac{2}{3} \qquad \text{Take } \sin^{-1} \text{ of each side}$$

$$\approx 0.72973 \qquad \text{Use a calculator}$$

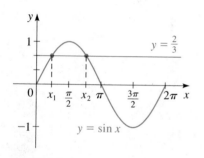

FIGURE 3

Note that since the range of $\sin^{-1}$ is the interval $[-\pi/2, \pi/2]$, the calculator gives us a number in that interval. From Figure 3 we see that the other solution lies between $\pi/2$ and π. (In fact, if we think of x as an angle in radian measure, then x_2 is the angle in the second quadrant whose reference angle is x_1.) Thus

$$x_2 = \pi - \sin^{-1} \frac{2}{3} \approx 2.41186$$

The solutions, correct to five decimal places are $x \approx 0.72973$ and $x \approx 2.41186$.
                                                                                       ■

If we are using inverse trigonometric functions to solve an equation, we must keep in mind that $\sin^{-1}$ and $\tan^{-1}$ give values in quadrants I and IV, and $\cos^{-1}$ gives values in quadrants I and II. To find other solutions, we must look at the quadrant where the trigonometric function in the equation can take on the value we need.

EXAMPLE 10 ■ Finding All Solutions

Find all solutions of the equation $4 \cos^2 x - 9 \cos x + 2 = 0$.

SOLUTION   We first find the solutions in the interval $[0, 2\pi)$, which is one complete period of the cosine function.

$$4\cos^2 x - 9\cos x + 2 = 0 \qquad \text{Given equation}$$

$$(\cos x - 2)(4\cos x - 1) = 0 \qquad \text{Factor}$$

$$\cos x - 2 = 0 \qquad \text{or} \qquad 4\cos x - 1 = 0$$

$$\cos x = 2 \qquad\qquad\qquad \cos x = \tfrac{1}{4}$$

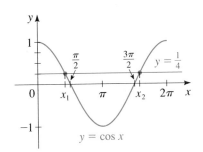

FIGURE 4

Since $\cos x$ cannot be larger than 1, the equation $\cos x = 2$ has no solution. From Figure 4 we see that the equation $\cos x = \tfrac{1}{4}$ has two solutions in the interval $[0, 2\pi)$. One of them is $x_1 = \cos^{-1}\tfrac{1}{4} \approx 1.31812$. Note that since the range of $\cos^{-1}$ is the interval $[0, \pi]$, the calculator gives us a number in that interval. The other solution is between $3\pi/2$ and 2π (the fourth quadrant), so its value is $x_2 = 2\pi - \cos^{-1}\tfrac{1}{4} \approx 4.96507$. Thus, all the solutions of the equation are of the form

$$x \approx 1.31812 + 2k\pi \qquad \text{or} \qquad x \approx 4.96507 + 2k\pi$$

for any integer k. ■

7.5   EXERCISES

1–30 ■ Find all solutions of the equation.

1. $2\cos x - 1 = 0$

2. $\sqrt{2}\sin x - 1 = 0$

3. $2\sin x - \sqrt{3} = 0$

4. $\tan x + 1 = 0$

5. $4\cos^2 x - 1 = 0$

6. $2\cos^2 x - 1 = 0$

7. $\sec^2 x - 2 = 0$

8. $\csc^2 x - 4 = 0$

9. $\cos x (2\sin x + 1) = 0$

10. $\sec x (2\cos x - \sqrt{2}) = 0$

11. $(\tan x + \sqrt{3})(\cos x + 2) = 0$

12. $(2\cos x + \sqrt{3})(2\sin x - 1) = 0$

13. $\cos x \sin x - 2\cos x = 0$

14. $\tan x \sin x + \sin x = 0$

15. $4\cos^2 x - 4\cos x + 1 = 0$

16. $2\sin^2 x - \sin x - 1 = 0$

17. $\sin^2 x = 2\sin x + 3$

18. $3\tan^3 x = \tan x$

19. $\sin^2 x = 4 - 2\cos^2 x$

20. $2\cos^2 x + \sin x = 1$

21. $2\sin 3x - 1 = 0$

22. $\sqrt{3}\sin 2x = \cos 2x$

23. $\cos\dfrac{x}{2} - 1 = 0$

24. $\csc 3x = \sin 3x$

25. $\tan^5 x - 9\tan x = 0$

26. $3\tan^3 x - 3\tan^2 x - \tan x + 1 = 0$

27. $4\sin x \cos x + 2\sin x - 2\cos x - 1 = 0$

28. $\sin 2x = 2\tan 2x$

29. $\cos^2 2x - \sin^2 2x = 0$

30. $\sec x - \tan x = \cos x$

31–38 ■ Find all solutions of the equation in the interval $[0, 2\pi)$.

31. $2\cos 3x = 1$

32. $3\csc^2 x = 4$

33. $2\sin x \tan x - \tan x = 1 - 2\sin x$

34. $\sec x \tan x - \cos x \cot x = \sin x$

35. $\tan x - 3\cot x = 0$

36. $2\sin^2 x - \cos x = 1$

37. $\tan 3x + 1 = \sec 3x$

38. $3\sec^2 x + 4\cos^2 x = 7$

39–46 ■ (a) Use a calculator to solve the equation in the interval $[0, 2\pi)$, correct to five decimal places.
(b) Find all solutions of the equation.

39. $\cos x = 0.4$

40. $2\tan x = 13$

41. $\sec x - 5 = 0$

42. $3 \sin x = 7 \cos x$

43. $5 \sin^2 x - 1 = 0$

44. $2 \sin 2x - \cos x = 0$

45. $3 \sin^2 x - 7 \sin x + 2 = 0$

46. $\tan^4 x - 13 \tan^2 x + 36 = 0$

47–50 ■ Sketch the graphs of f and g on the same axes, and find their points of intersection.

47. $f(x) = 3 \cos x + 1, \quad g(x) = \cos x - 1$

48. $f(x) = \sin 2x, \quad g(x) = 2 \sin 2x + 1$

49. $f(x) = \tan x, \quad g(x) = \sqrt{3}$

50. $f(x) = \sin x - 1, \quad g(x) = \cos x$

51–54 ■ Use an addition or subtraction formula to simplify the equation. Then find all solutions in the interval $[0, 2\pi)$.

51. $\cos x \cos 3x - \sin x \sin 3x = 0$

52. $\cos x \cos 2x + \sin x \sin 2x = \frac{1}{2}$

53. $\sin 2x \cos x + \cos 2x \sin x = \sqrt{3}/2$

54. $\sin 3x \cos x - \cos 3x \sin x = 0$

55–58 ■ Use a double- or half-angle formula to solve the equation in the interval $[0, 2\pi)$.

55. $\sin 2x + \cos x = 0$

56. $\tan \dfrac{x}{2} - \sin x = 0$

57. $\cos 2x + \cos x = 2$

58. $\tan x + \cot x = 4 \sin 2x$

59–62 ■ Solve the equation by first using a sum-to-product formula.

59. $\sin x + \sin 3x = 0$

60. $\cos 5x - \cos 7x = 0$

61. $\cos 4x + \cos 2x = \cos x$

62. $\sin 5x - \sin 3x = \cos 4x$

63–68 ■ Use a graphing device to find the solutions of the equation, correct to two decimal places.

63. $\sin 2x = x$

64. $\cos x = \dfrac{x}{3}$

65. $2^{\sin x} = x$

66. $\sin x = x^3$

67. $\dfrac{\cos x}{1 + x^2} = x^2$

68. $\cos x = \frac{1}{2}(e^x + e^{-x})$

DISCOVERY · DISCUSSION

69. Equations and Identities Which of the following statements is true?

    A. Every identity is an equation.

    B. Every equation is an identity.

Give examples to illustrate your answer. Write a short paragraph to explain the difference between an equation and an identity.

70. A Special Trigonometric Equation What makes the equation $\sin(\cos x) = 0$ different from all the other equations we've looked at in this section? Find all solutions of this equation.

7.6
TRIGONOMETRIC FORM OF COMPLEX NUMBERS; DeMOIVRE'S THEOREM

Complex numbers were introduced in Chapter 3 in order to solve certain algebraic equations. The applications of complex numbers go far beyond this initial use, however. Complex numbers are now used routinely in physics, electrical engineering, aerospace engineering, and many other fields. In this section we represent complex numbers using the functions sine and cosine. This will enable us to find the nth roots of complex numbers.

TRIGONOMETRIC FORM OF COMPLEX NUMBERS

Let $z = a + bi$ be a complex number, and in the complex plane let's draw the line segment joining the origin to the point $a + bi$ (see Figure 1). The length of this line segment is denoted by $r = |z| = \sqrt{a^2 + b^2}$. (Recall that $|z|$ is the modulus of z.) If θ is an angle in standard position whose terminal side coincides with this line segment, then by the definitions of sine and cosine (see Section 6.3)

$$a = r\cos\theta \qquad \text{and} \qquad b = r\sin\theta$$

so $z = r\cos\theta + ir\sin\theta = r(\cos\theta + i\sin\theta)$. We have shown the following.

Im

bi

$a + bi$

r

θ

0 a **Re**

FIGURE 1

> ### TRIGONOMETRIC FORM OF COMPLEX NUMBERS
>
> A complex number $z = a + bi$ has the **trigonometric form**
>
> $$z = r(\cos\theta + i\sin\theta)$$
>
> where $r = |z| = \sqrt{a^2 + b^2}$ and $\tan\theta = b/a$. The number r is the **modulus** of z, and θ is an **argument** of z.

The argument of z is not unique, but any two arguments of z differ by a multiple of 2π.

EXAMPLE 1 ■ **Writing Complex Numbers in Trigonometric Form**

Write each of the following complex numbers in trigonometric form.

(a) $1 + i$ (b) $-1 + \sqrt{3}\, i$ (c) $-4\sqrt{3} - 4i$ (d) $3 + 4i$

SOLUTION   These complex numbers are graphed in Figure 2, which helps us find their arguments.

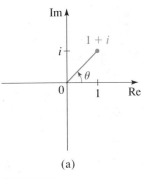

(a)

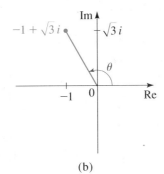

(b)

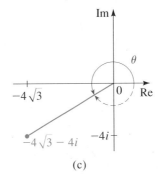

(c)

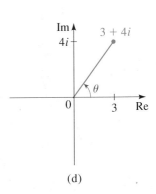

(d)

FIGURE 2

$\tan\theta = \frac{1}{1} = 1$

$\theta = \frac{\pi}{4}$

(a) An argument is $\theta = \pi/4$ and $r = \sqrt{1 + 1} = \sqrt{2}$. Thus

$$1 + i = \sqrt{2}\left(\cos\frac{\pi}{4} + i\sin\frac{\pi}{4}\right)$$

$$\tan \theta = \frac{\sqrt{3}}{-1} = -\sqrt{3}$$

$$\theta = \frac{2\pi}{3}$$

$$\tan \theta = \frac{-4}{-4\sqrt{3}} = \frac{1}{\sqrt{3}}$$

$$\theta = \frac{7\pi}{6}$$

$$\tan \theta = \frac{4}{3}$$

$$\theta = \tan^{-1}\frac{4}{3}$$

(b) An argument is $\theta = 2\pi/3$ and $r = \sqrt{1 + 3} = 2$. Thus

$$-1 + \sqrt{3}\, i = 2\left(\cos \frac{2\pi}{3} + i \sin \frac{2\pi}{3} \right)$$

(c) An argument is $\theta = 7\pi/6$ (or we could use $\theta = -5\pi/6$), and $r = \sqrt{48 + 16} = 8$. Thus

$$-4\sqrt{3} - 4i = 8\left(\cos \frac{7\pi}{6} + i \sin \frac{7\pi}{6} \right)$$

(d) An argument is $\theta = \tan^{-1}\frac{4}{3}$ and $r = \sqrt{3^2 + 4^2} = 5$. So

$$3 + 4i = 5\left[\cos\left(\tan^{-1}\tfrac{4}{3}\right) + i \sin\left(\tan^{-1}\tfrac{4}{3}\right) \right]$$ ∎

The addition formulas for sine and cosine discussed in Section 7.2 greatly simplify the multiplication and division of complex numbers in trigonometric form. The following theorem shows how.

MULTIPLICATION AND DIVISION OF COMPLEX NUMBERS

If the two complex numbers z_1 and z_2 have the trigonometric forms

$$z_1 = r_1(\cos \theta_1 + i \sin \theta_1) \qquad \text{and} \qquad z_2 = r_2(\cos \theta_2 + i \sin \theta_2)$$

then

$$z_1 z_2 = r_1 r_2 [\cos(\theta_1 + \theta_2) + i \sin(\theta_1 + \theta_2)] \qquad \text{Multiplication}$$

$$\frac{z_1}{z_2} = \frac{r_1}{r_2} [\cos(\theta_1 - \theta_2) + i \sin(\theta_1 - \theta_2)] \qquad (z_2 \neq 0) \quad \text{Division}$$

This theorem says: *To multiply two complex numbers, we multiply the moduli and add the arguments,* and *to divide two complex numbers, we divide the moduli and subtract the arguments.*

■ **Proof**   To prove the multiplication formula, we simply multiply the two complex numbers.

$$z_1 z_2 = r_1 r_2 (\cos \theta_1 + i \sin \theta_1)(\cos \theta_2 + i \sin \theta_2)$$

$$= r_1 r_2 [\cos \theta_1 \cos \theta_2 - \sin \theta_1 \sin \theta_2 + i(\sin \theta_1 \cos \theta_2 + \cos \theta_1 \sin \theta_2)]$$

$$= r_1 r_2 [\cos(\theta_1 + \theta_2) + i \sin(\theta_1 + \theta_2)]$$

In the last step we used the addition formulas for sine and cosine.

The proof of the division formula is left as an exercise.   □

EXAMPLE 2 ■ Multiplying and Dividing Complex Numbers

Let

$$z_1 = 2\left(\cos\frac{\pi}{4} + i\sin\frac{\pi}{4}\right) \quad \text{and} \quad z_2 = 5\left(\cos\frac{\pi}{3} + i\sin\frac{\pi}{3}\right)$$

Find (a) z_1z_2 and (b) z_1/z_2.

SOLUTION

(a) By the multiplication formula

$$z_1z_2 = (2)(5)\left[\cos\left(\frac{\pi}{4} + \frac{\pi}{3}\right) + i\sin\left(\frac{\pi}{4} + \frac{\pi}{3}\right)\right]$$

$$= 10\left(\cos\frac{7\pi}{12} + i\sin\frac{7\pi}{12}\right)$$

To approximate the answer we use a calculator in radian mode and get

$$z_1z_2 \approx 10(-0.2588 + 0.9659i) = -2.588 + 9.659i$$

(b) By the division formula

$$\frac{z_1}{z_2} = \frac{2}{5}\left[\cos\left(\frac{\pi}{4} - \frac{\pi}{3}\right) + i\sin\left(\frac{\pi}{4} - \frac{\pi}{3}\right)\right]$$

$$= \frac{2}{5}\left[\cos\left(-\frac{\pi}{12}\right) + i\sin\left(-\frac{\pi}{12}\right)\right]$$

$$= \frac{2}{5}\left(\cos\frac{\pi}{12} - i\sin\frac{\pi}{12}\right)$$

Using a calculator in radian mode gives the approximate answer

$$\frac{z_1}{z_2} \approx \frac{2}{5}(0.9659 - 0.2588i) = 0.3864 - 0.1035i \qquad ■$$

DeMOIVRE'S THEOREM

Repeated use of the multiplication formula gives a useful formula for raising a complex number to a power n for any positive integer n. Let z be a complex number written in trigonometric form

$$z = r(\cos\theta + i\sin\theta)$$

Then by the multiplication formula

$$z^2 = zz = r^2[\cos(\theta + \theta) + i\sin(\theta + \theta)]$$

$$= r^2(\cos 2\theta + i\sin 2\theta)$$

Now we multiply z^2 by z to get

$$z^3 = z^2z = r^3[\cos(2\theta + \theta) + i\sin(2\theta + \theta)]$$

$$= r^3(\cos 3\theta + i\sin 3\theta)$$

Repeating this argument, we see that for any positive integer n

$$z^n = r^n(\cos n\theta + i \sin n\theta)$$

A similar argument using the division formula shows that this also holds for negative integers. We have proved the following theorem.

DeMOIVRE'S THEOREM

If $z = r(\cos\theta + i\sin\theta)$, then for any integer n

$$z^n = r^n(\cos n\theta + i\sin n\theta)$$

This says that *to take the nth power of a complex number, we take the nth power of the modulus and multiply the argument by n.*

EXAMPLE 3 ■ **Finding a Power Using DeMoivre's Theorem**

Find $\left(\frac{1}{2} + \frac{1}{2}i\right)^{10}$.

SOLUTION Since $\frac{1}{2} + \frac{1}{2}i = \frac{1}{2}(1 + i)$, it follows from Example 1(a) that

$$\frac{1}{2} + \frac{1}{2}i = \frac{\sqrt{2}}{2}\left(\cos\frac{\pi}{4} + i\sin\frac{\pi}{4}\right)$$

So by DeMoivre's Theorem,

$$\left(\frac{1}{2} + \frac{1}{2}i\right)^{10} = \left(\frac{\sqrt{2}}{2}\right)^{10}\left(\cos\frac{10\pi}{4} + i\sin\frac{10\pi}{4}\right)$$

$$= \frac{2^5}{2^{10}}\left(\cos\frac{5\pi}{2} + i\sin\frac{5\pi}{2}\right) = \frac{1}{32}i$$ ■

nTH ROOTS OF COMPLEX NUMBERS

To find the nth roots of the complex number z, we need to find a complex number w such that

$$w^n = z$$

Let's write z in trigonometric form:

$$z = r(\cos\theta + i\sin\theta)$$

One nth root of z is

$$w = r^{1/n}\left(\cos\frac{\theta}{n} + i\sin\frac{\theta}{n}\right)$$

since by DeMoivre's Theorem, $w^n = z$. But the argument θ of z can be replaced by $\theta + 2k\pi$ for any integer k. Thus, other nth roots of z are

$$w = r^{1/n}\left[\cos\left(\frac{\theta + 2k\pi}{n}\right) + i\sin\left(\frac{\theta + 2k\pi}{n}\right)\right]$$

Since this expression gives a different value of w for $k = 0, 1, 2, \ldots, n - 1$, we have proved the following theorem.

---

nTH ROOTS OF COMPLEX NUMBERS

If $z = r(\cos\theta + i\sin\theta)$ and n is a positive integer, then z has the n distinct nth roots

$$w_k = r^{1/n}\left[\cos\left(\frac{\theta + 2k\pi}{n}\right) + i\sin\left(\frac{\theta + 2k\pi}{n}\right)\right]$$

for $k = 0, 1, 2, \ldots, n - 1$.

The following observations help us use the preceding formula.

---

1. The modulus of each nth root is $r^{1/n}$.

2. The argument of the first root is θ/n.

3. We repeatedly add $2\pi/n$ to get the argument of each successive root.

These observations show that, when graphed, the nth roots of z are spaced equally on the circle of radius $r^{1/n}$.

EXAMPLE 4 ■ Finding Roots of a Complex Number

Find the six sixth roots of $z = -64$, and graph these roots in the complex plane.

SOLUTION   In trigonometric form, $z = 64(\cos\pi + i\sin\pi)$. Applying the formula for nth roots with $n = 6$, we get

$$w_k = 64^{1/6}\left[\cos\left(\frac{\pi + 2k\pi}{6}\right) + i\sin\left(\frac{\pi + 2k\pi}{6}\right)\right]$$

for $k = 0, 1, 2, 3, 4, 5$. Thus, the six sixth roots of -64 are

We add $2\pi/6 = \pi/3$ to each argument to get the argument of the next root.

$$w_0 = 64^{1/6}\left(\cos\frac{\pi}{6} + i\sin\frac{\pi}{6}\right) = \sqrt{3} + i$$

$$w_1 = 64^{1/6}\left(\cos\frac{\pi}{2} + i\sin\frac{\pi}{2}\right) = 2i$$

$$w_2 = 64^{1/6}\left(\cos\frac{5\pi}{6} + i\sin\frac{5\pi}{6}\right) = -\sqrt{3} + i$$

$$w_3 = 64^{1/6}\left(\cos\frac{7\pi}{6} + i\sin\frac{7\pi}{6}\right) = -\sqrt{3} - i$$

$$w_4 = 64^{1/6}\left(\cos\frac{3\pi}{2} + i\sin\frac{3\pi}{2}\right) = -2i$$

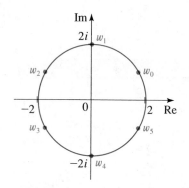

FIGURE 3

The six sixth roots of $z = -64$

and $$w_5 = 64^{1/6}\left(\cos\frac{11\pi}{6} + i\sin\frac{11\pi}{6}\right) = \sqrt{3} - i$$

All these points lie on the circle of radius 2, as shown in Figure 3. ∎

When finding roots of complex numbers, we sometimes write the argument θ of the complex number in degrees. In this case, the nth roots are obtained from the formula

$$w_k = r^{1/n}\left[\cos\left(\frac{\theta + 360°k}{n}\right) + i\sin\left(\frac{\theta + 360°k}{n}\right)\right]$$

for $k = 0, 1, 2, \ldots, n - 1$.

EXAMPLE 5 ■ **Finding Cube Roots of a Complex Number**

Find the three cube roots of $z = 2 + 2i$, and graph these roots in the complex plane.

SOLUTION   First we write z in trigonometric form using degrees. We have $r = \sqrt{2^2 + 2^2} = 2\sqrt{2}$ and $\theta = 45°$. Thus

$$z = 2\sqrt{2}\,(\cos 45° + i\sin 45°)$$

Applying the formula for nth roots (in degrees) with $n = 3$, we find the cube roots of z are of the form

$$(2\sqrt{2}\,)^{1/3} = (2^{3/2})^{1/3} = 2^{1/2} = \sqrt{2}$$

$$w_k = (2\sqrt{2}\,)^{1/3}\left[\cos\left(\frac{45° + 360°k}{3}\right) + i\sin\left(\frac{45° + 360°k}{3}\right)\right]$$

where $k = 0, 1, 2$. Thus, the three cube roots are

We add $360°/3 = 120°$ to each argument to get the argument of the next root.

$$w_0 = \sqrt{2}\,(\cos 15° + i\sin 15°) \approx 1.366 + 0.366i$$

$$w_1 = \sqrt{2}\,(\cos 135° + i\sin 135°) = -1 + i$$

$$w_2 = \sqrt{2}\,(\cos 255° + i\sin 255°) \approx -0.366 + 1.366i$$

The three cube roots of z are graphed in Figure 4. These roots are spaced equally on the circle of radius $\sqrt{2}$.

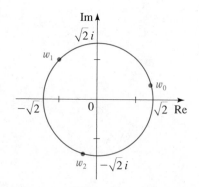

FIGURE 4

The three cube roots of $z = 2 + 2i$

EXAMPLE 6 ■ Solving an Equation Using the nth Roots Formula

Solve the equation $z^6 + 64 = 0$.

SOLUTION This equation can be written as $z^6 = -64$. Thus, the solutions are the sixth roots of -64, which we found in Example 4. ■

| 7.6 | **EXERCISES** |

1–24 ■ Write the complex number in trigonometric form with argument θ between 0 and 2π.

1. $1 + i$ **2.** $1 - \sqrt{3}\, i$ **3.** $\sqrt{2} - \sqrt{2}\, i$

4. $1 - i$ **5.** $2\sqrt{3} - 2i$ **6.** $-1 + i$

7. $-\sqrt{2}\, i$ **8.** $-3 - 3\sqrt{3}\, i$ **9.** $5 + 5i$

10. 4 **11.** $4\sqrt{3} - 4i$ **12.** $8i$

13. -20 **14.** $\sqrt{3} + i$ **15.** $3 + 4i$

16. $i(2 - 2i)$ **17.** $3i(1 + i)$ **18.** $2(1 - i)$

19. $4(\sqrt{3} + i)$ **20.** $-3 - 3i$ **21.** $2 + i$

22. $3 + \sqrt{3}\, i$ **23.** $\sqrt{2} + \sqrt{2}\, i$ **24.** $-\pi i$

25–30 ■ Find the product $z_1 z_2$ and the quotient z_1/z_2. Express your answer in trigonometric form.

25. $z_1 = \cos\dfrac{\pi}{4} + i\sin\dfrac{\pi}{4}, \quad z_2 = \cos\dfrac{3\pi}{4} + i\sin\dfrac{3\pi}{4}$

26. $z_1 = 3\left(\cos\dfrac{\pi}{6} + i\sin\dfrac{\pi}{6}\right),$

$z_2 = 5\left(\cos\dfrac{4\pi}{3} + i\sin\dfrac{4\pi}{3}\right)$

27. $z_1 = 7\left(\cos\dfrac{9\pi}{7} + i\sin\dfrac{9\pi}{7}\right),$

$z_2 = 2\left(\cos\dfrac{3\pi}{7} + i\sin\dfrac{3\pi}{7}\right)$

28. $z_1 = \sqrt{2}\,(\cos 75° + i\sin 75°),$

$z_2 = 3\sqrt{2}\,(\cos 60° + i\sin 60°)$

29. $z_1 = 4(\cos 200° + i\sin 200°),$

$z_2 = 25(\cos 150° + i\sin 150°)$

30. $z_1 = \frac{4}{5}(\cos 25° + i\sin 25°),$

$z_2 = \frac{1}{5}(\cos 155° + i\sin 155°)$

31–38 ■ Write z_1 and z_2 in trigonometric form, and then find the product $z_1 z_2$ and the quotients z_1/z_2 and $1/z_1$.

31. $z_1 = \sqrt{3} + i, \quad z_2 = 1 + \sqrt{3}\, i$

32. $z_1 = \sqrt{2} - \sqrt{2}\, i, \quad z_2 = 1 - i$

33. $z_1 = 2\sqrt{3} - 2i, \quad z_2 = -1 + i$

34. $z_1 = -\sqrt{2}\, i, \quad z_2 = -3 - 3\sqrt{3}\, i$

35. $z_1 = 5 + 5i, \quad z_2 = 4$

36. $z_1 = 4\sqrt{3} - 4i, \quad z_2 = 8i$

37. $z_1 = -20, \quad z_2 = \sqrt{3} + i$

38. $z_1 = 3 + 4i, \quad z_2 = 2 - 2i$

39–50 ■ Find the indicated power using DeMoivre's Theorem.

39. $(1 + i)^{20}$ **40.** $(1 - \sqrt{3}\, i)^5$

41. $(2\sqrt{3} + 2i)^5$ **42.** $(1 - i)^8$

43. $\left(\dfrac{\sqrt{2}}{2} + \dfrac{\sqrt{2}}{2}i\right)^{12}$ **44.** $(\sqrt{3} - i)^{-10}$

45. $(2 - 2i)^8$ **46.** $\left(-\dfrac{1}{2} - \dfrac{\sqrt{3}}{2}i\right)^{15}$

47. $(-1 - i)^7$ **48.** $(3 + \sqrt{3}\, i)^4$

49. $(2\sqrt{3} + 2i)^{-5}$ **50.** $(1 - i)^{-8}$

51–60 ■ Find the indicated roots, and sketch the roots in the complex plane.

51. The square roots of $4\sqrt{3} + 4i$

52. The cube roots of $4\sqrt{3} + 4i$

53. The fourth roots of $-81i$

54. The fifth roots of 32

55. The eighth roots of 1

56. The cube roots of $1 + i$

57. The cube roots of i

58. The fifth roots of i

59. The fourth roots of -1

60. The fifth roots of $-16 - 16\sqrt{3}\,i$

61–66 ■ Solve the equation.

61. $z^4 + 1 = 0$

62. $z^8 - i = 0$

63. $z^3 - 4\sqrt{3} - 4i = 0$

64. $z^6 - 1 = 0$

65. $z^3 + 1 = -i$

66. $z^3 - 1 = 0$

67. (a) Let $w = \cos\dfrac{2\pi}{n} + i\sin\dfrac{2\pi}{n}$ where n is a positive integer. Show that $1, w, w^2, w^3, \ldots, w^{n-1}$ are the n distinct nth roots of 1.

(b) If $z \neq 0$ is any complex number and $s^n = z$, show that the n distinct nth roots of z are $s, sw, sw^2, sw^3, \ldots, sw^{n-1}$.

DISCOVERY • DISCUSSION

68. Sums of Roots of Unity Find the exact values of all three cube roots of 1 (see Exercise 67) and then add them. Do the same for the fourth, fifth, sixth, and eighth roots of 1. What do you think is the sum of the nth roots of 1, for any n?

69. Products of Roots of Unity Find the product of the three cube roots of 1 (see Exercise 68). Do the same for the fourth, fifth, sixth, and eighth roots of 1. What do you think is the product of the nth roots of 1, for any n?

7.7 VECTORS

In applications of mathematics certain quantities are determined completely by their magnitude—for example, length, mass, area, temperature, and energy. We speak of a length of 5 m or a mass of 3 kg; only one number is needed to describe each of these quantities. Such a quantity is called a **scalar**.

On the other hand, to describe the displacement of an object, two numbers are required: the *magnitude* and the *direction* of the displacement. To describe the velocity of a moving object, we must specify both the *speed* and the *direction* of travel. Quantities such as displacement, velocity, acceleration, and force that involve magnitude as well as direction are called *directed quantities*. One way to represent such quantities mathematically is through the use of *vectors*.

GEOMETRIC DESCRIPTION OF VECTORS

A **vector** in the plane is a line segment with an assigned direction. We sketch a vector as shown in Figure 1 with an arrowhead to specify the direction. We denote this vector by $\overrightarrow{AB}$. The point A is the **initial point**, and B is the **terminal point** of the vector $\overrightarrow{AB}$. The length of the line segment AB is called the **magnitude** or **length** of the vector and is denoted by $|\overrightarrow{AB}|$. We use boldface letters to denote vectors. Thus, we write $\mathbf{u} = \overrightarrow{AB}$.

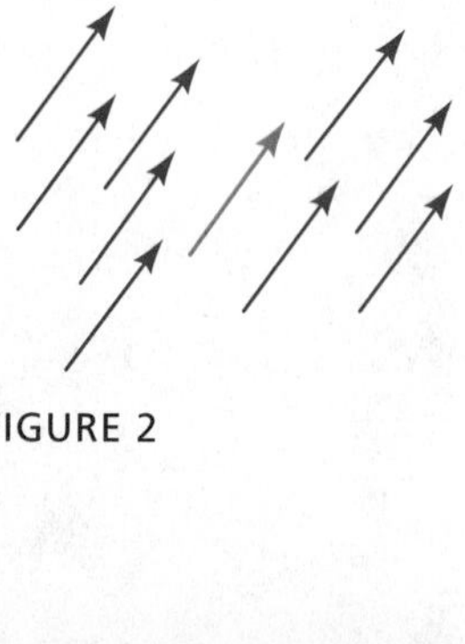

Two vectors are considered **equal** if they have equal magnitude and the same direction. Thus, all the vectors in Figure 2 are equal. This definition of equality makes sense if we think of a vector as representing a displacement. Two such displacements are the same if they have equal magnitudes and the same direction. So the vectors in Figure 2 can be thought of as the *same* displacement applied to objects in different locations in the plane.

If the displacement $\mathbf{u} = \overrightarrow{AB}$ is followed by the displacement $\mathbf{v} = \overrightarrow{BC}$, then the resulting displacement is $\overrightarrow{AC}$ as shown in Figure 3. In other words, the single dis-

$\mathbf{u} = \overrightarrow{AB}$

FIGURE 1

FIGURE 2

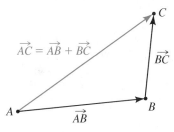

$$\vec{AC} = \vec{AB} + \vec{BC}$$

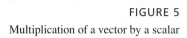

FIGURE 3

placement represented by the vector $\vec{AC}$ has the same effect as the other two displacements together. We call the vector $\vec{AC}$ the **sum** of the vectors $\vec{AB}$ and $\vec{BC}$ and we write $\vec{AC} = \vec{AB} + \vec{BC}$. (The **zero vector**, denoted by **0**, represents no displacement). Thus, to find the sum of any two vectors **u** and **v**, we sketch vectors equal to **u** and **v** with the initial point of one at the terminal point of the other [see Figure 4(a)]. If we draw **u** and **v** starting at the same point, then **u** + **v** is the vector that is the diagonal of the parallelogram formed by **u** and **v**, as shown in Figure 4(b).

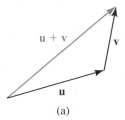

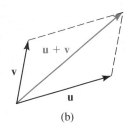

FIGURE 4

Addition of vectors

(a)   (b)

If a is a real number and **v** is a vector, we define a new vector a**v** as follows: The vector a**v** has magnitude $|a||\mathbf{v}|$ and has the same direction as **v** if $a > 0$, or the opposite direction if $a < 0$. If $a = 0$, then $a\mathbf{v} = \mathbf{0}$, the zero vector. This process is called **multiplication of a vector by a scalar**. Multiplication of a vector by a scalar has the effect of stretching or shrinking the vector. Figure 5 shows graphs of the vector a**v** for different values of a. We write the vector $(-1)\mathbf{v}$ as $-\mathbf{v}$. Thus, $-\mathbf{v}$ is the vector with the same length as **v** but with the opposite direction.

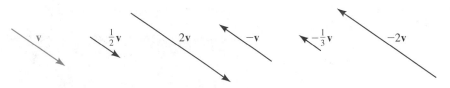

FIGURE 5

Multiplication of a vector by a scalar

The **difference** of two vectors **u** and **v** is defined by $\mathbf{u} - \mathbf{v} = \mathbf{u} + (-\mathbf{v})$. Figure 6 shows that the vector $\mathbf{u} - \mathbf{v}$ is the other diagonal of the parallelogram formed by **u** and **v**.

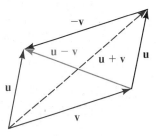

FIGURE 6

Subtraction of vectors

ANALYTIC DESCRIPTION OF VECTORS

So far we've been discussing vectors geometrically. We now give a method for describing vectors using coordinates (that is, analytically). Consider a vector **v** in the coordinate plane as in Figure 7(a). To go from the initial point of **v** to the terminal point, we move a units to the right and b units upward. We represent the vector **v** as an ordered pair of real numbers

$$\mathbf{v} = \langle a, b \rangle$$

where a is the **horizontal component** of **v** and b is the **vertical component** of **v**. We must remember that a vector represents a magnitude and a direction, not a particular arrow in the plane. Thus, again, the vector $\langle a, b \rangle$ has many different representations, depending on its initial point [see Figure 7(b)].

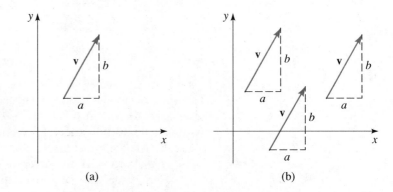

FIGURE 7   (a)   (b)

Using Figure 8, the relationship between a geometric representation of a vector and the analytic one can be stated as follows.

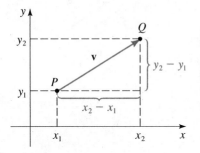

FIGURE 8

ANALYTIC FORM OF A VECTOR

If a vector **v** is represented in the plane with initial point $P(x_1, y_1)$ and terminal point $Q(x_2, y_2)$, then

$$\mathbf{v} = \langle x_2 - x_1, y_2 - y_1 \rangle$$

EXAMPLE 1 ■ Describing Vectors in Analytic Form

(a) Find the vector **u** with initial point $(-2, 5)$ and terminal point $(3, 7)$.
(b) If the vector $\mathbf{v} = \langle 3, 7 \rangle$ is sketched with initial point $(2, 4)$, what is its terminal point?
(c) Sketch representations of the vector $\mathbf{w} = \langle 2, 3 \rangle$ with initial points at $(0, 0)$, $(2, 2)$, $(-2, -1)$, and $(1, 4)$.

SOLUTION

(a) The desired vector is

$$\mathbf{u} = \langle 3 - (-2), 7 - 5 \rangle = \langle 5, 2 \rangle$$

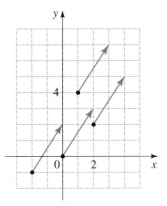

FIGURE 9

(b) Let the terminal point of **v** be (x, y). Then

$$\langle x - 2, y - 4 \rangle = \langle 3, 7 \rangle$$

So $x - 2 = 3$ and $y - 4 = 7$, or $x = 5$ and $y = 11$. The terminal point is $(5, 11)$.

(c) Representations of the vector **w** are sketched in Figure 9. ■

We now give analytic definitions of the various operations on vectors that we have described geometrically. Let's start with equality of vectors. We've said that two vectors are equal if they have equal magnitude and the same direction. For the vectors $\mathbf{u} = \langle a_1, b_1 \rangle$ and $\mathbf{v} = \langle a_2, b_2 \rangle$, this means that $a_1 = a_2$ and $b_1 = b_2$. In other words, two vectors are **equal** if and only if their corresponding components are equal. Thus, all the arrows in Figure 7(b) represent the same vector, as do all the arrows in Figure 9.

The length of a vector has the following meaning in terms of components.

MAGNITUDE OF A VECTOR

The **magnitude** or **length** of a vector $\mathbf{v} = \langle a, b \rangle$ is

$$|\mathbf{v}| = \sqrt{a^2 + b^2}$$

EXAMPLE 2 ■ Magnitudes of Vectors

Find the magnitude of each of the following vectors.

(a) $\mathbf{u} = \langle 2, -3 \rangle$ (b) $\mathbf{v} = \langle 5, 0 \rangle$ (c) $\mathbf{w} = \langle \frac{3}{5}, \frac{4}{5} \rangle$

SOLUTION

(a) $|\mathbf{u}| = \sqrt{2^2 + (-3)^2} = \sqrt{13}$

(b) $|\mathbf{v}| = \sqrt{5^2 + 0^2} = \sqrt{25} = 5$

(c) $|\mathbf{w}| = \sqrt{\left(\frac{3}{5}\right)^2 + \left(\frac{4}{5}\right)^2} = \sqrt{\frac{9}{25} + \frac{16}{25}} = 1$ ■

The following definitions of addition, subtraction, and scalar multiplication of vectors correspond to the geometric descriptions given earlier. Figure 10 shows how the analytic definition of addition corresponds to the geometric one.

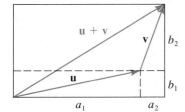

FIGURE 10

ALGEBRAIC OPERATIONS ON VECTORS

If $\mathbf{u} = \langle a_1, b_1 \rangle$ and $\mathbf{v} = \langle a_2, b_2 \rangle$, then

$$\mathbf{u} + \mathbf{v} = \langle a_1 + a_2, b_1 + b_2 \rangle$$

$$\mathbf{u} - \mathbf{v} = \langle a_1 - a_2, b_1 - b_2 \rangle$$

$$c\mathbf{u} = \langle ca_1, cb_1 \rangle, \qquad c \in \mathbb{R}$$

EXAMPLE 3 ■ Operations with Vectors

If $\mathbf{u} = \langle 2, -3 \rangle$ and $\mathbf{v} = \langle -1, 2 \rangle$, find $\mathbf{u} + \mathbf{v}$, $\mathbf{u} - \mathbf{v}$, $2\mathbf{u}$, $-3\mathbf{v}$, and $2\mathbf{u} + 3\mathbf{v}$.

SOLUTION By the definitions of the vector operations, we have

$$\mathbf{u} + \mathbf{v} = \langle 2, -3 \rangle + \langle -1, 2 \rangle = \langle 1, -1 \rangle$$

$$\mathbf{u} - \mathbf{v} = \langle 2, -3 \rangle - \langle -1, 2 \rangle = \langle 3, -5 \rangle$$

$$2\mathbf{u} = 2\langle 2, -3 \rangle = \langle 4, -6 \rangle$$

$$-3\mathbf{v} = -3\langle -1, 2 \rangle = \langle 3, -6 \rangle$$

$$2\mathbf{u} + 3\mathbf{v} = 2\langle 2, -3 \rangle + 3\langle -1, 2 \rangle = \langle 4, -6 \rangle + \langle -3, 6 \rangle = \langle 1, 0 \rangle \qquad ■$$

The following properties for vector operations follow easily from the definitions. The **zero vector** is the vector $\mathbf{0} = \langle 0, 0 \rangle$. It plays the same role for addition of vectors as the number 0 does for addition of real numbers.

PROPERTIES OF VECTORS

Vector addition	Multiplication by a scalar
$\mathbf{u} + \mathbf{v} = \mathbf{v} + \mathbf{u}$	$c(\mathbf{u} + \mathbf{v}) = c\mathbf{u} + c\mathbf{v}$
$\mathbf{u} + (\mathbf{v} + \mathbf{w}) = (\mathbf{u} + \mathbf{v}) + \mathbf{w}$	$(c + d)\mathbf{u} = c\mathbf{u} + d\mathbf{u}$
$\mathbf{u} + \mathbf{0} = \mathbf{u}$	$(cd)\mathbf{u} = c(d\mathbf{u}) = d(c\mathbf{u})$
$\mathbf{u} + (-\mathbf{u}) = \mathbf{0}$	$1\mathbf{u} = \mathbf{u}$
Length of a vector	$0\mathbf{u} = \mathbf{0}$
$|c\mathbf{u}| = |c||\mathbf{u}|$	$c\mathbf{0} = \mathbf{0}$

A vector of length 1 is called a **unit vector**. For instance, in Example 2(c), the vector $\mathbf{w} = \langle \frac{3}{5}, \frac{4}{5} \rangle$ is a unit vector. Two useful unit vectors are $\mathbf{i}$ and $\mathbf{j}$, defined by

$$\mathbf{i} = \langle 1, 0 \rangle \qquad \mathbf{j} = \langle 0, 1 \rangle$$

These vectors are special because any vector can be expressed in terms of them.

VECTORS IN TERMS OF i AND j

The vector $\mathbf{v} = \langle a, b \rangle$ can be expressed in terms of $\mathbf{i}$ and $\mathbf{j}$ by

$$\mathbf{v} = \langle a, b \rangle = a\mathbf{i} + b\mathbf{j}$$

EXAMPLE 4 ■ Vectors in Terms of i and j

(a) Write the vector $\mathbf{u} = \langle 5, -8 \rangle$ in terms of $\mathbf{i}$ and $\mathbf{j}$.

(b) If $\mathbf{u} = 3\mathbf{i} + 2\mathbf{j}$ and $\mathbf{v} = -\mathbf{i} + 6\mathbf{j}$, write $2\mathbf{u} + 5\mathbf{v}$ in terms of $\mathbf{i}$ and $\mathbf{j}$.

SOLUTION

(a) $\mathbf{u} = 5\mathbf{i} + (-8)\mathbf{j} = 5\mathbf{i} - 8\mathbf{j}$

(b) The properties of addition and scalar multiplication of vectors show that we can manipulate vectors in the same way as algebraic expressions. Thus

$$2\mathbf{u} + 5\mathbf{v} = 2(3\mathbf{i} + 2\mathbf{j}) + 5(-\mathbf{i} + 6\mathbf{j})$$
$$= (6\mathbf{i} + 4\mathbf{j}) + (-5\mathbf{i} + 30\mathbf{j})$$
$$= \mathbf{i} + 34\mathbf{j}$$ ∎

Let $\mathbf{v}$ be a vector in the plane with its initial point at the origin. The **direction** of $\mathbf{v}$ is θ, the smallest positive angle in standard position formed by the positive x-axis and $\mathbf{v}$ (see Figure 11). If we know the magnitude and direction of a vector, then Figure 11 shows that we can find the horizontal and vertical components of the vector.

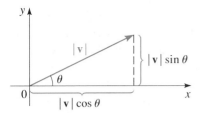

FIGURE 11

HORIZONTAL AND VERTICAL COMPONENTS OF A VECTOR

Let $\mathbf{v}$ be a vector with magnitude $|\mathbf{v}|$ and direction θ.
Then $\mathbf{v} = \langle a, b \rangle = a\mathbf{i} + b\mathbf{j}$, where

$$a = |\mathbf{v}| \cos\theta \quad \text{and} \quad b = |\mathbf{v}| \sin\theta$$

EXAMPLE 5 ■ Components and Direction of a Vector

(a) A vector $\mathbf{v}$ has length 8 and direction $\pi/3$. Find the horizontal and vertical components, and write $\mathbf{v}$ in terms of $\mathbf{i}$ and $\mathbf{j}$.

(b) Find the direction of the vector $\mathbf{u} = -\sqrt{3}\,\mathbf{i} + \mathbf{j}$.

SOLUTION

(a) We have $\mathbf{v} = \langle a, b \rangle$, where the components are given by

$$a = 8\cos\frac{\pi}{3} = 4 \quad \text{and} \quad b = 8\sin\frac{\pi}{3} = 4\sqrt{3}$$

Thus, $\mathbf{v} = \langle 4, 4\sqrt{3} \rangle = 4\mathbf{i} + 4\sqrt{3}\,\mathbf{j}$.

(b) From Figure 12 we see that the direction θ has the property that

$$\tan\theta = \frac{1}{-\sqrt{3}} = -\frac{\sqrt{3}}{3}$$

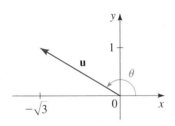

FIGURE 12

Thus, the reference angle for θ is $\pi/6$. Since the terminal point of the vector $\mathbf{u}$ is in quadrant II, it follows that $\theta = 5\pi/6$. ∎

APPLICATIONS TO VELOCITY AND FORCE

When applying vectors to navigation, the direction of a vector is usually given as a **bearing**, that is, as an acute angle measured from due north or due south. The bearing N 30° E, for example, indicates a direction that points 30° to the east of due north (see Figure 13).

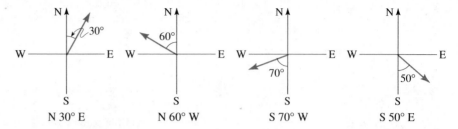

FIGURE 13   N 30° E   N 60° W   S 70° W   S 50° E

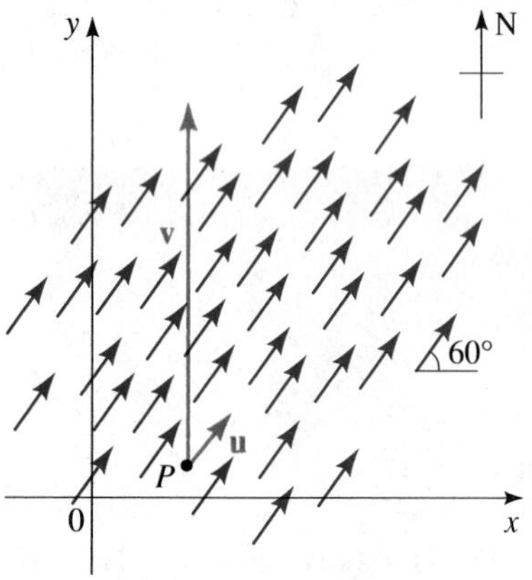

The **velocity** of a moving object is described by a vector whose direction is the direction of motion and whose magnitude is the speed. Figure 14 shows some vectors **u**, representing the velocity of wind flowing in the direction N 30° E, and a vector **v**, representing the velocity of an airplane flying through this wind at the point *P*. It's obvious from our experience that wind affects both the speed and the direction of an airplane. Figure 15 indicates that the true course of the plane (relative to the ground) is given by the vector **w** = **u** + **v**.

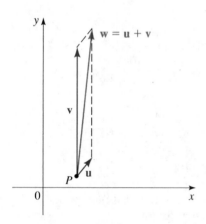

FIGURE 14

EXAMPLE 6 ■ The True Speed and Direction of an Airplane

An airplane heads due north at 300 mi/h. It experiences a 40 mi/h crosswind flowing in the direction N 30° E, as shown in Figure 14.

(a) Find the true velocity of the airplane as a vector.
(b) Find the true speed and direction of the airplane.

SOLUTION   The velocity of the airplane in still air is **v** = 0**i** + 300**j** = 300**j**. We must also write the velocity **u** of the wind as a vector. By the formulas for the component of a vector, we have

$$\mathbf{u} = (40 \cos 60°)\mathbf{i} + (40 \sin 60°)\mathbf{j}$$
$$= 20\mathbf{i} + 20\sqrt{3}\,\mathbf{j}$$
$$\approx 20\mathbf{i} + 34.64\mathbf{j}$$

(a) The true velocity of the airplane is given by the vector **w** = **u** + **v**.

$$\mathbf{w} = \mathbf{u} + \mathbf{v} = (20\mathbf{i} + 20\sqrt{3}\,\mathbf{j}) + (300\mathbf{j})$$
$$= 20\mathbf{i} + (20\sqrt{3} + 300)\mathbf{j}$$
$$\approx 20\mathbf{i} + 334.64\mathbf{j}$$

(b) The true speed of the airplane is given by the magnitude of **w**.

$$|\mathbf{w}| \approx \sqrt{(20)^2 + (334.64)^2} \approx 335.2 \text{ mi/h}$$

The direction of the airplane is the direction θ of the vector **w**. The angle θ has the property that $\tan \theta \approx 334.64/20 = 16.732$ and so $\theta \approx 86.6°$. Thus, the airplane is heading in the direction N 3.4° E.   ∎

FIGURE 15

EXAMPLE 7 ■ Calculating a Heading

A woman launches a boat from one shore of a straight river and wants to land at the point directly on the opposite shore. If the speed of the boat (in still water) is 10 mi/h and the river is flowing east at the rate of 5 mi/h, in what direction should she head the boat in order to arrive at the desired landing point?

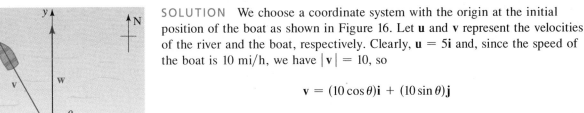

FIGURE 16

SOLUTION We choose a coordinate system with the origin at the initial position of the boat as shown in Figure 16. Let **u** and **v** represent the velocities of the river and the boat, respectively. Clearly, **u** = 5**i** and, since the speed of the boat is 10 mi/h, we have $|\mathbf{v}| = 10$, so

$$\mathbf{v} = (10\cos\theta)\mathbf{i} + (10\sin\theta)\mathbf{j}$$

where the angle θ is as shown in Figure 16. The true course of the boat is given by the vector **w** = **u** + **v**. We have

$$\mathbf{w} = \mathbf{u} + \mathbf{v} = 5\mathbf{i} + (10\cos\theta)\mathbf{i} + (10\sin\theta)\mathbf{j}$$

$$= (5 + 10\cos\theta)\mathbf{i} + (10\sin\theta)\mathbf{j}$$

Since the woman wants to land at a point directly across the river, her direction should have horizontal component 0. In other words, she should choose θ in such a way that

$$5 + 10\cos\theta = 0$$

$$\cos\theta = -\tfrac{1}{2}$$

$$\theta = 120°$$

Thus, she should head the boat in the direction $\theta = 120°$ (or N 30° W).   ■

Force is also represented by a vector. Intuitively, we can think of force as describing a push or a pull on an object, for example, a horizontal push of a book across a table or the downward pull of the earth's gravity on a ball. Force is measured in pounds (or in newtons, in the metric system). For instance, a man weighing 200 lb exerts a force of 200 lb downward on the ground. If several forces are acting on an object, the **resultant force** experienced by the object is the vector sum of these forces.

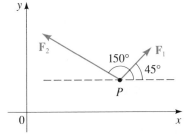

FIGURE 17

EXAMPLE 8 ■ Resultant Force

Two forces $\mathbf{F}_1$ and $\mathbf{F}_2$ with magnitudes 10 and 20 lb, respectively, act on an object at a point P as shown in Figure 17. Find the resultant force acting at P.

SOLUTION   We write $\mathbf{F}_1$ and $\mathbf{F}_2$ in terms of their components:

$$\mathbf{F}_1 = (10\cos 45°)\mathbf{i} + (10\sin 45°)\mathbf{j} = 10\,\frac{\sqrt{2}}{2}\,\mathbf{i} + 10\,\frac{\sqrt{2}}{2}\,\mathbf{j} = 5\sqrt{2}\,\mathbf{i} + 5\sqrt{2}\,\mathbf{j}$$

$$\mathbf{F}_2 = (20\cos 150°)\mathbf{i} + (20\sin 150°)\mathbf{j} = -20\,\frac{\sqrt{3}}{2}\,\mathbf{i} + 20\!\left(\frac{1}{2}\right)\!\mathbf{j}$$

$$= -10\sqrt{3}\,\mathbf{i} + 10\mathbf{j}$$

So the resultant force $\mathbf{F}$ is

$$\mathbf{F} = \mathbf{F}_1 + \mathbf{F}_2$$

$$= (5\sqrt{2}\,\mathbf{i} + 5\sqrt{2}\,\mathbf{j}) + (-10\sqrt{3}\,\mathbf{i} + 10\mathbf{j})$$

$$= (5\sqrt{2} - 10\sqrt{3}\,)\mathbf{i} + (5\sqrt{2} + 10)\mathbf{j}$$

$$\approx -10\mathbf{i} + 17\mathbf{j}$$

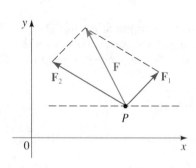

FIGURE 18

The resultant force $\mathbf{F}$ is shown in Figure 18.   ■

7.7   **EXERCISES**

1–6 ■ Sketch the vector indicated. (The vectors **u** and **v** are shown in the figure.)

1. $2\mathbf{u}$

2. $-\mathbf{v}$

3. $\mathbf{u} + \mathbf{v}$

4. $\mathbf{u} - \mathbf{v}$

5. $\mathbf{v} - 2\mathbf{u}$

6. $2\mathbf{u} + \mathbf{v}$

7–16 ■ Find the vector with initial point P and terminal point Q.

7.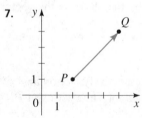

8.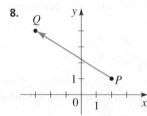

9.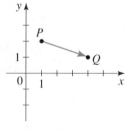

10.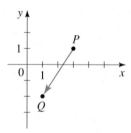

11. $P(3,2),\quad Q(8,9)$

12. $P(1,1),\quad Q(9,9)$

13. $P(5,3),\quad Q(1,0)$

14. $P(-1,3),\quad Q(-6,-1)$

15. $P(-1,-1),\quad Q(-1,1)$

16. $P(-8,-6),\quad Q(-1,-1)$

17–22 ■ Find $2\mathbf{u}$, $-3\mathbf{v}$, $\mathbf{u} + \mathbf{v}$, and $3\mathbf{u} - 4\mathbf{v}$ for the given vectors $\mathbf{u}$ and $\mathbf{v}$.

17. $\mathbf{u} = \langle 2, 7 \rangle$, $\mathbf{v} = \langle 3, 1 \rangle$

18. $\mathbf{u} = \langle -2, 5 \rangle$, $\mathbf{v} = \langle 2, -8 \rangle$

19. $\mathbf{u} = \langle 0, -1 \rangle$, $\mathbf{v} = \langle -2, 0 \rangle$

20. $\mathbf{u} = \mathbf{i}$, $\mathbf{v} = -2\mathbf{j}$

21. $\mathbf{u} = 2\mathbf{i}$, $\mathbf{v} = 3\mathbf{i} - 2\mathbf{j}$

22. $\mathbf{u} = \mathbf{i} + \mathbf{j}$, $\mathbf{v} = \mathbf{i} - \mathbf{j}$

23–26 ■ Find $|\mathbf{u}|$, $|\mathbf{v}|$, $|2\mathbf{u}|$, $\left|\frac{1}{2}\mathbf{v}\right|$, $|\mathbf{u} + \mathbf{v}|$, $|\mathbf{u} - \mathbf{v}|$, and $|\mathbf{u}| - |\mathbf{v}|$.

23. $\mathbf{u} = 2\mathbf{i} + \mathbf{j}$, $\mathbf{v} = 3\mathbf{i} - 2\mathbf{j}$

24. $\mathbf{u} = -2\mathbf{i} + 3\mathbf{j}$, $\mathbf{v} = \mathbf{i} - 2\mathbf{j}$

25. $\mathbf{u} = \langle 10, -1 \rangle$, $\mathbf{v} = \langle -2, -2 \rangle$

26. $\mathbf{u} = \langle -6, 6 \rangle$, $\mathbf{v} = \langle -2, -1 \rangle$

27–32 ■ Find the horizontal and vertical components of the vector with given length and direction, and write the vector as a linear combination of the vectors $\mathbf{i}$ and $\mathbf{j}$.

27. $|\mathbf{v}| = 40$, $\theta = 30°$ **28.** $|\mathbf{v}| = 50$, $\theta = 120°$

29. $|\mathbf{v}| = 1$, $\theta = 225°$ **30.** $|\mathbf{v}| = 800$, $\theta = 125°$

31. $|\mathbf{v}| = 4$, $\theta = 10°$ **32.** $|\mathbf{v}| = \sqrt{3}$, $\theta = 300°$

33. A man pushes a lawn mower with a force of 30 lb exerted at an angle of 30° to the ground. Find the horizontal and vertical components of the force.

34. A jet is flying in a direction N 20° E with a speed of 500 mi/h. Find the north and east components of the velocity.

35–40 ■ Find the magnitude and direction (in degrees) of the vector.

35. $\mathbf{v} = \langle 3, 4 \rangle$

36. $\mathbf{v} = \left\langle -\dfrac{\sqrt{2}}{2}, -\dfrac{\sqrt{2}}{2} \right\rangle$

37. $\mathbf{v} = \langle -12, 5 \rangle$

38. $\mathbf{v} = \langle 40, 9 \rangle$

39. $\mathbf{v} = \mathbf{i} + \sqrt{3}\,\mathbf{j}$

40. $\mathbf{v} = \mathbf{i} + \mathbf{j}$

41. A pilot heads his jet due east. The jet has a speed of 425 mi/h in still air. The wind is blowing due north with a speed of 40 mi/h.
(a) Find the true velocity of the jet as a vector.
(b) Find the true speed and direction of the jet.

42. A jet is flying through a wind that is blowing with a speed of 55 mi/h in the direction N 30° E. The jet has a speed of 765 mi/h in still air, and the pilot heads the jet in the direction N 45° E.
(a) Find the true velocity of the jet as a vector.
(b) Find the true speed and direction of the jet.

43. Find the true speed and direction of the jet in Exercise 42 if the pilot heads the plane in the direction N 30° W.

44. In what direction should the pilot in Exercise 42 head the plane in order for the true course to be due north?

45. A straight river flows east at a speed of 10 mi/h. A boater starts at the south shore of the river and heads in a direction 60° from the shore (see the figure). The motorboat has a speed of 20 mi/h in still water.
(a) Find the true velocity of the motorboat as a vector.
(b) Find the true speed and direction of the motorboat.

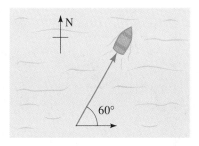

46. The boater in Exercise 45 wants to arrive at a point on the north shore of the river directly opposite the starting point. In what direction should the boat be headed?

47. A boat heads in the direction N 72° E. The speed of the boat in still water is 24 mi/h. The water is flowing directly south. It is observed that the true direction of the boat is directly east. Find the speed of the water and the true speed of the boat.

48. A woman walks due west on the deck of an ocean liner at 2 mi/h. The ocean liner is moving due north at a speed of 25 mi/h. Find the speed and direction of the woman relative to the surface of the water.

49–54 ■ The forces $\mathbf{F}_1$, $\mathbf{F}_2$, ..., $\mathbf{F}_n$ acting at the same point P are said to be in equilibrium if the resultant force is zero, that is, if $\mathbf{F}_1 + \mathbf{F}_2 + \cdots + \mathbf{F}_n = \mathbf{0}$. Find (a) the resultant forces acting at P, and (b) the additional force required (if any) in order for the forces to be in equilibrium.

49. $\mathbf{F}_1 = \langle 2, 5 \rangle, \quad \mathbf{F}_2 = \langle 3, -8 \rangle$

50. $\mathbf{F}_1 = \langle 3, -7 \rangle, \quad \mathbf{F}_2 = \langle 4, -2 \rangle, \quad \mathbf{F}_3 = \langle -7, 9 \rangle$

51. $\mathbf{F}_1 = 4\mathbf{i} - \mathbf{j}, \quad \mathbf{F}_2 = 3\mathbf{i} - 7\mathbf{j},$
$\mathbf{F}_3 = -8\mathbf{i} + 3\mathbf{j}, \quad \mathbf{F}_4 = \mathbf{i} + \mathbf{j}$

52. $\mathbf{F}_1 = \mathbf{i} - \mathbf{j}, \quad \mathbf{F}_2 = \mathbf{i} + \mathbf{j}, \quad \mathbf{F}_3 = -2\mathbf{i} + \mathbf{j}$

53.

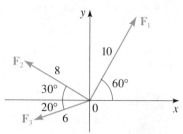

54.

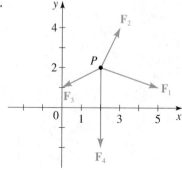

55. A 100-lb weight hangs from a string as shown in the figure. Find the tensions $\mathbf{T}_1$ and $\mathbf{T}_2$ in the string.

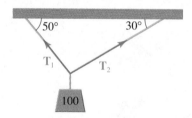

56. The cranes in the figure are lifting an object that weighs 18,278 lb. Find the tensions $\mathbf{T}_1$ and $\mathbf{T}_2$.

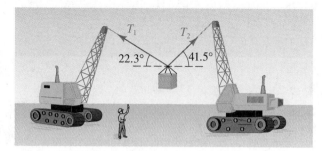

![DISCOVERY icon] **DISCOVERY · DISCUSSION**

57. Vectors that Form a Polygon   Suppose that n vectors can be placed head to tail in the plane so that they form a polygon. (The figure shows the case of a hexagon.) Explain why the sum of these vectors is $\mathbf{0}$.

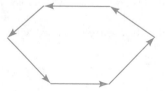

| 7 | REVIEW |

CONCEPT CHECK

1. (a) State the reciprocal identities.
   (b) State the Pythagorean identities.
   (c) State the even-odd identities.
   (d) State the cofunction identities.

2. Explain the difference between an equation and an identity.

3. How do you get started proving a trigonometric identity?

4. (a) State the addition formulas for sine, cosine, and tangent.
   (b) State the subtraction formulas for sine, cosine, and tangent.

5. (a) State the double-angle formulas for sine, cosine, and tangent.
(b) State the formulas for lowering powers.
(c) State the half-angle formulas.

6. (a) State the product-to-sum formulas.
(b) State the sum-to-product formulas.

7. (a) Define the inverse sine function $\sin^{-1}$. What are its domain and range?
(b) For what values of x is the equation $\sin(\sin^{-1}x) = x$ true?
(c) For what values of x is the equation $\sin^{-1}(\sin x) = x$ true?

8. (a) Define the inverse cosine function $\cos^{-1}$. What are its domain and range?
(b) For what values of x is the equation $\cos(\cos^{-1}x) = x$ true?
(c) For what values of x is the equation $\cos^{-1}(\cos x) = x$ true?

9. (a) Define the inverse tangent function $\tan^{-1}$. What are its domain and range?
(b) For what values of x is the equation $\tan(\tan^{-1}x) = x$ true?

(c) For what values of x is the equation $\tan^{-1}(\tan x) = x$ true?

10. What is the trigonometric form of a complex number z? What is the modulus of z? What is the argument of z?

11. (a) How do you multiply two complex numbers if they are given in trigonometric form?
(b) How do you divide two such numbers?

12. (a) State DeMoivre's Theorem.
(b) How do you find the nth roots of a complex number?

13. (a) What is the difference between a scalar and a vector?
(b) Draw a diagram to show how to add two vectors.
(c) Draw a diagram to show how to subtract two vectors.
(d) Draw a diagram to show how to multiply a vector by the scalars 2, $\frac{1}{2}$, -2, and $-\frac{1}{2}$.

14. If $\mathbf{u} = \langle a_1, b_1 \rangle$, $\mathbf{v} = \langle a_2, b_2 \rangle$ and c is a scalar, write expressions for $\mathbf{u} + \mathbf{v}$, $\mathbf{u} - \mathbf{v}$, $c\mathbf{u}$, and $|\mathbf{u}|$.

15. (a) If $\mathbf{v} = \langle a, b \rangle$, write $\mathbf{v}$ in terms of $\mathbf{i}$ and $\mathbf{j}$.
(b) Write the components of $\mathbf{v}$ in terms of the magnitude and direction of $\mathbf{v}$.

EXERCISES

1–22 ■ Verify the identity.

1. $\cos^2 x \csc x - \csc x = -\sin x$

2. $\dfrac{1}{1 - \sin^2 x} = 1 + \tan^2 x$

3. $\dfrac{\cos^2 x - \tan^2 x}{\sin^2 x} = \cot^2 x - \sec^2 x$

4. $\dfrac{1 + \sec x}{\sec x} = \dfrac{\sin^2 x}{1 - \cos x}$

5. $\dfrac{\cos^2 x}{1 - \sin x} = \dfrac{\cos x}{\sec x - \tan x}$

6. $(1 - \tan x)(1 - \cot x) = 2 - \sec x \csc x$

7. $\sin^2 x \cot^2 x + \cos^2 x \tan^2 x = 1$

8. $(\tan x + \cot x)^2 = \csc^2 x \sec^2 x$

9. $\dfrac{\sin 2x}{1 + \cos 2x} = \tan x$

10. $\dfrac{\cos(x + y)}{\cos x \sin y} = \cot y - \tan x$

11. $\tan \dfrac{x}{2} = \csc x - \cot x$

12. $\dfrac{\sin(x + y) + \sin(x - y)}{\cos(x + y) + \cos(x - y)} = \tan x$

13. $\sin(x + y)\sin(x - y) = \sin^2 x - \sin^2 y$

14. $\csc x - \tan \dfrac{x}{2} = \cot x$

15. $1 + \tan x \tan \dfrac{x}{2} = \sec x$

16. $\dfrac{\sin 3x + \cos 3x}{\cos x - \sin x} = 1 + 2 \sin 2x$

17. $\left(\cos \dfrac{x}{2} - \sin \dfrac{x}{2} \right)^2 = 1 - \sin x$

18. $\dfrac{\cos 3x - \cos 7x}{\sin 3x + \sin 7x} = \tan 2x$

19. $\dfrac{\sin 2x}{\sin x} - \dfrac{\cos 2x}{\cos x} = \sec x$

20. $(\cos x + \cos y)^2 + (\sin x - \sin y)^2 = 2 + 2\cos(x + y)$

21. $\tan\left(x + \dfrac{\pi}{4}\right) = \dfrac{1 + \tan x}{1 - \tan x}$

22. $\dfrac{\sec x - 1}{\sin x \sec x} = \tan \dfrac{x}{2}$

23–26 ■ (a) Graph f and g. (b) Do the graphs suggest that the equation $f(x) = g(x)$ is an identity? Prove your answer.

23. $f(x) = 1 - \left(\cos \dfrac{x}{2} - \sin \dfrac{x}{2}\right)^2$, $g(x) = \sin x$

24. $f(x) = \sin x + \cos x$, $g(x) = \sqrt{\sin^2 x + \cos^2 x}$

25. $f(x) = \tan x \tan \dfrac{x}{2}$, $g(x) = \dfrac{1}{\cos x}$

26. $f(x) = 1 - 8\sin^2 x + 8\sin^4 x$, $g(x) = \cos 4x$

27–28 ■ (a) Graph the function(s) and make a conjecture, and (b) prove your conjecture.

27. $f(x) = 2\sin^2 3x + \cos 6x$

28. $f(x) = \sin x \cot \dfrac{x}{2}$, $g(x) = \cos x$

29–44 ■ Solve the equation in the interval $[0, 2\pi)$.

29. $\cos x \sin x - \sin x = 0$

30. $\sin x - 2\sin^2 x = 0$

31. $2\sin^2 x - 5\sin x + 2 = 0$

32. $\sin x - \cos x - \tan x = -1$

33. $2\cos^2 x - 7\cos x + 3 = 0$

34. $4\sin^2 x + 2\cos^2 x = 3$

35. $\dfrac{1 - \cos x}{1 + \cos x} = 3$ **36.** $\sin x = \cos 2x$

37. $\tan^3 x + \tan^2 x - 3\tan x - 3 = 0$

38. $\cos 2x \csc^2 x = 2\cos 2x$

39. $\tan \frac{1}{2}x + 2\sin 2x = \csc x$

40. $\cos 3x + \cos 2x + \cos x = 0$

41. $\tan x + \sec x = \sqrt{3}$

42. $2\cos x - 3\tan x = 0$

43. $\cos x = x^2 - 1$

44. $e^{\sin x} = x$

45–54 ■ Find the exact value of the expression.

45. $\cos 15°$ **46.** $\sin \dfrac{5\pi}{12}$

47. $\tan \dfrac{\pi}{8}$ **48.** $2\sin \dfrac{\pi}{12} \cos \dfrac{\pi}{12}$

49. $\sin 5° \cos 40° + \cos 5° \sin 40°$

50. $\dfrac{\tan 66° - \tan 6°}{1 + \tan 66° \tan 6°}$

51. $\cos^2 \dfrac{\pi}{8} - \sin^2 \dfrac{\pi}{8}$ **52.** $\dfrac{1}{2}\cos \dfrac{\pi}{12} + \dfrac{\sqrt{3}}{2}\sin \dfrac{\pi}{12}$

53. $\cos 37.5° \cos 7.5°$ **54.** $\cos 67.5° + \cos 22.5°$

55–60 ■ Find the exact value of the expression given that $\sec x = \frac{3}{2}$, $\csc y = 3$, and x and y are in quadrant I.

55. $\sin(x + y)$ **56.** $\cos(x - y)$

57. $\tan(x + y)$ **58.** $\sin 2x$

59. $\cos \dfrac{y}{2}$ **60.** $\tan \dfrac{y}{2}$

61–68 ■ Find the exact value of the expression.

61. $\sin^{-1}(\sqrt{3}/2)$ **62.** $\tan^{-1}(\sqrt{3}/3)$

63. $\cos(\tan^{-1}\sqrt{3}\,)$ **64.** $\sin(\cos^{-1}(\sqrt{3}/2))$

65. $\tan(\sin^{-1}\frac{2}{5})$ **66.** $\sin(\cos^{-1}\frac{3}{8})$

67. $\cos(2\sin^{-1}\frac{1}{3})$ **68.** $\cos(\sin^{-1}\frac{5}{13} - \cos^{-1}\frac{4}{5})$

69–70 ■ Rewrite the expression as an algebraic function of x.

69. $\sin(\tan^{-1}x)$ **70.** $\sec(\sin^{-1}x)$

71–72 ■ Express θ in terms of x.

71.   **72.**

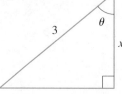

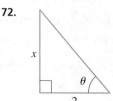

73. A satellite is in orbit at an altitude h above the earth. Compute the arc distance s between points P and Q, which represent the farthest points the satellite can "see." Express s as a function of h. The radius of the earth is 3960 mi. [*Hint:* First express s in terms of θ.]

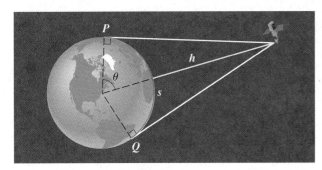

74. How high does the satellite in Exercise 73 have to be in order to see both Los Angeles and New York, 2450 mi apart?

75–80 ■ Write the complex number in trigonometric form with argument between 0 and 2π.

75. $4 + 4i$

76. $-10i$

77. $5 + 3i$

78. $1 + \sqrt{3}\,i$

79. $-1 + i$

80. -20

81–84 ■ Find the indicated power.

81. $(1 - \sqrt{3}\,i)^4$

82. $(1 + i)^8$

83. $(\sqrt{3} + i)^{-4}$

84. $\left(\dfrac{1}{2} + \dfrac{\sqrt{3}}{2}i\right)^{20}$

85–88 ■ Find the indicated roots.

85. The square roots of $-16i$

86. The cube roots of $4 + 4\sqrt{3}\,i$

87. The sixth roots of 1

88. The eighth roots of i

89–90 ■ Find $|\mathbf{u}|$, $\mathbf{u} + \mathbf{v}$, $\mathbf{u} - \mathbf{v}$, $2\mathbf{u}$, and $3\mathbf{u} - 2\mathbf{v}$.

89. $\mathbf{u} = \langle -2, 3 \rangle$, $\quad \mathbf{v} = \langle 8, 1 \rangle$

90. $\mathbf{u} = 2\mathbf{i} + \mathbf{j}$, $\quad \mathbf{v} = \mathbf{i} - 2\mathbf{j}$

91. Find the vector $\mathbf{u}$ with initial point $P(0, 3)$ and terminal point $Q(3, -1)$.

92. Find the vector $\mathbf{u}$ having length $|\mathbf{u}| = 20$ and direction $\theta = 60°$.

93. If the vector $5\mathbf{i} - 8\mathbf{j}$ is placed in the plane with its initial point at $P(5, 6)$, find its terminal point.

94. Find the direction of the vector $2\mathbf{i} - 5\mathbf{j}$.

95. Two tugboats are pulling a barge, as shown. One pulls with a force of 2.0×10^4 lb in the direction N 50° E and the other with a force of 3.4×10^4 lb in the direction S 75° E.

(a) Find the resultant force on the barge as a vector.

(b) Find the magnitude and direction of the resultant force.

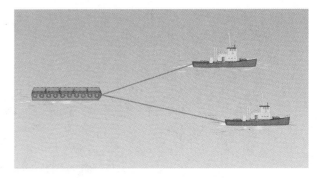

96. An airplane heads N 60° E at 600 mi/h in still air. A tail wind begins to blow in the direction N 30° E at 50 mi/h.

(a) Find the velocity of the airplane as a vector.

(b) Find the true speed and direction of the airplane.

1. Verify each identity.

(a) $\dfrac{\tan x}{1 - \cos x} = \csc x\,(1 + \sec x)$ (b) $\dfrac{2\tan x}{1 + \tan^2 x} = \sin 2x$

2. Solve each trigonometric equation in the interval $[0, 2\pi)$.

(a) $2\cos^2 x + 5\cos x + 2 = 0$ (b) $\sin 2x - \cos x = 0$

3. Find all solutions in the interval $[0, 2\pi)$, correct to five decimal places:

$$5\cos 2x = 2$$

4. Let $x = 2\sin\theta$, $-\pi/2 < \theta < \pi/2$. Simplify the expression

$$\frac{x}{\sqrt{4 - x^2}}$$

5. Find the exact value of each expression.

(a) $\sin 8° \cos 22° + \cos 8° \sin 22°$ (b) $\sin 75°$ (c) $\sin \dfrac{\pi}{12}$

6. For the angles α and β in the figures, find $\cos(\alpha + \beta)$.

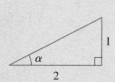

7. (a) Write $\sin 3x \cos 5x$ as a sum of trigonometric functions.

(b) Write $\sin 2x - \sin 5x$ as a product of trigonometric functions.

8. If $\sin\theta = -\frac{4}{5}$ and θ is in quadrant III, find $\tan(\theta/2)$.

9. Sketch the graphs of $y = \sin x$ and $y = \sin^{-1}x$, and specify the domain of each function.

10. Express θ in each figure in terms of x.

(a)   (b)

 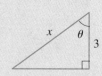

11. Find the exact value of $\cos(\tan^{-1}\frac{9}{40})$.

12. Let $z = 1 + \sqrt{3}\,i$.

(a) Write z in trigonometric form.

(b) Find the complex number z^9.

13. Let

$$z_1 = 4\left(\cos\frac{7\pi}{12} + i\sin\frac{7\pi}{12}\right)$$

$$z_2 = 2\left(\cos\frac{5\pi}{12} + i\sin\frac{5\pi}{12}\right)$$

Find $z_1 z_2$ and $\dfrac{z_1}{z_2}$.

14. Find the cube roots of $27i$, and sketch these roots in the complex plane.

15. Let $\mathbf{u}$ be the vector with initial point $P(3, -1)$ and terminal point $Q(-3, 9)$.
(a) Express $\mathbf{u}$ in terms of $\mathbf{i}$ and $\mathbf{j}$.
(b) Find the length of $\mathbf{u}$.

16. Let $\mathbf{u} = \langle 1, 3\rangle$ and $\mathbf{v} = \langle -6, 2\rangle$.
(a) Find $\mathbf{u} - 3\mathbf{v}$.
(b) Find $|\mathbf{u} + \mathbf{v}|$.

17. A river is flowing due east at 8 mi/h. A man heads his motorboat in a direction N 30° E in the river. The speed of the motorboat in still water is 12 mi/h.
(a) Express the true velocity of the motorboat as a vector.
(b) Find the true speed and direction of the motorboat.

FOCUS ON PROBLEM SOLVING

One way to show that a statement is true is to show that its opposite leads to a contradiction. This type of **indirect reasoning** is an important problem-solving tool. The two examples we give here are problems that were solved more than 2000 years ago but continue to have a profound influence in mathematics.

IRRATIONAL NUMBERS EXIST

The Pythagoreans were fascinated by the beauty of the natural numbers: 1, 2, 3, ... They hoped that the length of every interval constructed in geometry would be a ratio of two natural numbers so that the natural numbers would describe all of geometry.

But what is the length of the diagonal of a square of side 1? The Pythagoreans knew (by the Pythagorean Theorem) that this length is $\sqrt{2}$. The question was, Is $\sqrt{2}$ the ratio of two natural numbers? One of the Pythagoreans, Hippasus, is reputed to have proved, during a sea voyage, that $\sqrt{2}$ is in fact *not* rational. His fellow Pythagoreans were so upset by this discovery that they hurled him overboard! His proof, nevertheless, is a classic use of indirect reasoning. Here is the proof:

Suppose $\sqrt{2}$ is rational, so that

$$\sqrt{2} = \frac{a}{b}$$

where a and b are natural numbers with no factor in common. Then

$$a = \sqrt{2}\, b \qquad \text{Multiply by } b$$

$$a^2 = 2b^2 \qquad \text{Square each side}$$

This means that a^2 is an even number and so a is an even number, say $a = 2m$. So, from the preceding equation we get

$$(2m)^2 = 2b^2 \qquad \text{Substitute } a = 2m$$

$$4m^2 = 2b^2$$

$$2m^2 = b^2 \qquad \text{Divide by 2}$$

Thus, b is also even. So a and b have 2 as a common factor, and this contradicts our assumption that a and b have no factor in common. The assumption that $\sqrt{2}$ is rational therefore leads to a contradiction, and so $\sqrt{2}$ must be irrational.

THERE ARE INFINITELY MANY PRIMES

A prime number is one that has no factor other than 1 and itself. The first primes are

$$2,\ 3,\ 5,\ 7,\ 11,\ 13,\ 17,\ 19,\ 23,\ 29,\ 31,\ \ldots$$

How do we find the next prime? No formula is known that will produce the primes, and no one has found a pattern for the location of the primes. One thing that is known is that there are infinitely many primes. Euclid gave a proof for this fact over 2000 years ago by a brilliant use of indirect reasoning. Here is his famous proof:

Suppose there is only a finite number of primes. We list them as

$$p_1,\ p_2,\ p_3,\ \ldots,\ p_n$$

Then the number

$$N = p_1 \cdot p_2 \cdot p_3 \cdot \cdots \cdot p_n + 1$$

is not divisible by any prime (Why?) and so is itself prime. But N is not in our list of primes. (Why?) This contradicts our assumption that the list contains all the primes. Thus, there are infinitely many primes.

Eratosthenes (circa 276–195 B.C.) was a renowned Greek geographer, mathematician, and astronomer. He accurately calculated the circumference of the earth by an ingenious method. He is most famous, however, for his method for finding primes, now called the *sieve of Eratosthenes*. The method consists of listing the integers, beginning with 2 (the first prime), and then crossing out all the multiples of 2, which are not prime. The next number remaining on the list is 3 (the second prime), so we again cross out all multiples of it. The next remaining number is 5 (the third prime number), and we cross out all multiples of it, and so on. In this way all numbers that are not prime are crossed out, and the remaining numbers are the primes.

PROBLEMS

1. (a) Prove that $\sqrt{6}$ is an irrational number.
   (b) Prove that $\sqrt{2} + \sqrt{3}$ is an irrational number.

2. Prove that $\log_2 5$ is irrational.

3. Prove that it's possible to raise an irrational number to an irrational power and get a rational result. [*Hint:* The number $a = \sqrt{2}^{\sqrt{2}}$ is either rational or irrational. If a is rational, we are done. If a is irrational, consider $a^{\sqrt{2}}$.]

4. Each letter in the following multiplication represents a different digit. Find the value of each letter.

$$
\begin{array}{r}
ABCDE \\
\times\ 4 \\
\hline
EDCBA
\end{array}
$$

5. Prove that at any party there are two people who know the same number of people. (Assume that if person A knows person B, then B knows A. Assume also that everyone knows himself or herself.)

6. Of nine eggs, eight have exactly the same weight. The ninth egg weighs less than the other eight. How can you determine which is the lighter egg with exactly two weighings on a balance?

7. Of 12 similar coins, one is a counterfeit. It is not known whether the counterfeit coin is lighter or heavier than a genuine coin. Using a balance three times, how can the counterfeit be identified and in the process be determined to be lighter or heavier than a genuine coin?

8. Augustus DeMorgan, the famous 19th-century logician, once stated that he was x years old in the year x^2. He died at age 65. In what year did he die?

9. Find the number of solutions of the equation $\sin x = \dfrac{x}{100}$.

10. Three tangent circles of radius 10 cm are drawn. All centers lie on the line AB. The tangent AC to the right-hand circle is drawn, intersecting the middle circle at D and E. Find the length of the segment DE.

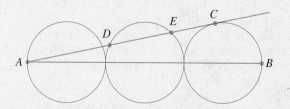

11. A ball of radius a is set inside a cone so that the surface of the cone is tangent to the ball. A larger ball of radius b fits inside the cone in such a way that it is tangent to both the ball of radius a and the sides of the cone. Express b in terms of a and the angle θ shown in the figure.

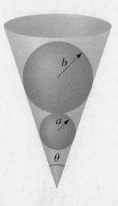

12. Points (m, n) in the coordinate plane, both of whose coordinates are integers, are called *lattice points*. Show that it's impossible for any equilateral triangle to have each of its vertices at a lattice point. [*Hint:* See Section 7.2, Exercise 42(b).]

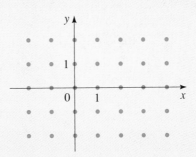

Lattice points in the plane

13. Find $\angle A + \angle B + \angle C$ in the figure.

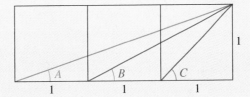

14. If $0 < \theta < \pi/2$ and $\sin 2\theta = a$, find $\sin \theta + \cos \theta$.

15. Show that $\pi/4 = \tan^{-1}\frac{1}{2} + \tan^{-1}\frac{1}{5} + \tan^{-1}\frac{1}{8}$.

   NOTE: This identity was used by Zacharias Dase in 1844 to calculate the decimal expansion of π to 200 places.

16. Evaluate $(\log_2 3)(\log_3 4)(\log_4 5) \cdots (\log_{31} 32)$.

17. Show that if $x > 0$ and $x \neq 1$, then

$$\frac{1}{\log_2 x} + \frac{1}{\log_3 x} + \frac{1}{\log_5 x} = \frac{1}{\log_{30} x}$$

18. A red ribbon is tied tightly around the earth at the equator. How much more ribbon would you need if you raised the ribbon 1 ft above the equator everywhere? (You don't need to know the radius of the earth to solve this problem.)

19. Find the area of the region between the two concentric circles shown in the figure.

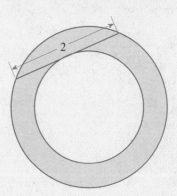

20. In a singles "knock-out" tennis tournament, a player is eliminated as soon as he or she loses a match. A prearranged schedule determines who plays whom initially, and the winner of each match advances to the next round, as shown in the following sample.

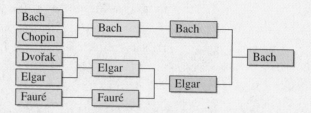

If an odd number of players participates in any round, then the round will have at least one *bye*, that is, a player who doesn't play but advances automatically to the next round. Suppose n players participate in the tournament.

(a) For what values of n is it possible to avoid a bye?

(b) Show that exactly $n - 1$ matches must be played, no matter how the schedule is arranged. [*Hint:* There's an easy way to see this. Consider the losers of matches.]

21. The ancient Egyptians considered rectangles whose perimeters and areas were numerically the same to be special. Find all rectangles whose perimeter (in feet) and whose area (in ft^2) are the same integer.

22. The Indian mathematician Bhaskara sketched the two figures shown here and wrote below them, "Behold!" Explain how his sketches prove the Pythagorean Theorem.

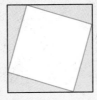

23. If the lengths of the sides of a right triangle, in increasing order, are a, b, and c, show that $a^3 + b^3 < c^3$.

8

SYSTEMS OF EQUATIONS AND INEQUALITIES

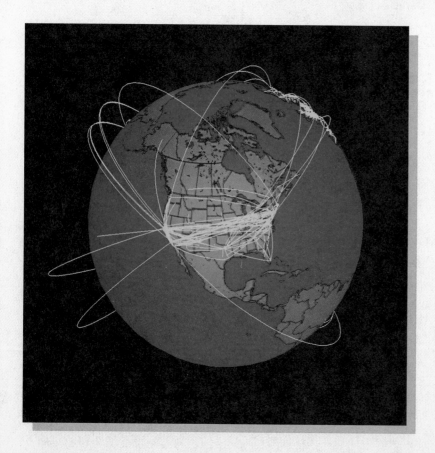

Systems of equations and inequalities can be used to determine how to allocate resources most effectively. These techniques are used, for example, to find the most efficient routing for a telephone call or an internet connection, or to determine the optimal configuration for an airline schedule.

Mathematics is the key and door to the sciences.

GALILEO GALILEI

Many of the problems to which we can apply the techniques of algebra give rise to sets of equations with several variables, rather than to just a single equation in a single variable. A set of equations with common variables is called a *system of equations*. In this chapter we develop techniques for finding simultaneous solutions of systems. We first consider pairs of equations with two unknowns, the simplest case of this situation. To help us solve *linear* equations in an arbitrary number of variables, we study the algebra of matrices and determinants. We also study systems of inequalities, and in the *Focus on Modeling* section we study linear programming, which is an optimization technique used widely in business and the social sciences.

8.1 | SYSTEMS OF EQUATIONS

In this section we study how to solve systems of two equations in two unknowns. Here is an example of such a system:

$$\begin{cases} 2x - y = 0 \\ 3x + 2y = 14 \end{cases}$$

To solve this system means to find values for x and y that satisfy *both* equations. A solution of this system is $x = 2$, $y = 4$ because

$$\begin{cases} 2(2) - (4) = 0 \\ 3(2) + 2(4) = 14 \end{cases}$$

We write the solution as the ordered pair $(2, 4)$. Graphically, the solutions of a system are the points where the graphs of the two equations intersect (see Figure 1).

To solve a system we may use the **substitution method** or the **elimination method**, or we may solve the system **graphically**. We study all three methods in this section.

In the substitution method we start with one equation in the system and solve for one variable in terms of the other variable. The following box describes the procedure.

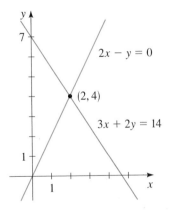

FIGURE 1

SUBSTITUTION METHOD

1. SOLVE FOR ONE OF THE VARIABLES. Use one of the equations to solve for one variable in terms of the other variable.

2. SUBSTITUTE. Substitute the expression you found in Step 1 into the other equation to get an equation in one variable, then solve for that variable.

3. BACK-SUBSTITUTE. Substitute the value you found in Step 2 back into the expression found in Step 1 to solve for the remaining variable.

EXAMPLE 1 ■ Substitution Method

Find all solutions of the following system.

$$\begin{cases} 2x + y = 1 \\ 3x + 4y = 14 \end{cases}$$

SOLUTION   We solve for y in the first equation.

$$y = 1 - 2x \qquad \text{Solve for } y \text{ in the first equation}$$

Now we substitute for y in the second equation and solve for x:

$$3x + 4(1 - 2x) = 14 \qquad \text{Substitute } y = 1 - 2x \text{ into the second equation}$$

$$3x + 4 - 8x = 14 \qquad \text{Expand}$$

$$-5x + 4 = 14 \qquad \text{Simplify}$$

$$-5x = 10 \qquad \text{Subtract 4}$$

$$x = -2 \qquad \text{Solve for } x$$

Next we back-substitute $x = -2$ into the equation $y = 1 - 2x$:

$$y = 1 - 2(-2) = 5 \qquad \text{Back-substitute}$$

Thus, $x = -2$ and $y = 5$, so the solution is the ordered pair $(-2, 5)$. Figure 2 shows that the graphs of the two equations are two lines intersecting at the point $(-2, 5)$.   ■

EXAMPLE 2 ■ Substitution Method

Find all solutions of the following system.

$$\begin{cases} x^2 - y = 2 \\ 2x - y = -1 \end{cases}$$

SOLUTION   We start by solving for y in the second equation.

$$y = 2x + 1 \qquad \text{Solve for } y \text{ in the second equation}$$

Next we substitute for y in the first equation and solve for x:

$$x^2 - (2x + 1) = 2 \qquad \text{Substitute } y = 2x + 1 \text{ into the first equation}$$

$$x^2 - 2x - 1 = 2 \qquad \text{Expand}$$

$$x^2 - 2x - 3 = 0 \qquad \text{Subtract 2}$$

$$(x + 1)(x - 3) = 0 \qquad \text{Factor}$$

$$x = -1 \qquad \text{or} \qquad x = 3 \qquad \text{Solve for } x$$

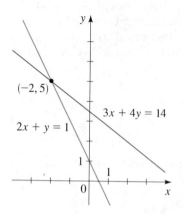

FIGURE 2

CHECK YOUR ANSWER

$x = -2, y = 5$:

$$\begin{cases} 2(-2) + 5 = 1 \\ 3(-2) + 4(5) = 14 \end{cases} \quad \checkmark$$

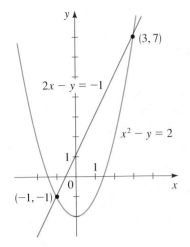

FIGURE 3

Now we back-substitute these values of x into the equation $y = 2x + 1$.

For $x = -1$: $y = 2(-1) + 1 = -1$ Back-substitute

For $x = 3$: $y = 2(3) + 1 = 7$ Back-substitute

So we have two solutions: $(-1, -1)$ and $(3, 7)$.

The graph of the first equation is a parabola, and the graph of the second equation is a line; Figure 3 shows that the graphs intersect at the two points $(-1, -1)$ and $(3, 7)$.

CHECK YOUR ANSWERS

$x = -1, y = -1$:
$$\begin{cases} (-1)^2 - (-1) = 2 \\ 2(-1) - (-1) = -1 \end{cases} \checkmark$$

$x = 3, y = 7$:
$$\begin{cases} 3^2 - 7 = 2 \\ 2(3) - 7 = -1 \end{cases} \checkmark$$

EXAMPLE 3 ■ Substitution Method

Find all solutions of the following system.

$$\begin{cases} x^2 + y^2 = 100 \\ 3x - y = 10 \end{cases}$$

SOLUTION   We start by solving for y in the second equation.

$$y = 3x - 10 \qquad \text{Solve for } y \text{ in the second equation}$$

Next we substitute for y in the first equation and solve for x:

$$x^2 + (3x - 10)^2 = 100 \qquad \begin{array}{l}\text{Substitute } y = 3x - 10 \\ \text{into the first equation}\end{array}$$

$$x^2 + (9x^2 - 60x + 100) = 100 \qquad \text{Expand}$$

$$10x^2 - 60x = 0 \qquad \text{Simplify}$$

$$10x(x - 6) = 0 \qquad \text{Factor}$$

$$x = 0 \quad \text{or} \quad x = 6 \qquad \text{Solve for } x$$

Now we back-substitute these values of x into the equation $y = 3x - 10$.

For $x = 0$: $y = 3(0) - 10 = -10$ Back-substitute

For $x = 6$: $y = 3(6) - 10 = 8$ Back-substitute

So we have two solutions: $(0, -10)$ and $(6, 8)$.

The graph of the first equation is a circle, and the graph of the second equation is a line; Figure 4 shows that the graphs intersect at the two points $(0, -10)$ and $(6, 8)$.

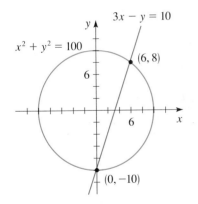

FIGURE 4

CHECK YOUR ANSWERS

$x = 0, y = -10$:
$$\begin{cases} (0)^2 + (-10)^2 = 100 \\ 3(0) - (-10) = 10 \end{cases} \checkmark$$

$x = 6, y = 8$:
$$\begin{cases} (6)^2 + (8)^2 = 36 + 64 = 100 \\ 3(6) - (8) = 18 - 8 = 10 \end{cases} \checkmark$$

To solve a system using the **elimination method**, we try to combine the equations using sums or differences so as to eliminate one of the variables.

ELIMINATION METHOD

1. ADJUST THE COEFFICIENTS.   Multiply one or more of the equations by appropriate numbers so that the coefficient of one of the variables in one equation is the negative of its coefficient in the other equation.

2. ADD THE EQUATIONS.   Add the two equations to eliminate one of the variables, then solve for the remaining variable.

3. BACK-SUBSTITUTE.   Substitute the value you found in Step 2 back into one of the original equations, and solve for the remaining variable.

EXAMPLE 4 ■ Elimination Method

Find all solutions of the following system.

$$\begin{cases} 3x^2 + 2y = 26 \\ 5x^2 + 7y = 3 \end{cases}$$

SOLUTION   We choose to eliminate the x-term, so we multiply the first equation by 5 and the second equation by -3. Then we add the two equations and solve for y.

$$\begin{cases} 15x^2 + 10y = 130 \qquad \text{First equation times 5} \\ -15x^2 - 21y = -9 \qquad \text{Second equation times } -3 \end{cases}$$

$$-11y = 121 \qquad \text{Add}$$
$$y = -11 \qquad \text{Solve for } y$$

Now we back-substitute $y = -11$ into one of the original equations, say $3x^2 + 2y = 26$, and solve for x:

$$3x^2 + 2(-11) = 26 \qquad \text{Back-substitute } y = -11 \text{ into the first equation}$$
$$3x^2 = 48 \qquad \text{Add 22}$$
$$x^2 = 16 \qquad \text{Divide by 3}$$
$$x = -4 \quad \text{or} \quad x = 4 \qquad \text{Solve for } x$$

So we have two solutions: $(-4, -11)$ and $(4, -11)$.

The graphs of both equations are parabolas; Figure 5 shows that the graphs intersect at the two points $(-4, -11)$ and $(4, -11)$.

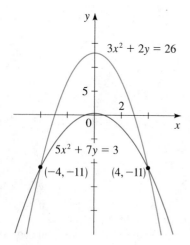

FIGURE 5

CHECK YOUR ANSWERS

$x = -4, y = -11$:

$$\begin{cases} 3(-4)^2 + 2(-11) = 26 \\ 5(-4)^2 + 7(-11) = 3 \end{cases} \checkmark$$

$x = 4, y = -11$:

$$\begin{cases} 3(4)^2 + 2(-11) = 26 \\ 5(4)^2 + 7(-11) = 3 \end{cases} \checkmark$$

■

EXAMPLE 5 ■ **Elimination Method**

Find all solutions of the following system.

$$\begin{cases} x^2 + 2y^2 = 11 \\ 3x^2 + 4y = 23 \end{cases}$$

SOLUTION   We choose to eliminate the x-term, so we multiply the first equation by -3 and then add the two equations.

$$\begin{cases} -3x^2 - 6y^2 = -33 \qquad \text{First equation times } -3 \\ \underline{3x^2 + 4y \;=\; 23} \qquad \text{Second equation} \end{cases}$$
$$ -6y^2 + 4y \;=\; -10 \qquad \text{Add}$$

Now we solve for y in the new equation:

$$-6y^2 + 4y + 10 = 0 \qquad \text{Add 10}$$
$$3y^2 - 2y - 5 = 0 \qquad \text{Divide by } -2$$
$$(y + 1)(3y - 5) = 0 \qquad \text{Factor}$$
$$y = -1 \quad \text{or} \quad y = \tfrac{5}{3} \qquad \text{Solve for } y$$

The solutions are the intersection points of an *ellipse* and a *parabola*, curves that we will study in Chapter 9.

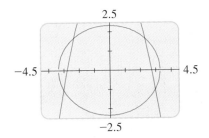

Now we back-substitute these values of y into one of the original equations, say $3x^2 + 4y = 23$, and solve for x.

If $y = -1$, we get

$$3x^2 + 4(-1) = 23 \qquad \text{Back-substitute } y = -1 \text{ into the second equation}$$
$$3x^2 = 27 \qquad \text{Add 4}$$
$$x^2 = 9 \qquad \text{Divide by 3}$$
$$x = -3 \quad \text{or} \quad x = 3 \qquad \text{Solve for } x$$

If $y = \tfrac{5}{3}$, we get

$$3x^2 + 4\left(\tfrac{5}{3}\right) = 23 \qquad \text{Back-substitute } y = \tfrac{5}{3} \text{ into the second equation}$$
$$3x^2 = 23 - \tfrac{20}{3} = \tfrac{49}{3} \qquad \text{Simplify}$$
$$x^2 = \tfrac{49}{9} \qquad \text{Divide by 3}$$
$$x = -\tfrac{7}{3} \quad \text{or} \quad x = \tfrac{7}{3} \qquad \text{Solve for } x$$

So we have four solutions:

$$(-3, -1), \quad (3, -1), \quad \left(-\tfrac{7}{3}, \tfrac{5}{3}\right), \quad \left(\tfrac{7}{3}, \tfrac{5}{3}\right)$$

You should check that each of the ordered pairs satisfies both of the original equations.   ■

EXAMPLE 6 ■ **An Application of Systems of Equations to Geometry**

A right triangle has an area of 120 ft^2 and a perimeter of 60 ft. Find the lengths of its sides.

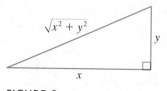

FIGURE 6

SOLUTION Let x and y be the lengths of the sides adjacent to the right angle. Then, by the Pythagorean Theorem, the hypotenuse has length $\sqrt{x^2 + y^2}$ (see Figure 6). Since the area is 120 ft^2, we have

$$\tfrac{1}{2}xy = 120 \qquad \text{Area of triangle}$$

$$xy = 240 \qquad \text{Simplify}$$

Also, since the perimeter is 60 ft,

$$x + y + \sqrt{x^2 + y^2} = 60 \qquad \text{Perimeter of triangle}$$

$$\sqrt{x^2 + y^2} = 60 - x - y \qquad \text{Isolate the square root on LHS}$$

$$\left(\sqrt{x^2 + y^2}\right)^2 = (60 - x - y)^2 \qquad \text{Square each side}$$

$$x^2 + y^2 = 3600 + x^2 + y^2 - 120x - 120y + 2xy \qquad \text{Expand}$$

$$120x + 120y = 3600 + 2xy \qquad \text{Combine like terms}$$

$$60x + 60y = 1800 + xy \qquad \text{Divide by 2}$$

Thus, we must solve the system

$$\begin{cases} xy = 240 \\ 60x + 60y = 1800 + xy \end{cases}$$

Substituting $xy = 240$ into the second equation, we get

$$60x + 60y = 1800 + 240 \qquad \text{Substitute } xy = 240$$

$$x + y = 34 \qquad \text{Divide by 60}$$

$$y = 34 - x \qquad \text{Solve for } y$$

Back-substituting this into the equation $xy = 240$ gives

$$x(34 - x) = 240 \qquad \text{Back-substitute } y = 34 - x$$

$$x^2 - 34x + 240 = 0 \qquad \text{Simplify}$$

$$(x - 24)(x - 10) = 0 \qquad \text{Factor}$$

So, either $x = 24$ or $x = 10$. Since $y = 240/x$, we get $y = 10$ or $y = 24$, respectively. In either case, the hypotenuse is

$$\sqrt{(10)^2 + (24)^2} = \sqrt{676} = 26$$

The lengths of the sides of the triangle are 10 ft, 24 ft, and 26 ft. ∎

CHECK YOUR ANSWER

Area of triangle $= \tfrac{1}{2}xy$

$$= \tfrac{1}{2} \cdot 10 \cdot 24$$

$$= 120 \text{ ft}^2$$

Perimeter $= 10 + 24 + 26$

$$= 60 \text{ ft} \qquad ✓$$

 ## USING GRAPHING DEVICES TO SOLVE SYSTEMS

Graphing devices are sometimes useful in solving systems of equations that involve just two variables. Note that with most graphing devices, any equation must first be expressed in terms of one or more functions of the form $y = f(x)$

before we can use the calculator to graph it. Not all equations can be readily expressed in this way, so not all systems can be solved by this technique.

EXAMPLE 7 ■ Solving a System of Equations with a Graphing Calculator

Find all solutions of the following system, correct to one decimal place.

$$\begin{cases} \dfrac{x^2}{12} + \dfrac{y^2}{7} = 1 \\ y = 3x^2 - 6x + \tfrac{1}{2} \end{cases}$$

SOLUTION The graph of the first equation has x-intercepts $x = \pm\sqrt{12} \approx \pm3.46$ and y-intercepts $y = \pm\sqrt{7} \approx \pm2.65$. We therefore select the viewing rectangle $[-4, 4]$ by $[-3, 3]$ to be sure that our graph contains the intercepts. Solving for y in terms of x, we get

$$\frac{y^2}{7} = 1 - \frac{x^2}{12} \qquad \text{Isolate } y\text{-term on LHS}$$

$$y^2 = 7\left(1 - \frac{x^2}{12}\right) \qquad \text{Solve for } y^2$$

$$y = \pm\sqrt{7\left(1 - \frac{x^2}{12}\right)} \qquad \text{Take square roots}$$

To graph the entire curve, we must graph both of the functions

$$y = \sqrt{7[1 - (x^2/12)]} \qquad \text{and} \qquad y = -\sqrt{7[1 - (x^2/12)]}$$

We graph the second equation in the system in the same viewing rectangle as the first (see Figure 7). The two equations in the system appear to intersect at three points. First, we zoom in to the two points above the x-axis. Their approximate coordinates are $(-0.3, 2.6)$ and $(2.2, 2.0)$. There also appears to be an intersection point in quadrant IV. However, when we zoom in, we see that the curves come close to each other here but don't intersect (see Figure 8). Thus, the system has only two solutions; correct to the nearest tenth, they are

$$(-0.3, 2.6) \qquad \text{and} \qquad (2.2, 2.0)$$

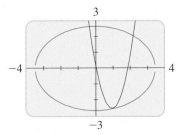

FIGURE 7

$\dfrac{x^2}{12} + \dfrac{y^2}{7} = 1, \quad y = 3x^2 - 6x + \tfrac{1}{2}$

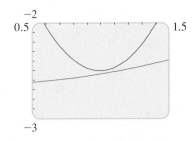

FIGURE 8

Zooming in on the system

8.1 EXERCISES

1–6 ■ Use the substitution method to find all solutions of the system of equations.

1. $\begin{cases} x + y = 8 \\ x - 3y = 0 \end{cases}$

2. $\begin{cases} 2x + y = 7 \\ 3x - y = 13 \end{cases}$

3. $\begin{cases} y = x^2 \\ y = x + 6 \end{cases}$

4. $\begin{cases} x^2 + y^2 = 25 \\ y = \frac{3}{4}x \end{cases}$

5. $\begin{cases} x^2 + y^2 = 8 \\ x + y = 0 \end{cases}$

6. $\begin{cases} x^2 + y = 9 \\ x - y + 3 = 0 \end{cases}$

7–12 ■ Use the elimination method to find all solutions of the system of equations.

7. $\begin{cases} 5x + 2y = 2 \\ 7x + 3y = 6 \end{cases}$

8. $\begin{cases} 4x - 3y = 10 \\ 9x + 4y = 1 \end{cases}$

9. $\begin{cases} x^2 - 2y = 1 \\ x^2 + 5y = 29 \end{cases}$

10. $\begin{cases} 3x^2 + 4y = 17 \\ 2x^2 + 5y = 2 \end{cases}$

11. $\begin{cases} 3x^2 - y^2 = 11 \\ x^2 + 4y^2 = 8 \end{cases}$

12. $\begin{cases} 2x^2 + 4y = 13 \\ x^2 - y^2 = \frac{7}{2} \end{cases}$

13–26 ■ Find all solutions of the system of equations.

13. $\begin{cases} y + x^2 = 4x \\ y + 4x = 16 \end{cases}$

14. $\begin{cases} x - y^2 = 0 \\ y - x^2 = 0 \end{cases}$

15. $\begin{cases} x - 2y = 2 \\ y^2 - x^2 = 2x + 4 \end{cases}$

16. $\begin{cases} y = 4 - x^2 \\ y = x^2 - 4 \end{cases}$

17. $\begin{cases} x - y = 4 \\ xy = 12 \end{cases}$

18. $\begin{cases} xy = 24 \\ 2x^2 - y^2 + 4 = 0 \end{cases}$

19. $\begin{cases} x^2 y = 16 \\ x^2 + 4y + 16 = 0 \end{cases}$

20. $\begin{cases} x + \sqrt{y} = 0 \\ y^2 - 4x^2 = 12 \end{cases}$

21. $\begin{cases} x^2 + y^2 = 9 \\ x^2 - y^2 = 1 \end{cases}$

22. $\begin{cases} x^2 + 2y^2 = 2 \\ 2x^2 - 3y = 15 \end{cases}$

23. $\begin{cases} 2x^2 - 8y^3 = 19 \\ 4x^2 + 16y^3 = 34 \end{cases}$

24. $\begin{cases} x^4 - y^3 = 17 \\ 3x^4 + 5y^3 = 53 \end{cases}$

25. $\begin{cases} \dfrac{2}{x} - \dfrac{3}{y} = 1 \\ -\dfrac{4}{x} + \dfrac{7}{y} = 1 \end{cases}$

26. $\begin{cases} \dfrac{4}{x^2} + \dfrac{6}{y^4} = \dfrac{7}{2} \\ \dfrac{1}{x^2} - \dfrac{2}{y^4} = 0 \end{cases}$

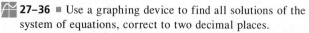 **27–36** ■ Use a graphing device to find all solutions of the system of equations, correct to two decimal places.

27. $\begin{cases} y = 2x + 6 \\ y = -x + 5 \end{cases}$

28. $\begin{cases} y = x^2 \\ y = x + 3 \end{cases}$

29. $\begin{cases} y = x^3 \\ y = 1 - x^4 \end{cases}$

30. $\begin{cases} y = x^2 - 4x \\ 2x - y = 2 \end{cases}$

31. $\begin{cases} x^2 + y^2 = 25 \\ x + 3y = 2 \end{cases}$

32. $\begin{cases} x^2 + y^2 = 17 \\ x^2 - 2x + y^2 = 13 \end{cases}$

33. $\begin{cases} \dfrac{x^2}{9} + \dfrac{y^2}{18} = 1 \\ y = -x^2 + 6x - 2 \end{cases}$

34. $\begin{cases} x^2 - y^2 = 3 \\ y = x^2 - 2x - 8 \end{cases}$

35. $\begin{cases} x^4 + 16y^4 = 32 \\ x^2 + 2x + y = 0 \end{cases}$

36. $\begin{cases} y = e^x + e^{-x} \\ y = 5 - x^2 \end{cases}$

37. A right triangle has a perimeter of 40 cm and an area of 60 cm². What are the lengths of its sides?

38. A rectangle has an area of 180 cm² and a perimeter of 54 cm. What are its dimensions?

39. A right triangle has an area of 54 in². The product of the lengths of the three sides is 1620 in³. What are the lengths of its sides?

40. A right triangle has an area of 84 ft² and a hypotenuse 25 ft long. What are the lengths of its other two sides?

41. The perimeter of a rectangle is 70 and its diagonal is 25. Find its length and width.

42. A circular piece of sheet metal has a diameter of 20 in. The edges are to be cut off to form a rectangle of area 160 in² (see the figure). What are the dimensions of the rectangle?

43. A hill is inclined so that its "slope" is $\frac{1}{2}$, as shown in the figure. We introduce a coordinate system with the origin at the base of the hill and with the scales on the axes measured in meters. A rocket is fired from the base of the hill in such a way that its trajectory is the parabola $y = -x^2 + 401x$. At what point does the rocket strike the hillside? How far is this point from the base of the hill (to the nearest cm)?

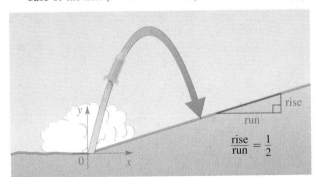

44. A rectangular piece of sheet metal with an area of 1200 in² is to be bent into a cylindrical length of stovepipe having a volume of 600 in³. What are the dimensions of the sheet metal?

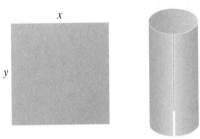

45. Find an equation for the line that passes through the points of intersection of the circles $x^2 + y^2 = 25$ and

$x^2 - 3x + y^2 + y = 30$. [*Hint:* Try to eliminate the x^2- and y^2-terms.]

46–49 ■ Find all solutions of the system.

46. $\begin{cases} x - y = 3 \\ x^3 - y^3 = 387 \end{cases}$ [*Hint:* Factor the left side of the second equation.]

47. $\begin{cases} x^2 + xy = 1 \\ xy + y^2 = 3 \end{cases}$ [*Hint:* Add the equations and factor the result.]

48. $\begin{cases} 2^x + 2^y = 10 \\ 4^x + 4^y = 68 \end{cases}$

49. $\begin{cases} \log x + \log y = \frac{3}{2} \\ 2 \log x - \log y = 0 \end{cases}$

DISCOVERY · DISCUSSION

50. Intersection of a Parabola and a Line On a sheet of graph paper, or using a graphing calculator, draw the parabola $y = x^2$. Then draw the graphs of the linear equation $y = x + k$ on the same coordinate plane for various values of k. Try to choose values of k so that the line and the parabola intersect at two points for some of your k's, and not for others. For what value of k is there exactly one intersection point? Use the results of your experiment to make a conjecture about the values of k for which the following system has two solutions, one solution, and no solution. Prove your conjecture.

$$\begin{cases} y = x^2 \\ y = x + k \end{cases}$$

8.2 PAIRS OF LINES

In Section 1.10 we saw that the graph of any equation of the form

$$Ax + By = C$$

is a line. In this section we study systems of two such equations:

$$\begin{cases} ax + by = c \\ dx + ey = f \end{cases}$$

We can use either the substitution method or the elimination method (described in Section 8.1) to solve such systems algebraically. But since the elimination method is usually easier for linear systems, we use elimination instead of substitution in our examples.

In general, there are three situations that can occur when we graph a system of two linear equations. The graphs may intersect at a single point (Figure 1), they may be parallel with no intersection point (Figure 2), or the two equations may simply be different equations for the same line (Figure 3). This means that the system can have one solution, no solution, or infinitely many solutions.

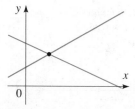

FIGURE 1
Linear system with one solution.
Lines intersect at a single point.

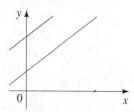

FIGURE 2
Linear system with no solution.
Lines are parallel—they do not intersect.

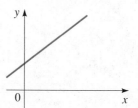

FIGURE 3
Linear system with infinitely many solutions.
Lines coincide—equations are for the same line.

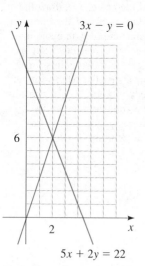

FIGURE 4

CHECK YOUR ANSWER

$x = 2$, $y = 6$:

$$\begin{cases} 3(2) - (6) = 0 \\ 5(2) + 2(6) = 22 \end{cases} \checkmark$$

EXAMPLE 1 ■ A Linear System with One Solution

Solve the following system and graph the lines.

$$\begin{cases} 3x - y = 0 \\ 5x + 2y = 22 \end{cases}$$

SOLUTION We eliminate y from the equations and solve for x.

$$\begin{cases} 6x - 2y = 0 & \text{First equation times 2} \\ \underline{5x + 2y = 22} \end{cases}$$
$$11x = 22 \qquad \text{Add}$$
$$x = \tfrac{22}{11} = 2 \qquad \text{Solve for } x$$

Now we back-substitute into the first equation and solve for y:

$$6(2) - 2y = 0 \qquad \text{Back-substitute } y = 2$$
$$-2y = -12 \qquad \text{Subtract } 6 \cdot 2 = 12$$
$$y = 6 \qquad \text{Solve for } y$$

The solution of the system is the ordered pair $(2, 6)$, that is,

$$x = 2, \qquad y = 6$$

The graph in Figure 4 shows that the lines in the system intersect at the point $(2, 6)$. ■

EXAMPLE 2 ■ A Pair of Linear Equations with No Solution

Solve the following system.

$$\begin{cases} 8x - 2y = 5 \\ -12x + 3y = 7 \end{cases}$$

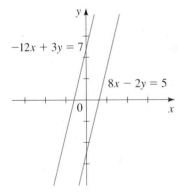

$-12x + 3y = 7$

$8x - 2y = 5$

FIGURE 5

SOLUTION   This time we try to find a suitable combination of the two equations to eliminate the variable y. Multiplying the first equation by 3 and the second by 2 gives

$$\begin{cases} 24x - 6y = 15 & \text{First equation times 3} \\ -24x + 6y = 14 & \text{Second equation times 2} \end{cases}$$
$$\quad\quad\quad\quad 0 = 29 \quad \text{Add}$$

Adding the two equations eliminates *both x and y* in this case, and we end up with $0 = 29$, which is obviously false. No matter what values we assign to x and y, we cannot make this statement true, so the system has *no solution*. Figure 5 shows that the lines in the system are parallel and do not intersect. ■

A system that has no solution, like the one in Example 2, is said to be **inconsistent**.

EXAMPLE 3 ■ A Pair of Linear Equations with Infinitely Many Solutions
Solve the following system.

$$\begin{cases} 3x - 6y = 12 \\ 4x - 8y = 16 \end{cases}$$

SOLUTION   We multiply the first equation by 4 and the second by 3 to prepare for subtracting the equations to eliminate x. The new equations are

$$\begin{cases} 12x - 24y = 48 & \text{First equation times 4} \\ 12x - 24y = 48 & \text{Second equation times 3} \end{cases}$$

We see that the two equations in the original system are simply different ways of expressing the equation of one single line. The coordinates of any point on this line give a solution of the system. Writing the equation in slope-intercept form, we have $y = \frac{1}{2}x - 2$. So, the solution of the system is

$$\left(x, \tfrac{1}{2}x - 2\right)$$

where x can be any real number. The system has infinitely many solutions. ■

APPLIED LINEAR SYSTEMS

Frequently, when we use equations to solve problems in the sciences or in other areas, we obtain systems like the ones we've been considering. The next two examples illustrate such situations.

EXAMPLE 4 ■ A Distance-Speed-Time Problem

A woman rows a boat upstream from one point on a river to another point 4 mi away in $1\frac{1}{2}$ hours. The return trip, traveling with the current, takes only 45 min. How fast does she row relative to the water, and at what speed is the current flowing?

SOLUTION   In this and any other problem that involves distance, time, and speed, we make use of the fundamental relationships between these quantities that are shown in the margin. We use these equations to translate the English sentences of the problem into mathematical statements. Since we are asked to find the rowing speed and the speed of the current, we give names to these quantities. Let

$$x = \text{rowing speed (in mi/h)}$$

$$y = \text{speed of the current (in mi/h)}$$

$$\text{speed} = \frac{\text{distance}}{\text{time}}$$

$$\text{distance} = \text{speed} \times \text{time}$$

$$\text{time} = \frac{\text{distance}}{\text{speed}}$$

When the woman is traveling with the current (downstream), she will be moving at a total of $x + y$ miles per hour, but upstream she moves at $x - y$ miles per hour, since the current decreases her net speed. The distance both upstream and downstream is 4 mi, so using the fact that distance = speed × time for both parts of the trip, we get the equations

$$4 = (x - y) \cdot \tfrac{3}{2} \qquad \text{and} \qquad 4 = (x + y) \cdot \tfrac{3}{4}$$

(*Note:* All times have been converted to hours, since we are expressing the speeds in miles per *hour.*) If we multiply the equations by 2 and 4, respectively, to clear the denominators, we get the system

$$\begin{cases} 3x - 3y = 8 & \text{First equation times 2} \\ 3x + 3y = 16 & \text{Second equation times 4} \end{cases}$$

$$\begin{array}{ll} 6x = 24 & \text{Add} \\ x = 4 & \text{Solve for } x \end{array}$$

CHECK YOUR ANSWER

Speed upstream is

$$\frac{\text{distance}}{\text{time}} = \frac{4 \text{ mi}}{1\frac{1}{2} \text{ h}} = 2\tfrac{2}{3} \text{ mi/h}$$

and this should equal

rowing speed − current flow

$$= 4 \text{ mi/h} - \tfrac{4}{3} \text{ mi/h}$$

$$= 2\tfrac{2}{3} \text{ mi/h} \qquad ✓$$

Speed downstream is

$$\frac{\text{distance}}{\text{time}} = \frac{4 \text{ mi}}{\frac{3}{4} \text{ h}} = 5\tfrac{1}{3} \text{ mi/h}$$

and this should equal

rowing speed + current flow

$$= 4 \text{ mi/h} + \tfrac{4}{3} \text{ mi/h} = 5\tfrac{1}{3} \text{ mi/h} \quad ✓$$

Back-substituting this value of x into the first equation (the second works just as well) and solving for y gives

$$\begin{array}{ll} 3(4) - 3y = 8 & \text{Back-substitute } x = 4 \\ -3y = 8 - 12 & \text{Subtract 12} \\ y = \tfrac{4}{3} & \text{Solve for } y \end{array}$$

The woman rows at 4 mi/h and the current flows at $1\frac{1}{3}$ mi/h.   ■

Many students find that the hardest part of solving a word problem is getting started. The following steps, adapted from the problem-solving principles on pages 122–124, should help you find your way to the equations that are needed to solve the problem.

SOLVING APPLIED PROBLEMS

1. GIVE NAMES TO THE VARIABLES. Assign letters to denote the variable quantities in the problem. Usually the last sentence of the problem tells you what is being asked for, so this is what the variable names will represent.

2. ORGANIZE THE GIVEN INFORMATION. If possible, draw a diagram or create a table that helps you see the relationship between the quantities involved in the problem.

3. TRANSLATE THE GIVEN INFORMATION INTO EQUATIONS. Translate the information about the variables given in the problem into mathematical equations. Remember that *an equation is just a sentence written using the symbols of mathematics.*

4. SOLVE THE EQUATIONS AND INTERPRET THE RESULTS. Solve the equations given by Step 3, and state in words what the solutions mean in terms of the original meanings of the variables.

EXAMPLE 5 ■ A Mixture Problem

A vintner wishes to fortify wine that contains 10% alcohol by adding some 70% alcohol solution to it. The resulting mixture is to have an alcoholic strength of 16% and is to fill 1000 one-liter bottles. How many liters (L) of the wine and of the alcohol solution should he use?

SOLUTION Let

Name the variables

$$x = \text{number of liters of wine to be used}$$

$$y = \text{number of liters of alcohol solution to be used}$$

To help us translate the information in the problem into equations, we organize the given data in a table.

Organize the information into a table

	Wine	Alcohol solution	Resulting mixture
Volume	x	y	1000
Percent alcohol	10%	70%	16%
Amount of alcohol	$(0.10)x$	$(0.70)y$	$(0.16)1000$

The volume of the mixture must be the total of the two volumes the vintner is adding together, so

Translate the information into equations

$$x + y = 1000$$

Similarly, the amount of alcohol in the mixture must be the total of the alcohol contributed by the wine and by the alcohol solution, which means that

$$(0.10)x + (0.70)y = (0.16)1000$$

$$(0.10)x + (0.70)y = 160 \qquad \text{Simplify}$$

$$x + 7y = 1600 \qquad \text{Multiply by 10 to clear decimals}$$

Thus, we must solve the system

$$\begin{cases} x + y = 1000 \\ x + 7y = 1600 \end{cases}$$

Subtracting the first equation from the second eliminates the variable x, and we get

Solve the equations

$$6y = 600 \qquad \text{Subtract equations}$$

$$y = 100 \qquad \text{Solve for } y$$

We now back-substitute $y = 100$ into the first equation and solve for x:

$$x + 100 = 1000 \qquad \text{Back-substitute } y = 100$$

$$x = 900 \qquad \text{Solve for } x$$

The vintner should use 900 L of wine and 100 L of the alcohol solution. ∎

8.2 EXERCISES

1–6 ■ Graph each pair of lines on a single set of axes. Determine whether the lines are parallel, and if they are not parallel, estimate the coordinates of their point of intersection from your graph.

1. $\begin{cases} x + y = 3 \\ 2x - y = 0 \end{cases}$

2. $\begin{cases} 3x + 2y = 3 \\ -x + 5y = 16 \end{cases}$

3. $\begin{cases} 2x + 3y = 12 \\ x - y = 1 \end{cases}$

4. $\begin{cases} 3x + 5y = 15 \\ x + \frac{5}{3}y = 10 \end{cases}$

5. $\begin{cases} 2x + 5y = 15 \\ 4x + 10y = 20 \end{cases}$

6. $\begin{cases} -4x + 14y = 28 \\ 10x - 35y = 70 \end{cases}$

7–28 ■ Solve the system. If a system has infinitely many solutions, express them in the form given in Example 3.

7. $\begin{cases} -x + y = 2 \\ 4x - 3y = -3 \end{cases}$

8. $\begin{cases} 4x - 3y = 28 \\ 9x - y = -6 \end{cases}$

9. $\begin{cases} x + 2y = 7 \\ 5x - y = 2 \end{cases}$

10. $\begin{cases} -4x + 12y = 0 \\ 12x + 4y = 160 \end{cases}$

11. $\begin{cases} \frac{1}{2}x + \frac{1}{3}y = 2 \\ \frac{1}{5}x - \frac{2}{3}y = 8 \end{cases}$

12. $\begin{cases} 0.2x - 0.2y = -1.8 \\ -0.3x + 0.5y = 3.3 \end{cases}$

13. $\begin{cases} 3x + 2y = 8 \\ x - 2y = 0 \end{cases}$

14. $\begin{cases} 4x + 2y = 16 \\ x - 5y = 70 \end{cases}$

15. $\begin{cases} x + 4y = 8 \\ 3x + 12y = 2 \end{cases}$

16. $\begin{cases} -3x + 5y = 2 \\ 9x - 15y = 6 \end{cases}$

17. $\begin{cases} 2x - 6y = 10 \\ -3x + 9y = -15 \end{cases}$

18. $\begin{cases} 2x - 3y = -8 \\ 14x - 21y = 3 \end{cases}$

19. $\begin{cases} 6x + 4y = 12 \\ 9x + 6y = 18 \end{cases}$

20. $\begin{cases} 25x - 75y = 100 \\ -10x + 30y = -40 \end{cases}$

21. $\begin{cases} 8s - 3t = -3 \\ 5s - 2t = -1 \end{cases}$

22. $\begin{cases} u - 30v = -5 \\ -3u + 80v = 5 \end{cases}$

23. $\begin{cases} \frac{1}{2}x + \frac{3}{5}y = 3 \\ \frac{5}{3}x + 2y = 10 \end{cases}$

24. $\begin{cases} \frac{3}{2}x - \frac{1}{3}y = -\frac{1}{2} \\ 2x - \frac{1}{2}y = -\frac{1}{2} \end{cases}$

25. $\begin{cases} \dfrac{2x-5}{3} + \dfrac{y-1}{6} = \dfrac{1}{2} \\ \dfrac{x}{5} + \dfrac{3y-6}{12} = 1 \end{cases}$

26. $\begin{cases} x - 3y = 4x - 6y - 10 \\ 2x = 12y + 10 \end{cases}$

27. $x - 2y = 2x + 2y = 1$

28. $x = 2x + y = 2y + 1$

29–32 ■ Find x and y in terms of a and b.

29. $\begin{cases} x + y = 0 \\ x + ay = 1 \end{cases}$ $(a \neq 1)$

30. $\begin{cases} ax + by = 0 \\ x + y = 1 \end{cases}$ $(a \neq b)$

31. $\begin{cases} ax + by = 1 \\ bx + ay = 1 \end{cases}$ $(a^2 - b^2 \neq 0)$

32. $\begin{cases} ax + by = 0 \\ a^2x + b^2y = 1 \end{cases}$ $(a \neq 0, b \neq 0, a \neq b)$

33. Find two numbers whose sum is 34 and whose difference is 10.

34. The sum of two numbers is twice their difference. The larger number is 6 more than twice the smaller. Find the numbers.

35. A man has 14 coins in his pocket, all of which are dimes and quarters. If the total value of his change is $2.75, how many dimes and how many quarters does he have?

36. The admission fee at an amusement park is $1.50 for children and $4.00 for adults. On a certain day, 2200 people entered the park, and the admission fees collected totaled $5050. How many children and how many adults were admitted?

37. A man flies a small airplane from Fargo to Bismarck, North Dakota—a distance of 180 mi. Because he is flying into a head wind, the trip takes him 2 hours.

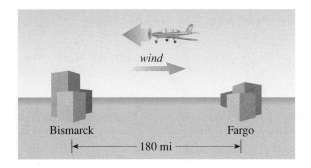

Bismarck   wind   Fargo

|← 180 mi →|

On the way back, the wind is still blowing at the same speed, so the return trip takes only 1 h 12 min. What is his speed in still air, and how fast is the wind blowing?

38. A boat on a river travels downstream between two points, 20 mi apart, in one hour. The return trip against the current takes $2\frac{1}{2}$ hours. What is the boat's speed, and how fast does the current in the river flow?

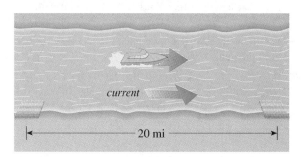

current

|← 20 mi →|

39. A woman keeps fit by bicycling and running every day. On Monday she spends $\frac{1}{2}$ hour at each activity, covering a total of $12\frac{1}{2}$ mi. On Tuesday, she runs for 12 min and cycles for 45 min, covering a total of 16 mi. Assuming her running and cycling speeds don't change from day to day, find these speeds.

40. A biologist has two brine solutions, one containing 5% salt and another containing 20% salt. How many milliliters of each solution should he mix to obtain 1 L of a solution that contains 14% salt?

41. A researcher performs an experiment to test a hypothesis that involves the nutrients niacin and retinol. She wishes to feed one of her groups of laboratory rats a diet that contains precisely 32 units of niacin and 22,000 units of retinol per day. She has two types of commercial pellet foods available. Food A contains 0.12 unit of niacin and 100 units of retinol per gram. Food B contains 0.20 unit of niacin and 50 units of retinol per gram. How many grams of each food should she feed this group of rats each day?

42. A customer in a coffee shop wishes to purchase a blend of two coffees: Kenyan, costing $3.50 a pound, and Sri Lankan, costing $5.60 a pound. He ends up buying 3 lb of such a blend, which costs him $11.55. How many pounds of each kind went into the mixture?

43. A chemist has two large containers of sulfuric acid solution, with different concentrations of acid in each container. Blending 300 mL of the first solution and 600 mL of the second gives a mixture that is 15% acid, whereas 100 mL of the first mixed with 500 mL of

the second gives a $12\frac{1}{2}\%$ acid mixture. What is the concentration of sulfuric acid in each of the original containers?

44. John and Mary leave their house at the same time and drive off in opposite directions. John drives at 60 mi/h and travels 35 mi farther than Mary, who drives at 40 mi/h. Mary's trip takes 15 min longer than John's. For what length of time does each of them drive?

45. The sum of the digits of a two-digit number is 7. When the digits are reversed, the number is increased by 27. Find the number.

46. Find the area of the triangle that lies in the first quadrant (with its base on the x-axis) and that is bounded by the lines $y = 2x - 4$ and $y = -4x + 20$.

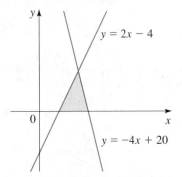

47–50 ■ Use a graphing device to graph both lines in the same viewing rectangle. (Note that you must solve for y in terms of x before graphing if you are using a graphing calculator.) Solve the system by zooming in to the point of intersection and then using the cursor to find its coordinates, correct to two decimal places.

47. $\begin{cases} 0.21x + 3.17y = 9.51 \\ 2.35x - 1.17y = 5.89 \end{cases}$

48. $\begin{cases} 18.72x - 14.91y = 12.33 \\ 6.21x - 12.92y = 17.82 \end{cases}$

49. $\begin{cases} 2371x - 6552y = 13,591 \\ 9815x + 992y = 618,555 \end{cases}$

50. $\begin{cases} -435x + 912y = 0 \\ 132x + 455y = 994 \end{cases}$

DISCOVERY · DISCUSSION

51. The Least Squares Line The *least squares* line or *regression* line is the line that best fits a set of points in the plane. We studied this line in *Principles of Modeling* (see page 212). Using calculus, it can be shown that the line that best fits the n data points (x_1, y_1), (x_2, y_2), ..., (x_n, y_n) is the line $y = ax + b$, where the coefficients a and b satisfy the following pair of linear equations. [The notation $\sum_{k=1}^{n} x_k$ stands for the sum of all the x's. See Section 10.1 for a complete description of sigma (Σ) notation.]

$$\left(\sum_{k=1}^{n} x_k\right) a + nb = \sum_{k=1}^{n} y_k$$

$$\left(\sum_{k=1}^{n} x_k^2\right) a + \left(\sum_{k=1}^{n} x_k\right) b = \sum_{k=1}^{n} x_k y_k$$

Use these equations to find the least squares line for the following data points.

$$(1, 3), \quad (2, 5), \quad (3, 6), \quad (5, 6), \quad (7, 9)$$

Sketch the points and your line to confirm that the line fits these points well. If you have a calculator that calculates regression lines, see whether the calculator gives you the same line as the formulas.

8.3   SYSTEMS OF LINEAR EQUATIONS

Linear equations
$$6x_1 - 3x_2 + \sqrt{5}\, x_3 = 1000$$
$$x + y + z = 2w - \tfrac{1}{2}$$

Nonlinear equations
$$x^2 + 3y - \sqrt{z} = 5$$
$$x_1 x_2 + 6x_3 = -6$$

A **linear equation in n variables** is an equation that can be put in the form

$$a_1 x_1 + a_2 x_2 + \cdots + a_n x_n = c$$

where $a_1, a_2, \ldots, a_n$ and c are real numbers, and $x_1, x_2, \ldots, x_n$ are the variables. If the number of variables is no more than three or four, we generally use x, y, z, and w instead of x_1, x_2, x_3, and x_4. Such equations are called *linear* because if we have just two variables, the equation is

$$a_1 x + a_2 y = c$$

which is the equation of a line. Each term of a linear equation is either a constant or a constant multiple of one of the variables.

To solve a system of linear equations, we adapt the elimination method introduced in Section 8.1 to obtain an **equivalent system** (that is, a system with the same solution as the original system), which is easy to solve. For a system of three equations, the idea is to eliminate all but one of the variables from one equation, then eliminate all but two variables from another equation. This allows us to use back-substitution to find the solution.

EXAMPLE 1 ■ Solving a System Using the Elimination Method

Solve the following system of linear equations.

$$\begin{cases} x - y + 3z = 4 \\ x + 2y - 2z = 10 \\ 3x - y + 5z = 14 \end{cases}$$

SOLUTION   We use elimination to obtain a system whose last equation involves only the variable z, and the next-to-last involves just y and z. Then we back-substitute to find first z, then y, and finally x.

We begin by eliminating the x-terms from the second and third equations. We multiply the first equation by -1 and add it to the second equation.

$$\begin{array}{ll} -x + y - 3z = -4 & \text{First equation times } -1 \\ \underline{x + 2y - 2z = 10} & \text{Second equation} \\ \quad\;\; 3y - 5z = 6 & \text{Add} \end{array}$$

The resulting equation does not contain the variable x. Similarly, we add -3 times the first equation to the third.

$$\begin{array}{ll} -3x + 3y - 9z = -12 & \text{First equation times } -3 \\ \underline{3x - y + 5z = 14} & \text{Third equation} \\ \quad\;\; 2y - 4z = 2 & \text{Add} \end{array}$$

This equation also does not contain the variable x. Now we have a simpler system that is equivalent to the original system.

$$\begin{cases} x - y + 3z = 4 \\ \quad\;\, 3y - 5z = 6 \qquad \text{New system} \\ \quad\;\, 2y - 4z = 2 \end{cases}$$

Since each term in the third equation has a common factor of 2, we multiply the equation by $\frac{1}{2}$.

$$\begin{cases} x - y + 3z = 4 \\ 3y - 5z = 6 \\ y - 2z = 1 \end{cases} \quad \text{Third equation times } \tfrac{1}{2}$$

We now eliminate the variable y from the second equation by adding -3 times the third equation to the second.

$$3y - 5z = 6$$
$$\underline{-3y + 6z = -3}$$
$$z = 3$$

$$\begin{cases} x - y + 3z = 4 \\ z = 3 \quad \text{Add } -3 \text{ times third equation to second} \\ y - 2z = 1 \end{cases}$$

To reach our goal, we interchange the second and third equations.

$$\begin{cases} x - y + 3z = 4 \\ y - 2z = 1 \quad \text{Interchange second and third equations} \\ z = 3 \end{cases}$$

Now we back-substitute $z = 3$ into the second equation and solve for y:

$$y - 2(3) = 1 \quad \text{Back-substitute } z = 3 \text{ into the second equation}$$

$$y = 7 \quad \text{Solve for } y$$

Finally, we back-substitute the values for y and z into the first equation and solve for x:

$$x - (7) + 3(3) = 4 \quad \text{Back-substitute } y = 7 \text{ and } z = 3 \text{ into the first equation}$$

$$x = 2 \quad \text{Solve for } x$$

Thus, the solution to the system is $x = 2$, $y = 7$, $z = 3$. The solution may be written as $(2, 7, 3)$.

CHECK YOUR ANSWER

$x = 2$, $y = 7$, $z = 3$:

$$\begin{cases} (2) - (7) + 3(3) = 4 \\ (2) + 2(7) - 2(3) = 10 \\ 3(2) - (7) + 5(3) = 14 \end{cases} \checkmark$$

■

If we examine the solution to Example 1, we see that the variables x, y, and z act simply as place-holders in our computations. It's only the coefficients of the variables and the constants that actually enter into the calculations. We will use this fact to simplify the notation for our computations. Instead of writing out the equations of a system in full, we write only the coefficients and constants in a rectangular array, called the **matrix form** of the system. The matrix form of the system of Example 1 is as follows.

System of linear equations   Matrix form

$$\begin{cases} x - y + 3z = 4 \\ x + 2y - 2z = 10 \\ 3x - y + 5z = 14 \end{cases} \qquad \begin{bmatrix} 1 & -1 & 3 & 4 \\ 1 & 2 & -2 & 10 \\ 3 & -1 & 5 & 14 \end{bmatrix}$$

Any rectangular array of numbers is called a **matrix**. The **rows** of a matrix are the horizontal lists of numbers in the array; the **columns** of a matrix are the vertical lists of numbers in the array. Each row of the matrix form of a system represents an equation. For example, the first row of the matrix in the preceding display is $\begin{bmatrix} 1 & -1 & 3 & 4 \end{bmatrix}$, and it represents the equation $x - y + 3z = 4$.

The operations we use on a system of equations (as in Example 1) correspond to operations on the matrix form of the system. These are called the **elementary row operations**.

ELEMENTARY ROW OPERATIONS

1. Add a multiple of one row to another.

2. Multiply a row by a nonzero constant.

3. Interchange two rows.

Note that performing any of these operations on the matrix form of a system does not change its solution. We use the following notation to describe the elementary row operations:

Symbol	Description
$R_i + kR_j \rightarrow R_i$	Change the ith row by adding k times row j to it, and then put the result back in row i.
kR_i	Multiply the ith row by k.
$R_i \leftrightarrow R_j$	Interchange the ith and jth rows.

In the next example we compare the two ways of writing systems of linear equations, using the system of Example 1.

EXAMPLE 2 ■ **Using the Matrix Form to Solve a System**

Solve the following system using its matrix form.

$$\begin{cases} x - y + 3z = 4 \\ x + 2y - 2z = 10 \\ 3x - y + 5z = 14 \end{cases}$$

SOLUTION   This is the system in Example 1. As before, our goal is to eliminate the x-term from the second equation and the x- and y-terms from the third equation. For comparison, we write both the full form of the system and the shorthand matrix form.

System

$$\begin{cases} x - y + 3z = 4 \\ x + 2y - 2z = 10 \\ 3x - y + 5z = 14 \end{cases}$$

Matrix form

$$\begin{bmatrix} 1 & -1 & 3 & 4 \\ 1 & 2 & -2 & 10 \\ 3 & -1 & 5 & 14 \end{bmatrix}$$

Add -1 times the first equation to the second.

Add -3 times the first to the third.

$$\begin{cases} x - y + 3z = 4 \\ 3y - 5z = 6 \\ 2y - 4z = 2 \end{cases}$$

$\xrightarrow{\begin{array}{c} R_2 - R_1 \to R_2 \\ R_3 - 3R_1 \to R_3 \end{array}}$

$$\begin{bmatrix} 1 & -1 & 3 & 4 \\ 0 & 3 & -5 & 6 \\ 0 & 2 & -4 & 2 \end{bmatrix}$$

Multiply the third equation by $\frac{1}{2}$.

$$\begin{cases} x - y + 3z = 4 \\ 3y - 5z = 6 \\ y - 2z = 1 \end{cases}$$

$\xrightarrow{\frac{1}{2}R_3}$

$$\begin{bmatrix} 1 & -1 & 3 & 4 \\ 0 & 3 & -5 & 6 \\ 0 & 1 & -2 & 1 \end{bmatrix}$$

Add -3 times the third equation to the second (to eliminate y from the second equation).

$$\begin{cases} x - y + 3z = 4 \\ z = 3 \\ y - 2z = 1 \end{cases}$$

$\xrightarrow{R_2 - 3R_3 \to R_2}$

$$\begin{bmatrix} 1 & -1 & 3 & 4 \\ 0 & 0 & 1 & 3 \\ 0 & 1 & -2 & 1 \end{bmatrix}$$

Interchange the second and third equations.

$$\begin{cases} x - y + 3z = 4 \\ y - 2z = 1 \\ z = 3 \end{cases}$$

$\xrightarrow{R_2 \leftrightarrow R_3}$

$$\begin{bmatrix} 1 & -1 & 3 & 4 \\ 0 & 1 & -2 & 1 \\ 0 & 0 & 1 & 3 \end{bmatrix}$$

At this point, we arrive at the solution $(2, 7, 3)$ by back-substitution, as in Example 1. ∎

In general, to solve a system of equations using a matrix, we use elementary row operations to arrive at a matrix in a certain form. This form is described in the following box.

ECHELON FORM AND REDUCED ECHELON FORM OF A MATRIX

A matrix is in **echelon form** if it satisfies the following conditions.

1. The first nonzero number in each row (reading from left to right) is 1. This is called the **leading entry**.

2. The leading entry in each row is to the right of the leading entry in the row immediately above it.

3. All rows consisting entirely of zeros are at the bottom of the matrix.

A matrix in echelon form is in **reduced echelon form** if it also satisfies the following condition.

4. Every number above and below each leading entry is a 0.

In the following matrices, the first matrix is in reduced echelon form, but the second one is just in echelon form. The third matrix is not in echelon form.

Reduced echelon form

$$\begin{bmatrix} 1 & 3 & 0 & 0 & 0 \\ 0 & 0 & 1 & 0 & -3 \\ 0 & 0 & 0 & 1 & \frac{1}{2} \\ 0 & 0 & 0 & 0 & 0 \end{bmatrix}$$

Echelon form

$$\begin{bmatrix} 1 & 3 & -6 & 10 & 0 \\ 0 & 0 & 1 & 4 & -3 \\ 0 & 0 & 0 & 1 & \frac{1}{2} \\ 0 & 0 & 0 & 0 & 0 \end{bmatrix}$$

Not in echelon form

$$\begin{bmatrix} 0 & 1 & -\frac{1}{2} & 0 & 7 \\ 1 & 0 & 3 & 4 & -5 \\ 0 & 0 & 0 & 1 & 0.4 \\ 0 & 1 & 1 & 0 & 0 \end{bmatrix}$$

To solve a system in matrix form, we use the elementary row operations to arrive at a matrix in echelon form or reduced echelon form. The technique of using elementary row operations to arrive at a matrix in echelon form is called **Gaussian elimination**, in honor of the German mathematician C. F. Gauss (see page 259). The process of arriving at a matrix in reduced echelon form is called **Gauss-Jordan elimination**.

EXAMPLE 3 ■ Solving a System Using Echelon Form

Solve the following system of linear equations using Gaussian elimination.

$$\begin{cases} 4x + 8y - 4z = 4 \\ 3x + 6y + 5z = -13 \\ -2x + y + 12z = -17 \end{cases}$$

SOLUTION We first write the system in matrix form.

$$\begin{bmatrix} 4 & 8 & -4 & 4 \\ 3 & 6 & 5 & -13 \\ -2 & 1 & 12 & -17 \end{bmatrix} \xrightarrow{\frac{1}{4}R_1} \begin{bmatrix} 1 & 2 & -1 & 1 \\ 3 & 6 & 5 & -13 \\ -2 & 1 & 12 & -17 \end{bmatrix} \xrightarrow[R_3 + 2R_1 \to R_3]{R_2 - 3R_1 \to R_2} \begin{bmatrix} 1 & 2 & -1 & 1 \\ 0 & 0 & 8 & -16 \\ 0 & 5 & 10 & -15 \end{bmatrix}$$

$$\xrightarrow{R_2 \leftrightarrow R_3} \begin{bmatrix} 1 & 2 & -1 & 1 \\ 0 & 5 & 10 & -15 \\ 0 & 0 & 8 & -16 \end{bmatrix} \xrightarrow[\frac{1}{8}R_3]{\frac{1}{5}R_2} \begin{bmatrix} 1 & 2 & -1 & 1 \\ 0 & 1 & 2 & -3 \\ 0 & 0 & 1 & -2 \end{bmatrix}$$

We now have an equivalent matrix in echelon form, and the corresponding system of equations is

$$\begin{cases} x + 2y - z = 1 \\ y + 2z = -3 \\ z = -2 \end{cases}$$

We now use back-substitution to solve the system.

$$y + 2(-2) = -3 \qquad \text{Back-substitute } z = -2 \text{ into the second equation}$$

$$y = 1 \qquad \text{Solve for } y$$

$$x + 2(1) - (-2) = 1 \qquad \text{Back-substitute } y = 1 \text{ and } z = -2 \text{ into the first equation}$$

$$x = -3 \qquad \text{Solve for } x$$

So the solution of the system is $(-3, 1, -2)$. ■

The advantage of using reduced echelon form is that we don't need to use back-substitution when solving a system in this form, as we see in the next example.

EXAMPLE 4 ■ Solving a System Using Reduced Echelon Form

Solve the following system of linear equations, using Gauss-Jordan elimination.

$$\begin{cases} 4x + 8y - 4z = 4 \\ 3x + 6y + 5z = -13 \\ -2x + y + 12z = -17 \end{cases}$$

SOLUTION   In Example 3 we used Gaussian elimination on the matrix form of this system of equations to arrive at an equivalent matrix in echelon form. We now continue using elementary row operations on the last matrix in Example 3 to arrive at an equivalent matrix in reduced echelon form.

$$\begin{bmatrix} 1 & 2 & -1 & 1 \\ 0 & 1 & 2 & -3 \\ 0 & 0 & 1 & -2 \end{bmatrix} \xrightarrow[R_1 + R_3 \rightarrow R_1]{R_2 - 2R_3 \rightarrow R_2} \begin{bmatrix} 1 & 2 & 0 & -1 \\ 0 & 1 & 0 & 1 \\ 0 & 0 & 1 & -2 \end{bmatrix} \xrightarrow{R_1 - 2R_2 \rightarrow R_1} \begin{bmatrix} 1 & 0 & 0 & -3 \\ 0 & 1 & 0 & 1 \\ 0 & 0 & 1 & -2 \end{bmatrix}$$

We now have an equivalent matrix in reduced echelon form, and the corresponding system of equations is

Since the system is in reduced echelon form, back-substitution is not required to get the solution.

$$\begin{cases} x = -3 \\ y = 1 \\ z = -2 \end{cases}$$

Hence we immediately arrive at the solution $(-3, 1, -2)$.   ■

INCONSISTENT AND DEPENDENT SYSTEMS

The systems of linear equations that we considered in Examples 1–4 had one solution for each of the unknowns. But as we saw in Section 8.2, a system of two linear equations in two variables can have one solution, no solution, or infinitely many solutions. The same cases arise when we study linear systems with more equations and more variables.

A system that has no solution is said to be **inconsistent**. If we use Gaussian elimination to change a system to echelon form, and if one of the equations we arrive at is false, then the system is inconsistent. (The false equation will always have the form $0 = c$, where c is not zero.)

A system that has infinitely many solutions is called **dependent**. If we use Gaussian elimination to change a system to echelon form and there are fewer nonzero rows than variables, then the system is dependent.

In the following box we give a procedure for solving a system in matrix form. A **leading variable** is a variable that corresponds to a leading entry in the echelon form of the matrix of a system.

When you study calculus or linear algebra, you will learn that the graph of a linear equation in three variables is a *plane* in a three-dimensional coordinate system. For a system of three equations in three variables, the following situations arise:

1. The three planes intersect in a single point.
   The system has a unique solution.

2. The three planes intersect in more than one point.
   The system has infinitely many solutions.

3. The three planes have no point in common.
   The system has no solution.

SOLVING A SYSTEM IN MATRIX FORM

Suppose the matrix form of a system of linear equations has been transformed by Gaussian elimination into echelon form.

1. If the echelon form contains a row that represents the equation $0 = c$, where c is not zero, then the system has *no solution.*

2. If each of the variables in the echelon form of the system is a leading variable, then the system has *exactly one solution,* which we find using back-substitution or Gauss-Jordan elimination.

3. If not all of the variables in the echelon form are leading variables, then the system has *infinitely many solutions,* which we find by using Gauss-Jordan elimination to convert to reduced echelon form and then expressing the leading variables in terms of the nonleading variables. The nonleading variables take on any real numbers as their values.

EXAMPLE 5 ■ **A System with No Solution**

Solve the following system.

$$\begin{cases} x - 3y + 2z = 12 \\ 2x - 5y + 5z = 14 \\ x - 2y + 3z = 20 \end{cases}$$

SOLUTION   We transform the system into echelon form.

$$\begin{bmatrix} 1 & -3 & 2 & 12 \\ 2 & -5 & 5 & 14 \\ 1 & -2 & 3 & 20 \end{bmatrix} \xrightarrow[R_3 - R_1 \to R_3]{R_2 - 2R_1 \to R_2} \begin{bmatrix} 1 & -3 & 2 & 12 \\ 0 & 1 & 1 & -10 \\ 0 & 1 & 1 & 8 \end{bmatrix}$$

$$\xrightarrow{R_3 - R_2 \to R_3} \begin{bmatrix} 1 & -3 & 2 & 12 \\ 0 & 1 & 1 & -10 \\ 0 & 0 & 0 & 18 \end{bmatrix} \xrightarrow{\frac{1}{18}R_3} \begin{bmatrix} 1 & -3 & 2 & 12 \\ 0 & 1 & 1 & -10 \\ 0 & 0 & 0 & 1 \end{bmatrix}$$

This last matrix is in echelon form, so we may stop the Gaussian elimination process. Now if we translate the last row back into equation form, we get $0x + 0y + 0z = 1$, or $0 = 1$, which is false. No matter what values we pick for x, y, and z, the last equation will never be a true statement. This means the system *has no solution.*   ■

EXAMPLE 6 ■ **A System with Infinitely Many Solutions**

Find the complete solution of the following system.

$$\begin{cases} -3x - 5y + 36z = 10 \\ -x \quad\quad + 7z = 5 \\ x + y - 10z = -4 \end{cases}$$

SOLUTION   We transform the system into echelon form.

$$\begin{bmatrix} -3 & -5 & 36 & 10 \\ -1 & 0 & 7 & 5 \\ 1 & 1 & -10 & -4 \end{bmatrix} \xrightarrow{R_1 \leftrightarrow R_3} \begin{bmatrix} 1 & 1 & -10 & -4 \\ -1 & 0 & 7 & 5 \\ -3 & -5 & 36 & 10 \end{bmatrix}$$

$$\xrightarrow[R_3 + 3R_1 \to R_3]{R_2 + R_1 \to R_2} \begin{bmatrix} 1 & 1 & -10 & -4 \\ 0 & 1 & -3 & 1 \\ 0 & -2 & 6 & -2 \end{bmatrix} \xrightarrow{R_3 + 2R_2 \to R_3} \begin{bmatrix} 1 & 1 & -10 & -4 \\ 0 & 1 & -3 & 1 \\ 0 & 0 & 0 & 0 \end{bmatrix}$$

$$\xrightarrow{R_1 - R_2 \to R_1} \begin{bmatrix} 1 & 0 & -7 & -5 \\ 0 & 1 & -3 & 1 \\ 0 & 0 & 0 & 0 \end{bmatrix}$$

The third row corresponds to the equation $0 = 0$. This equation is always true, no matter what values are used for x, y, and z. Since the equation adds no new information about the variables, we can drop it from the system. So the last matrix corresponds to the system

Leading variables $\left\{\begin{array}{l} x \quad\;\; - 7z = -5 \\ \quad\; y - 3z = \quad 1 \end{array}\right.$

We solve for the leading variables x and y in terms of the nonleading variable z:

$$x = 7z - 5 \qquad \text{Solve for } x \text{ in the first equation}$$

$$y = 3x + 1 \qquad \text{Solve for } y \text{ in the second equation}$$

To obtain the complete solution, we let z be any real number:

$$x = 7z - 5$$

$$y = 3z + 1$$

$$z = \text{any real number}$$

■

In Example 6, to get specific solutions we give a specific value to z. For example, if $z = 1$, then

$$x = 7(1) - 5 = 2$$

$$y = 3(1) + 1 = 4$$

Thus, $(2, 4, 1)$ is a solution to the system. We would get a different solution if we let $z = 2$, because we would then have

$$x = 7(2) - 5 = 9$$

$$y = 3(2) + 1 = 7$$

So $(9, 7, 2)$ is also a solution. There are infinitely many choices for z, so there are infinitely many solutions to the system.

EXAMPLE 7 ■ A System with Infinitely Many Solutions

Find the complete solution of the following system.

$$\begin{cases} x + 2y - 3z - 4w = 10 \\ x + 3y - 3z - 4w = 15 \\ 2x + 2y - 6z - 8w = 10 \end{cases}$$

SOLUTION   We transform the system into reduced echelon form.

$$\begin{bmatrix} 1 & 2 & -3 & -4 & 10 \\ 1 & 3 & -3 & -4 & 15 \\ 2 & 2 & -6 & -8 & 10 \end{bmatrix} \xrightarrow[\substack{R_3 - 2R_1 \rightarrow R_3}]{R_2 - R_1 \rightarrow R_2} \begin{bmatrix} 1 & 2 & -3 & -4 & 10 \\ 0 & 1 & 0 & 0 & 5 \\ 0 & -2 & 0 & 0 & -10 \end{bmatrix}$$

$$\xrightarrow{R_3 + 2R_2 \rightarrow R_3} \begin{bmatrix} 1 & 2 & -3 & -4 & 10 \\ 0 & 1 & 0 & 0 & 5 \\ 0 & 0 & 0 & 0 & 0 \end{bmatrix} \xrightarrow{R_1 - 2R_2 \rightarrow R_1} \begin{bmatrix} 1 & 0 & -3 & -4 & 0 \\ 0 & 1 & 0 & 0 & 5 \\ 0 & 0 & 0 & 0 & 0 \end{bmatrix}$$

This is in reduced echelon form. Since the last row represents the equation $0 = 0$, we may discard it. So the last matrix corresponds to the system

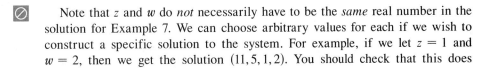

Leading variables $\begin{cases} x \qquad\ - 3z - 4w = 0 \\ \qquad y \qquad\qquad = 5 \end{cases}$

To obtain the complete solution, we solve for the leading variables x and y in terms of the nonleading variables z and w, and we let z and w be any real numbers. Thus, the complete solution is

$$x = 3z + 4w$$

$$y = 5$$

$$z = \text{any real number}$$

$$w = \text{any real number} \qquad\qquad\qquad ■$$

⊘   Note that z and w do *not* necessarily have to be the *same* real number in the solution for Example 7. We can choose arbitrary values for each if we wish to construct a specific solution to the system. For example, if we let $z = 1$ and $w = 2$, then we get the solution $(11, 5, 1, 2)$. You should check that this does indeed satisfy all three of the original equations in Example 7.

Examples 6 and 7 illustrate the following general fact: If a system in echelon form has n equations in m variables ($m > n$), then the complete solution will have $m - n$ nonleading variables. For instance, in Example 6 we arrived at two equations in the three variables x, y, and z with $3 - 2 = 1$ nonleading variable.

APPLICATIONS OF LINEAR SYSTEMS

Linear equations, often containing hundreds or even thousands of variables, occur frequently in the applications of algebra to the sciences and to other fields. For now, we consider an example that involves only three variables.

EXAMPLE 8 ■ Nutritional Analysis Using a System of Linear Equations

A nutritionist is performing an experiment on student volunteers. He wishes to feed one of his subjects a daily diet that consists of a combination of three commercial diet foods: MiniCal, SloStarve, and SlimQuick. For the experiment it's important that the subject consume exactly 500 mg of potassium, 75 g of protein, and 1150 units of vitamin D every day. The amounts of these nutrients in one ounce of each food are given in the following table.

	MiniCal	SloStarve	SlimQuick
Potassium (mg)	50	75	10
Protein (g)	5	10	3
Vitamin D (units)	90	100	50

How many ounces of each food should the subject eat every day to satisfy the nutrient requirements exactly?

SOLUTION   Let x, y, and z represent the number of ounces of MiniCal, SloStarve, and SlimQuick, respectively, that the subject should eat every day. This means that he will get $50x$ mg of potassium from MiniCal, $75y$ mg from SloStarve, and $10z$ mg from SlimQuick, for a total of $50x + 75y + 10z$ mg potassium in all. Since the potassium requirement is 500 mg, we get the first equation below. Similar reasoning for the protein and vitamin D requirements leads to the system

$$\begin{cases} 50x + 75y + 10z = 500 & \text{Potassium} \\ 5x + 10y + 3z = 75 & \text{Protein} \\ 90x + 100y + 50z = 1150 & \text{Vitamin D} \end{cases}$$

Dividing the first equation by 5 and the third one by 10 gives the system

$$\begin{cases} 10x + 15y + 2z = 100 \\ 5x + 10y + 3z = 75 \\ 9x + 10y + 5z = 115 \end{cases}$$

We solve this system using Gaussian elimination.

$$\begin{bmatrix} 10 & 15 & 2 & 100 \\ 5 & 10 & 3 & 75 \\ 9 & 10 & 5 & 115 \end{bmatrix} \xrightarrow{R_1 - R_3 \to R_1} \begin{bmatrix} 1 & 5 & -3 & -15 \\ 5 & 10 & 3 & 75 \\ 9 & 10 & 5 & 115 \end{bmatrix}$$

$$\xrightarrow[R_3 - 9R_1 \to R_3]{R_2 - 5R_1 \to R_2} \begin{bmatrix} 1 & 5 & -3 & -15 \\ 0 & -15 & 18 & 150 \\ 0 & -35 & 32 & 250 \end{bmatrix} \xrightarrow{-\frac{1}{3}R_2} \begin{bmatrix} 1 & 5 & -3 & -15 \\ 0 & 5 & -6 & -50 \\ 0 & -35 & 32 & 250 \end{bmatrix}$$

$$\xrightarrow{R_3 + 7R_2 \to R_3} \begin{bmatrix} 1 & 5 & -3 & -15 \\ 0 & 5 & -6 & -50 \\ 0 & 0 & -10 & -100 \end{bmatrix} \xrightarrow{-\frac{1}{10}R_3} \begin{bmatrix} 1 & 5 & -3 & -15 \\ 0 & 5 & -6 & -50 \\ 0 & 0 & 1 & 10 \end{bmatrix}$$

$x = 5$, $y = 2$, $z = 10$:

$$\begin{cases} 10(5) + 15(2) + 2(10) = 100 \\ 5(5) + 10(2) + 3(10) = 75 \\ 9(5) + 10(2) + 5(10) = 115 \checkmark \end{cases}$$

Now we back-substitute to get $z = 10$, $y = 2$, and $x = 5$. The subject should be fed 5 oz of MiniCal, 2 oz of SloStarve, and 10 oz of SlimQuick every day. ■

A more practical application might involve dozens of foods and nutrients, rather than just three. As you may imagine, such a problem would be almost impossible to solve without the assistance of a computer. Many graphing calculators, including the TI-82, TI-83, and TI-85, are capable of performing elementary row operations on matrices, so you can use them to solve more complicated systems.

8.3 EXERCISES

1–4 ■ State whether the equation or system of equations is linear.

1. $6x - 3y + 1000z - w = \sqrt{13}$

2. $x_1^2 + x_2^2 + x_3^2 = 36$

3. $e^2 x_1 + \pi x_2 - \sqrt{5} = x_3 - \frac{1}{2}x_4$

4. $\begin{cases} x - 3xy + 5y = 0 \\ 12x + 321y = 123 \end{cases}$

5–10 ■ (a) Determine whether the matrix is in echelon form.
(b) Determine whether the matrix is in reduced echelon form.
(c) Write the system of equations that corresponds to the matrix.

5. $\begin{bmatrix} 1 & 0 & -3 \\ 0 & 1 & 5 \end{bmatrix}$

6. $\begin{bmatrix} 1 & 3 & -3 \\ 0 & 1 & 5 \end{bmatrix}$

7. $\begin{bmatrix} 1 & 2 & 8 & 0 \\ 0 & 1 & 3 & 2 \\ 0 & 0 & 0 & 0 \end{bmatrix}$

8. $\begin{bmatrix} 1 & 0 & -7 & 0 \\ 0 & 1 & 3 & 0 \\ 0 & 0 & 0 & 1 \end{bmatrix}$

9. $\begin{bmatrix} 1 & 0 & 0 & 0 \\ 0 & 0 & 0 & 0 \\ 0 & 1 & 5 & 1 \end{bmatrix}$

10. $\begin{bmatrix} 1 & 0 & 0 & 1 \\ 0 & 1 & 0 & 2 \\ 0 & 0 & 1 & 3 \end{bmatrix}$

11–20 ■ Find the unique solution of the system of linear equations. Use Gaussian elimination or Gauss-Jordan elimination.

11. $\begin{cases} x - 2y + z = 1 \\ y + 2z = 5 \\ x + y + 3z = 8 \end{cases}$

12. $\begin{cases} x + y + 6z = 3 \\ x + y + 3z = 3 \\ x + 2y + 4z = 7 \end{cases}$

13. $\begin{cases} x + y + z = 2 \\ 2x - 3y + 2z = 4 \\ 4x + y - 3z = 1 \end{cases}$

14. $\begin{cases} x + y + z = 4 \\ -x + 2y + 3z = 17 \\ 2x - y = -7 \end{cases}$

15. $\begin{cases} x + 2y - z = -2 \\ x + z = 0 \\ 2x - y - z = -3 \end{cases}$

16. $\begin{cases} 2y + z = 4 \\ x + y = 4 \\ 3x + 3y - z = 10 \end{cases}$

17. $\begin{cases} x_1 + 2x_2 - x_3 = 9 \\ 2x_1 - x_3 = -2 \\ 3x_1 + 5x_2 + 2x_3 = 22 \end{cases}$

18. $\begin{cases} 2x_1 + x_2 = 7 \\ 2x_1 - x_2 + x_3 = 6 \\ 3x_1 - 2x_2 + 4x_3 = 11 \end{cases}$

19. $\begin{cases} 2x - 3y - z = 13 \\ -x + 2y - 5z = 6 \\ 5x - y - z = 49 \end{cases}$

20. $\begin{cases} 10x + 10y - 20z = 60 \\ 15x + 20y + 30z = -25 \\ -5x + 30y - 10z = 45 \end{cases}$

21–30 ■ Determine whether the system of linear equations is inconsistent or dependent. If it is dependent, find the complete solution.

21. $\begin{cases} x + y + z = 2 \\ y - 3z = 1 \\ 2x + y + 5z = 0 \end{cases}$

22. $\begin{cases} x + 3z = 3 \\ 2x + y - 2z = 5 \\ -y + 8z = 8 \end{cases}$

23. $\begin{cases} 2x - 3y - 9z = -5 \\ x + 3z = 2 \\ -3x + y - 4z = -3 \end{cases}$

24. $\begin{cases} x - 2y + 5z = 3 \\ -2x + 6y - 11z = 1 \\ 3x - 16y + 20z = -26 \end{cases}$

25. $\begin{cases} x - y + 3z = 3 \\ 4x - 8y + 32z = 24 \\ 2x - 3y + 11z = 4 \end{cases}$

26. $\begin{cases} -2x + 6y - 2z = -12 \\ x - 3y + 2z = 10 \\ -x + 3y + 2z = 6 \end{cases}$

27. $\begin{cases} x + 4y - 2z = -3 \\ 2x - y + 5z = 12 \\ 8x + 5y + 11z = 30 \end{cases}$ **28.** $\begin{cases} 3r + 2s - 3t = 10 \\ r - s - t = -5 \\ r + 4s - t = 20 \end{cases}$

29. $\begin{cases} 2x + y - 2z = 12 \\ -x - \frac{1}{2}y + z = -6 \\ 3x + \frac{3}{2}y - 3z = 18 \end{cases}$ **30.** $\begin{cases} y - 5z = 7 \\ 3x + 2y = 12 \\ 3x + 10z = 80 \end{cases}$

31–42 ■ Solve the system of linear equations.

31. $\begin{cases} 4x - 3y + z = -8 \\ -2x + y - 3z = -4 \\ x - y + 2z = 3 \end{cases}$

32. $\begin{cases} 2x - 3y + 5z = 14 \\ 4x - y - 2z = -17 \\ -x - y + z = 3 \end{cases}$

33. $\begin{cases} x + 2y - 3z = -5 \\ -2x - 4y - 6z = 10 \\ 3x + 7y - 2z = -13 \end{cases}$

34. $\begin{cases} 3x - y + 2z = -1 \\ 4x - 2y + z = -7 \\ -x + 3y - 2z = -1 \end{cases}$

35. $\begin{cases} -x + 2y + z - 3w = 3 \\ 3x - 4y + z + w = 9 \\ -x - y + z + w = 0 \\ 2x + y + 4z - 2w = 3 \end{cases}$

36. $\begin{cases} x + y - z - w = 6 \\ 2x + z - 3w = 8 \\ x - y + 4w = -10 \\ 3x + 5y - z - w = 20 \end{cases}$

37. $\begin{cases} x + y + 2z - w = -2 \\ 3y + z + 2w = 2 \\ x + y + 3w = 2 \\ -3x + z + 2w = 5 \end{cases}$

38. $\begin{cases} x - 3y + 2z + w = -2 \\ x - 2y - 2w = -10 \\ z + 5w = 15 \\ 3x + 2z + w = -3 \end{cases}$

39. $\begin{cases} x + z + w = 4 \\ y - z = -4 \\ x - 2y + 3z + w = 12 \\ 2x - 2z + 5w = -1 \end{cases}$

40. $\begin{cases} y - z + 2w = 0 \\ 3x + 2y + w = 0 \\ 2x + 4w = 12 \\ -2x - 2z + 5w = 6 \end{cases}$

41. $\begin{cases} x - y + w = 0 \\ 3x - z + 2w = 0 \\ x - 4y + z + 2w = 0 \end{cases}$

42. $\begin{cases} 2x - y + 2z + w = 5 \\ -x + y + 4z - w = 3 \\ 3x - 2y - z = 0 \end{cases}$

43. A doctor recommends that a patient take 50 mg each of niacin, riboflavin, and thiamin daily to alleviate a vitamin deficiency. Looking into his medicine chest at home, the patient finds three brands of vitamin pills. The amounts of the relevant vitamins per pill are given in the table.

	VitaMax	Vitron	VitaPlus
Niacin (mg)	5	10	15
Riboflavin (mg)	15	20	0
Thiamin (mg)	10	10	10

How many pills of each type should he take every day to fulfill the prescription?

44. A chemist has three containers of acid solution at various concentrations. The first is 10% acid, the second is 20%, and the third is 40%. How many milliliters of each should he mix together to make 100 mL of acid at 18% concentration, if he has to use four times as much of the 10% solution as the 40% solution?

45. Amanda, Bryce, and Corey enter a race in which they have to run, swim, and cycle over a marked course. Their average speeds are given in the table.

	Average Speed (mi/h)		
	Running	Swimming	Cycling
Amanda	10	4	20
Bryce	$7\frac{1}{2}$	6	15
Corey	15	3	40

Corey finishes first with a total time of 1 h 45 min. Amanda comes in second with a time of 2 h 30 min. Bryce finishes last with a time of 3 h. Find the distance (in miles) for each part of the race.

46. A small school has 100 students who occupy three classrooms: rooms A, B, and C. After the first period of the school day, half the students in room A move to room B, one-fifth of the students in room B move to room C, and one-third of the students in room C move to room A. Nevertheless, the total number of students in each room is the same for each period. How many students occupy each room?

47. A furniture factory makes wooden tables, chairs, and armoires. Each piece of furniture requires three production steps: cutting the wood, assembling, and finishing. The number of hours (h) of each operation required to make a piece of furniture is given in the table.

	Table	Chair	Armoire
Cutting (h)	$\frac{1}{2}$	1	1
Assembling (h)	$\frac{1}{2}$	$1\frac{1}{2}$	1
Finishing (h)	1	$1\frac{1}{2}$	2

The workers in the factory can provide 300 hours of cutting, 400 hours of assembling, and 590 hours of finishing each work week. How many tables, chairs, and armoires should be produced so that all available labor-hours are used? Or is this impossible?

48. A diagram of a section of the street network in a city is shown in the figure, where the arrows indicate one-way streets. The numbers on the diagram show how many cars enter or leave this section of the city via the indicated street in a certain one-hour period. The variables x, y, z, and w represent the number of cars that travel along the portions of First, Second, Avocado, and Birch

Streets during this period. Find x, y, z, and w, assuming that none of the cars involved in this problem stop or park on any of the streets shown in the figure.

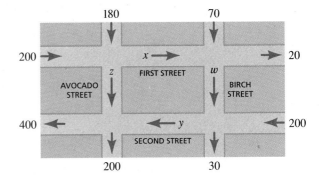

49. (a) Suppose that (x_0, y_0, z_0) and (x_1, y_1, z_1) are solutions of the system

$$\begin{cases} a_1 x + b_1 y + c_1 z = d_1 \\ a_2 x + b_2 y + c_2 z = d_2 \\ a_3 x + b_3 y + c_3 z = d_3 \end{cases}$$

Show that $\left(\dfrac{x_0 + x_1}{2}, \dfrac{y_0 + y_1}{2}, \dfrac{z_0 + z_1}{2}\right)$ is also a solution.

(b) Use the result of part (a) to prove that if the system has two different solutions, then it has infinitely many solutions.

▲ DISCOVERY · DISCUSSION

50. Polynomials Determined by a Set of Points We all know that two points uniquely determine a line $y = ax + b$ in the coordinate plane. Similarly, three points uniquely determine a quadratic (second-degree) polynomial $y = ax^2 + bx + c$, four points uniquely determine a cubic (third-degree) polynomial $y = ax^3 + bx^2 + cx + d$, and so on. (The only exceptions to this rule are if the three points actually lie on a line, or the four points lie on a quadratic or line, and so on.) For the following set of four points, find the line that contains the first two points, the quadratic that contains the first three points, and the cubic that contains all four points.

$$(0, 0), \quad (1, 12), \quad (3, 6), \quad (-1, -14)$$

Graph the points and functions in the same viewing rectangle using a graphing calculator.

8.4 THE ALGEBRA OF MATRICES

Up to this point we've been using matrices simply as a notational convenience. Matrices have many other uses in mathematics and the sciences, and for most of these applications a knowledge of matrix algebra is essential. Like numbers, matrices can be added, subtracted, multiplied, and divided. In this section we learn how to perform these algebraic operations on matrices.

Recall that a matrix is simply a rectangular array of numbers enclosed between brackets. For example, let A be the matrix

$$A = \begin{bmatrix} -1 & 4 & 7 & 0 \\ 0 & 2 & 13 & 14 \\ \frac{1}{2} & 22 & 8 & -2 \end{bmatrix}$$

The **dimension** of a matrix is a pair of numbers that indicates how many **rows** and **columns** a matrix has. The matrix A is a 3×4 matrix because it has 3 horizontal rows and 4 vertical columns. The individual numbers that make up a matrix are called its **entries**, and they are specified by their row and column position. In the matrix A, the number 13 is the $(2, 3)$ entry, since it is in the second row and the third column. If the name of a matrix is A, we will often use the symbol a_{ij} to denote the (i, j) entry of the matrix. Thus, for the preceding matrix, we have $a_{24} = 14$ and $a_{32} = 22$.

Two matrices are **equal** if they have the same dimension and their corresponding entries are equal. So

> $A = B$ if and only if both A and B have the same dimension $m \times n$, and $a_{ij} = b_{ij}$ for $i = 1, 2, \ldots, m$ and $j = 1, 2, \ldots, n$.

Equal matrices

$$\begin{bmatrix} \sqrt{4} & 2^2 & e^0 \\ 0.5 & 1 & 1-1 \end{bmatrix} = \begin{bmatrix} 2 & 4 & 1 \\ \frac{1}{2} & \frac{2}{2} & 0 \end{bmatrix}$$

Unequal matrices

$$\begin{bmatrix} 1 & 2 \\ 3 & 4 \\ 5 & 6 \end{bmatrix} \neq \begin{bmatrix} 1 & 3 & 5 \\ 2 & 4 & 6 \end{bmatrix}$$

ALGEBRAIC OPERATIONS ON MATRICES

Two matrices can be added or subtracted if they have the same dimension. (Otherwise, their sum or difference is undefined.) We add or subtract the matrices by adding or subtracting corresponding entries. To multiply a matrix by a number, we multiply every element of the matrix by that number. This is called the *scalar product*.

> ### SUM, DIFFERENCE, AND SCALAR PRODUCT OF MATRICES
>
> If A and B are matrices of the same dimension and if k is any real number, then
>
> 1. The **sum** $A + B$ is the matrix of the same dimension as A and B, and its (i, j) entry is $a_{ij} + b_{ij}$.
>
> 2. The **difference** $A - B$ is the matrix of the same dimension as A and B, and its (i, j) entry is $a_{ij} - b_{ij}$.
>
> 3. The **scalar product** kA is the matrix of the same dimension as A, and its (i, j) entry is ka_{ij}.

EXAMPLE 1 ■ Performing Algebraic Operations on Matrices

Let

$$A = \begin{bmatrix} 2 & -3 \\ 0 & 5 \\ 7 & -\frac{1}{2} \end{bmatrix} \qquad B = \begin{bmatrix} 1 & 0 \\ -3 & 1 \\ 2 & 2 \end{bmatrix}$$

$$C = \begin{bmatrix} 7 & -3 & 0 \\ 0 & 1 & 5 \end{bmatrix} \qquad D = \begin{bmatrix} 6 & 0 & -6 \\ 8 & 1 & 9 \end{bmatrix}$$

Julia Robinson (1919–1985) was born in St. Louis, Missouri, and grew up at Point Loma, California. Due to an illness, Robinson missed two years of school but later, with the aid of a tutor, she completed fifth, sixth, seventh, and eighth grades, all in one year. At San Diego State University, she found mathematics especially interesting after reading biographies of mathematicians in the book *Men of Mathematics* by E. T. Bell. She said, "I cannot overemphasize the importance of such books... in the intellectual life of a student." Robinson is famous for her work on Hilbert's tenth problem (page 578), which asks for a general procedure for determining whether an equation has integer solutions. Her ideas led to a complete answer to the problem. Interestingly, the answer involved certain properties of the Fibonacci numbers (page 687) discovered by the then-22-year-old Russian mathematician Yuri Matijasevič. As a result of her brilliant work on Hilbert's tenth problem, Robinson was offered a professorship at the University of California, Berkley, and became the first woman mathematician elected to the National Academy of Sciences. She also served as president of the American Mathematical Society.

Carry out each indicated operation, or explain why it cannot be performed.

(a) $A + B$ (b) $C - D$ (c) $C + A$ (d) $5A$

SOLUTION

(a) $A + B = \begin{bmatrix} 2 & -3 \\ 0 & 5 \\ 7 & -\frac{1}{2} \end{bmatrix} + \begin{bmatrix} 1 & 0 \\ -3 & 1 \\ 2 & 2 \end{bmatrix} = \begin{bmatrix} 3 & -3 \\ -3 & 6 \\ 9 & \frac{3}{2} \end{bmatrix}$

(b) $C - D = \begin{bmatrix} 7 & -3 & 0 \\ 0 & 1 & 5 \end{bmatrix} - \begin{bmatrix} 6 & 0 & -6 \\ 8 & 1 & 9 \end{bmatrix} = \begin{bmatrix} 1 & -3 & 6 \\ -8 & 0 & -4 \end{bmatrix}$

(c) $C + A$ is undefined because we can't add matrices of different dimensions.

(d) $5A = 5\begin{bmatrix} 2 & -3 \\ 0 & 5 \\ 7 & -\frac{1}{2} \end{bmatrix} = \begin{bmatrix} 10 & -15 \\ 0 & 25 \\ 35 & -\frac{5}{2} \end{bmatrix}$ ■

MATRIX MULTIPLICATION

Multiplication of two matrices is more difficult to describe than other matrix operations. We will see in later examples why taking the matrix product involves a rather complex procedure, which we now describe.

First, the product AB (or $A \cdot B$) of two matrices A and B is defined only when the number of columns in A is equal to the number of rows in B. This means that if we write their dimensions side by side, the two inner numbers must match:

Matrices:	A	B
Dimensions:	$m \times n$	$n \times k$
	↑	↑
	columns in A	rows in B

If the dimensions of A and B match in this fashion, then the product AB will have dimension $m \times k$. Before describing the procedure for obtaining the elements of AB, we define the *inner product* of a row of A and a column of B.

If $[a_1 \quad a_2 \quad \cdots \quad a_n]$ is a row of A, and if $\begin{bmatrix} b_1 \\ b_2 \\ \cdot \\ \cdot \\ \cdot \\ b_n \end{bmatrix}$ is a column of B, then

their **inner product** is the number $a_1b_1 + a_2b_2 + \cdots + a_nb_n$.

For example,

$$[2 \quad -1 \quad 0 \quad 4] \cdot \begin{bmatrix} 5 \\ 4 \\ -3 \\ \frac{1}{2} \end{bmatrix} = 2 \cdot 5 + (-1) \cdot 4 + 0 \cdot (-3) + 4 \cdot \tfrac{1}{2} = 8$$

We now define the **product** AB of two matrices.

THE PRODUCT OF TWO MATRICES

Suppose that A is an $m \times n$ matrix and B an $n \times k$ matrix. Then $C = AB$ is an $m \times k$ matrix, where c_{ij} is the inner product of the ith row of A and the jth column of B.

EXAMPLE 2 ■ Multiplying Matrices

Let

$$A = \begin{bmatrix} 1 & 3 \\ -1 & 0 \end{bmatrix} \quad \text{and} \quad B = \begin{bmatrix} -1 & 5 & 2 \\ 0 & 4 & 7 \end{bmatrix}$$

$$
\begin{array}{ccc}
A & \cdot & B \\
\downarrow & & \downarrow \\
2 \times 2 & & 2 \times 3
\end{array}
$$

Inner numbers match, so product is defined.
Outer numbers give dimension of product: 2×3.

Calculate, if possible, the products AB and BA.

SOLUTION   Since A has dimension 2×2 and B has dimension 2×3, the product AB is defined and has dimension 2×3. We can thus write

$$AB = \begin{bmatrix} 1 & 3 \\ -1 & 0 \end{bmatrix} \begin{bmatrix} -1 & 5 & 2 \\ 0 & 4 & 7 \end{bmatrix} = \begin{bmatrix} ? & ? & ? \\ ? & ? & ? \end{bmatrix}$$

where the question marks must be filled in using the rule defining the entries of a matrix product. The $(1, 1)$ entry will be the inner product of the first row of A and the first column of B:

$$\begin{bmatrix} 1 & 3 \\ -1 & 0 \end{bmatrix} \begin{bmatrix} -1 & 5 & 2 \\ 0 & 4 & 7 \end{bmatrix} \qquad 1 \cdot (-1) + 3 \cdot 0 = -1$$

Similarly, we calculate the remaining entries as follows.

Entry	Inner Product of:	Value	Product matrix
$(1,2)$	$\begin{bmatrix} 1 & 3 \\ -1 & 0 \end{bmatrix} \begin{bmatrix} -1 & 5 & 2 \\ 0 & 4 & 7 \end{bmatrix}$	$1 \cdot 5 + 3 \cdot 4 = 17$	$\begin{bmatrix} -1 & 17 & \\ & & \end{bmatrix}$
$(1,3)$	$\begin{bmatrix} 1 & 3 \\ -1 & 0 \end{bmatrix} \begin{bmatrix} -1 & 5 & 2 \\ 0 & 4 & 7 \end{bmatrix}$	$1 \cdot 2 + 3 \cdot 7 = 23$	$\begin{bmatrix} -1 & 17 & 23 \\ & & \end{bmatrix}$
$(2,1)$	$\begin{bmatrix} 1 & 3 \\ -1 & 0 \end{bmatrix} \begin{bmatrix} -1 & 5 & 2 \\ 0 & 4 & 7 \end{bmatrix}$	$(-1) \cdot (-1) + 0 \cdot 0 = 1$	$\begin{bmatrix} -1 & 17 & 23 \\ 1 & & \end{bmatrix}$
$(2,2)$	$\begin{bmatrix} 1 & 3 \\ -1 & 0 \end{bmatrix} \begin{bmatrix} -1 & 5 & 2 \\ 0 & 4 & 7 \end{bmatrix}$	$(-1) \cdot 5 + 0 \cdot 4 = -5$	$\begin{bmatrix} -1 & 17 & 23 \\ 1 & -5 & \end{bmatrix}$
$(2,3)$	$\begin{bmatrix} 1 & 3 \\ -1 & 0 \end{bmatrix} \begin{bmatrix} -1 & 5 & 2 \\ 0 & 4 & 7 \end{bmatrix}$	$(-1) \cdot 2 + 0 \cdot 7 = -2$	$\begin{bmatrix} -1 & 17 & 23 \\ 1 & -5 & -2 \end{bmatrix}$

Thus, we have

$$AB = \begin{bmatrix} -1 & 17 & 23 \\ 1 & -5 & -2 \end{bmatrix}$$

The product BA is not defined, however, because the dimensions are

$$2 \times 3 \quad \text{and} \quad 2 \times 2$$

The inner two numbers are not the same, so the rows and columns won't match up when we try to calculate the product. ■

⊘   The next example shows that even when both AB and BA are defined, they aren't necessarily equal. This result will prove that matrix multiplication is *not* commutative.

EXAMPLE 3 ■ **Matrix Multiplication Is Not Commutative**

Let

$$A = \begin{bmatrix} 5 & 7 \\ -3 & 0 \end{bmatrix} \quad \text{and} \quad B = \begin{bmatrix} 1 & 2 \\ 9 & -1 \end{bmatrix}$$

Calculate the products AB and BA.

SOLUTION   Since both matrices A and B have dimension 2×2, both products AB and BA are defined, and each product is also a 2×2 matrix.

$$AB = \begin{bmatrix} 5 & 7 \\ -3 & 0 \end{bmatrix} \begin{bmatrix} 1 & 2 \\ 9 & -1 \end{bmatrix} = \begin{bmatrix} 5 \cdot 1 + 7 \cdot 9 & 5 \cdot 2 + 7 \cdot (-1) \\ (-3) \cdot 1 + 0 \cdot 9 & (-3) \cdot 2 + 0 \cdot (-1) \end{bmatrix} = \begin{bmatrix} 68 & 3 \\ -3 & -6 \end{bmatrix}$$

$$BA = \begin{bmatrix} 1 & 2 \\ 9 & -1 \end{bmatrix} \begin{bmatrix} 5 & 7 \\ -3 & 0 \end{bmatrix} = \begin{bmatrix} 1 \cdot 5 + 2 \cdot (-3) & 1 \cdot 7 + 2 \cdot 0 \\ 9 \cdot 5 + (-1) \cdot (-3) & 9 \cdot 7 + (-1) \cdot 0 \end{bmatrix} = \begin{bmatrix} -1 & 7 \\ 48 & 63 \end{bmatrix}$$

This shows that, in general, $AB \neq BA$. In fact, in this example, AB and BA don't even have any entry in common.   ∎

Although matrix multiplication is not commutative, it does obey the Associative and Distributive Properties.

PROPERTIES OF MATRIX MULTIPLICATION

Let A, B, C, and D be matrices for which the following products are defined. Then

$$A(BC) = (AB)C \qquad \text{Associative Property}$$

$$A(B + C) = AB + AC$$
$$(B + C)D = BD + CD \qquad \text{Distributive Property}$$

The next two examples give some indication of why mathematicians chose to define the matrix product in such an apparently bizarre fashion.

EXAMPLE 4 ■ Writing a System of Equations as a Matrix Equation

Matrix equations are described in more detail on page 572.

Show that the matrix equation

$$\begin{bmatrix} 1 & -1 & 3 \\ 1 & 2 & -2 \\ 3 & -1 & 5 \end{bmatrix} \begin{bmatrix} x \\ y \\ z \end{bmatrix} = \begin{bmatrix} 4 \\ 10 \\ 14 \end{bmatrix}$$

is equivalent to the system of equations in Example 1 of Section 8.3.

SOLUTION   If we perform the matrix multiplication on the left side of the given equation, we get

$$\begin{bmatrix} x - y + 3z \\ x + 2y - 2z \\ 3x - y + 5z \end{bmatrix} = \begin{bmatrix} 4 \\ 10 \\ 14 \end{bmatrix}$$

Because two matrices are equal only if their corresponding entries are equal, this matrix equation means that

$$\begin{cases} x - y + 3z = 4 \\ x + 2y - 2z = 10 \\ 3x - y + 5z = 14 \end{cases}$$

This is exactly the system of equations we had in Example 1 of Section 8.3. ∎

The preceding example shows that our definition of matrix product allows us to express a system of linear equations as a single matrix equation in a natural way.

EXAMPLE 5 ■ Representing Demographic Data in Terms of Matrices

In a certain city the proportion of voters in each age group who are registered as Democrats, Republicans, or Independents is given by the following matrix.

	18–30	31–50	Over 50
Democrat	0.30	0.60	0.50
Republican	0.50	0.35	0.25
Independent	0.20	0.05	0.25

(Age across the top) $= A$

The next matrix gives the distribution, by age and sex, of the voting population of this city.

	Male	Female
18–30	5,000	6,000
31–50	10,000	12,000
Over 50	12,000	15,000

(Age down the side) $= B$

For the purpose of this problem, let's make the (highly unrealistic) assumption that within each age group, political preference is not related to gender. That is, the percentage of Democrat males in the 18–30 group, for example, is the same as the percentage of Democrat females in this group.

(a) Calculate the product AB.
(b) How many males are registered as Democrats in this city?
(c) How many females are registered as Republicans?

SOLUTION

(a) $$AB = \begin{bmatrix} 0.30 & 0.60 & 0.50 \\ 0.50 & 0.35 & 0.25 \\ 0.20 & 0.05 & 0.25 \end{bmatrix} \begin{bmatrix} 5{,}000 & 6{,}000 \\ 10{,}000 & 12{,}000 \\ 12{,}000 & 15{,}000 \end{bmatrix} = \begin{bmatrix} 13{,}500 & 16{,}500 \\ 9{,}000 & 10{,}950 \\ 4{,}500 & 5{,}550 \end{bmatrix}$$

Olga Taussky-Todd (1906–1995) was one of the world's leaders in developing applications of Matrix Theory. She has been described as "in love with anything matrices can do." She successfully applied matrices to the study of aerodynamics, a field used in the design of airplanes and rockets. Taussky-Todd was also famous for her work in Number Theory, a subject that deals with prime numbers and divisibility. Although Number Theory has often been called the least applicable branch of mathematics, it is now used in significant ways throughout the computer industry.

Taussky-Todd studied mathematics at a time when it was very uncommon for a young woman to want to be a mathematician. She said, "When I entered university I had no idea what it meant to study mathematics." But she became one of the most respected mathematicians of her time. She was for many years a professor of mathematics at Caltech in Pasadena.

(b) When we take the inner product of a row from A with a column from B, we are adding the number of people in each of the three age groups who belong to the category in question. For example, the $(2, 1)$ entry of AB (the 9,000) was obtained by taking the inner product of the Republican row from A with the Male column from B. This number is therefore the total number of male Republicans in this city. We can label the rows and columns of AB as follows.

$$\begin{array}{c} & \text{Male} \quad \text{Female} \\ \begin{array}{r} \text{Democrat} \\ \text{Republican} \\ \text{Independent} \end{array} & \begin{bmatrix} 13{,}500 & 16{,}500 \\ 9{,}000 & 10{,}950 \\ 4{,}500 & 5{,}550 \end{bmatrix} = AB \end{array}$$

There are 13,500 males registered as Democrats in this city.

(c) There are 10,950 females registered as Republicans. ∎

If we add the entries in the columns of matrix A in Example 5, we see that in each case the sum is 1. (Can you see why this has to be true, given what the matrix describes?) A matrix with this property is called **stochastic**. Stochastic matrices are studied extensively in statistics, where they arise frequently in situations like the one described in Example 5.

8.4 EXERCISES

1–21 ■ The matrices A, B, C, D, E, F, and G are defined as follows.

$$A = \begin{bmatrix} 2 & -5 \\ 0 & 7 \end{bmatrix} \qquad B = \begin{bmatrix} 3 & \frac{1}{2} & 5 \\ 1 & -1 & 3 \end{bmatrix} \qquad C = \begin{bmatrix} 2 & -\frac{5}{2} & 0 \\ 0 & 2 & -3 \end{bmatrix}$$

$$D = \begin{bmatrix} 7 & 3 \end{bmatrix} \qquad E = \begin{bmatrix} 1 \\ 2 \\ 0 \end{bmatrix} \qquad F = \begin{bmatrix} 1 & 0 & 0 \\ 0 & 1 & 0 \\ 0 & 0 & 1 \end{bmatrix}$$

$$G = \begin{bmatrix} 5 & -3 & 10 \\ 6 & 1 & 0 \\ -5 & 2 & 2 \end{bmatrix}$$

Carry out the indicated algebraic operation, or explain why it cannot be performed.

1. $B + C$

2. $B + F$

3. $C - B$

4. $5A$

5. $3B + 2C$

6. $C - 5A$

7. $2C - 6B$

8. DA

9. AD

10. BC

11. BF

12. GF

13. $(DA)B$

14. $D(AB)$

15. GE

16. A^2

17. A^3

18. $DB + DC$

19. B^2

20. F^2

21. $BF + FE$

22. What must be true about the dimensions of the matrices A and B if both products AB and BA are defined?

23–26 ■ Write the system of equations as a matrix equation (see Example 4).

23. $\begin{cases} 2x - 5y = 7 \\ 3x + 2y = 4 \end{cases}$

24. $\begin{cases} 6x - y + z = 12 \\ 2x + z = 7 \\ y - 2z = 4 \end{cases}$

25. $\begin{cases} 3x_1 + 2x_2 - x_3 + x_4 = 0 \\ x_1 - x_3 = 5 \\ 3x_2 + x_3 - x_4 = 4 \end{cases}$

26. $\begin{cases} x - y + z = 2 \\ 4x - 2y - z = 2 \\ x + y + 5z = 2 \\ -x - y - z = 2 \end{cases}$

27–28 ■ Solve for x and y.

27. $\begin{bmatrix} 6 & x \\ 1 & 0 \end{bmatrix} \begin{bmatrix} y & 2 \\ -1 & 2 \end{bmatrix} = \begin{bmatrix} 4 & 16 \\ 1 & 2 \end{bmatrix}$

28. $\begin{bmatrix} 2 & 5 \\ 1 & 3 \end{bmatrix} \begin{bmatrix} x \\ y \end{bmatrix} = \begin{bmatrix} -2 \\ -3 \end{bmatrix}$

29–32 ■ Solve the matrix equation for the unknown matrix X, or explain why there is no solution. Here

$$A = \begin{bmatrix} 4 & 6 \\ 1 & 3 \end{bmatrix} \qquad B = \begin{bmatrix} 2 & 5 \\ 3 & 7 \end{bmatrix}$$

$$C = \begin{bmatrix} 2 & 3 \\ 1 & 0 \\ 0 & 2 \end{bmatrix} \qquad D = \begin{bmatrix} 10 & 20 \\ 30 & 20 \\ 10 & 0 \end{bmatrix}$$

29. $2X - A = B$

30. $5(X - C) = D$

31. $3X + B = C$

32. $A + D = 3X$

33. A small fast-food chain has restaurants in Santa Monica, Long Beach, and Anaheim. Only hamburgers, hot dogs, and milk shakes are sold by this chain. On a certain day, sales were distributed according to the following matrix.

	Number of items sold		
	Santa Monica	Long Beach	Anaheim
Hamburgers	4000	1000	3500
Hot dogs	400	300	200
Milk shakes	700	500	9000

$= A$

The price of each item is given by the following matrix.

Hamburger	Hot dog	Milk shake
[$0.90	$0.80	$1.10]

$= B$

(a) Calculate the product BA.
(b) Interpret the entries in the product matrix BA.

34. A specialty-car manufacturer has plants in Auburn, Biloxi, and Chattanooga. Three models are produced, with daily production given in the following matrix.

	Cars produced each day		
	Model K	Model R	Model W
Auburn	12	10	0
Biloxi	4	4	20
Chattanooga	8	9	12

$= A$

Because of a wage increase, February profits are less than January profits. The profit per car is tabulated by model in the following matrix.

	January	February
Model K	$1000	$500
Model R	$2000	$1200
Model W	$1500	$1000

$= B$

(a) Calculate AB.
(b) Assuming all cars produced were sold, what was the daily profit in January from the Biloxi plant?
(c) What was the total daily profit (from all three plants) in February?

35. Let

$$A = \begin{bmatrix} 1 & 0 & 6 & -1 \\ 2 & \frac{1}{2} & 4 & 0 \end{bmatrix} \qquad C = \begin{bmatrix} 1 \\ 0 \\ -1 \\ -2 \end{bmatrix}$$

$$B = \begin{bmatrix} 1 & 7 & -9 & 2 \end{bmatrix}$$

Determine which of the following products are defined, and calculate the ones that are:

ABC	ACB	BAC
BCA	CAB	CBA

36. Let O represent the 2×2 **zero matrix**:

$$O = \begin{bmatrix} 0 & 0 \\ 0 & 0 \end{bmatrix}$$

If A and B are 2×2 matrices with $AB = O$, is it necessarily true that $A = O$ or $B = O$?

37. Let the matrix O be as in Exercise 36. Find a matrix $A \neq O$ such that $A^2 = O$.

38. Prove that if A and B are 2×2 matrices, then

$$(A + B)^2 = A^2 + AB + BA + B^2$$

39. If A and B are 2×2 matrices, is it necessarily true that

$$(A + B)^2 \stackrel{?}{=} A^2 + 2AB + B^2$$

◆ **DISCOVERY · DISCUSSION**

40. Powers of a Matrix   Let

$$A = \begin{bmatrix} 1 & 1 \\ 0 & 1 \end{bmatrix}$$

Calculate A^2, A^3, A^4, ... until you detect a pattern. Write a general formula for A^n.

41. Powers of a Matrix   Let $A = \begin{bmatrix} 1 & 1 \\ 1 & 1 \end{bmatrix}$. Calculate A^2, A^3, A^4, ... until you detect a pattern. Write a general formula for A^n.

42. Square Roots of Matrices   A **square root** of a matrix B is a matrix A with the property that $A^2 = B$. (This is the same definition as for the square root of a number.) Find as many square roots as you can of each matrix:

$$\begin{bmatrix} 4 & 0 \\ 0 & 9 \end{bmatrix} \qquad \begin{bmatrix} 1 & 5 \\ 0 & 9 \end{bmatrix}$$

[*Hint:* If $A = \begin{bmatrix} a & b \\ c & d \end{bmatrix}$, write the equations that a, b, c, and d would have to satisfy if A is the square root of the given matrix.]

8.5   INVERSES OF MATRICES AND MATRIX EQUATIONS

We have seen in the preceding section that matrices can, when the dimensions are appropriate, be added, subtracted, and multiplied. In this section we investigate division of matrices. With this operation we can solve equations that involve matrices.

First, we define *identity matrices,* which play the same role for matrix multiplication that the number 1 does for ordinary multiplication of numbers; that is, $1 \cdot a = a \cdot 1 = a$ for all numbers a. In the following definition the term **main diagonal** refers to the entries of a square matrix whose row and column numbers are the same. These entries stretch diagonally down the matrix, from top left to bottom right.

> The **identity matrix** I_n is the $n \times n$ matrix for which each main diagonal entry is a 1 and for which all other entries are 0.

Thus, the 2×2, 3×3, and 4×4 identity matrices are, respectively,

$$I_2 = \begin{bmatrix} 1 & 0 \\ 0 & 1 \end{bmatrix} \qquad I_3 = \begin{bmatrix} 1 & 0 & 0 \\ 0 & 1 & 0 \\ 0 & 0 & 1 \end{bmatrix} \qquad I_4 = \begin{bmatrix} 1 & 0 & 0 & 0 \\ 0 & 1 & 0 & 0 \\ 0 & 0 & 1 & 0 \\ 0 & 0 & 0 & 1 \end{bmatrix}$$

Identity matrices behave like the number 1 in the sense that

$$A \cdot I_n = A \qquad \text{and} \qquad I_n \cdot B = B$$

whenever these products are defined. Thus, multiplication by an identity of the appropriate size leaves a matrix unchanged. For example, we can verify by direct calculation that

$$\begin{bmatrix} 1 & 0 \\ 0 & 1 \end{bmatrix} \begin{bmatrix} 3 & 5 & 6 \\ -1 & 2 & 7 \end{bmatrix} = \begin{bmatrix} 3 & 5 & 6 \\ -1 & 2 & 7 \end{bmatrix}$$

or that

$$\begin{bmatrix} -1 & 7 & \frac{1}{2} \\ 12 & 1 & 3 \\ -2 & 0 & 7 \end{bmatrix} \begin{bmatrix} 1 & 0 & 0 \\ 0 & 1 & 0 \\ 0 & 0 & 1 \end{bmatrix} = \begin{bmatrix} -1 & 7 & \frac{1}{2} \\ 12 & 1 & 3 \\ -2 & 0 & 7 \end{bmatrix}$$

If A and B are $n \times n$ matrices, and if $AB = BA = I_n$, then we say that B is the *inverse* of A, and we write $B = A^{-1}$. The concept of the inverse of a matrix is analogous to that of the reciprocal of a real number.

INVERSE OF A MATRIX

Let A be a square $n \times n$ matrix. If there exists an $n \times n$ matrix A^{-1} with the property that

$$AA^{-1} = A^{-1}A = I_n$$

then we say that A^{-1} is the **inverse** of A.

Not every square matrix has an inverse. The following rule provides a simple way for calculating the inverse of a 2×2 matrix, when it exists. For larger matrices, there's a more general procedure for finding inverses, which we consider later in this section.

INVERSE OF A 2 × 2 MATRIX

If $A = \begin{bmatrix} a & b \\ c & d \end{bmatrix}$ then $A^{-1} = \dfrac{1}{ad - bc} \begin{bmatrix} d & -b \\ -c & a \end{bmatrix}$.

EXAMPLE 1 ■ Finding the Inverse of a 2 × 2 Matrix

Let

$$A = \begin{bmatrix} 4 & 5 \\ 2 & 3 \end{bmatrix}$$

Find A^{-1} and verify that $AA^{-1} = A^{-1}A = I_2$.

SOLUTION   Using the rule, we get

$$A^{-1} = \frac{1}{4 \cdot 3 - 5 \cdot 2} \begin{bmatrix} 3 & -5 \\ -2 & 4 \end{bmatrix} = \frac{1}{2} \begin{bmatrix} 3 & -5 \\ -2 & 4 \end{bmatrix} = \begin{bmatrix} \frac{3}{2} & -\frac{5}{2} \\ -1 & 2 \end{bmatrix}$$

To verify that this is indeed the inverse of A, we calculate AA^{-1} and $A^{-1}A$:

$$AA^{-1} = \begin{bmatrix} 4 & 5 \\ 2 & 3 \end{bmatrix} \begin{bmatrix} \frac{3}{2} & -\frac{5}{2} \\ -1 & 2 \end{bmatrix} = \begin{bmatrix} 4 \cdot \frac{3}{2} + 5(-1) & 4(-\frac{5}{2}) + 5 \cdot 2 \\ 2 \cdot \frac{3}{2} + 3(-1) & 2(-\frac{5}{2}) + 3 \cdot 2 \end{bmatrix} = \begin{bmatrix} 1 & 0 \\ 0 & 1 \end{bmatrix}$$

$$A^{-1}A = \begin{bmatrix} \frac{3}{2} & -\frac{5}{2} \\ -1 & 2 \end{bmatrix} \begin{bmatrix} 4 & 5 \\ 2 & 3 \end{bmatrix} = \begin{bmatrix} \frac{3}{2} \cdot 4 + (-\frac{5}{2})2 & \frac{3}{2} \cdot 5 + (-\frac{5}{2})3 \\ (-1)4 + 2 \cdot 2 & (-1)5 + 2 \cdot 3 \end{bmatrix} = \begin{bmatrix} 1 & 0 \\ 0 & 1 \end{bmatrix}$$

■

The quantity $ad - bc$ that appears in the rule for calculating the inverse is called the **determinant** of the matrix. If the determinant is 0, then the matrix does not have an inverse (since we cannot divide by 0). In the next section we will learn how to calculate the determinant of a square matrix of any size, and how to use determinants to solve systems of equations.

INVERSES OF $n \times n$ MATRICES

For 3×3 and larger square matrices, the following technique provides the most efficient way to calculate their inverses. If A is an $n \times n$ matrix, we begin by constructing the $n \times 2n$ matrix that has the entries of A on the left and of the identity matrix I_n on the right:

$$\left[\begin{array}{cccc|cccc} a_{11} & a_{12} & \cdots & a_{1n} & 1 & 0 & \cdots & 0 \\ a_{21} & a_{22} & \cdots & a_{2n} & 0 & 1 & \cdots & 0 \\ \vdots & \vdots & \ddots & \vdots & \vdots & \vdots & \ddots & \vdots \\ a_{n1} & a_{n2} & \cdots & a_{nn} & 0 & 0 & \cdots & 1 \end{array} \right]$$

We then use the elementary row operations on this new large matrix to change the left side into the identity matrix. The right side will be transformed automatically into A^{-1}. (We omit the proof of this fact.)

EXAMPLE 2 ■ Finding the Inverse of a 3 × 3 Matrix

Find the inverse of the matrix A, and verify that $AA^{-1} = A^{-1}A = I_3$.

$$A = \begin{bmatrix} 1 & -2 & -4 \\ 2 & -3 & -6 \\ -3 & 6 & 15 \end{bmatrix}$$

SOLUTION   We begin with the 3×6 matrix whose left half is A and whose right half is the identity matrix.

$$\left[\begin{array}{ccc|ccc} 1 & -2 & -4 & 1 & 0 & 0 \\ 2 & -3 & -6 & 0 & 1 & 0 \\ -3 & 6 & 15 & 0 & 0 & 1 \end{array} \right]$$

Arthur Cayley (1821–1895) was an English mathematician who invented matrices and developed Matrix Theory. He practiced law until the age of 42, but his primary interest from adolescence was mathematics, and he published almost 200 articles on the subject in his spare time. In 1863 he accepted the offer of a professorship in mathematics at Cambridge, where he taught until his death. Cayley's work on matrices was of purely theoretical interest in his day, but in the 20th century many of his results have found applications in physics, the social sciences, business, and other fields.

We then transform the left half of this new matrix into the identity matrix by performing the following sequence of elementary row operations on the *entire* new matrix:

$$\xrightarrow[\begin{array}{c} R_2 - 2R_1 \rightarrow R_2 \\ R_3 + 3R_1 \rightarrow R_3 \end{array}]{} \left[\begin{array}{ccc|ccc} 1 & -2 & -4 & 1 & 0 & 0 \\ 0 & 1 & 2 & -2 & 1 & 0 \\ 0 & 0 & 3 & 3 & 0 & 1 \end{array}\right]$$

$$\xrightarrow{\frac{1}{3}R_3} \left[\begin{array}{ccc|ccc} 1 & -2 & -4 & 1 & 0 & 0 \\ 0 & 1 & 2 & -2 & 1 & 0 \\ 0 & 0 & 1 & 1 & 0 & \frac{1}{3} \end{array}\right]$$

$$\xrightarrow{R_1 + 2R_2 \rightarrow R_1} \left[\begin{array}{ccc|ccc} 1 & 0 & 0 & -3 & 2 & 0 \\ 0 & 1 & 2 & -2 & 1 & 0 \\ 0 & 0 & 1 & 1 & 0 & \frac{1}{3} \end{array}\right]$$

$$\xrightarrow{R_2 - 2R_3 \rightarrow R_2} \left[\begin{array}{ccc|ccc} 1 & 0 & 0 & -3 & 2 & 0 \\ 0 & 1 & 0 & -4 & 1 & -\frac{2}{3} \\ 0 & 0 & 1 & 1 & 0 & \frac{1}{3} \end{array}\right]$$

We have now transformed the left half of this matrix into the identity matrix. (This means we've put the entire matrix in reduced echelon form.) Note that to do this in as systematic a fashion as possible, we first changed the elements below the main diagonal to zeros, just as we would if we were using Gaussian elimination. We then changed each main diagonal element to a 1 by multiplying by the appropriate constant(s). Finally, we completed the process by changing the remaining entries on the left side to zeros. The right half is now A^{-1}.

$$A^{-1} = \left[\begin{array}{ccc} -3 & 2 & 0 \\ -4 & 1 & -\frac{2}{3} \\ 1 & 0 & \frac{1}{3} \end{array}\right]$$

To verify this, we find the products AA^{-1} and $A^{-1}A$:

$$AA^{-1} = \left[\begin{array}{ccc} 1 & -2 & -4 \\ 2 & -3 & -6 \\ -3 & 6 & 15 \end{array}\right]\left[\begin{array}{ccc} -3 & 2 & 0 \\ -4 & 1 & -\frac{2}{3} \\ 1 & 0 & \frac{1}{3} \end{array}\right] = \left[\begin{array}{ccc} 1 & 0 & 0 \\ 0 & 1 & 0 \\ 0 & 0 & 1 \end{array}\right]$$

$$A^{-1}A = \left[\begin{array}{ccc} -3 & 2 & 0 \\ -4 & 1 & -\frac{2}{3} \\ 1 & 0 & \frac{1}{3} \end{array}\right]\left[\begin{array}{ccc} 1 & -2 & -4 \\ 2 & -3 & -6 \\ -3 & 6 & 15 \end{array}\right] = \left[\begin{array}{ccc} 1 & 0 & 0 \\ 0 & 1 & 0 \\ 0 & 0 & 1 \end{array}\right]$$

The next example shows that not every square matrix has an inverse.

EXAMPLE 3 ■ A Matrix that Does Not Have an Inverse

Try to find the inverse of the following matrix.

$$\begin{bmatrix} 2 & -3 & -7 \\ 1 & 2 & 7 \\ 1 & 1 & 4 \end{bmatrix}$$

SOLUTION We proceed as follows.

$$\begin{bmatrix} 2 & -3 & -7 & | & 1 & 0 & 0 \\ 1 & 2 & 7 & | & 0 & 1 & 0 \\ 1 & 1 & 4 & | & 0 & 0 & 1 \end{bmatrix} \xrightarrow{R_1 \leftrightarrow R_2} \begin{bmatrix} 1 & 2 & 7 & | & 0 & 1 & 0 \\ 2 & -3 & -7 & | & 1 & 0 & 0 \\ 1 & 1 & 4 & | & 0 & 0 & 1 \end{bmatrix}$$

$$\xrightarrow[R_3 - R_1 \to R_3]{R_2 - 2R_1 \to R_2} \begin{bmatrix} 1 & 2 & 7 & | & 0 & 1 & 0 \\ 0 & -7 & -21 & | & 1 & -2 & 0 \\ 0 & -1 & -3 & | & 0 & -1 & 1 \end{bmatrix}$$

$$\xrightarrow{-\frac{1}{7}R_2} \begin{bmatrix} 1 & 2 & 7 & | & 0 & 1 & 0 \\ 0 & 1 & 3 & | & -\frac{1}{7} & \frac{2}{7} & 0 \\ 0 & -1 & -3 & | & 0 & -1 & 1 \end{bmatrix}$$

$$\xrightarrow[R_1 - 2R_2 \to R_1]{R_3 + R_2 \to R_3} \begin{bmatrix} 1 & 0 & 1 & | & \frac{2}{7} & \frac{3}{7} & 0 \\ 0 & 1 & 3 & | & -\frac{1}{7} & \frac{2}{7} & 0 \\ 0 & 0 & 0 & | & -\frac{1}{7} & -\frac{5}{7} & 1 \end{bmatrix}$$

At this point, we would like to change the 0 in the $(3, 3)$ position of this matrix to a 1, without changing the zeros in the $(3, 1)$ and $(3, 2)$ positions. But there is no way to accomplish this, because no matter what multiple of rows 1 and/or 2 we add to row 3, we can't change the third zero in row 3 without changing the first or second as well. Thus, we cannot change the left half to the identity matrix. The original matrix doesn't have an inverse. ■

  If we encounter a row of zeros on the left when trying to find an inverse, as in Example 3, then the original matrix does not have an inverse.

MATRIX EQUATIONS

We saw in Section 8.4 that a system of linear equations can be written as a single matrix equation. For example, the system

$$\begin{cases} x - 2y - 4z = 7 \\ 2x - 3y - 6z = 5 \\ -3x + 6y + 15z = 0 \end{cases}$$

is equivalent to the matrix equation

$$\begin{bmatrix} 1 & -2 & -4 \\ 2 & -3 & -6 \\ -3 & 6 & 15 \end{bmatrix} \begin{bmatrix} x \\ y \\ z \end{bmatrix} = \begin{bmatrix} 7 \\ 5 \\ 0 \end{bmatrix}$$

If we let

$$A = \begin{bmatrix} 1 & -2 & -4 \\ 2 & -3 & -6 \\ -3 & 6 & 15 \end{bmatrix} \qquad X = \begin{bmatrix} x \\ y \\ z \end{bmatrix} \qquad B = \begin{bmatrix} 7 \\ 5 \\ 0 \end{bmatrix}$$

The matrix A is called the *coefficient matrix*.

then this matrix equation can be written as

$$AX = B$$

Solving the matrix equation $AX = B$ is very similar to solving the simple real-number equation

$$3x = 12$$

which we solve by multiplying each side by the reciprocal (or inverse) of 3:

$$\tfrac{1}{3}(3x) = \tfrac{1}{3}(12)$$

$$x = 4$$

We solve this matrix equation by multiplying each side by the inverse of A (provided this inverse exists):

$$AX = B$$
$$A^{-1}(AX) = A^{-1}B$$
$$(A^{-1}A)X = A^{-1}B$$
$$I_3 X = A^{-1}B$$
$$X = A^{-1}B$$

In Example 2 we showed that

$$A^{-1} = \begin{bmatrix} -3 & 2 & 0 \\ -4 & 1 & -\tfrac{2}{3} \\ 1 & 0 & \tfrac{1}{3} \end{bmatrix}$$

So, from $X = A^{-1}B$ we have

$$\begin{bmatrix} x \\ y \\ z \end{bmatrix} = \begin{bmatrix} -3 & 2 & 0 \\ -4 & 1 & -\tfrac{2}{3} \\ 1 & 0 & \tfrac{1}{3} \end{bmatrix} \begin{bmatrix} 7 \\ 5 \\ 0 \end{bmatrix} = \begin{bmatrix} -11 \\ -23 \\ 7 \end{bmatrix}$$

Thus, $x = -11$, $y = -23$, $z = 7$ is the solution of the original system.

SOLVING A MATRIX EQUATION

If A is a square $n \times n$ matrix that has an inverse A^{-1}, and if X is a variable matrix and B a known matrix, both with n rows, then the solution of the matrix equation $AX = B$ is given by

$$X = A^{-1}B$$

EXAMPLE 4 ■ Solving a System Using the Matrix Inverse

Solve the following system of equations.

$$\begin{cases} 2x - 5y = 15 \\ 3x - 6y = 36 \end{cases}$$

SOLUTION We first convert this to a matrix equation of the form $AX = B$:

$$\begin{bmatrix} 2 & -5 \\ 3 & -6 \end{bmatrix} \begin{bmatrix} x \\ y \end{bmatrix} = \begin{bmatrix} 15 \\ 36 \end{bmatrix}$$

Find A^{-1}.

Using the rule for calculating the inverse of a 2×2 matrix, we get

$$\begin{bmatrix} 2 & -5 \\ 3 & -6 \end{bmatrix}^{-1} = \frac{1}{2(-6) - (-5)3} \begin{bmatrix} -6 & -(-5) \\ -3 & 2 \end{bmatrix} = \frac{1}{3} \begin{bmatrix} -6 & 5 \\ -3 & 2 \end{bmatrix}$$

Multiplying each side of the matrix equation by this inverse matrix, we get

$X = A^{-1}B$

$$\begin{bmatrix} x \\ y \end{bmatrix} = \frac{1}{3} \begin{bmatrix} -6 & 5 \\ -3 & 2 \end{bmatrix} \begin{bmatrix} 15 \\ 36 \end{bmatrix} = \begin{bmatrix} 30 \\ 9 \end{bmatrix}$$

So $x = 30$ and $y = 9$. ∎

EXAMPLE 5 ■ An Application of Matrix Equations

A pet-store owner feeds his hamsters and gerbils different mixtures of three types of rodent food pellets, which we will call brands A, B, and C. He wishes to feed his animals the correct amount of each brand to satisfy their daily requirements for protein, fat, and carbohydrates exactly. Suppose that hamsters require 340 mg of protein, 280 mg of fat, and 440 mg of carbohydrates, and gerbils need 480 mg of protein, 360 mg of fat, and 680 mg of carbohydrates each day. The amount of each nutrient in one gram of each brand of food is given in the following table. How many grams of each food should the storekeeper feed his hamsters and gerbils daily to satisfy their nutrient requirements?

	Brand A	Brand B	Brand C
Protein (mg)	10	0	20
Fat (mg)	10	20	10
Carbohydrates (mg)	5	10	30

On the TI-82 and TI-83 calculators, matrices are stored in memory using names such as [A], [B], [C], ... To find the inverse of [A], we key in

[A] x^{-1} ENTER

SOLUTION   If we let x_1, x_2, and x_3 be the grams of brands A, B, and C, respectively, that the hamsters should eat, and if we let y_1, y_2, and y_3 be the corresponding amounts for the gerbils, then we want to solve the matrix equations

$$\begin{bmatrix} 10 & 0 & 20 \\ 10 & 20 & 10 \\ 5 & 10 & 30 \end{bmatrix} \begin{bmatrix} x_1 \\ x_2 \\ x_3 \end{bmatrix} = \begin{bmatrix} 340 \\ 280 \\ 440 \end{bmatrix} \qquad \text{Hamster equation}$$

and

$$\begin{bmatrix} 10 & 0 & 20 \\ 10 & 20 & 10 \\ 5 & 10 & 30 \end{bmatrix} \begin{bmatrix} y_1 \\ y_2 \\ y_3 \end{bmatrix} = \begin{bmatrix} 480 \\ 360 \\ 680 \end{bmatrix} \qquad \text{Gerbil equation}$$

Since the coefficient matrix on the left is the same in both of these equations, we can solve each one by multiplying each side by the inverse of this matrix. To find this inverse, we can use a calculator with matrix functions, and we find that

$$\begin{bmatrix} 10 & 0 & 20 \\ 10 & 20 & 10 \\ 5 & 10 & 30 \end{bmatrix}^{-1} = \begin{bmatrix} 0.10 & 0.04 & -0.08 \\ -0.05 & 0.04 & 0.02 \\ 0 & -0.02 & 0.04 \end{bmatrix} = \frac{1}{100}\begin{bmatrix} 10 & 4 & -8 \\ -5 & 4 & 2 \\ 0 & -2 & 4 \end{bmatrix}$$

We now multiply each side of our matrix equations by this inverse matrix, to get

$$\begin{bmatrix} x_1 \\ x_2 \\ x_3 \end{bmatrix} = \frac{1}{100}\begin{bmatrix} 10 & 4 & -8 \\ -5 & 4 & 2 \\ 0 & -2 & 4 \end{bmatrix}\begin{bmatrix} 340 \\ 280 \\ 440 \end{bmatrix} = \begin{bmatrix} 10 \\ 3 \\ 12 \end{bmatrix}$$

$$\begin{bmatrix} y_1 \\ y_2 \\ y_3 \end{bmatrix} = \frac{1}{100}\begin{bmatrix} 10 & 4 & -8 \\ -5 & 4 & 2 \\ 0 & -2 & 4 \end{bmatrix}\begin{bmatrix} 480 \\ 360 \\ 680 \end{bmatrix} = \begin{bmatrix} 8 \\ 4 \\ 20 \end{bmatrix}$$

We interpret these solution matrices as follows: Each hamster should be fed 10 g of brand A, 3 g of brand B, and 12 g of brand C, and each gerbil should be fed 8 g of brand A, 4 g of brand B, and 20 g of brand C daily.   ■

Since a lot of work is usually involved in finding the inverse of a 3 × 3 or larger matrix, the method used in Example 5 is really useful only when we are solving several systems of equations with the same coefficient matrix. However, if we have access to a calculator or computer program that calculates matrix inverses, then this becomes the preferred method in all circumstances. Most graphing calculators can compute matrix inverses.

8.5 EXERCISES

1–2 ■ Find the inverse of the matrix and verify that $A^{-1}A = AA^{-1} = I_2$ and $B^{-1}B = BB^{-1} = I_3$.

1. $A = \begin{bmatrix} 7 & 4 \\ 3 & 2 \end{bmatrix}$

2. $B = \begin{bmatrix} 1 & 3 & 2 \\ 0 & 2 & 2 \\ -2 & -1 & 0 \end{bmatrix}$

3–18 ■ Find the inverse of the matrix if it exists.

3. $\begin{bmatrix} 5 & 3 \\ 3 & 2 \end{bmatrix}$

4. $\begin{bmatrix} 3 & 4 \\ 7 & 9 \end{bmatrix}$

5. $\begin{bmatrix} 2 & 5 \\ -5 & -13 \end{bmatrix}$

6. $\begin{bmatrix} -7 & 4 \\ 8 & -5 \end{bmatrix}$

7. $\begin{bmatrix} 6 & -3 \\ -8 & 4 \end{bmatrix}$

8. $\begin{bmatrix} \frac{1}{2} & \frac{1}{3} \\ 5 & 4 \end{bmatrix}$

9. $\begin{bmatrix} 0.4 & -1.2 \\ 0.3 & 0.6 \end{bmatrix}$

10. $\begin{bmatrix} 4 & 2 & 3 \\ 3 & 3 & 2 \\ 1 & 0 & 1 \end{bmatrix}$

11. $\begin{bmatrix} 2 & 4 & 1 \\ -1 & 1 & -1 \\ 1 & 4 & 0 \end{bmatrix}$

12. $\begin{bmatrix} 5 & 7 & 4 \\ 3 & -1 & 3 \\ 6 & 7 & 5 \end{bmatrix}$

13. $\begin{bmatrix} 1 & 2 & 3 \\ 4 & 5 & -1 \\ 1 & -1 & -10 \end{bmatrix}$

14. $\begin{bmatrix} 2 & 1 & 0 \\ 1 & 1 & 4 \\ 2 & 1 & 2 \end{bmatrix}$

15. $\begin{bmatrix} 0 & -2 & 2 \\ 3 & 1 & 3 \\ 1 & -2 & 3 \end{bmatrix}$

16. $\begin{bmatrix} 3 & -2 & 0 \\ 5 & 1 & 1 \\ 2 & -2 & 0 \end{bmatrix}$

17. $\begin{bmatrix} 1 & 2 & 0 & 3 \\ 0 & 1 & 1 & 1 \\ 0 & 1 & 0 & 1 \\ 1 & 2 & 0 & 2 \end{bmatrix}$ **18.** $\begin{bmatrix} 1 & 0 & 1 & 0 \\ 0 & 1 & 0 & 1 \\ 1 & 1 & 1 & 0 \\ 1 & 1 & 1 & 1 \end{bmatrix}$

19–26 ■ Solve the system of equations by converting to a matrix equation and using the inverse of the coefficient matrix, as in Example 4. Use the inverses from Exercises 3–6, 11, 12, 15, and 17.

19. $\begin{cases} 5x + 3y = 4 \\ 3x + 2y = 0 \end{cases}$ **20.** $\begin{cases} 3x + 4y = 10 \\ 7x + 9y = 20 \end{cases}$

21. $\begin{cases} 2x + 5y = 2 \\ -5x - 13y = 20 \end{cases}$ **22.** $\begin{cases} -7x + 4y = 0 \\ 8x - 5y = 100 \end{cases}$

23. $\begin{cases} 2x + 4y + z = 7 \\ -x + y - z = 0 \\ x + 4y = -2 \end{cases}$

24. $\begin{cases} 5x + 7y + 4z = 1 \\ 3x - y + 3z = 1 \\ 6x + 7y + 5z = 1 \end{cases}$

25. $\begin{cases} -2y + 2z = 12 \\ 3x + y + 3z = -2 \\ x - 2y + 3z = 8 \end{cases}$

26. $\begin{cases} x + 2y + 3w = 0 \\ y + z + w = 1 \\ y + w = 2 \\ x + 2y + 2w = 3 \end{cases}$

27–28 ■ Solve the matrix equation by multiplying each side by the appropriate inverse matrix.

27. $\begin{bmatrix} 3 & -2 \\ -4 & 3 \end{bmatrix}\begin{bmatrix} x & y & z \\ u & v & w \end{bmatrix} = \begin{bmatrix} 1 & 0 & -1 \\ 2 & 1 & 3 \end{bmatrix}$

28. $\begin{bmatrix} 0 & -2 & 2 \\ 3 & 1 & 3 \\ 1 & -2 & 3 \end{bmatrix}\begin{bmatrix} x & u \\ y & v \\ z & w \end{bmatrix} = \begin{bmatrix} 3 & 6 \\ 6 & 12 \\ 0 & 0 \end{bmatrix}$

29. A nutritionist is studying the effects of the nutrients folic acid, choline, and inositol. He has three types of food available, and each type contains the following amounts of these nutrients per ounce:

	Type A	Type B	Type C
Folic acid (mg)	3	1	3
Choline (mg)	4	2	4
Inositol (mg)	3	2	4

(a) Find the inverse of the matrix

$$\begin{bmatrix} 3 & 1 & 3 \\ 4 & 2 & 4 \\ 3 & 2 & 4 \end{bmatrix}$$

and use it to solve the remaining parts of this problem.

(b) How many ounces of each food should the nutritionist feed his laboratory rats if he wants their daily diet to contain 10 mg of folic acid, 14 mg of choline, and 13 mg of inositol?

(c) How much of each food should be given to supply 9 mg of folic acid, 12 mg of choline, and 10 mg of inositol?

(d) Will any combination of these foods supply 2 mg of folic acid, 4 mg of choline, and 11 mg of inositol?

30. Refer to Exercise 29. Suppose it is found that food type C has been improperly labeled, and it actually contains 4 mg of folic acid, 6 mg of choline, and 5 mg of inositol per ounce. Would it still be possible to use matrix inversion to solve parts (b), (c), and (d) of Exercise 29? Why or why not?

31. An encyclopedia salesman works for a company that offers three different grades of bindings for its encyclopedias: standard, deluxe, and leather. For each set he sells, he earns a commission that is based on the set's binding grade. One week he sells one standard, one deluxe, and two leather sets and makes $675 in commission. The next week he sells two standard, one deluxe, and one leather set for a $600 commission. The third week he sells one standard, two deluxe, and one leather set, earning $625 in commission.

(a) Let x, y, and z represent the commission he earns on standard, deluxe, and leather sets, respectively. Translate the given information into a system of equations in x, y, and z.

(b) Express the system of equations you found in part (a) as a matrix equation of the form $AX = B$.

(c) Find the inverse of the coefficient matrix A and use it to solve the matrix equation in part (b). How much commission does the salesman earn on a set of encyclopedias in each grade of binding?

32–35 ■ Find the inverse of the matrix. For what value(s) of x, if any, does the matrix have no inverse?

32. $\begin{bmatrix} x & 1 \\ -1 & 1/x \end{bmatrix}$

33. $\begin{bmatrix} e^x & -e^{2x} \\ e^{2x} & e^{3x} \end{bmatrix}$

34. $\begin{bmatrix} 1 & e^x & 0 \\ e^x & -e^{2x} & 0 \\ 0 & 0 & 2 \end{bmatrix}$

35. $\begin{bmatrix} \sin x & \cos x \\ -\cos x & \sin x \end{bmatrix}$

36. Find the inverse of the following matrix, where $abcd \neq 0$:

$$\begin{bmatrix} a & 0 & 0 & 0 \\ 0 & b & 0 & 0 \\ 0 & 0 & c & 0 \\ 0 & 0 & 0 & d \end{bmatrix}$$

8.6 DETERMINANTS AND CRAMER'S RULE

If a matrix is **square** (that is, if it has the same number of rows as columns), then we can assign to it a number called its **determinant**. Determinants can be used to solve matrix equations, as we will see later in this section. They are also useful in determining whether a matrix has an inverse.

We denote the determinant of a square matrix A by the symbol $|A|$, and we begin by defining $|A|$ for the simplest case. If A is a 1×1 matrix, then it has only one entry, and we define its determinant to be the value of that entry; that is, if $A = [a]$, then $|A| = a$. If A is a 2×2 matrix,

$$A = \begin{bmatrix} a & b \\ c & d \end{bmatrix}$$

then the determinant of A is defined by

$$|A| = \begin{vmatrix} a & b \\ c & d \end{vmatrix} = ad - bc$$

EXAMPLE 1 ■ Determinant of a 2×2 Matrix

Evaluate $|A|$ for $A = \begin{bmatrix} 6 & -3 \\ 2 & 3 \end{bmatrix}$.

SOLUTION

$$\begin{vmatrix} 6 & -3 \\ 2 & 3 \end{vmatrix} = 6 \cdot 3 - (-3)2 = 18 - (-6) = 24$$

■

David Hilbert (1862–1943) was born in Königsberg, Germany, and became a professor at Göttingen University. He is considered by many to be the greatest mathematician of the 20th century. At the International Congress of Mathematicians held in Paris in 1900, Hilbert set the direction of mathematics for the about-to-dawn 20th century by posing 23 problems that he believed were of crucial importance. He said that "these are problems whose solutions we expect from the future." Most of Hilbert's problems have now been solved (see Julia Robinson, page 561, and Alan Turing, page 97), and these solutions have led to important new areas of mathematical research. But, as this century comes to a close, many important problems remain unsolved. In his work Hilbert emphasized structure, logic, and the foundations of mathematics. Part of his genius was the ability to see the most general possible statement of a problem. For instance, Euler proved that every whole number is the sum of four squares; Hilbert proved a similar statement for all powers of positive integers.

We can think of the evaluation of a 2×2 determinant as a "cross-product" operation. We take the product of the diagonal from top left to bottom right, and subtract the product from top right to bottom left.

To define the concept of determinant for an arbitrary $n \times n$ matrix, we must first introduce the following terminology.

Let A be an $n \times n$ matrix.

1. The **minor** M_{ij} of the element a_{ij} is the determinant of the matrix obtained by deleting the ith row and jth column of A.

2. The **cofactor** A_{ij} of the element a_{ij} is

$$A_{ij} = (-1)^{i+j} M_{ij}$$

For example, if A is the matrix

$$\begin{bmatrix} 2 & 3 & -1 \\ 0 & 2 & 4 \\ -2 & 5 & 6 \end{bmatrix}$$

then M_{12} is the determinant of the matrix obtained by deleting the first row and second column from A. Thus

$$M_{12} = \begin{vmatrix} 2 & 3 & -1 \\ 0 & 2 & 4 \\ -2 & 5 & 6 \end{vmatrix} = \begin{vmatrix} 0 & 4 \\ -2 & 6 \end{vmatrix} = 0(6) - 4(-2) = 8$$

So, $A_{12} = (-1)^{1+2} M_{12} = -8$. Similarly

$$M_{33} = \begin{vmatrix} 2 & 3 & -1 \\ 0 & 2 & 4 \\ -2 & 5 & 6 \end{vmatrix} = \begin{vmatrix} 2 & 3 \\ 0 & 2 \end{vmatrix} = 2 \cdot 2 - 3 \cdot 0 = 4$$

So, $A_{33} = (-1)^{3+3} M_{33} = 4$.

Note that the cofactor of a_{ij} is simply the minor of a_{ij} multiplied by either 1 or -1, depending on whether $i + j$ is even or odd. Thus, in a 3×3 matrix we obtain the cofactor of any element by prefixing its minor with the sign obtained from the following checkerboard pattern:

$$\begin{bmatrix} + & - & + \\ - & + & - \\ + & - & + \end{bmatrix}$$

We are now ready to define the determinant of any square matrix.

> ### THE DETERMINANT OF A SQUARE MATRIX
>
> If A is an $n \times n$ matrix, then the **determinant** of A is obtained by multiplying each element of the first row by its cofactor, and then adding the results. In symbols,
>
> $$|A| = \begin{vmatrix} a_{11} & a_{12} & \cdots & a_{1n} \\ a_{21} & a_{22} & \cdots & a_{2n} \\ \vdots & \vdots & \ddots & \vdots \\ a_{n1} & a_{n2} & \cdots & a_{nn} \end{vmatrix} = a_{11}A_{11} + a_{12}A_{12} + \cdots + a_{1n}A_{1n}$$

EXAMPLE 2 ■ Determinant of a 3 × 3 Matrix

Evaluate the determinant of the following matrix.

$$A = \begin{bmatrix} 2 & 3 & -1 \\ 0 & 2 & 4 \\ -2 & 5 & 6 \end{bmatrix}$$

SOLUTION

$$|A| = \begin{vmatrix} 2 & 3 & -1 \\ 0 & 2 & 4 \\ -2 & 5 & 6 \end{vmatrix} = 2\begin{vmatrix} 2 & 4 \\ 5 & 6 \end{vmatrix} - 3\begin{vmatrix} 0 & 4 \\ -2 & 6 \end{vmatrix} + (-1)\begin{vmatrix} 0 & 2 \\ -2 & 5 \end{vmatrix}$$

$$= 2(2 \cdot 6 - 4 \cdot 5) - 3[0 \cdot 6 - 4(-2)] - [0 \cdot 5 - 2(-2)]$$

$$= -16 - 24 - 4$$

$$= -44 \qquad ■$$

In our definition of the determinant, we used the cofactors of elements in the first row only. This is called **expanding the determinant by the first row**. In fact, *we can expand the determinant by any row or column in the same way, and obtain the same result in each case* (although we won't prove this). The next example illustrates this principle.

EXAMPLE 3 ■ Expanding a Determinant about a Row and a Column

Expand the determinant of the matrix A in Example 2 by the second row and by the third column, and show that the value obtained is the same in each case.

Emmy Noether (1882–1935) was one of the foremost mathematicians of the early 20th century. Her groundbreaking work in abstract algebra provided much of the foundation for this field, and her work in Invariant Theory was essential in the development of Einstein's theory of general relativity. Although women weren't allowed to study at German universities at that time, she audited courses unofficially and went on to receive a doctorate at Erlangen *summa cum laude,* despite the opposition of the academic senate, which declared that women students would "overthrow all academic order." She subsequently taught mathematics at Göttingen, Moscow, and Frankfurt. In 1933 she left Germany to escape Nazi persecution, accepting a position at Bryn Mawr College in suburban Philadelphia. She lectured there and at the Institute for Advanced Study in Princeton, New Jersey, until her untimely death in 1935.

SOLUTION Expanding by the second row, we get

$$|A| = \begin{vmatrix} 2 & 3 & -1 \\ 0 & 2 & 4 \\ -2 & 5 & 6 \end{vmatrix} = -0 \begin{vmatrix} 3 & -1 \\ 5 & 6 \end{vmatrix} + 2 \begin{vmatrix} 2 & -1 \\ -2 & 6 \end{vmatrix} - 4 \begin{vmatrix} 2 & 3 \\ -2 & 5 \end{vmatrix}$$

$$= 0 + 2[2 \cdot 6 - (-1)(-2)] - 4[2 \cdot 5 - 3(-2)]$$

$$= 0 + 20 - 64$$

$$= -44$$

Expanding by the third column gives

$$|A| = \begin{vmatrix} 2 & 3 & -1 \\ 0 & 2 & 4 \\ -2 & 5 & 6 \end{vmatrix} = -1 \begin{vmatrix} 0 & 2 \\ -2 & 5 \end{vmatrix} - 4 \begin{vmatrix} 2 & 3 \\ -2 & 5 \end{vmatrix} + 6 \begin{vmatrix} 2 & 3 \\ 0 & 2 \end{vmatrix}$$

$$= -[0 \cdot 5 - 2(-2)] - 4[2 \cdot 5 - 3(-2)] + 6(2 \cdot 2 - 3 \cdot 0)$$

$$= -4 - 64 + 24$$

$$= -44$$

In both cases, we obtain the same value for the determinant as when we expanded by the first row in Example 2. ∎

The following principle allows us to determine whether a square matrix has an inverse without actually calculating the inverse. This is one of the most important uses of the determinant in matrix algebra, and it is the reason for the name *determinant.*

INVERTIBILITY CRITERION

If A is a square matrix, then A has an inverse if and only if $|A| \neq 0$.

Although we won't prove this fact, we have already seen (in the preceding section) why it is true in the case of 2×2 matrices.

EXAMPLE 4 ■ **Using the Determinant to Show that a Matrix Is Not Invertible**

Show that the matrix A has no inverse.

$$A = \begin{bmatrix} 1 & 2 & 0 & 4 \\ 0 & 0 & 0 & 3 \\ 5 & 6 & 2 & 6 \\ 2 & 4 & 0 & 9 \end{bmatrix}$$

SOLUTION We begin by calculating the determinant of A. Since all but one of the elements of the second row is zero, we expand the determinant by the second row. If we do this, we see from the following equation that only the cofactor A_{24} will have to be calculated.

$$|A| = -0 \cdot A_{21} + 0 \cdot A_{22} - 0 \cdot A_{23} + 3 \cdot A_{24} = 3A_{24}$$

$$= 3 \begin{vmatrix} 1 & 2 & 0 \\ 5 & 6 & 2 \\ 2 & 4 & 0 \end{vmatrix} \qquad \text{Expand this by the third column}$$

$$= 3(-2) \begin{vmatrix} 1 & 2 \\ 2 & 4 \end{vmatrix}$$

$$= 3(-2)(1 \cdot 4 - 2 \cdot 2) = 0$$

Since the determinant of A is zero, A cannot have an inverse, by the Invertibility Criterion. ■

The preceding example shows that if we expand a determinant about a row or column that contains many zeros, our work is reduced considerably because we don't have to evaluate the cofactors of the elements that are zero. The following principle enables us in many cases to simplify the process of finding a determinant by introducing zeros into it without changing its value.

ROW AND COLUMN TRANSFORMATIONS OF A DETERMINANT

If A is a square matrix, and if the matrix B is obtained from A by adding a multiple of one row to another, or a multiple of one column to another, then $|A| = |B|$.

EXAMPLE 5 ■ Using Row and Column Transformations to Calculate a Determinant

Find the determinant of the matrix A. Does it have an inverse?

$$A = \begin{bmatrix} 8 & 2 & -1 & -4 \\ 3 & 5 & -3 & 11 \\ 24 & 6 & 1 & -12 \\ 2 & 2 & 7 & -1 \end{bmatrix}$$

SOLUTION If we add -3 times row 1 to row 3, we change all but one of the elements of row 3 to zeros:

$$\begin{bmatrix} 8 & 2 & -1 & -4 \\ 3 & 5 & -3 & 11 \\ 0 & 0 & 4 & 0 \\ 2 & 2 & 7 & -1 \end{bmatrix}$$

This new matrix has the same determinant as A, and if we expand its determinant by the third row, we get

$$|A| = 4 \begin{vmatrix} 8 & 2 & -4 \\ 3 & 5 & 11 \\ 2 & 2 & -1 \end{vmatrix}$$

Now, adding 2 times column 3 to column 1 in this determinant gives us

$$|A| = 4 \begin{vmatrix} 0 & 2 & -4 \\ 25 & 5 & 11 \\ 0 & 2 & -1 \end{vmatrix} \qquad \text{Expand this by the first column}$$

$$= 4(-25) \begin{vmatrix} 2 & -4 \\ 2 & -1 \end{vmatrix}$$

$$= 4(-25)[2(-1) - (-4)2] = -600$$

Since the determinant of A is not zero, A does have an inverse. ∎

CRAMER'S RULE

The solutions of linear equations can sometimes be expressed using determinants. To illustrate, let's try to solve the following pair of linear equations for the variable x.

$$\begin{cases} ax + by = r \\ cx + dy = s \end{cases}$$

To eliminate the variable y we multiply the first equation by d and the second by b, and subtract.

$$\begin{array}{c} adx + bdy = rd \\ \underline{bcx + bdy = bs} \\ adx - bcx = rd - bs \end{array}$$

Factoring the left-hand side, we get $(ad - bc)x = rd - bs$. Assuming that $ad - bc \neq 0$, we can now solve this equation for x, obtaining

$$x = \frac{rd - bs}{ad - bc}$$

The numerator and denominator of this fraction look like the determinants of 2×2 matrices. In fact, we can write the solution of the system as

$$x = \frac{\begin{vmatrix} r & b \\ s & d \end{vmatrix}}{\begin{vmatrix} a & b \\ c & d \end{vmatrix}} \qquad y = \frac{\begin{vmatrix} a & r \\ c & s \end{vmatrix}}{\begin{vmatrix} a & b \\ c & d \end{vmatrix}}$$

where the solution for y was obtained using the same sort of technique as the solution for x. Notice that the denominator in each case is the determinant of the coefficient matrix, which we will call D. The numerator in the solution for x is the determinant of the matrix obtained from D by replacing the first column, the coefficients of x, by r and s, respectively. Similarly, in the solution for y the numerator is the determinant of the matrix obtained from D by replacing the second column, the coefficients of y, by r and s. Thus, if we define

$$D = \begin{bmatrix} a & b \\ c & d \end{bmatrix} \qquad D_x = \begin{bmatrix} r & b \\ s & d \end{bmatrix} \qquad D_y = \begin{bmatrix} a & r \\ c & s \end{bmatrix}$$

then we can write the solution of the system as

$$x = \frac{|D_x|}{|D|} \qquad \text{and} \qquad y = \frac{|D_y|}{|D|}$$

This pair of formulas is known as **Cramer's Rule**, and it can be used to solve any pair of linear equations in two unknowns in which the determinant of the coefficient matrix is not zero.

EXAMPLE 6 ■ Using Cramer's Rule to Solve a System with Two Variables

Use Cramer's Rule to solve the following system.

$$\begin{cases} 2x + 6y = -1 \\ x + 8y = 2 \end{cases}$$

SOLUTION   For this system, we have

$$|D| = \begin{vmatrix} 2 & 6 \\ 1 & 8 \end{vmatrix} = 2 \cdot 8 - 6 \cdot 1 = 10$$

$$|D_x| = \begin{vmatrix} -1 & 6 \\ 2 & 8 \end{vmatrix} = (-1)8 - 6 \cdot 2 = -20$$

$$|D_y| = \begin{vmatrix} 2 & -1 \\ 1 & 2 \end{vmatrix} = 2 \cdot 2 - (-1)1 = 5$$

The solution is

$$x = \frac{|D_x|}{|D|} = \frac{-20}{10} = -2$$

$$y = \frac{|D_y|}{|D|} = \frac{5}{10} = \frac{1}{2}$$

■

Cramer's Rule can be extended to apply to any system of n linear equations in n variables in which the determinant of the coefficient matrix is not zero. As we saw in the preceding section, any such system can be written in matrix form as

$$\begin{bmatrix} a_{11} & a_{12} & \cdots & a_{1n} \\ a_{21} & a_{22} & \cdots & a_{2n} \\ \vdots & \vdots & \ddots & \vdots \\ a_{n1} & a_{n2} & \cdots & a_{nn} \end{bmatrix} \begin{bmatrix} x_1 \\ x_2 \\ \vdots \\ x_n \end{bmatrix} = \begin{bmatrix} b_1 \\ b_2 \\ \vdots \\ b_n \end{bmatrix}$$

By analogy with how we derived Cramer's Rule in the case of two equations in two unknowns, we let D be the coefficient matrix in this system, and we let D_{x_i} be the matrix obtained by replacing the ith column of D by the numbers b_1, $b_2, \ldots, b_n$ that appear to the right of the equal sign in the system. The solution of the system is then given by the following rule.

CRAMER'S RULE

If a system of n linear equations in the n variables $x_1, x_2, \ldots, x_n$ is equivalent to the matrix equation $DX = B$, and if $|D| \neq 0$, then its solutions are

$$x_1 = \frac{|D_{x_1}|}{|D|}, \quad x_2 = \frac{|D_{x_2}|}{|D|}, \quad \ldots, \quad x_n = \frac{|D_{x_n}|}{|D|}$$

where D_{x_i} is the matrix obtained by replacing the ith column of D by the $n \times 1$ matrix B.

EXAMPLE 7 ■ **Using Cramer's Rule to Solve a System with Three Variables**

Use Cramer's Rule to solve the following system.

$$\begin{cases} 2x - 3y + 4z = 1 \\ x \qquad + 6z = 0 \\ 3x - 2y \qquad = 5 \end{cases}$$

SOLUTION First, we evaluate the determinants that appear in Cramer's Rule.

$$|D| = \begin{vmatrix} 2 & -3 & 4 \\ 1 & 0 & 6 \\ 3 & -2 & 0 \end{vmatrix} = -38 \qquad |D_x| = \begin{vmatrix} 1 & -3 & 4 \\ 0 & 0 & 6 \\ 5 & -2 & 0 \end{vmatrix} = -78$$

$$|D_y| = \begin{vmatrix} 2 & 1 & 4 \\ 1 & 0 & 6 \\ 3 & 5 & 0 \end{vmatrix} = -22 \qquad |D_z| = \begin{vmatrix} 2 & -3 & 1 \\ 1 & 0 & 0 \\ 3 & -2 & 5 \end{vmatrix} = 13$$

Now we use Cramer's Rule to get the solution:

$$x = \frac{|D_x|}{|D|} = \frac{-78}{-38} = \frac{39}{19}$$

$$y = \frac{|D_y|}{|D|} = \frac{-22}{-38} = \frac{11}{19}$$

$$z = \frac{|D_z|}{|D|} = \frac{13}{-38} = -\frac{13}{38}$$ ∎

Solving the system in Example 7 using Gaussian elimination would involve matrices whose elements are fractions with fairly large denominators. Thus, in cases like Examples 6 and 7, Cramer's Rule provides an efficient method for solving systems of linear equations. But in systems with more than three equations, evaluating the various determinants involved is usually a long and tedious task. Moreover, the rule doesn't apply if $|D| = 0$ or if D is not a square matrix. So, Cramer's Rule is a useful alternative to Gaussian elimination, but only in some situations.

Most graphing calculators can be used to evaluate determinants.

8.6   **EXERCISES**

1–8 ■ Find the determinant of the matrix, if it exists.

1. $[3]$

2. $[0]$

3. $\begin{bmatrix} 4 & 5 \\ 0 & -1 \end{bmatrix}$

4. $\begin{bmatrix} -2 & 1 \\ 3 & -2 \end{bmatrix}$

5. $[2 \quad 5]$

6. $\begin{bmatrix} 3 \\ 0 \end{bmatrix}$

7. $\begin{bmatrix} \frac{1}{2} & \frac{1}{8} \\ 1 & \frac{1}{2} \end{bmatrix}$

8. $\begin{bmatrix} 2.2 & -1.4 \\ 0.5 & 1.0 \end{bmatrix}$

9–14 ■ Evaluate the minor and cofactor using the matrix A.

$$A = \begin{bmatrix} 1 & 0 & \frac{1}{2} \\ -3 & 5 & 2 \\ 0 & 0 & 4 \end{bmatrix}$$

9. M_{11}, A_{11}

10. M_{33}, A_{33}

11. M_{12}, A_{12}

12. M_{13}, A_{13}

13. M_{23}, A_{23}

14. M_{32}, A_{32}

15–20 ■ Find the determinant of the matrix. Determine whether the matrix has an inverse, but don't calculate the inverse.

15. $\begin{bmatrix} 1 & 3 & 7 \\ 2 & 0 & -1 \\ 0 & 2 & 6 \end{bmatrix}$

16. $\begin{bmatrix} -2 & -\frac{3}{2} & \frac{1}{2} \\ 2 & 4 & 0 \\ \frac{1}{2} & 2 & 1 \end{bmatrix}$

17. $\begin{bmatrix} 30 & 0 & 20 \\ 0 & -10 & -20 \\ 40 & 0 & 10 \end{bmatrix}$

18. $\begin{bmatrix} 1 & 2 & 5 \\ -2 & -3 & 2 \\ 3 & 5 & 3 \end{bmatrix}$

19. $\begin{bmatrix} 1 & 3 & 3 & 0 \\ 0 & 2 & 0 & 1 \\ -1 & 0 & 0 & 2 \\ 1 & 6 & 4 & 1 \end{bmatrix}$

20. $\begin{bmatrix} 1 & 2 & 0 & 2 \\ 3 & -4 & 0 & 4 \\ 0 & 1 & 6 & 0 \\ 1 & 0 & 2 & 0 \end{bmatrix}$

21–24 ■ Evaluate the determinant, using row or column operations whenever possible to simplify your work.

21. $\begin{vmatrix} 0 & 0 & 4 & 6 \\ 2 & 1 & 1 & 3 \\ 2 & 1 & 2 & 3 \\ 3 & 0 & 1 & 7 \end{vmatrix}$

22. $\begin{vmatrix} -2 & 3 & -1 & 7 \\ 4 & 6 & -2 & 3 \\ 7 & 7 & 0 & 5 \\ 3 & -12 & 4 & 0 \end{vmatrix}$

23. $\begin{vmatrix} 1 & 2 & 3 & 4 & 5 \\ 0 & 2 & 4 & 6 & 8 \\ 0 & 0 & 3 & 6 & 9 \\ 0 & 0 & 0 & 4 & 8 \\ 0 & 0 & 0 & 0 & 5 \end{vmatrix}$

24. $\begin{vmatrix} 2 & -1 & 6 & 4 \\ 7 & 2 & -2 & 5 \\ 4 & -2 & 10 & 8 \\ 6 & 1 & 1 & 4 \end{vmatrix}$

25. Let

$$B = \begin{bmatrix} 4 & 1 & 0 \\ -2 & -1 & 1 \\ 4 & 0 & 3 \end{bmatrix}$$

(a) Evaluate $|B|$ by expanding by the second row.

(b) Evaluate $|B|$ by expanding by the third column.

(c) Do your results in parts (a) and (b) agree?

26. Consider the system

$$\begin{cases} x + 2y + 6z = 5 \\ -3x - 6y + 5z = 8 \\ 2x + 6y + 9z = 7 \end{cases}$$

(a) Verify that $x = -1$, $y = 0$, $z = 1$ is a solution of the system.

(b) Find the determinant of the coefficient matrix.

(c) Without solving the system, determine whether there are any other solutions.

(d) Can Cramer's Rule be used to solve this system? Why or why not?

27–42 ■ Use Cramer's Rule to solve the system.

27. $\begin{cases} 2x - y = -9 \\ x + 2y = 8 \end{cases}$
28. $\begin{cases} 6x + 12y = 33 \\ 4x + 7y = 20 \end{cases}$

29. $\begin{cases} x - 6y = 3 \\ 3x + 2y = 1 \end{cases}$
30. $\begin{cases} \frac{1}{2}x + \frac{1}{3}y = 1 \\ \frac{1}{4}x - \frac{1}{6}y = -\frac{3}{2} \end{cases}$

31. $\begin{cases} 0.4x + 1.2y = 0.4 \\ 1.2x + 1.6y = 3.2 \end{cases}$
32. $\begin{cases} 10x - 17y = 21 \\ 20x - 31y = 39 \end{cases}$

33. $\begin{cases} x - y + 2z = 0 \\ 3x + z = 11 \\ -x + 2y = 0 \end{cases}$
34. $\begin{cases} 5x - 3y + z = 6 \\ 4y - 6z = 22 \\ 7x + 10y = -13 \end{cases}$

35. $\begin{cases} 2x_1 + 3x_2 - 5x_3 = 1 \\ x_1 + x_2 - x_3 = 2 \\ 2x_2 + x_3 = 8 \end{cases}$
36. $\begin{cases} -2a + c = 2 \\ a + 2b - c = 9 \\ 3a + 5b + 2c = 22 \end{cases}$

37. $\begin{cases} \frac{1}{3}x - \frac{1}{5}y + \frac{1}{2}z = \frac{7}{10} \\ -\frac{2}{3}x + \frac{2}{5}y + \frac{3}{2}z = \frac{11}{10} \\ x - \frac{4}{5}y + z = \frac{9}{5} \end{cases}$
38. $\begin{cases} 2x - y = 5 \\ 5x + 3z = 19 \\ 4y + 7z = 17 \end{cases}$

39. $\begin{cases} 3y + 5z = 4 \\ 2x - z = 10 \\ 4x + 7y = 0 \end{cases}$
40. $\begin{cases} 2x - 5y = 4 \\ x + y - z = 8 \\ 3x + 5z = 0 \end{cases}$

41. $\begin{cases} x + y + z + w = 0 \\ 2x + w = 0 \\ y - z = 0 \\ x + 2z = 1 \end{cases}$
42. $\begin{cases} x + y = 1 \\ y + z = 2 \\ z + w = 3 \\ w - x = 4 \end{cases}$

43. (a) Show that the equation

$$\begin{vmatrix} x_1 & y_1 & 1 \\ x_2 & y_2 & 1 \\ x & y & 1 \end{vmatrix} = 0$$

is an equation for the line that passes through the points (x_1, y_1) and (x_2, y_2).

(b) Use the result of part (a) to find an equation for the line that passes through the points $(20, 50)$ and $(-10, 25)$.

44. Evaluate the following determinant.

$$\begin{vmatrix} a & a & a & a & a \\ 0 & a & a & a & a \\ 0 & 0 & a & a & a \\ 0 & 0 & 0 & a & a \\ 0 & 0 & 0 & 0 & a \end{vmatrix}$$

45–48 ■ Solve for x.

45. $\begin{vmatrix} x & 12 & 13 \\ 0 & x - 1 & 23 \\ 0 & 0 & x - 2 \end{vmatrix} = 0$

46. $\begin{vmatrix} x & 1 & 1 \\ 1 & 1 & x \\ x & 1 & x \end{vmatrix} = 0$

47. $\begin{vmatrix} 1 & 0 & x \\ x^2 & 1 & 0 \\ x & 0 & 1 \end{vmatrix} = 0$
48. $\begin{vmatrix} a & b & x - a \\ x & x + b & x \\ 0 & 1 & 1 \end{vmatrix} = 0$

● **DISCOVERY · DISCUSSION**

49. Matrices with Determinant Zero Use the definition of the determinant and the elementary row and column operations to explain why matrices of the following types have determinant 0.

(a) A matrix with a row or column consisting entirely of zeros

(b) A matrix with two rows the same or two columns the same

(c) A matrix in which one row is a multiple of another row, or one column is a multiple of another column

50. Solving Linear Equations Suppose you had to solve a linear system with five equations and five variables without the assistance of a calculator or computer. Which method would you prefer: Cramer's Rule or Gaussian elimination? Write a short paragraph explaining the reasons for your answer.

8.7 SYSTEMS OF INEQUALITIES

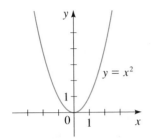

FIGURE 1

In this section we study systems of inequalities in two variables from a graphical point of view. First we consider the graph of a single inequality. We already know that the graph of $y = x^2$, for example, is the *parabola* in Figure 1. If we replace the equal sign by the symbol $\geq$, we obtain the *inequality*

$$y \geq x^2$$

Its graph consists of not just the parabola in Figure 1, but also every point whose y-coordinate is *larger* than x^2. We indicate the solution in Figure 2 by shading the points *above* the parabola.

Similarly, the graph of $y \leq x^2$ in Figure 3 consists of all points on and *below* the parabola, whereas the graphs of $y > x^2$ and $y < x^2$ don't include the points on the parabola itself, as indicated by the dashed curves in Figures 4 and 5.

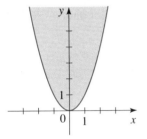

FIGURE 2
$y \geq x^2$

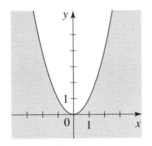

FIGURE 3
$y \leq x^2$

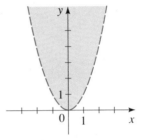

FIGURE 4
$y > x^2$

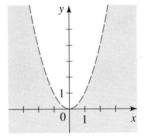

FIGURE 5
$y < x^2$

The graph of an inequality, in general, consists of a region in the plane whose boundary is the graph of the equation obtained by replacing the inequality sign ($\geq$, $\leq$, $>$, or $<$) with an equal sign. To determine which side of the graph gives the solution set of the inequality, we need to check only **test points**.

GRAPHING INEQUALITIES

To graph an inequality,

1. Graph the corresponding equation. (Use a dashed curve for $>$ or $<$, or a solid curve for $\leq$ or $\geq$.)

2. Test one point in each of the regions formed by the graph in Step 1. If the point satisfies the inequality, then all the points in that region satisfy the inequality. (In that case, shade the region to indicate it is part of the graph.) If the test point does not satisfy the inequality, then the region isn't part of the graph.

EXAMPLE 1 ■ Graphs of Nonlinear Inequalities

Graph each of the following inequalities.

(a) $x^2 + y^2 < 25$ (b) $x + 2y \geq 5$

SOLUTION

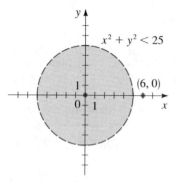

FIGURE 6

(a) The graph of $x^2 + y^2 = 25$ is a circle of radius 5 centered at the origin. The points on the circle itself do not satisfy the inequality because it is of the form $<$, so we graph the circle with a dashed curve, as shown in Figure 6.

   To determine whether the inside or the outside of the circle satisfies the inequality, we use the test points $(0,0)$ on the inside and $(6,0)$ on the outside. We do this by substituting the coordinates of each point into the inequality and determining whether the result satisfies the inequality. (Note that *any* point inside or outside the circle can serve as a test point. We have chosen these points for simplicity.)

Test point	$x^2 + y^2 < 25$	Conclusion
$(0,0)$	$0^2 + 0^2 = 0 < 25$	Part of graph
$(6,0)$	$6^2 + 0^2 = 36 \not< 25$	Not part of graph

Thus, the graph of $x^2 + y^2 < 25$ is the set of all points *inside* the circle (see Figure 6).

(b) The graph of $x + 2y = 5$ is the line shown in Figure 7. We use the test points $(0,0)$ and $(5,5)$ on opposite sides of the line.

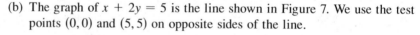

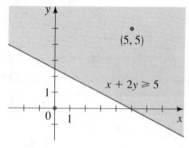

FIGURE 7

Test point	$x + 2y \geq 5$	Conclusion
$(0,0)$	$0 + 2(0) = 0 \not\geq 5$	Not part of graph
$(5,5)$	$5 + 2(5) = 15 \geq 5$	Part of graph

Our check shows that the points *above* the line satisfy the inequality.

   Alternatively, we could put the inequality into slope-intercept form and graph it directly:

$$x + 2y \geq 5$$

$$2y \geq -x + 5$$

$$y \geq -\tfrac{1}{2}x + \tfrac{5}{2}$$

From this form we see that the graph includes all points whose y-coordinates are *greater* than those on the line $y = -\tfrac{1}{2}x + \tfrac{5}{2}$; that is, the graph consists of the points *on or above* this line, as shown in Figure 7.   ■

EXAMPLE 2 ■ A System of Two Inequalities

Graph the solution set of the following pair of inequalities.

$$\begin{cases} x^2 + y^2 < 25 \\ x + 2y \geq 5 \end{cases}$$

SOLUTION These are the two inequalities of Example 1. In this example we wish to graph only those points that simultaneously satisfy both inequalities. The solution thus consists of the intersection of the graphs in Example 1. In Figure 8(a) we show the two regions on the same coordinate plane (in different colors), and in Figure 8(b) we show their intersection.

Vertices: The points $(-3, 4)$ and $(5, 0)$ in Figure 8(b) are the **vertices** of the solution set. They are obtained by solving the system of *equations*

$$\begin{cases} x^2 + y^2 = 25 \\ x + 2y = 5 \end{cases}$$

We solve this system of equations by substitution. Solving for x in the second equation gives $x = 5 - 2y$, and substituting this into the first equation gives

$$(5 - 2y)^2 + y^2 = 25 \qquad \text{Substitute } x = 5 - 2y$$

$$(25 - 20y + 4y^2) + y^2 = 25 \qquad \text{Expand}$$

$$-20y + 5y^2 = 0 \qquad \text{Simplify}$$

$$-5y(4 - y) = 0 \qquad \text{Factor}$$

Thus, $y = 0$ or $y = 4$. When $y = 0$, we have $x = 5 - 2(0) = 5$, and when $y = 4$, we have $x = 5 - 2(4) = -3$. So the points of intersection of these curves are $(5, 0)$ and $(-3, 4)$.

Note that in this case the vertices are not part of the solution set, since they don't satisfy the inequality $x^2 + y^2 < 25$ (and so they are graphed as open circles in the figure). They simply show where the "corners" of the solution set lie. ■

An inequality is **linear** if it can be put into one of the following forms:

$$ax + by \geq c \qquad ax + by \leq c \qquad ax + by > c \qquad ax + by < c$$

In the next example we graph the solution set of a system of linear inequalities.

EXAMPLE 3 ■ A System of Four Linear Inequalities

Graph the solution set of the following system, and label the vertices.

$$\begin{cases} x + 3y \leq 12 \\ x + y \leq 8 \\ x \geq 0 \\ y \geq 0 \end{cases}$$

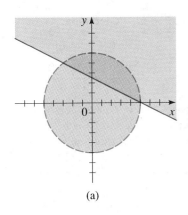

(a)

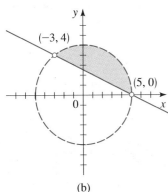

(b)

FIGURE 8

$$\begin{cases} x^2 + y^2 < 25 \\ x + 2y \geq 5 \end{cases}$$

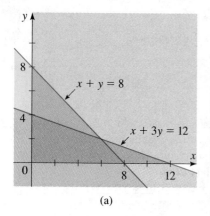

(a)

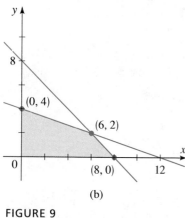

(b)

FIGURE 9

SOLUTION   In Figure 9 we first graph the lines given by the equations that correspond to each of the inequalities. To determine the graphs of the linear inequalities, we only need to check one test point. For simplicity we use the point $(0,0)$.

Inequality	Test point $(0,0)$	Conclusion
$x + 3y \leq 12$	$0 + 3(0) = 0 \leq 12$	Satisfies inequality
$x + y \leq 8$	$0 + 0 = 0 \leq 8$	Satisfies inequality

Since $(0,0)$ is below the line $x + 3y = 12$, our check shows that the region on or below the line must satisfy the inequality. Likewise, since $(0,0)$ is below the line $x + y = 8$, our check shows that the region on or below this line must satisfy the inequality. The inequalities $x \geq 0$ and $y \geq 0$ say that x and y are nonnegative. These regions are sketched in Figure 9(a), and the intersection—the solution set—is sketched in Figure 9(b).

Vertices: The coordinates of each vertex are obtained by simultaneously solving the equations of the lines that intersect at that vertex. From the system

$$\begin{cases} x + 3y = 12 \\ x + y = 8 \end{cases}$$

we get the vertex $(6, 2)$. The other two vertices are at the x- and y-intercepts of the corresponding lines: $(8, 0)$ and $(0, 4)$. In this case, all the vertices *are* part of the solution set.   ■

When a region in the plane can be covered by a (sufficiently large) circle, it's said to be **bounded**. A region that is not bounded is called **unbounded**. For example, the regions graphed in Figures 6, 8(b), and 9(b) are bounded, whereas those in Figures 2–5 and 7 are unbounded. An unbounded region cannot be "fenced in"—it extends infinitely far in at least one direction.

8.7   EXERCISES

1–12 ■ Graph the inequality.

1. $x \leq 2$

2. $y > -3$

3. $y > x$

4. $y < x + 2$

5. $y \geq 2x + 2$

6. $y < -x + 5$

7. $2x - y \leq 8$

8. $3x + 4y + 12 > 0$

9. $4x + 5y < 25$

10. $-x^2 + y \geq 10$

11. $y > x^2 + 1$

12. $x^2 + y^2 \geq 5$

13–34 ■ Graph the solution of the system of inequalities. In each case, find the coordinates of all vertices, and determine whether the solution set is bounded.

13. $\begin{cases} x + y \leq 4 \\ \quad\quad y \geq x \end{cases}$

14. $\begin{cases} 2x + 3y > 12 \\ 3x - y < 21 \end{cases}$

15. $\begin{cases} y < \frac{1}{4}x + 2 \\ y \geq 2x - 5 \end{cases}$

16. $\begin{cases} x - y > 0 \\ 4 + y \leq 2x \end{cases}$

17. $\begin{cases} x \geqslant 0 \\ y \geqslant 0 \\ 3x + 5y \leqslant 15 \\ 3x + 2y \leqslant 9 \end{cases}$

18. $\begin{cases} x > 2 \\ y < 12 \\ 2x - 4y > 8 \end{cases}$

19. $\begin{cases} y < 9 - x^2 \\ y \geqslant x + 3 \end{cases}$

20. $\begin{cases} y \geqslant x^2 \\ x + y \geqslant 6 \end{cases}$

21. $\begin{cases} x^2 + y^2 \leqslant 4 \\ x - y > 0 \end{cases}$

22. $\begin{cases} x > 0 \\ y > 0 \\ x + y < 10 \\ x^2 + y^2 > 9 \end{cases}$

23. $\begin{cases} x^2 - y \leqslant 0 \\ 2x^2 + y \leqslant 12 \end{cases}$

24. $\begin{cases} x^2 + y^2 < 9 \\ 2x + y^2 \geqslant 1 \end{cases}$

25. $\begin{cases} x + 2y \leqslant 14 \\ 3x - y \geqslant 0 \\ x - y \geqslant 2 \end{cases}$

26. $\begin{cases} y < x + 6 \\ 3x + 2y \geqslant 12 \\ x - 2y \leqslant 2 \end{cases}$

27. $\begin{cases} x \geqslant 0 \\ y \geqslant 0 \\ x \leqslant 5 \\ x + y \leqslant 7 \end{cases}$

28. $\begin{cases} x \geqslant 0 \\ y \geqslant 0 \\ y \leqslant 4 \\ 2x + y \leqslant 8 \end{cases}$

29. $\begin{cases} y > x + 1 \\ x + 2y \leqslant 12 \\ x + 1 > 0 \end{cases}$

30. $\begin{cases} x + y > 12 \\ y < \frac{1}{2}x - 6 \\ 3x + y < 6 \end{cases}$

31. $\begin{cases} x^2 + y^2 \leqslant 8 \\ x \geqslant 2 \\ y \geqslant 0 \end{cases}$

32. $\begin{cases} x^2 - y \geqslant 0 \\ x + y < 6 \\ x - y < 6 \end{cases}$

33. $\begin{cases} x^2 + y^2 < 9 \\ x + y > 0 \\ x \leqslant 0 \end{cases}$

34. $\begin{cases} y \geqslant x^3 \\ y \leqslant 2x + 4 \\ x + y \geqslant 0 \end{cases}$

35. A publishing company publishes a total of no more than 100 books every year. At least 20 of these are non-fiction, but the company always publishes at least as much fiction as nonfiction. Find a system of inequalities that describes the possible numbers of fiction and non-fiction books that the company can produce each year consistent with these policies. Graph the solution set.

36. A man and his daughter manufacture unfinished tables and chairs. Each table requires 3 hours of sawing and 1 hour of assembly. Each chair requires 2 hours of sawing and 2 hours of assembly. The two of them can do a total of up to 12 hours of sawing and 8 hours of assembly work each day. Find a system of inequalities that describes all possible combinations of tables and chairs that they can make daily. Graph the solution set.

8.8   PARTIAL FRACTIONS

To write a sum or difference of fractional expressions as a single fraction, we must bring them to a common denominator. For example,

$$\frac{1}{x - 1} + \frac{1}{2x + 1} = \frac{(2x + 1) + (x - 1)}{(x - 1)(2x + 1)} = \frac{3x}{2x^2 - x - 1}$$

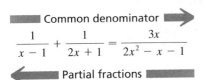

Common denominator ➡

$$\frac{1}{x - 1} + \frac{1}{2x + 1} = \frac{3x}{2x^2 - x - 1}$$

⬅ Partial fractions

But for some applications of algebra to calculus, we must know how to reverse this process—that is, we must know how to express a fraction such as $3x/(2x^2 - x - 1)$ as the sum of the simpler fractions $1/(x - 1)$ and $1/(2x + 1)$. These simpler fractions are called *partial fractions*; we learn how to find them in this section.

Let r be the rational function

$$r(x) = \frac{P(x)}{Q(x)}$$

where the degree of P is less than the degree of Q. It can be shown, using advanced algebra techniques, that every polynomial with real coefficients can be factored completely into linear and irreducible quadratic factors, that is, factors of the form $ax + b$ and $ax^2 + bx + c$, where a, b, and c are real numbers. For instance,

$$x^4 - 1 = (x^2 - 1)(x^2 + 1) = (x - 1)(x + 1)(x^2 + 1)$$

After we have completely factored the denominator Q of r, we will be able to express $r(x)$ as a sum of **partial fractions** of the form

$$\frac{A}{(ax + b)^i} \quad \text{and} \quad \frac{Ax + B}{(ax^2 + bx + c)^j}$$

This sum is called the **partial fraction decomposition** of r. We now explain the details in the four cases that occur.

CASE 1: The Denominator Is a Product of Distinct Linear Factors

This means that we can write

$$Q(x) = (a_1x + b_1)(a_2x + b_2)\cdots(a_nx + b_n)$$

with no factor repeated. In this case the partial fraction decomposition of r takes the form

$$r(x) = \frac{P(x)}{Q(x)} = \frac{A_1}{a_1x + b_1} + \frac{A_2}{a_2x + b_2} + \cdots + \frac{A_n}{a_nx + b_n}$$

where the constants $A_1, A_2, \ldots, A_n$ are determined as in the following example.

EXAMPLE 1 ■ Distinct Linear Factors

Find the partial fraction decomposition of $\dfrac{5x + 7}{x^3 + 2x^2 - x - 2}$.

SOLUTION   The denominator factors as follows.

$$x^3 + 2x^2 - x - 2 = x^2(x + 2) - (x + 2) = (x^2 - 1)(x + 2)$$
$$= (x - 1)(x + 1)(x + 2)$$

This gives us the partial fraction decomposition

$$\frac{5x + 7}{x^3 + 2x^2 - x - 2} = \frac{A}{x - 1} + \frac{B}{x + 1} + \frac{C}{x + 2}$$

Multiplying each side by the common denominator, $(x - 1)(x + 1)(x + 2)$, we get

$$5x + 7 = A(x + 1)(x + 2) + B(x - 1)(x + 2) + C(x - 1)(x + 1)$$
$$= A(x^2 + 3x + 2) + B(x^2 + x - 2) + C(x^2 - 1) \qquad \text{Expand}$$
$$= (A + B + C)x^2 + (3A + B)x + (2A - 2B - C) \qquad \text{Combine like terms}$$

If two polynomials are equal, then their coefficients are equal. Thus, since $5x + 7$ has no x^2-term, we have that $A + B + C = 0$. Similarly, by comparing the coefficients of x, we see that $3A + B = 5$, and by comparing constant terms, we get that $2A - 2B - C = 7$. This leads to the following system of linear equations for A, B, and C.

$$\begin{cases} A + B + C = 0 \\ 3A + B = 5 \\ 2A - 2B - C = 7 \end{cases}$$

We solve this system using the methods developed in Section 8.3.

$$\begin{bmatrix} 1 & 1 & 1 & 0 \\ 3 & 1 & 0 & 5 \\ 2 & -2 & -1 & 7 \end{bmatrix} \xrightarrow[R_3 - 2R_1 \to R_3]{R_2 - 3R_1 \to R_2} \begin{bmatrix} 1 & 1 & 1 & 0 \\ 0 & -2 & -3 & 5 \\ 0 & -4 & -3 & 7 \end{bmatrix} \xrightarrow[-R_2]{R_3 - 2R_2 \to R_3}$$

$$\begin{bmatrix} 1 & 1 & 1 & 0 \\ 0 & 2 & 3 & -5 \\ 0 & 0 & 3 & -3 \end{bmatrix} \xrightarrow[\frac{1}{3}R_3]{R_2 - R_3 \to R_2} \begin{bmatrix} 1 & 1 & 1 & 0 \\ 0 & 2 & 0 & -2 \\ 0 & 0 & 1 & -1 \end{bmatrix}$$

Thus, we see that $C = -1$, $B = -1$, and $A = 2$, so the required partial fraction decomposition is

$$\frac{5x + 7}{x^3 + 2x^2 - x - 2} = \frac{2}{x - 1} + \frac{-1}{x + 1} + \frac{-1}{x + 2} \qquad \blacksquare$$

The same method of attack works in each of the remaining cases. We set up the partial fraction decomposition with the unknown constants A, B, C, ... We then multiply each side of the resulting equation by the common denominator, simplify the right-hand side of the equation, and equate coefficients. This gives a set of linear equations that will always have a unique solution (provided the partial fraction decomposition has been set up correctly).

CASE 2: The Denominator Is a Product of Linear Factors, Some of Which Are Repeated

Suppose the complete factorization of $Q(x)$ contains the linear factor $ax + b$ repeated k times; that is, $(ax + b)^k$ is a factor of $Q(x)$. Then, corresponding to each such factor the partial fraction decomposition for $P(x)/Q(x)$ will contain

$$\frac{A_1}{ax + b} + \frac{A_2}{(ax + b)^2} + \cdots + \frac{A_k}{(ax + b)^k}$$

EXAMPLE 2 ■ Repeated Linear Factors

Find the partial fraction decomposition of $\dfrac{x^2 + 1}{x(x - 1)^3}$.

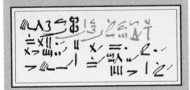

SOLUTION   Because the factor $x - 1$ is repeated three times in the denominator, the partial fraction decomposition has the form

$$\frac{x^2 + 1}{x(x - 1)^3} = \frac{A}{x} + \frac{B}{x - 1} + \frac{C}{(x - 1)^2} + \frac{D}{(x - 1)^3}$$

Multiplying each side by the common denominator, $x(x - 1)^3$, gives

$$x^2 + 1 = A(x - 1)^3 + Bx(x - 1)^2 + Cx(x - 1) + Dx$$

$$= A(x^3 - 3x^2 + 3x - 1) + B(x^3 - 2x^2 + x) + C(x^2 - x) + Dx \qquad \text{Expand}$$

$$= (A + B)x^3 + (-3A - 2B + C)x^2 + (3A + B - C + D)x - A \qquad \text{Combine like terms}$$

Equating coefficients, we get the equations

$$\begin{cases} A + B & = 0 \\ -3A - 2B + C & = 1 \\ 3A + B - C + D & = 0 \\ -A & = 1 \end{cases}$$

If we rearrange these equations by putting the last one in the first position, we can easily see (using substitution) that the solution to the system is $A = -1$, $B = 1$, $C = 0$, $D = 2$ and so the partial fraction decomposition is

$$\frac{x^2 + 1}{x(x - 1)^3} = \frac{-1}{x} + \frac{1}{x - 1} + \frac{2}{(x - 1)^3} \qquad \blacksquare$$

CASE 3: The Denominator Has Irreducible Quadratic Factors, None of Which Is Repeated

If the complete factorization of the denominator $Q(x)$ contains the quadratic factor $ax^2 + bx + c$ (which can't be factored further), then corresponding to this the partial fraction decomposition of $P(x)/Q(x)$ will have a term of the form

$$\frac{Ax + B}{ax^2 + bx + c}$$

EXAMPLE 3 ■ Distinct Quadratic Factors

Find the partial fraction decomposition of $\dfrac{2x^2 - x + 4}{x^3 + 4x}$.

SOLUTION   Since $x^3 + 4x = x(x^2 + 4)$, which can't be factored further, we write

$$\frac{2x^2 - x + 4}{x^3 + 4x} = \frac{A}{x} + \frac{Bx + C}{x^2 + 4}$$

Multiplying by $x(x^2 + 4)$, we get

$$2x^2 - x + 4 = A(x^2 + 4) + (Bx + C)x$$

$$= (A + B)x^2 + Cx + 4A$$

Equating coefficients gives us the equations

$$\begin{cases} A + B = 2 \\ \quad\quad C = -1 \\ \quad 4A = 4 \end{cases}$$

and so $A = 1$, $B = 1$, and $C = -1$. The required partial fraction decomposition is

$$\frac{2x^2 - x + 4}{x^3 + 4x} = \frac{1}{x} + \frac{x - 1}{x^2 + 4}$$

■

CASE 4: The Denominator Has a Repeated Irreducible Quadratic Factor

If the complete factorization of $Q(x)$ contains the factor $(ax^2 + bx + c)^k$, where $ax^2 + bx + c$ can't be factored further, then corresponding to this the partial fraction decomposition of $P(x)/Q(x)$ will have the terms

$$\frac{A_1 x + B_1}{ax^2 + bx + c} + \frac{A_2 x + B_2}{(ax^2 + bx + c)^2} + \cdots + \frac{A_k x + B_k}{(ax^2 + bx + c)^k}$$

EXAMPLE 4 ■ **Repeated Quadratic Factors**

Write out the form of the partial fraction decomposition of

$$\frac{x^5 - 3x^2 + 12x - 1}{x^3(x^2 + x + 1)(x^2 + 2)^3}$$

SOLUTION

$$\frac{x^5 - 3x^2 + 12x - 1}{x^3(x^2 + x + 1)(x^2 + 2)^3}$$

$$= \frac{A}{x} + \frac{B}{x^2} + \frac{C}{x^3} + \frac{Dx + E}{x^2 + x + 1} + \frac{Fx + G}{x^2 + 2} + \frac{Hx + I}{(x^2 + 2)^2} + \frac{Jx + K}{(x^2 + 2)^3}$$

■

In order to find the values of A, B, C, D, E, F, G, H, I, J, and K in Example 4, we would have to solve a system of 11 linear equations. Although certainly possible, this would involve a great deal of work!

The techniques we have described in this section apply only to rational functions $P(x)/Q(x)$ in which the degree of P is less than the degree of Q. If this isn't the case, we must first use long division to divide Q into P.

EXAMPLE 5 ■ **Using Long Division to Prepare for Partial Fractions**

Find the partial fraction decomposition of

$$\frac{2x^4 + 4x^3 - 2x^2 + x + 7}{x^3 + 2x^2 - x - 2}$$

SOLUTION   Since the degree of the numerator is larger than the degree of the denominator, we use long division to obtain

$$
\begin{array}{r}
2x \\
x^3 + 2x^2 - x - 2 \overline{)2x^4 + 4x^3 - 2x^2 + x + 7} \\
\underline{2x^4 + 4x^3 - 2x^2 - 4x} \\
5x + 7
\end{array}
$$

$$\frac{2x^4 + 4x^3 - 2x^2 + x + 7}{x^3 + 2x^2 - x - 2} = 2x + \frac{5x + 7}{x^3 + 2x^2 - x - 2}$$

The remainder term now satisfies the requirement that the degree of the numerator is less than the degree of the denominator. At this point we would proceed as in Example 1 to obtain the decomposition

$$\frac{2x^4 + 4x^3 - 2x^2 + x + 7}{x^3 + 2x^2 - x - 2} = 2x + \frac{2}{x - 1} + \frac{-1}{x + 1} + \frac{-1}{x + 2} \quad ■$$

8.8   **EXERCISES**

1–10 ■ Write out the form of the partial fraction decomposition of the function (as in Example 4). Do not determine the numerical values of the coefficients.

1. $\dfrac{1}{(x - 1)(x + 2)}$

2. $\dfrac{x}{x^2 + 3x - 4}$

3. $\dfrac{x^2 - 3x + 5}{(x - 2)^2(x + 4)}$

4. $\dfrac{1}{x^4 - x^3}$

5. $\dfrac{x^2}{(x - 3)(x^2 + 4)}$

6. $\dfrac{1}{x^4 - 1}$

7. $\dfrac{x^3 - 4x^2 + 2}{(x^2 + 1)(x^2 + 2)}$

8. $\dfrac{x^4 + x^2 + 1}{x^2(x^2 + 4)^2}$

9. $\dfrac{x^3 + x + 1}{x(2x - 5)^3(x^2 + 2x + 5)^2}$

10. $\dfrac{1}{(x^6 - 1)(x^4 - 1)}$

11–40 ■ Find the partial fraction decomposition of the rational function.

11. $\dfrac{5}{(x - 1)(x + 4)}$

12. $\dfrac{x + 6}{x(x + 3)}$

13. $\dfrac{12}{x^2 - 9}$

14. $\dfrac{x - 12}{x^2 - 4x}$

15. $\dfrac{4}{x^2 - 4}$

16. $\dfrac{2x + 1}{x^2 + x - 2}$

17. $\dfrac{x + 14}{x^2 - 2x - 8}$

18. $\dfrac{8x - 3}{2x^2 - x}$

19. $\dfrac{x}{8x^2 - 10x + 3}$

20. $\dfrac{7x - 3}{x^3 + 2x^2 - 3x}$

21. $\dfrac{9x^2 - 9x + 6}{2x^3 - x^2 - 8x + 4}$

22. $\dfrac{-3x^2 - 3x + 27}{(x + 2)(2x^2 + 3x - 9)}$

23. $\dfrac{x^2 + 1}{x^3 + x^2}$

24. $\dfrac{3x^2 + 5x - 13}{(3x + 2)(x^2 - 4x + 4)}$

25. $\dfrac{2x}{4x^2 + 12x + 9}$

26. $\dfrac{x - 4}{(2x - 5)^2}$

27. $\dfrac{4x^2 - x - 2}{x^4 + 2x^3}$

28. $\dfrac{x^3 - 2x^2 - 4x + 3}{x^4}$

29. $\dfrac{-10x^2 + 27x - 14}{(x - 1)^3(x + 2)}$

30. $\dfrac{-2x^2 + 5x - 1}{x^4 - 2x^3 + 2x - 1}$

31. $\dfrac{3x^3 + 22x^2 + 53x + 41}{(x + 2)^2(x + 3)^2}$

32. $\dfrac{3x^2 + 12x - 20}{x^4 - 8x^2 + 16}$

33. $\dfrac{x - 3}{x^3 + 3x}$

34. $\dfrac{3x^2 - 2x + 8}{x^3 - x^2 + 2x - 2}$

35. $\dfrac{2x^3 + 7x + 5}{(x^2 + x + 2)(x^2 + 1)}$

36. $\dfrac{x^2 + x + 1}{2x^4 + 3x^2 + 1}$

37. $\dfrac{x^4 + x^3 + x^2 - x + 1}{x(x^2 + 1)^2}$

38. $\dfrac{2x^2 - x + 8}{(x^2 + 4)^2}$

39. $\dfrac{x^5 - 2x^4 + x^3 + x + 5}{x^3 - 2x^2 + x - 2}$

40. $\dfrac{x^5 - 3x^4 + 3x^3 - 4x^2 + 4x + 12}{(x - 2)^2(x^2 + 2)}$

41. Determine A and B in terms of a and b:

$$\frac{ax + b}{x^2 - 1} = \frac{A}{x - 1} + \frac{B}{x + 1}$$

42. Determine A, B, C, and D in terms of a and b:

$$\frac{ax^3 + bx^2}{(x^2 + 1)^2} = \frac{Ax + B}{x^2 + 1} + \frac{Cx + D}{(x^2 + 1)^2}$$

---

8 REVIEW

CONCEPT CHECK

1. Suppose you are asked to solve a system of two equations (not necessarily linear) in two variables. Explain how you would solve them
  (a) by the substitution method
  (b) by the elimination method
  (c) graphically

2. Suppose you are asked to solve a system of two *linear* equations in two variables.
  (a) Would you prefer to use substitution or elimination?
  (b) How many solutions are possible? Draw diagrams to illustrate the possibilities.

3. Explain how Gaussian elimination works. Your explanation should include a discussion of matrices, elementary row operations, echelon form, back-substitution, and leading variables.

4. (a) What is meant by an inconsistent system?
  (b) What is meant by a dependent system? What is a leading variable?

5. Suppose you have used Gaussian elimination to transform a matrix that represents a system of linear equations into echelon form. How can you tell if the system has
  (a) no solution?
  (b) exactly one solution?
  (c) infinitely many solutions?

6. How can you tell if a matrix is in reduced echelon form? What is the advantage of using a matrix in reduced echelon form?

7. (a) What does it mean to say that A is a matrix with dimension $m \times n$?
  (b) If A and B have the same dimension and k is a real number, how do you find $A + B$, $A - B$, and kA?

8. (a) Under what conditions is it possible to find the product AB of two matrices?
  (b) If those conditions are satisfied, how do you calculate the product AB?

9. (a) What is the identity matrix I_n?
  (b) If A is a square $n \times n$ matrix, what is its inverse matrix?
  (c) Write a formula for the inverse of a 2×2 matrix.
  (d) Explain how you would find the inverse of a 3×3 matrix.

10. Suppose A is an $n \times n$ matrix.
  (a) What is the minor M_{ij} of the element a_{ij}?
  (b) What is the cofactor A_{ij}?
  (c) How do you find the determinant of A?
  (d) How can you tell if A has an inverse?

11. State Cramer's Rule for solving a system of linear equations in terms of determinants. Do you prefer to use Cramer's Rule or Gaussian elimination? Explain.

12. Explain the procedure for solving a system of inequalities.

13. Explain how to find the partial fraction decomposition of a fractional expression. Your explanation should discuss each of the four cases that arise.

EXERCISES

1–6 ■ Solve the system of equations and graph the lines.

1. $\begin{cases} 3x - y = 5 \\ 2x + y = 5 \end{cases}$

2. $\begin{cases} y = 2x + 6 \\ y = -x + 3 \end{cases}$

3. $\begin{cases} 2x - 7y = 28 \\ y = \frac{2}{7}x - 4 \end{cases}$

4. $\begin{cases} 6x - 8y = 15 \\ -\frac{3}{2}x + 2y = -4 \end{cases}$

5. $\begin{cases} 2x - y = 1 \\ x + 3y = 10 \\ 3x + 4y = 15 \end{cases}$

6. $\begin{cases} 2x + 5y = 9 \\ -x + 3y = 1 \\ 7x - 2y = 14 \end{cases}$

7–10 ■ Solve the system of equations.

7. $\begin{cases} y = x^2 + 2x \\ y = 6 + x \end{cases}$

8. $\begin{cases} x^2 + y^2 = 8 \\ y = x + 2 \end{cases}$

9. $\begin{cases} 3x + \dfrac{4}{y} = 6 \\ x - \dfrac{8}{y} = 4 \end{cases}$

10. $\begin{cases} x^2 + y^2 = 10 \\ x^2 + 2y^2 - 7y = 0 \end{cases}$

11–18 ■ Find the complete solution of the system using Gaussian elimination, or show that the system has no solution.

11. $\begin{cases} x + y + 2z = 6 \\ 2x \quad\quad + 5z = 12 \\ x + 2y + 3z = 9 \end{cases}$

12. $\begin{cases} x - 2y + 3z = 1 \\ x - 3y - z = 0 \\ 2x \quad\quad - 6z = 6 \end{cases}$

13. $\begin{cases} x - 2y + 3z = 1 \\ 2x - y + z = 3 \\ 2x - 7y + 11z = 2 \end{cases}$

14. $\begin{cases} x + y + z + w = 2 \\ 2x \quad\quad - 3z \quad\quad = 5 \\ x - 2y \quad\quad + 4w = 9 \\ x + y + 2z + 3w = 5 \end{cases}$

15. $\begin{cases} x - 3y + z = 4 \\ 4x - y + 15z = 5 \end{cases}$

16. $\begin{cases} 2x - 3y + 4z = 3 \\ 4x - 5y + 9z = 13 \\ 2x \quad\quad + 7z = 0 \end{cases}$

17. $\begin{cases} -x + 4y + z = 8 \\ 2x - 6y + z = -9 \\ x - 6y - 4z = -15 \end{cases}$

18. $\begin{cases} x \quad\quad - z + w = 2 \\ 2x + y \quad\quad - 2w = 12 \\ 3y + z + w = 4 \\ x + y - z \quad\quad = 10 \end{cases}$

19. A man invests his savings in two accounts, one paying 6% interest per year and the other paying 7%. He has twice as much invested in the 7% account as in the 6% account, and his annual interest income is $600. How much is invested in each account?

20. Find the values of a, b, and c if the parabola

$$y = ax^2 + bx + c$$

passes through the points $(1,0)$, $(-1,-4)$, and $(2,11)$.

21–32 ■ Let

$$A = [2 \quad 0 \;-1] \qquad\qquad B = \begin{bmatrix} 1 & 2 & 4 \\ -2 & 1 & 0 \end{bmatrix}$$

$$C = \begin{bmatrix} \frac{1}{2} & 3 \\ 2 & \frac{3}{2} \\ -2 & 1 \end{bmatrix} \qquad\qquad D = \begin{bmatrix} 1 & 4 \\ 0 & -1 \\ 2 & 0 \end{bmatrix}$$

$$E = \begin{bmatrix} 2 & -1 \\ -\frac{1}{2} & 1 \end{bmatrix} \qquad\qquad F = \begin{bmatrix} 4 & 0 & 2 \\ -1 & 1 & 0 \\ 7 & 5 & 0 \end{bmatrix}$$

$$G = [5]$$

Carry out the indicated operation, or explain why it cannot be performed.

21. $A + B$ **22.** $C - D$ **23.** $2C + 3D$

24. $5B - 2C$ **25.** GA **26.** AG

27. BC **28.** CB **29.** BF

30. FC **31.** $(C + D)E$ **32.** $F(2C - D)$

33–38 ■ Find the determinant and, if possible, the inverse of the matrix.

33. $\begin{bmatrix} 1 & 4 \\ 2 & 9 \end{bmatrix}$

34. $\begin{bmatrix} 2 & 2 \\ 1 & -3 \end{bmatrix}$

35. $\begin{bmatrix} 4 & -12 \\ -2 & 6 \end{bmatrix}$

36. $\begin{bmatrix} 2 & 4 & 0 \\ -1 & 1 & 2 \\ 0 & 3 & 2 \end{bmatrix}$

37. $\begin{bmatrix} 3 & 0 & 1 \\ 2 & -3 & 0 \\ 4 & -2 & 1 \end{bmatrix}$ **38.** $\begin{bmatrix} 1 & 0 & 0 & 1 \\ 0 & 2 & 0 & 2 \\ 0 & 0 & 3 & 3 \\ 0 & 0 & 0 & 4 \end{bmatrix}$

39–40 ■ Express the system of linear equations as a matrix equation. Then solve the matrix equation by multiplying each side by the inverse of the coefficient matrix.

39. $\begin{cases} 12x - 5y = 10 \\ 5x - 2y = 17 \end{cases}$

40. $\begin{cases} 2x + y + 5z = \frac{1}{3} \\ x + 2y + 2z = \frac{1}{4} \\ x \quad\quad + 3z = \frac{1}{6} \end{cases}$

41–44 ■ Solve the system using Cramer's Rule.

41. $\begin{cases} 2x + 7y = 13 \\ 6x + 16y = 30 \end{cases}$

42. $\begin{cases} 12x - 11y = 140 \\ 7x + 9y = 20 \end{cases}$

43. $\begin{cases} 2x - y + 5z = 0 \\ -x + 7y \quad\quad = 9 \\ 5x + 4y + 3z = -9 \end{cases}$

44. $\begin{cases} 3x + 4y - z = 10 \\ x \quad\quad - 4z = 20 \\ 2x + y + 5z = 30 \end{cases}$

45–48 ■ Graph the solution set of the system of inequalities. Find the coordinates of all vertices, and determine whether the solution set is bounded or unbounded.

45. $\begin{cases} x^2 + y^2 < 9 \\ x + y < 0 \end{cases}$

46. $\begin{cases} y - x^2 \geqslant 4 \\ y < 20 \end{cases}$

47. $\begin{cases} x \geqslant 0, \;\; y \geqslant 0 \\ x + 2y \leqslant 12 \\ y \leqslant x + 4 \end{cases}$

48. $\begin{cases} x \geqslant 4 \\ x + y \geqslant 24 \\ x \leqslant 2y + 12 \end{cases}$

49–50 ■ Solve for x, y, and z in terms of a, b, and c.

49. $\begin{cases} -x + y + z = a \\ x - y + z = b \\ x + y - z = c \end{cases}$

50. $\begin{cases} ax + by + cz = a - b + c \\ bx + by + cz = c \\ cx + cy + cz = c \end{cases}$ $(a \neq b, b \neq c, c \neq 0)$

51. For what values of k do the following three lines have a common point of intersection?

$$x + y = 12$$
$$kx - y = 0$$
$$y - x = 2k$$

52. For what value of k does the following system have infinitely many solutions?

$$\begin{cases} kx + y + z = 0 \\ x + 2y + kz = 0 \\ -x \quad\quad + 3z = 0 \end{cases}$$

53–56 ■ Use a graphing device to solve the system, correct to the nearest hundredth.

53. $\begin{cases} 0.32x + 0.43y = 0 \\ 7x - 12y = 341 \end{cases}$

54. $\begin{cases} \sqrt{12}\, x - 3\sqrt{2}\, y = 660 \\ 7137x + 3931y = 20{,}000 \end{cases}$

55. $\begin{cases} x - y^2 = 10 \\ x = \frac{1}{22}y + 12 \end{cases}$

56. $\begin{cases} y = 5^x + x \\ y = x^5 + 5 \end{cases}$

57–60 ■ Find the partial fraction decomposition of the rational function.

57. $\dfrac{3x + 1}{x^2 - 2x - 15}$

58. $\dfrac{8}{x^3 - 4x}$

59. $\dfrac{2x - 4}{x(x - 1)^2}$

60. $\dfrac{x + 6}{x^3 - 2x^2 + 4x - 8}$

1. In $2\frac{1}{2}$ hours an airplane travels 600 km against the wind. It takes 50 min to travel 300 km with the wind. Find the speed of the wind and the speed of the airplane in still air.

2–5 ■ Find all solutions of the system. Determine whether the system is linear or nonlinear. If it is linear, state whether it is inconsistent, dependent, or neither.

2. $\begin{cases} 3x - y = 10 \\ 2x + 5y = 1 \end{cases}$

3. $\begin{cases} x - y + 9z = -8 \\ x \quad\quad - 4z = 7 \\ 3x - y + z = 5 \end{cases}$

4. $\begin{cases} 2x - y + z = 0 \\ 3x + 2y - 3z = 1 \\ x - 4y + 5z = -1 \end{cases}$

5. $\begin{cases} 2x^2 + y^2 = 6 \\ 3x^2 - 4y = 11 \end{cases}$

6–13 ■ Let

$$A = \begin{bmatrix} 2 & 3 \\ 2 & 4 \end{bmatrix} \qquad B = \begin{bmatrix} 2 & 4 \\ -1 & 1 \\ 3 & 0 \end{bmatrix} \qquad C = \begin{bmatrix} 1 & 0 & 4 \\ -1 & 1 & 2 \\ 0 & 1 & 3 \end{bmatrix}$$

Carry out the indicated operation, or explain why it cannot be performed.

6. $A + B$ **7.** AB **8.** $BA - 3B$ **9.** CBA

10. A^{-1} **11.** B^{-1} **12.** $|B|$ **13.** $|C|$

14. Write a matrix equation equivalent to the following system.

$$\begin{cases} 4x - 3y = 10 \\ 3x - 2y = 30 \end{cases}$$

Find the inverse of the coefficient matrix, and use it to solve the system.

15. Solve using Cramer's Rule:

$$\begin{cases} 2x \quad\quad - z = 14 \\ 3x - y + 5z = 0 \\ 4x + 2y + 3z = -2 \end{cases}$$

16. Only one of the following matrices has an inverse. Find the determinant of each matrix, and use the determinants to identify the one that has an inverse. Then find the inverse.

$$A = \begin{bmatrix} 1 & 4 & 1 \\ 0 & 2 & 0 \\ 1 & 0 & 1 \end{bmatrix} \qquad B = \begin{bmatrix} 1 & 4 & 0 \\ 0 & 2 & 0 \\ -3 & 0 & 1 \end{bmatrix}$$

17. Graph the following system of inequalities, and state the coordinates of the vertices.

$$\begin{cases} x^2 - y + 5 \leq 0 \\ y \leq 5 + 2x \end{cases}$$

18. Find the partial fraction decomposition of the following rational function.

$$\frac{4x - 1}{(x - 1)^2(x + 2)}$$

19. Use a graphing calculator to find all solutions of the following system, correct to two decimal places.

$$\begin{cases} 2x^2 + y^2 = 16 \\ y = x^4 - 4x^3 + 6x^2 - 4x \end{cases}$$

FOCUS ON MODELING

Linear programming is a modeling technique used to determine the optimal allocation of resources in business, the military, and other areas of human endeavor. For example, a manufacturer who makes several different products from the same raw materials can use linear programming to determine how much of each product should be produced to maximize the profit. This modeling technique is probably the most important practical application of systems of linear inequalities. In 1975 Leonid Kantorovich and T. C. Koopmans won the Nobel Prize in economics for their work in the development of this subject.

Although linear programming can be applied to very complex problems with hundreds or even thousands of variables, we will consider only a few simple examples to which the graphical methods of Section 8.7 can be applied. (For large numbers of variables, a linear programming method based on matrices is used.) We introduce the technique with a typical problem.

EXAMPLE 1 ■ Manufacturing for Maximum Profit

A small shoe manufacturer makes two styles of shoes: oxfords and loafers. Two machines are used in the process: a cutting machine and a sewing machine. Each type of shoe requires 15 min per pair on the cutting machine. Oxfords require 10 min of sewing per pair, and loafers require 20 min of sewing per pair. Because the manufacturer can hire only one operator for each machine, each process is available for just 8 hours per day. If the profit is $15 on each pair of oxfords and $20 on each pair of loafers, how many pairs of each type should be produced per day for maximum profit?

SOLUTION First we organize the given information into a table. To be consistent, we convert all times to hours.

	Oxfords	Loafers	Time available
Time on cutting machine (h)	$\frac{1}{4}$	$\frac{1}{4}$	8
Time on sewing machine (h)	$\frac{1}{6}$	$\frac{1}{3}$	8
Profit	$15	$20	

We describe the model and solve the problem in four steps.

CHOOSING THE VARIABLES. To make a mathematical model we first give names to the variable quantities. For this problem we let

$$x = \text{number of pairs of oxfords made daily}$$
$$y = \text{number of pairs of loafers made daily}$$

FINDING THE OBJECTIVE FUNCTION. The objective of this problem is to determine which values for x and y give maximum profit. Since each pair of oxfords

Because loafers produce more profit, it would seem that it is best to manufacture only loafers. Surprisingly, this does not turn out to be the most profitable solution.

provides $15 profit and each pair of loafers $20, the total profit is given by

$$P = 15x + 20y$$

This function is called the *objective function.*

GRAPHING THE FEASIBLE REGION. The larger x and y are, the greater the profit. But we cannot choose arbitrarily large values for these variables because there are restrictions, or *constraints,* in the problem. Each restriction is an inequality in the variables.

In this problem the total number of cutting hours needed is $\frac{1}{4}x + \frac{1}{4}y$. Since only 8 hours are available on the cutting machine, we have

$$\tfrac{1}{4}x + \tfrac{1}{4}y \le 8$$

Similarly, by considering the amount of time needed and available on the sewing machine, we get

$$\tfrac{1}{6}x + \tfrac{1}{3}y \le 8$$

We cannot produce a negative number of shoes, so we also have

$$x \ge 0 \qquad \text{and} \qquad y \ge 0$$

Thus, x and y must satisfy the constraints

$$\begin{cases} \tfrac{1}{4}x + \tfrac{1}{4}y \le 8 \\ \tfrac{1}{6}x + \tfrac{1}{3}y \le 8 \\ \qquad\qquad x \ge 0 \\ \qquad\qquad y \ge 0 \end{cases}$$

If we multiply the first inequality by 4 and the second by 6, we obtain the simplified system

$$\begin{cases} x + y \le 32 \\ x + 2y \le 48 \\ \qquad x \ge 0 \\ \qquad y \ge 0 \end{cases}$$

The solution of this system (with vertices labeled) is sketched in Figure 1. The only values that satisfy the restrictions of the problem are the ones that correspond to points of the shaded region in Figure 1. This is called the *feasible region* for the problem.

FINDING MAXIMUM PROFIT. As x or y increases, profit will increase as well. Thus, it seems reasonable that the maximum profit will occur at a point on one of the outside edges of the feasible region, where it's impossible to increase x or

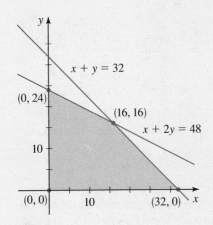

FIGURE 1

y without going outside the region. In fact, it can be shown that the maximum value will occur at a vertex. This means that we need to check the profit only at the vertices.

Vertex	$P = 15x + 20y$	
$(0, 0)$	0	
$(0, 24)$	$15(0) + 20(24) = \$480$	
$(16, 16)$	$15(16) + 20(16) = \$560$	$\leftarrow$ maximum profit
$(32, 0)$	$15(32) + 20(0) = \$480$	

The largest value of P occurs at the point $(16, 16)$, where $P = \$560$. Thus, the manufacturer should make 16 pairs of oxfords and 16 pairs of loafers, for a maximum daily profit of $560. ∎

All the linear programming problems that we will consider follow the pattern of Example 1. Each problem involves two variables, and certain restrictions, called **constraints**, described in the problem lead to a system of linear inequalities whose solution is called the **feasible region**. The function we are trying to maximize or minimize is called the **objective function**. This function will always attain its largest and smallest values at the **vertices** of the feasible region. We summarize this modeling technique in the following four steps.

---

1. CHOOSE THE VARIABLES.   Decide what variable quantities in the problem should be named x and y.

2. FIND THE OBJECTIVE FUNCTION.   Write an expression for the function we want to maximize or minimize.

3. GRAPH THE FEASIBLE REGION.   Express the constraints as a system of inequalities and graph the solution of this system (the feasible region).

4. FIND THE MAXIMUM OR MINIMUM.   Evaluate the objective function at the vertices of the feasible region to determine its maximum or minimum value.

EXAMPLE 2 ■ A Shipping Problem

A car dealer has warehouses in Millville and Trenton and dealerships in Camden and Atlantic City. Every car sold at the dealerships must be delivered from one of the warehouses. On a certain day the Camden dealers sell 10 cars, and the Atlantic City dealers sell 12. The Millville warehouse has 15 cars available, and the Trenton warehouse has 10. The cost of shipping one car is $50 from Millville to Camden, $40 from Millville to Atlantic City, $60 from Trenton to Camden, and $55 from Trenton to Atlantic City. How many cars should be moved from each warehouse to each dealership to fill the orders at minimum cost?

SOLUTION The first step is to organize the given information. Rather than construct a table, we draw a diagram to show the flow of cars from the warehouses to the dealerships (see Figure 2). The diagram shows the number of cars available at each warehouse or required at each dealership and the cost of shipping between these locations.

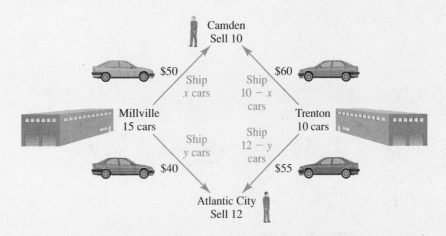

FIGURE 2

CHOOSING THE VARIABLES. The arrows in Figure 2 indicate four possible routes, so the problem seems to involve four variables. But we let

x = number of cars to be shipped from Millville to Camden

y = number of cars to be shipped from Millville to Atlantic City

To fill the orders, we must have

$10 - x$ = number of cars shipped from Trenton to Camden

$12 - y$ = number of cars shipped from Trenton to Atlantic City

So the only variables in the problem are x and y.

FINDING THE OBJECTIVE FUNCTION. The objective of this problem is to minimize cost. From Figure 2 we see that the total cost of shipping the cars is

$$C = 50x + 40y + 60(10 - x) + 55(12 - y)$$

$$= 50x + 40y + 600 - 60x + 660 - 55y$$

$$= 1260 - 10x - 15y$$

This is the objective function.

GRAPHING THE FEASIBLE REGION. We now derive the constraint inequalities that define the feasible region. First, the number of cars shipped on each route can't be negative, so we have

$$x \geq 0 \qquad\qquad y \geq 0$$
$$10 - x \geq 0 \qquad\qquad 12 - y \geq 0$$

Second, the total number of cars shipped from each warehouse can't exceed the number of cars available there, so

$$x + y \leq 15$$
$$(10 - x) + (12 - y) \leq 10$$

Simplifying the latter inequality, we get

$$22 - x - y \leq 10$$
$$-x - y \leq -12$$
$$x + y \geq 12$$

The inequalities $10 - x \geq 0$ and $12 - y \geq 0$ can be rewritten as $x \leq 10$ and $y \leq 12$, respectively. Thus, the feasible region is described by the constraints

$$\begin{cases} x + y \leq 15 \\ x + y \geq 12 \\ 0 \leq x \leq 10 \\ 0 \leq y \leq 12 \end{cases}$$

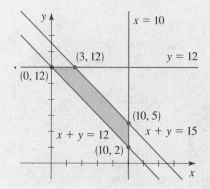

FIGURE 3

The feasible region is graphed in Figure 3.

FINDING MINIMUM COST. We check the value of the objective function at each vertex of the feasible region.

Vertex	$C = 1260 - 10x - 15y$	
$(0, 12)$	$1260 - 10(0) - 15(12) = \$1080$	
$(3, 12)$	$1260 - 10(3) - 15(12) = \$1050$	← minimum cost
$(10, 5)$	$1260 - 10(10) - 15(5) = \$1085$	
$(10, 2)$	$1260 - 10(10) - 15(2) = \$1130$	

The lowest cost is incurred at the point $(3, 12)$. Thus, the dealer should ship

3 cars from Millville to Camden

12 cars from Millville to Atlantic City

7 cars from Trenton to Camden

0 cars from Trenton to Atlantic City

In the 1940s mathematicians developed matrix methods for solving linear programming problems that involve more than two variables. These methods were first used by the Allies in World War II to solve supply problems similar to (but, of course much more complicated than) Example 2. Improving these matrix methods is an active and exciting area of current mathematical research.

PROBLEMS

1–4 ■ Find the maximum and minimum values of the given objective function on the indicated feasible region.

1. $M = 200 - x - y$

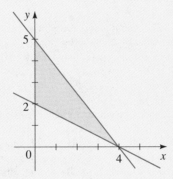

2. $N = \frac{1}{2}x + \frac{1}{4}y + 40$

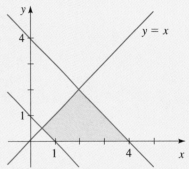

3. $P = 140 - x + 3y$

$$\begin{cases} x \geqslant 0, \quad y \geqslant 0 \\ 2x + y \leqslant 10 \\ 2x + 4y \leqslant 28 \end{cases}$$

4. $Q = 70x + 82y$

$$\begin{cases} x \geqslant 0, \quad y \geqslant 0 \\ x \leqslant 10, \quad y \leqslant 20 \\ x + y \geqslant 5 \\ x + 2y \leqslant 18 \end{cases}$$

5. A furniture manufacturer makes wooden tables and chairs. The production process involves two basic types of labor: carpentry and finishing. Making a table requires 2 hours of carpentry and 1 hour of finishing, whereas making a chair requires 3 hours of carpentry and $\frac{1}{2}$ hour of finishing. The profit is $35 per table and $20 per chair. The manufacturer's employees can supply a maximum of 108 hours of carpentry work and 20 hours of finishing work per day. How many tables and chairs should be made each day to maximize profit?

6. A housing contractor has subdivided a farm into 100 building lots. He has designed two types of homes for these lots: colonial and ranch style. A colonial requires $30,000 of capital and produces a profit of $4000 when sold. A ranch-style house requires $40,000 of capital and provides an $8000 profit. If he has $3.6 million of capital on hand, how many houses of each type should he build for maximum profit? Will any of the lots be left vacant?

7. A trucker is planning to haul citrus fruit from Florida to Montreal. Each crate of oranges is 4 ft^3 in volume and weighs 80 lb. Each crate of grapefruit has a volume of 6 ft^3 and weighs 100 lb. Her truck has a maximum capacity of 300 ft^3 and can carry no more than 5600 lb. Moreover, she is not permitted to carry more crates of grapefruit than crates of oranges. If she makes a profit of $2.50 on each crate of oranges and $4 on each crate of grapefruit, how many crates of each type of fruit should she carry for maximum profit?

8. A manufacturer of calculators produces two models: standard and scientific. Long-term demand for the two models mandates that the company manufacture at least 100 standard and 80 scientific calculators each day. However, because of limitations on production capacity, no more than 200 standard and 170 scientific calculators can be made daily. To satisfy a shipping contract, a total of at least 200 calculators must be shipped every day.
   (a) If the production cost is $5 for a standard calculator and $7 for a scientific one, how many of each model should be produced each day to minimize this cost?
   (b) If each standard calculator results in a $2 loss but each scientific one produces a $5 profit, how many of each model should be made each day to maximize profit?

9. An electronics discount chain has a sale on a certain brand of stereo. The chain has stores in Santa Monica and El Toro and warehouses in Long Beach and Pasadena. To satisfy rush orders, 15 sets must be shipped from the warehouses to the Santa Monica store, and 19 must be shipped to the El Toro store. The cost of shipping a set is $5 from Long Beach to Santa Monica, $6 from Long Beach to El Toro, $4 from Pasadena to Santa Monica, and $5.50 from Pasadena to El Toro. If the Long Beach warehouse has 24 sets and the Pasadena warehouse has 18 sets in stock, how many sets should be shipped from each warehouse to each store to fill the orders at a minimum shipping cost?

10. A man owns two building supply stores, one on the east side and one on the west side of a city. Two customers order some $\frac{1}{2}$-inch plywood. Customer A needs 50 sheets and customer B needs 70 sheets. The east-side store has 80 sheets and the west-side store has 45 sheets of this plywood in stock. Delivery cost per sheet is $0.50 from the east-side store to customer A, $0.60 from the east-side store to customer B, $0.40 from the west-side store to A, and $0.55 from the west-side store to B. How many sheets should be shipped from each store to each customer to minimize delivery cost?

11. A confectioner sells two types of nut mixtures. Each package of the standard mixture contains 100 g of cashews and 200 g of peanuts and sells for $1.95. Each package of the deluxe mixture contains 150 g of cashews and 50 g of peanuts and sells for $2.25. The confectioner has 15 kg of cashews and 20 kg of peanuts available. Based on past sales statistics, he wishes to have at least as many standard as deluxe packages available. How many bags of each type of mixture should he package to maximize his revenue?

12. A biologist wishes to feed laboratory rabbits a mixture of two types of foods. Type I contains 8 g of fat, 12 g of carbohydrate, and 2 g of protein per ounce, whereas type II contains 12 g of fat, 12 g of carbohydrate, and 1 g of protein per ounce. Type I costs $0.20 per ounce and type II costs $0.30 per ounce. Each rabbit is to receive a daily minimum of 24 g of fat, 36 g of carbohydrate, and 4 g of protein, but should get no more than 5 oz of food per day. How many ounces of each type of food should be fed to each rabbit daily to satisfy the dietary requirements at minimum cost?

13. A woman wishes to invest $12,000 in three types of bonds: municipal bonds paying 7% interest per year, bank investment certificates paying 8%, and high-risk bonds paying 12%. For tax reasons, she wants the amount invested in municipal bonds to be at least three times the amount invested in bank certificates. To keep her level of risk manageable, she will invest no more than $2000 in high-risk bonds. How much should she invest in each type of bond to maximize her annual interest yield? [*Hint:* Let $x =$ amount in municipal bonds and $y =$ amount in bank certificates. Then the amount in high-risk bonds will be $12,000 - x - y$.]

14. Refer to Exercise 13. Suppose the investor decides to increase to $3000 the maximum amount she will allow to be invested in high-risk bonds but leaves the other conditions unchanged. By how much will her maximum possible interest yield increase?

15. A small software company publishes computer games and educational and utility software. Their policy is to market a total of 36 new programs each year, with at least four of these being games. The number of utility programs published is never more than twice the number of educational programs. On average, the company can expect to make an annual profit of $5000 on each computer game, $8000 on each educational program, and $6000 on each utility program. How many of each type of program should they publish annually for maximum profit?

16. All parts of this problem refer to the following feasible region and objective function.

$$\begin{cases} x \geqslant 0 \\ x \geqslant y \\ x + 2y \leqslant 12 \\ x + y \leqslant 10 \end{cases}$$

$$P = x + 4y$$

(a) Graph the feasible region.
(b) On your graph from part (a), sketch the graphs of the linear equations obtained by setting P equal to 40, 36, 32, and 28.
(c) If we continue to decrease the value of P, at which vertex of the feasible region will these lines first touch the feasible region?
(d) Verify that the maximum value of P on the feasible region occurs at the vertex you chose in part (c).

9

TOPICS IN ANALYTIC GEOMETRY

The path of a projectile, such as a basketball, a missile, or a comet, is a conic section.

In studying the procedures of geometric thought we may hope to reach what is most essential in the human mind.

HENRI POINCARÉ

In this chapter we study the geometry of **conic sections** (or simply **conics**). Conic sections are the curves formed by the intersection of a plane with a pair of circular cones. These curves have four basic shapes, called **circles**, **ellipses**, **parabolas**, and **hyperbolas**, as illustrated in the figure.

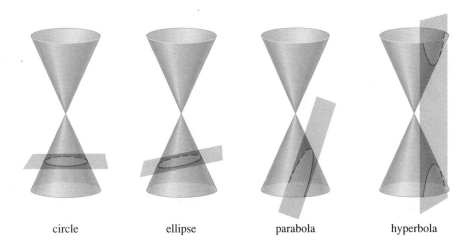

| circle | ellipse | parabola | hyperbola |

The ancient Greeks studied these curves because they considered the geometry of conic sections to be very beautiful. The mathematician Apollonius (262–190 B.C.) wrote a definitive eight-volume work on the subject. In more modern times, conics have been found to be useful as well as beautiful. Galileo discovered in 1590 that the path of a missile shot upward at an angle is a parabola. In 1609, Kepler found that the planets move in elliptical orbits around the sun. In 1668, Newton was the first to build a reflecting telescope, whose principle is based on the properties of parabolas and hyperbolas. In this century, many further applications of conic sections have been developed. One important application is the LORAN radio navigation system, which uses the intersection points of hyperbolas to pinpoint the location of ships and aircraft. Another application is the medical procedure of lithotripsy, a method of removing kidney stones without surgery; this method uses a property of ellipses. These and other applications of conic sections will be considered in this chapter.

In addition to studying conics, we will examine two other ways to describe points and curves in the Cartesian plane: polar coordinates and parametric equations. Both of these topics require a thorough understanding of trigonometry.

9.1 PARABOLAS

We saw in Section 2.5 that the graph of the equation $y = ax^2 + bx + c$ is a U-shaped curve called a *parabola* that opens either upward or downward, depending on whether the sign of a is positive or negative (see Figure 1). The lowest or highest point of the parabola is called the *vertex,* and the parabola is symmetric about its *axis.*

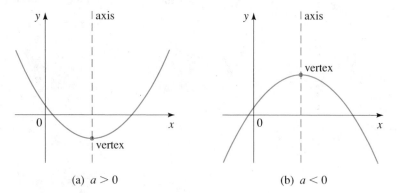

FIGURE 1
$y = ax^2 + bx + c$

(a) $a > 0$

(b) $a < 0$

In this section we study parabolas from a geometric rather than an algebraic point of view. We begin with the geometric definition of a parabola and show how this leads to the algebraic formula that we are already familiar with.

GEOMETRIC DEFINITION OF A PARABOLA

A **parabola** is the set of points in the plane equidistant from a fixed point F (called the **focus**) and a fixed line l (called the **directrix**).

This definition is illustrated in Figure 2. Note that the vertex V of the parabola lies halfway between the focus and the directrix and that the axis of symmetry is the line that runs through the focus perpendicular to the directrix.

In this section we restrict our attention to parabolas that are situated with the vertex at the origin and that have a vertical or horizontal axis of symmetry. (Parabolas in more general positions will be considered in Sections 9.4 and 9.5.) If the focus of such a parabola is the point $F(0, p)$, then the axis of symmetry must be vertical and the directrix has the equation $y = -p$. Figure 3 illustrates the case $p > 0$.

If $P(x, y)$ is any point on the parabola, then the distance from P to the focus F (using the Distance Formula) is

$$\sqrt{x^2 + (y - p)^2}$$

and the distance from P to the directrix is

$$|y - (-p)| = |y + p|$$

By the definition of a parabola, these two distances must be equal:

$$\sqrt{x^2 + (y - p)^2} = |y + p|$$
$$x^2 + (y - p)^2 = |y + p|^2 = (y + p)^2 \quad \text{Square both sides}$$

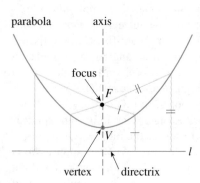

FIGURE 2

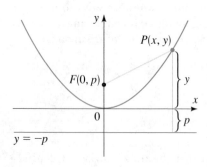

FIGURE 3

$$x^2 + y^2 - 2py + p^2 = y^2 + 2py + p^2 \qquad \text{Expand}$$

$$x^2 - 2py = 2py \qquad \text{Simplify}$$

$$x^2 = 4py$$

If $p > 0$, then the parabola opens upward, but if $p < 0$, it opens downward. When x is replaced by $-x$, the equation remains unchanged, so the graph is symmetric about the y-axis. We summarize what we have proved in the following box.

PARABOLA WITH VERTICAL AXIS

The graph of the equation

$$x^2 = 4py$$

is a parabola with the following properties.

VERTEX	$V(0,0)$
FOCUS	$F(0, p)$
DIRECTRIX	$y = -p$

The parabola opens upward if $p > 0$ or downward if $p < 0$.

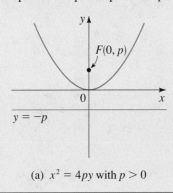

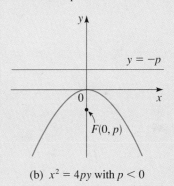

(a) $x^2 = 4py$ with $p > 0$ (b) $x^2 = 4py$ with $p < 0$

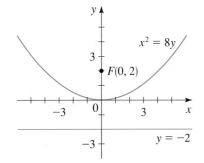

FIGURE 4

EXAMPLE 1 ■ Finding the Equation of a Parabola

Find the equation of the parabola with vertex $V(0,0)$ and focus $F(0,2)$, and sketch its graph.

SOLUTION Since the focus is $F(0,2)$, we conclude that $p = 2$ (and so the directrix is $y = -2$). Thus, the equation of the parabola is

$$x^2 = 4(2)y \qquad x^2 = 4py \text{ with } p = 2$$

$$x^2 = 8y$$

Since $p = 2 > 0$, the parabola opens upward. See Figure 4. ■

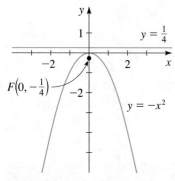

$y = \frac{1}{4}$

$F\left(0, -\frac{1}{4}\right)$

$y = -x^2$

FIGURE 5

EXAMPLE 2 ■ **Finding the Focus and Directrix of a Parabola from Its Equation**

Find the focus and directrix of the parabola $y = -x^2$, and sketch its graph.

SOLUTION   Comparing the equation $y = -x^2$ with the general equation $x^2 = 4py$, we see that $4p = -1$, so $p = -\frac{1}{4}$. Thus, the focus is $F\left(0, -\frac{1}{4}\right)$ and the directrix is $y = \frac{1}{4}$. The graph is shown in Figure 5.   ■

Reflecting the graph in Figure 3 about the diagonal line $y = x$ has the effect of interchanging the roles of x and y. This results in a parabola with horizontal axis. By the same method as before, we can prove the following properties.

PARABOLA WITH HORIZONTAL AXIS

The graph of the equation

$$y^2 = 4px$$

is a parabola with the following properties.

VERTEX	$V(0,0)$
FOCUS	$F(p,0)$
DIRECTRIX	$x = -p$

The parabola opens to the right if $p > 0$ or to the left if $p < 0$.

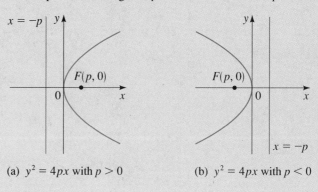

(a) $y^2 = 4px$ with $p > 0$ (b) $y^2 = 4px$ with $p < 0$

EXAMPLE 3 ■ **A Parabola with Horizontal Axis**

Find the focus and directrix of the parabola $6x + y^2 = 0$, and sketch its graph.

SOLUTION   We first write the equation as $y^2 = -6x$. Comparing this with the general equation $y^2 = 4px$, we see that $-6 = 4p$, so $p = -\frac{3}{2}$. Thus, the focus is $\left(-\frac{3}{2}, 0\right)$ and the directrix is $x = \frac{3}{2}$.

Since $p = -\frac{3}{2} < 0$, the parabola opens to the left. See Figure 6.   ■

We can use the coordinates of the focus to estimate the "width" of a parabola when sketching its graph. The line segment that runs through the focus perpendicular to the axis, with endpoints on the parabola, is called the **latus rectum**,

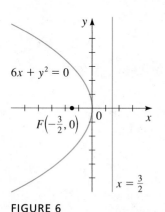

$6x + y^2 = 0$

$F\left(-\frac{3}{2}, 0\right)$

$x = \frac{3}{2}$

FIGURE 6

and its length is the **focal diameter** of the parabola. From Figure 7 we can see that the distance from an endpoint Q of the latus rectum to the directrix is $|2p|$. Thus, the distance from Q to the focus must be $|2p|$ as well (by the definition of a parabola), and so the focal diameter is $|4p|$. In the next example we use the focal diameter to determine the "width" of a parabola when graphing it.

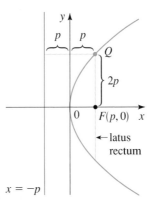

FIGURE 7

EXAMPLE 4 ■ Using the Focal Diameter to Sketch a Parabola

Find the focus, directrix, and focal diameter of the parabola $y = \frac{1}{2}x^2$, and sketch its graph.

SOLUTION   We first put the equation in the form $x^2 = 4py$.

$$y = \tfrac{1}{2}x^2$$
$$x^2 = 2y \qquad \text{Multiply each side by 2}$$

From this equation we see that $4p = 2$, so the focal diameter is 2. Solving for p gives $p = \frac{1}{2}$, so the focus is $(0, \frac{1}{2})$ and the directrix is $y = -\frac{1}{2}$. Since the focal diameter is 2, the latus rectum extends 1 unit to the left and 1 unit to the right of the focus. This enables us to sketch the graph in Figure 8.

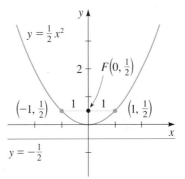

FIGURE 8

Parabolas have an important property that makes them useful as reflectors for lamps and telescopes. Light from a source placed at the focus of a surface with parabolic cross section will be reflected in such a way that it travels parallel to

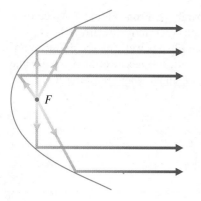

FIGURE 9

Parabolic reflector

the axis of the parabola (see Figure 9). Thus, a parabolic mirror reflects the light into a beam of parallel rays. Conversely, light approaching the reflector in rays parallel to its axis of symmetry is concentrated to the focus. This *reflection property,* which can be proved using calculus, is used in the construction of reflecting telescopes.

EXAMPLE 5 ■ Finding the Focal Point of a Searchlight Reflector

A searchlight has a parabolic reflector that forms a "bowl," which is 12 in. wide from rim to rim and 8 in. deep, as shown in Figure 10. If the filament of the lightbulb is located at the focus, how far from the vertex of the reflector is it?

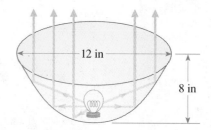

FIGURE 10

A parabolic reflector

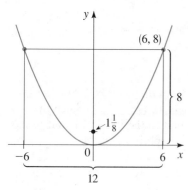

FIGURE 11

SOLUTION   We introduce a coordinate system and place a parabolic cross section of the reflector so that its vertex is at the origin and its axis is vertical (see Figure 11). Then the equation of this parabola has the form $x^2 = 4py$. From Figure 11 we see that the point $(6, 8)$ lies on the parabola. We use this to find p.

$$6^2 = 4p(8) \qquad \text{The point } (6, 8) \text{ satisfies the equation } x^2 = 4py$$

$$36 = 32p$$

$$p = \tfrac{9}{8}$$

The focus is $F(0, \tfrac{9}{8})$, so the distance between the vertex and the focus is $\tfrac{9}{8} = 1\tfrac{1}{8}$ in. Because the filament is positioned at the focus, it is located $1\tfrac{1}{8}$ in. from the vertex of the reflector. ■

In the next example we graph a family of parabolas, to show how changing the distance between the focus and the vertex affects the "width" of a parabola.

EXAMPLE 6 ■ A Family of Parabolas

(a) Find equations for the parabolas with vertex at the origin and foci $F_1(0, \tfrac{1}{8})$, $F_2(0, \tfrac{1}{2})$, $F_3(0, 1)$, and $F_4(0, 4)$.

(b) Draw the graphs of the parabolas in part (a). What do you conclude?

SOLUTION

(a) Since the foci are on the positive y-axis, the parabolas open upward and have equations of the form $x^2 = 4py$. This leads to the following equations.

Focus	p	Equation $x^2 = 4py$	Form of the equation for graphing calculator
$F_1\left(0, \frac{1}{8}\right)$	$p = \frac{1}{8}$	$x^2 = \frac{1}{2}y$	$y = 2x^2$
$F_2\left(0, \frac{1}{2}\right)$	$p = \frac{1}{2}$	$x^2 = 2y$	$y = 0.5x^2$
$F_3(0, 1)$	$p = 1$	$x^2 = 4y$	$y = 0.25x^2$
$F_4(0, 4)$	$p = 4$	$x^2 = 16y$	$y = 0.0625x^2$

(b) The graphs are drawn in Figure 12. We see that the closer the focus to the vertex, the narrower the parabola.

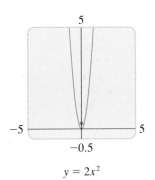

$y = 2x^2$

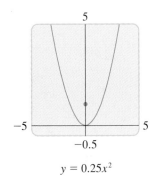

$y = 0.5x^2$

$y = 0.25x^2$

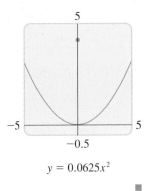

$y = 0.0625x^2$

FIGURE 12   A family of parabolas

9.1   EXERCISES

1–12 ■ Find the focus, directrix, and focal diameter of the parabola, and sketch its graph.

1. $y^2 = 4x$

2. $x^2 = y$

3. $x^2 = 9y$

4. $y^2 = 3x$

5. $y = 5x^2$

6. $y = -2x^2$

7. $x = -8y^2$

8. $x = \frac{1}{2}y^2$

9. $x^2 + 6y = 0$

10. $x - 7y^2 = 0$

11. $5x + 3y^2 = 0$

12. $8x^2 + 12y = 0$

13–24 ■ Find an equation for the parabola that has its vertex at the origin and satisfies the given condition(s).

13. Focus $F(0, 2)$

14. Focus $F\left(0, -\frac{1}{2}\right)$

15. Focus $F(-8, 0)$

16. Focus $F(5, 0)$

17. Directrix $x = 2$

18. Directrix $y = 6$

19. Directrix $y = -10$

20. Directrix $x = -\frac{1}{8}$

21. Focus on the positive x-axis, 2 units away from the directrix

22. Directrix has y-intercept 6

23. Opens upward with focus 5 units from the vertex

24. Focal diameter 8 and focus on the negative y-axis

25–32 ■ Use the sketch to find an equation of the parabola.

25.

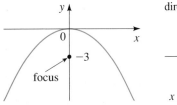

26.

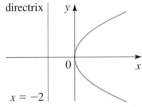

27.

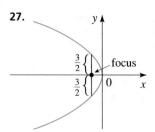

28.

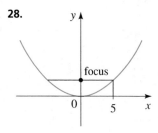

29.

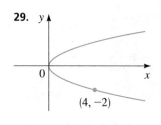

30.

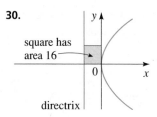

31.

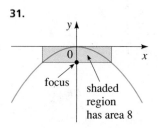

32.

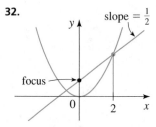

33. A lamp with a parabolic reflector is shown in the figure. The bulb is placed at the focus and the focal diameter is 12 cm.
(a) Find an equation of the parabola.
(b) Find the diameter $d(C, D)$ of the opening, 20 cm from the vertex.

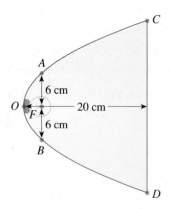

34. A reflector for a satellite dish is parabolic in cross section, with the receiver at the focus F. The reflector

is 1 ft deep and 20 ft wide from rim to rim (see the figure). How far is the receiver from the vertex of the parabolic reflector?

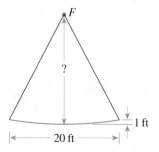

35. In a suspension bridge the shape of the suspension cables is parabolic. The bridge shown in the figure has towers that are 600 m apart, and the lowest point of the suspension cables is 150 m below the top of the towers. Find the equation of the parabolic part of the cables, placing the origin of the coordinate system at the vertex.

NOTE: This equation is used to find the length of cable needed in the construction of the bridge.

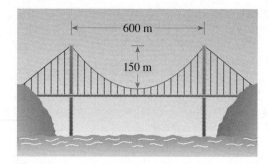

36. The Hale telescope at the Mount Palomar Observatory has a 200-inch mirror, as shown. The mirror is constructed in a parabolic shape that collects light from the stars and focuses it at the **prime focus**, that is, the focus of the parabola. The mirror is 3.79 in. deep at its center. Find the **focal length** of this parabolic mirror, that is, the distance from the vertex to the focus.

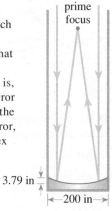

37. (a) Find equations for the family of parabolas with vertex at the origin and with directrixes $y = \frac{1}{2}$, $y = 1$, $y = 4$, and $y = 8$.

(b) Draw the graphs. What do you conclude?

DISCOVERY · DISCUSSION

38. Parabolas in the Real World Several examples of the uses of parabolas are given in the text. Find other situations in real life where parabolas occur. Consult a scientific encyclopedia in the reference section of your library, or search the Internet.

39. Light Cone from a Flashlight When a flashlight is held horizontally, a lighted area forms on the ground, as shown in the figure. Is it possible to angle the flashlight in such a way that the boundary of the lighted area is a parabola? Explain your answer.

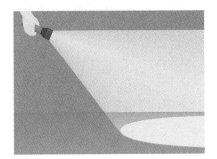

9.2 ELLIPSES

An ellipse is an oval curve that looks like an elongated circle. More precisely, we have the following definition.

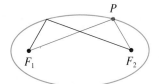

FIGURE 1

> **GEOMETRIC DEFINITION OF AN ELLIPSE**
>
> An **ellipse** is the set of all points in the plane the sum of whose distances from two fixed points F_1 and F_2 is a constant. (See Figure 1.) These two fixed points are the **foci** (plural of **focus**) of the ellipse.

The geometric definition suggests a simple method for drawing an ellipse. Place a sheet of paper on a drawing board and insert thumbtacks at the two points that are to be the foci of the ellipse. Attach the ends of a string to the tacks, as shown in Figure 2. With the point of a pencil, hold the string taut. Then carefully move the pencil around the foci, keeping the string taut at all times. The pencil will trace out an ellipse, since the sum of the distances from the point of the pencil to the foci will always equal the length of the string, which is constant.

If the string is only slightly longer than the distance between the foci, then the ellipse traced out will be elongated in shape as in Figure 3(a), but if the foci are

FIGURE 2

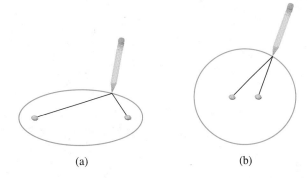

FIGURE 3   (a)   (b)

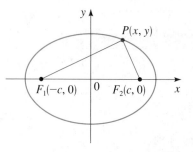

FIGURE 4

close together relative to the length of the string, the ellipse will be almost circular, as shown in Figure 3(b).

To obtain the simplest equation for an ellipse, we place the foci on the x-axis at $F_1(-c, 0)$ and $F_2(c, 0)$, so that the origin is halfway between them (see Figure 4). For later convenience, we let the sum of the distances from a point on the ellipse to the foci be $2a$. Then if $P(x, y)$ is any point on the ellipse, we have

$$d(P, F_1) + d(P, F_2) = 2a$$

so, from the Distance Formula,

$$\sqrt{(x + c)^2 + y^2} + \sqrt{(x - c)^2 + y^2} = 2a$$

or $$\sqrt{(x - c)^2 + y^2} = 2a - \sqrt{(x + c)^2 + y^2}$$

Squaring each side and multiplying, we get

$$x^2 - 2cx + c^2 + y^2 = 4a^2 - 4a\sqrt{(x + c)^2 + y^2} + (x^2 + 2cx + c^2 + y^2)$$

which simplifies to

$$4a\sqrt{(x + c)^2 + y^2} = 4a^2 + 4cx$$

Dividing each side by 4 and squaring again, we get

$$a^2[(x + c)^2 + y^2] = (a^2 + cx)^2$$
$$a^2x^2 + 2a^2cx + a^2c^2 + a^2y^2 = a^4 + 2a^2cx + c^2x^2$$
$$(a^2 - c^2)x^2 + a^2y^2 = a^2(a^2 - c^2)$$

Since the sum of the distances from P to the foci must be larger than the distance between the foci, we have that $2a > 2c$, or $a > c$. Thus, $a^2 - c^2 > 0$, and we can divide each side of the preceding equation by $a^2(a^2 - c^2)$ to get

$$\frac{x^2}{a^2} + \frac{y^2}{a^2 - c^2} = 1$$

For convenience, let $b^2 = a^2 - c^2$ (with $b > 0$). Since $b^2 < a^2$, it follows that $b < a$. The preceding equation then becomes

$$\frac{x^2}{a^2} + \frac{y^2}{b^2} = 1 \qquad \text{with } a > b$$

This is the equation of the ellipse. To graph it, we need to know the x- and y-intercepts. Setting $y = 0$, we get

$$\frac{x^2}{a^2} = 1$$

so $x^2 = a^2$, or $x = \pm a$. Thus, the ellipse crosses the x-axis at $(a, 0)$ and $(-a, 0)$. These points are called the **vertices** of the ellipse, and the segment that joins them is called the **major axis**. Its length is $2a$.

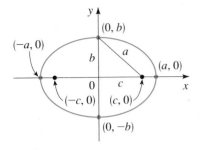

FIGURE 5

$\dfrac{x^2}{a^2} + \dfrac{y^2}{b^2} = 1$ with $a > b$

Similarly, if we set $x = 0$, we get $y = \pm b$, so the ellipse crosses the y-axis at $(0, b)$ and $(0, -b)$. The segment that joins these points is called the **minor axis**, and it has length $2b$. Note that $2a > 2b$, so the major axis is longer than the minor axis.

In Section 1.8 we studied several tests that detect symmetry in a graph. If we replace x by $-x$ or y by $-y$ in the ellipse equation, it remains unchanged. Thus, the ellipse is symmetric about both the x- and y-axes, and hence about the origin as well. For this reason, the origin is called the **center** of the ellipse. The complete graph is shown in Figure 5.

If the foci of the ellipse are placed on the y-axis at $(0, \pm c)$ rather than on the x-axis, then the roles of x and y are reversed in the preceding discussion, and we get a vertical ellipse. Thus, we have the following description of ellipses.

ELLIPSE WITH CENTER AT THE ORIGIN

The graph of the equation

$$\frac{x^2}{a^2} + \frac{y^2}{b^2} = 1 \quad \text{or} \quad \frac{x^2}{b^2} + \frac{y^2}{a^2} = 1 \quad \text{with } a > b > 0$$

is an **ellipse** with center at the origin and having the following properties.

GRAPH

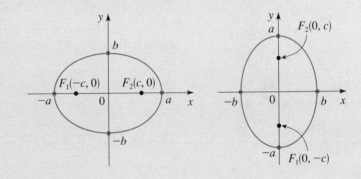

EQUATION	$\dfrac{x^2}{a^2} + \dfrac{y^2}{b^2} = 1$ $a > b > 0$	$\dfrac{x^2}{b^2} + \dfrac{y^2}{a^2} = 1$ $a > b > 0$
VERTICES	$(\pm a, 0)$	$(0, \pm a)$
MAJOR AXIS	Horizontal, length $2a$	Vertical, length $2a$
MINOR AXIS	Vertical, length $2b$	Horizontal, length $2b$
FOCI	$(\pm c, 0)$, $c^2 = a^2 - b^2$	$(0, \pm c)$, $c^2 = a^2 - b^2$

In the standard equation for an ellipse, a^2 is the *larger* denominator and b^2 is the *smaller*. To find c^2, we subtract: larger denominator minus smaller denominator.

EXAMPLE 1 ■ Sketching an Ellipse

Find the foci, vertices, and the lengths of the major and minor axes for the following ellipse, and sketch its graph.

$$\frac{x^2}{9} + \frac{y^2}{4} = 1$$

SOLUTION   Since the denominator of x^2 is larger, the ellipse has horizontal major axis. This gives $a^2 = 9$ and $b^2 = 4$, so $c^2 = a^2 - b^2 = 9 - 4 = 5$. Thus, $a = 3$, $b = 2$, and $c = \sqrt{5}$.

FOCI $(\pm\sqrt{5}, 0)$

VERTICES $(\pm 3, 0)$

LENGTH OF MAJOR AXIS   6

LENGTH OF MINOR AXIS   4

The graph is shown in Figure 6.   ■

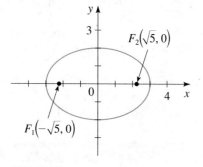

FIGURE 6
$\frac{x^2}{9} + \frac{y^2}{4} = 1$

EXAMPLE 2 ■ Finding the Equation of an Ellipse

The vertices of an ellipse are $(\pm 4, 0)$ and the foci are at $(\pm 2, 0)$. Find its equation and sketch the graph.

SOLUTION   Since the vertices are $(\pm 4, 0)$, we have $a = 4$. The foci are $(\pm 2, 0)$, so $c = 2$. To write the equation, we need to find b. Since $c^2 = a^2 - b^2$, we have

$$2^2 = 4^2 - b^2$$

$$b^2 = 16 - 4 = 12$$

Thus, the equation of the ellipse is

$$\frac{x^2}{16} + \frac{y^2}{12} = 1$$

The graph is shown in Figure 7.   ■

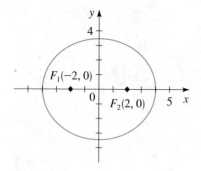

FIGURE 7
$\frac{x^2}{16} + \frac{y^2}{12} = 1$

EXAMPLE 3 ■ Finding the Foci of an Ellipse

Find the foci of the ellipse $16x^2 + 9y^2 = 144$, and sketch its graph.

SOLUTION   Dividing by 144, we get

$$\frac{x^2}{9} + \frac{y^2}{16} = 1$$

Since $16 > 9$, this is an ellipse with its foci on the y-axis, and with $a = 4$ and

$b = 3$. We have

$$c^2 = a^2 - b^2 = 16 - 9 = 7$$

$$c = \sqrt{7}$$

Thus, the foci are $(0, \pm\sqrt{7})$. The graph is shown in Figure 8.

ECCENTRICITIES OF THE ORBITS OF THE PLANETS

The orbits of the planets are ellipses with the sun at one focus. For most planets these ellipses have very small eccentricity, so they are nearly circular. However, Mercury and Pluto, the innermost and outermost known planets, have visibly elliptical orbits.

Planet	Eccentricity
Mercury	0.206
Venus	0.007
Earth	0.017
Mars	0.093
Jupiter	0.048
Saturn	0.056
Uranus	0.046
Neptune	0.010
Pluto	0.248

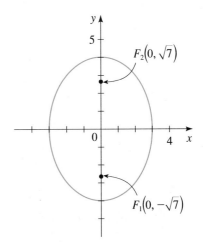

FIGURE 8
$16x^2 + 9y^2 = 144$

We saw earlier in this section (Figure 3) that if $2a$ is only slightly greater than $2c$, the ellipse is long and thin, whereas if $2a$ is much greater than $2c$, the ellipse is almost circular. We measure the deviation of an ellipse from being circular by the ratio of a and c.

DEFINITION OF ECCENTRICITY

For the ellipse $\dfrac{x^2}{a^2} + \dfrac{y^2}{b^2} = 1$ or $\dfrac{x^2}{b^2} + \dfrac{y^2}{a^2} = 1$ (with $a > b > 0$), the **eccentricity** e is the number

$$e = \frac{c}{a}$$

where $c = \sqrt{a^2 - b^2}$. The eccentricity of every ellipse satisfies $0 < e < 1$.

Thus, if e is close to 1, then c is almost equal to a, and the ellipse is elongated in shape, but if e is close to 0, then the ellipse is close to a circle in shape. The eccentricity is a measure of how "stretched" the ellipse is.

Johannes Kepler (1571–1630) was the first to give a correct description of the motion of the planets. The cosmology of his time postulated complicated systems of circles moving on circles to describe these motions. Kepler sought a simpler and more harmonious description. As official astronomer at the imperial court in Prague, he studied the astronomical observations of the Danish astronomer Tycho Brahe, whose data was at the time the most accurate available. After numerous attempts to find a theory, Kepler made the momentous discovery that the orbits of the planets are elliptical. His three great laws of planetary motion are

1. The orbit of each planet is an ellipse with the sun at one focus.
2. The line segment that joins the sun to a planet sweeps out equal areas in equal time (see the figure).
3. The square of the period of revolution of a planet is proportional to the cube of the length of the major axis of its orbit.

His formulation of these laws is perhaps the most impressive deduction from empirical data in the history of science.

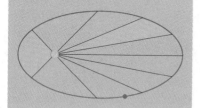

In Figure 9 we show a number of ellipses to demonstrate the effect of varying the eccentricity e.

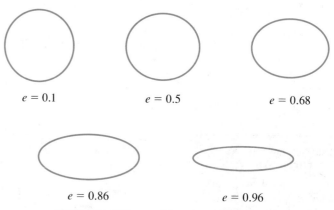

$e = 0.1$ $e = 0.5$ $e = 0.68$

$e = 0.86$ $e = 0.96$

FIGURE 9 Ellipses with various eccentricities

EXAMPLE 4 ■ Finding the Equation of an Ellipse from Its Eccentricity and Foci

Find the equation of the ellipse with foci $(0, \pm 8)$ and eccentricity $e = \frac{4}{5}$, and sketch its graph.

SOLUTION We are given $e = \frac{4}{5}$ and $c = 8$. Thus

$$\frac{4}{5} = \frac{8}{a} \qquad \text{Eccentricity } e = c/a$$

$$4a = 40 \qquad \text{Cross multiply}$$

$$a = 10$$

To find b, we use the fact that $c^2 = a^2 - b^2$.

$$8^2 = 10^2 - b^2$$

$$b^2 = 10^2 - 8^2 = 36$$

$$b = 6$$

Thus, the equation of the ellipse is

$$\frac{x^2}{36} + \frac{y^2}{100} = 1$$

Because the foci are on the y-axis, the ellipse is oriented vertically. To sketch the ellipse, we find the intercepts: The x-intercepts are ± 6 and the y-intercepts are ± 10. The graph is sketched in Figure 10.

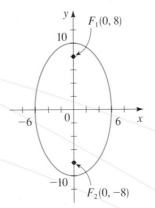

FIGURE 10

$$\frac{x^2}{36} + \frac{y^2}{100} = 1$$

Gravitational attraction causes the planets to move in elliptical orbits around the sun with the sun at one focus. This remarkable property was first observed by Johannes Kepler and was later deduced by Isaac Newton from his inverse square law of gravity, using calculus. The orbits of the planets have different eccentricities, but most are nearly circular (see the margin note on page 623).

Ellipses, like parabolas, have an interesting *reflection property* that leads to a number of practical applications. If a light source is placed at one focus of a reflecting surface with elliptical cross sections, then all the light will be reflected off the surface to the other focus, as shown in Figure 11. This principle, which works for sound waves as well as for light, is used in *lithotripsy*, a treatment for kidney stones. The patient is placed in a tub of water with elliptical cross sections in such a way that the kidney stone is accurately located at one focus. High-intensity sound waves generated at the other focus are reflected to the stone and destroy it with minimal damage to surrounding tissue. The patient is spared the trauma of surgery and recovers within days instead of weeks.

The reflection property of ellipses is also used in the construction of *whispering galleries*. Sound coming from one focus bounces off the walls and ceiling of an elliptical room and passes through the other focus. In these rooms, even quiet whispers spoken at one focus can be heard clearly at the other. Famous whispering galleries include the National Statuary Gallery of the U.S. Capitol in Washington, D.C., and the Mormon Tabernacle in Salt Lake City, Utah.

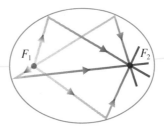

FIGURE 11

9.2   EXERCISES

1–14 ■ Find the vertices, foci, and eccentricity of the ellipse. Determine the lengths of the major and minor axes, and sketch the graph.

1. $\dfrac{x^2}{25} + \dfrac{y^2}{9} = 1$ **2.** $\dfrac{x^2}{16} + \dfrac{y^2}{25} = 1$

3. $9x^2 + 4y^2 = 36$ **4.** $4x^2 + 25y^2 = 100$

5. $x^2 + 4y^2 = 16$ **6.** $4x^2 + y^2 = 16$

7. $2x^2 + y^2 = 3$ **8.** $5x^2 + 6y^2 = 30$

9. $x^2 + 4y^2 = 1$ **10.** $9x^2 + 4y^2 = 1$

11. $\frac{1}{2}x^2 + \frac{1}{8}y^2 = \frac{1}{4}$ **12.** $x^2 = 4 - 2y^2$

13. $y^2 = 1 - 2x^2$ **14.** $20x^2 + 4y^2 = 5$

15–18 ■ Find an equation for the ellipse whose graph is shown.

15.

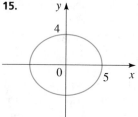

16.

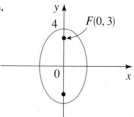

17.

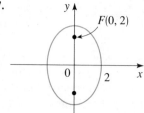

18.

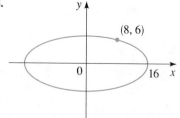

19–30 ■ Find an equation for the ellipse that satisfies the given conditions.

19. Foci $(\pm 4, 0)$, vertices $(\pm 5, 0)$

20. Foci $(0, \pm 3)$, vertices $(0, \pm 5)$

21. Length of major axis 4, length of minor axis 2, foci on y-axis

22. Length of major axis 6, length of minor axis 4, foci on x-axis

23. Foci $(0, \pm 2)$, length of minor axis 6

24. Foci $(\pm 5, 0)$, length of major axis 12

25. Endpoints of major axis $(\pm 10, 0)$, distance between foci 6

26. Endpoints of minor axis $(0, \pm 3)$, distance between foci 8

27. Length of major axis 10, foci on x-axis, ellipse passes through the point $\left(\sqrt{5}, 2\right)$

28. Eccentricity $\frac{1}{9}$, foci $(0, \pm 2)$

29. Eccentricity 0.8, foci $(\pm 1.5, 0)$

30. Eccentricity $\sqrt{3}/2$, foci on y-axis, length of major axis 4

31–32 ■ Find the intersection points of the pair of ellipses. Sketch the graphs of each pair of equations on the same coordinate axes and label the points of intersection.

31. $\begin{cases} 4x^2 + y^2 = 4 \\ 4x^2 + 9y^2 = 36 \end{cases}$

32. $\begin{cases} \dfrac{x^2}{16} + \dfrac{y^2}{9} = 1 \\ \dfrac{x^2}{9} + \dfrac{y^2}{16} = 1 \end{cases}$

33. The planets move around the sun in elliptical orbits with the sun at one focus. The point in the orbit at which the planet is closest to the sun is called **perihelion**, and the point at which it is farthest is called **aphelion**. These points are the vertices of the orbit. The earth's distance from the sun is 147,000,000 km at perihelion and 153,000,000 km at aphelion. Find an equation for the earth's orbit. (Place the origin at the center of the orbit with the sun on the *x*-axis.)

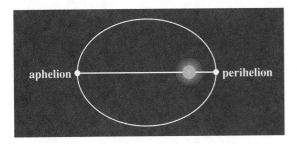

34. With an eccentricity of 0.25, Pluto's orbit is the most eccentric in the Solar System. The length of the minor axis of its orbit is approximately 10,000,000,000 km. Find the distance between Pluto and the sun at perihelion and at aphelion. (See Exercise 33.)

35. For an object in an elliptical orbit around the moon, the points in the orbit that are closest to and farthest from the center of the moon are called **perilune** and **apolune**, respectively. These are the vertices of the orbit. The center of the moon is at one of the foci of the orbit. The *Apollo 11* spacecraft was placed in a lunar orbit with perilune at 68 mi and apolune at 195 mi above the surface of the moon. Assuming the moon is a sphere of radius 1075 mi, find an equation for the orbit of *Apollo 11*. (Place the coordinate axes so that the origin is at the center of the orbit and the foci are located on the *x*-axis.)

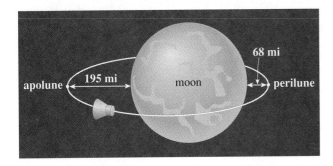

36. A carpenter wishes to construct an elliptical table top from a sheet of plywood, 4 ft by 8 ft. He will trace out the ellipse using the "thumbtack and string" method illustrated in Figures 2 and 3. What length of string should he use, and how far apart should the tacks be

located, if the ellipse is to be the largest possible that can be cut out of the plywood sheet?

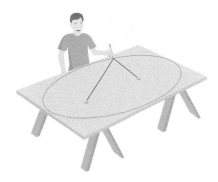

37. A "sunburst" window above a doorway is constructed in the shape of the top half of an ellipse, as shown in the figure. The window is 20 in. tall at its highest point and 80 in. wide at the bottom. Find the height of the window 25 in. from the center of the base.

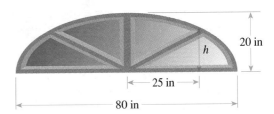

38. The **ancillary circle** of an ellipse is the circle with radius equal to half the length of the minor axis and center the same as the ellipse (see the figure). The ancillary circle is thus the largest circle that can fit within an ellipse.
   (a) Find an equation for the ancillary circle of the ellipse $x^2 + 4y^2 = 16$.
   (b) For the ellipse and ancillary circle of part (a), show that if (s, t) is a point on the ancillary circle, then $(2s, t)$ is a point on the ellipse.

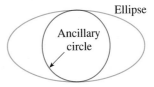

39. A **latus rectum** for an ellipse is a line segment perpendicular to the major axis at a focus, with endpoints on the ellipse, as shown. Show that the length of a latus rectum is $2b^2/a$ for the ellipse

$$\frac{x^2}{a^2} + \frac{y^2}{b^2} = 1 \quad \text{with } a > b$$

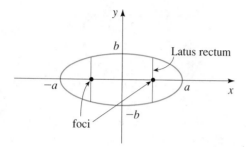

40. If $k > 0$, the following equation represents an ellipse:

$$\frac{x^2}{k} + \frac{y^2}{4 + k} = 1$$

Show that all the ellipses represented by this equation have the same foci, no matter what the value of k.

 41. Use a graphing device to draw each of the following ellipses by solving for y and graphing both solutions.

(a) $\dfrac{x^2}{25} + \dfrac{y^2}{20} = 1$ (b) $6x^2 + y^2 = 36$

 42. (a) Use a graphing device to sketch the top half (the portion in the first and second quadrants) of the family of ellipses $x^2 + ky^2 = 100$ for $k = 4, 10, 25,$ and 50.

(b) What do the members of this family of ellipses have in common? How do they differ?

 DISCOVERY · DISCUSSION

43. Drawing an Ellipse on a Blackboard   Try drawing an ellipse as accurately as possible on a blackboard. How would a piece of string and two friends help this process?

44. Light Cone from a Flashlight   A flashlight shines on a wall, as shown in the figure. What is the shape of the boundary of the lighted area? Explain your answer.

45. Is It an Ellipse?   A piece of paper is wrapped around a cylindrical bottle, and then a compass is used to draw a circle on the paper, as shown in the figure. When the paper is laid flat, is the shape drawn on the paper an ellipse? (You don't need to prove your answer, but you may want to do the experiment and see what you get.)

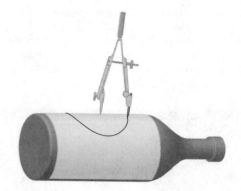

9.3 HYPERBOLAS

Although ellipses and hyperbolas have completely different shapes, their definitions and equations are similar. Instead of using the *sum* of distances from two fixed foci, as in the case of an ellipse, we use the *difference* to define a hyperbola.

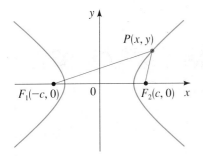

FIGURE 1

P is on the hyperbola if
$|d(P, F_1) - d(P, F_2)| = 2a$

GEOMETRIC DEFINITION OF A HYPERBOLA

A **hyperbola** is the set of all points in the plane, the difference of whose distances from two fixed points F_1 and F_2 is a constant. (See Figure 1.) These two fixed points are the **foci** of the hyperbola.

As in the case of the ellipse, we get the simplest equation for the hyperbola by placing the foci on the x-axis at $(\pm c, 0)$, as shown in Figure 1. From the definition, if $P(x, y)$ lies on the hyperbola, then either $d(P, F_1) - d(P, F_2)$ or $d(P, F_2) - d(P, F_1)$ must equal some positive constant, which we call $2a$. Thus, we have

$$d(P, F_1) - d(P, F_2) = \pm 2a$$

or

$$\sqrt{(x + c)^2 + y^2} - \sqrt{(x - c)^2 + y^2} = \pm 2a$$

Proceeding as we did in the case of the ellipse (Section 9.2), we simplify this to

$$(c^2 - a^2)x^2 - a^2 y^2 = a^2(c^2 - a^2)$$

From triangle PF_1F_2 in Figure 1, we see that $|d(P, F_1) - d(P, F_2)| < 2c$. It follows that $2a < 2c$, or $a < c$. Thus, $c^2 - a^2 > 0$, so we can set $b^2 = c^2 - a^2$. We then simplify the last displayed equation to get

$$\frac{x^2}{a^2} - \frac{y^2}{b^2} = 1$$

This is the *equation of the hyperbola.* If we replace x by $-x$ or y by $-y$ in this equation, it remains unchanged, so the hyperbola is symmetric about both the x- and y-axes and about the origin. The x-intercepts are $\pm a$, and the points $(a, 0)$ and $(-a, 0)$ are the **vertices** of the hyperbola. There is no y-intercept, because setting $x = 0$ in the equation of the hyperbola leads to $-y^2 = b^2$, which is impossible. Furthermore, the equation of the hyperbola implies that

$$\frac{x^2}{a^2} = \frac{y^2}{b^2} + 1 \geqslant 1$$

so $x^2/a^2 \geqslant 1$; thus, $x^2 \geqslant a^2$, and hence $x \geqslant a$ or $x \leqslant -a$. This means that the hyperbola consists of two parts, called its **branches**. The segment joining the two vertices on the separate branches is the **transverse axis** of the hyperbola, and the origin is called its **center**.

If we place the foci of the hyperbola on the y-axis rather than on the x-axis, then this has the effect of reversing the roles of x and y in the derivation of the equation of the hyperbola. This leads to a hyperbola with a vertical transverse axis.

The main properties of hyperbolas are listed in the following box.

HYPERBOLA WITH CENTER AT THE ORIGIN

The graph of the equation

$$\frac{x^2}{a^2} - \frac{y^2}{b^2} = 1 \quad \text{or} \quad \frac{y^2}{a^2} - \frac{x^2}{b^2} = 1 \quad \text{with } a > 0, \ b > 0$$

is a **hyperbola** with its center at the origin and with the following properties.

GRAPH

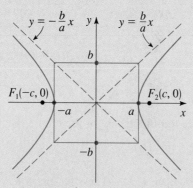

EQUATION	$\dfrac{x^2}{a^2} - \dfrac{y^2}{b^2} = 1$	$\dfrac{y^2}{a^2} - \dfrac{x^2}{b^2} = 1$
VERTICES	$(\pm a, 0)$	$(0, \pm a)$
TRANSVERSE AXIS	Horizontal, length $2a$	Vertical, length $2a$
ASYMPTOTES	$y = \pm \dfrac{b}{a} x$	$y = \pm \dfrac{a}{b} x$
FOCI	$(\pm c, 0), \quad c^2 = a^2 + b^2$	$(0, \pm c), \quad c^2 = a^2 + b^2$

The *asymptotes* mentioned in this box are lines that the hyperbola approaches for large values of x and y.

Asymptotes of rational functions are discussed in Section 3.5.

To find the asymptotes in the first case in the box, we solve the equation for y to get

$$y = \pm \frac{b}{a} \sqrt{x^2 - a^2}$$

$$= \pm \frac{b}{a} x \sqrt{1 - \frac{a^2}{x^2}}$$

As x gets large, a^2/x^2 gets closer to zero. In other words, as $x \to \infty$ we have $a^2/x^2 \to 0$. So, for large x the value of y can be approximated as $y = \pm(b/a)x$. This shows that these lines are asymptotes of the hyperbola.

Asymptotes are an essential aid for graphing a hyperbola; they help us determine its shape. A convenient way to find the asymptotes is to first plot the points $(a, 0)$, $(-a, 0)$, $(0, b)$, and $(0, -b)$. Then sketch horizontal and vertical segments through these points to construct a rectangle, as shown in Figure 2(a). We call this rectangle the **central box** of the hyperbola. The slopes of the diagonals of the central box are $\pm b/a$, so by extending them we obtain the asymptotes $y = \pm(b/a)x$, as sketched in part (b) of the figure. Finally, we plot the vertices and use the asymptotes as a guide in sketching the hyperbola shown in part (c). (A similar procedure applies to graphing a hyperbola that has a vertical transverse axis.)

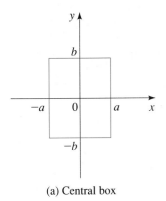

(a) Central box

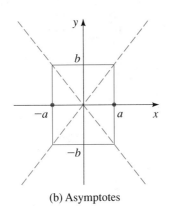

(b) Asymptotes

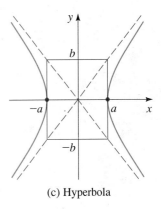

(c) Hyperbola

FIGURE 2

Steps in graphing the hyperbola $\dfrac{x^2}{a^2} - \dfrac{y^2}{b^2} = 1$

HOW TO SKETCH A HYPERBOLA

1. SKETCH THE CENTRAL BOX. This is the rectangle centered at the origin, with sides parallel to the axes, that crosses one axis at $\pm a$, the other at $\pm b$.

2. SKETCH THE ASYMPTOTES. These are the lines obtained by extending the diagonals of the central box.

3. PLOT THE VERTICES. These are the two x-intercepts or the two y-intercepts.

4. SKETCH THE HYPERBOLA. Start at a vertex and sketch a branch of the hyperbola, approaching the asymptotes. Sketch the other branch in the same way.

EXAMPLE 1 ■ A Hyperbola with Horizontal Transverse Axis

Find the vertices, foci, and asymptotes of the following hyperbola, and sketch its graph.

$$9x^2 - 16y^2 = 144$$

SOLUTION   First we divide both sides of the equation by 144 to put it into standard form:

$$\frac{x^2}{16} - \frac{y^2}{9} = 1$$

Because the x^2-term is positive, the hyperbola has a horizontal transverse axis; its vertices and foci are on the x-axis. Its vertices are $(\pm 4, 0)$. Since $a^2 = 16$ and $b^2 = 9$, we get $c = \sqrt{16 + 9} = 5$. Thus, the foci are $(\pm 5, 0)$. We have $a = 4$ and $b = 3$, so the asymptotes are $y = \pm \frac{3}{4}x$. After sketching the central box and asymptotes, we complete the sketch of the hyperbola as in Figure 3.

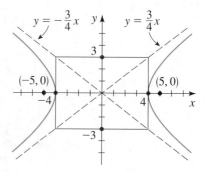

FIGURE 3
$9x^2 - 16y^2 = 144$

EXAMPLE 2 ■ A Hyperbola with Vertical Transverse Axis

Find the vertices, foci, and asymptotes of the following hyperbola, and sketch its graph.

$$x^2 - 9y^2 + 9 = 0$$

SOLUTION   We begin by writing the equation in the standard form for a hyperbola.

$$x^2 - 9y^2 = -9$$

$$y^2 - \frac{x^2}{9} = 1 \qquad \text{Divide by } -9$$

Because the y^2-term is positive, the hyperbola has a vertical transverse axis; its foci and vertices are on the y-axis. Its vertices are $(0, \pm 1)$. Since $a^2 = 1$ and $b^2 = 9$, we get $c = \sqrt{1 + 9} = \sqrt{10}$. Thus, the foci are $\left(0, \pm\sqrt{10}\right)$. We have $a = 1$ and $b = 3$, so the asymptotes are $y = \pm\frac{1}{3}x$. We sketch the central box and asymptotes, then complete the graph, as shown in Figure 4.

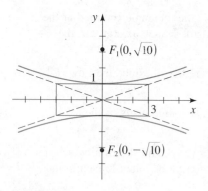

FIGURE 4
$x^2 - 9y^2 + 9 = 0$

EXAMPLE 3 ■ Finding the Equation of a Hyperbola from Its Vertices and Foci

Find the equation of the hyperbola with vertices $(\pm 3, 0)$ and foci $(\pm 4, 0)$. Sketch the graph.

SOLUTION   Since the vertices are on the x-axis, the hyperbola has a horizontal transverse axis. Its equation is of the form

$$\frac{x^2}{3^2} - \frac{y^2}{b^2} = 1$$

We have $a = 3$ and $c = 4$. To find b, we use the relation $a^2 + b^2 = c^2$:

$$3^2 + b^2 = 4^2$$
$$b^2 = 4^2 - 3^2 = 7$$
$$b = \sqrt{7}$$

Thus, the equation of the hyperbola is

$$\frac{x^2}{9} - \frac{y^2}{7} = 1$$

The graph is shown in Figure 5. ■

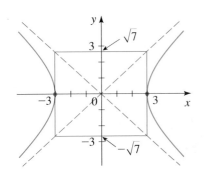

FIGURE 5
$$\frac{x^2}{9} - \frac{y^2}{7} = 1$$

EXAMPLE 4 ■ Finding the Equation of a Hyperbola from
                         Its Vertices and Asymptotes

Find the equation and the foci of the hyperbola with vertices $(0, \pm 2)$ and asymptotes $y = \pm 2x$. Sketch the graph.

SOLUTION   Since the vertices are on the y-axis, the hyperbola has a vertical transverse axis with $a = 2$. From the asymptote equation, we see that

$$\frac{a}{b} = 2$$

$$\frac{2}{b} = 2 \qquad \text{Since } a = 2$$

$$b = \frac{2}{2} = 1$$

Thus, the equation of the hyperbola is

$$\frac{y^2}{4} - x^2 = 1$$

To find the foci, we calculate $c^2 = a^2 + b^2 = 2^2 + 1^2 = 5$, so $c = \sqrt{5}$. Thus, the foci are $(0, \pm\sqrt{5})$. The graph is shown in Figure 6. ■

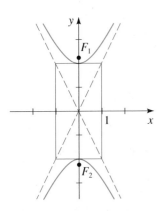

FIGURE 6
$$\frac{y^2}{4} - x^2 = 1$$

Like parabolas and ellipses, hyperbolas have an interesting *reflection property*. Light aimed at one focus of a hyperbolic mirror is reflected toward the other focus, as shown in Figure 7. This property is used in the construction of Cassegrain-type telescopes. A hyperbolic mirror is placed in the telescope tube so that light reflected from the primary parabolic reflector is aimed at one focus of the hyperbolic mirror. The light is then refocused at a more accessible point below the primary reflector (Figure 8).

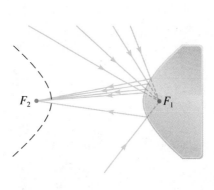

FIGURE 7

Reflection property of hyperbolas

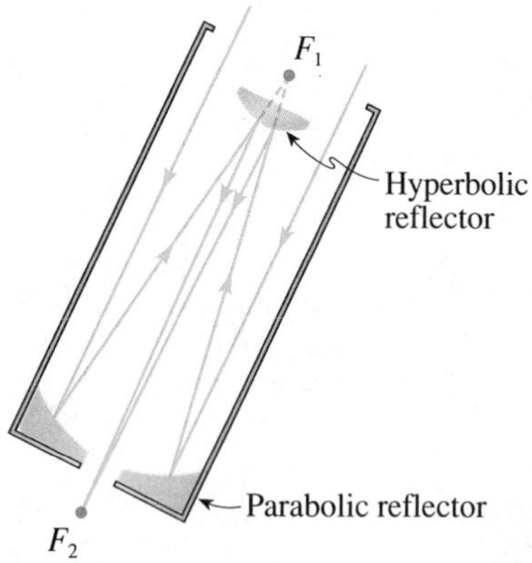

FIGURE 8

Cassegrain-type telescope

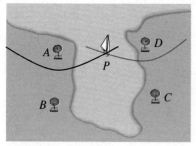

FIGURE 9

LORAN system for finding the location of a ship

In the LORAN (LOng RAnge Navigation) system, hyperbolas are used onboard a ship to determine its location. In Figure 9 radio stations at A and B transmit signals simultaneously for reception by the ship at P. The onboard computer converts the time difference in reception of these signals into a distance difference $d(P, A) - d(P, B)$. By the definition of a hyperbola, this locates the ship on one branch of a hyperbola with foci at A and B (sketched in black in the figure). The same procedure is carried out with two other radio stations at C and D, and this locates the ship on a second hyperbola (shown in red in the figure). (In practice, only three stations are needed because one station can be used as a focus for both hyperbolas.) The coordinates of the intersection point of these two hyperbolas, which can be calculated precisely by the computer, give the location of P.

9.3 **EXERCISES**

1–12 ■ Find the vertices, foci, and asymptotes of the hyperbola, and sketch its graph.

1. $\dfrac{x^2}{4} - \dfrac{y^2}{16} = 1$

2. $\dfrac{y^2}{9} - \dfrac{x^2}{16} = 1$

3. $y^2 - \dfrac{x^2}{25} = 1$

4. $\dfrac{x^2}{2} - y^2 = 1$

5. $x^2 - y^2 = 1$

6. $9x^2 - 4y^2 = 36$

7. $25y^2 - 9x^2 = 225$

8. $x^2 - y^2 + 4 = 0$

9. $x^2 - 4y^2 - 8 = 0$

10. $x^2 - 2y^2 = 3$

11. $4y^2 - x^2 = 1$

12. $9x^2 - 16y^2 = 1$

13–16 ■ Find the equation for the hyperbola whose graph is shown.

13.

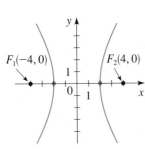

14.

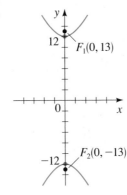

15.

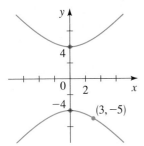

16.

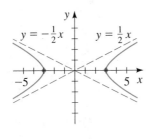

17–28 ■ Find an equation for the hyperbola that satisfies the given conditions.

17. Foci $(\pm 5, 0)$, vertices $(\pm 3, 0)$

18. Foci $(0, \pm 10)$, vertices $(0, \pm 8)$

19. Foci $(0, \pm 2)$, vertices $(0, \pm 1)$

20. Foci $(\pm 6, 0)$, vertices $(\pm 2, 0)$

21. Vertices $(\pm 1, 0)$, asymptotes $y = \pm 5x$

22. Vertices $(0, \pm 6)$, asymptotes $y = \pm \frac{1}{3}x$

23. Foci $(0, \pm 8)$, asymptotes $y = \pm \frac{1}{2}x$

24. Vertices $(0, \pm 6)$, hyperbola passes through $(-5, 9)$

25. Asymptotes $y = \pm x$, hyperbola passes through $(5, 3)$

26. Foci $(\pm 3, 0)$, hyperbola passes through $(4, 1)$

27. Foci $(\pm 5, 0)$, length of transverse axis 6

28. Foci $(0, \pm 1)$, length of transverse axis 1

29. (a) Show that the asymptotes of the hyperbola $x^2 - y^2 = 5$ are perpendicular to each other.

(b) Find an equation for the hyperbola with foci $(\pm c, 0)$ and with asymptotes perpendicular to each other.

30. The hyperbolas

$$\frac{x^2}{a^2} - \frac{y^2}{b^2} = 1 \quad \text{and} \quad \frac{x^2}{a^2} - \frac{y^2}{b^2} = -1$$

are said to be **conjugate** to each other.
(a) Show that the hyperbolas

$$x^2 - 4y^2 + 16 = 0 \quad \text{and} \quad 4y^2 - x^2 + 16 = 0$$

are conjugate to each other, and sketch their graphs on the same coordinate axes.
(b) What do the hyperbolas of part (a) have in common?
(c) Show that any pair of conjugate hyperbolas have the relationship you discovered in part (b).

31. In the derivation of the equation of the hyperbola at the beginning of this section, we said that the equation

$$\sqrt{(x + c)^2 + y^2} - \sqrt{(x - c)^2 + y^2} = \pm 2a$$

simplifies to

$$(c^2 - a^2)x^2 - a^2 y^2 = a^2(c^2 - a^2)$$

Supply the steps needed to show this.

32. (a) For the hyperbola

$$\frac{x^2}{9} - \frac{y^2}{16} = 1$$

determine the values of a, b, and c, and find the coordinates of the foci F_1 and F_2.
(b) Show that the point $P\left(5, \frac{16}{3}\right)$ lies on this hyperbola.
(c) Find $d(P, F_1)$ and $d(P, F_2)$.
(d) Verify that the difference between $d(P, F_1)$ and $d(P, F_2)$ is $2a$.

33. Refer to Figure 9 in the text. Suppose that the radio stations at A and B are 500 mi apart, and that the ship at P receives Station A's signal 2640 microseconds (μs) before it receives the signal from B.
(a) Assuming that radio signals travel at 980 ft/μs, find $d(P, A) - d(P, B)$.
(b) Find an equation for the branch of the hyperbola indicated in black in the figure. (Place A and B on the y-axis with the origin halfway between them. Use miles as the unit of distance.)
(c) If A is due north of B, and if P is due east of A, how far is P from A?

34. Some comets, such as Halley's comet, are a permanent part of the Solar System, traveling in elliptical orbits around the sun. Others pass through the Solar System only once, following a hyperbolic path with the sun at a focus. The figure shows the path of such a comet. Find an equation for the path, assuming that the closest the comet comes to the sun is 2×10^9 mi and that the path the comet was taking before it neared the Solar System is at a right angle to the path it continues on after leaving the Solar System.

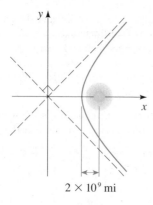

$$2 \times 10^9 \text{ mi}$$

35. Hyperbolas are called **confocal** if they have the same foci.

(a) Show that the hyperbolas

$$\frac{y^2}{k} - \frac{x^2}{16 - k} = 1 \qquad \text{with } 0 < k < 16$$

are confocal.

(b) Use a graphing device to draw the top branches of the family of hyperbolas in part (a) for $k = 1, 4, 8,$ and 12. How does the shape of the graph change as k increases?

 DISCOVERY · DISCUSSION

36. Hyperbolas in the Real World   Several examples of the uses of hyperbolas are given in the text. Find other situations in real life where hyperbolas occur. Consult a scientific encyclopedia in the reference section of your library, or search the Internet.

37. Light from a Lamp   The light from a lamp forms a lighted area on a wall, as shown in the figure. Why is the boundary of this lighted area a hyperbola? How can one hold a flashlight so that its beam forms a hyperbola on the ground?

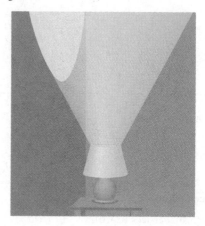

9.4   SHIFTED CONICS

In the preceding sections we studied parabolas with vertices at the origin and ellipses and hyperbolas with centers at the origin. We restricted ourselves to these cases because these equations have the simplest form. In this section we consider conics whose vertices and centers are not necessarily at the origin, and we determine how this affects their equations.

In Section 2.4 we studied transformations of functions that have the effect of shifting their graphs. In general, for any equation in x and y, if we replace x by $x - h$ or by $x + h$, the graph of the new equation is simply the old graph shifted horizontally; if y is replaced by $y - k$ or by $y + k$, the graph is shifted vertically. The following box gives the details.

> **SHIFTING GRAPHS OF EQUATIONS**
>
> If h and k are positive real numbers, then replacing x by $x - h$ or by $x + h$ and replacing y by $y - k$ or by $y + k$ has the following effect(s) on the graph of any equation in x and y.
>
> | Replacement | How the graph is shifted |
> |---|---|
> | **1.** x replaced by $x - h$ | Right h units |
> | **2.** x replaced by $x + h$ | Left h units |
> | **3.** y replaced by $y - k$ | Upward k units |
> | **4.** y replaced by $y + k$ | Downward k units |

For example, consider the ellipse with equation

$$\frac{x^2}{a^2} + \frac{y^2}{b^2} = 1$$

which is shown in Figure 1. If we shift it so that its center is at the point (h, k) instead of at the origin, then its equation becomes

$$\frac{(x - h)^2}{a^2} + \frac{(y - k)^2}{b^2} = 1$$

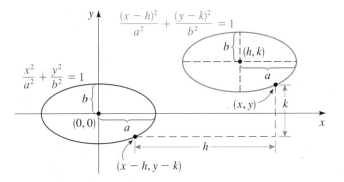

FIGURE 1

Shifted ellipse

EXAMPLE 1 ■ **Sketching the Graph of a Shifted Ellipse**

Sketch the graph of the ellipse

$$\frac{(x + 1)^2}{4} + \frac{(y - 2)^2}{9} = 1$$

and determine the coordinates of the foci.

SOLUTION The ellipse

$$\frac{(x + 1)^2}{4} + \frac{(y - 2)^2}{9} = 1 \qquad \text{(Shifted ellipse)}$$

is shifted so that its center is at $(-1, 2)$. It is obtained from the ellipse

$$\frac{x^2}{4} + \frac{y^2}{9} = 1 \qquad \text{(Ellipse with center at origin)}$$

by shifting it left 1 unit and upward 2 units. The endpoints of the minor and major axes of the unshifted ellipse are $(2, 0)$, $(-2, 0)$, $(0, 3)$, $(0, -3)$. We apply the required shifts to these points to obtain the corresponding points on the shifted ellipse:

$$(2, 0) \quad \rightarrow \quad (2 - 1, 0 + 2) = (1, 2)$$

$$(-2, 0) \quad \rightarrow \quad (-2 - 1, 0 + 2) = (-3, 2)$$

$$(0, 3) \quad \rightarrow \quad (0 - 1, 3 + 2) = (-1, 5)$$

$$(0, -3) \quad \rightarrow \quad (0 - 1, -3 + 2) = (-1, -1)$$

This helps us sketch the graph in Figure 2.

To find the foci of the shifted ellipse, we first find the foci of the ellipse with center at the origin. Since $a^2 = 9$ and $b^2 = 4$, we have $c^2 = 9 - 4 = 5$, so $c = \sqrt{5}$. So the foci are $(0, \pm\sqrt{5})$. Shifting left 1 unit and upward 2 units, we get

$$(0, \sqrt{5}) \quad \rightarrow \quad (0 - 1, \sqrt{5} + 2) = (-1, 2 + \sqrt{5})$$

$$(0, -\sqrt{5}) \quad \rightarrow \quad (0 - 1, -\sqrt{5} + 2) = (-1, 2 - \sqrt{5})$$

Thus, the foci of the shifted ellipse are

$$(-1, 2 + \sqrt{5}) \qquad \text{and} \qquad (-1, 2 - \sqrt{5}) \qquad\qquad ■$$

Applying shifts to parabolas and hyperbolas leads to the equations and their graphs shown in Figures 3 and 4.

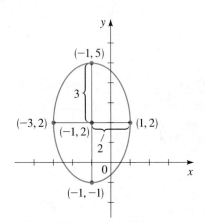

FIGURE 2

$$\frac{(x + 1)^2}{4} + \frac{(y - 2)^2}{9} = 1$$

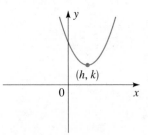

(a) $(x - h)^2 = 4p(y - k)$
$p > 0$

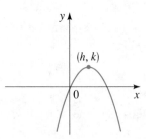

(b) $(x - h)^2 = 4p(y - k)$
$p < 0$

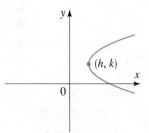

(c) $(y - k)^2 = 4p(x - h)$
$p > 0$

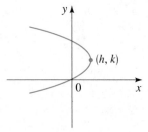

(d) $(y - k)^2 = 4p(x - h)$
$p < 0$

FIGURE 3   Shifted parabolas

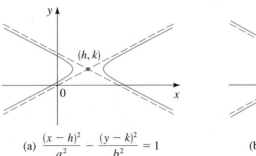

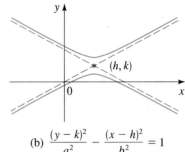

FIGURE 4

Shifted hyperbolas

(a) $\dfrac{(x-h)^2}{a^2} - \dfrac{(y-k)^2}{b^2} = 1$

(b) $\dfrac{(y-k)^2}{a^2} - \dfrac{(x-h)^2}{b^2} = 1$

EXAMPLE 2 ■ Graphing a Shifted Parabola

Determine the vertex, focus, and directrix and sketch the graph of the following parabola.

$$x^2 - 4x = 8y - 28$$

SOLUTION   We complete the square in x to put this equation into one of the forms in Figure 3.

$$x^2 - 4x + 4 = 8y - 28 + 4 \qquad \text{Add 4 to complete the square}$$

$$(x - 2)^2 = 8y - 24$$

$$(x - 2)^2 = 8(y - 3) \qquad \text{(Shifted parabola)}$$

This is a parabola that opens upwards with vertex at $(2, 3)$. It is obtained from the parabola

$$x^2 = 8y \qquad \text{(Parabola with vertex at origin)}$$

by shifting right 2 units and upward 3 units. Since $4p = 8$, we have $p = 2$, so the focus is 2 units above the vertex and the directrix is 2 units below the vertex. Thus, the focus is $(2, 5)$ and the directrix is $y = 1$. The graph is shown in Figure 5.   ■

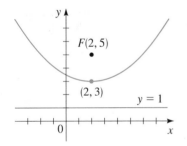

FIGURE 5

$x^2 - 4x = 8y - 28$

EXAMPLE 3 ■ Graphing a Shifted Hyperbola

Show that the following equation represents a hyperbola:

$$9x^2 - 72x - 16y^2 - 32y = 16$$

Find its center, vertices, foci, and asymptotes, and sketch its graph.

SOLUTION   We first complete the squares in both x and y:

$$9(x^2 - 8x \qquad) - 16(y^2 + 2y \qquad) = 16$$

$$9(x^2 - 8x + 16) - 16(y^2 + 2y + 1) = 16 + 9 \cdot 16 - 16 \cdot 1 \qquad \text{Complete the squares}$$

$$9(x - 4)^2 - 16(y + 1)^2 = 144 \qquad \text{Divide this by 144}$$

$$\frac{(x - 4)^2}{16} - \frac{(y + 1)^2}{9} = 1 \qquad \text{(Shifted hyperbola)}$$

This is a hyperbola with center $(4, -1)$ and with a horizontal transverse axis.

$$\text{CENTER} \quad (4, -1)$$

Its graph will have the same shape as the unshifted hyperbola

$$\frac{x^2}{16} - \frac{y^2}{9} = 1 \qquad \text{(Hyperbola with center at origin)}$$

Since $a^2 = 16$ and $b^2 = 9$, we have $a = 4$, $b = 3$, and $c = \sqrt{a^2 + b^2} = \sqrt{16 + 9} = 5$. Thus, the foci lie 5 units to the left and to the right of the center, and the vertices lie 4 units to either side of the center.

$$\text{FOCI} \quad (-1, -1) \text{ and } (9, -1)$$

$$\text{VERTICES} \quad (0, -1) \text{ and } (8, -1)$$

The asymptotes of the unshifted hyperbola are $y = \pm\frac{3}{4}x$, so the asymptotes of the shifted parabola are found as follows.

$$\text{ASYMPTOTES} \quad (y + 1) = \pm\frac{3}{4}(x - 4)$$

$$y + 1 = \pm\frac{3}{4}x \mp 3$$

$$y = \frac{3}{4}x - 4 \quad \text{and} \quad y = -\frac{3}{4}x + 2$$

To help us sketch the hyperbola, we draw the central box; it extends 4 units left and right from the center and 3 units upward and downward from the center. We then draw the asymptotes and complete the graph of the shifted hyperbola as shown in Figure 6.

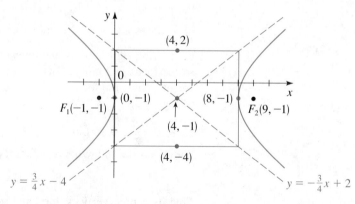

FIGURE 6
$9x^2 - 72x - 16y^2 - 32y = 16$

If we expand and simplify the equations of any of the shifted conics illustrated in Figures 1, 3, and 4, then we will always obtain an equation of the form

$$Ax^2 + Cy^2 + Dx + Ey + F = 0$$

where A and C are not both 0. Conversely, if we begin with an equation of this form, then we can complete the square in x and y to see which type of conic section the equation represents. In some cases, the graph of the equation turns out to be just a pair of lines, a single point, or there may be no graph at all. These cases are called **degenerate conics**. The next example illustrates such a case.

EXAMPLE 4 ■ An Equation that Leads to a Degenerate Conic

Sketch the graph of the equation

$$9x^2 - y^2 + 18x + 6y = 0$$

SOLUTION Because the coefficients of x^2 and y^2 are of opposite sign, this equation looks as if it should represent a hyperbola (like the equation of Example 3). To see whether this is in fact the case, we complete the squares:

$$9(x^2 + 2x \quad) - (y^2 - 6y \quad) = 0$$

$$9(x^2 + 2x + 1) - (y^2 - 6y + 9) = 0 + 9 - 9$$

$$9(x + 1)^2 - (y - 3)^2 = 0$$

$$(x + 1)^2 - \frac{(y - 3)^2}{9} = 0 \qquad \text{Divide by 9}$$

For this to fit the form of the equation of a hyperbola, we would need a non-zero constant to the right of the equal sign. In fact, further analysis shows that this is the equation of a pair of intersecting lines:

$$(y - 3)^2 = 9(x + 1)^2$$

$$y - 3 = \pm 3(x + 1) \qquad \text{Take the square roots}$$

$$y = 3(x + 1) + 3 \qquad \text{or} \qquad y = -3(x + 1) + 3$$

$$y = 3x + 6 \qquad\qquad\qquad y = -3x$$

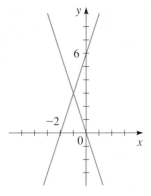

FIGURE 7
$9x^2 - y^2 + 18x + 6y = 0$

These lines are graphed in Figure 7.   ■

Because the equation in Example 4 looked at first glance like the equation of a hyperbola but, in fact, turned out to represent simply a pair of lines, we refer to its graph as a **degenerate hyperbola**. Degenerate ellipses and parabolas can also arise when we complete the square(s) in an equation that seems to represent a conic. For example, the equation

$$4x^2 + y^2 - 8x + 2y + 6 = 0$$

looks as if it should represent an ellipse, because the coefficients of x^2 and y^2 have the same sign. But completing the squares leads to

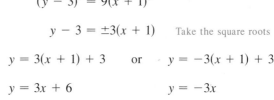

which has no solution at all (since the sum of two squares cannot be negative). This equation is therefore degenerate.

To summarize, we have the following theorem.

GENERAL EQUATION OF A SHIFTED CONIC

The graph of the equation

$$Ax^2 + Cy^2 + Dx + Ey + F = 0$$

where A and C are not both 0, is a conic or a degenerate conic. In the nondegenerate cases, the graph is

1. a parabola if A or C is 0.
2. an ellipse if A and C have the same sign (or a circle if $A = C$).
3. a hyperbola if A and C have opposite signs.

9.4   EXERCISES

1–4 ■ Find the center, foci, and vertices of the ellipse, and determine the lengths of the major and minor axes. Then sketch the graph.

1. $\dfrac{(x-2)^2}{9} + \dfrac{(y-1)^2}{4} = 1$

2. $\dfrac{(x-3)^2}{16} + (y+3)^2 = 1$

3. $\dfrac{x^2}{9} + \dfrac{(y+5)^2}{25} = 1$

4. $\dfrac{(x+2)^2}{4} + y^2 = 1$

5–8 ■ Find the vertex, focus, and directrix of the parabola, and sketch the graph.

5. $(x-3)^2 = 8(y+1)$ **6.** $(y+5)^2 = -6x + 12$

7. $-4\left(x + \tfrac{1}{2}\right)^2 = y$ **8.** $y^2 = 16x - 8$

9–12 ■ Find the center, foci, vertices, and asymptotes of the hyperbola. Then sketch the graph.

9. $\dfrac{(x+1)^2}{9} - \dfrac{(y-3)^2}{16} = 1$

10. $(x-8)^2 - (y+6)^2 = 1$

11. $y^2 - \dfrac{(x+1)^2}{4} = 1$

12. $\dfrac{(y-1)^2}{25} - (x+3)^2 = 1$

13–18 ■ Find an equation for the conic whose graph is shown.

13.

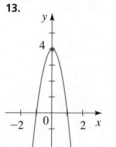

14.

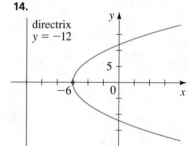

directrix
$y = -12$

15.

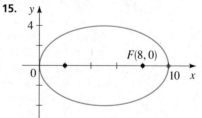

$F(8, 0)$

16.

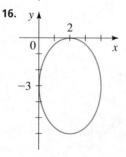

17.

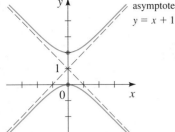

18.

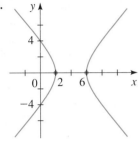

19–30 ■ Complete the square to determine whether the equation represents an ellipse, a parabola, a hyperbola, or a degenerate conic. Then sketch the graph of the equation. If the graph is an ellipse, find the center, foci, vertices, and lengths of the major and minor axes. If it is a parabola, find the vertex, focus, and directrix. If it is a hyperbola, find the center, foci, vertices, and asymptotes. If the equation has no graph, explain why.

19. $9x^2 - 36x + 4y^2 = 0$

20. $y^2 = 4(x + 2y)$

21. $x^2 - 4y^2 - 2x + 16y = 20$

22. $x^2 + 6x + 12y + 9 = 0$

23. $4x^2 + 25y^2 - 24x + 250y + 561 = 0$

24. $2x^2 + y^2 = 2y + 1$

25. $16x^2 - 9y^2 - 96x + 288 = 0$

26. $4x^2 - 4x - 8y + 9 = 0$

27. $x^2 + 16 = 4(y^2 + 2x)$

28. $x^2 - y^2 = 10(x - y) + 1$

29. $3x^2 + 4y^2 - 6x - 24y + 39 = 0$

30. $x^2 + 4y^2 + 20x - 40y + 300 = 0$

31. Determine what the value of F must be if the graph of the equation

$$4x^2 + y^2 + 4(x - 2y) + F = 0$$

is (a) an ellipse, (b) a single point, or (c) the empty set.

32. Find an equation for the ellipse that shares a vertex and a focus with the parabola $x^2 + y = 100$ and has its other focus at the origin.

 33. This exercise deals with **confocal parabolas**, that is, families of parabolas that have the same focus.
  (a) Draw the graphs of the family of parabolas

$$x^2 = 4p(y + p)$$

  for $p = -2, -\frac{3}{2}, -1, -\frac{1}{2}, \frac{1}{2}, 1, \frac{3}{2}, 2$.
  (b) Show that each parabola in this family has its focus at the origin.
  (c) Describe the effect on the graph of moving the vertex closer to the origin.

 DISCOVERY · DISCUSSION

34. A Family of Confocal Conics Conics that share a focus are called **confocal**. Consider the family of conics that have a focus at $(0, 1)$ and a vertex at the origin (see the figure).
  (a) Find equations of two different ellipses that have these properties.
  (b) Find equations of two different hyperbolas that have these properties.
  (c) Explain why there is only one parabola that satisfies these properties. Find its equation.
  (d) Sketch the conics you found in parts (a), (b), and (c) on the same coordinate axes (for the hyperbolas, sketch the top branches only).
  (e) How are the ellipses and hyperbolas related to the parabola?

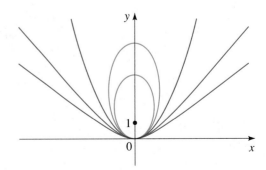

9.5   ROTATION OF AXES

In Section 9.4 we studied conics with equations of the form

$$Ax^2 + Cy^2 + Dx + Ey + F = 0$$

We saw that the graph is always an ellipse, parabola, or hyperbola with horizontal or vertical axes (except in the degenerate cases). In this section we study the most general second-degree equation

$$Ax^2 + Bxy + Cy^2 + Dx + Ey + F = 0$$

We will see that the graph of an equation of this form is also a conic. In fact, by rotating the coordinate axes through an appropriate angle, we can eliminate the term Bxy and then use our knowledge of conic sections to analyze the graph.

In Figure 1 the x- and y-axes have been rotated through an acute angle ϕ about the origin to produce a new pair of axes, which we call the X- and Y-axes. A point P that has coordinates (x, y) in the old system has coordinates (X, Y) in the new system. If we let r denote the distance of P from the origin and let θ be the angle that the segment OP makes with the new X-axis, then we can see from Figure 2 (by considering the two right triangles in the figure) that

$$X = r\cos\theta \qquad\qquad Y = r\sin\theta$$

$$x = r\cos(\theta + \phi) \qquad y = r\sin(\theta + \phi)$$

Using the addition formula for cosine, we see that

$$
\begin{aligned}
x &= r\cos(\theta + \phi) \\
&= r(\cos\theta\cos\phi - \sin\theta\sin\phi) \\
&= (r\cos\theta)\cos\phi - (r\sin\theta)\sin\phi \\
&= X\cos\phi - Y\sin\phi
\end{aligned}
$$

Similarly, we can apply the addition formula for sine to the expression for y to obtain $y = X\sin\phi + Y\cos\phi$. By treating these equations for x and y as a system of linear equations in the variables X and Y (see Exercise 27), we obtain expressions for X and Y in terms of x and y, as detailed in the following box.

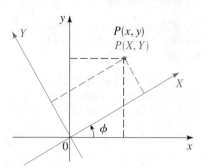

FIGURE 1

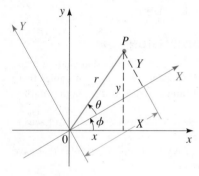

FIGURE 2

ROTATION OF AXES FORMULAS

Suppose the x- and y-axes in a coordinate plane are rotated through the acute angle ϕ to produce the X- and Y-axes, as shown in Figure 1. Then the coordinates (x, y) and (X, Y) of a point in the xy- and the XY-planes are related as follows:

$$x = X\cos\phi - Y\sin\phi \qquad X = x\cos\phi + y\sin\phi$$

$$y = X\sin\phi + Y\cos\phi \qquad Y = -x\sin\phi + y\cos\phi$$

EXAMPLE 1 ■ Rotation of Axes

If the coordinate axes are rotated through 30°, find the XY-coordinates of the point with xy-coordinates $(2, -4)$.

SOLUTION Using the Rotation of Axes Formulas with $x = 2$, $y = -4$, and $\phi = 30°$, we get

$$X = 2\cos 30° + (-4)\sin 30° = 2\left(\frac{\sqrt{3}}{2}\right) - 4\left(\frac{1}{2}\right) = \sqrt{3} - 2$$

$$Y = -2\sin 30° + (-4)\cos 30° = -2\left(\frac{1}{2}\right) - 4\left(\frac{\sqrt{3}}{2}\right) = -1 - 2\sqrt{3}$$

The XY-coordinates are $\left(-2 + \sqrt{3}, -1 - 2\sqrt{3}\right)$. ■

EXAMPLE 2 ■ Rotating a Hyperbola

Show by rotating the coordinate axes through 45° that the graph of the equation $xy = 2$ is a hyperbola.

SOLUTION We use the Rotation of Axes Formulas with $\phi = 45°$ to obtain

$$x = X\cos 45° - Y\sin 45° = \frac{X}{\sqrt{2}} - \frac{Y}{\sqrt{2}}$$

$$y = X\sin 45° + Y\cos 45° = \frac{X}{\sqrt{2}} + \frac{Y}{\sqrt{2}}$$

Substituting these expressions into the original equation gives

$$\left(\frac{X}{\sqrt{2}} - \frac{Y}{\sqrt{2}}\right)\left(\frac{X}{\sqrt{2}} + \frac{Y}{\sqrt{2}}\right) = 2$$

$$\frac{X^2}{2} - \frac{Y^2}{2} = 2$$

$$\frac{X^2}{4} - \frac{Y^2}{4} = 1$$

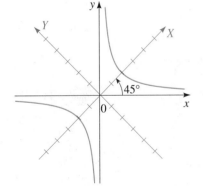

FIGURE 3

$xy = 2$

We recognize this as a hyperbola with vertices $(\pm 2, 0)$ in the XY-coordinate system. Its asymptotes are $Y = \pm X$, which correspond to the coordinate axes in the xy-system (see Figure 3). ■

The method of Example 2 can be used to transform any equation of the form

$$Ax^2 + Bxy + Cy^2 + Dx + Ey + F = 0$$

into an equation in X and Y that doesn't contain an XY-term by choosing an appropriate angle of rotation. To find the angle that works, we rotate the axes through an angle ϕ and substitute for x and y using the Rotation of Axes Formulas:

$$A(X\cos\phi - Y\sin\phi)^2 + B(X\cos\phi - Y\sin\phi)(X\sin\phi + Y\cos\phi)$$
$$+ C(X\sin\phi + Y\cos\phi)^2 + D(X\cos\phi - Y\sin\phi)$$
$$+ E(X\sin\phi + Y\cos\phi) + F = 0$$

If we expand this and collect like terms, we obtain an equation of the form

$$A'X^2 + B'XY + C'Y^2 + D'X + E'Y + F' = 0$$

where

$$A' = A\cos^2\phi + B\sin\phi\cos\phi + C\sin^2\phi$$
$$B' = 2(C - A)\sin\phi\cos\phi + B(\cos^2\phi - \sin^2\phi)$$
$$C' = A\sin^2\phi - B\sin\phi\cos\phi + C\cos^2\phi$$
$$D' = D\cos\phi + E\sin\phi$$
$$E' = -D\sin\phi + E\cos\phi$$
$$F' = F$$

To eliminate the XY-term, we would like to choose ϕ so that $B' = 0$, that is,

$$2(C - A)\sin\phi\cos\phi + B(\cos^2\phi - \sin^2\phi) = 0$$

$$(C - A)\sin 2\phi + B\cos 2\phi = 0 \qquad \text{Double-angle formulas for sine and cosine}$$

$$B\cos 2\phi = (A - C)\sin 2\phi$$

$$\cot 2\phi = \frac{A - C}{B} \qquad \text{Divide by } \sin 2\phi$$

The preceding calculation proves the following theorem.

SIMPLIFYING THE GENERAL CONIC EQUATION

To eliminate the xy-term in the general conic equation

$$Ax^2 + Bxy + Cy^2 + Dx + Ey + F = 0$$

rotate the axes through the acute angle ϕ that satisfies

$$\cot 2\phi = \frac{A - C}{B}$$

EXAMPLE 3 ■ Eliminating the xy-Term

Use a rotation of axes to eliminate the xy-term in the equation

$$6\sqrt{3}\, x^2 + 6xy + 4\sqrt{3}\, y^2 = 21\sqrt{3}$$

Identify and sketch the curve.

SOLUTION   To eliminate the xy-term, we rotate the axes through an angle ϕ that satisfies

$$\cot 2\phi = \frac{A - C}{B} = \frac{6\sqrt{3} - 4\sqrt{3}}{6} = \frac{\sqrt{3}}{3}$$

Thus, $2\phi = 60°$ and hence $\phi = 30°$. With this value of ϕ, we get

$$x = X\left(\frac{\sqrt{3}}{2}\right) - Y\left(\frac{1}{2}\right)$$

Rotation of Axes Formulas:

$$\cos\phi = \frac{\sqrt{3}}{2}, \sin\phi = \frac{1}{2}$$

$$y = X\left(\frac{1}{2}\right) + Y\left(\frac{\sqrt{3}}{2}\right)$$

Substituting these values for x and y into the given equation leads to

$$6\sqrt{3}\left(\frac{X\sqrt{3}}{2} - \frac{Y}{2}\right)^2 + 6\left(\frac{X\sqrt{3}}{2} - \frac{Y}{2}\right)\left(\frac{X}{2} + \frac{Y\sqrt{3}}{2}\right) + 4\sqrt{3}\left(\frac{X}{2} + \frac{Y\sqrt{3}}{2}\right)^2 = 21\sqrt{3}$$

Expanding and collecting like terms, we get

$$7\sqrt{3}\, X^2 + 3\sqrt{3}\, Y^2 = 21\sqrt{3}$$

$$\frac{X^2}{3} + \frac{Y^2}{7} = 1 \qquad \text{Divide by } 21\sqrt{3}$$

This is the equation of an ellipse in the XY-coordinate system. The foci lie on the Y-axis. Because $a^2 = 7$ and $b^2 = 3$, the length of the major axis is $2\sqrt{7}$, and the length of the minor axis is $2\sqrt{3}$. The ellipse is sketched in Figure 4. ■

In the preceding example we were able to determine ϕ without difficulty, since we remembered that $\cot 60° = \sqrt{3}/3$. In general, finding ϕ is not quite so easy. The next example illustrates how the following half-angle formulas, which are valid for $0 < \phi < \pi/2$, are useful in determining ϕ (see Section 7.3):

$$\cos\phi = \sqrt{\frac{1 + \cos 2\phi}{2}} \qquad \sin\phi = \sqrt{\frac{1 - \cos 2\phi}{2}}$$

EXAMPLE 4 ■ **Eliminating the xy-Term**

Use a rotation of axes to eliminate the xy-term in the equation

$$64x^2 + 96xy + 36y^2 - 15x + 20y - 25 = 0$$

Identify and sketch the curve.

SOLUTION   To eliminate the xy-term, we rotate the axes through an angle ϕ that satisfies

$$\cot 2\phi = \frac{A - C}{B} = \frac{64 - 36}{96} = \frac{7}{24}$$

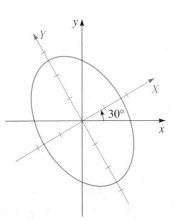

FIGURE 4
$6\sqrt{3}\, x^2 + 6xy + 4\sqrt{3}\, y^2 = 21\sqrt{3}$

FIGURE 5

In Figure 5 we sketch a triangle with $\cot 2\phi = \frac{7}{24}$. We see that

$$\cos 2\phi = \frac{7}{25}$$

so, using the half-angle formulas, we get

$$\cos\phi = \sqrt{\frac{1 + \frac{7}{25}}{2}} = \sqrt{\frac{16}{25}} = \frac{4}{5}$$

$$\sin\phi = \sqrt{\frac{1 - \frac{7}{25}}{2}} = \sqrt{\frac{9}{25}} = \frac{3}{5}$$

The Rotation of Axes Formulas then give

$$x = \tfrac{4}{5}X - \tfrac{3}{5}Y \quad \text{and} \quad y = \tfrac{3}{5}X + \tfrac{4}{5}Y$$

Substituting into the given equation, we have

$$64\left(\tfrac{4}{5}X - \tfrac{3}{5}Y\right)^2 + 96\left(\tfrac{4}{5}X - \tfrac{3}{5}Y\right)\left(\tfrac{3}{5}X + \tfrac{4}{5}Y\right)$$

$$+ 36\left(\tfrac{3}{5}X + \tfrac{4}{5}Y\right)^2 - 15\left(\tfrac{4}{5}X - \tfrac{3}{5}Y\right) + 20\left(\tfrac{3}{5}X + \tfrac{4}{5}Y\right) - 25 = 0$$

Expanding and collecting like terms, we get

$$100X^2 + 25Y - 25 = 0$$

$$-4X^2 = Y - 1 \qquad \text{Simplify}$$

We recognize this as the equation of a parabola that opens along the negative Y-axis and has vertex $(0, 1)$ in XY-coordinates. Since $4p = -4$, we have $p = -1$, so the focus is $(0, 0)$ and the directrix is $Y = 2$. Using

$$\phi = \cos^{-1}\tfrac{4}{5} \approx 37°$$

we sketch the graph in Figure 6. ∎

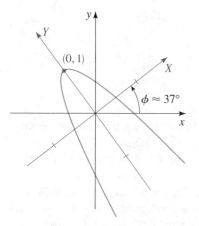

FIGURE 6
$64x^2 + 96xy + 36y^2 - 15x + 20y - 25 = 0$

In Examples 3 and 4 we were able to identify the type of conic by rotating the axes. The next theorem gives rules for identifying the type of conic directly from the equation, without rotating axes.

IDENTIFYING CONICS BY THE DISCRIMINANT

The graph of the equation

$$Ax^2 + Bxy + Cy^2 + Dx + Ey + F = 0$$

is either a conic or a degenerate conic. In the nondegenerate cases, the graph is

  1. a parabola if $B^2 - 4AC = 0$.

  2. an ellipse if $B^2 - 4AC < 0$.

  3. a hyperbola if $B^2 - 4AC > 0$.

The quantity $B^2 - 4AC$ is called the **discriminant** of the equation.

■ **Proof** If we rotate the axes through an angle ϕ, we get an equation of the form

$$A'X^2 + B'XY + C'Y^2 + D'X + E'Y + F' = 0$$

where A', B', C', ... are given by the formulas on page 646. A straightforward calculation shows that

$$(B')^2 - 4A'C' = B^2 - 4AC$$

Thus, the expression $B^2 - 4AC$ remains unchanged for any rotation. In particular, if we choose a rotation that eliminates the xy-term ($B' = 0$), we get

$$A'X^2 + C'Y^2 + D'X + E'Y + F' = 0$$

In this case, $B^2 - 4AC = -4A'C'$. So $B^2 - 4AC = 0$ if either A' or C' is zero; $B^2 - 4AC < 0$ if A' and C' have the same sign; and $B^2 - 4AC > 0$ if A' and C' have opposite signs. According to the box on page 642, these cases correspond to the graph of the last displayed equation being a parabola, an ellipse, or a hyperbola, respectively. ☐

In the proof we indicated that the discriminant is unchanged by any rotation; for this reason, the discriminant is said to be **invariant** under rotation.

EXAMPLE 5 ■ Identifying a Conic by the Discriminant

Use the discriminant to identify the graph of the following equation.

$$3x^2 + 5xy - 2y^2 + x - y + 4 = 0$$

SOLUTION Since $A = 3$, $B = 5$, and $C = -2$, the discriminant is

$$B^2 - 4AC = 5^2 - 4(3)(-2) = 49 > 0$$

So, the graph is a hyperbola. ■

9.5 EXERCISES

1–6 ■ Determine the XY-coordinates of the given point if the coordinate axes are rotated through the indicated angle.

1. $(1,1)$, $\phi = 45°$

2. $(-2,1)$, $\phi = 30°$

3. $(3, -\sqrt{3})$, $\phi = 60°$

4. $(2,0)$, $\phi = 15°$

5. $(0,2)$, $\phi = 55°$

6. $(\sqrt{2}, 4\sqrt{2})$, $\phi = 45°$

7–10 ■ Determine the equation of the given conic in XY-coordinates when the coordinate axes are rotated through the indicated angle.

7. $y = (x - 1)^2$, $\phi = 45°$

8. $x^2 - y^2 = 2y$, $\phi = \cos^{-1}\frac{3}{5}$

9. $x^2 + 2\sqrt{3}\,xy - y^2 = 4$, $\phi = 30°$

10. $xy = x + y$, $\phi = \pi/4$

11–24 ■ (a) Use the discriminant to determine whether the graph of the equation is a parabola, an ellipse, or a hyperbola. (b) Use a rotation of axes to eliminate the xy-term. (c) Sketch the graph.

11. $xy = 8$

12. $xy + 4 = 0$

13. $x^2 + 2xy + y^2 + x - y = 0$

14. $13x^2 + 6\sqrt{3}\,xy + 7y^2 = 16$

15. $x^2 + 2\sqrt{3}\,xy - y^2 + 2 = 0$

16. $21x^2 + 10\sqrt{3}\,xy + 31y^2 = 144$

17. $11x^2 - 24xy + 4y^2 + 20 = 0$

18. $25x^2 - 120xy + 144y^2 - 156x - 65y = 0$

19. $\sqrt{3}\,x^2 + 3xy = 3$

20. $153x^2 + 192xy + 97y^2 = 225$

21. $2\sqrt{3}\,x^2 - 6xy + \sqrt{3}\,x + 3y = 0$

22. $9x^2 - 24xy + 16y^2 = 100(x - y - 1)$

23. $52x^2 + 72xy + 73y^2 = 40x - 30y + 75$

24. $(7x + 24y)^2 = 600x - 175y + 25$

25. (a) Use rotation of axes to show that the following equation represents a hyperbola:

$$7x^2 + 48xy - 7y^2 - 200x - 150y + 600 = 0$$

 (b) Find the XY- and xy-coordinates of the center, vertices, and foci.

 (c) Find the equations of the asymptotes in XY- and xy-coordinates.

26. (a) Use rotation of axes to show that the following equation represents a parabola:

$$2\sqrt{2}(x + y)^2 = 7x + 9y$$

 (b) Find the XY- and xy-coordinates of the vertex and focus.

 (c) Find the equation of the directrix in XY- and xy-coordinates.

27. Solve the equations

$$x = X\cos\phi - Y\sin\phi$$
$$y = X\sin\phi + Y\cos\phi$$

for X and Y in terms of x and y. [*Hint:* To begin, multiply the first equation by $\cos\phi$ and the second by $\sin\phi$, and then add the two equations to solve for X.]

28. Show that the graph of the equation

$$\sqrt{x} + \sqrt{y} = 1$$

is a part of a parabola by rotating the axes through an angle of $45°$. [*Hint:* First convert the equation to one that does not involve radicals.]

29. Let Z, Z', and R be the matrices

$$Z = \begin{bmatrix} x \\ y \end{bmatrix} \qquad Z' = \begin{bmatrix} X \\ Y \end{bmatrix}$$

$$R = \begin{bmatrix} \cos\phi & -\sin\phi \\ \sin\phi & \cos\phi \end{bmatrix}$$

Show that the Rotation of Axes Formulas can be written as

$$Z = RZ' \qquad \text{and} \qquad Z' = R^{-1}Z$$

DISCOVERY · DISCUSSION

30. Algebraic Invariants A quantity is *invariant under rotation* if it does not change when the axes are rotated. It was shown in the text that for the general equation of a conic, the quantity $B^2 - 4AC$ is invariant under rotation. Use the formulas for A' and C' on page 646 to prove that the quantity $A + C$ is also invariant under rotation. Is the quantity F invariant under rotation?

31. Geometric Invariants Do you expect that the distance between two points is invariant under rotation? Prove your answer by comparing the distance $d(P, Q)$ and $d(P', Q')$ where P' and Q' are the images of P and Q under a rotation of axes.

9.6 POLAR COORDINATES

A coordinate system is a method for specifying the location of a point in the plane. Thus far we have been dealing with the rectangular (or Cartesian) coordinate system, which describes locations using a rectangular grid. Using rectangular coordinates is like describing a location in a city by saying that, for example, it's at the corner of 48th Street and 7th Avenue. But we might also describe this same location by saying that it's 3 miles northwest of City Hall. Instead of specifying the location with respect to a grid of streets and avenues, we can describe it by giving its distance and direction from a fixed reference point.

The *polar coordinate system* uses distances and directions to specify the locations of points in the plane. To set up this system, we first choose a fixed point O called the **origin** (or **pole**). We then draw a ray (half-line) starting at O, called the **polar axis**. This axis is usually drawn horizontally to the right of O and coincides with the positive x-axis in rectangular coordinates. Now let P be any point in the plane. Let r be the distance from P to the origin, and let θ be the angle between the polar axis and the segment OP, as shown in Figure 1. Then the ordered pair (r, θ) uniquely specifies the location of P. We write $P(r, \theta)$ and refer to r and θ as the **polar coordinates** of P. We use the convention that θ is positive if measured in a counterclockwise direction from the polar axis, or negative if measured in a clockwise direction. It's customary to use radian measure for θ. If $r = 0$, then $P = O$ no matter what value θ has, so $(0, \theta)$ represents the pole for any value of θ.

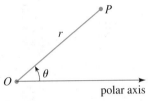

FIGURE 1

Because the angles $\theta + 2n\pi$ (for $n = \pm 1, \pm 2, \pm 3, \ldots$) all have the same terminal side as the angle θ, each point has infinitely many representations in polar coordinates. For example, $(2, \pi/3)$, $(2, 7\pi/3)$, and $(2, -5\pi/3)$ all represent the same point P, as shown in Figure 2.

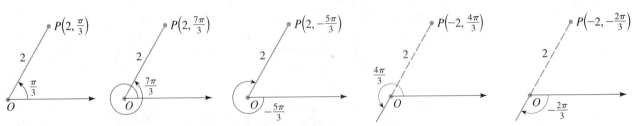

FIGURE 2

Moreover, we also allow r to take on negative values, with the understanding that if $-r$ is negative, then the point $P(-r, \theta)$ lies r units away from the origin in the direction *opposite* to that given by θ. Thus, the point P graphed in Figure 2 can also be described by the coordinates $(-2, 4\pi/3)$ or $(-2, -2\pi/3)$. By this convention, the coordinates (r, θ) and $(-r, \theta + \pi)$ represent the same point (see Figure 3). In fact, any point $P(r, \theta)$ can also be represented by

$$P(r, \theta + 2n\pi) \qquad \text{and} \qquad P(-r, \theta + (2n + 1)\pi)$$

for any integer n.

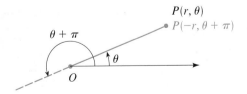

FIGURE 3

EXAMPLE 1 ■ Plotting Points in Polar Coordinates

Plot the points whose polar coordinates are given.

(a) $(1, 3\pi/4)$ (b) $(3, -\pi/6)$ (c) $(3, 3\pi)$ (d) $(-4, \pi/4)$

SOLUTION   The points are plotted in Figure 4. Note that the point in part (d) lies 4 units from the origin along the angle $5\pi/4$, because the given value of r is negative.

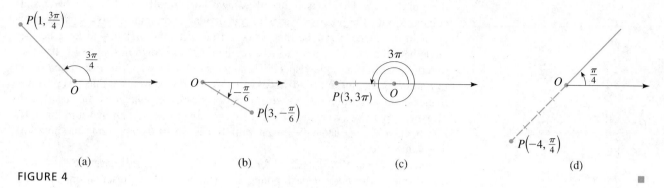

(a)   (b)   (c)   (d)

FIGURE 4

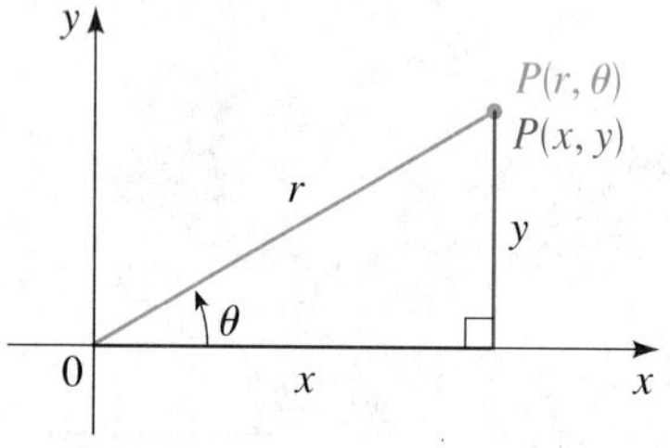

FIGURE 5

Situations often arise in which we need to consider polar and rectangular coordinates simultaneously. The connection between the two systems is illustrated in Figure 5, where the polar axis coincides with the positive x-axis. The formulas in the following box are obtained from the figure using the definitions of the trigonometric functions and the Pythagorean Theorem. (Although we have pictured the case where $r > 0$ and θ is acute, the formulas hold for any angle θ and for any value of r.)

> ### RELATIONSHIP BETWEEN POLAR AND RECTANGULAR COORDINATES
>
> **1.** To change from polar to rectangular coordinates, use the formulas
> $$x = r\cos\theta \quad \text{and} \quad y = r\sin\theta$$
> **2.** To change from rectangular to polar coordinates, use the formulas
> $$r^2 = x^2 + y^2 \quad \text{and} \quad \tan\theta = \frac{y}{x} \quad (x \neq 0)$$

EXAMPLE 2 ■ Converting Polar Coordinates to Rectangular Coordinates

Find rectangular coordinates for the point that has polar coordinates $(4, 2\pi/3)$.

SOLUTION   Since $r = 4$ and $\theta = 2\pi/3$, we have

$$x = r\cos\theta = 4\cos\frac{2\pi}{3} = 4 \cdot \left(-\frac{1}{2}\right) = -2$$

$$y = r\sin\theta = 4\sin\frac{2\pi}{3} = 4 \cdot \frac{\sqrt{3}}{2} = 2\sqrt{3}$$

Thus, the point has rectangular coordinates $(-2, 2\sqrt{3}\,)$.

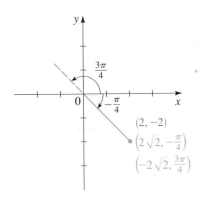

FIGURE 6

EXAMPLE 3 ■ Converting Rectangular Coordinates to Polar Coordinates

Find polar coordinates for the point that has rectangular coordinates $(2, -2)$.

SOLUTION Using $x = 2$, $y = -2$, we get

$$r^2 = x^2 + y^2 = 2^2 + (-2)^2 = 8$$

so $r = 2\sqrt{2}$ or $-2\sqrt{2}$. Also

$$\tan \theta = \frac{y}{x} = \frac{-2}{2} = -1$$

so $\theta = 3\pi/4$ or $-\pi/4$. Since the point $(2, -2)$ lies in quadrant IV (see Figure 6), we can represent it in polar coordinates as $\left(2\sqrt{2}, -\pi/4\right)$ or $\left(-2\sqrt{2}, 3\pi/4\right)$. ■

⊘ Note that the equations relating polar and rectangular coordinates do not uniquely determine r or θ. When we use these equations to find the polar coordinates of a point, we must be careful that the values we choose for r and θ give us a point in the correct quadrant, as we saw in Example 3.

GRAPHS OF POLAR EQUATIONS

The **graph of a polar equation** $r = f(\theta)$ consists of all points P that have at least one polar representation (r, θ) whose coordinates satisfy the equation. In the next two examples, we see that circles centered at the origin and lines that pass through the origin have particularly simple equations in polar coordinates.

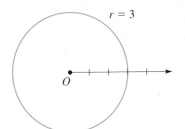

FIGURE 7

EXAMPLE 4 ■ Sketching the Graph of a Polar Equation

Sketch the graph of the equation $r = 3$.

SOLUTION The graph consists of all points whose r-coordinate is 3, that is, all points that are 3 units away from the origin. Therefore, this is the circle of radius 3 centered at the origin, as shown in Figure 7. ■

In general, the graph of the equation $r = a$ is a circle of radius $|a|$ centered at the origin. Squaring both sides of this equation, we get $r^2 = a^2$, and since $r^2 = x^2 + y^2$, the equivalent equation in rectangular coordinates is $x^2 + y^2 = a^2$.

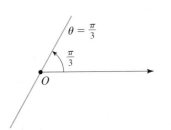

FIGURE 8

EXAMPLE 5 ■ Sketching the Graph of a Polar Equation

Sketch the graph of the equation $\theta = \pi/3$ and express the equation in rectangular coordinates.

SOLUTION The graph consists of all points whose θ-coordinate is $\pi/3$. This is the straight line that passes through the origin and makes an angle of $\pi/3$ with the polar axis (see Figure 8). Note that the points $(r, \pi/3)$ on the line with $r > 0$ lie in quadrant I, whereas those with $r < 0$ lie in quadrant III. If the point (x, y) lies on this line, then

$$\frac{y}{x} = \tan\theta = \tan\frac{\pi}{3} = \sqrt{3}$$

Thus, the rectangular equation of this line is $y = \sqrt{3}\,x$.   ∎

To sketch a polar curve whose graph isn't so obvious as the ones in the preceding examples, we rely on two techniques. One technique is to plot points calculated for sufficiently many values of θ and then join them in a continuous curve. This is what we did when we first learned to graph functions in rectangular coordinates. The other technique is to convert the polar equation into rectangular coordinates in the hope that the resulting equation is one that we recognize from our previous work. Both methods are used in the next example.

EXAMPLE 6 ■ Sketching the Graph of a Polar Equation

(a) Sketch the graph of the polar equation $r = 2\sin\theta$.

(b) Convert the equation of part (a) to rectangular coordinates.

SOLUTION

(a) We first use the equation to determine the polar coordinates of several points on the curve. The results are shown in the following table.

$\boldsymbol{\theta}$	0	$\pi/6$	$\pi/4$	$\pi/3$	$\pi/2$	$2\pi/3$	$3\pi/4$	$5\pi/6$	π
$r = 2\sin\theta$	0	1	$\sqrt{2}$	$\sqrt{3}$	2	$\sqrt{3}$	$\sqrt{2}$	1	0

We plot these points in Figure 9 and then join them to sketch the curve. The graph appears to be a circle. We have used values of θ only between 0 and π, since the same points (this time expressed with negative r-coordinates) would be obtained if we allowed θ to range from π to 2π.

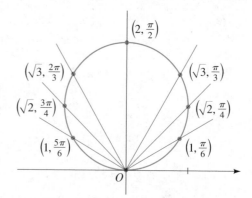

FIGURE 9
$r = 2\sin\theta$

(b) We multiply both sides of the equation by r to get

$$r^2 = 2r\sin\theta$$

Using the formulas that relate polar and rectangular coordinates, we replace r^2 by $x^2 + y^2$ and $r\sin\theta$ by y to obtain the rectangular equation

$$x^2 + y^2 = 2y$$

$$x^2 + y^2 - 2y = 0 \qquad \text{Subtract } 2y$$

$$x^2 + (y-1)^2 = 1 \qquad \text{Complete the square in } y$$

This is the equation of a circle of radius 1 centered at the point $(0, 1)$. ■

In general, the graphs of equations of the form

$$r = 2a\sin\theta \qquad \text{and} \qquad r = 2a\cos\theta$$

are circles with radius $|a|$ centered at the points with polar coordinates $(a, \pi/2)$ and $(a, 0)$, respectively.

EXAMPLE 7 ■ **Sketching the Graph of a Polar Equation**

(a) Sketch the graph of $r = 2 + 2\cos\theta$.
(b) Convert the equation of part (a) to rectangular coordinates.

SOLUTION

(a) Instead of plotting points as in Example 6, we sketch the graph of $r = 2 + 2\cos\theta$ in *rectangular* coordinates in Figure 10. We can think of this graph as a table of values that enables us to read at a glance the values of r that correspond to increasing values of θ. For instance, we see that as θ increases from 0 to $\pi/2$, r (the distance from O) decreases from 4 to 2, so we sketch the corresponding part of the polar graph in Figure 11(a). As θ increases from $\pi/2$ to π, Figure 10 shows that r decreases from 2 to 0, so we sketch the next part of the graph as in Figure 11(b). As θ increases from π to $3\pi/2$, r increases from 0 to 2, as shown in part (c). Finally, as θ increases from $3\pi/2$ to 2π, r increases from 2 to 4, as shown in part (d). If we let θ increase beyond 2π or decrease beyond 0, we would simply retrace our path. Combining the portions of the graph from parts (a) through (d) of Figure 11, we sketch the complete graph in part (e).

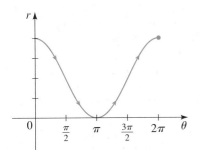

FIGURE 10
$r = 2 + 2\cos\theta$

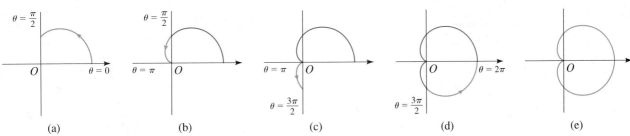

(a) (b) (c) (d) (e)

FIGURE 11 Steps in sketching $r = 2 + 2\cos\theta$

(b) We first multiply both sides of the equation by r:

$$r^2 = 2r + 2r\cos\theta$$

Using $r^2 = x^2 + y^2$ and $y = r\cos\theta$, we can convert two of the three terms in the equation into rectangular coordinates, but eliminating the remaining r requires more work:

$$x^2 + y^2 = 2r + 2y$$
$$x^2 + y^2 - 2y = 2r$$
$$(x^2 + y^2 - 2y)^2 = 4r^2 \qquad \text{Square both sides}$$
$$(x^2 + y^2 - 2y)^2 = 4(x^2 + y^2) \qquad \text{Since } r^2 = x^2 + y^2$$

In this case the rectangular equation is much more complicated than the polar equation, so the polar form is the more useful one. ■

The curve in Figure 11 is called a **cardioid** because it is heart-shaped. In general, the graph of any equation of the form

$$r = a(1 \pm \cos\theta) \qquad \text{or} \qquad r = a(1 \pm \sin\theta)$$

is a cardioid.

EXAMPLE 8 ■ Sketching the Graph of a Polar Equation

Sketch the curve $r = \cos 2\theta$.

SOLUTION   As in Example 7, we first sketch the graph of $r = \cos 2\theta$ in *rectangular* coordinates, as shown in Figure 12. As θ increases from 0 to $\pi/4$, Figure 12 shows that r decreases from 1 to 0 and so we draw the corresponding portion of the polar curve in Figure 13 (indicated by ①). As θ increases from $\pi/4$ to $\pi/2$, the value of r goes from 0 to -1. This means that the distance from the origin increases from 0 to 1, but instead of being in quadrant I, this portion of the polar curve (indicated by ②) lies on the opposite side of the origin in quadrant III. The remainder of the curve is drawn in a similar fashion, with the arrows and numbers indicating the order in which the portions are traced out. The resulting curve has four petals and is called a **four-leaved rose**.

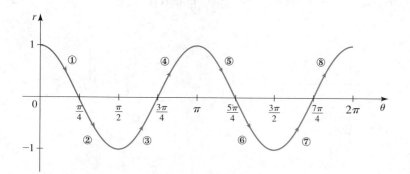

FIGURE 12

Graph of $r = \cos 2\theta$ sketched in rectangular coordinates

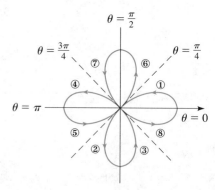

FIGURE 13

Four-leaved rose $r = \cos 2\theta$ sketched in polar coordinates   ■

In general, the graph of an equation of the form

$$r = a \cos n\theta \qquad \text{or} \qquad r = a \sin n\theta$$

is an **n-leaved rose** if n is odd or a $2n$-leaved rose if n is even (as in Example 8).

When sketching the graph of a polar equation, it's often helpful to take advantage of symmetry. We list three tests for symmetry; Figure 14 shows why these tests work.

TESTS FOR SYMMETRY

1. If a polar equation is unchanged when we replace θ by $-\theta$, then the graph is symmetric about the polar axis [Figure 14(a)].

2. If the equation is unchanged when we replace r by $-r$, then the graph is symmetric about the pole [Figure 14(b)].

3. If the equation is unchanged when we replace θ by $\pi - \theta$, the graph is symmetric about the vertical line $\theta = \pi/2$ (the y-axis) [Figure 14(c)].

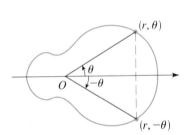

(a) Symmetry about the polar axis

FIGURE 14

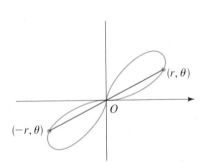

(b) Symmetry about the pole

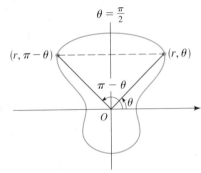

(c) Symmetry about the line $\theta = \frac{\pi}{2}$

The graphs in Figures 7, 11(e), and 13 are symmetric about the polar axis. The graph in Figure 13 is also symmetric about the pole. Figures 9 and 13 show graphs that are symmetric about $\theta = \pi/2$. Note that the four-leaved rose in Figure 13 meets all three tests for symmetry.

 GRAPHING POLAR EQUATIONS WITH GRAPHING DEVICES

Although it's useful to be able to sketch simple polar graphs by hand, we need a graphing calculator or computer when we are faced with a graph as complicated as the one in Figure 15. Fortunately, most graphing calculators are capable of graphing polar equations directly.

EXAMPLE 9 ■ Drawing the Graph of a Polar Equation

Graph the equation $r = \cos(2\theta/3)$.

SOLUTION We need to determine the domain for θ. So we ask ourselves: How many complete rotations are required before the graph starts to repeat

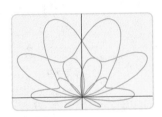

FIGURE 15
$r = \sin\theta + \sin^3(5\theta/2)$

itself? The graph repeats itself when the same value of r is obtained at θ and $\theta + 2n\pi$. Thus, we need to find an integer n, so that

$$\cos \frac{2(\theta + 2n\pi)}{3} = \cos \frac{2\theta}{3}$$

For this equality to hold, $4n\pi/3$ must be a multiple of 2π, and this first happens when $n = 3$. Therefore, we obtain the entire graph if we choose values of θ between $\theta = 0$ and $\theta = 0 + 2(3)\pi = 6\pi$. The graph is shown in Figure 16.

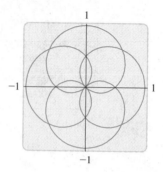

FIGURE 16

$r = \cos(2\theta/3)$

EXAMPLE 10 ■ A Family of Polar Equations

Graph the family of polar equations $r = 1 + c \sin \theta$ for $c = 3, 2.5, 2, 1.5, 1$. How does the shape of the graph change as c changes? (These curves are called **limaçons**, after the French word for snail, because of the shape of the curve for certain values of c.)

SOLUTION   Figure 17 shows computer-drawn graphs for the given values of c. For $c > 1$, the graph has an inner loop; the loop decreases in size as c decreases. When $c = 1$, the loop disappears and the graph becomes a cardioid (see Example 7).

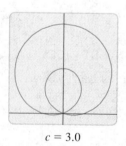 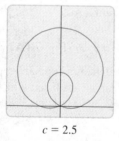 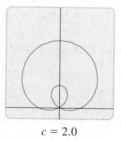 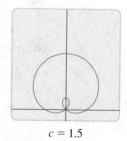 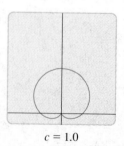

$c = 3.0$ $c = 2.5$ $c = 2.0$ $c = 1.5$ $c = 1.0$

FIGURE 17   A family of limaçons $r = 1 + c \sin \theta$ in the viewing rectangle $[-2.5, 2.5]$ by $[-0.5, 4.5]$

The following box gives a summary of some of the basic polar graphs used in calculus.

SOME COMMON POLAR CURVES

Circles and Spiral

$r = a$
circle

$r = a \sin \theta$
circle

$r = a \cos \theta$
circle

$r = a\theta$
spiral

Limaçons

$r = a \pm b \sin \theta$

$r = a \pm b \cos \theta$

$(a > 0, b > 0)$

Orientation depends on the trigonometric function (sine or cosine) and the sign of b.

$a < b$
limaçon with inner loop

$a = b$
cardioid

$a > b$
dimpled limaçon

$a \geq 2b$
convex limaçon

Roses

$r = a \sin n\theta$

$r = a \cos n\theta$

n-leaved if n is odd

$2n$-leaved if n is even

$r = a \cos 2\theta$
4-leaved rose

$r = a \cos 3\theta$
3-leaved rose

$r = a \cos 4\theta$
8-leaved rose

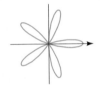

$r = a \cos 5\theta$
5-leaved rose

Lemniscates

Figure-eight-shaped curves

$r^2 = a^2 \sin 2\theta$
lemniscate

$r^2 = a^2 \cos 2\theta$
lemniscate

9.6 EXERCISES

1–6 ■ Plot the point that has the given polar coordinates. Then give two other polar coordinate representations of the point, one with $r < 0$ and the other with $r > 0$.

1. $(3, \pi/2)$ **2.** $(2, 3\pi/4)$ **3.** $(-1, 7\pi/6)$

4. $(-2, -\pi/3)$ **5.** $(-5, 0)$ **6.** $(3, 1)$

7–12 ■ Find the rectangular coordinates for the point whose polar coordinates are given.

7. $(4, \pi/6)$ **8.** $(6, 2\pi/3)$ **9.** $(\sqrt{2}, -\pi/4)$

10. $(-1, 5\pi/2)$ **11.** $(5, 5\pi)$ **12.** $(0, 13\pi)$

13–18 ■ Convert the rectangular coordinates to polar coordinates with $r > 0$ and $0 \leqslant \theta < 2\pi$.

13. $(-1, 1)$ **14.** $(3\sqrt{3}, -3)$ **15.** $(\sqrt{8}, \sqrt{8})$

16. $(-\sqrt{6}, -\sqrt{2})$ **17.** $(3, 4)$ **18.** $(1, -2)$

19–24 ■ Convert the equation to polar form.

19. $x = y$ **20.** $x^2 + y^2 = 9$ **21.** $y = x^2$

22. $y = 5$ **23.** $x = 4$ **24.** $x^2 - y^2 = 1$

25–38 ■ Convert the polar equation to rectangular coordinates.

25. $r = 7$ **26.** $\theta = \pi$

27. $r\cos\theta = 6$ **28.** $r = 6\cos\theta$

29. $r^2 = \tan\theta$ **30.** $r^2 = \sin 2\theta$

31. $r = \dfrac{1}{\sin\theta - \cos\theta}$ **32.** $r = \dfrac{1}{1 + \sin\theta}$

33. $r = 1 + \cos\theta$ **34.** $r = \dfrac{4}{1 + 2\sin\theta}$

35. $r = 2\sec\theta$ **36.** $r = 2 - \cos\theta$

37. $\sec\theta = 2$ **38.** $\cos 2\theta = 1$

39–60 ■ Sketch the graph of the polar equation.

39. $r = 3$ **40.** $r = -1$

41. $\theta = -\pi/2$ **42.** $\theta = 5\pi/6$

43. $r = 6\sin\theta$ **44.** $r = \cos\theta$

45. $r = -2\cos\theta$ **46.** $r = 2\sin\theta + 2\cos\theta$

47. $r = 2 - 2\cos\theta$ **48.** $r = 1 + \sin\theta$

49. $r = -3(1 + \sin\theta)$ **50.** $r = \cos\theta - 1$

51. $r = \theta, \quad \theta \geqslant 0$ (spiral)

52. $r\theta = 1, \quad \theta > 0$ (reciprocal spiral)

53. $r = \sin 2\theta$ (four-leaved rose)

54. $r = 2\cos 3\theta$ (three-leaved rose)

55. $r^2 = \cos 2\theta$ (lemniscate)

56. $r^2 = 4\sin 2\theta$ (lemniscate)

57. $r = 2 + \sin\theta$ (limaçon)

58. $r = 1 - 2\cos\theta$ (limaçon)

59. $r = 2 + \sec\theta$ (conchoid)

60. $r = \sin\theta\tan\theta$ (cissoid)

61–64 ■ Use a graphing device to graph the polar equation. Choose the domain of θ to make sure you produce the entire graph.

61. $r = \cos(\theta/2)$

62. $r = \sin(8\theta/5)$

63. $r = 1 + 2\sin(\theta/2)$ (nephroid)

64. $r = \sqrt{1 - 0.8\sin^2\theta}$ (hippopede)

65. Graph the family of polar equations $r = 1 + \sin n\theta$ for $n = 1, 2, 3, 4,$ and 5. How is the number of loops related to n?

66. Graph the family of polar equations $r = 1 + c\sin 2\theta$ for $c = 0.3, 0.6, 1, 1.5,$ and 2. How does the graph change as c increases?

67–70 ■ Match the polar equation with the graphs labeled I–IV. Give reasons for your answers.

67. $r = \sin(\theta/2)$ **68.** $r = 1/\sqrt{\theta}$

69. $r = \theta\sin\theta$ **70.** $r = 1 + 3\cos(3\theta)$

I

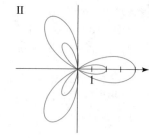

II

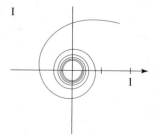

III

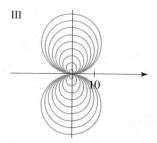

IV

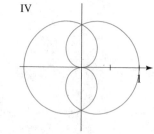

71. (a) Show that the distance between the points whose polar coordinates are (r_1, θ_1) and (r_2, θ_2) is

$$\sqrt{r_1^2 + r_2^2 - 2r_1 r_2\cos(\theta_2 - \theta_1)}$$

[*Hint:* Use the Law of Cosines.]

(b) Find the distance between the points whose polar coordinates are $(3, 3\pi/4)$ and $(-1, 7\pi/6)$.

72. Show that the graph of $r = a\cos\theta + b\sin\theta$ is a circle, and find its center and radius.

 DISCOVERY · DISCUSSION

73. **A Transformation of Polar Graphs** How are the graphs of $r = 1 + \sin(\theta - \pi/6)$ and $r = 1 + \sin(\theta - \pi/3)$ related to the graph of $r = 1 + \sin\theta$? In general, how is the graph of $r = f(\theta - \alpha)$ related to the graph of $r = f(\theta)$?

74. **Choosing a Convenient Coordinate System** Compare the polar equation of the circle $r = 2$ with its equation in rectangular coordinates. In which coordinate system is the equation simpler? Do the same for the equation of the four-leaved rose $r = \sin 2\theta$. Which coordinate system would you choose to use to study these curves?

75. **Choosing a Convenient Coordinate System** Compare the rectangular equation of the line $y = 2$ with its polar equation. In which coordinate system is the equation simpler? Which coordinate system would you choose to study lines?

9.7   POLAR EQUATIONS OF CONICS

Earlier in this chapter we defined a parabola in terms of a focus and directrix, but we defined the ellipse and hyperbola in terms of two foci. In this section we give a more unified treatment of all three types of conics in terms of a focus and directrix. If we place the focus at the origin, then a conic section has a simple polar equation. Moreover, in polar form, rotation of conics becomes a simple matter. Polar equations of ellipses are crucial in the derivation of Kepler's laws of motion (see page 624).

> **EQUIVALENT DESCRIPTION OF CONICS**
>
> Let F be a fixed point (the **focus**), ℓ a fixed line (the **directrix**), and e a fixed positive number (the **eccentricity**). The set of all points P such that the ratio of the distance from P to F to the distance from P to ℓ is the constant e is a conic. That is, the set of all points P such that
>
> $$\frac{d(P, F)}{d(P, \ell)} = e$$
>
> is a conic. The conic is a parabola if $e = 1$, an ellipse if $e < 1$, or a hyperbola if $e > 1$.

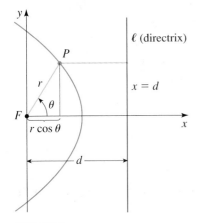

FIGURE 1

■ **Proof** If $e = 1$, then $d(P, F) = d(P, \ell)$, and so the given condition becomes the definition of a parabola as given in Section 9.1.

Now, suppose $e \neq 1$. Let's place the focus F at the origin and the directrix parallel to the y-axis and d units to the right. Thus, the directrix has equation $x = d$ and is perpendicular to the polar axis. If the point P has polar coordinates (r, θ), we see from Figure 1 that $d(P, F) = r$ and $d(P, \ell) = d - r\cos\theta$. Thus, the condition $d(P, F)/d(P, \ell) = e$, or $d(P, F) = e \cdot d(P, \ell)$, becomes

$$r = e(d - r\cos\theta)$$

If we square both sides of this polar equation and convert to rectangular coordinates, we get

$$x^2 + y^2 = e^2(d - x)^2$$

$$(1 - e^2)x^2 + 2de^2x + y^2 = e^2d^2 \qquad \text{Expand and simplify}$$

$$\left(x + \frac{e^2d}{1 - e^2}\right)^2 + \frac{y^2}{1 - e^2} = \frac{e^2d^2}{(1 - e^2)^2} \qquad \begin{array}{l}\text{Divide by } 1 - e^2 \text{ and} \\ \text{complete the square}\end{array}$$

If $e < 1$, then dividing both sides of this equation by $e^2d^2/(1 - e^2)^2$ gives an equation of the form

$$\frac{(x - h)^2}{a^2} + \frac{y^2}{b^2} = 1$$

where

$$h = \frac{-e^2d}{1 - e^2} \qquad a^2 = \frac{e^2d^2}{(1 - e^2)^2} \qquad b^2 = \frac{e^2d^2}{1 - e^2}$$

This is the equation of an ellipse with center $(h, 0)$. In Section 9.2 we found that the foci of an ellipse are a distance c from the center, where $c^2 = a^2 - b^2$. In our case,

$$c^2 = a^2 - b^2 = \frac{e^4d^2}{(1 - e^2)^2}$$

Thus, $c = e^2d/(1 - e^2) = -h$, which confirms that the focus defined in the theorem is the same as the focus defined in Section 9.2. It also follows that $e = c/a$.

If $e > 1$, a similar proof shows that the conic is a hyperbola with $e = c/a$, where $c^2 = a^2 + b^2$. $\qquad\qquad\qquad\qquad\qquad\qquad\qquad\qquad\qquad\qquad\qquad\quad\square$

In the proof we saw that the polar equation of the conic in Figure 1 is $r = e(d - r\cos\theta)$. Solving for r, we get

$$r = \frac{ed}{1 + e\cos\theta}$$

If the directrix is chosen to be to the *left* of the focus ($x = -d$), then we get the equation $r = ed/(1 - e\cos\theta)$. If the directrix is *parallel* to the polar axis ($y = d$ or $y = -d$), then we get $\sin\theta$ instead of $\cos\theta$ in the equation. These observations are summarized in the following box and in Figure 2.

POLAR EQUATIONS OF CONICS

A polar equation of the form

$$r = \frac{ed}{1 \pm e \cos \theta} \quad \text{or} \quad r = \frac{ed}{1 \pm e \sin \theta}$$

represents a conic with eccentricity e. The conic is

1. a parabola if $e = 1$.

2. an ellipse if $e < 1$.

3. a hyperbola if $e > 1$.

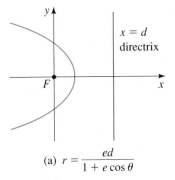

(a) $r = \dfrac{ed}{1 + e \cos \theta}$

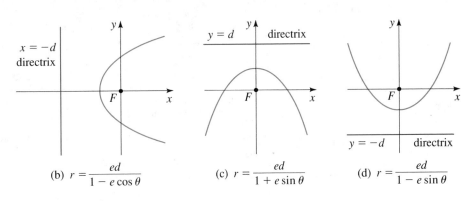

(b) $r = \dfrac{ed}{1 - e \cos \theta}$

(c) $r = \dfrac{ed}{1 + e \sin \theta}$

(d) $r = \dfrac{ed}{1 - e \sin \theta}$

FIGURE 2

EXAMPLE 1 ■ **Finding a Polar Equation for a Conic**

Find a polar equation for the parabola that has its focus at the origin and whose directrix is the line $y = -6$.

SOLUTION   Using $e = 1$ and $d = 6$, and using part (d) of Figure 2, we see that the polar equation of the parabola is

$$r = \frac{6}{1 - \sin \theta}$$ ■

EXAMPLE 2 ■ **Identifying and Sketching a Conic**

A conic is given by the polar equation

$$r = \frac{10}{3 - 2 \cos \theta}$$

Identify the conic and sketch its graph.

SOLUTION   Dividing the numerator and denominator by 3, we write the equation as

$$r = \frac{\frac{10}{3}}{1 - \frac{2}{3} \cos \theta}$$

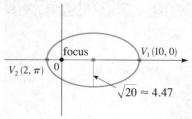

FIGURE 3

$$r = \frac{10}{3 - 2\cos\theta}$$

We see that the equation represents an ellipse with $e = \frac{2}{3}$ and with major axis parallel to the polar axis (because of the $\cos\theta$ in the equation). To find the endpoints of the major axis, we let $\theta = 0$ and $\theta = \pi$ in the equation. This gives the points $V_1(10, 0)$ and $V_2(2, \pi)$. The distance between these two points is 12, thus, $2a = 12$ and so $a = 6$. The center of the ellipse is at $C(4, 0)$, the midpoint of $V_1 V_2$. To sketch the graph, we need to find b. We have $c = ae = 6\left(\frac{2}{3}\right) = 4$, so

$$b^2 = a^2 - c^2 = 6^2 - 4^2 = 20$$

and so $b = \sqrt{20} \approx 4.47$. With this information, we can sketch the graph in Figure 3. ∎

EXAMPLE 3 ■ Identifying and Sketching a Conic

A conic is given by the polar equation

$$r = \frac{12}{2 + 4\sin\theta}$$

Identify the conic and sketch its graph.

SOLUTION Dividing numerator and denominator by 2, we write the equation as

$$r = \frac{6}{1 + 2\sin\theta}$$

We see that the equation represents a hyperbola with $e = 2$ and with transverse axis perpendicular to the polar axis (because of the $\sin\theta$ in the equation). The vertices occur when $\theta = \pi/2$ and $\theta = 3\pi/2$, so they are $V_1(2, \pi/2)$ and $V_2(-6, 3\pi/2) = V_2(6, \pi/2)$. The x-intercepts occur when $\theta = 0$ and $\theta = \pi$; they are $(6, 0)$ and $(6, \pi)$. The x-intercepts help us sketch the lower branch of the hyperbola (Figure 4).

The distance between the two vertices is 4; thus, $2a = 4$ and so $a = 2$. To find b, we first find $c = ae = 2 \cdot 2 = 4$, so

$$b^2 = c^2 - a^2 = 4^2 - 2^2 = 12$$

and so $b = \sqrt{12} \approx 3.46$. Knowing a and b helps us sketch the asymptotes and complete the graph in Figure 4. ∎

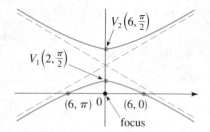

FIGURE 4

$$r = \frac{12}{2 + 4\sin\theta}$$

When we are rotating conic sections, it is much more convenient to use polar equations than Cartesian equations. We simply use the fact that the graph of $r = f(\theta - \alpha)$ is the graph of $r = f(\theta)$ rotated counterclockwise about the origin through an angle α (see Exercise 73 in Section 9.6).

 EXAMPLE 4 ■ Rotating an Ellipse

Suppose the ellipse of Example 2 is rotated through an angle $\pi/4$ about the origin. Find a polar equation for the resulting ellipse, and draw its graph.

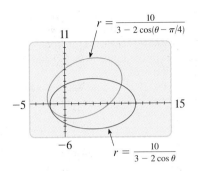

FIGURE 5

SOLUTION We get the equation of the rotated ellipse by replacing θ with $\theta - \pi/4$ in the equation given in Example 2. So the new equation is

$$r = \frac{10}{3 - 2\cos(\theta - \pi/4)}$$

We use this equation to graph the rotated ellipse in Figure 5. Notice that the ellipse has been rotated about its left focus. ∎

In Figure 6 we use a computer to sketch a number of conics to demonstrate the effect of varying the eccentricity e. Notice that when e is close to 0, the ellipse is nearly circular and becomes more elongated as e increases. When $e = 1$, of course, the conic is a parabola. As e increases beyond 1, the conic is an ever steeper hyperbola.

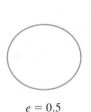

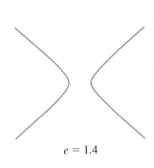

$e = 0.5$ $e = 0.86$ $e = 1$ $e = 1.4$ $e = 4$

FIGURE 6

9.7 EXERCISES

1–8 ■ Write a polar equation of a conic that has its focus at the origin and satisfies the given conditions.

1. Ellipse, eccentricity $\frac{2}{3}$, directrix $x = 3$

2. Hyperbola, eccentricity $\frac{4}{3}$, directrix $x = -3$

3. Parabola, directrix $y = 2$

4. Ellipse, eccentricity $\frac{1}{2}$, directrix $y = -4$

5. Hyperbola, eccentricity 4, directrix $r = 5 \sec \theta$

6. Ellipse, eccentricity 0.6, directrix $r = 2 \csc \theta$

7. Parabola, vertex at $(5, \pi/2)$

8. Ellipse, eccentricity 0.4, vertex at $(2, 0)$

9–16 ■ (a) Find the eccentricity and identify the conic.
(b) Sketch the conic and label the vertices.

9. $r = \dfrac{4}{1 + 3\cos\theta}$

10. $r = \dfrac{8}{3 + 3\cos\theta}$

11. $r = \dfrac{2}{1 - \cos\theta}$

12. $r = \dfrac{10}{3 - 2\sin\theta}$

13. $r = \dfrac{6}{2 + \sin\theta}$

14. $r = \dfrac{5}{2 - 3\sin\theta}$

15. $r = \dfrac{7}{2 - 5\sin\theta}$

16. $r = \dfrac{8}{3 + \cos\theta}$

17. (a) Find the eccentricity and directrix of the conic $r = 1/(4 - 3\cos\theta)$ and graph the conic and its directrix.
(b) If this conic is rotated about the origin through an angle $\pi/3$, write the resulting equation and draw its graph.

18. Graph the parabola $r = 5/(2 + 2\sin\theta)$ and its directrix. Also graph the curve obtained by rotating this parabola about its focus through an angle $\pi/6$.

19. Graph the conics $r = e/(1 - e\cos\theta)$ with $e = 0.4, 0.6$, 0.8, and 1.0 on a common screen. How does the value of e affect the shape of the curve?

20. (a) Graph the conics $r = ed/(1 + e\sin\theta)$ for $e = 1$ and various values of d. How does the value of d affect the shape of the conic?

   (b) Graph these conics for $d = 1$ and various values of e. How does the value of e affect the shape of the conic?

21. (a) Show that the polar equation of an ellipse with directrix $x = -d$ can be written in the form

$$r = \frac{a(1 - e^2)}{1 - e\cos\theta}$$

   (b) Find an approximate polar equation for the elliptical orbit of the earth around the sun (at one focus) given that the eccentricity is about 0.017 and the length of the major axis is about 2.99×10^8 km.

22. (a) The planets move around the sun in elliptical orbits with the sun at one focus. The positions of a planet that are closest to, and farthest from, the sun are called its **perihelion** and **aphelion**, respectively. Use Exercise 21(a) to show that the perihelion distance from a planet to the sun is $a(1 - e)$ and the aphelion distance is $a(1 + e)$.

   (b) Use the data of Exercise 21(b) to find the distances from the earth to the sun at perihelion and at aphelion.

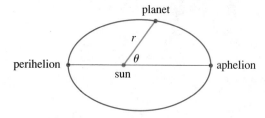

23. The distance from the planet Pluto to the sun is 4.43×10^9 km at perihelion and 7.37×10^9 km at aphelion. Use Exercise 22 to find the eccentricity of Pluto's orbit.

◢ DISCOVERY · DISCUSSION

24. **Distance to a Focus** When we found polar equations for the conics, we placed one focus at the pole. It's easy to find the distance from that focus to any point on the conic. Explain how the polar equation gives us this distance.

25. **Polar Equations of Orbits** When a satellite orbits the earth, its path is an ellipse with one focus at the center of the earth. Why do scientists use polar (rather than rectangular) coordinates to track the position of satellites? Your answer to Exercise 24 is relevant here.

9.8   PARAMETRIC EQUATIONS

So far we've described a curve by giving an equation that the coordinates of all points on the curve must satisfy. For example, we know that the equation $y = x^2$ represents a parabola in rectangular coordinates and that $r = \sin\theta$ represents a circle in polar coordinates. We now study another method for describing a curve in the plane, which in many situations turns out to be more useful and natural than either rectangular or polar equations. In this method, the x- and y-coordinates of points on the curve are given separately as functions of an additional variable t, called the **parameter**:

$$x = f(t) \qquad y = g(t)$$

These are called **parametric equations** for the curve. Substituting a value of t into each equation determines the coordinates of a point (x, y). As t varies, the point $(x, y) = (f(t), g(t))$ varies and traces out the curve. If we think of t as representing time, then as t increases, we can imagine a particle at $(x, y) = (f(t), g(t))$ moving along the curve.

EXAMPLE 1 ■ Sketching a Parametric Curve

Sketch the curve defined by the parametric equations

$$x = t^2 - 3t \qquad y = t - 1$$

Eliminate the parameter t to obtain a single equation for the curve in the variables x and y.

SOLUTION   For every value of t, we get a point on the curve. For example, if $t = 0$, then $x = 0$ and $y = -1$, so the corresponding point is $(0, -1)$. In Figure 1 we plot the points (x, y) determined by the values of t shown in the following table.

t	x	y
-2	10	-3
-1	4	-2
0	0	-1
1	-2	0
2	-2	1
3	0	2
4	4	3
5	10	4

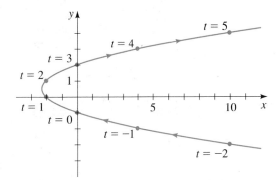

FIGURE 1

   As t increases, a particle whose position is given by the parametric equations moves along the curve in the direction of the arrows. The curve seems to be a parabola. We can confirm this by eliminating the parameter t from the parametric equations and reducing them to a single equation as follows. First we solve for t in the second equation to get $t = y + 1$. Substituting this into the first equation, we get

$$x = (y + 1)^2 - 3(y + 1) = y^2 - y - 2$$

The curve is the parabola $x = y^2 - y - 2$.   ■

   Notice that we would obtain the same graph as in Example 1 from the parametrization

$$x = t^2 - t - 2 \qquad y = t$$

because the points on this curve also satisfy the equation $x = y^2 - y - 2$. But the same value of t produces different points on the curve in these two parametrizations. For example, when $t = 0$, the particle that traces out the curve in Figure 1 is at $(0, -1)$, whereas in the parametrization of the preceding equations, the particle is already at $(-2, 0)$ when $t = 0$. Thus, *a parametrization contains more information than just the curve being parametrized; it also indicates how the curve is being traced out.*

EXAMPLE 2 ■ Eliminating the Parameter

Describe and graph the curve represented by the parametric equations

$$x = \cos t \qquad y = \sin t \qquad 0 \leq t \leq 2\pi$$

SOLUTION To identify the curve, we eliminate the parameter. Since $\cos^2 t + \sin^2 t = 1$ and since $x = \cos t$ and $y = \sin t$ for every point (x, y) on the curve, we have

$$x^2 + y^2 = (\cos t)^2 + (\sin t)^2 = 1$$

This means that all points on the curve satisfy the equation $x^2 + y^2 = 1$, so the graph is a circle of radius 1 centered at the origin. As t increases from 0 to 2π, the point given by the parametric equations starts at $(1, 0)$ and moves counterclockwise once around the circle, as shown in Figure 2. Notice that the parameter t can be interpreted as the angle shown in the figure. ■

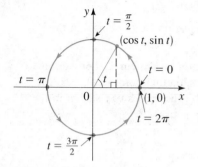

FIGURE 2

EXAMPLE 3 ■ Finding Parametric Equations for a Graph

Find parametric equations for the line of slope 3 that passes through the point $(2, 6)$.

SOLUTION Let's start at the point $(2, 6)$ and move up and to the right along this line. Because the line has slope 3, for every 1 unit we move to the right, we must move up 3 units. In other words, if we increase the x-coordinate by t units, we must correspondingly increase the y-coordinate by $3t$ units. This leads to the parametric equations

$$x = 2 + t \qquad y = 6 + 3t$$

To confirm that these equations give the desired line, we eliminate the parameter. We solve for t in the first equation and substitute into the second to get

$$y = 6 + 3(x - 2) = 3x$$

Thus, the slope-intercept form of the equation of this line is $y = 3x$, which is a line of slope 3 that does pass through $(2, 6)$ as required. The graph is shown in Figure 3. ■

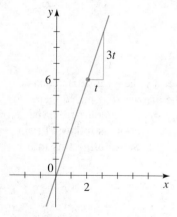

FIGURE 3

EXAMPLE 4 ■ Sketching a Parametric Curve

Sketch the curve with parametric equations

$$x = \sin t \qquad y = 2 - \cos^2 t$$

SOLUTION To eliminate the parameter, we first use the trigonometric identity $\cos^2 t = 1 - \sin^2 t$ to change the second equation:

$$y = 2 - \cos^2 t = 2 - (1 - \sin^2 t) = 1 + \sin^2 t$$

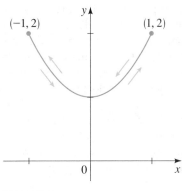

FIGURE 4

Now we can substitute $\sin t = x$ from the first equation to get

$$y = 1 + x^2$$

and so the point (x, y) moves along the parabola $y = 1 + x^2$. However, since $-1 \le \sin t \le 1$, we have $-1 \le x \le 1$, so the parametric equations represent only the part of the parabola between $x = -1$ and $x = 1$. Since $\sin t$ is periodic, the point $(x, y) = (\sin t, 2 - \cos^2 t)$ moves back and forth infinitely often along the parabola between the points $(-1, 2)$ and $(1, 2)$ as shown in Figure 4. ∎

EXAMPLE 5 ■ Parametric Equations for the Cycloid

As a circle rolls along a straight line, the curve traced out by a fixed point P on the circumference of the circle is called a **cycloid** (see Figure 5). If the circle has radius a and rolls along the x-axis, with one position of the point P being at the origin, find parametric equations for the cycloid.

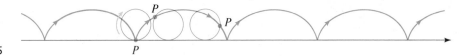

FIGURE 5

SOLUTION   Figure 6 shows the circle and the point P after the circle has rolled through an angle θ (in radians). The distance $d(O, T)$ that the circle has rolled must be the same as the length of the arc PT, which, by the arc length formula, is $a\theta$ (see Section 6.1). This means that the center of the circle is $C(a\theta, a)$.

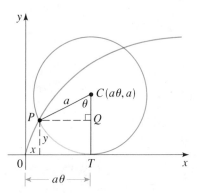

FIGURE 6

Let the coordinates of P be (x, y). Then from Figure 6 (which illustrates the case $0 < \theta < \pi/2$), we see that

$$x = d(O, T) - d(P, Q) = a\theta - a \sin \theta = a(\theta - \sin \theta)$$
$$y = d(T, C) - d(Q, C) = a - a \cos \theta = a(1 - \cos \theta)$$

so parametric equations for the cycloid are

$$x = a(\theta - \sin \theta) \qquad y = a(1 - \cos \theta)$$

∎

The cycloid has a number of interesting physical properties. It is the "curve of quickest descent" in the following sense. Let's choose two points P and Q that are not directly above each other, and join them with a wire. Suppose we allow a bead to slide down the wire under the influence of gravity (ignoring friction). Of all possible shapes that the wire can be bent into, the bead will slide from P to Q the fastest when the shape is half of an arch of an inverted cycloid (see Figure 7). The cycloid is also the "curve of equal descent" in the sense that no matter where we place a bead B on a cycloid-shaped wire, it takes the same time to slide to the bottom (see Figure 8). These rather surprising properties of the cycloid were proved (using calculus) in the 17th century by several mathematicians and physicists, including Johann Bernoulli, Blaise Pascal, and Christiaan Huygens.

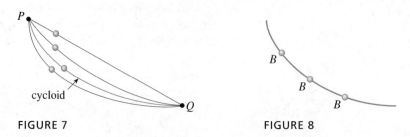

FIGURE 7   FIGURE 8

 ## USING GRAPHING DEVICES TO GRAPH PARAMETRIC CURVES

Most graphing calculators and computer graphing programs can be used to graph parametric equations. Such devices are particularly useful when sketching complicated curves like the one shown in Figure 9.

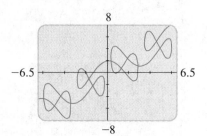

FIGURE 9
$x = t + 2\sin 2t,\ y = t + 2\cos 5t$

EXAMPLE 6 ■ Graphing Parametric Curves

Use a graphing device to draw the following parametric curves. Discuss their similarities and differences.

(a) $x = \sin 2t$
    $y = 2\cos t$

(b) $x = \sin 3t$
    $y = 2\cos t$

SOLUTION   In both parts (a) and (b), the graph will lie inside the rectangle given by $-1 \leqslant x \leqslant 1$, $-2 \leqslant y \leqslant 2$, since both the sine and the cosine of any number will be between -1 and 1. Thus, we may use the viewing rectangle $[-1.5, 1.5]$ by $[-2.5, 2.5]$.

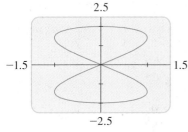

(a) $x = \sin 2t$, $y = 2 \cos t$

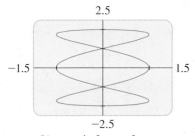

(b) $x = \sin 3t$, $y = 2 \cos t$

FIGURE 10

(a) Since $2 \cos t$ is periodic with period 2π (see Section 5.3), and since $\sin 2t$ has period π, letting t vary over the interval $0 \leqslant t \leqslant 2\pi$ gives us the complete graph, which is shown in Figure 10(a).

(b) Again, letting t take on values between 0 and 2π gives the complete graph shown in Figure 10(b).

Both graphs are *closed curves*, which means they form loops with the same starting and ending point; also, both graphs cross over themselves. However, the graph in Figure 10(a) has two loops, like a figure eight, whereas the graph in Figure 10(b) has three loops. ∎

The curves graphed in Example 6 are called Lissajous figures. A **Lissajous figure** is the graph of a pair of parametric equations of the form

$$x = A \sin \omega_1 t \qquad y = B \cos \omega_2 t$$

where A, B, ω_1, and ω_2 are real constants. Since $\sin \omega_1 t$ and $\cos \omega_2 t$ are both between -1 and 1, a Lissajous figure will lie inside the rectangle determined by $-A \leqslant x \leqslant A$, $-B \leqslant y \leqslant B$. This fact can be used to choose a viewing rectangle when graphing a Lissajous figure, as in Example 6.

Recall from Section 9.6 that rectangular coordinates (x, y) and polar coordinates (r, θ) are related by the equations $x = r \cos \theta$, $y = r \sin \theta$. Thus, we can graph the polar equation $r = f(\theta)$ by changing it to parametric form as follows:

$$x = r \cos \theta = f(\theta) \cos \theta$$
$$y = r \sin \theta = f(\theta) \sin \theta \qquad \text{Since } r = f(\theta)$$

Replacing θ by the standard parametric variable t, we have the following result.

POLAR EQUATIONS IN PARAMETRIC FORM

The graph of the polar equation $r = f(\theta)$ is the same as the graph of the parametric equations

$$x = f(t) \cos t \qquad y = f(t) \sin t$$

EXAMPLE 7 ■ Parametric Form of a Polar Equation

Consider the polar equation $r = \ln \theta$, $1 \leqslant \theta \leqslant 10\pi$.

(a) Express the equation in parametric form.
(b) Draw a graph of the parametric equations from part (a).

SOLUTION

(a) The given polar equation is equivalent to the parametric equation

$$x = \ln t \cos t \qquad y = \ln t \sin t$$

(b) Since $\ln 10\pi \approx 3.45$, we use the viewing rectangle $[-3.5, 3.5]$ by

$[-3.5, 3.5]$, and we let t vary from 1 to $10\pi \approx 31.42$. The resulting graph shown in Figure 11 is a *logarithmic spiral*.

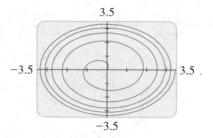

3.5

−3.5   3.5 .

−3.5

FIGURE 11

$x = \ln t \cos t, \; y = \ln t \sin t$

| 9.8 | EXERCISES |

1–22 ■ (a) Sketch the curve represented by the parametric equations. (b) Find a rectangular-coordinate equation for the curve by eliminating the parameter.

1. $x = 2t, \quad y = t + 6$

2. $x = 6t - 4, \quad y = 3t, \quad t \geq 0$

3. $x = t^2, \quad y = t - 2, \quad 2 \leq t \leq 4$

4. $x = 2t + 1, \quad y = \left(t + \frac{1}{2}\right)^2$

5. $x = \sqrt{t}, \quad y = 1 - t$ 　　**6.** $x = t^2, \quad y = t^4 + 1$

7. $x = \dfrac{1}{t}, \quad y = t + 1$ 　　**8.** $x = t + 1, \quad y = \dfrac{t}{t + 1}$

9. $x = 4t^2, \quad y = 8t^3$ 　　**10.** $x = |t|, \quad y = |1 - |t||$

11. $x = 2 \sin t, \quad y = 2 \cos t, \quad 0 \leq t \leq \pi$

12. $x = 2 \cos t, \quad y = 3 \sin t, \quad 0 \leq t \leq 2\pi$

13. $x = \sin^2 t, \quad y = \sin^4 t$ 　　**14.** $x = \sin^2 t, \quad y = \cos t$

15. $x = \cos t, \quad y = \cos 2t$ 　　**16.** $x = \cos 2t, \quad y = \sin 2t$

17. $x = \sec t, \quad y = \tan t, \quad 0 \leq t < \pi/2$

18. $x = \cot t, \quad y = \csc t, \quad 0 < t < \pi$

19. $x = e^t, \quad y = e^{-t}$

20. $x = e^{2t}, \quad y = e^t, \quad t \geq 0$

21. $x = \cos^2 t, \quad y = \sin^2 t$

22. $x = \cos^3 t, \quad y = \sin^3 t, \quad 0 \leq t \leq 2\pi$

23–26 ■ Find parametric equations for the line with the given properties.

23. Slope $\frac{1}{2}$, passing through $(4, -1)$

24. Slope -2, passing through $(-10, -20)$

25. Passing through $(6, 7)$ and $(7, 8)$

26. Passing through $(12, 7)$ and the origin

27. Find parametric equations for the circle $x^2 + y^2 = a^2$.

28. Find parametric equations for the ellipse

$$\frac{x^2}{a^2} + \frac{y^2}{b^2} = 1$$

29. Show by eliminating the parameter θ that the following parametric equations represent a hyperbola:

$$x = a \tan \theta \qquad y = b \sec \theta$$

30. Show that the following parametric equations represent a part of the hyperbola of Exercise 29:

$$x = a \sqrt{t} \qquad y = b \sqrt{t + 1}$$

31–34 ■ Sketch the curve given by the parametric equations.

31. $x = t \cos t, \quad y = t \sin t, \quad t \geq 0$

32. $x = \sin t, \quad y = \sin 2t$

33. $x = \dfrac{3t}{1 + t^3}, \quad y = \dfrac{3t^2}{1 + t^3}$

34. $x = \cot t, \quad y = 2 \sin^2 t, \quad 0 < t < \pi$

35. If a projectile is fired with an initial speed of v_0 ft/s at an angle α above the horizontal, then its position after t seconds is given by the parametric equations

$$x = (v_0 \cos \alpha)t \qquad y = (v_0 \sin \alpha)t - 16t^2$$

(where x and y are measured in feet). Show that the

path of the projectile is a parabola by eliminating the parameter t.

36. Referring to Exercise 35, suppose a gun fires a bullet into the air with an initial speed of 2048 ft/s at an angle of 30° to the horizontal.
 (a) After how many seconds will the bullet hit the ground?
 (b) How far from the gun will the bullet hit the ground?
 (c) What is the maximum height attained by the bullet?

 37–42 ■ Use a graphing device to draw the curve represented by the parametric equations.

37. $x = \sin t, \quad y = 2 \cos 3t$

38. $x = 2 \sin t, \quad y = \cos 4t$

39. $x = 3 \sin 5t, \quad y = 5 \cos 3t$

40. $x = \sin 4t, \quad y = \cos 3t$

41. $x = \sin(\cos t), \quad y = \cos(t^{3/2}), \quad 0 \le t \le 2\pi$

42. $x = 2 \cos t + \cos 2t, \quad y = 2 \sin t - \sin 2t$

 43–46 ■ (a) Express the polar equation in parametric form. (b) Use a graphing device to graph the parametric equations you found in part (a).

43. $r = e^{\theta/12}, \quad 0 \le \theta \le 4\pi$

44. $r = \sin\theta + 2\cos\theta$

45. $r = \dfrac{4}{2 - \cos\theta}$

46. $r = 2^{\sin\theta}$

47–50 ■ Match the parametric equations with the graphs labeled I–IV. Give reasons for your answers.

47. $x = t^3 - 2t, \quad y = t^2 - t$

48. $x = \sin 3t, \quad y = \sin 4t$

49. $x = t + \sin 2t, \quad y = t + \sin 3t$

50. $x = \sin(t + \sin t), \quad y = \cos(t + \cos t)$

I
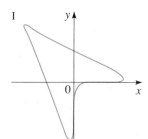

II
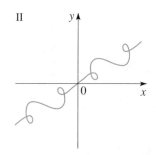

III
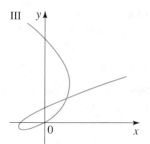

IV
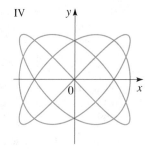

51. In Example 5 suppose the point P that traces out the curve lies not on the edge of the circle, but rather at a fixed point inside the rim, at a distance b from the center (with $b < a$). The curve traced out by P is called a **curtate cycloid** (or **trochoid**). Show that parametric equations for the curtate cycloid are

$$x = a\theta - b\sin\theta \qquad y = a - b\cos\theta$$

Sketch the graph.

52. In Exercise 51 if the point P lies *outside* the circle at a distance b from the center (with $b > a$), then the curve traced out by P is called a **prolate cycloid**. Show that parametric equations for the prolate cycloid are the same as the equations for the curtate cycloid, and sketch the graph for the case where $a = 1$ and $b = 2$.

53. A circle C of radius b rolls on the inside of a larger circle of radius a centered at the origin. Let P be a fixed point on the smaller circle, with initial position at the point $(a, 0)$, as shown in the figure. The curve traced out by P is called a **hypocycloid**.

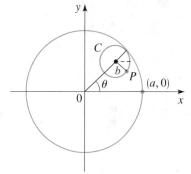

(a) Show that parametric equations for the hypocycloid are

$$x = (a - b)\cos\theta + b\cos\left(\frac{a - b}{b}\theta\right)$$

$$y = (a - b)\sin\theta - b\sin\left(\frac{a - b}{b}\theta\right)$$

(b) If $a = 4b$, the hypocycloid is called an **astroid**. Show that in this case the parametric equations can be reduced to

$$x = a\cos^3\theta \qquad y = a\sin^3\theta$$

Sketch the curve and eliminate the parameter to obtain an equation for the astroid in rectangular coordinates.

54. If the circle C of Exercise 53 rolls on the *outside* of the larger circle, the curve traced out by P is called an **epicycloid**. Find parametric equations for the epicycloid.

55. In the figure, the circle of radius a is stationary and, for every θ, the point P is the midpoint of the segment QR. The curve traced out by P for $0 < \theta < \pi$ is called the **longbow curve**. Find parametric equations for this curve.

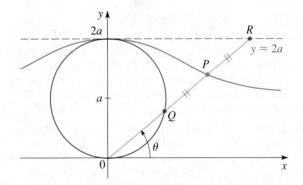

56. A string is wound around a circle and then unwound while being held taut. The curve traced out by the point P at the end of the string is called the **involute** of the circle, as shown in the figure. If the circle has radius a and is centered at the origin, and if the initial position of P is $(a, 0)$, show that parametric equations for the involute in terms of the parameter θ are

$$x = a(\cos\theta + \theta\sin\theta) \qquad y = a(\sin\theta - \theta\cos\theta)$$

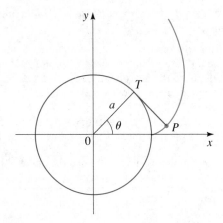

57. Eliminate the parameter θ in the parametric equations for the cycloid (Example 5) to obtain a rectangular coordinate equation for the section of the curve given by $0 \le \theta \le \pi$.

◆ DISCOVERY · DISCUSSION

58. More Information in Parametric Equations In this section we stated that parametric equations contain more information than just the shape of a curve. Write a short paragraph explaining this statement. Use the following example and your answers to the following questions in your explanation.

The position of a particle is given by the parametric equations

$$x = \sin t \qquad y = \cos t$$

where t represents time. We know that the shape of the path of the particle is a circle.

(a) How long does it take the particle to go once around the circle? Find parametric equations if the particle moves twice as fast around the circle.

(b) Does the particle travel clockwise or counterclockwise around the circle? Find parametric equations if the particle moves in the opposite direction around the circle.

CONCEPT CHECK

1. (a) Give the geometric definition of a parabola. What are the focus and directrix of the parabola?

(b) Sketch the parabola $x^2 = 4py$ for the case $p > 0$. Identify on your diagram the vertex, focus, and directrix. What happens if $p < 0$?

(c) Sketch the parabola $y^2 = 4px$, together with its vertex, focus, and directrix, for the case $p > 0$. What happens if $p < 0$?

2. (a) Give the geometric definition of an ellipse. What are the foci of the ellipse?

(b) For the ellipse with equation

$$\frac{x^2}{a^2} + \frac{y^2}{b^2} = 1$$

where $a > b > 0$, what are the coordinates of the vertices and the foci? What are the major and minor axes? Illustrate with a graph.

(c) Give an expression for the eccentricity of the ellipse in part (b).

(d) State the equation of an ellipse with foci on the y-axis.

3. (a) Give the geometric definition of a hyperbola. What are the foci of the hyperbola?

(b) For the hyperbola with equation

$$\frac{x^2}{a^2} - \frac{y^2}{b^2} = 1$$

what are the coordinates of the vertices and foci? What are the equations of the asymptotes? What is the transverse axis? Illustrate with a graph.

(c) State the equation of a hyperbola with foci on the y-axis.

(d) What steps would you take to sketch a hyperbola with a given equation?

4. Suppose h and k are positive numbers. What is the effect on the graph of an equation in x and y if

(a) x is replaced by $x - h$? By $x + h$?

(b) y is replaced by $y - k$? By $y + k$?

5. How can you tell whether the following nondegenerate conic is a parabola, an ellipse, or a hyperbola?

$$Ax^2 + Cy^2 + Dx + Ey + F = 0$$

6. Suppose the x- and y-axes are rotated through an acute angle ϕ to produce the X- and Y-axes. Write equations that relate the coordinates (x, y) and (X, Y) of a point in the xy-plane and XY-plane, respectively.

7. (a) How do you eliminate the xy-term in this equation?

$$Ax^2 + Bxy + Cy^2 + Dx + Ey + F = 0$$

(b) What is the discriminant of the conic in part (a)? How can you use the discriminant to determine whether the conic is a parabola, an ellipse, or a hyperbola?

8. (a) Describe how polar coordinates represent the position of a point in the plane.

(b) What equations do you use to change from polar to rectangular coordinates?

(c) What equations do you use to change from rectangular to polar coordinates?

9. (a) Write polar equations that represent a conic with eccentricity e.

(b) For what values of e is the conic an ellipse? A hyperbola? A parabola?

10. How do you sketch a curve with parametric equations $x = f(t)$, $y = g(t)$?

EXERCISES

1–4 ■ Find the vertex, focus, and directrix of the parabola, and sketch the graph.

1. $x^2 + 8y = 0$

2. $2x - y^2 = 0$

3. $x - y^2 + 4y - 2 = 0$

4. $2x^2 + 6x + 5y + 10 = 0$

5–8 ■ Find the center, vertices, foci, and the lengths of the major and minor axes of the ellipse, and sketch the graph.

5. $x^2 + 4y^2 = 16$

6. $9x^2 + 4y^2 = 1$

7. $4x^2 + 9y^2 = 36y$

8. $2x^2 + y^2 = 2 + 4(x - y)$

9–12 ■ Find the center, vertices, foci, and asymptotes of the hyperbola, and sketch the graph.

9. $x^2 - 2y^2 = 16$

10. $x^2 - 4y^2 + 16 = 0$

11. $9y^2 + 18y = x^2 + 6x + 18$

12. $y^2 = x^2 + 6y$

13–18 ■ Find an equation for the conic whose graph is shown.

13.

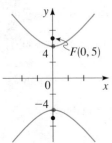

14.

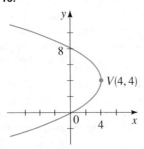

15.

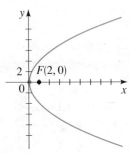

16.

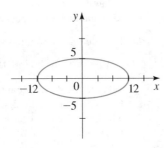

17.

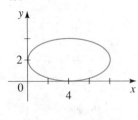

18.

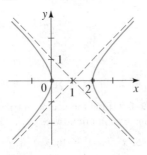

19–30 ■ Determine the type of curve represented by the equation. Find the foci and vertices (if any), and sketch the graph.

19. $\dfrac{x^2}{12} + y = 1$

20. $\dfrac{x^2}{12} + \dfrac{y^2}{144} = \dfrac{y}{12}$

21. $x^2 - y^2 + 144 = 0$

22. $x^2 + 6x = 9y^2$

23. $4x^2 + y^2 = 8(x + y)$

24. $3x^2 - 6(x + y) = 10$

25. $x = y^2 - 16y$

26. $2x^2 + 4 = 4x + y^2$

27. $2x^2 - 12x + y^2 + 6y + 26 = 0$

28. $36x^2 - 4y^2 - 36x - 8y = 31$

29. $9x^2 + 8y^2 - 15x + 8y + 27 = 0$

30. $x^2 + 4y^2 = 4x + 8$

31–38 ■ Find an equation for the conic section with the given properties.

31. The parabola with focus $F(0, 1)$ and directrix $y = -1$

32. The ellipse with center $C(0, 4)$, foci $F_1(0, 0)$ and $F_2(0, 8)$, and major axis of length 10

33. The hyperbola with vertices $V(0, \pm 2)$ and asymptotes $y = \pm \frac{1}{2} x$

34. The hyperbola with center $C(2, 4)$, foci $F_1(2, 1)$ and $F_2(2, 7)$, and vertices $V_1(2, 6)$ and $V_2(2, 2)$

35. The ellipse with foci $F_1(1, 1)$ and $F_2(1, 3)$, and with one vertex on the x-axis

36. The parabola with vertex $V(5, 5)$ and directrix the y-axis

37. The ellipse with vertices $V_1(7, 12)$ and $V_2(7, -8)$, and passing through the point $P(1, 8)$

38. The parabola with vertex $V(-1, 0)$ and horizontal axis of symmetry, and crossing the y-axis at $y = 2$

39. A cannon fires a cannonball as shown in the figure. The path of the cannonball is a parabola with vertex at the highest point of the path. If the cannonball lands 1600 ft from the cannon and the highest point it reaches is 3200 ft above the ground, find an equation for the path of the cannonball. Place the origin at the location of the cannon.

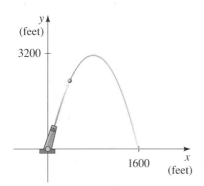

40. A satellite is in an elliptical orbit around the earth with the center of the earth at one focus. The height of the satellite above the earth varies between 140 mi and 440 mi. Assume the earth is a sphere with radius 3960 mi. Find an equation for the path of the satellite with the origin at the center of the earth.

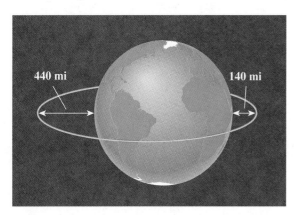

41. The path of the earth around the sun is an ellipse with the sun at one focus. The ellipse has major axis 186,000,000 mi and eccentricity 0.017. Find the distance between the earth and the sun when the earth is (a) closest to the sun and (b) furthest from the sun.

42. A ship is located 40 mi from a straight shoreline. LORAN stations A and B are located on the shoreline, 300 mi apart. From the LORAN signals, the captain determines that his ship is 80 mi closer to A than to B. Find the location of the ship. (Place A and B on the y-axis with the x-axis halfway between them. Find the x- and y-coordinates of the ship.)

 43. (a) Draw graphs of the following family of ellipses for $k = 1, 2, 4,$ and 8.

$$\frac{x^2}{16 + k^2} + \frac{y^2}{k^2} = 1$$

    (b) Prove that all the ellipses in part (a) have the same foci.

 44. (a) Draw graphs of the following family of parabolas for $k = \frac{1}{2}, 1, 2,$ and 4.

$$y = kx^2$$

    (b) Find the foci of the parabolas in part (a).
    (c) How does the location of the focus change as k increases?

45–48 ■ (a) Use the discriminant to determine whether the graph of the equation is a parabola, an ellipse, or a hyperbola. (b) Use a rotation of axes to eliminate the xy-term. (c) Sketch the graph.

45. $x^2 + 4xy + y^2 = 1$

46. $5x^2 - 6xy + 5y^2 - 8x + 8y - 8 = 0$

47. $7x^2 - 6\sqrt{3}\,xy + 13y^2 - 4\sqrt{3}\,x - 4y = 0$

48. $9x^2 + 24xy + 16y^2 = 25$

49–56 ■ (a) Sketch the graph of the polar equation. (b) Express the equation in rectangular coordinates.

49. $r = 3 + 3\cos\theta$ **50.** $r = 3\sin\theta$

51. $r = 2\sin 2\theta$ **52.** $r = 4\cos 3\theta$

53. $r^2 = \sec 2\theta$ **54.** $r^2 = 4\sin 2\theta$

55. $r = \sin\theta + \cos\theta$ **56.** $r = \dfrac{4}{2 + \cos\theta}$

 57–60 ■ Use a graphing device to graph the polar equation. Choose the domain of θ to make sure you produce the entire graph.

57. $r = \cos(\theta/3)$ **58.** $r = \sin(9\theta/4)$

59. $r = 1 + 4\cos(\theta/3)$ **60.** $r = \theta\sin\theta$

61–64 ■ (a) Find the eccentricity and identify the conic. (b) Sketch the conic and label the vertices.

61. $r = \dfrac{1}{1 - \cos\theta}$ **62.** $r = \dfrac{2}{1 + 2\sin\theta}$

63. $r = \dfrac{4}{1 + 2\sin\theta}$ **64.** $r = \dfrac{12}{1 - 4\cos\theta}$

65–68 ■ Sketch the parametric curve, and eliminate the parameter to find an equation in rectangular coordinates that all points on the curve satisfy.

65. $x = 1 - t^2, \quad y = 1 + t$

66. $x = t^2 - 1, \quad y = t^2 + 1$

67. $x = 1 + \cos t, \quad y = 1 - \sin t, \quad 0 \le t \le \pi/2$

68. $x = \dfrac{1}{t} + 2, \quad y = \dfrac{2}{t^2}, \quad 0 < t \le 2$

69–70 ■ Use a graphing device to draw the parametric curve.

69. $x = \cos 2t, \quad y = \sin 3t$

70. $x = \sin(t + \cos 2t), \quad y = \cos(t + \sin 3t)$

71. The curves C, D, E, and F are defined parametrically as follows, where the parameter t takes on all real values unless otherwise stated:

$$C: \quad x = t, \quad y = t^2$$
$$D: \quad x = \sqrt{t}, \quad y = t, \quad t \ge 0$$
$$E: \quad x = \sin t, \quad y = 1 - \cos^2 t$$
$$F: \quad x = e^t, \quad y = e^{2t}$$

(a) Show that the points on all four of these curves satisfy the same rectangular coordinate equation.

(b) Draw the graph of each curve and explain how the curves differ from one another.

72. In the figure the point P is the midpoint of the segment QR and $0 \le \theta < \pi/2$. Using θ as the parameter, find a parametric representation for the curve traced out by P.

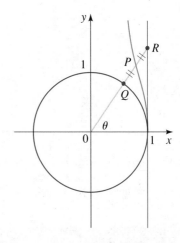

1. Find the focus and directrix of the parabola $x^2 = -12y$, and sketch its graph.

2. Find the vertices, foci, and the lengths of the major and minor axes for the ellipse $\dfrac{x^2}{16} + \dfrac{y^2}{4} = 1$. Then sketch its graph.

3. Find the vertices, foci, and asymptotes of the hyperbola $\dfrac{y^2}{9} - \dfrac{x^2}{16} = 1$. Then sketch its graph.

4–6 ■ Find an equation for the conic whose graph is shown.

4.

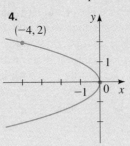

5.

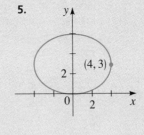

6.

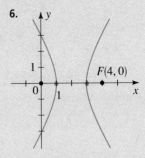

7–9 ■ Sketch the graph of the equation.

7. $16x^2 + 36y^2 - 96x + 36y + 9 = 0$ 8. $9x^2 - 8y^2 + 36x + 64y = 92$

9. $2x + y^2 + 8y + 8 = 0$

10. Find an equation for the hyperbola with foci $(0, \pm5)$ and with asymptotes $y = \pm\frac{3}{4}x$.

11. Find an equation for the parabola with focus $(2, 4)$ and directrix the x-axis.

12. A parabolic reflector for a car headlight forms a bowl shape that is 6 in. wide at its opening and 3 in. deep, as shown in the figure at the left. How far from the vertex should the filament of the bulb be placed if it is to be located at the focus?

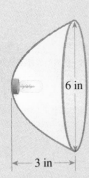

13. (a) Use the discriminant to determine whether the graph of this equation is a parabola, an ellipse, or a hyperbola:

$$5x^2 + 4xy + 2y^2 = 18$$

(b) Use rotation of axes to eliminate the xy-term in the equation.
(c) Sketch the graph of the equation.
(d) Find the coordinates of the vertices of this conic (in the xy-coordinate system).

14. Graph the polar equation $r = 2 + \cos\theta$.

15. Convert the polar equation $r = 2\cos\theta - 4\sin\theta$ to rectangular coordinates, and identify the graph.

16. (a) Sketch the graph of the parametric curve

$$x = 3\sin\theta + 3 \qquad y = 2\cos\theta \qquad 0 \le \theta \le \pi$$

(b) Eliminate the parameter θ in part (a) to obtain an equation for this curve in rectangular coordinates.

FOCUS ON MODELING

Modeling motion is one of the most important ideas in both classical and modern physics. Much of Isaac Newton's work dealt with creating a mathematical model for how objects move and interact—this was the main reason for his invention of calculus (see page 226). Albert Einstein developed his Special Theory of Relativity in the early 1900s to refine Newton's laws of motion.

In this section we use coordinate geometry to model the motion of a projectile, such as a ball thrown upward into the air, a bullet fired from a gun, or any other sort of missile. A similar model was created by Galileo (see the margin), but we have the advantage of using our modern mathematical notation to make describing the model much easier than it was for Galileo!

Suppose that we fire a projectile into the air from ground level, with an initial speed v_0 and at an angle θ upward from the ground. The initial *velocity* of the projectile is a vector (see Section 7.7) with horizontal component $v_0 \cos \theta$ and vertical component $v_0 \sin \theta$, as shown in Figure 1.

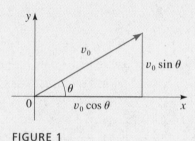

FIGURE 1

If there were no gravity (and no air resistance), the projectile would just keep moving indefinitely at the same speed and in the same direction. Since distance = speed × time, the projectile's position at time t would therefore be given by the following parametric equations (assuming the origin of our coordinate system is placed at the initial location of the projectile):

$$x = (v_0 \cos \theta)t \qquad y = (v_0 \sin \theta)t \qquad \text{No gravity}$$

But, of course, we know that gravity will pull the projectile back to ground level. Using calculus it can be shown that the effect of gravity can be accounted for by subtracting $\frac{1}{2}gt^2$ from the vertical position of the projectile. In this expression, g is the gravitational acceleration: $g \approx 32 \text{ ft/s}^2 \approx 9.8 \text{ m/s}^2$. Thus, we have the following parametric equations for the path of the projectile:

$$x = (v_0 \cos \theta)t \qquad y = (v_0 \sin \theta)t - \frac{1}{2}gt^2 \qquad \text{With gravity}$$

Galileo Galilei (1564–1642) was born in Pisa, Italy. He studied medicine, but later abandoned this in favor of science and mathematics. At the age of 25 he demonstrated that light objects fall at the same rate as heavier ones, by dropping cannonballs of various sizes from the Leaning Tower of Pisa. This contradicted the then-accepted view of Aristotle that heavier objects fall more quickly. He also showed that the distance an object falls is proportional to the square of the time it has been falling, and from this was able to prove that the path of a projectile is a parabola.

Galileo constructed the first telescope, and using it, discovered the moons of Jupiter. His advocacy of the Copernican view that the earth revolves around the sun (rather than being stationary) led to his being called before the Inquisition. By then an old man, he was forced to recant his views, but he is said to have muttered

(continued)

under his breath "the earth nevertheless does move." Galileo revolutionized science by expressing scientific principles in the language of mathematics. He said, "The great book of nature is written in mathematical symbols."

EXAMPLE ■ The Path of a Cannonball

Find parametric equations that model the path of a cannonball fired into the air with an initial speed of 150.0 m/s at a 30° angle of elevation. Sketch the path of the cannonball. How far from the cannon does the ball hit the ground, and for how long is it in the air?

SOLUTION Substituting the given initial speed and angle into the general parametric equations of the path of a projectile, we get

$$x = (150.0 \cos 30°)t \qquad y = (150.0 \sin 30°)t - \tfrac{1}{2}(9.8)t^2 \qquad \text{Substitute } v_0 = 150.0, \theta = 30°$$

$$x = 129.9t \qquad y = 75.0t - 4.9t^2 \qquad \text{Simplify}$$

This path is graphed in Figure 2.

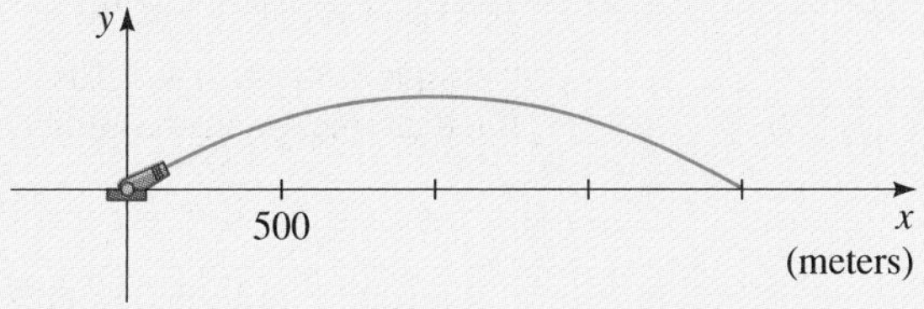

FIGURE 2
Path of a cannonball

How can we tell where and when the cannonball hits the ground? Since ground level corresponds to $y = 0$, we substitute this value for y and solve for t.

$$0 = 75.0t - 4.9t^2 \qquad \text{Set } y = 0$$

$$0 = t(75.0 - 4.9t) \qquad \text{Factor}$$

$$t = 0 \quad \text{or} \quad t = \frac{75.0}{4.9} \approx 15.3 \qquad \text{Solve for } t$$

The first solution, $t = 0$, is the time when the cannon was fired; the second solution means that the cannonball hits the ground after 15.3 s of flight. To see *where* this happens, we substitute this value into the equation for x, the horizontal location of the cannonball.

$$x = 129.9(15.3) \approx 1987.5 \text{ m}$$

The cannonball travels almost 2 km before hitting the ground. ■

Figure 3 shows the paths of several projectiles, all fired with the same initial speed but at different angles. From the graphs we see that if the firing angle is too high or too low, the projectile doesn't travel very far.

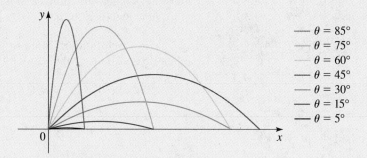

FIGURE 3
Paths of projectiles

Let's try to find the optimal firing angle—the angle that shoots the projectile as far as possible. We'll go through the same steps as we did in the preceding example, but we'll use the general parametric equations instead. First, we solve for the time when the projectile hits the ground by substituting $y = 0$.

$$0 = (v_0 \sin \theta)t - \tfrac{1}{2}gt^2 \qquad \text{Substitute } y = 0$$

$$0 = t\left(v_0 \sin \theta - \tfrac{1}{2}gt\right) \qquad \text{Factor}$$

$$0 = v_0 \sin \theta - \tfrac{1}{2}gt \qquad \text{Set second factor equal to 0}$$

$$t = \frac{2v_0 \sin \theta}{g} \qquad \text{Solve for } t$$

Now we substitute this into the equation for x to see how far the projectile has traveled horizontally when it hits the ground.

$$x = (v_0 \cos \theta)t \qquad \text{Parametric equation for } x$$

$$= (v_0 \cos \theta)\left(\frac{2v_0 \sin \theta}{g}\right) \qquad \text{Substitute } t = (2v_0 \sin \theta)/g$$

$$= \frac{2v_0^2 \sin \theta \cos \theta}{g} \qquad \text{Simplify}$$

$$= \frac{v_0^2 \sin 2\theta}{g} \qquad \text{Use identity } \sin 2\theta = 2 \sin \theta \cos \theta$$

We want to choose θ so that x is as large as possible. The largest value that the sine of any angle can have is 1, the sine of 90°. Thus, we want $2\theta = 90°$, or $\theta = 45°$. So to send the projectile as far as possible, shoot it up at an angle of 45°.

From the last equation in the preceding display, we see that it will then travel a distance $x = v_0^2/g$.

PROBLEMS

1. From the graphs in Figure 3, it seems that the paths of projectiles are parabolas that open downward. Eliminate the parameter t from the general parametric equations to verify that these are indeed parabolas.

2. Suppose a baseball is thrown at 30 ft/s at a 60° angle to the horizontal, from a height of 4 ft above the ground. Find parametric equations for the path of the baseball, and sketch its graph. How far does the baseball travel, and when does it hit the ground?

3. Suppose that a rocket is fired at an angle of 5° from the vertical, with an initial speed of 1000 ft/s.
   (a) Find the length of time the rocket is in the air.
   (b) Find the greatest height it reaches.
   (c) Find the horizontal distance it has traveled when it hits the ground.
   (d) Sketch a graph of the rocket's path.

4. The initial speed of a missile is 330 m/s. At what angle should it be fired so that it hits a target 10 km away? You should find that there are two possible angles. Sketch graphs of the missile paths for both angles. For which angle is the target hit sooner?

5. Find the maximum height reached by a projectile as a function of its initial speed v_0 and its firing angle θ.

6. Suppose that a projectile is fired into a headwind that pushes it back so as to reduce its horizontal speed by a constant amount w. Find parametric equations for the path of the projectile.

 7. Using the general parametric equations you derived in Problem 6, draw graphs of the path of a projectile with initial speed $v_0 = 32$ ft/s, fired into a headwind of $w = 24$ ft/s, for the angles $\theta = 5°$, 15°, 30°, 40°, 45°, 55°, 60°, and 75°. Is it still true that the greatest range is attained when firing at 45°? Draw some more graphs for different angles, and use these graphs to estimate the optimal firing angle.

10

SEQUENCES AND SERIES

Sequences and series abound in nature; the Fibonacci sequence, for example, is hidden in the intricate structure of the nautilus.

A mathematician, like a painter or a poet, is a maker of patterns.

G. H. HARDY

In this chapter we study sequences and series of numbers. Roughly speaking, a *sequence* is a list of numbers written in a specific order, and a *series* is what one gets by adding the numbers in a sequence. Sequences and series have many theoretical and practical uses. Among other applications, we consider how series are used to calculate the value of an annuity.

Section 10.5 introduces a special kind of proof called *mathematical induction*. In Section 10.6 we use mathematical induction to prove a formula for expanding $(a + b)^n$ for any natural number n.

10.1 SEQUENCES AND SUMMATION NOTATION

A *sequence* is a set of numbers written in a specific order:

$$a_1, a_2, a_3, a_4, \ldots, a_n, \ldots$$

The number a_1 is called the *first term*, a_2 is the *second term*, and in general a_n is the *nth term*. Since for every natural number n there is a corresponding number a_n, we can define a sequence as a function.

> ### DEFINITION OF A SEQUENCE
>
> A **sequence** is a function f whose domain is the set of natural numbers. The values $f(1), f(2), f(3), \ldots$ are called the **terms** of the sequence.

We usually write a_n instead of the function notation $f(n)$ for the value of the function at the number n.

Here is a simple example of a sequence:

$$2, 4, 6, 8, 10, \ldots$$

The dots indicate that the sequence continues indefinitely. We can write a sequence in this way when it's clear what the subsequent terms of the sequence are. This sequence consists of even numbers. To be more accurate, however, we need to specify a procedure for finding *all* the terms of the sequence. This can be done by giving a formula for the *n*th term a_n of the sequence. In this case,

Another way to write this sequence is to use function notation:

$$a(n) = 2n$$

so $a(1) = 2, a(2) = 4, a(3) = 6, \ldots$

$$a_n = 2n$$

and the sequence can be written as

$$
\begin{array}{cccccc}
2, & 4, & 6, & 8, & \ldots, & 2n, & \ldots \\
\uparrow & \uparrow & \uparrow & \uparrow & & \uparrow & \\
\text{1st} & \text{2nd} & \text{3rd} & \text{4th} & & \text{nth} & \\
\text{term} & \text{term} & \text{term} & \text{term} & & \text{term} &
\end{array}
$$

Notice how the formula $a_n = 2n$ gives all the terms of the sequence. For instance, substituting 1, 2, 3, and 4 for n gives the first four terms:

$$a_1 = 2 \cdot 1 = 2 \qquad a_2 = 2 \cdot 2 = 4$$

$$a_3 = 2 \cdot 3 = 6 \qquad a_4 = 2 \cdot 4 = 8$$

To find the 103rd term of this sequence, we use $n = 103$ to get

$$a_{103} = 2 \cdot 103 = 206$$

EXAMPLE 1 ■ Finding the Terms of a Sequence

Find the first five terms and the 100th term of the sequence defined by each formula.

(a) $a_n = 2n - 1$ (b) $c_n = n^2 - 1$ (c) $t_n = \dfrac{n}{n + 1}$ (d) $r_n = \dfrac{(-1)^n}{2^n}$

SOLUTION   To find the first five terms, we substitute $n = 1, 2, 3, 4,$ and 5 in the formula for the nth term. To find the 100th term, we substitute $n = 100$. This gives the following.

nth term	First five terms	100th term
(a) $2n - 1$	1, 3, 5, 7, 9	199
(b) $n^2 - 1$	0, 3, 8, 15, 24	9999
(c) $\dfrac{n}{n + 1}$	$\dfrac{1}{2}, \dfrac{2}{3}, \dfrac{3}{4}, \dfrac{4}{5}, \dfrac{5}{6}$	$\dfrac{100}{101}$
(d) $\dfrac{(-1)^n}{2^n}$	$-\dfrac{1}{2}, \dfrac{1}{4}, -\dfrac{1}{8}, \dfrac{1}{16}, -\dfrac{1}{32}$	$\dfrac{1}{2^{100}}$

■

In Example 1(d) the presence of $(-1)^n$ in the sequence has the effect of making successive terms alternately negative and positive.

It is often useful to picture a sequence by sketching its graph. Since a sequence is a function whose domain is the natural numbers, we can draw its graph in the Cartesian plane. For instance, the graph of the sequence

$$1, \frac{1}{2}, \frac{1}{3}, \frac{1}{4}, \frac{1}{5}, \frac{1}{6}, \dots, \frac{1}{n}, \dots$$

is shown in Figure 1. Compare this to the graph of

$$1, -\frac{1}{2}, \frac{1}{3}, -\frac{1}{4}, \frac{1}{5}, -\frac{1}{6}, \dots, \frac{(-1)^{n+1}}{n}, \dots$$

shown in Figure 2. The graph of every sequence consists of isolated points that are *not* connected.

Some sequences do not have simple defining formulas like those of the preceding example. The nth term of a sequence may depend on some or all of the terms preceding it. A sequence defined in this way is called **recursive**. Here are two examples.

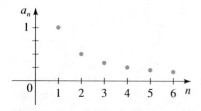

FIGURE 1

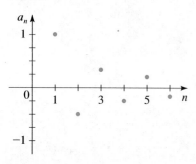

FIGURE 2

EXAMPLE 2 ■ Finding the Terms of a Recursive Sequence

Find the first five terms of the sequence defined recursively by

$$a_n = 3(a_{n-1} + 2)$$

and $a_1 = 1$.

SOLUTION   The defining formula for this sequence is recursive. It allows us to find the *n*th term a_n if we know the preceding term a_{n-1}. Thus, we can find the second term from the first term, the third term from the second term, the fourth term from the third term, and so on. Since we are given the first term $a_1 = 1$, we can proceed as follows.

$$a_2 = 3(a_1 + 2) = 3(1 + 2) = 9$$
$$a_3 = 3(a_2 + 2) = 3(9 + 2) = 33$$
$$a_4 = 3(a_3 + 2) = 3(33 + 2) = 105$$
$$a_5 = 3(a_4 + 2) = 3(105 + 2) = 321$$

Thus, the first five terms of this sequence are

$$1, 9, 33, 105, 321, \ldots$$ ■

Note that in order to find the 100th term of the sequence in Example 2 we must first find all 99 preceding terms.

EXAMPLE 3 ■ The Fibonacci Sequence

Find the first 11 terms of the sequence defined recursively by

$$F_n = F_{n-1} + F_{n-2}$$

with $F_1 = 1$ and $F_2 = 1$.

SOLUTION   To find F_n we need to find the two preceding terms F_{n-1} and F_{n-2}. Since we are given F_1 and F_2, we proceed as follows.

$$F_3 = F_2 + F_1 = 1 + 1 = 2$$
$$F_4 = F_3 + F_2 = 2 + 1 = 3$$
$$F_5 = F_4 + F_3 = 3 + 2 = 5$$

It's clear what is happening here. Each term is simply the sum of the two terms that precede it, so we can easily write down as many terms as we please. Here are the first 11 terms:

$$1, 1, 2, 3, 5, 8, 13, 21, 34, 55, 89, \ldots$$ ■

The sequence in Example 3 is called the **Fibonacci sequence**, named after the 13th-century Italian mathematician who used it to solve a problem about the breeding of rabbits (see Exercise 56). The sequence also occurs in numerous other applications in nature. (See Figures 3 and 4.) In fact, so many phenomena behave

8

5

3

2

1

1

FIGURE 3

The Fibonacci sequence in the branching of a tree

like the Fibonacci sequence that one mathematical journal, the *Fibonacci Quarterly,* is devoted entirely to its properties.

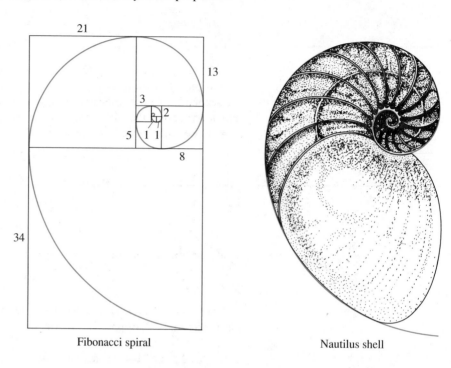

FIGURE 4   Fibonacci spiral   Nautilus shell

The sequences we have considered so far are defined by either a formula or a recursive procedure. But not all sequences can be defined in this way. For example, there is no known formula that produces the sequence of prime numbers:

$$2, 3, 5, 7, 11, 13, 17, 19, 23, \ldots$$

A *prime number* is a natural number with no factor other than 1 and itself. By convention, 1 is not considered prime.

If we let a_n be the nth digit in the decimal expansion of the number π, we get the sequence

$$3, 1, 4, 1, 5, 9, 2, 6, 5, 4, \ldots$$

Again, no simple formula can be given for finding the terms of this sequence.

Finding patterns is a very important part of mathematics. Consider a sequence

$$1, 4, 9, 16, \ldots$$

Can you detect a pattern in these numbers? In other words, can you define a sequence whose first four terms are these numbers? The answer to this question seems easy; these numbers are the squares of the numbers 1, 2, 3, 4. Thus, the sequence we are looking for is defined by $a_n = n^2$. However, this is not the *only* sequence whose first four terms are 1, 4, 9, 16. In other words, the answer to our problem is not unique (see Exercise 57). In the next example we are interested in finding an *obvious* sequence whose first few terms agree with the given ones.

EXAMPLE 4 ■ Finding the *n*th Term of a Sequence

Find the *n*th term of a sequence whose first several terms are given.

(a) $\frac{1}{2}, \frac{3}{4}, \frac{5}{6}, \frac{7}{8}, \ldots$ (b) $-2, 4, -8, 16, -32, \ldots$

SOLUTION

(a) We notice that the numerators of these fractions are the odd numbers and the denominators are the even numbers. Even numbers are of the form $2n$, and odd numbers are of the form $2n - 1$ (an odd number differs from an even number by 1). So, a sequence that has these numbers for its first four terms is given by

$$a_n = \frac{2n - 1}{2n}$$

(b) These numbers are powers of 2 and they alternate in sign, so a sequence that agrees with these terms is given by

$$a_n = (-1)^n 2^n$$

You should check that these formulas do indeed generate the given terms.   ■

Fibonacci (1175–1250) was born in Pisa, Italy, and educated in North Africa. He traveled widely in the Mediterranean area and learned the various methods then in use for writing numbers. On returning to Pisa in 1202, Fibonacci advocated the use of the Hindu-Arabic decimal system, the one we use today, over the Roman numeral system used in Europe in his time. His most famous book *Liber Abaci* is devoted to expounding the advantages of the Hindu-Arabic numerals. In fact, multiplication and division were so complicated using Roman numerals that it required a college degree to master these skills. Interestingly, in 1299 the city of Florence outlawed the use of the decimal system for merchants and other businesses, allowing numbers to be written using only Roman numerals or words. One can only speculate about the reasons for this law.

THE PARTIAL SUMS OF A SEQUENCE

In calculus we are often interested in adding the terms of a sequence. This leads to the following definition.

THE PARTIAL SUMS OF A SEQUENCE

For the sequence

$$a_1, \quad a_2, \quad a_3, \quad a_4, \quad \ldots, \quad a_n, \quad \ldots$$

the **partial sums** are

$$S_1 = a_1$$
$$S_2 = a_1 + a_2$$
$$S_3 = a_1 + a_2 + a_3$$
$$S_4 = a_1 + a_2 + a_3 + a_4$$
$$\vdots$$
$$S_n = a_1 + a_2 + a_3 + \cdots + a_n$$
$$\vdots$$

S_1 is called the **first partial sum**, S_2 is the **second partial sum**, and so on. S_n is called the ***n*th partial sum**. The sequence $S_1, S_2, S_3, \ldots, S_n, \ldots$ is called the **sequence of partial sums**.

EXAMPLE 5 ■ Finding the Partial Sums of a Sequence

Find the first four partial sums and the nth partial sum of the sequence given by $a_n = 1/2^n$.

SOLUTION The terms of the sequence are

$$\frac{1}{2}, \frac{1}{4}, \frac{1}{8}, \cdots$$

The first four partial sums are

$$S_1 = \frac{1}{2} \qquad\qquad = \frac{1}{2}$$

$$S_2 = \frac{1}{2} + \frac{1}{4} \qquad\qquad = \frac{3}{4}$$

$$S_3 = \frac{1}{2} + \frac{1}{4} + \frac{1}{8} \qquad = \frac{7}{8}$$

$$S_4 = \frac{1}{2} + \frac{1}{4} + \frac{1}{8} + \frac{1}{16} = \frac{15}{16}$$

Notice that in the value of each partial sum the denominator is a power of 2 and the numerator is one less than the denominator. In general, the nth partial sum is

$$S_n = \frac{2^n - 1}{2^n} = 1 - \frac{1}{2^n}$$

The first five terms of a_n and S_n are graphed in Figure 5. ■

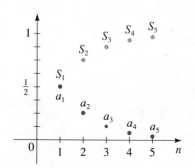

FIGURE 5

Graphs of the sequence a_n and the sequence of partial sums S_n

EXAMPLE 6 ■ Finding the Partial Sums of a Sequence

Find the first four partial sums and the nth partial sum of the sequence given by

$$a_n = \frac{1}{n} - \frac{1}{n+1}$$

SOLUTION The first four partial sums are

$$S_1 = \left(1 - \frac{1}{2}\right) \qquad\qquad\qquad\qquad = 1 - \frac{1}{2}$$

$$S_2 = \left(1 - \frac{1}{2}\right) + \left(\frac{1}{2} - \frac{1}{3}\right) \qquad\qquad = 1 - \frac{1}{3}$$

$$S_3 = \left(1 - \frac{1}{2}\right) + \left(\frac{1}{2} - \frac{1}{3}\right) + \left(\frac{1}{3} - \frac{1}{4}\right) \qquad = 1 - \frac{1}{4}$$

$$S_4 = \left(1 - \frac{1}{2}\right) + \left(\frac{1}{2} - \frac{1}{3}\right) + \left(\frac{1}{3} - \frac{1}{4}\right) + \left(\frac{1}{4} - \frac{1}{5}\right) = 1 - \frac{1}{5}$$

Do we detect a pattern here? Of course. The nth partial sum is

$$S_n = 1 - \frac{1}{n+1}$$

SIGMA NOTATION

Given a sequence

$$a_1, a_2, a_3, a_4, \ldots$$

we can write the sum of the first n terms using **summation notation**, or **sigma notation**. This notation derives its name from the Greek letter Σ (capital sigma, corresponding to our S for "sum"). Sigma notation is used as follows:

$$\sum_{k=1}^{n} a_k = a_1 + a_2 + a_3 + a_4 + \cdots + a_n$$

The left side of this expression is read "The sum of a_k from $k = 1$ to $k = n$." The letter k is called the **index of summation**, or the **summation variable**, and the idea is to replace k in the expression after the sigma by the integers $1, 2, 3, \ldots, n$, and add the resulting expressions, arriving at the right side of the equation.

EXAMPLE 7 ■ Sigma Notation

Find each sum.

(a) $\displaystyle\sum_{k=1}^{5} k^2$ (b) $\displaystyle\sum_{j=3}^{5} \frac{1}{j}$ (c) $\displaystyle\sum_{i=5}^{10} i$ (d) $\displaystyle\sum_{i=1}^{6} 2$

SOLUTION

(a) $\displaystyle\sum_{k=1}^{5} k^2 = 1^2 + 2^2 + 3^2 + 4^2 + 5^2 = 55$

(b) $\displaystyle\sum_{j=3}^{5} \frac{1}{j} = \frac{1}{3} + \frac{1}{4} + \frac{1}{5} = \frac{47}{60}$

(c) $\displaystyle\sum_{i=5}^{10} i = 5 + 6 + 7 + 8 + 9 + 10 = 45$

(d) $\displaystyle\sum_{i=1}^{6} 2 = 2 + 2 + 2 + 2 + 2 + 2 = 12$

EXAMPLE 8 ■ Writing Sums in Sigma Notation

Write each sum using sigma notation.

(a) $1^3 + 2^3 + 3^3 + 4^3 + 5^3 + 6^3 + 7^3$

(b) $\sqrt{3} + \sqrt{4} + \sqrt{5} + \cdots + \sqrt{77}$

The ancient Greeks considered a line segment to be divided into the **golden ratio** if the ratio of the shorter part to the longer part is the same as the ratio of the longer part to the whole segment.

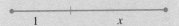

Thus, the segment shown is divided into the golden ratio if

$$\frac{1}{x} = \frac{x}{1+x}$$

This leads to a quadratic equation whose positive solution is

$$x = \frac{1 + \sqrt{5}}{2} \approx 1.618$$

This ratio occurs naturally in many places. For instance, psychological experiments show that the most pleasing shape of rectangle is one whose sides are in golden ratio. The ancient Greeks agreed with this and built their temples in this ratio.

The golden ratio is related to the Fibonacci numbers (see page 687). In fact, it can be shown using calculus* that the ratio of two successive Fibonacci numbers

$$\frac{F_{n+1}}{F_n}$$

gets closer to the golden ratio the larger the value of n. Try finding this ratio for $n = 10$.

*James Stewart, *Calculus, Third Edition* (Pacific Grove, CA: Brooks/Cole, 1995) p. 607.

SOLUTION

(a) We can write

$$1^3 + 2^3 + 3^3 + 4^3 + 5^3 + 6^3 + 7^3 = \sum_{k=1}^{7} k^3$$

(b) A natural way to write this sum is

$$\sqrt{3} + \sqrt{4} + \sqrt{5} + \cdots + \sqrt{77} = \sum_{k=3}^{77} \sqrt{k}$$

However, there is no unique way of writing a sum in sigma notation. We also could write this sum as

$$\sqrt{3} + \sqrt{4} + \sqrt{5} + \cdots + \sqrt{77} = \sum_{k=0}^{74} \sqrt{k + 3}$$

or $$\sqrt{3} + \sqrt{4} + \sqrt{5} + \cdots + \sqrt{77} = \sum_{k=1}^{75} \sqrt{k + 2}$$ ∎

The following properties of sums are natural consequences of properties of the real numbers.

PROPERTIES OF SUMS

Let $a_1, a_2, a_3, a_4, \ldots$ and $b_1, b_2, b_3, b_4, \ldots$ be sequences. Then for every positive integer n and any real number c,

1. $\displaystyle\sum_{k=1}^{n} (a_k + b_k) = \sum_{k=1}^{n} a_k + \sum_{k=1}^{n} b_k$

2. $\displaystyle\sum_{k=1}^{n} (a_k - b_k) = \sum_{k=1}^{n} a_k - \sum_{k=1}^{n} b_k$

3. $\displaystyle\sum_{k=1}^{n} ca_k = c\left(\sum_{k=1}^{n} a_k\right)$

■ **Proof**   To prove Property 1, we write out the left side of the equation to get

$$\sum_{k=1}^{n} (a_k + b_k) = (a_1 + b_1) + (a_2 + b_2) + (a_3 + b_3) + \cdots + (a_n + b_n)$$

Because addition is commutative and associative, we can rearrange the terms on the right side to read

$$\sum_{k=1}^{n} (a_k + b_k) = (a_1 + a_2 + a_3 + \cdots + a_n) + (b_1 + b_2 + b_3 + \cdots + b_n)$$

Rewriting the right side using sigma notation gives Property 1. Property 2 is proved in a similar manner. To prove Property 3, we use the Distributive Property:

$$\sum_{k=1}^{n} ca_k = ca_1 + ca_2 + ca_3 + \cdots + ca_n$$

$$= c(a_1 + a_2 + a_3 + \cdots + a_n) = c\left(\sum_{k=1}^{n} a_k\right) \qquad \square$$

10.1 EXERCISES

1–10 ■ Find the first four terms and the 1000th term of the sequence.

1. $a_n = n + 1$

2. $a_n = 2n + 3$

3. $a_n = \dfrac{1}{n + 1}$

4. $a_n = n^2 + 1$

5. $a_n = \dfrac{(-1)^n}{n^2}$

6. $a_n = \dfrac{1}{n^2}$

7. $a_n = 1 + (-1)^n$

8. $a_n = (-1)^{n+1} \dfrac{n}{n + 1}$

9. $a_n = n^n$

10. $a_n = 3$

11–16 ■ Find the first five terms of the given recursively defined sequence.

11. $a_n = 2(a_{n-1} - 2)$ and $a_1 = 3$

12. $a_n = \dfrac{a_{n-1}}{2}$ and $a_1 = -8$

13. $a_n = 2a_{n-1} + 1$ and $a_1 = 1$

14. $a_n = \dfrac{1}{1 + a_{n-1}}$ and $a_1 = 1$

15. $a_n = a_{n-1} + a_{n-2}$ and $a_1 = 1, a_2 = 2$

16. $a_n = a_{n-1} + a_{n-2} + a_{n-3}$ and $a_1 = a_2 = a_3 = 1$

17–24 ■ Find the nth term of a sequence whose first several terms are given.

17. $2, 4, 8, 16, \ldots$

18. $-\frac{1}{3}, \frac{1}{9}, -\frac{1}{27}, \frac{1}{81}, \ldots$

19. $1, 4, 7, 10, \ldots$

20. $5, -25, 125, -625, \ldots$

21. $1, \frac{3}{4}, \frac{5}{9}, \frac{7}{16}, \frac{9}{25}, \ldots$

22. $\frac{3}{4}, \frac{4}{5}, \frac{5}{6}, \frac{6}{7}, \ldots$

23. $0, 2, 0, 2, 0, 2, \ldots$

24. $1, \frac{1}{2}, 3, \frac{1}{4}, 5, \frac{1}{6}, \ldots$

25–28 ■ Find the first six partial sums $S_1, S_2, S_3, S_4, S_5, S_6$ of the sequence.

25. $1, 3, 5, 7, \ldots$

26. $1^2, 2^2, 3^2, 4^2, \ldots$

27. $\dfrac{1}{3}, \dfrac{1}{3^2}, \dfrac{1}{3^3}, \dfrac{1}{3^4}, \ldots$

28. $-1, 1, -1, 1, \ldots$

29–32 ■ Find the first four partial sums and the nth partial sum of the sequence a_n.

29. $a_n = \dfrac{2}{3^n}$

30. $a_n = \dfrac{1}{n + 1} - \dfrac{1}{n + 2}$

31. $a_n = \sqrt{n} - \sqrt{n + 1}$

32. $a_k = \log\left(\dfrac{k}{k + 1}\right)$ [*Hint:* Use a property of logarithms to write the kth term as a difference.]

33–40 ■ Find the sum.

33. $\displaystyle\sum_{k=1}^{4} k$

34. $\displaystyle\sum_{k=1}^{4} k^2$

35. $\displaystyle\sum_{k=1}^{3} \dfrac{1}{k}$

36. $\displaystyle\sum_{j=1}^{100} (-1)^j$

37. $\displaystyle\sum_{i=1}^{8} [1 + (-1)^i]$

38. $\displaystyle\sum_{i=4}^{12} 10$

39. $\displaystyle\sum_{k=1}^{5} 2^{k-1}$

40. $\displaystyle\sum_{i=1}^{3} i2^i$

41–46 ■ Write the sum without using sigma notation.

41. $\displaystyle\sum_{k=1}^{5} \sqrt{k}$

42. $\displaystyle\sum_{i=0}^{4} \dfrac{2i - 1}{2i + 1}$

43. $\displaystyle\sum_{k=0}^{6} \sqrt{k+4}$ **44.** $\displaystyle\sum_{k=6}^{9} k(k+3)$

45. $\displaystyle\sum_{k=3}^{100} x^k$ **46.** $\displaystyle\sum_{j=1}^{n} (-1)^{j+1} x^j$

47–54 ■ Write the sum using sigma notation.

47. $1 + 2 + 3 + 4 + \cdots + 100$

48. $2 + 4 + 6 + \cdots + 20$

49. $1^2 + 2^2 + 3^2 + \cdots + 10^2$

50. $\dfrac{1}{2\ln 2} - \dfrac{1}{3\ln 3} + \dfrac{1}{4\ln 4} - \dfrac{1}{5\ln 5} + \cdots + \dfrac{1}{100\ln 100}$

51. $\dfrac{1}{1\cdot 2} + \dfrac{1}{2\cdot 3} + \dfrac{1}{3\cdot 4} + \cdots + \dfrac{1}{999\cdot 1000}$

52. $\dfrac{\sqrt{1}}{1^2} + \dfrac{\sqrt{2}}{2^2} + \dfrac{\sqrt{3}}{3^2} + \cdots + \dfrac{\sqrt{n}}{n^2}$

53. $1 + x + x^2 + x^3 + \cdots + x^{100}$

54. $1 - 2x + 3x^2 - 4x^3 + 5x^4 + \cdots - 100x^{99}$

55. Find a formula for the nth term of the sequence

$$\sqrt{2},\ \sqrt{2\sqrt{2}},\ \sqrt{2\sqrt{2\sqrt{2}}},\ \sqrt{2\sqrt{2\sqrt{2\sqrt{2}}}},\ \ldots$$

[*Hint:* Write each term as a power of 2.]

56. Fibonacci posed the following problem: Suppose that rabbits live forever and that every month each pair produces a new pair that becomes productive at age 2 months. If we start with one newborn pair, how many pairs of rabbits will we have in the nth month? Show that the answer is F_n, where F_n is the nth term of the Fibonacci sequence.

DISCOVERY • DISCUSSION

57. Different Sequences that Start the Same
(a) Show that the first four terms of the sequence $a_n = n^2$ are

$$1, 4, 9, 16, \ldots$$

(b) Show that the first four terms of the sequence $a_n = n^2 + (n-1)(n-2)(n-3)(n-4)$ are also

$$1, 4, 9, 16, \ldots$$

(c) Find a sequence whose first six terms are the same as those of $a_n = n^2$ but whose succeeding terms differ from this sequence.

(d) Find two different sequences that begin

$$2, 4, 8, 16, \ldots$$

58. A Recursive Sequence Find the first 40 terms of the sequence defined by

$$a_{n+1} = \begin{cases} \dfrac{a_n}{2} & \text{if } a_n \text{ is an even number} \\ 3a_n + 1 & \text{if } a_n \text{ is an odd number} \end{cases}$$

and $a_1 = 11$. Do the same if $a_1 = 25$. Make a conjecture about this type of sequence.

59. A Different Type of Recursion Find the first ten terms of the sequence defined by

$$a_n = a_{n-a_{n-1}} + a_{n-a_{n-2}}$$

with $a_1 = 1$ and $a_2 = 1$. How is this recursive sequence different from the others in this section?

10.2   **ARITHMETIC SEQUENCES**

Perhaps the simplest way to generate a sequence is to start with a number a and add to it a fixed constant d, over and over again.

DEFINITION OF AN ARITHMETIC SEQUENCE

An **arithmetic sequence** is a sequence of the form

$$a,\ a+d,\ a+2d,\ a+3d,\ a+4d,\ \ldots$$

The number a is the **first term**, and d is the **common difference** of the sequence. The **nth term** of an arithmetic sequence is given by

$$a_n = a + (n-1)d$$

The number d is called the common difference because any two consecutive terms of an arithmetic sequence differ by d.

EXAMPLE 1 ■ Arithmetic Sequences

(a) If $a = 2$ and $d = 3$, then we have the arithmetic sequence

$$2, \ 2 + 3, \ 2 + 6, \ 2 + 9, \ \ldots$$

or $$2, 5, 8, 11, \ldots$$

Any two consecutive terms of this sequence differ by $d = 3$. The nth term is $a_n = 2 + 3(n - 1)$.

(b) Consider the arithmetic sequence

$$9, 4, -1, -6, -11, \ldots$$

Here the common difference is $d = -5$. The terms of an arithmetic sequence decrease if the common difference is negative. The nth term is $a_n = 9 - 5(n - 1)$. ■

An arithmetic sequence is determined completely by the first term a and the common difference d. Thus, if we know the first two terms of an arithmetic sequence, then we can find a formula for the nth term, as the next example shows.

EXAMPLE 2 ■ Finding Terms of an Arithmetic Sequence

Find the first six terms and the 300th term of the arithmetic sequence

$$13, 7, \ldots$$

SOLUTION   Since the first term is 13, we have $a = 13$. The common difference is $d = 7 - 13 = -6$. Thus, the nth term of this sequence is

$$a_n = 13 - 6(n - 1)$$

From this we find the first six terms:

$$13, 7, 1, -5, -11, -17, \ldots$$

The 300th term is $a_{300} = 13 - 6(299) = -1781$. ■

The next example shows that an arithmetic sequence is determined completely by *any* two of its terms.

EXAMPLE 3 ■ Finding Terms of an Arithmetic Sequence

The 11th term of an arithmetic sequence is 52, and the 19th term is 92. Find the 1000th term.

SOLUTION   To find the nth term of this sequence, we need to find a and d in the formula

$$a_n = a + (n - 1)d$$

From this formula, we get

$$a_{11} = a + (11 - 1)d = a + 10d$$

$$a_{19} = a + (19 - 1)d = a + 18d$$

Since $a_{11} = 52$ and $a_{19} = 92$, we get the two equations:

$$\begin{cases} 52 = a + 10d \\ 92 = a + 18d \end{cases}$$

Solving this system for a and d, we get $a = 2$ and $d = 5$. (Verify this.) Thus, the nth term of this sequence is

$$a_n = 2 + 5(n - 1)$$

The 1000th term is $a_{1000} = 2 + 5(999) = 4997$. ∎

PARTIAL SUMS OF ARITHMETIC SEQUENCES

Suppose we want to find the sum of the numbers 1, 2, 3, 4, ..., 100, that is,

$$\sum_{k=1}^{100} k$$

When the famous mathematician C. F. Gauss was a schoolboy, his teacher asked the class this question and expected that it would keep the students busy for a long time. But Gauss answered the question almost immediately. His idea was this: Since we are adding numbers that are produced according to a fixed pattern, there must also be a pattern (or formula) for finding the sum. He started by writing the numbers from 1 to 100 and below them the same numbers in reverse order. Writing S for the sum and adding corresponding terms gives

$$\begin{array}{rcccccccc} S = & 1 + & 2 + & 3 + \cdots + & 98 + & 99 + & 100 \\ S = & 100 + & 99 + & 98 + \cdots + & 3 + & 2 + & 1 \\ \hline 2S = & 101 + & 101 + & 101 + \cdots + & 101 + & 101 + & 101 \end{array}$$

It follows that $2S = 100(101) = 10,100$ and so $S = 5050$.

Of course, the sequence of natural numbers 1, 2, 3, ... is an arithmetic sequence (with $a = 1$ and $d = 1$), and the method used for summing the first 100 terms of this sequence can be used to find a formula for the nth partial sum of any arithmetic sequence. We want to find the sum of the first n terms of the arithmetic sequence whose terms are $a_k = a + (k - 1)d$; that is, we want to find

$$S_n = \sum_{k=1}^{n} [a + (k - 1)d]$$

$$= a + (a + d) + (a + 2d) + (a + 3d) + \cdots + [a + (n - 1)d]$$

Using Gauss's method, we write

$$S_n = \quad a \quad + \quad (a + d) \quad + \cdots + [a + (n - 2)d] + [a + (n - 1)d]$$

$$S_n = [a + (n - 1)d] + [a + (n - 2)d] + \cdots + \quad (a + d) \quad + \quad a$$

$$\overline{2S_n = [2a + (n - 1)d] + [2a + (n - 1)d] + \cdots + [2a + (n - 1)d] + [2a + (n - 1)d]}$$

There are n identical terms on the right side of this equation, so

$$2S_n = n[2a + (n - 1)d]$$

$$S_n = \frac{n}{2}[2a + (n - 1)d]$$

Notice that $a_n = a + (n - 1)d$ is the nth term of this sequence. So, we can write

$$S_n = \frac{n}{2}[a + a + (n - 1)d] = n\left(\frac{a + a_n}{2}\right)$$

This last formula says that the sum of the first n terms of an arithmetic sequence is the average of the first and nth terms multiplied by n, the number of terms in the sequence. We now summarize this result.

PARTIAL SUMS OF AN ARITHMETIC SEQUENCE

The **nth partial sum**

$$S_n = a + (a + d) + (a + 2d) + (a + 3d) + \cdots + [a + (n - 1)d]$$

of the first n terms of an arithmetic sequence is given by either of the following formulas.

1. $S_n = \dfrac{n}{2}[2a + (n - 1)d]$

2. $S_n = n\left(\dfrac{a + a_n}{2}\right)$

EXAMPLE 4 ■ Finding a Partial Sum of an Arithmetic Sequence

Find the sum of the first 50 odd numbers.

SOLUTION   The odd numbers form an arithmetic sequence with $a = 1$ and $d = 2$. The nth term is $a_n = 1 + 2(n - 1) = 2n - 1$, so the 50th odd number is $a_{50} = 2(50) - 1 = 99$. Substituting in Formula 2 for the partial sum of an arithmetic sequence, we get

$$S_{50} = 50\left(\frac{a + a_{50}}{2}\right) = 50\left(\frac{1 + 99}{2}\right) = 50 \cdot 50 = 2500$$

■

EXAMPLE 5 ■ Finding a Partial Sum of an Arithmetic Sequence

Find the sum of the first 40 terms of the arithmetic sequence

$$3, 7, 11, 15, \ldots$$

SOLUTION   For this arithmetic sequence, $a = 3$ and $d = 4$. Using Formula 1 for the partial sum of an arithmetic sequence, we get

$$S_{40} = \tfrac{40}{2}[2(3) + (40 - 1)4] = 20(6 + 156) = 3240 \qquad ■$$

EXAMPLE 6 ■ Finding the Seating Capacity of an Amphitheater

An amphitheater has 50 rows of seats with 30 seats in the first row, 32 in the second, 34 in the third, and so on. Find the total number of seats.

SOLUTION   The numbers of seats in the rows form an arithmetic sequence with $a = 30$ and $d = 2$. Since there are 50 rows, the total number of seats is the sum

$$S_{50} = \tfrac{50}{2}[2(30) + 49(2)] \qquad \text{Formula 1}$$

$$= 3950$$

Thus, there are 3950 seats in the amphitheater.   ■

EXAMPLE 7 ■ Finding the Number of Terms in a Partial Sum

How many terms of the arithmetic sequence 5, 7, 9, ... must be added to get 572?

SOLUTION   We are asked to find n when $S_n = 572$. Substituting $a = 5$, $d = 2$, and $S_n = 572$ in Formula 1 for the partial sum of an arithmetic sequence, we get

$$572 = \frac{n}{2}[2 \cdot 5 + (n - 1)2] \qquad \text{Formula 1}$$

$$572 = 5n + n(n - 1)$$

$$0 = n^2 + 4n - 572$$

$$0 = (n - 22)(n + 26)$$

This gives $n = 22$ or $n = -26$. But since n is the *number* of terms in this partial sum, we must have $n = 22$.   ■

10.2   EXERCISES

1–6 ■ Determine whether the sequence is arithmetic. If it is arithmetic, find the common difference.

1. $5, 8, 11, 14, \ldots$

2. $3, 6, 9, 13, \ldots$

3. $2, 4, 8, 16, \ldots$

4. $2, 4, 6, 8, \ldots$

5. $3, \frac{3}{2}, 0, -\frac{3}{2}, \ldots$

6. $\ln 2, \ln 4, \ln 8, \ln 16, \ldots$

7–16 ■ Determine the common difference, the fifth term, the nth term and the 100th term of the arithmetic sequence.

7. 2, 5, 8, 11, …

8. 1, 5, 9, 13, …

9. 4, 9, 14, 19, …

10. 11, 8, 5, 2, …

11. −12, −8, −4, 0, …

12. $\frac{7}{6}, \frac{5}{3}, \frac{13}{6}, \frac{8}{3}, \dots$

13. 25, 26.5, 28, 29.5, …

14. 15, 12.3, 9.6, 6.9, …

15. 2, 2 + s, 2 + 2s, 2 + 3s, …

16. −t, −t + 3, −t + 6, −t + 9, …

17. The tenth term of an arithmetic sequence is $\frac{55}{2}$, and the second term is $\frac{7}{2}$. Find the first term.

18. The 12th term of an arithmetic sequence is 32, and the fifth term is 18. Find the 20th term.

19. The 100th term of an arithmetic sequence is 98, and the common difference is 2. Find the first three terms.

20. The 20th term of an arithmetic sequence is 101, and the common difference is 3. Find a formula for the nth term.

21. Which term of the arithmetic sequence 1, 4, 7, … is 88?

22. The first term of an arithmetic sequence is 1, and the common difference is 4. Is 11,937 a term of this sequence? If so, which term is it?

23–28 ■ Find the partial sum S_n of the arithmetic sequence that satisfies the given conditions.

23. $a = 1$, $d = 2$, $n = 10$

24. $a = 3$, $d = 2$, $n = 12$

25. $a = 4$, $d = 2$, $n = 20$

26. $a = 100$, $d = -5$, $n = 8$

27. $a_1 = 55$, $d = 12$, $n = 10$

28. $a_2 = 8$, $a_5 = 9.5$, $n = 15$

29–34 ■ A partial sum of an arithmetic sequence is given. Find the sum.

29. $1 + 5 + 9 + \dots + 401$

30. $-3 + \left(-\frac{3}{2}\right) + 0 + \frac{3}{2} + 3 + \dots + 30$

31. $0.7 + 2.7 + 4.7 + \dots + 56.7$

32. $-10 - 9.9 - 9.8 - \dots - 0.1$

33. $\displaystyle\sum_{k=0}^{10} (3 + 0.25k)$

34. $\displaystyle\sum_{n=0}^{20} (1 - 2n)$

35. The purchase value of an office computer is \$12,500. Its annual depreciation is \$1875. Find the value of the computer after 6 years.

36. Telephone poles are stored in a pile with 25 poles in the first layer, 24 in the second, and so on. If there are 12 layers, how many telephone poles does the pile contain?

37. A man gets a job with a salary of \$30,000 a year. He is promised a \$2300 raise each subsequent year. Find his total earnings for a 10-year period.

38. In a drive-in theater there are spaces for 20 cars in the first parking row, 22 in the second, 24 in the third, and so on. If there are 21 rows in the theater, find the number of cars that can be parked.

39. An architect designs a theater with 15 seats in the first row, 18 in the second, 21 in the third, and so on. If the theater is to have a seating capacity of 870, how many rows must the architect use in his design?

40. An arithmetic sequence has first term $a = 5$ and common difference $d = 2$. How many terms of this sequence must be added to get 2700?

41. When an object is allowed to fall freely near the surface of the earth, the gravitational pull is such that the object falls 16 ft in the first second, 48 ft in the next second, 80 ft in the next second, and so on.
(a) Find the total distance a ball falls in 6 s.
(b) Find a formula for the total distance a ball falls in n seconds.

42. In the well-known song "The Twelve Days of Christmas," a person gives his sweetheart k gifts on the kth day for each of the 12 days of Christmas. The person also repeats each gift identically on each subsequent day. Thus, on the 12th day the sweetheart receives a gift for the first day, 2 gifts for the second, 3 gifts for the third, and so on. Show that the number of gifts received on the 12th day is a partial sum of an arithmetic sequence. Find this sum.

43. Show that a right triangle whose sides are in arithmetic progression is similar to a 3–4–5 triangle.

44. Find the product of the numbers
$$10^{1/10},\ 10^{2/10},\ 10^{3/10},\ 10^{4/10},\ \ldots,\ 10^{19/10}$$

45. A sequence is **harmonic** if the reciprocals of the terms of the sequence form an arithmetic sequence. Determine whether the following sequence is harmonic:
$$1, \tfrac{3}{5}, \tfrac{3}{7}, \tfrac{1}{3}, \ldots$$

46. The **harmonic mean** of two numbers is the reciprocal of the average of the reciprocals of the two numbers. Find the harmonic mean of 3 and 5.

 DISCOVERY · DISCUSSION

47. Arithmetic Means The **arithmetic mean** (or average) of two numbers a and b is
$$m = \frac{a + b}{2}$$

Note that m is the same distance from a as from b, so a, m, b is an arithmetic sequence. In general, if $m_1, m_2, \ldots, m_k$ are equally spaced between a and b so that
$$a, m_1, m_2, \ldots, m_k, b$$
is an arithmetic sequence, then $m_1, m_2, \ldots, m_k$ are called k arithmetic means between a and b.

(a) Insert two arithmetic means between 10 and 18.

(b) Insert three arithmetic means between 10 and 18.

(c) Suppose a doctor wants to increase a patient's dosage of a certain medicine from 100 mg to 300 mg per day in five equal steps. How many arithmetic means must be inserted between 100 and 300 to give the progression of daily doses, and what are these means?

10.3   GEOMETRIC SEQUENCES

Another simple way of generating a sequence is to start with a number a and repeatedly multiply by a fixed nonzero constant r.

> **DEFINITION OF A GEOMETRIC SEQUENCE**
>
> A **geometric sequence** is a sequence of the form
> $$a,\ ar,\ ar^2,\ ar^3,\ ar^4,\ \ldots$$
> The number a is the **first term**, and r is the **common ratio** of the sequence. The **nth term** of a geometric sequence is given by
> $$a_n = ar^{n-1}$$

The number r is called the common ratio because the ratio of any two consecutive terms of the sequence is r.

EXAMPLE 1 ■ Geometric Sequences

(a) If $a = 3$ and $r = 2$, then we have the geometric sequence
$$3,\ 3 \cdot 2,\ 3 \cdot 2^2,\ 3 \cdot 2^3,\ 3 \cdot 2^4,\ \ldots$$
or
$$3, 6, 12, 24, 48, \ldots$$
Notice that the ratio of any two consecutive terms is $r = 2$. The nth term is $a_n = 3(2)^{n-1}$.

(b) The sequence
$$2,\ -10,\ 50,\ -250,\ 1250,\ \ldots$$

is a geometric sequence with $a = 2$ and $r = -5$. When r is negative, the terms of the sequence alternate in sign. The nth term is $a_n = 2(-5)^{n-1}$.

(c) The sequence

$$1, \tfrac{1}{3}, \tfrac{1}{9}, \tfrac{1}{27}, \tfrac{1}{81}, \ldots$$

is a geometric sequence with $a = 1$ and $r = \tfrac{1}{3}$. The nth term is $a_n = 1\left(\tfrac{1}{3}\right)^{n-1}$.

If $0 < r < 1$, then the terms of the sequence decrease, but if $r > 1$, then the terms increase. (What happens if $r = 1$?) ∎

Geometric sequences occur naturally. Here is a simple example: Suppose a ball has elasticity such that when it is dropped it bounces up one-third of the distance it has fallen. If this ball is dropped from a height of 2 m, then it bounces up to a height of $2\left(\tfrac{1}{3}\right) = \tfrac{2}{3}$ m. On its second bounce, it returns to a height of $\left(\tfrac{2}{3}\right)\left(\tfrac{1}{3}\right) = \tfrac{2}{9}$ m, and so on (see Figure 1). Thus, the height h_n that the ball reaches on its nth bounce is given by the geometric sequence

$$h_n = \tfrac{2}{3}\left(\tfrac{1}{3}\right)^{n-1} = 2\left(\tfrac{1}{3}\right)^{n}$$

We can find the nth term of a geometric sequence if we know any two terms, as the following examples show.

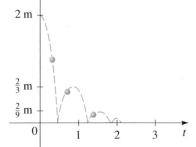

FIGURE 1

EXAMPLE 2 ■ Finding Terms of a Geometric Sequence

Find the eighth term of the geometric sequence 5, 15, 45, ...

SOLUTION To find a formula for the nth term of this sequence, we need to find a and r. Clearly, $a = 5$. To find r, we find the ratio of any two consecutive terms. For instance, $r = \tfrac{45}{15} = 3$. Thus

$$a_n = 5(3)^{n-1}$$

The eighth term is $a_8 = 5(3)^{8-1} = 5(3)^7 = 10{,}935$. ∎

EXAMPLE 3 ■ Finding Terms of a Geometric Sequence

The third term of a geometric series is $\tfrac{63}{4}$, and the sixth term is $\tfrac{1701}{32}$. Find the fifth term.

SOLUTION Since this series is geometric, its nth term is given by the formula $a_n = ar^{n-1}$. Thus

$$a_3 = ar^{3-1} = ar^2$$

$$a_6 = ar^{6-1} = ar^5$$

From the values we are given for these two terms, we get the following system of equations:

$$\begin{cases} \tfrac{63}{4} = ar^2 \\ \tfrac{1701}{32} = ar^5 \end{cases}$$

One way to solve this system is to divide:

$$\frac{ar^5}{ar^2} = \frac{\frac{1701}{32}}{\frac{63}{4}}$$

$$r^3 = \frac{27}{8} \qquad \text{Simplify}$$

$$r = \frac{3}{2} \qquad \text{Take the cube root of each side}$$

Substituting for r in the first equation, $\frac{63}{4} = ar^2$, gives

$$\frac{63}{4} = a\left(\frac{3}{2}\right)^2$$

$$a = 7$$

It follows that the nth term of this sequence is

$$a_n = 7\left(\frac{3}{2}\right)^{n-1}$$

and so the fifth term is

$$a_5 = 7\left(\frac{3}{2}\right)^{5-1} = 7\left(\frac{3}{2}\right)^4 = \frac{567}{16} \qquad \blacksquare$$

PARTIAL SUMS OF GEOMETRIC SEQUENCES

For the geometric sequence a, ar, ar^2, ar^3, ar^4, ..., ar^{n-1}, ..., the nth partial sum is

$$S_n = \sum_{k=1}^{n} ar^{k-1} = a + ar + ar^2 + ar^3 + ar^4 + \cdots + ar^{n-1}$$

To find a formula for S_n, we multiply S_n by r and subtract from S_n to get

$$S_n = a + ar + ar^2 + ar^3 + ar^4 + \cdots + ar^{n-1}$$
$$rS_n = \qquad ar + ar^2 + ar^3 + ar^4 + \cdots + ar^{n-1} + ar^n$$
$$\overline{}$$
$$S_n - rS_n = a - ar^n$$

So,

$$S_n(1 - r) = a(1 - r^n)$$

$$S_n = \frac{a(1 - r^n)}{1 - r} \qquad (r \neq 1)$$

We summarize this result.

PARTIAL SUMS OF A GEOMETRIC SEQUENCE

The *nth partial sum*

$$S_n = a + ar + ar^2 + ar^3 + ar^4 + \cdots + ar^{n-1} \qquad (r \neq 1)$$

of the first n terms of a geometric sequence is given by

$$S_n = a\frac{1 - r^n}{1 - r}$$

EXAMPLE 4 ■ Finding a Partial Sum of a Geometric Sequence

Find the sum of the first five terms of the geometric sequence

$$1, 0.7, 0.49, 0.343, \ldots$$

SOLUTION The required sum is the sum of the first five terms of a geometric sequence with $a = 1$ and $r = 0.7$. Using the formula for S_n with $n = 5$, we get

$$S_5 = 1 \frac{1 - (0.7)^5}{1 - 0.7} = 2.7731$$

Thus, the sum of the first five terms of this sequence is 2.7731. ■

EXAMPLE 5 ■ Finding a Partial Sum of a Geometric Sequence

Find the sum $\displaystyle\sum_{k=1}^{5} 7\left(-\frac{2}{3}\right)^k$.

SOLUTION The given sum is the fifth partial sum of a geometric sequence with first term $a = 7\left(-\frac{2}{3}\right) = -\frac{14}{3}$ and common ratio $r = -\frac{2}{3}$. Thus, by the formula for S_n, we have

$$S_5 = -\frac{14}{3} \cdot \frac{1 - \left(-\frac{2}{3}\right)^5}{1 - \left(-\frac{2}{3}\right)} = -\frac{14}{3} \cdot \frac{1 + \frac{32}{243}}{\frac{5}{3}} = -\frac{770}{243}$$ ■

WHAT IS AN INFINITE SERIES?

An expression of the form

$$a_1 + a_2 + a_3 + a_4 + \cdots$$

is called an **infinite series**. The dots mean that we are to continue the addition indefinitely. What meaning can we attach to the sum of infinitely many numbers? It seems at first that it is not possible to add infinitely many numbers and arrive at a finite number. But consider the following problem. You have a cake and you want to eat it by first eating half the cake, then eating half of what remains, then again eating half of what remains. This process can continue indefinitely because at each stage some of the cake remains. (See Figure 2.)

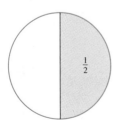
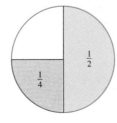
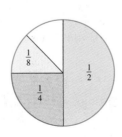
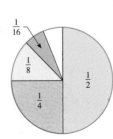
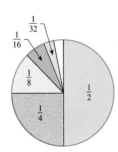

FIGURE 2

Does this mean that it's impossible to eat all of the cake? Of course not. Let's write down what you have eaten from this cake:

$$\frac{1}{2} + \frac{1}{4} + \frac{1}{8} + \frac{1}{16} + \cdots + \frac{1}{2^n} + \cdots$$

This is an infinite series, and we note two things about it: First, from Figure 2 it's clear that no matter how many terms of this series we add, the total will never exceed 1. Second, the more terms of this series we add, the closer the sum is to 1 (see Figure 2). This suggests that the number 1 can be written as the sum of infinitely many smaller numbers:

$$1 = \frac{1}{2} + \frac{1}{4} + \frac{1}{8} + \frac{1}{16} + \cdots + \frac{1}{2^n} + \cdots$$

To make this more precise, let's look at the partial sums of this series:

$$S_1 = \frac{1}{2} \qquad\qquad\qquad\qquad = \frac{1}{2}$$

$$S_2 = \frac{1}{2} + \frac{1}{4} \qquad\qquad\qquad = \frac{3}{4}$$

$$S_3 = \frac{1}{2} + \frac{1}{4} + \frac{1}{8} \qquad\qquad = \frac{7}{8}$$

$$S_4 = \frac{1}{2} + \frac{1}{4} + \frac{1}{8} + \frac{1}{16} = \frac{15}{16}$$

$$\vdots$$

and, in general (see Example 5 of Section 10.1),

$$S_n = 1 - \frac{1}{2^n}$$

As n gets larger and larger, we are adding more and more of the terms of this series. Intuitively, as n gets larger, S_n gets closer to the sum of the series. Now notice that as n gets large, $1/2^n$ gets closer and closer to 0. Thus, S_n gets close to $1 - 0 = 1$. Using the notation of Section 3.5, we can write

$$S_n \to 1 \qquad \text{as} \qquad n \to \infty$$

In general, if S_n gets close to a finite number S as n gets large, we say that S is the **sum of the infinite series.**

INFINITE GEOMETRIC SERIES

We call an infinite series of the form

$$a + ar + ar^2 + ar^3 + ar^4 + \cdots + ar^{n-1} + \cdots$$

Here is another way to arrive at the formula for the sum of an infinite geometric series:

$$S = a + ar + ar^2 + ar^3 + \cdots$$
$$= a + r(a + ar + ar^2 + \cdots)$$
$$= a + rS$$

Solve the equation $S = a + rS$ for S to get

$$S - rS = a$$
$$(1 - r)S = a$$
$$S = \frac{a}{1 - r}$$

an **infinite geometric series**. We can apply the reasoning used earlier to find the sum of an infinite geometric series. The nth partial sum of such a series is given by the formula

$$S_n = a\,\frac{1 - r^n}{1 - r} \qquad (r \neq 1)$$

It can be shown that if $|r| < 1$, then r^n gets close to 0 as n gets large (you can easily convince yourself of this using a calculator). It follows that S_n gets close to $a/(1 - r)$ as n gets large, or

$$S_n \to \frac{a}{1 - r} \qquad \text{as} \qquad n \to \infty$$

Thus, the sum of this infinite geometric series is $a/(1 - r)$.

SUM OF AN INFINITE GEOMETRIC SERIES

If $|r| < 1$, then the infinite geometric series

$$a + ar + ar^2 + ar^3 + ar^4 + \cdots + ar^{n-1} + \cdots$$

has the sum

$$S = \frac{a}{1 - r}$$

EXAMPLE 6 ■ Finding the Sum of an Infinite Geometric Series

Find the sum of the infinite geometric series

$$2 + \frac{2}{5} + \frac{2}{25} + \frac{2}{125} + \cdots + \frac{2}{5^n} + \cdots$$

SOLUTION We use the formula for the sum of an infinite geometric series. In this case, $a = 2$ and $r = \frac{1}{5}$. Thus, the sum of this infinite series is

$$S = \frac{2}{1 - \frac{1}{5}} = \frac{5}{2}$$

EXAMPLE 7 ■ Writing a Repeated Decimal as a Fraction

Find the fraction that represents the rational number $2.3\overline{51}$.

SOLUTION This repeating decimal can be written as a series:

$$\frac{23}{10} + \frac{51}{1000} + \frac{51}{100,000} + \frac{51}{10,000,000} + \frac{51}{1,000,000,000} + \cdots$$

After the first term, the terms of this series form an infinite geometric series with

$$a = \frac{51}{1000} \qquad \text{and} \qquad r = \frac{1}{100}$$

Thus, the sum of this part of the series is

$$S = \frac{\frac{51}{1000}}{1 - \frac{1}{100}} = \frac{\frac{51}{1000}}{\frac{99}{100}} = \frac{51}{1000} \cdot \frac{100}{99} = \frac{51}{990}$$

So,

$$2.3\overline{51} = \frac{23}{10} + \frac{51}{990} = \frac{2328}{990} = \frac{388}{165}$$

$\blacksquare$

10.3   EXERCISES

1–6 $\blacksquare$ Determine whether the sequence is geometric. If it is geometric, find the common ratio.

1. 2, 4, 8, 16, . . .

2. 2, 6, 18, 36, . . .

3. 3, $\frac{3}{2}$, $\frac{3}{4}$, $\frac{3}{8}$, . . .

4. 27, −9, 3, −1, . . .

5. $\frac{1}{2}$, $\frac{1}{3}$, $\frac{1}{4}$, $\frac{1}{5}$, . . .

6. e^2, e^4, e^6, e^8, . . .

7–16 $\blacksquare$ Determine the common ratio, the fifth term, and the nth term of the geometric sequence.

7. 2, 6, 18, 54, . . .

8. 7, $\frac{14}{3}$, $\frac{28}{9}$, $\frac{56}{27}$, . . .

9. 0.3, −0.09, 0.027, −0.0081, . . .

10. 1, $\sqrt{2}$, 2, 2$\sqrt{2}$, . . .

11. 144, −12, 1, −$\frac{1}{12}$, . . .

12. −8, −2, −$\frac{1}{2}$, −$\frac{1}{8}$, . . .

13. 3, $3^{5/3}$, $3^{7/3}$, 27, . . .

14. t, $\dfrac{t^2}{2}$, $\dfrac{t^3}{4}$, $\dfrac{t^4}{8}$, . . .

15. 1, $s^{2/7}$, $s^{4/7}$, $s^{6/7}$, . . .

16. 5, 5^{c+1}, 5^{2c+1}, 5^{3c+1}, . . .

17. The first term of a geometric sequence is 8, and the second term is 4. Find the fifth term.

18. The first term of a geometric sequence is 3, and the third term is $\frac{4}{3}$. Find the fifth term.

19. The common ratio in a geometric sequence is $\frac{2}{5}$, and the fourth term is $\frac{5}{2}$. Find the third term.

20. The common ratio in a geometric sequence is $\frac{3}{2}$, and the fifth term is 1. Find the first three terms.

21. Which term of the geometric sequence 2, 6, 18, . . . is 118,098?

22. The second and the fifth terms of a geometric sequence are 10 and 1250, respectively. Is 31,250 a term of this sequence? If so, which term is it?

23–26 $\blacksquare$ Find the partial sum S_n of the geometric sequence that satisfies the given conditions.

23. $a = 5$, $r = 2$, $n = 6$

24. $a = \frac{2}{3}$, $r = \frac{1}{3}$, $n = 4$

25. $a_3 = 28$, $a_6 = 224$, $n = 6$

26. $a_2 = 0.12$, $a_5 = 0.00096$, $n = 4$

27–30 $\blacksquare$ Find the sum.

27. $1 + 3 + 9 + \cdots + 2187$

28. $1 - \frac{1}{2} + \frac{1}{4} - \frac{1}{8} + \cdots - \frac{1}{512}$

29. $\displaystyle\sum_{k=0}^{10} 3\left(\frac{1}{2}\right)^k$

30. $\displaystyle\sum_{j=0}^{5} 7\left(\frac{3}{2}\right)^j$

31. A ball is dropped from a height of 80 ft. The elasticity of this ball is such that it rebounds three-fourths of the distance it has fallen. How high does the ball rebound on the fifth bounce? Find a formula for how high the ball rebounds on the nth bounce.

32. A culture initially has 5000 bacteria, and its size increases by 8% every hour. How many bacteria are present at the end of 5 hours? Find a formula for the number of bacteria present after n hours.

33. A truck radiator holds 5 gal and is filled with water. A gallon of water is removed from the radiator and replaced with a gallon of antifreeze; then, a gallon of the mixture is removed from the radiator and again replaced by a gallon of antifreeze. This process is repeated indefinitely. How much water remains in the tank after this process is repeated 3 times? 5 times? n times?

34. A very patient woman wishes to become a billionaire. She decides to follow a simple scheme: She puts aside 1 cent the first day, 2 cents the second day, 4 cents the third day, and so on, doubling the number of cents each day. How much money will she have at the end of 30 days? How many days will it take this woman to realize her wish?

35. A ball is dropped from a height of 9 ft. The elasticity of the ball is such that it always bounces upward one-third the distance it has fallen.
(a) Find the total distance the ball has traveled at the instant it hits the ground the fifth time.
(b) Find a formula for the total distance the ball has traveled at the instant it hits the ground the nth time.

36. The following is a well-known children's rhyme:

> As I was going to St. Ives
>
> I met a man with seven wives;
>
> Every wife had seven sacks;
>
> Every sack had seven cats;
>
> Every cat had seven kits;
>
> Kits, cats, sacks, and wives,
>
> How many were going to St. Ives?

Assuming that the entire group is actually going to St. Ives, show that the answer to the question in the rhyme is a partial sum of a geometric sequence, and find the sum.

37–44 ■ Find the sum of the infinite geometric series.

37. $1 + \dfrac{1}{3} + \dfrac{1}{9} + \dfrac{1}{27} + \cdots$

38. $1 - \dfrac{1}{2} + \dfrac{1}{4} - \dfrac{1}{8} + \cdots$

39. $1 - \dfrac{1}{3} + \dfrac{1}{9} - \dfrac{1}{27} + \cdots$

40. $\dfrac{2}{5} + \dfrac{4}{25} + \dfrac{8}{125} + \cdots$

41. $\dfrac{1}{3^6} + \dfrac{1}{3^8} + \dfrac{1}{3^{10}} + \dfrac{1}{3^{12}} + \cdots$

42. $3 - \dfrac{3}{2} + \dfrac{3}{4} - \dfrac{3}{8} + \cdots$

43. $-\dfrac{100}{9} + \dfrac{10}{3} - 1 + \dfrac{3}{10} - \cdots$

44. $\dfrac{1}{\sqrt{2}} + \dfrac{1}{2} + \dfrac{1}{2\sqrt{2}} + \dfrac{1}{4} + \cdots$

45–50 ■ Express the repeating decimal as a fraction.

45. $0.777\ldots$

46. $0.2\overline{53}$

47. $0.030303\ldots$

48. $2.11\overline{25}$

49. $0.\overline{112}$

50. $0.123123123\ldots$

51. A certain ball rebounds to half the height from which it is dropped. Use an infinite geometric series to approximate the total distance the ball travels, after being dropped from 1 m above the ground, until it comes to rest.

52. If the ball in Exercise 51 is dropped from a height of 8 ft, then 1 s is required for its first complete bounce—from the instant it first touches the ground until it next touches the ground. Each subsequent complete bounce requires $1/\sqrt{2}$ as long as the preceding complete bounce. Use an infinite geometric series to estimate the time period from the instant the ball first touches the ground until it stops bouncing.

53. The midpoints of the sides of a square of side 1 are joined to form a new square. This procedure is repeated for each new square. (See the figure.)
(a) Find the sum of the areas of all the squares.
(b) Find the sum of the perimeters of all the squares.

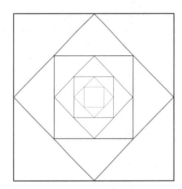

54. A circular disk of radius R is cut out of paper, as shown in figure (a). Two disks of radius $\tfrac{1}{2}R$ are cut out of paper and placed on top of the first disk, as in figure (b), and then four disks of radius $\tfrac{1}{4}R$ are placed on these two disks [figure (c)]. Assuming that this process can be repeated indefinitely, find the total area of all the disks.

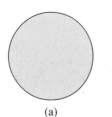 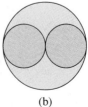 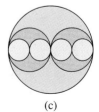

(a)   (b)   (c)

55. A yellow square of side 1 is divided into nine smaller squares, and the middle square is colored blue as shown in the figure. Each of the smaller yellow squares is in turn divided into nine squares, and each middle square is colored blue. If this process is continued indefinitely, what is the total area colored blue?

56. If the numbers $a_1, a_2, \ldots, a_n$ form a geometric sequence, then $a_2, a_3, \ldots, a_{n-1}$ are **geometric means** between a_1 and a_n. Insert three geometric means between 5 and 80.

57. Find the sum of the first ten terms of the sequence

$$a + b, \ a^2 + 2b, \ a^3 + 3b, \ a^4 + 4b, \ \ldots$$

 DISCOVERY · DISCUSSION

58. Arithmetic or Geometric?   The first four terms of a sequence are given. Determine whether these terms can be the terms of an arithmetic sequence, a geometric sequence, or neither. Find the next term if the sequence is arithmetic or geometric.

(a) $5, -3, 5, -3, \ldots$ (b) $\frac{1}{3}, 1, \frac{5}{3}, \frac{7}{3}, \ldots$

(c) $\sqrt{3}, 3, 3\sqrt{3}, 9, \ldots$
(d) $1, -1, 1, -1, \ldots$
(e) $2, -1, \frac{1}{2}, 2, \ldots$
(f) $x - 1, \ x, \ x + 1, \ x + 2, \ \ldots$
(g) $-3, -\frac{3}{2}, 0, \frac{3}{2}, \ldots$
(h) $\sqrt{5}, \sqrt[3]{5}, \sqrt[6]{5}, 1, \ldots$

59. Reciprocals of a Geometric Sequence   If $a_1, a_2, a_3, \ldots$ is a geometric sequence with common ratio r, show that the sequence

$$\frac{1}{a_1}, \frac{1}{a_2}, \frac{1}{a_3}, \ldots$$

is also a geometric sequence, and find the common ratio.

60. Logarithms of a Geometric Sequence   If $a_1, a_2, a_3, \ldots$ is a geometric sequence with common ratio $r > 0$ and $a_1 > 0$, show that the sequence

$$\log a_1, \ \log a_2, \ \log a_3, \ \ldots$$

is an arithmetic sequence, and find the common difference.

61. Exponentials of an Arithmetic Sequence   If $a_1, a_2, a_3, \ldots$ is an arithmetic sequence with common difference d, show that the sequence

$$10^{a_1}, \ 10^{a_2}, \ 10^{a_3}, \ \ldots$$

is a geometric sequence, and find the common ratio.

10.4   **ANNUITIES AND INSTALLMENT BUYING**

Many financial transactions involve payments that are made at regular intervals. For example, if you deposit $100 each month in an interest-bearing account, what will the value of the account be at the end of 5 years? If you borrow $100,000 to buy a house, how much will the monthly payments be in order to pay off the loan in 30 years? Each of these questions involves the sum of a sequence of numbers; we use the results of the preceding section to answer them here.

 THE AMOUNT OF AN ANNUITY

An **annuity** is a sum of money that is paid in regular equal payments. Although the word *annuity* suggests annual (or yearly) payments, they can be made semi-annually, quarterly, monthly, or at some other regular interval. Payments are usually made at the end of the payment interval. The **amount of an annuity** is the sum of all the individual payments from the time of the first payment until the last payment is made, together with all the interest. We denote this sum by A_f (the subscript f here is used to denote *final* amount).

EXAMPLE 1 ■ Calculating the Amount of an Annuity

An investor deposits $400 every December 15 and June 15 for 10 years into an account that earns interest at the rate of 8% per year, compounded semi-annually. How much will be in the account immediately after the last payment?

SOLUTION   We need to find the amount of an annuity consisting of 20 semi-annual payments of $400 each. Since the interest rate is 8% per year, com-pounded semiannually, the interest rate per time period is $i = 0.04$. The first payment is in the account for 19 time periods, the second for 18 time periods, and so on. The last payment receives no interest. The situation can be illus-trated by the time line in Figure 1.

⊘ When using interest rates in calculators, remember to convert percentages to decimals. For example, 8% is 0.08.

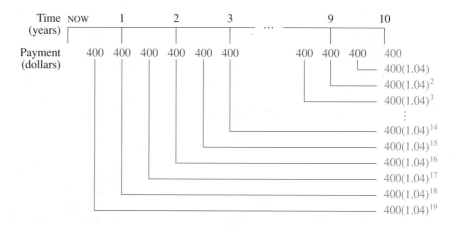

FIGURE 1

The amount A_f of the annuity is the sum of these 20 amounts. Thus

$$A_f = 400 + 400(1.04) + 400(1.04)^2 + \cdots + 400(1.04)^{19}$$

But this is a geometric series with $a = 400$, $r = 1.04$, and $n = 20$, so

$$A_f = 400\ \frac{1 - (1.04)^{20}}{1 - 1.04} \approx 11{,}911.23$$

Thus, the amount of the annuity after the last payment is $11,911.23.   ■

In general, the regular annuity payment is called the **periodic rent** and is denoted by R. We also let i denote the interest rate per time period and n the num-ber of payments. *We always assume that the time period in which interest is com-pounded is equal to the time between payments.* By the same reasoning as in Example 1, we see that the amount A_f of an annuity is

$$A_f = R + R(1 + i) + R(1 + i)^2 + \cdots + R(1 + i)^{n-1}$$

Since this is the nth partial sum of a geometric sequence with $a = R$ and

$r = 1 + i$, the formula for the partial sum gives

$$A_f = R \frac{1 - (1 + i)^n}{1 - (1 + i)} = R \frac{1 - (1 + i)^n}{-i} = R \frac{(1 + i)^n - 1}{i}$$

AMOUNT OF AN ANNUITY

The amount A_f of an annuity consisting of n regular equal payments of size R with interest rate i per time period is given by

$$A_f = R \frac{(1 + i)^n - 1}{i}$$

EXAMPLE 2 ■ Calculating the Amount of an Annuity

How much money should be invested every month at 12% per year, compounded monthly, in order to have $4000 in 18 months?

SOLUTION   In this problem $i = 0.12/12 = 0.01$, $A_f = 4000$, and $n = 18$. We need to find the amount R of each payment. By the formula for the amount of an annuity,

$$4000 = R \frac{(1 + 0.01)^{18} - 1}{0.01}$$

Solving for R, we get

$$R = \frac{4000(0.01)}{(1 + 0.01)^{18} - 1} \approx \frac{40}{1.196147 - 1} \approx 203.928$$

Thus, the monthly investment should be $203.93.   ■

THE PRESENT VALUE OF AN ANNUITY

If you were to receive $10,000 five years from now, it would be worth much less than getting $10,000 right now. This is because of the interest you could accumulate during the next five years if you invested the money now. What smaller amount would you be willing to accept *now* instead of receiving $10,000 in five years? This is the amount of money that, together with interest, would be worth $10,000 in five years. The amount we are looking for here is called the *discounted value,* or *present value.* If the interest rate is 8% per year, compounded quarterly, then the interest per time period is $i = 0.08/4 = 0.02$, and there are $4 \times 5 = 20$ time periods. If we let PV denote the present value, then by the formula for compound interest (Section 4.2) we have

$$10,000 = PV(1 + i)^n = PV(1 + 0.02)^{20}$$

so

$$PV = 10,000(1 + 0.02)^{-20} \approx 6729.713$$

Thus, in this situation, the present value of $10,000 is $6729.71. This reasoning leads to a general formula for present value:

$$PV = A(1 + i)^{-n}$$

Similarly, the **present value of an annuity** is the amount A_p that must be invested now at the interest rate i per time period in order to provide n payments, each of amount R. Clearly, A_p is the sum of the present values of each of the individual payments (see Exercise 22). Another way of finding A_p is to note that A_p is the present value of A_f:

$$A_p = A_f(1 + i)^{-n} = R\,\frac{(1 + i)^n - 1}{i}(1 + i)^{-n} = R\,\frac{1 - (1 + i)^{-n}}{i}$$

THE PRESENT VALUE OF AN ANNUITY

The **present value** A_p of an annuity consisting of n regular equal payments of size R and interest rate i per time period is given by

$$A_p = R\,\frac{1 - (1 + i)^{-n}}{i}$$

EXAMPLE 3 ■ Calculating the Present Value of an Annuity

A person wins $10,000,000 in the California lottery, and the amount is paid in yearly installments of half a million dollars each for 20 years. What is the present value of his winnings? Assume that he can earn 10% interest, compounded annually.

SOLUTION   Since the amount won is paid as an annuity, we need to find its present value. Here $i = 0.1$, $R = \$500,000$, and $n = 20$. Thus

$$A_p = 500,000\,\frac{1 - (1 + 0.1)^{-20}}{0.1} \approx 4{,}256{,}781.859$$

This means that the winner really won only $4,256,781.86 if it were paid immediately. ■

INSTALLMENT BUYING

When you buy a house or a car by installment, the payments you make are an annuity whose present value is the amount of the loan.

EXAMPLE 4 ■ The Amount of a Loan

A student wishes to buy a car. He can afford to pay $200 per month but has no money for a down payment. If he can make these payments for four years and if the interest rate is 12%, what purchase price can he afford for a car?

SOLUTION   The payments the student makes constitute an annuity whose present value is the price of the car (which is also the amount of the loan, in this case). Here we have $i = 0.12/12 = 0.01$, $R = 200$, $n = 12 \times 4 = 48$, so

$$A_p = R\,\frac{1 - (1 + i)^{-n}}{i} = 200\,\frac{1 - (1 + 0.01)^{-48}}{0.01} \approx 7594.792$$

Thus, the student can buy a car worth $7594.79.   ■

When a bank makes a loan that is to be repaid with regular equal payments R, then the payments form an annuity whose present value A_p is the amount of the loan. So, to find the size of the payments, we solve for R in the formula for the amount of an annuity. This gives the following formula for R.

INSTALLMENT BUYING

If a loan A_p is to be repaid in n regular equal payments with interest rate i per time period, then the size R of each payment is given by

$$R = \frac{iA_p}{1 - (1 + i)^{-n}}$$

EXAMPLE 5 ■ Calculating Monthly Mortgage Payments

A couple borrows $100,000 at 9% interest as a mortgage loan on a house. They expect to make monthly payments for 30 years to repay the loan. What is the size of each payment?

SOLUTION The mortgage payments form an annuity whose present value is $A_p = \$100{,}000$. Also, $i = 0.09/12 = 0.0075$, and $n = 12 \times 30 = 360$. We are looking for the amount R of each payment. From the formula for installment buying, we get

$$R = \frac{iA_p}{1 - (1 + i)^{-n}}$$

$$= \frac{(0.0075)(100{,}000)}{1 - (1 + 0.0075)^{-360}} \approx 804.623$$

Thus, the monthly payments are $804.62.   ■

We now give an example that illustrates the use of graphing devices in solving problems related to installment buying.

 EXAMPLE 6 ■ Calculating the Interest Rate from the Size
                             of Monthly Payments

A car dealer sells a new car for $18,000. He offers to sell the car for payments of $405 per month for 5 years. What interest rate is this car dealer charging?

SOLUTION The payments form an annuity with present value $A_p = \$18{,}000$, $R = 405$, and $n = 12 \times 5 = 60$. To find the interest rate, we must solve for i

in the equation

$$R = \frac{iA_p}{1 - (1 + i)^{-n}}$$

A little experimentation will convince you that it's not possible to algebraically solve this equation for i. So, to find i we use a graphing device to graph R as a function of the interest rate x, and we then use the graph to find the interest rate corresponding to the value of R we want ($405 in this case). Since $i = x/12$, we graph the function

$$R(x) = \frac{\frac{x}{12}(18{,}000)}{1 - \left(1 + \frac{x}{12}\right)^{-60}}$$

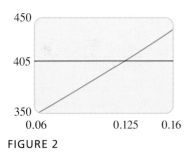

FIGURE 2

in the viewing rectangle $[0.06, 0.16] \times [350, 450]$, as shown in Figure 2. We also graph the horizontal line $R(x) = 405$ in the same viewing rectangle. Then, by moving the cursor to the point of intersection of the two graphs, we find that the corresponding x-value is approximately 0.125. Thus, the interest rate is about $12\frac{1}{2}\%$. ∎

10.4 EXERCISES

1. Find the amount of an annuity that consists of 10 annual payments of $1000 each into an account that pays 6% interest per year.

2. Find the amount of an annuity that consists of 24 monthly payments of $500 each into an account that pays 8% interest per year, compounded monthly.

3. Find the amount of an annuity that consists of 20 annual payments of $5000 each into an account that pays interest of 12% per year.

4. Find the amount of an annuity that consists of 20 semi-annual payments of $500 each into an account that pays 6% interest per year, compounded semiannually.

5. Find the amount of an annuity that consists of 16 quarterly payments of $300 each into an account that pays 8% interest per year, compounded quarterly.

6. How much money should be invested every quarter at 10% per year, compounded quarterly, in order to have $5000 in 2 years?

7. How much money should be invested monthly at 6% per year, compounded monthly, in order to have $2000 in 8 months?

8. What is the present value of an annuity that consists of 20 semiannual payments of $1000 at the interest rate of 9% per year, compounded semiannually?

9. How much money must be invested now at 9% per year, compounded semiannually, to fund an annuity of 20 payments of $200 each, paid every 6 months, the first payment being 6 months from now?

10. A 55-year-old man deposits $50,000 to fund an annuity with an insurance company. The money will be invested at 8% per year, compounded semiannually. He is to draw semiannual payments until he reaches age 65. What is the amount of each payment?

11. A woman wants to borrow $12,000 in order to buy a car. She wants to repay the loan by monthly installments for 4 years. If the interest rate on this loan is $10\frac{1}{2}\%$ per year, compounded monthly, what is the amount of each payment?

12. What is the monthly payment on a 30-year mortgage of $80,000 at 9% interest? What is the monthly payment on this same mortgage if it is to be repaid over a 15-year period?

13. What is the monthly payment on a 30-year mortgage of $100,000 at 8% interest per year, compounded monthly? What is the total amount paid on this loan over the 30-year period?

14. A couple can afford to make a monthly mortgage payment of $650. If the mortgage rate is 9% and the couple intends to secure a 30-year mortgage, how much can they borrow?

15. A couple secures a 30-year loan of $100,000 at $9\frac{3}{4}$% per year, compounded monthly, to buy a house.
(a) What is the amount of the monthly payment?
(b) What is the total amount that will be paid by the couple over the 30-year period?
(c) If, instead of taking the loan, the couple deposits the monthly payments in an account that pays $9\frac{3}{4}$% interest per year, compounded monthly, how much will be in the account at the end of the 30-year period?

16. Jane agrees to buy a car by making a down payment of $2000 and payments of $220 per month for 3 years. If the interest rate is 8% per year, compounded monthly, what is the actual purchase price of the car?

17. Mike buys a ring for his fiancee by paying $30 per month for one year. If the interest rate is 10% per year, compounded monthly, what is the price of the ring?

18. Janet decides to buy a $12,500 car by paying $420 a month for 3 years. Assuming that interest is compounded monthly, what interest rate is she paying on the car loan?

19. John buys a stereo system that is advertised for $640. He agrees to pay $32 per month for 2 years. Assuming that interest is compounded monthly, what interest rate is he paying?

20. A woman purchases a $2000 diamond ring by paying a down payment of $200 and monthly installments of $88 for 2 years. Assuming that interest is compounded monthly, what interest rate is she paying?

21. An item at a department store is priced at $189.99 and can be bought by making 20 payments of $10.50. Find the interest rate, assuming that interest is compounded monthly.

22. (a) Draw a time line as in Example 1 to show that the present value of an annuity is the sum of the present values of each of the payments, that is,

$$A_p = \frac{R}{1+i} + \frac{R}{(1+i)^2} + \frac{R}{(1+i)^3} + \cdots + \frac{R}{(1+i)^n}$$

(b) Use part (a) to derive the formula for A_p given in the text.

23. An **annuity in perpetuity** is one that continues forever. Such annuities are useful in setting up scholarship funds to ensure that the award continues.
(a) Draw a time line (as in Example 1) to show that in order to set up an annuity in perpetuity of amount R per time period, the amount to be invested now is

$$A_p = \frac{R}{1+i} + \frac{R}{(1+i)^2} + \frac{R}{(1+i)^3} + \cdots + \frac{R}{(1+i)^n} + \cdots$$

where i is the interest rate per time period.
(b) Find the sum of the infinite series in part (a) to show that

$$A_p = \frac{R}{i}$$

24. How much money must be invested now at 10% per year, compounded annually, to provide an annuity in perpetuity of $5000 per year? The first payment is due in one year. (Refer to Exercise 23.)

25. How much money must be invested now at 8% per year, compounded quarterly, in order to provide an annuity in perpetuity of $3000 per year? The first payment is due in one year. (Refer to Exercise 23.)

10.5   MATHEMATICAL INDUCTION

There are two aspects to mathematics—discovery and proof—and both are of equal importance. It is necessary to discover something before attempting to prove it, and we can only be certain of its truth once it has been proved. In this section we examine the relationship between these two parts of mathematics more closely.

CONJECTURE AND PROOF

Consider the polynomial

$$p(n) = n^2 - n + 41$$

Here are some values of $p(n)$:

$p(1) = 41 \quad p(2) = 43$
$p(3) = 47 \quad p(4) = 53$
$p(5) = 61 \quad p(6) = 71$
$p(7) = 83 \quad p(8) = 97$

All the values so far are prime numbers. In fact, if you keep going, you will find that $p(n)$ is prime for all natural numbers up to $n = 40$. It may seem reasonable at this point to conjecture that $p(n)$ is prime for *every* natural number n. But our conjecture would be too hasty, because it is easily seen that $p(41)$ is *not* prime. This illustrates that we cannot be certain of the truth of a statement no matter how many special cases we check. We need a convincing argument—a *proof*—to determine the truth of a statement.

Let's try a simple experiment. We add up more and more of the odd numbers as follows:

$$1 = 1$$
$$1 + 3 = 4$$
$$1 + 3 + 5 = 9$$
$$1 + 3 + 5 + 7 = 16$$
$$1 + 3 + 5 + 7 + 9 = 25$$

What do you notice about the numbers on the right side of these equations? They are in fact all perfect squares. These equations say the following:

The sum of the first 1 odd number is 1^2.

The sum of the first 2 odd numbers is 2^2.

The sum of the first 3 odd numbers is 3^2.

The sum of the first 4 odd numbers is 4^2.

The sum of the first 5 odd numbers is 5^2.

This leads naturally to the following question: Is it true that for every natural number n, the sum of the first n odd numbers is n^2? Could this remarkable property be true? We could try a few more numbers and find that the pattern persists for the first 6, 7, 8, 9, and 10 odd numbers. At this point, we feel quite sure that this is always true, so we make a conjecture:

The sum of the first n odd numbers is n^2.

Since we know that the nth odd number is $2n - 1$, we can write this statement more precisely as

$$1 + 3 + 5 + \cdots + (2n - 1) = n^2$$

It's important to realize that this is still a conjecture. We cannot conclude by checking a finite number of cases that a property is true for all numbers (there are infinitely many). To see this more clearly, suppose someone tells us that he has added up the first trillion odd numbers and found out that they do *not* add up to 1 trillion squared. What would you tell this person? It would be silly to tell him that you're sure it's true because you've already checked the first five cases. You could, however, take out paper and pencil and start checking it yourself, but this task would probably take the rest of your life. The tragedy would be that after completing this task you would still not be sure of the truth of the conjecture. Do you see why?

Herein lies the power of mathematical proof. A **proof** is a clear argument that demonstrates the truth of a statement beyond doubt. We consider here a special kind of proof called *mathematical induction* that will help us prove statements like the one we have just considered.

MATHEMATICAL INDUCTION

We consider a special kind of proof called **mathematical induction**. Here is how it works: Suppose we have a statement that says something about all natural numbers n. Let's call this statement P. For example, we could consider the statement

P: For every natural number n, the sum of the first n odd numbers is n^2.

Since this statement is about *all* natural numbers, it contains infinitely many statements; we will call them $P(1)$, $P(2)$, ...

$P(1)$:   The sum of the first 1 odd number is 1^2.

$P(2)$:   The sum of the first 2 odd numbers is 2^2.

$P(3)$:   The sum of the first 3 odd numbers is 3^2.

$\vdots$ $\vdots$

How can we prove all of these statements at once? Mathematical induction is a clever way of doing just that.

The crux of the idea is this: Suppose we can prove that whenever one of these statements is true, then the one following it in the list is also true. In other words,

For every k, if $P(k)$ is true, then $P(k + 1)$ is true.

This is called the **induction step** because it leads us from the truth of one statement to the next. Now, suppose that we can also prove that

$P(1)$ is true.

The induction step now leads us through the following chain of statements:

$P(1)$ is true, so $P(2)$ is true.

$P(2)$ is true, so $P(3)$ is true.

$P(3)$ is true, so $P(4)$ is true.

$\vdots$ $\vdots$

So we see that if both of these statements are proved, then statement P is proved for all n. We summarize this important method of proof.

PRINCIPLE OF MATHEMATICAL INDUCTION

For each natural number n, let $P(n)$ be a statement depending on n.
Suppose that the following two conditions are satisfied.

1. $P(1)$ is true.

2. For every natural number k, if $P(k)$ is true, then $P(k + 1)$ is true.

Then $P(n)$ is true for all natural numbers n.

To apply this principle, there are two steps:

Step 1 Prove that $P(1)$ is true.

Step 2 Assume that $P(k)$ is true and use this assumption to prove that $P(k + 1)$ is true.

Notice that in Step 2 we do not prove that $P(k)$ is true. We only show that *if* $P(k)$ is true, *then $P(k + 1)$ is also true. The assumption that $P(k)$ is true is called the **induction hypothesis**.

FOOTIES AT THE INDUCTION STEP

IF I WERE ON RUNG NUMBER K THEN IT WOULD BE EASY TO GET TO THE NEXT RUNG.

SURE, AND THEN YOU COULD CLIMB THE WHOLE LADDER.

BUT ONE HAS TO MANAGE TO REACH THE FIRST RUNG TO BEGIN WITH.

We now use mathematical induction to prove that the conjecture we made at the beginning of this section is true.

EXAMPLE 1 ■ A Proof by Mathematical Induction

Prove that for all natural numbers n,

$$1 + 3 + 5 + \cdots + (2n - 1) = n^2$$

SOLUTION   Let $P(n)$ denote the statement $1 + 3 + 5 + \cdots + (2n - 1) = n^2$.

Step 1 We need to show that $P(1)$ is true. But $P(1)$ is simply the statement that $1 = 1^2$, which is of course true.

Step 2 We assume that $P(k)$ is true. Thus, our induction hypothesis is

$$1 + 3 + 5 + \cdots + (2k - 1) = k^2$$

We want to use this to show that $P(k + 1)$ is true, that is,

$$1 + 3 + 5 + \cdots + (2k - 1) + [2(k + 1) - 1] = (k + 1)^2$$

[Note that we get $P(k + 1)$ by substituting $k + 1$ for each n in the statement $P(n)$.] We start with the left side and use the induction hypothesis

Blaise Pascal (1623–1662) is considered one of the most versatile minds in modern history. He was a writer and philosopher as well as a gifted mathematician and physicist. Among his contributions that appear in this book are the theory of probability, Pascal's triangle, and the Principle of Mathematical Induction.

Pascal's father, himself a mathematician, believed that his son should not study mathematics until he was 15 or 16. But at age 12, Blaise insisted on learning geometry, and proved most of its elementary theorems himself. At 19, he invented the first mechanical adding machine. In 1647, after writing a major treatise on the conic sections, he abruptly abandoned mathematics because he felt that his intense studies were contributing to his ill health. He devoted himself instead to frivolous recreations such as gambling, but this only served to pique his interest in probability. In 1654, he miraculously survived a carriage accident in which his horses ran off a bridge. Taking this to be a sign from God, he entered a monastery, where he pursued theology and philosophy, writing his famous *Pensées*. He also continued his

(continued)

to obtain the right side of the equation:

$$1 + 3 + 5 + \cdots + (2k - 1) + [2(k + 1) - 1]$$

$$= [1 + 3 + 5 + \cdots + (2k - 1)] + [2(k + 1) - 1]$$

Group the first k terms

$$= k^2 + [2(k + 1) - 1] \qquad \text{Induction hypothesis}$$

$$= k^2 + [2k + 2 - 1] \qquad \text{Distributive Property}$$

$$= k^2 + 2k + 1 \qquad \text{Simplify}$$

$$= (k + 1)^2 \qquad \text{Factor}$$

Thus, $P(k + 1)$ follows from $P(k)$ and this completes the induction step.

Having proved Steps 1 and 2, we conclude by the Principle of Mathematical Induction that $P(n)$ is true for all natural numbers n. ∎

EXAMPLE 2 ■ A Proof by Mathematical Induction

Prove that for every natural number n,

$$1 + 2 + 3 + \cdots + n = \frac{n(n + 1)}{2}$$

SOLUTION   Let $P(n)$ be the statement $1 + 2 + 3 + \cdots + n = n(n + 1)/2$. We want to show that $P(n)$ is true for all natural numbers n.

Step 1 We need to show that $P(1)$ is true. But $P(1)$ says that

$$1 = \frac{1(1 + 1)}{2}$$

and this statement is clearly true.

Step 2 Assume that $P(k)$ is true. Thus, our induction hypothesis is

$$1 + 2 + 3 + \cdots + k = \frac{k(k + 1)}{2}$$

We want to use this to show that $P(k + 1)$ is true, that is,

$$1 + 2 + 3 + \cdots + k + (k + 1) = \frac{(k + 1)[(k + 1) + 1]}{2}$$

To show this we start with the left side and use the induction hypothesis to obtain the right side:

$$1 + 2 + 3 + \cdots + k + (k + 1)$$

$$= [1 + 2 + 3 + \cdots + k] + (k + 1) \qquad \text{Group the first } k \text{ terms}$$

$$= \frac{k(k + 1)}{2} + (k + 1) \qquad \text{Induction hypothesis}$$

$$= (k + 1)\left(\frac{k}{2} + 1\right)$$ Factor $k + 1$

$$= (k + 1)\left(\frac{k + 2}{2}\right)$$ Common denominator

$$= \frac{(k + 1)[(k + 1) + 1]}{2}$$ Write $k + 2$ as $k + 1 + 1$

Thus, $P(k + 1)$ follows from $P(k)$ and this completes the induction step.

Having proved Steps 1 and 2, we conclude by the Principle of Mathematical Induction that $P(n)$ is true for all natural numbers n. ∎

It might happen that a statement $P(n)$ is false for the first few natural numbers, but true from some number on. For example, we may want to prove that $P(n)$ is true for $n \geq 5$. Notice that if we prove that $P(5)$ is true, then this fact, together with the induction step, would imply the truth of $P(5)$, $P(6)$, $P(7)$, ... The next example illustrates this point.

EXAMPLE 3 ■ Proving an Inequality by Mathematical Induction

Prove that $4n < 2^n$ for all $n \geq 5$.

SOLUTION Let $P(n)$ denote the statement $4n < 2^n$.

Step 1 $P(5)$ is the statement that $4 \cdot 5 < 2^5$, or $20 < 32$, which is true.

Step 2 Assume that $P(k)$ is true. Thus, our induction hypothesis is

$$4k < 2^k$$

We want to use this to show that $P(k + 1)$ is true, that is,

$$4(k + 1) < 2^{k+1}$$

To show this we start with the left side of the inequality and use the induction hypothesis to show that it is less than the right side. For $k \geq 5$, we have

$$4(k + 1) = 4k + 4$$
$$< 2^k + 4 \qquad \text{Induction hypothesis}$$
$$< 2^k + 4k \qquad 4 < 4k$$
$$< 2^k + 2^k \qquad \text{Induction hypothesis}$$
$$= 2 \cdot 2^k$$
$$= 2^{k+1} \qquad \text{Property of exponents}$$

Thus, $P(k + 1)$ follows from $P(k)$ and this completes the induction step.

Having proved Steps 1 and 2, we conclude by the Principle of Mathematical Induction that $P(n)$ is true for all natural numbers n. ∎

10.5   EXERCISES

1–12 ■ Use mathematical induction to prove that the formula is true for all natural numbers n.

1. $2 + 4 + 6 + \cdots + 2n = n(n + 1)$

2. $1 + 4 + 7 + \cdots + 3(n - 2) = \dfrac{n(3n - 1)}{2}$

3. $5 + 8 + 11 + \cdots + (3n + 2) = \dfrac{n(3n + 7)}{2}$

4. $1^2 + 2^2 + 3^2 + \cdots + n^2 = \dfrac{n(n + 1)(2n + 1)}{6}$

5. $1 \cdot 2 + 2 \cdot 3 + 3 \cdot 4 + \cdots + n(n + 1)$
$= \dfrac{n(n + 1)(n + 2)}{3}$

6. $1 \cdot 3 + 2 \cdot 4 + 3 \cdot 5 + \cdots + n(n + 2)$
$= \dfrac{n(n + 1)(2n + 7)}{6}$

7. $1^3 + 2^3 + 3^3 + \cdots + n^3 = \dfrac{n^2(n + 1)^2}{4}$

8. $1^3 + 3^3 + 5^3 + \cdots + (2n - 1)^3 = n^2(2n^2 - 1)$

9. $2^3 + 4^3 + 6^3 + \cdots + (2n)^3 = 2n^2(n + 1)^2$

10. $\dfrac{1}{1 \cdot 2} + \dfrac{1}{2 \cdot 3} + \dfrac{1}{3 \cdot 4} + \cdots + \dfrac{1}{n(n + 1)} = \dfrac{n}{(n + 1)}$

11. $1 \cdot 2 + 2 \cdot 2^2 + 3 \cdot 2^3 + 4 \cdot 2^4 + \cdots + n \cdot 2^n$
$= 2[1 + (n - 1)2^n]$

12. $1 + 2 + 2^2 + \cdots + 2^{n-1} = 2^n - 1.$

13. Show that $n^2 + n$ is divisible by 2 for all natural numbers n.

14. Show that $5^n - 1$ is divisible by 4 for all natural numbers n.

15. Show that $n^2 - n + 41$ is odd for all natural numbers n.

16. Show that $n^3 - n + 3$ is divisible by 3 for all natural numbers n.

17. Show that $8^n - 3^n$ is divisible by 5 for all natural numbers n.

18. Show that $3^{2n} - 1$ is divisible by 8 for every natural number n.

19. Prove that $n < 2^n$ for all natural numbers n.

20. Prove that $(n + 1)^2 < 2n^2$ for all natural numbers $n \geq 3$.

21. Prove that if $x > -1$, then $(1 + x)^n \geq 1 + nx$ for all natural numbers n.

22. Show that $100n \leq n^2$ for all $n \geq 100$.

23. Let $a_{n+1} = 3a_n$ and $a_1 = 5$. Show that $a_n = 5 \cdot 3^{n-1}$ for all natural numbers n.

24. A sequence is defined recursively by $a_{n+1} = 3a_n - 8$ and $a_1 = 4$. Find an explicit formula for a_n and then use mathematical induction to prove that the formula you found is true.

25. Show that $x - y$ is a factor of $x^n - y^n$ for all natural numbers n.
[*Hint:* $x^{k+1} - y^{k+1} = x^k(x - y) + (x^k - y^k)y$]

26. Show that $x + y$ is a factor of $x^{2n-1} + y^{2n-1}$ for all natural numbers n.

27–31 ■ F_n denotes the nth term of the Fibonacci sequence discussed in Section 10.1. Use mathematical induction to prove the statement.

27. F_{3n} is even for all natural numbers n.

28. $F_1 + F_2 + F_3 + \cdots + F_n = F_{n+2} - 1$

29. $F_1^2 + F_2^2 + F_3^2 + \cdots + F_n^2 = F_n F_{n+1}$

30. If $a_{n+2} = a_{n+1} \cdot a_n$ and $a_1 = a_2 = 2$, then $a_n = 2^{F_n}$ for all natural numbers n.

31. For all $n \geq 2$,

$$\begin{bmatrix} 1 & 1 \\ 1 & 0 \end{bmatrix}^n = \begin{bmatrix} F_{n+1} & F_n \\ F_n & F_{n-1} \end{bmatrix}$$

32. Let a_n be the nth term of the sequence defined recursively by

$$a_{n+1} = \dfrac{1}{1 + a_n}$$

and $a_1 = 1$. Find a formula for a_n in terms of the Fibonacci numbers F_n. Prove that the formula you found is valid for all natural numbers n.

33. Let F_n be the nth term of the Fibonacci sequence. Find and prove an inequality relating n and F_n for natural numbers n.

34. Find and prove an inequality relating $100n$ and n^3.

DISCOVERY · DISCUSSION

35. True or False?   Determine whether each of the following statements is true or false. If you think the statement is true, prove it. If you think it is false, give an example where it fails.

(a) $p(n) = n^2 - n + 11$ is prime for all n.

(b) $n^2 > n$ for all $n \geq 2$.

(c) $2^{2n+1} + 1$ is divisible by 3 for all $n \geq 1$.

(d) $n^3 \geq (n + 1)^2$ for all $n \geq 2$.

(e) $n^3 - n$ is divisible by 3 for all $n \geq 2$.

(f) $n^3 - 6n^2 + 11n$ is divisible by 6 for all $n \geq 1$.

36. All Cats are Black?   What is wrong with the following "proof" by mathematical induction that all cats are black? Let $P(n)$ denote the statement: In any group of n cats, if one is black, then they are all black.

Step 1   The statement is clearly true for $n = 1$.

Step 2   Suppose that $P(k)$ is true. We show that $P(k + 1)$ is true. Suppose we have a group of $k + 1$ cats, one of whom is black; call this cat "Midnight." Remove some other cat (call it

"Sparky") from the group. We are left with k cats, one of whom (Midnight) is black, so by the induction hypothesis, all k of these are black. Now put Sparky back in the group and take out Midnight. We again have a group of k cats, all of whom—except possible Sparky—are black. Then by the induction hypothesis, Sparky must be black, too. So all $k + 1$ cats in the original group are black.

Thus, by induction $P(n)$ is true for all n. Since everyone has seen at least one black cat, it follows that all cats are black.

Midnight   Sparky

10.6   **THE BINOMIAL THEOREM**

An expression of the form $a + b$ is called a **binomial**. Although in principle it's easy to raise $a + b$ to any power, raising it to a very high power would be a tedious task. In this section we find a formula that gives the expansion of $(a + b)^n$ for any natural number n. We discover this formula by finding a pattern for the successive powers of $a + b$. So, we first look at some special cases:

$$(a + b)^1 = a + b$$

$$(a + b)^2 = a^2 + 2ab + b^2$$

$$(a + b)^3 = a^3 + 3a^2b + 3ab^2 + b^3$$

$$(a + b)^4 = a^4 + 4a^3b + 6a^2b^2 + 4ab^3 + b^4$$

$$(a + b)^5 = a^5 + 5a^4b + 10a^3b^2 + 10a^2b^3 + 5ab^4 + b^5$$

$$\vdots$$

The following simple patterns emerge for the expansion of $(a + b)^n$:

1. There are $n + 1$ terms, the first being a^n and the last b^n.

2. The exponents of a decrease by 1 from term to term while the exponents of b increase by 1.

3. The sum of the exponents of a and b in each term is n.

Pascal's triangle appears in this Chinese document by Chu Shi-kie, dated 1303. The title reads "The Old Method Chart of the Seven Multiplying Squares." The triangle was rediscovered by Pascal (see page 718).

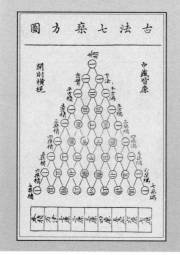

For instance, notice how the exponents of a and b behave in the expansion of $(a + b)^5$.

The exponents of a decrease:

$$(a + b)^5 = a^{⑤} + 5a^{④}b^1 + 10a^{③}b^2 + 10a^{②}b^3 + 5a^{①}b^4 + b^5$$

The exponents of b increase:

$$(a + b)^5 = a^5 + 5a^4b^{①} + 10a^3b^{②} + 10a^2b^{③} + 5a^1b^{④} + b^{⑤}$$

With these observations we can write the form of the expansion of $(a + b)^n$ for any natural number n. For example, writing a question mark for the missing coefficients, we have

$$(a + b)^8 = a^8 + ?a^7b + ?a^6b^2 + ?a^5b^3 + ?a^4b^4 + ?a^3b^5 + ?a^2b^6 + ?ab^7 + b^8$$

To complete the expansion, we need to determine these coefficients. To find a pattern, let's write the coefficients in the expansion of $(a + b)^n$ for the first few values of n in a triangular array as shown in the following array, which is called **Pascal's triangle**.

$$
\begin{array}{cccccccccccc}
(a + b)^0 & & & & & 1 & & & & \\
(a + b)^1 & & & & 1 & & 1 & & & \\
(a + b)^2 & & & 1 & & 2 & & 1 & & \\
(a + b)^3 & & 1 & & 3 & & 3 & & 1 & \\
(a + b)^4 & 1 & & 4 & & 6 & & 4 & & 1 \\
(a + b)^5 & 1 & 5 & & 10 & & 10 & & 5 & & 1 \\
\end{array}
$$

The row corresponding to $(a + b)^0$ is called the zeroth row and is included to show the symmetry of the array. The key observation about Pascal's triangle is the following property.

KEY PROPERTY OF PASCAL'S TRIANGLE

Every entry (other than a 1) is the sum of the two entries diagonally above it.

From this property it's easy to find any row of Pascal's triangle from the row above it. For instance, we find the sixth and seventh rows, starting with the fifth row:

$$
\begin{array}{ccccccccccccc}
(a + b)^5 & & & 1 & & 5 & & 10 & & 10 & & 5 & & 1 \\
(a + b)^6 & & 1 & & 6 & & 15 & & 20 & & 15 & & 6 & & 1 \\
(a + b)^7 & 1 & & 7 & & 21 & & 35 & & 35 & & 21 & & 7 & & 1 \\
\end{array}
$$

To see why this property holds, let's consider the following expansions:

$$(a + b)^5 = a^5 + 5a^4b + 10a^3b^2 + \boxed{10a^2b^3} + \boxed{5ab^4} + b^5$$

$$(a + b)^6 = a^6 + 6a^5b + 15a^4b^2 + 20a^3b^3 + \boxed{15a^2b^4} + 6ab^5 + b^6$$

We arrive at the expansion of $(a + b)^6$ by multiplying $(a + b)^5$ by $(a + b)$. Now notice, for instance, that the circled term in the expansion of $(a + b)^6$ is obtained via this multiplication from the two circled terms above it. We get this term when the two terms above it are multiplied by b and a, respectively. Thus, its coefficient is the sum of the coefficients of these two terms. This is the observation we will use at the end of this section in proving the Binomial Theorem.

Having found these patterns, we can now easily obtain the expansion of any binomial, at least to relatively small powers.

EXAMPLE 1 ■ Expanding a Binomial Using Pascal's Triangle

Find the expansion of $(a + b)^7$ using Pascal's triangle.

SOLUTION   The first term in the expansion is a^7, and the last term is b^7. Using the fact that the exponent of a decreases by 1 from term to term and that of b increases by 1 from term to term, we have

$$(a + b)^7 = a^7 + ?a^6b + ?a^5b^2 + ?a^4b^3 + ?a^3b^4 + ?a^2b^5 + ?ab^6 + b^7$$

The appropriate coefficients appear in the seventh row of Pascal's triangle. Thus

$$(a + b)^7 = a^7 + 7a^6b + 21a^5b^2 + 35a^4b^3 + 35a^3b^4 + 21a^2b^5 + 7ab^6 + b^7$$

■

EXAMPLE 2 ■ Expanding a Binomial Using Pascal's Triangle

Use Pascal's triangle to expand $(2 - 3x)^5$.

SOLUTION   We find the expansion of $(a + b)^5$ and then substitute 2 for a and $-3x$ for b. Using Pascal's triangle for the coefficients, we get

$$(a + b)^5 = a^5 + 5a^4b + 10a^3b^2 + 10a^2b^3 + 5ab^4 + b^5$$

Substituting $a = 2$ and $b = -3x$ gives

$$(2 - 3x)^5 = (2)^5 + 5(2)^4(-3x) + 10(2)^3(-3x)^2 + 10(2)^2(-3x)^3 + 5(2)(-3x)^4 + (-3x)^5$$

$$= 32 - 240x + 720x^2 - 1080x^3 + 810x^4 - 243x^5$$

■

THE BINOMIAL COEFFICIENTS AND PASCAL'S TRIANGLE

Although Pascal's triangle is useful in finding the binomial expansion for reasonably small values of n, it isn't practical for finding $(a + b)^n$ for large values of n. The reason is that the method we use for finding the successive rows of Pascal's triangle is recursive. Thus, to find the 100th row of this triangle, we must first find the preceding 99 rows.

We need to examine the pattern in the coefficients more carefully to develop a formula that will allow us to calculate directly any coefficient in the binomial expansion. Such a formula exists, and the rest of this section is devoted to finding and proving it. However, to state this formula we need some notation.

The product of the first n natural numbers is denoted by $n!$ and is called **n factorial**:

$$4! = 1 \cdot 2 \cdot 3 \cdot 4 = 24$$
$$7! = 1 \cdot 2 \cdot 3 \cdot 4 \cdot 5 \cdot 6 \cdot 7 = 5040$$
$$10! = 1 \cdot 2 \cdot 3 \cdot 4 \cdot 5 \cdot 6 \cdot 7 \cdot 8 \cdot 9 \cdot 10$$
$$= 3{,}628{,}800$$

$$\boxed{n! = 1 \cdot 2 \cdot 3 \cdot \ \cdots \ \cdot (n-1) \cdot n}$$

We also define $0!$ as follows:

$$\boxed{0! = 1}$$

This definition of $0!$ makes many formulas involving factorials shorter and easier to write.

THE BINOMIAL COEFFICIENT

Let n and r be nonnegative integers with $r \leq n$. The **binomial coefficient** is denoted by $\binom{n}{r}$ and is defined by

$$\binom{n}{r} = \frac{n!}{r!\,(n-r)!}$$

EXAMPLE 3 ■ Calculating Binomial Coefficients

(a) $\displaystyle \binom{9}{4} = \frac{9!}{4!\,(9-4)!} = \frac{9!}{4!\,5!} = \frac{1 \cdot 2 \cdot 3 \cdot 4 \cdot 5 \cdot 6 \cdot 7 \cdot 8 \cdot 9}{(1 \cdot 2 \cdot 3 \cdot 4)(1 \cdot 2 \cdot 3 \cdot 4 \cdot 5)}$

$$= \frac{6 \cdot 7 \cdot 8 \cdot 9}{1 \cdot 2 \cdot 3 \cdot 4} = 126$$

(b) $\displaystyle \binom{100}{3} = \frac{100!}{3!\,(100-3)!} = \frac{1 \cdot 2 \cdot 3 \cdot \cdots \cdot 97 \cdot 98 \cdot 99 \cdot 100}{(1 \cdot 2 \cdot 3)(1 \cdot 2 \cdot 3 \cdot \cdots \cdot 97)}$

$$= \frac{98 \cdot 99 \cdot 100}{1 \cdot 2 \cdot 3} = 161{,}700$$

(c) $\displaystyle \binom{100}{97} = \frac{100!}{97!\,(100-97)!} = \frac{1 \cdot 2 \cdot 3 \cdot \cdots \cdot 97 \cdot 98 \cdot 99 \cdot 100}{(1 \cdot 2 \cdot 3 \cdot \cdots \cdot 97)(1 \cdot 2 \cdot 3)}$

$$= \frac{98 \cdot 99 \cdot 100}{1 \cdot 2 \cdot 3} = 161{,}700 \qquad ■$$

Although the binomial coefficient $\binom{n}{r}$ is defined in terms of a fraction, all the results of Example 3 are natural numbers. In fact, $\binom{n}{r}$ is always a natural number (see Exercise 50). Notice that the binomial coefficients in parts (b) and (c) of

Example 3 are equal. This is a special case of the relation

$$\binom{n}{r} = \binom{n}{n-r}$$

which you are asked to prove in Exercise 48.

To see the connection between the binomial coefficients and the binomial expansion of $(a+b)^n$, let's calculate the following binomial coefficients:

$$\binom{5}{2} = \frac{5!}{2!\,(5-2)!} = 10$$

$$\binom{5}{0} = 1 \qquad \binom{5}{1} = 5 \qquad \binom{5}{2} = 10 \qquad \binom{5}{3} = 10 \qquad \binom{5}{4} = 5 \qquad \binom{5}{5} = 1$$

These are precisely the entries in the fifth row of Pascal's triangle. In fact, we can write Pascal's triangle as follows.

$$\binom{0}{0}$$

$$\binom{1}{0} \qquad \binom{1}{1}$$

$$\binom{2}{0} \qquad \binom{2}{1} \qquad \binom{2}{2}$$

$$\binom{3}{0} \qquad \binom{3}{1} \qquad \binom{3}{2} \qquad \binom{3}{3}$$

$$\binom{4}{0} \qquad \binom{4}{1} \qquad \binom{4}{2} \qquad \binom{4}{3} \qquad \binom{4}{4}$$

$$\binom{5}{0} \qquad \binom{5}{1} \qquad \binom{5}{2} \qquad \binom{5}{3} \qquad \binom{5}{4} \qquad \binom{5}{5}$$

$$\cdot \qquad \cdot \qquad \cdot \qquad \cdot \qquad \cdot \qquad \cdot \qquad \cdot$$

$$\binom{n}{0} \qquad \binom{n}{1} \qquad \binom{n}{2} \qquad \cdot \qquad \cdot \qquad \cdot \qquad \binom{n}{n-1} \qquad \binom{n}{n}$$

To show that this pattern holds, we need to show that any entry in this version of Pascal's triangle is the sum of the two entries diagonally above it. In other words, we need to show that each entry satisfies the key property of Pascal's triangle. We now state this property in terms of the binomial coefficients.

KEY PROPERTY OF THE BINOMIAL COEFFICIENTS

For any nonnegative integers r and k with $r \leq k$,

$$\binom{k}{r-1} + \binom{k}{r} = \binom{k+1}{r}$$

Notice that the two terms on the left side of this equation are adjacent entries in the kth row of Pascal's triangle and the term on the right side is the entry diagonally below them, in the $(k + 1)$st row. Thus, this equation is a restatement of the key property of Pascal's triangle in terms of the binomial coefficients. A proof of this formula is outlined in Exercise 49.

THE BINOMIAL THEOREM

We are now ready to state the Binomial Theorem.

THE BINOMIAL THEOREM

$$(a + b)^n = \binom{n}{0}a^n + \binom{n}{1}a^{n-1}b + \binom{n}{2}a^{n-2}b^2 + \cdots + \binom{n}{n-1}ab^{n-1} + \binom{n}{n}b^n$$

We prove this theorem at the end of this section. First, we show some of its applications.

EXAMPLE 4 ■ Expanding a Binomial Using the Binomial Theorem

Use the Binomial Theorem to expand $(x + y)^4$.

SOLUTION By the Binomial Theorem,

$$(x + y)^4 = \binom{4}{0}x^4 + \binom{4}{1}x^3y + \binom{4}{2}x^2y^2 + \binom{4}{3}xy^3 + \binom{4}{4}y^4$$

Verify that

$$\binom{4}{0} = 1 \qquad \binom{4}{1} = 4 \qquad \binom{4}{2} = 6 \qquad \binom{4}{3} = 4 \qquad \binom{4}{4} = 1$$

It follows that

$$(x + y)^4 = x^4 + 4x^3y + 6x^2y^2 + 4xy^3 + y^4 \qquad ■$$

EXAMPLE 5 ■ Expanding a Binomial Using the Binomial Theorem

Use the Binomial Theorem to expand $(\sqrt{x} - 1)^8$.

SOLUTION We first find the expansion of $(a + b)^8$ and then substitute $\sqrt{x}$ for a and -1 for b. Using the Binomial Theorem, we have

$$(a + b)^8 = \binom{8}{0}a^8 + \binom{8}{1}a^7b + \binom{8}{2}a^6b^2 + \binom{8}{3}a^5b^3 + \binom{8}{4}a^4b^4$$

$$+ \binom{8}{5}a^3b^5 + \binom{8}{6}a^2b^6 + \binom{8}{7}ab^7 + \binom{8}{8}b^8$$

Verify that

$$\binom{8}{0} = 1 \qquad \binom{8}{1} = 8 \qquad \binom{8}{2} = 28 \qquad \binom{8}{3} = 56 \qquad \binom{8}{4} = 70$$

$$\binom{8}{5} = 56 \qquad \binom{8}{6} = 28 \qquad \binom{8}{7} = 8 \qquad \binom{8}{8} = 1$$

So

$$(a + b)^8 = a^8 + 8a^7b + 28a^6b^2 + 56a^5b^3 + 70a^4b^4 + 56a^3b^5$$
$$+ 28a^2b^6 + 8ab^7 + b^8$$

Performing the substitutions $a = x^{1/2}$ and $b = -1$ gives

$$(\sqrt{x} - 1)^8 = (x^{1/2})^8 + 8(x^{1/2})^7(-1) + 28(x^{1/2})^6(-1)^2 + 56(x^{1/2})^5(-1)^3$$
$$+ 70(x^{1/2})^4(-1)^4 + 56(x^{1/2})^3(-1)^5 + 28(x^{1/2})^2(-1)^6$$
$$+ 8(x^{1/2})(-1)^7 + (-1)^8$$

This simplifies to

$$(\sqrt{x} - 1)^8 = x^4 - 8x^{7/2} + 28x^3 - 56x^{5/2} + 70x^2 - 56x^{3/2} + 28x - 8x^{1/2} + 1$$

∎

The Binomial Theorem can be used to find a particular term of a binomial expansion without having to find the entire expansion.

GENERAL TERM OF THE BINOMIAL EXPANSION

The term that contains a^r in the expansion of $(a + b)^n$ is

$$\binom{n}{n - r} a^r b^{n-r}$$

Recall that $\binom{n}{n - r} = \binom{n}{r}$.

EXAMPLE 6 ■ Finding a Particular Term in a Binomial Expansion

Find the term that contains x^5 in the expansion of $(2x + y)^{20}$.

SOLUTION The term that contains x^5 is given by the formula for the general term with $a = 2x$, $b = y$, $n = 20$, and $r = 5$. So, this term is

$$\binom{20}{15} a^5 b^{15} = \frac{20!}{15!\,(20 - 5)!}(2x)^5 y^{15} = \frac{20!}{15!\,5!} 32x^5y^{15} = 496{,}128x^5y^{15}$$

∎

EXAMPLE 7 ■ Finding a Particular Term in a Binomial Expansion

Find the coefficient of x^8 in the expansion of $\left(x^2 + \dfrac{1}{x} \right)^{10}$.

SOLUTION Both x^2 and $1/x$ are powers of x, so the power of x in each term of the expansion is determined by both terms of the binomial. To find the required coefficient we first find the general term in the expansion. By the formula, we have $a = x^2$, $b = 1/x$, and $n = 10$, so the general term is

$$\binom{10}{10-r}(x^2)^r\left(\frac{1}{x}\right)^{10-r} = \binom{10}{10-r}x^{2r}(x^{-1})^{10-r} = \binom{10}{10-r}x^{3r-10}$$

Thus, the term that contains x^8 is the term in which

$$3r - 10 = 8$$

$$r = 6$$

So the required coefficient is

$$\binom{10}{10-6} = \binom{10}{4} = 210$$

∎

PROOF OF THE BINOMIAL THEOREM

We now give a proof of the Binomial Theorem using mathematical induction.

■ **Proof** Let $P(n)$ denote the statement

$$(a + b)^n = \binom{n}{0}a^n + \binom{n}{1}a^{n-1}b + \binom{n}{2}a^{n-2}b^2 + \cdots + \binom{n}{n-1}ab^{n-1} + \binom{n}{n}b^n$$

Step 1 We show that $P(1)$ is true. But $P(1)$ is just the statement

$$(a + b)^1 = \binom{1}{0}a^1 + \binom{1}{1}b^1 = 1a + 1b = a + b$$

which is certainly true.

Step 2 We assume that $P(k)$ is true and show that $P(k + 1)$ is true. The statement $P(k)$ reads

$$(a + b)^k = \binom{k}{0}a^k + \binom{k}{1}a^{k-1}b + \binom{k}{2}a^{k-2}b^2 + \cdots + \binom{k}{k-1}ab^{k-1} + \binom{k}{k}b^k$$

Multiplying each side of this equation by $(a + b)$ and collecting like terms gives the following.

$$(a + b)^{k+1} = (a + b)\left[\binom{k}{0}a^k + \binom{k}{1}a^{k-1}b + \binom{k}{2}a^{k-2}b^2 + \cdots + \binom{k}{k-1}ab^{k-1} + \binom{k}{k}b^k\right]$$

$$= a\left[\binom{k}{0}a^k + \binom{k}{1}a^{k-1}b + \binom{k}{2}a^{k-2}b^2 + \cdots + \binom{k}{k-1}ab^{k-1} + \binom{k}{k}b^k\right]$$

$$+ b\left[\binom{k}{0}a^k + \binom{k}{1}a^{k-1}b + \binom{k}{2}a^{k-2}b^2 + \cdots + \binom{k}{k-1}ab^{k-1} + \binom{k}{k}b^k\right]$$

$$= \binom{k}{0}a^{k+1} + \binom{k}{1}a^k b + \binom{k}{2}a^{k-1}b^2 + \cdots + \binom{k}{k-1}a^2 b^{k-1} + \binom{k}{k}ab^k$$

$$+ \binom{k}{0}a^k b + \binom{k}{1}a^{k-1}b^2 + \binom{k}{2}a^{k-2}b^3 + \cdots + \binom{k}{k-1}ab^k + \binom{k}{k}b^{k+1}$$

$$= \binom{k}{0}a^{k+1} + \left[\binom{k}{0} + \binom{k}{1}\right]a^k b + \left[\binom{k}{1} + \binom{k}{2}\right]a^{k-1}b^2 + \cdots + \left[\binom{k}{k-1} + \binom{k}{k}\right]ab^k + \binom{k}{k}b^{k+1}$$

Using the key property of the binomial coefficients, we can write each of the expressions in square brackets as a single binomial coefficient. Also, writing the first and last coefficients as $\binom{k+1}{0}$ and $\binom{k+1}{k+1}$ (these are equal to 1 by Exercise 46) gives

$$(a + b)^{k+1} = \binom{k+1}{0}a^{k+1} + \binom{k+1}{1}a^k b + \binom{k+1}{2}a^{k-1}b^2 + \cdots + \binom{k+1}{k}ab^k + \binom{k+1}{k+1}b^{k+1}$$

But this last equation is precisely $P(k + 1)$, and this completes the induction step.

Having proved Steps 1 and 2, we conclude by the Principle of Mathematical Induction that the theorem is true for all natural numbers n. $\square$

10.6 EXERCISES

1–12 ■ Use Pascal's triangle to expand the expression.

1. $(x + y)^6$

2. $(2x + 1)^4$

3. $\left(x + \dfrac{1}{x}\right)^4$

4. $(x - y)^5$

5. $(x - 1)^5$

6. $(\sqrt{a} + \sqrt{b})^6$

7. $(x^2 y - 1)^5$

8. $(1 + \sqrt{2})^6$

9. $(2x - 3y)^3$

10. $(1 + x^3)^3$

11. $\left(\dfrac{1}{x} - \sqrt{x}\right)^5$

12. $\left(2 + \dfrac{x}{2}\right)^5$

13–20 ■ Evaluate the expression.

13. $\binom{6}{4}$

14. $\binom{8}{3}$

15. $\binom{100}{98}$

16. $\binom{10}{5}$

17. $\binom{3}{1}\binom{4}{2}$

18. $\binom{5}{2}\binom{5}{3}$

19. $\binom{5}{0} + \binom{5}{1} + \binom{5}{2} + \binom{5}{3} + \binom{5}{4} + \binom{5}{5}$

20. $\binom{5}{0} - \binom{5}{1} + \binom{5}{2} - \binom{5}{3} + \binom{5}{4} - \binom{5}{5}$

21–24 ■ Use the Binomial Theorem to expand the expression.

21. $(x + 2y)^4$

22. $(1 - x)^5$

23. $\left(1 + \dfrac{1}{x}\right)^6$

24. $(2A + B^2)^4$

25. Find the first three terms in the expansion of $(x + 2y)^{20}$.

26. Find the first four terms in the expansion of $(x^{1/2} + 1)^{30}$.

27. Find the last two terms in the expansion of $(a^{2/3} + a^{1/3})^{25}$.

28. Find the first three terms in the expansion of

$$\left(x + \dfrac{1}{x}\right)^{40}$$

29. Find the middle term in the expansion of $(x^2 + 1)^{18}$.

30. Find the fifth term in the expansion of $(ab - 1)^{20}$.

31. Find the 24th term in the expansion of $(a + b)^{25}$.

32. Find the 28th term in the expansion of $(A - B)^{30}$.

33. Find the 100th term in the expansion of $(1 + y)^{100}$.

34. Find the second term in the expansion of

$$\left(x^2 - \frac{1}{x} \right)^{25}$$

35. Find the term containing x^4 in the expansion of $(x + 2y)^{10}$.

36. Find the term containing y^3 in the expansion of $(\sqrt{2} + y)^{12}$.

37. Find the term containing b^8 in the expansion of $(a + b^2)^{12}$.

38. Find the term that does not contain x in the expansion of

$$\left(8x + \frac{1}{2x} \right)^{8}$$

39–42 ■ Factor using the Binomial Theorem.

39. $x^4 + 4x^3y + 6x^2y^2 + 4xy^3 + y^4$

40. $(x - 1)^5 + 5(x - 1)^4 + 10(x - 1)^3 + 10(x - 1)^2$ $+ 5(x - 1) + 1$

41. $8a^3 + 12a^2b + 6ab^2 + b^3$

42. $x^8 + 4x^6y + 6x^4y^2 + 4x^2y^3 + y^4$

43–44 ■ Simplify using the Binomial Theorem.

43. $\dfrac{(x + h)^3 - x^3}{h}$ **44.** $\dfrac{(x + h)^4 - x^4}{h}$

45. Show that $(1.01)^{100} > 2$.
[*Hint:* Note that $(1.01)^{100} = (1 + 0.01)^{100}$ and use the Binomial Theorem to show that the sum of the first two terms of the expansion is greater than 2.]

46. Show that $\dbinom{n}{0} = 1$ and $\dbinom{n}{n} = 1$.

47. Show that $\dbinom{n}{1} = \dbinom{n}{n - 1} = n$.

48. Show that $\dbinom{n}{r} = \dbinom{n}{n - r}$ for $0 \le r \le n$.

49. In this exercise we prove the identity

$$\binom{n}{r - 1} + \binom{n}{r} = \binom{n + 1}{r}$$

(a) Write the left side of this equation as the sum of two fractions.

(b) Show that a common denominator of the expression you found in part (a) is $r!\,(n - r + 1)!$.

(c) Add the two fractions using the common denominator in part (b), simplify the numerator, and note that the resulting expression is equal to the right side of the equation.

50. Prove that $\dbinom{n}{r}$ is an integer for all n and for $0 \le r \le n$. [*Suggestion:* Use induction to show that the statement is true for all n, and use Exercise 49 for the induction step.]

■ **DISCOVERY · DISCUSSION**

51. Powers of Factorials   Which is larger, $(100!)^{101}$ or $(101!)^{100}$?   [*Hint:* Try factoring the expressions. Do they have any common factors?]

52. Sums of Binomial Coefficients   Add each of the first five rows of Pascal's triangle, as indicated. Do you see a pattern?

$$1 + 1 = ?$$
$$1 + 2 + 1 = ?$$
$$1 + 3 + 3 + 1 = ?$$
$$1 + 4 + 6 + 4 + 1 = ?$$
$$1 + 5 + 10 + 10 + 5 + 1 = ?$$

Based on the pattern you have found, find the sum of the nth row:

$$\binom{n}{0} + \binom{n}{1} + \binom{n}{2} + \cdots + \binom{n}{n}$$

Prove your result by expanding $(1 + 1)^n$ using the Binomial Theorem.

53. Alternating Sums of Binomial Coefficients   Find the sum

$$\binom{n}{0} - \binom{n}{1} + \binom{n}{2} - \cdots + (-1)^n \binom{n}{n}$$

by finding a pattern as in Exercise 52. Prove your result by expanding $(1 - 1)^n$ using the Binomial Theorem.

10   REVIEW

CONCEPT CHECK

1. (a) What is a sequence?
   (b) What is an arithmetic sequence? Write an expression for the nth term of an arithmetic sequence.
   (c) What is a geometric sequence? Write an expression for the nth term of a geometric sequence.

2. (a) What is a recursive sequence?
   (b) What is the Fibonacci sequence?

3. (a) What is meant by the partial sums of a sequence?
   (b) If an arithmetic sequence has first term a and common difference d, write an expression for the sum of its first n terms.
   (c) If a geometric sequence has first term a and common ratio r, write an expression for the sum of its first n terms.
   (d) Write an expression for the sum of an infinite geometric series with first term a and common ratio r. For what values of r is your formula valid?

4. (a) Explain the meaning of the symbol $\sum\limits_{k=1}^{n} a_k$.
   (b) Write an expression for each of the following.
   (i) $\sum\limits_{k=1}^{n} (a_k + b_k)$ (ii) $\sum\limits_{k=1}^{n} (a_k - b_k)$ (iii) $\sum\limits_{k=1}^{n} ca_k$

5. Write an expression for the amount A_f of an annuity consisting of n regular equal payments of size R with interest rate i per time period.

6. State the Principle of Mathematical Induction.

7. Write the first five rows of Pascal's Triangle. How are the entries related to each other?

8. (a) What does the symbol $n!$ mean?
   (b) Write an expression for the binomial coefficient $\binom{n}{r}$.
   (c) State the Binomial Theorem.
   (d) Write the term that contains a^r in the expansion of $(a + b)^n$.

EXERCISES

1–6 ■ Find the first four terms as well as the tenth term of the sequence with the given nth term.

1. $a_n = \dfrac{n^2}{n + 1}$

2. $a_n = (-1)^n \dfrac{2^n}{n}$

3. $a_n = \dfrac{(-1)^n + 1}{n^3}$

4. $a_n = \dfrac{n(n + 1)}{2}$

5. $a_n = \dfrac{(2n)!}{2^n n!}$

6. $a_n = \binom{n + 1}{2}$

7–10 ■ A sequence is defined recursively. Find the first seven terms of the sequence.

7. $a_n = a_{n-1} + 2n - 1$, $a_1 = 1$

8. $a_n = \dfrac{a_{n-1}}{n}$, $a_1 = 1$

9. $a_n = a_{n-1} + 2a_{n-2}$, $a_1 = 1, a_2 = 3$

10. $a_n = \sqrt{3a_{n-1}}$, $a_1 = \sqrt{3}$

11–18 ■ The first four terms of a sequence are given. Determine whether the given terms can be the terms of an arithmetic sequence, a geometric sequence, or neither. If the sequence is arithmetic or geometric, find the fifth term.

11. $5, 5.5, 6, 6.5, \ldots$

12. $1, -\frac{3}{2}, 2, -\frac{5}{2}, \ldots$

13. $\sqrt{2}, 2\sqrt{2}, 3\sqrt{2}, 4\sqrt{2}, \ldots$

14. $\sqrt{2}, 2, 2\sqrt{2}, 4, \ldots$

15. $t - 3, t - 2, t - 1, t, \ldots$

16. $t^3, t^2, t, 1, \ldots$

17. $\dfrac{3}{4}, \dfrac{1}{2}, \dfrac{1}{3}, \dfrac{2}{9}, \ldots$

18. $a, 1, \dfrac{1}{a}, \dfrac{1}{a^2}, \ldots$

19. Show that $3, 6i, -12, -24i, \ldots$ is a geometric sequence, and find the common ratio.

20. Find the nth term of the geometric sequence $2, 2 + 2i,$ $4i, -4 + 4i, -8, \ldots$.

21. The sixth term of an arithmetic sequence is 17, and the fourth term is 11. Find the second term.

22. The 20th term of an arithmetic sequence is 96, and the common difference is 5. Find the nth term.

23. The third term of a geometric sequence is 9, and the common ratio is $\frac{3}{2}$. Find the fifth term.

24. The second term of a geometric sequence is 10, and the fifth term is $\frac{1250}{27}$. Find the nth term.

25. The frequencies of musical notes (measured in cycles per second) form a geometric sequence. Middle C has a frequency of 256, and C, an octave higher, has a frequency of 512. Find the frequency of C two octaves below middle C.

26. A person has two parents, four grandparents, eight great-grandparents, and so on. How many ancestors does a person have 15 generations back?

27. A certain type of bacteria divides every 5 s. If three of these bacteria are put into a petri dish, how many bacteria are in the dish at the end of 1 min?

28. If $a_1, a_2, a_3, \ldots$ and $b_1, b_2, b_3, \ldots$ are arithmetic sequences, show that $a_1 + b_1, a_2 + b_2, a_3 + b_3, \ldots$ is also an arithmetic sequence.

29. If $a_1, a_2, a_3, \ldots$ and $b_1, b_2, b_3, \ldots$ are geometric sequences, show that $a_1b_1, a_2b_2, a_3b_3, \ldots$ is also a geometric sequence.

30. (a) If $a_1, a_2, a_3, \ldots$ is an arithmetic sequence, is the sequence $a_1 + 2, a_2 + 2, a_3 + 2, \ldots$ arithmetic?
(b) If $a_1, a_2, a_3, \ldots$ is a geometric sequence, is the sequence $5a_1, 5a_2, 5a_3, \ldots$ geometric?

31. Find the values of x for which the sequence 6, x, 12, $\ldots$ is
(a) arithmetic   (b) geometric

32. Find the values of x and y for which the sequence 2, x, y, 17, $\ldots$ is
(a) arithmetic   (b) geometric

33–36 ■ Find the sum.

33. $\displaystyle\sum_{k=3}^{6} (k + 1)^2$

34. $\displaystyle\sum_{i=1}^{4} \frac{2i}{2i - 1}$

35. $\displaystyle\sum_{k=1}^{6} (k + 1)2^{k-1}$

36. $\displaystyle\sum_{m=1}^{5} 3^{m-2}$

37–40 ■ Write the sum without using sigma notation. Do not evaluate.

37. $\displaystyle\sum_{k=1}^{10} (k - 1)^2$

38. $\displaystyle\sum_{j=2}^{100} \frac{1}{j - 1}$

39. $\displaystyle\sum_{k=1}^{50} \frac{3^k}{2^{k+1}}$

40. $\displaystyle\sum_{n=1}^{10} n^2 2^n$

41–44 ■ Write the sum using sigma notation. Do not evaluate.

41. $3 + 6 + 9 + 12 + \cdots + 99$

42. $1^2 + 2^2 + 3^2 + \cdots + 100^2$

43. $1 \cdot 2^3 + 2 \cdot 2^4 + 3 \cdot 2^5 + 4 \cdot 2^6 + \cdots + 100 \cdot 2^{102}$

44. $\dfrac{1}{1 \cdot 2} + \dfrac{1}{2 \cdot 3} + \dfrac{1}{3 \cdot 4} + \cdots + \dfrac{1}{999 \cdot 1000}$

45–50 ■ Determine whether the expression is a partial sum of an arithmetic or geometric sequence. Then find the sum.

45. $1 + 0.9 + (0.9)^2 + \cdots + (0.9)^5$

46. $3 + 3.7 + 4.4 + \cdots + 10$

47. $\sqrt{5} + 2\sqrt{5} + 3\sqrt{5} + \cdots + 100\sqrt{5}$

48. $\frac{1}{3} + \frac{2}{3} + 1 + \frac{4}{3} + \cdots + 33$

49. $\displaystyle\sum_{n=0}^{6} 3(-4)^n$

50. $\displaystyle\sum_{k=0}^{8} 7(5)^{k/2}$

51. The first term of an arithmetic sequence is $a = 7$, and the common difference is $d = 3$. How many terms of this sequence must be added to obtain 325?

52. The sum of the first three terms of a geometric series is 52, and the common ratio is $r = 3$. Find the first term.

53. Refer to Exercise 26. What is the total number of a person's ancestors in 15 generations?

54. Find the amount of an annuity consisting of 16 annual payments of $1000 each into an account that pays 8% interest per year, compounded annually.

55. How much money should be invested every quarter at 12% per year, compounded quarterly, in order to have $10,000 in one year?

56. What are the monthly payments on a mortgage of $60,000 at 9% interest if the loan is to be repaid in
(a) 30 years? (b) 15 years?

57–60 ■ Find the sum of the infinite geometric series.

57. $1 - \frac{2}{5} + \frac{4}{25} - \frac{8}{125} + \cdots$

58. $0.1 + 0.01 + 0.001 + 0.0001 + \cdots$

59. $1 + \dfrac{1}{3^{1/2}} + \dfrac{1}{3} + \dfrac{1}{3^{3/2}} + \cdots$

60. $a + ab^2 + ab^4 + ab^6 + \cdots$

61–63 ■ Use mathematical induction to prove that the formula is true for all natural numbers n.

61. $1 + 4 + 7 + \cdots + (3n - 2) = \dfrac{n(3n - 1)}{2}$

62. $\dfrac{1}{1 \cdot 3} + \dfrac{1}{3 \cdot 5} + \dfrac{1}{5 \cdot 7} + \cdots + \dfrac{1}{(2n - 1)(2n + 1)}$

$$= \dfrac{n}{2n + 1}$$

63. $\left(1 + \dfrac{1}{1}\right)\left(1 + \dfrac{1}{2}\right)\left(1 + \dfrac{1}{3}\right)\cdots\left(1 + \dfrac{1}{n}\right) = n + 1$

64. Show that $7^n - 1$ is divisible by 6 for all natural numbers n.

65. Let $a_{n+1} = 3a_n + 4$ and $a_1 = 4$. Show that $a_n = 2 \cdot 3^n - 2$ for all natural numbers n.

66. Prove that the Fibonacci number F_{4n} is divisible by 3 for all natural numbers n.

67. Find and prove an inequality that relates 2^n and $n!$.

68–71 ■ Evaluate the expression.

68. $\dbinom{5}{2}\dbinom{5}{3}$

69. $\dbinom{10}{2} + \dbinom{10}{6}$

70. $\displaystyle\sum_{k=0}^{5} \dbinom{5}{k}$

71. $\displaystyle\sum_{k=0}^{8} \dbinom{8}{k}\dbinom{8}{8 - k}$

72–73 ■ Expand the expression.

72. $(1 - x^2)^6$

73. $(2x + y)^4$

74. Find the 20th term in the expansion of $(a + b)^{22}$.

75. Find the first three terms in the expansion of $(b^{-2/3} + b^{1/3})^{20}$.

76. Find the term containing A^6 in the expansion of $(A + 3B)^{10}$.

1. Find the first four terms and the tenth term of the sequence whose nth term is

$$a_n = n^2 - 1$$

2. Find the common difference, the fourth term, and the nth term in the arithmetic sequence 80, 76, 72, ...

3. The first term of a geometric sequence is 25, and the fourth term is $\frac{1}{5}$. Find the common ratio and the fifth term.

4. Determine whether each of the following statements is true or false. If it is true, prove it. If it is false, give an example where it fails.
   (a) If $a_1, a_2, a_3, \ldots$ is an arithmetic sequence, then the sequence $a_1^2, a_2^2, a_3^2, \ldots$ is also arithmetic.
   (b) If $a_1, a_2, a_3, \ldots$ is a geometric sequence, then the sequence $a_1^2, a_2^2, a_3^2, \ldots$ is also geometric.

5. (a) Write the formula for the nth partial sum of an arithmetic sequence.
   (b) The first term of an arithmetic sequence is 10, and the tenth term is 2. Find the partial sum of the first ten terms.
   (c) Find the common difference and the 100th term of the sequence in part (b).

6. (a) Write the formula for the nth partial sum of a geometric sequence.
   (b) Find the sum

$$\frac{1}{3} + \frac{2}{3^2} + \frac{2^2}{3^3} + \frac{2^3}{3^4} + \cdots + \frac{2^9}{3^{10}}$$

7. Find the sum of the infinite geometric series $1 + \dfrac{1}{2^{1/2}} + \dfrac{1}{2} + \dfrac{1}{2^{3/2}} + \cdots$.

8. Use mathematical induction to prove that, for all natural numbers n,

$$1^2 + 2^2 + 3^2 + \cdots + n^2 = \frac{n(n + 1)(2n + 1)}{6}$$

9. Write the expression without using sigma notation, and then find the sum.
   (a) $\displaystyle\sum_{n=1}^{5} (1 - n^2)$ (b) $\displaystyle\sum_{n=3}^{6} (-1)^n 2^{n-2}$

10. Expand $(2x + y^2)^5$.

11. A sequence is defined recursively by

$$a_{n+2} = (a_n)^2 - a_{n+1}$$

If $a_1 = 1$ and $a_2 = 1$, find a_5.

FOCUS ON PROBLEM SOLVING

The solutions to many of the problems of mathematics involve **finding patterns**. The algebraic formulas we've found in this book are compact ways of describing patterns. For example, the familiar equation $(a + b)^2 = a^2 + 2ab + b^2$ gives the pattern for squaring the sum of two numbers. Another pattern we have encountered is the pattern for the sum of the first n natural numbers:

$$1 + 2 + 3 + \cdots + n = \frac{n(n + 1)}{2}$$

How do we discover patterns? In many cases, a good way to start is to experiment with the problem. To prove that a pattern always holds, we can often use mathematical induction, but other proofs are possible. A geometrical method is used in the problem we give here. It is attributed to the 11th-century mathematician Abu Bekr Mohammed ibn Al Husain Al Karchi.

THE GNOMONS OF AL KARCHI

We prove the beautiful formula

$$1^3 + 2^3 + 3^3 + \cdots + n^3 = (1 + 2 + 3 + \cdots + n)^2$$

But first, let's see how this formula was discovered. The sum of the first n natural numbers is called a *triangular number*. The name "triangular" comes from the following figures:

| 1 | $1 + 2 = 3$ | $1 + 2 + 3 = 6$ | $1 + 2 + 3 + 4 = 10$ | $1 + 2 + 3 + 4 + 5 = 15$ |

The first few triangular numbers are

$$1, 3, 6, 10, 15, \ldots$$

Now, let's look at the sums of the cubes:

$$1^3 = 1$$
$$1^3 + 2^3 = 9$$
$$1^3 + 2^3 + 3^3 = 36$$
$$1^3 + 2^3 + 3^3 + 4^3 = 100$$
$$1^3 + 2^3 + 3^3 + 4^3 + 5^3 = 225$$

We get the sequence

$$1, 9, 36, 100, 225, \ldots$$

It doesn't take long to notice that these are the squares of the triangular numbers. It appears that the sum of the first n cubes equals the square of the sum of the first n numbers.

To show that this pattern always holds, Al Karchi sketches the following diagram:

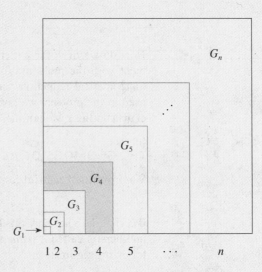

Each region G_n in the shape of an inverted **L** is called a *gnomon*. The area of gnomon G_4 is

$$\text{area} = 4^2 + 2[4 \times (1 + 2 + 3)] = 64$$

Similarly, the areas of gnomons G_1, G_2, G_3, G_4, G_5, ... are 1, 8, 27, 64, 125, ... These are the cubes of the natural numbers. The pattern persists here also, because the area of the nth gnomon G_n is n^3, as the following calculation shows:

$$n^2 + 2(n \times [1 + 2 + \cdots + (n - 1)]) = n^2 + 2n\,\frac{(n - 1)n}{2}$$
$$= n^2 + n^3 - n^2 = n^3$$

Now comes Al Karchi's punchline: The first n gnomons form a square of side $1 + 2 + 3 + \cdots + n$ and so the sum of the areas of these gnomons equals the area of the square, that is, $1^3 + 2^3 + \cdots + n^3 = (1 + 2 + \cdots + n)^2$.

▨ PROBLEMS

1. Use the diagram at the left to find and prove a formula for the sum of the first n odd numbers.

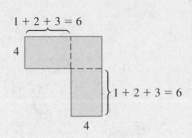

$1 + 2 + 3 = 6$

4

$1 + 2 + 3 = 6$

4

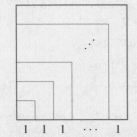

$1 \quad 1 \quad 1 \quad \cdots \quad 1$

2. Use the figure to find a simple formula for $F_1^2 + F_2^2 + \cdots + F_n^2$, where F_k is the kth Fibonacci number.

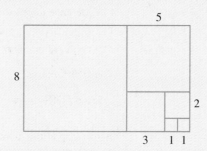

3. Show that every perfect cube is a difference of two squares. [*Hint:* Use the formula for the sum of cubes that we have just proved.]

4. (a) Find the product $\left(1 - \dfrac{1}{2}\right)\left(1 - \dfrac{1}{3}\right)\left(1 - \dfrac{1}{4}\right) \cdots \left(1 - \dfrac{1}{200}\right)$.

(b) Find the product $\left(1 - \dfrac{1}{4}\right)\left(1 - \dfrac{1}{9}\right)\left(1 - \dfrac{1}{16}\right) \cdots \left(1 - \dfrac{1}{n^2}\right)$.

5. An $n \times n$ *magic square* consists of the numbers from 1 to n^2 arranged in a square in such a way that the sum of the numbers in any row, column, or diagonal is the same. Let's call this constant sum S. Find a formula for S for an $n \times n$ magic square.

6. The ancient Greeks studied the triangular numbers that we described on page 735. In a similar way, we can define other *polygonal numbers,* such as the square and pentagonal numbers.

Square numbers   1   4   9   16   25

Pentagonal numbers   1   5   12   22   35

To find a pattern for such numbers, we construct the *difference table* by taking differences of successive terms in the sequence, then taking differences of the differences, and so on. For example, the triangular numbers give the following difference table:

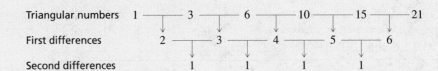

Triangular numbers   1 —— 3 —— 6 —— 10 —— 15 —— 21

First differences   2 —— 3 —— 4 —— 5 —— 6

Second differences   1   1   1   1

We stop at this point because we have obtained a constant sequence. Working backward from the bottom row, we can easily find more of the first differences, and from these, more of the triangular numbers.

(a) Construct the difference tables for the square numbers and the pentagonal numbers. Use the pattern to find the tenth pentagonal number.

(b) Using the patterns you've observed so far, what do you think the second differences would be for the *hexagonal numbers*? Use this, together with the fact that the first two hexagonal numbers are 1 and 6, to construct the difference table that gives the first eight hexagonal numbers.

(c) Sketch a dot pattern like those shown on page 737 to illustrate the first four hexagonal numbers.

7. Choose *n* different points on the circumference of a circle, and then connect them with line segments. We are interested in the number of regions into which these segments divide the circle. Let's see what happens for *n* = 1, 2, 3, and 4.

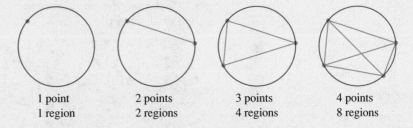

| 1 point | 2 points | 3 points | 4 points |
| 1 region | 2 regions | 4 regions | 8 regions |

(a) How many regions result from five points on the circle?

(b) Based on the pattern you have observed, how many regions do you think would arise from *n* different points on the circle?

(c) Draw a diagram of the regions that result from six points on the circle. (Do this very carefully.) Does this result fit with the pattern you conjectured in part (b)?

(d) Take the sequence of numbers of regions for *n* = 1, 2, 3, 4, 5, 6, and construct the difference table (see Problem 6). Use the difference table to determine the number of regions for *n* = 7, 8, 9, and 10.

8. Starting with an equilateral triangle of side 1, we successively construct new figures with more and more sides as shown. The *snowflake curve* is the result of repeating this process indefinitely.

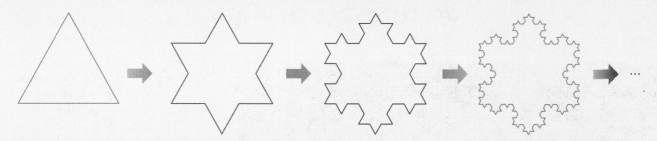

(a) Find the area enclosed by the snowflake curve. [*Hint:* This area is the sum of an infinite series.]

(b) Observe that the length of the snowflake curve is infinite.

Parts (a) and (b) show that if your lawn has the shape of a snowflake curve, you could easily mow it, because it has finite area, but you could never put a fence along its boundary.

9. Prove that the number of people who have shaken hands an odd number of times is an even number.

10. Let p be a prime number.

(a) Show that $\binom{p}{k}$ is divisible by p for $1 \le k \le p$.

(b) Use mathematical induction to prove that $n^p - n$ is divisible by p for all n. [*Hint:* For the induction step, use the Binomial Theorem to expand $(n + 1)^p$, and then use part (a).]

11. Consider a 6×6 grid as shown in the figure.

(a) Find the number of squares of all sizes in this grid. Generalize your result to an $n \times n$ grid.

(b) Find the number of rectangles of all sizes in this grid. Generalize your result to an $n \times n$ grid.

COUNTING AND PROBABILITY

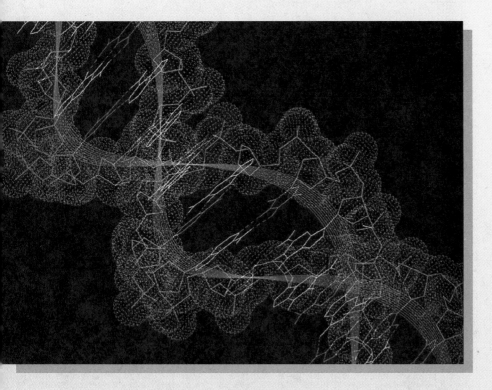

Probability is the mathematical study of chance and random processes. The laws of probability are essential for understanding such fields as genetics, opinion polls, games of chance, and many others.

When it is not in our power to determine what is true, we ought to follow what is most probable.

RENÉ DESCARTES

Many questions in mathematics involve *counting*. For example, in how many ways can a committee of two men and three women be chosen from a group of 35 men and 40 women? How many different license plates can be made using three letters followed by three digits? How many different poker hands are possible?

Closely related to the problem of counting is that of *probability*. We consider questions such as these: If a committee of five people is chosen randomly from a group of 35 men and 40 women, what are the chances that no women will be chosen for the committee? What is the likelihood of getting a straight flush in a poker game? In studying probability we give precise mathematical meaning to phrases such as "what are the chances...?" and "what is the likelihood...?"

11.1 COUNTING PRINCIPLES

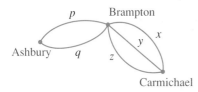

Suppose that three towns, Ashbury, Brampton, and Carmichael, are located in such a way that two roads connect Ashbury to Brampton and three roads connect Brampton to Carmichael. How many different routes can one take to travel from Ashbury to Carmichael via Brampton? The key idea in answering this question is to consider the problem in stages. At the first stage—from Ashbury to Brampton—there are two choices to make. For each one of these choices, there are three choices to make at the second stage—from Brampton to Carmichael. Thus, the number of different routes is $2 \times 3 = 6$. These routes are conveniently enumerated by a *tree diagram* as in Figure 1.

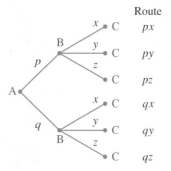

FIGURE 1
Tree diagram

The method used to solve this problem leads to the following principle.

FUNDAMENTAL COUNTING PRINCIPLE

Suppose that two events occur in order. If the first can occur in m ways and the second in n ways (after the first has occurred), then the two events can occur in order in $m \times n$ ways.

There is an immediate consequence of this principle for any number of events: If $E_1, E_2, \ldots, E_k$ are events that occur in order and if E_1 can occur in n_1 ways, E_2 in n_2 ways, and so on, then the events can occur in order in $n_1 \times n_2 \times \cdots \times n_k$ ways.

EXAMPLE 1 ■ Using the Fundamental Counting Principle

An ice-cream store offers three types of cones and 31 flavors. How many different single-scoop ice-cream cones is it possible to buy at this store?

SOLUTION   There are two choices to make: type of cone and flavor of ice cream. At the first stage we choose a type of cone, and at the second stage we choose a flavor. We can think of the different stages as boxes:

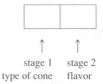

stage 1   stage 2
type of cone   flavor

The first box can be filled in three ways and the second in 31 ways:

stage 1   stage 2

Thus, by the Fundamental Counting Principle, there are $3 \times 31 = 93$ ways of choosing a single-scoop ice-cream cone at this store.   ■

EXAMPLE 2 ■ Using the Fundamental Counting Principle

In a certain state, automobile license plates display three letters followed by three digits. How many such plates are possible if repetition of the letters

(a) is allowed?   (b) is not allowed?

SOLUTION

(a) There are six choices to be made, one choice for each letter or digit on the license plate. As in the preceding example, we sketch a box for each stage:

letters   digits

At the first stage, we choose a letter (from 26 possible choices); at the second stage, another letter (again from 26 choices); at the third stage, another letter (26 choices); at the fourth stage, a digit (from 10 possible choices); at the fifth stage, a digit (again from 10 choices); and at the sixth

stage, another digit (10 choices). By the Fundamental Counting Principle, the number of possible license plates is

$$26 \times 26 \times 26 \times 10 \times 10 \times 10 = 17{,}576{,}000$$

(b) If repetition of letters is not allowed, then we can arrange the choices as follows:

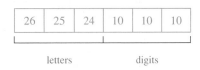

26	25	24	10	10	10

  letters   digits

At the first stage, we have 26 letters to choose from, but once the first letter is chosen, there are only 25 letters to choose from at the second stage. Once the first two letters are chosen, 24 letters are left to choose from for the third stage. The digits are chosen as before. Thus, the number of possible license plates in this case is

$$26 \times 25 \times 24 \times 10 \times 10 \times 10 = 15{,}600{,}000 \qquad \blacksquare$$

EXAMPLE 3 ■ **Using Factorial Notation**

In how many different ways can a race with six runners be completed? Assume there is no tie.

Factorial notation is explained on page 724.

SOLUTION   There are six possible choices for first place, five choices for second place (since only five runners are left after first place has been decided), four choices for third place, and so on. So, by the Fundamental Counting Principle, the number of different ways this race can be completed is

$$6 \times 5 \times 4 \times 3 \times 2 \times 1 = 6! = 720 \qquad \blacksquare$$

11.1   **EXERCISES**

1. A vendor sells ice cream from a cart on the boardwalk. He offers vanilla, chocolate, strawberry, and pistachio ice cream, served on either a waffle, sugar, or plain cone. How many different single-scoop ice-cream cones can you buy from this vendor?

2. How many three-letter "words" (strings of letters) can be formed using the 26 letters of the alphabet if repetition of letters
(a) is allowed?   (b) is not allowed?

3. How many three-letter "words" (strings of letters) can be formed using the letters *WXYZ* if repetition of letters
(a) is allowed?   (b) is not allowed?

4. Eight horses are entered in a race.
 (a) How many different orders are possible for completing the race?
 (b) In how many different ways can first, second, and third places be decided? (Assume there is no tie.)

5. A multiple-choice test has five questions with four choices for each question. In how many different ways can the test be completed?

6. Telephone numbers consist of seven digits; the first digit cannot be 0 or 1. How many telephone numbers are possible?

7. In how many different ways can a race with five runners be completed? (Assume there is no tie.)

8. In how many ways can five people be seated in a row of five seats?

9. A restaurant offers six different main courses, eight types of drinks, and three kinds of desserts. How many different meals consisting of a main course, a drink, and a dessert does the restaurant offer?

10. In how many ways can five different mathematics books be placed next to each other on a shelf?

11. Towns A, B, C, and D are located in such a way that there are four roads from A to B, five roads from B to C, and six roads from C to D. How many routes are there from town A to town D via towns B and C?

12. In a family of four children, how many different boy-girl birth-order combinations are possible? (The birth orders *BBBG* and *BBGB* are different.)

13. A coin is flipped five times, and the resulting sequence of heads and tails is recorded. How many such sequences are possible?

14. A red die and a white die are rolled, and the numbers showing are recorded. How many different outcomes are possible? (The singular form of the word *dice* is *die*.)

15. A red die, a blue die, and a white die are rolled, and the numbers showing are recorded. How many different outcomes are possible?

16. Two cards are chosen in order from a deck. In how many ways can this be done if
    (a) the first card must be a spade and the second must be a heart?
    (b) both cards must be spades?

17. A girl has five skirts, eight blouses, and 12 pairs of shoes. How many different skirt-blouse-shoe outfits can she wear? (Assume that each item matches all the others, so she is willing to wear any combination.)

18. A company's employee ID number system consists of one letter followed by three digits. How many different ID numbers are possible with this system?

19. A company has 2844 employees. Each employee is to be given an ID number that consists of one letter followed by two digits. Is it possible to give each employee a different ID number using this scheme? Explain.

20. An all-star baseball team has a roster of seven pitchers and three catchers. How many pitcher-catcher pairs can the manager select from this roster?

21. Standard automobile license plates in California display a nonzero digit, followed by three letters, followed by three digits. How many different standard plates are possible in this system?

22. A combination lock has 60 different positions. In order to open the lock, the dial is turned to a certain number in the clockwise direction, then to a number in the counterclockwise direction, and finally to a third number in the clockwise direction. If successive numbers in the combination cannot be the same, how many different combinations are possible?

23. A true-false test contains ten questions. In how many different ways can this test be completed?

24. An automobile dealer offers five models. Each model comes in a choice of four colors, three types of stereo equipment, with or without air conditioning, and with or without a sunroof. In how many different ways can a customer order an auto from this dealer?

25. The registrar at a certain university classifies students according to major, minor, year (1, 2, 3, 4), and sex (M, F). Each student must choose one major and either one or no minor from the 32 fields taught at this university. How many different student classifications are possible?

26. How many monograms consisting of three initials are possible?

27. A state has registered 8 million automobiles. In order to simplify the license plate system, a state employee suggests that each plate display only two letters followed by three digits. Will this system create enough different license plates for all the vehicles registered?

28. A state license plate design has six places. Each plate begins with a fixed number of letters, and the remaining places are filled with digits. (For example, one letter followed by five digits, two letters followed by four digits, and so on.) The state has registered 17 million vehicles.
(a) The state decides to change to a system consisting of one letter followed by five digits. Will this design allow for enough different plates to accommodate all the vehicles registered?
(b) Find a system that will be sufficient if the smallest possible number of letters is to be used.

29. In how many ways can a president, vice president, and secretary be chosen from a class of 30 students?

30. In how many ways can a president, vice president, and secretary be chosen from a class of 20 females and 30 males if the president must be a female and the vice president a male?

31. A senate subcommittee consists of ten Democrats and seven Republicans. In how many ways can a chairman, vice chairman, and secretary be chosen if the chairman must be a Democrat and the vice chairman must be a Republican?

32. Social Security numbers consist of nine digits, with the first digit between 0 and 6, inclusive. How many Social Security numbers are possible?

33. Five-letter "words" are formed using the letters *A*, *B*, *C*, *D*, *E*, *F*, *G*. How many such words are possible for each of the following conditions?
(a) No condition is imposed.
(b) No letter can be repeated in a word.
(c) Each word must begin with the letter *A*.
(d) The letter *C* must be in the middle.
(e) The middle letter must be a vowel.

34. How many five-letter palindromes are possible? (A *palindrome* is a string of letters that reads the same backward and forward, such as the string *XCZCX*.)

35. A certain computer programming language allows names of variables to consist of two characters, the first being any letter and the second any letter or digit. How many names of variables are possible?

36. How many different three-character code words consisting of letters or digits are possible for each of the following code designs?
(a) The first entry must be a letter.
(b) The first entry cannot be zero.

37. In how many ways can four men and four women be seated in a row of eight seats for each of the following situations?
(a) The women are to be seated together, and the men are to be seated together.
(b) They are to be seated alternately by gender.

38. In how many ways can five different mathematics books be placed on a shelf if the two algebra books are to be placed next to each other?

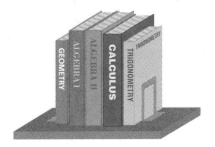

39. Eight mathematics books and three chemistry books are to be placed on a shelf. In how many ways can this be done if the mathematics books are to be next to each other and the chemistry books are next to each other?

40. Three-digit numbers are formed using the digits 2, 4, 5, and 7, with repetition of digits allowed. How many such numbers can be formed if
(a) the numbers are less than 700?
(b) the numbers are even?
(c) the numbers are divisible by 5?

41. How many three-digit odd numbers can be formed using the digits 1, 2, 4, and 6 if repetition of digits is not allowed?

DISCOVERY · DISCUSSION

42. Pairs of Initials Explain why in any group of 677 people, at least two people must have the same pair of initials.

43. Area Codes Until recently, telephone area codes in the United States, Canada, and the Caribbean islands were chosen according to the following rules: (*i*) The first digit *cannot* be a 0 or a 1, and (*ii*) the second digit *must* be a 0 or a 1. But in 1995, the second rule was abandoned when the area code 360 was introduced in parts of western Washington State. Since then, many other new area codes that violate Rule (*ii*) have come into use, although Rule (*i*) still remains in effect.

(a) How many area code + telephone number combinations were possible under the old rules? (See Exercise 6 for a description of local telephone numbers.)

(b) How many area code + telephone number combinations are now possible under the new rules?

(c) Why do you think it was necessary to make this change?

(d) Look at the area code listings in a recent telephone directory and find five or more area codes that violate Rule (*ii*).

11.2 PERMUTATIONS AND COMBINATIONS

In this section we single out two important special cases of the multiplication principle—permutations and combinations.

PERMUTATIONS

Permutations of three colored squares

A **permutation** of a set of distinct objects is an ordering of these objects. For example, some of the permutations of the letters *ABCDWXYZ* are

XAYBZWCD   *ZAYBCDWX*   *DBWAZXYC*   *YDXAWCZB*

How many such permutations are possible? Since there are eight choices for the first position, seven for the second (after the first has been chosen), six for the third (after the first two have been chosen), and so on, the Fundamental Counting Principle tells us that the number of possible permutations is

$$8 \times 7 \times 6 \times 5 \times 4 \times 3 \times 2 \times 1 = 40{,}320$$

This same reasoning with 8 replaced by *n* leads to the following observation.

> The number of permutations of *n* objects is *n*!.

How many permutations consisting of five letters can be made from these same eight letters? Some of these permutations are

XYZWC   *AZDWX*   *AZXYB*   *WDXZB*

Again, there are eight choices for the first position, seven for the second, six for the third, five for the fourth, and four for the fifth. By the Fundamental Counting Principle, the number of such permutations is

$$8 \times 7 \times 6 \times 5 \times 4 = 6720$$

In general, if a set has *n* elements, then the number of ways of ordering *r* elements from the set is denoted by $P(n, r)$ and is called **the number of permutations of *n* objects taken *r* at a time**.

We have just shown that $P(8, 5) = 6720$. The same reasoning used to find $P(8, 5)$ will help us to find a general formula for $P(n, r)$. Indeed, there are n objects and r positions to place them in. Thus, there are n choices for the first position, $n - 1$ choices for the second, $n - 2$ choices for the third, and so on. The last position can be filled in $n - r + 1$ ways. By the Fundamental Counting Principle,

$$P(n, r) = n(n - 1)(n - 2) \cdots (n - r + 1)$$

This formula can be written more compactly using factorial notation:

$$P(n, r) = n(n - 1)(n - 2) \cdots (n - r + 1)$$

$$= \frac{n(n - 1)(n - 2) \cdots (n - r + 1)(n - r) \cdots 3 \cdot 2 \cdot 1}{(n - r) \cdots 3 \cdot 2 \cdot 1} = \frac{n!}{(n - r)!}$$

PERMUTATIONS OF n OBJECTS TAKEN r AT A TIME

The number of permutations of n objects taken r at a time is

$$P(n, r) = \frac{n!}{(n - r)!}$$

EXAMPLE 1 ■ Finding the Number of Permutations

A club has nine members. In how many ways can a president, vice president, and secretary be chosen from the members of this club?

SOLUTION   We need the number of ways of selecting three members *in order* for the positions of president, vice president, and secretary from the nine club members. This number is

$$P(9, 3) = \frac{9!}{(9 - 3)!} = \frac{9!}{6!} = 9 \times 8 \times 7 = 504$$

■

EXAMPLE 2 ■ Finding the Number of Permutations

From 20 raffle tickets in a hat, four tickets are to be selected in order. The holder of the first ticket wins a car, the second a motorcycle, the third a bicycle, and the fourth a skateboard. In how many different ways can these prizes be awarded?

SOLUTION   The order in which the tickets are chosen determines who wins each prize. So, we need to find the number of ways of selecting four objects *in order* from 20 objects (the tickets). This number is

$$P(20, 4) = \frac{20!}{(20 - 4)!} = \frac{20!}{16!} = 20 \times 19 \times 18 \times 17 = 116{,}280$$

■

DISTINGUISHABLE PERMUTATIONS

If we have a collection of ten balls, each a different color, then the number of permutations of these balls is $P(10, 10) = 10!$. If all ten balls are red, then we have just one distinguishable permutation because all the ways of ordering these balls look the same. In general, when considering a set of objects, some of which are of the same kind, then two permutations are **distinguishable** if one cannot be obtained from the other by interchanging the positions of elements of the same kind. For example, if we have ten balls, of which six are red and the other four are each a different color, then how many distinguishable permutations are possible? The key point here is that balls of the same color are not distinguishable. So each rearrangement of the red balls, keeping all the other balls fixed, gives essentially the same permutation. Since there are 6! rearrangements of the red balls for each fixed position of the other balls, the total number of distinguishable permutations is 10!/6!. The same type of reasoning gives the following general rule.

DISTINGUISHABLE PERMUTATIONS

If a set of n objects consists of k different kinds of objects with n_1 objects of the first kind, n_2 objects of the second kind, n_3 objects of the third kind, and so on, where $n_1 + n_2 + \cdots + n_k = n$, then the number of distinguishable permutations of these objects is

$$\frac{n!}{n_1!\, n_2!\, n_3! \cdots n_k!}$$

EXAMPLE 3 ■ Finding the Number of Distinguishable Permutations

Find the number of different ways of placing 15 balls in a row given that 4 are red, 3 are yellow, 6 are black, and 2 are blue.

SOLUTION   We want to find the number of distinguishable permutations of these balls. By the formula, this number is

$$\frac{15!}{4!\, 3!\, 6!\, 2!} = 6{,}306{,}300$$

■

Suppose we have 15 wooden balls in a row and four colors of paint: red, yellow, black, and blue. In how many different ways can the 15 balls be painted in such a way that we have 4 red, 3 yellow, 6 black, and 2 blue balls? A little thought will show that this number is exactly the same as that calculated in Example 3. This way of looking at the problem is somewhat different, however. Here we think of the number of ways to **partition** the balls into four groups, each containing 4, 3, 6, and 2 balls to be painted red, yellow, black, and blue, respectively. The next example shows how this reasoning is used.

EXAMPLE 4 ■ Finding the Number of Partitions

Fourteen construction workers are to be assigned to three different tasks. Seven workers are needed for mixing cement, five for laying bricks, and two for carrying the bricks to the brick layers. In how many different ways can the workers be assigned to these tasks?

SOLUTION   We need to partition the workers into three groups containing 7, 5, and 2 workers, respectively. This number is

$$\frac{14!}{7!\,5!\,2!} = 72{,}072$$

■

Ronald Graham was born in Taft, California, in 1935. He has been described as the world's leading mathematician in the field of combinatorics, the branch of mathematics that deals with counting. For many years Graham was head of the Mathematical Studies Center at Bell Laboratories in Murray Hill, New Jersey. There he solved numerous problems that arose in the telephone industry. During the *Apollo* program, NASA needed to evaluate mission schedules so that the three astronauts aboard the spacecraft could find the time to perform all the necessary tasks. The number of ways to allot these tasks was astronomical—too vast for even a computer to sort out. Graham, using his knowledge of combinatorics, was able to reassure NASA that there were easy ways of solving their problem that were not too far from the theoretically best possible solution. Besides being a prolific mathematician, Graham is an accomplished juggler and has been president of the International Jugglers Association.

COMBINATIONS

When finding permutations, we are interested in the number of ways of ordering elements of a set. In many counting problems, however, order is *not* important. For example, a poker hand is the same hand, regardless of how it is ordered. A poker player interested in the number of possible hands wants to know the number of ways of drawing five cards from 52 cards, without regard to the order in which the cards of a given hand are dealt. In this section we develop a formula for counting in situations such as this, where order doesn't matter.

A **combination** of r elements of a set is any subset of r elements from the set (without regard to order). If the set has n elements, then the number of combinations of r elements is denoted by $C(n, r)$ and is called the **number of combinations of n elements taken r at a time**.

For example, consider a set with the four elements A, B, C, and D. The combinations of these four elements taken three at a time are

ABC	*ABD*	*ACD*	*BCD*

The permutations of these elements taken three at a time are

ABC	*ABD*	*ACD*	*BCD*
ACB	*ADB*	*ADC*	*BDC*
BAC	*BAD*	*CAD*	*CBD*
BCA	*BDA*	*CDA*	*CDB*
CAB	*DAB*	*DAC*	*DBC*
CBA	*DBA*	*DCA*	*DCB*

We notice that the number of combinations is a lot fewer than the number of permutations. In fact, each combination of three elements generates 3! permutations. So

$$C(4, 3) = \frac{P(4, 3)}{3!} = \frac{4!}{3!\,(4-3)!} = 4$$

In general, each combination of r objects gives rise to $r!$ permutations of these objects. Thus

$$C(n, r) = \frac{P(n, r)}{r!} = \frac{n!}{r!(n-r)!}$$

In Section 10.6 we denoted $C(n, r)$ by $\binom{n}{r}$, but it is customary to use the notation $C(n, r)$ in the context of counting. For an explanation of why these are the same, see Exercise 66.

COMBINATIONS OF n OBJECTS TAKEN r AT A TIME

The number of combinations of n objects taken r at a time is

$$C(n, r) = \frac{n!}{r!(n-r)!}$$

The key difference between permutations and combinations is *order*. If we are interested in ordered arrangements, then we are counting permutations; but if we are concerned with subsets without regard to order, then we are counting combinations. Compare Examples 5 and 6 below (where order doesn't matter) with Examples 1 and 2 (where order does matter).

EXAMPLE 5 ■ Finding the Number of Combinations

A club has nine members. In how many ways can a committee of three be chosen from the members of this club?

SOLUTION We need the number of ways of choosing three of the nine members. Order is not important here, because the committee is the same no matter how its members are ordered. So, we want the number of combinations of nine objects (the club members) taken three at a time. This number is

$$C(9, 3) = \frac{9!}{3!(9-3)!} = \frac{9!}{3!\,6!} = \frac{9 \times 8 \times 7}{3 \times 2 \times 1} = 84$$
■

EXAMPLE 6 ■ Finding the Number of Combinations

From 20 raffle tickets in a hat, four tickets are to be chosen at random. The holders of the winning tickets are to be awarded free trips to the Bahamas. In how many ways can the four winners be chosen?

SOLUTION We need to find the number of ways of choosing four winners from 20 entries. The order in which the tickets are chosen doesn't matter, because the same prize is awarded to each of the four winners. So, we want the number of combinations of 20 objects (the tickets) taken four at a time. This number is

$$C(20, 4) = \frac{20!}{4!(20-4)!} = \frac{20!}{4!\,16!} = \frac{20 \times 19 \times 18 \times 17}{4 \times 3 \times 2 \times 1} = 4845$$
■

If a set S has n elements, then $C(n, k)$ is the number of ways of choosing k elements from S, that is, the number of k-element subsets of S. Thus, the number of subsets of S of all possible sizes is given by the sum

$$C(n,0) + C(n,1) + C(n,2) + \cdots + C(n,n) = 2^n$$

(See Section 10.6, Exercise 52, where this sum is discussed.)

> A set with n elements has 2^n subsets.

EXAMPLE 7 ■ Finding the Number of Subsets of a Set

A pizza parlor offers the basic cheese pizza and a choice of 16 toppings. How many different kinds of pizza can be ordered at this pizza parlor?

SOLUTION   We need the number of possible subsets of the 16 toppings (including the empty set, which corresponds to a plain cheese pizza). Thus, $2^{16} = 65,536$ different pizzas can be ordered.   ■

The crucial step in solving counting problems is deciding whether to use permutations, combinations, or the Fundamental Counting Principle.

GUIDELINES FOR USING PERMUTATIONS AND COMBINATIONS

When we want to find the number of ways of picking r objects from n objects, we need to ask ourselves: Does the order of the objects matter?

If the order matters, we use permutations.

If the order doesn't matter, we use combinations.

EXAMPLE 8 ■ A Problem Involving Permutations and Combinations

A class of 20 students is going to choose a committee of seven, consisting of a chairman, a vice chairman, a secretary, and four other members. In how many ways can this committee be chosen?

SOLUTION   In choosing the three officers, order is important. So, the number of ways of choosing them is

$$P(20,3) = 6840$$

Next, we need to choose four other students from the 17 remaining. Since order doesn't matter in this case, the number of ways of doing this is

$$C(17,4) = 2380$$

We could have first chosen the four unordered members of the committee—in $C(20,4)$ ways—and then the three officers from the remaining 16 members, in $P(16,3)$ ways. Check to see that this gives the same answer.

Thus, by the Fundamental Counting Principle, the number of ways of choosing this committee is

$$P(20,3) \times C(17,4) = 6840 \times 2380 = 16,279,200$$ ■

11.2   EXERCISES

1–6 ■ Evaluate the expression.

1. $P(8, 3)$ **2.** $P(9, 2)$ **3.** $P(11, 4)$

4. $P(10, 5)$ **5.** $P(100, 1)$ **6.** $P(99, 3)$

7. In how many different ways can a president, vice president, and secretary be chosen from a class of 15 students?

8. In how many different ways can first, second, and third prizes be awarded in a game with eight contestants?

9. In how many different ways can six of ten people be seated in a row of six chairs?

10. In how many different ways can six people be seated in a row of six chairs?

11. How many three-letter "words" can be made from the letters *FGHIJK*? (Letters may not be repeated.)

12. How many permutations are possible from the letters of the word *LOVE*?

13. How many different three-digit whole numbers can be formed using the digits 1, 3, 5, and 7 if no repetition of digits is allowed?

14. A pianist plans to play eight pieces at a recital. In how many ways can she arrange these pieces in the program?

15. In how many different ways can a race with nine runners be completed, assuming there is no tie?

16. A ship carries five signal flags of different colors. How many different signals can be sent by hoisting exactly three of the five flags on the ship's flagpole in different orders?

17. In how many ways can first, second, and third prizes be awarded in a contest with 1000 contestants?

18. In how many ways can a president, vice president, secretary, and treasurer be chosen from a class of 30 students?

19. In how many ways can five students be seated in a row of five chairs if Jack insists on sitting in the first chair?

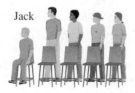

Jack

20. In how many ways can the students in Exercise 19 be seated if Jack insists on sitting in the middle chair?

21–24 ■ Find the number of distinguishable permutations of the given letters.

21. *AAABBC*

22. *AAABBBCCC*

23. *AABCD*

24. *ABCDDDEE*

25. In how many ways can two blue marbles and four red marbles be arranged in a row?

26. In how many different ways can five red balls, two white balls, and seven blue balls be arranged in a row?

27. In how many different ways can four pennies, three nickels, two dimes, and three quarters be arranged in a row?

28. In how many different ways can the letters of the word *ELEEMOSYNARY* be arranged?

29. A man bought three vanilla ice-cream cones, two chocolate cones, four strawberry cones, and five butterscotch cones for his 14 children. In how many ways can he distribute the cones among his children?

30. When seven students take a trip, they find a hotel with three rooms available—a room for one person, a room for two people, and a room for three people. In how many different ways can the students be assigned to these rooms? (One student has to sleep in the car.)

31. Eight workers are cleaning a large house. Five are needed to clean windows, two to clean the carpets, and one to clean the rest of the house. In how many different ways can these tasks be assigned to the eight workers?

32. A jogger jogs every morning to his health club, which is eight blocks east and five blocks north of his home. He always takes a route that is as short as possible, but he likes to vary the path he follows (see the figure). How many different paths can he take?   [*Hint:* The path shown can be thought of as *ENNEEENENEENE*, where *E* is East and *N* is North.]

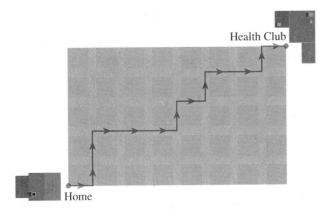

Health Club

Home

33–38 ■ Evaluate the expression.

33. $C(8, 3)$ **34.** $C(9, 2)$ **35.** $C(11, 4)$

36. $C(10, 5)$ **37.** $C(100, 1)$ **38.** $C(99, 3)$

39. In how many ways can three books be chosen from a group of six?

40. In how many ways can three pizza toppings be chosen from 12 available toppings?

41. In how many ways can six people be chosen from a group of ten?

42. In how many ways can a committee of three members be chosen from a club of 25 members?

43. How many five-card hands can be dealt from a deck of 52 cards?

44. How many seven-card hands can be picked from a deck of 52 cards?

45. A student must answer seven of the ten questions on an exam. In how many ways can she choose the seven questions?

46. A pizza parlor offers a choice of 16 different toppings. How many three-topping pizzas are possible?

47. A violinist has practiced 12 pieces. In how many ways can he choose eight of these pieces for a recital?

48. If a woman has eight skirts, in how many ways can she choose five of these to take on a weekend trip?

49. In how many ways can seven students from a class of 30 be chosen for a field trip?

50. In how many ways can the seven students in Exercise 49 be chosen if Jack must go on the field trip?

51. In how many ways can the seven students in Exercise 49 be chosen if Jack is not allowed to go on the field trip?

52. In the 6/49 lottery game, a player picks six numbers from 1 to 49. How many different choices does the player have?

53. In the California Lotto game, a player chooses six numbers from 1 to 53. It costs $1 to play this game. How much would it cost to buy every possible combination of six numbers to ensure picking the winning six numbers?

YOU CAN WIN $1,000,000									
●●●LOTTO●●●									
1	2	3	4	5	6	7	8	9	10
11	12	13	14	15	16	17	18	19	20
21	22	23	24	25	26	27	28	29	30
31	32	33	34	35	36	37	38	39	40
41	42	43	44	45	46	47	48	49	50
51	52	53							

54. A class has 20 students, of which 12 are females and 8 are males. In how many ways can a committee of five students be picked from this class under each of the following conditions?
   (a) No restriction is placed on the number of males or females on the committee.
   (b) No males are to be included on the committee.
   (c) The committee must have three females and two males.

55. A set has eight elements.
   (a) How many subsets containing five elements does this set have?
   (b) How many subsets does this set have?

56. A travel agency has limited numbers of eight different free brochures about Australia. The agent tells you to take any that you like, but no more than one of any kind. How many different ways can you choose brochures (including not choosing any)?

57. A hamburger chain gives their customers a choice of ten different hamburger toppings. In how many different ways can a customer order a hamburger?

58. Each of 20 shoppers in a shopping mall chooses to enter or not to enter the Dressfastic clothing store. How many different outcomes of their decisions are possible?

59. From a group of ten male and ten female tennis players, two men and two women are to face each other in a men-versus-women doubles match. In how many different ways can this match be arranged?

60. A school dance committee is to consist of two freshmen, three sophomores, four juniors, and five seniors. If six freshmen, eight sophomores, twelve juniors, and ten seniors are eligible to be on the committee, in how many ways can the committee be chosen?

61. A group of 22 aspiring thespians contains ten men and twelve women. For the next play the director wants to choose a leading man, a leading lady, a supporting male role, a supporting female role, and eight extras—three women and five men. In how many ways can the cast be chosen?

62. A hockey team has 20 players of which twelve play forward, six play defense, and two are goalies. In how many ways can the coach pick a starting lineup consisting of three forwards, two defense players, and one goalie?

63. A pizza parlor offers four sizes of pizza (small, medium, large, and colossus), two types of crust (thick and thin), and 14 different toppings. How many different pizzas can be made with these choices?

DISCOVERY · DISCUSSION

64. Complementary Combinations Without performing any calculations, explain in words why the number of ways of choosing two objects from ten objects is the same as the number of ways of choosing eight objects from ten objects. In general, explain why $C(n, r) = C(n, n - r)$.

65. An Identity Involving Combinations Kevin has ten different marbles, and he wants to give three of them to Luke and two to Mark. How many ways can he choose to do this? There are two ways of analyzing this problem: He could first pick three for Luke and then two for Mark, or he could first pick two for Mark and then three for Luke. Explain how these two viewpoints show that $C(10, 3) \cdot C(7, 2) = C(10, 2) \cdot C(8, 3)$. In general, explain why

$$C(n, r) \cdot C(n - r, k) = C(n, k) \cdot C(n - k, r)$$

66. Why Is $\binom{n}{r}$ the Same as $C(n, r)$? This exercise explains why the binomial coefficients $\binom{n}{r}$ that appear in the expansion of $(x + y)^n$ are the same as $C(n, r)$, the number of ways of choosing r objects from n objects. First, note that expanding a binomial using only the Distributive Property gives

$$(x + y)^2 = (x + y)(x + y)$$
$$= (x + y)x + (x + y)y$$
$$= xx + xy + yx + yy$$
$$(x + y)^3 = (x + y)(xx + xy + yx + yy)$$
$$= xxx + xxy + xyx + xyy + yxx$$
$$+ yxy + yyx + yyy$$

(a) Expand $(x + y)^5$ using only the Distributive Property.

(b) Write all the terms that represent $x^2 y^3$ together. These are all the terms that contain two x's and three y's.

(c) Note that the two x's appear in all possible positions. Conclude that the number of terms that represent $x^2 y^3$ is $C(5, 2)$.

(d) In general, explain why $\binom{n}{r}$ in the Binomial Theorem is the same as $C(n, r)$.

11.3 PROBABILITY

If you roll a pair of dice, what are the chances of rolling a double six? What is the likelihood of winning a state lottery? The subject of probability was invented to give precise answers to questions like these. It is now an indispensable tool for making decisions in such diverse areas as business, manufacturing, psychology,

The mathematical theory of probability began in 1654 in a series of letters between Pascal (see page 718) and Fermat (see page 458). Their correspondence was prompted by a question raised by the experienced gambler the Chevalier de Méré. The Chevalier was interested in the equitable distribution of the stakes of an interrupted gambling game (see Problem 3, page 775).

genetics, and in many of the sciences. Probability is used to determine the effectiveness of new medicines, assess a fair price for an insurance policy, decide on the likelihood of a candidate winning an election, determine the opinion of many people on a certain topic (without interviewing everyone), and answer many other questions that involve a measure of uncertainty.

To discuss probability, we begin by defining some terms. An **experiment** is a process, such as tossing a coin or rolling a die, that gives definite results, called the **outcomes** of the experiment. For tossing a coin, the possible outcomes are "heads" and "tails"; for rolling a die, the outcomes are 1, 2, 3, 4, 5, and 6. The **sample space** of an experiment is the set of all possible outcomes. If we let H stand for heads and T for tails, then the sample space of the coin-tossing experiment is

$$S = \{H, T\}$$

The sample space for rolling a die is

$$S = \{1, 2, 3, 4, 5, 6\}$$

We will be concerned only with experiments for which all the outcomes are "equally likely." We already have an intuitive feeling for what this means. When tossing a perfectly balanced coin, heads and tails are equally likely outcomes in the sense that if this experiment is repeated many times, we expect that about half the results will be heads and half will be tails.

In any given experiment we are often concerned with a particular set of outcomes. We might be interested in a die showing an even number or in picking an ace from a deck of cards. Any particular set of outcomes is a subset of the sample space. This leads to the following definition.

DEFINITION OF AN EVENT

If S is the sample space of an experiment, then an **event** is any subset of the sample space.

EXAMPLE 1 ■ Events in a Sample Space

If an experiment consists of tossing a coin three times and recording the results in order, the sample space is

$$S = \{HHH, HHT, HTH, THH, TTH, THT, HTT, TTT\}$$

The event E of showing "exactly two heads" is the subset of S that consists of all outcomes with two heads. Thus

$$E = \{HHT, HTH, THH\}$$

The event F of showing "at least two heads" is

$$F = \{HHH, HHT, HTH, THH\}$$

and the event of showing "no heads" is $G = \{TTT\}$. ■

We are now ready to define the notion of probability. Intuitively, we know that rolling a die may result in any of six equally likely outcomes, so the chance of any particular outcome occurring is $\frac{1}{6}$. What is the chance of showing an even number? Of the six equally likely outcomes possible, three are even numbers. So, it's reasonable to say that the chance of showing an even number is $\frac{3}{6} = \frac{1}{2}$. This reasoning is the intuitive basis for the following definition of probability.

DEFINITION OF PROBABILITY

Let S be the sample space of an experiment and E an event. The probability of E, written $P(E)$, is

$$P(E) = \frac{n(E)}{n(S)} = \frac{\text{number of elements in } E}{\text{number of elements in } S}$$

Notice that $0 \leq n(E) \leq n(S)$, so the probability $P(E)$ of an event is a number between 0 and 1, that is,

$$0 \leq P(E) \leq 1$$

The closer the probability of an event is to 1, the more likely the event is to happen; the closer to 0, the less likely. If $P(E) = 1$, then E is called the **certain event** and if $P(E) = 0$, then E is called the **impossible event**.

EXAMPLE 2 ■ Finding the Probability of an Event

A coin is tossed three times and the results are recorded. What is the probability of getting exactly two heads? At least two heads? No heads?

SOLUTION   By the results of Example 1, the sample space S of this experiment contains eight outcomes and the event E of getting "exactly two heads" contains three outcomes, $\{HHT, HTH, THH\}$, so by the definition of probability,

$$P(E) = \frac{n(E)}{n(S)} \quad \frac{3}{8}$$

Similarly, the event F of getting "at least two heads" has four outcomes, $\{HHH, HHT, HTH, THH\}$, and so

$$P(F) = \frac{n(F)}{n(S)} = \frac{4}{8} = \frac{1}{2}$$

The event G of getting "no heads" has one element, so

$$P(G) = \frac{n(G)}{n(S)} \quad \frac{1}{8}$$

■

To find the probability of an event, we do not need to list all the elements in the sample space and the event. All we do need is the *number* of elements in these

Persi Diaconis (b. 1945) is currently professor of statistics at Stanford University in California. He was born in New York City into a musical family and studied violin until the age of 14. At that time he left home to become a magician. He was a magician (apprentice and master) for ten years. Magic is still his major passion, and if there were a professorship for magic, he would certainly qualify for such a post! His interest in card tricks led him to a study of probability and statistics. He is now one of the leading statisticians in the world. With his background he approaches mathematics with an undeniable flair. He says "Statistics is the physics of numbers. Numbers seem to arise in the world in an orderly fashion. When we examine the world, the same regularities seem to appear again and again." Among his many original contributions to mathematics is a probabilistic study of the perfect card shuffle.

sets. The counting techniques we've learned in the preceding sections will be very useful here.

EXAMPLE 3 ■ Finding the Probability of an Event

A five-card poker hand is drawn from a standard deck of 52 cards. What is the probability that all five cards are spades?

SOLUTION　The experiment here consists of choosing five cards from the deck, and the sample space S consists of all possible five-card hands. Thus, the number of elements in the sample space is

$$n(S) = C(52, 5) = \frac{52!}{5!\,(52 - 5)!} = 2{,}598{,}960$$

The event E we are interested in consists of choosing five spades. Since the deck contains only 13 spades, the number of ways of choosing five spades is

$$n(E) = C(13, 5) = \frac{13!}{5!\,(13 - 5)!} = 1287$$

Thus, the probability of drawing five spades is

It's traditional to write probabilities without a leading zero before the decimal point.

$$P(E) = \frac{n(E)}{n(S)} = \frac{1287}{2{,}598{,}960} \approx .0005 \qquad ■$$

What does the answer to Example 3 tell us? Since $.0005 = \frac{1}{2000}$, this means that if you play poker many, many times, on average you will be dealt a hand consisting of only spades about once in every 2000 hands.

EXAMPLE 4 ■ Finding the Probability of an Event

A bag contains 20 tennis balls, of which four are defective. If two balls are selected at random from the bag, what is the probability that both are defective?

SOLUTION　The experiment consists of choosing two balls from 20, so the number of elements in the sample space S is $C(20, 2)$. Since there are four defective balls, the number of ways of picking two defective balls is $C(4, 2)$. Thus, the probability of the event E of picking two defective balls is

$$P(E) = \frac{n(E)}{n(S)} = \frac{C(4, 2)}{C(20, 2)} = \frac{6}{190} \approx .032 \qquad ■$$

The **complement** of an event E is the set of outcomes in the sample space that are not in E. We denote the complement of an event E by E'. We can calculate the probability of E' using the definition and the fact that $n(E') = n(S) - n(E)$:

$$P(E') = \frac{n(E')}{n(S)} = \frac{n(S) - n(E)}{n(S)} = \frac{n(S)}{n(S)} - \frac{n(E)}{n(S)} = 1 - P(E)$$

> ### PROBABILITY OF THE COMPLEMENT OF AN EVENT
>
> Let S be the sample space of an experiment and E an event. Then
>
> $$P(E') = 1 - P(E)$$

This is an extremely useful result, since it is often difficult to calculate the probability of an event E but easy to find the probability of E', from which $P(E)$ can be calculated immediately using this formula.

EXAMPLE 5 ■ Finding the Probability of the Complement of an Event

An urn contains 10 red balls and 15 blue balls. Six balls are drawn at random from the urn. What is the probability that at least one ball is red?

SOLUTION Let E be the event that at least one red ball is drawn. It is tedious to count all the possible ways in which one or more of the balls drawn are red. So, let's consider E', the complement of this event—namely, that none of the balls chosen is red. The number of ways of choosing 6 blue balls from the 15 blue balls is $C(15,6)$; the number of ways of choosing 6 balls from the 25 balls is $C(25,6)$. Thus

$$P(E') = \frac{n(E')}{n(S)} = \frac{C(15,6)}{C(25,6)} = \frac{5005}{177{,}100} = \frac{13}{460}$$

By the formula for the complement of an event, we have

Since

$$P(E') = 1 - P(E)$$

we have

$$P(E) = 1 - P(E')$$

$$P(E) = 1 - P(E') = 1 - \frac{13}{460} = \frac{447}{460} \approx .97 \qquad ■$$

MUTUALLY EXCLUSIVE EVENTS

Two events that have no outcome in common are said to be **mutually exclusive** (see Figure 1). For example, in drawing a card from a deck, the events

E: The card is an ace

F: The card is a queen

are mutually exclusive, because a card cannot be both an ace and a queen.

If E and F are mutually exclusive events, what is the probability that E or F occurs? The word *or* indicates that we want the probability of the *union* of these events, that is, $E \cup F$. Since E and F have no element in common,

$$n(E \cup F) = n(E) + n(F)$$

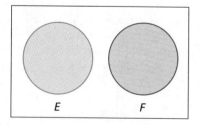

FIGURE 1

Thus

$$P(E \cup F) = \frac{n(E \cup F)}{n(S)} = \frac{n(E) + n(F)}{n(S)} = \frac{n(E)}{n(S)} + \frac{n(F)}{n(S)} = P(E) + P(F)$$

We have proved the following formula.

> ## PROBABILITY OF THE UNION OF MUTUALLY EXCLUSIVE EVENTS
>
> If E and F are mutually exclusive events in a sample space S, then the probability of E *or* F is
>
> $$P(E \cup F) = P(E) + P(F)$$

There is a natural extension of this formula for any number of mutually exclusive events: If $E_1, E_2, \ldots, E_n$ are pairwise mutually exclusive, then

$$P(E_1 \cup E_2 \cup \cdots \cup E_n) = P(E_1) + P(E_2) + \cdots + P(E_n)$$

EXAMPLE 6 ■ **The Probability of Mutually Exclusive Events**

A card is drawn at random from a standard deck of 52 cards. What is the probability that the card is either a seven or a face card?

SOLUTION Let E and F denote the following events.

$$E: \quad \text{The card is a seven}$$

$$F: \quad \text{The card is a face card}$$

Since a card cannot be both a seven and a face card, the events are mutually exclusive. We want the probability of E *or* F; in other words, the probability of $E \cup F$. By the formula,

$$P(E \cup F) = P(E) + P(F) = \frac{4}{52} + \frac{12}{52} = \frac{4}{13}$$ ■

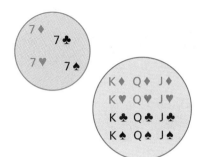

THE PROBABILITY OF THE UNION OF TWO EVENTS

If two events E and F are not mutually exclusive, then they share outcomes in common. The situation is described graphically in Figure 2. The overlap of the two sets is their intersection, that is, $E \cap F$. Again, we are interested in the event E *or* F, so we must count the elements in $E \cup F$. If we simply added the number of elements in E to the number of elements in F, then we would be counting the elements in the overlap twice—once in E and once in F. So, to get the correct total, we must subtract the number of elements in $E \cap F$. Thus

$$n(E \cup F) = n(E) + n(F) - n(E \cap F)$$

Using the formula for probability, we get

$$P(E \cup F) = \frac{n(E \cup F)}{n(S)} = \frac{n(E) + n(F) - n(E \cap F)}{n(S)}$$

$$= \frac{n(E)}{n(S)} + \frac{n(F)}{n(S)} - \frac{n(E \cap F)}{n(S)}$$

$$= P(E) + P(F) - P(E \cap F)$$

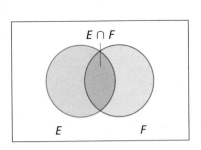

FIGURE 2

We have proved the following.

> ### PROBABILITY OF THE UNION OF TWO EVENTS
>
> If E and F are events in a sample space S, then the probability of E *or* F is
>
> $$P(E \cup F) = P(E) + P(F) - P(E \cap F)$$

EXAMPLE 7 ■ The Probability of the Union of Events

What is the probability that a card drawn at random from a standard 52-card deck is either a face card or a spade?

SOLUTION   We let E and F denote the following events:

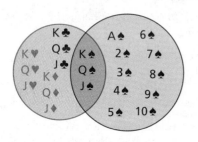

$$E: \quad \text{The card is a face card}$$

$$F: \quad \text{The card is a spade}$$

There are 12 face cards and 13 spades in a 52-card deck, so

$$P(E) = \frac{12}{52} \quad \text{and} \quad P(F) = \frac{13}{52}$$

Since there are 3 cards that are both face cards and spades,

$$P(E \cap F) = \frac{3}{52}$$

Thus, by the formula for the probability of the union of two events, we have

$$P(E \cup F) = P(E) + P(F) - P(E \cap F)$$

$$= \frac{12}{52} + \frac{13}{52} - \frac{3}{52} = \frac{11}{26}$$ ■

THE INTERSECTION OF INDEPENDENT EVENTS

We have considered the probability of events joined by the word *or*, that is, the union of events. Now we study the probability of events joined by the word *and*—in other words, the intersection of events.

When the occurrence of one event does not affect the probability of another event, we say that the events are **independent**. For instance, if a balanced coin is tossed, the probability of showing heads on the second toss is $\frac{1}{2}$, regardless of the outcome of the first toss. So, any two tosses of a coin are independent.

> **PROBABILITY OF THE INTERSECTION OF INDEPENDENT EVENTS**
>
> If E and F are independent events in a sample space S, then
>
> $$P(E \cap F) = P(E)\,P(F)$$

EXAMPLE 8 ■ The Probability of Independent Events

A jar contains five red balls and four black balls. A ball is drawn at random from the jar and then replaced; then another ball is picked. What is the probability that both balls are red?

SOLUTION The events are independent. The probability that the first ball is red is $\frac{5}{9}$. The probability that the second is red is also $\frac{5}{9}$. Thus, the probability that both balls are red is

$$\frac{5}{9} \times \frac{5}{9} = \frac{25}{81} \approx .31$$

■

EXAMPLE 9 ■ The Birthday Problem

What is the probability that in a class of 35 students, at least two have the same birthday?

SOLUTION It's reasonable to assume that the 35 birthdays are independent and that each day of the 365 days in a year is equally likely as a date of birth. (We ignore February 29.)

Let E be the event that two of the students have the same birthday. It is tedious to list all the possible ways in which at least two of the students have matching birthdays. So, we consider the complementary event E', that is, that *no* two students have the same birthday. To find this probability, we consider the students one at a time. The probability that the first student has a birthday is 1, the probability that the second has a birthday different from the first is $\frac{364}{365}$, the probability that the third has a birthday different from the first two is $\frac{363}{365}$, and so on. Thus

$$P(E') = 1 \cdot \frac{364}{365} \cdot \frac{363}{365} \cdot \frac{362}{365} \cdot \,\cdots\, \cdot \frac{331}{365} \approx .186$$

So

$$P(E) = 1 - P(E') \approx 1 - .186 = .814$$

■

Number of people in a group	Probability that at least two have the same birthday
5	.02714
10	.11695
15	.25290
20	.41144
22	.47569
23	.50730
24	.53834
25	.56870
30	.70631
35	.81438
40	.89123
50	.97037

Most people find it very surprising that the probability in Example 9 is so high. For this reason, this problem is sometimes called the "birthday paradox." The table in the margin gives the probability that two people in a group will share the same birthday for groups of various sizes.

11.3 EXERCISES

1. An experiment consists of tossing a coin twice.
   (a) Find the sample space.
   (b) Find the probability of getting heads exactly two times.
   (c) Find the probability of getting heads at least one time.
   (d) Find the probability of getting heads exactly one time.

2. An experiment consists of tossing a coin and rolling a die.
   (a) Find the sample space.
   (b) Find the probability of getting heads and an even number.
   (c) Find the probability of getting heads and a number greater than 4.
   (d) Find the probability of getting tails and an odd number.

3–4 ■ A die is rolled. Find the probability of the given event.

3. (a) The number showing is a six.
   (b) The number showing is an even number.
   (c) The number showing is greater than 5.

4. (a) The number showing is a two or a three.
   (b) The number showing is an odd number.
   (c) The number showing is a number divisible by 3.

5–6 ■ A card is drawn randomly from a standard 52-card deck. Find the probability of the given event.

5. (a) The card drawn is a king.
   (b) The card drawn is a face card.
   (c) The card drawn is not a face card.

6. (b) The card drawn is a heart.
   (b) The card drawn is either a heart or a spade.
   (c) The card drawn is a heart, a diamond, or a spade.

7–8 ■ A ball is drawn randomly from a jar that contains five red balls, two white balls, and one yellow ball. Find the probability of the given event.

7. (a) A red ball is drawn.
   (b) The ball drawn is not yellow.
   (c) A black ball is drawn.

8. (a) Neither a white nor yellow ball is drawn.
   (b) A red, white, or yellow ball is drawn.
   (c) The ball drawn is not white.

9. A drawer contains an unorganized collection of 18 socks—three pairs are red, two pairs are white, and four pairs are black.
   (a) If one sock is drawn at random from the drawer, what is the probability that it is red?
   (b) Once a sock is drawn and discovered to be red, what is the probability of drawing another red sock to make a matching pair?

10. A child's game has a spinner as shown in the figure. Find the probability of the given event.
   (a) The spinner stops on an even number.
   (b) The spinner stops on an odd number or a number greater than 3.

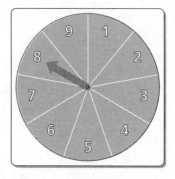

11. A letter is chosen at random from the word *EXTRATERRESTRIAL*. Find the probability of the given event.
   (a) The letter *T* is chosen.
   (b) The letter chosen is a vowel.
   (c) The letter chosen is a consonant.

12–15 ■ A poker hand, consisting of five cards, is dealt from a standard deck of 52 cards. Find the probability that the hand contains the cards described.

12. Five hearts

13. Five cards of the same suit

14. Five face cards

15. An ace, king, queen, jack, and 10 of the same suit (royal flush)

16. A pair of dice is rolled, and the numbers showing are observed.
   (a) List the sample space of this experiment.
   (b) Find the probability of getting a sum of 7.

(c) Find the probability of getting a sum of 9.

(d) Find the probability that the two dice show doubles (the same number).

(e) Find the probability that the two dice show different numbers.

(f) Find the probability of getting a sum of 9 or higher.

17. A couple intends to have four children. Assume that having a boy or a girl is an equally likely event.

(a) List the sample space of this experiment.

(b) Find the probability that the couple has only boys.

(c) Find the probability that the couple has two boys and two girls.

(d) Find the probability that the couple has four children of the same sex.

(e) Find the probability that the couple has at least two girls.

18. What is the probability that a 13-card bridge hand consists of all cards from the same suit?

19. An American roulette wheel has 38 slots: two slots are numbered 0 and 00, and the remaining slots are numbered from 1 to 36. Find the probability that the ball lands in an odd-numbered slot.

20. A toddler has wooden blocks showing the letters C, E, F, H, N, and R. Find the probability that the child arranges the letters in the indicated order.

(a) In the order FRENCH

(b) In alphabetical order

21. In the 6/49 lottery game, a player selects six numbers from 1 to 49. What is the probability of picking the six winning numbers?

22. The president of a large company selects six employees to receive a special bonus. He claims that the six employees are chosen randomly from among the 30 employees, of which 19 are women and 11 are men. What is the probability that no woman is chosen?

23. An exam has ten true-false questions. A student who has not studied answers all ten questions by just guessing. Find the probability that the student correctly answers the given number of questions.

(a) All ten questions

(b) Exactly seven questions

24. To control the quality of their product, the Bright-Light Company inspects three light bulbs out of each batch of ten bulbs manufactured. If a defective bulb is found, the batch is discarded. Suppose that a batch contains two defective bulbs. What is the probability that the batch will be discarded?

25. An often-quoted example of an event of extremely low probability is that a monkey types the Shakespearean play *Hamlet* by randomly striking keys on a typewriter. Assume that the typewriter has 48 keys (including the space bar) and that the monkey is equally likely to hit any key.

(a) Find the probability that such a monkey will actually correctly type just the title of the play as his first word.

(b) What is the probability that the monkey will type the phrase "To be or not to be" as his first words?

26. A monkey is trained to arrange wooden blocks in a straight line. He is then given six blocks showing the letters A, E, H, L, M, T. What is the probability that he will arrange them to spell the word *HAMLET*?

27. A monkey is trained to arrange wooden blocks in a straight line. He is then given 11 blocks showing the letters A, B, B, I, I, L, O, P, R, T, Y. What is the probability that the monkey will arrange the blocks to spell the word *PROBABILITY*?

28. Eight horses are entered in a race. You randomly predict a particular order for the horses to complete the race. What is the probability that your prediction is correct?

29. Many genetic traits are controlled by two genes, one dominant and one recessive. In Gregor Mendel's original experiments with peas, the genes controlling the height of the plant are denoted by T (tall) and t (short). The gene T is dominant, so a plant with the genotype (genetic makeup) TT or Tt is tall, whereas one with genotype tt is short. By a statistical analysis of the offspring in his experiments, Mendel concluded that offspring inherit one gene from each parent, and each possible combination of the two genes is equally likely.

If each parent has the genotype Tt, then the following chart gives the possible genotypes of the offspring:

		Parent 2 T	t
Parent 1	T	TT	Tt
	t	Tt	tt

Find the probability that a given offspring of these parents will be (a) tall or (b) short.

30. Refer to Exercise 29. Make a chart of the possible genotypes of the offspring if one parent has genotype Tt and the other tt. Find the probability that a given offspring will be (a) tall or (b) short.

31–32 ■ Determine whether the events E and F in the given experiment are mutually exclusive.

31. The experiment consists of selecting a person at random.
   (a) E: The person is male
      F: The person is female
   (b) E: The person is tall
      F: The person is blond

32. The experiment consists of choosing at random a student from your class.
   (a) E: The student is female
      F: The student wears glasses
   (b) E: The student has long hair
      F: The student is male

33–34 ■ A die is rolled and the number showing is observed. Determine whether the events E and F are mutually exclusive. Then find the probability of the event E ∪ F.

33. (a) E: The number is even
      F: The number is odd
   (b) E: The number is even
      F: The number is greater than 4

34. (a) E: The number is greater than 3
      F: The number is less than 5
   (b) E: The number is divisible by 3
      F: The number is less than 3

35–36 ■ A card is drawn at random from a standard 52-card deck. Determine whether the events E and F are mutually exclusive. Then find the probability of the event E ∪ F.

35. (a) E: The card is a face card
      F: The card is a spade
   (b) E: The card is a heart
      F: The card is a spade

36. (a) E: The card is a club
      F: The card is a king
   (b) E: The card is an ace
      F: The card is a spade

37–38 ■ Refer to the spinner shown in the figure. Find the probability of the given event.

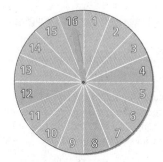

37. (a) The spinner stops on red.
   (b) The spinner stops on an even number.
   (c) The spinner stops on red or an even number.

38. (a) The spinner stops on blue.
   (b) The spinner stops on an odd number.
   (c) The spinner stops on blue or an odd number.

39. An American roulette wheel has 38 slots: two of the slots are numbered 0 and 00, and the rest are numbered from 1 to 36. Find the probability that the ball lands in an odd-numbered slot or in a slot with a number higher than 31.

40. A monkey is trained to arrange wooden blocks in a straight line. He is then given blocks with the letters A, E, H, L, M, T. What is the probability that he will arrange them to spell one of the words *HAMLET* or *THELMA*?

41. A committee of five is chosen randomly from a group of six males and eight females. What is the probability that the committee includes either all males or all females?

42. In the 6/49 lottery game a player selects six numbers from 1 to 49. What is the probability of selecting at least five of the six winning numbers?

43. A jar contains six red marbles numbered 1 to 6 and ten blue marbles numbered 1 to 10. A marble is drawn at random from the jar. Find the probability that the given event occurs.
(a) The marble is red.
(b) The marble is odd-numbered.
(c) The marble is red or odd-numbered.
(d) The marble is blue or even-numbered.

44. A coin is tossed twice. Let E be the event "the first toss shows heads" and F the event "the second toss shows heads."
(a) Are the events E and F independent?
(b) Find the probability of showing heads on both tosses.

45. A die is rolled twice. Let E be the event "the first roll shows a six" and F the event "the second roll shows a six."
(a) Are the events E and F independent?
(b) Find the probability of showing a six on both rolls.

46–47 ■ Spinners A and B shown in the figure are spun at the same time.

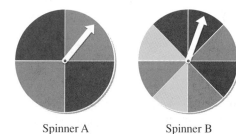

Spinner A   Spinner B

46. (a) Are the events "spinner A stops on red" and "spinner B stops on yellow" independent?
(b) Find the probability that spinner A stops on red and spinner B stops on yellow.

47. (a) Find the probability that both spinners stop on purple.
(b) Find the probability that both spinners stop on blue.

48. A die is rolled twice. What is the probability of showing a one on both rolls?

49. A die is rolled twice. What is the probability of showing a one on the first roll and an even number on the second roll?

50. A card is drawn from a deck and replaced, and then a second card is drawn.
(a) What is the probability that both cards are aces?
(b) What is the probability that the first is an ace and the second a spade?

51. A roulette wheel has 38 slots: Two slots are numbered 0 and 00, and the rest are numbered 1 to 36. A player places a bet on a number between 1 and 36 and wins if a ball thrown into the spinning roulette wheel lands in the slot with the same number. Find the probability of winning on two consecutive spins of the roulette wheel.

52. A researcher claims that she has taught a monkey to spell the word *MONKEY* using the six wooden letters *E, O, K, M, N, Y*. If the monkey has not actually learned anything and is merely arranging the blocks randomly, what is the probability that he will spell the word correctly three consecutive times?

53. What is the probability of rolling "snake eyes" (double ones) three times in a row with a pair of dice?

54. In the 6/49 lottery game, a player selects six numbers from 1 to 49 and wins if he selects the winning six numbers. What is the probability of winning the lottery two times in a row?

55. Jar A contains three red balls and four white balls. Jar B contains five red balls and two white balls. Which one of the following ways of randomly selecting balls gives the greatest probability of drawing two red balls?
(i) Draw two balls from jar B.
(ii) Draw one ball from each jar.
(iii) Put all the balls in one jar, and then draw two balls.

56. A slot machine has three wheels: Each wheel has 11 positions—a bar and the digits 0, 1, 2, . . . , 9. When the handle is pulled, the three wheels spin independently

before coming to rest. Find the probability that the wheels stop on each of the following positions.
(a) Three bars
(b) The same number on each wheel
(c) At least one bar

57. Find the probability that in a group of eight students at least two people have the same birthday.

58. What is the probability that in a group of six students at least two have birthdays in the same month?

 DISCOVERY · DISCUSSION

59. **The "Second Son" Paradox** Mrs. Smith says, "I have two children—the older one is named William." Mrs.

Jones replies, "One of my two children is also named William." For each woman, list the sample space for the genders of her children, and calculate the probability that her other child is also a son. Explain why these two probabilities are different.

60. **The "Oldest Son or Daughter" Phenomenon** Poll your class to determine how many of your male classmates are the oldest sons in their families, and how many of your female classmates are the oldest daughters in their families. You will most likely find that they form a majority of the class. Explain why a randomly selected individual has a high probability of being the oldest son or daughter in his or her family.

11.4   EXPECTED VALUE

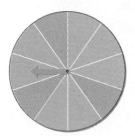

FIGURE 1

In the game shown in Figure 1, you pay $1 to spin the arrow. If the arrow stops in a red region, you get $3 (the dollar you paid plus $2); otherwise, you lose the dollar you paid. If you play this game many times, how much would you expect to win? Or lose? To answer these questions, let's consider the probabilities of winning and losing. Since three of the regions are red, the probability of winning is $\frac{3}{10} = .3$ and that of losing is $\frac{7}{10} = .7$. Remember, this means that if you play this game many times, you expect to win "on average" three out of ten times. So, suppose you play the game 1000 times. Then you would expect to win 300 times and lose 700 times. Since we win $2 or lose $1 in each game, our expected payoff in 1000 games is

$$2(300) + (-1)(700) = -100$$

So the average expected return per game is $\frac{-100}{1000} = -0.1$. In other words, we expect to lose, on average, 10 cents per game. Another way to view this average is to divide each side of the preceding equation by 1000. Writing E for the result, we get

$$E = \frac{2(300) + (-1)(700)}{1000}$$

$$= 2\left(\frac{300}{1000}\right) + (-1)\frac{700}{1000}$$

$$= 2(.3) + (-1)(.7)$$

Thus, the expected return, or *expected value*, per game is

$$E = a_1 p_1 + a_2 p_2$$

where a_1 is the payoff that occurs with probability p_1 and a_2 is the payoff that occurs with probability p_2. This example leads us to the following definition of expected value.

> **DEFINITION OF EXPECTED VALUE**
>
> A game gives payoffs of $a_1, a_2, \ldots, a_n$ with probabilities $p_1, p_2, \ldots, p_n$. The **expected value** (or **expectation**) E of this game is
>
> $$E = a_1 p_1 + a_2 p_2 + \cdots + a_n p_n$$

The expected value is an average expectation per game if the game is played many times. In general, E need not be one of the possible payoffs. In the preceding example the expected value is -10 cents, but it's impossible to lose exactly 10 cents in any given trial of the game.

EXAMPLE 1 ■ Finding an Expected Value

A die is rolled, and you receive \$1 for each point that shows. What is your expectation?

SOLUTION Each face of the die has probability $\frac{1}{6}$ of showing. So you get \$1 with probability $\frac{1}{6}$, \$2 with probability $\frac{1}{6}$, \$3 with probability $\frac{1}{6}$, and so on. Thus, the expected value is

$$E = 1\left(\frac{1}{6}\right) + 2\left(\frac{1}{6}\right) + 3\left(\frac{1}{6}\right) + 4\left(\frac{1}{6}\right) + 5\left(\frac{1}{6}\right) + 6\left(\frac{1}{6}\right) = \frac{21}{6} = 3.5$$

This means that if you play this game many times, you will make, on average, \$3.50 per game. ■

EXAMPLE 2 ■ Finding an Expected Value

In Monte Carlo, the game of roulette is played on a wheel with slots numbered 0, 1, 2, $\ldots$, 36. The wheel is spun, and a ball dropped in the wheel is equally likely to end up in any one of the slots. To play the game, you bet \$1 on any number other than zero. (For example, you may bet \$1 on number 23.) If the ball stops in your slot, you get \$36 (the \$1 you bet plus \$35). Find the expected value of this game.

SOLUTION You gain \$35 with probability $\frac{1}{37}$, and you lose \$1 with probability $\frac{36}{37}$. Thus

$$E = (35)\frac{1}{37} + (-1)\frac{36}{37} \approx -0.027$$

In other words, if you play this game many times, you would expect to lose 2.7 cents on every dollar you bet (on average). Consequently, the house expects to gain 2.7 cents on every dollar that is bet. This expected value is what makes gambling very profitable for the gaming house and very unprofitable for the gambler. ■

11.4 EXERCISES

1–10 ■ Find the expected value (or expectation) of the game described.

1. Mike wins $2 if a coin toss shows heads and $1 if it shows tails.

2. Jane wins $10 if a die roll shows a six, and she loses $1 otherwise.

3. The game consists of drawing a card from a deck. You win $100 if you draw the ace of spades or lose $1 if you draw any other card.

4. Tim wins $3 if a coin toss shows heads or $2 if it shows tails.

5. Carol wins $3 if a die roll shows a six, and she wins $0.50 otherwise.

6. A coin is tossed twice. Albert wins $2 for each heads and must pay $1 for each tails.

7. A die is rolled. Tom wins $2 if the die shows an even number and he pays $2 otherwise.

8. A card is drawn from a deck. You win $104 if the card is an ace, $26 if it is a face card, and $13 if it is the 8 of clubs.

9. A bag contains two silver dollars and eight slugs. You pay 50 cents to reach into the bag and take a coin, which you get to keep.

10. A bag contains eight white balls and two black balls. John picks two balls at random from the bag, and he wins $5 if he does not pick a black ball.

11. In the game of roulette as played in Las Vegas, the wheel has 38 slots: Two slots are numbered 0 and 00, and the rest are numbered 1 to 36. A $1 bet on any number other than 0 or 00 wins $36 ($35 plus the $1 bet). Find the expected value of this game.

12. A sweepstakes offers a first prize of $1,000,000, second prize of $100,000, and third prize of $10,000. Suppose that two million people enter the contest and three names are drawn randomly for the three prizes.
   (a) Find the expected winnings for a person participating in this contest.
   (b) Is it worth paying a dollar to enter this sweepstakes?

13. A box contains 100 envelopes. Ten envelopes contain $10 each, ten contain $5 each, two are "unlucky," and the rest are empty. A player draws an envelope from the box and keeps whatever is in it. If a person draws an

unlucky envelope, however, he must pay $100. What is the expectation of a person playing this game?

14. A safe containing one million dollars is locked with a combination lock. You pay $1 for one guess at the six-digit combination. If you open the lock, you get to keep the million dollars. What is your expectation?

15. An investor buys 1000 shares of a risky stock for $5 a share. She estimates that the probability the stock will rise in value to $20 a share is .1 and the probability that it will fall to $1 a share is .9. If the only criterion for her decision to buy this stock was the expected value of her profit, did she make a wise investment?

16. A slot machine has three wheels, and each wheel has 11 positions—the digits 0, 1, 2, . . . , 9 and the picture of a watermelon. When a quarter is placed in the machine and the handle is pulled, the three wheels spin independently and come to rest. When three watermelons show, the payout is $5; otherwise, nothing is paid. What is the expected value of this game?

17. In a 6/49 lottery game, a player pays $1 and selects six numbers from 1 to 49. Any player who has chosen the six winning numbers wins $1,000,000. Assuming this is the only way to win, what is the expected value of this game?

18. A bag contains two silver dollars and six slugs. A game consists of reaching into the bag and drawing a coin, which you get to keep. Determine the "fair price" of playing this game, that is, the price at which the player can be expected to break even if he plays the game many times (in other words, the price at which his expectation is zero).

19. A game consists of drawing a card from a deck. You win $13 if you draw an ace. What is a "fair price" to pay to play this game? (See Exercise 18.)

◆ DISCOVERY · DISCUSSION

20. The Expected Value of a Sweepstakes Contest   A magazine clearinghouse holds a sweepstakes contest to sell subscriptions. If you return the winning number, you win $1,000,000. You have a 1-in-20-million chance of winning, but your only cost to enter the contest is a first-class stamp to mail the entry. Use the current price of a first-class stamp to calculate your expected net winnings if you enter this contest. Is it worth entering the sweepstakes?

11 REVIEW

CONCEPT CHECK

1. What does the Fundamental Counting Principle say?

2. (a) What is a permutation of a set of distinct objects?
   (b) How many permutations are there of n objects?
   (c) How many permutations are there of n objects taken r at a time?
   (d) What is the number of distinguishable permutations of n objects if there are k different kinds of objects with n_1 objects of the first kind, n_2 objects of the second kind, and so on?

3. (a) What is a combination of r elements of a set?
   (b) How many combinations are there of n elements taken r at a time?
   (c) How many subsets does a set with n elements have?

4. In solving a problem involving picking r objects from n objects, how do you know whether to use permutations or combinations?

5. (a) What is meant by the sample space of an experiment?
   (b) What is an event?
   (c) Define the probability of an event E in a sample space S.
   (d) What is the probability of the complement of E?

6. (a) What are mutually exclusive events?
   (b) If E and F are mutually exclusive events, what is the probability of the union of E and F? What if E and F are not mutually exclusive?

7. (a) What are independent events?
   (b) If E and F are independent events, what is the probability of the intersection of E and F?

8. Suppose that a game gives payoffs of $a_1, a_2, \ldots, a_n$ with probabilities $p_1, p_2, \ldots, p_n$. What is the expected value of the game?

EXERCISES

1. A coin is tossed, a die is rolled, and a card is drawn from a deck. How many possible outcomes does this experiment have?

2. How many three-digit numbers can be formed using the digits 1, 2, 3, 4, 5, 6 if repetition of digits
   (a) is allowed?
   (b) is not allowed?

3. (a) How many different two-element subsets does the set $\{A, E, I, O, U\}$ have?
   (b) How many different two-letter "words" can be made using the letters from the set in part (a)?

4. An airline company overbooks a particular flight and seven passengers are "bumped" from the flight. If 120 passengers are booked on this flight, in how many ways can the airline choose the seven passengers to be bumped?

5. A quiz has ten true-false questions. How many different ways is it possible to earn a score of exactly 70% on this quiz?

6. A test has ten true-false questions and five multiple-choice questions with four choices for each. In how many ways can this test be completed?

7. If you must answer only eight of ten questions on a test, how many ways do you have of choosing the questions you will omit?

8. An ice-cream store offers 15 flavors of ice cream. The specialty is a banana split with four scoops of ice cream. If each scoop must be a different flavor, how many different banana splits may be ordered?

9. A company uses a different three-letter security code for each of its employees. What is the maximum number of codes this security system can generate?

10. A group of students determines that they can stand in a row for their class picture in 120 different ways. How many students are in this class?

11. A coin is tossed ten times. In how many different ways can the result be three heads and seven tails?

12. The Yukon Territory in Canada uses a license-plate system for automobiles that consists of two letters

followed by three numbers. Explain how we can know that fewer than 700,000 autos are licensed in the Yukon.

13. A group of friends have one tennis court. They find that there are ten different ways in which two of them can play a singles game on this court. How many friends are in this group?

14. A pizza parlor advertises that they prepare 2048 different types of pizza. How many toppings does this parlor offer?

15. In Morse code, each letter is represented by a sequence of dots and dashes, with repetition allowed. How many letters can be represented using Morse code if three or fewer symbols are used?

16. The genetic code is based on the four nucleotides adenine (A), cytosine (C), guanine (G), and thymine (T). These are connected in long strings to form DNA molecules. For example, a sequence in the DNA may look like CAGTGGTACC.... The code uses "words," all the same length, that are composed of the nucleotides A, C, G, and T. It is known that at least 20 different words exist. What is the minimum word length necessary to generate 20 words?

17. Given 16 subjects from which to choose, in how many ways can a student select fields of study as follows?
(a) A major and a minor
(b) A major, a first minor, and a second minor
(c) A major and two minors

18. (a) How many three-digit numbers can be formed using the digits 0, 1, ..., 9? (Remember, a three-digit number cannot have 0 as the leftmost digit.)
(b) If a number is chosen randomly from the set {0, 1, 2, ..., 1000}, what is the probability that the number chosen is a three-digit number?

19–20 ■ An **anagram** of a word is a permutation of the letters of that word. For example, anagrams of the word *triangle* include *griantle, integral,* and *tenalgir.*

19. How many anagrams of the word *TRIANGLE* are possible?

20. How many anagrams are possible from the word *MISSISSIPPI*?

21. A shelf has ten books: two mysteries, four romance novels, and four mathematics textbooks. If you select a book at random to take to the beach, what is the probability that it turns out to be a mathematics text?

22. A jar contains ten red balls labeled 0, 1, 2, ..., 9 and five white balls labeled 0, 1, 2, 3, 4. If a ball is drawn from the jar, find the probability of the given event.
(a) The ball is red.
(b) The ball is even-numbered.
(c) The ball is white and odd-numbered.
(d) The ball is red or odd-numbered.

23. A coin is tossed three times in a row, and the outcomes of each toss are observed.
(a) Find the sample space for this experiment.
(b) Find the probability of getting three heads.
(c) Find the probability of getting two or more heads.
(d) Find the probability of getting tails on the first toss.

24. A die is rolled and a card is selected from a standard 52-card deck. What is the probability that both the die and the card show a six?

25. Find the probability that the indicated card is drawn at random from a 52-card deck.
(a) An ace
(b) An ace or a jack
(c) An ace or a spade
(d) A red ace

26. A card is drawn from a 52-card deck, a die is rolled, and a coin is tossed. Find the probability of each of the indicated outcomes.
(a) The ace of spades, a six, and heads
(b) A spade, a six, and heads
(c) A face card, a number greater than 3, and heads

27. Two dice are rolled. Find the probability of each outcome.
(a) The dice show the same number.
(b) The dice show different numbers.

28. Four cards are dealt from a standard 52-card deck. Find the probability that the cards are
(a) all kings   (b) all spades   (c) all the same color

29. In the "numbers game" lottery, a player picks a three-digit number (from 000 to 999), and if the number is selected in the drawing, the player wins $500. If another number with the same digits (in any order) is drawn, the player wins $50. John plays the number 159.
(a) What is the probability that he will win $500?
(b) What is the probability that he will win $50?

30. In a television game show, a contestant is given five cards with a different digit on each and is asked to arrange them to match the price of a brand-new car. If she gets the price right, she wins the car. What is the probability that she wins, assuming that she knows the first digit but must guess the remaining four?

31. Two dice are rolled. John gets $5 if they show the same number, or he pays $1 if they show different numbers. What is the expected value of this game?

32. Three dice are rolled. John gets $5 if they all show the same number; he pays $1 otherwise. What is the expected value of this game?

33. Mary will win $1,000,000 if she can name the 13 original states in the order in which they ratified the U.S. Constitution. Mary has no knowledge of this order, so she makes a guess. What is her expectation?

34. A pizza parlor offers 12 different toppings, one of which is anchovies. If a pizza is ordered at random, what is the probability that anchovies is used as one of the toppings?

35. A drawer contains an unorganized collection of 50 socks—20 are red and 30 are blue. Suppose the lights go out so Kathy can't distinguish the color of the socks.
(a) What is the minimum number of socks Kathy must take out of the drawer to be sure of getting a matching pair?
(b) If two socks are taken at random from the drawer, what is the probability that they make a matching pair?

36. A volleyball team has nine players. In how many ways can a starting lineup be chosen if it consists of two forward players and three defense players?

37. Zip codes consist of five digits.
(a) How many different zip codes are possible?
(b) How many different zip codes can be read when the envelope is turned upside down? (An upside down 9 is a 6; and 0, 1, and 8 are the same when read upside down.)
(c) What is the probability that a randomly chosen zip code can be read upside down?
(d) How many zip codes read the same upside down as right side up?

38. In the Zip+4 postal code system, zip codes consist of nine digits.
(a) How many different Zip+4 codes are possible?
(b) How many Zip+4 codes are palindromes? (A *palindrome* is a number that reads the same from left to right as right to left.)
(c) What is the probability that a randomly chosen Zip+4 code is a palindrome?

39. Let $N = 3,600,000$. (Note that $N = 2^7 3^2 5^5$.)
(a) How many divisors does N have?
(b) How many even divisors does N have?
(c) How many divisors of N are multiples of 6?
(d) What is the probability that a randomly chosen divisor of N is even?

40. The U.S. Senate has two senators from each of the 50 states. In how many ways can a committee of five senators be chosen if no state is to have two members on the committee?

1. How many five-letter "words" can be made from the letters *A, B, C, D, E, F, G, H, I, J* if repetition (a) is allowed or (b) is not allowed?

2. A restaurant offers five main courses, three types of desserts, and four kinds of drinks. In how many ways can a customer order a meal consisting of one choice from each category?

3. A board of directors consisting of eight members is to be chosen from a pool of 30 candidates. The board is to have a chairman, a treasurer, a secretary, and five other members. In how many ways can the board of directors be chosen?

4. A commuter must travel from Ajax to Barrie and back every day. Four roads join the two cities. The commuter likes to vary the trip as much as possible, so he always leaves and returns by different roads. In how many different ways can he make the round-trip?

5. A pizza parlor offers four sizes of pizza and 14 different toppings. A customer may choose any number of toppings (or no topping at all). How many different pizzas does this parlor offer?

6. An *anagram* of a word is a rearrangement of the letters of the word.
   (a) How many anagrams of the word *LOVE* are possible?
   (b) How many different anagrams of the word *KISSES* are possible?

7. Three people are chosen at random from a group of five men and ten women. What is the probability that all three are men?

8. Two dice are rolled. What is the probability of getting doubles?

9. One card is drawn from a deck. Find the probability of the given event.
   (a) The card is red.
   (b) The card is a king.
   (c) The card is a red king.

10. A jar contains five red balls, numbered 1 to 5, and eight white balls, numbered 1 to 8. A ball is chosen at random from the jar. Find the probability of the given event.
    (a) The ball is red.
    (b) The ball is even-numbered.
    (c) The ball is red or even-numbered.

11. You are to draw one card from a deck. If it is an ace, you win $10; if it is a face card, you win $1; otherwise, you lose $0.50. What is the expected value of this game?

12. In a group of four students, what is the probability that at least two have the same astrological sign?

FOCUS ON MODELING

A good way to familiarize ourselves with a fact is to experiment with it. For instance, to convince ourselves that the earth is a sphere (which was considered a major paradox at one time), we could go up in a space shuttle to see that it is so; to see whether a given equation is an identity, we might try some special cases to make sure there are no obvious counterexamples. In problems involving probability, we can perform an experiment many times and use the results to estimate the probability in question. In fact, we often model the experiment on a computer, thereby making it feasible to perform the experiment a large number of times. This technique is called the **Monte Carlo method**, named after the famous gambling casino in Monaco.

EXAMPLE 1 ■ The Contestant's Dilemma

In a TV game show, a contestant chooses one of three doors. Behind one of them is a valuable prize—the other two doors have nothing behind them. After the contestant has made her choice, the host opens one of the other two doors, one that he knows does not conceal a prize, and then gives her the opportunity to change her choice.

Should the contestant switch, stay, or does it matter? In other words, by switching doors, does she increase, decrease, or leave unchanged her probability of winning? At first, it may seem that switching doors doesn't make any difference. After all, two doors are left—one with the prize and one without—so it seems reasonable that the contestant has an equal chance of winning or losing. But if you play this game many times, you will find that by switching doors you actually win about $\frac{2}{3}$ of the time.

The authors modeled this game on a computer and found that in one million games the simulated contestant (who always switches) won 667,049 times—very close to $\frac{2}{3}$ of the time. Thus, it seems that switching doors does make a difference: Switching increases the contestant's chances of winning. This experiment forces us to reexamine our reasoning. Here is why switching doors is the correct strategy:

1. When the contestant first made her choice, she had a $\frac{1}{3}$ chance of winning. If she doesn't switch, no matter what the host does, her probability of winning remains $\frac{1}{3}$.

2. If the contestant decides to switch, she will switch to the winning door if she had initially chosen a losing one, or to a losing door if she had initially chosen the winning one. Since the probability of having initially selected a losing door is $\frac{2}{3}$, by switching the probability of winning then becomes $\frac{2}{3}$.

We conclude that the contestant should switch, because her probability of winning is $\frac{2}{3}$ if she switches and $\frac{1}{3}$ if she doesn't. Put simply, there is a much

Contestant: "I choose door number 2."

Contestant: "Oh no, what should I do?"

greater chance that she had initially chosen a losing door (since there are more
of these), so she should switch.

■

An experiment can be modeled using any computer language or programmable calculator that has a random-number generator. This is a command or
function (usually called $\boxed{\text{RND}}$ or $\boxed{\text{RAND}}$) that returns a randomly chosen number x
with $0 \leq x < 1$. In the next example we see how to use this to model a simple
experiment.

EXAMPLE 2 ■ Monte Carlo Model of a Coin Toss

When a balanced coin is tossed, each outcome—"heads" or "tails"—has
probability $\frac{1}{2}$. This doesn't mean that if we toss a coin several times, we will
necessarily get exactly half heads and half tails. We would expect, however,
the proportion of heads and of tails to get closer and closer to $\frac{1}{2}$ as the number
of tosses increases. To test this hypothesis, we could toss a coin a very large
number of times and keep track of the results. But this is a very tedious
process, so we will use the Monte Carlo method to model this process.

To model a coin toss with a calculator or computer, we use the random-
number generator to get a random number x such that $0 \leq x < 1$. Because the
number is chosen randomly, the probability that it lies in the first half of this
interval $\left(0 \leq x < \frac{1}{2}\right)$ is the same as the probability that it lies in the second
half $\left(\frac{1}{2} \leq x < 1\right)$. Thus, we could model the outcome "heads" by the event that
$0 \leq x < \frac{1}{2}$ and the outcome "tails" by the event that $\frac{1}{2} \leq x < 1$.

An easier way to keep track of heads and tails is to note that if $0 \leq x < 1$,
then $0 \leq 2x < 2$, and so $[\![2x]\!]$, the integer part of $2x$, is either 0 or 1, each with
probability $\frac{1}{2}$. (On most programmable calculators, the function $\boxed{\text{INT}}$ gives the
integer part of a number.) Thus, we could model "heads" with the outcome "0"
and "tails" with the outcome "1" when we take the integer part of $2x$. The
program in the margin models 100 tosses of a coin on the TI-82 calculator. The
graph in Figure 1 shows what proportion p of the tosses have come up "heads"
after n tosses. As you can see, this proportion settles down near .5 as the number of tosses increases—just as we hypothesized.

■

```
PROGRAM: HEADTAIL
:0→J:0→K
:For(N,1,100)
:rand→X
:int(2X)→Y
:J+(1−Y)→J
:K+Y→K
:END
:Disp "HEADS=",J
:Disp "TAILS=",K
```

In general, if a process has n equally likely outcomes, then we can model the
process using a random-number generator as follows: If our program or calculator produces the random number x, with $0 \leq x < 1$, then the integer part of
nx will be a random choice from the n integers $0, 1, 2, \ldots, n - 1$. Thus, we can
use the outcomes $0, 1, 2, \ldots, n - 1$ as models for the outcomes of the actual
experiment.

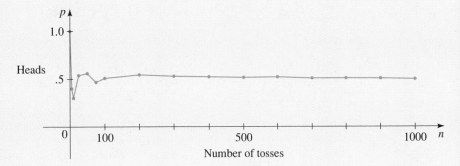

FIGURE 1
Relative frequency of "heads"

PROBLEMS

1. In a game show like the one described in Example 1, a prize is concealed behind one of ten doors. After the contestant chooses a door, the host opens eight losing doors and then gives the contestant the opportunity to switch to the other unopened door.

(a) Play this game with a friend 30 or more times, using the strategy of switching doors each time. Count the number of times you win, and estimate the probability of winning with this strategy.

(b) Calculate the probability of winning with the "switching" strategy using reasoning similar to that in Example 1. Compare with your result from part (a).

2. A couple intend to have two children. What is the probability that they will have one child of each sex? The French mathematician D'Alembert analyzed this problem (incorrectly) by reasoning that three outcomes are possible: two boys, or two girls, or one child of each sex. He concluded that the probability of having one of each sex is $\frac{1}{3}$, mistakenly assuming that the three outcomes are "equally likely."

(a) Model this problem with a pair of coins (using "heads" for boys and "tails" for girls), or write a program to model the problem. Perform the experiment 40 or more times, counting the number of boy-girl combinations. Estimate the probability of having one child of each sex.

(b) Calculate the correct probability of having one child of each sex, and compare this with your result from part (a).

3. A game between two players consists of tossing a coin. Player A gets a point if the coin shows heads, and player B gets a point if it shows tails. The first player to get six points wins an $8000 jackpot. As it happens, the police raid the place when player A has five points and B has three points. After everyone has calmed down, how should the jackpot be divided between the two players? In other words, what is the probability of A winning (and that of B winning) if the game were to continue?

The French mathematicians Pascal and Fermat corresponded about this problem, and both came to the same correct conclusion (though by very different reasonings). Their friend Roberval disagreed with both of them. He argued that player A has probability $\frac{3}{4}$ of winning since, he claimed, the game can end in the four ways *H, TH, TTH, TTT*, and in three of these, A wins. Roberval was wrong.

(a) Continue the game from the point at which it was interrupted, using either a coin or a modeling program. Perform this experiment 80 or more times, and estimate the probability that player A wins.

(b) Calculate the probability that player A wins. Compare with your estimate from part (a).

4. In the World Series, the top teams in the National League and the American League play a best-of-seven series; that is, they play until one team has won four games. (No tie is allowed, so this results in a maximum of seven games.) Suppose the teams are evenly matched, so that the probability that either team wins a given game is $\frac{1}{2}$.

(a) Use a coin or a modeling program to model a World Series, where "heads" represents a win by Team A and "tails" a win by Team B. Perform this experiment at least 80 times, keeping track of how many games are needed to decide each series. Estimate the probability that an evenly matched series will end in four games. Do the same for five, six, and seven games.

(b) What is the probability that the series will end in four games? Five games? Six games? Seven games? Compare with your estimates from part (a).

(c) Find the expected value for the number of games until the series ends.
[*Hint:* This will be $P(\text{four games}) \times 4 + P(\text{five}) \times 5 + P(\text{six}) \times 6 + P(\text{seven}) \times 7$.]

5. In this problem we use the Monte Carlo method to estimate the value of π. The circle in the figure has radius 1 so its area is π, and the square has area 4. If we choose a point at random from the square, the probability that it lies inside the circle will be

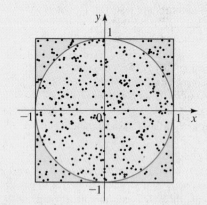

$$\frac{\text{area of circle}}{\text{area of square}} = \frac{\pi}{4}$$

The Monte Carlo method involves choosing many points inside the square. Then we have

$$\frac{\text{number of hits inside circle}}{\text{number of hits inside square}} \approx \frac{\pi}{4}$$

so 4 times this ratio will give us an approximation for π.

```
PROGRAM: PI
:0→P
:For(N,1,1000)
:rand→X:rand→Y
:P+((X²+Y²)<1)→P
:End
:Disp "PI IS APPROX",4*P/N
```

To implement this method, we use a random-function generator to obtain the coordinates (x, y) of a random point in the square, and then check to see if it lies inside the circle (that is, we check if $x^2 + y^2 < 1$). Note that we only need to use points in the first quadrant, since the ratio of areas is the same in each quadrant. The program in the margin shows a way of doing this on the TI-82 calculator for 1000 randomly selected points.

Carry out this Monte Carlo simulation for as many points as you can. How do your results compare with the actual value of π? Do you think this is a reasonable way to get a good approximation for π?

6. Choose two numbers at random from the interval $[0, 1)$. What is the probability that the sum of the two numbers is less than 1?
   (a) Use a Monte Carlo model to estimate the probability.
   (b) Calculate the exact value of the probability. [*Hint:* Call the numbers x and y. Choosing these two numbers is the same as choosing an ordered pair (x, y) in the unit square $\{(x, y) \mid 0 \le x < 1, 0 \le y < 1\}$. What proportion of the points in this square corresponds to $x + y$ being less than 1?]

ANSWERS TO ODD-NUMBERED EXERCISES AND CHAPTER TESTS

CHAPTER 1

Section 1.1 ■ page 12

1. Commutative Property for addition
3. Associative Property for addition
5. Distributive Property   7. Distributive Property
9. $3x + 3y$ 11. $8m$ 13. $-8y$ 15. $-5x + 10y$
17. (a) $\frac{7}{13}$ (b) $\frac{5}{18}$ 19. (a) $\frac{4}{9}$ (b) $\frac{15}{2}$
21. (a) False   (b) True   23. (a) False   (b) True
25. (a) $x > 0$ (b) $t < 4$ (c) $a \geq \pi$
(d) $-5 < x < \frac{1}{3}$ (e) $|p - 3| \leq 5$
27. (a) $\{1, 2, 3, 4, 5, 6, 8\}$ (b) $\{2, 4, 6\}$
29. (a) $\{1, 2, 3, 4, 5, 6, 7, 8, 9, 10\}$ (b) $\varnothing$
31. (a) $\{x \mid x \leq 5\}$ (b) $\{x \mid -1 < x < 4\}$
33. $-3 < x < 0$ 35. $2 \leq x < 8$

37. $x \geq 2$ 39. $(-\infty, 1]$

41. $(-2, 1]$ 43. $(-1, \infty)$

45.   47.

49.

51. (a) 100   (b) 73   53. (a) 2   (b) -1
55. $3|x + 3|$ 57. $\frac{1}{2}|x - 5|$ 59. $x^2 + 9$
61. (a) 15   (b) 24   (c) $\frac{67}{40}$
63. (a) $\frac{7}{9}$ (b) $\frac{13}{45}$ (c) $\frac{19}{33}$

Section 1.2 ■ page 24

1. (a) 16   (b) -16 (c) 1
3. (a) $\frac{625}{8}$ (b) 100,000   (c) 4096
5. (a) $\frac{2}{3}$ (b) 4   (c) $\frac{1}{2}$ 7. (a) 6   (b) 4   (c) $\frac{3}{5}$
9. (a) $\frac{3}{2}$ (b) $\frac{9}{4}$ (c) $\frac{125}{512}$ 11. $\sqrt[3]{4}$ 13. $2\sqrt{5}$
15. t^5 17. $6x^7y^5$ 19. $16x^{10}$ 21. $4/b^2$
23. $64r^7s$ 25. $648y^7$ 27. $\dfrac{x^3}{y}$ 29. $\dfrac{y^2z^9}{x^5}$
31. $\dfrac{s^3}{q^7r^6}$ 33. $x^{13/15}$ 35. $16b^{9/10}$ 37. $\dfrac{1}{c^{2/3}d}$
39. $y^{1/2}$ 41. $\dfrac{32x^{12}}{y^{16/15}}$ 43. $\dfrac{x^{15}}{y^{15/2}}$ 45. $\dfrac{4a^2}{3b^{1/3}}$
47. $\dfrac{3t^{25/6}}{s^{1/2}}$ 49. $|x|$ 51. $x\sqrt[3]{y}$ 53. $ab\sqrt[5]{ab^2}$
55. $2|x|$ 57. (a) $\dfrac{\sqrt{6}}{6}$ (b) $\dfrac{\sqrt{3xy}}{3y}$ (c) $\dfrac{\sqrt{15}}{10}$
59. (a) $\dfrac{\sqrt[3]{x^2}}{x}$ (b) $\dfrac{\sqrt[5]{x^3}}{x}$ (c) $\dfrac{\sqrt[7]{x^4}}{x}$
61. (a) 6.93×10^7 (b) 2.8536×10^{-5} (c) 1.2954×10^8
63. (a) 319,000   (b) 0.00000002670
(c) 710,000,000,000,000
65. (a) 5.9×10^{12} mi   (b) 4×10^{-13} cm
(c) 3.3×10^{19} molecules
67. 1.3×10^{-20} 69. 1.429×10^{19} 71. 7.4×10^{-14}
73. $8\frac{1}{3}$ min   75. 41.3 mi

Section 1.3 ■ page 34

1. $6x + 6$ 3. $3x^2 - 2x + 6$ 5. $x^3 + 3x^2 - 6x + 11$
7. $-t^4 + t^3 - t^2 - 10t + 5$ 9. $x^{3/2} - x$

11. $y^{7/3} - y^{1/3}$ **13.** $21t^2 - 29t + 10$
15. $3x^2 + 5xy - 2y^2$ **17.** $1 - 4y + 4y^2$
19. $2x^3 - 7x^2 + 7x - 5$ **21.** $x^3 + x^2 - 2x$
23. $30y^4 + y^5 - y^6$ **25.** $4x^4 + 12x^2y^2 + 9y^4$
27. $x^4 - a^4$ **29.** $1 + 3a^3 + 3a^6 + a^9$ **31.** $a - 1/b^2$
33. $x^5 + x^4 - 3x^3 + 3x - 2$ **35.** $1 - x^{2/3} + x^{4/3} - x^2$
37. $1 - 2b^2 + b^4$ **39.** $3x^4y^4 + 7x^3y^5 - 6x^2y^3 - 14xy^4$
41. $2x(1 + 6x^2)$ **43.** $3y^3(2y - 5)$ **45.** $(x + 6)(x + 1)$
47. $(x - 4)(x + 2)$ **49.** $(y - 3)(y - 5)$
51. $(2x + 3)(x + 1)$ **53.** $9(x - 2)(x + 2)$
55. $(3x + 2)(2x - 3)$ **57.** $3(x - 1)(x + 2)$
59. $y^4(y + 2)^3(y + 1)^2$ **61.** $(a + 1)(a - 1)(b + 2)(b - 2)$
63. $(t + 1)(t^2 - t + 1)$ **65.** $(2t - 3)^2$ **67.** $x(x + 1)^2$
69. $(2x + y)^2$ **71.** $x^2(x + 3)(x - 1)$
73. $(2x - 5)(4x^2 + 10x + 25)$ **75.** $(x^2 + 2)(x + 1)(x - 1)$
77. $(y + 2)(y - 2)(y - 3)$ **79.** $(2x^2 + 1)(x + 2)$
81. $(x + y)(x - y)(x^2 + xy + y^2)(x^2 - xy + y^2)$
83. $x^{1/2}(x + 1)(x - 1)$ **85.** $x^{-3/2}(x + 1)^2$
87. $(x^2 + 3)(x^2 + 1)^{-1/2}$
89. $(a + 2)(a - 2)(a + 1)(a - 1)$
91. $(x^2 - x + 2)(x^2 + x + 2)$
95. $(a + b + c)(a - b + c)(a + b - c)(-a + b + c)$

Section 1.4 ■ page 42

1. $\dfrac{x + 1}{x + 3}$ **3.** $\dfrac{-y}{y + 1}$ **5.** $\dfrac{x(2x + 3)}{2x - 3}$ **7.** $\dfrac{1}{t^2 + 9}$
9. $\dfrac{x + 4}{x + 1}$ **11.** $\dfrac{(2x + 1)(2x - 1)}{(x + 5)^2}$ **13.** $x^2(x + 1)$
15. $\dfrac{x}{yz}$ **17.** $\dfrac{3x + 7}{(x - 3)(x + 5)}$ **19.** $\dfrac{1}{(x + 1)(x + 2)}$
21. $\dfrac{3x + 2}{(x + 1)^2}$ **23.** $\dfrac{u^2 + 3u + 1}{u + 1}$ **25.** $\dfrac{2x + 1}{x^2(x + 1)}$
27. $\dfrac{2x + 7}{(x + 3)(x + 4)}$ **29.** $\dfrac{x - 2}{(x + 3)(x - 3)}$ **31.** $\dfrac{5x - 6}{x(x - 1)}$
33. $\dfrac{-5}{(x + 1)(x + 2)(x - 3)}$ **35.** $-xy$ **37.** $\dfrac{c}{c - 2}$
39. $\dfrac{3x + 7}{x^2 + 2x - 1}$ **41.** $\dfrac{y - x}{xy}$ **43.** $\dfrac{-1}{a(a + h)}$
45. $\dfrac{-3}{(2 + x)(2 + x + h)}$ **47.** $\dfrac{1}{\sqrt{1 - x^2}}$
49. $\dfrac{x + 2}{(x + 1)^{3/2}}$ **51.** $\dfrac{2x + 3}{(x + 1)^{4/3}}$ **53.** $\dfrac{3 - \sqrt{5}}{2}$
55. $\dfrac{2(\sqrt{7} - \sqrt{2})}{5}$ **57.** $\dfrac{-4}{3(1 + \sqrt{5})}$ **59.** $\dfrac{r - 2}{5(\sqrt{r} - \sqrt{2})}$
61. $\dfrac{1}{\sqrt{x^2 + 1} + x}$ **63.** True   **65.** False   **67.** False
69. True   **71.** False

73. (a) $\dfrac{R_1 R_2}{R_1 + R_2}$ **(b)** $\frac{20}{3} \approx 6.7$ ohms

Section 1.5 ■ page 56

1. (a) Yes   **(b)** No   **3. (a)** No   **(b)** Yes
5. $x = 4$ **7.** $w = -3$ **9.** $y = 12$ **11.** $x = -\frac{3}{4}$
13. $x = -\frac{1}{3}$ **15.** $t = \frac{8}{11}$ **17.** $t = -2$ **19.** $x = -\frac{4}{9}$
21. $x = \frac{29}{2}$ **23.** No solution   **25.** $x = -4$
27. $x = \pm 3\sqrt{2}$ **29.** $-2, 3$ **31.** 2
33. $x = 2 \pm \sqrt{2}$ **35.** $x = \frac{3}{2}$ or $x = -\frac{1}{2}$ **37.** $-2, 4$
39. $-6 \pm 3\sqrt{7}$ **41.** $\dfrac{-3 \pm 2\sqrt{6}}{3}$ **43.** $-\frac{9}{2}, \frac{1}{2}$
45. $\dfrac{-5 \pm \sqrt{13}}{2}$ **47.** $\dfrac{\sqrt{5} \pm 1}{2}$ **49.** $-50, 100$
51. -4 **53.** ± 2 **55.** $-1, 0, 2$ **57.** 4 **59.** 4
61. 21 **63.** $-1, 0, 3$ **65.** $\pm 2\sqrt{2}, \pm 3\sqrt{3}$
67. $27, 729$ **69.** $-\frac{1}{2}$ **71.** ± 1.5 **73.** $3.99, 4.01$
75. $x \approx 5.06$ **77.** $1.200, 1.250$ **79.** $R = \dfrac{PV}{nT}$
81. $h = \dfrac{A - 2lw}{2l + 2w}$ **83.** $x = \dfrac{2d - b}{a - 2c}$
85. $r = \pm\sqrt{\dfrac{3V}{\pi h}}$ **87.** $b = \pm\sqrt{c^2 - a^2}$
89. $i = 100(-1 \pm \sqrt{A/P})$ **91.** 2 **93.** 1
95. $k = \pm 20$

Section 1.6 ■ page 67

1. $0.14x$ **3.** $50w$ **5.** $3s + 15$ **7.** 714
9. $111, 112, 113$ **11.** 19 and 36
13. \$9000 at $4\frac{1}{2}$% and \$3000 at 4%   **15.** $7\frac{1}{2}$%
17. 45 ft   **19.** 4 in.   **21.** 25 ft by 35 ft
23. 13 in. by 13 in.   **25.** $1\frac{1}{2}$ h   **27.** 450 mi
29. 4 h   **31.** 50 mi/h (or 240 mi/h)   **33.** 200 mL
35. 18 g   **37.** 0.6 L   **39.** 37 min 20 s   **41.** 3 h
43. 3 h   **45.** 200 m
47. (a) After 1 s and $1\frac{1}{2}$ s   **(b)** Never   **(c)** 25 ft
(d) After $1\frac{1}{4}$ s   **(e)** After $2\frac{1}{2}$ s
49. (a) After 17 yr, on Jan. 1, 2014
(b) After 18.612 yr, on Aug. 12, 2015
51. 215,000 mi   **53.** 49 ft, 168 ft, and 175 ft
55. 8 h   **57.** 7 yr   **59.** 300 mi   **61.** 35 yd
63. 4.55 ft

Section 1.7 ■ page 78

1. $\{-1, 0, \frac{1}{2}, \sqrt{2}, 2\}$ **3.** $\{-1, 2\}$
5. $(-\infty, 4]$ **7.** $(-\infty, -5)$

9. $(4, \infty)$

11. $(-\infty, 2]$

13. $\left(-\infty, -\frac{1}{2}\right)$

15. $[1, \infty)$

17. $[-1, \infty)$

19. $(-\infty, -1]$

21. $(-\infty, 2) \cup (5, \infty)$

23. $[-3, 6]$

25. $(-\infty, -1] \cup \left[\frac{1}{2}, \infty\right)$

27. $(-\infty, -3) \cup (6, \infty)$

29. $(-2, 2)$

31. $[-2, 0] \cup [2, \infty)$

33. $(-\infty, -1) \cup [3, \infty)$

35. $\left(-\infty, -\frac{3}{2}\right)$

37. $(-\infty, 5) \cup [16, \infty)$

39. $(-2, 0) \cup (2, \infty)$

41. $(-\infty, -2] \cup [2, \infty)$

43. $\left(0, \frac{3}{4}\right] \cup (1, \infty)$

45. $\left[-8, -\frac{5}{2}\right)$

47. $(-2, 2)$

49. $[2, 8]$

51. $(-7, -3)$

53. $[1.3, 1.7]$

55. $(-6.001, -5.999)$

57. $(2, 6)$

59. $(0, 1]$

61. $(-\infty, 0) \cup \left(\frac{1}{4}, \infty\right)$

63. $(-\infty, -1) \cup (1, \infty)$

65. $68 \le F \le 86$

67. (a) $T = 20 - \dfrac{h}{100}$　(b) $20\,°C$ to $-30\,°C$

69. 0 s to 3 s　**71.** Distances greater than 30 m

73. $x \ge \dfrac{c(a + b)}{ab}$

Section 1.8 ■ page 92

1.
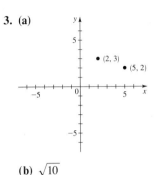

3. (a)
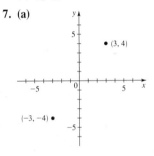

(b) $\sqrt{10}$
(c) $\left(\frac{7}{2}, \frac{5}{2}\right)$

5. (a)
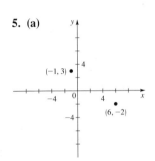

(b) $\sqrt{74}$
(c) $\left(\frac{5}{2}, \frac{1}{2}\right)$

7. (a)
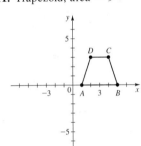

(b) 10
(c) $(0, 0)$

9. 24

11. Trapezoid, area $= 9$

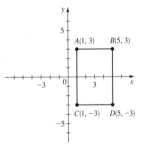

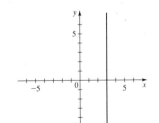

13.

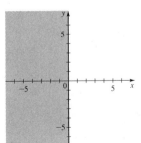

15.

17.

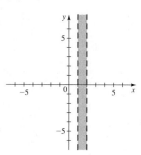

19.

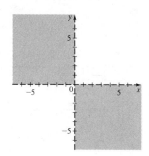

21.

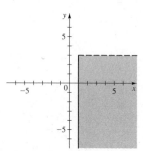

23.

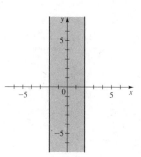

25. $A(6,7)$ **29. (b)** 10

31. (a)

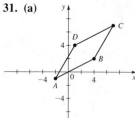

33. (a) $(8,5)$
   (b) $(a + 3, b + 2)$
   (c) $A'(-2,1)$, $B'(0,4)$,
        $C'(5,3)$

(b) $\left(\frac{5}{2},3\right)$, $\left(\frac{5}{2},3\right)$

35. x-intercept 0,
   y-intercept 0,
   symmetry about origin

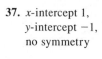

37. x-intercept 1,
   y-intercept -1,
   no symmetry

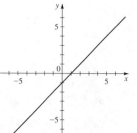

39. x-intercept $\frac{5}{3}$,
   y-intercept -5,
   no symmetry

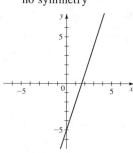

41. x-intercepts ± 1,
   y-intercept 1,
   symmetry about y-axis

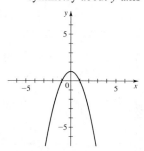

43. x-intercepts ± 3,
   y-intercept -9,
   symmetry about y-axis

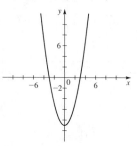

45. x-intercept 0,
   y-intercept 0,
   no symmetry

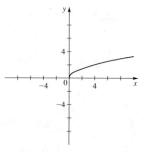

47. x-intercept 0,
   y-intercept 0,
   symmetry about y-axis

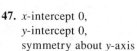
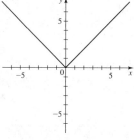

49. x-intercepts ± 4,
   y-intercept 4,
   symmetry about y-axis

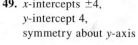
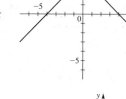

51. x-intercept 0,
   y-intercept 0,
   symmetry about origin

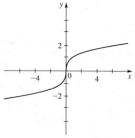

53. Symmetry about *y*-axis   **55.** Symmetry about origin
57. Symmetry about origin
59.

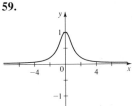

61.

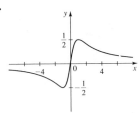

63. $(x - 2)^2 + (y + 1)^2 = 9$ **65.** $x^2 + y^2 = 65$
67. $(x - 3)^2 + (y + 1)^2 = 32$
69. $(x + 2)^2 + (y - 2)^2 = 4$ **71.** $(1, -2), 2$
73. $(2, -5), 4$ **75.** $\left(-\frac{1}{4}, 0\right), \frac{1}{4}$
77.

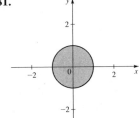

79.

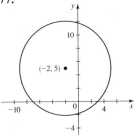

81.

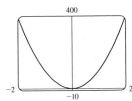

83. 12π

85. $a^2 + b^2 > 4c, \left(-\frac{1}{2}a, -\frac{1}{2}b\right), \frac{1}{2}\sqrt{a^2 + b^2 - 4c}$

Section 1.9 ■ page 101

1. (c)   **3.** (c)   **5.** (c)
7.

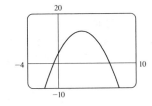

9.

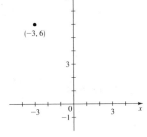

11.

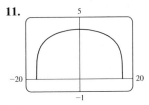

13.

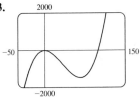

15.

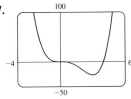

17.

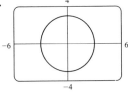

19.

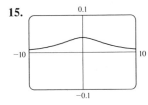

21.

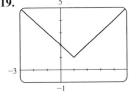

23.

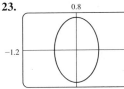

25. 3.0, 4.0
27. 1.0, 2.0, 3.0
29. 1.62
31. $-1.0, 0, 1.0$
33. $[-2.0, 5.0]$
35. $(-\infty, 1.0] \cup [2.0, 3.0]$

37. $(-1.0, 0) \cup (1.0, \infty)$ **39.** $(-\infty, 0)$

Section 1.10 ■ page 112

1. 4   **3.** $-\frac{9}{2}$ **5.** 1   **7.** $-2, \frac{1}{2}, 3, -\frac{1}{4}$
9. $x + y - 4 = 0$ **11.** $3x - 2y - 6 = 0$
13. $x - y + 1 = 0$ **15.** $2x - 3y + 19 = 0$
17. $5x + y - 11 = 0$ **19.** $3x - y - 2 = 0$
21. $3x - y - 3 = 0$ **23.** $y = 5$
25. $x + 2y + 11 = 0$ **27.** $x = -1$
29. $5x - 2y + 1 = 0$ **31.** $x - y + 6 = 0$
33. (a)   (b) $3x - 2y + 8 = 0$

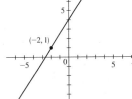

35. All lines pass through $(3, 2)$

37. $-1, 3$

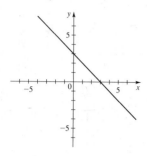

39. $-\frac{1}{3}, 0$

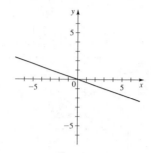

41. $\frac{3}{2}, 3$

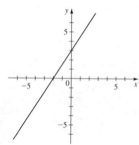

43. $0, 4$

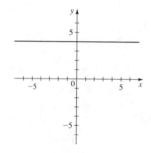

45. $\frac{3}{4}, -3$

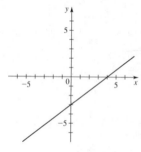

47. $-\frac{3}{4}, \frac{1}{4}$

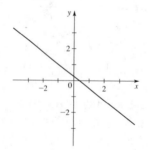

53. $x - y - 3 = 0$ **55.** (b) $4x - 3y - 24 = 0$

57. 16,667 ft

59. (a)

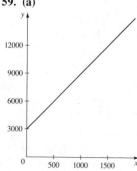

(b) The slope represents production cost per toaster; the y-intercept represents monthly fixed cost.

61. (a) $t = \frac{5}{24}n + 45$ (b) $76\,°F$

63. (a) $P = 0.434d + 15$, where P is pressure in lb/in² and d is depth in feet   (b) 196 ft

65. (a) $C = \frac{1}{4}d + 260$ (b) \$635

(c) The slope represents cost per mile.

(d) The y-intercept represents annual fixed cost.

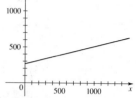

67. (a)

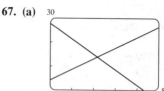

(b) $(21.82, 13.82)$

(c) Price \$21.82, amount 13.82

Chapter 1 Review ■ page 117

1. Commutative Property for addition

3. Distributive Property

5. $-1 < x \le 3$

7. $(2, \infty)$

9. 6   **11.** $\frac{1}{6}$ **13.** 11   **15.** 4   **17.** $12x^5y^4$

19. $9x^3$ **21.** x^2y^2 **23.** $\dfrac{x(2 - \sqrt{x})}{4 - x}$ **25.** $\dfrac{4r^{5/2}}{s^7}$

27. 7.825×10^{10} **29.** 1.65×10^{-32}

31. $3xy^2(4xy^2 - y^3 + 3x^2)$ **33.** $(x - 2)(x + 5)$

35. $(4t + 3)(t - 4)$ **37.** $(5 - 4t)(5 + 4t)$

39. $x^{-1/2}(x - 1)^2$ **41.** $y(a + b)(a - b)$

43. $2x^3(x - 3)(x - 6)$ **45.** $6x^2 - 21x + 3$

47. $4a^4 - 4a^2b + b^2$ **49.** $2x^{3/2} + x - x^{1/2}$

51. $\dfrac{x - 3}{2x + 3}$ **53.** $\dfrac{x + 1}{x - 4}$ **55.** $\dfrac{1}{x + 1}$ **57.** $-\dfrac{1}{2x}$

59. $\dfrac{1}{\sqrt{x + h} + \sqrt{x}}$ **61.** 4   **63.** $\frac{15}{2}$ **65.** -30

67. $2, 7$ **69.** $-1, \frac{1}{2}$ **71.** $\dfrac{-2 \pm \sqrt{7}}{3}$ **73.** $\dfrac{3 \pm \sqrt{6}}{3}$

75. 20 lb raisins, 30 lb nuts   **77.** 4:00 P.M.

79. 1 h 50 min

81. $(-3, \infty)$

83. $\left[\frac{10}{3}, \infty\right)$

85. $[-4, -1)$

87. $(-\infty, -2) \cup (2, 4]$

89. $(-\infty, -1] \cup [0, \infty)$

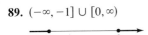

91. (a)

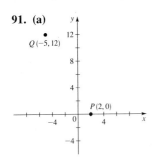

(b) $\sqrt{193}$ **(c)** $\left(-\frac{3}{2}, 6\right)$

(d) $y = -\frac{12}{7}x + \frac{24}{7}$ **(e)** $(x - 2)^2 + y^2 = 193$

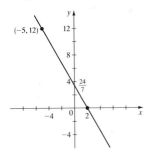   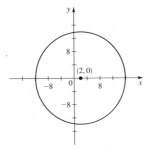

93.

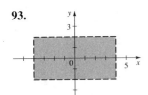

95. B **97.** $(x + 5)^2 + (y + 1)^2 = 26$

99. Circle, center $(-1, 3)$, radius 1

101. No graph   **103.** $2x - 3y - 16 = 0$

105. $3x + y - 12 = 0$ **107.** $x + 5y = 0$

109. $x^2 + y^2 = 169$, $5x - 12y + 169 = 0$

111. No symmetry   **113.** No symmetry

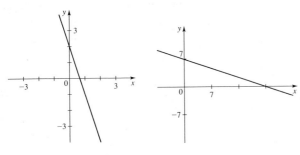

115. No symmetry

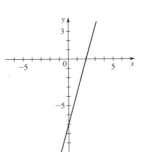

117. Symmetry about y-axis

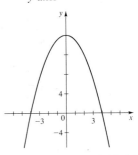

119. No symmetry

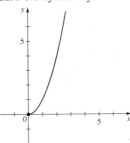

121. $x = -1.53, -0.35, 1.88$

123. $x \leqslant 1.03$

Chapter 1 Test ■ page 120

1. (a)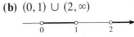

(b) $(-\infty, 5)$; $[-2, 1]$ **(c)** 53

2. (a) 81   **(b)** $5^6 = 15{,}625$ **(c)** 2   **(d)** $\frac{1}{8}$

3. (a) $8\sqrt{2}$ **(b)** $54a^6b^{14}$ **(c)** $\dfrac{8x^{1/4}}{y}$

(d) $\dfrac{x + 2}{x - 2}$ **(e)** $\dfrac{1}{x - 2}$

4. (a) 3.25×10^{11} **(b)** 8.931×10^{-6}

5. (a) $2x^2 - 7x - 15$ **(b)** $x - y$

(c) $9t^2 + 24t + 16$

6. (a) $(3x - 5)(3x + 5)$ **(b)** $(3x + 5)(2x - 1)$

(c) $(x^2 - 3)(x - 4)$ **(d)** $x(x + 3)(x^2 - 3x + 9)$

(e) $3x^{-1/2}(x - 1)(x - 2)$

7. $\dfrac{x(\sqrt{x} + 2)}{x - 4}$ **8.** $\dfrac{V}{2h(2h + 1)}$

9. (a) -10 **(b)** 1   **(c)** $-3, 4$ **(d)** No solution

(e) $0, \frac{1}{2}, 1$ **(f)** $0, -3$

10. 120 mi   **11.** 50 ft by 120 ft

12. (a) $(-\infty, 3]$ **(b)** $(0, 1) \cup (2, \infty)$

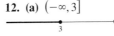   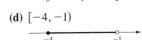

(c) $(1, 5)$ **(d)** $[-4, -1)$

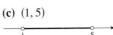

13. 41 °F to 50 °F

14. (a) 20   **(b)** $(-1, -4)$ **(c)** $4x + 3y + 16 = 0$
(d) $3x - 4y - 13 = 0$ **(e)** $(x + 1)^2 + (y + 4)^2 = 100$

15. (a)   **(b)**

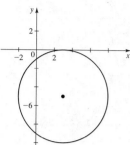

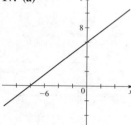

$(x - 3)^2 + (y + 5)^2 = 25$ $\left(x + \frac{3}{2}\right)^2 + \left(y + \frac{5}{2}\right)^2 = 0$

16. (a) $x + 3y - 7 = 0$ **(b)** $4x - y + 12 = 0$

17. (a)   **(b)**

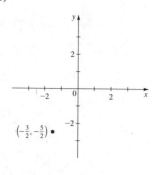

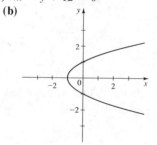

PRINCIPLES OF PROBLEM SOLVING ■ page 126

1. 37.5 mi/h   **3.** 150 mi   **5.** 2   **7.** 57 min
9. No, not necessarily   **11.** The same amount
13. (a) Uncle   **(b)** Father   **(d)** No   **15.** 8.49
21. 7   **23.** $\frac{11}{2}$

CHAPTER 2

Section 2.1 ■ page 137

1. $f(x) = 3x + 1$ **3.** $f(x) = (x + 2)^2$
5. Divide by 3, then subtract 5
7. Square, multiply by 2, then subtract 3
9.

11. 3, -3, 2, $2\sqrt{5} + 1$, $2a + 1$, $-2a + 1$, $2(a + b) + 1$

13. $-\frac{1}{3}$, -3, $\dfrac{1 - \pi}{1 + \pi}$, $\dfrac{1 - a}{1 + a}$, $\dfrac{2 - a}{a}$, $\dfrac{1 + a}{1 - a}$

15. -4, 10, $3\sqrt{2}$, $5 + 7\sqrt{2}$, $2x^2 - 3x - 4$, $2x^2 + 7x + 1$, $4x^2 + 6x - 8$, $8x^2 + 6x - 4$

17. $3a + 2$, $3a + 3h + 4$, $3(a + h) + 2$, 3

19. 5, 10, 5, 0

21. $3 - 5a + 4a^2$, $6 - 5a - 5h + 4a^2 + 4h^2$,
$3 - 5a - 5h + 4a^2 + 8ah + 4h^2$, $-5 + 8a + 4h$

23. $15 + 8h$, $8x + 8h - 1$, 8

25. $\dfrac{1}{2 + h}$, $\dfrac{1}{x + h}$, $\dfrac{-1}{x(x + h)}$ **27.** $(-\infty, \infty)$, $(-\infty, \infty)$

29. $[-1, 5]$, $[-2, 10]$ **31.** $(-\infty, \infty)$, $(-\infty, 2]$

33. $\left[\frac{5}{2}, \infty\right)$, $[0, \infty)$ **35.** $\{x \mid x \neq 3\}$ **37.** $\{x \mid x \neq \pm 1\}$

39. $(-\infty, \infty)$ **41.** $[5, \infty)$ **43.** $(-\infty, \infty)$

45. $[-2, 3) \cup (3, \infty)$ **47.** $(10, \infty)$ **49.** $[0, 1]$

51. $(-\infty, 0] \cup [6, \infty)$ **53.** $[0, \pi)$

Section 2.2 ■ page 148

1. (a) 2, 0, 2, 3   **(b)** $[-3, 3]$, $[0, 3]$
(c) Increasing on $[0, 3]$; decreasing on $[-3, 0]$
3. (a) 3, 2, -2, 1, 0   **(b)** $[-4, 4]$, $[-2, 3]$
(c) Increasing on $[0, 2]$; decreasing on $[-4, 0]$, $(2, 4]$
5. (a) $f(0)$ **(b)** $g(-3)$ **(c)** -2, 2
7. (a) Yes   **(b)** No   **(c)** Yes   **(d)** No
9. Function, domain $[-3, 2]$, range $[-2, 2]$
11. Not a function

13. (a)

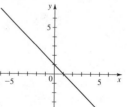

(b) $(-\infty, \infty)$
(c) Decreasing on $(-\infty, \infty)$

15. (a)

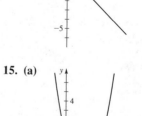
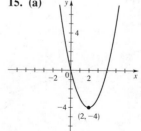

(b) $(-\infty, \infty)$
(c) Increasing on $[2, \infty)$, decreasing on $(-\infty, 2]$

17. (a)

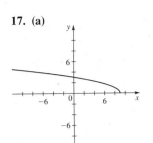

(b) $(-\infty, 9]$
(c) Decreasing on $(-\infty, 9]$

19. (a)

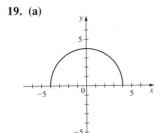

(b) $[-4, 4]$
(c) Increasing on $[-4, 0]$, decreasing on $[0, 4]$

21.

23.

25.

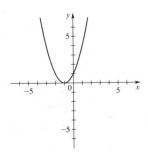

27.

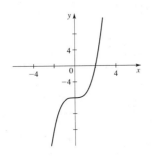

29.

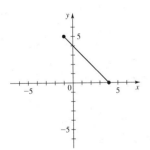

31.

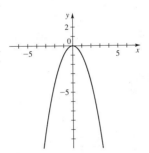

33.

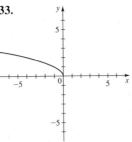

35.

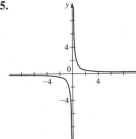

37.

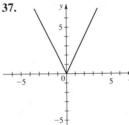

39.

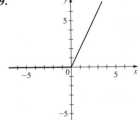

41.

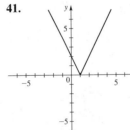

43.

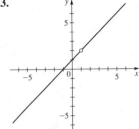

45. (a)

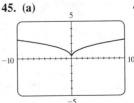

(b) Increasing on $[0, \infty)$, decreasing on $(-\infty, 0]$

47. (a)

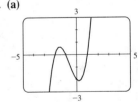

(b) Increasing on $(-\infty, -1.55]$, $[0.22, \infty)$; decreasing on $[-1.55, 0.22]$

49. (a)

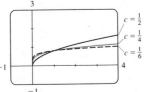

(b)

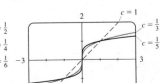

(c)

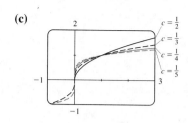

(d) Graphs of even roots are similar to $\sqrt{x}$; graphs of odd roots are similar to $\sqrt[3]{x}$. As c increases, the graph of $y = \sqrt[c]{x}$ becomes steeper near 0 and flatter when $x > 1$.

51. (a)

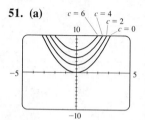

(b)

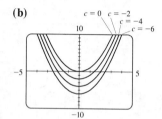

(c) If $c > 0$, then the graph of $f(x) = x^2 + c$ is the same as the graph of $y = x^2$ shifted upward c units. If $c < 0$, then the graph of $f(x) = x^2 + c$ is the same as the graph of $y = x^2$ shifted downward c units.

53. (a)

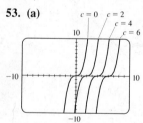

(b)

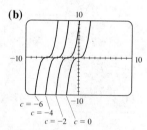

(c) If $c > 0$, then the graph of $f(x) = (x - c)^3$ is the same as the graph of $y = x^3$ shifted right c units. If $c < 0$, then the graph of $f(x) = (x - c)^3$ is the same as the graph of $y = x^3$ shifted left c units.

55. $g(x) = x^3/10$

57. (a)

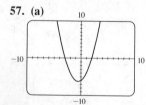

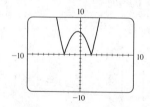

(b) To obtain the graph of g, reflect in the x-axis the part of the graph of f that is below the x-axis.

59.

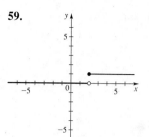

61.

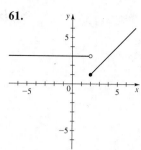

63.

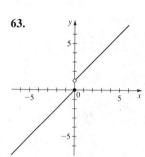

65.

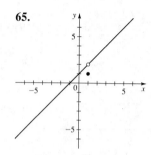

67.

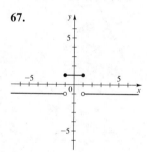

69.

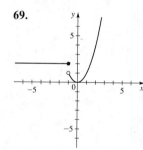

71.

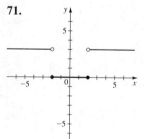

73.

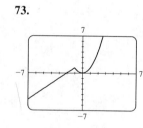

75.

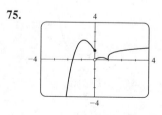

77.

$$C(x) = \begin{cases} 2 & 0 < x \leq 1 \\ 2.2 & 1 < x \leq 1.1 \\ 2.4 & 1.1 < x \leq 1.2 \\ \vdots \\ 4.0 & 1.9 < x < 2.0 \end{cases}$$

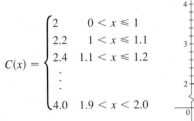

79. $f(x) = -\frac{7}{6}x - \frac{4}{3}$, $-2 \leq x \leq 4$ **81.** $f(x) = 1 - \sqrt{-x}$

Section 2.3 ■ page 158

1. This person's weight increases as he grows, then continues to increase; the person then goes on a crash diet (possibly) at age 30, then gains weight again, the weight gain eventually leveling off.

3.

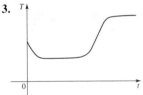

5.

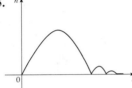

7.

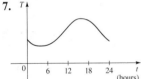

9. (a) 500 MW   **(b)** Between 3:00 A.M. and 4:00 A.M.
(c) Just before noon

11.

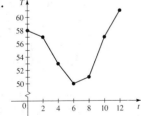

13. $A = 10x - x^2$, $0 < x < 10$
15. $A = (\sqrt{3}x^2)/4$, $x > 0$ **17.** $r = \sqrt{A/\pi}$, $A > 0$
19. $A = x^2 + (48/x)$, $x > 0$
21. $A = 15x - (\pi + 4)x^2/8$, $x > 0$
23. $A = 2x(1200 - x)$, $0 < x < 1200$
25. $d = 25t$, $t \geq 0$

27.

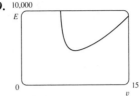

29.

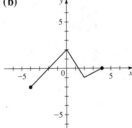

31. $R = kt$ **33.** $v = k/z$ **35.** $y = ks/t$
37. $z = k\sqrt{y}$ **39.** $y = 18x$ **41.** $M = 15x/y$
43. $W = 360/r^2$ **45. (a)** $F = kx$ **(b)** 8   **(c)** 32 N
47. (a) $C = \frac{1}{8}pn$ **(b)** \$57,500
49. (a) $R = kL/d^2$, $k = \frac{7}{2400} = 0.00291\overline{6}$ **(b)** $\approx$137 Ω

Section 2.4 ■ page 172

1. (a) Shift downward 4 units   **(b)** Shift right 4 units
3. (a) Stretch vertically by a factor of 3
(b) Shrink vertically by a factor of $\frac{1}{3}$
5. (a) Reflect in the x-axis and shift upward 5 units
(b) Reflect in the y-axis and shift upward 5 units
7. (a) Shift right 2 units and downward 3 units
(b) Shift right 3 units and stretch vertically by a factor of 2
9. (a)

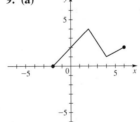

(b)

(c)

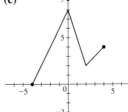

(d)

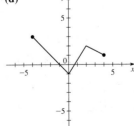

(e)

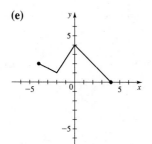

(f)

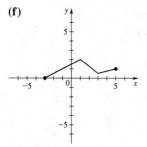

17.

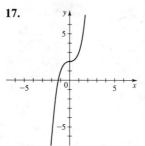

19.

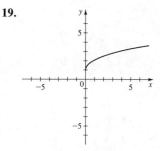

11. (a)

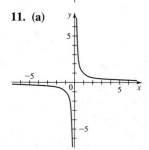

21.

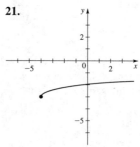

23.

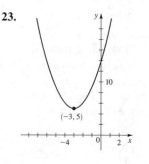

(b) (i)

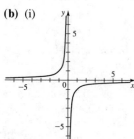

(ii)

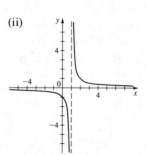

25.

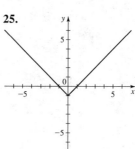

27.

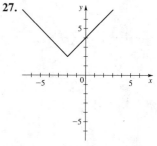

(iii)

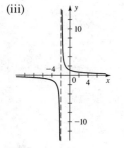

(iv)

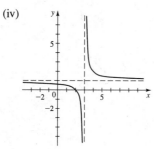

29.

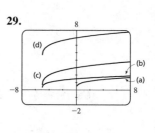

For part (b) shift the graph in (a) left 5 units; for part (c) shift the graph in (a) left 5 units and stretch vertically by a factor of 2; for part (d) shift the graph in (a) left 5 units, stretch vertically by a factor of 2, and then shift upward 4 units.

13.

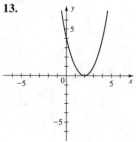

15.

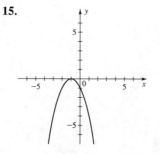

31.

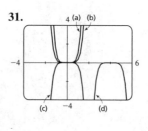

For part (b) shrink the graph in (a) vertically by a factor of $\frac{1}{3}$; for part (c) shrink the graph in (a) vertically by a factor of $\frac{1}{3}$ and reflect in the x-axis; for part (d) shift the graph in (a) right 4 units, shrink vertically by a factor of $\frac{1}{3}$, and then reflect in the x-axis.

33. To obtain the graph of g, reflect in the x-axis the part of the graph of f that is below the x-axis.

35. (a) 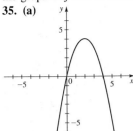   **(b)**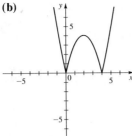

37. (a)    **(b)**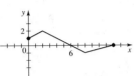

39.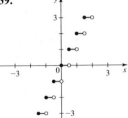

41.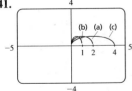

For part (b) shrink the graph in (a) horizontally by a factor of $\frac{1}{2}$; for part (c) stretch the graph in (a) horizontally by a factor of 2.

43. Even 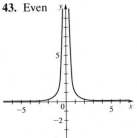   **45.** Neither

47. Odd 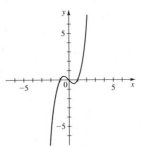   **49.** Neither

Section 2.5 ■ page 184

1. Vertex $(0, -8)$
x-intercepts $\pm 2\sqrt{2}$
y-intercept -8
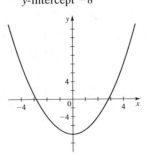

3. Vertex $(0, -2)$
no x-intercept
y-intercept -2
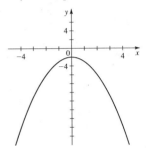

5. Vertex $\left(\frac{3}{2}, -\frac{9}{2}\right)$
x-intercepts $0, 3$
y-intercept 0
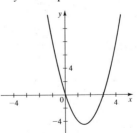

7. Vertex $(-3, -1)$
x-intercepts $-4, -2$
y-intercept 8
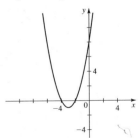

9. Vertex $(-1, 1)$
no x-intercept
y-intercept 3
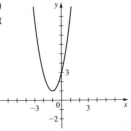

11. Vertex $(5,7)$
no x-intercept
y-intercept 57

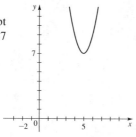

13. (a) $f(x) = -(x - 1)^2 + 1$
(b)

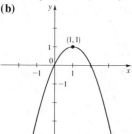

(c) Maximum $f(1) = 1$

15. (a) $(x + 1)^2 - 2$
(b)

(c) Minimum $f(-1) = -2$

17. (a) $f(x) = -\left(x + \frac{3}{2}\right)^2 + \frac{21}{4}$
(b)

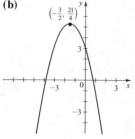

(c) Maximum $f\left(-\frac{3}{2}\right) = \frac{21}{4}$

19. (a) $g(x) = 3(x - 2)^2 + 1$
(b)

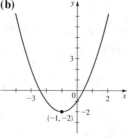

(c) Minimum $g(2) = 1$

21. (a) $h(x) = -\left(x + \frac{1}{2}\right)^2 + \frac{5}{4}$
(b)

(c) Maximum $h\left(-\frac{1}{2}\right) = \frac{5}{4}$

23. Minimum $f\left(-\frac{1}{2}\right) = \frac{3}{4}$
25. Maximum $f(-3.5) = 185.75$
27. Minimum $f(0.6) = 15.64$
29. Minimum $h(-2) = -8$
31. $f(x) = 2x^2 - 4x$ **33.** $(-\infty, \infty), (-\infty, 1]$ **35.** 25 ft

37. $50, -50$ **39.** $-12, -12$ **41.** 600 ft by 1200 ft
43. 14,062.5 ft^2 **45. (a)** -4.01 **(b)** -4.011025
47. Local maximum ≈ 0.38 when $x \approx -0.58$;
local minimum ≈ -0.38 when $x \approx 0.58$
49. Local maximum 0 when $x = 0$;
local minimum ≈ -13.61 when $x \approx -1.71$;
local minimum ≈ -73.32 when $x \approx 3.21$
51. Local maximum ≈ 5.66 when $x \approx 4.00$
53. Local maximum ≈ 0.38 when $x \approx -1.73$;
local minimum ≈ -0.38 when $x \approx 1.73$
55. 7.5 mi/h
57. Height ≈ 1.44 ft, length and width of base ≈ 2.88 ft
59. Width ≈ 8.40 ft, height of rectangular part ≈ 4.20 ft
61. Width ≈ 3.27 ft, height ≈ 5.33 ft
63. (b) 9.23, 13.00   **65.** $f(x) = x^2 + x - 3$

Section 2.6 ■ page 193

1. $(f + g)(x) = x^2 + 5, (-\infty, \infty)$;
$(f - g)(x) = x^2 - 2x - 5, (-\infty, \infty)$;
$(fg)(x) = x^3 + 4x^2 - 5x, (-\infty, \infty)$;
$(f/g)(x) = (x^2 - x)/(x + 5), (-\infty, -5) \cup (-5, \infty)$
3. $(f + g)(x) = \sqrt{1 + x} + \sqrt{1 - x}, [-1, 1]$;
$(f - g)(x) = \sqrt{1 + x} - \sqrt{1 - x}, [-1, 1]$;
$(fg)(x) = \sqrt{1 - x^2}, [-1, 1]$;
$(f/g)(x) = \sqrt{(1 + x)/(1 - x)}, [-1, 1)$
5. $(f + g)(x) = 8/(x(x + 4)), x \neq -4, x \neq 0$;
$(f - g)(x) = 4(x + 2)/(x(x + 4)), x \neq -4, x \neq 0$;
$(fg)(x) = -4/(x(x + 4)), x \neq -4, x \neq 0$;
$(f/g)(x) = -(x + 4)/x, x \neq -4, x \neq 0$
7. $[-3, -1) \cup (-1, 1) \cup (1, 4]$
9.

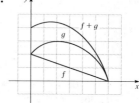

11.

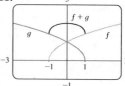

13.

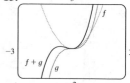

15. (a) 1   **(b)** -23 **17. (a)** -11 **(b)** -119
19. (a) $-3x^2 + 1$ **(b)** $-9x^2 + 30x - 23$
21. 4   **23.** 5   **25.** 4

27. $(f \circ g)(x) = 8x + 1$, $(-\infty, \infty)$;
$(g \circ f)(x) = 8x + 11$, $(-\infty, \infty)$;
$(f \circ f)(x) = 4x + 9$, $(-\infty, \infty)$;
$(g \circ g)(x) = 16x - 5$, $(-\infty, \infty)$

29. $(f \circ g)(x) = 3(6x^2 + 7x + 2)$, $(-\infty, \infty)$;
$(g \circ f)(x) = 6x^2 - 3x + 2$, $(-\infty, \infty)$;
$(f \circ f)(x) = 8x^4 - 8x^3 + x$, $(-\infty, \infty)$;
$(g \circ g)(x) = 9x + 8$, $(-\infty, \infty)$

31. $(f \circ g)(x) = \sqrt{x^2 - 1}$, $(-\infty, -1] \cup [1, \infty)$;
$(g \circ f)(x) = x - 1$, $[1, \infty)$;
$(f \circ f)(x) = \sqrt{\sqrt{x - 1} - 1}$, $[2, \infty)$;
$(g \circ g)(x) = x^4$, $(-\infty, \infty)$

33. $(f \circ g)(x) = -\frac{1}{2}(x + 1)$, $x \neq -1$;
$(g \circ f)(x) = (2 - x)/x$, $x \neq 0$, $x \neq 1$;
$(f \circ f)(x) = (x - 1)/(2 - x)$, $x \neq 1$, $x \neq 2$;
$(g \circ g)(x) = -1/x$, $x \neq -1$, $x \neq 0$

35. $(f \circ g)(x) = \sqrt[3]{1 - \sqrt{x}}$, $[0, \infty)$;
$(g \circ f)(x) = 1 - \sqrt[6]{x}$, $[0, \infty)$; $(f \circ f)(x) = \sqrt[9]{x}$, $(-\infty, \infty)$;
$(g \circ g)(x) = 1 - \sqrt{1 - \sqrt{x}}$, $[0, 1]$

37. $(f \circ g)(x) = (3x - 4)/(3x - 2)$, $x \neq 2$, $x \neq \frac{2}{3}$;
$(g \circ f)(x) = -(x + 2)/(3x)$, $x \neq -\frac{1}{2}$, $x \neq 0$;
$(f \circ f)(x) = (5x + 4)/(4x + 5)$, $x \neq -\frac{1}{2}$, $x \neq -\frac{5}{4}$;
$(g \circ g)(x) = x/(4 - x)$, $x \neq 2$, $x \neq 4$

39. $(f \circ g \circ h)(x) = \sqrt{x - 1} - 1$

41. $(f \circ g \circ h)(x) = (\sqrt{x} - 5)^4 + 1$

43. $g(x) = x - 9$, $f(x) = x^5$

45. $g(x) = x^2$, $f(x) = x/(x + 4)$

47. $g(x) = 1 - x^3$, $f(x) = |x|$

49. $h(x) = x^2$, $g(x) = x + 1$, $f(x) = 1/x$

51. $h(x) = \sqrt[3]{x}$, $g(x) = 4 + x$, $f(x) = x^9$

53. $A(t) = 3600\pi t^2$

55. (a) $s = f(d) = \sqrt{1 + d^2}$
(b) $d = g(t) = 350t$
(c) $s = f(g(t)) = \sqrt{1 + 122{,}500t^2}$

Section 2.7 ■ page 201

1. No   **3.** Yes   **5.** No   **7.** Yes   **9.** Yes
11. No   **13.** (a) 2   (b) 3   **15.** $f^{-1}(3) = 1$
23. $f^{-1}(x) = \frac{1}{2}(x - 1)$
25. $f^{-1}(x) = \frac{1}{4}(x - 7)$
27. $f^{-1}(x) = (1/x) - 2$, $x > 0$
29. $f^{-1}(x) = (5x - 1)/(2x + 3)$
31. $f^{-1}(x) = \frac{1}{5}(x^2 - 2)$, $x \geq 0$
33. $f^{-1}(x) = \sqrt{4 - x}$
35. $f^{-1}(x) = (x - 4)^3$
37. $f^{-1}(x) = x^2 - 2x$, $x \geq 1$
39. $f^{-1}(x) = \sqrt[4]{x}$

41. (a) 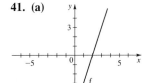   (b)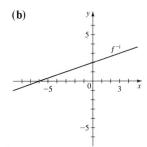

(c) $f^{-1}(x) = \frac{1}{3}(x + 6)$

43. (a) 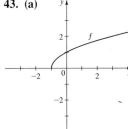   (b)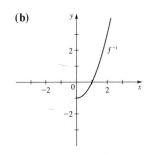

(c) $f^{-1}(x) = x^2 - 1$, $x \geq 0$

45. Not one-to-one   **47.** One-to-one

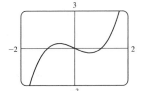   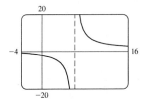

Chapter 2 Review ■ page 204

1. $1, 3, 7, a^2 - a + 1, a^2 + a + 1, x^2 + x + 1$,
$4x^2 - 2x + 1, 2x^2 - 2x$
3. (a) $-1, 2$ (b) $[-4, 5]$ (c) $[-4, 4]$
(d) Increasing on $[-4, -2]$ and $[-1, 4]$;
decreasing on $[-2, -1]$ and $[4, 5]$ (e) No
5. Domain $[-3, \infty)$, range $[0, \infty)$ **7.** $(-\infty, \infty)$
9. $[-4, \infty)$ **11.** $\{x \mid x \neq -2, -1, 0\}$
13. $(-\infty, -1] \cup [1, 4]$
15.

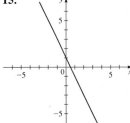

17.

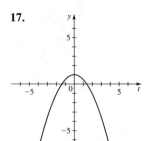

19.

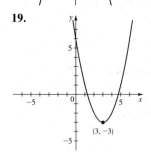

(3, −3)

21.

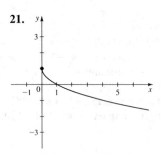

23.

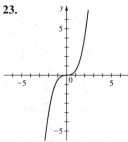

25.

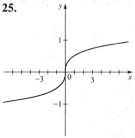

27.

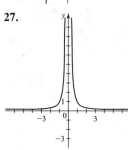

29.

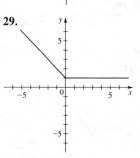

31.

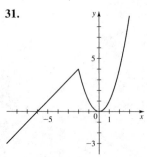

33. (iii)

35.

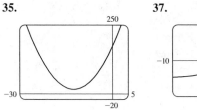

37.

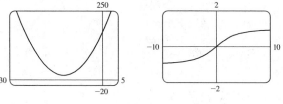

39. $[-2.1, 0.2] \cup [1.9, \infty)$

41. (a) Shift upward 8 units **(b)** Shift left 8 units
(c) Stretch vertically by a factor 2, then shift upward 1 unit
(d) Shift right 2 units and downward 2 units
(e) Reflect in y-axis
(f) Reflect in y-axis, then in x-axis
(g) Reflect in x-axis **(h)** Reflect in line $y = x$
43. (a) Neither **(b)** Odd **(c)** Even **(d)** Neither
45. $f(x) = (x + 2)^2 - 3$ **47.** $g(-1) = -7$
49. Local maximum ≈ 3.79 when $x \approx 0.46$;
local minimum ≈ 2.81 when $x \approx -0.46$
51. (a) $(f + g)(x) = x^2 - 6x + 6$
(b) $(f - g)(x) = x^2 - 2$
(c) $(fg)(x) = -3x^3 + 13x^2 - 18x + 8$
(d) $(f/g)(x) = (x^2 - 3x + 2)/(4 - 3x)$
(e) $(f \circ g)(x) = 9x^2 - 15x + 6$
(f) $(g \circ f)(x) = -3x^2 + 9x - 2$
53. $(f \circ g)(x) = -3x^2 + 6x - 1, (-\infty, \infty)$;
$(g \circ f)(x) = -9x^2 + 12x - 3, (-\infty, \infty)$;
$(f \circ f)(x) = 9x - 4, (-\infty, \infty)$;
$(g \circ g)(x) = -x^4 + 4x^3 - 6x^2 + 4x, (-\infty, \infty)$
55. $(f \circ g \circ h)(x) = 1 + \sqrt{x}$ **57.** Yes **59.** No
61. No **63.** $f^{-1}(x) = \frac{1}{3}(x + 2)$
65. (a), (b)

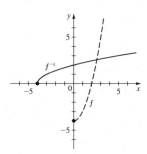

f^{-1}

f

(c) $f^{-1}(x) = \sqrt{x + 4}$
67. $M = 8z$
69. (a) $I = k/d^2$ **(b)** 64,000 **(c)** 160 candles
71. $A = b\sqrt{4 - b}$
73. (a) $A(x) = (x^2/16) + (\sqrt{3}/36)(10 - x)^2$,
$0 \le x \le 10$ **(b)** $40\sqrt{3}/(9 + 4\sqrt{3}) \approx 4.35$ m
75. Square with side 10 m **77.** $ad \ne bc$

Chapter 2 Test ■ page 208

1. (a), (b) are graphs of functions, (a) is one-to-one

2. $[0, 1) \cup (1, \infty)$

3. (a)   (b)

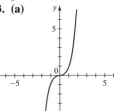
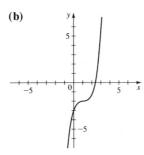

4. (a) Shift left 3 units, reflect in x-axis, then shift upward 2 units

(b) Reflect in y-axis, then in x-axis

5. (a) $f(x) = 2(x - 2)^2 + 5$

(b) $f(2) = 5$

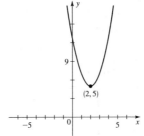

(2, 5)

6. (a) $-3, 3$ (b)

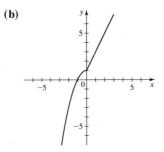

7. (a) $(f \circ g)(x) = 4x^2 - 8x + 2$

(b) $(g \circ f)(x) = 2x^2 + 4x - 5$ (c) 2   (d) 11

(e) $(g \circ g \circ g)(x) = 8x - 21$

8. (a) $f^{-1}(x) = 3 - x^2, x \geqslant 0$ (b)

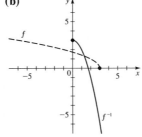

f

f^{-1}

9. (a) $A(x) = 400x - 2x^2$ (b) 100 ft

10. (a)

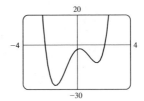

(b) No

(c) Local minimum ≈ -27.18 when $x \approx -1.61$; local maximum ≈ -2.55 when $x \approx 0.18$; local minimum ≈ -11.93 when $x \approx 1.43$

(d) $[-27.18, \infty)$

(e) Increasing on $[-1.61, 0.18] \cup [1.43, \infty)$; decreasing on $(-\infty, -1.61] \cup [0.18, 1.43]$

PRINCIPLES OF MODELING ■ page 216

1. (a)

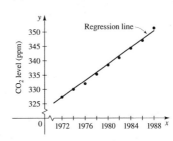

Regression line

CO_2 level (ppm)

1972 1976 1980 1984 1988

(b) $y = 1.4775x - 2586.9$

(c) 353.3 ppm

3. (a)

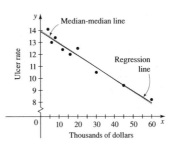

Median-median line

Ulcer rate

Regression line

Thousands of dollars

(b) $y = -0.0995x + 13.9$, x in thousands of dollars

(c) $y = -0.1026x + 14.0$, x in thousands of dollars

(d) 11.4 per 100 population

(e) 5.8 per 100 population

5. (a)

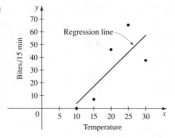

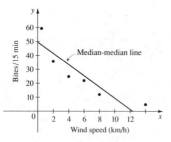

(b) $y = 2.6768x - 22.19$
(c) $y = -4.0103x + 49.5$
(d) 0.77 is the correlation coefficient for the temperature-bites data, so a linear model is not appropriate; -0.89 is the correlation coefficient for the wind/bites data, so a linear model is more suitable in this case.

7. (a)

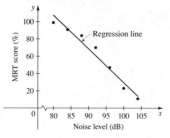

(b) $y = -3.9018x + 419.7$
(c) The correlation coefficient is -0.98, so a linear model is appropriate.
(d) 53%

9. (a)

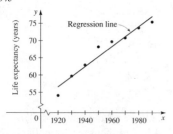

(b) $y = 0.29083x - 501.8$
(c) 79.9 years

CHAPTER 3

Section 3.1 ■ page 230

1.

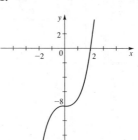

3.

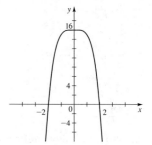

5.

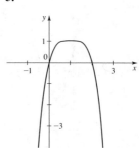

7.

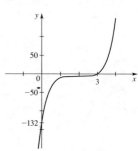

9.

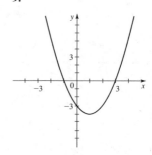

11.

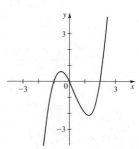

13.

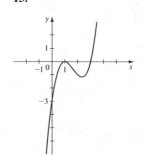

15.

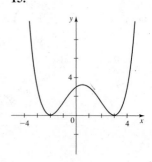

17.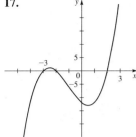

19.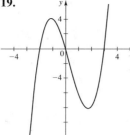

21.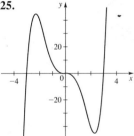

23.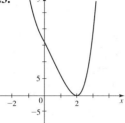

25.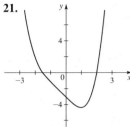

27. III   **29.** IV

31.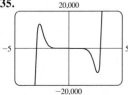

$y \to \infty$ as $x \to \infty$
$y \to -\infty$ as $x \to -\infty$

33.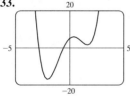

$y \to \infty$ as $x \to \pm\infty$

35.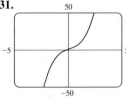

$y \to \infty$ as $x \to \infty$
$y \to -\infty$ as $x \to -\infty$

37.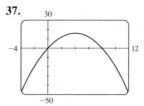

x-intercepts 0, 8
y-intercept 0
local maximum $(4, 16)$

39.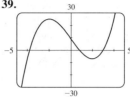

x-intercepts -3.79, 0.79, 3
y-intercept 9
local maximum $(-2, 25)$
local minimum $(2, -7)$

41.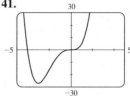

x-intercepts -4, 0
y-intercept 0
local minimum $(-3, -27)$

43.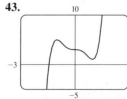

x-intercept -1.42
y-intercept 3
local maximum $(-1, 5)$
local minimum $(1, 1)$

45. Local maximum $(0.75, 6.13)$
47. Local maximum $(-0.33, 0.19)$,
local minimum $(1.00, -1.00)$
49. Local maximum $(0, 4)$;
local minima $(-1.58, -2.25)$, $(1.58, -2.25)$
51. No extrema
53. Local maximum: $(0.44, 0.33)$;
local minima $(1.09, -1.15)$, $(-1.12, -3.36)$
55.

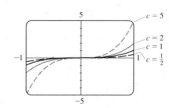

Increasing the value of c stretches the graph vertically.

57.

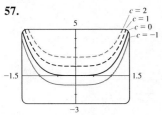

Increasing the value of c moves the graph up.

59.

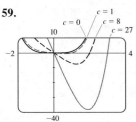

Increasing the value of c causes a deeper dip in the graph in the fourth quadrant and moves the positive x-intercept to the right.

61. (a)

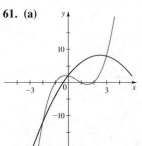

(b) Three
(c) $(0, 2), (3, 8), (-2, -12)$

63. (g) $P(x) = P_O(x) + P_E(x)$, where
$P_O(x) = x^5 + 6x^3 - 2x$ and $P_E(x) = -x^2 + 5$

65.

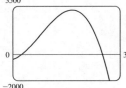

(a) 26 blenders
(b) Maximum profit is $3,276.22 when producing 166 blenders

67. (a) Local maximum $(1.8, 2.1)$,
local minimum $(3.5, -0.6)$
(b) Local maximum $(1.8, 7.1)$, local minimum $(3.5, 4.4)$

Section 3.2 ■ page 245

In answers 1–11, the first polynomial given is the quotient and the second is the remainder.

1. $x^2 + 2, -3$ **3.** $x^2 - 3x + 1, -1$
5. $x^4 + x^3 + 4x^2 + 4x + 4, -2$ **7.** $x + 2, 8x - 1$
9. $3x + 1, 7x - 5$ **11.** $2x^2 + 4x, 1$ **13.** -3
15. 12 **17.** -483 **19.** $\frac{7}{3}$ **21.** $\pm 1, \pm 3$
23. $\pm 1, \pm 2, \pm 3, \pm 6, \pm \frac{1}{2}, \pm \frac{3}{2}$ **25.** $-2, 2, 3$
27. $-1, 2, 3$ **29.** $1, 2, 4$ **31.** -1 **33.** $\pm 1, \pm 2$
35. $1, -1, -2, -4$ **37.** $\pm 2, \pm \frac{3}{2}$ **39.** $-\frac{3}{2}, \frac{1}{2}, 1$

41. $-1, \pm \frac{1}{2}$ **43.** $-1, \pm 2, 3$ **47.** $x^3 - 3x^2 - x + 3$
49. $x^4 - 8x^3 + 14x^2 + 8x - 15$
51. 1 positive, 2 or 0 negative; 3 or 1 real
53. 0 positive, 4, 2, or 0 negative; 4, 2, or 0 real
55. 5, 3, or 1 positive, 2 or 0 negative; 7, 5, 3, or 1 real
61. $3, -2$ **63.** $3, -1$ **65.** $-2, \frac{1}{2}, \pm 1$
67. $\pm \frac{1}{2}, \pm \sqrt{5}$ **69.** $-2, 1, 3, 4$ **71.** $-2, 2, 3$
73. $-\frac{3}{2}, -1, 1, 4$ **75.** 5.15 **77.** 0.67
79. $-1.00, 1.50$ **81.** -1.50 **83.** 11.3 ft
85. (a) It began to snow again. **(b)** No
(c) Just before midnight on Saturday night
87. 2.76 m **89.** 88 in. (or 3.21 in.)

Section 3.3 ■ page 254

1. Real part 3, imaginary part -5
3. Real part 0, imaginary part 6
5. Real part $\sqrt{2}$, imaginary part $\sqrt{3}$
7. $9 + i$ **9.** $12 + i$ **11.** $-19 + 4i$ **13.** $-4 + 8i$
15. $30 + 10i$ **17.** $-33 - 56i$ **19.** $-i$ **21.** $\frac{8}{5} + \frac{1}{5}i$
23. $-5 + 12i$ **25.** $-4 + 2i$ **27.** $-i$ **29.** 1
31. $5i$ **33.** -6 **35.** $(3 + \sqrt{5}) + (3 - \sqrt{5})i$ **37.** 2
39. $-i\sqrt{2}$
41. 3 **43.** $\sqrt{29}$

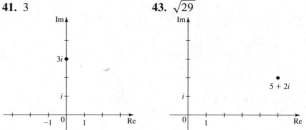

45. 2 **47.** 1

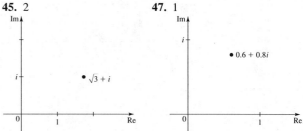

49. **51.**

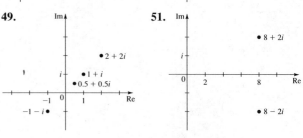

53.

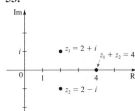

55.

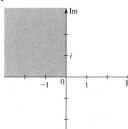

57.

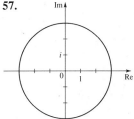

59.

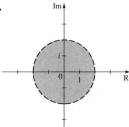

61.

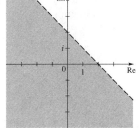

41. $(x - 2)(x + 1 + i\sqrt{2})(x + 1 - i\sqrt{2})$
43. $(x - 2)(x + 1)(x - 3i)(x + 3i)$
45. 0 positive, 2 or 0 negative; 4 or 2 imaginary
47. 1 positive, 2 or 0 negative; 4 or 2 imaginary
49. (a) 4 real　**(b)** 2 real, 2 imaginary
(c) 4 imaginary

Section 3.5 ■ page 274

1. x-intercept 6, y-intercept -6
3. x-intercept 0, y-intercept 0
5. x-intercept 3, y-intercept 3
vertical $x = 2$; horizontal $y = 2$
7. Vertical $x = -3$; horizontal $y = 0$
9. Vertical $x = 3$, $x = -2$; horizontal $y = 1$
11. Horizontal $y = 0$
13. Vertical $x = 1$; slant $y = x + 1$
15. Vertical $x = -3$

17.

19.

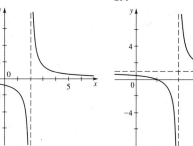

21.

23.

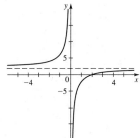

25.

27.

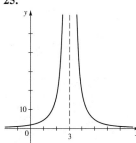

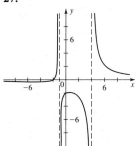

Section 3.4 ■ page 261

1. $\pm 4i$　　**3.** $-1 \pm i$　　**5.** $-2 \pm 2i$　　**7.** $\dfrac{5 \pm i\sqrt{23}}{6}$

9. $4 \pm i$　　**11.** $\dfrac{-3 \pm i\sqrt{3}}{2}$　　**13.** $\pm 1, \pm i$

15. $-2, 1 \pm i\sqrt{3}$　　**17.** $\pm 2, \pm 2i$　　**19.** $\pm 3, \dfrac{\pm 3 \pm 3\sqrt{3}\,i}{2}$

21. $\pm i\sqrt{5}$　　**23.** $x^3 - 2x^2 + x - 2$
25. $x^4 - 4x^3 + 10x^2 - 12x + 5$
27. $6x^4 - 12x^3 + 18x^2 - 12x + 12$　　**29.** $-2, \pm 2i$
31. $1, \dfrac{1 \pm i\sqrt{3}}{2}$　　**33.** $2, \dfrac{1 \pm i\sqrt{3}}{2}$　　**35.** $-2, 1, \pm 3i$

37. $(x + 3)\left(x - \dfrac{3 + 3\sqrt{3}\,i}{2}\right)\left(x - \dfrac{3 - 3\sqrt{3}\,i}{2}\right)$

39. $(x + 2)(x - 1)(x - 1 + i\sqrt{3})(x - 1 - i\sqrt{3}) \cdot$
$\left(x + \dfrac{1 + i\sqrt{3}}{2}\right)\left(x + \dfrac{1 - i\sqrt{3}}{2}\right)$

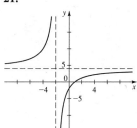

29.

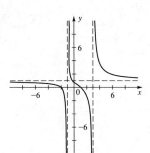

31.

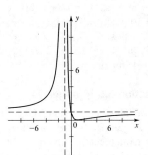

33.

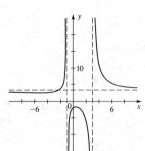

35.

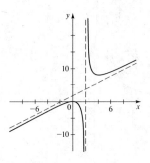

37.

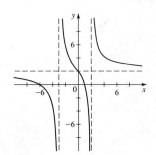

39.

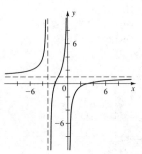

41.

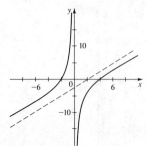

43.

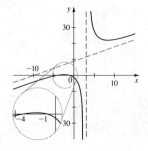

45.

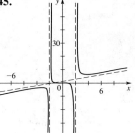

47.

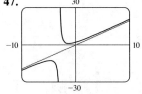

vertical $x = -3$

49.

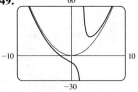

vertical $x = 2$
vertical $x = -1.5$
x-intercepts 0, 2.5
y-intercept 0
local maximum $(-3.9, -10.4)$
local minimum $(0.9, -0.6)$
end behavior: $y = x - 4$

51.

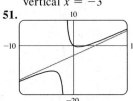

53.

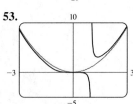

vertical $x = 1$
x-intercept 0
y-intercept 0
local minimum $(1.4, 3.1)$
end behavior: $y = x^2$

57.

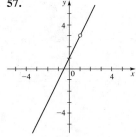

59.

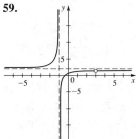

61. (a)

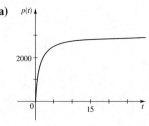

(b) It levels off at 3000.

63. (a) 2.50 mg/L   **(b)** It decreases to 0.   **(c)** 16.61 h

65.

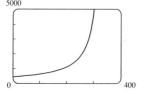

If the speed of the train approaches the speed of sound, then the pitch increases indefinitely (a sonic boom).

Chapter 3 Review ■ page 277

1.

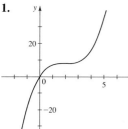

3.

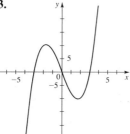

5.

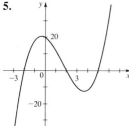

7.

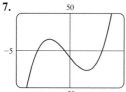

x-intercepts: $-3, -0.5, 3$
y-intercept: -9
local maximum: $(-1.9, 15.1)$
local minimum: $(1.6, -27.1)$
$y \to \infty$ as $x \to \infty$
$y \to -\infty$ as $x \to -\infty$

9.

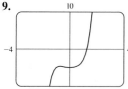

x-intercept: 1.3
y-intercept: -5
local maximum: $(-0.7, -4.7)$
local minimum: $(0, -5)$
$y \to \infty$ as $x \to \infty$
$y \to -\infty$ as $x \to -\infty$

In answers 11–17, the first polynomial given is the quotient and the second is the remainder.

11. $x^2 + 2x + 7, 10$ **13.** $x - 3, -9$

15. $x^3 - 5x^2 + 4, -5$

17. $x^3 + (\sqrt{3} + 1)x^2 + (\sqrt{3} + 1)x + \sqrt{3}, 2$

19. 3   **23.** 8   **25.** $\pm 1, \pm 2, \pm 3, \pm 6, \pm 9, \pm 18$

27. $-3 - 9i$ **29.** $19 + 40i$ **31.** $(-5 - 12i)/13$

33. i **35.** $(2 + 2\sqrt{3}) + (2 - 2\sqrt{3})i$

37. $4x^3 - 18x^2 + 14x + 12$

39. No; since the complex conjugates of imaginary zeros will also be zeros, the polynomial would have 8 zeros, contradicting the requirement that it have degree 4.

41. $-3, 1, 5$ **43.** $-1 \pm 2i, -2$ (multiplicity 2)

45. $\pm 2, 1$ (multiplicity 3)   **47.** $\pm 2, \pm 1 \pm i\sqrt{3}$

49. $1, 3, \dfrac{-1 \pm i\sqrt{7}}{2}$

51. $x = -0.5, 3$ **53.** $x \approx -0.24, 4.24$

55.

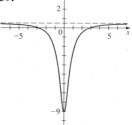

57.

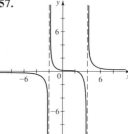

59.

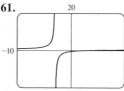

61.

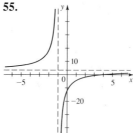

x-intercept 3
y-intercept -0.5
vertical $x = -3$
horizontal $y = 0.5$
no local extrema

63.

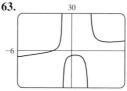

x-intercept -2
y-intercept -4
vertical $x = -1, x = 2$
slant $y = x + 1$
local maximum
$(0.425, -3.599)$
local minimum $(4.216, 7.175)$

Chapter 3 Test ▪ page 280

1.

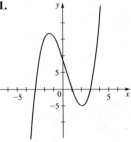

2. quotient: $x^3 + 2x^2 + 2$
remainder: 9
3. (a) $\pm\frac{1}{2}$, ±1, $\pm\frac{3}{2}$, ±3
(b) $2\left(x - \frac{1}{2}\right)(x - 1)(x + 1)(x - 3)$ **(c)** $-1, \frac{1}{2}, 1, 3$
4. (a) $\frac{6}{13} - \frac{22}{13}i$ **(b)** $2 - 11i$ **(c)** i
5. $-1, 2, -1 \pm i$
6. (a) For P and Q: by Rational Zeros Theorem;
For R: by Descartes' Rule   **(b)** No; by Descartes' Rule
(c) Two (one positive, one negative); by Descartes' Rule
(d) Only possible rational zeroes are 1 and -1, neither of
which is a zero
7. $x^4 + 2x^2 + 8x + 5$
8. (a) 4, 2, or 0 positive real root(s); 0 negative real root
9. (a) r, u **(b)** s **(c)** s **(d)**

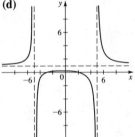

10.

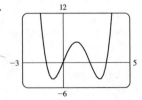

x-intercepts: $-1.24, 0, 2, 3.24$
local maximum: $(1, 5)$
local minima: $(-0.73, -4)$, $(2.73, -4)$

FOCUS ON PROBLEM SOLVING ▪ page 283

1. Not traversable   **3.** Traversable
5. Yes; tour is possible for odd n.   **7.** Inside
9. -36 **11.** No   **13.** $\frac{2}{3}$

15.

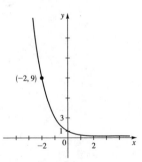

CHAPTER 4

Section 4.1 ▪ page 292

1.

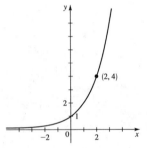

3.

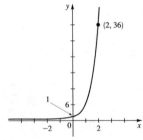

5.

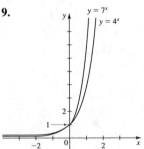

7.

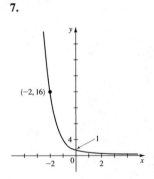

9.

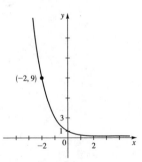

11. $f(x) = 3^x$ **13.** $f(x) = \left(\frac{1}{4}\right)^x$
15. III   **17.** I   **19.** II

21. $\mathbb{R}$, $(-\infty, 0)$, $y = 0$

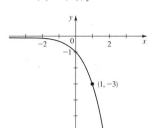

23. $\mathbb{R}$, $(-3, \infty)$, $y = -3$

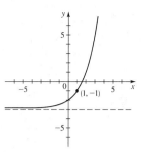

39. (a)

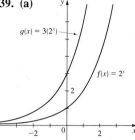

(b) The graph of g is steeper than that of f.

25. $\mathbb{R}$, $(4, \infty)$, $y = 4$

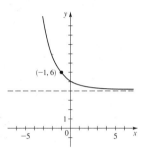

27. $\mathbb{R}$, $(0, \infty)$, $y = 0$

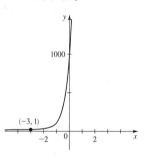

43. (a)
(i)

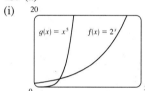

(ii)

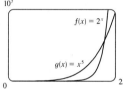

(iii)

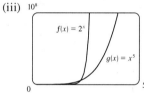

The graph of f ultimately increases much more quickly than g.

(b) 1.2, 22.4

29. $\mathbb{R}$, $(-\infty, 0)$, $y = 0$

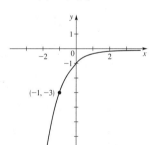

31. $\mathbb{R}$, $(0, \infty)$, $y = 0$

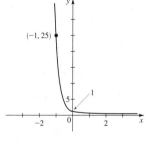

45.

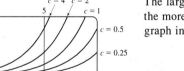

The larger the value of c, the more rapidly the graph increases.

33. $\mathbb{R}$, $(-\infty, 5)$, $y = 5$

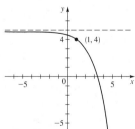

35. $\mathbb{R}$, $[1, \infty)$, no asymptote

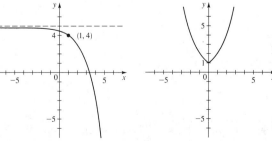

47.

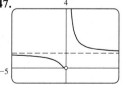

Vertical asymptote $x = 0$, horizontal asymptote $y = 1$
49. Local minimum $\approx (0.37, 0.69)$
51. (a) Increasing on $(-\infty, 0.50]$, decreasing on $[0.50, \infty)$
(b) $(0, 1.78]$

37. $y = 3(2^x)$

Section 4.2 ■ page 301

1.

x	$f(x) = 3e^x$
-2	0.4
-1.5	0.7
-1	1.1
-0.5	1.8
0	3
0.5	4.9
1	8.2
1.5	13.4
2	22.2

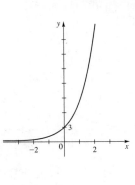

3. $\mathbb{R}, (-\infty, 0), y = 0$

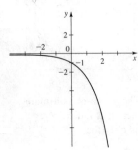

5. $\mathbb{R}, (-1, \infty), y = -1$

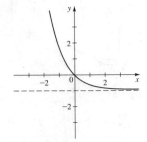

7. $\mathbb{R}, (0, \infty), y = 0$

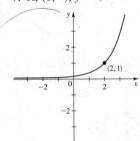

9. (a) \$16,288.95
(b) \$26,532.98
(c) \$43,219.42

11. (a) \$4,615.87 **(b)** \$4,658.91 **(c)** \$4,697.04
(d) \$4,703.11 **(e)** \$4,704.68 **(f)** \$4,704.93
(g) \$4,704.94
13. (i) **15.** \$7,678.96
17. (a) 45% **(b)** 500 **(c)** 4744
19. (a) $n(t) = 18{,}000e^{0.08t}$ **(b)** 34,137
(c)

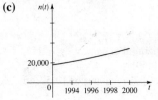

21. (a) 233 million **(b)** 181 million
23. 5.87 billion **25. (a)** About 500 **(b)** 361,000
27. (a) 6 g **(b)** 1 g
29. (a) 2.7 lb **(b)** 4.9 lb
(c) **(d)** 15 lb

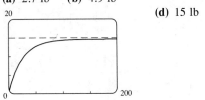

31. (a) 5164 **(b)**

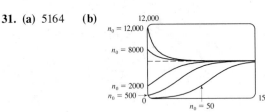

(c) 6000
33.

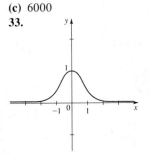

35.

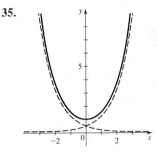

41.

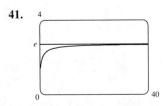

43. (a)

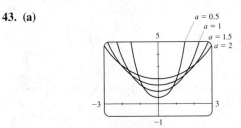

(b) The larger the value of a, the less steep the graph and the higher its minimum value.
45. Local maximum $\approx (1.00, 0.37)$

Section 4.3 ■ page 311

1. (a) $2^5 = 32$ (b) $5^0 = 1$
3. (a) $4^{1/2} = 2$ (b) $2^{-4} = \frac{1}{16}$
5. (a) $e^x = 5$ (b) $e^5 = y$
7. (a) $\log_2 8 = 3$ (b) $\log_{10} 0.001 = -3$
9. (a) $\log_4 0.125 = -\frac{3}{2}$ (b) $\log_7 343 = 3$
11. (a) $\ln 2 = x$ (b) $\ln y = 3$
13. (a) 4 (b) 3 (c) 1 **15.** (a) 2 (b) 2 (c) 10
17. (a) -3 (b) $\frac{1}{2}$ (c) -1
19. (a) 37 (b) 8 (c) $\sqrt{5}$
21. (a) $-\frac{2}{3}$ (b) 4 (c) -1 **23.** (a) 32 (b) 4
25. (a) 100 (b) 25 **27.** (a) 2 (b) 4
29. (a) 0.3010 (b) 1.5465 (c) -0.1761
31. (a) 1.6094 (b) 3.2308 (c) 1.0051
33. $y = \log_5 x$ **35.** $y = \log_9 x$
37. II **39.** III **41.** VI
43.

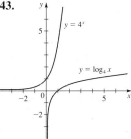

45. $(4, \infty)$, $\mathbb{R}$, $x = 4$ **47.** $(-\infty, 0)$, $\mathbb{R}$, $x = 0$

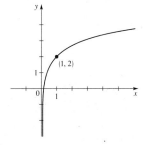

49. $(0, \infty)$, $\mathbb{R}$, $x = 0$ **51.** $(0, \infty)$, $\mathbb{R}$, $x = 0$

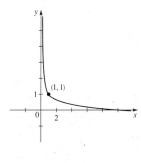

53. $(0, \infty)$, $[0, \infty)$, $x = 0$

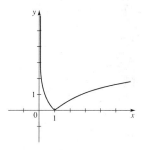

55. $\left(-\frac{2}{5}, \infty\right)$ **57.** $(-\infty, -1) \cup (1, \infty)$ **59.** $(0, 2)$

61.

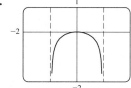

domain $= (-1, 1)$
vertical asymptotes
$x = 1$, $x = -1$
local maximum $(0, 0)$

63.

domain $= (0, \infty)$
vertical asymptote $x = 0$
no maximum or minimum

65.

domain $= (0, \infty)$
vertical asymptote $x = 0$
horizontal asymptote $y = 0$
local maximum
$\approx (2.72, 0.37)$

67. The graph of f grows more slowly than g.
69. (a)

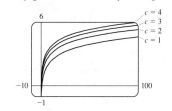

(b) The graph of $f(x) = \log(cx)$ is the graph of $f(x) = \log x$ shifted upward $\log c$ units.
71. (a) $(1, \infty)$ (b) $f^{-1}(x) = 10^{2^x}$
73. (a) $f^{-1}(x) = \log_2\left(\dfrac{x}{1 - x}\right)$ (b) $(0, 1)$

Section 4.4 ■ page 317

1. $\log_2 x + \log_2(x - 1)$ **3.** $23 \log 7$
5. $\log_2 A + 2 \log_2 B$ **7.** $\log_3 x + \frac{1}{2} \log_3 y$
9. $\frac{1}{3} \log_5(x^2 + 1)$ **11.** $\frac{1}{2}(\ln a + \ln b)$
13. $3 \log x + 4 \log y - 6 \log z$
15. $\log_2 x + \log_2(x^2 + 1) - \frac{1}{2} \log_2(x^2 - 1)$
17. $\ln x + \frac{1}{2}(\ln y - \ln z)$ **19.** $\frac{1}{4} \log(x^2 + y^2)$
21. $\frac{1}{2}[\log(x^2 + 4) - \log(x^2 + 1) - 2 \log(x^3 - 7)]$
23. $\frac{1}{2} \ln x + 4 \ln z - \frac{1}{3} \ln(y^2 + 6y + 17)$
25. $\frac{3}{2}$ **27.** 1   **29.** 3   **31.** $\ln 8$ **33.** 16

35. $\log_3 160$ **37.** $\log_2(AB/C^2)$ **39.** $\log\left[\dfrac{x^4(x - 1)^2}{\sqrt[3]{x^2 + 1}}\right]$
41. $\ln[5x^2(x^2 + 5)^3]$
43. $\log[\sqrt[3]{2x + 1}\,\sqrt{(x - 4)/(x^4 - x^2 - 1)}\,]$
45. 2.807355   **47.** 2.182658   **49.** 0.655407
51. 4.165458
53.

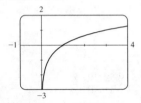

Section 4.5 ■ page 323

1. 1.7227   **3.** −0.5850   **5.** 1.2040   **7.** 0.0767
9. 0.2524   **11.** 1.9349   **13.** −43.0677
15. 2.1492   **17.** 6.2126   **19.** −2.9469
21. −2.4423   **23.** 14.0055   **25.** ±1   **27.** $0, \frac{4}{3}$
29. $\ln 2 \approx 0.6931, 0$ **31.** $\frac{1}{2} \ln 3 \approx 0.5493$
33. $e^{10} \approx 22026$ **35.** 0.01   **37.** $\frac{95}{3}$
39. $3 - e^2 \approx -4.3891$ **41.** 5   **43.** 5   **45.** $\frac{13}{12}$
47. 6   **49.** $\frac{3}{2}$ **51.** $1/\sqrt{5} \approx 0.4472$ **53.** 13 days
55. (a) $t = -\frac{5}{13} \ln\left(1 - \frac{13}{60}I\right)$ (b) 0.218 s
57. 2.21   **59.** 0.00, 1.14   **61.** −0.57   **63.** 0.36
65. $2 < x < 4$ or $7 < x < 9$ **67.** $\log 2 < x < \log 5$
69. 101, 1.1   **71.** $\log_2 3 \approx 1.58$

Section 4.6 ■ page 334

1. (a) $12,870.19   (b) 8.24 yr   **3.** 5 yr
5. 8.15 yr   **7.** 8.30%
9. (a) 500   (b) 45%   (c) 1929   (d) 6.66 h
11. (a) $n(t) = 112,000e^{0.04t}$ (b) About 142,000   (c) 2008
13. (a) 20,000   (b) $n(t) = 20,000e^{0.1096t}$
(c) About 48,000   (d) 2004
15. (a) $n(t) = 8600e^{0.1508t}$ (b) About 11,600   (c) 4.6 h
17. (a) 2029   (b) 2049   **19.** 22.85 h
21. (a) $n(t) = 10e^{-0.0231t}$ (b) 1.6 g   (c) 70 yr

23. 16 yr   **25.** 149 h   **27.** 3560 yr
29. (a) 210°F   (b) 153°F   (c) 28 min
31. (a) 137°F   (b) 116 min
33. (a) 2.3   (b) 3.5   (c) 8.3
35. (a) 10^{-3} M   (b) 3.2×10^{-7} M   **37.** $4.8 \leq$ pH ≤ 6.4
39. $\log 20 \approx 1.3$ **41.** Twice as intense   **43.** 8.2
45. 6.3×10^{-3} W/m²   **47.** (b) 106 dB

Chapter 4 Review ■ page 338

1. $\mathbb{R}, (0, \infty), y = 0$ **3.** $\mathbb{R}, (-\infty, 5), y = 5$

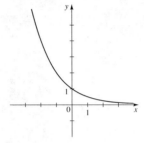   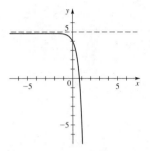

5. $(1, \infty), \mathbb{R}, x = 1$ **7.** $(0, \infty), \mathbb{R}, x = 0$

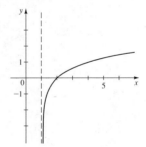   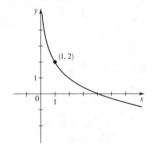

9. $\mathbb{R}, (-1, \infty), y = -1$ **11.** $(0, \infty), \mathbb{R}, x = 0$

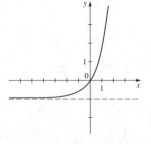   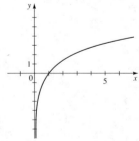

13. $\left(-\infty, \frac{1}{2}\right)$ **15.** $2^{10} = 1024$ **17.** $10^y = x$
19. $\log_2 64 = 6$ **21.** $\log 74 = x$ **23.** 7   **25.** 45
27. 6   **29.** −3   **31.** $\frac{1}{2}$ **33.** 2   **35.** 92   **37.** $\frac{2}{3}$
39. $\log A + 2 \log B + 3 \log C$
41. $\frac{1}{2}[\ln(x^2 - 1) - \ln(x^2 + 1)]$

43. $2 \log_5 x + \frac{3}{2} \log_5(1 - 5x) - \frac{1}{2} \log_5(x^3 - x)$

45. $\log 96$ **47.** $\log_2 \left[\dfrac{(x - y)^{3/2}}{(x^2 + y^2)^2} \right]$ **49.** $\log \left(\dfrac{x^2 - 4}{\sqrt{x^2 + 4}} \right)$

51. -15 **53.** $\frac{1}{3}(5 - \log_5 26) \approx 0.99$

55. $\frac{4}{3} \ln 10 \approx 3.07$ **57.** 3   **59.** $-4, 2$ **61.** 0.430618

63. 2.303600

65.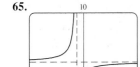

vertical asymptote
$x = -2$
horizontal asymptote
$y = 2.72$
no maximum or minimum

67.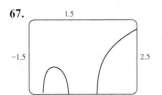

vertical asymptotes
$x = -1, x = 0, x = 1$
local maximum
$\approx (-0.58, -0.41)$

69. 2.42   **71.** $0.16 < x < 3.15$

73. Increasing on $(-\infty, 0]$ and $[1.10, \infty)$, decreasing on $[0, 1.10]$

75. 1.953445   **77.** $\log_4 258$

79. (a) $\$16,081.15$ (b) $\$16,178.18$ (c) $\$16,197.64$
(d) $\$16,198.31$

81. (a) $n(t) = 30e^{0.15t}$ (b) 55   (c) 19 yr

83. (a) 9.97 mg   (b) 1.39×10^5 yr

85. (a) $n(t) = 150e^{-0.0004359t}$ (b) 97.0 mg   (c) 2520 yr

87. (a) $n(t) = 1500e^{0.1515t}$ (b) 7940   **89.** 7.9, basic

91. 8.0

Chapter 4 Test ■ page 341

1.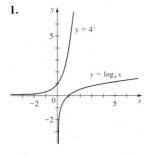

2.

$(-2, \infty)$, $\mathbb{R}$, $x = -2$

3. (a) $\frac{3}{2}$ (b) 3   (c) $\frac{2}{3}$ (d) 2

4. $\frac{1}{2}[\log(x^2 - 1) - 3 \log x - 5 \log(y^2 + 1)]$

5. $\ln \left[\dfrac{x\sqrt{3 - x^4}}{(x^2 + 1)^2} \right]$

6. (a) 4.32   (b) 0.77   (c) 5.39   (d) 2

7. (a) $n(t) = 1000e^{2.07944t}$ (b) 22,627   (c) 1.3 h

(d)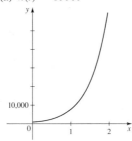

8. 8.33 yr   **9.** $\left(-4, \frac{8}{5}\right)$

10. (a) 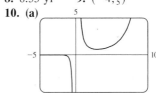   (b) $x = 0, y = 0$

(c) Local minimum $\approx (3.00, 0.74)$
(d) $(-\infty, 0) \cup [0.74, \infty)$ (e) $-0.85, 0.96, 9.92$

FOCUS ON MODELING ■ page 347

1. (a)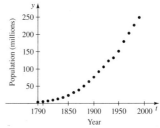

(b) $y = ab^t$, where $a = 4.041807 \times 10^{-16}$ and $b = 1.021003194$, and y is the population in millions in the year t.
(c) 457.9 million   (d) 221.2 million   (e) No

3. (a) Yes
(b) Yes, the scatter plot appears linear.

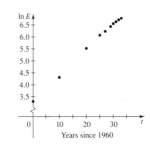

(c) $\ln E = 3.30161 + 0.10769t$, where t is years since 1960 and E is expenditure in billions of dollars.
(d) $E = 27.15633e^{0.10769t}$
(e) 1310.9 billion dollars
5. (a) $y = ab^t$, where $a = 301.813054$, $b = 0.819745$, and t is the number of years since 1970
(b) $y = at^4 + bt^3 + ct^2 + dt + e$, where $a = -0.002430$, $b = 0.135159$, $c = -2.014322$, $d = -4.055294$, $e = 199.092227$, and t is the number of years since 1970
(c) From the graphs we see that the fourth-degree polynomial is a better model.
(d) 202.8, 27.8; 184.0, 43.5

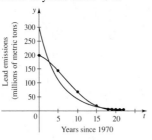

Lead emissions (millions of metric tons)
Years since 1970

CHAPTER 5

Section 5.1 ■ page 358

5. $P\left(\frac{3}{5}, \frac{4}{5}\right)$ **7.** $P\left(\frac{2}{3}, -\sqrt{5}/3\right)$ **9.** $P\left(\sqrt{2}/3, -\sqrt{7}/3\right)$
11. Quadrant I   **13.** Quadrant II

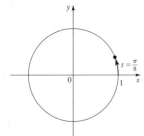

15. Quadrant III   **17.** $\left(\sqrt{2}/2, \sqrt{2}/2\right)$
19. $\left(\frac{1}{2}, -\sqrt{3}/2\right)$
21. $(-1, 0)$

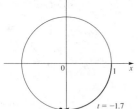

23. $\left(-\frac{1}{2}, \sqrt{3}/2\right)$ **25.** $\left(-\sqrt{2}/2, -\sqrt{2}/2\right)$
27. (a) $\left(-\frac{3}{5}, \frac{4}{5}\right)$ (b) $\left(\frac{3}{5}, -\frac{4}{5}\right)$ (c) $\left(-\frac{3}{5}, -\frac{4}{5}\right)$
(d) $\left(\frac{3}{5}, \frac{4}{5}\right)$

29. (a) $\pi/4$ (b) $\pi/3$ (c) $\pi/3$ (d) $\pi/6$
31. (a) $\pi/5$ (b) $\pi/6$ (c) $\pi/3$ (d) $\pi/6$
33. (a) $\pi/4$ (b) $\left(-\sqrt{2}/2, \sqrt{2}/2\right)$
35. (a) $\pi/3$ (b) $\left(-\frac{1}{2}, -\sqrt{3}/2\right)$
37. (a) $\pi/4$ (b) $\left(-\sqrt{2}/2, -\sqrt{2}/2\right)$
39. (a) $\pi/6$ (b) $\left(-\sqrt{3}/2, -\frac{1}{2}\right)$
41. (a) $\pi/3$ (b) $\left(\frac{1}{2}, \sqrt{3}/2\right)$
43. (a) $\pi/3$ (b) $\left(-\frac{1}{2}, -\sqrt{3}/2\right)$ **45.** $(0.5, 0.8)$
47. $(0.5, -0.9)$

Section 5.2 ■ page 366

1. (a) 0   (b) 1   **3.** (a) 0   (b) -1
5. (a) 1   (b) -1 **7.** (a) 0   (b) 0
9. (a) $\frac{1}{2}$ (b) 2   **11.** (a) $\frac{1}{2}$ (b) $\frac{1}{2}$
13. (a) $\sqrt{3}/3$ (b) $-\sqrt{3}/3$ **15.** (a) 2   (b) $-2\sqrt{3}/3$
17. (a) $\sqrt{2}/2$ (b) $\sqrt{2}$ **19.** (a) -1 (b) -1
21. $\sin 0 = 0$, $\cos 0 = 1$, $\tan 0 = 0$, $\sec 0 = 1$, others undefined
23. $\sin \pi = 0$, $\cos \pi = -1$, $\tan \pi = 0$, $\sec \pi = -1$, others undefined
25. $\frac{4}{5}, \frac{3}{5}, \frac{4}{3}$ **27.** $-\sqrt{13}/7, \frac{6}{7}, -\sqrt{13}/6$ **29.** $\frac{9}{41}, \frac{40}{41}, \frac{9}{40}$
31. $-\frac{12}{13}, -\frac{5}{13}, \frac{12}{5}$ **33.** (a) 0.8   (b) 0.84147
35. (a) 0.9   (b) 0.93204   **37.** (a) 1   (b) 1.02964
39. (a) -0.6 (b) -0.57482 **41.** Positive
43. Negative   **45.** II   **47.** II
49. $\sin t = \sqrt{1 - \cos^2 t}$ **51.** $\tan t = (\sin t)/\sqrt{1 - \sin^2 t}$
53. $\sec t = -\sqrt{1 + \tan^2 t}$ **55.** $\tan t = \sqrt{\sec^2 t - 1}$
57. $\tan^2 t = (\sin^2 t)/(1 - \sin^2 t)$
59. $\cos t = -\frac{4}{5}$, $\tan t = -\frac{3}{4}$, $\csc t = \frac{5}{3}$, $\sec t = -\frac{5}{4}$, $\cot t = -\frac{4}{3}$
61. $\sin t = -\frac{3}{5}$, $\cos t = \frac{4}{5}$, $\csc t = -\frac{5}{3}$, $\sec t = \frac{5}{4}$, $\cot t = -\frac{4}{3}$
63. $\sin t = -\sqrt{3}/2$, $\cos t = \frac{1}{2}$, $\tan t = -\sqrt{3}$, $\csc t = -2\sqrt{3}/3$, $\cot t = -\sqrt{3}/3$
65. $\cos t = -\sqrt{15}/4$, $\tan t = \sqrt{15}/15$, $\csc t = -4$, $\sec t = -4\sqrt{15}/15$, $\cot t = \sqrt{15}$
67. Odd   **69.** Odd   **71.** Even   **73.** Neither

Section 5.3 ■ page 380

1.

3.

5.

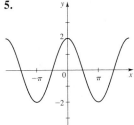

7.

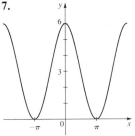

21. 2, 2π, $\pi/6$

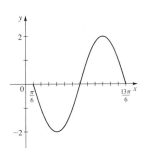

23. 5, $2\pi/3$, $\pi/12$

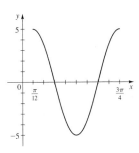

9.

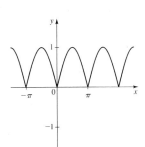

11. 3, $2\pi/3$

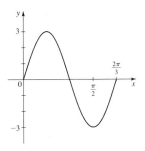

25. 2, 3π, $\pi/4$

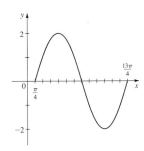

27. 3, 2, $-\frac{1}{2}$

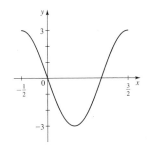

13. 10, 4π

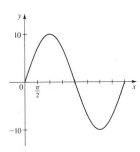

15. 1, 6π

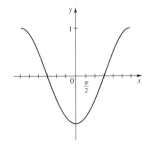

29. $\frac{1}{2}$, π, $\pi/6$

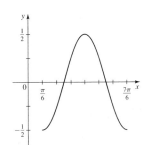

31. 1, $2\pi/3$, $-\pi/3$

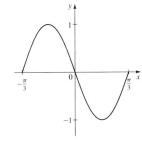

17. 3, $\frac{2}{3}$

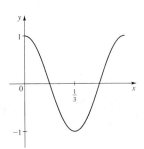

19. 1, 2π, $\pi/2$

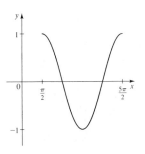

33. (a) 4, 2π, 0   **(b)** $y = 4\sin x$

35. (a) 3, 4π, 0   **(b)** $y = 3\sin\frac{1}{2}x$

37. (a) $\frac{1}{2}$, π, $-\pi/3$ **(b)** $y = -\frac{1}{2}\cos 2(x + \pi/3)$

39.   **41.**

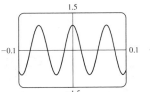

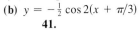

43.

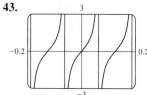

45.

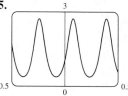

47.

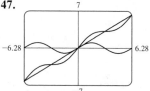

49.

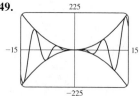

$y = x^2 \sin x$ is a sine curve that lies between the graphs of $y = x^2$ and $y = -x^2$

51.

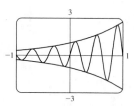

$y = e^x \sin 5\pi x$ is a sine curve that lies between the graphs of $y = e^x$ and $y = -e^x$

53.

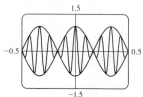

$y = \cos 3\pi x \cos 21\pi x$ is a cosine curve that lies between the graphs of $y = \cos 3\pi x$ and $y = -\cos 3\pi x$

55. (a)

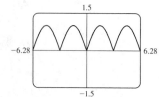

(b) Period π **(c)** Even

57. (a)

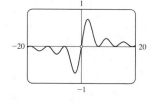

(b) Period 2π **(c)** Neither

59. (a)

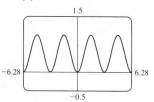

(b) Period π
(c) Even

61. Maximum value 1.76 when $x \approx 0.94$, minimum value -1.76 when $x \approx -0.94$ (The same maximum and minimum values occur at infinitely many other values of x.)
63. Maximum value 3.00 when $x \approx 1.57$, minimum value -1.00 when $x \approx -1.57$ (The same maximum and minimum values occur at infinitely many other values of x.)
65. 1.16   **67.** 0.34, 2.80
69. (a) Odd   **(b)** 0, $\pm 2\pi$, $\pm 4\pi$, $\pm 6\pi$, . . .
(c)

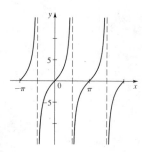

(d) $f(x)$ approaches 0
(e) $f(x)$ approaches 0

Section 5.4 ■ page 388

1. π **3.** π

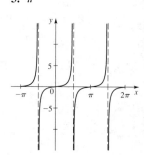

5. π **7.** 2π

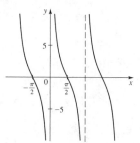

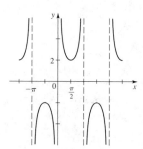

9. 2π

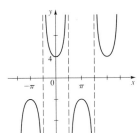

11. π

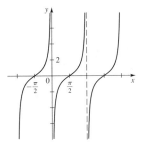

25. π

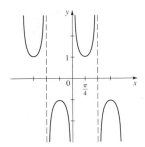

27. $\pi/3$

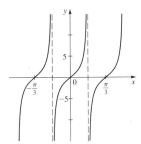

13. 2π

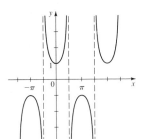

15. π

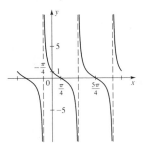

29. $2\pi/3$

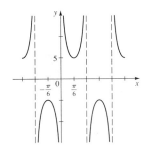

31. $\pi/2$

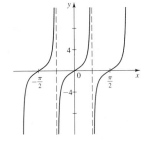

17. 2π

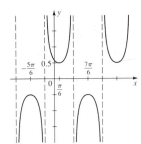

19. $\pi/2$

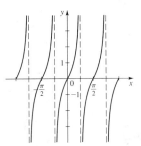

33. $\pi/2$

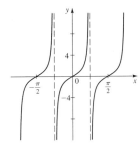

35. $\pi/2$

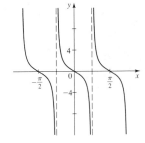

21. 1

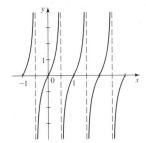

23. π

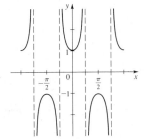

37. 2

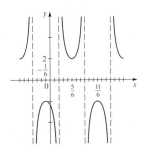

39. $2\pi/3$

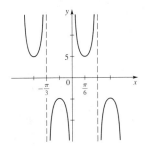

41. $3\pi/2$

43. 2

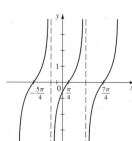

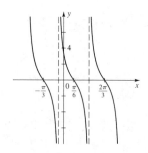

45. $\pi/2$

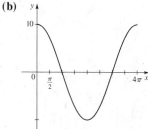

Chapter 5 Review ■ page 390

1. (b) $\sqrt{3}/2, \frac{1}{2}, \sqrt{3}$

3. (a) $\pi/3$ (b) $\left(\frac{1}{2}, \sqrt{3}/2\right)$ (c) $\sin t = \sqrt{3}/2$, $\cos t = \frac{1}{2}$, $\tan t = \sqrt{3}$, $\csc t = 2\sqrt{3}/3$, $\sec t = 2$, $\cot t = \sqrt{3}/3$

5. (a) $\pi/4$ (b) $\left(-\sqrt{2}/2, \sqrt{2}/2\right)$
(c) $\sin t = \sqrt{2}/2$, $\cos t = -\sqrt{2}/2$, $\tan t = -1$, $\csc t = \sqrt{2}$, $\sec t = -\sqrt{2}$, $\cot t = -1$

7. (a) $\sqrt{2}/2$ (b) $-\sqrt{2}/2$

9. (a) 0.89121   (b) 0.45360   **11.** (a) 0   (b) Undefined

13. (a) Undefined   (b) 0

15. (a) 0.41421   (b) 2.41421   **17.** $(\sin t)/(1 - \sin^2 t)$

19. $(\sin t)/\sqrt{1 - \sin^2 t}$

21. $\tan t = -\frac{5}{12}$, $\csc t = \frac{13}{5}$, $\sec t = -\frac{13}{12}$, $\cot t = -\frac{12}{5}$

23. $\sin t = 2\sqrt{5}/5$, $\cos t = -\sqrt{5}/5$, $\tan t = -2$, $\sec t = -\sqrt{5}$

25. $\left(16 - \sqrt{17}\right)/4$ **27.** 3

29. (a) $10, 4\pi, 0$

(b)

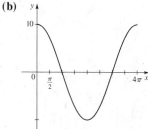

31. (a) $1, 4\pi, 0$

(b)

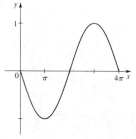

33. (a) $3, \pi, 1$
(b)

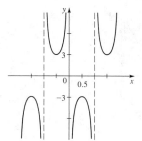

35. (a) $1, 4, -\frac{1}{3}$
(b)

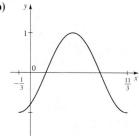

37. $y = 5 \sin 4x$

39. $y = \frac{1}{2} \sin 2\pi\left(x + \frac{1}{3}\right)$

41. π

43. π

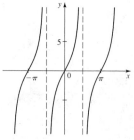

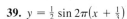

45. π

47. 2π

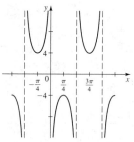

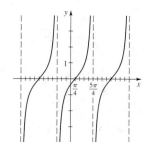

49. (a)

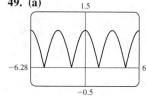

51. (a)

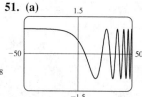

(b) Period π
(c) Even

(b) Not periodic
(c) Neither

53. (a)

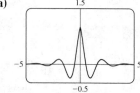

(b) Not periodic
(c) Even

55.

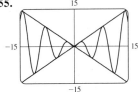

$y = x \sin x$ is a sine function whose graph lies between those of $y = x$ and $y = -x$

57.

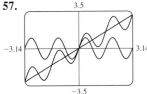

The graphs are related by graphical addition.

59. 1.76, −1.76   **61.** 0.30, 2.84
63. (a) Odd   **(b)** 0, ±π, ±2π, . . .
(c)

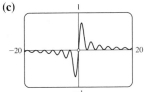

(d) $f(x)$ approaches 0
(e) $f(x)$ approaches 0

Chapter 5 Test ■ page 392

1. $y = -\frac{5}{13}$ **2. (a)** $\frac{4}{5}$ **(b)** $-\frac{3}{5}$ **(c)** $\frac{4}{3}$ **(d)** $\frac{5}{3}$
3. $\tan t = -(\sin t)/\sqrt{1 - \sin^2 t}$ **4.** $-\frac{2}{15}$
5. (a) 2, 2π/3, 0   **(b)**

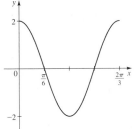

6. (a) 1, 4π, π/3

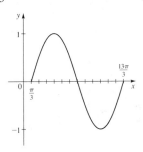

7. π

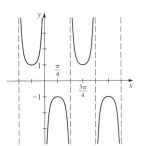

8. π/2

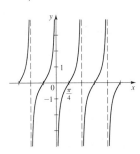

9. $y = 2 \sin 2(x + \pi/3)$
10. (a)

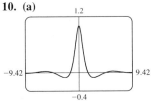

(b) Even
(c) Minimum value −0.11
when $x \approx \pm 2.54$,
maximum value 1
when $x = 0$

11. (a)

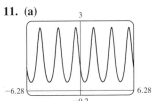

(b) Neither
(c) Minimum value 0.37,
maximum value 2.72
(Both values are attained
at infinitely many values
of x.)

12. (a)

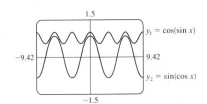

(b) y_1 has period π, y_2 has period 2π
(c) $\sin(\cos x) < \cos(\sin x)$

FOCUS ON MODELING ■ page 401

1. $d(t) = 5 \sin 5\pi t$

3. $y = 21 \sin\left(\dfrac{\pi}{6} t\right)$

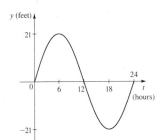

5. $y = 5\cos 2\pi t$ **7.** $y = 11 + 10\sin\left(\dfrac{\pi t}{10}\right)$

9. $y = 3.8 + 0.2\sin\left(\dfrac{\pi}{5}\,t\right)$ **11.** $E(t) = 155\cos 120\pi t$

13. (a) $45V$ **(b)** 40 **(c)** 40 **(d)** $V(t) = 45\cos 80\pi t$

15. $f(t) = e^{-0.9t}\sin \pi t$ **17.** $c = \frac{1}{3}\ln 4 \approx 0.46$

CHAPTER 6

Section 6.1 ■ page 413

1. $2\pi/9 \approx 0.698$ rad   **3.** $2\pi/5 \approx 1.257$ rad
5. $\pi/4 \approx 0.785$ rad   **7.** $17\pi/4 \approx 13.352$ rad
9. $\pi/5 \approx 0.628$ rad   **11.** $-630°$
13. $360°/\pi \approx 114.6°$ **15.** $40°$ **17.** $36°$
19. $660°, 1020°, -60°, -420°$
21. $11\pi/4, 19\pi/4, -5\pi/4, -13\pi/4$
23. $7\pi/4, 15\pi/4, -9\pi/4, -17\pi/4$
25. Yes   **27.** Yes   **29.** Yes
31. $13°$ **33.** $63°$ **35.** $280°$ **37.** $2\pi/5$
39. π **41.** $\pi/4$ **43.** $55\pi/9 \approx 19.2$ **45.** 4
47. 4 mi   **49.** 2 rad $\approx 114.6°$ **51.** $36/\pi \approx 11.459$ m
53. $330\pi \approx 1037$ mi   **55.** 1.6 million mi   **57.** 1.15 mi
59. 50 m^2 **61.** 4 m   **63.** 6 cm^2

Section 6.2 ■ page 422

1. $\sin\theta = \frac{4}{5}$, $\cos\theta = \frac{3}{5}$, $\tan\theta = \frac{4}{3}$, $\csc\theta = \frac{5}{4}$, $\sec\theta = \frac{5}{3}$, $\cot\theta = \frac{3}{4}$
3. $\sin\theta = \frac{4}{7}$, $\cos\theta = \sqrt{33}/7$, $\tan\theta = 4\sqrt{33}/33$, $\csc\theta = \frac{7}{4}$, $\sec\theta = 7\sqrt{33}/33$, $\cot\theta = \sqrt{33}/4$
5. (a) $2\sqrt{13}/13, 2\sqrt{13}/13$ **(b)** $\frac{2}{3}, \frac{2}{3}$ **(c)** $\sqrt{13}/3, \sqrt{13}/3$
7. $\frac{25}{2}$ **9.** $13\sqrt{3}/2$ **11.** 16.51658
13. $x = 28\cos\theta$, $y = 28\sin\theta$
15. $\cos\theta = \frac{4}{5}$, $\tan\theta = \frac{3}{4}$, $\csc\theta = \frac{5}{3}$, $\sec\theta = \frac{5}{4}$, $\cot\theta = \frac{4}{3}$

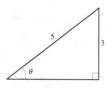

17. $\sin\theta = \sqrt{2}/2$, $\cos\theta = \sqrt{2}/2$,
$\tan\theta = 1$, $\csc\theta = \sqrt{2}$,
$\sec\theta = \sqrt{2}$

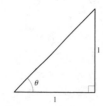

19. $\sin\theta = 4\sqrt{3}/7$, $\cos\theta = \frac{1}{7}$, $\tan\theta = 4\sqrt{3}$,
$\csc\theta = 7\sqrt{3}/12$, $\cot\theta = \sqrt{3}/12$

21. $(1 + \sqrt{3})/2$ **23.** 1 **25.** $\frac{1}{2}$
27.   **29.**

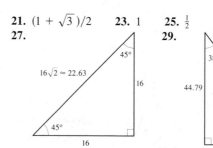

31. $\sin\theta \approx 0.45$, $\cos\theta \approx 0.89$, $\tan\theta \approx 0.50$, $\csc\theta \approx 2.24$, $\sec\theta \approx 1.12$, $\cot\theta \approx 2.00$
33. 1026 ft   **35. (a)** 2100 mi   **(b)** No   **37.** 19 ft
39. $38.7°$ **41.** 345 ft   **43.** 415 ft, 152 ft
45. 2570 ft   **47.** 5808 ft   **49.** 91.7 million mi
51. 3960 mi   **53.** 230.9   **55.** 63.7
57. $a = \sin\theta$, $b = \tan\theta$, $c = \sec\theta$, $d = \cos\theta$

Section 6.3 ■ page 433

1. (a) $45°$ **(b)** $35°$ **(c)** $1°$
3. (a) $25°$ **(b)** $85°$ **(c)** $15°$
5. (a) $\pi/3$ **(b)** $\pi/4$ **(c)** $\pi - 3 \approx 0.14$
7. $\frac{1}{2}$ **9.** $\sqrt{2}/2$ **11.** $-\sqrt{3}/2$ **13.** 1 **15.** $-\sqrt{3}/2$
17. $\sqrt{3}/3$ **19.** $\sqrt{3}/2$ **21.** -1 **23.** $\frac{1}{2}$ **25.** 2
27. -1 **29.** undefined   **31.** III   **33.** IV
35. $\tan\theta = -\sqrt{1 - \cos^2\theta}/\cos\theta$ **37.** $\cos\theta = \sqrt{1 - \sin^2\theta}$
39. $\sec\theta = -\sqrt{1 + \tan^2\theta}$
41. $\cos\theta = -\frac{4}{5}$, $\tan\theta = -\frac{3}{4}$, $\csc\theta = \frac{5}{3}$, $\sec\theta = -\frac{5}{4}$, $\cot\theta = -\frac{4}{3}$
43. $\sin\theta = -\frac{3}{5}$, $\cos\theta = \frac{4}{5}$, $\csc\theta = -\frac{5}{3}$, $\sec\theta = \frac{5}{4}$, $\cot\theta = -\frac{4}{3}$
45. $\sin\theta = \frac{1}{2}$, $\cos\theta = \sqrt{3}/2$, $\tan\theta = \sqrt{3}/3$, $\sec\theta = 2\sqrt{3}/3$, $\cot\theta = \sqrt{3}$
47. $\sin\theta = 3\sqrt{5}/7$, $\tan\theta = -3\sqrt{5}/2$, $\csc\theta = 7\sqrt{5}/15$, $\sec\theta = -\frac{7}{2}$, $\cot\theta = -2\sqrt{5}/15$
49. (a) $\sqrt{3}/2, \sqrt{3}$ **(b)** $\frac{1}{2}, \sqrt{3}/4$ **(c)** $\frac{3}{4}, 0.88967$
51. 19.1 **53.** $66.1°$

Section 6.4 ■ page 439

1. 21.5   **3.** 134.6   **5.** 44°
7. $\angle C = 114°$, $a \approx 51$, $b \approx 24$
9. $\angle C = 62°$, $a \approx 200$, $b \approx 242$

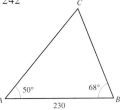

11. $\angle B = 85°$, $a \approx 5$,   **13.** $\angle A = 100°$, $a \approx 89$,
$c \approx 9$ $c \approx 71$

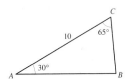 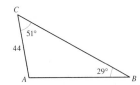

15. $\angle B \approx 30°$, $\angle C \approx 40°$, $c \approx 19$ **17.** No solution
19. $\angle A_1 \approx 125°$, $\angle C_1 \approx 30°$, $a_1 \approx 49$;
$\angle A_2 \approx 5°$, $\angle C_2 \approx 150°$, $a_2 \approx 5.6$
21. No solution   **23.** 219 ft
25. (a) 1018 mi   (b) 1017 mi   **27.** 155 m
29. (a) $d\,\dfrac{\sin \alpha}{\sin(\beta - \alpha)}$ (c) 2350 ft.   **31.** 48.2°

Section 6.5 ■ page 446

1. 13   **3.** 29.89°   **5.** 15
7. $\angle A \approx 39.4°$, $\angle B \approx 20.6°$, $c \approx 24.6$
9. $\angle A \approx 48°$, $\angle B \approx 79°$, $c \approx 3.2$
11. $\angle A \approx 50°$, $\angle B \approx 73°$, $\angle C \approx 57°$
13. $\angle A_1 \approx 83.6°$, $\angle C_1 \approx 56.4°$, $a_1 \approx 193$;
$\angle A_2 \approx 16.4°$, $\angle C_2 \approx 123.6$, $a_2 \approx 54.9$
15. No such triangle   **17.** 2   **19.** 25.4   **21.** 84.6°
23. 24.3   **25.** 2.30 mi   **27.** 23.1 mi   **29.** 2179 mi
31. 96°   **33.** 211 ft   **35.** 3835 ft   **37.** 3.85 cm^2
39. 14.3 m   **41.** $165,554

Chapter 6 Review ■ page 449

1. (a) $7\pi/18 \approx 1.22$ (b) $7\pi/3 \approx 7.33$
(c) $-4\pi/3 \approx -4.19$ (d) $-2\pi/9 \approx -0.70$
3. (a) 630°   (b) $-60°$ (c) 315°   (d) $(378/\pi)° \approx 120.3°$
5. 8 m   **7.** 82 ft   **9.** 0.619 rad $\approx 35.4°$
11. 18,151 ft^2
13. $\sin \theta = 5/\sqrt{74}$, $\cos \theta = 7/\sqrt{74}$, $\tan \theta = \frac{5}{7}$,
$\csc \theta = \sqrt{74}/5$, $\sec \theta = \sqrt{74}/7$, $\cot \theta = \frac{7}{5}$
15. $x \approx 3.83$, $y \approx 3.21$ **17.** $x \approx 2.92$, $y \approx 3.11$

19.

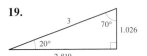

21. $a = \cot \theta$, $b = \csc \theta$
23. 48 m   **25.** 1076 mi
27. $-\sqrt{2}/2$ **29.** 1
31. $-\sqrt{3}/3$ **33.** $-\sqrt{2}/2$ **35.** $2\sqrt{3}/3$ **37.** $-\sqrt{3}$
39. $\sin \theta = \frac{12}{13}$, $\cos \theta = -\frac{5}{13}$, $\tan \theta = -\frac{12}{5}$, $\csc \theta = \frac{13}{12}$,
$\sec \theta = -\frac{13}{5}$, $\cot \theta = -\frac{5}{12}$
41. 60°   **43.** $\tan \theta = -\sqrt{1 - \cos^2\theta}/\cos \theta$
45. $\tan^2\theta = \sin^2\theta/(1 - \sin^2\theta)$
47. $\sin \theta = \sqrt{7}/4$, $\cos \theta = \frac{3}{4}$, $\csc \theta = 4\sqrt{7}/7$,
$\cot \theta = 3\sqrt{7}/7$
49. $\cos \theta = -\frac{4}{5}$, $\tan \theta = -\frac{3}{4}$, $\csc \theta = \frac{5}{3}$, $\sec \theta = -\frac{5}{4}$,
$\cot \theta = -\frac{4}{3}$
51. $-\sqrt{5}/5$ **53.** 1   **55.** 5.32   **57.** 148.07
59. 77.82   **61.** 77.3 mi   **63.** 3.9 mi   **65.** 119.2 m
67. 14.98

Chapter 6 Test ■ page 453

1. $5\pi/3$, $-\pi/10$ **2.** 150°, 137.5°
3. (a) $\sqrt{2}/2$ (b) $\sqrt{3}/3$ (c) 2   (d) 1
4. $(26 + 6\sqrt{13})/39$ **5.** $a = 24 \sin \theta$, $b = 24 \cos \theta$
6. $(4 - 3\sqrt{2})/4$ **7.** $-\frac{13}{12}$ **8.** $\tan \theta = -\sqrt{\sec^2\theta - 1}$
9. 19.6 ft   **10.** 9.1   **11.** 250.5   **12.** 8.4   **13.** 19.5
14. 15.3 m^2 **15.** 24.3 m   **16.** 129.9°   **17.** 44.9
18. 554 ft

FOCUS ON PROBLEM SOLVING ■ page 457

1. Tetrahedron 4, 6, 4; octahedron 8, 12, 6; cube 6, 12, 8;
dodecahedron 12, 30, 20; icosahedron 20, 30, 12
3. Hexagons   **5.** $P + 2\pi$ **7.** $2\left(\dfrac{\pi}{3} - \dfrac{\sqrt{3}}{4}\right) \approx 1.2$
9. 0   **11.** $\sqrt{a^2 + (b + c)^2}$
15. (a) $13 = 2^2 + 3^2$, $41 = 4^2 + 5^2$
(c) $533 = 2^2 + 23^2 = 7^2 + 22^2$

CHAPTER 7

Section 7.1 ■ page 466

1. $\sin x$ **3.** 1   **5.** $\sec u$ **7.** $\tan x$ **9.** $\sin y$
11. $\sin^2 x$ **13.** $\sec x$ **15.** $2 \sec u$ **17.** $\cos^2 x$
19. $\cos \theta$ **83.** $\tan \theta$ **85.** $\tan \theta$ **87.** $3 \cos \theta$
93.   Yes

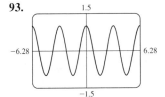

95.

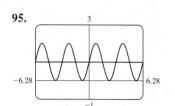

No

Section 7.2 ■ page 473

1. $\left(\sqrt{6} - \sqrt{2}\right)/4$ **3.** $-2 - \sqrt{3}$ **5.** $\left(\sqrt{6} - \sqrt{2}\right)/4$

7. $\sqrt{2}/2$ **9.** $\sqrt{3}$ **33.** $2 \sin\left(x + \dfrac{5\pi}{6}\right)$

35. $5\sqrt{2} \sin\left(2x + \dfrac{7\pi}{4}\right)$

37. $f(x) = \sqrt{2} \sin\left(x + \dfrac{\pi}{4}\right)$

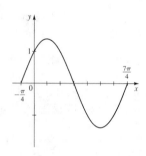

41. $\tan \gamma = \dfrac{17}{6}$

43. (a)

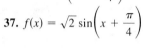

$$\sin^2\left(x + \frac{\pi}{4}\right) + \sin^2\left(x - \frac{\pi}{4}\right) = 1$$

45. (a)

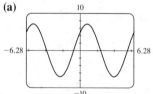

(b) $k = 5\sqrt{2}$, $\theta = \pi/4$

Section 7.3 ■ page 482

1. $\dfrac{120}{169}, \dfrac{119}{169}, \dfrac{120}{119}$ **3.** $-\dfrac{24}{25}, -\dfrac{7}{25}, \dfrac{24}{7}$ **5.** $\dfrac{24}{25}, \dfrac{7}{25}, \dfrac{24}{7}$

7. $\dfrac{1}{2}\left(\dfrac{3}{4} - \cos 2x + \dfrac{1}{4} \cos 4x\right)$

9. $\dfrac{1}{32}\left(\dfrac{3}{4} - \cos 4x + \dfrac{1}{4} \cos 8x\right)$

11. $\dfrac{1}{16}\left(1 - \cos 2x - \cos 4x + \cos 2x \cos 4x\right)$

13. $\dfrac{1}{2}\sqrt{2 - \sqrt{3}}$ **15.** $\dfrac{1}{2}\sqrt{2 + \sqrt{2}}$ **17.** $\dfrac{1}{2}\sqrt{2 - \sqrt{3}}$

19. (a) $\sin 36°$ **(b)** $\sin 6\theta$

21. (a) $\cos 68°$ **(b)** $\cos 10\theta$ **23. (a)** $\tan 4°$ **(b)** $\tan 2\theta$

25. $\sqrt{10}/10, 3\sqrt{10}/10, \dfrac{1}{3}$

27. $\sqrt{(3 + 2\sqrt{2})/6}, \sqrt{(3 - 2\sqrt{2})/6}, 3 + 2\sqrt{2}$

29. $\sqrt{6}/6, -\sqrt{30}/6, -\sqrt{5}/5$

31. $\dfrac{1}{2}(\sin 5x - \sin x)$ **33.** $\dfrac{3}{2}(\cos 11x + \cos 3x)$

35. $2 \sin 4x \cos x$ **37.** $2 \sin 5x \sin x$

39. $-2 \cos \dfrac{9}{2}x \sin \dfrac{5}{2}x$ **41.** $\left(\sqrt{2} + \sqrt{3}\right)/2$

43. $\dfrac{1}{4}\left(\sqrt{2} - 1\right)$ **45.** $\sqrt{2}/2$

69. (a)

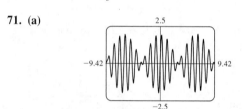

$\dfrac{\sin 3x}{\sin x} - \dfrac{\cos 3x}{\cos x} = 2$

71. (a)

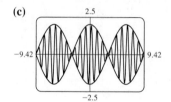

(c)

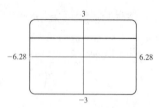

The graph of $y = f(x)$ lies between the two other graphs.

75. (a) $P(t) = 8t^4 - 8t^2 + 1$
(b) $Q(t) = 16t^5 - 20t^3 + 5t$

Section 7.4 ■ page 491

1. (a) $\pi/6$ **(b)** $\pi/3$ **(c)** Not defined

3. (a) $\pi/4$ **(b)** $\pi/4$ **(c)** $-\pi/4$

5. (a) $\pi/2$ **(b)** 0 **(c)** π

7. (a) $\pi/6$ **(b)** $-\pi/6$ **(c)** Not defined

9. (a) 0.87696 **(b)** 2.09601 **11.** $\dfrac{1}{3}$ **13.** 10

15. $\pi/3$ **17.** $-\pi/6$ **19.** $-\pi/3$ **21.** $\sqrt{3}/3$ **23.** $\dfrac{1}{2}$

25. $\pi/3$ **27.** $\dfrac{4}{5}$ **29.** $\dfrac{12}{13}$ **31.** $\dfrac{13}{5}$ **33.** $\sqrt{5}/5$

35. $\dfrac{24}{25}$ **37.** 1 **39.** $\sqrt{1 - x^2}$ **41.** $x/\sqrt{1 - x^2}$

43. $\dfrac{1 - x^2}{1 + x^2}$ **45.** 0 **47.** $\theta = \tan^{-1}\dfrac{50}{s}$

49. **(b)** 17.1°, 29.7°, 19.7°

51. **(a)**

Conjecture: $y = \pi/2$ for $-1 \le x \le 1$

53. **(a)** 0.28   **(b)** $(-3 + \sqrt{17})/4$

Section 7.5 ■ page 499

1. $\dfrac{\pi}{3} + 2k\pi, \dfrac{5\pi}{3} + 2k\pi$ **3.** $\dfrac{\pi}{3} + 2k\pi, \dfrac{2\pi}{3} + 2k\pi$

5. $\dfrac{\pi}{3} + 2k\pi, \dfrac{2\pi}{3} + 2k\pi, \dfrac{4\pi}{3} + 2k\pi, \dfrac{5\pi}{3} + 2k\pi$

7. $\dfrac{(2k+1)\pi}{4}$ **9.** $\dfrac{\pi}{2} + k\pi, \dfrac{7\pi}{6} + 2k\pi, \dfrac{11\pi}{6} + 2k\pi$

11. $-\dfrac{\pi}{3} + k\pi$ **13.** $\dfrac{\pi}{2} + k\pi$

15. $\dfrac{\pi}{3} + 2k\pi, \dfrac{5\pi}{3} + 2k\pi$ **17.** $\dfrac{3\pi}{2} + 2k\pi$

19. No solution   **21.** $\dfrac{\pi}{18} + \dfrac{2k\pi}{3}, \dfrac{5\pi}{18} + \dfrac{2k\pi}{3}$

23. $4k\pi$ **25.** $\dfrac{k\pi}{3}$

27. $\dfrac{\pi}{6} + 2k\pi, \dfrac{2\pi}{3} + 2k\pi, \dfrac{5\pi}{6} + 2k\pi, \dfrac{4\pi}{3} + 2k\pi$

29. $\dfrac{\pi}{8} + \dfrac{k\pi}{2}, \dfrac{3\pi}{8} + \dfrac{k\pi}{2}$

31. $\dfrac{\pi}{9}, \dfrac{5\pi}{9}, \dfrac{7\pi}{9}, \dfrac{11\pi}{9}, \dfrac{13\pi}{9}, \dfrac{17\pi}{9}$

33. $\dfrac{\pi}{6}, \dfrac{3\pi}{4}, \dfrac{5\pi}{6}, \dfrac{7\pi}{4}$

35. $\dfrac{\pi}{3}, \dfrac{2\pi}{3}, \dfrac{4\pi}{3}, \dfrac{5\pi}{3}$

37. $0, \dfrac{2\pi}{3}, \dfrac{4\pi}{3}$

39. **(a)** 1.15928, 5.12391
(b) 1.15928 + $2k\pi$, 5.12391 + $2k\pi$
41. **(a)** 1.36944, 4.91375
(b) 1.36944 + $2k\pi$, 4.91375 + $2k\pi$
43. **(a)** 0.46365, 2.67795, 3.60524, 5.81954
(b) 0.46365 + $k\pi$, 2.67795 + $k\pi$
45. **(a)** 0.33984, 2.80176
(b) 0.33984 + $2k\pi$, 2.80176 + $2k\pi$

47. $((2k+1)\pi, -2)$ **49.** $\left(\dfrac{\pi}{3} + k\pi, \sqrt{3}\right)$

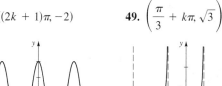

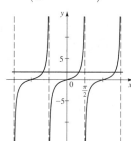

51. $\dfrac{\pi}{8}, \dfrac{3\pi}{8}, \dfrac{5\pi}{8}, \dfrac{7\pi}{8}, \dfrac{9\pi}{8}, \dfrac{11\pi}{8}, \dfrac{13\pi}{8}, \dfrac{15\pi}{8}$

53. $\dfrac{\pi}{9}, \dfrac{2\pi}{9}, \dfrac{7\pi}{9}, \dfrac{8\pi}{9}, \dfrac{13\pi}{9}, \dfrac{14\pi}{9}$

55. $\dfrac{\pi}{2}, \dfrac{7\pi}{6}, \dfrac{3\pi}{2}, \dfrac{11\pi}{6}$ **57.** 0   **59.** $\dfrac{k\pi}{2}$

61. $\dfrac{\pi}{9} + \dfrac{2k\pi}{3}, \dfrac{\pi}{2} + k\pi, \dfrac{5\pi}{9} + \dfrac{2k\pi}{3}$

63. 0, ±0.95   **65.** 1.92   **67.** ±0.71

Section 7.6 ■ page 507

1. $\sqrt{2}\left(\cos\dfrac{\pi}{4} + i\sin\dfrac{\pi}{4}\right)$ **3.** $2\left(\cos\dfrac{7\pi}{4} + i\sin\dfrac{7\pi}{4}\right)$

5. $4\left(\cos\dfrac{11\pi}{6} + i\sin\dfrac{11\pi}{6}\right)$ **7.** $\sqrt{2}\left(\cos\dfrac{3\pi}{2} + i\sin\dfrac{3\pi}{2}\right)$

9. $5\sqrt{2}\left(\cos\dfrac{\pi}{4} + i\sin\dfrac{\pi}{4}\right)$

11. $8\left(\cos\dfrac{11\pi}{6} + i\sin\dfrac{11\pi}{6}\right)$ **13.** $20(\cos\pi + i\sin\pi)$

15. $5\left[\cos\left(\tan^{-1}\tfrac{4}{3}\right) + i\sin\left(\tan^{-1}\tfrac{4}{3}\right)\right]$

17. $3\sqrt{2}\left(\cos\dfrac{3\pi}{4} + i\sin\dfrac{3\pi}{4}\right)$

19. $8\left(\cos\dfrac{\pi}{6} + i\sin\dfrac{\pi}{6}\right)$

21. $\sqrt{5}\left[\cos\left(\tan^{-1}\tfrac{1}{2}\right) + i\sin\left(\tan^{-1}\tfrac{1}{2}\right)\right]$

23. $2\left(\cos\dfrac{\pi}{4} + i\sin\dfrac{\pi}{4}\right)$

25. $z_1z_2 = \cos\pi + i\sin\pi; \dfrac{z_1}{z_2} = \cos\left(\dfrac{\pi}{2}\right) - i\sin\left(\dfrac{\pi}{2}\right)$

27. $z_1z_2 = 14\left(\cos\dfrac{12\pi}{7} + i\sin\dfrac{12\pi}{7}\right)$

$\dfrac{z_1}{z_2} = \dfrac{7}{2}\left(\cos\dfrac{6\pi}{7} + i\sin\dfrac{6\pi}{7}\right)$

29. $z_1 z_2 = 100(\cos 350° + i \sin 350°)$

$\dfrac{z_1}{z_2} = \dfrac{4}{25}(\cos 50° + i \sin 50°)$

31. $z_1 = 2\left(\cos \dfrac{\pi}{6} + i \sin \dfrac{\pi}{6}\right)$

$z_2 = 2\left(\cos \dfrac{\pi}{3} + i \sin \dfrac{\pi}{3}\right)$

$z_1 z_2 = 4\left(\cos \dfrac{\pi}{2} + i \sin \dfrac{\pi}{2}\right)$

$\dfrac{z_1}{z_2} = \cos \dfrac{\pi}{6} - i \sin \dfrac{\pi}{6}$

$\dfrac{1}{z_1} = \dfrac{1}{2}\left(\cos \dfrac{\pi}{6} - i \sin \dfrac{\pi}{6}\right)$

33. $z_1 = 4\left(\cos \dfrac{11\pi}{6} + i \sin \dfrac{11\pi}{6}\right)$

$z_2 = \sqrt{2}\left(\cos \dfrac{3\pi}{4} + i \sin \dfrac{3\pi}{4}\right)$

$z_1 z_2 = 4\sqrt{2}\left(\cos \dfrac{7\pi}{12} + i \sin \dfrac{7\pi}{12}\right)$

$\dfrac{z_1}{z_2} = 2\sqrt{2}\left(\cos \dfrac{13\pi}{12} + i \operatorname{sn} \dfrac{13\pi}{12}\right)$

$\dfrac{1}{z_1} = \dfrac{1}{4}\left(\cos \dfrac{11\pi}{6} - i \sin \dfrac{11\pi}{6}\right)$

35. $z_1 = 5\sqrt{2}\left(\cos \dfrac{\pi}{4} + i \sin \dfrac{\pi}{4}\right)$

$z_2 = 4(\cos 0 + i \sin 0)$

$z_1 z_2 = 20\sqrt{2}\left(\cos \dfrac{\pi}{4} + i \sin \dfrac{\pi}{4}\right)$

$\dfrac{z_1}{z_2} = \dfrac{5\sqrt{2}}{4}\left(\cos \dfrac{\pi}{4} + i \sin \dfrac{\pi}{4}\right)$

$\dfrac{1}{z_1} = \dfrac{\sqrt{2}}{10}\left(\cos \dfrac{\pi}{4} - i \sin \dfrac{\pi}{4}\right)$

37. $z_1 = 20(\cos \pi + i \sin \pi)$

$z_2 = 2\left(\cos \dfrac{\pi}{6} + i \sin \dfrac{\pi}{6}\right)$

$z_1 z_2 = 40\left(\cos \dfrac{7\pi}{6} + i \sin \dfrac{7\pi}{6}\right)$

$\dfrac{z_1}{z_2} = 10\left(\cos \dfrac{5\pi}{6} + i \sin \dfrac{5\pi}{6}\right)$

$\dfrac{1}{z_1} = \dfrac{1}{20}(\cos \pi - i \sin \pi)$

39. -1024　　　**41.** $512(-\sqrt{3} + i)$　　**43.** -1

45. 4096　　**47.** $8(-1 + i)$　　**49.** $\dfrac{1}{2048}(-\sqrt{3} - i)$

51. $2\sqrt{2}\left(\cos \dfrac{\pi}{12} + i \sin \dfrac{\pi}{12}\right),$

$2\sqrt{2}\left(\cos \dfrac{13\pi}{12} + i \sin \dfrac{13\pi}{12}\right)$

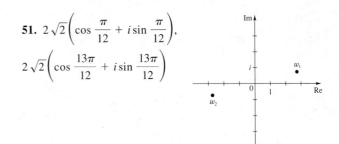

53. $3\left(\cos \dfrac{3\pi}{8} + i \sin \dfrac{3\pi}{8}\right),\ 3\left(\cos \dfrac{7\pi}{8} + i \sin \dfrac{7\pi}{8}\right),$

$3\left(\cos \dfrac{11\pi}{8} + i \sin \dfrac{11\pi}{8}\right),\ 3\left(\cos \dfrac{15\pi}{8} + i \sin \dfrac{15\pi}{8}\right)$

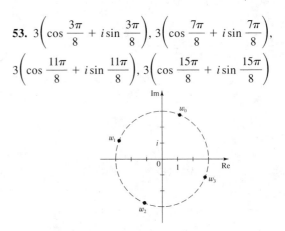

55. $\pm 1,\ \pm i,\ \pm \dfrac{\sqrt{2}}{2} \pm \dfrac{\sqrt{2}}{2}i$

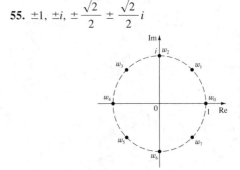

57. $\dfrac{\sqrt{3}}{2} + \dfrac{1}{2}i,\ -\dfrac{\sqrt{3}}{2} + \dfrac{1}{2}i,\ -i$

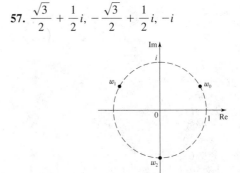

59. $\pm \dfrac{\sqrt{2}}{2} \pm \dfrac{\sqrt{2}}{2} i$

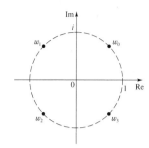

61. $\pm \dfrac{\sqrt{2}}{2} \pm \dfrac{\sqrt{2}}{2} i$

63. $2\left(\cos \dfrac{\pi}{18} + i\sin \dfrac{\pi}{18}\right)$, $2\left(\cos \dfrac{13\pi}{18} + i\sin \dfrac{13\pi}{18}\right)$,

$2\left(\cos \dfrac{25\pi}{18} + i\sin \dfrac{25\pi}{18}\right)$

65. $2^{1/6}\left(\cos \dfrac{5\pi}{12} + i\sin \dfrac{5\pi}{12}\right)$, $2^{1/6}\left(\cos \dfrac{13\pi}{12} + i\sin \dfrac{13\pi}{12}\right)$,

$2^{1/6}\left(\cos \dfrac{21\pi}{12} + i\sin \dfrac{21\pi}{12}\right)$

Section 7.7 ■ page 516

1.

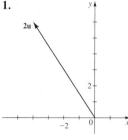

3.

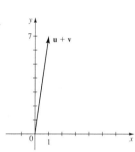

5.

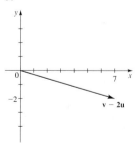

7. $\langle 3, 3\rangle$ **9.** $\langle 3, -1\rangle$ **11.** $\langle 5, 7\rangle$ **13.** $\langle -4, -3\rangle$
15. $\langle 0, 2\rangle$ **17.** $\langle 4, 14\rangle, \langle -9, -3\rangle, \langle 5, 8\rangle, \langle -6, 17\rangle$
19. $\langle 0, -2\rangle, \langle 6, 0\rangle, \langle -2, -1\rangle, \langle 8, -3\rangle$
21. $4\mathbf{i}, -9\mathbf{i} + 6\mathbf{j}, 5\mathbf{i} - 2\mathbf{j}, -6\mathbf{i} + 8\mathbf{j}$
23. $\sqrt{5}, \sqrt{13}, 2\sqrt{5}, \frac{1}{2}\sqrt{13}, \sqrt{26}, \sqrt{10}, \sqrt{5} - \sqrt{13}$

25. $\sqrt{101}, 2\sqrt{2}, 2\sqrt{101}, \sqrt{2}, \sqrt{73}, \sqrt{145}, \sqrt{101} - 2\sqrt{2}$
27. $20\sqrt{3}\,\mathbf{i} + 20\mathbf{j}$ **29.** $-\dfrac{\sqrt{2}}{2}\mathbf{i} - \dfrac{\sqrt{2}}{2}\mathbf{j}$
31. $4\cos 10°\mathbf{i} + 4\sin 10°\mathbf{j} \approx 3.94\mathbf{i} + 0.69\mathbf{j}$
33. $15\sqrt{3}, -15$ **35.** $5, 53.13°$ **37.** $13, 157.38°$
39. $2, 60°$
41. (a) $425\mathbf{i} + 40\mathbf{j}$ (b) 427 mi/h, N $84.6°$ E
43. 794 mi/h, N $26.6°$ W
45. (a) $20\mathbf{i} + 17.32\mathbf{j}$ (b) 26.5 mi/h, N $49.1°$ E
47. 7.4 mi/h, 22.8 mi/h   **49.** (a) $\langle 5, -3\rangle$ (b) $\langle -5, 3\rangle$
51. (a) $-4\mathbf{j}$ (b) $4\mathbf{j}$
53. (a) $\langle -7.57, 10.61\rangle$ (b) $\langle 7.57, -10.61\rangle$
55. $\mathbf{T}_1 \approx -56.5\mathbf{i} + 67.4\mathbf{j}, \mathbf{T}_2 \approx 56.5\mathbf{i} + 32.6\mathbf{j}$

Chapter 7 Review ■ page 519

23. (a)

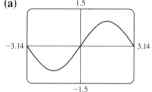

(b) Yes

25. (a)

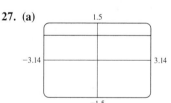

(b) No

27. (a)

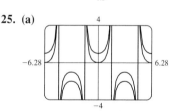

$2\sin^2 3x + \cos 6x = 1$

29. $0, \pi$ **31.** $\dfrac{\pi}{6}, \dfrac{5\pi}{6}$ **33.** $\dfrac{\pi}{3}, \dfrac{5\pi}{3}$ **35.** $\dfrac{2\pi}{3}, \dfrac{4\pi}{3}$
37. $\dfrac{\pi}{3}, \dfrac{2\pi}{3}, \dfrac{3\pi}{4}, \dfrac{4\pi}{3}, \dfrac{5\pi}{3}, \dfrac{7\pi}{4}$
39. $\dfrac{\pi}{6}, \dfrac{\pi}{2}, \dfrac{5\pi}{6}, \dfrac{7\pi}{6}, \dfrac{3\pi}{2}, \dfrac{11\pi}{6}$ **41.** $\dfrac{\pi}{6}$ **43.** 1.18
45. $\frac{1}{2}\sqrt{2 + \sqrt{3}}$ **47.** $\sqrt{2} - 1$ **49.** $\sqrt{2}/2$
51. $\sqrt{2}/2$ **53.** $\dfrac{\sqrt{2} + \sqrt{3}}{4}$ **55.** $2\dfrac{\sqrt{10} + 1}{9}$
57. $\frac{2}{3}(\sqrt{2} + \sqrt{5})$ **59.** $\sqrt{(3 + 2\sqrt{2})/6}$ **61.** $\pi/3$
63. $\frac{1}{2}$ **65.** $2/\sqrt{21}$ **67.** $\frac{7}{9}$ **69.** $x/\sqrt{1 + x^2}$

71. $\theta = \cos^{-1}\dfrac{x}{3}$ **73.** $s = 7920\cos^{-1}\left(\dfrac{3960}{h+3960}\right)$

75. $4\sqrt{2}\left(\cos\dfrac{\pi}{4} + i\sin\dfrac{\pi}{4}\right)$

77. $\sqrt{34}\left[\cos(\tan^{-1}\tfrac{3}{5}) + i\sin(\tan^{-1}\tfrac{3}{5})\right]$

79. $\sqrt{2}\left(\cos\dfrac{3\pi}{4} + i\sin\dfrac{3\pi}{4}\right)$ **81.** $8(-1 + i\sqrt{3})$

83. $-\tfrac{1}{32}(1 + i\sqrt{3})$ **85.** $\pm 2\sqrt{2}\,(1 - i)$

87. $\pm 1,\ \pm\tfrac{1}{2} \pm \dfrac{\sqrt{3}}{2}i$

89. $\sqrt{13}$, $\langle 6,4\rangle$, $\langle -10,2\rangle$, $\langle -4,6\rangle$, $\langle -22,7\rangle$

91. $3i - 4j$ **93.** $(10,-2)$

95. (a) $(4.8i + 0.4j) \times 10^4$
(b) 4.8×10^4 lb, N $85.2°$ E

Chapter 7 Test ■ page 522

2. (a) $\dfrac{2\pi}{3}, \dfrac{4\pi}{3}$ **(b)** $\dfrac{\pi}{6}, \dfrac{\pi}{2}, \dfrac{5\pi}{6}, \dfrac{3\pi}{2}$

3. $0.57964, 2.56195, 3.72123, 5.70355$ **4.** $\tan\theta$

5. (a) $\tfrac{1}{2}$ **(b)** $\dfrac{\sqrt{2}+\sqrt{6}}{4}$ **(c)** $\tfrac{1}{2}\sqrt{2-\sqrt{3}}$

6. $(10 - 2\sqrt{5})/15$

7. (a) $\tfrac{1}{2}(\sin 8x - \sin 2x)$ **(b)** $-2\cos\tfrac{7}{2}x\sin\tfrac{3}{2}x$ **8.** -2

9.

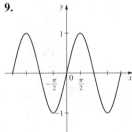
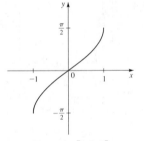

Domain $\mathbb{R}$ Domain $[-1,1]$

10. (a) $\theta = \tan^{-1}\dfrac{x}{4}$ **(b)** $\theta = \cos^{-1}\dfrac{3}{x}$ **11.** $\tfrac{40}{41}$

12. (a) $2\left(\cos\dfrac{\pi}{3} + i\sin\dfrac{\pi}{3}\right)$ **(b)** -512

13. $-8, \sqrt{3} + i$

14. $-3i,\ 3\left(\pm\dfrac{\sqrt{3}}{2} + \dfrac{1}{2}i\right)$

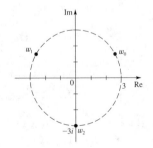

15. (a) $-6i + 10j$ **(b)** $2\sqrt{34}$
16. (a) $\langle 19, -3\rangle$ **(b)** $5\sqrt{2}$
17. (a) $14i + 6\sqrt{3}j$ **(b)** 17.4 mi/h, N $53.4°$ E

FOCUS ON PROBLEM SOLVING ■ page 525

5. *Hint:* Try proof by contradiction.
7. *Hint:* First divide the coins into groups of four.

9. 63 **11.** $b = \dfrac{a[1 + \sin(\theta/2)]}{1 - \sin(\theta/2)}$ **13.** $\pi/2$ **19.** π

21. 3 ft by 6 ft, 4 ft by 4 ft

CHAPTER 8

Section 8.1 ■ page 538

1. $(6,2)$ **3.** $(-2,4), (3,9)$ **5.** $(2,-2), (-2,2)$
7. $(-6,16)$ **9.** $(-3,4), (3,4)$
11. $(-2,-1), (-2,1), (2,-1), (2,1)$ **13.** $(4,0)$
15. $(-2,-2)$ **17.** $(6,2), (-2,-6)$
19. No solution
21. $(\sqrt{5},2), (\sqrt{5},-2), (-\sqrt{5},2), (-\sqrt{5},-2)$
23. $(3, -\tfrac{1}{2}), (-3, -\tfrac{1}{2})$ **25.** $(\tfrac{1}{5}, \tfrac{1}{3})$
27. $(-0.33, 5.33)$
29. $(-1.38, -2.63), (0.82, 0.55)$
31. $(-4.51, 2.17), (4.91, -0.97)$
33. $(1.23, 3.87), (-0.35, -4.21)$
35. $(-2.30, -0.70), (0.48, -1.19)$
37. 8 cm, 15 cm, 17 cm
39. 9 in., 12 in., 15 in.
41. $15, 20$ **43.** $(400.50, 200.25), 447.77$ m
45. $y = 3x + 5$
47. $\left(-\tfrac{1}{2}, -\tfrac{3}{2}\right), \left(\tfrac{1}{2}, \tfrac{3}{2}\right)$
49. $(\sqrt{10}, 10)$

Section 8.2 ■ page 544

1. $(1,2)$

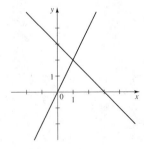

3. $(3, 2)$　　　　　　　**5.** Parallel

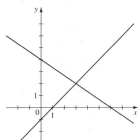　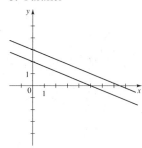

7. $(3, 5)$　　**9.** $(1, 3)$　　**11.** $(10, -9)$　　**13.** $(2, 1)$
15. No solution　　**17.** $\left(x, \frac{1}{3}x - \frac{5}{3}\right)$　　**19.** $\left(x, 3 - \frac{3}{2}x\right)$
21. $(-3, -7)$　　**23.** $\left(x, 5 - \frac{5}{6}x\right)$　　**25.** $\left(\frac{5}{2}, 4\right)$

27. $\left(\frac{2}{3}, -\frac{1}{6}\right)$　　**29.** $\left(-\dfrac{1}{a-1}, \dfrac{1}{a-1}\right)$

31. $\left(\dfrac{1}{a+b}, \dfrac{1}{a+b}\right)$　　**33.** $22, 12$

35. 5 dimes, 9 quarters
37. Plane's speed 120 mi/h, wind speed 30 mi/h
39. Run 5 mi/h, cycle 20 mi/h
41. 200 g of A, 40 g of B　　**43.** 25%, 10%　　**45.** 25
47. $(3.87, 2.74)$　　**49.** $(61.00, 20.00)$

Section 8.3 ■ page 557

1. Linear　　**3.** Linear

5. (a) Yes　(b) Yes　(c) $\begin{cases} x = -3 \\ y = 5 \end{cases}$

7. (a) Yes　(b) No　(c) $\begin{cases} x + 2y + 8z = 0 \\ y + 3z = 0 \end{cases}$

9. (a) No　(b) No　(c) $\begin{cases} x = 0 \\ 0 = 0 \\ y + 5z = 1 \end{cases}$

11. $(1, 1, 2)$　　**13.** $(1, 0, 1)$　　**15.** $(-1, 0, 1)$
17. $(-1, 5, 0)$　　**19.** $(10, 3, -2)$　　**21.** No solution
23. $x = 2 - 3z, y = 3 - 5z, z =$ any number
25. No solution
27. $x = -2z + 5, y = z - 2, z =$ any number
29. $x = -\frac{1}{2}y + z + 6, y =$ any number, $z =$ any number
31. $(-2, 1, 3)$　　**33.** $(-9, 2, 0)$
35. $(0, -3, 0, -3)$　　**37.** $(-1, 0, 0, 1)$
39. $x = \frac{7}{4} - \frac{7}{4}w, y = -\frac{7}{4} + \frac{3}{4}w, z = \frac{9}{4} + \frac{3}{4}w,$
$w =$ any number
41. $x = \frac{1}{3}z - \frac{2}{3}w, y = \frac{1}{3}z + \frac{1}{3}w, z =$ any number,
$w =$ any number
43. 2 VitaMax, 1 Vitron, 2 VitaPlus
45. 5 mile run, 2 mile swim, 30 mile cycle.
47. Impossible

Section 8.4 ■ page 566

1. $\begin{bmatrix} 5 & -2 & 5 \\ 1 & 1 & 0 \end{bmatrix}$　**3.** $\begin{bmatrix} -1 & -3 & -5 \\ -1 & 3 & -6 \end{bmatrix}$　**5.** $\begin{bmatrix} 13 & -\frac{7}{2} & 15 \\ 3 & 1 & 3 \end{bmatrix}$

7. $\begin{bmatrix} -14 & -8 & -30 \\ -6 & 10 & -24 \end{bmatrix}$　**9.** Impossible　**11.** $\begin{bmatrix} 3 & \frac{1}{2} & 5 \\ 1 & -1 & 3 \end{bmatrix}$

13. $\begin{bmatrix} 28 & 21 & 28 \end{bmatrix}$　**15.** $\begin{bmatrix} -1 \\ 8 \\ -1 \end{bmatrix}$　**17.** $\begin{bmatrix} 8 & -335 \\ 0 & 343 \end{bmatrix}$

19. Impossible　**21.** Impossible

23. $\begin{bmatrix} 2 & -5 \\ 3 & 2 \end{bmatrix} \begin{bmatrix} x \\ y \end{bmatrix} = \begin{bmatrix} 7 \\ 4 \end{bmatrix}$

25. $\begin{bmatrix} 3 & 2 & -1 & 1 \\ 1 & 0 & -1 & 0 \\ 0 & 3 & 1 & -1 \end{bmatrix} \begin{bmatrix} x_1 \\ x_2 \\ x_3 \\ x_4 \end{bmatrix} = \begin{bmatrix} 0 \\ 5 \\ 4 \end{bmatrix}$　**27.** $x = 2, y = 1$

29. $\begin{bmatrix} 3 & \frac{11}{2} \\ 2 & 5 \end{bmatrix}$　　**31.** Impossible

33. (a) $\begin{bmatrix} 4{,}690 & 1{,}690 & 13{,}210 \end{bmatrix}$
(b) Total revenue in Santa Monica, Long Beach, and
Anaheim, respectively
35. Only ACB is defined.　$\begin{bmatrix} -3 & -21 & 27 & -6 \\ -2 & -14 & 18 & -4 \end{bmatrix}$
37. There are infinitely possible answers.
One example is $A = \begin{bmatrix} 0 & 1 \\ 0 & 0 \end{bmatrix}$
39. No.

Section 8.5 ■ page 575

1. $\begin{bmatrix} 1 & -2 \\ -\frac{3}{2} & \frac{7}{2} \end{bmatrix}$　**3.** $\begin{bmatrix} 2 & -3 \\ -3 & 5 \end{bmatrix}$　**5.** $\begin{bmatrix} 13 & 5 \\ -5 & -2 \end{bmatrix}$

7. No inverse　**9.** $\begin{bmatrix} 1 & 2 \\ -\frac{1}{2} & \frac{2}{3} \end{bmatrix}$　**11.** $\begin{bmatrix} -4 & -4 & 5 \\ 1 & 1 & -1 \\ 5 & 4 & -6 \end{bmatrix}$

13. No inverse　**15.** $\begin{bmatrix} -\frac{9}{2} & -1 & 4 \\ 3 & 1 & -3 \\ \frac{7}{2} & 1 & -3 \end{bmatrix}$

17. $\begin{bmatrix} 0 & 0 & -2 & 1 \\ -1 & 0 & 1 & 1 \\ 0 & 1 & -1 & 0 \\ 1 & 0 & 0 & -1 \end{bmatrix}$　**19.** $x = 8, y = -12$

21. $x = 126, y = -50$　　**23.** $x = -38, y = 9, z = 47$

25. $x = -20, y = 10, z = 16$　　**27.** $\begin{bmatrix} 7 & 2 & 3 \\ 10 & 3 & 5 \end{bmatrix}$

29. (a) $\begin{bmatrix} 0 & 1 & -1 \\ -2 & \frac{3}{2} & 0 \\ 1 & -\frac{3}{2} & 1 \end{bmatrix}$ **(b)** 1 oz A, 1 oz B, 2 oz C

(c) 2 oz A, 0 oz B, 1 oz C   **(d)** No

31. (a) $\begin{cases} x + y + 2z = 675 \\ 2x + y + z = 600 \\ x + 2y + z = 625 \end{cases}$

(b) $\begin{bmatrix} 1 & 1 & 2 \\ 2 & 1 & 1 \\ 1 & 2 & 1 \end{bmatrix} \begin{bmatrix} x \\ y \\ z \end{bmatrix} = \begin{bmatrix} 675 \\ 600 \\ 625 \end{bmatrix}$

(c) $A^{-1} = \begin{bmatrix} -\frac{1}{4} & \frac{3}{4} & -\frac{1}{4} \\ -\frac{1}{4} & -\frac{1}{4} & \frac{3}{4} \\ \frac{3}{4} & -\frac{1}{4} & -\frac{1}{4} \end{bmatrix}$

He earns \$125 on a standard set, \$150 on a deluxe set, and \$200 on a leather-bound set.

33. $\frac{1}{2} \begin{bmatrix} e^{-x} & e^{-2x} \\ -e^{-2x} & e^{-3x} \end{bmatrix}$

35. $\begin{bmatrix} \sin x & -\cos x \\ \cos x & \sin x \end{bmatrix}$

Section 8.6 ■ page 585

1. 3   **3.** -4 **5.** Does not exist   **7.** $\frac{1}{8}$ **9.** 20, 20
11. $-12, 12$ **13.** 0, 0   **15.** -6, has an inverse
17. 5000, has an inverse   **19.** -4, has an inverse
21. -18 **23.** 120
25. (a) -2 **(b)** -2 **(c)** Yes
27. $(-2, 5)$ **29.** $(0.6, -0.4)$ **31.** $(4, -1)$
33. $(4, 2, -1)$ **35.** $(1, 3, 2)$ **37.** $(0, -1, 1)$
39. $\left(\frac{189}{29}, -\frac{108}{29}, \frac{88}{29}\right)$ **41.** $\left(\frac{1}{2}, \frac{1}{4}, \frac{1}{4}, -1\right)$
43. (b) $5x - 6y = -200$ **45.** 0, 1, 2   **47.** 1, -1

Section 8.7 ■ page 590

1.

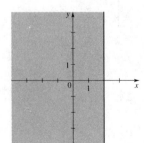

3.

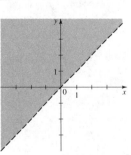

5.

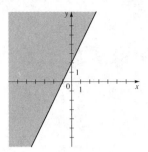

7.

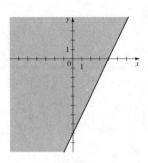

9.

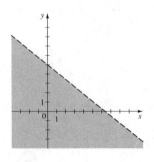

11.

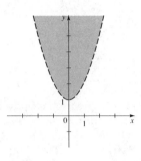

13.

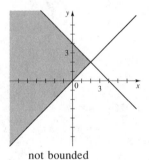

not bounded

15.

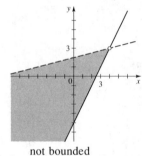

not bounded

17.

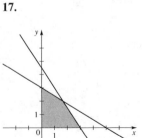

bounded

19.

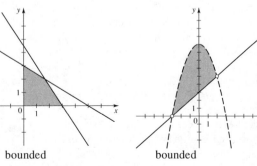

bounded

21.

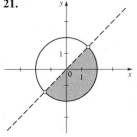

bounded

23.

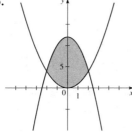

bounded

35. x = number of fiction books
y = number of nonfiction books

$$\begin{cases} x + y \leq 100 \\ 20 \leq y, \ x \geq y \\ x \geq 0, \ y \geq 0 \end{cases}$$

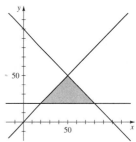

25.

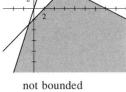

not bounded

27.

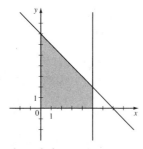

bounded

Section 8.8 ■ page 596

1. $\dfrac{A}{x - 1} + \dfrac{B}{x + 2}$ **3.** $\dfrac{A}{x - 2} + \dfrac{B}{(x - 2)^2} + \dfrac{C}{x + 4}$

5. $\dfrac{A}{x - 3} + \dfrac{Bx + C}{x^2 + 4}$ **7.** $\dfrac{Ax + B}{x^2 + 1} + \dfrac{Cx + D}{x^2 + 2}$

9. $\dfrac{A}{x} + \dfrac{B}{2x - 5} + \dfrac{C}{(2x - 5)^2} + \dfrac{D}{(2x - 5)^3}$

$+ \dfrac{Ex + F}{x^2 + 2x + 5} + \dfrac{Gx + H}{(x^2 + 2x + 5)^2}$

11. $\dfrac{1}{x - 1} - \dfrac{1}{x + 4}$ **13.** $\dfrac{2}{x - 3} - \dfrac{2}{x + 3}$

15. $\dfrac{1}{x - 2} - \dfrac{1}{x + 2}$ **17.** $\dfrac{3}{x - 4} - \dfrac{2}{x + 2}$

19. $\dfrac{-\frac{1}{2}}{2x - 1} + \dfrac{\frac{3}{2}}{4x - 3}$

21. $\dfrac{2}{x - 2} + \dfrac{3}{x + 2} - \dfrac{1}{2x - 1}$

23. $\dfrac{2}{x + 1} - \dfrac{1}{x} + \dfrac{1}{x^2}$ **25.** $\dfrac{1}{2x + 3} - \dfrac{3}{(2x + 3)^2}$

27. $\dfrac{2}{x} - \dfrac{1}{x^3} - \dfrac{2}{x + 2}$

29. $\dfrac{4}{x + 2} - \dfrac{4}{x - 1} + \dfrac{2}{(x - 1)^2} + \dfrac{1}{(x - 1)^3}$

31. $\dfrac{3}{x + 2} - \dfrac{1}{(x + 2)^2} - \dfrac{1}{(x + 3)^2}$

33. $\dfrac{x + 1}{x^2 + 3} - \dfrac{1}{x}$ **35.** $\dfrac{2x - 5}{x^2 + x + 2} + \dfrac{5}{x^2 + 1}$

37. $\dfrac{1}{x^2 + 1} - \dfrac{x + 2}{(x^2 + 1)^2} + \dfrac{1}{x}$

39. $x^2 + \dfrac{3}{x - 2} - \dfrac{x + 1}{x^2 + 1}$

41. $A = \dfrac{a + b}{2}, \ B = \dfrac{a - b}{2}$

29.

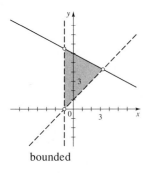

bounded

31.

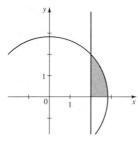

bounded

33.

bounded

Chapter 8 Review ■ page 598

1. $(2, 1)$

3. $x =$ any number
$y = \frac{2}{7}x - 4$

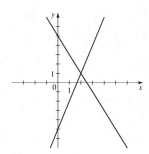

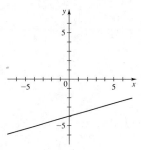

5. No solution

7. $(-3, 3), (2, 8)$
9. $\left(\frac{16}{7}, -\frac{14}{3}\right)$
11. $(1, 1, 2)$
13. No solution
15. $x = -4z + 1,$
$y = -z - 1,$
$z =$ any number
17. $x = 6 - 5z,$
$y = \frac{1}{2}(7 - 3z),$
$z =$ any number

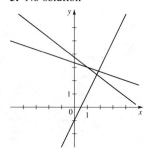

19. \$3000 at 6%, \$6000 at 7% **21.** Impossible

23. $\begin{bmatrix} 4 & 18 \\ 4 & 0 \\ 2 & 2 \end{bmatrix}$ **25.** $[10 \quad 0 \quad -5]$ **27.** $\begin{bmatrix} -\frac{7}{2} & 10 \\ 1 & -\frac{9}{2} \end{bmatrix}$

29. $\begin{bmatrix} 30 & 22 & 2 \\ -9 & 1 & -4 \end{bmatrix}$ **31.** $\begin{bmatrix} -\frac{1}{2} & \frac{11}{2} \\ \frac{15}{4} & -\frac{3}{2} \\ -\frac{1}{2} & 1 \end{bmatrix}$ **33.** $1, \begin{bmatrix} 9 & -4 \\ -2 & 1 \end{bmatrix}$

35. 0, no inverse **37.** $-1, \begin{bmatrix} 3 & 2 & -3 \\ 2 & 1 & -2 \\ -8 & -6 & 9 \end{bmatrix}$ **39.** $(65, 154)$

41. $\left(\frac{1}{5}, \frac{9}{5}\right)$ **43.** $\left(-\frac{87}{26}, \frac{21}{26}, \frac{3}{2}\right)$
45. **47.**

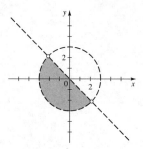

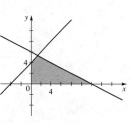

bounded bounded

49. $x = \dfrac{b + c}{2}, y = \dfrac{a + c}{2}, z = \dfrac{a + b}{2}$ **51.** $2, 3$
53. $(21.41, -15.93)$ **55.** $(11.94, -1.39), (12.07, 1.44)$
57. $\dfrac{2}{x - 5} + \dfrac{1}{x + 3}$ **59.** $\dfrac{-4}{x} + \dfrac{4}{x - 1} + \dfrac{-2}{(x - 1)^2}$

Chapter 8 Test ■ page 600

1. Wind speed 60 km/h, airplane 300 km/h
2. $(3, -1)$; linear, neither inconsistent nor dependent
3. No solution; linear, inconsistent
4. $x = \frac{1}{7}(z + 1), y = \frac{1}{7}(9z + 2), z =$ any number;
linear, dependent
5. $(\pm 1, -2), \left(\pm\frac{5}{3}, -\frac{2}{3}\right)$; nonlinear
6. Incompatible dimensions
7. Incompatible dimensions
8. $\begin{bmatrix} 6 & 10 \\ 3 & -2 \\ -3 & 9 \end{bmatrix}$ **9.** $\begin{bmatrix} 36 & 58 \\ 0 & -3 \\ 18 & 28 \end{bmatrix}$ **10.** $\begin{bmatrix} 2 & -\frac{3}{2} \\ -1 & 1 \end{bmatrix}$
11. B is not square **12.** B is not square **13.** -3
14. $(70, 90)$ **15.** $(5, -5, -4)$
16. $|A| = 0, |B| = 2, B^{-1} = \begin{bmatrix} 1 & -2 & 0 \\ 0 & \frac{1}{2} & 0 \\ 3 & -6 & 1 \end{bmatrix}$

17.

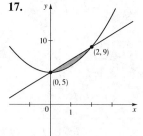

18. $\dfrac{1}{x - 1} + \dfrac{1}{(x - 1)^2} - \dfrac{1}{x + 2}$
19. $(-0.49, 3.93), (2.34, 2.24)$

FOCUS ON MODELING ■ page 607

1. $198, 195$
3.

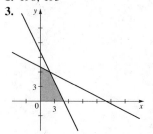

maximum 161
minimum 135

5. 3 tables, 34 chairs

7. 30 grapefruit crates, 30 orange crates

9. 15 Pasadena to Santa Monica, 3 Pasadena to El Toro,
0 Long Beach to Santa Monica, 16 Long Beach to El Toro

11. 90 standard, 40 deluxe

13. $7500 in municipal bonds, $2500 in bank certificates,
$2000 in high-risk bonds

15. 4 games, 32 educational, 0 utility

CHAPTER 9

Section 9.1 ■ page 617

Order of answers: focus; directrix; focal diameter

1. $F(1,0)$; $x = -1$; 4   **3.** $F(0, \frac{9}{4})$; $y = -\frac{9}{4}$; 9

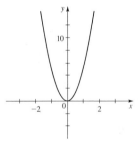

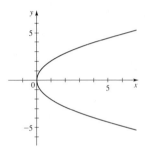

5. $F(0, \frac{1}{20})$; $y = -\frac{1}{20}$; $\frac{1}{5}$ **7.** $F(-\frac{1}{32}, 0)$; $x = \frac{1}{32}$; $\frac{1}{8}$

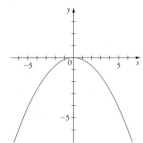

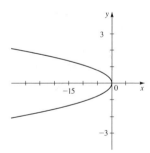

9. $F(0, -\frac{3}{2})$; $y = \frac{3}{2}$; 6   **11.** $F(-\frac{5}{12}, 0)$; $x = \frac{5}{12}$; $\frac{5}{3}$

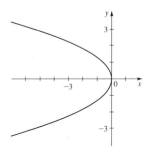

13. $x^2 = 8y$ **15.** $y^2 = -32x$

17. $y^2 = -8x$ **19.** $x^2 = 40y$

21. $y^2 = 4x$ **23.** $x^2 = 20y$

25. $x^2 = -12y$ **27.** $y^2 = -3x$

29. $x = y^2$ **31.** $x^2 = -4\sqrt{2}\, y$

33. (a) $y^2 = 12x$ (b) $8\sqrt{15} \approx 31$ cm

35. $x^2 = 600y$

37. (a) $x^2 = -4py$, $p = \frac{1}{2}$, 1, 4, and 8

(b) The closer the directrix to the vertex, the steeper the
parabola.

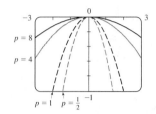

Section 9.2 ■ page 626

*Order of answers: vertices; foci; eccentricity; major axis
and minor axis*

1. $V(\pm 5, 0)$; $F(\pm 4, 0)$;   **3.** $V(0, \pm 3)$; $F(0, \pm\sqrt{5}\,)$;
$\frac{4}{5}$; 10, 6 $\sqrt{5}/3$; 6, 4

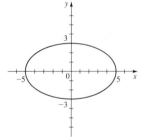

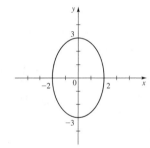

5. $V(\pm 4, 0)$; $F(\pm 2\sqrt{3}, 0)$;   **7.** $V(0, \pm\sqrt{3}\,)$; $F(0, \pm\sqrt{3/2}\,)$;
$\sqrt{3}/2$; 8, 4 $1/\sqrt{2}$; $2\sqrt{3}$, $\sqrt{6}$

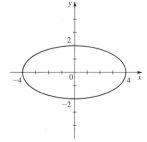

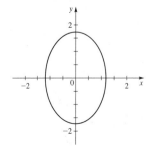

9. $V(\pm 1, 0)$; $F(\pm\sqrt{3}/2, 0)$; $\sqrt{3}/2$; 2, 1

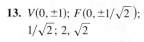

11. $V(0, \pm\sqrt{2})$; $F(0, \pm\sqrt{3/2})$; $\sqrt{3}/2$; $2\sqrt{2}$, $\sqrt{2}$

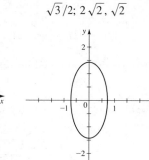

13. $V(0, \pm 1)$; $F(0, \pm 1/\sqrt{2})$; $1/\sqrt{2}$; 2, $\sqrt{2}$

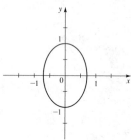

15. $\dfrac{x^2}{25} + \dfrac{y^2}{16} = 1$ **17.** $\dfrac{x^2}{4} + \dfrac{y^2}{8} = 1$

19. $\dfrac{x^2}{25} + \dfrac{y^2}{9} = 1$ **21.** $x^2 + \dfrac{y^2}{4} = 1$

23. $\dfrac{x^2}{9} + \dfrac{y^2}{13} = 1$ **25.** $\dfrac{x^2}{100} + \dfrac{y^2}{91} = 1$

27. $\dfrac{x^2}{25} + \dfrac{y^2}{5} = 1$ **29.** $\dfrac{64x^2}{225} + \dfrac{64y^2}{81} = 1$

31. $(0, \pm 2)$

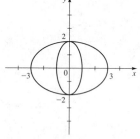

33. $\dfrac{x^2}{2.2500 \times 10^{16}} + \dfrac{y^2}{2.2497 \times 10^{16}} = 1$

35. $\dfrac{x^2}{1{,}455{,}642} + \dfrac{y^2}{1{,}451{,}610} = 1$

37. $5\sqrt{39}/2 \approx 15.6$ in.

41. (a)

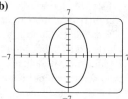

(b)

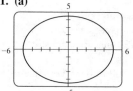

Section 9.3 ■ page 634

Order of answers: vertices; foci; asymptotes

1. $V(\pm 2, 0)$; $F(\pm 2\sqrt{5}, 0)$; $y = \pm 2x$

3. $V(0, \pm 1)$; $F(0, \pm\sqrt{26})$; $y = \pm\frac{1}{5}x$

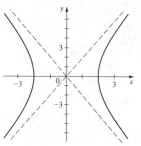

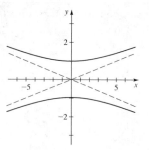

5. $V(\pm 1, 0)$; $F(\pm\sqrt{2}, 0)$; $y = \pm x$

7. $V(0, \pm 3)$; $F(0, \pm\sqrt{34})$; $y = \pm\frac{3}{5}x$

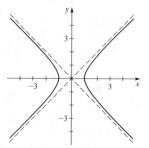

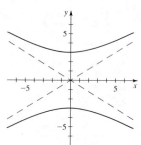

9. $V(\pm 2\sqrt{2}, 0)$; $F(\pm\sqrt{10}, 0)$; $y = \pm\frac{1}{2}x$

11. $V(0, \pm\frac{1}{2})$; $F(0, \pm\sqrt{5}/2)$; $y = \pm\frac{1}{2}x$

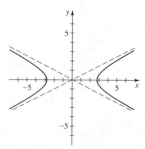

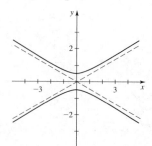

13. $\dfrac{x^2}{4} - \dfrac{y^2}{12} = 1$ **15.** $\dfrac{y^2}{16} - \dfrac{x^2}{16} = 1$

17. $\dfrac{x^2}{9} - \dfrac{y^2}{16} = 1$ **19.** $y^2 - \dfrac{x^2}{3} = 1$

21. $x^2 - \dfrac{y^2}{25} = 1$ **23.** $\dfrac{5y^2}{64} - \dfrac{5x^2}{256} = 1$

25. $\dfrac{x^2}{16} - \dfrac{y^2}{16} = 1$ **27.** $\dfrac{x^2}{9} - \dfrac{y^2}{16} = 1$

29. (b) $x^2 - y^2 = c^2/2$

33. (a) 490 mi   (b) $\dfrac{y^2}{60{,}025} - \dfrac{x^2}{2475} = 1$ (c) 10.1 mi

35. (b)

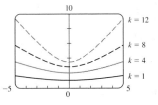

Section 9.4 ■ page 642

1. Center $C(2, 1)$;
foci $F(2 \pm \sqrt{5}, 1)$;
vertices $V_1(-1, 1)$,
$V_2(5, 1)$; major axis 6,
minor axis 4

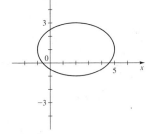

3. Center $C(0, -5)$;
foci $F_1(0, -1)$, $F_2(0, -9)$;
vertices $V_1(0, 0)$, $V_2(0, -10)$;
major axis 10, minor axis 6

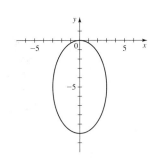

5. Vertex $V(3, -1)$;
focus $F(3, 1)$;
directrix $y = -3$

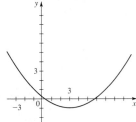

7. Vertex $V\left(-\frac{1}{2}, 0\right)$;
focus $F\left(-\frac{1}{2}, -\frac{1}{16}\right)$;
directrix $y = \frac{1}{16}$

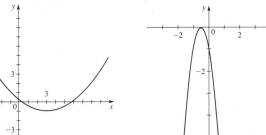

9. Center $C(-1, 3)$;
foci $F_1(-6, 3)$, $F_2(4, 3)$;
vertices $V_1(-4, 3)$, $V_2(2, 3)$;
asymptotes
$y = \pm \frac{4}{3}(x + 1) + 3$

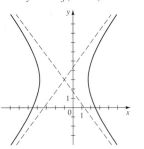

11. Center $C(-1, 0)$;
foci $F(-1, \pm\sqrt{5})$;
vertices $V(-1, \pm 1)$;
asymptotes
$y = \pm \frac{1}{2}(x + 1)$

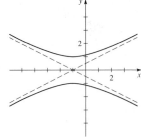

13. $x^2 = -\frac{1}{4}(y - 4)$ **15.** $\dfrac{(x - 5)^2}{25} + \dfrac{y^2}{16} = 1$

17. $(y - 1)^2 - x^2 = 1$

19. Ellipse; $C(2, 0)$;
$F(2, \pm\sqrt{5})$; $V(2, \pm 3)$;
major axis 6,
minor axis 4

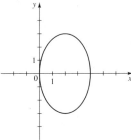

21. Hyperbola; $C(1, 2)$;
$F_1\left(-\frac{3}{2}, 2\right)$, $F_2\left(\frac{7}{2}, 2\right)$;
$V(1 \pm \sqrt{5}, 2)$;
asymptotes
$y = \pm \frac{1}{2}(x - 1) + 2$

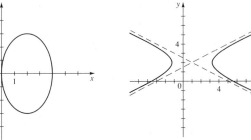

23. Ellipse; $C(3, -5)$;
$F(3 \pm \sqrt{21}, -5)$;
$V_1(-2, -5)$, $V_1(8, -5)$;
major axis 10,
minor axis 4

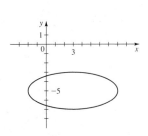

25. Hyperbola; $C(3, 0)$;
$F(3, \pm 5)$; $V(3, \pm 4)$;
asymptotes
$y = \pm \frac{4}{3}(x - 3)$

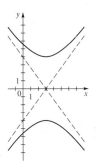

27. Degenerate conic
(pair of lines),
$y = \pm \frac{1}{2}(x - 4)$

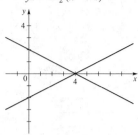

29. Point $(1, 3)$

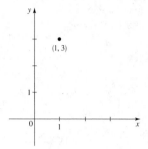

31. (a) $F < 17$ **(b)** $F = 17$ **(c)** $F > 17$
33. (a)

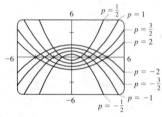

(c) The parabolas become narrower.

Section 9.5 ■ page 649

1. $(\sqrt{2}, 0)$ **3.** $(0, -2\sqrt{3})$ **5.** $(1.6383, 1.1472)$
7. $X^2 + Y^2 - 2XY - 3\sqrt{2}X + \sqrt{2}Y + 2 = 0$
9. $X^2 - Y^2 = 2$
11. (a) Hyperbola   **(b)** $X^2 - Y^2 = 16$
(c) $\phi = 45°$

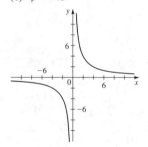

13. (a) Parabola   **(b)** $Y = \sqrt{2}X^2$
(c) $\phi = 45°$

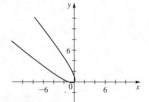

15. (a) Hyperbola
(b) $Y^2 - X^2 = 1$
(c) $\phi = 30°$

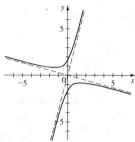

17. (a) Hyperbola
(b) $\dfrac{X^2}{4} - Y^2 = 1$
(c) $\phi \approx 53°$

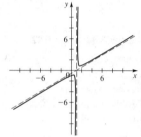

19. (a) Hyperbola
(b) $3X^2 - Y^2 = 2\sqrt{3}$
(c) $\phi = 30°$

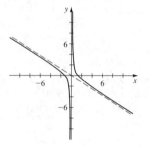

21. (a) Hyperbola
(b) $(X - 1)^2 - 3Y^2 = 1$
(c) $\phi = 60°$

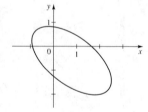

23. (a) Ellipse
(b) $X^2 + \dfrac{(Y + 1)^2}{4} = 1$
(c) $\phi \approx 53°$
See graph at right.

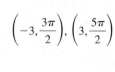

25. (a) $(X - 5)^2 - Y^2 = 1$
(b) XY-coordinates: $C(5, 0)$; $V_1(6, 0)$, $V_2(4, 0)$; $F(5 \pm \sqrt{2}, 0)$;
xy-coordinates: $C(4, 3)$; $V_1\left(\frac{24}{5}, \frac{18}{5}\right)$, $V_2\left(\frac{16}{5}, \frac{12}{5}\right)$;
$F\left(4 \pm \frac{4}{5}\sqrt{2}, 3 \pm \frac{3}{5}\sqrt{2}\right)$
(c) $Y = \pm(X - 5)$; $7x - y - 25 = 0$, $x + 7y - 25 = 0$

Section 9.6 ■ page 659

1.

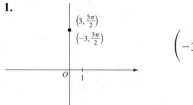

$$\left(-3, \frac{3\pi}{2}\right), \left(3, \frac{5\pi}{2}\right)$$

3.

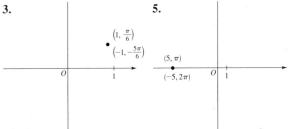

$\left(-1, -\dfrac{5\pi}{6}\right), \left(1, \dfrac{\pi}{6}\right)$

5.

$(-5, 2\pi), (5, \pi)$

7. $\left(2\sqrt{3}, 2\right)$　**9.** $(1, -1)$　**11.** $(-5, 0)$

13. $\left(\sqrt{2}, \dfrac{3\pi}{4}\right)$　**15.** $\left(4, \dfrac{\pi}{4}\right)$

17. $\left(5, \tan^{-1}\tfrac{4}{3}\right)$　**19.** $\theta = \dfrac{\pi}{4}$

21. $r = \tan\theta\sec\theta$　**23.** $r = 4\sec\theta$

25. $x^2 + y^2 = 49$　**27.** $x = 6$

29. $x^2 + y^2 = \dfrac{y}{x}$　**31.** $y - x = 1$

33. $x^2 + y^2 = (x^2 + y^2 - x)^2$

35. $x = 2$　**37.** $y = \pm\sqrt{3}\, x$

39.

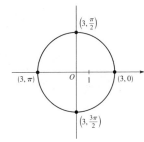

41.

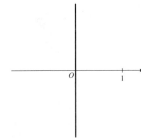

43.

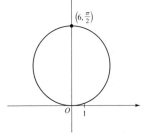

45.

47.

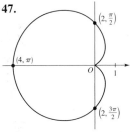

49.

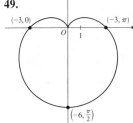

51.

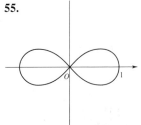

53.

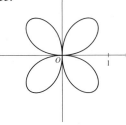

55.

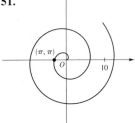

57.

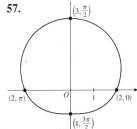

59.

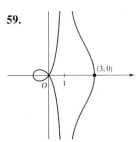

61. $0 \le \theta \le 4\pi$

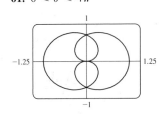

63. $0 \le \theta \le 4\pi$

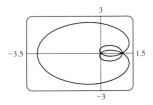

65.

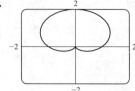 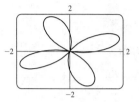

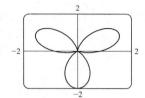 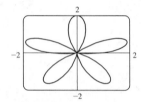

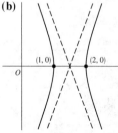

The graph of $r = 1 + \sin n\theta$ has n loops.
67. IV   **69.** III   **71. (b)** $\sqrt{10 + 6\cos(5\pi/12)} \approx 3.40$

Section 9.7 ■ page 665

1. $r = 6/(3 + 2\cos\theta)$ **3.** $r = 2/(1 + \sin\theta)$
5. $r = 20/(1 + 4\cos\theta)$ **7.** $r = 10/(1 + \sin\theta)$
9. (a) 3, Hyperbola   **11. (a)** 1, Parabola
(b)   **(b)**

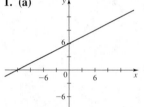

13. (a) $\frac{1}{2}$, Ellipse   **15. (a)** $\frac{5}{2}$, Hyperbola
(b)   **(b)**

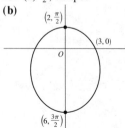 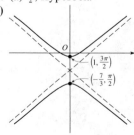

17. (a) $e = \frac{3}{4}$,
   directrix $x = -\frac{1}{3}$
(b) $r = \dfrac{1}{4 - 3\cos\left(\theta - \dfrac{\pi}{3}\right)}$

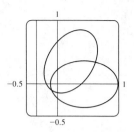

19. The ellipse is nearly circular when e is close to 0 and becomes more elongated as $e \to 1^-$. At $e = 1$, the curve becomes a parabola.

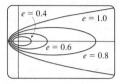

21. (b) $r = (1.49 \times 10^8)/(1 - 0.017\cos\theta)$ **23.** 0.25

Section 9.8 ■ page 672

1. (a)   **3. (a)**

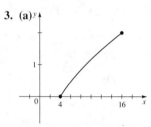

(b) $x - 2y + 12 = 0$ **(b)** $x = (y + 2)^2$
5. (a)   **7. (a)**

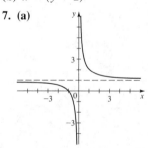

(b) $x = \sqrt{1 - y}$ **(b)** $y = \dfrac{1}{x} + 1$
9. (a)   **(b)** $x^3 = y^2$

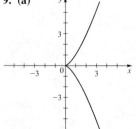

11. (a)

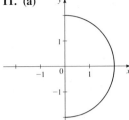

(b) $x^2 + y^2 = 4$

13. (a)

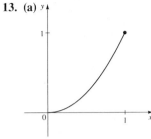

(b) $y = x^2$

15. (a)

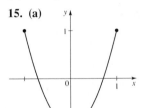

(b) $y = 2x^2 - 1$

17. (a)

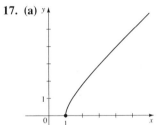

(b) $x^2 - y^2 = 1$

19. (a)

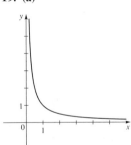

(b) $xy = 1$

21. (a)

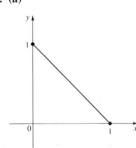

(b) $x + y = 1$

23. $x = 4 + t,\ y = -1 + \frac{1}{2}t$

25. $x = 6 + t,\ y = 7 + t$

27. $x = a\cos t,\ y = a\sin t$

31.

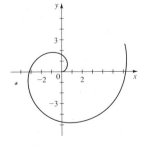

33.

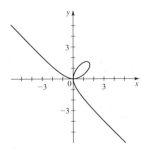

37.

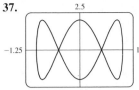

39.

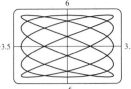

41.

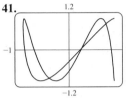

43. (a) $x = e^{t/12}\cos t,\ y = e^{t/12}\sin t$

(b)

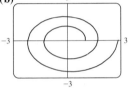

45. (a) $x = \dfrac{4\cos t}{2 - \cos t},\ y = \dfrac{4\sin t}{2 - \cos t}$

(b)

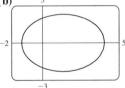

47. III   **49.** II

51.

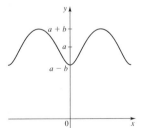

53. (b) $x^{2/3} + y^{2/3} = a^{2/3}$

55. $x = a(\sin\theta\cos\theta + \cot\theta),\ y = a(1 + \sin^2\theta)$

57. $y = a - a\cos\left(\dfrac{x + \sqrt{2ay - y^2}}{a}\right)$

Chapter 9 Review ■ page 675

1. $V(0,0)$; $F(0,-2)$; $y = 2$

3. $V(-2,2)$; $F(-\frac{7}{4},2)$; $x = -\frac{9}{4}$

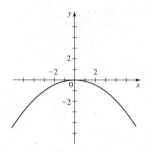

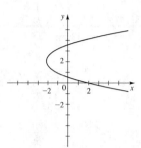

19. Parabola; $F(0,-2)$; $V(0,1)$

21. Hyperbola; $F(0,\pm 12\sqrt{2}\,)$; $V(0,\pm 12)$

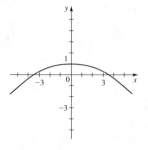

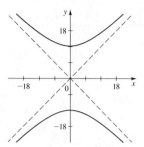

5. $C(0,0)$; $V(\pm 4,0)$; $F(\pm 2\sqrt{3},0)$; axes 8, 4

7. $C(0,2)$; $V(\pm 3,2)$; $F(\pm\sqrt{5},2)$; axes 6, 4

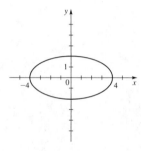

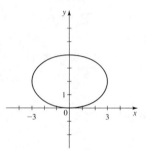

23. Ellipse; $F(1, 4 \pm \sqrt{15}\,)$; $V(1, 4 \pm 2\sqrt{5}\,)$

25. Parabola; $F(-\frac{255}{4},8)$; $V(-64,8)$

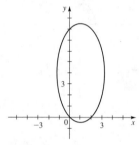

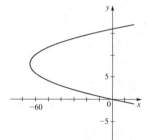

9. $C(0,0)$; $V(\pm 4,0)$; $F(\pm 2\sqrt{6},0)$; asymptotes $y = \pm \dfrac{1}{\sqrt{2}}x$

11. $C(-3,-1)$; $V(-3,-1 \pm \sqrt{2}\,)$; $F(-3,-1 \pm 2\sqrt{5}\,)$; asymptotes $y = \frac{1}{3}x$, $y = -\frac{1}{3}x - 2$

27. Ellipse; $F(3, -3 \pm 1/\sqrt{2}\,)$; $V_1(3,-4)$, $V_2(3,-2)$

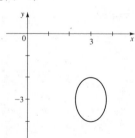

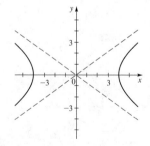

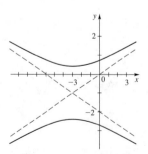

13. $y^2 = 8x$

15. $\dfrac{y^2}{16} - \dfrac{x^2}{9} = 1$

17. $\dfrac{(x-4)^2}{16} + \dfrac{(y-2)^2}{4} = 1$

29. Has no graph   **31.** $x^2 = 4y$ **33.** $\dfrac{y^2}{4} - \dfrac{x^2}{16} = 1$

35. $\dfrac{(x-1)^2}{3} + \dfrac{(y-2)^2}{4} = 1$

37. $\dfrac{4(x-7)^2}{225} + \dfrac{(y-2)^2}{100} = 1$

39. $(x-800)^2 = -200(y - 3200)$

41. **(a)** 91,419,000 mi   **(b)** 94,581,000 mi

43. (a)

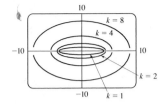

45. (a) Hyperbola
(b) $3X^2 - Y^2 = 1$
(c) $\phi = 45°$

47. (a) Ellipse
(b) $(X - 1)^2 + 4Y^2 = 1$
(c) $\phi = 30°$

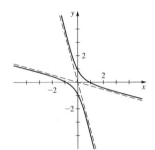

49. (a)

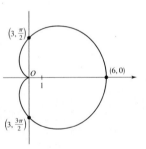

(b) $(x^2 + y^2 - 3x)^2 = 9(x^2 + y^2)$

51. (a)

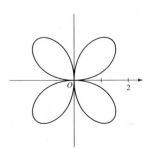

(b) $(x^2 + y^2)^3 = 16x^2y^2$

53. (a)

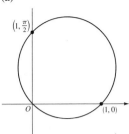

(b) $x^2 - y^2 = 1$
57. $0 \leqslant \theta \leqslant 6\pi$

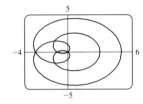

61. (a) $e = 1$, parabola
(b)

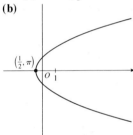

65. $x = 2y - y^2$

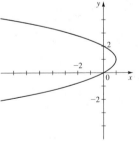

55. (a)

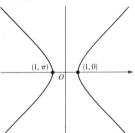

(b) $x^2 + y^2 = x + y$
59. $0 \leqslant \theta \leqslant 6\pi$

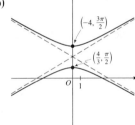

63. (a) $e = 2$, hyperbola
(b)

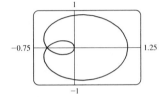

67. $(x - 1)^2 + (y - 1)^2 = 1$

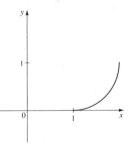

69.

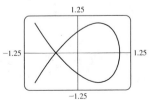

71. (a) $y = x^2$
(b)

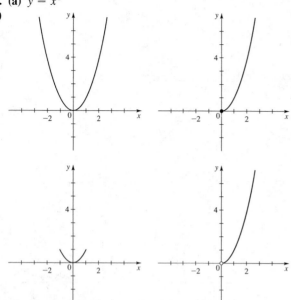

The curves are different parts of the parabola $y = x^2$.

Chapter 9 Test ■ page 679

1. $F(0, -3)$, $y = 3$

2. $V(\pm 4, 0)$; $F(\pm 2\sqrt{3}, 0)$;
8, 4

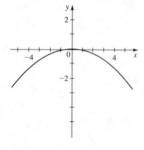

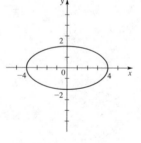

3. $V(0, \pm 3)$; $F(0, \pm 5)$; $y = \pm \frac{3}{4}x$

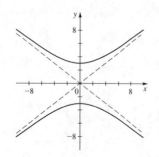

4. $y^2 = -x$ **5.** $\dfrac{x^2}{16} + \dfrac{(y-3)^2}{9} = 1$

6. $(x-2)^2 - \dfrac{y^2}{3} = 1$

7. $\dfrac{(x-3)^2}{9} + \dfrac{\left(y+\frac{1}{2}\right)^2}{4} = 1$

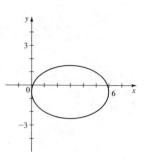

8. $9(x+2)^2 - 8(y-4)^2 = 0$

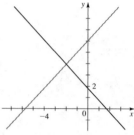

9. $(y+4)^2 = -2(x-4)$

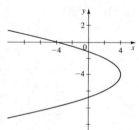

10. $\dfrac{y^2}{9} - \dfrac{x^2}{16} = 1$ **11.** $x^2 - 4x - 8y + 20 = 0$

12. $\frac{3}{4}$ in.

13. (a) Ellipse   **(b)** $\dfrac{X^2}{3} + \dfrac{Y^2}{18} = 1$

(c) $\phi \approx 27°$

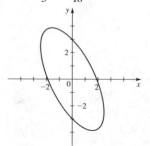

(d) $\left(-3\sqrt{2/5}, 6\sqrt{2/5}\right)$, $\left(3\sqrt{2/5}, -6\sqrt{2/5}\right)$

14.

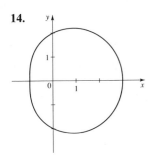

15. $(x - 1)^2 + (y + 2)^2 = 5$, circle

16. (a)

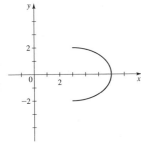

(b) $\dfrac{(x - 3)^2}{9} + \dfrac{y^2}{4} = 1$

FOCUS ON MODELING ■ page 683

1. $y = -\left(\dfrac{g}{2v_0^2 \cos^2\theta}\right)x^2 + (\tan\theta)x$

3. (a) 5.45 s   **(b)** 118.7 ft   **(c)** 5426.5 ft
(d)

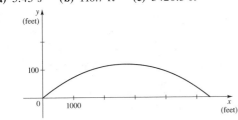

5. $\dfrac{v_0^2 \sin^2\theta}{2g}$ **7.** No, $\theta \approx 23°$

CHAPTER 10

Section 10.1 ■ page 693

1. 2, 3, 4, 5; 1001   **3.** $\frac{1}{2}, \frac{1}{3}, \frac{1}{4}, \frac{1}{5}; \frac{1}{1001}$
5. $-1, \frac{1}{4}, -\frac{1}{9}, \frac{1}{16}; \frac{1}{1,000,000}$ **7.** 0, 2, 0, 2; 2
9. 1, 4, 27, 256; 1000^{1000} **11.** 3, 2, 0, −4, −12
13. 1, 3, 7, 15, 31   **15.** 1, 2, 3, 5, 8   **17.** 2^n

19. $3n - 2$ **21.** $(2n - 1)/n^2$ **23.** $1 + (-1)^n$
25. 1, 4, 9, 16, 25, 36   **27.** $\frac{1}{3}, \frac{4}{9}, \frac{13}{27}, \frac{40}{81}, \frac{121}{243}, \frac{364}{729}$

29. $\frac{2}{3}, \frac{8}{9}, \frac{26}{27}, \frac{80}{81}; S_n = 1 - \dfrac{1}{3^n}$

31. $1 - \sqrt{2}, 1 - \sqrt{3}, -1, 1 - \sqrt{5}; S_n = 1 - \sqrt{n + 1}$
33. 10   **35.** $\frac{11}{6}$ **37.** 8   **39.** 31
41. $\sqrt{1} + \sqrt{2} + \sqrt{3} + \sqrt{4} + \sqrt{5}$
43. $\sqrt{4} + \sqrt{5} + \sqrt{6} + \sqrt{7} + \sqrt{8} + \sqrt{9} + \sqrt{10}$

45. $x^3 + x^4 + \cdots + x^{100}$ **47.** $\displaystyle\sum_{k=1}^{100} k$ **49.** $\displaystyle\sum_{k=1}^{10} k^2$

51. $\displaystyle\sum_{k=1}^{999} \dfrac{1}{k(k + 1)}$ **53.** $\displaystyle\sum_{k=0}^{100} x^k$ **55.** $2^{(2^n-1)/2^n}$

Section 10.2 ■ page 698

1. Arithmetic, 3   **3.** Not arithmetic   **5.** Arithmetic, $-\frac{3}{2}$
7. 3, 14, $2 + 3(n - 1)$, 299   **9.** 5, 24, $4 + 5(n - 1)$, 499
11. 4, 4, $-12 + 4(n - 1)$, 384
13. 1.5, 31, $25 + 1.5(n - 1)$, 173.5
15. s, $2 + 4s$, $2 + (n - 1)s$, $2 + 99s$ **17.** $\frac{1}{2}$
19. $-100, -98, -96$ **21.** 30th   **23.** 100   **25.** 460
27. 1090   **29.** 20,301   **31.** 832.3   **33.** 46.75
35. \$1250   **37.** \$403,500   **39.** 20
41. (a) 576 ft   **(b)** $16n^2$ ft   **45.** Yes

Section 10.3 ■ page 706

1. Geometric, 2   **3.** Geometric, $\frac{1}{2}$ **5.** Not geometric
7. 3, 162, $2 \cdot 3^{n-1}$ **9.** -0.3, 0.00243, $(0.3)(-0.3)^{n-1}$
11. $-\frac{1}{12}, \frac{1}{144}, 144\left(-\frac{1}{12}\right)^{n-1}$ **13.** $3^{2/3}, 3^{11/3}, 3^{(2n+1)/3}$
15. $s^{2/7}, s^{8/7}, s^{2(n-1)/7}$ **17.** $\frac{1}{2}$ **19.** $\frac{25}{4}$ **21.** 11th
23. 315   **25.** 441   **27.** 3280   **29.** $\frac{6141}{1024}$
31. 19 ft, $80\left(\frac{3}{4}\right)^n$ **33.** $\frac{64}{25}, \frac{1024}{625}, 5\left(\frac{4}{5}\right)^n$
35. (a) $17\frac{8}{9}$ ft   **(b)** $18 - \left(\frac{1}{3}\right)^{n-3}$ **37.** $\frac{3}{2}$ **39.** $\frac{3}{4}$ **41.** $\frac{1}{648}$
43. $-\frac{1000}{117}$ **45.** $\frac{7}{9}$ **47.** $\frac{1}{33}$ **49.** $\frac{112}{999}$ **51.** 3 m
53. (a) 2   **(b)** $8 + 4\sqrt{2}$ **55.** 1
57. $a \cdot \dfrac{1 - a^9}{1 - a} + 55b$

Section 10.4 ■ page 713

1. \$13,180.79   **3.** \$360,262.21   **5.** \$5,591.79
7. \$245.66   **9.** \$2,601.59   **11.** \$307.24
13. \$733.76, \$264,153.60
15. (a) \$859.15 **(b)** \$309,294.00 **(c)** \$1,841,519.29
17. \$341.24   **19.** 18.16%   **21.** 11.68%   **25.** \$36,393.56

Section 10.6 ■ page 729

1. $x^6 + 6x^5y + 15x^4y^2 + 20x^3y^3 + 15x^2y^4 + 6xy^5 + y^6$

3. $x^4 + 4x^2 + 6 + \dfrac{4}{x^2} + \dfrac{1}{x^4}$

5. $x^5 - 5x^4 + 10x^3 - 10x^2 + 5x - 1$

7. $x^{10}y^5 - 5x^8y^4 + 10x^6y^3 - 10x^4y^2 + 5x^2y - 1$

9. $8x^3 - 36x^2y + 54xy^2 - 27y^3$

11. $\dfrac{1}{x^5} - \dfrac{5}{x^{7/2}} + \dfrac{10}{x^2} - \dfrac{10}{x^{1/2}} + 5x - x^{5/2}$

13. 15 **15.** 4950 **17.** 18 **19.** 32

21. $x^4 + 8x^3y + 24x^2y^2 + 32xy^3 + 16y^4$

23. $1 + \dfrac{6}{x} + \dfrac{15}{x^2} + \dfrac{20}{x^3} + \dfrac{15}{x^4} + \dfrac{6}{x^5} + \dfrac{1}{x^6}$

25. $x^{20} + 40x^{19}y + 760x^{18}y^2$ **27.** $25a^{26/3} + a^{25/3}$

29. $48,620x^{18}$ **31.** $300a^2b^{23}$ **33.** $100y^{99}$

35. $13,440x^4y^6$ **37.** $495a^8b^8$ **39.** $(x + y)^4$

41. $(2a + b)^3$ **43.** $3x^2 + 3xh + h^2$

Chapter 10 Review ■ page 731

1. $\frac{1}{2}, \frac{4}{3}, \frac{9}{4}, \frac{16}{5}; \frac{100}{11}$ **3.** $0, \frac{1}{4}, 0, \frac{1}{32}; \frac{1}{500}$

5. $1, 3, 15, 105; 654,729,075$ **7.** $1, 4, 9, 16, 25, 36, 49$

9. $1, 3, 5, 11, 21, 43, 85$ **11.** Arithmetic, 7

13 Arithmetic, $5\sqrt{2}$ **15.** Arithmetic, $t + 1$

17. Geometric, $\frac{4}{27}$ **19.** $2i$ **21.** 5 **23.** $\frac{81}{4}$

25. 64 **27.** 12,288 **31. (a)** 9 **(b)** $\pm 6\sqrt{2}$

33. 126 **35.** 384 **37.** $0^2 + 1^2 + 2^2 + \cdots + 9^2$

39. $\dfrac{3}{2^2} + \dfrac{3^2}{2^3} + \dfrac{3^3}{2^4} + \cdots + \dfrac{3^{50}}{2^{51}}$ **41.** $\sum\limits_{k=1}^{33} 3k$

43. $\sum\limits_{k=1}^{100} k2^{k+2}$ **45.** Geometric; 4.68559

47. Arithmetic, $5050\sqrt{5}$ **49.** Geometric, 9831

51. 13 **53.** 65,534 **55.** \$2390.27 **57.** $\frac{5}{7}$

59. $\frac{1}{2}(3 + \sqrt{3})$ **67.** $2^n < n!$ for $n \geq 4$ **69.** 255

71. 12,870 **73.** $16x^4 + 32x^3y + 24x^2y^2 + 8xy^3 + y^4$

75. $b^{-40/3} + 20b^{-37/3} + 190b^{-34/3}$

Chapter 10 Test ■ page 734

1. $0, 3, 8, 15; 99$ **2.** $-4; 68, a_n = 80 - 4(n - 1)$

3. $\frac{1}{5}, \frac{1}{25}$ **4. (a)** False **(b)** True

5. (a) $S_n = \dfrac{n}{2}[2a + (n - 1)d]$ or $S_n = n\left(\dfrac{a + a_n}{2}\right)$

(b) 60 **(c)** $-\frac{8}{9}, -78$

6. (a) $S_n = \dfrac{a(1 - r^n)}{1 - r}$ **(b)** $58,025/59,049$

7. $2 + \sqrt{2}$

9. (a) $(1 - 1^2) + (1 - 2^2) + (1 - 3^2) + (1 - 4^2)$
$+ (1 - 5^2) = -50$

(b) 10

10. $32x^5 + 80x^4y^2 + 80x^3y^4 + 40x^2y^6 + 10xy^8 + y^{10}$

11. -1

FOCUS ON PROBLEM SOLVING ■ page 736

1. $1 + 3 + 5 + \cdots + (2n - 1) = n^2$ **5.** $\dfrac{n(n^2 + 1)}{2}$

7. (a) 16 **(c)** 31 regions **(d)** 57, 99, 163, 256

11. (a) $91; \dfrac{n(n + 1)(2n + 1)}{6}$ **(b)** $441; \left[\dfrac{n(n + 1)}{2}\right]^2$

CHAPTER 11

Section 11.1 ■ page 743

1. 12 **3. (a)** 64 **(b)** 24 **5.** 1024 **7.** 120

9. 144 **11.** 120 **13.** 32 **15.** 216 **17.** 480

19. No **21.** 158,184,000 **23.** 1024 **25.** 8192

27. No **29.** 24,360 **31.** 1050

33. (a) 16,807 **(b)** 2520 **(c)** 2401 **(d)** 2401

(e) 4802

35. 936 **37. (a)** 1152 **(b)** 1152 **39.** 483,840

41. 6

Section 11.2 ■ page 752

1. 336 **3.** 7920 **5.** 100 **7.** 2730 **9.** 151,200

11. 120 **13.** 24 **15.** 362,880 **17.** 997,002,000

19. 24 **21.** 60 **23.** 60 **25.** 15 **27.** 277,200

29. 2,522,520 **31.** 168 **33.** 56 **35.** 330

37. 100 **39.** 20 **41.** 210 **43.** 2,598,960

45. 120 **47.** 495 **49.** 2,035,800 **51.** 1,560,780

53. \$22,957,480 **55. (a)** 56 **(b)** 256 **57.** 1024

59. 2025 **61.** 79,833,600 **63.** 131,072

Section 11.3 ■ page 762

1. (a) $S = \{HH, HT, TH, TT\}$ **(b)** $\frac{1}{4}$ **(c)** $\frac{3}{4}$ **(d)** $\frac{1}{2}$

3. (a) $\frac{1}{6}$ **(b)** $\frac{1}{2}$ **(c)** $\frac{1}{6}$ **5. (a)** $\frac{1}{13}$ **(b)** $\frac{3}{13}$ **(c)** $\frac{10}{13}$

7. (a) $\frac{5}{8}$ **(b)** $\frac{7}{8}$ **(c)** 0 **9. (a)** $\frac{1}{3}$ **(b)** $\frac{5}{17}$

11. (a) $\frac{3}{16}$ **(b)** $\frac{3}{8}$ **(c)** $\frac{5}{8}$

13. $4C(13, 5)/C(52, 5) \approx .00198$

15. $4/C(52, 5) \approx 1.53908 \times 10^{-6}$

17. (b) $\frac{1}{16}$ **(c)** $\frac{3}{8}$ **(d)** $\frac{1}{8}$ **(e)** $\frac{11}{16}$ **19.** $\frac{9}{19}$

21. $1/C(49, 6) \approx 7.15 \times 10^{-8}$ **23. (a)** $\frac{1}{1024}$ **(b)** $\frac{15}{128}$

25. (a) $1/48^6 \approx 8.18 \times 10^{-11}$ **(b)** $1/48^{18} \approx 5.47 \times 10^{-31}$

27. $4/11! \approx 1.00 \times 10^{-7}$ **29. (a)** $\frac{3}{4}$ **(b)** $\frac{1}{4}$

31. (a) Yes **(b)** No

33. (a) Mutually exclusive; 1

(b) Not mutually exclusive; $\frac{2}{3}$

35. (a) Not mutually exclusive; $\frac{11}{26}$

(b) Mutually exclusive; $\frac{1}{2}$

37. (a) $\frac{3}{4}$ **(b)** $\frac{1}{2}$ **(c)** 1 **39.** $\frac{21}{38}$ **41.** $\frac{31}{1001}$

43. (a) $\frac{3}{8}$ **(b)** $\frac{1}{2}$ **(c)** $\frac{11}{16}$ **(d)** $\frac{13}{16}$

45. (a) Yes **(b)** $\frac{1}{36}$ **47. (a)** $\frac{1}{16}$ **(b)** $\frac{1}{32}$

49. $\frac{1}{12}$ **51.** $\frac{1}{1444}$ **53.** $\frac{1}{36^3} \approx 2.14 \times 10^{-5}$

55. (i)   **57.** $1 - \dfrac{P(365, 8)}{365^8} \approx 0.07434$

Section 11.4 ■ page 768

1. \$1.50   **3.** \$0.94   **5.** \$0.92   **7.** 0   **9.** −\$0.30
11. −\$0.0526   **13.** −\$0.50
15. No, she should expect to lose \$2.10 per stock.
17. −\$0.93   **19.** \$1

Chapter 11 Review ■ page 769

1. 624   **3.** (a) 10   (b) 20   **5.** 120   **7.** 45
9. 17,576   **11.** 120   **13.** 5   **15.** 14
17. (a) 240   (b) 3360   (c) 1680
19. 40,320   **21.** $\frac{2}{5}$
23. (a)
$S = \{HHH, HHT, HTH, HTT, THH, THT, TTH, TTT\}$
(b) $\frac{1}{8}$ (c) $\frac{1}{2}$ (d) $\frac{1}{2}$

25. (a) $\frac{1}{13}$ (b) $\frac{2}{13}$ (c) $\frac{4}{13}$ (d) $\frac{1}{26}$
27. (a) $\frac{1}{6}$ (b) $\frac{5}{6}$ **29.** (a) $\frac{1}{1000}$ (b) $\frac{1}{200}$ **31.** 0
33. \$0.00016   **35.** (a) 3   (b) .51
37. (a) 10^5 (b) 5^5 (c) $\frac{1}{32}$ (d) 75
39. (a) 144   (b) 126   (c) 84   (d) $\frac{7}{8}$

Chapter 11 Test ■ page 772

1. (a) 10^5 (b) 30,240   **2.** 60
3. $30 \cdot 29 \cdot 28 \cdot C(27, 5) = 1,966,582,800$
4. 12   **5.** $4 \cdot 2^{14} = 65,536$
6. (a) $4! = 24$ (b) $6!/3! = 120$
7. $C(5, 3)/C(15, 3) \approx .022$ **8.** $\frac{1}{6}$
9. (a) $\frac{1}{2}$ (b) $\frac{1}{13}$ (c) $\frac{1}{26}$
10. (a) $\frac{5}{13}$ (b) $\frac{6}{13}$ (c) $\frac{9}{13}$
11. \$0.65   **12.** $1 - 1 \cdot \frac{11}{12} \cdot \frac{10}{12} \cdot \frac{9}{12} \approx .427$

FOCUS ON MODELING ■ page 775

1. (b) $\frac{9}{10}$ **3.** (b) $\frac{7}{8}$

INDEX

Abbreviations for units, xx
Abscissa, 80
Absolute value
  of a complex number, 253
  properties of, 11
  of a real number, 10
Absolute-value function, 146
Absolute-value inequality, 77
  properties of, 77
Addition
  of complex numbers, 250
  of functions, 187
  of matrices, 560
  of polynomials, 27
  of rational expressions, 38
  of real numbers, 6
  of vectors, 509, 511, 512
Addition formulas, 468, 474
Additive identity, 6
Adleman, Leonard, 32
Ahmes, 593
Al Karchi, Abu Bekr, 735
Al-Khowarizmi, 46
Algebraic expression, 26
Algebraic operations on vectors, 511
Alternating current (ac) generator, 403
Ambiguous case, 436, 437
Amount (A_f) of an annuity, 708, 710
Amplitude, 372, 394
Amplitude modulation (AM), 400
Anagram, 772
Analogy, used in problem solving, 123
Analytic form of a vector, 510

Analytic Geometry, Fundamental
  Principle of, 84
Ancillary circle of an ellipse, 627
Angle(s), 407
  coterminal, 409
  definition of, 407
  of depression, 419
  of elevation, 419
  of inclination, 419
  initial side of, 407
  measure of, 407
  negative, 407
  positive, 407
  quadrantal, 428
  reference, 428
  in standard position, 709
  terminal side of, 407
Annual percentage yield (APY), 326
Annuity, 708
  amount (A_f) of, 708, 710
  in perpetuity, 714
  present value (A_p) of, 710, 711
Aphelion, 626, 666
Apollonius, 611
Apolune, 627
Appel, Kenneth, 456
Applied functions, 152
  steps in setting up, 153
Applied linear systems, 541, 555
  steps in solving, 543
Applied maximum and minimum
  problems, 180
Apollo 11, 627

Arc of a circle, length of, 411
Archimedes, 354, 615
Archimedes' principle, 127
Arccosine function, 487
Arcsine function, 485
Arctangent function, 489
Area
  of a sector, 412
  of a triangle, 432, 433, 444
Area codes, 746
Areas, formulas for, *inside front cover*
Argument of a complex number, 501
Aristarchus of Samos, 418
Arithmetic mean(s), 700
Arithmetic sequence, 694
  common difference, 694
  first term, 694
  nth partial sum of, 697
  nth term, 694
  partial sums of, 696, 697
Arrow diagram
  for a function, 133
  for a composition of functions, 190
Arrow notation for end behavior, 265
Ars Magna, 256
Associative properties, 5
  for matrices, 564
Astroid, 674
Astronomical unit (AU), 345
Asymptote(s), 265
  horizontal, 263, 265
  of a hyperbola, 630
  oblique, 271

Asymptote(s) *(continued)*
  of a rational function, 268
  slant, 270, 271
  vertical, 263, 265
Axes, coordinate, 81
  rotation of, 644
Axis
  imaginary, 252
  polar, 651
  real, 252
Axis of symmetry of a parabola, 612

Babylonian process for finding square
      roots, 128
Back substitution, 531
Bacterial growth, 297
Base
  of an exponent, 14
  of an exponential function, 289
  of a logarithm, 304
  of a power, 14
Baseline, 438
Bearing, 513
Beats, 483
Bernoulli, Johann, 670
Bhaskara, 529
Binomial, 26, 721
Binomial coefficient(s), 723, 724
  compared with $C(n, r)$, 750, 754
  key property, 725
Binomial expansion, general term of, 727
Binomial Theorem, 721, 726
  proof of, 728
Birthday paradox, 761
Book on the Games of Chance, 256
Borwein, Jonathan and Peter, 355
Bounded region, 590
Bounds for roots, upper and lower, 241
Brahe, Tycho, 625
Branches of a hyperbola, 629

$C(n, r)$, 749, 750
  compared with binomial coefficients,
      750, 754
Calculations and significant digits, xviii
Calculator, graphing, xviii, 95
Calculator, scientific, xviii
  entering scientific notation on, 19
  notation for matrices, 574
  notation for trigonometric ratios, 421
Cancellation of common factors, 36
Cardano, Gerolamo, 256
Cardioid, 656
Carrier signal, 400
Cartesian coordinate system, 81

Cartesian plane, 81
Cassegrain-type telescope, 634
Catenary, 298
Cayley, Arthur, 571
Center
  of a circle, 87
  of an ellipse, 621
  of a hyperbola, 629
Central box of a hyperbola, 631
Certain event, 756
Change of base formula, 316
Chevalier de Méré, 755
Chudnovsky, David and Gregory, 354
Chui-chang suan-shu [*Nine Chapters on
      the Mathematical Art*], 71
Chu Shi-kie, 722
Circle
  as a conic section, 611
  equation of, 67
  involute of, 674
  polar equation of, 659
  unit, 351
Circular arc, length of, 411
Circular functions, 360
Circular sector, area of, 412
Cissoid, 660
Closed interval, 9
Coefficient matrix, 573
Coefficient of a polynomial, 229
  complex, 262
  constant, 221
  leading, 221
  real, 262
Cofactor (A_{ij}) of a matrix element, 578
Cofunction identities, 461
Coin toss, Monte Carlo method, 774
Column transformation of a determinant,
      581
Combination, 749
  $C(n, r)$, 749, 750
  guidelines for using, 751
Combining functions, 187
Comet, path of, 632
Common difference of an arithmetic
      sequence, 694
Common factor, 30
Common logarithm, 308
Common ratio of a geometric sequence,
      700
Commutative properties, 5
Complement of an event, 757
  probability of, 758
Complete Factorization Theorem, 257
Completing the square, 49, 50
  to graph a quadratic function, 176

Complex coefficient, 262
Complex conjugate, 251
Complex number(s), 248, 249
  absolute value of, 253
  addition of, 250
  argument, 501
  arithmetic operations on, 250
  division of, 250, 502
  equality of, 249
  graph of, 252
  imaginary part, 249
  modulus of, 253, 501
  multiplication of, 250, 502
  nth root of, 504, 505
  pure imaginary, 249
  real part, 249
  subtraction of, 250
  trigonometric form of, 501
Complex plane, 252
Complex root of a quadratic equation, 255
Components of a vector, 510
Composite function, 189
Composition
  of functions, 189
  of trigonometric functions, 381
Compound fraction, 39
Compound-fraction form of a rational
      function, 268
Compound interest, 195, 295, 296, 325
Conchoid, 660
Confocal conics, 643
Confocal hyperbolas, 636
Confocal parabolas, 643
Conic section(s), 611
  confocal, 643
  degenerate, 640
  discriminant of equation for, 648
  focus-directrix definition of, 661
  general equation, 646
  polar equations of, 661, 663
  shifted, 636, 642
Conjecture, 715
Conjugate, complex, 251
Conjugate hyperbolas, 635
Conjugate radical, 41
Conjugate Roots Theorem, 260
  proof of, 260
Constant, 26
Constant coefficient, 221
Constant function, 139
Constant of proportionality, 155
Constraints, 603, 604
Contestant's dilemma, 773, 774
Continuously compounded interest, 297,
      325

Coordinate(s)
  polar, 651
  rectangular, 80
  relationship between polar and
    rectangular, 652
Coordinate axes, 81
  rotation of, 644
Coordinate line, 4
Coordinate plane, 81
Coordinate system, polar, 651
Coordinate system, rectangular, 81
  used for locating street addresses, 83
Correlation coefficient, 213
Cosecant curve, 387
Cosecant function
  definition of, 360, 427
  graph of, 382, 383, 384
  inverse, 490
  period of, 387
  periodic property of, 382
Cosine curve, 372, 373
  shifted, 374
Cosine function
  definition of, 360, 427
  graph of, 368, 370
  inverse, 487
  period of, 369
  periodic property of, 369
  values of, 370
Cotangent curve, 385
Cotangent function
  definition of, 360, 427
  graph of, 382, 383, 384
  period of, 385
  periodic property of, 382
Coterminal angles, 409
Counting principles, 741
Cramer's Rule, 582, 583, 584
Cube(s), 455, 456
  perfect, 31
  sum of, 31
  difference of, 31
Curtate cycloid, 673
Cycle, 393
Cycloid, 669, 670
  curtate, 673
  prolate, 673

D'Alembert, Jean Le Rond, 775
Damped harmonic motion, 399
Damping constant, 399
Dase, Zacharias, 527
Decibel scale, 334
Decomposition of a partial fraction, 592
  distinct linear factors, 593

irreducible quadratic factors, 594
  repeated irreducible quadratic factors, 595
  repeated linear factors, 593
Decreasing function, 142, 143
Degenerate conic, 640
Degenerate hyperbola, 641
Degree (angle measure), 407
  relationship to radian measure, 408
Degree of a polynomial, 26, 221
  even or odd degree, 225
Demand equation, 111
DeMoivre's Theorem, 503, 504
DeMorgan, Augustus, 350
Denominator, 7
  rationalizing, 23, 41
  least common, 7, 38
Dependent system, 552
Dependent variable, 132
Depression, angle of, 419
Descartes' Rule of Signs, 240, 241
Descartes, René, 88, 220, 240, 740
Determinant of a matrix, 570, 577
  expanding, 579
  transformation of, 581
Diaconis, Persi, 756
Difference
  of cubes, 31
  of functions, 187
  of squares, 31
  of two vectors, 509
Difference quotient, 134
Difference table, 738
Dimension of a matrix, 560
Diophantus, 3, 458
Directed quantity, 508
Direction of a vector, 513
Directrix
  of a conic, 661
  of a parabola, 612
Direct variation, 154, 155
Discounted value of an annuity, 710
Discriminant
  of a quadratic equation, 52
  of the general conic equation, 648
Display window, 95
Distance
  between two points on a line, 11
  between two points in a plane, 82
  to the sun, trigonometric methods for
    finding, 424
Distance formula, 82
Distance-rate-time problems, 63, 542
Distinguishable permutation, 748
Distributive property, 5
  for matrices, 564

Dividend of a polynomial, 233, 234
Division
  of complex numbers, 250, 502
  of polynomials, 233
  of real numbers, 6
Division Algorithm for polynomials, 234
Divisor, 7
  of a polynomial, 233, 234
Dodecahedron, 455, 456
Domain(s)
  of a function, 132
  of trigonometric functions, 361
  of a variable, 26
Doppler effect, 276, 404
Double-angle formulas, 475
Double zero, 258
Doubling period, 298
Drawing a diagram in problem solving, 59,
    123, 281

e (the number), 294
Earthquake, magnitude of, 332, 333
Eccentricity
  of a conic, 661, 665
  of an ellipse, 623
  of the orbit of a planet, 623
Echelon form of a matrix, 550
Edge of a network, 281
Einstein, Albert, 126, 680
Electrical resistance, 273
Elementary row operation, 549
  notation for, 549
Element of a set, 8
Elevation, angle of, 419
Elimination method, 531, 534
Ellipse, 101, 414, 535, 619
  ancillary circle of, 627
  axes of, 620
  center of, 621
  as a conic section, 611
  discriminant of equation for, 648
  eccentricity of, 623
  foci of, 619
  geometric definition, 619
  latus rectum of, 627
  major axis of, 620
  minor axis of, 621
  polar equation of, 663
  properties of, 621
  reflection property of, 625
  shifted, 637
  vertices of, 620
Empty set ($\emptyset$), 8
End behavior of a polynomial function,
    224, 225

End behavior of a rational function, 263, 264

Entry (a_{ij}) of a matrix, 560

Epicycloid, 674

Equality
  of complex numbers, 249
  of matrices, 560
  properties of, 45
  of vectors, 508, 511

Equation(s), 44
  of a circle, 87
  of a conic, 646
  of an ellipse, 619
  equivalent, 45
  exponential, 318
  graph of, 84
  of a hyperbola, 629
  identity, 45
  of a line, 104, 105, 106
  linear, 46, 106, 539, 540, 546
  logarithmic, 321
  matrix, 572
  of a parabola, 612
  parametric, 666
  polar, 653
  quadratic, 48
  involving radicals, 53
  root of, 45
  shifted conic, 642
  shifting the graph of, 637
  solution(s) of, 45, 49
  solving graphically, 99, 100
  system of, 531
  system of linear, 539
  trigonometric, 493
  variables of, 44

Equilibrium point, 111

Equivalent equations, 45

Equivalent inequalities, 72

Equivalent systems, 547

Eratosthenes, 414, 525

Establishing subgoals in problem solving, 123

Euclid, 49

Euler, Leonhard, 249, 294

Euler's Formula, 456

Even function, 170, 171

Even-odd identities, 461

Even-odd properties of trigonometric functions, 364

Event(s), 755
  certain, 756
  complement of, 757
  impossible, 756
  independent, 760

intersection of, 760
  mutually exclusive, 758
  union of, 759

Everest, Sir George, 438

Expansion
  of a binomial, 727
  of a determinant, 579
  of an expression, 30

Expectation, 767

Expected value, 766, 767

Experiment, 755

Experimenting in problem solving, 773

Exponent(s)
  fractional, 22
  integer, 14
  Laws of, 15, 17
  negative, 15
  of a power, 14
  rational, 22
  zero, 15

Exponential decay formula, 329

Exponential equation(s), 318
  steps in solving, 319

Exponential form of a logarithm, 304

Exponential function, 287, 289
  natural, 294
  as a mathematical model, 342

Exponential growth, 297, 327

Exponential growth formula, 298, 327

Exponential model, 342

Exponential notation, 14

Expression
  algebraic, 26
  fractional, 36
  rational, 36

Extraneous solution, 53
  in a trigonometric equation, 496

Extrema, local, 228, 229

Extreme values of a quadratic function, 175, 177, 179

Factor(s)
  cancellation of, 36
  of an exponent, 14
  of a number, 32
  of a polynomial, 30, 257

Factored form of a rational function, 268

Factorial, 724

Factoring, 30, 32
  expressions involving fractional exponents, 33
  formulas for, 31
  by grouping, 34
  a higher-degree equation, 54
  a nonlinear inequality, 73

a polynomial, 30
  a quadratic, 48

Factoring formulas, 31

Factor Theorem, 237

Family
  of conics, 643
  of functions, 144, 150
  of parabolas, 230, 616, 643
  of polar equations, 658
  of power functions, 144

Feasible region, 603, 604

Fermat, Pierre de, 249, 458, 755, 776

Fermat's Last Theorem, 458, 459

Fibonacci, 689

Fibonacci numbers, 691

Fibonacci sequence, 687

First term
  of an arithmetic sequence, 694
  of a geometric sequence, 700
  of a sequence, 685

Focal diameter of a parabola, 615

Focal length of a parabola, 618

Focus (foci)
  of a conic, 661
  of an ellipse, 619
  of a hyperbola, 629
  of a parabola, 612

FOIL method, 27

Force, 515

Four-color problem, 456

Four-leaved rose, 656

Fourier, Joseph B., 286, 378, 476

Fraction(s), 7
  compound, 39
  properties of, 7

Fractional exponent(s), 22
  factoring an expression involving, 33
  solving an equation involving, 55

Fractional expression(s), 36
  adding and subtracting, 38
  multiplying and dividing, 37
  simplifying, 36

Frequency, 394

Frequency modulation (FM), 400

Function(s), 131
  absolute-value, 146
  adding and subtracting, 187
  algebra of, 187
  algebraic representation of, 136, 137
  applied, 152
  arrow diagram for, 133, 190
  circular, 360
  combining, 187
  composite, 189
  constant, 139

decreasing, 142, 143
definition of, 132
dependent variable in, 132
difference of, 187
dividing, 187
domain of, 132
even, 170, 171
exponential, 287, 289
extreme values of, 174, 175, 177, 179
family of, 144
four ways to represent, 136, 137
graph of, 139, 368
graphical addition of, 188
greatest integer, 146
increasing, 142, 143
independent variable in, 132
inverse, 197
linear, 139
local maximum of, 182
local minimum of, 182
logarithmic, 304
machine diagram for, 133, 190
as a mathematical model, 152
maximum value of, 177, 179
minimum value of, 176, 177, 179
multiplying, 187
natural exponential, 294
numerical representation of, 136, 137
objective, 603, 604
odd, 171
one-to-one, 195
periodic, 369
piecewise-defined, 145
polynomial, 221
power, 144
product of, 187
quadratic, 175
quotient of, 187
range of, 132
rational, 221, 262
square root, 142
squaring, 133
step, 151
sum of, 187
transformations of, 162
trigonometric (of angles), 426, 427
trigonometric (of real numbers), 360
value of, 132
verbal representation of, 136, 137
visual representation of, 136, 137
Fundamental Counting Principle, 741
Fundamental identities, 364, 365, 430, 431, 461
Fundamental Principle of Analytic
    Geometry, 84

Fundamental Theorem of Algebra, 257

Galilei, Galileo, 530, 611, 680
Galois, Evariste, 234
Gateway Arch, 298
Gauss, Carl Friedrich, 257, 259, 551, 696
Gaussian elimination, 551
Gauss-Jordan elimination, 551
General equation of a line, 106
General equation of a shifted conic, 642, 646
General term of the binomial expansion, 727
Geometric mean, 708
Geometric sequence, 700
    common ratio, 700
    first term, 700
    nth partial sum of, 702
    nth term, 700
    partial sums of, 702
Gnomons of Al Karchi, 735, 736
Golden ratio, 691
Googol, 313
Googolplex, 313
Grad, 415
Graham, Ronald, 749
Graph(s)
    of an absolute-value function, 146
    of a circle, 87
    of a complex number, 252
    of an equation, 84
    of a function, 139
    horizontal shifting of, 164, 165
    horizontal shrinking and stretching of, 169
    of an inequality, 587
    of a line, 106
    of a linear equation in two variables, 140
    of a linear equation in three variables, 553
    local maximum and minimum values of, 182
    of a logarithmic function, 305
    of a piecewise-defined function, 145
    of a polar equation, 653
    of a polynomial, 221, 226
    of a power function, 144
    of a rational function, 262
    of the square root function, 142
    of the squaring function, 140
    of a trigonometric function, 368, 382
    reflecting, 166
    shifted, 637
    transformations of, 162

vertical shifting of, 162, 163
vertical shrinking and stretching of, 167
Graphical addition of functions, 188
Graphical solution to a nonlinear system,
    531, 536
Graphing calculator, xviii, 95
    avoiding extraneous lines in graphs, 98, 99
    calculating interest rates, 712
    comparing functions, 291
    comparing graphs of exponential and
        power functions, 291
    comparing rates of population growth, 299
    finding extreme values, 182
    finding increasing and decreasing
        intervals, 144
    graphing a circle, 98
    graphing an cubic equation, 96
    graphing an equation, 95
    graphing a family of parabolas, 616
    graphing a family of polar equations, 658
    graphing a family of polynomials, 230
    graphing a family of power functions, 144
    graphing a function, 143, 144
    graphing a logarithmic function, 310, 317
    graphing a parametric curve, 670
    graphing a piecewise-defined function, 147
    graphing a polar equation, 657
    graphing a polynomial, 227
    graphing a rational function, 271
    graphing a trigonometric function, 376
    instructions for using to draw graphs, 95
    rotating an ellipse, 664
    setting axes scales, 98
    solving an equation, 99
    solving an exponential equation, 320
    solving an inequality, 99
    solving a linear system, 111
    solving a nonlinear system, 536
    solving a polynomial equation, 243
Gravity, Newton's Law of, 47
Greatest integer function, 146
Great Trigonometric Survey of India, 438
Grouping, factoring by, 34

Haken, Wolfgang, 456
Hale-Bopp comet, 632
Half-angle formulas, 475, 477, 478

Half-life, 329
  of a radioactive element, 331
Halley's comet, 632, 636
Hardy, G. H., 684
Harmonic mean, 700
Harmonic motion, 394
  damped, 399
  simple, 394
Harmonic sequence, 700
Heron's Formula, 444
Hexagonal numbers, 738
Hilbert, David, 561, 578
Hipparchus, 417
Hippopede, 660
Horizontal asymptote, 263, 265, 268
Horizontal component of a vector, 510, 513
Horizontal line, 106
Horizontal Line Test, 196
Horizontal shift of a graph, 165
Horizontal shrinking and stretching of a graph, 169
Hours of daylight, mathematical model for, 397
Huygens, Christiaan, 670
Hyperbola, 101, 628
  asymptotes of, 630
  branches of, 629
  center of, 629
  central box of, 631
  confocal, 636
  as a conic section, 611
  conjugate, 635
  degenerate, 641
  discriminant of equation for, 648
  foci of, 629
  geometric definition of, 629
  polar equation of, 663
  properties of, 630
  reflection property of, 633
  shifted, 638, 639
  steps for sketching graph of, 631
  transverse axis of, 629
  vertices of, 629
Hyperbolic cosine and sine functions, 303
Hyperbolic reflector, 634
Hypocycloid, 673

Icosahedron, 455, 456
Identity(-ies), 45
  addition, 468, 474
  cofunction, 461
  double-angle, 475
  even-odd, 461
  fundamental, 364, 365, 430, 431, 461

half-angle, 475, 477, 478
  product-to-sum, 480
  Pythagorean, 365, 431, 461
  reciprocal, 364, 431, 461
  subtraction, 468
  sum-to-product, 481
  trigonometric, 364
Identity matrix, 568
Image of x under f, 132
Imaginary axis, 252
Imaginary number, 249
Imaginary part of a complex number, 249
Imaginary zero of a polynomial, 261
Impossible event, 756
Impossible museum tour problem, 282
Inclination, angle of, 419
Inconsistent system, 552
Increasing function, 142, 143
Independent events, 760
  probability formula for intersection of, 761
Independent variable, 132
Index of summation, 691
Indirect reasoning in problem solving, 124, 524
Induction, mathematical, 714
  principle of, 716
  steps in, 717
Induction hypothesis, 717
Induction step, 716
Inequalities, system of, 587
Inequality(-ies), 71, 587
  absolute-value, 77
  checking test points for, 587
  equivalent, 72
  graphing, steps in, 587
  linear, 73, 589
  nonlinear, 73
  quadratic, 74
  rules for, 72
  solution(s) of, 71
  solving algebraically, 71
  solving graphically, 99, 100, 101, 587
Infinite geometric series, 704, 705
  sum of, 705
Infinite series, 703
  sum of, 704
Infinity (∞), 9
Initial point of a vector, 508
Initial side of an angle, 407
Inner product, 561, 562
Installment buying, 711, 712
Intensity level(s) on decibel scale (β), 334
  table of, 334
Integer(s), 3

Integer exponent, 14
Intercepts of a graph, finding, 86
Interest
  compound, 195, 295, 296, 325
  continuously compounded, 297, 325
  simple, 325, 326
Intersection of independent events, 760
  probability formula for, 761
Intersection of sets, 8
Interval(s), 8
  notation for, 9
Introducing notation in problem solving, 59
Introducing something extra in problem solving, 123, 281, 282, 464
Invariant
  algebraic, 650
  geometric, 650
  under rotation, 649, 650
Inverse
  of a real number, 6
  of a matrix, 569
  of a 2×2 matrix, 569
  of an $n \times n$ matrix, 570
Inverse cosecant function, 490
Inverse cosine function, 487
Inverse cotangent function, 490, 491
Inverse function(s), 197
  graph of, 200
  property of, 198
  steps for finding, 199
Inverse secant function, 490
Inverse sine function, 485
Inverse tangent function, 488, 489
Inverse trigonometric functions, 484
Inverse variation, 156
Invertibility criterion for a matrix, 580
Involute of a circle, 674
Irrational number(s), 3
  proof of existence, 524

Joint variation, 157
Jordan curve, 283

Kanada, Yasumasa, 355
Kantorovich, T. C., 602
Karmarkar, Narendra, 604
Kepler, Johannes, 346, 611, 624, 625
Kepler's laws of planetary motion, 162, 624
Klein, Felix, 406
Knuth, Donald, 143
Königsberg bridge problem, 281
Koopmans, T. C., 602
Kovalevsky, Sonya, 165

Lattice points, 527
Latus rectum
    of an ellipse, 627
    of a parabola, 614
Law of Cosines, 442
Law of Gravity, 47
Law of Sines, 435
Laws of Exponents, 15, 17
Laws of Logarithms, 313
Leading coefficient of a polynomial, 221
Leading entry of a matrix row, 550
Leading variable of a system, 552
Learning curve, 324
Least common denominator, 7, 38
Least squares line, 212, 546
Lemniscate, polar equation of, 659
Length
    of a circular arc, 411
    of a vector, 508, 511, 512
Like terms of a polynomial, 27
Lilavati [*The Beautiful*], 529
Limaçon, 658
    polar equation of, 659
Limiting behavior, 13, 26, 44
Line(s), 102
    of best fit, 211
    equation of, 106
    horizontal, 106
    least squares, 212
    parallel, 107
    perpendicular, 107, 108
    point-slope equation for, 104
    regression, 212
    rise and run of, 102
    slope-intercept equation for, 105
    slope of, 103
    two-intercept equation for, 114
    vertical, 106
Linear depreciation, 114
Linear equation, 46, 106
    applications of, 110
    in n variables, 546
    solving, 46
    in three variables, 553
Linear function, 139
    as a mathematical model, 210
Linear inequality, 73, 589
Linear programming, 602
    uses in industry, 604
Linear systems, applications of, 555
Line of sight, 419
Lissajous figure, 671
Lithotripsy, 625
Local extrema of a polynomial, 228, 229
Local maximum, 182, 227

Local minimum, 182, 227, 228
Logarithm(s)
    base of, 304
    change of base, 315, 316
    common, 308
    definition of, 304
    exponential form, 304
    Laws of, 313
    natural, 308, 309
    properties of, 308, 310
Logarithmic equation, 321
    steps in solving, 321
Logarithmic form of an exponent, 304
Logarithmic function, 304
    graph of, 305
Logarithmic scale, 331
Logarithmic spiral, 672
Log-log plot, 345
Logistic growth formula, 300, 303
Longbow curve, 674
Long division of polynomials, 233
LORAN system, 634
Lower bound for roots, 241
Lowering powers, formula for, 477

Machine diagram
    for a function, 133
    for a composition of functions, 190
Magic square, 737
Magnitude
    of an earthquake, 332, 333
    of a vector, 508, 511
Main diagonal of a matrix, 568
Major axis of an ellipse, 620
Making a table in problem solving, 59
Mathematical induction, 714, 716
    principle of, 716
    used in problem solving, 124
Mathematical model(s), 152, 210, 342, 393
    exponential function as, 342
    linear function as, 210
    linear programming, 602
    Monte Carlo method, 737
    parametric equations as, 680
    polynomial function as, 346
    power function as, 344
Matijasevič, Yuri, 561
Matrix (matrices), 549
    addition of, 560
    algebraic operations on, 560
    calculator notation for, 574
    column of, 549, 560
    determinant of, 570, 577
    difference of, 560
    dimension of, 560

    echelon form of, 550
    elementary row operations on, 549
    entry (a_{ij}), 560
    equality of, 560
    identity, 568
    inverse of, 568, 569
    leading entry of, 550
    main diagonal of, 568
    multiplication of, 561
    product of, 562
    reduced echelon form, 550
    row of, 549, 560
    scalar product of, 560
    square, 577
    square root of, 568
    stochastic, 566
    subtraction of, 560
    sum of, 560
    zero, 567
Matrix equation, 572
    solution of, 573
Matrix form of a system, 548
    guidelines for using to solve a system, 553
Matrix multiplication, 561
    properties of, 564
Maximum and minimum problems, 180
Maximum value
    of a quadratic function, 177, 179
    local, 182
McCormack, Thomas J., 130
Mean
    arithmetic, 700
    harmonic, 700
Measure of an angle, 407
    in degrees, 407
    in radians, 408
Median, 214
Median-median line, 214
Mendel, Gregor, 763
Mercator projection, 459
Mersenne numbers, 527
Method of elimination, 531, 534
Method of substitution, 531
Midpoint formula, 83
Minimum value
    of a quadratic function, 176, 177, 179
    local, 182
Minor axis of an ellipse, 621
Minor of an element (M_{ij}), 578
Mixture problems, 61, 543
Model, mathematical, 152, 210
Modulus of a complex number, 253, 501
Monomial, 26, 221
Monte Carlo method, 773

Motion of a projectile, parametric model
    for, 680
Mount Everest, 439
Multiplication
    of complex numbers, 250
    of matrices, 561, 564
    of polynomials, 27
    of powers with the same base, 14
    properties of, 42
    of a vector by a scalar, 509, 511, 512
Multiplicative identity, 6
Multiplicity of a zero, 258
Mutually exclusive events, 758
    probability formula for the union of, 759

$n!$ (n factorial), 724
nth partial sum, 689, 697, 702
nth power of a real number, 14
nth root(s), 19, 20
    of a complex number, 504, 505
    properties of, 20
nth term
    of an arithmetic sequence, 694
    of a geometric sequence, 700
    of a sequence, 685
Napier, John, 314
Natural exponential function, 294
Natural logarithm, 308, 309
    properties of, 310
Natural number(s), 3
Nautical mile, 414
Negative angle, 407
Negative exponent, 15
Negative of a real number, 6
    properties of, 6
Nephroid, 660
Network, 281
    traversable, 282
Network theory, 281
Newton, Sir Isaac, 226, 611, 625, 680
Newton's law of cooling, 330
Newton's law of gravity, 47
Newton's laws of motion, 632
Nickel, Laura, 527
n-leaved rose, 657, 659
Noble, Ben, 460
Noether, Emmy, 580
Noll, Curt, 527
Nominal annual interest rate, 296
Nonlinear equation, 46
Nonlinear inequality, 73
    solving, 75
Number(s)
    complex, 248, 249
    imaginary, 249

integer, 3
irrational, 3
natural, 3
prime, 525, 688
pure imaginary, 249
rational, 3
real, 3
Numerator, 7
    rationalizing, 41
Numerical methods, 363, 417
Nutritional content, using a system of
    linear equations to analyze, 556

Objective function, 603, 604
Oblique asymptote, 271
Oblique triangle, 435
Octahedron, 455
Odd function, 171
Oldest son or daughter phenomenon, 766
One-to-one function, 195
Open interval, 8
Orbit of a planet, eccentricity of, 623
Order of a vertex, 281
Ordinate, 80
Origin, 4, 80
    symmetry with respect to, 89, 90
Outcome of an experiment, 755

$P(n, r)$, 746, 747
π (the number), 4
    estimate of value of, 354, 776
Palindrome, 745
Parabola, 85, 140, 175, 535, 611, 612
    axis of symmetry of, 612
    confocal, 643
    as a conic section, 611
    directrix of, 612
    discriminant of equation for, 648
    family of, 616
    focal diameter of, 615
    focal length of, 618
    focus of, 612
    geometric definition, 612
    with horizontal axis, 614
    latus rectum of, 614
    polar equation of, 663
    prime focus of, 618
    properties of, 613, 614
    reflection property of, 615, 616
    shifted, 636, 638
    vertex of, 612
    with vertical axis, 613
Parabolic reflector, 634
Parallax, 425
Parallel lines, 107

Parameter, 666
    for a family of functions, 144
Parametric equations, 666
    graphing, 670
    as a mathematical model for motion of a
        projectile, 680
Parametric form of a polar equation, 671
Partial fractions, 591, 592
    decomposition of, 592
Partial sums
    of a sequence, 689
    of an arithmetic sequence, 696, 697
    of a geometric sequence, 702
Partition, 748
Pascal, Blaise, 670, 718, 755, 776
Pascal's triangle, 722
Patterns, used in problem solving, 122, 735
Pentagonal numbers, 737
Perfect cube, 31
Perfect square, 31
Perihelion, 626, 666
Perilune, 627
Periodic properties of the trigonometric
    functions, 369, 382
Periodic behavior, modeling of, 393
Periodic function, 369, 378, 382
Periodic rent, 709
Period of a function, 369, 378, 382, 394
Permutation(s), 746
    distinguishable, 748
    guidelines for using, 751
    $P(n, r)$, 746, 747
Perpendicular lines, 107, 108
Phase shift, 374
pH scale, 332
Piecewise-defined function, 145
Planetary orbit
    aphelion, 626, 666
    eccentricity of, 623
    elliptical shape of, 625
    Kepler's laws of, 162, 624
    perihelion, 626, 666
Planets, power model for periods of, 345
Platonic solid, 455
Poincaré, Henri, 610
Point-slope form of an equation for a line,
    104
Polar axis, 651
Polar coordinates, 650, 651
    relationship to rectangular coordinates,
        652
Polar coordinate system, 651
Polar curves, table of equations for, 659
Polar equations, 653
    of conics, 661, 663

family of, 658
graph of, 653
graphing with a graphing calculator, 657
in parametric form, 671
symmetry of graphs, 657
Pole, 651
Polya, George, 123
Polygon, regular, 455
Polygonal number(s), 737
Polyhedron, regular, 455
Polynomial(s), 26, 221
    addition of, 27
    coefficient(s) of, 221
    degree of, 26, 221
    division of, 233
    end behavior of graph, 224
    factoring, 30
    factors of, 30, 257
    families of, 230
    graph of, 225, 226
    like terms of, 27
    local extrema of, 228, 229
    multiplication of, 27
    product of, 27
    quadratic, 30
    rational zeros of, 238
    real zeros of, 233
    second-degree, 30
    steps in graphing, 225
    subtraction of, 27
    synthetic division of, 235, 236
    Tchebycheff, 484
    zeros of, 223, 237
Polynomial equation, 54
    roots of, 223, 237
Polynomial function, 259
    as a mathematical model, 346
Polynomial model, 346
Positive angle, 407
Positive square root, 19
Power, nth, 14
Power function, 144
    as a mathematical model, 344
Power model, 344
Present value (A_p) of an annuity, 710, 711
Present value of a sum, 301
Prime focus of a parabola, 618
Prime number(s), 525, 527, 688
Principal, 295
Principal nth root, 20
Principle of Mathematical Induction, 716
Principle of Substitution, 29
Principles of problem solving, 122
    analogy, 123

drawing a diagram, 59, 123, 281
establishing subgoals, 123
experimenting, 773
indirect reasoning, 124, 524
introducing notation, 59
introducing something extra, 123, 281, 282, 464
making a table, 59
mathematical induction, 124
patterns, finding and recognizing, 122, 735
recognizing the familiar, 122
taking cases, 123, 455
working backward, 123
Principal square root, 19
    of a negative number, 251, 252
Probability, 754
    of the complement of an event, 758
    definition of, 756
    of intersection of independent events, 761
    notation for, 757
    of union of mutually exclusive events, 759
    of union of two events, 760
Problem solving
    with equations, 58
    guidelines for word problems, 59
    with linear equations, 110
    principles of, 122
Product
    of functions, 187
    inner, 562
    of matrices, 562
    of polynomials, 27
    scalar, 560
    sign of, 74
Product formulas, 29
Product-to-sum formulas, 480
Projectile, finding the path of, 64
Projection Laws, 448
Prolate cycloid, 673
Proof, 715
    of the Binomial Theorem, 728
    by contradiction, 124
    by induction, 716
Pure imaginary number, 249
Pythagoras, 59
Pythagorean identities, 365, 431, 461
Pythagoreans, 59, 524
Pythagorean Theorem, 59

Quadrant, 81
Quadrantal angle, 428
Quadratic equation, 48
    roots of, 51

solution of, 49
Quadratic formula, 51
Quadratic function, 175
    extreme values of, 175, 177, 179
    graph of, 175
    maximum value of, 177, 179
    minimum value of, 176, 177, 179
    standard form of, 177
Quadratic polynomial, factoring, 30
Quotient
    of functions, 187
    of polynomials, 234
    of real numbers, 7
    sign of, 74

Radian, 408
Radian measure, relationship to degree measure, 408
Radical, 19
Radicals, conjugate, 41
Radioactive decay, 328
Radioactive element, half-life of, 329, 331
Radioactive waste, 329
Radiocarbon dating, 322
Radio signals, 400
Radius, 87
Ramanujan, Srinivasa, 284
Range of a function, 132
Rational exponent, 22
Rational expression, 36. *See also* Fractional expression(s).
    limiting behavior of, 44
Rational function, 221, 262
    asymptotes of, 268
    compound-fraction form, 268
    factored form, 268
    graphing with a graphing calculator, 271
    steps in graphing, 267
Rationalizing a denominator, 23, 41
Rationalizing a numerator, 41
Rational number(s), 3, 9
    repeating decimal, 4
Rational zero of a polynomial, 238
    steps in finding, 239
Rational Zeros Theorem, 238
Real axis, 252
Real coefficient, 262
Real line, 4
Real number(s), 3
    decimal representation of, 4
    properties of, 5
Real part of a complex number, 249
Real zero of a polynomial, 233
Reciprocal identities, 365, 431, 461
Reciprocal relations, 363, 417

Rectangular coordinate system, 81
  relationship to polar coordinates, 652
Recognizing the familiar in problem
    solving, 122
Recursive sequence, 686
Reduced echelon form, 550
Reduction formulas, 368, 389
Reference angle, 428
Reference number, 355
  steps in finding, 356
Reflection of a graph, 166
Reflection property
  of an ellipse, 625
  of a hyperbola, 633
  of a parabola, 615, 616
Region
  bounded, 590
  feasible, 603, 604
  unbounded, 590
Regression line, 212, 546
Regular polygon, 455
Regular polyhedra, classifying, 455
Regular polyhedron, 455
Relativity, theory of, 26, 161
Remainder, 234
Remainder Theorem, 236
Representations of a function, 136, 137
Resistance, electrical, 273
Resultant force, 515
Reversible operation, 465
Rhind papyrus, 593
Richter, Charles, 332
Richter scale, 332
Rise between points of a line, 102
Rivest, Ted, 32
Robinson, Julia, 561
Root(s)
  bounds for, 241
  complex, 255
  of an equation, 45
  nth, 19, 20, 504, 505
  number of, 52
  of a polynomial equation, 223, 237
  of a quadratic equation, 51
  square, 19, 251, 252
Root-mean-square (rms) voltage, 403
Roots of unity, 508
Rose, 659
  four-leaved, 656, 659
  n-leaved, 657, 659
Rotation of axes, 644
  formulas for, 644
Rounding to significant digits, xix
Row and column transformations of a
    determinant, 581
Run between points of a line, 102

Salt Lake City addresses, locating by
    coordinates, 83
Sample space of an experiment, 755
Scalar, 508
Scalar multiplication, 509, 511, 512
Scalar product, 560
Scatter plot, 211
Scientific calculator, xviii
Scientific notation, 18
  calculator notation for, 19
Secant curve, 387
Secant function
  definition of, 360, 427
  graph of, 382, 383, 384
  inverse, 490
  period of, 387
  periodic property of, 382
Second son paradox, 766
Sector of a circle, area of, 412
Semiperimeter, 444
Sequence, 685
  arithmetic, 694
  common ratio of a geometric, 700
  Fibonacci, 687
  first partial sum of, 689
  first term of, 685, 694, 700
  geometric, 700
  geometric mean of, 708
  harmonic, 700
  nth partial sum, 689, 697, 702
  nth term of, 685, 694, 700
  partial sums of, 689, 702
  of partial sums, 689
  recursively defined, 686
  term of, 685
Series, 685
  infinite, 703
  infinite geometric, 704, 705
  sum of, 704, 705
Set(s), 8
  element of, 8
  empty ($\varnothing$), 8
  intersection of, 8
  union of, 8
Set-builder notation, 8
Shamir, Adi, 32
Shanks, William, 354
Shifted conic, 636
  general equation of, 642, 646
Shifting a graph, 163, 165, 637
Shrinking a graph, 167, 169
Sieve of Eratosthenes, 525
Sigma (Σ) notation, 691
Sign, variation in, 240
Significant digits, xviii
  rules for calculating, xix

Sign of a product or quotient, 74
Signs of the trigonometric functions, 362,
    427
Simple harmonic motion, 394
Simple interest, 325, 326
Sine curve, 372, 373
  shifted, 374
Sine function
  definition of, 360, 427
  graph of, 368, 370
  inverse, 485
  period of, 369
  periodic property of, 369
  values of, 370
Sinusoidal curve, 381
Slant asymptote, 270, 271
Slope-intercept form of an equation for a
    line, 105
Slope of a line, 103
Snowflake curve, 739
Solids, volumes of, 57
Solution(s)
  of an equation, 45
  of an equation involving absolute value,
    55
  of an equation involving fractional
    exponents, 55
  of an equation involving higher powers,
    54
  extraneous, 52
  of an inequality, 71
  number of, 52
  of a quadratic equation, 49
  of a quadratic-type equation, 54
  of a system, 531
Solving a triangle, 418, 435
Sophocles, 2
Sound production, mathematical model
    for, 396
Special Product Formulas, 29
Spiral
  logarithmic, 672
  polar equation of, 659
Spring constant, 161
Square, perfect, 31
Square matrix, 577
  determinant of, 579
Square numbers, 737
Square root(s)
  Babylonian process for finding, 128
  of a matrix, 568
  of a negative number, 251, 252
  principal, 19, 251, 252
Square root function, 142
Squaring function, 133
Standard position of an angle, 409

Step function, 151
Stochastic matrix, 566
Stretching a graph, 167, 169
Substitution method, 531
Subtraction
  of complex numbers, 250
  of functions, 187
  of matrices, 560
  of polynomials, 27
  of real numbers, 6
  of vectors, 509, 511
Subtraction formulas, 468
Sum(s)
  alternating, 730
  of binomial coefficients, 730
  of combinations, 750, 751
  of cubes, 31
  of functions, 187
  of an infinite series, 704
  of an infinite geometric series, 705
  partial, 689, 696, 697
  of vectors, 509
Sum formulas for sine and cosine, 472
Summary point, 214
Summation (sigma) notation, 691
Summation variable, 691
Sums, properties of, 692
Sum-to-product formulas, 481
Supply equation, 111
Surveying, methods of, 438
Symmetry, 89
  tests for, 90, 657
Synthetic division, 235, 236
System of equations, 531
  dependent, 552
  equivalent, 547
  inconsistent, 541
  leading variable(s) of, 552
  linear, 531, 539, 546
  matrix form, 548
  solutions of, 531
  steps in solving, 553
System of inequalities, 587

Taking cases in problem solving, 123, 455
Tangent curve, 385
Tangent function
  definition of, 360, 427
  graph of, 382, 383, 384
  inverse, 488, 489
  period of, 385
  periodic property of, 382
Tangent line, 226
Taussky-Todd, Olga, 565
Taylor, Brook, 363
Tchebycheff, P. L., 484

Tchebycheff polynomial, 484
Term
  of an arithmetic sequence, 694
  of a geometric sequence, 700
  of a polynomial, 26
  of a sequence, 685
Terminal point of a vector, 508
Terminal point on the unit circle, 352, 353
  guidelines for finding, 356
Terminal side of an angle, 407
Terminal velocity, 303
Test point, 587
Tests for symmetry, 90
Test value, 74
Tetrahedron, 455
Thales of Miletus, 420
Theodolite, 438
Transformation(s)
  of a determinant, 581
  of a function, 162
  of a polar graph, 661
  of a trigonometric function, 370, 385, 387
Transverse axis of a hyperbola, 629
Traversable network, 282
Tree diagram, 741
Triangle(s)
  area of, 432, 433, 444
  oblique, 435
  solving, 418, 435
  special, 416
Triangular numbers, 735
Triangulation, 438
Trigonometric equation, 493
  solving with a calculator, 498
Trigonometric expression(s), simplifying, 461
  of the form $A \sin x + B \cos x$, 471
Trigonometric form of a complex number, 501
Trigonometric function(s), 359
  of angles, 426, 427
  definitions of, 360
  domains of, 360
  evaluating, 362, 427, 429
  even-odd properties of, 364
  graphs of, 368
  inverse, 484
  as a model for periodic behavior, 393
  of real numbers, 359, 360, 426
  signs of, 362, 427
  steps in evaluating at any angle, 429
  values of, 361
Trigonometric graphs, 368
Trigonometric identities, 364, 365, 430, 431, 461

Trigonometric identities, proving, 462, 463
Trigonometric ratios, 415
  table of special values for, 417
Trigonometric substitution, 465
Trigonometry of right triangles,
  relationship to trigonometric functions
  of real numbers, 360, 426
Trinomial, 26
Trochoid, 673
Truncated pyramid, volume of, 128
Tsu Ch'ung-chih, 354
Turing, Alan, 97
Two-intercept form of an equation for a
  line, 114

Unbounded region, 590
Union of events, 759
  probability formula for, 760
Union of mutually exclusive events, 758
  probability formula for, 759
Union of sets, 8
Unit circle, 351
Unit vector, 512
  terminal point on, 352, 353
Upper and Lower Bounds Theorem, 241
Upper bound for roots, 241

Value of a function, 132
Variable, 26, 44
  dependent, 132
  domain of, 26
  independent, 132
Variable star brightness, mathematical
  model for, 396
Variation, 154
  direct, 154, 155
  inverse, 156
  joint, 157
Variation in length of daylight,
  mathematical model for, 398
Variation in sign, 240
Vector(s), 508
  addition of, 509, 511, 512
  algebraic operations on, 511
  analytic form, 510
  components of, 510, 513
  difference of, 509, 511
  direction of, 513
  equality of, 508, 511
  horizontal component of, 510, 513
  i and **j** components of, 512
  initial point, 508
  length of, 508, 511
u magnit de of, 508, 511
  multiplication by a scalar, 509, 511, 5

Vector(s) *(continued)*
   sum of, 509, 511
   terminal point of, 508
   unit, 512
   vertical component of, 510, 513
   zero, 509, 512
Velocity, 514, 680
   terminal, 303
Vertex (vertices)
   of an ellipse, 620
   of a feasible region, 604
   of a hyperbola, 629
   in a network, 281
   of a parabola, 612
   of a polyhedron, 455
   of the graph of a system of inequalities,
      589
Vertical asymptote, 263, 265, 268
Vertical component of a vector, 510, 513
Vertical line, 106
Vertical Line Test, 141
Vertical shift of a graph, 162, 163
Vertical shrinking and stretching of a
      graph, 167

Vibrating string, mathematical model for,
      400
Viète, François, 51
Viète's relations, 284
Viewing rectangle(s), 95
   comparison of, 95
   selection of, 97
Viewing screen, 95
Volume(s)
   formulas for, *inside front cover*
   of solids, 57
   of a truncated pyramid, 128

Water wave motion, mathematical model
      for, 400
Whispering gallery, 625
Wiles, Andrew, 459
Word problems, guidelines for solving, 59
Working backward in problem solving, 123
World population, 327, 342
   exponential model for, 342
   polynomial model for, 346
World series, estimating number of games
      in, 776

x-axis, 80
   symmetry with respect to, 89, 90
x-coordinate, 80
x-intercept, 86

y-axis, 80
   symmetry with respect to, 89, 90
y-coordinate, 80
y-intercept, 86

Zero
   complex, 258
   double, 258
   multiplicity of, 258
   of a polynomial, 233, 237
   rational, 238
Zero exponent, 15
Zero matrix, 567
Zero-Product Property, 48
Zeros Theorem, 258
Zero vector, 509, 512

Turn the graphing calculator into a powerful tool for your success!

Explorations in Precalculus Using the TI-82/TI-83: With Appendix Notes for the TI-85

by Deborah J. Cochener and Bonnie M. Hodge,
both of Austin Peay State University

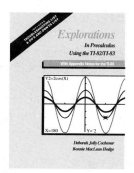

$20.95 Single Copy Price. 432 pages. Spiral bound. 8 1/2 x 11.
ISBN: 0-534-34227-2. © 1997. Published by Brooks/Cole.

Learn to use the graphing calculator to develop problem-solving and critical-thinking skills that will help improve your performance in your precalculus course!

Designed to help you succeed in your precalculus course, this unique and student-friendly workbook improves both your understanding and retention of precalculus concepts—using the graphing calculator. By integrating technology into mathematics, the authors help you develop problem-solving and critical-thinking skills.

To guide you in your explorations, you'll find:
- hands-on applications with solutions
- correlation charts that relate course topics to the workbook units
- key charts (specific to the TI-82, TI-83, and TI-85) that show which units introduce keys on the calculator
- a Troubleshooting Section to help you avoid common errors

I think this is an excellent workbook. . . . It is definitely easier to follow than the book that comes with the calculator. The examples are clear and the keystrokes are easy to follow. . . . The correlation charts are extremely helpful.
Terry Teegarden, San Diego Mesa College

I particularly like the writing style. It is clear and easy to follow. The correlation charts and the introduction of the keys is really nice. I like the summary questions, which provide the opportunity for students to write and collect their thoughts about a given topic.
Gladys Crates, Chattanooga State Technical Community College

Topics

This text contains 50 units divided into the following subsections:
  Textbook Correlation Charts
    Basic Calculator Operations
    Graphically Solving Equations and Inequalities
    Graphing and Applications of Equations in Two Variables
    Trigonometric Funtions
    Conic Sections
    Miscellaneous (Matrices, Combinatorics and Probability, and Sequences and Series)
    Stat Plots
    Programming
  Troubleshooting
  Calculator Menus

Order your copy today!

To receive your copy of *Explorations in Precalculus Using the TI-82/TI-83: With Appendix Notes for the TI-85,* simply mail in the order form attached.

ORDER FORM

To order, simply fill out this coupon and return it to Brooks/Cole along with your check, money order, or credit card information.

_____Yes! I would like to order *Explorations in Precalculus Using the TI-82/TI-83: With Appendix Notes for the TI-85*, by Deborah J. Cochener and Bonnie M. Hodge, ISBN: 0-534-34227-2 for $20.95.

Residents of: AL, AZ, CA, CT, CO, FL, GA, IL, IN, KS, KY, LA, MA, MD, MI, MN, MO, NC, NJ, NY, OH, PA, RI, SC, TN, TX, UT, VA, WA, WI must add appropriate state sales tax.

Subtotal _____
Tax _____
Handling $4.00
Total _____

Payment Options

_____ Check or money order enclosed

or bill my ____VISA ____MasterCard ____American Express

Card Number: _____

Expiration Date: _____

Signature: _____

Note: Credit card billing and shipping addresses must be the same.

Please ship my order to: (Please print.)

Name _____

Street Address_____

City _____ State _____ Zip+4_____

Telephone (   )_____ e-mail _____

Mail to:

Brooks/Cole Publishing Company
Dept. 8BCMA136
511 Forest Lodge Road, Pacific Grove, California 93950-5098
Phone: (408) 373-0728; Fax: (408) 375-6414

I(T)P **International Thomson Publishing Education Group**

10/97

P.S. If this book is appropriate for a course that you teach, please send your request for a complimentary review copy, on department letterhead, to the address listed above. Prices subject to change without notice.

SEQUENCES AND SERIES

Arithmetic

$$a, a + d, a + 2d, a + 3d, a + 4d, \ldots$$

$$a_n = a + (n - 1)d$$

$$S_n = \sum_{k=1}^{n} a_k = \frac{n}{2}[2a + (n - 1)d] = n\left(\frac{a + a_n}{2}\right)$$

Geometric

$$a, ar, ar^2, ar^3, ar^4, \ldots \qquad a_n = ar^{n-1}$$

$$S_n = \sum_{k=1}^{n} a_k = a\frac{1 - r^n}{1 - r}$$

If $|r| < 1$, then the sum of an infinite geometric series is

$$S = \frac{a}{1 - r}$$

THE BINOMIAL THEOREM

$$(a + b)^n = \binom{n}{0}a^n + \binom{n}{1}a^{n-1}b + \cdots + \binom{n}{n-1}ab^{n-1} + \binom{n}{n}b^n$$

FINANCE

Compound interest

$$A = P\left(1 + \frac{r}{n}\right)^{nt}$$

where A is the amount after t years, P is the principal, r is the interest rate, and the interest is compounded n times per year.

Amount of an annuity

$$A_f = R\frac{(1 + i)^n - 1}{i}$$

where A_f is the final amount, i is the interest rate per time period, and there are n payments of size R.

Present value of an annuity

$$A_p = R\frac{1 - (1 + i)^{-n}}{i}$$

where A_p is the present value, i is the interest rate per time period, and there are n payments of size R.

Installment buying

$$R = \frac{iA_p}{1 - (1 + i)^{-n}}$$

where R is the size of each payment, i is the interest rate per time period, A_p is the amount of the loan, and n is the number of payments.

COUNTING

Fundamental Counting Principle

Suppose that two events occur in order. If the first can occur in m ways and the second can occur in n ways (after the first has occurred), then the two events can occur in order in $m \times n$ ways.

The number of **permutations** of n objects taken r at a time is

$$P(n, r) = \frac{n!}{(n - r)!}$$

The number of **combinations** of n objects taken r at a time is

$$C(n, r) = \frac{n!}{r!\,(n - r)!}$$

The number of **subsets** of a set with n elements is 2^n.

The number of **distinct permutations** of n elements, with n_i elements of the ith kind (where $n_1 + n_2 + \cdots + n_k = n$), is

$$\frac{n!}{n_1!\,n_2!\cdots n_k!}$$

PROBABILITY

If S is a sample space consisting of equally likely outcomes, and E is an event in F, then the probability of E is

$$P(E) = \frac{n(E)}{n(F)} = \frac{\text{number of elements in } E}{\text{number of elements in } F}$$

Complement of an event:

$$P(E') = 1 - P(E)$$

Union of two events:

$$P(E \cup F) = P(E) + P(F) - P(E \cap F)$$

Intersection of two independent events:

$$P(E \cap F) = P(E)P(F)$$

If a game gives payoffs of $a_1, a_2, \ldots, a_n$ with probabilities $p_1, p_2, \ldots, p_n$, respectively, then the **expected value** is

$$E = a_1p_1 + a_2p_2 + \ldots + a_np_n$$

ANGLE MEASUREMENT

π radians $= 180°$

$1° = \dfrac{\pi}{180}$ rad 1 rad $= \dfrac{180°}{\pi}$

$s = r\theta$ $A = \frac{1}{2}r^2\theta$ (θ in radians)

To convert from degrees to radians, multiply by $\dfrac{\pi}{180}$.

To convert from radians to degrees, multiply by $\dfrac{180}{\pi}$.

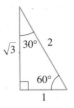

TRIGONOMETRIC FUNCTIONS OF REAL NUMBERS

$\sin t = y$ $\csc t = \dfrac{1}{y}$

$\cos t = x$ $\sec t = \dfrac{1}{x}$

$\tan t = \dfrac{y}{x}$ $\cot t = \dfrac{x}{y}$

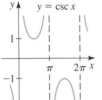

TRIGONOMETRIC FUNCTIONS OF ANGLES

$\sin \theta = \dfrac{y}{r}$ $\csc \theta = \dfrac{r}{y}$

$\cos \theta = \dfrac{x}{r}$ $\sec \theta = \dfrac{r}{x}$

$\tan \theta = \dfrac{y}{x}$ $\cot \theta = \dfrac{x}{y}$

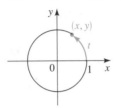

RIGHT ANGLE TRIGONOMETRY

$\sin \theta = \dfrac{\text{opp}}{\text{hyp}}$ $\csc \theta = \dfrac{\text{hyp}}{\text{opp}}$

$\cos \theta = \dfrac{\text{adj}}{\text{hyp}}$ $\sec \theta = \dfrac{\text{hyp}}{\text{adj}}$

$\tan \theta = \dfrac{\text{opp}}{\text{adj}}$ $\cot \theta = \dfrac{\text{adj}}{\text{opp}}$

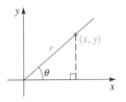

SPECIAL VALUES OF THE TRIGONOMETRIC FUNCTIONS

θ	radians	$\sin \theta$	$\cos \theta$	$\tan \theta$
0°	0	0	1	0
30°	$\pi/6$	$1/2$	$\sqrt{3}/2$	$\sqrt{3}/3$
45°	$\pi/4$	$\sqrt{2}/2$	$\sqrt{2}/2$	1
60°	$\pi/3$	$\sqrt{3}/2$	$1/2$	$\sqrt{3}$
90°	$\pi/2$	1	0	—
180°	π	0	−1	0
270°	$3\pi/2$	−1	0	—

SPECIAL TRIANGLES

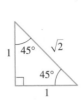
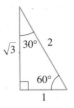

GRAPHS OF THE TRIGONOMETRIC FUNCTIONS

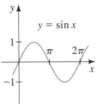
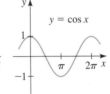
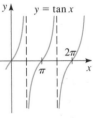

SINE AND COSINE CURVES

$y = a \sin k(x - b)$ $(k > 0)$ $y = a \cos k(x - b)$ $(k > 0)$

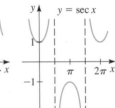
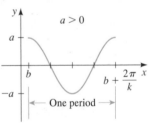
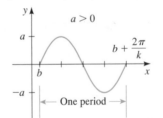
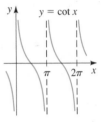

amplitude: $|a|$ period: $2\pi/k$ phase shift: b

GRAPHS OF THE INVERSE TRIGONOMETRIC FUNCTIONS

$y = \sin^{-1} x$ $y = \cos^{-1} x$ $y = \tan^{-1} x$

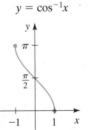
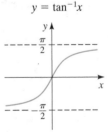